W9-CKS-284

The Firefly Encyclopedia of Astronomy

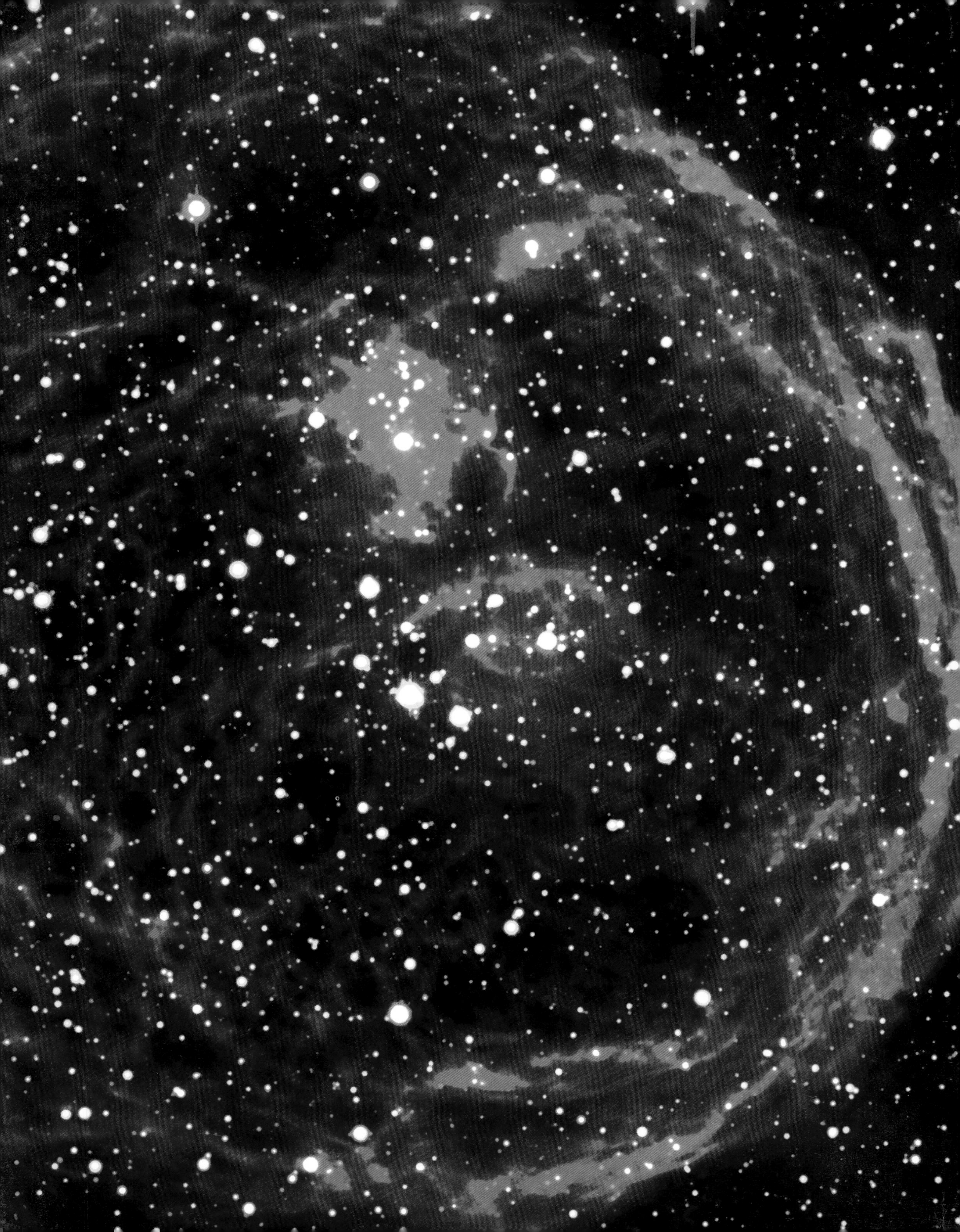

The Firefly Encyclopedia of Astronomy

Edited by Paul Murdin & Margaret Penston

Firefly Books

Published by Firefly Books Ltd. 2004

Publisher Cataloging-in-Publication Data (U.S.)

The Firefly encyclopedia of astronomy / edited by Paul Murdin and Margaret Penston. —1st ed.
[480] p. : col. ill., photos. (chiefly col.) ; cm.
Includes bibliographical references.
Summary: Reference to the whole of modern astronomy, including results of leading research in the field.
ISBN 1-55297-797-8
1. Astronomy.
I. Murdin, Paul. II. Penston, Margaret. III. Title.
522 21 QB44.3.F57 2004

National Library of Canada Cataloguing in Publication

The Firefly encyclopedia of astronomy / edited by Paul Murdin and Margaret Penston ; artworks by James Symonds; introduction by Sir Martin Rees.
Includes bibliographical references and index.
ISBN 1-55297-797-8
1. Astronomy—Encyclopedias.
I. Murdin, Paul II. Penston, Margaret
QB14.F575 2004 520'.3 C2004-900136-1

Published in the United States in 2004 by
Firefly Books (U.S.) Inc.
P.O. Box 1338, Elliott Station
Buffalo, New York 14205

Published in Canada in 2004 by
Firefly Books Ltd.
66 Leek Crescent
Richmond Hill, Ontario L4B 1H1

Printed in Singapore

Produced by Canopus Publishing Limited
27 Queen Square, Bristol, BS1 4ND, UK
www.canopusbooks.com

Project director: Robin Rees
Editors: Paul Murdin, Margaret Penston
Astronomy consultants: Terence Dickinson, Chris Lintott, Patrick Moore
Project editor: Julian Brigstocke
Production editors: Seth Burgess, Lindsey Coles
Commissioning editor (practical articles): Pam Spence
Art Director: Hariet Athay
Designers: Paul Cook, Kevin Lowry
Picture research: Robin Rees
Artist: James Symonds

Authors of practical astronomy features:

Bob Argyle	Nick Hewitt
Mel Bartels	Andrew Hollis
Graham Boots	Guy Hurst
Owen Brazell	Taichi Kato
Robin Chassagne	Chris Lintott
David Crawford	David Malin
Ron Dantowitz	John McNally
Terence Dickinson	Patrick Moore
Tim Doyle	Arto Oksanen
Alan Dyer	Jay Pasachoff
Stephen Edberg	Margaret Penston
Robert Evans	Nick Quinn
Rick Fienberg	Robert Reeves
Jeff Foust	Jonathan Shanklin
Maurice Gavin	Giovanni Sostero
Douglas George	Robert Steele
Monica Grady	Nik Szymanek
Arne Henden	Stephen Tonkin

Photos in preliminary pages

Facing title page: N70 nebula in the Large Magellanic Cloud (European Southern Observatory)
Facing preface: Orion Nebula (Nik Szymanek)

Acknowledgements

This work is derived with permission from the four-volume *Encyclopedia of Astronomy and Astrophysics*, published in 2001 in print and on-line editions by the Nature Publishing Group and Institute of Physics Publishing.

We would like to thank all the contributors to *Encyclopedia of Astronomy and Astrophysics* for their enthusiasm for the original project and for this, its daughter. We hope they are pleased with the way that their thoughts and words have made up this book. They will provide inspiration to a wider readership. A list of these contributors can be found on page 467.

Paul Murdin would like to thank Mr Andrew Ritchie and Dr Leonard Shapiro and the highly skilled staff of the cardiac units of the Addenbrooks and Papworth Hospitals in Cambridge, without whom he could not have brought this book to completion.

Foreword

This encyclopedia is witness to a crescendo of discovery in astronomy. It describes the farthest reaches of the solar system, the planets that have been discovered around other stars, the most distant galaxies (so far away that their light set out 10 billion years ago), and the first few seconds of cosmic history disclosed by the properties of the 'afterglow' of the Big Bang. It describes what astronomers know of huge black holes, some of them the same mass as a billion suns, and cosmic explosions that disgorge more light in a few seconds than the Sun emits in its lifetime.

As the subject advances, new problems come into focus. We have learned, for instance, that most stuff in the universe isn't ordinary atoms at all – it consists of dark particles of quite unknown nature and of an energy latent in space. The key properties of our universe and its constituents were imprinted when the universe was compressed to enormous density. We need scientific breakthroughs to think properly about these unknown components of matter, and the extreme conditions in which they play a significant role.

Astronomers address these issues with elaborate instruments. Historical astronomical artifacts, such as astrolabes, clocks and telescopes, have been and still are monuments to human ingenuity. As a bonus, they often have great aesthetic appeal. They are part of the heritage of astronomy and it is good to see that they are celebrated in this encyclopedia. Present-day astronomers are heirs to a long tradition, but technical advances are happening at an accelerating pace. The power of astronomy's instruments has leapt in the last four decades by as much as in the previous four centuries. We can soon expect new generations of telescopes – huge, ground-based optical telescopes too big to put into buildings, radio arrays spread across whole deserts, and mirrors in space linked over 5 million kilometers – by far the largest human construct ever made.

Astronomers have always been a widely distributed, international community. The community is now broader and the links are now stronger. The largest instruments in future will be 'world facilities' and there will be intense competition to access them. This doesn't mean that astronomy will become more elitist and exclusive – quite the reverse, in fact. Telescopes small enough to be built by amateurs are not only providing exquisite views of the astronomical components of the universe, they are yielding ever-greater scientific dividends about asteroids, about black holes, about white dwarfs and about supernovae. Enthusiasts with computers, wherever they live, can, via the World Wide Web, enter a 'virtual observatory,' downloading data from the world's largest telescopes, now and in the future. Professionals are working to make the repositories of almost all the information known about astronomical bodies easier to access. Everybody can participate actively in the exploration of the universe.

What will be the most exciting discoveries in the future? Past experience suggests that these will be the completely unexpected ones. But I will hazard a guess that within a few years dark matter will be much less of a mystery, and cosmologists will have convincingly pinned down the theory of the universe. There will be exciting news from space probes to other planets – the results of searches for life on Mars, on Titan and under the frozen oceans of Europa. There will be detailed information about the planets orbiting thousands of other stars. Some will surely contain Earth-like planets, and we will learn from them essential properties of our own solar system and of our own planet – we will be looking into space but thinking about ourselves.

Will any of those planets harbor life? And will we know about it? The origins – of planets, of stars, of galaxies, of life and of the cosmos itself – will be the focus of upcoming research. How, from a simple beginning, did atoms assemble on at least one planet around at least one star into at least one creature able to think about these mysteries? This question will challenge and fascinate us for a millennium.

Sir Martin Rees
Astronomer Royal

Preface

Photons of light can travel for years, centuries, or even millennia, across the unfathomable vastness of space. When they reach our small world, they enter our eyes, activate our optic nerves and fire our imagination. In response to these small triggers, we can paint a mental picture of the source of the photons – the planets, stars and galaxies in the night sky. We can begin to wrestle with the concepts of modern astronomy, which sometimes seem more like science fiction than science: pulsars, quasars, black holes, dark matter and the ultimate fate of the universe.

Because astronomy creates these extraordinary images in our minds as well as in print or on the Web, and because it is something everyone can enjoy directly with the unaided eye or with a telescope, it is a science with a huge popular following. There are a few lucky professionals, but they are supported by a small army of unpaid enthusiasts who observe the night sky on a regular basis.

There are few sciences where professionals work so closely with amateurs, and this encyclopedia is evidence of that proximity. In non-technical language, we synthesize reports from 800 of the world's leading astronomers (listed at the back of the book) to make the latest professional research available to everyone. (In what other field would leading-edge research be of such general interest?) And these articles are linked to features on practical astronomy (printed on a yellow background) by amateurs whose work crosses the border into professional territory. We hope that the combination of theory and practice will make the picture of the universe expressed in this encyclopedia more vivid, more realistic, more understandable – more like it actually is.

Smaller entries explain key concepts, and illustrate others by reference to individual planets, stars, nebulae and galaxies. We sketch out the world community of astronomy, past and present, with entries about all the significant astronomical satellites, telescopes and professional facilities, and about important astronomers from all over the world.

We have tried to make a popularly accessible astronomy encyclopedia that sails as close to the direction of the wind of research as possible. If the balance of the topics in this book is different from the balance in other amateur astronomy encyclopedias, that is because it is derived from professional astronomy as it is now, and looks forward. Assisted by open access to the mammoth professional *Encyclopedia of Astronomy and Astrophysics* (Nature Publishing Group/IOP Publishing 2001), we have set out a landscape into which amateur astronomy might travel further in the future. The increasing sophistication of astronomical education and the development of the technology available to amateurs have put amateur astronomy now where professional astronomy was not very long ago.

So here is the subject matter of astronomy, and a road map through it. We hope that we can share our excitement for modern astronomy and what you can do yourself with cheap and modest, or elaborate and expensive, equipment. Add your imagination, and you can reap the reward of constructing those elusive pictures, knowing that they are as accurate as anyone else can see without actually being there.

Paul Murdin and Margaret Penston

A note on how to use the book

Within an article, cross references are indicated in SMALL CAPITAL letters. These take the reader to related articles. Metric units are used throughout other than for larger amateur telescopes and historical professional telescopes. A conversion table is included at the back of the book. Acronyms and abbreviations are sometimes used to refer to well-known astronomical organizations, space missions and observatories. There is a list of common acronyms and abbreviations at the back of the encyclopedia, which the reader may refer to to find out what these stand for and where to look them up. Photographs are credited *in situ* in the practical features where the photography is often key to the discussion. All other credits are listed at the back of the book.

A

Abastumani Astrophysical Observatory

The Abastumani Astrophysical Observatory is located in the southwest part of the Republic of Georgia, 250 km west of the capital, Tbilisi, on the top of Mount Kanobili at 1700 m. Its main facilities are 125-cm and 70-cm telescopes.

Abell cluster

George Ogden Abell (1927–83) constructed a catalog, published in 1958, that contained 2712 of the richest GALAXY CLUSTERS AND GROUPS in the northern sky. The catalog was later extended to the southern sky. The total sample of 4076 cluster candidates over the whole sky has revolutionized the study of the large-scale structure in the universe. The *Abell Catalog of Rich Galaxy Clusters* has formed the basis of the first quantitative studies of the densest components of the large-scale structure inthe local universe.

The reality of the cluster candidates in Abell's catalog has been the subject of some debate. Spectroscopic observations of large numbers of galaxies in the directions of the Abell clusters showed convincingly that only a small fraction of the rich clusters are the result of superpositions of galaxies that lie by chance along the line of sight. That is, a very large fraction of the rich cluster candidates in the catalog made by Abell (or, including the southern clusters, by Abell, Corwin and Olowin) represent compact, localized peaks in the spatial distribution of galaxies, mostly with a REDSHIFT of less than 0.2, and held together by gravity.

◄◄ ***Abell cluster*** *The Abell 1060 cluster of galaxies in the constellation Hydra is framed between two nearby bright stars. The galaxies, which are collections of billions of stars, are seen as fuzzy objects beyond the hard, circular images of faint stars in our Milky Way.*

Aberration

(1) The phenomenon, discovered in 1729 by JAMES BRADLEY, by which a star is apparently displaced from its mean position on the CELESTIAL SPHERE due to the combined effects of the velocity of the Earth in its orbit around the Sun and the finite velocity of light. During the course of a year a star will appear to trace out a tiny ELLIPSE in the sky about its mean position, with maximum displacement about 20.5 arcsec.

Bradley was in fact trying to measure stellar PARALLAX. The displacement due to aberration is much greater than that due to parallax (the annual parallax of the nearest star is 0.77 arcsec). Diurnal aberration, a smaller but still significant effect, is due to the speed of rotation of the Earth on its axis.

(2) In optical systems, such as lenses and curved mirrors, aberration refers to the inability of the system to produce a perfect image, unlike that produced by a plane mirror. The main aberrations are CHROMATIC and SPHERICAL ABERRATION, COMA and ASTIGMATISM. The quality of an image is also affected by SCINTILLATION (twinkling) and ATMOSPHERIC REFRACTION.

Abetti, Antonio (1846–1928) and Abetti, Giorgio (1882–1982)

Antonio Abetti was born in San Pietro di Gorizia, Italy. A civil engineer, he turned to astronomy and became director of the forerunner of the ARCETRI ASTROPHYSICAL OBSERVATORY (1893–1921), and professor of astronomy at the University of Florence, Italy. His main interests were positional astronomy, COMETS, the observation of minor planets, and star occultations. In 1874 he observed the transit of VENUS across the Sun's disk through a SPECTROSCOPE. His son, Giorgio, succeeded him as director of the Arcetri Observatory (1921–53). Giorgio took up solar physics at the MOUNT WILSON OBSERVATORY in the USA, and on his return to Italy established a SPECTROHELIOGRAPH in Arcetri, about 180° in longitude away from Mount Wilson, to give nearly 24-hour coverage of solar phenomena. He discovered the radial motion of gases in sunspots (the Evershed–Abetti effect). He was well known as a popularizer of astronomy.

Absolute magnitude *See* MAGNITUDE AND PHOTOMETRY

Absolute zero

The lowest possible temperature. According to the kinetic theory of gases, this is the temperature at which all motion of atoms and molecules ceases. This temperature is equivalent to –273.16 °C, and defines the zero of the Kelvin, or absolute, temperature scale.

Absorption nebula *See* DARK NEBULA

Absorption spectrum

A pattern of dark lines or bands superimposed on a CONTINUOUS SPECTRUM. When a continuous spectrum of radiation (a broad range of wavelengths) passes through a material medium (for example, a cool, low-pressure gas), selective absorption occurs at certain wavelengths. This gives rise to a series of dips in intensity (absorption lines) that, in the visible region of the spectrum, appear as dark lines against the bright background of the 'rainbow' band of colors that comprises the continuous spectrum.

Photons of ELECTROMAGNETIC RADIATION may be absorbed through radiative excitation. This is a process that occurs when an electron in one of the lower energy levels of an ATOM or ION absorbs a PHOTON with energy precisely equal to the difference in energy between that level and one of the higher permitted levels. As a result, the electron jumps (makes an 'upward transition') from the lower to the higher level.

The prominence (strength) of any particular line depends on the number of atoms of the appropriate chemical ELEMENT that have electrons residing in the energy level from which the relevant upward transition takes place (the degree of excitation). That in turn depends on the abundance of the particular chemical element (the relative proportion of that element in the absorbing medium) and on other factors, in particular the temperature (the higher the temperature, the greater the proportion of electrons in excited states).

As well as producing absorption through electronic transitions (as in atoms and ions), MOLECULES may also absorb (or emit) radiation by changing their states of vibration (their constituent atoms vibrate relative to each other) or rotation (a molecule, having a physical shape, can rotate about a particular axis). Molecular absorption spectra are complex, their various lines often merging into broader bands.

See also EMISSION SPECTRUM.

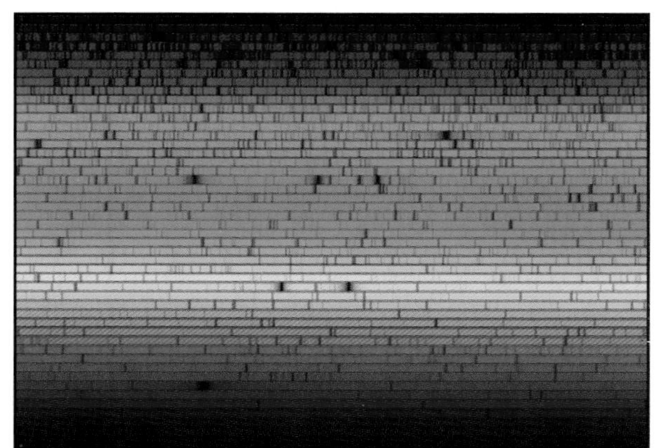

◄ ***Absorption spectrum*** *The rainbow spectrum of the Sun (arranged in sections, left to right and top to bottom in this échelle spectrogram) shows dark lines caused by atoms in the Sun's atmosphere absorbing radiation at specific frequencies.*

Abundance *See* COSMIC ABUNDANCE OF ELEMENTS

Acceleration

Rate of change of velocity. This is expressed in units of meters per second per second (m s^{-2}), feet per second per second (ft s^{-2}), or other equivalent units. Thus, an acceleration of 10 m s^{-2} would imply an increase in velocity of 10 meters per second in one second.

Accretion

The process by which a celestial body increases its mass by aggregating smaller objects that collide with it. Several types of object grow by accretion. In BINARY STARS in which mass transfer is taking place, one member grows at the expense of the other. BLACK HOLES, including supermassive black holes that are believed to be present in active galactic nuclei (*see* ACTIVE GALAXY), also increase their mass by accretion. In both of these cases matter is accumulated via a disk of material orbiting the accreting object (an accretion disk). The process is particularly important in the formation of planets. Dust grains in a PROTOPLANETARY DISK (or proplyd) around a young star collide and coalesce, gradually building up larger objects, which in turn collide and merge.

Accretion disk *See* ACCRETION

Achernar

The star Alpha Eridani, a bluish-white MAIN SEQUENCE star of spectral type B3Vp (*see* STELLAR SPECTRUM: CLASSIFICATION). Observations with the interferometer on the VLT show this rapidly rotating star to be highly flattened; its equatorial radius is more than 50% greater than its polar radius, and this poses a challenge for theorists.

Achilles

The first TROJAN ASTEROID to be discovered, by Max WOLF in 1906. Its designation is (588) Achilles. It belongs to the largest group of Trojans, which orbit ahead of Jupiter around the L_4 LAGRANGIAN POINT. Achilles has an estimated diameter of 147 km, and orbits the Sun at a mean distance of 5.17 AU over a period of 11.78 years. The INCLINATION is 10° and the ECCENTRICITY 0.15. Like most Trojans, Achilles is D-type: it has a reddish reflectance spectrum that indicates a carbon-rich surface. *See also* ASTEROID.

Achondrite

One of the two main divisions of STONY METEORITES, the other being CHONDRITES. Achondrites have a coarser crystalline structure than chondrites. They also lack chondrules, which are millimeter-sized grains of silicate. Achondrites contain hardly any nickel-iron or sulfide. They are further classified principally by their calcium content.

Achondrites form the largest group of differentiated meteorites. These consist of once-molten rock that has separated into regions of different composition. Basaltic achondrites have all the characteristics of igneous terrestrial rocks, such as basalt, that were once molten. These are thought to be fragments from the crust of a parent body that underwent at least partial DIFFERENTIATION. The prime candidate for this is the asteroid (4) VESTA. The largest known single achondritic mass is a 1070-kg fragment of the Norton County meteorite, found in Kansas, USA, in 1948. All known lunar meteorites also fall into this category. Their composition is consistent with them having been ejected from the surface of the Moon by impact. All of the so-called SNC METEORITES are in this category too. They are believed to have been ejected from the surface of Mars.

Achromatic lens *See* CHROMATIC ABERRATION

Active galactic nucleus *See* ACTIVE GALAXY

Active galaxy

An active galaxy is a galaxy that is exceptionally bright. It emits its energy from within a few light years of its central nucleus. It does not generate the energy by processes normally found in galaxies. The energy is not starlight, SUPERNOVA explosions, or any of the ordinary consequences of the evolution of stars and interstellar matter.

A typical active galaxy exhibits all or most of the following:

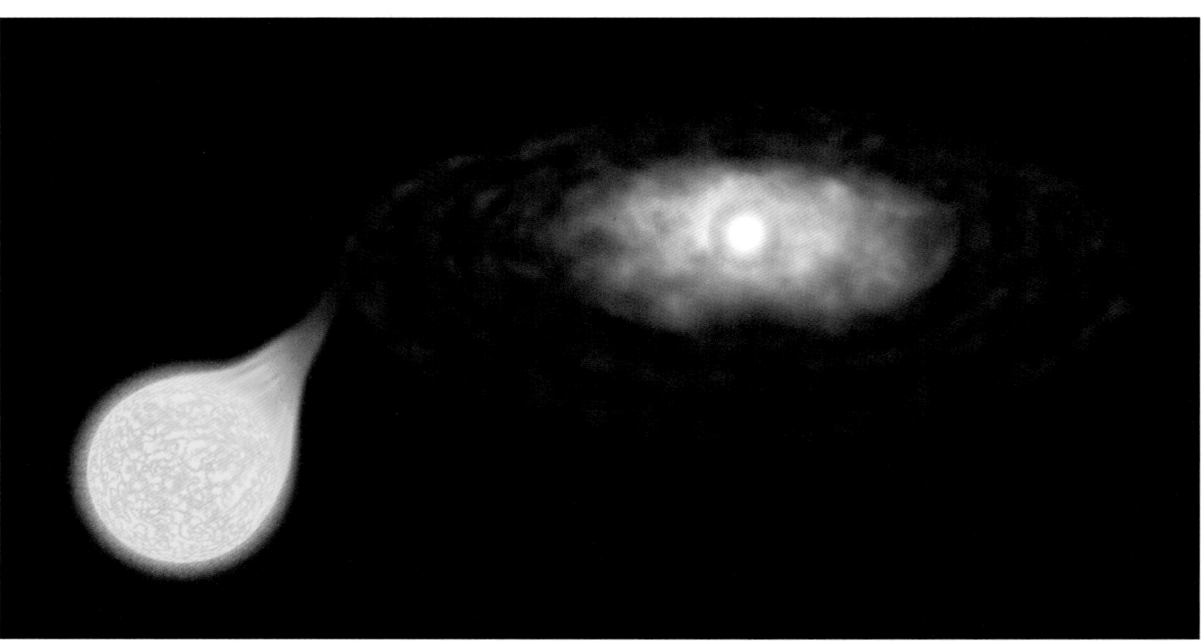

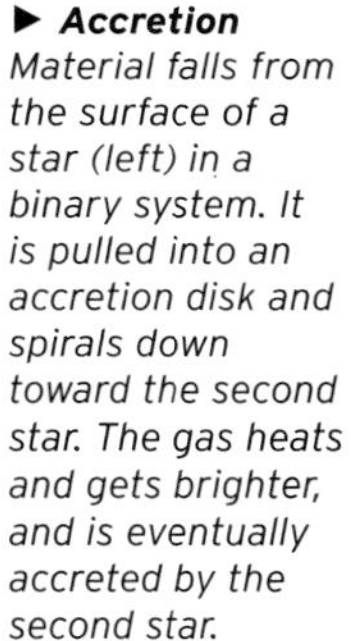

► ***Accretion*** *Material falls from the surface of a star (left) in a binary system. It is pulled into an accretion disk and spirals down toward the second star. The gas heats and gets brighter, and is eventually accreted by the second star.*

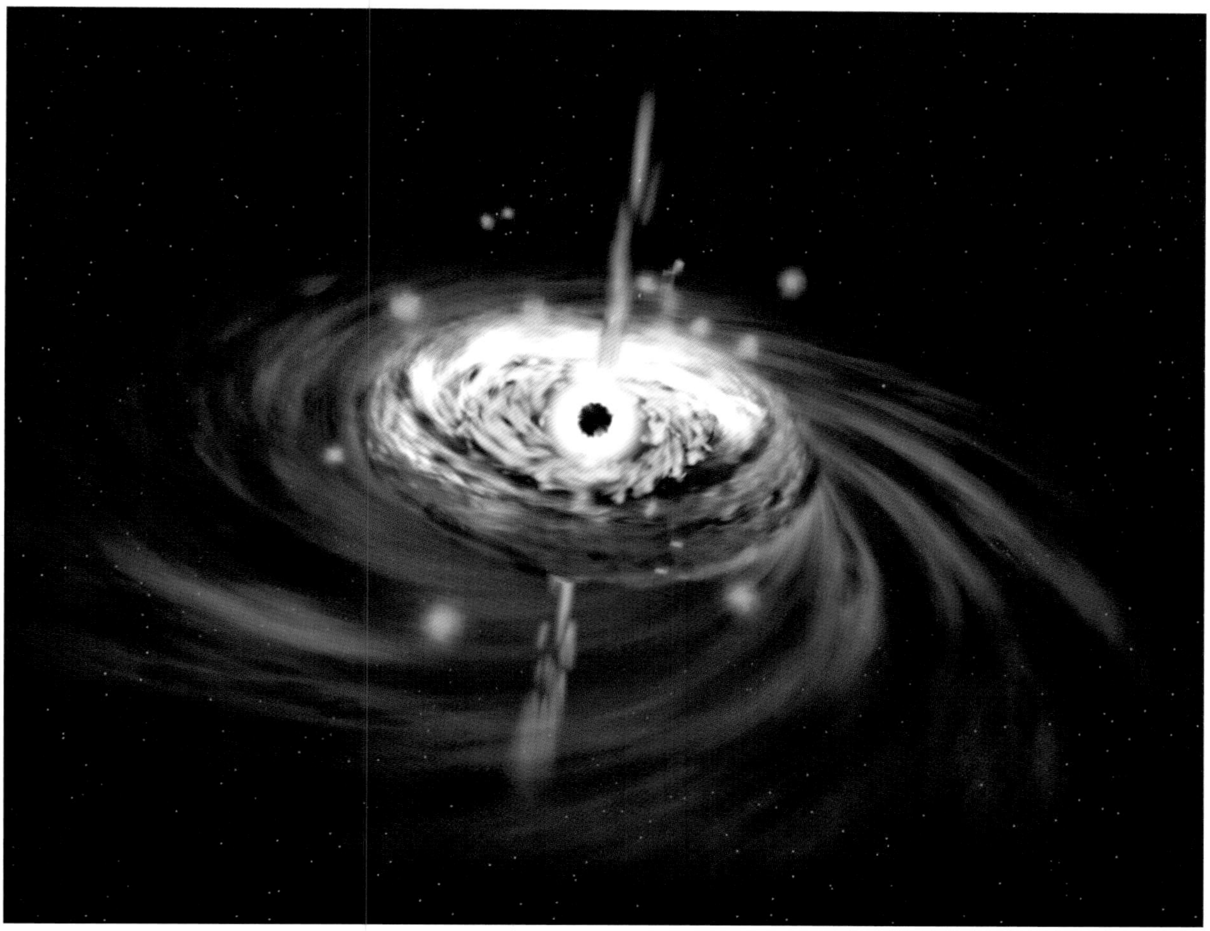

◀ ***Active galactic nucleus***
Interstellar material and disrupted stars spiral through an accretion disk onto a central black hole. Light and X-rays from the vicinity of the black hole illuminate the inner edge of the accretion disk and the high-speed material at its center. Material rejected by the black hole squirts in jets along the axis of the accretion disk, passing through low-density, low-velocity clouds that have strayed from the accretion path.

- a much higher output of X-ray, ultraviolet (UV), infrared and radio radiation than a normal galaxy;
- a very bright and very compact central core, known as an active galactic nucleus (AGN). In many cases the AGN varies rapidly;
- narrow jets of matter shooting from the central core;
- large-scale clouds of radio-emitting material;
- in some cases, broad lines in its EMISSION SPECTRUM. This indicates material that is moving at high speeds.

Luminosity, variability, jets, clouds, high speeds – these are why active galaxies are called active. They include the following principal types, which have proved to be fundamentally similar, all of them containing a central BLACK HOLE feeding on infalling gas:

Seyfert galaxies

The first known active galaxies were the SPIRAL GALAXIES identified as a group by Carl Seyfert (1911–60) in 1943. He was the first to notice their unusually bright and concentrated nuclei, and the strong emission lines in their spectra. All Seyfert galaxies have narrow emission lines from relatively slow-moving gas, but so-called Type 1 Seyfert galaxies also have emission lines that are unusually wide. This indicates that the gas that emitted them has velocities of up to several thousands of kilometers per second. Type 2 Seyfert galaxies have only narrow emission lines. Seyfert galaxies have strong X-ray emissions. They are often distorted by a nearby companion galaxy, and this is a clue to why an ordinary galaxy has turned into a Seyfert galaxy.

Radio galaxies and quasars

Almost two decades after Seyfert's discovery, the first radio surveys of the sky found vast numbers of galaxies that emit radio waves. Some of these RADIO GALAXIES were at first thought to be optical stars, but they proved to be distant galaxies with nuclei so luminous that they outshone the galaxy in which they resided. They became known as quasi-stellar radio sources, or QUASARS.

BL Lacertae objects and blazars

One quasar, BL LACERTAE (BL Lac), had been listed as a VARIABLE STAR since 1929. Observations in the late 1960s showed it to have highly variable radio emission as well. In the 1970s it was found to be outside our Galaxy and not a star. Along with quasars such as 3C273 and 3C279, BL Lacertae is known as a BLAZAR. It is the original BL Lacertae object.

Unification theory

The unification model of AGN says that all of these active galaxy types are fundamentally similar. They only appear to be different because we are viewing them from different angles.

At the center of an active galaxy is a supermassive black hole, of millions to billions of solar masses. This black hole is the source of the galaxy's energy. The black hole rotates and is surrounded by an ACCRETION disk of gas orbiting at high speeds. Gas falls from the accretion disk into the black hole. The gas is then compressed, heats, and emits X-rays. The X-rays change in intensity because the flow of gas into

Observing active galaxies

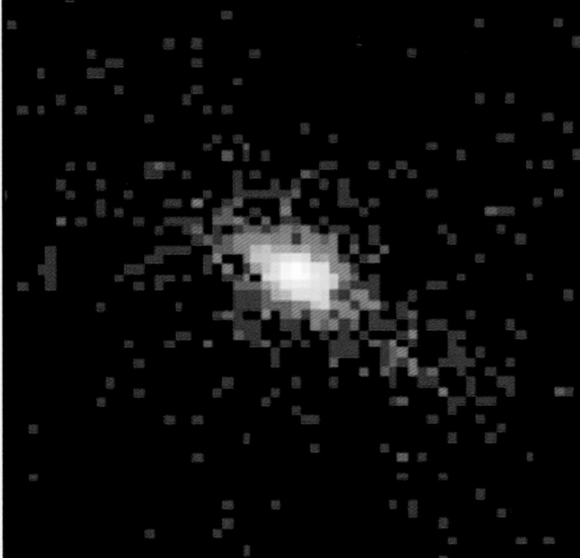

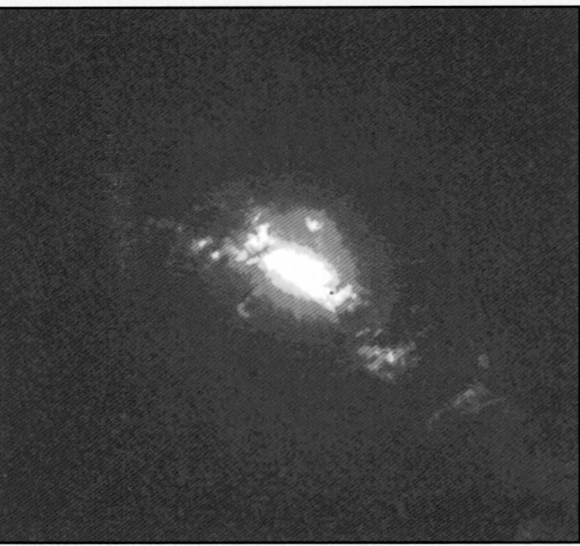

► **Galaxy NGC 4151** *Images by the Chandra X-ray Observatory (top) and the HST (bottom). This active galaxy became an object for amateur observation in the late 1970s.*

Amateurs began to take an interest in observing the brighter active galactic nuclei (AGN) (*see* ACTIVE GALAXY) quite soon after their discovery in the late 1960s, although some had been observed as VARIABLE STARS before then. Leading amateurs have thus been observing the brighter AGN for more than three decades, monitoring changes in their brightness. Many recreational observers have also tracked down these faint specks visually (especially 3C273 in Virgo), and have marveled at them, although less at their appearance than their extraordinary nature.

Coordinated programs for amateurs to monitor AGN are run by AAVSO, the BAA and *The Astronomer* organization. They are monitored as if variable stars.

The potential for serious photographic work was realized in the late 1980s. Photography enables astronomers to add several exposures to reveal AGN that were too faint for most visual observers. But this was a limited success owing to the difficulties of photographic patrolling. It was not until the advent of the CCD and its availability to the amateur that the next boost to AGN observation was given. There is no doubt that professionals are interested in observations by amateurs: several international astronomers have given their support and encouragement.

The broad aims of amateur observations of AGN are:

- to monitor apparent changes in the brightness of AGN;
- to measure such changes and build up as accurate a light curve in visible wavelengths as possible;
- to alert professional astronomers to unusual activity and anticipated activity of interest, and to provide routine observations on a regular basis.

More than 20 AGN have been chosen for the coordinated programs of observations that are available to the well-equipped amateur. Although most are relatively faint, several are accessible to modest telescopes visually, and all should be quite straightforward targets for CCD users. There are still more than a dozen that can be observed by the owner of a 16-in or larger telescope.

Observers are encouraged to make observations as regularly as skies and conditions allow. Almost all AGN on the program are plotted in the star atlas *Uranometria 2000*, although some have alternative designations to their more popular terms. However, the use of coordinates should allow them to be identified easily.

Observing methods

Once located, the AGN's brightness can be estimated or measured visually, photographically or with a CCD. Each method has its merits and drawbacks. The advantages of taking a visual approach are that several objects can be observed during the session; the results are rapidly obtained; and the equipment is relatively inexpensive, although realistically a larger aperture is needed for such faint objects. However, the telescope need not be equatorially mounted. Set against these advantages are the problems of finding relatively faint objects from less than ideal sites, with problematic sky conditions (due to light pollution, moonlight, high cirrus, and so forth). Experience in variable star work is essential. Inevitably, some observer bias occurs, although this can be compensated for, as with other variable star observations. Even if the AGN itself poses no great problem, the fainter comparison stars may well be difficult.

Photography would seem to reduce some of these barriers. Fainter magnitudes can be reached, so more active galaxies become available and the comparisons more easily accessed. Personal bias is reduced and the results are easily displayed and reproduced. However, an equatorial or driven telescope is essential, which increases the financial commitment. Standardization and formal measurement become necessary

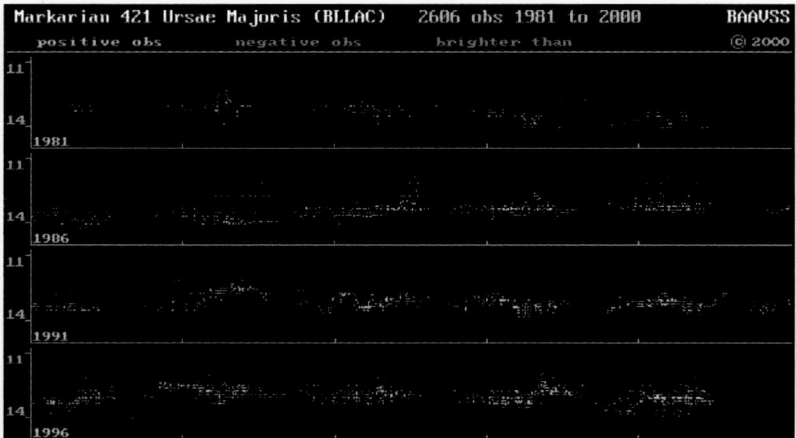

▼ **Markarian 421** *A BL Lacertae object that has been observed by amateurs for more than two decades. The magnitude is shown on the left-hand axis, and each of the panels shows five years of activity. The brightest outbursts, in 1982 and 1992, reached magnitude 12.*

Best 10 target AGN										
Object	**Right ascension**			**Declination**		**Constellation**	**Type**	**Magnitude or range**	***Uranometria 2000* chart no.**
	h	**m**	**s**	**°**	**'**				
3C66A	00	22	39.0	+43	05	Andromeda	BL Lac	14–16.3	62
3C120 (BW TAU)	04	33	00.0	+05	21	Taurus	BL Lac	13.7–14.6	178
OJ287	08	54	48.9	+20	06	Cancer	BL Lac	12.4–16	142
Markarian 421	11	04	27.0	+38	12	Ursa Major	BL Lac	13.6–14	106
NGC 4151	12	10	00.0	+39	24	Canes Venatici	Seyfert	11.1	74
W Comae	12	21	31.6	+28	13	Coma Berenices	BL Lac	11.5–16	148
Markarian 205	12	21	44.0	+75	18	Draco	BL Lac	14.5	9
3C273	12	29	6.8	+02	03	Virgo	Quasar	12.2	238
3C279	12	56	11.0	-05	47	Virgo	Quasar	17.7	284
BL Lacertae	22	02	00.0	+42	16	Lacerta	BL Lac	12.5–15	87

to achieve the accurate results that the time and the level of financial outlay deserve. This requires the observer to use a standard film. (Kodak 103aB was recommended in the past, but it is difficult to obtain and other emulsions need to be tested.) Also, filtered images are desirable, and microdensitometric measures are ideal for accurate estimates of magnitude. Add to this the laborious process of photography and darkroom work, and it is perhaps understandable that the photographic programs of the late 1980s never really came to fruition. The chance of discovering a nova or supernova is often adequate incentive to be involved in patrol work. Monitoring an already known object is perhaps less enticing.

CCD observations

The CCD revolution seems be the answer to this. Images made with CCDs can go very deep, even with modest apertures. Many AGN are available to the owner of a 10-in telescope equipped with CCD. Finding objects with smaller CCD chips was once notoriously difficult, but the 'Go To' facilities of many modern telescopes mean that the field can be found confidently, the AGN identified from charts and imaged quite rapidly, and a measurement made.

Computer software is becoming more sophisticated, and with care it should be possible to carry out photometry accurate to about 0.1 magnitudes. The data can be stored on disk easily and retrieved at leisure. It is possible to make rapid comparisons with a master image to show any significant changes in brightness. This happy state of affairs is tempered by the need to spend a fair amount of money on equipment. As well as a driven telescope, a CCD camera, computer and software are also needed. So it may cost the observer several thousand dollars to achieve the first meaningful CCD image of an AGN.

Not all CCDs have the same sensitivities. Most have a red bias, which distorts the results; also, different cameras have different profiles, so for photometry to be really useful to the professionals it is increasingly clear that filtration is needed. Ultimately it may be acceptable to use just B and V filters, or even one of these. However, filtered images are not needed for simple monitoring without photometry. The observations can still be useful even if no filters are available to the observer. At present the majority of observations are visual, with some CCD imaging – but with more CCD users looking for serious projects, the emphasis may change.

AGN offer a wonderful challenge to the serious amateur observer who wishes to contribute, in his or her small way, to an area of science in which the professionals welcome help, as long as it is provided carefully. While they offer little in the way of spectacular views, AGN feed the imagination in a way that is equally satisfying to the enquiring mind. The techniques are improving, the quality of results will become more refined, and the study of the deepest of deep-sky objects is about to take off.

Nick Hewitt is director of the Deep Sky Section of the British Astronomical Association. He earns his living as a doctor in Northamptonshire, UK.

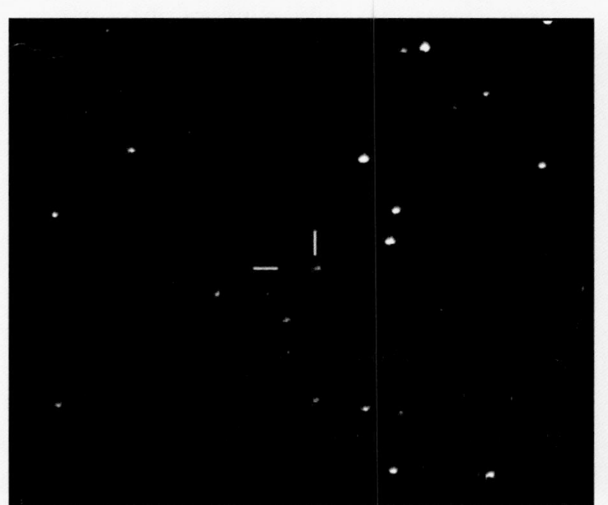

◄ ***OJ287***
(Left) On April 21, 2000, the AGN OJ 287 was faint at magnitude 16. (Right) On December 28, 2000, it had brightened to magnitude 14.5. To see the change in brightness, compare the indicated AGN with the other stars in the chain of three to the right.

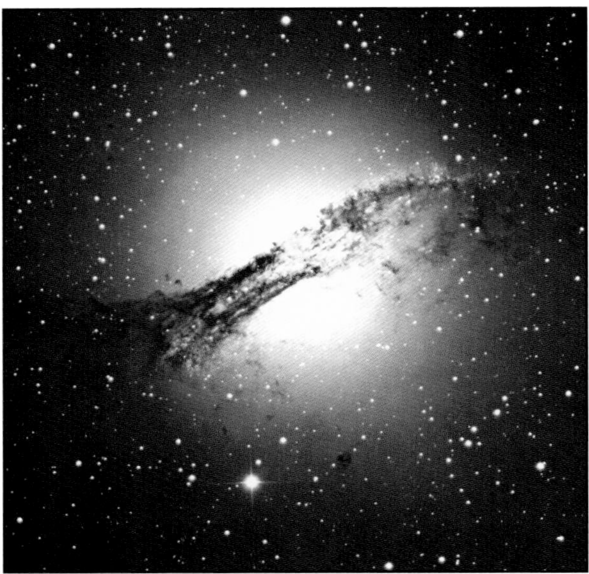

▶ ***Centaurus A*** *(NGC 5128) An elliptical galaxy crossed by a prominent dust lane, which, some astronomers think, is a spiral galaxy in collision with the elliptical, feeding the active galactic nucleus. Centaurus A is a strong source of radio waves, gamma-rays and X-rays, as well as visible and infrared radiation.*

the black hole is not uniform. The X-rays, and friction, heat the accretion disk, producing UV and optical light.

An opaque, rotating torus (doughnut) of dust and thick gas, with a radius of a few parsecs, is aligned with the rotation of the black hole and the accretion disk. The dust in the torus absorbs the X-rays and is warmed. The energy is then re-radiated as infrared emission. The dust may obscure the inner parts of the AGN, depending on the viewing angle. When viewed down the axis of the dust disk, the inner parts are revealed. These include the full blaze of light from the nucleus (as seen in a quasar or blazar), and the rapidly moving material closest to the black hole (Seyfert Type 1 galaxy). However, if we view an active galaxy edge on, the inner parts are obscured from direct view (as in a Seyfert Type 2 galaxy).

The properties of the inner zones can be inferred indirectly even in edge-on galaxies. Light from the inner parts may radiate along the axis of the accretion disk and torus, and encounter galactic material. It may be reflected into our line of sight, giving us a faint view of the inner parts of the AGN. This indirect view is revealed in light that is polarized (*see* POLARIZATION) by the reflection, or by energy from the black hole that is absorbed and re-radiated (reprocessed). Because the X-rays and light radiated from the black hole area right at the nucleus may vary, the reflected or reprocessed light also varies, in sympathy but with a time delay due to the time it takes for the light to travel from the nucleus to the material of the 'mirror.'

In observational campaigns on some particularly active AGN, space satellites have observed X-rays, and ground-based telescopes have observed light, infrared radiation, and spectra. Time delays in reverberation are used to measure the dimensions of the material and its distribution (*see* REVERBERATION MAPPING).

AGN may also have equal and opposite high-speed JETS of material that have been generated from accretion disk material whose infall has been deflected by the radiation from the black hole. The jets shoot out along the axes of the accretion disk. Emission from the jets is strongly beamed by relativistic effects in the direction of the jet's motion. This is what happens in a blazar. The jets and the accretion disk structure contribute to the X-ray emission.

The broad emission lines seen in many AGN, particularly in Seyfert Type 1 galaxies, are produced in dense gas clouds near the black hole. These clouds move at high speeds in the black hole's gravity. The narrow emission lines arise in low-density clouds that are driven outwards by radiation from the central source, and are therefore pushed more gently along the system axis.

Radio emission from AGN

About 10% of AGN are strong radio sources. Some are double radio sources separated by millions of parsecs (megaparsecs), each 'lobe' having a weak central core. In other systems the core is the strongest component. The core-dominated sources may eject patches of material that seem to move across the line of sight. The speed of the patches is apparently faster than the speed of light, but this cannot be. In fact, the patches are moving at a little less than the speed of light, but toward us (*see* SUPERLUMINAL MOTION). The double sources often have one-sided, straight jets that extend from their compact core toward one of the extended lobes. The one-sided nature of the jets can be understood if an approaching jet and a jet receding in the opposite direction are respectively boosted and dimmed by relativistic effects. In core-dominated sources we are looking down the relativistic jet, while in double sources the jet is projected onto the plane of the sky.

Influence of host galaxies

AGN account for only a fraction of all galaxies. It is, however, difficult to identify AGN that have very low luminosities. For example, the emission-line spectra of many galaxies are too weak for Seyfert galaxies, but too strong for normal galaxies. Because they have Low Ionization Nuclear Emission line Regions, they are called LINER galaxies. It is possible that most galaxies are active, to a greater or lesser degree.

To be a really active AGN, the two most obvious requirements are a massive black hole and enough fuel close to the nucleus to 'feed' it. A massive black hole probably forms when a dense star cluster collapses in the center of a galaxy. Massive black holes may be present in many or even most galaxies. The fact that most galaxies are not AGN must then be attributable to a lack of fuel. The black holes in the most powerful quasars require tens of solar masses per year. Gas might be fed to the black hole by the influence of a bar (which causes the gas to have a non-circular motion), or by a disturbance caused by another galaxy passing nearby.

AGN are very compact, and the largest relevant scales for most of the interesting and extreme AGN phenomena are of the order of 10 parsecs or even less. Typical galaxies are too far away for this to be distinguishable. During the next decade, observations of AGN will benefit from new and planned space observatories, and from new ground-based techniques such as INTERFEROMETRY.

Further reading: Peterson B M (1997) *An Introduction to Active Galactic Nuclei* Cambridge University Press.

Active optics

A system that enables the shape and relative positions of the principal optical elements of a telescope to be adjusted to compensate for deformations. Sources of deformation include: flexure of the PRIMARY MIRROR or telescope structure as the orientation of the telescope changes, thermal effects induced by changing temperatures, buffeting by wind, and errors during the manufacture of the mirrors. In active optics the shape of the primary mirror is controlled by a system of actuators that applies forces to the rear of the mirror. The

◀ ***Active region*** *An X-ray color image of the Sun from NASA's TRACE satellite. Hot plasma explodes off the Sun's photosphere and travels along loops defined by the Sun's magnetic field.*

primary mirror in such a system has to be thin relative to its diameter in order to respond effectively to these forces. For example, the primary mirror in each of the four telescopes that comprise the EUROPEAN SOUTHERN OBSERVATORY's VLT has a diameter of 8.2 m, weighs 23 tonnes, and is only 175 mm thick. The shape of each mirror is controlled by 150 actuators. *Contrast* ADAPTIVE OPTICS.

Active region

A localized region of the SUN's surface and atmosphere that displays most or all of the following phenomena: sunspots, faculae, plages, filaments, prominences, flares, and bright condensations in the CORONA. The underlying feature of a typical active region is an area of concentrated magnetic fields, and its overall diameter may be several hundred thousand kilometers. Localized fields of up to 0.4 T occur within sunspots. An active region begins to form when magnetic flux tubes (bundles of magnetic field lines) erupt through the PHOTOSPHERE from below to create loop structures. Typically, an active region will grow to its maximum extent in about ten days and may persist for several months, declining slowly as its underlying magnetic fields dissipate. Sunspots within an active region usually disappear within two to four weeks. *See also* SUNSPOT, FLARE AND ACTIVE REGION.

Adams, John Couch (1819–92)

Born in Cornwall, England, Adams became a fellow and mathematical tutor at Cambridge University. He developed a procedure for numerically integrating differential equations and, inspired by Mary SOMERVILLE, deduced mathematically the existence and location of the planet NEPTUNE from its perturbations on the planet URANUS. Neptune was discovered by Johann GALLE in Berlin in 1846 using Urbain LEVERRIER's independently computed position. Soon afterwards Adams claimed priority. It transpired that, with a letter of introduction from James CHALLIS in Cambridge, he had applied to George AIRY, the ASTRONOMER ROYAL, for assistance. Unfortunately, Adams failed to secure an interview with Airy, and nothing further happened. Challis himself had begun a somewhat half-hearted search for Neptune based on Adams' calculations, but he was overtaken by Galle. Adams was appointed professor of astronomy at Cambridge in 1858, and was director of the Cambridge Observatory from 1861.

Adams, Walter Sydney (1876–1956)

Born in Antioch, Syria, to American missionary parents, he worked under George HALE at YERKES OBSERVATORY at the University of Chicago. Adams accompanied Hale to California to set up the MOUNT WILSON OBSERVATORY, and became its director when Hale retired. Adams helped design the 200-in telescope for the PALOMAR OBSERVATORY. His method of SPECTROSCOPIC PARALLAXES, a technique using spectra to give indications of stellar luminosities, made it possible to measure the distances to more distant stars than those whose PARALLAX could be measured by the trigonometric method.

Adaptive optics

A system that compensates for distortions in incoming light waves as they pass through the Earth's atmosphere. Light waves from a distant source, such as a star, arrive at the top

► ***Airglow*** *This hand-held 35-mm photograph was taken from the Space Shuttle Endeavor. The thin, greenish band stretching along the Earth's horizon is airglow: light emitted from the atmosphere from a layer about 30 km thick and 100 km in altitude. Brighter patches of green superimposed on the airglow are the aurora borealis.*

of the atmosphere as straight parallel wavefronts, rather like waves advancing across the surface of the ocean. These are then distorted by the atmosphere and arrive as 'corrugated' wavefronts at the PRIMARY MIRROR of a telescope. Because all parts of such a wavefront cannot be focused at the same point, the image quality is degraded. This severely limits the resolutions that can be achieved by large, ground-based telescopes. By monitoring a suitable bright star, the adaptive optics technique senses and counteracts these wavefront distortions. If no natural star is available, an artificial 'star' may be produced by a powerful sodium laser beam. This stimulates the emission of light by sodium atoms in a layer that exists at an altitude of about 100 km, and generates a star-like point of light that may be positioned conveniently close to the object that is being studied by the telescope.

A small deformable mirror is located behind the FOCUS of the primary at or near the image of the pupil. Because the distortions induced by the atmosphere are continually changing, the system has to sense and respond to them as rapidly as possible, some tens or even hundreds of times per second. The field of view over which the image can be corrected is very small.

In principle, adaptive optics systems should allow telescopes to achieve a resolution (*see* RESOLVING POWER) approaching the theoretical limits. A system of this kind has been in operation at near-infrared wavelengths on the ESO's 3.6-m New Technology Telescope since 1992. Adaptive optics systems are now being developed on all of the world's largest telescopes (8 m and larger), and are expected to enable them to achieve theoretical (DIFFRACTION-limited) resolution at near-infrared wavelengths.

Not to be confused with ACTIVE OPTICS.

► **Air Force Maui Optical Station** *The sturdy, squat 3.6-m Advanced Electro-Optical System telescope is built to track and image fast-moving artificial satellites, to identify what they are and what their purpose is.*

Advanced Satellite for Cosmology and Astrophysics *See* ASCA

Aerolite *See* STONY METEORITE

Aerosol

A solid particle or liquid drop, free-floating in a gas. Describes the non-gaseous components of a planetary ATMOSPHERE, such as dust and clouds of ice crystals or water drops.

Air Force Maui Optical Station

Located on the island of Maui, Hawaii, this US Air Force station is used to track and image Earth-orbiting satellites and astronomical objects such as ASTEROIDS. Civilian astronomers have limited access to the 3.6-m Advanced Electro-Optical System telescope.

Airglow

The non-thermal radiation emitted by the Earth's atmosphere. In its simplest form, it arises from the de-excitation of

MOLECULES, ATOMS OR IONS excited by solar ultraviolet (UV) PHOTONS or energetic particles. The resulting emitted photons are mostly located in the visible and UV ranges of the spectrum. By contrast, the thermal emission of the Earth, heated by sunlight and subsequently radiating to space, takes place in the infrared portion of the spectrum, at wavelengths longer than 7 μm. Spectroscopic techniques show the presence of the green line of atomic oxygen at 558 nm, the red oxygen lines at 630–636 nm, as well as the sodium D line.

Air mass

The length of the sight line to a star through the atmosphere, in units such that the air mass to a star at the ZENITH is unity. In a horizontally layered atmosphere (that is, over an area not much affected by the curvature of the Earth), the air mass is given by the formula $AM = \sec(z)$, where z is the angle of the sight line from the zenith (zenith angle).

Airy, George Biddell (1801–92)

A brilliant mathematician, Airy became the seventh ASTRONOMER ROYAL at Greenwich, London, in 1835 after a brief period as Lucasian Professor at Cambridge University. His output was prodigious, and he published nearly 400 scientific papers and 150 reports on various issues, including the gauge of railways, spectacles to correct astigmatism and methods to correct for compass readings in ships made of iron. His work on optics is recognized by the use of the term AIRY DISK for the resolution element of a telescope due to DIFFRACTION at its APERTURE, which he studied. As Astronomer Royal he saw that the ROYAL OBSERVATORY AT GREENWICH was re-equipped with modern instruments, and that its work was carried out punctiliously by its many human 'calculators' and observers.

He himself determined the density of the Earth, the mass of the planet Jupiter and its rotation period. He calculated the orbits of comets and cataloged stars. He made many contributions to the prediction of the motion of the Moon, and he analyzed transits of the planet Venus and eclipses. As a result of the accuracy of the observations made under Airy at Greenwich, and his practical exploitation of the railway telegraph to distribute it, Greenwich Mean Time was established in 1880 as the official time service throughout Britain, and in 1884 it became the basis for timekeeping throughout the world.

Airy disk

The bright spot at the center of the DIFFRACTION pattern that is formed when a point source, such as a star, is imaged at the FOCUS of a telescope. When the light from a distant star is brought to a focus, INTERFERENCE effects between different parts of each wavefront result in the formation of a diffraction pattern. For a point source, the resulting image (assuming a perfect optical system) consists of a central spot of light surrounded by a series of light and dark fringes, or rings. According to theory, 84% of the light energy is concentrated into the central spot (the Airy disk), the diameter of which depends on the APERTURE of the telescope and the wavelength of the light. The Airy disk is named for George AIRY.

Akebono

A Japanese satellite launched in 1989 to study the Earth's AURORAE. Named from the Japanese for the 'rising Sun.' Auroral image capability was lost in early 1995, but other instruments continue to operate. Also known as EXOS-D.

al-Battani, Abu Abdullah (c.868 – c.929)

Born in Harran in present-day Syria, al-Battani (also known as Albategnius) is best known for determining the solar year as being 365 days, 5 hours, 46 minutes and 24 seconds – an extremely accurate value that was used in the Gregorian CALENDAR reform of the Julian Calendar. Al-Battani also determined the true and mean orbit of the Sun, proving the variation of the apparent angular diameter of the Sun (an indication of the variable distance between Sun and Earth).

Albedo

A measure of the reflectivity of a material or object. For bodies in the solar system, albedo is the proportion of sunlight falling on them that is reflected away. It is measured on a scale from 0 (a perfectly absorbing black surface) to 1 (a perfectly reflecting white surface).

Of the major planets, Mercury has the lowest geometrical albedo at 0.11 (comparable to the Moon's 0.12), while Venus, by virtue of its blanket of highly reflective clouds, has the highest at 0.65. The highest measured albedo is that of Saturn's satellite Enceladus, to which some sources assign a value of 1. The particles that make up the rings of Neptune appear to have the lowest albedo in the solar system, with a value probably close to the theoretical lower limit of 0.

Albireo

The star Beta Cygni, which is said to represent the eye of the swan in the ancient constellation figure. With the unaided eye it is seen as a single star, but with good binoculars or a small telescope it appears as a beautiful double, comprising an orange star ($Beta^1$ Cyg) with a bluish-white companion ($Beta^2$ Cyg).

Alcor *See* MIZAR AND ALCOR

Aldebaran

The star Alpha Tauri. Aldebaran appears within the OPEN CLUSTER the HYADES although it is not actually a member of the cluster, which is much more distant. The name derives from the Arabic *al Dabaran*, 'the follower' – presumably of the PLEIADES. Aldebaran was one of the four 'royal stars' or 'guardians of the sky' of the Persian astronomers/astrologers *c.*3000 BC. Although not the brightest stars, they were carefully chosen, apparently to mark the seasons, as they are approximately 6 h apart in RIGHT ASCENSION. Aldebaran (ancient Persian name Tascheter) was prominent in the evening sky in March and was associated with the vernal EQUINOX. The other stars were REGULUS (summer SOLSTICE), ANTARES (autumnal equinox) and Fomalhaut (winter solstice).

Alfonso X (1221–84)

Also known as Alfonso X of Castile, Alfonso the Astronomer and Alfonso the Wise. Born in Burgos, Spain, he was the king of León and Castile (from 1252). Among the works he commissioned was a table of planetary positions, known as the *Alfonsine Tables*.

Alfvén, Hannes Olof Gösta (1908–95)

A Swedish physicist and Nobel prizewinner (1970), he developed plasma physics and applied the theory to a range of physical and astrophysical phenomena, including charged particle beams in accelerators, and interplanetary and magnetospheric physics. He founded magnetohydrodynamics, the branch of physics that shows how magnetic and electrical phenomena act on gases that are hot and in motion. In 1937 he postulated the galactic magnetic field as the 'bottle' that contained cosmic rays, and in 1950 he identified non-thermal radiation from astronomical sources as SYNCHROTRON RADIATION, which is produced by fast-moving electrons in the presence of magnetic fields. This is the theoretical basis for most of RADIO ASTRONOMY. Alfvén held positions in Uppsala and Stockholm, Sweden and California.

Alfvén waves, named for him, are electromagnetic waves that propagate through a highly conducting medium, such as the ionized gas of the Sun.

Algol

Algol, also known as Beta Persei, is the prototype of a class of ECLIPSING BINARY STARS that undergo dramatic brightness changes during the eclipses. Like other Algols, Algol itself consists of a hot, unevolved B star primary and a cool, evolved K SUBGIANT secondary. When the hot primary is eclipsed by the larger, cool secondary, the overall brightness of the system drops for a few hours from its out-of-eclipse brightness of 2.1 magnitudes to about 3.4 magnitudes. These eclipses, repeating every 2.87 days, are easily seen with the naked eye, and it is obvious that the ancients observed them given that Algol is invariably associated with something evil or demonic. The name Algol comes from the Arabic *Ra's al Ghul,* meaning 'the head of the demon.' The Greeks saw Algol as the blinking eye of Medusa, whose severed head was held aloft by the hero Perseus.

According to stellar evolution theory, the more massive star should evolve more rapidly, using up its fuel more quickly than its lighter companion. This gave rise to the Algol Paradox. The explanation is that, in this close binary system, the cooler star was once the more massive star. As it expanded to become a RED GIANT, matter flowed onto the companion reversing the ratio of the masses.

A third component in the system is Algol C, a late A or early F star orbiting AB with a period of 1.86 years.

Algonquin Radio Observatory

A Canadian RADIO ASTRONOMY observatory in Ontario. The main instrument is a 46-m fully steerable dish that began operation in May 1966. Today it is mainly used in international VLBI programs to maintain the International Celestial Reference Frame (*see* ASTROMETRY).

Allegheny Observatory

At the Allegheny Observatory, University of Pittsburgh, USA, the main telescope is the Thaw 30-in REFRACTOR, completed in 1912. In 1985 the original lens was replaced by one optimized for working at the red end of the spectrum. With its long FOCAL LENGTH (*f*/18), the telescope is a major contributor of accurate PARALLAXES, stellar masses and the astrometric search for extrasolar planets (*see* EXOPLANET AND BROWN DWARF). Founded in 1869, the observatory established the standard US time service, often called Rail-Road Time.

Allende meteorite

The largest known fall of a CARBONACEOUS CHONDRITE. On February 8, 1969, an exceptionally brilliant fireball was seen in the skies of northern Mexico, followed by a series of sonic booms. The next day a METEORITE was found in the village of Pueblito de Allende. This was the first of hundreds of fragments, totaling 2 tonnes, subsequently collected from the surrounding area. It has been estimated that the parent body weighed as much as 30 tonnes.

Almagest

An astronomical work compiled by PTOLEMY of Alexandria in about AD 140. The purpose of the *Almagest* was a comprehensive empirical and mathematical deduction of models to explain the motions and phenomena of the Sun, Moon, the five known planets and the fixed stars. The work laid the methodological foundation for Islamic and European mathematical astronomy of the Middle Ages and Renaissance.

Almanac

An annual publication that tabulates daily positional and related data for the Sun, Moon, planets and other celestial bodies, and lists events and phenomena such as lunar phases, times of sunrise and sunset, and eclipses. Almanacs of this kind are used primarily by astronomers and navigators. Since 1981, one of the most important almanacs for astronomers, the *Astronomical Almanac*, has been prepared jointly by the Nautical Almanac Office at the US Naval Observatory and HM Nautical Almanac Office in the UK. This unification brought together two series that had been published separately in the UK (since 1767) and in the USA (since 1855).

Alpha Centauri A, B and C

The star Alpha Centauri is also known (especially to navigators) as Rigil Kentaurus or Rigil Kent. It is a BINARY STAR, the components being a yellow G2V star, Alpha Cen A, of magnitude –0.01; and an orange K1V star, Alpha Cen B, of magnitude 1.35. Their separation is currently 19.1 arcsec; the period is 81.2 years. There is also a much more distant third member of the multiple star system, PROXIMA CENTAURI.

Alpha particle

A term that is sometimes used to describe a HELIUM nucleus – a positively charged particle that consists of two PROTONS and two NEUTRONS, bound together. Alpha particles, which were discovered in 1898 by Ernest Rutherford (1871–1937), are emitted by atomic nuclei that are undergoing alpha radioactivity. During this process, an unstable heavy nucleus spontaneously emits an alpha particle and transmutes (changes) into a nucleus of a different (lighter) element. Studies of the way in which alpha particles were scattered by atoms in thin sheets of mica and gold led Rutherford, in 1910, to the discovery of the atomic nucleus.

Alpha particles are found in the SOLAR WIND and are produced in collision events in high-energy particle accelerators. An important nuclear reaction, called the triple-alpha reaction, that welds helium nuclei (alpha particles) together to form carbon nuclei, occurs in RED GIANT STARS.

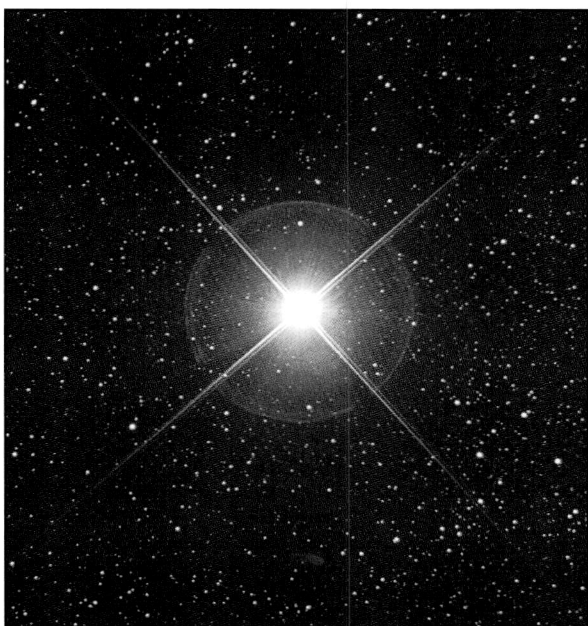

◀ ***Alpha Centauri*** *The image of the star is large because the star's bright light has scattered in the emulsion. The bright cross, called diffraction spikes, is caused by the support structure that holds the telescope's secondary mirror in the light beam.*

al-Sufi, Abu al-Rahman (903–86)

A Persian astronomer who published works on astronomical instruments. He revised PTOLEMY's star catalog as the *Book of the Constellations of the Fixed Stars*, which included improved magnitudes, stars named by reference to constellation figures (much copied, for example, by ALFONSO X), and nebulae (*see* NEBULAE AND INTERSTELLAR MATTER), including the ANDROMEDA GALAXY (M31).

Altair

The star Alpha Aquilae, a white star of apparent magnitude 0.76, spectral type A7IV-V (*see* STELLAR SPECTRUM: CLASSIFICATION). Believed to be about 1.3 times the diameter of the Sun, Altair has an exceptionally high rate of axial spin, rotating in a period of about 6.5 hours. As a result, the diameter at its equator is at least 14% greater than at its poles.

Altazimuth mounting

A telescope mounting that has axes in the horizontal and vertical planes to enable the telescope to be swiveled in ALTITUDE (perpendicular to the horizon) and AZIMUTH (parallel to the horizon). This type of mounting is simpler, cheaper to construct, and less prone to flexure and changes of balance than the EQUATORIAL MOUNTING. In recent years, altazimuth mountings in which the drive rates for the two axes are computer controlled have become the only way large optical telescopes over, say, 2-m aperture, are constructed. Altazimuth mountings are also widely used for large radio dishes.

Altitude

The angle between the HORIZON and a star, measured in a direction perpendicular to the plane of the horizon. Together with a value of AZIMUTH, this specifies the position in the sky of a celestial body at a particular instant, as seen by an observer at a particular point on the Earth's surface. Due to the rotation of the Earth, the altitude of a star continually changes. For example, an observer in the northern hemisphere will see a star rise in the east, reach its maximum altitude when due south (culmination), and set in the west. Observers located at the north or south poles, however, will see stars move parallel to the horizon, so maintaining constant altitude.

Altitude also refers to the height of an object (for example, an artificial satellite or atmospheric feature) above mean sea level.

Alvarez, Luis Walter (1911–88)

A physicist and astronomer, born in San Francisco, USA. He was a professor at the University of California, and won the Nobel prize in 1968 for his discoveries in particle physics. Alvarez and his son discovered globally distributed iridium at the Cretaceous/Tertiary boundary in the Earth's geological record. From this, he proposed that the change of climate and the disappearance of the dinosaurs at that time (some 65 million years ago), resulted from a cataclysmic collision of an asteroid with the Earth, asteroids being relatively rich in iridium. The crater of the collision has since been identified as the CHICXULUB BASIN.

Ambartsumian, Viktor Amazaspovich (1908–96)

An astrophysicist, born in Armenia, who became professor at the University of Leningrad and at Erevan, and founded the BYURAKAN ASTROPHYSICAL OBSERVATORY. He applied an original and creative scientific imagination to a range of observational phenomena in astronomy. He suggested that T TAURI STARS are very young, and that loose stellar associations are dissociating. In other words, they are too sparsely populated to be gravitationally bound and may be moving away from a common origin. He made the connection between the various energetic phenomena in galaxies (for example, jets, high-speed motions and bright ultraviolet luminosities) as pointing to energetic sources at the centers of galaxies, now identified as massive black holes.

AM CVn star

Ultrashort period variable star (periods 5–65 min), probably a binary star, named for the prototype. The light from each of these rare stars (12 known as of July 2003) comes from an accretion disk around a WHITE DWARF star, with the donor star too faint to be seen. The accretion disk's spectrum is helium with no hydrogen, so the donor star is presumably also of this composition. If so, it is also a white dwarf star

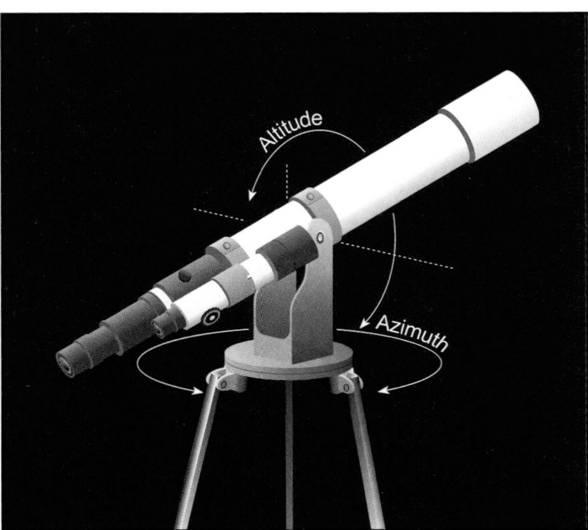

◀ ***Altazimuth mounting*** *Enables this refracting telescope to rotate around vertical and horizontal axes.*

originating in some way that interrupted its development to a composition of carbon and oxygen. One member of this class is the X-ray star RX J0806.2+1527, with period 321.5 s, the shortest binary star period known. Close binary stars like this are expected to be sources of gravitational waves (*see* GRAVITATION, GRAVITATIONAL LENSING AND GRAVITATIONAL WAVES) and to merge in the future as SUPERNOVAE of Type Ia. *See also* VARIABLE STAR: CLOSE BINARY STAR.

American Association of Variable Star Observers (AAVSO)

The largest organization of variable star observers in the world, with over 1000 members in more than 40 countries. Its database contains over 9 million observations of variable stars. The organization's headquarters are in Cambridge, Massachusetts. The association was founded in 1911 by amateur observer William Tyler Olcott (1873–1936). Inspired by a talk at the 1909 meeting of the American Association for the Advancement of Science, Olcott began to send his observations of variations in stellar brightness to Edward PICKERING at Harvard College Observatory.

American Astronomical Society (AAS)

Founded in 1899, the American Astronomical Society is a non-profit scientific association created to promote the advancement of astronomy and closely related branches of science. Its membership consists primarily of professional researchers in the astronomical sciences, but also includes educators, students and others interested in the advancement of astronomical research. About 85% of members are from North America. The AAS operates in five major areas: publications, meetings, education, employment and public policy. Its headquarters are in Washington, DC.

The society's research journals, *The Astrophysical Journal* and *The Astronomical Journal*, appear in both print and electronic formats. *The Bulletin of the American Astronomical Society* reports the latest institutional developments, and documents the content of the meetings held by the society and its five divisions. AAS meetings, held twice a year, provide a forum for the presentation of scientific results.

AM Herculis

AM Herculis (AM Her) is the prototype and brightest of a class of short-period CATACLYSMIC BINARIES, in which a strongly magnetic WHITE DWARF accretes gas from a companion. The magnetic field is sufficiently strong to affect qualitatively the binary's most fundamental characteristics. It prevents the formation of an ACCRETION disk and forces mass transfer to occur entirely through one or more gas streams. The flow is channeled near the white dwarf into a near-radial funnel that liberates its accretion energy in a shock above the magnetic pole. The rotation of the white dwarf becomes synchronized to the orbital motion of the companion, such that all basic phenomena are orchestrated on a common period.

Because the gas arrives at the white dwarf surface under virtual free-fall, the post-shock plasma temperature is high and AM Her-type binaries are strong X-ray emitters. Most of the approximately 50 known systems have been discovered in X-ray surveys. *See also* VARIABLE STAR: CLOSE BINARY STAR.

Analemma

The figure-eight shape traced out by the position of the Sun at the same time each day (usually midday) over the course of a year from the same location. An analemma is an illustration of the seasonal changes caused by the tilt of the Earth's axis, and of the EQUATION OF TIME caused by the Earth's elliptical orbit around the Sun.

Andromeda *See* CONSTELLATION

Andromeda Galaxy (Andromeda Nebula)

The Andromeda Galaxy is the closest SPIRAL GALAXY to the Milky Way, just visible to the naked eye on a dark night as

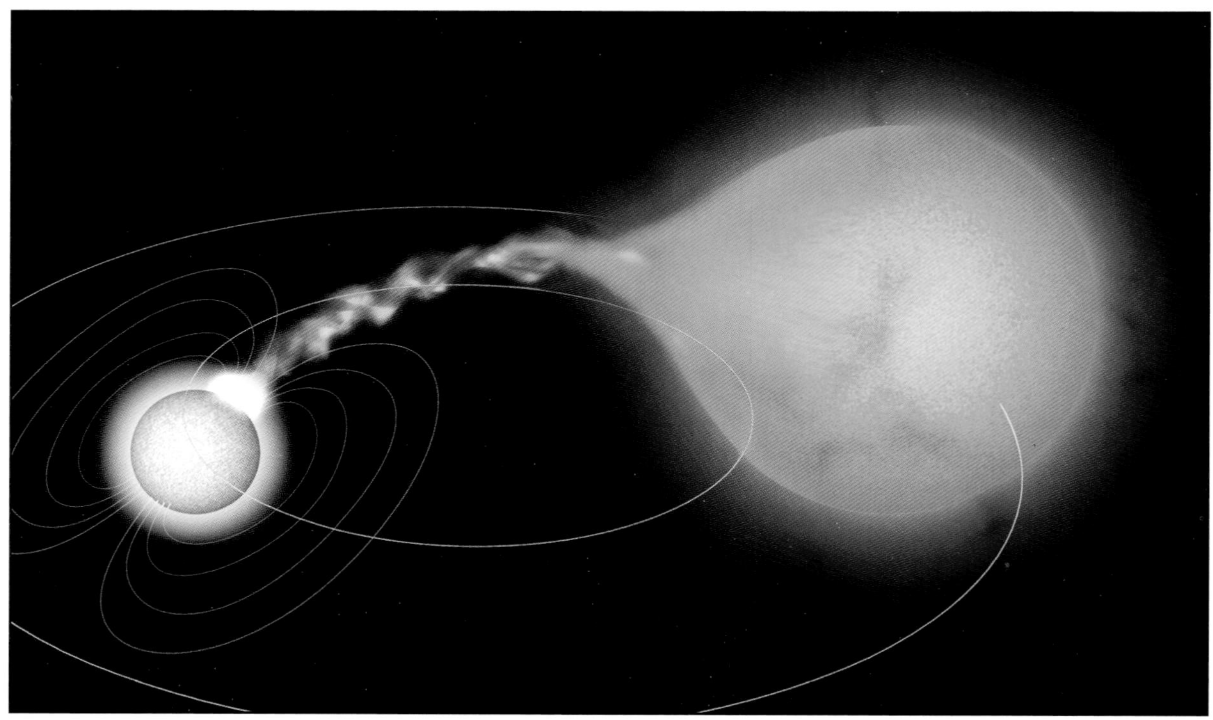

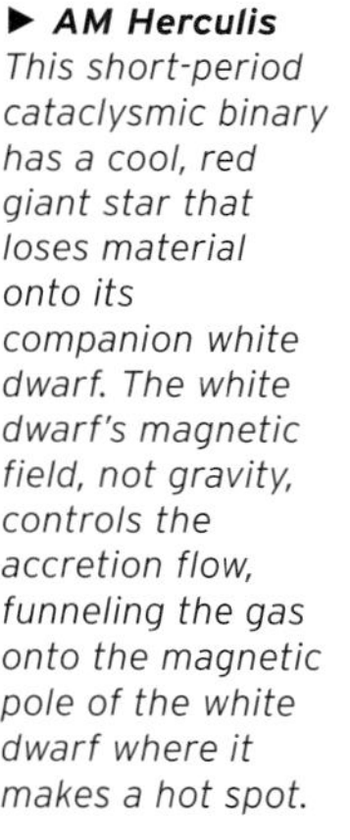

▶ ***AM Herculis*** *This short-period cataclysmic binary has a cool, red giant star that loses material onto its companion white dwarf. The white dwarf's magnetic field, not gravity, controls the accretion flow, funneling the gas onto the magnetic pole of the white dwarf where it makes a hot spot.*

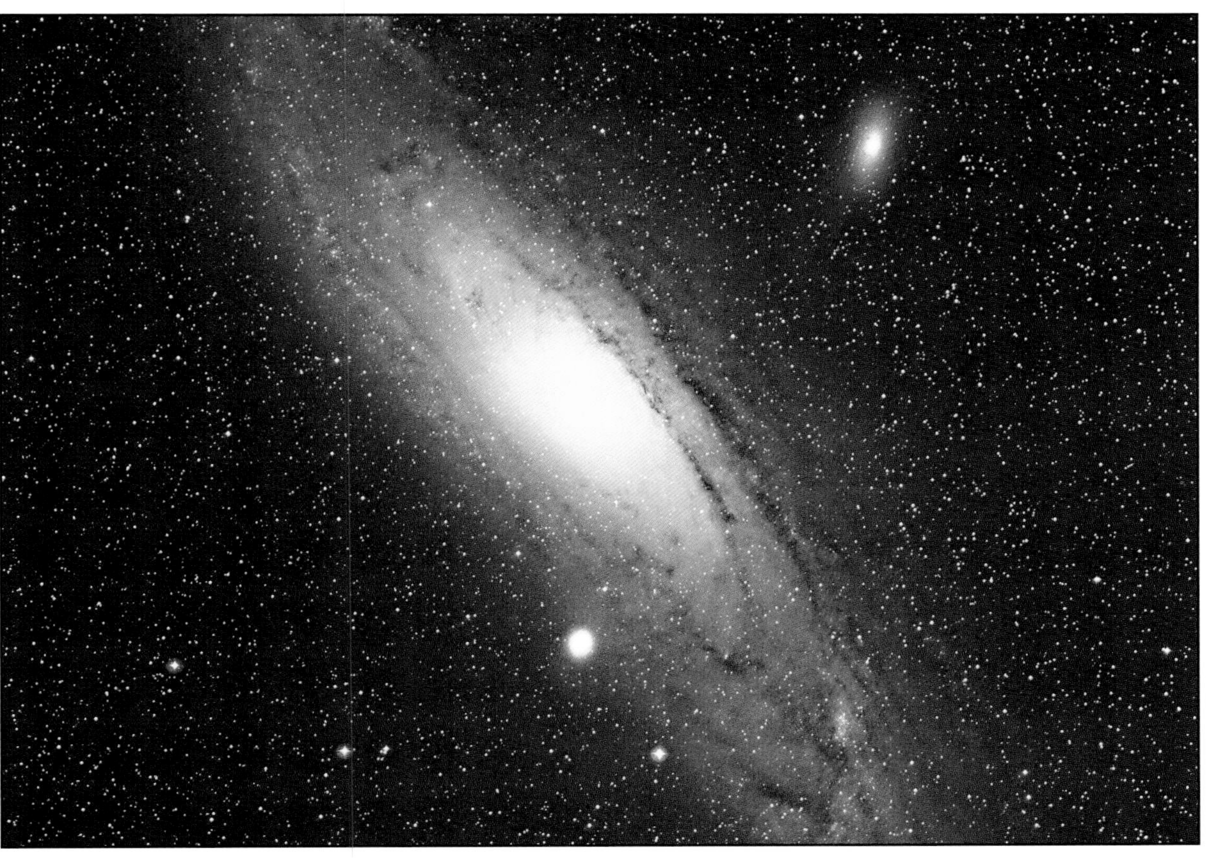

◀ ***Andromeda Galaxy*** *(M31) A spiral galaxy tilted at an angle, with the lower left-hand side of the galaxy furthest from us. The image includes the companion galaxies M32 at lower center and M110 at upper right.*

a faint smudge of light in the constellation Andromeda. The earliest records of the Andromeda Nebula, as it is still often referred to, date back to AD 964, to the *Book of the Constellations of the Fixed Stars* published by the Persian astronomer AL-SUFI. The object is number 31 in the famous MESSIER CATALOG, and is therefore often referred to as M31. Another alias is NGC 224, since the galaxy is number 224 in the NEW GENERAL CATALOGUE.

Modern measurements based on CEPHEID VARIABLES place M31 at a distance of 740 kiloparsecs or about 2.4 million l.y. It is the dominant member of the Local Group, and as such it fills an important role in studies of galaxy structure, evolution and dynamics, stellar populations, star formation and interstellar medium. Crucial historical developments include Edwin HUBBLE's work in the 1920s on the distance to M31, which proved that galaxies outside our own Milky Way exist, and Wilhelm BAADE's work in the 1940s on stellar populations, which led to the concepts of old (population II) and young (population I) stars.

Anglo-Australian Observatory (AAO)

The Anglo-Australian Observatory operates two optical telescopes at Siding Spring Mountain: the 3.9-m Anglo-Australian Telescope and the 1.2-m UK SCHMIDT TELESCOPE. These are located beside the spectacular Warrumbungle National Park outside Coonabarabran in New South Wales, 450 km northwest of Sydney.

The UK Schmidt Telescope is a special-purpose survey telescope. Its initial task was the first detailed photographic survey in blue light of the southern skies. Other major projects have since been undertaken, and more are in progress.

The AAO has pioneered the use of optical fibers in astronomy. The so-called 2dF (2-degree-field) facility uses optical fibers to enable 400 objects to be observed at once.

David Malin (1941–), an astronomer at the AAO, has developed techniques to make astronomical color photographs from plates taken in three separate colors. These beautiful images have earned recognition as among the finest in the world.

Angstrom

A unit of length frequently used to describe the wavelength of light. It is named after the Swedish physicist Anders Jonas Ångström (1814–74), a founder of SPECTROSCOPY. It is equal to 10^{-10} meters (that is, one ten-thousand-millionth of a meter) and is denoted by the symbol Å. Most of the visible spectrum lies in the wavelength range 3900–7500 Å (390–750 nm).

Angular momentum

A property of rotating bodies or systems of masses that depends on the distribution and velocities of the masses about the axis of rotation. The angular momentum of a single particle of mass m, moving in a circular orbit of radius R, at velocity v, is given by mvR. Angular momentum is a conserved quantity. In other words, the total angular momentum of a system is constant. Thus if a large, slowly rotating gas cloud contracts, its rotational velocity must increase to conserve angular momentum. As it contracts, therefore, it spins more rapidly. Appropriate SI units for angular momentum are kilogram meters squared per second (kg m^2 s^{-1}). The conservation of angular momentum is of central importance to such questions as star formation, and the origin of planetary systems.

Annual parallax

See PARALLAX

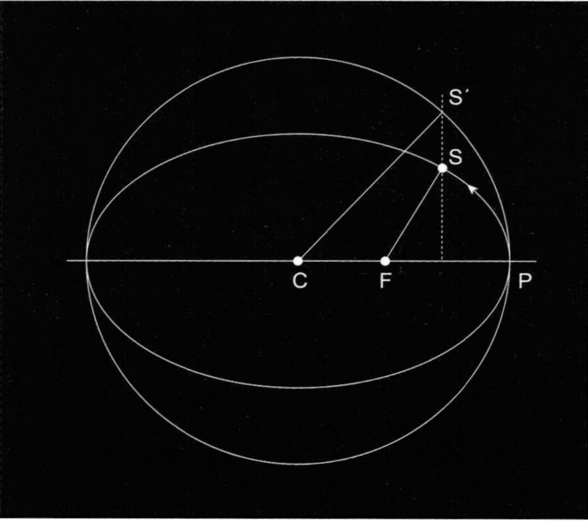

▶ ***Anomaly*** *S is a body (for example, a planet) in orbit around another at the focus, F, of the orbital ellipse. The angle PFS is called the 'true anomaly.' The angle PCS' is the 'eccentric anomaly.'*

Annular eclipse *See* ECLIPSE

Anomaly

For a particle moving in an orbit that is a CONIC SECTION, the true anomaly is the angle between the point of closest approach to a FOCUS (the pericenter, P), the focus, F, and the position of the particle, S, measured in the direction of motion of the particle. In the case of a planet moving around the Sun in an elliptical orbit, its true anomaly at any particular instant is the angle between its PERIHELION position, the Sun (which lies at one focus of the ELLIPSE), and the position of the planet at that instant.

The mean anomaly is the angle between P, F, and the position of a hypothetical particle moving at the mean velocity of a particle with the same orbital period. The eccentric anomaly is the angle between P, the center of the circumscribing circle of the ellipse, C, and the projection of the position of S onto that circle.

For an ellipse, mean anomaly equals true anomaly at opposite ends of the major axis. For a circular orbit, mean and true anomaly are identical at all times. The concept of mean anomaly is useful for calculating the time taken for a body to move along a segment of its orbit.

Antapex

The point on the CELESTIAL SPHERE directly opposite the solar APEX (the point toward which the Sun is moving). The antapex is located in the CONSTELLATION of Columba, at about right ascension 6h and declination –30°.

Antares

The star Alpha Scorpii, which is said to represent the heart of the scorpion. The name derives from 'rival to MARS,' that planet being of similar brightness and color at favorable oppositions. It is a striking red supergiant star, of spectral type M1Ib and apparent magnitude 1.06. Antares is a slow, irregular, pulsating variable, its magnitude varying by ±0.14. There is a binary companion of magnitude 5.3 at separation 2.9 arcsec; the binary period is estimated to be of the order of 1000 years.

Anthropic principle

Early humans started with the anthropocentric view that we are located at the center of the universe and are the focus of the laws of nature. Then Nicolaus COPERNICUS revealed that the Earth moves around the Sun, and Isaac NEWTON's mechanistic view of the world conceived the universe as a giant machine, oblivious to the presence of humanity. However, in the last 40 years or so an anthropic view has developed. This says that, in some respects, the universe has to be the way it is because otherwise it could not produce life and we would not be here speculating about it. Although the term 'anthropic' derives from the Greek word for 'man,' this is a misnomer since most of the arguments pertain to life in general rather than humans in particular.

Why is the universe as big as it is?

The mechanistic answer is that, at any particular time, the size of the observable universe is the distance traveled by light since the BIG BANG. Since the universe's present age is about 10 000 million years, its present size is about 10 000 million l.y. But there is no compelling reason for this; it just happens to be that old.

Robert DICKE first gave another answer to this question in 1961. In order for life to exist, carbon must exist, or at least some form of chemistry. Carbon is produced inside stars and this takes about 10 000 million years. After this time the star can explode as a SUPERNOVA, scattering the newly baked elements throughout space, where they may eventually become part of life-evolving planets. But the universe cannot be much older than 10 000 million years, otherwise all the material would have been processed into stellar remnants such as WHITE DWARFS, neutron stars (*see* NEUTRON STAR AND PULSAR) and BLACK HOLES. This suggests that life can only exist when the universe is aged about 10 000 million years.

This startling conclusion reverses the mechanistic answer. The hugeness of the universe, which seems at first to point to human insignificance, is actually a consequence of our existence. This is not to say that the universe itself could not exist with a different size, only that we would not be aware of it if its size were different.

Triple-alpha

The most sensitive example of the anthropic principle is associated with the 'triple-alpha' reaction. A star makes carbon by first combining two ALPHA PARTICLES to make a beryllium nucleus, and then adding a third alpha particle to form a carbon nucleus. Beryllium is unstable and it used to be thought that it would decay before the third alpha particle could combine with it. In this case it was difficult to understand why there was any carbon in the universe at all. Fred HOYLE realized that there must be a resonance (an enhanced interaction rate) in the second step which helps the carbon to form before the beryllium disappears; that is, the carbon nucleus must have a state with energy just above the sum of the energies of beryllium and helium. There is, however, no similar favorably placed resonance in oxygen, otherwise almost all of the carbon would be transmuted into oxygen. The resonance was rapidly found in laboratory experiments, so this might be regarded as the first confirmed anthropic prediction. Indeed, the coincidence is so precise that Hoyle concluded that the universe has to be a 'put-up job.' There are other coincidences with similar results.

Many universes?

Do anthropic arguments come under the heading of physics or metaphysics? Anthropic coincidences may reflect the existence of a 'beneficent being' who created the universe

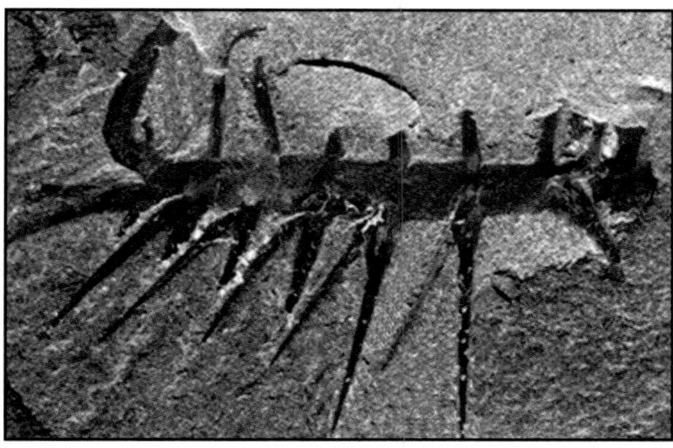

◀ ***Anthropic principle*** *Hallucigenia is a creature from the Burgess Shale fossil site, in the Canadian Rocky Mountains, whose biological form is now extinct. Human beings and Hallucigenia are both carbon-based, and the anthropic principle infers from this that the universe is about 10 000 million years old.*

with the specific intention of producing us. Such an interpretation is logically possible and appeals to theologians, but not to most physicists.

From a physical point of view, the anthropic 'explanation' of the various scientific coincidences can be criticized on a number of grounds. First, the anthropic arguments are mainly *post hoc* – the triple-alpha resonance is the only successful prediction based on the anthropic principle. Second, the anthropic argument is unduly anthropocentric in that it assumes conditions that are specifically associated with human-type life (for example, carbon chemistry). Third, the anthropic principle does not predict exact values for the constants, only order-of-magnitude relationships between them, so it is not a complete explanation.

The last two objections might be met with the possibility that there are many universes (the multiverse), all with different randomly distributed constants of physics. In this case, we necessarily reside in one of the small proportion of universes that satisfy the anthropic constraints, with scientific constants that have the values required for life of some sort, including ours.

Invoking lots of extra universes might seem rather uneconomical, but there are several physical contexts in which this makes sense, especially the theory of the INFLATIONARY UNIVERSE, which explains the origin of the Big Bang. This proposes that at very early times quantum fluctuations cause tiny regions to expand rapidly. Each region becomes a 'bubble,' within one of which our entire visible universe is contained. In principle, as stressed by Andrei Linde (1948–), there could be different values for the constants within each bubble. In only a small fraction of them would life develop.

Further reading: Barrow J D and Tipler F J (1986) *The Anthropic Cosmological Principle* Oxford University Press.

Antimatter

Matter composed of ELEMENTARY PARTICLES that have the same masses as their ordinary matter equivalents but which have opposite, or mirror-image, values of other quantities, such as charge. For example, electrons and positrons (anti-electrons) have the same mass but equal and opposite charge. Antiparticles are found naturally among COSMIC RAYS and can be produced in high-energy particle accelerators. Because of the equivalence of mass and energy, high-energy gamma-ray photons can transform into particle-antiparticle pairs (for example, an electron and a positron). When a particle and its antiparticle collide, they annihilate each other, transforming their mass into energy, normally in the form of gamma-ray photons.

Any antiparticles that are produced in our locality are quickly annihilated by colliding with particles of ordinary matter. All the bulk matter in our part of the universe is composed of conventional matter, and it is widely believed that the universe is dominated by matter rather than antimatter.

Antlia *See* CONSTELLATION

Apache Point Observatory

The principal projects at Apache Point Observatory at Sunspot, New Mexico, USA, are the 3.5-m telescope, the SLOAN DIGITAL SKY SURVEY and New Mexico State University's 1-m telescope. The 3.5-m telescope construction incorporates many innovations, including remote observing. The Sloan Digital Sky Survey is generating a three-dimensional map of a large volume of the northern night sky using an array of sensitive instruments coupled to a 2.5-m telescope.

Apastron

The farthest point in the relative orbit of one component of a binary system (*see* BINARY STAR) from the other component. The term is also used for the most distant position of a planet orbiting a star other than the Sun. The closest point in such an orbit is called periastron. *See also* APHELION.

Aperture

The clear diameter of the PRIMARY MIRROR of a REFLECTOR, or the OBJECTIVE LENS of a REFRACTOR. A telescope is characterized by its aperture; for example, a telescope with an aperture of 2 m would be referred to as a '2-m telescope.'

Aperture synthesis

A technique, also known as Earth-rotation synthesis, that enables the outputs of two or more RADIO TELESCOPES to be combined to give the same resolution as a single radio dish of aperture equivalent to the maximum separation of the radio dishes.

A single observation with a conventional RADIO INTERFEROMETER produces high resolution only in a direction parallel to the line joining the two antennae, or dishes. If two radio telescopes (A and B) are set up at opposite ends of a track and continue to observe a particular radio source for 12 hours then, during that time, the rotation of the Earth will change the orientation of the baseline relative to the source. Dish A is then moved a little closer to B to enable another 'ring' to be filled in and the process repeated until the entire 'dish' has been synthesized.

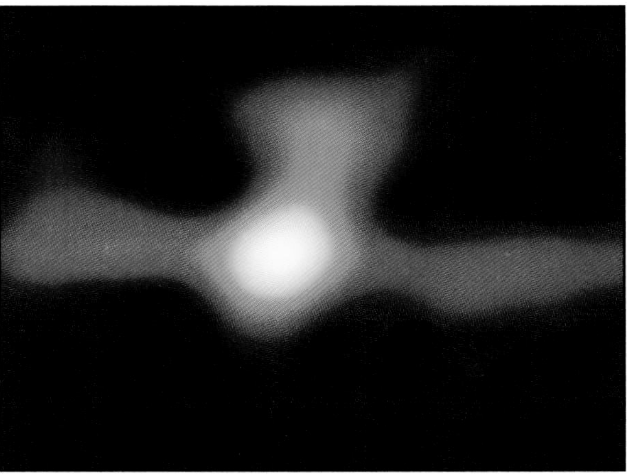

◀ ***Antimatter*** *This galactic cloud of antimatter at the center of the Milky Way is glowing in gamma-rays produced by annihilating antimatter particles.*

► ***Aperture synthesis*** *The VLA in New Mexico comprises 27 dishes, each 25 m in diameter, set out along a Y-shaped track. It can simulate a single dish 36 km in diameter.*

A synthesis array with only two dishes would take a very long time to build up the effect of a single large dish. The process may be speeded up by using more than two dishes, some of which may be fixed and others movable. In some cases, the dishes making up the array can be set out along more than one track. The largest complete Earth-rotation synthesis system is the VLA, located in the New Mexican desert close to Socorro. Usually, the outputs from the various dishes are linked together by cables, but the Multi-Element Radio-Linked Interferometer Network (MERLIN) in the UK uses radio links to connect dishes separated by distances of up to 230 km. Because it links a number of widely separated individual dishes it cannot simulate, and attain the sensitivity of, a complete dish. Nevertheless, it can detect details as small as 0.01 arcsec at its shortest operating wavelengths.

Apex

The direction of motion of the Sun relative to the stars in the local part of the Galaxy is known as the solar apex. The group motions of stars appear to diverge from a point in the constellation Hercules, and to converge toward a point in the constellation Columba. The solar motion is equal and opposite to this group motion. The observations indicate that the Sun is moving at a velocity of about 19.5 km s^{-1} toward a point at right ascension 18h and declination +30° in Hercules, and away from the corresponding point in Columba (the solar ANTAPEX). Apex can also refer to the point on the CELESTIAL SPHERE toward which the Earth appears to be moving as a result of its orbital motion around the Sun.

Aphelion

The point in the elliptical orbit of a planet or other object around the Sun at which it is farthest from the Sun. The Earth, for example, reaches aphelion in July, when it is just over 152 million km from the Sun. The Earth's PERIHELION passage is in early January. The word derives from Helios, the Greek god of the Sun. *See also* APOAPSIS.

Apoapsis

The point in an elliptical ORBIT at which the orbiting body is furthest from the body it is orbiting. The prefix 'apo-' or 'ap-' may be attached to various words or roots depending on the body being orbited: for example, apastron for an orbit around a star; apojove for an orbit around Jupiter.

See also APHELION; APOGEE.

Apochromatic lens *See* CHROMATIC ABERRATION

Apogee

The point in its orbit around the Earth at which the Moon or an orbiting satellite is farthest away.

Apollo

The US program to land humans on the Moon. It included 11 manned missions, from October 1968 to December 1972, with three missions restricted to a lunar flyby or orbital survey (Apollos 8, 10 and 13), and six landings (Apollos 11, 12, 14, 15, 16 and 17). Some 385 kg of lunar soil and rock samples were returned, and these provided evidence that the Moon is about the same age as the Earth and probably originated from material derived from Earth during a gigantic impact event. Astronauts deployed a variety of surface experiments, including seismometers, laser reflectors for refining the Earth-Moon distance, magnetometers, ion detectors, and solar wind spectrometers. Orbital surveys were conducted from the command modules (most notably on Apollos 15–17), which carried mapping and panoramic cameras and various spectrometers. Data on the distribution of lunar gravity variations were obtained by studies of two subsatellites deployed from Apollos 15 and 16.

Apollonius of Perga (200–100 BC)

A Greek geometer from Perga, which is now in Turkey. He proposed that the planets revolved around the Sun, and the Sun revolved around the Earth. Apollonius left no astronomical works but is believed to have invented the geometric system of epicycles and eccentric circles, by which perfect circular motions were combined into the actual elliptical orbits of the planets.

Apparent magnitude *See* MAGNITUDE AND PHOTOMETRY

Apparent solar time

This refers to TIME reckoned by the apparent position of the Sun in the sky. The local apparent solar time (or local apparent time) is defined to be the local HOUR ANGLE of the Sun (the angle between the observer's MERIDIAN and the Sun, measured westwards from the meridian) plus 12 hours. Thus, when the Sun is crossing the meridian at noon its hour angle is zero, and the local apparent solar time is 0 h + 12 h = 12 h; when the hour angle of the Sun has increased to 6 hours, the local apparent solar time is 6 h + 12 h = 18 h, and so on. Apparent solar time is the time that is displayed on a SUNDIAL. Compared to a regular clock, such as a pendulum, it is fast or slow by up to 15 min. This effect is known as the EQUATION OF TIME.

Apparition

The appearance of a celestial body at a time when it is well placed for observation. The term is used especially for objects in the solar system whose orbits are such that they are unobservable for periods of time. For the planets Mercury and Venus, morning and evening apparitions occur when they are at greatest ELONGATION (their maximum angular separation from the Sun), and for long-period comets apparition occurs when their orbits bring them into the inner solar system.

Appulse

An apparently close approach between two objects as seen on the CELESTIAL SPHERE, as their motions bring them close to the same line of sight of an observer on Earth. Examples are when the Moon or a planet narrowly misses occulting a star (a so-called occultation).

Apus, Aquarius, Aquila, Ara *See* CONSTELLATION

Arago, Dominique François Jean (1786–1853)

A French scientist and statesman. Arago became director of the PARIS OBSERVATORY, where he invited Urbain LEVERRIER to determine why Uranus was deviating from its predicted orbit, a study which led to the discovery of Neptune. Arago defined the MERIDIAN of Paris, used by French sailors until it was supplanted by the 'prime' meridian of Greenwich in 1884. He made measurements along the Paris meridian of the circumference of the Earth. A 'virtual monument' to Arago by the Dutch sculptor Jan Dibbets runs through Paris, in the form of 135 brass circles engraved with his name and located along the meridian line.

Arcetri Astrophysical Observatory

The Arcetri Astrophysical Observatory, founded in 1872, is located close to the villa in Florence, Italy, where GALILEO GALILEI spent the last 11 years of his life. Under the directorship of Giorgio ABETTI (from 1921 to 1953) it became the growth point of Italian astrophysics, with emphasis on solar physics.

▲ ***Apollo 11*** *Edwin (Buzz) Aldrin planting the Stars and Stripes on the lunar surface during the first moonwalk on June 20, 1969.*

Archeo-astronomy *See* PREHISTORIC ASTRONOMY

Arc minute (arcmin), arc second (arcsec)

Units of angular measure. One arcsec (") is one-sixtieth of 1 arcmin ('), which is one-sixtieth of a degree.

Arcturus

The star Alpha Boötis, a golden-yellow giant star of spectral type K2IIIp (*see* STELLAR SPECTRUM: CLASSIFICATION). Its high velocity through space will carry it past the solar system in a few thousand years' time. It will be a spectacular visitor, being about 21 times the Sun's diameter and 98 times its luminosity, having an absolute magnitude of –0.3.

Arecibo Observatory

Located in the limestone hills of northwestern Puerto Rico, Arecibo Observatory is operated by the National Astronomy and Ionosphere Center, managed by Cornell University for the US National Science Foundation.

With its 305-m diameter dish, which was built into a hillside and completed in 1963, Arecibo's radio-radar telescope has the largest collecting area of any radio antenna in the world (8 hectares or 18 acres). The Arecibo system operates at FREQUENCIES between 50 MHz and 10 GHz (WAVELENGTHS between 6 m and 3 cm).

Arecibo has been used for solar system exploration, including detailed mapping of the surfaces of the Moon by RADAR ASTRONOMY, Venus and a handful of near-Earth asteroids.

On November 16, 1974, Arecibo sent the first radio message to extraterrestrials. The transmission contained representations of the fundamental chemicals of life, the formula for DNA, a crude diagram of our solar system, and simple pictures of a human being and the Arecibo telescope. The telescope is still used for SETI research today (*see* EXOBIOLOGY AND SETI).

Argelander, Friedrich Wilhelm August (1799–1875)

An astronomer, born in Memel, East Prussia. He was director of the Bonn Observatory, Germany, where he organized a survey of the position of all 324 198 stars of the northern hemisphere above magnitude 9. These were published as the star charts and catalogs of the *Bonner Durchmusterung* (Bonn Survey). His assistant, Eduard Schönfeld, made the

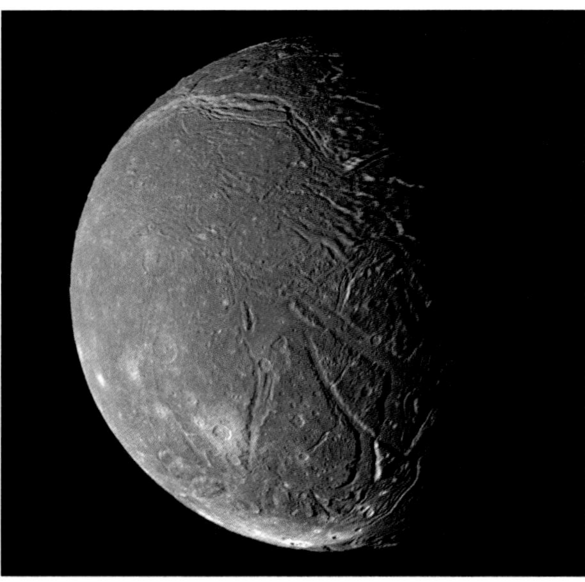

► ***Ariel*** *This satellite of Uranus was viewed by NASA's Voyager 2 spacecraft from a distance of 103 000 km on January 24, 1986. Kachina Chasmata is the valley at the top of the picture, the Korrigan Chasma complex is at the bottom. The crater with bright ejecta (lower left) is Melusine.*

extension into the southern sky. It is staggering to think about this project, carried out in the days before information technology, to observe by eye half a million stars twice each, to calculate their brightnesses and positions, to chart them and to print them by hand.

Argo Navis *See* CONSTELLATION

Ariel

(1) A mid-sized satellite of URANUS, discovered by William LASSELL in 1851. Its diameter is 1160 km and it orbits at a distance of 191 000 km. Of Uranus's five mid-sized satellites, Ariel has the brightest surface, and, with the exception of MIRANDA, has been the most geologically active. An older, cratered surface is criss-crossed by spectacular, deep, steep-sided canyons whose broad floors themselves show signs of further activity, with volcanic flooding and sinuous, winding troughs reminiscent of the Moon's sinuous rilles. The longest of these canyons is the Kachina Chasmata, which measures 622 km. Such activity must have been caused by tidal heating. At present Ariel lacks any orbital resonance with neighboring satellites that could cause tidal heating, but this may have been different in the past. When the molten icy interior eventually solidified, it would have expanded, splitting the crust to form the canyons.

(2) A series of six UK-led scientific satellites launched between 1962 and 1979. The first four studied the IONOSPHERE, while the last two were X-RAY ASTRONOMY satellites. Ariel 3 was the first all-British satellite. The UK–US Ariel 5, launched in October 1974, operated for more than eight years and discovered dozens of new X-ray sources, including a number of transient sources such as Cygnus X-1. Ariel 6 suffered technical problems and produced few results.

Aries *See* CONSTELLATION

Aristarchus of Samos (310–230 BC)

A Greek philosopher who, according to Archimedes, hypothesized that the fixed stars are stationary and that the Earth orbits the Sun, which is at the center of the universe. Since the stars did not show motions that reflected the motion of the Earth, the stars must be at great distances. The Babylonian astronomer Seleucus (358–280 BC) was the only one to take up this concept until Nicolaus COPERNICUS in 1543, who knew of Aristarchus's theory. Aristarchus also used geometric methods to measure the size of the Sun, Moon and Earth.

Aristotle of Stagira (384–322 BC)

Aristotle was a Greek philosopher and scientist. His lectures, compiled into 150 volumes, included *Physics*, *Metaphysics* and *De Caelo et Mundo* (On the Heavens and Earth). In this he accepted the heavenly spheres proposed by the astronomer EUDOXUS. Aristotle thought the Earth to be spherical and imagined a perfectly spherical, unchanging universe of spheres centered on the Earth and carrying the planets. Aristotle followed Eudoxus in suggesting that there were four elements: earth (solid), fire (energy), water (liquid) and air (gas). Each element could be hot, wet, dry or cold. Aristotle added a fifth element called ETHER, which he believed to be the main constituent of the celestial bodies. Aristotle's view of the universe was hierarchical, and he made a sharp distinction between the eternal and immutable heavens, and the sublunar, imperfect world of change. The motion of celestial objects was therefore always natural, circular and continuous.

Arizona Crater *See* METEOR CRATER

Armagh Observatory

The Armagh Observatory in Northern Ireland is a modern astronomical research institute with a rich heritage, founded in 1790 by Archbishop Richard Robinson.

Armillary sphere

A device of ancient origin that was used to measure or describe celestial positions. In essence, it is a model of the CELESTIAL SPHERE consisting of a set of rings. Each ring represents a great circle, such as the celestial equator or ecliptic, that revolves within a framework of fixed circles. These represent, for example, the observer's horizon and meridian. When used for observation open sights were mounted on a pivoted ring, and the position of the object of interest was read off from graduated scales on the various rings.

Armillary spheres were used to take positional measurements in ancient and medieval times by observers such as HIPPARCHUS, PTOLEMY and Tycho BRAHE. They were widely used from the Middle Ages onwards to teach astronomy.

Arrhenius, Svante August (1859–1927)

Swedish chemist and Nobel prizewinner (1903). In *Worlds in the Making* (1908) he suggested that life on Earth had begun when spores from space reached the Earth on a METEORITE. This is known as the PANSPERMIA hypothesis.

Art and astronomy

There is a long tradition of the use of astronomical references in art. Among the intents of fine art is to comment by metaphor or analogy on human activities. Such activities may be set against or associated with astronomical phenomena, either for their own interest or as a comment.

In paintings of the crucifixion, for example, images of eclipses are often shown. These are literal representations of the sudden darkness that the scriptures say occurred when Jesus died. They are also metaphors for the extinguishing of an important person, and examples of the

◀ ***Art and astronomy*** *Vincent van Gogh's* Starry Night *shows a French hamlet under a night sky with a crescent Moon, Venus (possibly the large white star left of center), and a swirling Milky Way. Van Gogh painted the picture in June 1889 from naked eye observations he had made while staying at Saint Rémy. The picture expresses the connection between humankind and the cosmos, as well as van Gogh's psychological turmoil.*

'pathetic fallacy,' in which the natural world is identified as reacting to the human activities that happen in it.

Likewise, the Florentine painter Giotto di Bondone (*c.*1267–1337) – who was commissioned in 1303 to decorate the interior of a chapel belonging to a wealthy Paduan merchant – painted 38 religious scenes, one of which was *The Adoration of the Magi*. The worship of the infant Jesus in a stable was set under the Star of Bethlehem, a symbol of change and a portentous celestial mark of the birth. Giotto depicted the star as a blazing comet. He was inspired by COMET HALLEY, which had returned to PERIHELION in 1301 and which he had undoubtedly seen. It was fitting that the spacecraft sent into the nucleus of Comet Halley during the return of 1986 should have been named GIOTTO.

On a more individual level, Vincent van Gogh (1853–90) painted *Starry Night* (1889). It shows a church and houses with lit windows in a village in Provence, France, among a landscape of hills and olive and cypress trees. Above, there is a starry sky, with crescent moon and swirling Milky Way. The skyscape appears to be a free rendering of the night sky at the time of its painting. Whatever its natural realism, the dizzying lights that overarch the human landscape represent a disturbing cosmic background to the smaller human events below, including, one must suppose, van Gogh's developing madness, which culminated in his suicide the following year.

The same Milky Way runs across the night sky in Adam Elsheimer's (1578–1610) *Flight into Egypt* (1609). Again, it is a cosmic background to the events, shown small, that are taking place on Earth as the Holy Family travels on donkey, fleeing from the massacre of the infants by Herod. Elsheimer's is the first representation of the Milky Way as a mass of individual stars.

Astronomical instruments are often shown in portraits. They are decorations that reflect the interests, attainments and education of the sitter(s), and may or may not be directly associated with the subject. *The Ambassadors* (1533) is a full-length portrait of two wealthy, educated and powerful young men: Jean de Dinteville, the French ambassador to England, and Georges de Selve, Bishop of Lavaur. The artist, Hans Holbein the Younger (1497/8–1543), has depicted the two men standing on either side of a table bearing a collection of globes, books, and astronomical and musical instruments. A sundial on the table is identical to that featured in Holbein's 1528 portrait of the astronomer Nicholas Kratzer (1487–1550?). The plausible guess is that the sundial was designed or made by Kratzer, and its image was re-used and incorporated by Holbein into *The Ambassadors* as flattery.

Astronomical references in the illustrated books of the English romantic poet William Blake (1759–1827) are not favorable to science. *God as an Architect* (1794), an illustration from *The Ancient of Days*, depicts God as an old man, kneeling on stormy clouds against a fiery Sun, coldly measuring the cosmos with dividers. Blake depicted Isaac NEWTON as a misguided Greek-god-like hero whose gaze was directed only downwards at sterile geometrical diagrams that he also measured with dividers. It is curious that this scornful depiction of Newton is often seen displayed as a poster on the walls of university physics departments.

By contrast to Blake's pessimism, his contemporary and fellow countryman Samuel Palmer (1805–81) shows bucolic scenes in an idealized England – for example, in *Coming from Evening Church* (1830) – under a benign, evening sky with a glowing full Moon.

Space art

Space art is a form of scientific illustration, akin to botanical illustration. It is intended mostly to convey ideas about space and astronomy. Since the subject matter is not readily seen directly, space artists extrapolate what is known about a subject into an illustration, much as an illustrator would do when reconstructing a dinosaur from fossils or a historical site from an archeological dig. Space artists might show what a far-off planet, double-star system or galaxy would look like if we were there in person, or how an activity in space might be implemented, such as mining an asteroid or colonizing Mars.

Space art may also incorporate fantasy elements into its genre – placing fish, flowers and so on into a space scene for dramatic effect. An early example of this form of art was *The Sower of Systems* (1902), a representation by George Frederick Watts (1817–1904) of the chaos that existed before planets were formed. An expressionist image of Pierre LAPLACE's nebular hypothesis is combined with the vague form of a striding, gray-cloaked figure who appears to be sowing a field.

The pioneer space artist in the modern school was Chesley Bonestell (1888–1986), who illustrated books by space-travel experts such as Wernher von Braun (1912–77) and Willy Ley (1906–69). Modern space art is distinguished by a naturalistic method of illustration, creating detailed landscapes reminiscent of those of the American Romantic landscape painters. The subject matter and the material available both favor digital methods of work, but more traditional techniques are also common. About 120 space artists form the International Association of Astronomical Artists, founded in 1982. Modern practitioners include David Hardy (UK), Lynette Cooke (USA), Georgii Poplavski, Andrei Sokolov, and cosmonauts Vladimir Dzhanibekov and Alexei Leonov (Russia).

There is endless scope for space art. In the future it will be fascinating to look back at some of the speculations of the twentieth century; for example, will a lunar base really be like the graceful, domed structure visualized by Bonestell?

Artificial satellite

A man-made object placed in orbit around the Earth or some other celestial body. The first artificial Earth satellite was Sputnik 1, launched by the then Soviet Union on October 4, 1957. Spherical in shape, and with a mass of 84 kg, it entered an orbit with a PERIGEE altitude of 229 km, an APOGEE altitude of 947 km, and a period of 96 min. Artificial satellites are used for a variety of roles, including astronomical observation; studying the Earth's magnetosphere and space environment; monitoring the atmosphere, weather, oceans and surface of the Earth; geodesy (studying the shape and gravitational field of the Earth); communications and military reconnaissance.

All satellites follow elliptical (or circular) orbits around their parent body. The shape and orientation (the ORBITAL ELEMENTS) of a satellite's orbit change under the action of various perturbing forces. For example, the tenuous outermost layers of the Earth's atmosphere exert a small but finite frictional drag on the motion of a satellite; this causes its perigee altitude to decrease (tending to make the orbit more circular), and leads eventually to the satellite spiraling inwards and burning up in the denser atmospheric layers. Drag forces of this kind caused Sputnik 1 to re-enter the atmosphere and burn up 92 days after it had been launched.

Gravitational perturbations exerted, for example, by the Earth's equatorial bulge, or by the Moon and Sun, cause the planes of satellite orbits to undergo PRECESSION (slowly rotate around the Earth). If the orbital inclination (the tilt of the orbital plane relative to the Earth's equator) is carefully chosen, the orbit can be made to precess at a rate that ensures that its orientation relative to the direction of the Sun remains the same while the Earth revolves around the Sun. An orbit of this kind is called Sun-synchronous.

If a satellite is placed in a circular orbit at an altitude of just less than 36 000 km above the Earth's equator, its orbital period will be exactly equal to the rotation period of the Earth. Because the satellite then remains continuously above the same point on the equator (and remains stationary in the sky when viewed from the Earth's surface), it is said to be geostationary. The geostationary orbit is used extensively by communications and meteorological satellites.

Artificial satellites, observing

See practical astronomy feature overleaf.

ASCA (Advanced Satellite for Cosmology and Astrophysics)

Japanese–US X-RAY satellite (also known as Astro-D, *see*

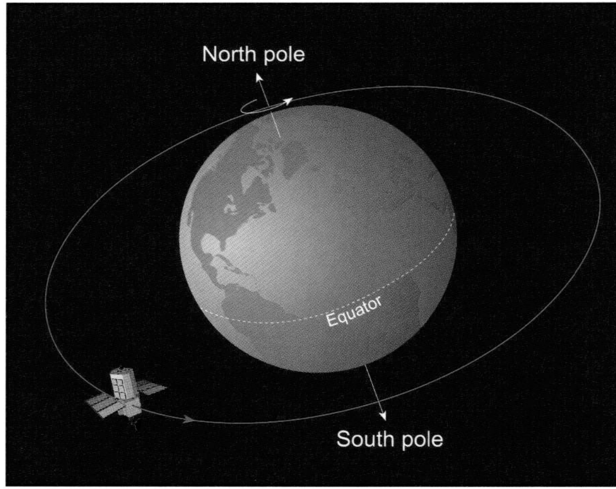

▶ ***Geostationary artificial satellite*** *Orbits around the equator in 24 hours, staying above the same place on Earth.*

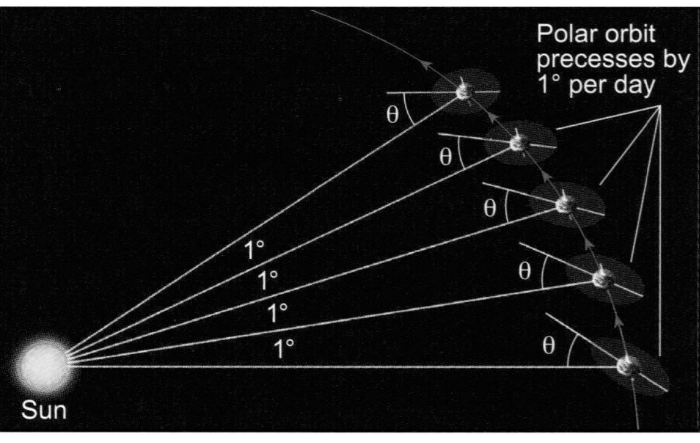

▶ ***Sun-synchronous artificial satellite*** *Orbits over the Earth's poles. As the Earth moves in its orbit around the Sun at 1° each day, the satellite's orbit precesses by the same amount. The orientation of the orbit relative to the Sun, θ, remains constant.*

ASTRO). It was launched in February 1993 and re-entered the atmosphere over the Pacific Ocean on March 2, 2001. During its eight-year lifetime it provided broadband imaging over the 1–12 keV energy range. It carried four US conical grazing telescopes and the first two CCD-based imaging spectrometers to be used on an X-ray observatory. The satellite found evidence for the creation of COSMIC RAYS during SUPERNOVAE, and played a major role in studies of X-ray binaries and gamma-ray bursters. It is also known as Asuka, which means 'flying bird.'

Ashen light

An apparent slight brightening of the night side of VENUS occasionally seen when the planet is observed as a thin crescent, near inferior conjunction, when it is nearest the Earth. The ashen light may first have been observed by Giambattista Riccioli in 1643, though the first unequivocal description was by William Derham in 1715. The ashen light, like other observational phenomena on Venus long reported by amateur observers (such as cusp caps and the Schröter effect), has been a controversial subject: it has never been satisfactorily imaged, for example. It has been likened in appearance to EARTHSHINE though it is much more feeble, and of course it cannot have a comparable cause.

There are two possible physical explanations for the ashen light. First, it may be similar to the Earth's AIRGLOW, a faint but persistent illumination of the Earth's upper atmosphere caused by the recombination of atoms and molecules that have been ionized by solar ultraviolet radiation. On recombination, the atoms and molecules emit a weak radiation at optical wavelengths. The effect may be greater on Venus because the flux of solar radiation is stronger closer to the Sun.

Alternatively, the ashen light may be the result of the refraction of sunlight in Venus's thick atmosphere. In theory, the atmosphere is so dense that 'super-refraction' can occur: light could be refracted all the way around the planet. Some people maintain that Venus's ashen light is a wholly subjective phenomenon – that the observer 'sees' the whole disk as illuminated when only the crescent is visible, particularly when cusp caps (brightenings of the cusps, the tips of the crescent) are pronounced.

Association of Universities for Research in Astronomy (AURA)

The Association of Universities for Research in Astronomy Inc. is a consortium of educational and other non-profit institutions that manages the NATIONAL OPTICAL ASTRONOMY OBSERVATORY (NOAO) and the GEMINI OBSERVATORY for the US National Science Foundation. It manages the SPACE TELESCOPE SCIENCE INSTITUTE (Baltimore, Maryland) for NASA.

NOAO operates ground-based observatories for night-time astronomy at KITT PEAK NATIONAL OBSERVATORY (Arizona), and CERRO TOLOLO INTER-AMERICAN OBSERVATORY (Chile); and for solar research at Kitt Peak and the NATIONAL SOLAR OBSERVATORY (New Mexico).

Asterism

A readily recognizable group or arrangement of stars (usually bright) that are not necessarily members of a single constellation. Well-known examples are the Big Dipper (part of the constellation Ursa Major), the False Cross, the Summer Triangle and the Square of Pegasus, all of which comprise stars from more than one constellation. The term is also used occasionally to denote a close group of faint stars that appear to be, but in fact are not, members of a cluster.

▲ ***Asterism*** *The Big Dipper is one of the best-known asterisms in the northern night sky.*

Asteroid

Asteroids are also known as minor planets. They are small objects orbiting the Sun, mostly in the region between MARS and JUPITER, but also elsewhere in the solar system.

Any small body orbiting the Sun that shows no cometary activity is called an asteroid. Asteroids range in size from 950 km in diameter down to a few centimeters, at which size they are called METEOROIDS. Fewer than 1000 asteroids are larger than 30 km across, and of these about 200 asteroids are larger than 100 km. The total mass of all asteroids is estimated to be about one-thousandth of the Earth's mass.

In 1766 an empirical formula for the distances of the planets from the Sun was popularized by Johann Bode (1747–1826). This became known as BODE'S LAW or the Titus–Bode Law. Bode's law was a good fit to all the then-known planets (Mercury to Saturn), and proved to fit Uranus when it was discovered in 1781. However, it predicted a planet in the gap between Mars and Jupiter, which bolstered the belief in a missing planet at 2.8 AU, which was first expressed by Johannes KEPLER.

Discovery of asteroids

In September 1800 Baron Franz von Zach (1754–1832) organized a concerted search for the presumed planet by several astronomers known as the 'celestial police.' On January 1, 1801, before they could even start work, a Sicilian monk, Father Guiseppe Piazzi (1746–1826), discovered a moving stellar object while he was constructing a star catalog. But Piazzi lost the object, his observations interrupted by illness. The German mathematician Carl Friedrich GAUSS developed a new method to compute orbits, which enabled the new object, (1) CERES, to be recovered. Further objects were discovered by Heinrich OLBERS ((2) PALLAS and (4) VESTA) and Karl Harding (1765–1834) ((3) JUNO). One hundred new 'minor planets' had been discovered by 1868, 200 by 1879, and 300 by 1890.

The size and reflectivity of an asteroid are determined by comparing the sunlight and solar heat absorbed by the asteroid with the infrared that it reradiates and the light that it reflects. Asteroids generally have low ALBEDO.

Observing artificial satellites

Speeding across the skies at up to 28 000 km h^{-1}, artificial satellites are surprisingly visible as they go about their business hundreds or even thousands of kilometers above the Earth. They appear as bright stars cruising across the celestial sphere. Aiming a telescope at them, however, can resolve large structures in low-Earth orbit. It is possible to use telescopes to track and image satellites at high resolution, even during the daytime.

One problem is predicting and following satellite motion. Most satellites move too quickly to keep them centered in an eyepiece's tiny field of view. Try it manually and you are in for a bad case of orbital whiplash. Attempts at tracking by hand also introduce considerable telescope vibration, which is fatal to the sub-arcsec resolution desired for discerning fine details. At the Clay Center Observatory at Dexter and Southfield Schools, Brookline, Massachusetts, USA, our telescope mount excels at moving along trajectories that would be impossible to pursue by hand.

Equally important, a method for predicting a target's location and speed is required to guide the mount along the computed path. Being unable to find suitable existing software, we created it ourselves. After eight years of algorithms, coding, testing and repetition of this cycle, the software is now complete. The final product, C-Sat, is a Windows-based program that is so powerful it enables the user to lock onto satellites thousands of kilometers away, even during the daytime.

The software requires accurate details of time and location, which can be set either manually or via an inexpensive Global Positioning System (GPS) receiver. Once C-Sat has determined the precise location and time, the computer automatically scans the space above the telescope and displays an ever-changing list of all the satellites visible from your location at that moment. Selecting a target from the list causes the telescope to move to a pre-computed 'rendezvous' point, and a countdown begins. At zero, the satellite crosses through the telescope's field of view and the mount's drive system matches the target's angular speed, resulting in the spacecraft remaining centered in the eyepiece.

Although hundreds of satellites can be tracked each night, most of them appear only as pinpoints even at ×1000 magnification. Basic trigonometry illustrates the reason for this: a satellite measuring 2 m that is 400 km away subtends only 1 arcsec. While most objects in orbit appear star-like, there are some notable exceptions, for example, the Space Shuttle, which is 37 m long, can subtend some 40 arcsec when passing overhead at an orbit of 200 km. This is roughly the angular size of Jupiter.

International Space Station (ISS)

Another favorite target is the ISS. It is one of the largest objects in orbit, and is, at the time of writing, still under construction. The ISS is also easily visible; its high-inclination orbit yields favorable flyovers for more than 90% of the Earth's population. Because of its growing size and frequent visitors (for example, resupply vehicles), the ISS is an especially fascinating object to observe. The Clay Center Observatory's 12-in guide scope easily resolves the solar panels and details on the ISS modules, and our 25-in *f*/9.6 main telescope can resolve the windows, booms and antennae on each ISS module. We can even detect a single astronaut on a spacewalk.

Among the other interesting targets are the hundreds of burned-out rocket stages. Many of them look like little pencils, sometimes tumbling end over end as they fly through space. Occasionally satellites can be seen spinning as well. We recently imaged a defunct Iridium satellite at high resolution, and could see it spinning twice per second – a good indication that it was indeed out of order.

There are also, of course, more than a few top-secret spy

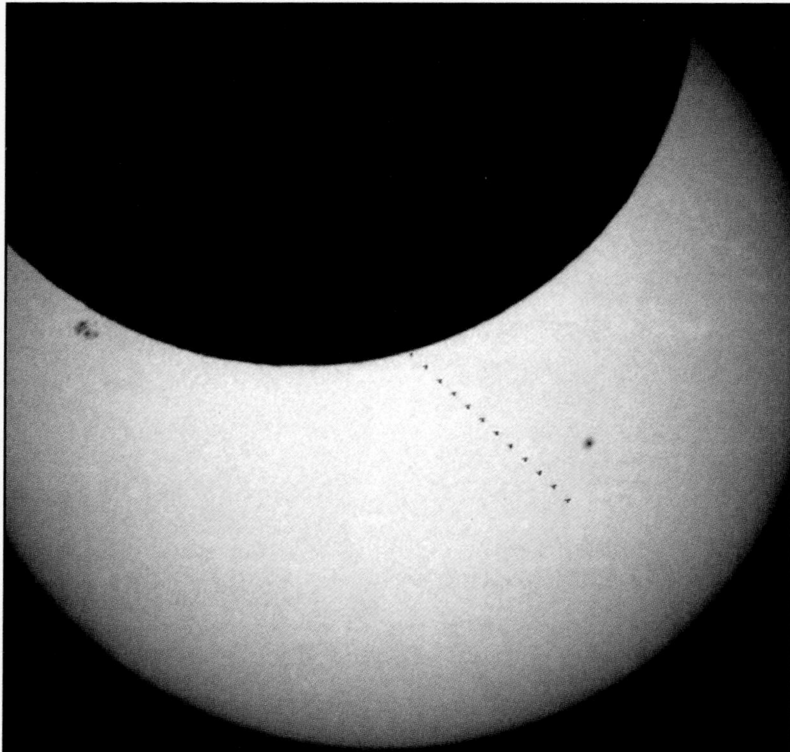

► ***Transit of the ISS*** *A multiple exposure recorded 12 silhouettes of the ISS as it passed under the Moon (circular dark shadow) and in front of the Sun (with sunspots) during the partial solar eclipse on December 25, 2000.*

►► ***Mir space station*** *Its long trail above trees was captured with a fixed camera using a 30-s exposure through a 50-mm lens on Ektachrome 200 film.*

satellites in orbit, from several governments. Some (optical telescopes or radar imagers, perhaps) appear surprisingly large, about 15–20 m long with solar panels. Some even appear to have large dish antennae more than 12 m across.

Viewing satellites with the eye is enjoyable, but it is also satisfying to image them. The challenge is to take sub-arcsec resolution images of objects measuring a few meters across that are hundreds of kilometers away and traveling at 28 000 km h^{-1}. This has to be done in a few optimal seconds during each pass, through sea-level atmospheric turbulence, often during the daytime.

To record big, bright satellites such as the Space Shuttle and ISS, we use a commercially available monochrome closed-circuit television (CCTV) video camera on our 25-in *f*/9.6 RITCHEY–CHRÉTIEN TELESCOPE. The camera's 512 × 768 array of 9 µm pixels offers excellent resolution. When the skies are especially stable we use a ×3 apochromatic BARLOW LENS to yield more magnification and better sampling. When the telescope locks onto a passing satellite, the target is kept centered on the tiny CCD chip with real-time closed-loop feedback. The computer automatically senses the satellite's position on the CCD, and continually updates the telescope velocities and accelerations to keep the target centered on the chip. There is also a joystick in case the telescope operator wishes to make manual adjustments to the track. The video output is recorded on a professional-grade video deck.

Atmospheric turbulence obscures most detail, even in the best conditions. Since a laser ADAPTIVE OPTICS system was beyond our price range, we adopted a poor man's solution to 'untwinkling' the spacecraft images: videotape. By recording 30 high-resolution frames per second, we could view each image by stepping through the video in slow motion. Out of the thousands of frames recorded on the tape, a few flawless frames show spectacular detail, with almost no atmospheric distortion at all. Digitizing the best video frames and combining them increases the signal-to-noise ratio of the resulting image, which is then suitable for processing.

Satellites transiting the lunar disk are quite common, but one of the more unusual events we observed was the solar double transit of the Space Shuttle Columbia on October 22, 1995. On this date the Shuttle was due to transit the Sun on two consecutive orbits as seen from a single location in Attleboro, Massachusetts. Using a GPS receiver, we drove to the precise location and set up two Meade 8-in LX200 telescopes to observe the event in white and hydrogen-alpha light. At the duly appointed time, the Shuttle's shadow raced over us at more than 25 000 km h^{-1}. Columbia became visible in both telescopes as a black dot zipping across the solar disk. Ninety minutes later we watched another half-second solar transit from the same location.

Another fascinating sight is the changing color of orbital sunset playing across the ISS as it slips into the Earth's shadow. Upon entering the PENUMBRA, the ISS shifts from brilliant white to gold to a lovely iridescent orange in just a few seconds. Once inside the UMBRA, the spacecraft remains visible by refracted sunlight for several more seconds, before fading to the deepest red imaginable and disappearing completely.

In addition to our ongoing studies of the ISS, we have succeeded in making three-dimensional images of satellites by combining images taken several seconds apart. Images of low-altitude targets show enough PARALLAX to measure the dimensions of satellites with accuracies in the centimeter range. When the Shuttle flights resume after their suspension due to the Columbia disaster in 2003, it will be a real challenge to image the Shuttle's payload bay lights while the spacecraft is otherwise hidden in the Earth's shadow.

Ron Dantowitz is director of the Clay Center Observatory at Dexter and Southfield Schools, Massachusetts, USA.

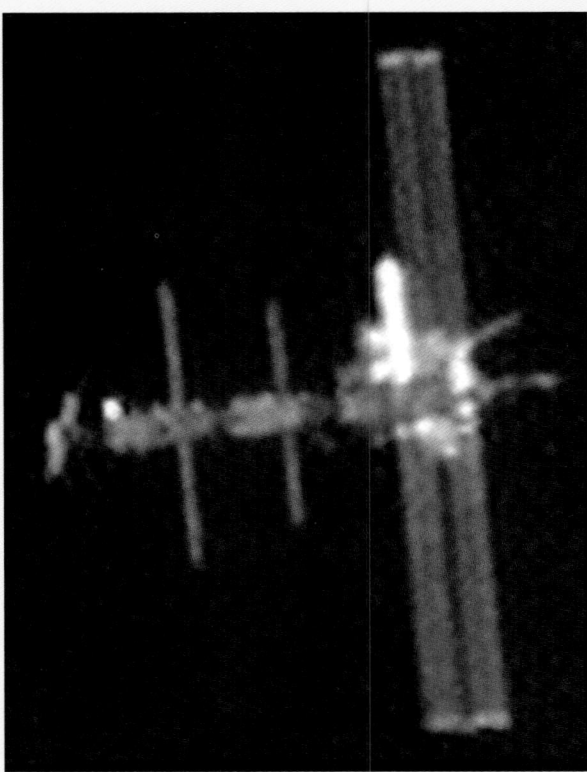

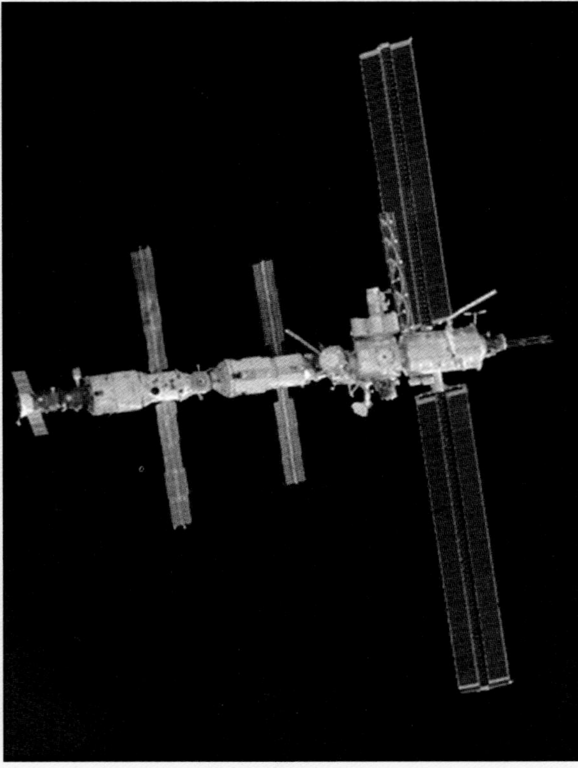

◄◄ ***ISS*** *Daytime pass at 420 km on September 10, 2002, at 13.22 UT. Imaged using a 12-in Schmidt–Cassegrain telescope.*

◄ ***ISS*** *Graphic for comparison.*

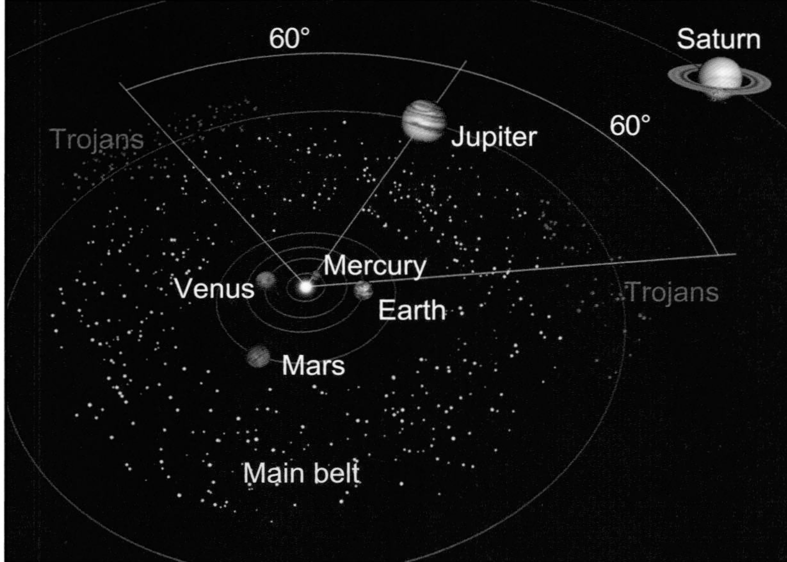

▲ ***Main asteroid belt*** *between the orbits of Mars and Jupiter. Some asteroids, the Trojans, have found their way into the orbit of Jupiter, following or preceding it by 60°.*

The asteroid belt

Most asteroids orbit between Mars and Jupiter, in the main asteroid belt (2.1–3.3 AU). Out of about 30 000 known objects, 8500 asteroids have precisely determined orbits. Each of these asteroids is designated with a number that indicates its position in the catalog maintained by the Minor Planet Center in Cambridge, Massachusetts, USA. It may also have a name proposed by the discoverer, in the form '(1234) Name.' Only preliminary orbits are available for the other 21 500 asteroids. Each is temporarily identified by its year of discovery and by two letters indicating the date it was first observed. For example, (1998) BC identifies the third asteroid (C) discovered during the second half of January (B) in 1998. Spacecraft use radar, or in a few cases remote sensing, to determine, by regular light curves, the rotational period of about 700 asteroids. Several are known from RADAR ASTRONOMY or space imaging to be orbited by a satellite asteroid.

The structure of the asteroid belt is determined by gravitational interactions, principally with JUPITER. Jupiter causes a depletion of objects in correspondence to the RESONANCES (where the orbital period of an asteroid is an exact integer ratio of Jupiter's orbital period). For example, according to Kepler's third law an asteroid 2.5 AU from the Sun would complete exactly three orbits around the Sun for every one orbit by Jupiter, which is at a distance of 5.2 AU. Those breaks in the uniformity of the asteroid belt are called Kirkwood gaps after the name of the discoverer.

Some main-belt asteroids that have similar orbits and are usually of the same type form HIRAYAMA FAMILIES (such as the Themis, Eos, Koronis, Flora and Karin families or groups). They are the result of the catastrophic disruption of one asteroid by collision with another. Larger asteroids are more regular in shape than the smaller asteroids, which may be fragments from collisions, or which may never have grown large enough to settle into a spherical shape and form a differentiated CORE or mantle structure.

Formation of asteroids

In general, asteroids are believed to be the remnants, either fragments or 'survivors,' of the swarm of PLANETESIMALS from which the TERRESTRIAL PLANETS were formed. The asteroid belt coincides with the boundary separating the inner solar system (where the water in the PROTOSOLAR NEBULA vaporized) from the outer regions (where water condensation occurred). Forming in the cold outer zone, Jupiter accreted its large mass of otherwise gaseous material, and its consequent gravitational perturbations continuously interrupted the condensation of a single planetesimal in the asteroid zone. This left many asteroidal bodies, whose interactions and impacts created the present asteroid population.

The records of primordial chemical and physical processes can still be found frozen on the surface of asteroids. The chemical composition of asteroids, determined by SPECTROSCOPY, has been classified by the letters S, C, M, D, F, P, V, G, E, B and A (the initials of the words used to describe an asteroid's spectrum, such as stony, carbonaceous or metallic). A small number of asteroids show spectra typical of rocky planets and satellites (A- and V-type). The latter are probably members of the Vesta family, fragments of (4) Vesta, whose surface composition resembles that of the Moon. Most of the asteroids belong to the low-albedo C-type, their spectra matching those of the carbonaceous chondritic meteorites (*see* CHONDRITE). The variety of other types is in part due to 'space weathering' of the surfaces by collisions, solar radiation and micrometeoroid impacts. The composition of main-belt asteroids varies with distance from the Sun, a result of the formation processes.

Space investigations

The GALILEO MISSION approached two asteroids, (951) GASPRA (October 1992) and (243) IDA (August 1993), offering a close look at the surfaces of asteroids for the first time. It discovered around Ida the first asteroid satellite, named Dactyl. Both Gaspra and Ida have an elongated, irregular shape, while Dactyl has a rather smooth, regular shape, like the Martian moons (PHOBOS AND DEIMOS), which may be captured asteroids. Because of Dactyl, Ida's mass was determined, with its density consistent with a bulk chondritic composition.

The NEAR mission targeted (253) MATHILDE (flyby in June 1997) and (433) EROS (flyby in December 1998, survey from orbit February 2000 to 2001; it impacted the surface of Eros on February 14, 2001). The flyby of Mathilde enabled its mass and density to be determined, suggesting that it is a loosely assembled froth of chondrite material. NEAR's year-long orbital inspection of Eros is the most complete study

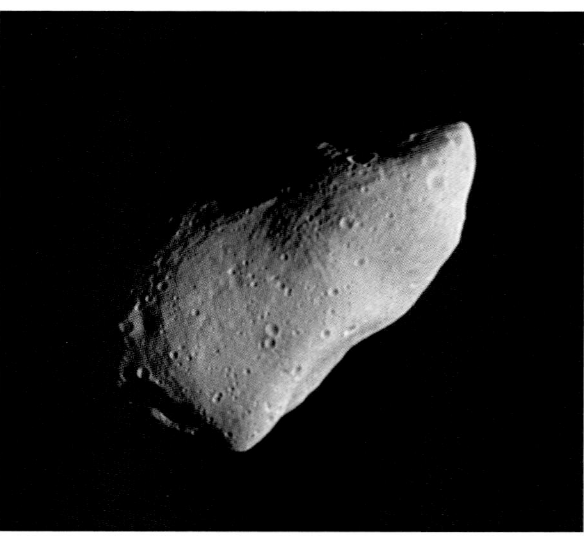

▶ ***Gaspra*** *This image of asteroid (951) Gaspra is a mosaic of two images taken by the Galileo spacecraft moments before closest approach on October 29, 1991. The illuminated portion measures about 18 km from lower left to upper right.*

was the Copernican system (*see* Nicolaus COPERNICUS). This demanded the accurate measurement of planetary positions relative to the stars in order to develop CELESTIAL MECHANICS, as by Tycho BRAHE. This basic science was applied in celestial navigation, calculating one's position on Earth from the observed star positions at any moment. This in turn sparked even more accurate measurements, such as by John FLAMSTEED of the ROYAL OBSERVATORY, GREENWICH.

The Copernican hypothesis also implied that the stars were at great distances from the Earth (many thousands or millions of times further than the Sun), presumably in orbit around each other and the Galaxy. The cyclic motion of the Earth around the Sun, and of the solar system through space in its orbit around our Galaxy, causes the stars to shift their positions relative to each other (*see* PARALLAX). The parallax of a star is a measure of its distance; the greater the parallax the closer is the star. Because of the great distances of the stars, the parallaxes and proper motions of even close stars are small, and it is difficult to measure them.

In 1718 Edmond HALLEY showed from measurements by Hipparchus, Ptolemy, Tycho and his own contemporaries like Flamsteed, that Aldebaran, Sirius and Arcturus moved relative to the bulk of the other stars, with a proper motion. The motions at the level of 1 arcsec per year built up cumulatively over 1000 years to something measurable. But as astrometrists' accuracy increased from Tycho's 1 arcmin to Flamsteed's 10 arcsec, no parallax was measured. It was not until the nineteenth century that Friedrich BESSEL, Thomas HENDERSON and Wilhelm STRUVE measured the first three stellar parallaxes, at the 1 arcsec level. The third dimension, the distance of celestial objects in space, remains one of the biggest astronomical challenges. Only distances within our solar system can be measured with precision by radar and laser ranging.

Modern astrometry

Isaac NEWTON recognized that the fundamental reference frame in which to measure motions should be an INERTIAL REFERENCE FRAME, fixed in space. It is difficult to establish such a frame relative to virtual points and lines. The objective of modern astrometry is to measure the motions of stars relative to reference stars that are stationary on the sky. QUASARS form such a stationary background of fixed 'stars' at immense distances. Those brighter quasars that are also radio 'stars' can be measured with a radio interferometer, which observes the angle between radio stars as they pass across the telescope beam, carried by the Earth's rotation. The positions of the quasars form the International Celestial Reference Frame (ICRF), agreed by astronomers to be their current best effort to define an inertial frame in practice. Likewise, the positions of optical stars can be measured with a meridian circle – an optical telescope that moves only along the meridian (the great circle from the southern horizon through the point overhead to the northern horizon). The angle of the stars above the horizon and the time interval between transits of successive stars across the meridian as the Earth rotates constitute the measurements of position.

Stars and quasars can also be linked together by moving a freely pointing telescope from one star to the other. The HIPPARCOS satellite did this, spinning in space and measuring the separation of stars by timing transits of the stars in its telescopes.

These observations are possible for the brighter stars and quasars. The position of the fainter stars in our Galaxy can be determined relative to the quasars by taking images with a telescope used like a camera. The images are recorded on photographic emulsion or electronic detectors such as a charge-coupled device (CCD). Such telescopes (astrographs) must have a large field of view to allow for simultaneous observation of a sufficiently large number of reference stars. Historically the most important examples of astrographs that resulted in the *Astrographic Catalogue* were built around 1900 with a focal length of 3.7 m for the Carte du Ciel project (*see* Paul HENRY).

Distant quasars

Nearby stars

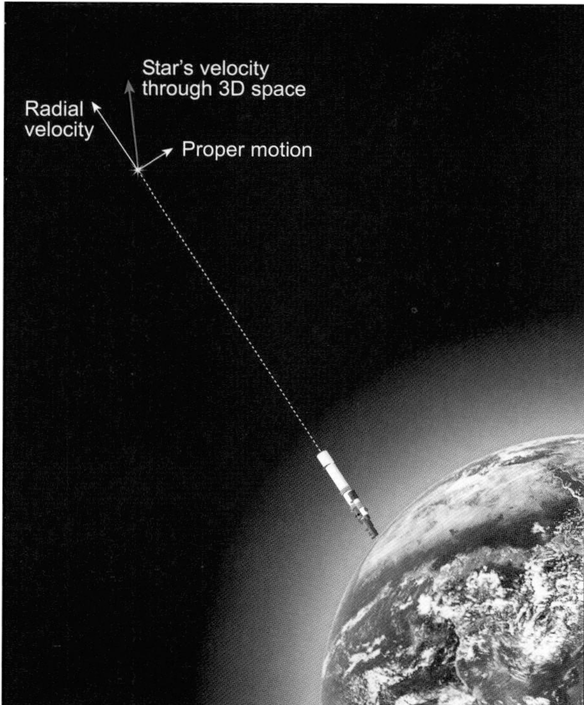

◄◄ ***Star positions*** *The positions of nearby stars are measured by rotating a telescope from one star to another. The positions of distant quasars are measured in the same way. These provide an unmoving inertial network against which small shifts in the positions of the stars are determined.*

◄ ***Stellar velocity*** *A star's velocity through space is the sum of its velocity away from the Earth and its proper motion across the line of sight.*

Astrometric measurements

A modern astrometric telescope with a high-quality wide field of view of several degrees requires a complex optical design, which cannot be achieved with a two-lens refractor or a two-mirror telescope. Multi-lens astrographs are similar to telephoto lenses. For larger apertures (about 0.7 m or more) that can reach a large enough number of stars as faint as quasars, catadioptic systems are often used (reflecting telescopes with additional lenses). Examples are the SCHMIDT TELESCOPE and large reflectors with 'field correctors' to correct optical ABERRATIONS. The number of stars recorded on very wide-angle photographs may be huge, and MEASURING MACHINES such as APM, Galaxy and Cosmos are used automatically to seek them out and measure their positions.

Very precise angular measurements in the order of 1 milliarcsec can be achieved by ground-based optical telescopes, astrometric satellites and radio telescopes. A pair of candles 5 cm apart are separated by 1 milliarcsec when seen from the other side of the world at a distance of 10 000 km.

Currently the most accurate positions of stars across the sky are obtained from VERY LONG BASELINE INTERFEROMETRY (VLBI) radio observations for several hundred compact, extragalactic sources. Some sources change their structure at the 0.1 milliarcsec level, and this is a confusing problem. VLBI is also an expensive technique (in terms of time) and observing is slow, one object at a time. The most accurate astrometry of all is obtained by optical interferometers, and several groups of astronomers are developing these techniques. For example, observers using the Navy Prototype Optical Interferometer at the LOWELL OBSERVATORY predict the ability to separate binary stars with separations of only 70 microarcsec under ideal conditions.

Parallaxes on the 0.5 milliarcsec level are achievable by traditional ground-based, long-focus telescopes with CCDs. However, ground-based telescopes perform at the limits of the atmosphere and next-generation space missions are likely to improve on this by a factor of ten.

The Hipparcos astrometric catalog is the most accurate, giving positions of 118 000 stars on the milliarcsec level. A serendipitous additional outcome of the Hipparcos project produced the most accurate positions for 2.5 million stars of magnitude 9 to 12. These are published in the *Tycho-2 Catalogue* (0.002–0.007 arcsec).

Further reading: Kovalevsky J (1995) *Modern Astrometry* Springer.

Astronomer Royal

An honorary title, established in England by King Charles II for the first director of the ROYAL OBSERVATORY, GREENWICH, founded in 1675. The association of the office with the Royal Observatory was broken after Astronomer Royal Sir Richard Woolley retired in 1971. The title of Astronomer Royal for Scotland was created in 1834 and formally linked with the post of regius professor of astronomy at the University of Edinburgh. The title was in abeyance between 1990 and 1995, during which time it was decided to sever the traditional link with the University of Edinburgh.

Astronomical Society of Australia (ASA)

The Astronomical Society of Australia was formed in 1966, publishes an international journal, *Publications of the Astronomical Society of Australia*, and holds an annual four-day scientific meeting.

Astronomical Society of the Pacific (ASP)

The Astronomical Society of the Pacific has worked since 1890 to advance astronomy in the community. It is the world's largest organization for general astronomy. Its annual meetings bring together professional and amateur astronomers, historians, teachers, students and the public.

Its publications include the professional journal *Publications of the ASP*; the *ASP Conference Series*, which covers over 250 astronomy conferences; and the non-technical magazine *Mercury*. The quarterly teacher's newsletter *The Universe in the Classroom* reaches thousands of schools worldwide. The ASP also produces teaching resources, including books and slides. Its Project ASTRO develops partnerships between astronomers and teachers across North America and the wider world.

Astronomical unit (AU)

The semi-major axis of the elliptical orbit of the Earth. The astronomical unit is also described as the 'mean' distance between the Earth and the Sun, where 'mean' refers to the average of the maximum (APHELION) and minimum (PERIHELION) distances between the Earth and the Sun. The currently accepted value for the AU is 149 597 870 km.

Astronomische Gesellschaft (AG)

The AG was founded in 1863 in Leipzig, Germany, as an international society (currrently 800 members) dedicated to the 'advancement of science by means of supporting projects which require systematic cooperations of many people.'

Astronomische Nederlands Satelliet (ANS)

The first Dutch satellite. Launched in August 1974, it operated until 1977. It carried X-RAY and ultraviolet instruments, and discovered X-rays from cool dwarf stars (like the prototype, UV Ceti). The satellite also discovered the first X-ray burster.

Astronomy

The science of the universe, its constituent bodies and phenomena. Arguably the oldest of the sciences, its beginnings go back to the dawn of recorded history, to the time when mankind first began seriously to study the apparent motions of the Sun, Moon and planets, and to identify patterns and cycles in their behavior. Astronomy provided the basis for calendars, timekeeping and navigation, but it is now essentially a pure science.

Astronomers Royal, England

John Flamsteed	1675–1719
Edmond Halley	1720–1742
James Bradley	1742–1762
Nathaniel Bliss	1762–1764
Nevil Maskelyne	1765–1811
John Pond	1811–1835
George Biddell Airy	1835–1881
William Christie	1881–1910
Frank Watson Dyson	1910–1933
Harold Spencer Jones	1933–1955
Richard Woolley	1956–1971
Martin Ryle	1972–1982
Francis Graham-Smith	1982–1990
Arnold Wolfendale	1991–1995
Martin Rees	1995–

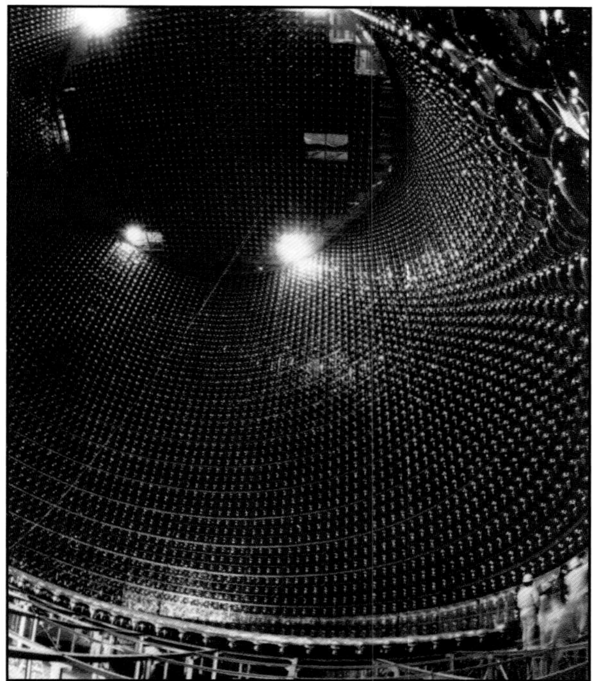

◀ ***Super-Kamiokande*** *Located 1000 m below ground in the Mozumi mine in Japan, Super-Kamiokande contains 50 000 tonnes of pure water and is lined with 11 200 photo-multiplier tubes to detect neutrinos.*

The fundamental types of observation that astronomers carry out include ASTROMETRY (the measurement of positions and motions), PHOTOMETRY (the measurement of brightness and brightness changes), imaging, and SPECTROSCOPY (the detailed analysis of the spectrum of radiation arriving from an astronomical source).

See also ASTROPHYSICS.

Astronomy on the Web

See practical astronomy feature overleaf.

Astroparticle physics

This is an area of research that lies on the boundary of particle physics and astrophysics. Because most interest in the field is in high-energy phenomena, COSMOLOGY (particularly the early universe) is at the core of the research activity.

Neutrinos in astrophysics

Astroparticle physics was developed in the 1970s to examine the proposition that NEUTRINOS, if massive, could have a profound effect on the overall mass density of the universe because they were produced in prodigious numbers in the BIG BANG. In the Standard Model of particle physics (the reference model used by physicists, even though it is known to be incorrect in some details), neutrinos have exactly zero mass, while laboratory experiments show only that the mass is undetectable. Variations of the model can produce neutrinos with mass, which, even if very slight, could dominate the dynamics of the universe.

There are three known types, or 'flavors,' of neutrino. In the early universe, even massless neutrinos played a significant role. If more than the three known flavors were present, this would change the calculations of the abundances of the elements created in the Big Bang (deuterium, helium and lithium). Because the calculations agree so well with observations, this places strong constraints on the particle content of the universe about 2 min after the Big Bang.

Establishing a cosmological limit of three neutrino flavors (which has since been confirmed by accelerator experiments) developed the credibility of the particle-astrophysics connection.

Astroparticle physics has placed constraints on a multitude of particle properties. Recent observations of solar neutrinos have established that the three flavors change from one to another (they 'oscillate'). These experiments provide the first positive indication of a neutrino mass.

The proton

Astroparticle physics is also concerned with the lifetime of the PROTON. Stars, planets and people could not exist if its lifetime were too short. Astroparticle physics calculates that the lifetime of the proton is theoretically at least 10^{32} years. Compare this with the current age of the universe of about 1.4×10^{10} years. Thus, protons decay very rarely. On the other hand, there are a lot of protons to stare at. Since the late 1970s none has been seen to decay, and their lifetime is empirically established at over 10^{30} years.

Grand unified theories

A major motivating force in the development of astroparticle physics is the search for GRAND UNIFIED THEORIES, or GUTs. GUTs attempt to describe three FUNDAMENTAL FORCES in nature (electromagnetism, and the weak and strong nuclear forces) in a single unified theory. GUTs possess the ingredients to solve a long-standing problem in cosmology, namely why the universe is 'asymmetric,' made of matter rather than ANTIMATTER, when matter and antimatter are symmetric and equally likely.

The study of GUTs led to one of the major recent developments in cosmology by introducing the notion of 'inflation' (a period of extremely rapid expansion in the very early universe). Although the early models of inflation were not successful, the study of other exotic phenomena, such as the production of 'cosmic strings,' 'domain walls' and 'textures,' thrived. As theories of unification were pushed to include gravity, even the number of spatial dimensions was no longer held sacred. In string theory the universe is ten-dimensional spacetime, and in extensions of string theory, called M-theory or brane-theory (referring to the membranes that separate domains), the universe is eleven dimensional.

Inflation

Astroparticle physics addresses other problems that occur in the standard Big Bang model. Although the density of the universe is uncertain, it is known that its value is close to the critical value of 1 in certain units. As the universe evolves, the density changes – from an initial value extremely close to 1, to values much larger than 1 if the universe is closed, and to values much smaller than 1 if the universe is open. Since the evolution of the universe is basically a gravitational phenomenon, one might expect that the density evolves on a typical gravitational timescale. The timescale, called the Planck time, is 10^{-44} s. However, the density has remained near 1 for more than 10^{10} years (3×10^{17} s), thus outlasting its welcome by over 61 orders of magnitude. Inflation theory requires that the density should be exactly 1, to which, observationally, it is known to be near.

Another problem with cosmology is related to the near-perfect isotropy of the COSMIC MICROWAVE BACKGROUND (CMB). Its temperature is the same in all directions to extreme accuracy of about 10 parts per million. In standard cosmology

Astronomy on the Web

In the early to mid 1990s, as the World Wide Web began to evolve from a tool for collaboration among research scientists to a global medium for communications and commerce, backyard astronomers jumped in with both feet. Today the Web has become as integral to the hobby of astronomy as the telescope itself. Amateurs use the Web to find information, buy products, interact with other hobbyists, plan and conduct observations and to collaborate with professionals in doing astronomical research.

Virtually every observatory, space mission, astronomy department, research project and individual professional astronomer has a website accessible to the public. This means that people interested in astronomy are no longer dependent on science journalists to tell them when something interesting happens. Anyone can now view press releases and images, and even read research papers, on the Web.

Traditional sources of hobbyist information, such as publishers, telescope manufacturers and equipment dealers (both retail and mail order), also maintain websites. These offer such resources as tips for beginners, telescope reviews and comparisons, interactive sky charts, directories of astronomy clubs and other organizations, and, for magazines, searchable archives of back issues. Because the Web enables instant publication, sources that might otherwise appear only monthly or quarterly now provide updates on astronomical discoveries, events and products as soon as information becomes available. Some astronomy books, especially those dealing with new developments in the science or the hobby, have online editions that, like the *Encyclopedia of Astronomy and Astrophysics* (at www.ency-astro.com), get revised regularly.

Several universities and other accredited educational institutions offer astronomy degree programs online. Lectures are supplied as downloadable videos, discussions with the professor and fellow students are conducted via e-mail or chat and exams are administered via interactive forms. But it is not necessary to enroll in a degree program to learn astronomy at the college level. Most university professors post lecture notes and other supplementary materials online, where anyone can review them. Instruction in practical aspects of amateur astronomy, especially astrophotography and image processing, is also available on the Web, courtesy of several equipment and software vendors, and knowledgable amateurs.

It is even possible to get cosmic questions answered online by submitting them to websites such as 'Ask the Astronomer' and 'Ask the Space Scientist.' These are maintained by professional astronomers enthusiastic about public outreach, sometimes on their own and sometimes with support from NASA or other funding agencies.

Since the late 1980s desktop-planetarium programs have been available for personal computers. These simulate the appearance of the night sky as seen from any location on Earth, on any date, at any time. Most such products are now integrated with the Web. For example, they can be automatically updated with the orbital elements of newly discovered comets and asteroids, which are available online from the IAU.

When amateur astronomers become interested in a new product, they invariably turn to the Web to learn more about it and to find a vendor offering it for sale at a good price. Most manufacturers post their products' specifications online, while reviews and customer opinions can be found on a variety of websites, including those produced by the major astronomy magazines. Many amateurs routinely buy, sell or trade used astronomy books, telescopes and other products using the popular auction site 'eBay' and various classified advertising sites.

The Web has been a boon to astronomy vendors, most of which are small businesses with very limited marketing budgets. Retailers that once served a small, local clientele now ship telescopes and accessories all over the globe, having gained access to new customers by publishing their product catalogs and taking orders online.

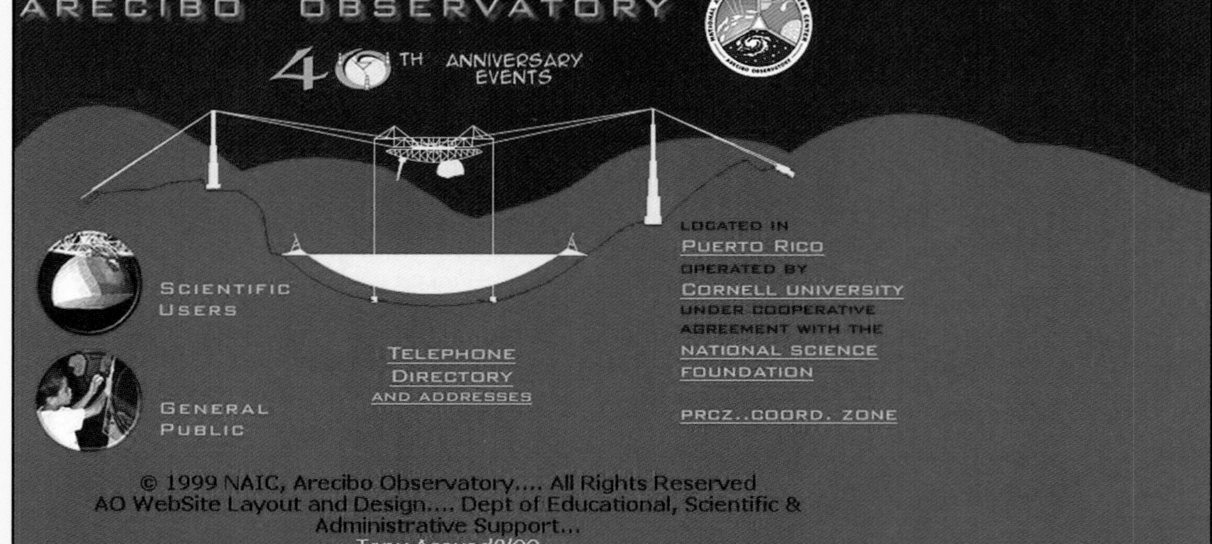

► ***Arecibo home page*** *at www.naic.edu*

Communicating your results

Even before the Web, amateurs used commercial online services such as CompuServe and America Online to communicate with fellow hobbyists. Now much of this traffic has migrated to newsgroups such as 'sci.astro.amateur' and various subject-specific mailing lists and Yahoo! groups, all of which can be accessed via the Web.

Consumer digital cameras are well suited to astrophotography, with or without a telescope. More and more hobbyists are shooting the sky with these devices and getting impressive results. Several websites now invite amateurs to submit their digital astronomical images for posting in online galleries. Some of these sites are automated, so that images appear on the Web only moments after being submitted. The morning after a meteor shower or other widely observed transient celestial event, dozens or even hundreds of pictures from skygazers around the world can be perused online.

Most astronomy clubs post membership information, the agendas of upcoming meetings and the minutes of past meetings on their websites. Amateur astronomers routinely use the Web when planning skygazing sessions. First and foremost, they check the weather forecast. They might also use any of several online almanacs to determine the times of sunrise/set and moonrise/set. They can also print out finder charts for all of the items on their observing list by using interactive sky charts or the Digitized Sky Survey.

Do you think you've found a new comet or asteroid? With utilities available on the website of the Central Bureau for Astronomical Telegrams and the Minor Planet Center, you can find out if any known solar system objects are currently at the position of your suspect object.

A new innovation is the ability to control telescopes and electronic cameras via the Internet using off-the-shelf hardware and software. Some amateurs with remote observing sites are operating their own equipment this way. Others are collaborating with hobbyists in different latitudes and longitudes to give each other access to the entire celestial sphere. Commercial ventures are in the works that will enable anyone, anywhere, to take control of large telescopes and sensitive cameras at dark-sky sites for a modest fee. Armed with high-quality telescopes, sensitive electronic detectors and powerful computers, amateur astronomers are more capable of providing useful data to their professional counterparts than ever before.

The international Virtual Observatory project, which aims to provide access to much of the world's astronomical data via a single Web-based search-and-retrieval interface, will include both amateur and professional observations. Amateurs will be invited to submit their images to be archived and cataloged in the Virtual Observatory database so that researchers can, for example, search for prediscovery observations of supernovae in amateur astrophotos. Amateurs, who generally have more time to pore over images than professionals, will also be able to browse through the database.

Like many technologies, the Web has a downside: some of the information online is wrong or, worse, deliberately misleading. This is inevitable in a medium where almost anyone can publish whatever he or she wants without restriction. So when you find something on the Web, consider the source. Reliable astronomy information generally comes from universities, space agencies, publishers, vendors, clubs and well-known individuals. Much of the rest is debunked on sites like 'Bad Astronomy,' (www.badastronomy.com) where a professional astronomer and educator confronts popular misconceptions and misinformation found on less trustworthy websites.

A sign that people crave not just information but reliable information, is that astronomy magazines like *Sky & Telescope* have thrived and grown since the advent of the Web, not withered as some of the Web's early advocates predicted.

Rick Fienberg is Editor-in-Chief of *Sky & Telescope* magazine. He was formerly president of Sky Publishing Corporation.

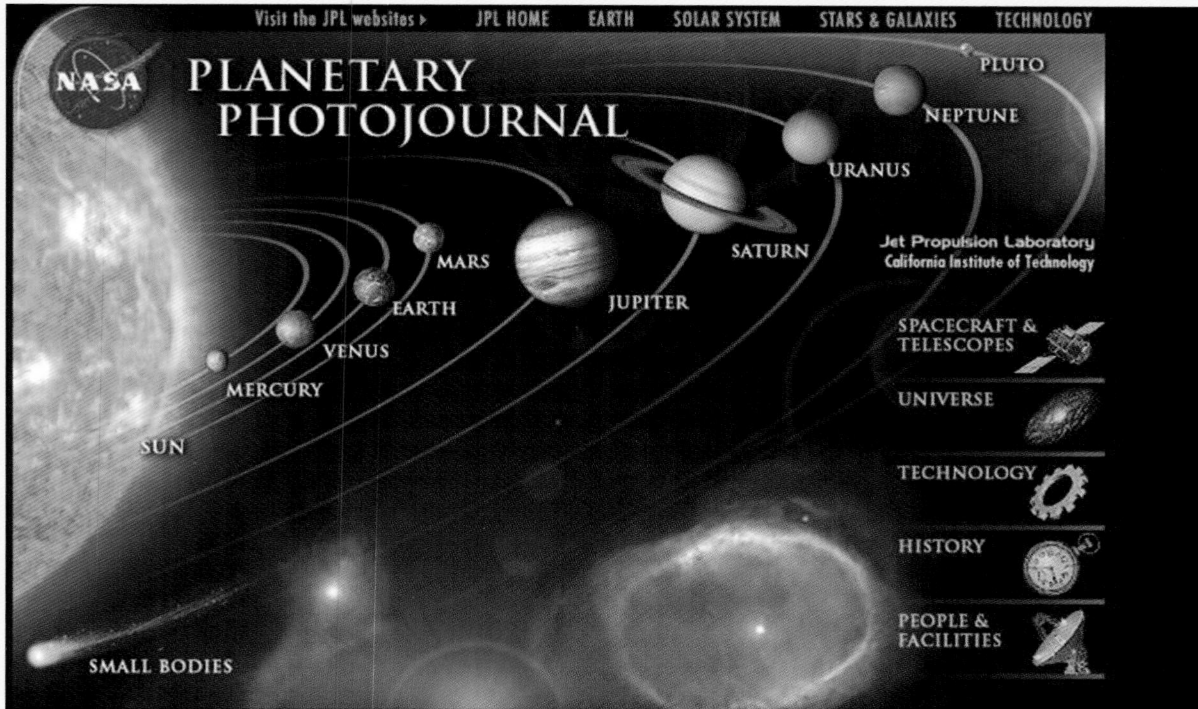

◀ **NASA's site for planetary images** at photojournal.jpl.nasa.gov

▶ ***William Huggins*** *in his observatory in Tulse Hill, near London, UK. He has his hand on a spectroscope, attached to the 15-in refracting telescope. Huggins and his spectroscopic investigations were two of the foundation stones of astrophysics.*

there is no explanation for this isotropy. But in the inflationary theory of the universe there was initially a period of rapid (exponential) expansion. Because our entire visible universe, and more, originated from a small region, everywhere can be traced back to this one initial state, and isotropy of the CMB is expected. Its residual perturbations were caused by quantum fluctuations during inflation. The development of the particle physics associated with the theory of the INFLATIONARY UNIVERSE has been a focus of astroparticle physics since the early 1980s.

Dark-matter particles

Astroparticle physics is particularly interested in DARK MATTER. The issue of the neutrino masses focused attention on whether other stable particles were created in the Big Bang that may have cosmological significance, as components of dark matter.

Although neutrinos are too light for this, other calculations suggest different possibilities. For example, a step on the way to GUTs, known as 'supersymmetry,' predicts a stable massive particle, the neutralino. In principle, the most direct way of searching for such dark matter particles is by accelerator, since it may be possible to produce and detect the actual particle, establishing its identity. However, these searches are typically designed to search for a very specific candidate with the predicted properties. Other searches are more general, looking for something other than the dark-matter candidate itself. For example, the Sun and the Earth orbit through dark matter distributed in our Galaxy. If a dark-matter particle streams through the Sun, it may become trapped. The dark-matter particles in the Sun then annihilate and are a source of high-energy neutrinos. Neutrino detectors (the same ones as used to measure neutrino oscillations) are searching for this flux of neutrinos.

Further reading: Borner G (1993) *The Early Universe* Springer; Kolb E W and Turner M S (1990) *The Early Universe* Addison-Wesley.

Astrophotography

See practical astronomy features overleaf.

Astrophysics

Branch of ASTRONOMY concerned with the physical properties of, and physical processes operating in, stars and other celestial bodies. It may be argued that the subject of astrophysics really began with the development of spectroscopy in the nineteenth century. Today astrophysics has transformed COSMOLOGY, the study of the evolution of the universe, from an almost purely speculative activity to a modern science capable of predictions that can be tested.

Observations of astrophysical significance date back to the proof by Tycho BRAHE that the supernova of 1572 lay beyond the orbit of the Moon and also to the invention of the telescope in the early seventeenth century, when GALILEO GALILEI in 1610 discovered that lunar topography is like the Earth's (the first step toward lunar geology), and that the Sun had spots.

Astroseismology

The study of the internal structure and dynamics of the Sun through the analysis of its oscillations is called HELIOSEISMOLOGY. Astroseismology (also 'asteroseismology') is the study of the oscillations of stars in general. Like the oscillations of a musical instrument, a star will resonate in different ways and at different frequencies; each distinct oscillation is called a mode. The oscillation causes the surface of the star to expand and contract, and this may be seen directly as a variation in radial velocity (*see* DOPPLER EFFECT) or inferred from as a variation in light output, both observables being at a frequency corresponding to each mode of oscillation. Thus the observed light or radial-velocity curve may be quite complex, being the sum of all the modes of oscillation happening at once.

Each mode is an independent probe of the star's interior, and astroseismology is about using this information to determine (or constrain) the interior structure of a star. In the same way, one may infer details about the construction of a musical instrument from the way it plays: brass instruments sound different from strings.

Even if a star pulsates in only one or two modes, this can reveal the mean density. In 1982 Norman Simon showed how details of the interior structure and composition of Cepheids and Delta Scuti stars could be inferred from their light variations. More precise determinations of the internal structure of a star are possible if many pulsation modes can be identified, because each mode is the result of sound waves that travel through different regions of the body of the star. The WHITE DWARFS PG 1159-035 and GD 358 have more than 100 modes detected in each star, the largest number in any star (except for the Sun). Next best are Delta Scuti stars, some of which have 20 modes detected.

Donald Winget predicted in 1981 that white dwarf stars with helium atmospheres would oscillate, just like hydrogen-atmosphere white dwarfs. This was the first example of a class of pulsating star theoretically predicted before discovery.

After seeing the importance of solar pulsations in determining the Sun's structure, people started trying to find similar pulsations in other solar-type stars, with Procyon and ALPHA CENTAURI being two of the favorite targets (because they are stars very like the Sun and they are bright). The oscillations of the Sun hardly affect its brightness, and none of these searches in other stars has been successful.

The analysis of complex oscillations with many modes is made easier by observing from many sites, so there are no gaps in the observations due to the star being below the horizon, confusion by daylight, or interference by weather or equipment availability. Multilongitude, multisite networks have been set up. The Delta Scuti Network and the WHOLE EARTH TELESCOPE are two of the longest running and best known collaborations.

Atacama Large Millimeter Array (ALMA)

The Atacama Large Millimeter Array will be the largest APERTURE SYNTHESIS telescope ever constructed, operating at millimeter and submillimeter wavelengths. It is to be built on the plateau close to Cerro Chajnantor in the eastern Atacama Desert of northern Chile, at an altitude of 5000 m. The project is an international collaboration between Europe and North America and should be ready for scientific use in 2009/10. ALMA will have 64 antennae with a diameter of 12 m, and will operate at wavelengths from 10 to 0.3 mm, allowing it to map at high resolution the molecular gas and dust in nearby star-forming regions, and star-forming galaxies at high redshift.

Atmosphere

Atmospheres are the external gaseous envelopes that surround planets, larger satellites or stars.

The atmosphere of a telluric (Earth-like) planet is a negligible fraction of its mass, but plays an essential role as the interface between its surface and the Sun. The GIANT PLANETS are mostly gaseous, and their atmospheres may constitute their major component.

Planetary atmospheres are diverse, with surface pressures ranging from nanobars to hundreds of bars; surface temperatures ranging from 40 to over 700 K; and different chemical compositions, from the lightest, hydrogen-dominated, to the densest, dominated by carbon dioxide. In spite of these differences there are striking similarities in the interaction of the atmospheres with solar radiation. Condensation takes place on almost all planets, generating cloud structures and promoting seasonal effects.

All nine planets of the solar system have an atmosphere except Mercury, which is too small and too hot. Within the TERRESTRIAL PLANETS, Venus has an atmosphere with a surface pressure of almost 100 bar and a surface temperature of 730 K; while at the surface of Mars, the pressure is only 6 mbar and the mean temperature ranges from around 210 to 230 K; Earth lies between these two extremes. Venus has a thick, white-yellow cloud layer composed of sulfuric acid; the blue planet Earth is surrounded by white clouds; and Mars has a thin, transparent atmosphere with occasional white clouds. All have similar chemical atmospheric composition, dominated by nitrogen and carbon dioxide. In the case of Earth, carbon dioxide is trapped under the oceans in the form of carbonates, and there are large amounts of oxygen, due to the presence of life.

Jupiter, Saturn, Uranus and Neptune also show striking differences in their visual appearance. Although they are all composed mostly of hydrogen and helium, their clouds, composed of various minor constituents, are very different. Pluto, the ninth planet, is a small, icy body surrounded by a very tenuous atmosphere of nitrogen, and shows more similarities with the satellites of the outer solar system (like Triton, Neptune's largest satellite) than with any other planet. Titan, Saturn's largest satellite, has a nitrogen-dominated atmosphere with a surface pressure comparable to that of Earth. Io, the Galilean satellite closest to Jupiter, has a very tenuous atmosphere of sulfur dioxide generated by active volcanism.

The density of an atmosphere generally decreases with height, because the dense lower regions have to support the less dense higher regions. The density structure of a planet can be studied when the planet covers a star in an occultation (*see* ECLIPSE AND OCCULTATION).

The temperature of an atmosphere is a result of its interaction with the incoming solar flux. In the case of the telluric (Earth-like) atmospheres, a small fraction of the solar radiation is absorbed by gas and/or aerosols in the atmosphere itself. Near the surface, the atmosphere is heated by the infrared radiation of the surface, itself heated by the Sun. CONVECTION transports the energy toward higher levels (and creates clouds); this part of the atmosphere is called the troposphere. Above it, in the stratosphere, the energy is transferred by radiation. At even higher altitudes, the solar flux is energetic enough to ionize all atoms. This region, called the ionosphere, strongly interacts with the magnetic field, when present, to produce the AURORA.

The giant planets have no surface to absorb the visible solar flux, which is all absorbed in the troposphere. In addition, with the exception of Uranus, all giant planets exhibit an internal source of energy due to settling of the body of the planet under gravity or radioactive heating.

The greenhouse effect can strongly heat the lower troposphere. Its mechanism is as follows: the surface, heated by the visible solar flux, mostly emits in the infrared range. Carbon dioxide, water and methane can absorb this radiation. The lower atmosphere is thus heated. It radiates back to the surface to heat it even more, and the process spirals. The extreme case is Venus, which now has a surface temperature of 730 K. In the absence of the greenhouse effect its equilibrium temperature would be 229 K. Both Earth and Mars also present a significant greenhouse effect, which causes an increase of 33 K for Earth and 4 K for Mars.

The Sun's radiation falls directly on the surface near the equator of all planets (except Uranus). Its heat is redistributed by thermal winds. The effect of the planet's ROTATION then determines the general circulation, through the Coriolis force. Poleward winds are deflected in the direction of the planet's rotation. A typical example on Earth is the cyclonic motions (in the northern hemisphere) around low-pressure centers at midlatitudes. Similar examples exist in the southern hemisphere of Jupiter.

With the space era, we can now directly measure planetary atmospheres. Some space missions have been especially fruitful: VIKING on Mars, VOYAGER toward the giant planets, and, more recently, GALILEO toward Jupiter. After the success of the MARS PATHFINDER and MARS GLOBAL SURVEYOR spacecraft, several probes will be sent to Mars in forthcoming years. The atmosphere of Venus was extensively explored by the VENERA and PIONEER MISSIONS, and by Galileo on its flyby toward Jupiter. The CASSINI/HUYGENS MISSION includes the descent of the Huygens probe into Titan's atmosphere and the four-year monitoring of the whole Saturnian system by the Cassini orbiter.

Further reading: Atreya S K, Pollack J B and Matthews M S (eds) (1989) *Origin and Evolution of Planetary and Satellite Atmospheres* University of Arizona Press; Shirley J H and Fairbridge R W (eds) (1997) *Encyclopedia of Planetary Sciences* Chapman and Hall.

Widefield astrophotography

Widefield astrophotography is generally defined as celestial photography using a consumer-grade camera or a specialized astrograph without shooting through a telescope. As practiced by the amateur astronomer, widefield astrophotography uses readily available 35-mm and medium-format cameras to photograph celestial phenomena using the various lenses available for these cameras. Any camera capable of time exposures can be used for celestial photography. Older-model, fully mechanical 35-mm single-lens reflex (SLR) cameras are preferred because they can operate over long time exposures without batteries, unlike modern electronic cameras. The versatility of interchangeable lenses and the wide selection of available films make the 35-mm camera ideal for widefield astrophotography.

Astrophotography can be accomplished by using a fixed camera mounted on a tripod, or by mounting the camera on top of an equatorially mounted clock-driven telescope to allow the camera to follow the moving stars. Each mounting configuration produces different results. Exposures longer than 15 s with a fixed camera will show the stars as streaks because of the natural movement of the stars around the celestial pole. By mounting the camera on top of an equatorially driven telescope to follow the motion of the stars, it will record the star images as pinpoints instead.

Over the past decade, film manufacturers have improved both the sensitivity and the long-exposure reciprocity-law-failure characteristics of their medium- to high-speed color films. Reciprocity law failure makes the film appear to lose sensitivity over long time-exposures. Previously, this effect rendered many commercial films impractical for astrophotography because the film seemingly stopped recording greater image density with increased exposure time. As a result of recent film improvements, amateur astrophotographers are now able to record celestial scenes that were impossible to photograph just a generation ago. By using such consumer films as Kodak Ektachrome E200 or Fuji Provia 400 color slide film, or Kodak Supra 400 color negative film, with a 20-s time exposure, the astrophotographer can record celestial objects that are dimmer than the naked eye can detect.

Color slide film is the best way to start because the slide is the finished image. Images from color negative films have to be printed on photographic paper, and most automated photo-processing machines will misinterpret the black sky background as a badly underexposed image and render a poor print.

Fixed-camera astrophotography can be used to record a variety of interesting celestial phenomena with exposures short enough to minimize the trailing of stars. With wide-angle lenses, exposures of up to 30 s will reveal little star trailing. Exposures of between 15 and 20 s through a normal focal-length lens will also remain trail free. The increased image scale of telephoto lenses will magnify the apparent

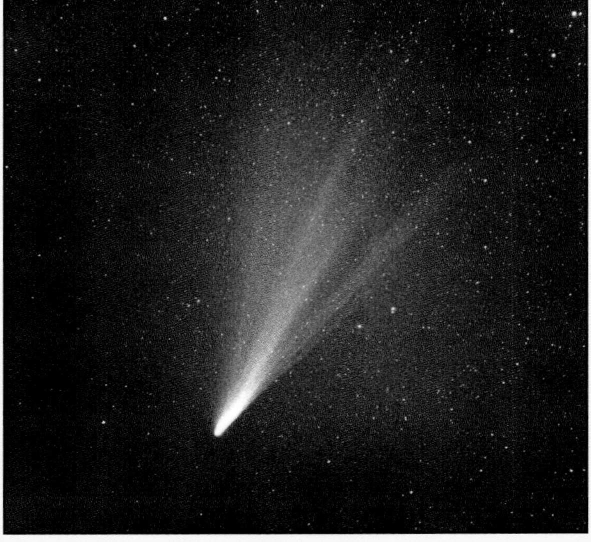

► *Comet West* Captured with a 4-min exposure using an 85-mm lens and Kodak Tri-X film.

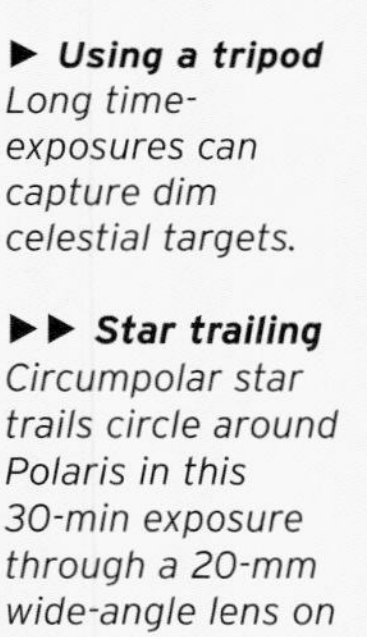

► ***Using a tripod*** *Long time-exposures can capture dim celestial targets.*

►► ***Star trailing*** *Circumpolar star trails circle around Polaris in this 30-min exposure through a 20-mm wide-angle lens on Ektachrome 200 film.*

movement of the stars in the sky, thus exposures through telephoto lenses must be limited to less than 10 s to record untrailed stars. Targets for fixed cameras using short exposures include constellation star patterns, comets, meteors, aurorae, the passage of Earth satellites and long-exposure circumpolar star trails.

To take sky images with a fixed camera, first set the focus at infinity and open the lens aperture to its widest setting. After composing the image, begin the exposure by locking the shutter open using a cable release to minimize camera vibration. Exposure times will be limited by the brightness of the sky to a point known as the sky fog limit. In urban areas near city lights, exposures with 800 ISO film will be fogged by light pollution after about 30 s. Away from the city, under dark skies, exposures up to 10 min or more are possible. To find out how long you can expose from a given location, bracket your initial exposure times from 30 s to 16 min, doubling the length of each succeeding exposure. The point where the background sky brightness becomes objectionable in the resulting images is the sky fog limit for the film and exposure used.

Using a telescope as a mount

Mounting the camera on top of a clock-driven telescope to arrest the apparent motion of the stars allows longer exposures without star trailing, and thus reveals even dimmer celestial objects. This technique, which is sometimes called piggyback astrophotography, requires a telescope and mount that are sufficiently sturdy to carry both the camera and the camera counterweights that are used to maintain the telescope's balance on its mount. Most equatorially mounted telescopes with an aperture of 8 in or larger can carry a piggyback camera.

It is also necessary to ensure that the telescope accurately follows the movement of the stars. This can be done by aligning a bright star on the crosshairs of a guiding eyepiece and using the telescope's electric slow-motion controls to keep the star centered on the crosshairs during the exposure. Targets for widefield astrophotography using the piggyback technique include constellations, the Milky Way, and telephoto images of nebulae and star clusters.

A valuable technique to improve the quality of star images with piggyback astrophotography is to close the lens down one or two f-stops. This eliminates much of the inherent CHROMATIC ABERRATION and vignetting present in most wide-angle and normal lenses. At the widest aperture, chromatic aberration will render stars near the edge of the image field as elongated, out-of-focus blobs. With the lens stopped down, dimmer and smaller stars will remain in focus. While it may seem counterproductive to reduce the lens aperture on nearly invisible celestial targets, the longer exposure times made possible by tracking the stars during the exposure will more than offset the light loss from reducing the aperture. Telephoto lenses suffer less chromatic aberration than shorter focal-length lenses and can generally be used at their widest aperture.

Robert Reeves lives in San Antonio, Texas, and is the author of *Wide-Field Astrophotography* (Willmann-Bell, 1994).

◄ ***Piggybacking*** *Positioning a camera on an equatorially mounted telescope allows the camera to follow the movement of the stars across the sky and render stars as pinpoints instead of streaks.*

◄ ***Cygnus*** *The North America Nebula (upper left) and the Butterfly Nebula (lower right) can be seen in this image of the northern part of Cygnus. Captured using a 20-min exposure through an 8-in lens on Ektachrome E200 film.*

▶ Lunar eclipse *Image by Nik Szymanek, taken with a Meade 10-in LX200 at f/6.3 with Kodak Elitechrome 100 ISO film.*

Astrophotography with telescopes

The modern amateur astronomer is capable of taking beautiful pictures of the night sky, often with surprisingly humble equipment. Most people start out in astrophotography by holding a camera up to the telescope's eyepiece and taking snapshots of the Moon, Sun (with proper filtration) and brighter planets. This is known as the afocal projection method and is now very popular with digital camera users. The telescope is focused in the normal way and then the camera lens is held as close as possible to the eyepiece. A single lens reflex (SLR) camera or digital camera work best because the camera lens also needs to be focused. The limitation is that it only works well with bright objects, as short exposures must be used. Improved results are obtained if the camera and lens are placed on a sturdy tripod and then aligned with the eyepiece. Better still is to use a purpose-built adjustable camera bracket that will hold the camera securely in place behind the eyepiece. As stated, this method works well with both types of camera: film and digital. For long exposure photography of deep-sky objects however, film is still the preferred medium. Best results are obtained if exposure times are greatly extended, typically to half an hour or longer.

Unfortunately modern SLR cameras are not really suitable for deep-exposure astrophotography as they rely on batteries to keep the shutter open. Astrophotographic exposures of faint objects can be quite long, and these will drain the camera's batteries very quickly. The best cameras for this work are those that were manufactured in the late 1970s, such as Nikon, Pentax and Canon. These are rugged, manual cameras that can be purchased for reasonable prices from most second-hand camera shops.

Attaching the camera to a modern telescope allows detailed photography but means that the drive system has to be more accurate. The camera is usually attached by placing an adapter between the rear cell of the telescope focuser and the camera body. The Moon is a good target for telescopic photography because, due to its brightness, exposure times can be quite short. Capturing a sequence of the Moon's changing phases can be rewarding, as can the eerie spectacle of a total lunar eclipse.

For all types of astronomical photography it is a good idea to bracket the exposures. This means using a series of varying exposure times around the likely correct value. By doing this, a good result will usually be attained. Also, by noting the exposure time that gives the best result, a more educated guess can be made the next time the same object is photographed under similar conditions. Because the Moon is a large, bright object, the SLR's exposure meter can be used to give a good starting point for approximate exposures. The Sun is also a good photographic target, but great care must be taken when observing the Sun as even glancing at it can damage the eyes. Care must also be taken with equipment because the heat from the Sun can damage lenses if they are pointed directly at it for too long. The easiest method of solar photography is to project an image of the Sun onto a piece of white card, and then photograph that image. In this way, it is possible to photograph sunspots on a daily basis (*see* SUN, OBSERVING).

Full-aperture and off-axis solar filters give good views of sunspots, as do dedicated solar-observing telescopes. Advanced amateurs also use hydrogen-alpha filters to

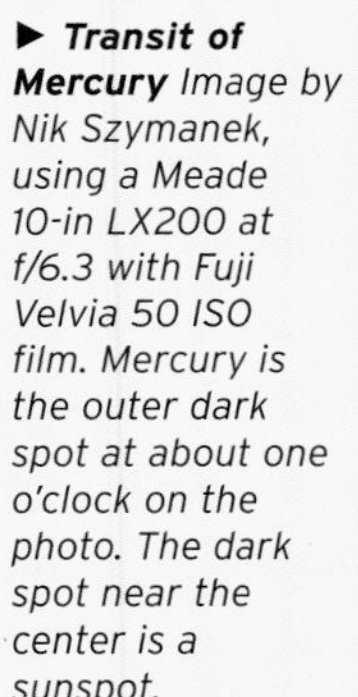

▶ Transit of Mercury *Image by Nik Szymanek, using a Meade 10-in LX200 at f/6.3 with Fuji Velvia 50 ISO film. Mercury is the outer dark spot at about one o'clock on the photo. The dark spot near the center is a sunspot.*

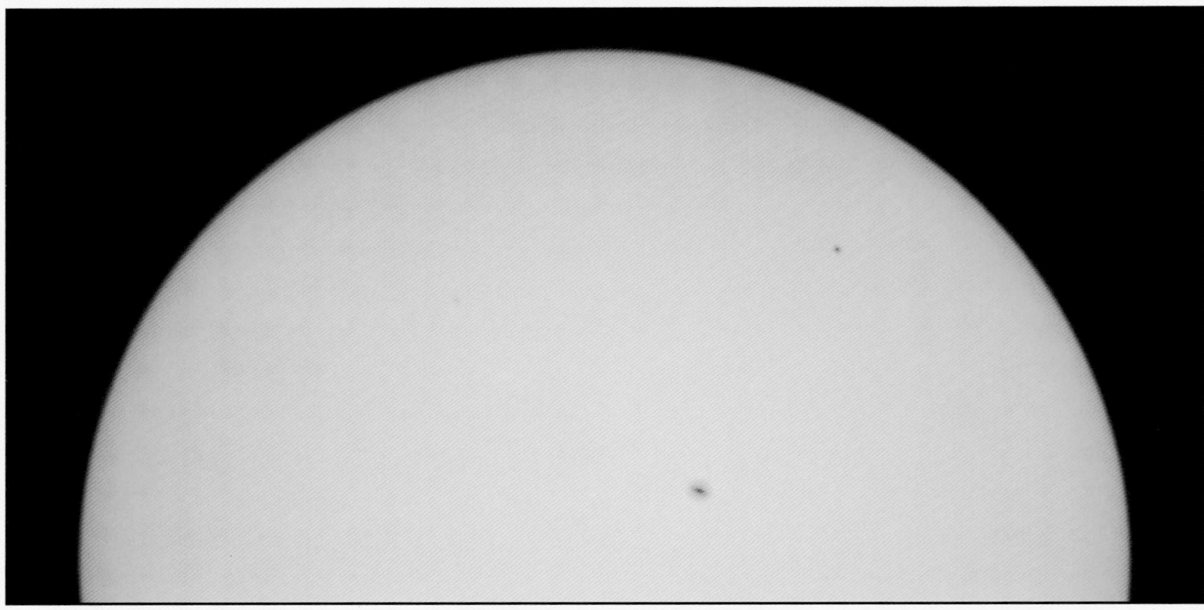

◀ ***Lagoon Nebula***
A diffuse emission nebula in Sagittarius. From Hawaii, Nik Szymanek used a Pentax SDHF 75-mm f/6.6 apochromatic refractor, Kodak Elitechrome 200 ISO film, and an SBIG ST-4 autoguider for this 20-min exposure.

photograph the Sun at a very specific wavelength, chosen to show the dramatic prominences that are visible around the edge of the Sun. Although expensive, these filters produce stunning results, especially when the Sun is in an active phase. As with the Moon, the brightness of the subject means that photographic exposures can be kept short.

To photograph the planets, which have very small angular diameters, one option is to use eyepiece projection. Here an eyepiece is placed between the telescope and camera to project a highly magnified image of the planet onto the film. Although capable of producing good results, it is usually quite difficult due to the unsteadiness of the Earth's atmosphere (poor SEEING) and the difficulty of focusing on such a faint image (*see* FOCUS).

Planetary photography has recently been revolutionized by the introduction of inexpensive webcams and this is the preferred choice of equipment for discerning amateur photographers.

Deep-sky photography

Deep-sky photography of galaxies and nebulae is rewarding but requires very good tracking from a polar-aligned equatorial drive. Although most modern telescopes are equipped with accurate drives, during longer exposures it is necessary to make minor corrections to the tracking of the telescope. Traditionally this has been done in one of two ways. A separate telescope can be attached to the mount and used as a guide telescope to monitor any deviations from perfect tracking performance. This method works well, because bright stars that are suitable for guiding are usually easy to find. The guidescope needs to be fixed rigidly, otherwise flexure can occur and this will degrade the guiding. The second method involves the use of an off-axis guider, attached between the telescope and camera, which diverts the light from a guide star using a small prism or mirror. This works well because the guiding is carried out through the main optics of the telescope so no flexure occurs, but it can be difficult in practice because the choice of suitable guide stars is limited. A modern option is to use a CCD autoguider with the guidescope or off-axis guider. This will deliver the best tracking of all and alleviates the need for an observer to peer through a guiding eyepiece during long exposures.

Focusing on deep-sky objects is quite demanding and is best achieved by using a moderately bright star and racking the focus back and forth until good focus is attained. If the camera's focusing screen can be changed, a good option is to install a high-transmission screen such as those manufactured by Beattie. These feature very bright views and minimal clutter because no 'daytime' focusing aids (split image, microprism and so on) are necessary. Auxiliary equipment that utilizes the precise knife-edge method to achieve exceptionally fine focus can also be purchased from astronomical suppliers.

Traditionally, fast films have been used for deep-sky astrophotography (400–1600 ISO), but one of the best recent films to emerge is Kodak Elitechrome 200 ISO slide film. Several characteristics, such as fine grain, excellent reciprocity characteristics and good red-sensitivity, make it an excellent performer. Also, digitally manipulating astrophotographs is very popular among modern astrophotographers. Film can be scanned and then digitally enhanced to remove light pollution, aircraft and satellite trails, and images can be contrast- and color-enhanced in what is effectively a digital darkroom.

Nik Szymanek drives a District Line train on London's Underground railway and vacations on La Palma, Canary Islands and Hawaii to pursue his astrophotography hobby, as well as observing from his observatory in Essex, UK.

► ***Atmospheric refraction*** *A ray of light from a star, B, is incident on the top of the atmosphere at an angle, i, to the vertical, and refracted to an angle, r, entering the telescope, A. The telescope looks back along the sight line and puts the star at C. The shift in position (i - r) is atmospheric refraction. In practice, atmospheric refraction does not occur at one place, the top of the atmosphere. It is progressive throughout the depth of the atmosphere.*

Atmospheric extinction

The absorption of starlight by gases and AEROSOLS in the atmosphere. In general, the extinction:

- increases away from the ZENITH toward the horizon;
- is smaller from high altitude sites;
- is larger for shorter wavelength light (more absorption for blue than for red – this is why sunsets are red).

The amount of extinction in magnitudes is proportional to the AIR MASS to the star.

Atmospheric pressure

The pressure (the force per unit area) exerted on the surface of a planet by the weight of gas contained in a column extending vertically upwards to the limit of the atmosphere. At the surface of the Earth, the average value of atmospheric pressure, in SI units, is 1.013×10^5 pascal (Pa), also referred to as 1 atmosphere (atm). By way of comparison, the atmospheric pressures at the surfaces of Venus and Mars are, respectively, 90 and 0.006 atm.

Atmospheric refraction

Rays of light entering the Earth's atmosphere (passing from the vacuum of space to the medium of air) are bent, or refracted, so that the apparent position of a star is displaced by a small amount toward the ZENITH. In other words, the apparent ALTITUDE of the star is increased. Close to the zenith the effect is small and is proportional to the tangent of the zenith distance. However, at the horizon (90° from the zenith) the apparent altitude is increased by about 35 arcmin. This means, for example, that at the moment the lower edge of the Sun appears to touch the horizon at sunset, the whole disk of the Sun is actually below the horizon.

Atom

The smallest particle of a chemical element that retains the properties of that element. Each atom consists of a compact nucleus, in which all but a tiny fraction of its total mass resides, surrounded by a cloud of ELECTRONS (lightweight particles with negative electrical charge).

The nucleus consists of a number of PROTONS (massive particles with positive electrical charge) and a number of NEUTRONS (particles of similar mass to the proton but with zero electrical charge), all of these particles being bound together by the strong nuclear interaction. The surrounding cloud of negatively charged electrons is held around the positively charged nucleus by the electromagnetic force. A complete (neutral) atom contains the same number of protons as electrons and, because it has equal numbers of positively and negatively charged particles, has zero net charge. An atom that has lost one or more of its electrons and which, therefore, has a net positive charge, is called a positive ion. An atom that has acquired one or more extra electrons and which, therefore, has a net negative charge, is called a negative ion.

An atom is characterized by its atomic number (denoted by the symbol Z), which is the number of protons in its nucleus, and by its mass number (A), which is the total number of protons and neutrons contained in its nucleus. Atomic number defines the chemical element of which a particular atom is an example. Thus, every hydrogen atom has $Z = 1$, every helium atom has $Z = 2$, and so on. Nuclei of the same chemical element may, however, contain different numbers of neutrons. Atoms that have the same number of protons but different numbers of neutrons in their nuclei, and which, therefore, have the same atomic number but different mass numbers, are called ISOTOPES. For example, 'heavy hydrogen,' or deuterium, denoted by ^2H (or ^2D) is an isotope of hydrogen that has one proton and one neutron, rather than just a single proton, in its nucleus.

The average of the various naturally occurring isotopes of a particular element, weighted according to their relative abundances, is the atomic weight of that element.

Electrons can be visualized as points, localized in space, orbiting the nuclei of various atoms. This is the Bohr model of the atom, proposed in 1913 by the Danish physicist Niels Bohr (1885–1962). Quantum mechanics has shown that the electrons are more like clouds surrounding a nucleus, representing regions where the probability of finding electrons is highest. Nevertheless, the Bohr model continues to provide a good basis for describing the way in which atoms emit and absorb radiation.

See also SPECTRUM.

Atomic time *See* INTERNATIONAL ATOMIC TIME; TIME

Auriga *See* CONSTELLATION

Aurora

An aurora (plural auroras or aurorae) is an extended source of light of different forms and colors that is observable in the atmosphere at high latitudes and sometimes at mid-latitudes. Its brightness may reach the intensity of full moonlight.

The region of maximum occurrence of aurorae is known as the 'auroral zone.' The auroral oval (one in each hemisphere) is the location where most of the aurorae are observable. It is a belt surrounding each of Earth's magnetic poles, which are located in northern Canada and a point diametrically opposite on the Antarctic coast. The belt lies around 70° magnetic latitude and has a mean width of about 5° in quiet conditions. Aurorae occur more frequently in

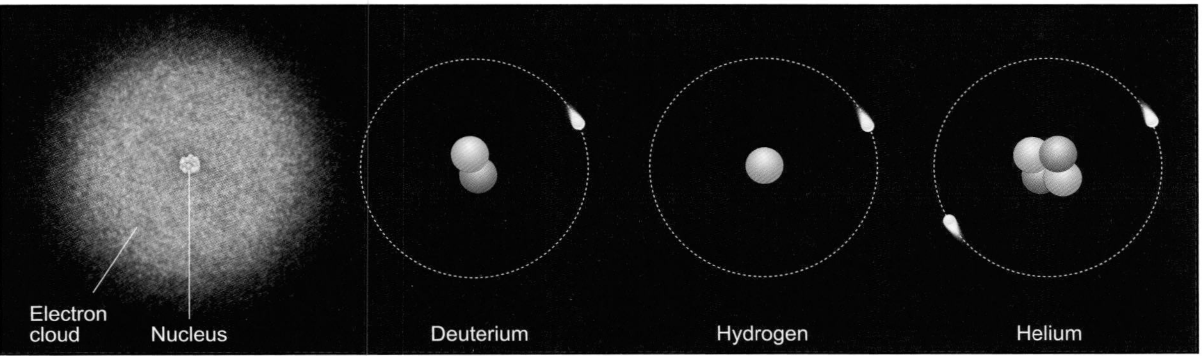

◀ ***Atom***
(Left to right) A quantum mechanical picture of an atom as a nucleus and electron cloud. Three Bohr models of atoms: deuterium, hydrogen and helium. Protons are red, neutrons are blue and electrons are yellow.

the band of the auroral oval than within its central part. This latter region is named the POLAR CAP, and aurorae here have a weaker intensity and a smaller rate of change with time than in the auroral oval. To see aurorae it is not necessary to travel all the way to the poles: Alaska, Canada and Scandinavia are centers for 'auroral tourism.'

Aurorae are visible on other planets. The auroral ovals of Jupiter and Saturn have been imaged by the HST.

Aurorae are generated by particles of solar origin interacting with the ATOMS and MOLECULES of the Earth's upper ATMOSPHERE through direct collisions and chemical processes. The aurorae are 80–300 km high, depending on the characteristics of the particles (ELECTRONS and PROTONS) that initiate them. There are differences between electron and proton aurorae because electrons and protons propagate differently in the upper atmosphere. Proton aurorae are of weak intensity and diffuse. The height of occurrence is around 100 km. The proton auroral oval is shifted slightly toward the equator with respect to the electron auroral oval.

The occurrence and intensity of aurorae follow solar activity. Like sunspots, aurorae have an 11-year solar cycle with, however, a time lag of one to two years. Records of aurorae were used to reconstruct solar activity before the first instrumental observations, or when observations were intermittent. There were far fewer aurorae from 1645 to 1715, confirming the MAUNDER MINIMUM in sunspot numbers. Solar activity resumed in the early eighteenth century, and some intense aurorae were observed at midlatitudes in 1716 and 1726.

In the first few centuries AD, some aurorae were recorded by the Chinese, Greeks and Romans. However, people living in high-latitude regions were more accustomed to their occurrence and they named them the Northern Lights.

The first modern observations of aurorae started in the second half of the sixteenth century with Tycho BRAHE from his observatory at Uraniborg. Sometime later the occurrence of aurorae decreased, especially at midlatitudes. Pierre Gassendi (1592–1655) described several aurorae, observed in the early seventeenth century, and suggested the name 'aurora borealis' (northern dawn).

Edmond HALLEY and later Jean Jacques d'Ortous de Mairan (1678–1771) were asked to study the origin of aurorae by the Royal Society in London and the Académie Royale des Sciences (Royal Academy of Sciences) in Paris, respectively. Mairan suggested in 1733 that the auroral light was generated by a solar fluid impinging upon the Earth's atmosphere. Even though approximate, this was the first hypothesis based on a relationship between the Sun and the Earth. In the second edition of his report, Mairan felt there was a link between sunspots and aurorae as well as their variations in seasonal occurrence. The relationship between the occurrence of aurorae and magnetic perturbations was demonstrated experimentally in 1741 by Olof Hiorter (1696–1750), who observed the tiny perturbations experienced by a compass needle (provided by Anders CELSIUS). Mairan suspected that aurorae could occur in the southern hemisphere and addressed a note to Don Antonion de Ulloa (1716–95), a Spanish naval officer, who confirmed observations of a polar light when rounding Cape Horn in 1745. Captain James Cook (1728–79), exploring the high southern latitudes, confirmed the occurrence of the 'aurora australis' in 1770. In the second half of the eighteenth century, Pehr Wilhelm Wargentin (1717–83) concluded from various observations that aurorae can be seen from different places at the same time and extend horizontally as a belt surrounding the north pole.

Observations of aurorae

The modern means of observing aurorae are spectrometers, which analyze their spectrum; photometers, which study the behavior of one or several emissions lines; cameras, which record the aurora forms as a function of time; and radar. These instruments can be used on the ground or from orbiting platforms, where they provide a global survey of aurorae. However, before instruments were used to study aurorae, observations were made visually. In order to allow comparisons of observations made by different authors, a nomenclature to describe auroral characteristics (intensity, form and color) was adopted internationally. It is still in use.

An auroral SPECTRUM is made of many lines from atomic and molecular constituents of the atmosphere. The spectral line of the greatest intensity is due to the emission of atomic oxygen at 557.7 nm. This is frequently referred to as the 'green line.' Auroral brightness is scaled to the intensity of the green line through the International Brightness Coefficient (IBC), as shown in the table, p. 44. The unit used is the Rayleigh (R), which corresponds to a million photons emitted over a sphere and received per square centimeter per second. This scale was proposed by Michael Seaton in 1954 and formalized by Donald Hunten in 1955.

Visual observations and images recorded by cameras reveal several typical forms. They are accurately described as arcs, bands, rayed or not, diffuse or pulsating surfaces, coronae and flaming. The description of auroral forms is based on an international nomenclature (*see* table, p. 44).

Auroral colors are usually green, red or blue, depending on the atmospheric species involved in the excitation processes. Aurorae of intensity IBC I or II appear colorless. Yellow aurorae are due to the blending of red and green aurorae. Most of the auroral displays are green due to the oxygen emission line at 557.7 nm. However, many other

► **A mixture of green and red** *hues characterized the aurora of September 7, 2002. Using a 6-megapixel digital camera, Terence Dickinson took this photo with a 15-s exposure from Ontario, Canada.*

Observing aurorae

Aurorae range from barely perceptible, nondescript glows near the northern horizon to majestic displays of activity dominated by vivid red and green shafts and arches that fill the heavens like the vaultings of a ghostly cathedral. About once a decade a truly great auroral display occurs that ranks second only to a total eclipse of the Sun as a naked-eye celestial spectacle of nature (*see* ECLIPSE AND OCCULATATION).

The stupendous aurora on the night of March 12/13, 1989, for example, was seen as far south as Guatemala and Turkey. That night, anyone in North America or Europe who happened to be outside was treated to a sky filled with dancing streamers and curtains of auroral light in shades of yellow, green, purple and intense cherry red, rarely seen outside the Arctic. So brilliant was the display that it even overpowered the nighttime illumination of some of the world's largest cities, including Chicago, Detroit and Baltimore.

In rural areas from coast to coast in North America, the 1989 spectacle was unrivaled since the great aurorae of the late 1950s. Once there was no Moon in the sky after midnight, auroral light alone was at times as bright as the full Moon, enveloping the entire sky in a shimmering, multicolored natural kaleidoscope that streamed down from the zenith like the underside of an umbrella, a phenomenon known as the aurora's corona.

Although the March 1989 aurora was exceptional for its intensity and duration (more than 36 hours), other displays in 1990, 1991, 2001 and 2002 were nearly as impressive. In general, the strongest aurorae coincide with the maximum of the 11-year cycle of solar activity, though a major burst can occur unpredictably. For example, on October 28, 2003, a flare on the Sun of record-breaking intensity occurred a year after the period of solar maximum had subsided.

The first detailed aurora report is credited to the Greek scholar Anaxagoras (500–428 BC), who wrote in 467 BC that 'there was seen in the heavens a fiery body of vast size, as if it had been a flaming cloud not resting in one place but moving along with intricate and regular motions.' Aurorae are rare sights from Greece, where only two or three displays are seen in a decade. In Rome, aurorae are equally rare. There, Pliny the Elder recorded the phenomena in these words: 'We sometimes see ... a flame in the sky which seems to descend to the Earth on showers of blood.'

► **An extreme wide-angle lens** *provides a near-overhead view of the corona of a brilliant aurora on November 5, 2001, from Ontario, Canada. This 30-s exposure was taken by Terence Dickinson on Ektachrome 200 film.*

Red is the most common color seen at more southerly latitudes, since red auroral curtains reach the highest altitudes because of the emissions from atomic oxygen. Because of its altitude, the light can be seen from much farther south, although (because of the curvature of the Earth) it is always close to the horizon.

Until recently, the only way to ensure that you did not miss the next aurora was to be standing outside, looking up. That's still a good plan, but thanks to constant monitoring of the Earth's near-space environment by satellites, it is easy to get an accurate overview of global auroral activity on the Web at www.sec.noaa.gov/pmap. This website displays a graphic showing the intensity of the auroral oval within the past two hours, as gathered by the instruments onboard the NOAA Polar-orbiting Operational Environmental Satellite (POES), which continually monitors the power flux carried by the protons and electrons that produce aurorae in the atmosphere. The red color code indicates sufficient activity for the aurora to be visible. If red color covers, or is close to, where you live, there is a good chance that a visible aurora is occurring outside right now (if it is night, of course).

The auroral oval is centered on the magnetic pole in each hemisphere. The maximum brightness of the aurora is in the oval, not at the magnetic poles, as the graphic on p. 43 displays. Typical altitude of the aurora is 100–300 km, though during intense displays it can extend above 800 km. The colors of an aurora depend on its height and the gases involved. Aurorae appear at altitudes where the atmosphere is thin enough for the electric current passing through to make it glow, in roughly the same way as the gas in a fluorescent lamp. Blue and red emissions come from the glow of nitrogen atoms. Green, the most common auroral

◀ ***A typical greenish aurora*** *hugs the northern horizon. Taken with a 6-megapixel digital camera from Ontario, Canada, by Terence Dickinson.*

color, comes from oxygen. In the thinner atmosphere, from 250–1000 km up, a different reaction with oxygen produces red aurorae. Provided the aurora is bright enough to exceed the color threshold of human vision, the strongest light emissions are blue, green or red. These, of course, are the three primary colors. Thus, various combinations of light from different heights will mix to give the fantastic variety of glorious auroral hues.

Photographing aurorae

Photographing the aurora is one of the easiest and most satisfying activities for the amateur astronomer. The most important ingredient is the aurora itself. If you live in Canada or the northern tier of the USA or Europe, you are near enough to the auroral oval to have a reasonable chance of seeing a couple of displays a year.

A standard 35-mm camera, tripod-mounted and loaded with 400-speed color film was for decades the standard equipment for aurora photography. However, the early years of the twenty-first century have seen the rapid replacement of traditional film cameras with digital cameras. In the case of aurora photography, this is generally a good thing. On the plus side, tripod-mounted digital cameras set at 400- speed and exposed for 10 to 20 s at *f*/2.8 will produce aurora portraits at least as well as film cameras. On the minus side, some digital cameras are incapable of taking exposures longer than 8 s, or even 4 s. This is not long enough to capture faint detail.

On the other hand, the top-of-the-line 5–8 megapixel digital cameras available from 2003 can capture brighter and more detailed aurora images, which are superior to anything being done with film. Images recorded with these cameras have brought recording the appearance of aurorae to a new level of accuracy and beauty.

Reporting observations

Observers interested in reporting auroral activity as part of a global network can become involved through the Auroral Activity Observation Network. Report forms and information are available at the website of Solar Terrestrial Dispatch (www.spacew.com). Software that can access real-time aurora predictions and information, called STD Aurora Monitor, was reviewed in *Sky & Telescope*, June 2001. Updated software information is available from Solar Terrestrial Dispatch.

Terence Dickinson is editor of the Canadian astronomy magazine *SkyNews* and author of several guidebooks for amateur astronomers.

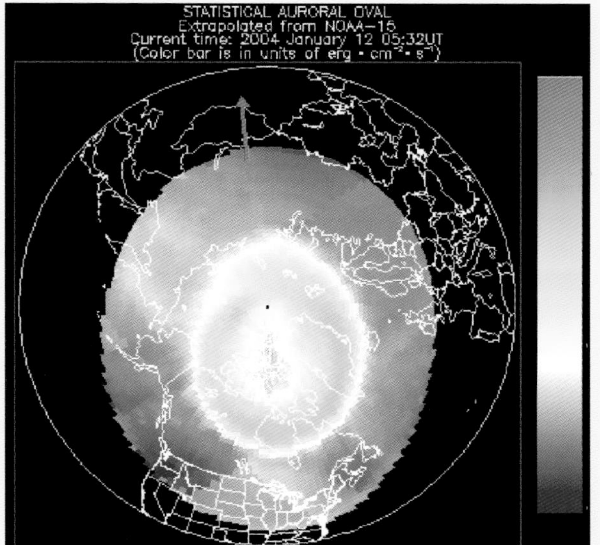

◀ ***The NOAA website*** *allows you to see whether any auroral activity is occurring. At the moment there is none. However, if red color covers, or is close to, where you live, there is a good chance that a visible aurora is occurring outside right now (if it is night).*

Auroral brightness

IBC	Intensity (kR)	Equivalent source
I	1	Milky Way
II	10	Thin moonlit cirrus clouds
III	100	Moonlit cumulus clouds
IV	1000	Full moonlight

IBC = International Brightness Coefficient, R = Rayleigh

Nomenclature of auroral forms

Homogeneous arc (HA)	Regular form in height, direction and intensity.
Homogeneous band (HB)	Similar to HA, but variable along length.
Rays arc or band (RA or RB)	Narrow with luminosity quasi-vertically structured.
Diffuse surface (DS)	Uniform emitting area.
Pulsating surface (PS)	As DS, with intensity variable in time (several seconds).
Pulsating arc (PA)	As PS for an arc.
Corona (C)	Ray system viewed parallel to the magnetic field lines.
Flaming (F)	Forms of variable intensity moving toward zenith.

lines from either atoms or molecular or ionized species exist. The blue color originates from the band at 391.4 nm due to ionized molecular nitrogen.

Auroral intensity changes rapidly. Pulsating surfaces of auroral light change on timescales of 10 s, but certain homogeneous forms are much faster (0.1 s). Changes in time may be due to bulk motion of the light, as for the flaming-type aurorae, which appear to move upward. Apparent horizontal movements are found in arc or band aurorae with velocities as great as 100 km s^{-1} for the strongest magnetic storms. The rapidity of the changes with time is certainly the most striking and spectacular feature of the auroral phenomenon.

Further reading: Bone N (1997) *The Aurora* Wiley; Brekke A and Egeland A (1983) *The Northern Light* Springer; Vallance Jones A (1974) *Aurora* Reidel.

Australia Telescope National Facility (ATNF)

The ATNF, created in 1989, operates the Australia Telescope Compact Array, which is the only radio synthesis interferometer array in the southern hemisphere. The array, opened in 1988, consists of six 22-m diameter antennae, which together simulate a single antenna 6 km in diameter. It is located near the town of Narrabri, about 580 km northwest of Sydney. The ATNF also operates the 64-m diameter radio telescope at the PARKES OBSERVATORY (opened in 1961) and a third telescope near the town of Coonabarabran, midway between Narrabri and Parkes, where there is a 22-m diameter antenna used for millimeter wave astronomy (*see* MILLIMETER AND SUBMILLIMETER ASTRONOMY). Antennae from all three observatories are used together and with those of other institutions for VLBI.

Autoguider

An instrument that senses the position of a star at the focus of a telescope and, if the telescope drifts, provides an error signal to adjust the position of the telescope to track the star accurately. In a space satellite, two such devices (then called 'star trackers') serve to position the telescope to the desired direction, and stabilize the satellite against motion induced by magnetic fields, the solar wind or other perturbations.

Averted vision

A technique for seeing faint objects, in which the observer uses peripheral vision; light then falls on the more sensitive part of the retina. An object that is invisible when stared at directly may swing into view if the observer looks slightly to one side of it. *See also* BINOCULAR ASTRONOMY.

Axis

A reference line in space or drawn through a body. Usually this is a line about which a body has some degree of symmetry; for example, a line passing through the center of a sphere or which runs centrally through the length of a cylinder. For an ELLIPSE the major axis is a line passing through the center and crossing the greatest diameter of the figure.

The axis of rotation is a real or imaginary line passing through a body, around which that body rotates.

Azimuth

The angle, measured parallel to the horizon in a clockwise direction from north, between the MERIDIAN and a celestial body. It is the angle between the north point of the horizon and a point on the horizon vertically below the celestial body.

Azimuth may take any value between 0° and 360°. Together with a value of ALTITUDE, this specifies the position in the sky of a celestial body as seen at a particular instant from a particular point on the Earth's surface.

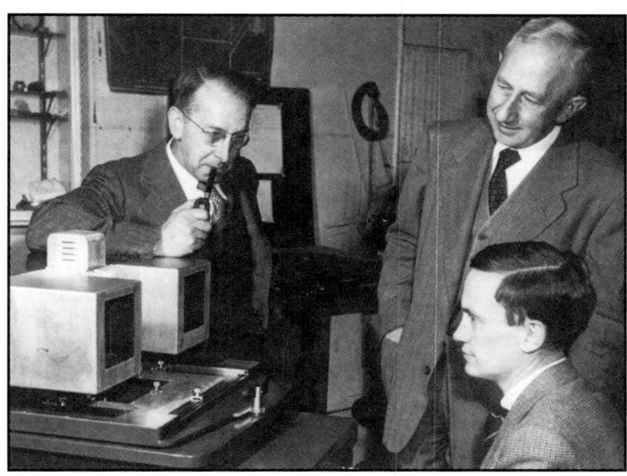

◀ ***Walter Baade*** *(right), Laurits Eichner (left), and Martin Schwarzschild (sitting). They are looking at the astrophotometer created by Eichner. Taken at Mount Palomar Observatory, USA, in 1950.*

Baade, Wilhelm Heinrich Walter (1893–1960)

German-born astronomer who worked at the Hamburg Observatory and moved to California in 1931. He and Fritz ZWICKY proposed in 1934 that SUPERNOVAE could produce COSMIC RAYS and neutron stars (*see* NEUTRON STAR AND PULSAR). Baade made extensive studies of the CRAB NEBULA, identifying it as the supernova of 1054. In blackouts in the Los Angeles area during World War II, Baade, confined as an alien to MOUNT WILSON OBSERVATORY, used the 100-in telescope to resolve stars in the central region of the ANDROMEDA GALAXY for the first time. This led to his definition of two stellar populations (*see* POPULATIONS I and II), to the realization that there were two kinds of CEPHEID VARIABLE star, and from there to a doubling of the assumed scale of the universe. With Rudolph MINKOWSKI, he identified and took SPECTRA of optical counterparts of many of the first-discovered radio sources, including CYGNUS A and CASSIOPEIA A.

Babcock, Harold Delos (1882–1968) and Babcock, Horace Welcome (1912–2003)

Harold was one of the first staff members at the MOUNT WILSON OBSERVATORY in the USA. He worked on precise laboratory studies of atomic SPECTRA, improved the precision of the wavelengths of some 22 000 lines in the solar spectrum, and, with his son Horace, measured the distribution of magnetic fields over the solar surface. They ruled excellent large gratings (the key component used to disperse light in a spectrograph), including those in the Coude spectrographs of the 100-in and 200-in telescopes (*see* COUDE TELESCOPE).

Horace is best known for his pioneering work on the magnetic fields of stars. In 1947 he detected the ZEEMAN EFFECT in the spectrum of 78 Virginis, and went on to investigate the effect in many more stars. Horace worked with his father on observational astronomy and also built many pioneering astronomical instruments. Horace was director of the Mount Wilson and PALOMAR OBSERVATORIES from 1964 to 1978.

Background radiation

See COSMIC MICROWAVE BACKGROUND

Bailly, Jean Sylvain (1736–93)

French astronomer and politician, born in Paris. Bailly studied COMET HALLEY and the satellites of JUPITER, and wrote *L'Histoire de l'astronomie*. He was guillotined following an incident when, as mayor of Paris during the French revolution in 1789, he let the National Guard fire on republican crowds.

Baily's beads

Brilliant points of light seen at the edge of the Moon's disk during a total eclipse of the Sun, just at the onset and termination of totality. They are caused by sunlight passing through the valleys and indentations at the edge of the Moon's visible disk. The English astronomer Francis Baily (1774–1844) drew attention to the phenomenon during the solar eclipse of 1836. *See also* ECLIPSE AND OCCULTATION; ECLIPSES, OBSERVING.

Balmer lines

See HYDROGEN SPECTRUM

Banneker, Benjamin (1731–1806)

The first black American scientist, born in Maryland of a family of former slaves. As a young man, Banneker took apart a borrowed pocket watch and made an enlarged wooden replica that worked for more than 50 years. At the age of 58 he began to study astronomy and was soon predicting eclipses. He compiled this information, as well as other astronomical and tidal calculations, weather predictions, proverbs, poems and essays, into *Benjamin Banneker's Almanac* (1792–1802). This almanac was cited by opponents of slavery (including Thomas Jefferson) as evidence of African-Americans' abilities.

Bappu, Manali Kallat Vainu (1927–82)

Born in India of an astronomer father, Bappu got his PhD in the USA in 1952 at Harvard University. He co-discovered Comet Bappu–Bok–Newkirk and, while working with Olin WILSON at the MOUNT WILSON OBSERVATORY, discovered the Wilson–Bappu effect (an indicator of the absolute magnitude and hence the distances of late-type DWARF STARS).

Barium star

See CH STAR AND BARIUM STAR

Barlow lens

An additional lens, named for Peter Barlow (1776–1862), that increases the effective FOCAL LENGTH and MAGNIFICATION of a telescope. It is a negative diverging lens (either concave on both sides or, more usually, plano-concave: flat on one side and concave on the other) that is placed in the converging cone of light a short distance in front of the focal plane of the OBJECTIVE LENS or PRIMARY MIRROR. By decreasing the angle at which the light rays converge, it increases the effective focal length of the telescope and enables a given eyepiece to achieve a higher magnification. The amplification factor (the ratio of the effective focal length of the objective with the Barlow lens in place, to the focal length of the objective alone) of a Barlow lens of focal length f_b, placed a distance, d, inside the focal plane of an objective is $f_b/(f_b - d)$. For example, if a Barlow lens with a focal length of 100 mm is placed 50 mm inside the FOCUS of the objective, the effective focal length will be increased by a factor of 100/(100 – 50), which equals 2.

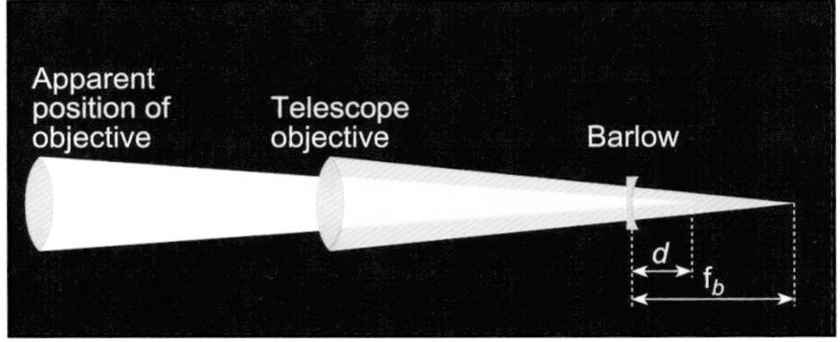

◀ ***Barlow lens*** *This is a popular way to increase the power of a telescope's eyepiece. A ×2 Barlow lens will double the power of an existing eyepiece.*

Barnard, Edward Emerson (1857–1923)

An American astronomer, born in Tennessee into an impoverished family. From a young start as a photographer's assistant, he made his own telescope and became an exceptional observer. A wealthy patron of astronomy awarded $200 each time a new COMET was found. Barnard discovered eight comets, which earned him enough money to build a house and to gain an education at Vanderbilt University, Tennessee. He joined the initial staff of the LICK OBSERVATORY, then in 1895 he moved to the new YERKES OBSERVATORY. He discovered BARNARD'S STAR, Jupiter's satellite Amalthea, and (using widefield photography) dark clouds, globules and nebulae in the Milky Way, including BARNARD'S LOOP around Orion.

Barnard's Loop (Sh2-276)

A large EMISSION NEBULA in the CONSTELLATION Orion, in the form of an arc centered on the Sword of Orion. The emission is thought to be caused by radiation pressure from the hot stars in the Sword region ionizing interstellar MATTER. It was discovered by Edward BARNARD.

Barnard's Star

A noteworthy star in the constellation Ophiuchus, it lies about 34° south of VEGA and 28° west of ALTAIR. Despite being faint (apparent magnitude 9.54 and visible only with powerful binoculars or a small telescope) it is interesting for several reasons. With a PARALLAX of 0.549 arcsec (distance 5.9 l.y.), only the triple system of ALPHA CENTAURI is closer to us. Discovered in 1916 by Edward BARNARD, this star still has the largest known PROPER MOTION, which is 10.36 arcsec per year. This is equivalent to moving the distance of the Moon's diameter every 174 years. It is a red SUBDWARF star of spectral type sdM4, and has very low intrinsic luminosity (absolute magnitude 13.2). A study of its observed motion by Peter VAN DE KAMP suggested the presence of a faint, low-mass, close companion, but this has not been confirmed.

Barred spiral galaxy

A galaxy in which the SPIRAL ARMS emerge from the ends of a bar (or elongated ellipsoid of stars) that straddles the nucleus, rather than from the nucleus itself. Barred spiral galaxies, denoted by 'SB' in the Hubble classification scheme (*see* GALAXY), are classified according to the size of the bulge in the center of the galaxy, the tightness of the spiral pattern and the degree of clumpiness in the arms. It is possible that our own MILKY WAY GALAXY is a barred spiral.

Barycenter

The center of mass of the Earth-Moon system or any other set of bodies (for example, the solar system). If two bodies move under the influence of their mutual gravitational attractions, they will describe ORBITS around their center of mass. The relative distances of the two bodies from this point will be inversely proportional to their masses. For example, if one body is twice the mass of the other, the center of mass will lie one-third of the way from the more massive body to the less massive one (their distances are in the ratio 1:2).

In the case of the Earth and Moon, the ratio of masses is approximately 81:1. The barycenter is therefore located at a distance from the Earth's center equal to 1/82 of the distance to the center of the Moon. This is about 4700 km from the center of the Earth, which is in fact located within the body of the Earth.

Baryon

Subatomic particle that is composed of three QUARKS and that is acted on by the strong nuclear force. This class of particle includes nucleons (PROTONS and NEUTRONS), which are the constituents of atomic nuclei; and a variety of heavier but short-lived particles called hyperons. MATTER that is composed of baryons is known as 'baryonic matter.'

Basalt

Volcanic igneous rock produced by the cooling of lava. Its mineral content is primarily plagioclase and pyroxene.

Basin

A huge, bowl-shaped, often multi-ringed crater on the surface of any of the rocky planets, caused by the impact of a massive METEOROID. Basins were produced during the early history of the SOLAR SYSTEM. Many basins have since been filled with lava, such as the maria on the lunar surface (*see* MARE).

Bayer, Johann (1572–1625)

German lawyer and astronomer whose catalog and atlas *Uranometria* (1603) depicts the positions of double the 1000 stars in Tycho BRAHE's catalog, on which it is based. He designated the brighter stars by the letters of the Greek alphabet (usually in brightness order), followed by Latin letters. Bayer's lettering system is still used.

Becklin–Neugebauer object

One of the most powerful infrared sources known, located in the KLEINMANN–LOW NEBULA behind the ORION NEBULA, and discovered by Eric Becklin and Gerry Neugebauer in 1967. It is thought to be a very young and massive B star embedded in a dense, expanding dust envelope that absorbs the star's visible light and re-emits it in the infrared.

Beehive Cluster (M44) *See* MESSIER CATALOG

Beijing Astronomical Observatory (BAO)

In 2001 the BAO was designated as the headquarters of the National Astronomical Observatories of China (NAOC), to link China's main observatories. It has numerous optical, radio and millimeter telescopes including the new 6.67 m × 6.05 m large-area multi-object spectroscopic telescope (LAMOST) at the Xinglong station. The programs of the still independent PURPLE MOUNTAIN OBSERVATORY and SHANGHAI ASTRONOMICAL OBSERVATORY are coordinated by the NAOC.

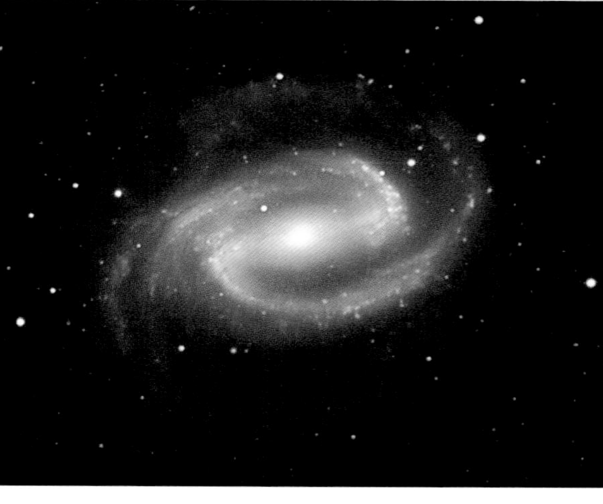

▶ ***Barred spiral galaxy*** *NGC 1300 in the southern constellation of Fornax shows the 'bar' extending more or less straight out on either side of the nucleus.*

Belgium, Royal Observatory of

The observatory was founded in 1826, and installed in its present site in Brussels in 1891. It maintains the Sunspot Index Data Center (SIDC), which was founded in 1981, monitors solar activity and manages a large planetarium.

Bell-Burnell, Susan Jocelyn (1943–)

Known as Joceyln Bell, a radio astronomer born in Northern Ireland, UK. Bell-Burnell studied at Glasgow and Cambridge, where in 1967 she was a research student working with Antony HEWISH on interplanetary SCINTILLATION. She noticed an unusually regular signal, shown to be bursts of radio energy at a constant interval of just over a second. This was the first pulsar (*see* NEUTRON STAR AND PULSAR). Hewish was awarded the Nobel prize for this discovery.

BepiColombo mission

BepiColombo is ESA's Cornerstone mission to MERCURY, to be launched in 2011/12 on a journey lasting up to three-and-a-half years. It is ESA's first mission in cooperation with Japan. The mission is named for the Italian astronomer Giuseppe (Bepi) Colombo (1920–84), who studied Mercury in detail and first explained the peculiar rotation of Mercury which rotates on its axis three times for each two orbits around the Sun. Colombo also worked on satellite orbits and interplanetary traveling.

The only probe so far to have visited Mercury is NASA's Mariner 10 (*see* MARINER MISSIONS). It flew past the planet three times in 1974–5 and returned close-up images of the planet but was only able to map about half of the surface. Mercury is too close to the Sun to be safely imaged with the HST and so the rest of Mercury's surface is still a mystery to us.

BepiColombo will consist of two orbiters and a lander. One orbiter will map the planet, a second will investigate its magnetosphere and a third will land on Mercury to study the surface. All of ESA's previous interplanetary missions have been to the relatively cold outer parts of the solar system. BepiColombo will have very different problems; it must brake against the Sun's gravity rather than accelerate away from it and it will have to withstand very high temperatures when facing the Sun and cold temperatures when in the shadow of the planet.

BeppoSAX

Italian-Dutch gamma-ray and X-ray astronomy satellite that was launched in April 1996. It was named in honor of Italian physicist Giuseppe Occhialini (1907–93), whose nickname was Beppo. The satellite was used for spectroscopic and time-variability studies of X-ray sources in the energy band 1–200 keV, including an all-sky monitoring investigation of X-ray NOVAE in the 2–30 keV energy range. In 1997 it played a major role in the identification of gamma-ray bursters by pinpointing X-ray emissions from these events, so enabling confirmation of their cosmological distances. BeppoSAX was switched off on May 2, 2002, having made nearly 1500 observations and discovering more than 50 gamma-ray bursts. *See also* GAMMA-RAY ASTRONOMY; X-RAY ASTRONOMY.

Besançon Observatory

Founded in Besançon, France, in 1878 to provide an astronomically derived time-reference service for the region's watch-making industry. Today the observatory continues its research and service activities in the field of time and frequency metrology. It has also developed active research groups in stellar and galactic evolution and in the chemical physics of astrophysical media.

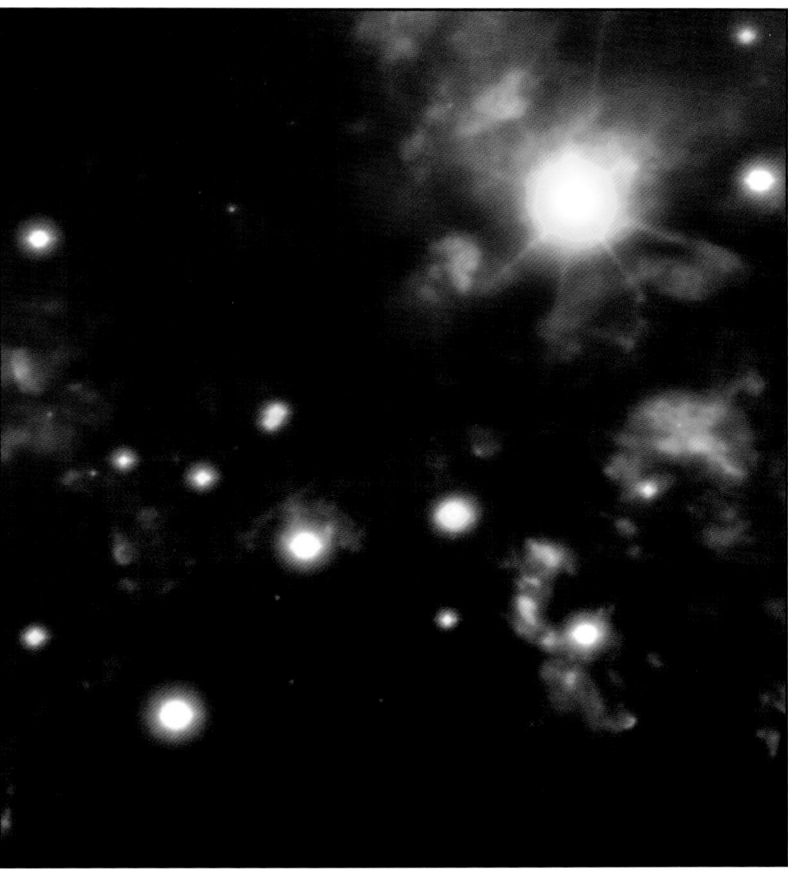

▲ ***Becklin–Neugebauer object*** *The brightest star in this image of the Orion star-forming region, taken in the near-infrared using the VLT.*

Bessel, Friedrich Wilhelm (1784–1846)

A German astronomer, geodesist and mathematician. At the age of 26 he became director of the new Königsberg Observatory, Germany, where he remained for the rest of his life. In 1838 he published the first recognizably accurate stellar PARALLAX (10 l.y. for 61 Cygni) (but *see* Friedrich STRUVE's and Thomas HENDERSON's claims to priority). Bessel discovered the orbital deflections of SIRIUS and Procyon, due to their then-unseen WHITE DWARF companions.

Be star

A Be star (pronounced 'bee-ee' star) is a non-supergiant B-type star whose spectrum emits or has emitted one or more Balmer lines (*see* HYDROGEN SPECTRUM). Be is the notation for the spectral classification of such a star (*see* STELLAR SPECTRUM: CLASSIFICATION). Classical Be stars are believed to have acquired the circumstellar material that produces the Balmer emission through ejection of matter from their photospheres instead of from mass ACCRETION. As a group, Be stars are rapidly rotating. Spectral analysis, and, more recently, interferometric imaging, suggests that the circumstellar material is in the form of a disk positioned around the equator of the star. Be-shell stars are classical Be stars that also exhibit narrow absorption lines, as well as central narrow components to the Balmer emission lines that extend below the continuum. The current consensus is that the Be-shell stars are classical Be stars viewed nearly equator on. *See also* DELTA SCORPII.

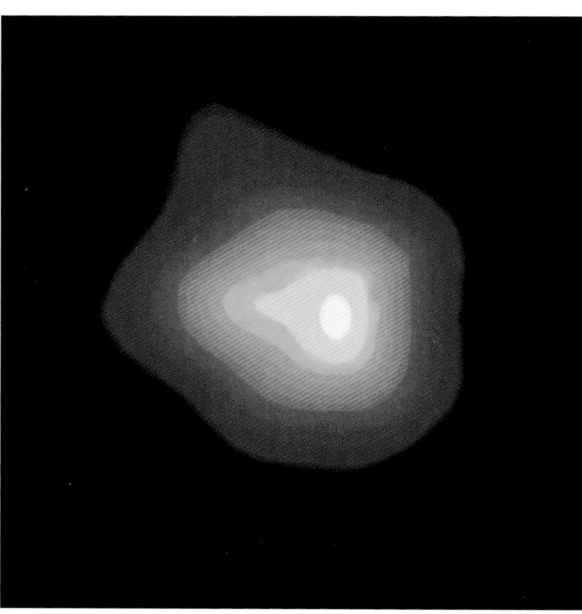

► ***Betelgeuse*** *The star is so big and sufficiently close that modern instruments can see detail on its disk. Radio images from the VLA show it had a bright spot in 1998 – its spots come and go from year to year. The VLA image also shows giant plumes propelling gas from its surface up into its atmosphere. If Betelgeuse were at the center of our solar system, its atmosphere would extend beyond Saturn.*

Beta Lyrae

Prototype of a class of close BINARY STAR in which mass transfers rapidly from the more massive star onto the companion. In the case of Beta Lyrae itself, the primary is a B8.5 supergiant. The secondary is hidden by a thick torus (doughnut) of material forming an ACCRETION disk around it. The system is an ECLIPSING BINARY. The orbital period of about 12.9 days is increasing at the rate of 19 s per year due to the transfer of material.

Beta particle

An energetic ELECTRON or POSITRON emitted during a type of radioactive decay, called beta decay.

Beta Pictoris

A MAIN SEQUENCE star of apparent magnitude 3.85, found to be a very strong emitter of infrared radiation by the INFRARED ASTRONOMY SATELLITE survey in 1983. Extensive studies have revealed the presence of an edge-on disk of dust surrounding the star, which is possible raw material for the formation of a planetary system. This is particularly interesting because Beta Pictoris is similar in size to the Sun (1.4 solar diameters), although hotter (spectral type A3V).

Betelgeuse

The star Alpha Orionis, a red supergiant of spectral type M2Ib. At a distance of 427 l.y. (PARALLAX 0.008 arcsec), its apparent brightness arises from its high intrinsic luminosity, which is about 14 000 times that of the Sun. It was the first star to have its diameter reliably determined – by Albert MICHELSON using an INTERFEROMETER attached to the 100-in MOUNT WILSON telescope. The observed angular diameter of 0.047 arcsec is equivalent to an actual diameter more than double the size of the orbit of Mars. Betelgeuse is a semi-regular pulsating VARIABLE STAR, its magnitude varying by ±0.15 over a period averaging about 6.4 years. It is also a radio source and an emitter of strong infrared radiation. Observations by IRAS indicated that three concentric shells, ejected from the surface of the star in the past, are the sources of the long-wavelength radiation. There are several very faint optical companions.

Bevis, John (1695–1771)

English doctor and amateur astronomer. Bevis is the only person to have telescopically observed (through patchy cloud from Greenwich) the occultation of one planet by another: namely the occultation of Mercury by Venus on May 28, 1737 (*see* ECLIPSE AND OCCULTATION). The next chance, in theory, is an occultation of Neptune by Mercury in 2067. Bevis erected an observatory in Stoke Newington, England, from which he observed star positions and created a star atlas, called *Uranographia Britannica* (1750). The printer went bankrupt and fewer than 20 proof copies survive. The complementary catalog was never published. Bevis discovered the CRAB NEBULA, apparently in 1731, and illustrated it on the atlas. Charles MESSIER had access to a copy, which he referred to as 'the English Atlas.'

Bielids

A periodic METEOR SHOWER that produced November METEOR STORMS after the break-up of COMET BIELA in the 1850s. The shower, also known as the Andromedids, had its RADIANT in the constellation Andromeda. The comet, which had a period of 6.6 years, was last seen in 1852. It should have made a favorable return in 1872, but instead there was a meteor storm, peaking at 6000 per hour. The shower returned again in 1885 with a spectacular display that touched 75 000 meteors per hour. Weaker activity was seen in 1899 and 1904, after which gravitational perturbations by Jupiter shifted the Bielid stream away from Earth's orbit. It may return again early in the twenty-second century.

Big Bang

This theory asserts that the UNIVERSE originated a finite time ago by expanding from an infinitely compressed state. According to this model, space, TIME and MATTER originated together, and the universe has been expanding ever since. Key stages in the history of the Big Bang universe are summarized below.

According to the standard Big Bang model, the very early universe consisted of an extremely high-temperature mixture of radiation (PHOTONS) and particles. In accordance with the equivalence of mass and energy (matter and ANTIMATTER) inherent in Albert Einstein's theory of SPECIAL RELATIVITY, collisions between high-energy photons would have transformed radiation energy into particles of matter. To be consistent with the conservation laws of particle physics, a collision between sufficiently energetic photons creates a particle-antiparticle pair (for example, an ELECTRON and a POSITRON, or a QUARK and an antiquark). Conversely, a collision between a particle and its antiparticle results in the annihilation of the particle-antiparticle pair and their conversion into photons.

As the universe expanded and cooled, photon energies quickly declined below the thresholds at which they could continue to create particle-antiparticle pairs. After that, particles and antiparticles rapidly collided and annihilated each other. BARYONS and antibaryons (particles such as PROTONS, NEUTRONS and the quarks of which they are composed) underwent mutual annihilation about one-millionth of a second after the initial event, when the temperature dropped below 10^{13} K. Less-massive electrons and positrons experienced a similar event a few seconds later, when the temperature dropped to about 5×10^9 K. Had there been exactly the same number of particles and antiparticles, virtually no matter would have remained to form stars and galaxies. Because the destruction of each particle-antiparticle pair produced a pair of photons, and the ratio of photons to baryons in the present-day universe is nearly a billion to one, it follows that the excess of particles over antiparticles in the early universe must also have been about a billion to one.

About 100 s after the beginning of time, when the temperature had dropped below 10^9 K, nuclear reactions combined some of the protons and virtually all of the neutrons into nuclei of HELIUM, together with small quantities of other light elements and ISOTOPES, such as deuterium, helium-3 and lithium. At the end of this phase there were about 11 HYDROGEN nuclei (protons) for every helium nucleus. Because helium nuclei are four times heavier than hydrogen nuclei, this corresponds to a ratio, by mass, of hydrogen to helium of about 73:27. The observed relative abundances of the lightest elements closely match the predictions of the Big Bang theory (*see* COSMIC ABUNDANCE OF ELEMENTS).

After that, the expanding universe consisted of a mixture of atomic nuclei, electrons and photons in thermal equilibrium (frequent collisions between photons and particles ensured that matter and radiation shared a common temperature). A few hundred thousand years after the initial event, when the temperature everywhere had declined to a few thousand degrees kelvin, nuclei were able to capture electrons to form complete neutral atoms. This process mopped up the free electrons that had been responsible for scattering photons, thereby making space opaque to ELECTROMAGNETIC RADIATION. The close connection between matter and radiation was broken (this event is called the 'decoupling' of matter and radiation), space became essentially transparent, and the primordial radiation content of the universe was free to spread ever more thinly through the expanding volume of space. The dilute, REDSHIFTED remnant of the primordial radiation is detectable today as the COSMIC MICROWAVE BACKGROUND radiation.

Some time later, probably within the first billion years after the decoupling, clumps of matter aggregated to form galaxies, clusters of galaxies, superclusters and the other large-scale features of the universe. How, when and in what order these various structures appeared is a matter of debate.

The Big Bang theory is consistent with the observed expansion of the universe, the age of the oldest stars, the relative abundance of the lighter chemical elements, and the existence and properties of the cosmic microwave background radiation. Astronomers' attentions are now focused on the properties of the Big Bang itself (*see* ASTROPARTICLE PHYSICS; INFLATIONARY UNIVERSE).

See also UNIVERSE: COSMOLOGICAL THEORY.

Big Bear Solar Observatory (BBSO)

The Big Bear Solar Observatory is located at the end of a causeway in a mountain lake in California more than 2 km above sea level. The site has more than 300 sunny days a year, and atmospheric properties caused by the lake make for very sharp images. The BBSO is operated by the New Jersey Institute of Technology. It is the only university observatory in the USA that makes high-resolution observations of the Sun.

Big Dipper

The well-known, pan-shaped ASTERISM that forms part of the constellation of Ursa Major.

◄ ***Big Bang*** *The beginning of everything – space, time and matter.*

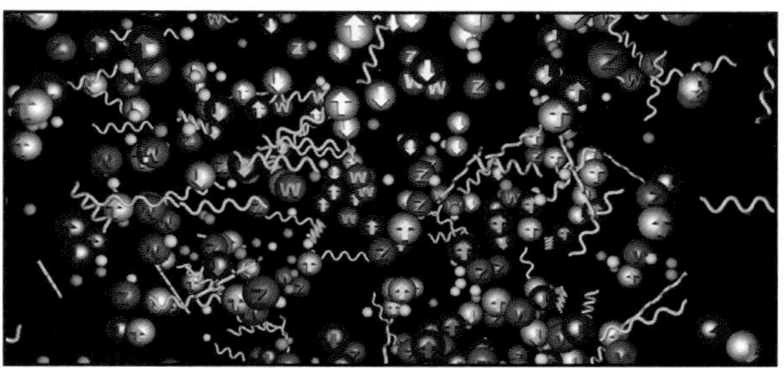

The very early universe was a high-temperature mix of radiation (wavy lines) and subatomic particles such as quarks (gray).

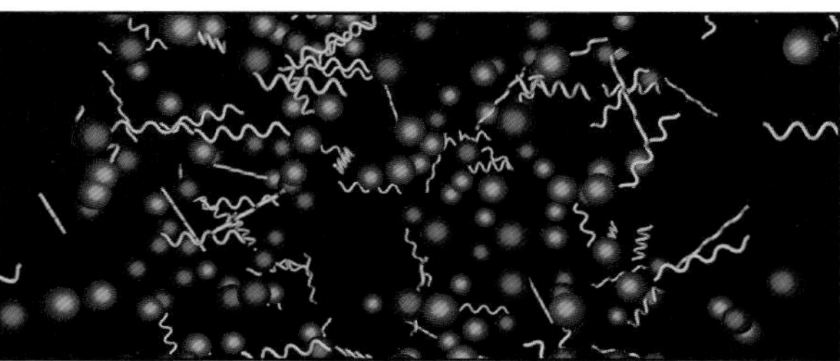

10^{-4} s after the Big Bang the temperature of the universe is 10^{12} K. Quarks combine to form protons (red) and neutrons (blue) – the building blocks of atoms.

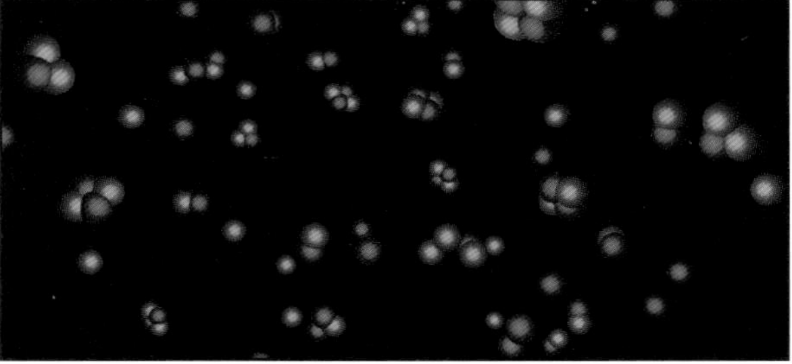

100 s after the Big Bang the temperature has reduced to 10^9 K. Protons and neutrons combine to form the nuclei of helium.

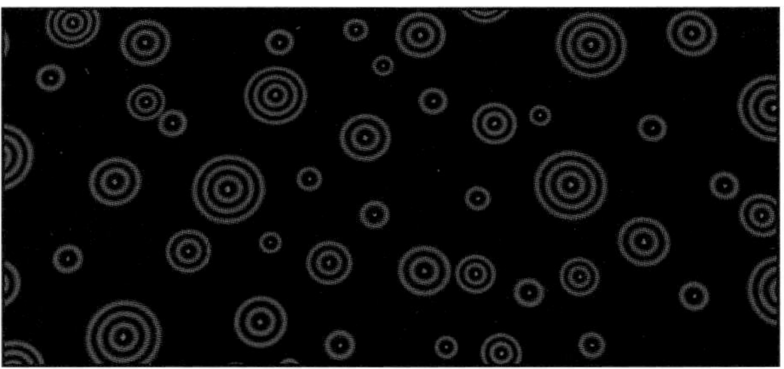

300 000 years later the temperature has fallen to 4000 K. Electrons stick to nuclei, and the first atoms form.

10–15 billion years later the universe is as we know it today, populated with stars and galaxies.

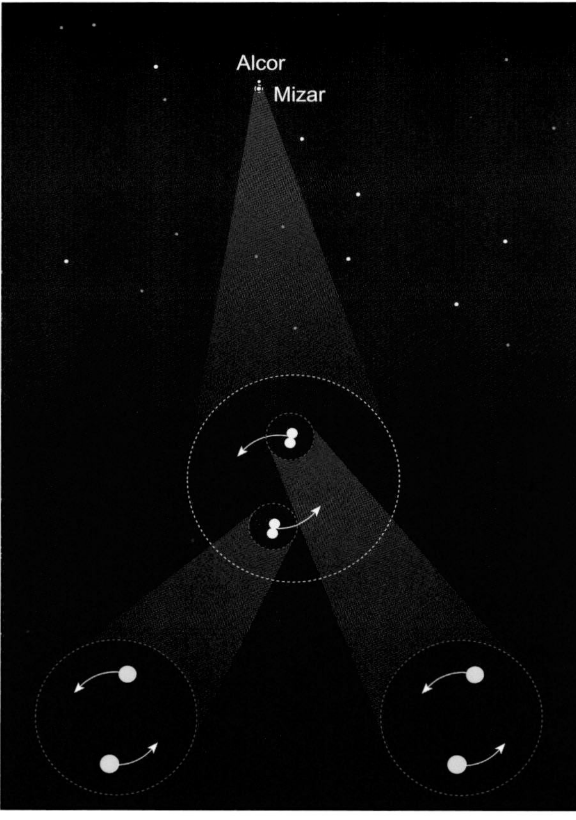

▶ ***Binary star*** *Alcor and Mizar are easily seen as a binary star in the Big Dipper. A small telescope shows that Mizar is itself a binary, and spectroscopic observations reveal that each of those stars is in turn a binary, as shown in the diagram.*

Binary star (or binary system)

Binary stars are pairs of stars in a gravitationally bound, periodic ORBIT around their common center of mass. At least 50% of all stars are known to be parts of binary systems, but this could reach 90% if we take into account the undiscovered ones. In most cases, the two stars of a binary formed out of a single gas cloud, so they are the same age and have the same initial chemical composition. A few systems arise from gravitational-capture processes between previously single stars, or via star exchange between a pair and a single star or two pairs. This occurs most often in environments like the centers of GLOBULAR CLUSTERS, where stars are very close together.

Binary stars are of interest and importance for several reasons. First, they provide the only accurate measurements of stellar masses – the primary determinant of their life cycle (*see* STAR and HERTZSPRUNG–RUSSELL DIAGRAM). Second, a variety of astronomical processes and phenomena occur only in binary systems, especially ones where the stars interact at some time in their lives (called close binary stars); the opposite is 'wide'. Third, possible orbits for stable, habitable planets around a star are greatly restricted when the star is part of a binary (*see* THREE-BODY PROBLEM).

Binary stars are classified in many different ways (*see* DOUBLE STAR):

- common proper motion (CPM) pairs, where the stars are seen as separate dots of light and move across the sky together, but the orbit period is too long for orbital motion to have been seen;
- visual binaries, where two dots of light are resolved and seen to orbit each other;
- astrometric binaries, where a single image is seen, but it is elongated and the orientation changes with the orbit period, or a single image moves periodically in the sky;
- spectrum binaries, where the spectrogram of a single image shows two sets of lines characteristic of two stars of different spectral types;
- spectroscopic binaries (SBs), where the spectrogram of a single image shows one or two sets of lines whose radial velocity varies periodically. These are called single-lined and double-lined SBs;
- ECLIPSING BINARIES, where one star passes in front of the other from our vantage point, so that the total light we see coming from the system is temporarily dimmed.

Many systems can be studied by more than one of these methods. ALGOL, for instance, is both a spectroscopic and an eclipsing binary with a period of about three days, while 70 Ophiuchi is both a visual and a spectroscopic binary with a period of 88 years.

The range in orbital periods of binary stars where both components are normal MAIN SEQUENCE stars can be from a few hours to several million years. Binaries with WHITE DWARF, neutron star (*see* NEUTRON STAR AND PULSAR) or BLACK HOLE components can have much shorter periods, although the systems will be short lived.

Close binaries may interact with each other physically during the later stages of their evolution. As one component swells up to become a red giant it reaches a radius (the ROCHE LOBE) where the gravitational pull of the companion exceeds that of the star itself. Material may then spill from one star to the other, either directly onto the surface as in AM HERCULIS stars, or via an ACCRETION disk as in BETA LYRAE. *See also* VARIABLE STAR: CLOSE BINARY STAR.

Binocular astronomy

See practical astronomy feature overleaf.

Biot, Jean-Baptiste (1774–1862)

A French physicist and astronomer, and professor at the Collège de France. Biot headed a commission from the French Academy of Sciences to study the two or three thousand pieces of the l'Aigle METEORITE, and he succeeded in convincing the academy that meteorites fell from space, a conclusion it had previously scorned.

BiSON

The telescopes of the Birmingham Solar Oscillation Network (BiSON) are located in east and west Australia, South Africa, Chile, California, and Tenerife, Canary Islands. They continuously monitor the radial oscillations of the SUN using very narrow absorption lines of potassium in the solar atmosphere. Absorption cells of potassium vapor are used in the optical path of the telescopes to measure the precise wavelength of solar features, and from this the Doppler motion of the Sun's surface can be detected to a precision of 1 cm s^{-1}. These velocities are analyzed to detect oscillations in the Sun, which give quantitative clues to the temperature and density of the Sun's interior.

Black-body radiation

An ideal emitter or absorber of radiation is called a black body. A perfect black body will absorb all radiation that falls on it, and will emit radiation that has a smooth, continuous SPECTRUM (known as the black-body or Planck spectrum), determined only by its temperature. According to the Wien displacement law, the wavelength at which the maximum quantity of radiation is emitted is inversely proportional to its absolute temperature. Stars are not perfect black bodies, but their radiation can be described in terms of black-body properties. *See also* ELECTROMAGNETIC RADIATION.

Black drop

An optical phenomenon observed at the beginning and end of TRANSITS of Venus across the Sun's disk. The dark silhouette of Venus, having just wholly passed onto the Sun's disk at one LIMB, or being just about to make contact with the opposite limb, is seen to be 'connected' to the limb by a narrow, dark 'bridge.' This bridge is the so-called black drop. It was probably first observed at the transit of 1761 by Mikhail Lomonsov, who concluded that it signified the presence of an ATMOSPHERE. It has to be allowed for when timings are made of transits. However, the black drop effect has been reported at transits of Mercury (which has almost no atmosphere) and, in 2003, in a transit of Mercury witnessed from space (outside Earth's atmosphere).

Black dwarf

Hypothetical final state of a WHITE DWARF, which cools slowly until it is invisible. The universe is not yet old enough for any white dwarfs to have reached this state.

Black Eye Galaxy (M64) *See* MESSIER CATALOG

Black hole

A black hole is a compact region of space that contains a large quantity of MATTER and whose gravitational field is so powerful that no material object, light or signal of any kind can escape from it.

In 1783 John MICHELL and in 1798 Pierre LAPLACE pointed out that if a body was sufficiently massive and compact, the force of gravity at its surface would be high, and its ESCAPE VELOCITY possibly greater than the speed of light. According to eighteenth-century theory, if LIGHT was made of particles it would not be able to leave the body's surface. It would be black.

In the modern language of GENERAL RELATIVITY, SPACETIME near a black hole has such a strong curvature that light and orbiting material particles cannot leave the volume of space within a critical distance known as the Schwarzschild radius. The boundary of the black hole (known as the EVENT HORIZON) would be a sphere having this radius. Events taking place within the event horizon cannot signal their occurrence to the outside world. For a mass the same as the Sun, the Schwarzschild radius is only 3 km, and for a mass the size of the Earth it is only millimeters.

In general relativity black holes have mass, a rotation, and the theoretical possibility of being electrically charged, but no other discernible properties. However, the phenomena surrounding an event horizon can be very complex because black holes are powerful energy sources if matter falls into them. A rotating black hole is a particularly efficient source.

When QUANTUM MECHANICS is added to general relativity (a step on the way to a GRAND UNIFIED THEORY), theory permits black holes to radiate, with PHOTONS being made in pairs in the spacetime around the black hole, according to Stephen HAWKING. The radiation evaporates the mass of the black hole. For astronomically sized black holes the evaporation rate is insignificant, but if tiny primordial black holes with masses less than an ASTEROID were created at the time of the BIG BANG, these mini black holes would have radiated all their mass since then, and disappeared.

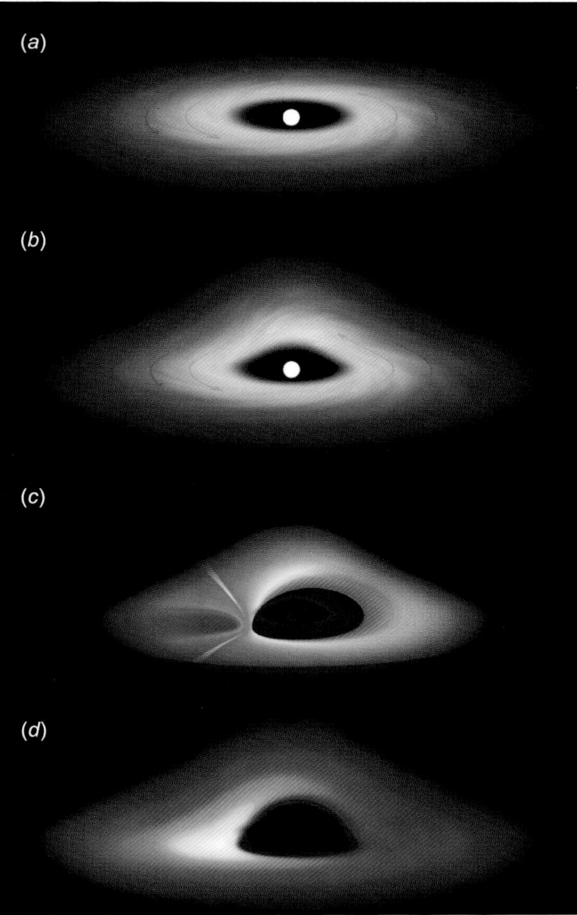

◄ ***Black hole***
(a) Under Kepler's and Newton's laws, gas in an accretion disk is symmetrical and orbits in an ellipse. (b) If the gas orbits a black hole, the gravitational lens effect of the black hole raises the image of the back of the elliptical disk. (c) The gas near a black hole orbits so close that the Doppler shifts of the rotation speed are all redshifted and the overall color is reddened. (d) The gas beams its light in the direction it travels, so the approaching (blueshifted) gas is brighter than the receding (redshifted) gas. These relativistic effects have been seen in supermassive black holes by the Chandra X-ray, XMM-Newton and ASCA satellites.

Nature makes black holes in at least two ways:
- by supernova explosions in stars;
- in the nuclei of galaxies.

Isolated black holes are DARK MATTER and difficult to identify. Accreting black holes release gravitational potential energy; this heats the accreting material, which is then visible as X-ray BINARY STARS and AGN.

Black holes in X-ray binary stars

X-ray binary stars are close binaries that interact over short periods (minutes to days). One component is an (almost) normal star that transfers material onto its compact companion via an ACCRETION disk. Matter spirals in and falls onto the surface of the compact companion, becoming hot enough (10 million K) to radiate X-rays. To release this power, the compact star must be a WHITE DWARF, neutron star (*see* NEUTRON STAR AND PULSAR) or black hole. To determine which, the compact star's mass must be calculated from the normal star's orbital motion around it, by application of Kepler's laws (*see* CELESTIAL MECHANICS). White dwarfs have a maximum mass of 1.5 solar masses, and theory suggests that the maximum mass of a neutron star is 3.2 times the mass of the Sun (no neutron star has been observed that exceeds twice the Sun's mass). If the mass of the compact object exceeds 3.2 times the solar mass then it can only be a black hole.

To apply this method in practice, astronomers must have a clear view of the normal companion, and it must be normal enough for its mass to be inferred. CYGNUS X-1 was the first black hole found this way, in 1971. The normal companion

Binocular astronomy

It is often supposed that anyone who means to take a serious interest in astronomy must acquire a large and expensive telescope. Nothing could be further from the truth! A great deal can be done with the naked eye alone. Small, cheap telescopes on sale in camera shops are most unsatisfactory; the most that can be said for them is that they are slightly better than nothing at all. Luckily there is an alternative: binoculars.

As with telescopes, the larger the APERTURE, the greater the light grasp, but this is not the only important parameter. Binoculars are described as $M \times A$, where M is the magnification and A is the aperture (in millimeters). With increased magnification, the field of view becomes smaller and the binoculars become heavier. If only one pair is to be obtained, then something around 7 × 50 is a good choice. The magnification is adequate, the field is pleasingly large and the binoculars are light enough to be hand-held without any awkward shake. Once the magnification exceeds around ×12, hand-holding is a major problem, and the only solution is to use some form of mounting, usually a tripod.

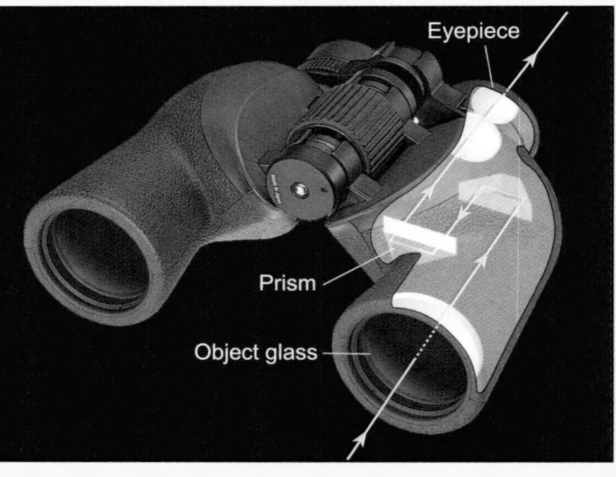

► *Binoculars fold the light path by prisms so as to make the parallel pair of telescopes more compact and to turn the image upright. The eyepieces are focused together by rotating the knurled central adjustment.*

Types of binoculars

The following notes may be useful as a general guide:

3 × 20: These binoculars are light and can be slipped into a pocket. They have their uses, but the low magnification means that they are decidedly limited and probably not to be recommended.

7 × 50: Binoculars of this kind are common. They will show considerable detail of the MOON, and almost endless star-fields. Because of their large field of view, they are ideal for looking at loose star clusters of the PLEIADES type.

8.5 × 50: Also very suitable. The light grasp is adequate, and the binoculars can be comfortably hand-held. The increased power means that more detail becomes available, though the field of view is noticeably reduced.

11 × 80: These, too, are common, although they are not cheap. The field of view is reasonably large and the light grasp is excellent, but they are rather too heavy to be held steadily.

12 × 50: These binoculars are reaching the limit of comfortable hand-holding. One solution is to sit down and jam one's elbows into one's body. This works fairly well, but is by no means ideal.

It is, of course, possible to be more ambitious and use a really powerful pair of binoculars. Some COMET hunters and observers of VARIABLE STARS often have binoculars with objective lenses up to 15 cm across. The cost is high, and the mounting is all important, so binoculars with objective lenses of this kind will only be found in the hands of serious researchers.

In short: if only one pair of binoculars is to be obtained, it is wise to choose a magnification between 7 and 10, with an aperture of around 40 to 60 mm.

Observing with binoculars

First, a word of warning: never use binoculars to look directly at the Sun, even with a dark filter over the eyepiece. If this is done, the Sun's light and heat will be focused on the eyes, with tragic results. In principle, the Sun should be avoided completely with binoculars.

The Moon is a different matter, because it sends us virtually no heat. With a low power and a full Moon you may dazzle yourself, but that is all. In any case, during full Moon is the worst time to observe because there are virtually no shadows on the lunar surface. The sunlight is coming straight down, so to speak, and little can be seen apart from the broad, dark plains still known as seas (*see* MARE). The best views are obtained when the Moon is in the CRESCENT, half or GIBBOUS (three-quarter) stage. Look along the TERMINATOR (the

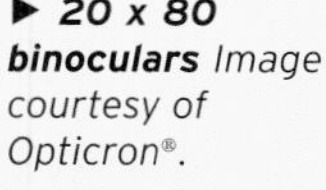

► **20 x 80 binoculars** *Image courtesy of Opticron®.*

►► **10 x 50 binoculars** *Image courtesy of Opticron®.*

◀ ***Moon*** *Near full Moon, the streaks radiating from the crater Tycho at the bottom of the image are spectacular. Image by Jamie Cooper.*

boundary between the sunlit and the night hemispheres) and the CRATERS and mountains will be beautifully displayed. The view alters from night to night, or even from hour to hour, as the angle of illumination changes. A crater that is prominent when on the terminator with its floor wholly or partially filled with shadow may be hard to identify when the Sun is high above it. Near full, the scene is dominated by the bright streaks or rays that are associated with a few of the craters, notably the TYCHO crater in the Moon's southern uplands. It does not take long to learn one's way around the lunar surface, and binoculars are ideal for this purpose.

The situation is not so good from the point of view of the planetary enthusiast because of the relatively low magnification. Mars, for example, will appear as nothing more than a small, red disk. Venus, which is closer to the Sun than we are, shows phases similar to those of the Moon. During the crescent stage, either in the west after sunset or in the east before dawn, the changing phase is easy to follow even with a magnification of 7. Venus is so bright that it can often be seen even when the Sun is well above the horizon; remember, however, not to sweep around for it, because one might look at the Sun by mistake. Similar restrictions apply when searching for the elusive planet Mercury. Binoculars often show it, either soon after sunset or soon before sunrise. Again, never sweep for it with binoculars unless the Sun is completely below the horizon.

Jupiter is more interesting, because binoculars will show all four of the large satellites: IO, EUROPA, GANYMEDE and Callisto. They look like small stars close to Jupiter, and their movements from night to night are easy to follow. Saturn's rings are below binocular range, though with a magnification of 10 or more it is possible to see that there is something unusual about the planet's shape. The two other planets, Uranus and Neptune, are available to binoculars, but they look like stars, and you will need a star map in order to locate them.

A bright comet is an ideal binocular object, and in recent years we have seen two: COMET HYAKUTAKE in 1996 and COMET HALE–BOPP in 1997. These were both superb, with long tails and gleaming heads. We cannot tell when the next brilliant COMET will appear, but there are plenty of faint ones, and the binocular owner armed with a good star map will be able to identify them.

It is wrong to say that all stars look white. Some of them are colorless, but others are blue, yellow, orange or red. Only really brilliant stars show obvious color as seen with the naked eye, and these are mainly orange or red. For instance, nobody can overlook the lovely warm glow of BETELGEUSE in the constellation of Orion. With fainter stars, the colors do not show up, and it is here that binoculars come into their own. Consider Mu Cephei, in the far north of the sky, not too far from the Pole Star. It is so dim that it is difficult to see without optical aid. Turn binoculars toward it, however, and you will see why it has been nicknamed the 'Garnet Star.' It has even been likened to a glowing coal.

BINARY STARS are also on view. One of the best examples is ALBIREO, in Cygnus. With the naked eye it looks commonplace enough, but binoculars show a golden-yellow primary together with a vivid blue companion. Many binary stars are within binocular range, and there are also VARIABLE STARS, many of which are orange or red. These are stars that fluctuate in brightness over short periods. Binoculars make it possible to estimate their magnitudes (*see* MAGNITUDE AND PHOTOMETRY), and to make really useful observations.

Star clusters are glorious; look, for example, in the Pleiades (or Seven Sisters) in Taurus, the Beehive in the Crab, and (from the southern part of the world) the Jewel Box cluster in the Southern Cross. Binoculars also show a myriad of stars in the Milky Way, which stretches across the sky from horizon to horizon.

Gaseous nebulae (*see* NEBULAE AND INTERSTELLAR MATTER), in which stars are being formed, are also to be seen. Look at the Great Nebula in Orion, close to the three stars of the Hunter's Belt. Powerful binoculars even show the four stars of the TRAPEZIUM, which are silhouetted against the nebulosity and are responsible for making it shine. Then there are a few galaxies, vast external star systems so remote that light takes millions of years to reach us. The Great Spiral in Andromeda, the nearest of the really large galaxies, is an easy binocular object.

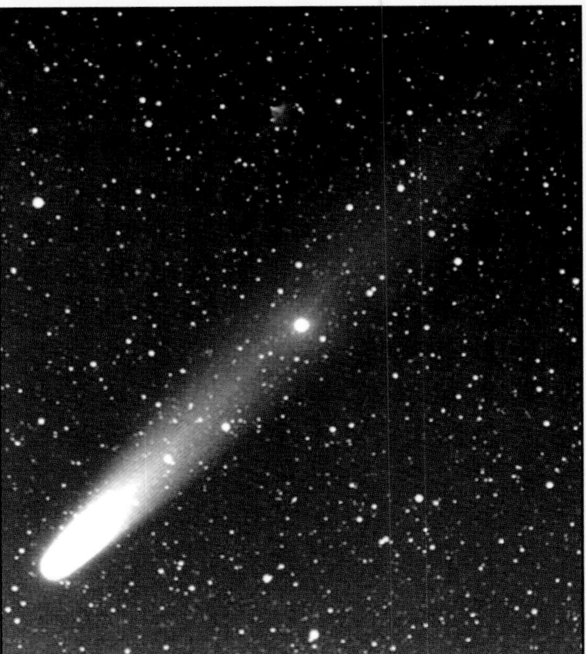

◀ ***Comet Hyakutake*** *seen in 1996. An ideal binocular object. Image by David Hanon.*

Sir Patrick Moore has presented the BBC's *Sky at Night* non-stop since 1957. He is the author of over 100 books, the composer of three operettas and a Fellow of the Royal Society. He was one of the pre-Apollo Moon mappers and developer of the British Astronomical Association's lunar section.

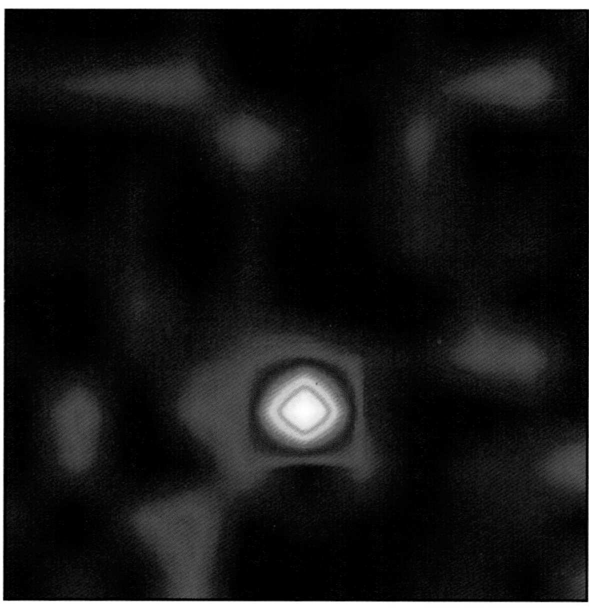

▶ ***Cygnus X-1*** *The first focused high-energy X-ray image of the Cygnus X-1 binary star system.*

is a blue supergiant and bright enough to see, but its unusual evolutionary history means that there is considerable uncertainty about its mass (12–20 solar masses). Nevertheless the compact object in Cygnus X-1 is almost certainly a black hole, with mass greater than 4 solar masses.

For other X-ray binary stars with more normal companions, the star's light is outshone by light produced as a by-product of the X-radiation. It is difficult to measure the normal companion, but all-sky monitors on space-borne X-ray observatories have led to the discovery of 'X-ray transients.' These undergo rare, dramatic X-ray outbursts, which are followed by an extended period of X-ray inactivity (typically decades) during which the normal companions are visible. Mass determinations have been made for 10 of the two dozen such stars known (there are 200–1000 altogether in the Galaxy). The compact star in V404 Cygni is 12 solar masses, which is clearly into the black hole range.

The X-rays and mass measurements from some X-ray binaries are the main evidence that they contain black holes. Some X-ray binary stars, including GRS1915+105, XTE J1550-564 and GRO J1655-40, have quasiperiodic oscillations (QPOs). QPOs are caused by blobs of heated material orbiting rapidly in the accretion disk, near to the black hole. QPOs are potentially important as a phenomenon that is influenced by general relativistic properties (mass and spin) of the black hole.

Black holes in X-ray binaries

Source	Black hole mass (solar masses)
V404 Cyg	12 ± 2
G2000+25	10 ± 4
N Oph 77	6 ± 2
N Mus 91	6 +5/-2
A0620-00	10 ± 5
J0422+32	0 ± 5
J1655-40	6.9 ± 1
4U1543-47	5.0 ± 2.5
Cygnus X-1	>4

Supermassive black holes

These are the energy sources in AGN (the generic name for for quasars, Seyfert galaxies, radio galaxies and BL Lacertae objects; *see* ACTIVE GALAXY). The key observations are as follows:

- AGN have prodigious luminosities – perhaps 10 times the luminosity of the brightest galaxies. Yet they are tiny, because they vary on timescales of hours;
- AGN contain very fast-moving material, with speeds typically 2000–10 000 km s^{-1}, orbiting around objects that are both massive and compact;
- some AGN have radio JETS that are long, narrow and straight, with blobs that move at speeds close to the speed of light (*see* SUPERLUMINAL MOTION). The outer tip of the jet is perhaps 10 million l.y. from the nucleus, and must have been emitted more than 10 million years before the inner end of the jet near the nucleus. Whatever is causing the jet, the AGN engine can remember the ejection direction with precision for 10 million years. The natural explanation is a single rotating body that acts as a stable gyroscope, with the jet aligned with the rotation axis and the axis of the disk of accreting material. The axis is fixed in space, unless it precesses because of the gravitational field of a companion (this phenomenon is sometimes manifest as a corkscrewing jet);
- recent X-ray observations by the ADVANCED SATELLITE FOR COSMOLOGY AND ASTROPHYSICS (ASCA) and other X-ray satellites show enormous orbital motions in the nuclei of many Seyfert galaxies, up to 100 000 km s^{-1}, or 0.3 times the speed of light. The X-ray emission shows effects of SPECIAL RELATIVITY and GENERAL RELATIVITY such as GRAVITATIONAL REDSHIFT and Doppler beaming in the inner zones of the accretion disks as close as a few Schwarzschild radii to the black hole.

Messier 87

High-resolution images confirm that disks of dust and ionized gas orbit the nuclei of many giant elliptical galaxies. The disk of Messier 87 measures 500 l.y. across and approaches as close as 15 l.y. to the nucleus. Its rotation axis is closely

Supermassive black holes

Galaxy	Hubble type*	Mass (million solar masses)	Nuclear activity
Milky Way	Sbc	3	exceedingly weak
M31	Sb	30	exceedingly weak
M32	E	30	completely inactive
NGC 3115	S0/†	1000	completely inactive
NGC 4594	Sa/†	1000	moderately active
NGC 3377	E	80	
NGC 3379	E	100	
NGC 4342	S0	300	
NGC 4486B	E	600	
M87	E	3000	giant AGN elliptical
NGC 4374	E	1000	giant AGN elliptical
NGC 4261	E	500	giant AGN elliptical
NGC 7052	E	300	giant AGN elliptical
NGC 6251	E	600	giant AGN elliptical
NGC 4945	Scd/†	1	
NGC 4258	Sbc	40	moderately active
NGC 1068	Sb	10	Seyfert galaxy

* *See* GALAXY
† *edge-on*

aligned with its optical and radio jet. The disk is orbiting around an object of mass 3 billion solar masses. Furthermore, this object is dark matter – much darker than any known stars – and its density exceeds 1 million solar masses per cubic light year. (Compare this with the value in the solar neighborhood of less than 1 solar mass per cubic light year.)

NGC 4258
The Seyfert galaxy NGC 4258 has radio emission (22 GHz microwave maser emission) from water in orbit in an edge-on nuclear disk. The VLA mapped the disk with exquisite angular resolution. The masers trace out a slightly warped, very thin disk (with an inner radius of 0.5 l.y., and an outer radius of 1 l.y.). There is a 36 million solar mass object within the disk, of density 100 000 million solar masses per cubic light year.

Normal galaxies
Several billion years after the Big Bang, most large galaxies were fitted with black holes as standard equipment, and quasars were 10 000 times more numerous than they are now. Now that most quasars have switched off, dim or dead black hole engines are hiding in virtually every galaxy with a substantial bulge component. Indeed, the bigger the bulge, the more massive the black hole (this must be a clue to the formation of black holes). Only when galaxies collide do the starving black holes feed on disturbed gas and reactivate.

It is easier to find black holes in quasars, but they can be seen more easily in normal galaxies, because, as astronomer Alan Dressler has said, we do not 'have a searchlight in our eyes.' The main evidence for black holes in nearby normal galaxies is found by analyzing the motions of the stars in their nuclei.

NGC 3115
The determination of the mass of the black hole in NGC 3115 is particularly clear. It is a prototypical S0 galaxy (an armless disk galaxy), very symmetrical, and almost exactly edge-on. It has a tiny, dense, central star cluster that has accumulated around a black hole of mass 1 billion solar masses.

Andromeda galaxy
M31 is the nearest comparable galaxy to our own, but it has a double nucleus. It rotates rapidly, and stars close to the center move quickly. The ANDROMEDA GALAXY contains a central dark mass of 30 million solar masses at the position of the cluster of blue stars embedded in the fainter nucleus.

Off-center black holes are the result of galaxy mergers (*see* GALAXIES, COLLIDING). Most large galaxies contain black holes, and mergers produce binary black holes. How much offset we see (whether we see two black holes, one, or none at all) depends on the number of mergers and how the orbits of binary black holes decay. NGC 4486B also has a double nucleus containing a black hole.

Milky Way
Our Galaxy contains the exceedingly compact radio source Sagittarius A* (pronounced 'A-star') (*see* SAGITTARIUS A), about 4 AU in dimension. The small size is impressive, but Sgr A* is a feeble AGN. The MILKY WAY GALAXY contains so much dust that infrared detectors are necessary to see the stars near Sgr A*. Repeated images of the stars showed them to be in very fast motion, at 0.03 l.y. revolving around the galactic center in a human lifetime! The mass that makes them do this is remarkably accurately determined at 2.9 ± 0.4 million solar masses. The case for a black hole in our own Galaxy is now very compelling.

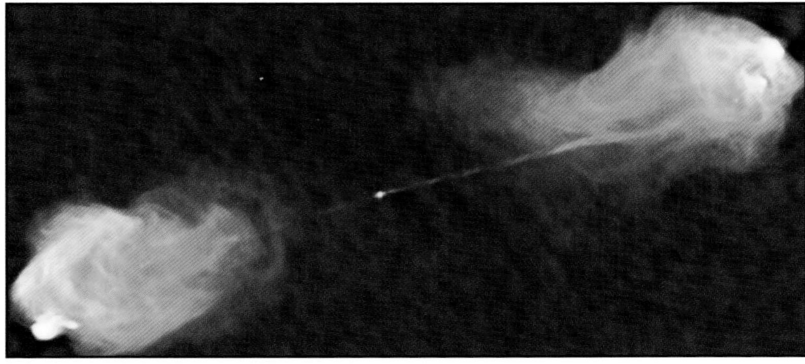

▲ ***Radio jets*** *Cygnus A has two symmetrical radio jets emanating from its central black hole. After traveling 250 000 l.y. these jets collide with the intergalactic medium to make puffy clouds or 'radio lobes.'*

◀ ***Globular clusters*** *M15 (left) and M31-G1 (right) have recently been found to contain black holes that are estimated to be 4000 and 20 000 times more massive than our Sun.*

Globular clusters
The HUBBLE SPACE TELESCOPE has observed the motions of stars in the cores of the GLOBULAR CLUSTERS M15 and M31-G1. These suggest that there are compact objects that have a mass 4000 and 20 000 times the mass of the Sun respectively. No other evidence directly identifies these massive objects as black holes, but if they are, they are a missing link between stellar-sized black holes formed by the collapse of a single star, and supermassive black holes formed in the centers of galaxies. Indeed the discovery suggests that black holes in star systems grow from some initial seed to a size that corresponds to the size of the parent star system.

Further reading: Charles P (1998) *Theory of Black Hole Accretion Disks* Cambridge University Press; Ho L C (1999) *Observational Evidence for Black Holes in the Universe* Kluwer; Van Paradijs J and McClintock J E (1995) *X-Ray Binaries* Cambridge University Press.

Blazar
The phenomenon seen in both BL LACERTAE objects and optically violently variable QUASARS when a beam of non-thermal emission produced in a jet in the core of an active galaxy is beamed directly toward the observer.

Blink comparator
A device, also known as a blink microscope, that highlights any differences between two images of a region of sky, obtained at different times. The light paths from the two images are arranged so that, when viewed through an eyepiece, they coincide exactly. The images are then illuminated alternately, switching from one to the other at a frequency of once or twice per second. Objects that have

changed in brightness fluctuate in size, while objects that have changed position appear to jump to and fro in the alternating illumination. This greatly facilitates the discovery and identification of objects such as variable stars, novae, supernovae, asteroids and comets. Virtual blink comparators are used in digital-image analysis systems for CCD data.

BL Lacertae

BL Lacertae (BL Lac or S4 2200+420) was originally thought to be a VARIABLE STAR. It was known to be variable in the optical from as early as 1929, and it has a stellar appearance. In the 1970s, however, BL Lac was found to be an extragalactic object with a REDSHIFT of $z = 0.069$, based on the detection of very weak emission lines. It is understood to be an unusual type of ACTIVE GALAXY and is the prototype of the class of active galactic nuclei (AGN) known as BL Lac objects. Along with some QUASARS such as 3C273 and 3C279, they are known as BLAZARS. These objects are characterized as the bright nucleus in an underlying elliptical galaxy with strongly polarized optical emission and large variability at all wavelengths. Their nonthermal radio-gamma ray continua are thought to be emitted by a relativistic JET oriented close to the line of sight. BL Lac objects make up a small subset of AGN, with about 350 known at present.

Radio observations show BL Lac to exhibit SUPERLUMINAL MOTION, indicative of material being ejected at relativistic velocities from the nucleus. BL Lac is also a strong and variable X-RAY and gamma-ray source. The rapid variability timescales and high luminosities observed at these energies indicate that the gamma-rays are produced in a very compact region of the jet.

BL Lac objects, as a class, are characterized by optical SPECTRA that are featureless or that have extremely weak lines. This property makes it difficult to determine their distances. BL Lac itself is no exception. However, strong, broad and variable emission lines (for example, the HYDROGEN SPECTRUM) have been seen in the spectrum of BL Lac. Current thinking is that the lines appear when the variable-continuum emission is low. Furthermore, the presence of emission lines indicates the presence of a radiation field external to the jet, which may play an important role in the jet energetics.

Blue moon

An expression in common use to imply that an event occurs only rarely. Originally it may have referred to the fact that the Moon does appear blue on occasion, due to particular atmospheric conditions arising as a result of volcanoes or forest fires. Some ascribe the expression to the unusual occurrence of two full Moons in the same calendar month.

▶ ***Blink comparator*** *Clyde Tombaugh, the discoverer of Pluto, using the Zeiss Comparator at the Lowell Observatory c.1930.*

Blue stars at high galactic latitudes

Most of the blue stars we see in the night sky are hot MAIN SEQUENCE stars associated with the Milky Way, which have formed rather recently from the thin layer of dust and gas in the plane of our Galaxy. Because these stars evolve rapidly, we do not see them at high galactic latitudes, where there is no interstellar material from which they can form. Hence blue stars at high galactic latitudes are rare and must have some other explanation. It was this rarity and the lack of a theoretical explanation for them that led Milton HUMASON and Fritz ZWICKY to undertake the first survey for such stars (the HZ survey), with the 18-in Schmidt telescope on Mount Palomar in 1938. The HZ and subsequent surveys for blue stars at high galactic latitudes have paid off handsomely. These stars tell us not about the early stages of stellar evolution, but about the final WHITE DWARF state and the rapid phases of stellar evolution that immediately precede it. Later studies have also shown that most of the faintest blue 'stars' at high galactic latitudes are in fact interesting, extragalactic objects.

Blue straggler

Allan SANDAGE was the first to note the existence of blue straggler stars when he constructed a color–magnitude (HERTZSPRUNG–RUSSELL) diagram for the GLOBULAR CLUSTER Messier 3 as part of his PhD thesis in 1952. This classic diagram demonstrated beyond doubt that at least 24 stars in the cluster lie near the hydrogen-burning MAIN SEQUENCE, but that they are hotter and more luminous than stars at the main sequence turn-off point. The cluster is about 13 billion years old and yet the blue stragglers cannot be more than 1 to 2 billion years old.

In a globular cluster all stars are considered to be the same age. Stars spend most of their lives on the main sequence, and as massive stars evolve fastest the age of the cluster can be found by the main sequence turn-off point. No stars more massive than this should remain in the cluster. It is clear that blue stragglers do not simply lie along the extensions of cluster main sequences, but are also found considerably redward of the zero-age main sequence. This is an important observational constraint that must be explained by any viable theory of blue stragglers.

Over a thousand blue stragglers are now known. They have also been found in open clusters and in the general field of the Milky Way and other galaxies. In clusters they are found to be more massive than turn-off stars, and more concentrated to the center than red giant or HORIZONTAL-BRANCH STARS. Of the various theories about blue stragglers, the most plausible says that they are formed by the collision and merger of two stars. A core-hydrogen-rich star has its evolutionary clock reset during the merger, and so can appear to be younger than its neighbors.

Bode's Galaxy (M81) *See* MESSIER CATALOG

Bode's law

A numerical sequence that is roughly in proportion to the distances of the major planets from the Sun. It is named after Johann Bode (1747–1826), who announced it in 1772, but it is more properly referred to as the Titius–Bode law, because

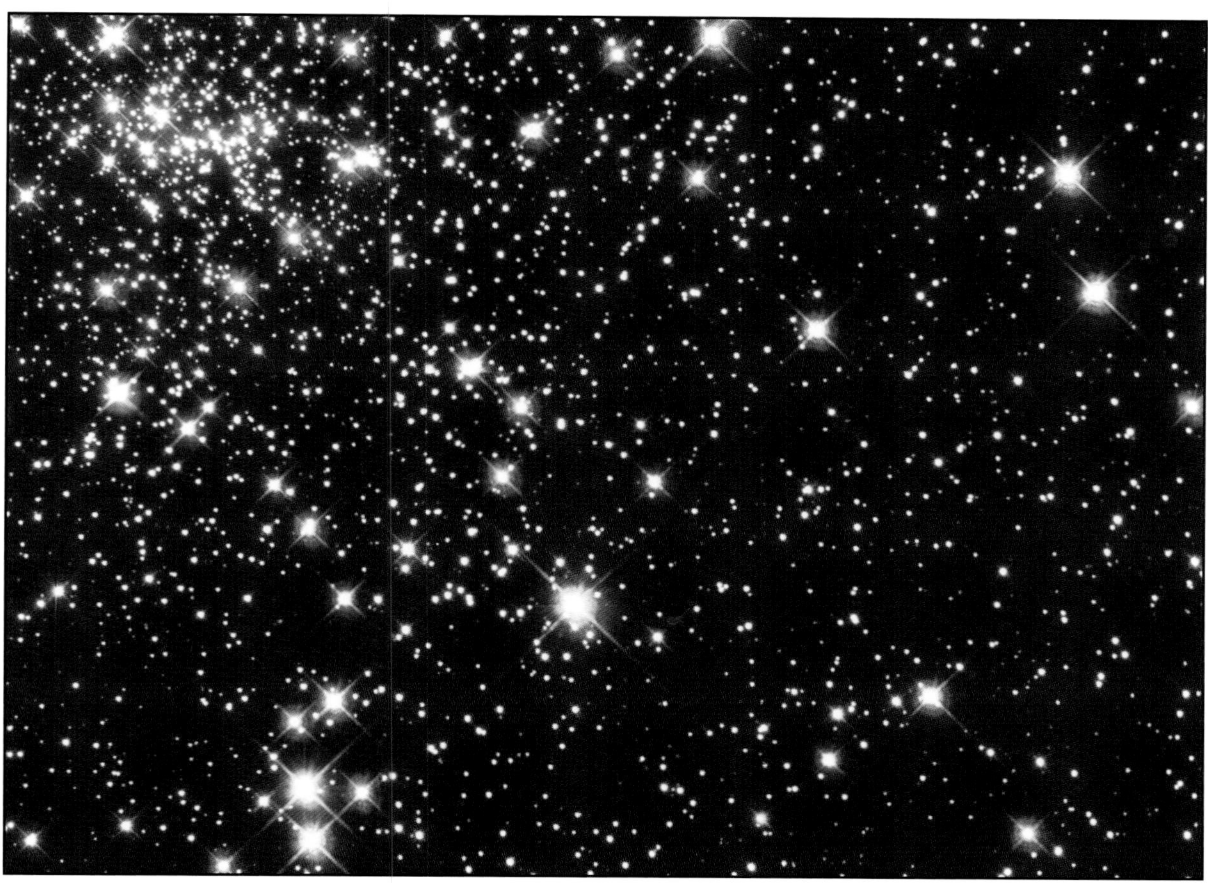

◀ ***Blue straggler*** *In the dense central part of globular cluster NGC 6397, stars collided to form new stars that are more massive, bluer and younger looking than the original ones. These are known as blue stragglers.*

it was first put forward by Johann TITIUS six years previously. The law starts with the sequence 0, 3, 6, 12, 24…, then adds 4 and divides by 10 to obtain 0.4, 0.7, 1, 1.6, 2.8…. This is a good approximation for the distances from the Sun of all of the planets, measured in ASTRONOMICAL UNITS. Bode made his announcement before the asteroids or the planets beyond Saturn had been discovered. When Uranus was discovered in 1781 and its distance from the Sun was seen to fit the numerical pattern, astronomers considered more seriously the apparent gap between Mars and Jupiter, and began searching for a missing planet. The first asteroid to be discovered, CERES, in 1801, was by no means a major planet, but it did orbit at the predicted distance from the Sun. This predictive power gave the Titius–Bode relationship the status of a 'law,' though it was transgressed by the subsequent discoveries of Neptune and Pluto at distances that broke the pattern.

Bok, Bart Jan (1906–83)

A Dutch astronomer who worked at Harvard University, Massachusetts, and directed the MOUNT STROMLO Observatory (near Canberra, Australia). Bok's investigations of interstellar gas and dust led to studies of star formation, and he became known for his work on DARK NEBULAE, small, spherical, dense, dark nebulae which are now called Bok GLOBULES.

Bolide

A bright METEOR (a FIREBALL) that produces audible sounds. Sonic effects are generated once the incoming METEOROID has penetrated to an altitude of around 30 km, and so will be heard some minutes after the fireball is seen. An initial sonic boom indicates that the meteoroid has exploded. If large enough, the resulting fragments will create their own sonic booms, and an object that undergoes multiple fragmentation may generate a continuous, thunder-like rumbling. The fragments of a meteoroid that produce a bolide may well survive their atmospheric passage to fall as METEORITES. The ALLENDE METEORITE, which underwent multiple fragmentation, was preceded by a bolide.

Bologna Astronomical Observatory

Astronomy in Bologna, Italy, has a long history. Official records of systematic teaching of astronomy at the Bologna 'Studio' date back to 1297. Records of observing work can be traced to the beginning of the seventeenth century, when Gian Domenico Cassini built a huge sundial in the major church of the city (*see* CASSINI DYNASTY). The present institute manages a 1.5-m telescope in Loiano.

Bologna Institute of Radio Astronomy

The institute operates three radio telescopes. The Croce del Nord radio telescope in Medicina is the largest transit instrument in the world, with its T-shaped arms of 564 m and 640 m respectively. Since it was built in the early 1960s, it has produced two important radio source catalogs (known as B2 and B3), with data on more than 30 000 objects. There are also two identical single-dish antennae, each 32 m in diameter, located in Medicina and Noto.

Bolometer

An instrument for PHOTOMETRY that measures the energy received from a radiating source, such as a star. The intention would be to measure the entire SPECTRUM of radiation, but in practice, although the most suitable materials available as detectors can cope with a wide spectral region, they cannot cope with everything.

Bolton, John Gatenby (1922–93)

Australian radio astronomer, born in England, who invented the radio interferometric equivalent of Lloyd's mirror. One antenna of the INTERFEROMETER was real and stood on a cliff overlooking the sea. The other antenna was virtual, being the reflection of the real antenna in the water, analogous to the mirror in Lloyd's mirror. Bolton used it to identify a radio source, Taurus A, with an optical object, the CRAB NEBULA. This was the first time a radio source was identified using a previously known object.

Bond, William Cranch (1789–1859) and Bond, George Phillips (1825–65)

William, an American businessman, instrument maker and astronomer, became the first (unpaid) director of the Harvard College Observatory, with its magnificent 15-in telescope. William identified Saturn's innermost Crêpe Ring. Independently of William LASSELL, he and his son George discovered HYPERION, the seventh satellite of Saturn. George (the observatory's second director) correctly realized that Saturn's rings could not be solid, as was later shown by James Clerk MAXWELL.

Boötes *See* CONSTELLATION

Boulliau, Ismael (1605-94)

A French mathematician, librarian and priest. In 1645 he published *Astronomia Philolaica* (Philolaic Astronomy) which, on the basis of elliptical orbits for planets and Kepler's laws, suggested that the force that attracted planets to the Sun should vary inversely as the square of the distance, correcting Johannes KEPLER's formulation. This was the most significant advance in the theory of gravitation between Kepler and Isaac NEWTON. It was acknowledged by Newton in his *Principia*.

Bouvard, Alexis (1767–1843)

A French astronomer who started life as a shepherd boy in Chamonix, became assistant to Pierre-Simon LAPLACE, and eventually director of the PARIS OBSERVATORY. He attempted to produce an accurate orbit of Uranus, calculating the perturbations of other planets according to corrected tables of planetary positions by Jean Delambre (1742–1822) that were published in 1792. Bouvard still could not make all the observations fit. He published his new orbit of Uranus in 1821 but noted: 'I leave to the future the task of discovering whether the difficulty of reconciling [the data] is connected with the ancient observations, or whether it depends on some foreign and unperceived cause which may have been acting upon the planet.' Urbain LEVERRIER and John Couch ADAMS showed that the latter, Neptune, was the reason.

► ***Ira Bowen*** *(left) and Robert Millikan with spectroscope.*

Bowen, Ira Sprague (1898–1973)

An American astronomer, also known as Ike Bowen, who worked at the California Institute of Technology. In 1927 his investigation of the ultraviolet spectra of highly ionized atoms led him to identify mysterious spectral lines of gaseous nebulae not as an unknown element that was prematurely called nebulium, but as forbidden lines of ionized oxygen and nitrogen. Bowen oversaw the completion of the 200-in Hale and 48-in SCHMIDT TELESCOPE.

Bradley, James (1693–1762)

English astronomer who, in 1742, became third ASTRONOMER ROYAL. With Samuel Molyneux (1689–1728) he discovered the ABERRATION of light (a large and unsuspected apparent motion of the stars that was, up to that time, an uncontrolled error in star observations). The explanation for the discovery provided the first observational proof of the Copernican system in which the Earth moved around the Sun, and showed that the speed of light was a constant of physics. Bradley also discovered that the INCLINATION of the Earth's axis to the ECLIPTIC is not constant (*see* NUTATION).

Brahe, Tycho (1546–1601)

Tycho (Latinized from Tyge, pronounced 'Teeko') Brahe was born into the highest stratum of Danish society, and many of his achievements were made possible by this. Strangely, before he was two years old he was kidnapped by his uncle and raised by him. This seemed to influence Tycho to cultivate learning rather than the aristocratic martial pursuits of his birth brothers. He attended the University of Copenhagen (1559–61), and thereafter the Lutheran universities of Germany, primarily Leipzig (1562–5) and Rostock (1566–7). Ostensibly he was studying humanities and law, but in fact he was becoming increasingly interested in alchemy and astronomy. At Rostock in December 1566 Tycho lost the bridge of his nose in a duel with another Danish noble student. Thereafter he wore a prosthesis of gold and silver, blended to a flesh color.

Recalled by his ailing father, Tycho returned to Denmark at the end of 1570 and settled at his ancestral home, Kundstrup Manor in Skaane (now southern Sweden), and then at nearby Herrevad Abbey. Around 1571 he fell in love with Kirsten Jørgensdatter, with whom he remained for the rest of his life, though they could never marry because she was a commoner.

In 1576 King Frederick II offered Tycho the isolated island of Hven in the strait off Copenhagen, and the funds to found and maintain an observatory. There Tycho constructed a castle/scientific institution, which he named Uraniborg (Urania's castle). Every element of the structure was devoted to Tycho's dual interests of astronomy and alchemy. The facilities also included a printing press and paper mill for

◀ ***Tycho Brahe*** *demonstrating a celestial globe to the Emperor Rudolf II in Prague. An imagined scene painted by Eduard Ender in 1855.*

publishing his findings. Later Tycho also founded a nearby subterranean observatory called Stjerneborg (the castle of the stars), which housed his largest and most sophisticated instruments.

After the death of King Frederick, in 1588, the generous royal funding diminished and stopped altogether when Frederick's successor, King Christian IV, came of age in 1596. Because of this, Tycho was obliged to close Uraniborg and withdraw to northern Germany. By the middle of 1599 Tycho had secured another patron and became imperial mathematician to the Holy Roman Emperor Rudolf II. He settled in Benatky Castle, north of Prague, where he began trying to construct a 'New Uraniborg.' However, by early 1601 the emperor had him move into the city to be more accessible for giving astrological advice.

The two most celebrated of Tycho's observational findings involved the nova of 1572 and the comet of 1577. Both, observed early in his astronomical career, had profound cosmological implications.

On November 11, 1572, Tycho noticed a new, intensely bright star in the constellation of Cassiopeia. The new star (nova stella), was in fact a SUPERNOVA. It is now known as TYCHO'S SUPERNOVA and is one of only a handful known in the Milky Way. For about two weeks it was bright enough to be visible during the daytime. Tycho's observations, published in a modest book entitled *De Nova Stella* (About the New Star) (1573), revealed something of extreme importance: the star displayed no diurnal PARALLAX, and therefore lay beyond the distance of the Moon. According to Aristotelian cosmology, the realm beyond the Moon consisted entirely of the fifth element, ETHER, which by its nature was incapable of change. However, the new star had come into being and passed away. This single inconsistency became an important element in the reconsideration of Aristotelian cosmology that took place during the latter half of the sixteenth century, although Tycho regarded the 'nova' as a special creation of God.

As well as being privileged to observe a supernova, in November 1577 Tycho first sighted one of the most spectacular comets in recorded history. He eventually concluded from the comet's motion and its imperceptibly small diurnal parallax that the comet was moving through space somewhere near the 'sphere of Venus.' This result was of great significance, because by the late sixteenth century many had come to the conclusion that Aristotle's ether spheres were hard, impenetrable and crystalline. According to Tycho the comet had moved through those spheres. He later claimed to have proved that the spheres had no reality.

Ever since the publication of Nicolaus COPERNICUS's *De Revolutionibus* (1543) astronomers had been in a quandary. The motion of the Earth in the heliocentric system was regarded by most as 'offending the principles of physics and of Holy Scripture,' as Tycho put it. From his observations of the comet of 1577 and of Mars, Tycho proposed a new cosmological system. In the TYCHONIC SYSTEM, the Earth is at rest in the center of the universe while the Sun, with the planets orbiting around it, orbits the Earth. Tycho's compromise cosmology was widely adopted and remained a serious alternative to the Copernican system well into the seventeenth century, long after the demise of the PTOLEMAIC SYSTEM.

Tycho improved instrument design in every respect: from the details of sights and divisions, to materials, to the type of instruments he made. His working instruments fell into three families.

Greatest in number were the AZIMUTH quadrants, a design series that culminated in the revolving azimuth quadrant (1586) and the great steel sextant (1588) in Stjerneborg.

Tycho had less success with another traditional class of instruments, the armillary family. Although armillaries offered the advantage of being able to measure directly in equatorial or ecliptic coordinates, with their many rings they were fussy to make and use. Tycho progressively simplified the design, and finally abandoned the ring arrangement altogether. His great equatorial armillary (1585) in Stjerneborg consisted only of a very large DECLINATION circle pivoting around a polar axis, with differences in RIGHT ASCENSION being measured by means of a large fixed equatorial arc (*see* ARMILLARY SPHERE).

Tycho's signature instrument (the one for which he most vigorously claimed priority) was his astronomical sextant, which reached its mature form in 1582. It was designed to replace the unsatisfactory traditional astronomical radius, or Jacob's staff, for measuring angular separation in any plane. In the absence of accurate clocks, Tycho did not measure positions in absolute coordinates but determined relative positions by triangulating from the positions of reference stars, for which the sextant was essential.

Accuracy of observation was important for Tycho to prove his cosmological theories, and various analyses have shown that during his mature observing career Tycho consistently achieved his goal of 1 arcmin accuracy. Tycho's dying words, 'Let me not seem to have lived in vain,' were inscribed on the final page of his observing log by an assistant, who became his successor. His name was Johannes KEPLER.

Breccia

A type of rock composed of small, shattered, crushed and sometimes melted angular fragments cemented together in a finer-grained material. The shock from the impact of even a meter-sized METEOROID or ASTEROID on a planetary surface may be sufficient to compact and weld surface material into breccias. They are thought to make up the surfaces of all rocky bodies that have not been subsequently modified by volcanic or other geological activity. For example, lunar highland rock, sampled by the crews of Apollo missions, consists of breccias. The fact that many chondritic and achondritic METEORITES are breccias shows that they were once part of larger bodies that suffered impacts.

Bremsstrahlung *See* FREE-FREE RADIATION

Brera Astronomical Observatory

The observatory, based in Milan and Merate, Italy, is a world leader in X-ray mirror technology. The observatory dates back to 1764 when it was founded in Milan by Ruggiero Boscovich, who became the first director in 1770. In the 1920s a 102-cm REFLECTING TELESCOPE became operational at the nearby observing station in Merate.

Brightness

In astronomy, brightness is used to denote the perceived quantity of light from a star or galaxy, as it appears at whatever distance it lies. It is measured as a magnitude (*see* MAGNITUDE AND PHOTOMETRY). 'Surface brightness' refers to the brightness of an extended source, for example a nebula or galaxy. It is measured in magnitudes per square arcsec.

Brisbane, Thomas Makdougall (1773–1860)

Scottish soldier and astronomer who, after a distinguished army career, became governor of New South Wales, Australia, in 1821. Continuing his early interest in astronomy, he built an observatory at Parramatta in New South Wales. Here he made the first observations of 7385 stars in the southern hemisphere since Nicholas LACAILLE in 1752. He is possibly the only astronomer to have a major city named for him (Brisbane, the capital of Queensland, Australia).

British Astronomical Association (BAA)

The British Astronomical Association was founded in 1890 and has 3000 members. It is open to anyone interested in astronomy and aims to bring together like-minded amateur astronomers, to collect their observations and to organize their observational efforts under the guidance of experienced section directors. The BAA has a library, and papers and astronomical observations are circulated by journal, webpage, handbook, circulars, memoirs and section publications.

Brouwer, Dirk (1902–66)

A Dutch celestial mechanician, he became director of the Yale University Observatory (*see* CELESTIAL MECHANICS). He determined the mass of Titan from its influence on other saturnian moons, and developed methods for finding orbits of comets, asteroids and planets, redetermining the astronomical constants. He was among the first to use electronic computers for astronomical computations, determining the orbits of the first ARTIFICIAL SATELLITES, and gaining increased knowledge of the figure of the Earth. He was editor of the *Astronomical Journal*.

Brown, Robert Hanbury (1916–2002)

British-Australian physicist who, after war-time service, became a radio astronomer. Brown invented the intensity INTERFEROMETER, despite being told by colleagues (who understood less quantum mechanics than he did) that it was impossible. The intensity interferometer created wave-like interference in optical sensors after the incoming light had been detected. Brown used the interferometer, constructed at Narrabri in Australia, to determine the angular diameters of the brighter stars.

Brown dwarf

A star with mass between about 0.01 and 0.08 solar masses (10–80 Jupiter masses), whose core temperature does not rise high enough for nuclear reactions to take place. The surface temperature of brown dwarfs is in the region of 2500 K, so they are are picked up in surveys at infrared wavelengths. Heat is generated by the release of gravitational energy as the stars slowly contract. *See also* EXOPLANET AND BROWN DWARF; *contrast* PLANET.

Bruno, Giordano (1548–1600)

An Italian philosopher. Born as Filippo, he took the name Giordano upon becoming a Dominican. He published *Cena de le Ceneri* (The Ash Wednesday Supper), in which he defended the heliocentric theory of Nicolaus COPERNICUS; and *De l'Infinito, Universo e Mondi* (On Infinity, the Universe

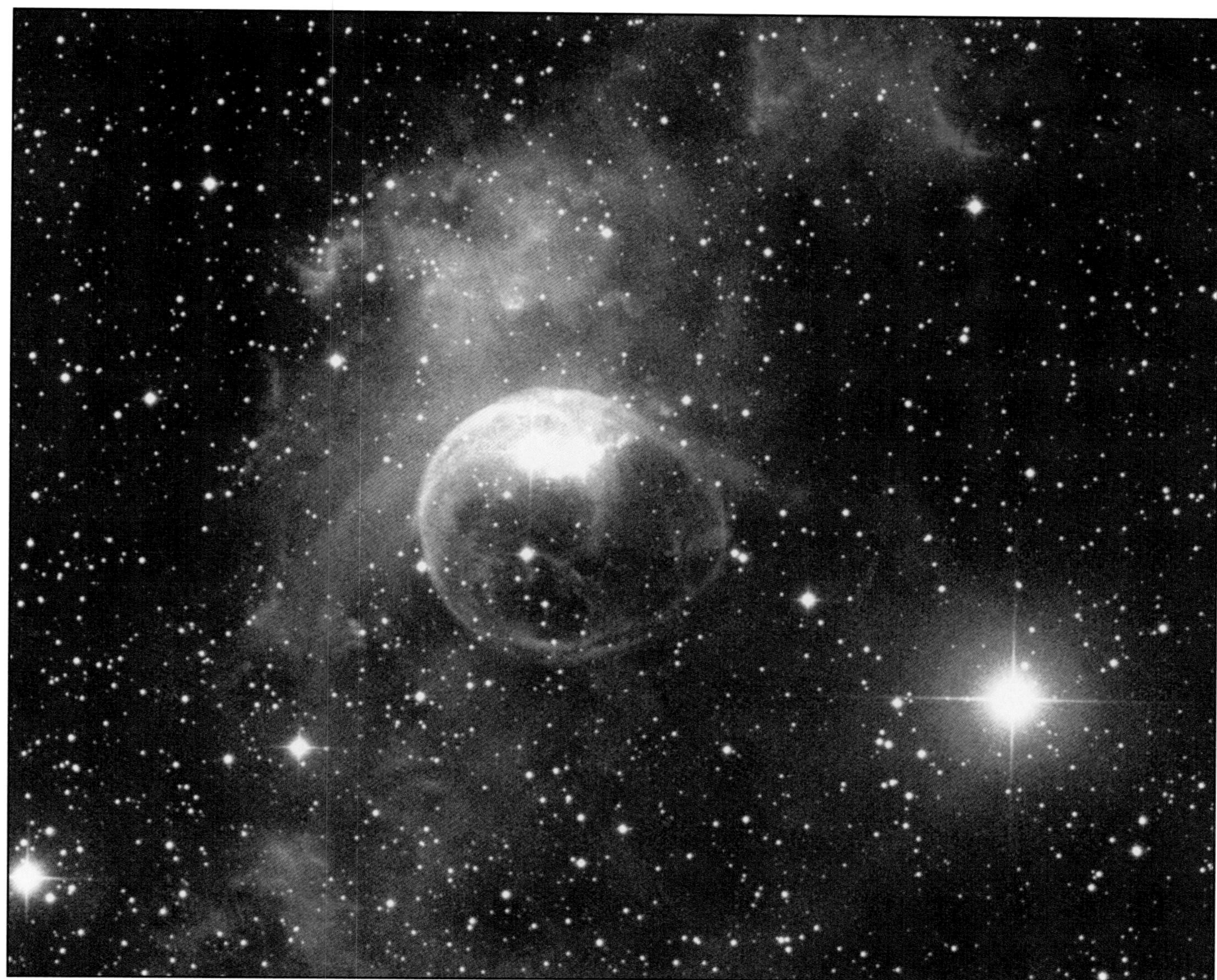

and Its Worlds), in which he argued that the UNIVERSE was infinite, that it contained an infinite number of worlds and that these were all inhabited by intelligent beings. He was burned at the stake, and it is often maintained that this was because of these beliefs.

Bubble Nebula (NGC 7635)

A faint, diffuse EMISSION NEBULA in the constellation Cassiopeia, appearing as a bubble with a diameter of 3 arcmin surrounding a magnitude 7 hot blue star. The bubble is a spherical shell of gas about 10 l.y. across, apparently shed by the massive central star and ionized by starlight. It is part of a larger complex of similar bubbles.

Burnham, Sherburne Wesley (1838–1921)

DOUBLE STAR observer at YERKES OBSERVATORY. His *General Catalogue of Double Stars* (1906) listed 13 665 stars in the northern hemisphere, including more than 1000 that he had discovered and observed himself.

Butterfly Cluster (M6) *See* MESSIER CATALOG

Butterfly diagram

Plot of the solar latitude of sunspots with time, showing that during each solar cycle sunspots first appear at high latitudes and progressively appear at lower latitudes as the cycle progresses. *See also* SUNSPOT CYCLE.

Byurakan Astrophysical Observatory

Located 40 km north of Yerevan in the former Soviet republic of Armenia, the observatory was founded in 1946 through the initiative of astronomer Viktor AMBARTSUMIAN.

▲ ***Bubble Nebula*** *NGC 7635 in Cassiopeia contains many extremely hot stars of the Wolf–Rayet type, which often have intense stellar winds blasting off from their surfaces. The bubble is the glowing interface between these winds and the interstellar medium.*

◄◄ **California Nebula** *(NGC 1499) Six times the width of the full Moon, but of very low surface brightness, the nebula glows red in the characteristic color of ionized hydrogen.*

Caelum *See* CONSTELLATION

Calar Alto Observatory

The Calar Alto Observatory, or Centro Astronomico Hispano-Aleman, is located at an altitude of 2168 m in the Sierra de los Filabres, in southern Spain. It is operated jointly by the Max-Planck Institute for Astronomy (MPIA) in Heidelberg, Germany, and Spain's National Astronomy Commission. The MPIA provides four telescopes of diameters 3.5 m, 2.2 m, 1.23 m and 0.8 m. The latter is a SCHMIDT TELESCOPE. Spain's National Astronomical Observatory independently runs a 1.5-m telescope.

Caldwell Catalog

A list of 109 deep-sky objects compiled by Patrick Caldwell MOORE in 1995 and first published in *Sky & Telescope* magazine. The list contains clusters, galaxies and nebulae that are visible with a small telescope and are of interest to amateur astronomers. These are objects, many of which are in the NEW GENERAL CATALOGUE, that did not quite make it into Charles MESSIER's more famous list. About half of the list is in each hemisphere.

Calendar

A calendar is a system of organizing units of time for the purpose of reckoning time over extended periods. The day is the smallest calendrical unit of time. Some calendars replicate astronomical cycles according to fixed rules, others are regulated by astronomical observations.

The periods of the REVOLUTION of the Sun and Moon, and the daily ROTATION of the Earth, have been the natural bases for the development of calendars. Unfortunately, none of the resulting periods (year, lunation or day) are exact multiples of the other. Thus, calendars based on lunar MONTHS require adjustments to remain close to the year based on the apparent motion of the Sun.

The seasons and some religious holidays are derived from the motions of natural bodies. Some are defined in terms of the vernal EQUINOX. A virtual ecclesiastical equinox has been defined so that Easter can be readily determined.

The Roman republican calendar was based on an older lunar calendar, and had 12 months. This civil calendar was a day longer than the tropical year, which measures exactly one complete orbit of the Earth around the Sun. By the first century BC, the new year fell almost two months after the vernal equinox, its original definition.

Julius Caesar (100–44 BC) saw the value of adopting the Egyptian solar calendar and, on the advice of the Alexandrine astronomer Sosigenes, promulgated what is now called the Julian calendar in 46 BC. This was an entirely solar calendar based on a year of 365.25 days, but fixed at 365. To account for the difference, an extra day was added every fourth year (a leap year).

The Julian year was some 11 min longer than the tropical year, and by the sixteenth century the vernal equinox was occurring 10 days early. To rectify this, in 1582 Pope Gregory introduced a new calendar and removed 10 days from the calendar that year. The Gregorian calendar, which is now used internationally, has 365 days per year and a leap day is inserted every four years, except in centuries that are not divisible by 400. Thus the years 1900, 2100 and 2200 are not leap years, but 2000 was a leap year.

The Gregorian calendar year is 365.2425 days long, compared with the current tropical year of 365.24219 days. The difference of 0.00031 days produces an error of only 1 day every 3000 years.

The Islamic calendar had a common origin with the Jewish and Christian calendars except that it dropped the link to the tropical year. Twelve lunar months of 29 and 30 days alternate. Being strictly lunar (not lunisolar), no months are intercalated to keep the lunar months in alignment with the seasons (although an intercalated day may be added to a given year). So religious observances such as Ramadan fall in different seasons over the years, cycling through all the seasons in about 33 years. Like the Jewish calendar, the Islamic calendar has seven days, each beginning at sundown. Most Islamic countries now use the Gregorian calendar for civil purposes, but the Islamic era takes as its starting point the Prophet Mohammed's flight to Medina in AD 622.

The foundation of the Christian era (Anno Domini, AD) was suggested by Dionysius Exiguus in the sixth century. The English adopted the Christian era for ecclesiastical purposes in the seventh century, but it was not generally accepted in Europe until the eleventh century.

There are non-astronomical features to calendars, such as the seven-day week. The origin of the seven-day week is uncertain, with explanations ranging from weather cycles to biblical and Talmudic texts. The Jewish seven-day week, adopted by early Christians, did not become legally binding in the Roman empire until the time of Constantine (AD 321).

Further reading: Richards E G (1999) *Mapping Time: The Calendar and its History* Oxford University Press.

California Nebula (NGC 1499)

An EMISSION NEBULA in the constellation Perseus whose shape resembles the US state of California. The nebula is large but of low surface brightness. It is illuminated by the star Xi Persei.

Callisto *See* JUPITER

Caloris Basin *See* MERCURY

Caltech Submillimeter Observatory (CSO)

The Caltech Submillimeter Observatory operates a Leighton 10.4-m diameter telescope on the summit of Mauna Kea, Hawaii, USA at an elevation of 4200 m. In partnership with the University of Texas (USA) and the University of Hawaii, CSO provides the international astronomical community with a state-of-the-art observing facility at submillimeter wavelengths (300–1000 μm). It studies mainly molecular and atomic lines, and continuum emission from dust in star-forming regions in the interstellar medium of the Milky Way and distant galaxies.

Camelopardalis *See* CONSTELLATION

Camera and filter

Readers of this article are using a natural camera, the human eye. It has a simple lens, but phenomenal signal-processing power compensates for its limitations.

A telescope is a big camera. Whether a mirror or a lens, the first optical element is called the objective. A telescope system is one or a series of cameras, the last of which has a detector at its focus. Often only the last camera is called the camera. The camera will likely be demountable from the telescope, and an observatory may have a number of cameras or instruments that can be attached to the telescope for different astronomical investigations.

▶ ***Camera*** *This generic telescope and instrument has a camera as its final system. The 'processor' might be a complex optical system (such as a diffraction grating), or simply a filter.*

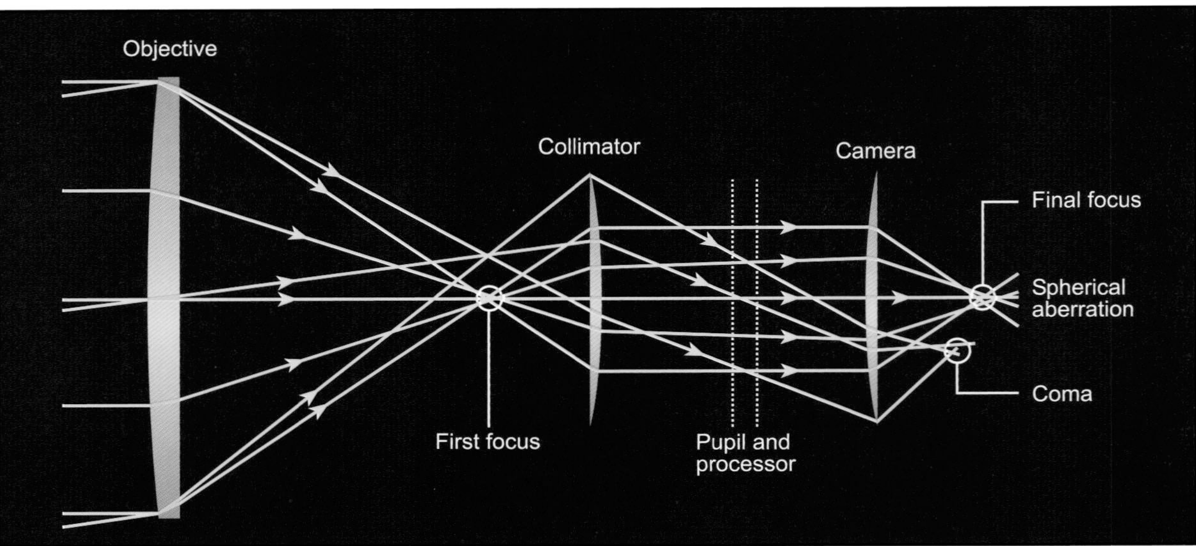

A camera consists of optical elements that gather and focus light onto a focal surface. Reflecting and transmitting elements are used in combination (CATADIOPTRIC systems) in sophisticated camera designs (for example, MAKSUTOV and SCHMIDT TELESCOPES). Optical designers try to make the focal surface flat but it might be curved, as in Schmidt cameras. The difficulties caused by a curved focal surface may be worth the effort if there is a compensating gain: for example, a wide field.

A detector is accurately located at the focal surface. This records the image, and if the focal surface is curved must match the curvature and retain good focus. Typical astronomical cameras use as detectors the retina, photographic film or a charge-coupled device (CCD). Several CCDs may be used in a mosaic, possibly mounted on a curve (*see* CCDS AND OTHER DETECTORS).

The camera is enclosed in a light-tight box, and a shutter controls the time interval over which light is gathered by the detector (the exposure). The shutter must operate quickly compared to the exposure time, so that the scan of the shutter across the detector produces negligible changes in exposure time in different areas of the detector.

An optical device is usually inserted into the light path through the camera. This will analyze the light in some way. The inserted device is called a processor. It is mostly the processor that distinguishes one instrument from another.

The simplest processor is a filter that restricts the bandwidth of light that falls on the detector. Other processors include:

- a very flat, close pair of windows (an INTERFEROMETER);
- a polarizer, such as a Wollaston prism (a polarimeter);
- a DIFFRACTION GRATING or prism (a spectrometer);
- a grating on a mirror (or 'reflectance grating') that reflects the light straight back onto the collimator, which doubles as the camera lens (a Littrow spectrograph).

In an accurate camera design all the optical elements must be taken into account, even a plane-parallel glass filter. Any change to the elements, such as replacement, change of separation or angle, will alter the optical performance.

The figure shows the basic optical design of all telescope/camera systems. The rays in the figure are traced accurately, but the horizontal scale has been contracted to show the separation of rays more clearly. In practice, even if the optical components are perfectly made, the rays do not all come to a sharp focus. The faults are called ABERRATIONS, and the principal types are spherical aberration, coma, astigmatism and chromatic aberration. The optical design work consists of minimizing the aberrations.

The objective in the figure is shown focusing beams of light (each consisting of parallel rays) from two stars onto the first focal surface. The rays cross here and diverge. A collimator lens (*see* COLLIMATION) changes the diverging cone to a parallel, collimated beam. The short-focal-length collimator lens produces a beam that is smaller in diameter than the objective. An instrument/processor placed here needs smaller, more practical optical components.

The objective and its support structure limit the portion of the enormous parallel beam from a star that can enter the telescope. The collimator lens produces an image of the entrance aperture. This image is called a pupil. If a black plate with a hole is located there, it transmits the beam but blocks stray light. It is called an aperture stop, or a Lyot stop.

In the figure a marginal ray from the edge of the field intersects the collimator lens far from its center, resulting in such a large blur of the pupil and of the final image. If a positive lens were located at the first focus it would move the rays closer to the center of the collimator lens and would move the pupil closer to the collimator lens, but would not change the location of the final focus. Such a lens at an intermediate focus is called a field lens.

Filters

Filters are used in astronomy to restrict the wavelength range of the light that is being measured, providing spectral information. The first filters were used with different photographic emulsions to define the passbands.

Some filters work by selective absorption and transmittance: these are the colored-glass (Schott or Corning), or gelatin (Kodak) filters. Others are interference filters.

An interference filter comprises a multilayer of thin films that selectively transmits or reflects light of particular wavelengths.They fall into two broad types: (1) bandpass filters (usually narrow bands) and edge filters, which transmit above or below a certain wavelength; (2) dichroic beamsplitters – edge filters that divide a light beam into wavelengths above and below the edge wavelength. The Fabry–Perot interferometer is a tunable interference filter.

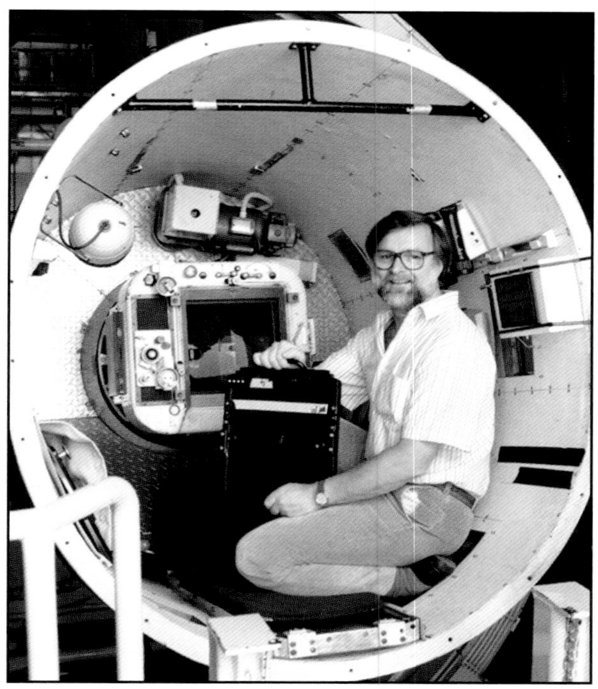

◀ ***Prime focus camera*** *of the Anglo-Australian Telescope. Astrophotographer David Malin is holding the 14-in photographic plate holder, which fits at the focal plane behind his right hand. The telescope has been parked nearly horizontal, and its primary is 12 m beyond the focal plane. Malin is kneeling on the backrest of the observer's chair, which is more nearly vertical when the telescope is pointing up to a star.*

Colored glasses are produced in one of two ways: by ionic coloration or by absorption and scattering from a suspension of colloidal particles in the glass, controlled in size by heat treatment.

Ionic glasses are made by dissolving particular salts (such as cobalt or nickel oxide) in glass. The UG (violet) and BG (blue) glasses have spectral passbands of half-width between 100 and 200 nm, but most also transmit red light beyond 700 nm. This 'red leak' can activate panchromatic detectors, such as CCDs. Polished copper-sulfate crystals or a copper-sulfate solution can be used as an ultraviolet-transmitting blue filter to block red leaks, but S8612 glass is almost as good. The WG, GG (sulfur and cadmium sulfide), OG (cadmium selenide) and RG (gold) filters absorb light blueward of a quite sharply defined wavelength, ranging from 400 to 800 nm in steps of 20 nm. Colored glass filters may be used in combination with each other and with the particular sensitivity of the detector.

It is difficult to make and handle many of the BG and all of the UG glasses in good-quality large sizes, because the required bandpass is produced with only a thin piece of the glass (1 mm or less), and they tarnish easily and are brittle.

Computer-controlled, thin-film coating technology is used to make interference filters by coating a glass or fused silica substrate with alternating layers of $\lambda/4$ reflective and $\lambda/2$ transmitting dielectric materials (λ is the wavelength of peak transmission). The filters are sandwiched between two colored blocking glasses, altogether 5 to 10 mm thick. The coatings are hygroscopic, so the edges are sealed with epoxy into an aluminum cylinder.

The bandpass of an interference filter shifts blueward as the incident angle deviates from perpendicular. Typically a 5° tilt shifts the central wavelength of a 656 nm filter by about 0.7 nm. The filter can be blue-tuned by tilting it, but if the filter is used in a converging beam the bandpass will both broaden and shift blueward, depending on the convergence. Temperature changes also shift the bandpass, at about 0.003% of the peak wavelength per degree Celsius.

A rugate filter is an interference coating in which the refractive index of the coating varies continuously through its thickness. Multiple passbands can be produced, for instance to transmit two emission lines of interest, or to reject certain wavelengths, such as the strongest night-sky emission lines.

Canada-France-Hawaii Telescope (CFHT)

The Canada-France-Hawaii Telescope is a 3.6-m optical-infrared telescope located on the summit of Mauna Kea on the island of Hawaii. The facility was dedicated in September 1979. Time is allocated to the partners in proportion to their financial support, with Canada and France each receiving 42.5% of the time. The telescope has interchangeable top-ends supporting prime-focus (*f*/4), Cassegrain (*f*/8 and *f*/35, the latter for infrared use), and Coude foci with a large complement of imaging and spectroscopic instruments. The CFHT pioneered much of the methodology used to enhance image quality on ground-based telescopes, beginning with the dome environment. More recently, CFHT has implemented a widefield-imaging plan, and about half the time is assigned to its large format camera Mega Prime.

Canal (martian)

Elusive linear feature on Mars. They were first given prominence by Giovanni SCHIAPARELLI on a map he drew of the planet in 1877. He gave the features the Italian name *canali*, which can mean 'channels' or 'canals,' but was translated as the latter in English reports. From his first claimed sighting of canals in 1894, Percival LOWELL began to believe that they were artificial waterways built by Martians. This idea, propounded by Lowell in books such as *Mars and its Canals* and depicted by him in maps of Mars showing an extensive and intricate canal network, fired the public imagination and infused literature, inspiring such stories as H G Wells's *The War of the Worlds*. Images returned by spacecraft from the late 1960s showed no canals.

Cancer, Canes Venatici, Canis Major, Canis Minor *See* CONSTELLATION

Cannon, Annie Jump (1863–1941)

An American astronomer. Cannon studied at Wellesley College, Massachusetts, USA, and in 1896 joined Harvard College Observatory, USA, as an assistant to Edward PICKERING, who had conceived a long-term project to obtain and classify stellar SPECTRA. Following on from Williamina FLEMING, who had developed a system containing 22 classes, Cannon used her own examination of bright southern-hemisphere stars to divide them into the spectral classes O, B, A, F, G, K, M and so on. She also reclassified the previously classified stars. The resulting HENRY DRAPER CATALOG listed the spectral classes of nearly 400 000 stars. Cannon also published catalogs of VARIABLE STARS, including 300 she herself discovered.

See also STELLAR SPECTRUM: CLASSIFICATION.

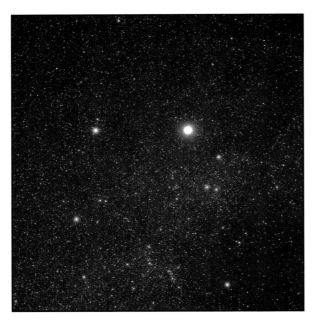

► ***Capella*** *The brightest star in this photograph of the constellation Auriga. Capella appears slightly yellow to the naked eye or with binoculars or a small telescope. Professional telescopes reveal that it is a pair.*

Canopus

The star Alpha Carinae. It was named for the chief pilot of the fleet of Menelaus, who died in Egypt *c.*1193 BC on the return from the Trojan War, the star being prominent at the time. It is the second brightest star, having an apparent magnitude of –0.62. It is a cream giant, of spectral type F0Ib (*see* STELLAR SPECTRUM: CLASSIFICATION), at a distance of 326 l.y. Its PARALLAX is 0.010 arcsec, and its absolute magnitude –5.6.

Capella

Capella, the Goat Star in the constellation Auriga, is the sixth brightest star in the sky. It is a nearby (12.9 parsecs), multiple-star system (*see* DOUBLE STARS). The system consists of a close pair of nearly identical yellow giants of spectral type G8III and G1III (*see* STELLAR SPECTRUM: CLASSIFICATION) circulated at a distance by a looser binary of RED DWARF stars. Capella is a member of the HYADES moving group, and the spectroscopically determined masses of the giants (2.6 and 2.5 times the solar mass respectively) indicate an evolutionary age of 600 million years, similar to the cluster turn-off time of the Hyades cluster.

Capella was the first 'normal' extrasolar star detected in X-rays and is a bright far-ultraviolet emission-line source. The high 'activity' of the system has made it a popular target for spacecraft observatories, beginning with COPERNICUS in the mid 1970s, and extending in the modern era to the Japanese X-ray spectroscopy satellite ASCA and the Italian-Dutch high-energy observatory BEPPOSAX.

Cape York meteorite

An iron METEORITE that fell at Cape York in northern Greenland. Its fragments total 58 tonnes, making it the second largest meteorite known (*see* HOBA METEORITE). Local people had used the largest fragment (31 tonnes) as a source of iron before the explorer Robert Peary visited the region in 1894. He shipped this fragment, known locally as Ahnighito, and two other fragments to New York. Ahnighito, now kept at the American Museum of Natural History in New York, is the world's largest meteorite on public display. Two further fragments (Savik 1, weighing 3.4 tonnes, and Agpalilik, 20 tonnes) were moved to Denmark in the twentieth century.

Capricornus *See* CONSTELLATION

Carbonaceous chondrite

A METEORITE that has a higher carbon content than other classes of meteorite. Carbonaceous chondrites account for about 3% of all known CHONDRITES. They are classified according to the proportion and size of the chondrules (millimeter-sized grains) they contain. One rare subclass lacks chondrules, so is therefore an ACHONDRITE. Carbonaceous chondrites are similar in composition to the Sun (with the exception of volatiles), and the material from which they are formed is believed to have condensed from the PROTOSOLAR NEBULA early in the history of the SOLAR SYSTEM, making them vital pieces of evidence in the study of cosmogony. The carbon is in the form of organic compounds, including amino acids (which are important biological precursors). Carbonaceous chondrites may therefore hold clues to the origin of life. The ALLENDE METEORITE in Mexico is the largest known example, with an estimated total mass before fragmentation of 2 tonnes.

Carbon star

A cool giant star that shows strong bands of carbon molecules in its SPECTRUM. This type of star was first recognized in 1868 by Father Angelo SECCHI. The presence of these bands is controlled by the chemistry of the stable carbon monoxide molecule. If the abundance of oxygen exceeds that of carbon, the remaining oxygen (after forming carbon monoxide) is available to form the oxide molecules seen in spectral type M (*see* STELLAR SPECTRUM: CLASSIFICATION). However, if there is more carbon than oxygen then the remaining carbon is available to form carbon molecules. The modern system of classifying carbon stars was established by Philip Keenan (1908–2000) in 1993. He subdivided them into three classes: C-R, C-N and C-H, corresponding to the old RN and CH classifications. Carbon stars are very cool and very luminous, and many have been found in other galaxies such as the Magellanic Clouds (*see* LOCAL GROUP OF GALAXIES; MAGELLANIC CLOUDS). Absolute magnitudes in the infrared K band are as bright as $M_K = -9$. The masses of carbon stars are in the range of one to six times the mass of the Sun.

Carina *See* CONSTELLATION

Carlsberg Meridian Telescope

An 18-cm diameter REFRACTOR, formerly known as the Carlsberg Automatic Meridian Circle. It is part of the ROQUE DE LOS MUCHACHOS OBSERVATORY on La Palma, one of the Canary Islands, Spain, and is dedicated to carrying out high-precision optical ASTROMETRY.

Carnegie Observatories

The Carnegie Observatories were founded in 1902 by George HALE. Their first facility was the MOUNT WILSON OBSERVATORY, located in the San Gabriel Mountains above Pasadena, California, USA. Originally a solar observatory, it moved into stellar, galactic and extragalactic research with the construction of the 60-in telescope and the 100-in Hale Telescope, each of which was the largest in the world at the time of its construction. Carnegie ran the Mount Wilson and PALOMAR observatories from the 1940s through the 1970s, in partnership with the California Institute of Technology.

Today the offices of the observatories remain in Pasadena, but their main observational facility is the LAS CAMPANAS

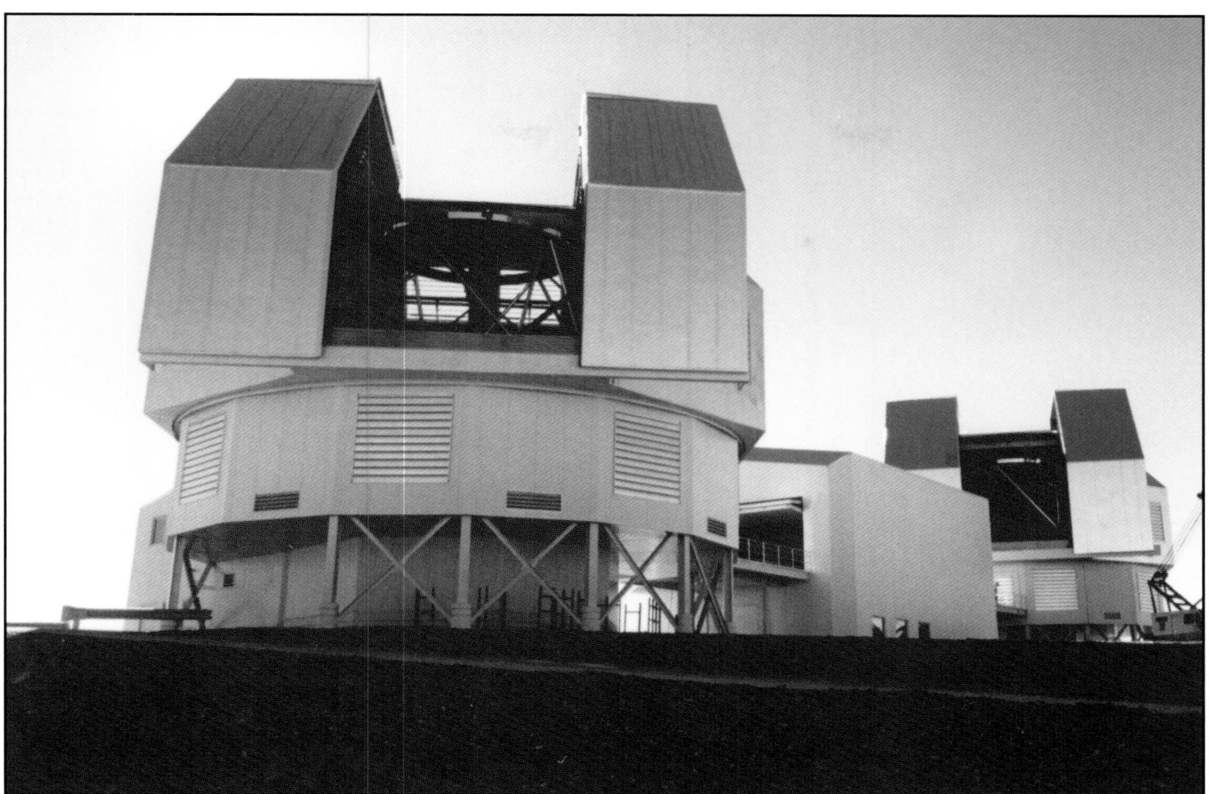

◀ **Carnegie Observatories** *The twin 6.5-m Magellan Telescopes at Las Campanas Observatory in Chile. Each telescope weighs 150 tonnes and stands 15 m high.*

OBSERVATORY, located in the southern fringes of Chile's Atacama Desert. There, Carnegie operates the 1-m Swope and 2.5-m du Pont telescopes. The twin 6.5-m Magellan Telescopes, built in partnership with Harvard, the Massachusetts Institute of Technology, and the universities of Michigan and Arizona, were named the Walter Baade and Landon Clay telescopes, and they began operations in 2000 and 2002 respectively.

The Carnegie Observatories have played a central role in the history of twentieth-century astronomy. Major contributions of Carnegie astronomers include Harlow SHAPLEY's use of globular clusters to establish the structure of our Galaxy, Edwin HUBBLE's discovery of the expansion of the universe, Wilhelm BAADE's work on stellar populations, and Allan SANDAGE's pursuit of the Hubble constant.

Carrington, Richard Christopher (1826–75)

An English amateur astronomer, he was the first person to observe a solar flare (1859): 'two patches of intensely bright and white light.' This was confirmed by Richard Hodgson and by the correlated appearance of AURORAE, seen as far south as Cuba, and magnetic disturbances, observed at the Kew Observatory in England. Carrington noted the connections but cautioned that 'one swallow does not make a summer.' Only about 50 flares have been observed in the white-light spectral range, as used by Carrington's visual observations with an 11.5-cm refractor.

Carter National Observatory

The Carter National Observatory is situated in the Botanic Gardens in Wellington, New Zealand. Opened in 1941, the observatory is equipped with a 41-cm REFLECTING TELESCOPE, a historic 23-cm Cooke photo-visual REFRACTOR, and a 36-seat Zeiss PLANETARIUM.

Cartwheel Galaxy

A peculiar galaxy in the constellation of Sculptor that derives its name from its wheel-like appearance. The Cartwheel has a ring-shaped 'rim' that is 150 000 l.y. in diameter. This contains billions of recently formed stars and is dominated by massive clusters of bright blue stars and H II regions (*see* NEBULAE AND INTERSTELLAR MATTER). The nucleus, or 'hub,' of the galaxy contains a predominantly older population of stars and is surrounded by faint arms, or 'spokes,' that stretch out toward the ring. The galaxy's striking appearance is the result of a collision (*see* GALAXIES, COLLIDING). A few hundred million years ago, a smaller galaxy plunged straight through the Cartwheel, which at that time is presumed to have been a normal SPIRAL GALAXY. The resulting shock spread out through its gaseous disk like a ripple on a pond, pushing gas and dust before it and compressing it into a ring. The compressed gas clouds collapsed and fragmented, thereby initiating the dramatic burst of star formation that is evident today in the ring. Located at a distance of some 500 million l.y., the Cartwheel is an example of a STARBURST GALAXY (one that is undergoing a major bout of star formation).

Cassegrain telescope

A type of REFLECTING TELESCOPE based on a design by 'Monsieur Cassegrain' (possibly Laurent Cassegrain (1629–93) or Guillaume Cassegrain). It utilizes a concave paraboloidal primary mirror and a convex ellipsoidal secondary mirror that is located a short distance inside the focus of the primary. The converging cone of light from the primary is reflected by the secondary, back down the telescope tube and through a hole in the center of the primary to the eyepiece. Because the curved surface of the secondary causes rays of light to converge at a narrower angle than the rays reflected directly from the primary, it increases the effective focal length of

► ***Cassegrain telescope*** *Cutaway image showing the folded light path between two mirrors. This version has a conveniently angled eyepiece.*

the instrument, so enabling an instrument of long effective focal length to be contained within a relatively short tube. Although the FOCAL RATIO of a Cassegrain primary mirror is typically in the region of *f*/3 to *f*/4, the effect of the secondary mirror enables these instruments to operate at focal ratios in the region of *f*/10 to *f*/30.

The standard Cassegrain design produces sharply focused images only in the central parts of its field of view. Images away from the center (off-axis images) are subject to various distortions, or ABERRATIONS, such as COMA. Most large modern telescopes are of the Cassegrain design or one of its variants.

Cassini dynasty

Gian Domenico Cassini (Cassini I, 1625–1712) was an Italian-born French astronomer. Attracted to ASTROLOGY in his youth, he became professor at Bologna University, Italy, during which time he conducted hydrological studies for the Pope to mitigate flooding of the River Po.

In 1669 Cassini moved to France and set up the PARIS OBSERVATORY, remaining director for the rest of his career. He pushed continually for the observatory to acquire the latest technology, and his powerful new telescopes were used to make important findings about the solar system. He saw spots on Mars and measured the planet's rotation period. He also measured the rotation of Jupiter and mapped its spots, bands and flattening at the poles. He timed the revolution of Jupiter's satellites, which allowed the Danish astronomer Ole RØMER to compute the speed of light while he was working in Paris.

Cassini made the best map of the Moon until the invention of photography. He observed comets and wrote about planetary and satellite orbits. And he was the first to record (in 1683) scientific observations of the ZODIACAL LIGHT, although the phenomenon had been known since the Middle Ages, and possibly since classical times. In 1684 he discovered four more satellites of Saturn (IAPETUS, RHEA, TETHYS and DIONE), in addition to those previously discovered by Christiaan HUYGENS. In 1675 he discovered that Saturn's rings are split largely into two parts by a narrow gap, which has since been known as the Cassini Division. The Cassini spacecraft, launched in 1997 and scheduled to arrive at Saturn in 2004 to begin exploring the planet and its moons, is named for him (*see* CASSINI/HUYGENS MISSION).

Jacques Cassini (Cassini II, 1677–1756) succeeded his father as director of the Paris Observatory. He measured the proper motion of ARCTURUS. Cassini II fought continually and unsuccessfully to defend the work of his father and to reconcile observations with the Cartesian theory of vortices.

The second son of Jacques, César-François Cassini (Cassini III, 1714–84), became director of the Paris Observatory in 1771. Here he concentrated on geodesy and made the first modern map of France, which, for navigational purposes, would accurately place the French ports relative to Paris. To everyone's surprise, France was some 20% smaller than previously believed, and the King commented that France had lost more territory to the astronomers than to its enemies.

Jean-Dominique Cassini (Cassini IV, 1748–1845) succeeded his father as director of the Paris Observatory in 1784. He persuaded King Louis XVI to restore the observatory, but the French Revolution occurred and Cassini, a monarchist, bitterly opposed the involvement of the new revolutionary government in observatory affairs and lost his position in 1793. Cassini IV spent his old age in his chateau, writing polemics justifying his own position and defending the scientific reputation of the family. The Paris Observatory had been in the charge of the Cassini dynasty for 120 years.

▼ ***Cassini/ Huygens mission*** *A complicated trajectory is needed to get gravity-assisted acceleration from Venus, Earth and Jupiter to speed up the flight to Saturn.*

Cassini/Huygens mission

The Cassini/Huygens mission is a planetary mission designed to explore in detail the Saturnian system. The mission is a joint undertaking between NASA and the ESA. It was successfully launched from Cape Canaveral, Florida, on October 15, 1997 for a seven-year interplanetary journey, reaching SATURN in July 2004.

The 5.6-tonne spacecraft was too heavy to be launched on a direct trajectory to Saturn, and its journey includes several planetary encounters that are used for gravity-assist maneuvers to accelerate the spacecraft to the next planet. The Cassini/Huygens spacecraft comprises the main craft (the Saturn Orbiter), and the Titan Probe (Huygens). The aim of the mission is to carry out a detailed study of the planet Saturn, its rings, satellites and magnetosphere. Titan, Saturn's largest moon, is a prime target. The mission will also study the relationships between the rings and the satellites, the interactions between the magnetospheric plasma and the satellites, and the rings and the atmosphere of Titan. Cassini will study each part of the Saturnian system and the system as a whole.

Cassiopeia *See* CONSTELLATION

Cassiopeia A (Cass A)

The brightest cosmic radio source in the sky. It is a SUPERNOVA remnant – an expanding cloud of gas that was ejected in a catastrophic stellar explosion. From the present size and

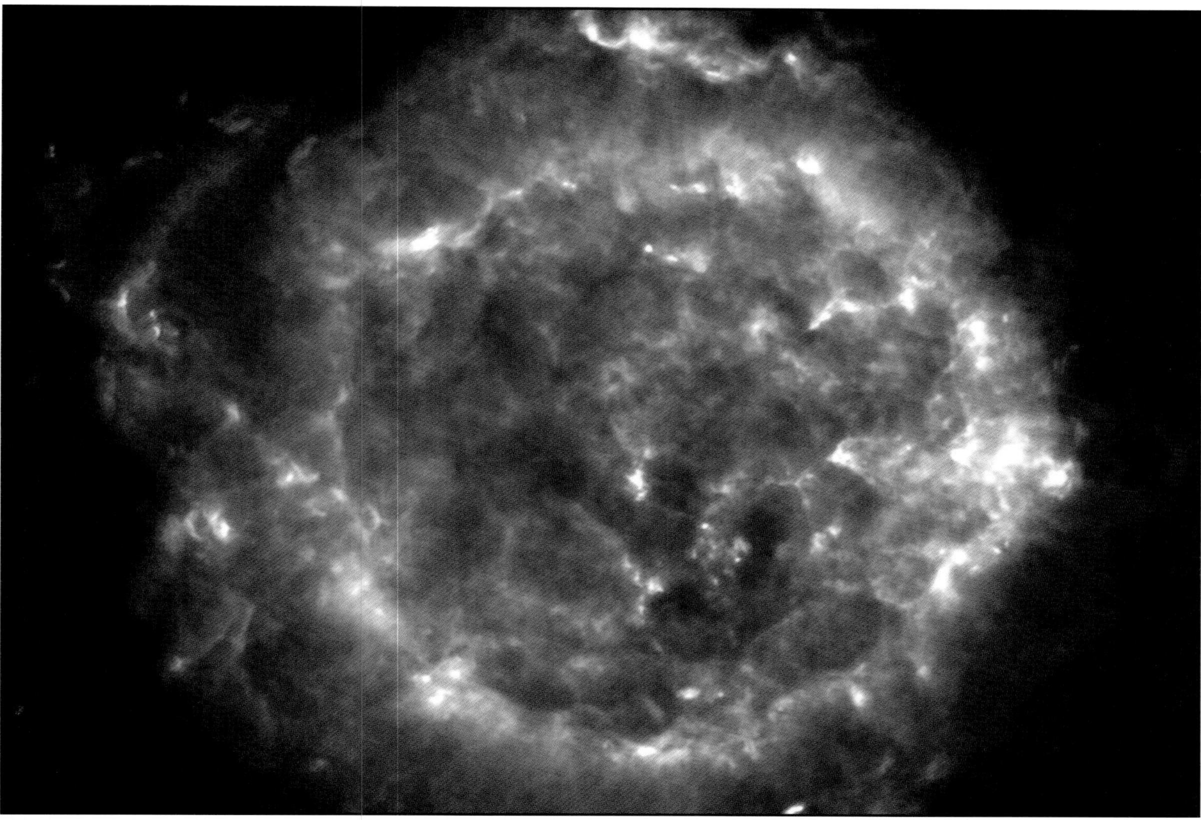

◀ ***Cassiopeia A*** *This radio image, created by the VLA telescope in Mexico, shows the shell of gas pushed out by the exploding supernova fragments.*

rate of expansion of the cloud, the supernova must have occurred around the year 1660. At its estimated distance of about 10 000 l.y., a typical supernova should be brighter than the brightest naked-eye star, perhaps even comparable to the planet Venus, but no bright 'new star' was recorded in Cassiopeia at that time. For whatever reason, the supernova that produced Cassiopeia A must have been of unusually low optical luminosity. Although not conspicuous in visible light (only a few faint wisps of nebulosity can be seen), Cassiopeia A radiates across the ELECTROMAGNETIC SPECTRUM from X-ray to radio wavelengths.

Castor

The star Alpha Geminorum, apparent magnitude 1.58. It is a complex multiple-star system (*see* DOUBLE STARS). Observed with a moderately powerful telescope it is seen to be double, consisting of a pair of white stars: Alpha Gem A (apparent magnitude 1.9, spectral type A1V), and Alpha Gem B (apparent magnitude 3 and spectral type A2V) (*see* STELLAR SPECTRUM: CLASSIFICATION). Their current separation is 3.1 arcsec at position angle 76°; the period is 420 years. Each of these component stars is itself a spectroscopic binary, the period of Alpha Gem A being 2.9 days and that of Alpha Gem B 9.2 days.

The system has another more distant component, Alpha Gem C (the ECLIPSING BINARY STAR YY Geminorum), which is 72.5 arcsec distant at position angle 164°. Its components are a pair of RED DWARFS of spectral type M1, their combined magnitude varying between 9.2 and 9.6 in a period of 19.6 hours. Although part of the gravitational system, this component is so far from the main pair (about 1000 AU) that its orbital period must be of the order of a million years.

This sextuple star system is situated 52 l.y. away, and has a PARALLAX of 0.063 arcsec. The combined absolute magnitude of the principal components is 0.6.

Cataclysmic binary (or variable) star

The Latin word NOVA, meaning 'new,' was applied historically to the sudden appearance of a bright star that had not been previously recorded. Post-outburst observations of Nova Ophiuchi 1848 showed that after a few years it faded to magnitude 13.5, which demonstrated that whatever the cause of the dramatic nova explosion, it had not destroyed the star altogether. The realization that some novae have repeated eruptions (for example, Nova Corona Borealis in 1866 and 1946) underlined the fact that the process is cataclysmic, rather than catastrophic. The cataclysmic outbursts are triggered from the nuclear detonation of a build up of HYDROGEN transferred on to a WHITE DWARF star from a nearby companion star. Cataclysmic binaries are thus nova-like variable binary stars with major, repeated outbursts.

See also VARIABLE STAR: CLOSE BINARY STAR.

Catadioptric optics

An optical system with mirrors and lenses.

Catoptric optics

An optical system with mirrors only.

Cat's Eye Nebula (NGC 6543)

A PLANETARY NEBULA in the constellation Draco, named for its oval shape and greenish color. It is about 20 arcsec in diameter, and of magnitude 9. The central star is magnitude 10. William HUGGINS's spectroscopic observation of the Cat's Eye in 1864, which revealed a SPECTRUM consisting of three bright lines, proved that not all nebulae consist of stars.

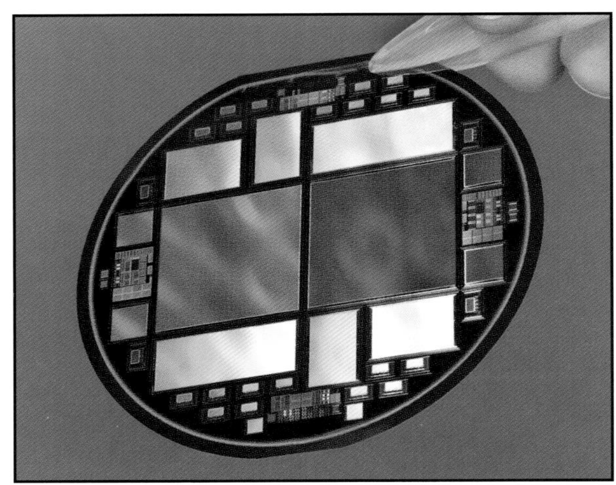

► ***CCD*** *This CCD imager was prepared at the Lawrence Berkeley National Laboratory, California, USA, for use in the giant Keck telescopes in Hawaii. The light-sensitive areas are central on the device, with read-out electronics in the small packages around the edges.*

Cavendish, Henry (1731–1810)

English chemist and physicist. Cavendish used a sensitive torsion balance (the Cavendish balance), which was made to oscillate under the gravitational force of attraction of large masses, to measure the value of the GRAVITATIONAL CONSTANT, *G*. He used this value to determine the mass of the Earth.

CCD imaging

See practical astronomy feature overleaf.

CCDs and other detectors

Charge-coupled devices (CCDs) are the detectors of choice for most modern optical/infrared astronomical instruments. They range in size up to 10 000 × 10 000 pixels. Modern CCDs have low noise, can distinguish 100 000 brightness levels, and can detect 90% of the incident light over their operating wavelengths. The images are readily calibrated.

Invented by Willard Boyle and George Smith at Bell Labs in 1969, CCDs were first used in astronomy in the late 1970s. The CCD consists of a two-dimensional array of metal-oxide-semiconductor capacitors in which the gate of each capacitor is connected to buses bonded to the front of the capacitors. External positive voltages are applied through the gates, which are made of transparent polysilicon. The voltages form potential wells in the silicon under the CCD gates. When light in an image is incident on the semiconductors, through the gates, it creates ELECTRONS via the photoelectric effect. This fundamental property makes the CCD linear, that is the number of electrons generated is proportional to the light accumulated in the image. The electrons collect in the potential wells under the gates. Each potential well is surrounded by more negative voltages, creating a barrier to stop the charge spreading and smearing the image.

CCDs are sensitive to all PHOTONS that are more energetic than the semiconductor's bandgap energy. CCDs of silicon, the most commonly used material, detect wavelengths shorter than 1.1 μm, that is the whole of the optical region and the near infrared, known as the OPTIR SPECTRUM.

CCD operation

Typical CCD pixel sizes are 5–30 μm in a two-dimensional array. The array consists of columns of pixels with a single serial shift register at their base. In operation, the CCD is first 'cleared,' to remove unwanted charge acquired by the device before exposure. Charge is shifted from every pixel out of the device without recording the data. The shutter is then opened, and an image is integrated on the CCD for the exposure time. Photoelectrons are collected in the potential wells in the array of pixels.

After sufficient charge has built up on the CCD, the shutter is closed and the charge is repeatedly shifted (or charge-coupled), one row of pixels at a time in all columns at once, at each stage shifting the bottom row into the serial register. The charge on the serial register is itself shifted along the register and read out into an amplifier at its end. External electronics and a computer convert this output sequence of analog voltages into a row of a digital image. The next row of charge is shifted into the register, for read out, until the whole image has been read. The performance required of the electronics is challenging. If the charge coupling in a column is inefficient, this shows as a dead column in the image, which must be removed from the final image by software interpolation.

Front-illuminated CCDs are exposed to photons through the transparent gate structure. This, however, absorbs almost all blue and ultraviolet (UV) light. Blue-sensitive CCDs are back illuminated, that is they are exposed to photons on the backside, opposite the gate structure. These devices must be very thin (less than 20 μm) for photoelectrons to be collected at the potential minima under the frontside gates.

Image processing

If the image is faint and the integration time is more than a few seconds, then the charge that it creates is less than the silicon generates itself by thermal noise – the dark signal. This signal can be reduced by cooling. Most professional astronomical CCDs are cooled to about –100 °C by liquid nitrogen. Many smaller CCDs, used for applications where exposure times are just a few seconds, rely on thermoelectric coolers that operate at about –40 °C. The average value of the dark current and any bias in the electronic conversion of charge to digital data must be subtracted from the image.

Because of manufacturing irregularities, each pixel of a CCD has an individual sensitivity that must be calibrated by exposing the CCD to a 'flat field' of uniform light. Repeated exposures of the same field shifted slightly in position on the pixel array also help reduce the fixed pattern of noise that remains after calibration.

COSMIC RAYS passing through the semiconductor during the exposure may ionize the silicon. Electrons accumulate in the potential wells and show as spuriously bright pixels in the image. Of course, in repeated exposures the pattern of cosmic rays is different. The brightness and the randomness of the cosmic ray events can be exploited to remove them from the final image.

Photography

CCDs have now virtually replaced photography, bringing its long, productive partnership with astronomy almost to a close. Practical photography appeared in 1837. By 1890 it had largely displaced the eye as the main astronomical detector, and it made possible the new science of ASTROPHYSICS.

Photography remains useful as a detector because of its almost unlimited sensitive area, high resolution and signal to noise, excellent uniformity, ready availability and low cost. It is also an efficient, stable, compact, readily viewed recording medium. However, the detection mechanism is nonlinear, and photographic images can be rendered quantitative only in complex measuring machines, with considerable effort and residual uncertainty.

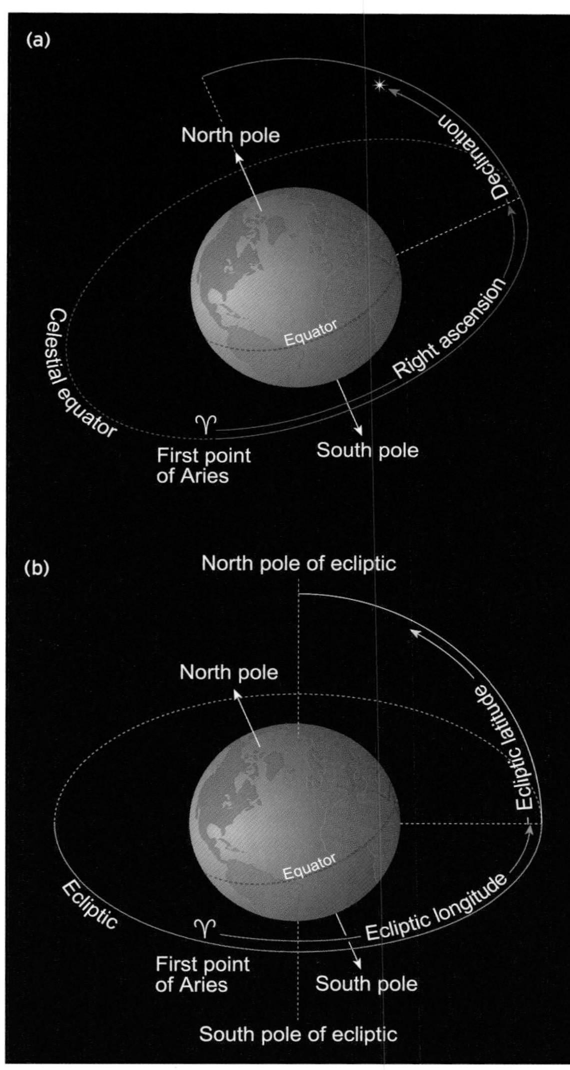

◀ ***Celestial coordinates***
(a) Equatorial coordinates based on the Earth's equator.
(b) Ecliptic coordinates based on the plane of the Earth's orbit.

The light-sensitive layer of all common photographic materials is a thin coating of gelatin. Throughout the gelatin are minute crystals of silver halides (silver bromide or bromo-iodide for astronomical emulsion types). These may be treated with dyestuffs to extend or enhance spectral sensitivity.

The aim of astronomical emulsions is to detect faint sources. At first, the main goal of astronomical photography was to develop the most sensitive detector. This interested C E Kenneth Mees (1882–1960), founding director of the Eastman Kodak research laboratories, who developed extremely sensitive Kodak spectroscopic plates that were widely used for astronomy from the 1930s until about 1990. They were low resolution, with high granularity. This suited long-focus telescopes with large plate scales, but it was not the ideal combination for recording images where the object of interest was fainter than the natural background, such as a galaxy on the background of night-sky AIRGLOW. Fast SCHMIDT TELESCOPES emphasized this problem. In the late 1960s, fast, grainy materials were displaced by emulsions designed for faint object detection, not high speed. Older emulsions, such as Kodak Type 103a, had a detective quantum efficiency (DQE) of 0.1% at best, but newer types, such as Type IIIa, reached DQE of 3–5%, and the recently developed Tech Pan is even higher.

In the early 1970s various treatments of the emulsions (such as baking, and infusing in nitrogen or hydrogen gases) were found to improve long-exposure performance, mainly by eliminating inefficiencies in the light detection process that appear at the low photon arrival rates of astronomy. These treatments are known as 'hypersensitization' (hypering). Many astronomers became skilled at the complex processes during the Schmidt telescope surveys of the 1970s. Schmidt telescopes still use photography today, as it is the most efficient recorder of their widefield images.

Superconducting tunnel junctions (STJs)

In both photography and CCDs, the absorption of a photon creates an effect which indicates that one photon has been detected. In STJs, the number of electrons created in the superconductor (niobium or tantalum) indicates the energy of the incident photon. This new technology holds the prospect of carrying out simultaneous SPECTROSCOPY and photometry of astrophysical objects, particularly those emitting X-ray and UV photons. Currently the STJ detectors are limited in practice to small format arrays (3 × 3 and 6 × 6 pixels), but they have promise as future detectors for both ground-based and space-based astronomy.

Further reading: Theuwissen A J P (1995) *Solid-State Imaging with Charge-Coupled Devices* Kluwer; Malin D F (1988) *Astrophotography* Springer.

Celestial coordinates

Celestial coordinates are the means of specifying the angular location of a celestial object within a reference frame. Many different coordinate systems are used for convenience within a particular context. The term 'celestial coordinate frame' is used for a frame that does not rotate with the Earth (or other planet).

In general, an object moves with respect to the coordinate frame, and the frame moves through space. Therefore it is necessary to specify the time to which the coordinates refer (EPOCH) and the time for which the coordinate frame is defined (EQUINOX). The times may be the same but are often different. The origin of the coordinate frame may be, for example, the center of the Sun (HELIOCENTRIC) or of the Earth (GEOCENTRIC). Positions referred to these origins are often called the 'true' and 'apparent' places, respectively. The positions of the fixed stars are usually given in terms of their RIGHT ASCENSION (measured from the FIRST POINT OF ARIES) and DECLINATION, but they may also be given in terms of their ecliptic or galactic coordinates. PRECESSION constantly changes geocentric positions, and for accurate positions other effects such as nutation, aberration, parallax, proper motion and refraction have to be taken into account.

Celestial equator

The great circle on the CELESTIAL SPHERE obtained by its intersection with the plane of the Earth's equator. The Sun, traveling along the ECLIPTIC, crosses the celestial equator twice a year. At these times, known as the EQUINOXES, day and night are everywhere of equal length.

Celestial latitude

The angular distance between the ECLIPTIC and a celestial body measured in a direction perpendicular to the ecliptic, and taking values between 0° and ±90°. The position of a celestial body may be expressed in terms of celestial latitude (denoted by β) and CELESTIAL LONGITUDE.

CCD imaging

The introduction of the charge-coupled device (*see* CCDS AND OTHER DETECTORS) to the world of amateur astronomy in 1990 was nothing short of a revelation. Although small and of low quality compared to today's devices, the early CCD detectors revolutionized the way amateur astronomers were able to image the night sky. Until then it had been necessary to use photographic film, with all its vagaries and deficiencies, to record the faint light emitted from galaxies, nebulae and so on. Once the CCD arrived it not only brought incredible sensitivity but also a means to carry out serious scientific research, such as ASTROMETRY and PHOTOMETRY, using the modest equipment available to the amateur astronomer.

The CCD detector consists of an array of light-sensitive pixels (or photosites) that convert the incoming light to an electric charge. The charge is then measured and stored on a computer. This is a linear procedure, which means that, in principle, doubling the exposure time doubles the signal strength (unlike photographic film).

An image in an astronomical camera (*see* CAMERA AND FILTER) is built up and converted to a digital form. The computer can process this data and extract a wealth of information from it. The digital nature of the information means that very precise photometric measurements can be carried out, and amateur astronomers are able to contribute accurate data of events concerning variable stars, novae and supernovae, asteroids and even GRB afterglows. Astrometry, the precise positional measurement of astronomical objects, is also much simpler using CCDs.

Unlike photographs, CCD images have to be calibrated using dark frames and flat-field frames. For example, suppose an image of a galaxy is taken with a 10-min exposure. During the exposure, PHOTONS from the galaxy are recorded by the CCD, and at the end these are read out to the computer.

► ***Horsehead Nebula*** *(B33) Imaged by the Prime Focus CCD Camera on the 4.2-m William Herschel Telescope at La Palma. Image courtesy of Isaac Newton Group of Telescopes and Nik Szymanek.*

During this exposure the CCD generates its own noise, called dark current, which is caused by the random motion of ELECTRONS within the detector itself. This component can be retarded by cooling the CCD chip, not as much as in professional astronomy by liquid nitrogen but by a small thermocooler Indeed, reducing the temperature of the CCD by 8 °C will lead to 50% less dark current.

To remove the effects of the dark current from the galaxy image, it is necessary to take a second 10-min exposure, but this time with the telescope capped so no signal from the sky is recorded. This dark frame is then subtracted from the first image to remove the noise component. A second calibration frame, the flat field, is produced by taking a CCD

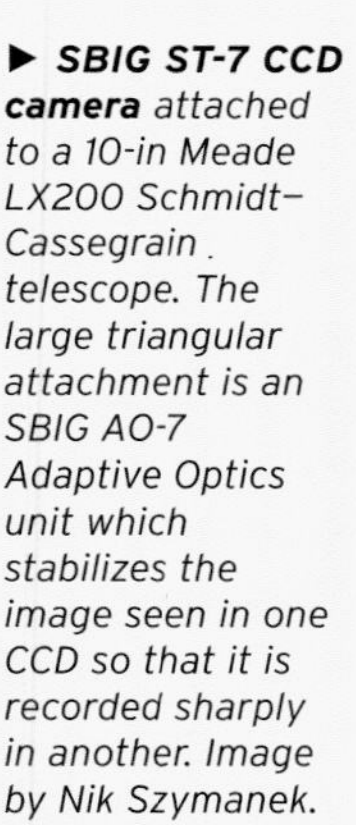

► ***SBIG ST-7 CCD camera*** *attached to a 10-in Meade LX200 Schmidt–Cassegrain telescope. The large triangular attachment is an SBIG AO-7 Adaptive Optics unit which stabilizes the image seen in one CCD so that it is recorded sharply in another. Image by Nik Szymanek.*

◀ ***Trifid Nebula*** *(M20) in Sagittarius, imaged with a 10-in Meade LX200 f/10 Schmidt–Cassegrain telescope, an SBIG ST-8 CCD camera, CFW-8 filter wheel and AO-7 Adaptive Optics unit. LRGB image by Robert Dalby and Nik Szymanek from La Palma, Canary Islands.*

image of the twilight sky or of an evenly illuminated surface. This image will record all the irregularities in the optical path such as vignetting, or the uneven illumination of the CCD detector caused by out-of-focus dust. The flat field is then divided into the dark subtracted frame to produce a calibrated image.

The ability of the CCD to collect light is termed 'quantum efficiency' (QE). Modern CCD detectors usually have a QE of about 50%, which means that half of all the incoming photons are recorded. This compares very well with the QE of photographic film, which is generally only 2–3%. In most cases, the quantum efficiency of CCD depends on the wavelength of light being imaged, and it is quite common for the QE to fall to only 20% at blue wavelengths.

Most CCDs image in monochrome, so to produce color pictures of the night sky it is necessary to use color filters. A true-color image can be synthesized by taking three consecutive exposures through red, green and blue (RGB) filters. Because the CCD detector is generally much more sensitive to red and infrared wavelengths of light, the consecutive filtered exposures have to be biased to produce the correct color balance. This usually means increasing the exposures of the green and blue images to compensate for the CCD's reduced QE at these wavelengths.

Another popular method of color imaging is to include a fourth, unfiltered exposure (the luminance frame), which consists of a high-resolution, deep exposure. When combined with the color information contained in the RGB exposures, a far superior LRGB image is produced. Some CCD cameras use a matrix of microscopic color filters to produce a color image with one exposure, which dispenses with the need for costly color filters. Although much easier to use, they generally produce inferior results to the RGB method.

Other filters can also be employed. A popular choice is the hydrogen-alpha filter, which can produce excellent images of deep-sky objects. The narrow bandwidth of this filter allows imaging under very light-polluted skies, and also in skies brightened by moonlight.

CCDs were physically very small when first introduced to the amateur market, and this meant that it was quite difficult to place the object of interest onto the detector for imaging. Widefield imaging is possible using standard photographic lenses, and CCD images can be stitched together into a mosaic.

The general trend has been for CCD detectors to get larger, although unfortunately prices have not fallen much and large detectors are still very expensive. Entry-level CCDs can be purchased for about $250, with prices rising for bigger and more sensitive cameras. For color imaging, it will be necessary to add a filter set costing about $70, as well as a filter holder or a dedicated filter wheel, the latter operated by a computer.

Many innovative amateur astronomers are turning to consumer devices such as digital cameras and webcams to produce excellent images of the sky. These devices contain CCD detectors that are mass-produced and therefore comparatively cheap. But they are not cooled so can only produce short exposures before the dark current becomes obtrusive. The short exposures obtained can, however, be co-added to produce a strong image. Dedicated amateurs are using cheap computer webcams to produce lunar and planetary images that rival those taken with vastly more expensive CCDs. Webcams act more like camcorders and take up to 30 frames per second. Once the data is secured, sophisticated software is used to extract and co-add the best frames to produce a high-resolution image.

Widefield CCD imaging

Compared to the generous field of view given by 35-mm film frames, the tiny CCD field initially seemed suited only to planetary and small deep-sky objects. As interest grew, however, and CCD cameras became more widespread, the physical size of the CCD detector increased, which made the option of widefield imaging possible. Amateurs using smaller CCDs soon realized that they could combine them with short-focal-length telescopes and camera lenses to create widefield vistas of deep-sky targets.

When using a CCD camera, the best results are obtained if the FOCAL LENGTH of the telescope is matched to the physical size of the CCD. This is important both in terms of the detector area and the size of the individual light-sensitive pixels that make up the CCD array. The detector area will dictate the amount of sky recorded by the CCD at a given focal length. To calculate the FIELD OF VIEW in arcminutes of a given telescope/CCD combination, the following formula is useful: field of view = 3438 × CCD size/focal length.

For example, if a 254-mm *f*/10 SCHMIDT–CASSEGRAIN TELESCOPE with a focal length of 2500 mm is used with an SBIG ST-9 CCD camera (10.2 × 10.2 mm), the area of the sky covered will be given by: field of view = 3438 × 10.2/2500, where the figure 3438 is the prefactor which provides an answer in arcminutes; in this example, an area of sky 14 arcmin across. This view is generally adequate for galaxies and smaller nebulae. If the telescope is used with an *f*/6.3 focal reducer, the field of view will increase to 22 arcmin. By reducing the focal length, using standard photographic lenses, correspondingly larger areas of the night sky can be imaged.

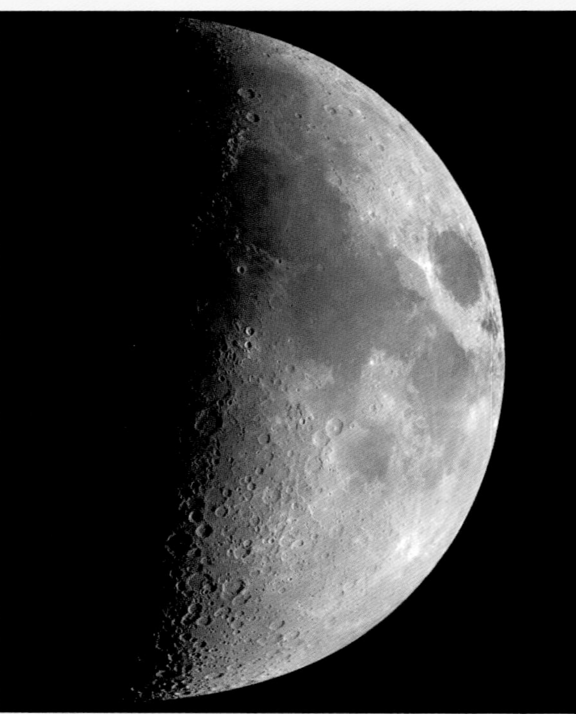

► ***Moon*** *Composite from 12 frames by Nik Szymanek and students at the University of Hertfordshire, UK, using a Meade 16-in LX200 at f/10 and an SBIG ST-9 CCD camera.*

► ***Rosette Nebula*** *(NGC 2237–39, 2246) in Monoceros. This photo, taken by Nik Szymanek, is a 2-image composite based on 30 x 5-min exposures, using a Pentax SDHF 75-mm f/6.6 apochromatic refractor, Starlight Xpress SXV-H9 CCD camera, Astronomik 10-nm Hydrogen Alpha filter and an SBIG ST-4 autoguider.*

Using equipment like this, the modern amateur astronomer can produce high-quality widefield portraits of large deep-sky objects.

It is also necessary to consider the resolution of an imaging system. In practice, this means choosing a combination of focal length and CCD pixel size to deliver optimum resolution. If the focal length is so short that the image of a STAR is smaller than an individual pixel, information will be lost and the star will appear to be unnaturally 'blocky.' This is known as undersampling. Generally, a good match of focal length and CCD pixel size is one that produces a resolution of about 1 arcsec per pixel. This can be calculated using the following formula: resolution = 206 265 × pixel size/ focal length, where the prefactor now gives a solution in arcseconds.

Thus, to obtain a resolution of 1 arcsec per pixel with the SBIG ST-9 CCD camera (which has 20-μm pixels), a focal length of 4120 mm will be needed.

Most modern CCD cameras feature pixels that are between 6 and 10 μm in size. Assuming a pixel size of 6 μm, the above equation suggests a focal length of 1230 mm for optimum resolution. This implies that even using 6-μm pixels, the resolution will produce undersampled images with the typical lenses that are used to make widefield images. However, several factors will conspire to degrade the sought-after resolution of 1 arcsec per pixel. The worst offender is the unsteadiness of the atmosphere, which will generally only allow seeing of 2 arcsec, even on better nights in a home-observing situation. Other factors, like poor telescope drives, errors in focusing and wind vibration also degrade the final image (*see* FOCUS).

However, there is no denying that widefield color CCD images can be truly spectacular. Targets like the NORTH AMERICA NEBULA and the CALIFORNIA NEBULA make excellent subjects for camera lenses.

Several manufacturers are now producing adapters that allow camera lenses to be coupled to filter wheels. This is

a perfect way to produce stunning tricolor images. Such imaging equipment can easily be attached to telescopes fitted with drive systems, and the short focal lengths involved mean that guiding is far less critical than at longer focal lengths. The fast f-ratio of the lens (typically less than *f*/5.6) allows shorter exposure times than do traditional telescopes. Many modern amateurs choose to image wide fields using hydrogen-alpha filters. These produce beautiful 'photographic' results and are relatively unaffected by light pollution and bright moonlight.

Other options for creating widefield pictures include building a mosaic of images. Most graphic-manipulation software can produce a widefield image out of smaller components. If the CCD detector is not capable of imaging the target in one attempt, then several (sometimes dozens) of separate images can be stitched together. This method works particularly well when attempting to build high-resolution images of the Moon. The blending capabilities of modern software can easily cope with the lines and borders at the edges of multiple images, enabling a seamless mosaic to be produced.

Of course, there is no substitute for a large-format CCD detector. Unfortunately, large CCD chips are still expensive, particularly the new breed of thinned, back-illuminated detectors that offer unprecedented sensitivity.

Nik Szymanek drives a District Line train on London's Underground railway and vacations on La Palma, Canary Islands and Hawaii, USA to pursue his astrophotography hobby, as well as observing from his observatory in Essex, UK.

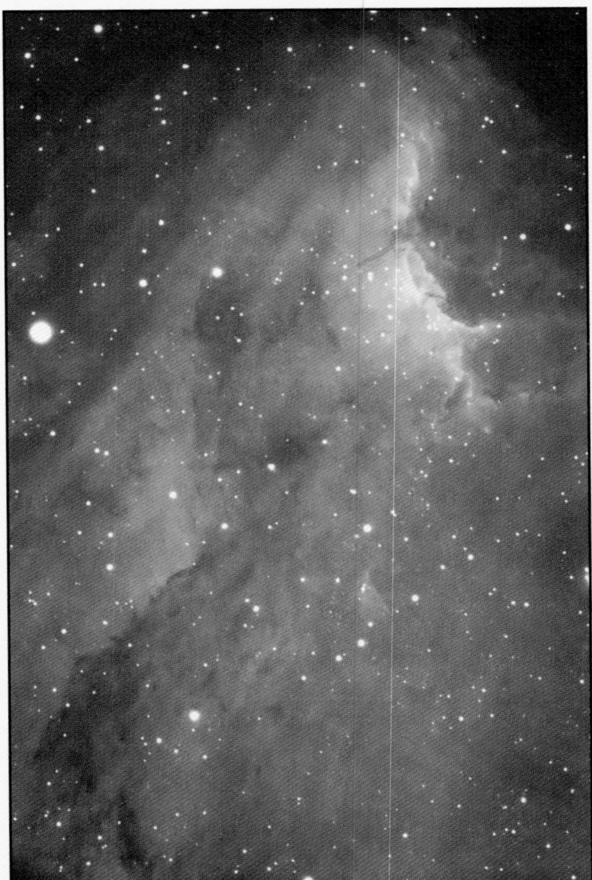

▲ ***Orion Nebula*** *(M42) Imaged by Nik Szymanek. This LRGB image is based on 9 x 5-min luminance exposures.*

◀ ***IC 1396*** *in Cepheus by Nik Szymanek. The LRGB image is based on 19 x 5-min luminance exposures.*

◀◀ ***Pelican Nebula*** *(IC 5070) in Cygnus. The photo, by Nik Szymanek, is an LRGB image based on 21 x 5-min luminance exposures. All images were taken with the equipment listed for the Rosette Nebula on p. 74.*

Celestial longitude

The angle between the great circle passing through the poles of the ECLIPTIC and the vernal EQUINOX, and the great circle passing through these poles and a celestial object, measured in an anticlockwise (eastward) direction from the vernal equinox. The angle is analogous to longitude on the Earth. Planetary positions are sometimes expressed in terms of celestial longitude (denoted by λ) and CELESTIAL LATITUDE.

Celestial mechanics

The study of moving objects in the universe. Nowadays the term is generally taken to mean the theory of motion of the SOLAR SYSTEM and multiple systems of stars. Celestial mechanics is primarily concerned with the consequences of the force of gravity, but additional forces can operate, such as the tidal interactions of planets and satellites. Celestial mechanics was originally considered to be the 'pure' form of theoretical astronomy, but the rise of modern ASTROPHYSICS made it appear somewhat archaic. It has revived in recent years with the realization that its deterministic equations of motion can give rise to chaotic solutions of great interest, with new applications to the solar system and exoplanets (*see* EXOPLANET AND BROWN DWARF).

The subject started in the seventeenth century, when Johannes KEPLER derived his three laws of planetary motion, empirically based on the precise observations of Mars's orbit made by Tycho BRAHE.

Kepler's laws apply to any system of two objects in close gravitational interaction. They apply to each planet in the solar system, paired with the central massive Sun, because the effects of other planets can be ignored to a certain degree of accuracy.

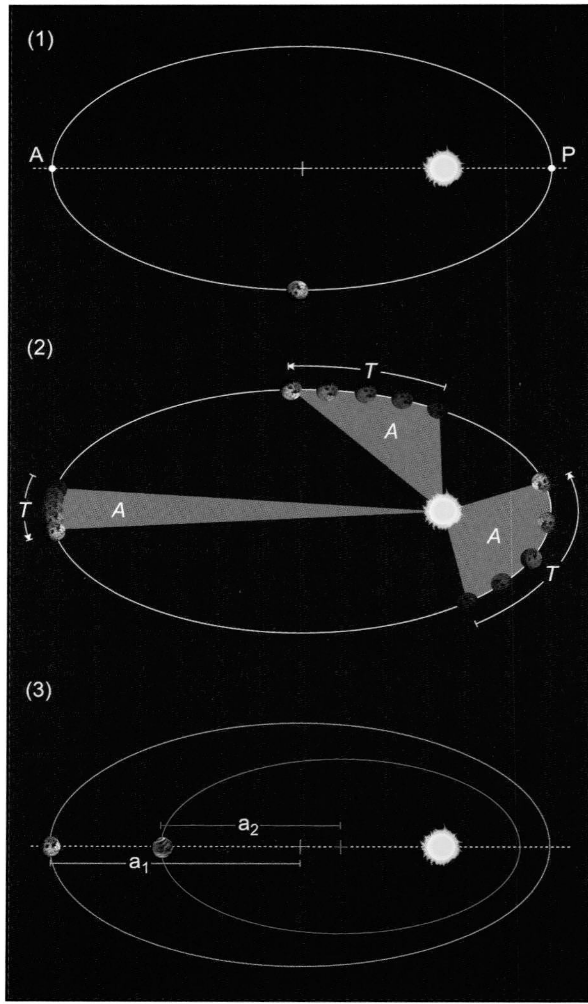

▶ Kepler's laws *(1) The path of each planet around the Sun is an ellipse, with the Sun at one of the foci. A is aphelion, P is perihelion. (2) Each planet moves in such a way along its orbit that a line drawn from the Sun to the planet sweeps equal areas* (A) *in equal amounts of time* (T)*. (3) The ratio of the cube of the semimajor axis of orbit to the square of the orbital period is the same for all planets.* a_1 *and* a_2 *represent the semimajor axes for two planets.*

Kepler's first law

The content of Kepler's first law is purely geometric. At its most fundamental it shows that, although planets move in three-dimensional space, they ORBIT in a plane. Kepler identified the exact shape of the orbit (an ELLIPSE) and the role of the Sun. It occupies one of the foci of the ellipse, the other focus and the center of the ellipse playing no part.

A set of orbital elements determines the orbit, of which two define the orientation of the trajectory plane. The semimajor axis a and the ECCENTRICITY give the extent and the shape of the ellipse; the direction of the major axis of the ellipse is determined by a further constant. The final elements are the starting time of an orbit, chosen as the time at which it passes PERIHELION, and the period.

Kepler's second law

Kepler's first law gives all possible positions of the planet, but no time indications. The second law implies that the velocity of a planet in its orbit decreases with increasing distance to the Sun. A planet moves faster when at perihelion, and slower at APHELION. The second law gives the time dependence, crucial to predict planetary positions.

Kepler's third law

The third law is different from the previous two. The first and second laws are about one single object orbiting the Sun, but the third gives a relation valid for all the objects doing so. The physical reason for this law is the inverse square law of gravitation, discovered by Isaac NEWTON. This shows that the ratio in the third law is determined by the mass of the Sun (more exactly the total mass of the system, including the mass of the planet, which is in practice negligible).

Kepler's third law offers the unique possibility to measure masses in the universe, when extended to BINARY STAR systems or ACCRETION disks around BLACK HOLES.

As Newton showed, it is possible to derive Kepler's laws from physical principles:

- if the two bodies are isolated, the total momentum of the system is constant;
- if the force between the two bodies is central and always directed toward the center of mass of the system, the angular momentum is constant;
- if the gravitational interaction force derives from a potential energy, the total mechanical energy of the system remains constant. The semimajor axis depends only on the mechanical energy;
- the gravitational force varies as the inverse of the square of the distance.

Beyond Kepler's laws

What happens when more than two bodies interact? The impact by COMET SHOEMAKER–LEVY 9 on Jupiter shows that the solar system is not only governed by the Sun's gravity: the comet's repetitive elliptical trajectory around the Sun was altered by the pull of Jupiter. In general the solar system has a hierarchical structure, each subsystem consisting of a central mass with one or more orbiting smaller masses. The planets orbit the Sun but most also have their own small satellites. When considering the motion of a satellite orbiting close to a planet, the perturbations from the Sun can usually be ignored. But the Sun perturbs the outer satellites. Likewise, planetary orbits themselves are not fixed because each planet is perturbed by others, especially by the more massive ones. Therefore, although at any given instant the planet is always following an elliptical path (its osculating orbit), the shape and orientation of that ellipse are continually changing owing to perturbations from the other planets.

Smaller objects in the solar system, such as grains of dust, are also perturbed by non-gravitational forces, particularly by the effects of solar radiation. For example, the POYNTING–ROBERTSON EFFECT is caused by the anisotropic emission of radiation from the dust grain (the side facing the Sun behaves differently from the side facing space). It results in orbits becoming more circular and smaller. This effect is responsible for the transport of dust grains from the asteroid belt into the inner solar system (therefore producing many meteors).

One of the outstanding successes of celestial mechanics was the explanation by John Couch ADAMS and Urbain LEVERRIER for discrepancies in the motion of the outer planets, leading to the discovery of Neptune. Further improvements in precision led to the even more momentous discovery by Albert EINSTEIN that a discrepancy in the motion of Mercury was due to a small breakdown of Newton's theory of gravity, and that his own general theory of relativity was better.

Celestial mechanics gained importance in the space age. There was a practical need to predict how satellites orbited the Earth. Complicated orbital tours were designed to send spacecraft to other planets (such as the GALILEO and CASSINI/HUYGENS MISSIONS), using the gravity-assist technique whereby a spacecraft boosts its orbit through a close flyby of a nearby planet or satellite.

The fact that motion in the solar system arises from simple mathematical laws led to the belief that it would be possible to predict the past and future state of the entire universe. Pierre-Simon LAPLACE optimistically held this view. However, at the end of the nineteenth century Jules POINCARÉ gave the first indication that deterministic systems could have unpredictable solutions.

Modern numerical simulations show that the planetary orbits, especially those of the inner planets, are technically chaotic: an error as small as a centimeter in a planet's position today propagates exponentially, and it becomes impossible to predict its location in 10 million to 100 million years. The finding of unstable exoplanetary systems will make it possible to discover why Earth's orbit has been stable for the 4000 million years that have resulted in our own evolution.

Further reading: Dreyer J L E (1953) *A History of Astronomy from Thales to Kepler* Dover; Murray C D and Dermott S F (1999) *Solar System Dynamics* Cambridge University Press.

Celestial Observation Satellite (COS-B)

European GAMMA-RAY ASTRONOMY satellite launched in 1975. It also carried a small X-ray detector. COS-B remained in operation until April 1982. It conducted a sky survey and detected some 25 sources of gamma-rays, including the CRAB PULSAR and VELA PULSAR.

Celestial pole

Either point on the CELESTIAL SPHERE at which the projected axis of the Earth intersects the sphere. The north celestial pole is thus vertically above the terrestrial north pole, and the south celestial pole is above the terrestrial south pole. At the present time a star, Polaris (the Pole Star), is close to the north celestial pole. The position of the south celestial pole at the present time is close to the point defined by an extension of the long axis of the Southern Cross and the perpendicular bisector of the Pointers. Due to PRECESSION the direction of the Earth's axis changes, and the position of the celestial poles changes on the sky as a result. The altitude of the celestial pole is equal to the latitude of an observer on Earth. Due to the rotation of the Earth, stars appear to move in circles centered on the celestial poles.

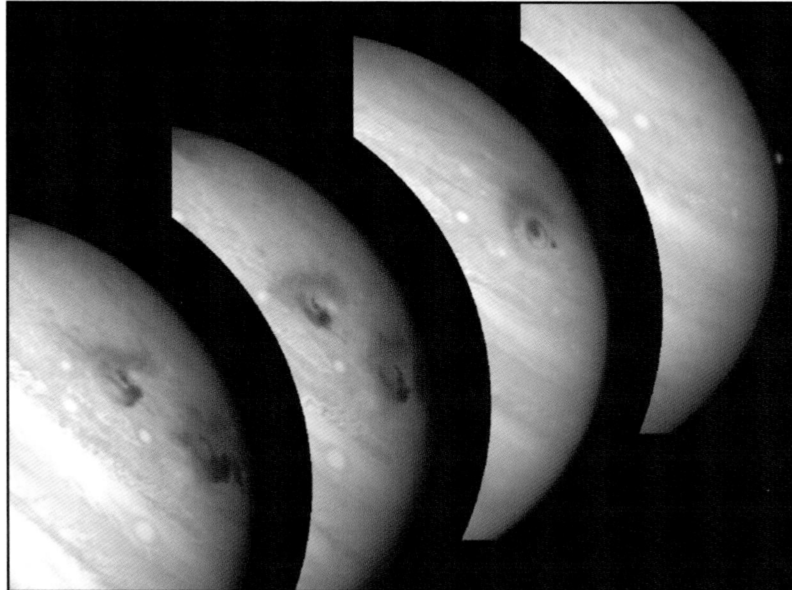

▲ ***Celestial mechanics in action*** *Comet Shoemaker–Levy 9 impacts Jupiter in this time-lapse sequence, which shows two fragments of the comet striking the planet in 1994. The impacts occurred within a few minutes of the time calculated.*

Celestial sphere

An imaginary sphere of very large radius, centered on the Earth, and to which the stars are considered to be fixed for the purposes of the measurement of CELESTIAL COORDINATES. Due to the ROTATION of the Earth on its axis, the celestial sphere appears to rotate around the Earth once a day, and it is convenient to imagine that the Earth is stationary and the sphere rotates. The sphere is therefore considered to rotate about an axis joining the north and south CELESTIAL POLES in an east to west direction (that is, clockwise) at a rate of 15° per hour (of SIDEREAL TIME). The plane of the Earth's equator, extended outwards to the sphere, marks the CELESTIAL EQUATOR. At any instant, an observer at a particular point on the Earth's surface can see only one half of the sphere. If the observer is located at one of the poles, one hemisphere is permanently below the horizon; if at the equator, then each part of the sphere is visible at some time.

Celsius, Anders (1701–44)

Swedish mathematician and astronomer. As professor of astronomy at Uppsala, Sweden, he participated in Pierre de Maupertuis' expedition in 1736 to the most northern part of Sweden to measure the length of a degree of latitude along a MERIDIAN, close to the pole, and compare the result with measurements near the equator. The findings confirmed Isaac NEWTON's opinion that the shape of the Earth is an ellipsoid flattened at the poles. Celsius founded the modern observatory in Uppsala in 1741, and made geographical measures, and meteorological and astronomical observations. He realized that AURORAE have magnetic causes by correlating the dip-angle of a compass needle with auroral activity. Celsius published catalogs of the brightnesses of 300 stars employing a photometer that attenuated a star's light until its extinction by using a number of glass plates. He constructed the eponymous Celsius thermometer for his meteorological observations.

Centaur
Any of a group of objects in the outer solar system whose orbits are mostly confined between the orbits of Jupiter and Neptune. The first to be discovered, in 1977, was CHIRON, which was originally given the asteroidal designation (2060) Chiron. In 1989, however, seven years before PERIHELION, it was found to have developed a COMA and was given the cometary designation 95P/Chiron. Another object, (5145) Pholus, was found in 1992, and by December 2002 the total had reached 126. It may well be that the Centaurs are objects that have been perturbed inward from the Kuiper Belt (*see* OORT CLOUD AND KUIPER BELT). This is supported by their reflectance spectra, which show them to be mostly dark red objects, and by their estimated sizes, which have a distribution consistent with that of Kuiper Belt objects. The largest known Centaur discovered to date is 2002 TC302, with a diameter of about 500 km.

Centaurus *See* CONSTELLATION

Centaurus A
An ACTIVE GALAXY, known by its radio name as the brightest radio galaxy in Centaurus.

Center for Astrophysical Research in Antarctica
The Center for Astrophysical Research in Antarctica is operated by the US National Science Foundation's Science and Technology Center and has its headquarters at the University of Chicago. The astrophysical observatory is located at the south pole. Instruments deployed there take advantage of the cold, dry and stable conditions that make the Antarctic Plateau the best site on Earth for observations at infrared and submillimeter wavelengths. They allow small telescopes at the south pole to outperform much larger telescopes at temperate sites. Observations are made at microwave, submillimeter and infrared wavelengths.

Center for High Angular Resolution Astronomy (CHARA)
Georgia State University's Center for High Angular Resolution Astronomy operates the CHARA Array at the MOUNT WILSON OBSERVATORY, California, USA. This optical/infrared interferometric array consists of six 1-m aperture telescopes configured in a Y shape with a maximum baseline of 350 m. Its limiting resolution of 200 microarcsec makes the instrument a powerful tool for studying fundamental properties and surface features of stars.

Cepheid variable star
Cepheid variable stars (prototype DELTA CEPHEI) have distinctive LIGHT CURVES that rise rapidly in brightness to a maximum, then fall more gradually to minimum light. This cycle repeats consistently, with individual periods ranging from a few days to many hundreds of days, depending on the intrinsic brightness of the star. The stars are pulsating variable stars, with the variability arising from the change in surface area and temperature as the stars expand and contract.

The period over which this cycling occurs for Cepheids is tightly coupled to the luminosities of the stars, and is known as the period-luminosity law. This law was discovered in 1912 by the Harvard astronomer Henrietta LEAVITT. She was analyzing the periods and apparent magnitudes of a selection of Cepheid variables all at a common distance in the LARGE and SMALL MAGELLANIC CLOUDS. What Leavitt found was to set in motion one of the most important and long-lasting astronomical activities of the twentieth century – the calibration of the size of the universe, which is continuing to this day (*see* DISTANCE).

Cepheids are supergiant stars, and as such they are among the most luminous objects in those galaxies that are today producing new populations of stars. According to the theory of the evolution of the STARS, the immediate precursors to Cepheids are the massive, young O- and B-type stars. These hot (blue) MAIN SEQUENCE stars have formed recently, but because of their high mass and prodigious energy output they soon evolve away from the hydrogen-core-burning region to cooler surface temperatures and redder colors. In doing so, these enormous stars briefly pass through a zone in which their outer atmospheres are unstable to periodic radial oscillations. A narrow range of temperature defines the so-called 'instability strip' in the HERTZSPRUNG–RUSSELL DIAGRAM, where Cepheid pulsation happens.

See also VARIABLE STAR: PULSATING.

Cepheus *See* CONSTELLATION

Ceres
The first ASTEROID to be discovered, by Giuseppe PIAZZI on January 1, 1801, and so designated (1) Ceres. Ceres is by far the largest of the asteroids, with a diameter of 933 km, and accounts for over one-quarter of the mass of the entire main-belt asteroid population. It orbits the Sun at a mean distance of 2.77 AU in a period of 4.6 years; it rotates in 9.08 h, and has a density of 2.7 g cm^{-3}. It is a G-type asteroid, with a reflectance spectrum similar to that of CARBONACEOUS CHONDRITES.

Cerro Tololo Inter-American Observatory
The Cerro Tololo observatory is operated by AURA under a cooperative agreement with the US National Science Foundation as part of the National Optical Astronomy Observatories. The observatory is about 2200 m above sea level, and 500 km north of Santiago in Chile. Its headquarters are located at La Serena. On site are various optical telescopes and one radio telescope. The largest instrument is the 4-m Victor M Blanco Reflecting Telescope, whose twin is located at KITT PEAK OBSERVATORY in Arizona, USA. This telescope has been in operation since the early 1970s. Other instruments at the site include a 1.5-m RITCHEY–CHRÉTIEN TELESCOPE, the YALO 1-m Ritchey–Chrétien telescope, a 92-cm reflector, a 60/90-cm Curtis/Schmidt telescope, the 60-cm Lowell Telescope, and El Enano (The Dwarf), the smallest telescope on Tololo. The 1.2-m radio telescope of the University of Chile is also at the site.

Cetus *See* CONSTELLATION

Cetus A (M77) *See* MESSIER CATALOG

Challis, James (1803–82)

British astronomer who became Plumian professor and director of the Observatory at Cambridge. Challis was persuaded by John Couch ADAMS to search for Neptune, but having no up-to-date chart and little confidence in the prediction, he set out to laboriously observe all the stars in the area twice to find one that moved. He observed Neptune, but only once, in July 1846. Because he did not re-observe the field of stars with any urgency, he lost priority to Johann GALLE and Heinrich d'Arrest (1822–75) in Berlin.

Chamaeleon *See* CONSTELLATION

Chandrasekhar, Subrahmanyan (1910–95)

A theoretical astrophysicist, born in India. In 1983 he won the Nobel prize with William FOWLER 'for his theoretical studies of the physical processes of importance to the structure and evolution of the stars.' At Cambridge, UK and Copenhagen he developed the theory of WHITE DWARF stars, showing that quantum mechanical degeneracy pressure cannot stabilize a massive star and that white dwarfs have a maximum mass (the Chandrasekhar limit). At the University of Chicago and YERKES OBSERVATORY he investigated and wrote several seminal text books. He edited the *Astrophysical Journal* for nearly 20 years. NASA's X-ray astronomy satellite was renamed CHANDRA in his honor. *See also* DEGENERATE MATTER.

Chandra X-ray Observatory

The third of NASA's Great Observatories, named after Nobel prize-winning astrophysicist Subrahmanyan CHANDRASEKHAR. It was previously called the Advanced X-ray Astrophysics Facility (AXAF). The observatory, which was launched from the Space Shuttle in July 1999, was designed for high-resolution imaging, with a resolving power 10 times greater than that of previous X-RAY telescopes. It carries a nested array of four mirrors. The largest of these is 1.2 m in diameter and has a focal length of 10 m. It also carries a CCD imaging spectrometer. Chandra and the ESA mission XMM-NEWTON (which was launched in the same year), are providing us with X-ray views of previously invisible X-ray sources, including BLACK HOLES, SUPERNOVAE and interstellar gases. We are also seeing completely new views of familiar objects, such as the planets.

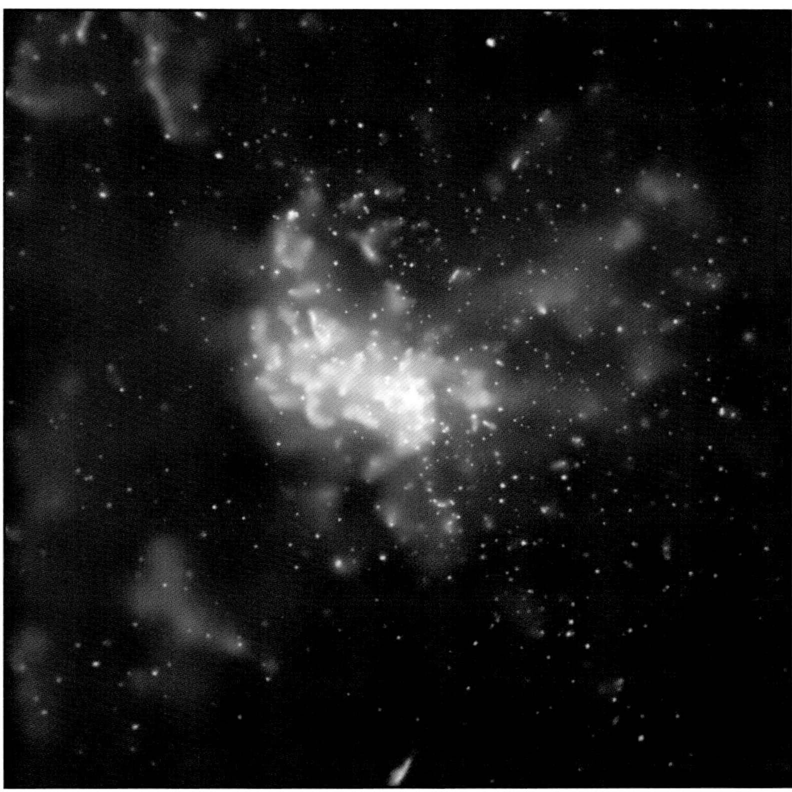

▲ ***Chandra X-ray Observatory*** *This view of Sagittarius A, the massive black hole at the center of the Milky Way, was created from a long X-ray exposure taken by Chandra. The reddish lobes extending for dozens of light years on either side of the black hole indicate that enormous explosions have occurred during the last 10 000 years.*

Chaos

1) Literally 'disorder,' as in chaotic terrain, which is a region of jumbled rock on a PLANETARY SURFACE with no particular alignment, scale length or scale height. Chaotic terrain on MARS resulted when subsurface ice melted and the surface slumped. Similarly, chaotic terrain on EUROPA arose from sudden melting of the ice on the surface, causing blocks of the ice layer to rotate and tilt.

2) In mathematical physics, chaos describes behavior that is predictable in the short term but that, in the long term, depends so much on the initial state that the long term cannot be calculated. Weather can be predicted one week or one day ahead. However, since the flapping of the wings of every butterfly cannot be taken into account at the outset, even hurricanes cannot be predicted a year or a century ahead. Likewise, in the N-BODY PROBLEM, as in the solar system, the multiple gravitational interactions between the planets cause chaotic behavior of their orbits, which cannot be predicted beyond tens of millions of years. *See also* CELESTIAL MECHANICS.

Charge-coupled device *See* CCDS AND OTHER DETECTORS

Charon *See* PLUTO

CH Cygni

CH Cygni is one of the most enigmatic symbiotic stars (*see* VARIABLE STAR: CLOSE BINARY STAR). Once known as a semiregular variable, the system began an unusual series of eruptions in 1964. CH Cyg displays a bewildering array of photometric variations, from short-timescale fluctuations at optical and X-ray wavelengths to much longer timescale variations of between 100 days and 32 years in the optical and near infrared. The current picture begins with an interacting BINARY STAR composed of an M-type giant and a hot WHITE DWARF. The white dwarf accretes material lost by the giant. Irregular pulsations of the giant modulate the ACCRETION rate onto the white dwarf, which varies the activity level.

Chicxulub basin

A terrestrial impact feature about 170 km in diameter centered near the coastal town of Puerto Chicxulub on the northern coast of Mexico's Yucatán Peninsula. The structure is well preserved but not apparent on the surface, being buried beneath about 1 km of sediment. Its presence was suspected when arc-like formations showed up in plots of surface-

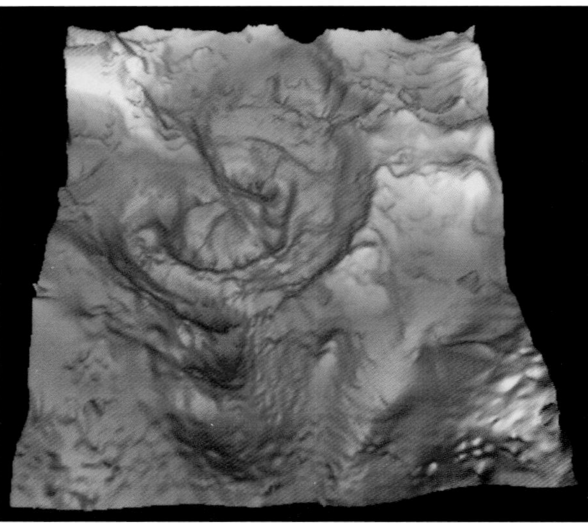

► ***Chicxulub basin*** *The basin is buried beneath 1 km of sediment, but this image, made from gravity and magnetic-field data from the region, shows the ring structure of the crater that lies below the surface. The image is an oblique view, looking north, about 300 km long.*

gravity variations. Supporting evidence came in the form of microtektites (microscopic glass beads formed from impact-melted surface material) found in Haiti. Confirmation was provided by offshore drilling carried out in a search for new oilfields. The Chicxulub basin is 65 million years old, and its size is consistent with the impact of a 10-km diameter ASTEROID. Such an impact would have had catastrophic consequences for life and the environment. It seems likely that this event was responsible for the mass extinctions that occurred at the end of the Cretaceous period 65 million years ago, leaving its mark as a sedimentary layer enriched in iridium at what is known to stratigraphists as the K/T boundary.

CHIPS mission

The Cosmic Hot Interstellar Plasma Spectrometer (CHIPS) is the first NASA-funded University-Class Explorer (UNEX) mission. It was carried into space in January 2003 aboard a dedicated spacecraft called CHIPSat. The CHIPS spectrometer is a survey instrument for ULTRAVIOLET ASTRONOMY studying the million-degree gas in the INTERSTELLAR MEDIUM. It will record the spectrum of the hot PLASMA that exists within roughly 100 l.y. of the solar system, probably the remnant of one or more SUPERNOVA explosions. CHIPS is expected to be operational for one year, first surveying the sky at low resolution and then re-examining in greater detail regions of particular interest. As a UNEX mission, CHIPS was developed primarily as a training device for young scientists and engineers, but is one that can also obtain actual and valuable science data. CHIPSat was one of two NASA science research satellites launched on a single rocket in January 2003. The primary cargo was the Earth Observation satellite ICESat (Ice, Cloud, and Land Elevation Satellite).

Chiron

The first of the population of minor planets called CENTAURS to be discovered (in 1977). (2060) Chiron orbits the Sun between Saturn and Uranus on a 50-year orbit, ranging in heliocentric distance from 8.5 AU to just under 19 AU. Chiron is also known as comet 95P/Chiron.

Chondrite

This is the most common type of METEORITE to fall to Earth from space. Chondrites are so called because most of them contain chondrules (silicate beads). Chondrules are sometimes found in large numbers and have always fascinated scientists. The inventor of the petrographic microscope, Henry Sorby (1826–1908), described them as droplets of fiery rain. They are essentially glassy beads made by a violent but brief heating event that caused dust grains to form millimeter-sized melt droplets. However, the cause of the heating remains a mystery. The most important feature of chondrites is that, with the exception of a few highly volatile elements, they have the same composition as the SUN. They are also extremely ancient rocks, having formation ages comparable with the age of the solar system (4600 million years). Finally, they have unique textures that suggest the accumulation of diverse components with little or no subsequent alteration. *Contrast* ACHONDRITES. *See also* CARBONACEOUS CHONDRITE.

Christie, William Henry Mahoney (1845–1922)

British astronomer who worked at the ROYAL OBSERVATORY, GREENWICH as chief assistant before succeeding George AIRY as ASTRONOMER ROYAL in 1881. Christie turned the Royal Observatory toward ASTROPHYSICS as opposed to positional astronomy. He installed large collecting-area telescopes; for example, the 28-in refractor, which is still in operation, used originally to observe DOUBLE STARS. Under his administration the Greenwich meridian was given the status of prime meridian of the world at the start of the international time-zone system at the Washington Conference of 1884. He expeditiously measured the positions of all the stars in the large Greenwich Zone (northern region) of the *Astrographic Catalogue*. Christie founded and first edited *The Observatory* magazine, which still preserves a human face for British astronomy.

Chromatic aberration

An optical defect whereby light rays of different wavelengths are focused at different points along the optical axis of a lens. Because the refractive index of glass is a function of wavelength, shorter-wavelength light is refracted (deflected) to a greater extent than longer-wavelength light. Since white light consists of a mixture of wavelengths, the image of a star, formed by a simple lens, consists, in effect, of a series of different-colored images spread out along the optical axis. The human eye is most sensitive to the yellow-green part of the SPECTRUM, so the best visual image is normally obtained when the eyepiece is adjusted to put this wavelength region into sharpest focus.

The effects of chromatic aberration can be considerably reduced by means of compound OBJECTIVE LENSES that have two or more appropriately shaped components, each made of glass with different dispersions. An achromatic lens (or achromatic doublet), which has two components, can bring two wavelengths to the same focus and reduce the spread in focal lengths for the other wavelengths by a factor of about 10. A three-element apochromat reduces the spread even further.

Because the laws of reflection are independent of wavelength, chromatic aberration does not affect the mirrors that are used in REFLECTING TELESCOPES.

Chromosphere

The layer of a star's atmosphere between the PHOTOSPHERE and the CORONA, typically 2000 km thick and containing mainly HYDROGEN and HELIUM gas. During a total solar eclipse the SUN's chromosphere can be seen as a pinkish, narrow ring around the LIMB of the Sun, often with protruding structures called PROMINENCES. The pink color is largely due to the presence of hydrogen-alpha (Hα) emissions. When there is not an eclipse, the chromosphere may be observed using a SPECTROHELIOGRAPH (or spectrohelioscope), which uses narrowband filters to trace the distribution of the gas.

CH star and barium star

The classical barium (or Ba II) stars are RED GIANT STARS whose spectra show strong absorption lines of barium, strontium and certain other heavy elements, as well as strong features due to carbon molecules. They were recognized in 1951 by Philip Keenan (1908–2000) and William Bidelman. CH stars were recognized even earlier by Keenan, in 1942. Like barium stars they have similar enhancements of some elements, but also very strong bands of the CH molecule, and, by contrast, weak lines of the ordinary metals such as iron. CH stars are found in the Galactic halo and in several GLOBULAR CLUSTERS. The cluster OMEGA CENTAURI contains the richest population of CH stars, with about six. All this implies that CH stars are old population stars, the POPULATION II equivalents of Ba II stars. CH stars and Ba II stars were crucial in establishing the existence of neutron-capture reactions in stellar interiors, which are responsible for the synthesis of heavy elements. More recently, the Ba II and CH stars have provided strong evidence for mass transfer in binary systems (*see* BINARY STAR) that drastically alters the surface composition of a companion star.

Cigar Galaxy (M82) *See* MESSIER CATALOG

Circinus *See* CONSTELLATION

Circumpolar star

A star that never sets, as seen from a particular location. These stars rotate once a day around the pole of the northern or southern sky as the Earth rotates. The altitude of the CELESTIAL POLE is equal to the latitude. Thus from a location at latitude 52°N, stars with a north polar distance of less than 52° (with DECLINATIONS of between +38° and +90°) are circumpolar and will be seen to circle around the north celestial pole. Observed from the Earth's geographical poles, all stars in the particular hemisphere are circumpolar, whereas from the equator none are circumpolar. Circumpolar stars transit the MERIDIAN twice a day, at upper and lower CULMINATION alternately.

▲ ***Circumpolar star*** *Star trails around the south celestial pole, seen at the AAO.*

Clark family

Alvan Clark (1804–87) was an American astronomer and telescope maker. With his sons George Bassett Clark (1827–91) and Alvan Graham Clark (1832–97) he founded Alvan Clark & Sons, makers of optical lenses for telescopes. This was America's first significant contribution to the making of astronomical instruments. While testing a lens on a newly manufactured telescope in 1862, Alvan Graham Clark discovered the WHITE DWARF companion of SIRIUS, Sirius B.

Classification of stellar spectrum

See STELLAR SPECTRUM: CLASSIFICATION.

Clementine

A joint project between the US Strategic Defense Initiative Organization and NASA, launched in January 1994. It was designed to test military sensors and spacecraft components in the space environment. It was also intended to make scientific observations of the Moon and the near-Earth asteroid (1620) Geographos. Clementine obtained global ultraviolet and infrared images of the lunar surface, laser-ranging altimetry coverage from 60°S to 60°N, and charged-particle measurements. It also obtained the first data to indicate the presence of water ice inside permanently shaded craters at the lunar poles. Towards the end of the lunar-mapping phase of the mission, Clementine suffered an onboard malfunction that left the spacecraft spinning out of control. This prevented it from performing the planned close flyby of Geographos.

Climate

Earth's climate may be defined as the global physical condition, averaged over a period of time (typically decades or longer), of the Earth's atmosphere, oceans and ice sheets. It is the presence of a relatively dense atmosphere that maintains the planet with a habitable climate. Without the blanketing of infrared energy radiated from Earth's surface and lower atmosphere, conditions would be well below freezing almost everywhere around the globe.

Close binary star *See* VARIABLE STAR: CLOSE BINARY STAR

Clown Face Nebula (NGC 2392)

A PLANETARY NEBULA in the constellation Gemini, also known as the Eskimo Nebula. It is 13 arcsec in diameter and of

▶ Clown Face Nebula *(NGC 2392) Also known as the Eskimo Nebula. The inner filaments have been ejected by a strong wind of particles from the central star. The outer filaments (the Eskimo's fur hood) were ejected by the star at a previous stage of its evolution.*

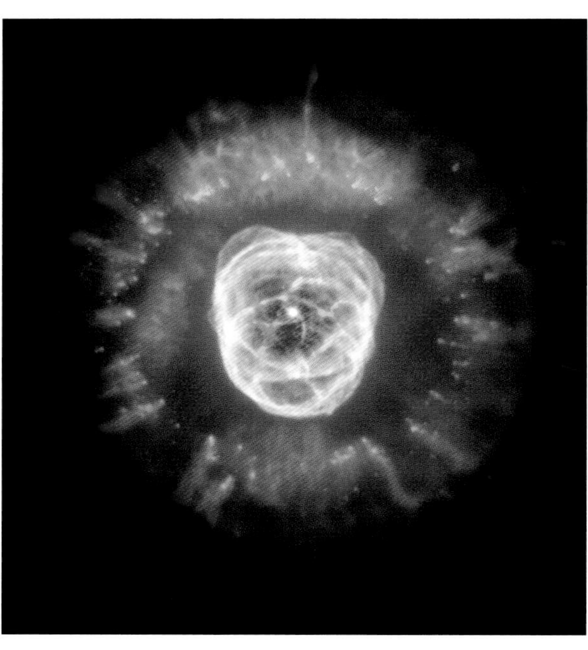

magnitude 9, with a magnitude 10 central star. The blue-green nebula's hazy outer regions are thought to resemble an Eskimo's hood or clown's ruff.

Cluster
Two ESA missions to send four identical spacecraft to explore the interaction between Earth's MAGNETOSPHERE and the SOLAR WIND. The four original satellites of Cluster I were lost in the failed maiden launch of the Ariane 5 rocket in 1996. Cluster II, replicating the original mission, was launched from Baikonur in August 2000 by two Russian Soyuz rockets, each carrying two identical spacecraft. The four satellites have joined a number of spacecraft from many countries (including SOHO and ULYSSES) to study the Sun and solar wind, and to provide the first three-dimensional measurements of large- and small-scale phenomena in the near-Earth environment.

Cluster of galaxies *See* GALAXY CLUSTER AND GROUP

Cluster of stars
See GLOBULAR CLUSTER; OPEN CLUSTER

Clusters of stars, observing
See practical astronomy feature overleaf.

CNES
CNES, the Centre National d'Études Spatiales, is the French space agency.

Coalsack
A large and prominent DARK NEBULA in the CONSTELLATION Crux, encroaching into neighboring Centaurus and Musca. It is centered approximately on position RIGHT ASCENSION 12h 50m, DECLINATION –63°. Coalsack is easily visible to the naked eye against the bright background of the Milky Way, and measures nearly 7° by 5°.

Coelostat *See* SIDEROSTAT/HELIOSTAT/COELOSTAT

Colliding galaxies *See* GALAXIES, COLLIDING

Collimation
The process whereby the components of an optical system, such as a telescope, are positioned and aligned to give optimum performance. For a REFRACTOR, each lens should be centered on, and perpendicular to, the optical axis of the telescope. REFLECTORS, for example the CASSEGRAIN TELESCOPE, are collimated by ensuring that the axis of the PRIMARY MIRROR passes through the center of the secondary mirror and that the centers of the eyepiece and the secondary mirror lie on this same axis. Where light from the primary is reflected through a right angle (as in the NEWTONIAN TELESCOPE), the system is collimated when the axis of the primary mirror and the axis of the eyepiece intersect at the center of the diagonal flat mirror and are inclined to its surface by the same angle.

A collimated beam of light consists of a bundle of parallel rays. A collimator is a device (lens or mirror) that produces a collimated beam. This may be achieved, in principle, by arranging for light to pass through a small aperture or pinhole placed at the focus of a lens. After passing through the lens, the emerging beam of light should consist of parallel rays.

See also SPECTROSCOPE AND SPECTROGRAPH.

Color index
A simple measure of the unevenness of the spectral distribution of light to indicate the COLORS OF STARS. Color index is determined from the ratio of the intensity of light from the star in just two regions of the spectrum, for example the blue (B) region and the yellow (V), or the yellow and the red (R). In mathematics, the logarithm of the ratio of two numbers is the difference between the logarithms of the numbers. Stellar magnitudes are logarithms of the intensity of starlight, so the color index is the difference between the magnitudes of starlight determined with a B and a V filter, or a V and an R filter. These color indexes are denoted (B-V) or (V-R), respectively. Color index is principally related to a star's temperature and its interstellar reddening, but also to other factors, like its composition, its REDSHIFT and its age.

Color-magnitude diagram
See HERTZSPRUNG–RUSSELL DIAGRAM

Colors of stars
See practical astronomy feature, p. 86.

Columba *See* CONSTELLATION

Coma
(1) The roughly spherical, temporary atmosphere of gas and dust expelled from a COMET'S nucleus as a result of solar heating when the comet is sufficiently close to the Sun. The nucleus and coma are together known as the comet's head. The coma, which is typically drawn into a teardrop shape by the SOLAR WIND, contains neutral and ionized gas molecules that shine by fluorescence and reflected sunlight. The coma typically develops at 3–4 AU from the Sun and can grow to between 100 000 and 1 000 000 km, reaching its maximum extent just after the comet's PERIHELION passage.

(2) The distortion of off-axis images produced by lenses or mirrors. Rays of light that enter a lens, or reflect from a mirror, at an angle to its optical axis (the line passing perpendicularly through the center of the lens or mirror) form pear-shaped images. The magnitude of the distortion increases with increasing distance from the optical axis. This

◀ ***Coalsack*** *This dark nebula is prominent against the bright band of the Milky Way at lower left. The red patch at upper right is the Carina Nebula.*

particular optical ABERRATION derives its name from the comet-like appearance of such images.

Coma Berenices *See* CONSTELLATION

Coma cluster

This well-studied cluster of galaxies lies close to the north galactic pole in the constellation Coma Berenices. Cataloged as Abell 1656 (*see* ABELL CLUSTER), it is one of the richest nearby clusters, having 650 confirmed member galaxies and probably as many as 2000 in total. The mean REDSHIFT of cluster members is 6900 km s^{-1}. As in most clusters, the galaxies in the core of Coma are mostly ELLIPTICALS and LENTICULARS while the outskirts have a higher proportion of SPIRALS. The total mass of the Coma cluster has been estimated from observations of the gravitational effects on the galaxies and on the hot X-ray gas in the cluster. Assuming the cluster is in dynamical equilibrium, the total mass is found to be much greater than the total observed mass. This result, first obtained by Fritz ZWICKY in 1933, remains one of the strongest pieces of evidence that the universe is predominantly composed of some form of unseen DARK MATTER.

Combustion

The process whereby a substance combines with oxygen and produces heat. Burning is a familiar example of this process. The energy required to propel chemical rockets is provided by the combustion of fuel with an oxidant at very high temperatures. A common oxidant is liquid oxygen. Others include hydrogen peroxide and nitrogen tetroxide.

Comet

Comets are small bodies of the SOLAR SYSTEM that follow highly eccentric ORBITS that may be inclined at any angle to the ECLIPTIC. They have a solid part, the nucleus, which is up to several tens of kilometers in size and composed of both ices and solid material. When comets come close to the Sun they become active and develop a bright, dusty atmosphere (*see* COMA; GAS and DUST TAILS).

For a long time comets were interpreted superstitiously. ARISTOTLE considered comets as atmospheric phenomena. Tycho BRAHE accurately measured the PARALLAX of the comet of 1577 and established that it was further than the Moon, demonstrating that it was astronomical.

Isaac NEWTON computed that the comet of 1680 was following a parabolic orbit around the Sun, visiting the Sun only once as far as he could tell. Edmond HALLEY, following Newton, computed the orbits of 24 comets, of which three, the comets that appeared in 1531, 1607 and 1682, had similar orbits. He postulated that it was the same comet, revisiting every 76 years, and predicted its next appearance in 1758 or early 1759. This happened after his death, when COMET HALLEY (or 1P/Halley) was recovered by Johann Palitzch (1723–88), an amateur astronomer. This prediction was one of the major successes of CELESTIAL MECHANICS.

Tails

The main tails of comets point away from the Sun. Occasionally a short antitail can be seen apparently pointing toward the Sun, but it is an effect of perspective. Heinrich OLBERS (1812) and Friedrich BESSEL (1836) proposed that comet tails were solid particles (dust). François Arago (1786–1853) measured the POLARIZATION of light emitted by the tails of the comet of 1819, and established that it was mainly sunlight reflected by the dust. The dust of comet tails undergoes a repulsive force from the Sun, identified by Svante ARRHENIUS and Karl SCHWARZSCHILD as solar radiation pressure. The tails have a curve – the locus of the dust particles

Observing star clusters

Some clusters of stars (*see* GLOBULAR CLUSTER; OPEN CLUSTER) were known in antiquity, visible to the naked eye. They include the PLEIADES, HYADES, and Praesepe (the Beehive) (*see* MESSIER CATALOG). PTOLEMY additionally identified a not-quite-resolved, sparse star cluster around the star Lambda Orionis as the 'mistiness on Orion's head.' Binoculars such as 8 × 30 and 10 × 50 offer fine views of these open clusters (*see* BINOCULAR ASTRONOMY). Many larger objects can even be difficult to fit into a binocular view, although the Hyades are just about right.

The Hyades is a V-shaped arrangement of stars centered on ALDEBARAN, and it is the closest cluster of stars to us. Indeed, the Sun is closer to the center of the Hyades than some of its members. As a result, the stars are so spread out that neither Charles MESSIER nor the Herschels (*see* HERSCHEL FAMILY) thought to include it in their catalogs of clusters and nebulae. In binoculars, the stars appear more numerous and more crowded.

Praesepe is the twin of the Hyades, but four times further away. In 1609, GALILEO GALILEI, with his primitive telescope, counted more than 40 glittering stars, and was astonished and delighted by them. Modern binoculars show a better view than Galileo's, and the view with a wide-field telescope is breathtaking.

The Pleiades cluster is visible in even relatively hazy skies, but offers a test as to how many individual stars can be seen with the naked eye. There are 'seven sisters' in the Pleiades of mythology, but usually people see six stars. In 1579, well before the invention of the telescope, Johannes KEPLER's teacher Michael Mästlin drew a chart showing eleven, and managed to map their positions in places that correspond to the telescopic view (which is reproduced in *Norton's Star Atlas*). Curiously, the Pleiades can be disappointing in a large telescope simply because the area of the cluster of about 1.5° cannot be fitted into the field of view.

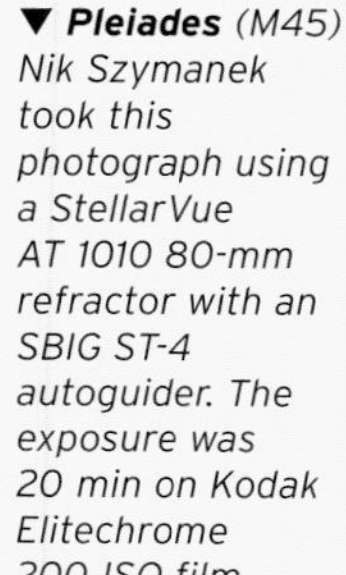

▼ ***Pleiades*** *(M45) Nik Szymanek took this photograph using a StellarVue AT 1010 80-mm refractor with an SBIG ST-4 autoguider. The exposure was 20 min on Kodak Elitechrome 200 ISO film.*

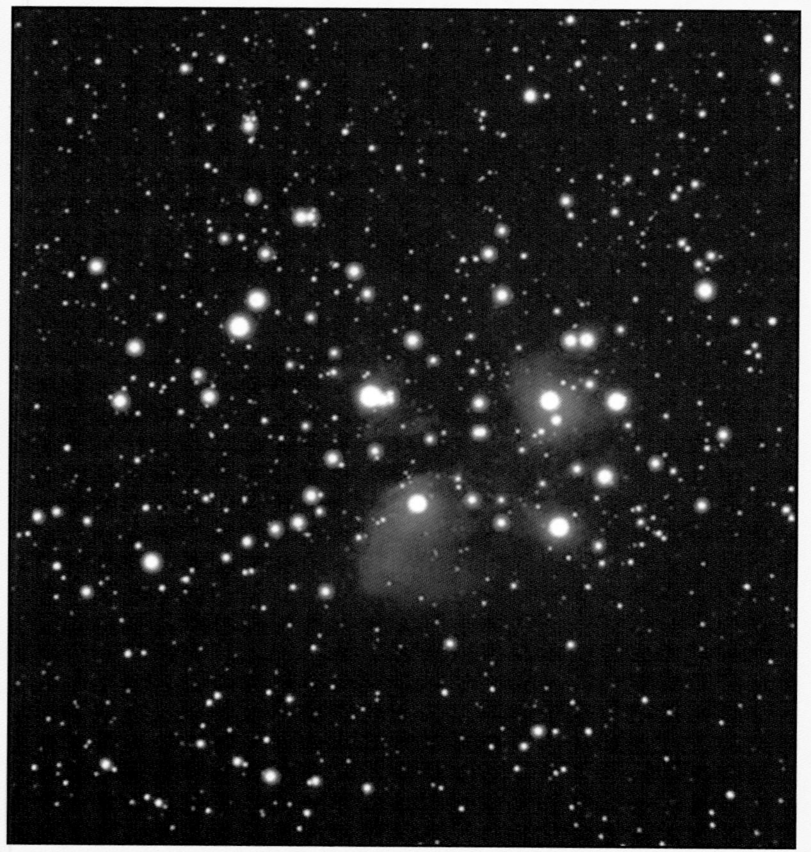

The stars h and Chi Persei are the two brightest stars in a striking pair of star clusters in Perseus, known as the DOUBLE CLUSTER, NGC 869 and 884. They are visible as hazy patches to the naked eye. Whether viewed through binoculars or a wide-field telescope, the Double Cluster remains a favorite for amateur astronomers. The stars are bright enough and close enough together that their colors appear very striking, partially as an effect of contrast.

Two clusters of stars in the southern hemisphere also have the names of stars but are clearly not point sources, namely OMEGA CENTAURI and 47 Tucanae. Although Omega Centauri was previously included in Ptolemy's star catalog as a star, Edmond HALLEY was the first, after having observed it telescopically in 1677, to list it as a non-stellar object in his catalog of six nebulae. Halley came across the prominent northern globular cluster M13 in the constellation Hercules in 1714. Halley noted that it was visible to the naked eye when the 'sky is serene and the Moon absent.' Omega Centauri is the finest cluster of stars in the whole sky when seen through a telescope; it has many more stars than can be counted by eye; many tens of thousands are recorded on photographs. From the United States, it is low in the spring and summer sky and visible only from the southern states; from Europe, it is never visible – excuses (if you need one) for a trip to Australia, southern Africa or South America!

Messier compiled early accidental discoveries of star clusters, as well as his own, into his catalog published in 1784. Messier wanted to list objects that look similar to comets, and therefore, from a comet observer's viewpoint, were to be avoided. The second item in Messier's list, M2, is a globular cluster in the constellation Aquarius. This cluster gives the impression of a three-dimensional ball of stars and looks particularly comet-like. However, soon Messier was including the looser star groups, or open clusters. The first such objects in Messier's list were M6 and M7, both located in the constellation Scorpius.

The number of known clusters of stars was greatly expanded by William Herschel in the course of his systematic surveys of the sky. Herschel was able, in some cases, to resolve what Messier had seen as nebulae into individual stars. Messier described the appearance of M98 as 'nebulous, without stars, round and clear.' To Herschel it was 'the form of a solid ball, consisting of small stars, quite compressed into a blaze of light, with a great number of small ones surrounding it.' It was Herschel who, in 1784, first used the term 'globular cluster,' now recognized as distinct from 'open cluster.' Open clusters are usually visible near to the Milky Way, and although globular clusters are more widely distributed, more than half are in the direction of the Galactic Center and in the constellations of Sagittarius, Ophiuchus, Scorpius and Centaurus – more reasons to be in the southern hemisphere for stargazing.

Classification of star clusters

Herschel noted in 1789 that 'almost all the nebulae and clusters of stars that I have seen, the number of which is not less than three and twenty hundred [2300], are more condensed and brighter in the middle.' In this statement, before the differences were realized, he was lumping together galaxies and clusters, as well as true nebulae. But the globular clusters in particular are 'clusters of stars of nearly equal

size (that is, brightness), which are scattered evenly at equal distances from the middle (that is, the clusters are circular in shape), but with an increasing accumulation towards the center.' In 1927, Harlow SHAPLEY and Helen HOGG classified globular clusters according to how concentrated the stars appear toward the center, with Class 1 being the most highly concentrated, and Class 12 the least.

Clusters with many stars are called 'rich.' Rich clusters such as those of Per Collinder (1890–1974) and Robert TRUMPLER offer even the binocular observer much potential enjoyment. The use of CCDs with a basic 50-mm lens can give pleasing images of these larger objects, but some are extremely obscured clusters buried deep behind or within the obscuring dust of our galaxy.

The most common classification scheme for open clusters is the Trumpler Classification System. It describes the three major attributes of an open cluster: concentration, the range in brightness of the component stars and richness.

Trumpler Classification System

Concentration of space

- I - Detached, strong concentration toward the center
- II - Detached, weak concentration toward the center
- III - Detached, no concentration toward the center
- IV - Not well detached from surrounding star field

Range in brightness

- 1 - Small range
- 2 - Moderate range
- 3 - Large range

Richness

- p - Poor (less than 50 stars)
- m - Moderately rich (50–100 stars)
- r - Rich (more than 100 stars)

Equipment and resources for the observer

Small amateur telescopes (50 mm) can detect the fuzzy outlines of many clusters. Medium-sized telescopes (100–200 mm) can resolve the clusters containing the brightest stars. Large telescopes (10–20 in and larger) are required to resolve clusters where the faintest stars constitute the majority of the cluster stars.

The NEW GENERAL CATALOGUE and INDEX CATALOGUE by John DREYER in 1888 was based on the surveys by William and John Herschel. It constitutes the major source of deep-sky star clusters used by astronomers. Finding clusters is sometimes not trivial. Some of the fainter and more distant clusters have incorrectly cataloged positions. Arizona deep-sky observer Brian Erdmann organizes a Pro-Am, Web-based attempt to identify all the NGC and IC nebulae.

One can select target sources from Norwegian astronomer Mikkel Steine's webpages, or go back to the historical compilations by the Reverend Thomas William Webb (1807–85) (*see* WEBB SOCIETY), who published the first edition of his *Celestial Objects for Common Telescopes* in 1859. Although this contains objects of all types, clusters featured prominently as so many were found to be within the range of amateur telescopes. Sherburne BURNHAM's *Celestial Handbook* was inspired by Webb. To this day, the Webb Society promotes the observation of deep-sky objects and, in Volume 3 of its *Deep-Sky Observer's Handbook* series, an entire book was devoted to drawings and descriptions of star clusters made by amateur astronomers. Curious chains of stars are quite common in open clusters, and the challenge is to be able to convey accurately on paper the (in truth) haphazard arrangement seen in the telescope.

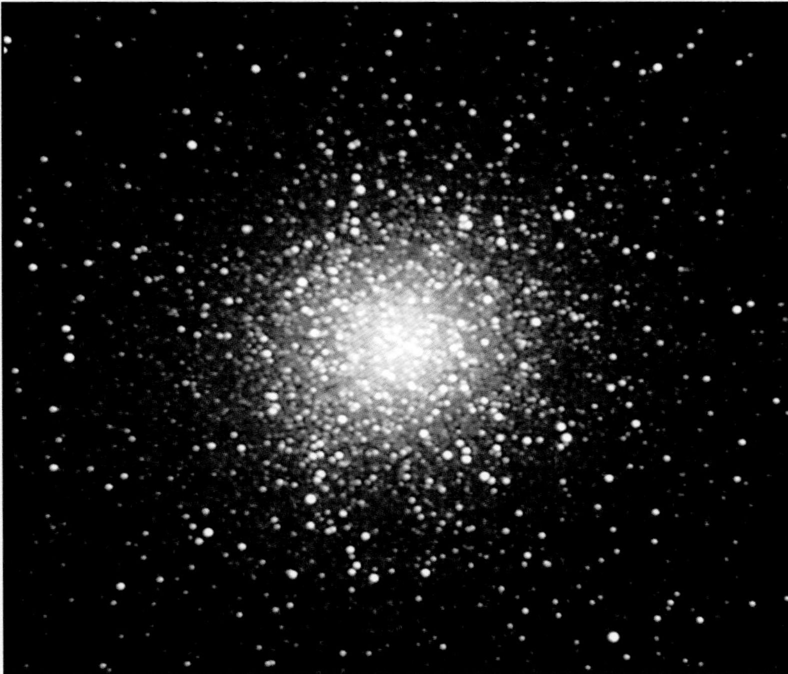

▲ ***Globular cluster M13*** *as imaged on July 11, 2002, by Chris Peterson using a Meade 12-in LX200 at f/7.6 and an SBIG ST8i camera. This is a single 10-min exposure with a non-linear contrast stretch applied to enhance the dynamic range. Such clusters require good optics to resolve their many individual stars.*

In any case, faint clusters offer a fruitful area of extended observation with larger telescopes. Objects in the lists of Ruprecht, Dolizde and Berkeley offer real challenges. The faintest clusters include the Terzan clusters and Palomar clusters, which are barely detectable with even the larger amateur telescopes. These clusters vary in richness and density of stars. Some are extremely sparse, such as Palomar 5, which is currently undergoing disruption due to tidal forces in our galaxy.

The brighter stars in the nearer clusters show distinctive colors when viewed through good telescopes. Discovered by Nicholas LACAILLE in 1751 on his trip to the Cape of Good Hope, NGC 4755 is an open cluster worth observing for its colors. John Herschel named it the Jewel Box because of the startling contrast between its brightest star, Kappa Crucis, and the rest, and its visual resemblance to jewels spread on black velvet. The brilliant red color of Kappa is easily seen through a modest telescope. The remaining stars, which give no particular impression of color, are measured at more or less blue.

A similar cluster is NGC 3324. The bright red star is a red supergiant, swelling up from its youthful state like the other blue DWARF STARS as the hydrogen in its center is exhausted. The brighter stars in globular clusters like Omega Centauri are all RED GIANT STARS, all of about the same brightness. The HORIZONTAL-BRANCH STARS show as individual blue stars by contrast with the red giants. With a good telescope, one can see their color directly and experience the HERTZSPRUNG–RUSSELL DIAGRAM with the heart and soul.

Guy Hurst was formerly President of the British Astronomical Association and is a retired bank manager. His special astronomical interests are novae and clusters.

Colors of stars

The colors of the stars are real and easy to photograph if you know how. They are not seen by the unaided eye as stars are both intrinsically faint and at the same time minute, single points of light, almost always observed by a more or less dark-adapted eye. However, the colors of the stars are an important and quantifiable observational property that is indicative of their surface temperature. This, in turn, is a function of a star's mass and its evolutionary status. These ideas are brought together in the well-known color-magnitude diagram, a cornerstone of stellar astrophysics (*see* HERTZSPRUNG–RUSSELL DIAGRAM).

With a very few exceptions (such as ANTARES and BETELGEUSE which are orange-yellow) most of the stars that appear bright, such as those defining the constellations, are in reality as blue as the midday sky. This is not recorded in the long history of non-telescopic stargazing as the eye is not very good at distinguishing delicate shades of blue when light levels are low and when the size of the apparent object is small. Even with a telescope, colors are usually only noted by observers of double stars, who sometimes comment on the contrasting colors of the adjacent objects.

The range of colors found among the stars, from pale blue to deep orange-yellow, is known to photographers in the concept of color temperature. For example, they use filters to adjust the color of tungsten lighting when it is shot on a film designed for daylight use. More broadly, the star colors are known as black-body colors and are the visible part of the radiation from any object that is at a temperature above about 1000 K (*see* BLACK-BODY RADIATION). The universe is full of such thermal radiation and much of it is invisible, originating from bodies much cooler than 1000 K. The ubiquitous COSMIC MICROWAVE BACKGROUND is a remnant of a time when the universe was unimaginably hot, but it now corresponds to a temperature of just 2.73 K.

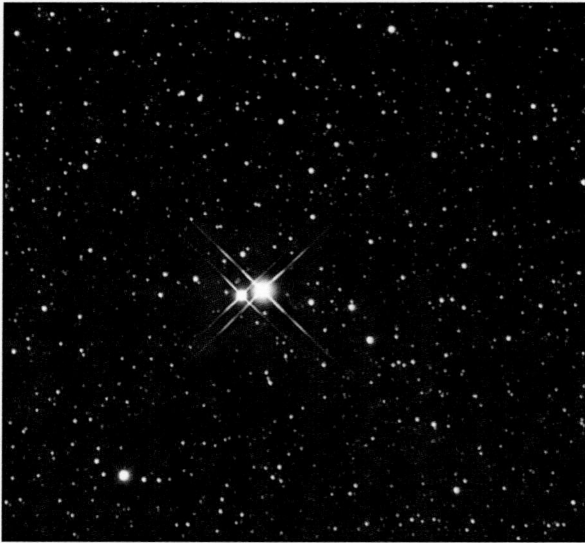

► ***Albireo*** *(Beta Cygni) pairs a vivid blue secondary with a golden primary star. Although best viewed using a telescope, color contrast can be easily seen with a good pair of binoculars. Image by Yuuji Kitahara.*

The reality of star colors can be demonstrated in several ways. One of the simplest is to look at a bright star through a defocused telescope; normal practice when, for example, manually focusing a telescope using the Foucault knife-edge technique (*see* Léon FOUCAULT). Although it can be difficult to distinguish between randomly chosen O and B stars, it is easier to identify cooler F and G stars, and the differences between the still-cooler K and M stars are still more obvious.

A more convincing demonstration is a simple photographic experiment that involves making a star-trail photograph. From a dark site beneath a clear night sky, a time exposure made with a camera loaded with color film captures the

► ***Orion*** *Image by Akira Fujii.*

►► ***The colors of Orion*** *A step-focus image of the colors of the stars in Orion, made on daylight color film by defocusing the camera lens in a series of steps during a normal star-trail exposure. Image by David Malin.*

long, narrow trails made by stars as they move across the sky due to the rotation of the Earth. If the lens is regularly defocused in a series of steps every five minutes or so during the exposure, the stepwise spreading of the star images on the film effectively changes the exposure time. At some stage, the colors of all the stars on the film are revealed.

The realization that some stars are much bluer than others came in the early 1890s when the relative magnitudes of stars in the early photographic catalogs were compared with stellar magnitudes obtained visually. It was gradually appreciated that the eye and the early photographic plates saw things differently. The plates of the time were sensitive only to blue and ultraviolet light, while the eye's maximum sensitivity lies in the yellow-green part of the SPECTRUM. This difference between the blue (B) photographic magnitudes and the visual (V) magnitudes was eventually understood as a way of measuring the color of a star and is the origin of the still widely used B-V color index (*see* COLOR INDEX).

Of course, the eye is no longer used for stellar PHOTOMETRY. Now the B-V color index system uses optical filters with well-defined characteristics to make two measurements in two well-separated color bands, nowadays using an electronic detector. Only after two-color systems were established was it realized that the difference in energy in the two bands amounted to measures of the slope of the black-body radiation distribution. In the visible part of the spectrum, this is seen as a color shift. Stars are not perfect black bodies, but allowances can be made for this. Thus, measuring temperatures using the colors of stars became fundamental to understanding and classifying them.

The zero-point (white point) of the B-V color index is now defined as an A0 star, which happens to be the color index of SIRIUS, the brightest star in the sky. But A0 stars have a surface temperature of 10 000 K, a temperature that we would see as quite blue if the star was as big as the SUN in the sky. Regrettably, this does not prevent astronomers from describing stars like Sirius as white and stars cooler than Sirius as yellow, or even red. If there are some truly red stars, they are intrinsically extremely faint; the famous, bright, so-called red giant stars such as Betelgeuse and Antares are in fact several hundred degrees hotter (and therefore bluer) than most domestic tungsten-filament lamps. Though light from these lamps undoubtedly has a warm color, not even an astronomer would describe it as red.

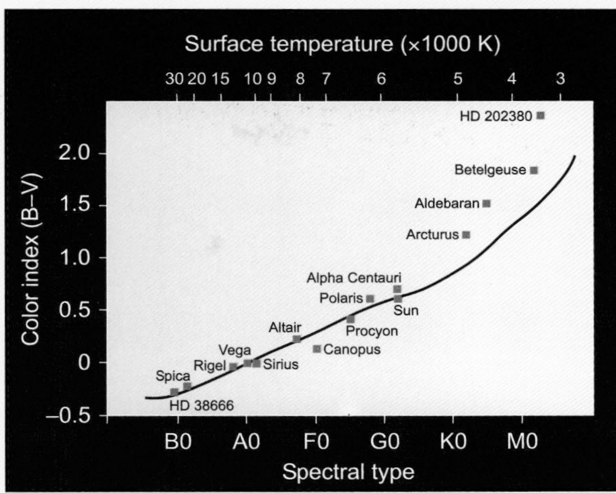

◀ ***Color-temperature diagram*** *This diagram shows a few well-known stars and the main sequence line, plotted according to surface temperature and B-V color index. The background color represents the color of each star.*

It has already been noted that the colors of the stars are most evident when observed in close double stars, so that the contrast between the two stars making up the pair is evident. This contrast can lead to some strange effects; red stars like Antares or Alpha Herculis have companions that many observers have recorded as greenish. The archetypal double star, providing a beautiful color contrast is ALBIREO (Beta Cygni), which pairs a vivid blue secondary with golden primary. Separated by 35 arcsec, it is best seen in a small telescope, but there are plenty of other targets regardless of what equipment one is using.

An investigation into historical records of the colors of stars is an interesting exercise; the discrimination between colors seems much greater, with many stars recorded as redder than we observe them today. It was believed for a long while that these observations reflected evolutionary changes in the stars, but the timescales involved are just too short and it seems we must accept that perceptions (or, perhaps more accurately, the language used to record them) have altered with time.

Colors can also cause problems for observers, and particularly for observers attempting to measure the brightness of objects. For example, variable star observers attempting comparison of the variable with nearby stars may need to take into account the eye's sensitivity to different colors. As if this wasn't difficult enough, this sensitivity will change between individual observers, causing problems when data sets are combined. This is also the reason for the difficulties described above in combining visual with photographic results. Often a set of standard filters is employed to improve the situation, although this doesn't completely eliminate the problems.

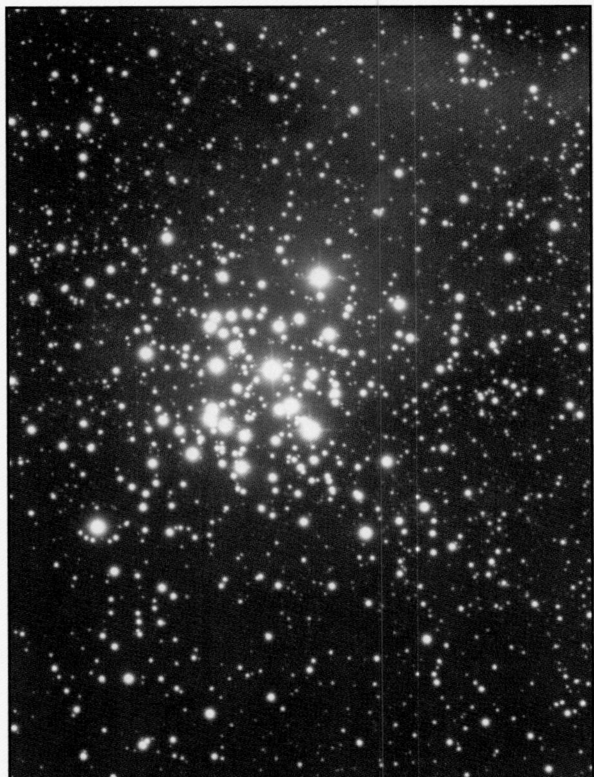

◀ ***NGC 3293*** *This brilliant cluster in Carina shows the beautiful contrast between the bright orange star and the bright blue stars which surround it.*

David Malin has pioneered novel astronomical imaging techniques and has used some of the world's finest telescopes to reveal cosmic landscapes of brilliant color and unexpected form.

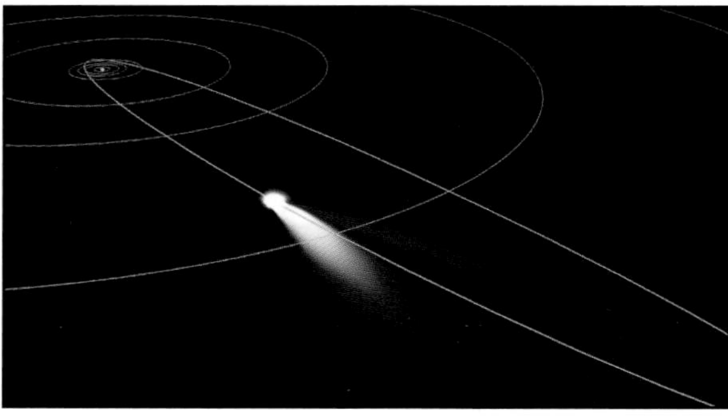

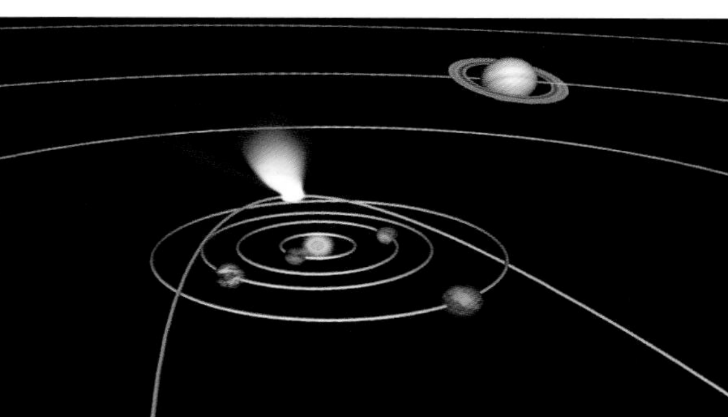

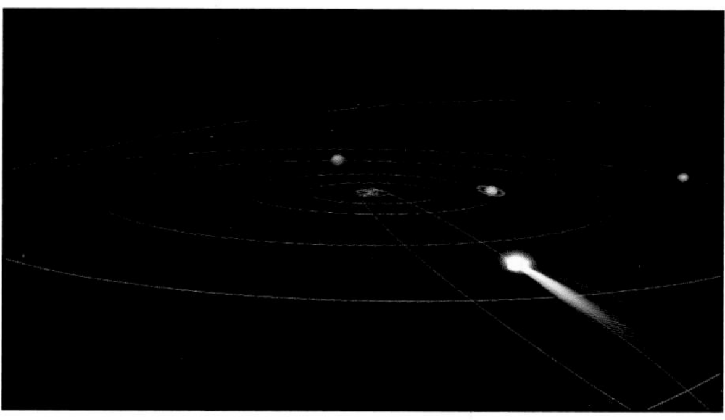

▶ ***Cometary orbit*** *As a comet sweeps into and out of the solar system in a parabolic orbit, the pressure of solar radiation causes its dust tail to point away from the Sun at all times.*

let loose from the comet as it proceeds in its orbit and pushed back radially from the Sun according to their size and mass (known as the Finson–Probstein model). In 1866 Giovanni SCHIAPARELLI linked the PERSEID meteor stream with dust shed by COMET SWIFT–TUTTLE.

Comets also have a second, faint, straight tail that points at an angle from the Sun. It is not reflected sunlight but has a SPECTRUM of various ions. In the 1950s Ludwig Biermann (1907–86) and Hannes ALFVÉN hypothesized that the Sun had an out-flowing SOLAR WIND, composed of high-velocity ions that follow the spiral trajectory of its magnetic field. The solar wind catches ionic material from the comet to make the second, ionic tail. Spacecraft later confirmed the existence of this new component of the interplanetary medium.

The spectra of comets show not only reflected sunlight, but also the EMISSION SPECTRA of radicals (fragments of molecules). Many of these were observed first in comets and only later in the laboratory. In the 1930s and 1940s Karl Wurm and Pol Swings (1906–83) showed that the observed radicals derived from parent vapors such as water and methane.

The nature of comets

In the early 1950s Fred WHIPPLE proposed the 'dirty snowball' model for comets. As opposed to the 'sand bank' model, which suggests that the nucleus of a comet is a loose assembly of dust grains, the dirty snowball model suggests that it is made of dust grains cemented together by ices, such as water, ammonia and methane. In the vacuum of space, the ices sublimate to gas directly from solid and are released from the comet as it warms on approaching the Sun. The vaporization of the ices lets loose the dust grains, and these are dragged from the nucleus by the gas, which expands with a velocity of about 1 km s^{-1} at 1 AU. The grains, of circumstellar origin, are silicates and carbon material. The latter is responsible for the release of some of the organic gas molecules observed in the coma. With the release of the grains, the comet develops its reflective coma and increases in brightness, often in bursts as puffs of dust are released.

The brightness of comets is measured in the same way as stars and is defined as the integrated brightness of the comet's entire coma that is visible above the sky background, determined with the smallest possible instrument needed to easily detect the comet. A large database is maintained by the *International Comet Quarterly*. The absolute magnitude of a comet is the magnitude that the comet would have at 1 AU from the Earth and from the Sun, and serves to compare different comets in their various orbits.

Spacecraft exploration

In 1985 the US INTERNATIONAL SUN-EARTH EXPLORER probe explored the plasma tail of COMET GIACOBINI–ZINNER. In 1986 GIOTTO and the two craft on the VEGA MISSION surveyed 1P/Halley from a short distance (Giotto passed within 600 km of the nucleus). Two Japanese probes also investigated its plasma tail. In the future, the ESA rendezvous mission ROSETTA will survey the nucleus of a comet, attempting a landing and following it as activity develops on its approach to the Sun.

Cometary nuclei are irregular, often elongated, with densities in the range 0.5–1.2 g cm^{-3} (water ice has a density of 1 g cm^{-3}). This suggests a highly porous but solid material. Cometary nuclei rotate with periods from hours to days. They typically have an ALBEDO of about 0.04, putting them among the darkest objects in the solar system. Their surface

◀ ***Cometary nucleus*** *Fountains of vaporized icy material break out from the interior of the comet and are swept back by the motion of the comet through the solar wind. The vaporized material liberates dust for the comet's coma and tail. The comet itself is as black as coal, covered with a crust of tar. (Artist's concept.)*

is probably covered by a crust of dust that insulates the ices below from heating and SUBLIMATION, except where fountains of gas are expelled by increasing pressure below. The comet may fragment (like COMET BIELA) or the surface evolves into a thick and solid dust crust, causing the comet's activity to decrease from one return to the next, perhaps eventually becoming an extinct comet, as CHIRON may be.

The main cometary volatile is water (about 80% by number of molecules). Other gases include carbon monoxide, carbon dioxide, methanol, methane, ammonia, hydrogen cyanide and hydrogen sulfide (hence the description of comets as being the smelliest places in the solar system).

The sublimation of water causes cometary activity at distances from the Sun less than 4 AU (the average temperature of the comet's surface is more than 140 K). However, many comets have been active at larger distances: C/1995 O1 (COMET HALE–BOPP) at 7 AU, 29P/Schwassmann–Wachmann 1 at 6 AU, 95P/Chiron at 10 AU, and 1P/Halley at 15 AU. This is probably due to the sublimation of carbon monoxide.

Comets deviate from the simply computed orbit due to perturbations by the GIANT PLANETS (especially Jupiter); this was first pointed out for Comet D/1770 L1 (Lexell). Other deviations are due to non-gravitational forces, such as the rocket effect that occurs when matter is ejected from the comet's nucleus in a preferential direction. These perturbations may lead to slight, or spectacularly chaotic, changes in the orbits. A comet may even be captured into temporary orbit around Jupiter and make a few revolutions around the planet before being ejected. In extreme cases it may be disrupted by a close approach to the giant planet, or may collide with it, as with COMET SHOEMAKER–LEVY 9 in 1994.

Sungrazing comets

The Kreutz sungrazing comets pass through the solar atmosphere within about one solar radius (696 000 km) of the solar photosphere. In 1888 Heinrich Kreutz (1854–1907) established that they are all small (about 10 m) fragments of a single comet that disrupted during previous orbits. The comets all approach the Sun from below the ECLIPTIC plane and in a retrograde orbit. Some have been seen approaching the Sun, but not leaving, that is, they have completely melted. The Kreutz family is the only known group of sungrazing comets, but it consists of subgroups that broke up from larger fragments at different times (one of them was possibly the Great Comet of February 1106). Before 1979 only about 10 sungrazing comets had been identified. Over the next decade 16 were discovered with space-based solar observatories (the P78-1 and SOLAR MAXIMUM MISSION spacecraft). The total number has increased dramatically owing to discoveries made, in particular by amateurs, using the coronagraphs on the SOHO spacecraft.

Source of comets

The source of comets remains speculative. In 1932 Ernst ÖPIK first discussed the idea of a distant reservoir of comets, situated on the verge of the solar system. Jan OORT showed in 1950, from data gathered by Erik Sinding and Adrianus van Woerkom, that many LONG-PERIOD COMETS have their aphelia at more than 20 000 AU (*see* APHELION). This spherical

Imaging comets

Photography

Even as electronic imagers become ubiquitous, the photography of comets with film still retains importance after more than a century of development, especially for wide-angle imaging. Simple astrophotography methods and advanced techniques will both yield beautiful images for casual photographers and experienced astronomers.

The principle equipment requirements for cometary photography are a camera whose shutter can be opened for long periods – one with a 'B' (for bulb) setting, a 'fast' (low f-number) lens, a cable release to open and close that shutter without introducing vibrations and a stand or support for the camera.

In the simplest approach to taking comet photographs, a tripod is used to hold and aim the camera. More elaborate efforts use an EQUATORIAL MOUNTING, which counteracts the rotation of the Earth, following the comet as it moves across the sky with Earth's diurnal rotation. Such a mount can even be used to track the comet as it moves with respect to the background stars.

Bright comets that are easily seen by the naked eye can be photographed using fast (large ISO value, from ISO 800 to 3200) films and a camera mounted on a tripod. The camera should be aimed so that the comet and tail are oriented along the long side of the rectangular field of view, with the head sufficiently off-center so the tail fits in the field of view. Different lens FOCAL LENGTHS and exposure durations affect the length of the recorded comet, and thus the overall filling of the frame. Foreground objects or landscape may require a different orientation if included (to beautiful effect) and may be illuminated by the camera's flash, local ambient lighting or moonlight. The scene will determine whether a portrait or landscape orientation (in which the long side of the image is horizontal) is more favorable.

The camera lens should be set at infinity focus and at its lowest f-number, with the shutter on the B setting. Using the cable release, the plunger that opens the shutter should be gently pressed for a 10-s exposure. Additional exposures of 20 s, 40 s, 1 min and 2 min should be taken. Many cable releases can be locked in the open position to minimize fatigue during longer exposures.

Depending on the duration of the exposure and the DECLINATION of the comet, greater or lesser amounts of trailing will be apparent. Exposures made near the CELESTIAL EQUATOR with a lens of 50-mm focal length show virtually undetectable trailing (due to the Earth's rotation) for durations of 10 s or less. Closer to the CELESTIAL POLES and with wider-angle lenses, the exposure duration with undetectable trailing increases.

A properly polar-aligned equatorial mount with clock drive can track the sky for minutes to hours at a time with negligible trailing. Comet exposures can be 5–20 min in duration, or longer, permitting more of the tail to be photographed, fainter comets to be recorded and different lens focal lengths or telescopes to be used. Reducing the APERTURE of a camera lens by a photographic stop or two (to reduce ABERRATIONS in the corners) can improve picture quality, as can using somewhat slower speed emulsions (as low as ISO 200 or 400) with their finer grain and higher resolution.

Since comets move with respect to the background stars, a separate GUIDE TELESCOPE should be used to keep the photographic instrument centered on the comet's central condensation for exposures exceeding 60 s. This will provide the ultimate in high-resolution imaging, both of the tail in wide-angle pictures and of the head of the comet at higher magnifications.

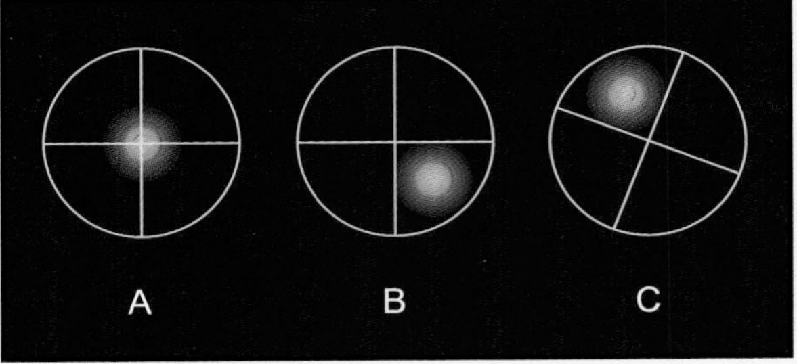

*▶ **To guide on a comet** with an auxiliary telescope: (A) Center the crosshairs on the coma's central condensation. (B) If there is no apparent central condensation in the coma, guiding may be easier with the coma at a tangent to the crosshairs. It is generally easiest to have the crosshairs oriented north–south and east–west. (C) Alternatively, they can be oriented so the comet's motion on the sky is 45° from the directions of the crosshairs.*

For faint comets and for high-magnification photographs of the comet's head, a full-size telescope, equatorially mounted, will be necessary for photography. With faint comets, the telescope provides the light-gathering power necessary. For high-resolution pictures, the telescope provides light-gathering power and a long focal length for magnification. (A teleconverter or EYEPIECE projection can be used if the telescope's own focal length is too short.) Careful guiding through an auxiliary guide telescope is a necessity for these types of photography.

Stephen Edberg has been observing comets since the 1960s. He was Coordinator for Amateur Observations for the International Halley Watch.

CCD imaging

Amateur astronomy has taken great advantage from the introduction of modern digital imaging techniques, and their application to cometary science makes no exception.

*▶ **Foreground objects,** including the distant horizon, generally force the frame to be vertical. To show the comet best with non-astronomical objects, a different lens with a different focal length, magnification and field of view might have to be used for the best composition. This photograph of Hale–Bopp was taken by Jamie Cooper using a Canon EOS 1000 FN with a 35-mm lens. Exposure was 50 s on Fuji Superia 800 ISO film.*

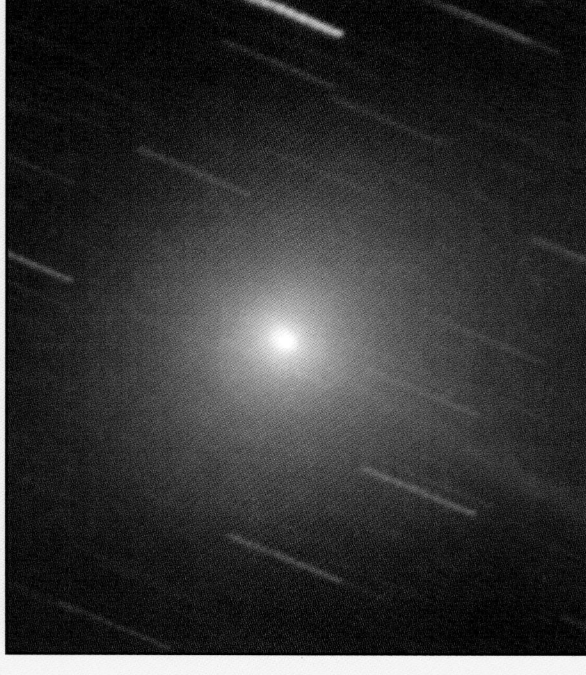

◀◀ ***Hale–Bopp*** *imaged by Nik Szymanek using a Brandon 94-mm f/7 aperture refractor, an SBIG ST-4 autoguider, and Kodak Ektachrome 100 film in a 20-min exposure.*

◀ ***False color picture*** *of the inner coma of C/2002 C1 (Ikeya–Zhang), imaged on May 1, 2002, with the 12-in, f/2.8 Baker–Schmidt camera of the Remanzacco Observatory, Italy. The telescope was fitted with an unfiltered Hi-Sis 24 CCD (chip Kodak KAF400). Ten single 3-min exposures were co-added, and were aligned on the false nucleus of the comet. The trails are background stars.*

Thanks to the characteristics of CCDs, interesting images of comets are routinely obtained, even under light-polluted suburban skies.

Considering that most of the time we are dealing with moderately faint and extended objects, the ideal telescope for the CCD imaging of comets generally has a good field of view and a suitable focal ratio (*f*/4 to *f*/7 are recommended). Its aperture will not be a major concern because of the high quantum efficiency of CCDs; a 20-cm NEWTONIAN TELESCOPE, or a SCHMIDT–CASSEGRAIN TELESCOPE of similar diameter fitted with a suitable focal reducer, will be good enough to image comets down to magnitude 15. This size of telescope is acceptably portable, which will be an additional advantage for imaging faint targets from dark and remote observing sites.

One important issue to consider is the need to match the focal length of a given telescope with the CCD's pixel size. Considering that the average SEEING normally found at an amateur's observing site is of the order of few arcseconds, a compromise needs to be found between the RESOLVING POWER and the sensitivity of the setup; for this reason, it is advisable to obtain images with a scale close to 2–3 arcsec per pixel. Because of their relatively high motion relative to background stars, one is generally prevented from obtaining cometary exposures in excess of a few minutes. In order to reach a good signal-to-noise ratio, one should therefore co-add a number of single exposures, realigning them onto the moving comet by means of suitable software, which is available in different image-processing packages.

If the comet is close to Earth and bright enough, it is worth investigating its innermost regions by means of dedicated image processing routines (like the so-called rotational-gradient algorithms). These are capable of enhancing subtle details in the inner cometary's comae, such as hoods and JETS.

If a big comet has to be imaged, one should consider a different approach. Generally, these objects span a few degrees in the sky, and the telescope and CCD combination frequently provides a field of view far smaller than the size of the target. In this instance, it is advisable to couple the CCD with a common telephoto lens (focal lengths ranging from 80 to 200 mm) by means of a suitable adapter. In order to cut down the CHROMATIC ABERRATION, a filter that cuts out the infrared will produce sharper images.

Further reading: Edberg S J and Levy D H (1994) *Observing Comets, Asteroids, Meteors, and the Zodiacal Light* Cambridge University Press; Merlin J C and Martinez P (eds), Dunlop S (translator) *The Observer's Guide to Astronomy, Vol. 1* Cambridge University Press.

Giovanni Sostero works during the day at the synchrotron 'Elettra' in Trieste, Italy, while at night he enjoys cometary imaging.

◀ ***A dusty parabolic envelope*** *is readily visible in the outer coma of C/2002 F1 (Utsunomiya) after a rotational-gradient treatment via software. Imaged on April 14, 2002, with the Remanzacco Observatory (Italy) equipment described above; 16 exposures of 30 s each were averaged. Image processing thanks to M Facchini.*

Discovering comets

The traditional ways in which COMETS were discovered by amateur astronomers may be things of the past. The development of increasing numbers of professional all-sky survey programs, many specifically designed to spot moving or changing objects, means that the future prospects for amateur astronomers of discovering comets visually are bleak. In the near future, the professional surveys are likely to cover most of the dark sky, perhaps down to a magnitude of 24. In addition, the near-Sun area will be covered by solar coronagraphs similar to those of the SOHO satellite, and the rest of the sky by Lyman-alpha imagers (*see* LYMAN SERIES). On SOHO, the coronagraphs from the LASCO (Large Angle and Spectrometric Coronagraph) instrument reach magnitude 9 out to around 9° ELONGATION. Over the rest of the sky, the SWAN (Solar Wind Anisotropies) Lyman alpha imager detects the huge hydrogen COMA of comets down to around magnitude 12.

However, all is not lost. There will still be scope for amateurs using photographic or CCD techniques to discover that an object that was previously unknown or one that had been incorrectly logged as an asteroid, is, in fact, cometary. It is also possible that there will be a few holes in the professional scans, perhaps in the 'twilight zone' between around 18° and 60° from the Sun. This area may remain the preserve of the dedicated amateur.

Some comets undergo sudden outbursts, and systematic search programs may miss these. Another area in which amateurs currently contribute and excel is the using of professional survey images to discover comets. Any comets found will not bear the name of the discoverer, just the name of the satellite or telescope from which the data originated.

▼ ***Comet Bennett*** *was discovered in 1969 by John Bennett from Pretoria, South Africa. Photo by Akira Fujii.*

Coronagraph images from the SOHO satellite exemplify this amateur contribution. The two LASCO coronagraphs take images roughly every half hour, and these are usually available in real time on the Internet. Over the past few years, amateurs have identified virtually every comet that has appeared on the images, either by recognizing the comet from its form, or by putting a sequence of images together and finding an unknown moving object. In addition, amateurs have scanned archival images and identified many objects missed by the professional team. This has resulted in the discovery of several new groups of Sun-approaching comets, and has also encouraged the professionals to hand over the search task to the amateurs.

Comet discovery has been a traditional goal of the amateur observer. Because a comet is named after its discoverer, finding a comet creates instant fame, particularly if the object is destined to become prominently visible, such as COMET HALE–BOPP (C/1995 O1) and COMET HYAKUTAKE (C/1996 B2). Some discoveries are made by accident such as, for example, C/1995 O1. Others are the result of a dedicated search, as in the case of C/1996 B2. Searches like these often take several hundred hours of hunting before a new comet is discovered. Instruments used have ranged from simple binoculars (*see* BINOCULAR ASTRONOMY) to large REFLECTORS, and observations have involved detectors that range from the naked eye to photography to CCD cameras.

A successful strategy for discovering a comet requires a systematic approach, clear skies and an element of luck. Good dark skies are essential, as light pollution or atmospheric obscuration will hide the faint COMA of a prospective new comet (*see* DARK SKIES AND GOOD OUTDOOR LIGHTING). Discovery is more likely in the early-morning sky for several reasons. First, there are fewer competitors because most observers would rather stay up late than get up early. Second, skies are darker due to the fainter natural AIRGLOW (hence the saying, 'the darkest skies are just before dawn'), and because there are fewer artificial lights. Third, comets in a direct orbit emerge rapidly into the morning sky after solar CONJUNCTION.

Discovery may take a long time. The average time it takes to find one is about 300 hours of searching per comet, although SOHO images produce one comet every 100 hours. An accidental discovery may involve virtually no searching, but in some cases observers have sought for a thousand hours without finding anything.

A key starting point is learning what comets look like in general and how they look through the specific instrument that will be used to search the skies. One can reasonably spend a year or more following a variety of already-discovered comets before even starting to search for new comets.

Large binoculars (for example, 25 × 100 mm) are one of the most common instruments now used to discover comets. This is because they give a wide FIELD OF VIEW and a slightly fainter limiting magnitude as both eyes are used. A pair of binoculars or a telescope with an ALTAZIMUTH MOUNTING (for example, a DOBSONIAN TELESCOPE) enables a simple scan pattern to be used. This means that one starts scanning in the morning in AZIMUTH, at a high elevation above the sunrise point, and then works down toward the horizon. In the evening, the reverse pattern is used. If a suspect object is found, the first step is to make sure that it is not an already-reported comet or a nebula. Comets can be eliminated by

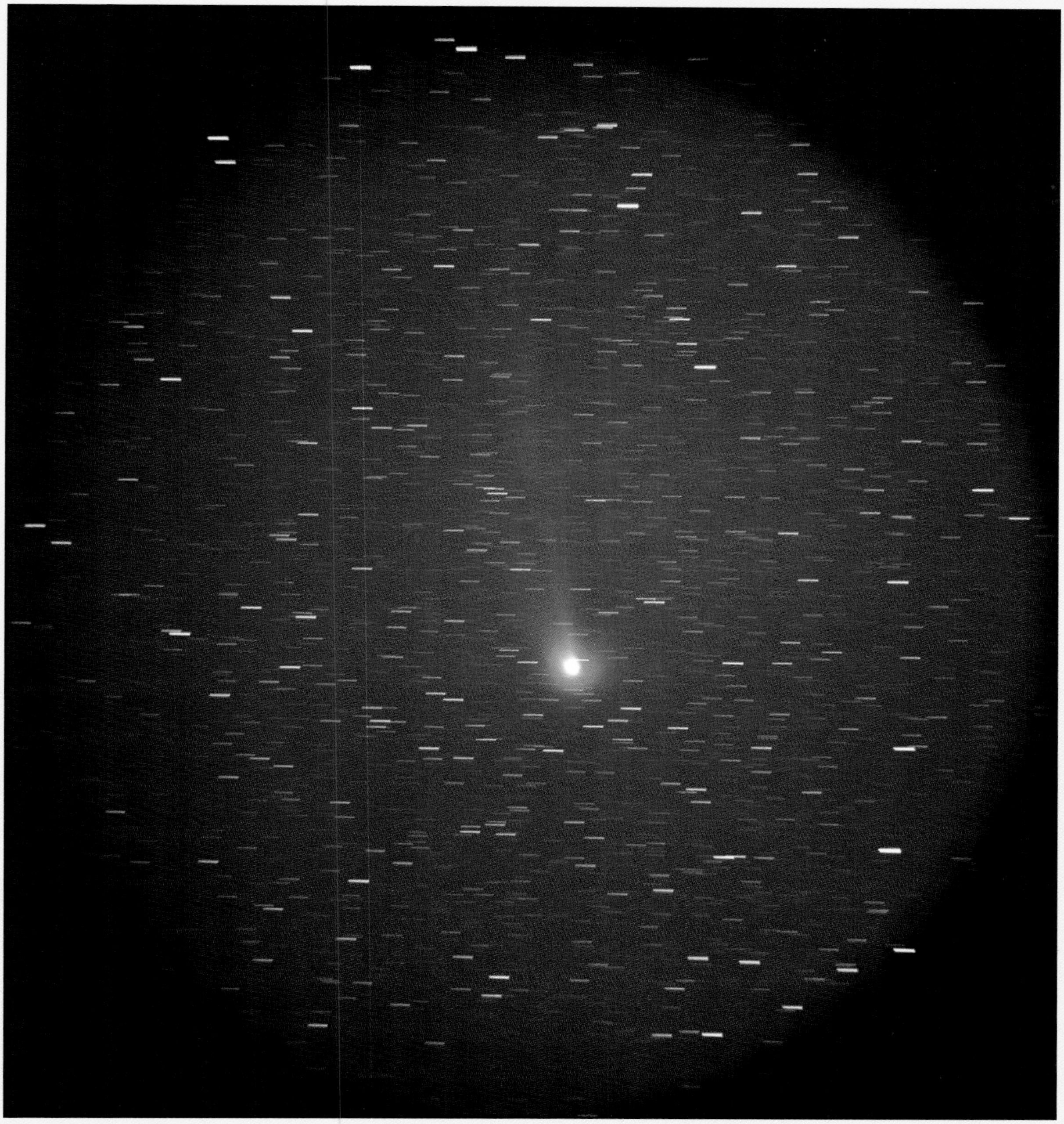

◀ ***Comet Swift–Tuttle*** *was discovered by Lewis Swift on July 16, 1862, and independently by Horace Tuttle three days later. It has a period of around 130 years. Photo by Akira Fujii.*

referring to the website of the IAU and nebulae by using one of the many computer sky-display programs.

The next step is to check the object's motion. A comet should move, and the direction of motion is an important requirement of any reported discovery.

Photographic discovery is still possible. One approach is to take a widefield image, covering a larger area, and thus a larger number of possible comets. The other approach is to take a narrowfield image down to a fainter limiting magnitude, thus detecting fainter comets that may not be visible in the widefield images. Either option requires a well set-up EQUATORIAL MOUNTING and a suitably fast film. It is important to take at least two separate exposures to avoid film flaws, which may appear similar to a comet.

A discovery technique that is often more likely to lead to success involves using a CCD camera. The camera can be attached to a telescope to facilitate going down to faint limiting magnitudes. Alternatively, an ordinary photographic lens can be used with the camera, with the image covering a wide field. The latter may be the most effective technique for surveying the twilight zone, which is too bright for visual searches by amateurs and which is not covered by the professional observers (*see* TWILIGHT GLOW). Observers in the southern hemisphere have an advantage because no major professional search programs are operating there.

Recent discoveries, observable comets and other resources are listed on the websites of the *International Comet Quarterly* and of the BAA; these pages also give guidelines for reporting a possible discovery.

Jonathan Shanklin works for the British Antarctic Survey and frequently travels to Antarctica.

reservoir is now known as the Oort Cloud and is supposed to extend between 20 000 and 200 000 AU. At almost the same period, Kenneth EDGEWORTH and Gerard KUIPER proposed that SHORT-PERIOD COMETS, with their low-inclination orbits (the so-called Jupiter-family comets) come from another reservoir, a disk of comets extending beyond Neptune, now named the Edgeworth–Kuiper Belt (*see* OORT CLOUD AND KUIPER BELT). Among the 1416 comets listed in the 2001 edition of *The Catalogue of Cometary Orbits* by Brian Marsden and Gareth Williams, 250 are short-period comets (with periods of less than 200 years), of which 213 have periods less than 20 years. For the 1166 others, 260 are long period (periods greater than 200 years), 732 have parabolic orbits (the eccentricity of their orbits cannot be distinguished from 1), and 174 are hyperbolic (*see* HYPERBOLA). The largest ECCENTRICITY of the hyperbolic orbit of a comet is 1.057, being that of Comet C/1980 E1 (Bowell). Gravitational or non-gravitational perturbations have changed the orbits of the hyperbolic comets from elliptical and none come from outside the solar system.

In 1992 TRANS-NEPTUNIAN OBJECTS (TNO) were identified. These are thought to be members of the Edgeworth–Kuiper Belt. Transitional objects have also been identified. These are objects that have orbits intermediate between TNO and short-period comets (such as the CENTAURS), or objects that are physically intermediate between comets and asteroids (such as dormant comets, extinct comets and active asteroids, like Chiron). It seems that we now have most of the pieces of the puzzle that will allow us to understand the interrelations between all the small bodies of the solar system and their role in its formation.

The solar system resulted from the collapse of an interstellar cloud, forming a growing proto-Sun surrounded by a disk. Comets were formed as PLANETESIMALS from the accretion of dust and gas in this PROTOSOLAR NEBULA, and they retain primordial matter, which has remained essentially unprocessed since that time. Comets that were formed inside Neptune's orbit were not in stable orbits. They were ejected through perturbations by giant planets, either to interstellar space or to distant orbits to form the Oort Cloud. Further occasional perturbations by nearby stars of Oort Cloud comets re-inject some of them into the inner solar system as dynamically new comets. These may evolve into Halley-type shorter-period comets. Such comets display a large range of orbital periods, and the INCLINATION of their orbital planes relative to the ecliptic is random. Comets that were formed beyond Neptune remain on relatively stable orbits in the Edgeworth–Kuiper Belt. Further orbital evolution, however, could make them evolve into short-period orbits that maintain their original low inclination relative to the ecliptic: the Jupiter-family comets.

It is likely that cometary collisions with Earth, which were much more frequent early in the history of the solar system, played an important role in the history of our planet. Part, though likely not all, of the ocean water could have been brought by comets. As well as water, complex organic molecules (and especially prebiotic organic molecules) could also have acted as seeds for the development of life on Earth.

Further reading: Crovisier J and Encrenaz T (2000) *Comet Science: The Study of Remnants from the Birth of the Solar System* Cambridge University Press; Marsden B G and Williams G V (1999) *Catalogue of Cometary Orbits 1999* Central Bureau for Astronomical Telegrams & Minor Planet Center; Yeomans D K (1991) *Comets: A Chronological History of Observation, Science, Myth, and Folklore* Wiley Science Editions.

Comet Arend–Roland (C/1956 R1)

A LONG-PERIOD COMET discovered jointly by Sylvain Arend (1902–92) and Georges Roland in November 1956. It reached PERIHELION (0.32 AU) on April 8, 1957, and passed closest to the Earth (0.57 AU) on April 21. The comet attained a peak brightness of magnitude 0, and developed a DUST TAIL that

Cometary designations

The IAU's Central Bureau for Astronomical Telegrams is responsible for naming comets, computing their orbits and maintaining a catalog.

Comets are called by the name(s) of the discoverer(s), or more rarely by the names of other people, the orbit computer, or something associated with the comet (such as COMET HALLEY, Comet SOHO and Comet CHIRON). The names can be ambiguous, because some prolific discoverers have found several comets.

Comets are now cataloged by IAU rules that date from 1995. The designation has a letter-prefix: C/ for comets with periods greater than 200 years; P/ for the so-called SHORT-PERIOD COMETS, which have periods of less than 200 years; more rarely, X/ for comets whose orbit cannot be evaluated; and D/ ('defunct') for uncertain objects that can no longer be observed (for example, Comet 3D/Biela).

For short-period comets that have been observed at more than one return, the serial number of the comet is listed before the prefix P/, followed by the name of the discoverer. For example, 1P/Halley, the first known periodic comet, was 'discovered' by Halley. If the discoverer has found several periodic comets they are numbered in order; for example, 100P/Hartley 1, 103P/Hartley 2, 110P/Hartley 3. As of early 2003 there were 155 numbered short-period comets.

Other comets have a prefix followed by a designation indicating the time of discovery (the year, a letter identifying the half-month of discovery and a serial number within this half-month) and discoverer. Thus C/1996 B2 (Hyakutake) is a LONG-PERIOD COMET, the second comet to be discovered in the second half of January 1996, and discovered by the Japanese amateur Yuji Hyakutake.

Before 1995 comets were given two designations. The provisional designation consisted of the year of discovery, followed by a lower-case letter giving the order of discovery within that year; thus 1975n West was the 14th comet to be discovered in 1975. The definitive designation, given several years later, was the year of PERIHELION passage, followed by a Roman numeral giving the order of perihelion passage within the year; thus 1975n became 1976 VI West. This method was abandoned as impractical because of the increasing number of comets being discovered, and the fact that perihelion passage for some comets discovered a long time after their perihelion was not definitive.

was 30° long at its greatest extent. But the comet was most notable for its antitail, which at best was half the length of the main tail and appeared as a pronounced spike extending from the head in the opposite direction to the tail. Antitails are an effect of perspective: when the Earth is close to the comet's orbital plane and the comet is heading in the general direction of the Earth, the curvature of the dust tail makes the tail visible on either side of the head. The INCLINATION was 120° and the ECCENTRICITY given as 1, making the APHELION distance and period indeterminately large (*see* CELESTIAL MECHANICS).

Cometary globule

A small DARK NEBULA that, with its 'head' and 'tail,' bears a superficial resemblance to a COMET. Cometary globules are concentrations of matter within a nebula that are situated near a young star with a strong stellar wind. The wind ionizes gases in the head, causing it to glow, and drives away gases in the surrounding nebula, leaving a dark tail of nebular material in the 'shadow' of the head, which may be several light years long.

Comet Bennett (C/1969 Y1)

A LONG-PERIOD COMET discovered by Jack Bennett (1919–90) in December 1969. It passed closest to Earth on March 26, 1970, having reached PERIHELION (0.54 AU) on March 20, and was then at its brightest, at magnitude 0. The comet was best placed for observation in April, when its tail reached around 11°, some estimates putting it at 25°. Jets of material emitted by the nucleus spiraled out as the nucleus rotated. Bennett was one of the first comets for which a vast hydrogen cloud was detected, by ultraviolet observations made from Earth orbit. The cloud surrounded the head at 13 million km, around 10 times the Sun's diameter. The ECCENTRICITY was 0.995, INCLINATION 90°, and the period about 1700 years.

Comet Biela (3D/Biela)

A COMET that disintegrated in the nineteenth century and gave rise to the briefly active BIELID meteor shower; the D in the prefix indicates that it is 'defunct.' The comet was discovered in 1772 by Jacques Montaigne (1716–85?), and recovered by Jean-Louis Pons (1761–1831) in 1805, but on neither occasion was it observed for long enough for an ORBIT to be computed and its period to be determined. This was accomplished by Wilhelm von Biela (1782–1836) when he recovered the comet in February 1826. It thus became the third known periodic comet. At its return in 1845/6 the comet was observed to have split into two well-separated nuclei. Both nuclei were seen to return separately in 1852, differing more in brightness and now over 2 million km apart. That was the comet's final apparition. There was no sign of either nucleus on what should have been the next favorable return in 1865/6, but in 1872 a METEOR STORM occurred, suggesting that the comet had completely disintegrated. There were further notable displays of what became known as the Bielid METEOR SHOWER at intervals corresponding to the comet's period of just over 6.6 years.

Comet Borelly (19P/Borelly)

Discovered in 1904 by Alphonse Borelly (1842–1926), this SHORT-PERIOD COMET has an orbital period of 6.8 years. It has never yet been visible to the naked eye, reaching a maximum brightness of 7.5 magnitude at its return in 1987. In 1999, when the NASA satellite DEEP SPACE 1 had successfully

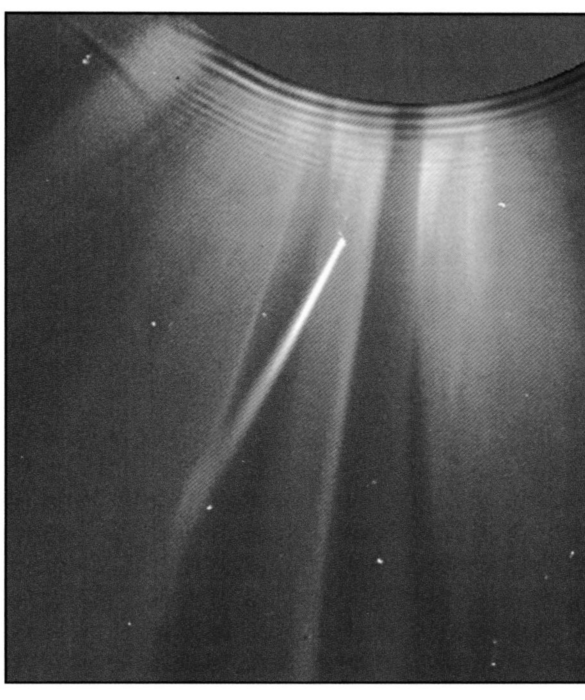

◀ ***Sungrazing comet*** *A comet approaches the Sun through its corona, as imaged by the SOHO satellite.*

completed its primary objectives the mission was extended to flyby Comet Borelly. Passing within 3700 km of the 8-km wide cometary nucleus the spacecraft returned some of the best data ever received from a comet, showing a terrain of rugged mountains and fault structures and jets of dust and gas shooting out from all sides as the solid nucleus rotated.

Comet Churyumov–Gerasimenko (67P/Churyumov–Gerasimenko)

Discovered in 1969 by Klim Churyumov and Svetlana Gerasimenko at Kiev, this SHORT-PERIOD COMET orbits the Sun with a period of 6.57 years. It was selected as the new target for the ESA ROSETTA mission after a delayed launch date put the original target of Comet 46P/Wirtanen out of reach. The nucleus of 67P is estimated to be about 5 km in diameter. Several encounters with Jupiter over its lifetime have resulted in the current orbit with PERIHELION 1.28 AU and APHELION 5.7 AU. The comet normally reaches a magnitude around 12 and is unusually active for a short-period comet. During the 2002/3 apparition the tail was as long as 10 arcmin.

Comet Donati (C/1858 L1)

A LONG-PERIOD COMET discovered by Giovanni Donati (1826–73) in June 1858. It reached its greatest MAGNITUDE of –1 shortly before PERIHELION (0.58 AU) on September 20, and was closest to the Earth (0.5 AU) on October 9. Its appearance was impressive, with a prominent, curved DUST TAIL stretching an estimated 60°, and two thin GAS TAILS. Concentric, sharply defined shells of material ahead of the nucleus suggested that material was being ejected mainly from one vent when the rotation of the nucleus carried the vent around to the sunward side. From the separation of the shells, the ROTATION period was estimated to be 4.6 hours. The eccentricity was 0.996, INCLINATION 117°, and period roughly 2000 years.

Comet Encke (2P/Encke)

The SHORT-PERIOD COMET with the shortest known period, at around 3.3 years. It was discovered independently in 1786,

▲ ***Comet Hale–Bopp*** *seen over the Armagh Observatory, Ireland.*

1795, 1805 and again in 1818. In 1821 Johann ENCKE calculated the orbit of the 1818 comet and predicted its return to PERIHELION in 1821 to an accuracy of one day. Two years later he established this comet's identity with those observed in previous years, and Encke's Comet, as it was called, thus became the second known periodic comet. Its early discovery and short period give it the record for the most observed returns to perihelion (0.33 AU), the 60th occurring in 2000. The comet seems to have been in a stable orbit for several thousand years, which poses the problem of why the historical record contains no sightings of it before 1786. The comet is now quite faint (magnitude 6) and shows only a short tail, presumably having shed most of its dust and gas. The orbit shows signs of slow decay, the present period being 3.28 years, and its APHELION distance is 4.06 AU. Debris from the comet forms a METEOR STREAM, which the Earth intersects twice a year, producing the Taurid shower in November and the daytime Beta Taurid shower in June.

Comet family

A group of COMETS with broadly similar orbital elements. The distribution of aphelia of members of the group peaks at a value corresponding to the distance of a particular major planet from the Sun (*see* APHELION). Perturbations by that planet have been responsible for changing the diverse original orbits of the comets into those they now occupy. It was once thought that all the MAJOR PLANETS possessed comet families. Only Jupiter's comet family is now recognized as real, those of Saturn, Uranus and Neptune being explained in terms of orbital resonances with Jupiter. Jupiter's comet family itself contains over two-thirds of all the officially designated SHORT-PERIOD COMETS. The orbits of Jupiter's comet family are continually decaying, and members are eventually ejected from the solar system by a gravitational 'slingshot,' or are tidally disrupted. Fresh captures from the Kuiper Belt replenish the stock (*see* OORT CLOUD AND KUIPER BELT).

Comet Giacobini–Zinner (21P/Giacobini–Zinner)

A SHORT-PERIOD COMET discovered independently in 1900 by Michel Giacobini (1873–1938) and in 1913 by Ernst Zinner (1886–1970). With a period of 6.61 years it is, like most short-period comets, a member of Jupiter's COMET FAMILY. Short outbursts in brightness have been seen on occasion. These outbursts may be linked to storm-level activity in the COMET'S associated METEOR SHOWER, known as the GIACOBINIDS, as occurred in 1946. The PERIHELION distance is 1.03 AU, APHELION distance 5.99 AU. In 1985 the comet was the first to be visited by a space probe when the INTERNATIONAL COMETARY EXPLORER passed within 7800 km of its nucleus. A shock front was detected where the ionized species in the COMA encountered the SOLAR WIND.

Comet Hale–Bopp (C/1995 O1)

Comet Hale–Bopp was discovered simultaneously by two American amateur astronomers, Alan Hale (1958–) and Thomas Bopp (1949–), on July 23, 1995. At that time it was at 7.1 AU from the Sun and had already a total visual magnitude of 10, which is more than 100 times brighter than COMET HALLEY at the same distance. It passed PERIHELION on April 1, 1997, at 0.91 AU from the Sun. The orbit of Comet Hale–Bopp is inclined by 89° to the ECLIPTIC. Its orbital period was about 4200 years before entering the inner solar system, and changed to 2400 years after gravitational perturbation by Jupiter. It is thus a LONG-PERIOD COMET, presumably coming from the Oort Cloud (*see* OORT CLOUD AND KUIPER BELT).

Comet Hale–Bopp was one of the brightest comets ever recorded, reaching a total visual magnitude of −1 around perihelion, and visible as a naked-eye object in the northern hemisphere for more than two months. The early detection of the comet enabled astronomers to witness its evolution over a large span of HELIOCENTRIC distances and to see the 'turning on' of the SUBLIMATION of various molecules as the comet approached the Sun. At large distances the more volatile substances such as carbon monoxide were observed. Water showed up at only 4.8 AU, with more exotic molecules observed closer to the Sun. Cometary activity is driven by the sublimation processes.

One of the highlights of the observations was the detection of a thin, straight tail of sodium atoms in addition to the GAS and DUST TAILS.

Comet Halley (1P/Halley)

Comet Halley is probably the most famous COMET on record. It is the only bright comet, easily visible with the unaided eye, that returns so often, namely with a periodicity that is never very different from 76 years. Since such a periodicity is roughly comparable to the duration of a human life, grandparents talk about it to their grandchildren and oral tradition has established its fame.

Until the 1682 passage of Comet Halley no comet was known to be periodic, and the shape of the cometary trajectories was unknown. However, a couple of years earlier Isaac NEWTON established for the first time that a PARABOLA, one of the possible solutions given by his new theory of gravitation, fitted the observed trajectory of the Great Comet of 1680. In 1703 Edmond HALLEY computed the parabolic orbits for the well-observed historical comets and noticed that three of them coincided. He concluded that they were passages of the same single comet and successfully predicted its return again in 1758.

More than 30 of Comet Halley's previous passages have since been computed backwards, enabling comparison with historical records. Famously the comet is seen on the Bayeux Tapestry, which depicts the events surrounding the Norman invasion of England in 1066.

The present 76-year orbit of Comet Halley is extremely elongated with ECCENTRICITY 0.967 (PERIHELION distance 0.387 AU, APHELION almost 35 AU). It orbits the Sun in a RETROGRADE direction with an INCLINATION of 17.8° to the ECLIPTIC plane. Its period is intermediate between the SHORT-PERIOD COMETS in the Kuiper Belt and the LONG-PERIOD COMETS originating in the Oort Cloud (*see* OORT CLOUD AND KUIPER BELT). Because of its retrograde orbit it is generally believed to have been captured from the Oort Cloud, having its orbit drastically changed by close encounter with a planet.

During its 1986 passage, Comet Halley became the first comet to be photographed from nearby, with the flyby of the ESA spacecraft GIOTTO. This revealed that the nucleus is black and irregular, with a diameter of approximately 10 km. The thick, black crust, assumed to be made of silicate grains and clearly containing a large amount of carbon, is pierced with many openings that allow vaporized snows to escape from beneath.

All comets are short lived because they lose material at each perihelion passage, eventually fading into invisibility. Comet Halley is no exception.

Comet Hyakutake (C/1996 B2)

Comet C/1996 B2 (Hyakutake) was discovered on January 30, 1996, by the Japanese amateur astronomer Yuji Hyakutake (1950–2002). It should not to be mistaken for another Comet Hyakutake, C/1995 Y1, which was discovered by the same observer five weeks before and was much less spectacular. Comet C/1996 B2 made a close approach to the Earth on March 25, 1996, at only 0.102 AU. Such approaches are rare but not exceptional (18 comets have approached within 0.1 AU of the Earth since 1700). Hyakutake passed PERIHELION on May 1,1996, at 0.23 AU from the Sun. This body is a LONG-PERIOD COMET (about 9000 years) with an orbital plane inclined by 125° to the ECLIPTIC; it is presumably coming from the Oort Cloud (*see* OORT CLOUD AND KUIPER BELT).

Comet Ikeya–Seki (C/1965 S1)

Comet Ikeya–Seki belongs to an exceptional class of COMETS, the Kreutz sungrazing group, so called because its members pass extremely close to the Sun at PERIHELION. Ikeya–Seki is one of the brightest members of the family, discovered by Japanese amateur astronomers Kaoru Ikeya (1944–) and Tsuotomu Seki (1930–). It is certainly the most brilliant and best observed in our epoch. Only the larger members of this group (including Ikeya–Seki) survive the close proximity to the Sun. At this time, the 'dust' is vaporized owing to the very intense heat, so that heavy, metallic elements, which are not seen at larger distances, are set free and can be detected. Numerous emissions due to such elements were indeed photographed and identified in detail for the first time in the optical SPECTRA of Comet Ikeya–Seki. Similarly, the first infrared multiband observations were made of this comet, and used to derive further information about the cometary dust.

◀ ***Comet Halley*** *viewed by the Giotto satellite. Compare this image with the artist's impression of a comet in the main article on comets.*

Comet IRAS–Araki–Alcock (C/1983 H1)

A LONG-PERIOD COMET discovered independently in 1983 by Genichi Araki (1954–), George Alcock (1912–2000) and IRAS. It was closest to the Earth on May 11, at just 0.03 AU (the closest approach of any comet since Lexell's Comet of 1770). At this time C/1983 H1 was a large diffuse object of magnitude 2, moving rapidly across the sky. There was no discernible tail, just a diffuse COMA 2° or 3° across, and more extended on the sunward side of the nucleus. Radar observations showed the nucleus to have a diameter of 9.3 km. PERIHELION (0.99 AU) was on May 21. The period is approximately 1000 years, the ECCENTRICITY is 0.99, and the INCLINATION 73°.

Comet Kohoutek (C/1973 E1)

A LONG-PERIOD COMET discovered by Czech astronomer Lubos Kohoutek (1935–) in March 1973. At that time it was near the orbit of Jupiter, and very bright for a COMET at that distance. This led to predictions that it would be exceptionally bright at PERIHELION, which attracted great interest and made it the target of an international observing effort. However, the comet failed to brighten significantly as it approached the Sun. Following its perihelion passage (0.14 AU) on December 28, it appeared in the evening sky in January 1974 at magnitude 4 with a tail 25° long. It is possible that Kohoutek was making its first visit from the Oort Cloud to the inner solar system (*see* OORT CLOUD AND KUIPER BELT). This would account for its brightness at the distance of Jupiter, as it would then have been a pristine object outgassing the most volatile of its constituents. The INCLINATION is 14°, and the ECCENTRICITY is quoted as 1, meaning that its period and APHELION distance are almost indeterminately large.

Comet Shoemaker–Levy 9

Comet Shoemaker–Levy 9 was discovered in March 1993 by Carolyn and Eugene SHOEMAKER and by David Levy (1948–), as a trail of about 20 individual fragments. Its trajectory showed that it was a JUPITER-family comet that had been disrupted by tidal forces in July 1992 at its previous perijove passage (*see* PERIAPSIS). It would collide with Jupiter at its next perijove passage.

A collision of this type is rare; such an event might be expected once every few hundred years. One took place about three centuries ago. In 1690 Jean-Dominique Cassini reported an unusual feature on the disk of Jupiter, which was remarkably like the stains observed after the collision of Comet Shoemaker–Levy 9 (*see* CASSINI DYNASTY).

The fragments of Comet Shoemaker–Levy 9 entered the Jovian atmosphere between July 16 and July 22, 1994, at a

▲ ***Cone Nebula***
Part of a vast cloud of gas and dust. Within and around the cloud are many recently formed stars, one of them at the apex of the cone. The cone is the wake of material formed by the interaction of the outflowing gas cloud and the new star, like the downstream wake of a rock in a river.

latitude of –44°, within a few minutes of the predicted times, and at various longitudes depending on Jupiter's rotation. The impacts took place just behind the LIMB, and rotated into direct view from the Earth about 10 min later. The Galileo spacecraft, then at a distance of 1.6 AU from Jupiter, was able to view the impacts directly (*see* GALILEO MISSION TO JUPITER). The effects of the collisions were also observed by the HST, the INTERNATIONAL ULTRAVIOLET EXPLORER, ROSAT and various ground-based telescopes. The explosions resulted in a huge temperature increase in the stratosphere of Jupiter, the comet was vaporized, and new molecules were detected from below Jupiter's cloud tops. Changes also took place in Jupiter's magnetosphere.

Each impact showed three main phases: entry, explosion (about 1 min later) and splash (lasting from 6 to 15 min).

The entry of the fragments was accompanied by a flash of light, caused by a METEOR STORM. The fragments ranged between 150 m and 600 m in diameter. The explosion that they caused ejected pieces of comet to an altitude of 3000 km. The temperature of the fireball resulting from the explosions started at over 10 000 K, and decreased to about 2000 K after some 15 s. The fireball increased in size from 15 km 10 s after the explosion, to 100 km after 40 s. Hot gases and dust sprayed up the inclined tunnel that was momentarily formed by the passage of each fragment and fell back to Jupiter in off-center concentric rings, making the splash.

The observed impact times yielded a solution for the comet's orbit (the best orbital fit ever obtained), which shed light on the comet's pre-break-up history. Its progenitor had been captured by Jupiter in 1929, plus or minus nine years. Before capture, the comet was in a low-ECCENTRICITY, low-INCLINATION, heliocentric orbit, probably inside Jupiter's orbit. This was consistent with a group of Jupiter-family comets called the quasi-Hildas.

The length of the chain of fragments made it possible to say something about the break-up. Two extreme models for comets have been proposed: one, a solid body subject to sequential cracking; the other, a loose agglomeration of rubble. The break-up of Comet Shoemaker–Levy 9 had features of both, so it was presumably an intermediate case. The progenitor was 1.5 km in diameter with a density of 0.5 g cm^{-3}. The break-up of the comet triggered activity in the fragments, which released dust continuously from July 1992 to July 1994, creating a COMA around each fragment.

A spectacular result, during the splash phase of impacts L, Q1 and Q2, was the ground-based detection of the elements hydrogen, helium, sulfur, silicon, magnesium, aluminum, iron, potassium, calcium, sodium, manganese, chromium and, for the first time, lithium. All of these elements, absent from the jovian outer atmosphere, undoubtedly came from the comet. However, the chemical composition of the comet is difficult to retrieve.

Was Comet Shoemaker–Levy 9 really a comet or an asteroid? Although no definite conclusion can be drawn on the basis of its chemical composition, several arguments strongly favor the cometary origin: the small size, the activity level (weak but real), the low density and low tensile force, and the silicate signature. It was most probably a small, very common Jupiter-family comet, which would never have caught astronomers' attention without its premature and spectacular death.

Further reading: Noll K S, Weaver H A and Feldman P D (eds) (1996) *The Collision of Comet Shoemaker–Levy 9 and Jupiter* Cambridge University Press; Spencer J R and Mitton J (eds) (1995) *The Great Comet Crash* Cambridge University Press.

Comet Swift–Tuttle (109P/Swift–Tuttle)

A SHORT-PERIOD COMET discovered in 1862 independently by several observers, the first of whom were Lewis Swift (1820–1913) and Horace Tuttle (1837–1923). Calculations indicated that the period was around 120 years, but searches in the early 1980s failed to find it. It was recovered in September 1992 as a 'new' COMET reported by Tsuruhiko Kiuchi (1954–), reaching magnitude 5 and with a 7° tail. PERIHELION (0.96 AU) was on December 12. The nucleus showed one particularly active jet, observations of which indicated a ROTATION period of 2.9 days. The comet was subsequently identified with Kegler's Comet of 1737, and with two comets recorded in Chinese annals from 68 BC and AD 188. On its present ORBIT the comet has an estimated period of 135 years, its ECCENTRICITY is 0.96, and its INCLINATION 113°. Debris from the comet forms the PERSEID meteor stream.

Comet Tempel–Tuttle (55P/Tempel–Tuttle)

A SHORT-PERIOD COMET discovered by Ernst Tempel (1821–89) in December 1865 and by Horace Tuttle (1837–1923) in January 1866. Its period was calculated to be around 33 years. In 1867 Giovanni SCHIAPARELLI showed a very close match between the orbital elements of the comet and those of the LEONID meteor stream, which had produced a spectacular METEOR STORM in November 1866; this was the first successful demonstration of a link between comets and meteors. Though not seen as expected in 1899 and 1932, it was recovered in

1965 but only as a distant object of magnitude 16. At the 1998 apparition it reached magnitude 5; as on previous occasions, it showed no tail. The comet's best apparition was in 1366 when it is calculated to have passed just 0.023 AU from the Earth. This was the third-closest approach of any recorded comet. The PERIHELION distance is 0.98 AU, the ECCENTRICITY is 0.90, and the INCLINATION 163°.

Compton Gamma Ray Observatory (CGRO)

The second of NASA's four 'great observatories.' The Compton Gamma Ray Observatory was named in honor of American Nobel prize–winning physicist Arthur Compton (1892–1962), and was launched from the Space Shuttle in April 1991. CGRO studied the sources and astrophysical processes that produce gamma radiation. Four instruments provided simultaneous observations over the energy range from 0.1 MeV to 30 GeV. A survey of the entire sky produced a catalog of gamma-ray sources, which included pulsars, X-ray binary stars and AGN. The Burst and Transient Source Experiment (BATSE) recorded more than 2000 GRBs. It helped to locate GRBs and discover what happens during these short-lived events. The CGRO made a controlled re-entry into Earth's atmosphere in June 2000. *See also* GAMMA-RAY ASTRONOMY.

Cone Nebula

A dark, tapering nebula in the constellation Monoceros. It is sometimes called the Conus Nebula, and is likened to the Madonna and Child. It forms part of the nebulosity surrounding the OPEN CLUSTER NGC 2264.

Conic section

The curve obtained by cutting a right circular cone with a plane that does not pass through the APEX of that cone. If the plane makes an angle relative to the base that is less than the angle of slope of the side of the cone, an ELLIPSE is obtained. If the plane is parallel to the base, the special case of a circle arises. A plane parallel to the side produces a PARABOLA. A plane making an angle with the base greater than that made by the side produces the HYPERBOLA.

Conic sections are important in astronomy and astronautics because they represent the various forms of orbit that may be followed by a body moving in a gravitational field of a point-like mass.

Conjunction

The position of a planet in its ORBIT when it and the Earth are aligned, so that it is on the same side of the sky as the Sun and has an ELONGATION of 0°. The term is also used for the time at which this alignment occurs. An INFERIOR PLANET (whose orbit lies inside that of the Earth) comes to conjunction at two points in its orbit: when it is between the Earth and the Sun (inferior conjunction) and when it is behind the Sun (superior conjunction). Conjunction is the least favorable time for observing a planet, as it is lost in the glare of the Sun in the daytime, and SUPERIOR PLANETS (whose orbits lie outside the Earth's) are then at their most distant. The inferior planets, though, are at their brightest either side of inferior conjunction. Venus, for example, attains its greatest brilliancy around 36 days either side of inferior conjunction. The term is also applied to two or more planets, or to the Moon and one or more planets, when they appear close together in the sky.

When a superior planet has an elongation of 180° it is said to be at OPPOSITION.

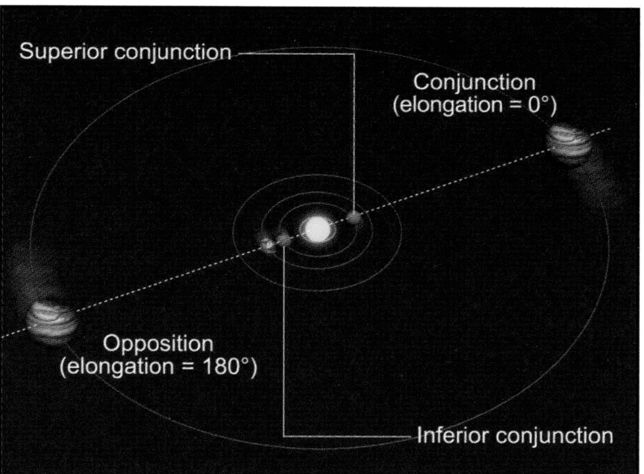

◄ ***Conjunction*** *In this simplified diagram of the inner solar system, the Earth and Sun are shown once, Jupiter and Venus twice, all of them on a line. When an outer planet (like Jupiter) appears away from the Sun, it is said to be at opposition. When it appears near the Sun, it is at conjunction. An inner planet (like Venus) could be at conjunction beyond the Sun (superior conjunction) or between the Earth and Sun (inferior conjunction).*

Constellation

A region of the sky containing a number of bright (naked-eye) stars that have been described as forming a definite pattern. Over the centuries different civilizations identified star patterns with mythical heroes, fabled creatures, actual animals and birds and other objects. In 1922 the IAU adopted the 88 modern constellations, and in 1930 it defined their boundaries precisely. These constellations are described in the table overleaf.

Some of our constellations were first established between 20 000 and 16 000 BC, most clearly Ursa Major. The stars in this constellation were identified with a bear in many cultures in both North America and northern Eurasia. The common cultural heritage of peoples in Alaska and Siberia must have pre-dated the opening up of the Bering Strait and the disappearance of the land-bridge that joined their two

◄ ***Constellation*** *Key map to the constellations from Petrus Apianus's sumptuous book* Astronomicum Caesareum, *published in 1540. Dedicated to the Holy Roman Emperor Charles V, it contained multilayer paper disks arranged to find planetary positions and other astronomical data and to explain the use of astronomical instruments.*

Constellations

Constellation	Common name	Genitive	Abbr.	Area (sq. deg.)	Location	Alpha star*
Andromeda	Andromeda	Andromedae	And	722	N	Alpheratz
Antlia	Air Pump	Antliae	Ant	239	S	
Apus	Bird of Paradise	Apodis	Aps	206	S	
Aquarius	Water Carrier	Aquarii	Aqr	980	SZ	Sadalmelik
Aquila	Eagle	Aquilae	Aql	652	E	Altair
Ara	Altar	Arae	Ara	237	S	
Aries	Ram	Arietis	Ari	441	NZ	Hamal
Auriga	Charioteer	Aurigae	Aur	657	N	Capella
Boötes	Herdsman	Boötis	Boo	907	N	Arcturus
Caelum	Chisel	Caeli	Cae	125	S	
Camelopardalis	Giraffe	Camelopardalis	Cam	757	N	
Cancer	Crab	Cancri	Cnc	506	NZ	Acubens
Canes Venatici	Hunting Dogs	Canum Venaticorum	CVn	465	N	Cor Caroli
Canis Major	Big Dog	Canis Majoris	CMa	380	S	Sirius
Canis Minor	Little Dog	Canis Minoris	CMi	183	N	Procyon
Capricornus	Sea Goat	Capricorni	Cap	414	SZ	Algedi
Carina	Keel	Carinae	Car	494	S	Canopus
Cassiopeia	Cassiopeia	Cassiopeiae	Cas	598	N	Schedar
Centaurus	Centaur	Centauri	Cen	1060	S	Rigil Kent
Cepheus	Cepheus	Cephei	Cep	588	S	Alderamin
Cetus	Whale	Ceti	Cet	1231	S	Menkar
Chamaeleon	Chameleon	Chamaeleontis	Cha	132	S	
Circinus	Pair of Compasses	Circini	Cir	93	S	
Columba	Dove	Columbae	Col	270	S	Phact
Coma Berenices	Berenice's Hair	Comae Berenices	Com	386	N	Diadem
Corona Australis	Southern Crown	Coronae Australis	CrA	128	S	
Corona Borealis	Northern Crown	Coronae Borealis	CrB	179	N	Alphecca
Corvus	Crow	Corvi	Crv	184	S	Alchiba
Crater	Cup	Crateris	Crt	282	S	Alkes
Crux	Southern Cross	Crucis	Cru	68	S	Acrux
Cygnus	Swan	Cygni	Cyg	804	N	Deneb
Delphinus	Dolphin	Delphini	Del	189	N	Sualocin
Dorado	Swordfish	Doradus	Dor	179	S	
Draco	Dragon	Draconis	Dra	1083	N	Thuban
Equuleus	Little Horse	Equulei	Equ	72	N	Kitalpha
Eridanus	River Eridanus	Eridani	Eri	1138	S	Achernar
Fornax	Furnace	Fornacis	For	398	S	
Gemini	Twins	Geminorum	Gem	514	NZ	Castor
Grus	Crane	Gruis	Gru	366	S	Al Na'ir
Hercules	Hercules	Herculis	Her	1225	N	Ras Algethi
Horologium	Clock	Horologii	Hor	249	S	
Hydra	Sea Serpent	Hydrae	Hya	1303	S	Alphard
Hydrus	Water Snake	Hydri	Hyi	243	S	
Indus	Indian	Indi	Ind	294	S	

** brightest star*
sq. deg. = square degrees, N = northern hemisphere, S = southern hemisphere, E = equatorial, Z = zodiac

continents, around 15 000 years ago.

Forty-five constellations were described in the *Phaenomena*, a poem by Aratus of Soli (*c*.315 – *c*.245 BC) written in about 275 BC. This work was itself based on earlier work, now lost, by EUDOXUS.

Aratus's system of constellations was invented by people who lived close to latitude 36°N. This is known because Aratus did not describe any constellations around the south pole, and the empty zone had a radius of 36°. This zone was below the horizon of people who lived at latitude 36°N. The constellation-free zone was not centered on the position of the south celestial pole at the time of Aratus, but on the position at 2000 BC, having moved by PRECESSION. The date and place correspond to the Babylonians and their Sumerian ancestors, or to the Minoans of Crete. By the age of Homer (eighth century BC), constellations were interwoven with Greek mythology. The mythologization of the constellations was virtually complete in the third century BC. A major step in the process by which they acquired their present names was the transfer of mythology from the Greeks to the Romans.

In the second century AD, PTOLEMY in the ALMAGEST described 48 of the 88 modern constellations. The remaining 40

Constellations						
Constellation	**Common name**	**Genitive**	**Abbr.**	**Area (sq. deg.)**	**Location**	**Alpha star***
Lacerta	Lizard	Lacertae	Lac	201	N	
Leo	Lion	Leonis	Leo	947	NZ	Regulus
Leo Minor	Little Lion	Leonis Minoris	LMi	232	N	
Lepus	Hare	Leporis	Lep	290	S	Arneb
Libra	Scales	Librae	Lib	538	SZ	Zuben el Genubi
Lupus	Wolf	Lupi	Lup	334	S	Men
Lynx	Lynx	Lyncis	Lyn	545	N	
Lyra	Lyre	Lyrae	Lyr	286	N	Vega
Mensa	Table (Mountain)	Mensae	Men	153	S	
Microscopium	Microscope	Microscopii	Mic	210	S	
Monoceros	Unicorn	Monocerotis	Mon	482	S	
Musca	Fly	Muscae	Mus	138	S	
Norma	Square (Carpenter's)	Normae	Nor	165	S	
Octans	Octant	Octantis	Oct	291	S	
Ophiucus	Serpent Bearer	Ophiuchi	Oph	948	E	Ras Alhague
Orion	Orion, the Hunter	Orionis	Ori	594	E	Betelgeuse
Pavo	Peacock	Pavonis	Pav	378	S	Joo Tseo
Pegasus	Winged Horse	Pegasi	Peg	1121	N	Markab
Perseus	Perseus	Persei	Per	615	N	Marfak
Phoenix	Phoenix	Phoenicis	Phe	469	S	Ankaa
Pictor	Easel	Pictoris	Pic	247	SZ	
Pisces	Fish	Piscium	Psc	889	N	El Rischa
Pisces Austrinus	Southern Fish	Piscis Austrini	PsA	245	S	Fomalhaut
Puppis	Stern	Puppis	Pup	673	S	
Pyxis	Mariner's Compass	Pyxidis	Pyx	221	S	
Reticulum	Net	Reticuli	Ret	114	S	
Sagitta	Arrow	Sagittae	Sge	80	N	
Sagittarius	Archer	Sagittarii	Sgr	867	SZ	Rukbat
Scorpius	Scorpion	Scorpii	Sco	497	SZ	Antares
Sculptor	Sculptor	Sculptoris	Scl	475	S	
Scutum	Shield	Scuti	Sct	109	S	
Serpens	Serpent	Serpentis	Ser	637	E	Unuck al Hai
Sextans	Sextant	Sextantis	Sex	314	S	
Taurus	Bull	Tauri	Tau	797	NZ	Aldebaran
Telescopium	Telescope	Telescopii	Tel	252	S	
Triangulum	Triangle	Trianguli	Tri	132	N	Ras al Muthallath
Triangulum Australe	Southern Triangle	Trianguli Australis	TrA	110	S	Atria
Tucana	Toucan	Tucanae	Tuc	295	S	
Ursa Major	Great Bear (Big Dipper)	Ursae Majoris	UMa	1280	N	Dubhe
Ursa Minor	Little Bear (Little Dipper)	Ursae Minoris	UMi	256	N	Polaris
Vela	Sails	Velorum	Vel	500	S	
Virgo	Virgin	Virginis	Vir	1294	EZ	Spica
Volans	Flying Fish	Volantis	Vol	141	S	
Vulpecula	Fox	Vulpeculae	Vul	268	N	

** brightest star*
sq. deg. = square degrees, N = northern hemisphere, S = southern hemisphere, E = equatorial, Z = zodiac

constellations fall into two categories: minor constellations that fill in gaps, and southern constellations. In the 1590s two Dutch navigators, Pieter Keyser and Frederick de Houtman, invented 12 far-southern constellations, which Johann BAYER included in his *Uranometria* star atlas of 1603. Among these were Tucana and Volans, representing southern-hemisphere wildlife. Petrus Plancius (1552–1622), a Dutch theologian and astronomer, added four constellations in 1613, drawing on the Bible for inspiration: Camelopardalis (named for the Old Testament camel that carried Rebecca to Isaac, although it is now represented as a giraffe); Columba (originally Columba Noachi, named for Noah's dove); Monoceros (since a unicorn is mentioned in the Old Testament); and Crux (from stars that were formerly part of Centaurus).

Johannes HEVELIUS added seven constellations in 1687. They included Lacerta, since, he reasoned wryly, only a lizard could wriggle into the small space available; and Lynx, because the eyes of a lynx were needed to see any stars at all. Nicholas LACAILLE, on his chart of 1752, filled in the less populous regions of the southern skies. He also broke down the huge, ancient constellation of Argo into the smaller groups Carina,

Puppis and Vela. In an effort to inject an air of modernity he included such names as Antlia Pneumatica and Fornax Chemica, labels that have survived in truncated form.

Many astronomers attempted to honor patrons with constellations, but only one survives. Hevelius's constellation Scutum was originally Scutum Sobiescianum, named after Poland's King John III Sobieski. Others never found favor and were included on some charts and ignored on others. The IAU, founded in 1919, took charge of the chaotic situation. Under Eugène Delporte (1882–1953) it standardized the constellations to the official modern system, abbreviating some over-elaborate names and rendering many constellations obsolete.

Stars are named with a letter or number and the genitive case of the Latin constellation name or a three (or four) letter abbreviation; see table.

Alongside the official names of the constellations, common names are in use in various languages. Gemini, for example, is known as the Twins in English, Gémaux in French, Zwillinge in German and Gemelli in Italian.

Further reading: Allen R H (1963) *Star Names: Their Lore and Meaning* Dover; Condos T (1997) *Star Myths of the Greeks and Romans: A Sourcebook* Phanes Press; Delporte E (1930) *Délimitation Scientifique des Constellations (Tables et Cartes)* Cambridge University Press.

Continuous spectrum

A continuous distribution of ELECTROMAGNETIC RADIATION spread over a wide range of wavelengths. In the visible region, a continuous spectrum corresponds to an unbroken rainbow-band of colors. A continuous spectrum, or continuum radiation, is emitted by a hot, dense, opaque body (solid, liquid or gaseous). Examples of continuum radiation include BLACK-BODY RADIATION, thermal radiation and SYNCHROTRON RADIATION.

The spectrum of a typical star consists of a continuous spectrum on which a pattern of dark absorption lines and/or bright emission lines has been superimposed.

Convection

The motion of a liquid or gas caused by the buoyancy of lower material. The pressure in the lower layers of a liquid or gas standing in equilibrium under the force of gravity is higher than the pressure in the upper layers, because the lower layers support the upper layers. If the liquid or gas is heated from below, the lower layers become warm, expand, become buoyant, and rise by convection. In the upper layers they cool and fall, completing a convective cycle. In a large amount of material, convection occurs in numerous convection cells. The convective cycle can also be initiated by the cooling of the upper layers.

Convection mixes cream from the bottom of a cup of coffee (as the surface of the coffee cools); it causes the clouds in the Earth's atmosphere (as air is warmed by the Earth's surface); and it causes GRANULATION on the Sun's surface (as solar material is warmed by heat generated from the nuclear core of the Sun below). The granulation, the clouds in the sky, and the clouds in coffee are all examples of convection cells. Convection does not generally occur in the ocean, because ocean water is heated from sunlight above. Nor does convection occur in the air inside the International Space Station, because there is effectively no gravity. For this reason fans must be used to drive hot, exhaled, oxygen-depleted air from the noses of sleeping astronauts, rather than relying on convection to replenish the oxygen content of the in-breathed air, as happens to people sleeping in terrestrial beds.

Co-orbital satellites

Two or more satellites that orbit at the same mean distance from their parent planet. The term was coined to describe two different types of orbit-sharing among Saturn's family of satellites, the only place where co-orbital behavior has been observed.

The first type of co-orbital behavior is shown by Epimetheus and Janus, two irregular-shaped bodies each of approximate diameter 100 km orbiting between the planet's F and G rings. They have slightly different ECCENTRICITIES and INCLINATIONS, their mean distances from Saturn differ by only about 50 km, and their periods by a little more than a minute. When one satellite catches up to the other one, gravitational interaction causes a small energy transfer between them and they trade places, the satellite in the lower orbit moving to the higher orbit, and vice versa.

In the second version of the phenomenon, minor satellites share the orbit of a larger satellite. Telesto and Calypso, roughly 30 km in size, have exactly the same period and mean distance from Saturn as the larger Tethys, and they orbit 60° ahead of and 60° behind Tethys respectively, at the L_4 and L_5 LAGRANGIAN POINTS of Tethys's orbit. Similarly, Helene, which also measures approximately 30 km, orbits at the L_4 Lagrangian point in the orbit of DIONE.

Copenhagen University Astronomical Observatory (CUAO)

CUAO was founded as a university observatory when the Astronomical Round Tower was built in central Copenhagen, Denmark, in 1637. In 1861 all astronomical activities were transferred to a new observatory just outside the center of Copenhagen. The Brorfelde Observatory, 60 km west of Copenhagen, was part of CUAO for about 40 years from the early 1950s. Staff and workshops from there have now been transferred to Copenhagen. The main astronomical facilities are a 1.54-m telescope and a specialized 0.5-m photometric telescope, both at La Silla Observatory, Chile (*see* EUROPEAN SOUTHERN OBSERVATORY); and the CARLSBERG MERIDIAN TELESCOPE at La Palma, Canary Islands.

Copernican system

See NICOLAUS COPERNICUS

Copernicus, Nicolaus (1473–1543)

Copernicus was a Polish astronomer whose *De Revolutionibus Orbium Coelestium* (On the Revolutions of Celestial Spheres), completed in 1543, was the final achievement under the agenda of the ancient Greek astronomers and – by its claim that the Earth orbits the Sun – the start of the development of the theory of dynamics that replaced it.

Mainstream Greek astronomy conceived the universe to be bound by the spherical heavens, in the midst of which was the spherical (and motionless) Earth. Astronomers attempted to 'save the appearances' (to reproduce the observed movements) of the planets by means of calculations based on geometrical models centered on the Earth; these models comprised circles on whose circumferences the center of another circle, or the planet in question, moved. In the second century AD, PTOLEMY in his *Almagest* and elsewhere had developed models that came near to saving the appearances. However, in these models Ptolemy had found it necessary to introduce non-uniform motion, namely motion that appeared uniform when viewed from an off-center 'equant' point that we recognize as analogous to the empty focus in a Keplerian ellipse. Another limitation of the models was their ad hoc character. Ptolemy used devices that had no rationale other than that they worked. Copernicus, however, developed geometrical models that more than matched Ptolemy's in predictive accuracy and provided an elegant and convincing picture of the cosmos.

Copernicus was born in Toru on the Vistula on February 19, 1473. In 1496 he studied law at Bologna, the leading university for the subject. In 1497 and (twice) in 1500 he made astronomical observations that he was later to use in his major publication, and about 1500 he is reported to have lectured in Rome on astronomy. In 1501, he started to study medicine at Padua for two years. Back in Poland, he became physician and administrative assistant to his bishop uncle. His interest in astronomy took second place to his duties as a canon. Furthermore, he lived in troubled times and his medical skills were much in demand.

But by 1539 his work *De Revolutionibus* was essentially complete, and in May of that year he allowed sight of it to a young visitor, Georg Joachim Rheticus (1514–74), professor of mathematics at the University of Wittenberg, Germany. By this time, despite Copernicus's isolation, word of his astronomical work had spread in the German-speaking world, and even as far as Rome, and Rheticus's curiosity had been aroused. Copernicus allowed Rheticus to summarize his treatise in a *Narratio Prima*, or First Account, which appeared in 1540. In October 1541 Rheticus returned to Wittenberg to resume his teaching, and he either took Copernicus's manuscript with him or had it sent to him soon thereafter. In May 1542 Rheticus was able to deliver the fair copy to the publisher. He was appointed to a post at Leipzig and the task of seeing the book through the press then passed to a Lutheran clergyman, Andreas Osiander (1498–1552), who took it upon himself to forestall criticism of Copernicus by adding an unsigned preface to say that the motion of the Earth was not being proposed as a truth of nature but merely for purposes of calculation. The misleading preface was taken to be the author's until Johannes KEPLER revealed Osiander's role after the turn of the century.

Late in 1542, Copernicus suffered a cerebral hemorrhage and paralysis of his right side, but he lingered on until May 24 of the following year. We are told he received the final pages of *De Revolutionibus* on the day of his death; but if so, he would hardly have been aware of it.

For the remainder of the century Copernicus's treatise was mined by mathematical astronomers for the ingenuity of its planetary models. Not until 1596 did Kepler publish an unabashedly heliocentric cosmology, and not until 1610 did GALILEO GALILEI's telescopic observations lead him to embark on a reform of physics that would make it possible for Earth-dwellers to accept the notion that they inhabited a spinning, orbiting planet without feeling any sensation of movement.

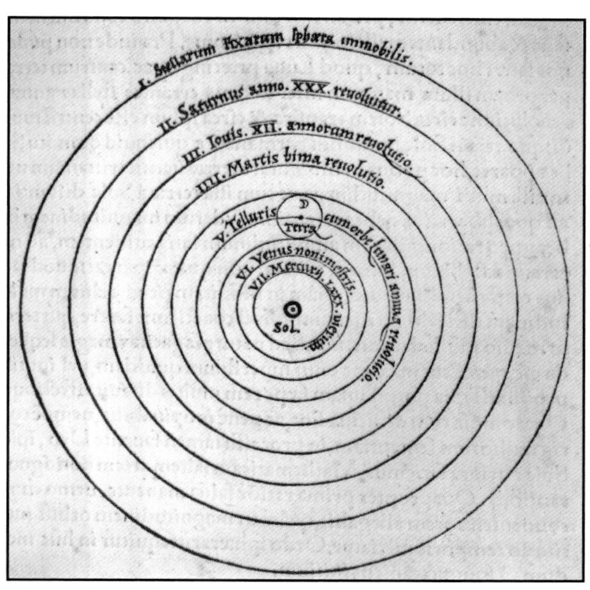

◄ ***Copernican system*** *Diagram of the heliocentric universe from Copernicus's* De Revolutionibus Orbium Coelestium *of 1543.*

Copernicus crater

One of the MOON's most conspicuous craters, with a diameter of 93 km. Named for the Polish astronomer Nicolaus COPERNICUS, it is a younger feature of the Moon, being created by an impact an estimated 1 billion years ago. Like other young craters, it is surrounded by a system of bright RAYS formed by EJECTA from the impact. The rays from Copernicus extend for more than 600 km over the neighboring Mare IMBRIUM and Oceanus Procellarum. Samples collected by astronauts during the APOLLO 12 mission were originally thought to be ejecta from Copernicus, but they were subsequently dated to 850 million years. The crater itself has terraced walls, caused by the partial collapse of sections of the inner rim, and several central peaks.

Copernicus (OAO-3)

A NASA observatory for ULTRAVIOLET ASTRONOMY that was launched in August 1972. The observatory, which operated for nine years, carried an 80-cm CASSEGRAIN TELESCOPE and spectrograph that covered the range 75–300 nm, giving wide spectral coverage and very high resolution. It was used to study the interstellar medium and outer CORONAS of stars. Copernicus also carried a British X-ray instrument that discovered pulsating X-ray sources (known as 'slow rotators') and variations in X-ray output from the galaxy CENTAURUS A.

See also ORBITING ASTRONOMICAL OBSERVATORY.

Cor Caroli

The star Alpha Canum Venaticorum, with apparent magnitude 2.89, is the brightest in this constellation of faint stars below the handle of the Big Dipper. It was formerly thought to be a binary system (*see* BINARY STAR), but more accurate PARALLAX determinations by the HIPPARCOS satellite have confirmed that it is an OPTICAL DOUBLE, the component stars having a SEPARATION of 19.4 arcsec. The main component, Alpha^2 CVn is the type star for the ACVn VARIABLE STARS.

Cor Caroli means 'Heart of Charles,' from the belief of Royalists in England that the star shone with exceptional brightness the night before King Charles II returned to London in 1660 following the re-establishment of the monarchy. But its variability (magnitude ±0.015) is too slight for there to be any scientific foundation for this legend.

Core

The central part of a differentiated planetary body (*see* DIFFERENTIATION). In the inner solar system cores are metallic, while in the outer solar system they are rocky. The giant planets also have rocky cores.

Coriolis force

Named after Gaspard de Coriolis (1792–1843), the Coriolis force is the apparent force acting on a moving body as observed from a moving frame of reference. Air flowing to a low-pressure region (or from a high-pressure region) is caused to rotate over the surface of the Earth by the Coriolis force generated by Earth's rotation. This causes anticlockwise-rotating low-pressure systems and clockwise-rotating high-pressure systems in the northern hemisphere, and reverse rotations in the southern hemisphere.

Corona

The name given to the extended outer atmosphere of the SUN. It extends to heights of more than a solar radius above the solar photosphere. The corona is a tenuous gas at temperatures in excess of 1 000 000 K, and it emits strongly in the extreme ultraviolet and X-rays. The corona can be seen in visible light as a faint halo around the Sun during a total solar eclipse. The shape changes markedly during the solar cycle. At solar minimum the corona is elongated along the equatorial direction, and polar plumes are seen. At solar maximum the disk is surrounded by coronal loops and streamers, which give the corona a rounder shape. An instrument called a coronagraph provides an alternative to the Moon by blocking out the bright light of the photosphere at times other than during an eclipse. Scattered light is usually still a problem, but a coronagraph works well in space, such as on the SOHO spacecraft. Huge clouds of PLASMA (called coronal mass ejections or CMEs) are ejected from the Sun into space approximately once a day. Occasionally this material can engulf the Earth, generating GEOMAGNETIC STORMS.

Corona Australis, Corona Borealis

See CONSTELLATION

Coronal cavity

Coronal cavities on the Sun, first observed during the eclipse of 1898, are regions of low emission surrounding PROMINENCES. They separate the prominence from the rest of the CORONA.

Coronal hole

Coronal holes are regions in the solar CORONA where the density is low, so that the magnetic fields that thread through the region open freely into interplanetary space. Coronal holes can be seen during total solar eclipses, or using coronagraphs such as those on the SOHO spacecraft. During times of low solar activity, coronal holes extend out from the north and south polar caps of the Sun. During more active periods, coronal holes can exist at all solar latitudes, but as the Sun rotates, the outgoing magnetic field gets twisted and the hole persists for only a few months. Ionized atoms and electrons flow along the open magnetic fields in coronal holes and escape at high speeds into interplanetary and even interstellar space, constituting the most extensive part of the SOLAR WIND. When, in 5 billion years, the Sun eventually becomes an ultraviolet-emitting WHITE DWARF, the solar wind material will show as a PLANETARY NEBULA, and the hourglass shape of the solar wind will be revealed, caused by today's polar coronal holes.

Coronal mass ejection

See CORONA

Coronas

Two Russian/Soviet solar physics and astronomy satellites. Coronas-I was launched in December 1992. Coronas-F was launched in July 2001. They are designed to observe the solar atmosphere from near-Earth orbit and to observe solar activity and MAGNETOSPHERIC solar effects. Instruments include X-ray spectrometers, multilayer imaging telescopes, coronagraphs and detectors for HELIOSEISMOLOGY.

COROT mission

COROT, an acronym for Convection Rotation and planetary Transits, is a French-led CNES/ESA mission to search for planets around other stars (*see* EXOPLANET AND BROWN DWARF). It will look for tiny dips in the brightness of a star as a planet passes in front of it. Jupiter-mass gas giants have been detected from the ground, but COROT will be the first spacecraft capable of detecting rocky planets, smaller than the gas giants but several times larger than the Earth. Such planets would represent a new, as yet undiscovered, class of object and astronomers expect to find 10–40 of them in the 2.5-year mission, together with tens of new gas giants. COROT is a 30-cm diameter space telescope scheduled for launch in late 2005. While it is monitoring a star COROT will also look for 'starquakes' in a technique known as ASTROSEISMOLOGY. The exact nature of the ripples allows astronomers to calculate the star's precise mass, age and chemical composition.

Corvus

See CONSTELLATION

Cosmic abundance of elements

The relative proportions of the different chemical elements in the universe. The usually quoted 'cosmic' abundance figures are derived from spectroscopic analysis of the SUN supplemented by chemical analyses of chondritic meteorites and terrestrial and lunar rocks. Abundances may be expressed as the ratio of the number of atoms of that element to the number of atoms of hydrogen (or other element such as silicon). They may also be expressed in terms of the relative proportions by mass of the various elements.

Although there are about 11 times as many hydrogen as HELIUM atoms in the universe, an atom of helium has about four times the mass of a hydrogen atom so the relative proportions by mass are greater than the relative proportions by number of atoms. Likewise for other elements.

Hydrogen and helium are by far the most abundant elements in the universe, so the composition of stars is often discussed in terms of the mass fractions of hydrogen (X), of helium (Y), and of all the other elements combined, usually referred to as 'metals' or 'heavy elements' (Z). For the Sun, the mass fractions are $X = 0.75$, $Y = 0.25$, $Z = 0.02$. A more general value for the solar system is $X = 0.71$, $Y = 0.27$, $Z = 0.02$.

The relative proportions of the lightest chemical elements (hydrogen, helium, lithium) and their ISOTOPES are widely assumed to have been determined in the BIG BANG. The heavier elements (up to iron) have been synthesized inside stars, and elements heavier than iron in SUPERNOVAE. *See also* ELEMENTS, FORMATION OF.

Cosmic Background Explorer (COBE)

NASA satellite designed to survey the COSMIC MICROWAVE BACKGROUND radiation at infrared and millimeter wavelengths across the entire sky. It was launched in November 1989, and operated until December 1993.

It determined the temperature of the cosmic background radiation to be 2.73 K (precisely as predicted by BIG BANG theories of the universe), and detected small variations (one part in 100 000) in the temperature. This was seen as evidence for a 'lumpy' universe in which galaxies could form.

Cosmic microwave background (CMB)

The cosmic microwave background is the BLACK-BODY RADIATION left over from the fireball of the BIG BANG. This radiation was emitted before stars, galaxies or quasars existed and is like the thermal emission from an object with a temperature of 2.7 K.

The SPECTRUM of radiation from a black body, or the amount of power at different frequencies or wavelengths, depends only on its temperature and is given by a theoretical function developed by Max Planck (1858–1947). The measured spectrum of the CMB is black body to within 50 parts per million. The temperature of the black body that best matches the sky is 2.725 ± 0.002 K, one of the most accurately known cosmological parameters. The amount of energy carried by the CMB is larger than any other cosmic radiation field. Most of its energy is at wavelengths near 1 mm.

Origin of the cosmic microwave background

George GAMOW, Ralph Alpher (*c.*1926–) and Robert Herman (1914–97) predicted the existence of the CMB in 1947 from their model for the formation of the elements during the Big Bang. In the early universe, radiation and matter interacted strongly. The temperature of the radiation was equal to the temperature of the matter, which started off very hot and cooled as the universe expanded. The temperature of the radiation became frozen when the matter and the radiation decoupled, and the universe became transparent. This happened about 379 000 years after the Big Bang, during the recombination epoch. At this time, at a temperature of perhaps 10 000 K, the ELECTRONS combined with the PROTONS in the universe, producing neutral HYDROGEN. Whereas before this time the free electrons had scrambled the radiation into a black body, the radiation now propagated unhindered through the universe, to become the CMB. The expansion of the universe has produced a REDSHIFT that preserved the black-body character of the radiation, but reduced its temperature. As an approximate illustration of this, since the universe was one-thousandth of its current size the wavelengths of the PHOTONS have increased by a factor of 1000 and the temperature deduced from the photon energy is 1000 times higher.

The properties of the CMB rule out almost all cosmological models except for those starting with a hot Big Bang. Less than 60 parts per million of the energy in the CMB was generated later than two months after the creation of the universe. Virtually all of it came from the universe when it was in a very hot, dense state,and there has been nothing comparable to the Big Bang in the subsequent history of the universe.

Discovery

Gamow, Alpher and Herman's cosmological model for the formation of the elements failed because it could not create elements heavier than lithium (*see* ELEMENTS, FORMATION OF). The model's predictions about the CMB were thus neglected, even though the development of radar during World War II and other advances in RADIO ASTRONOMY provided the tools necessary for its detection.

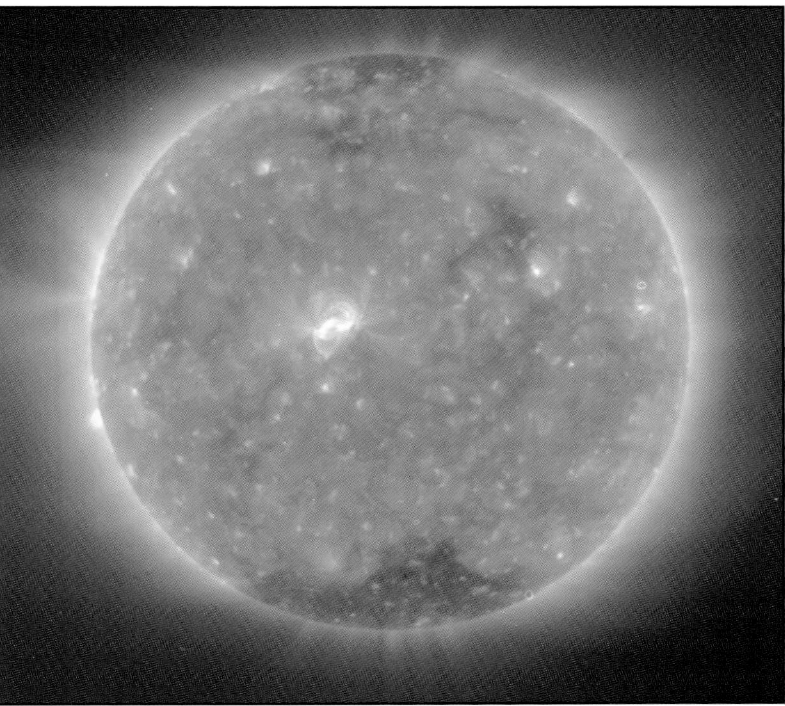

▲ ***Coronal holes*** *show as dark areas at the Sun's poles.*

Arno PENZIAS and Robert WILSON discovered the CMB in 1965, as they systematically tried to identify all the sources of noise in a very sensitive antenna used by Bell Labs for early communication satellite experiments. In every direction of sky they saw excess noise that they could not account for. It was the CMB. Many groups measured the intensity of the CMB at different wavelengths, and they quickly showed that the spectrum was black body.

One test for the cosmological status of the CMB was early satisfied. The CMB should appear brighter or hotter on one side of the sky and fainter or cooler on the opposite side. This dipole pattern would be caused through the DOPPLER EFFECT by the motion of our solar system relative to the observable universe (the Sun's motion through our GALAXY and the Local Group of galaxies, and the Local Group's motion relative to the CMB) (*see* LOCAL GROUP OF GALAXIES; MAGELLANIC CLOUDS). The pattern was detected in 1976. The velocity of the solar system is 369 ± 3 km s^{-1} relative to the observable universe. The velocity of the solar system relative to the center of mass of the Local Group is less well known, but the resulting velocity of the Local Group relative to the observable universe is 600 ± 45 km s^{-1}.

Isotropy

A second test for the cosmological status of the CMB was that, after the dipole pattern was subtracted out of the observed temperature map, the CMB was isotropic. It should appear the same in any direction, because there is no reason to think that any part of the universe is different from any other. Since the pattern produced on the sky by a Doppler shift can be calculated, the dipole pattern due to the motion of the solar system can be removed from the observations.

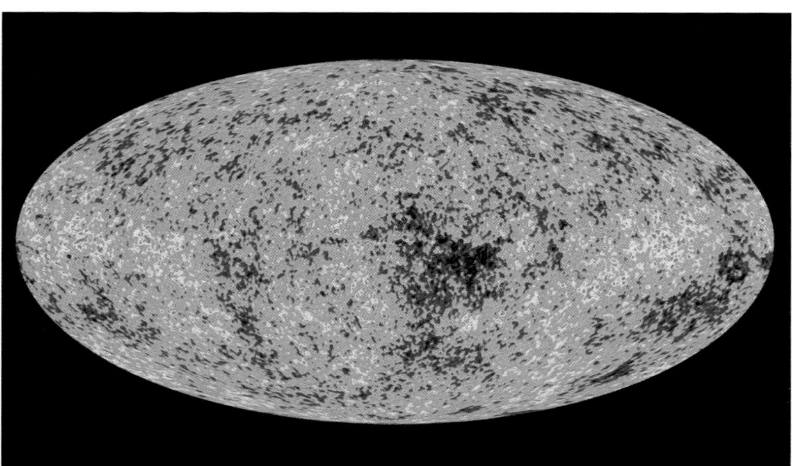

▲ Cosmic microwave background *Measurements by the WMAP satellite depict the slightly cooler-than-average regions of the universe (blue shades) that eventually became seeds for giant clusters of galaxies. The slightly warmer areas (red and yellow shades) would give rise to the huge voids seen today between galaxy clusters. The black areas indicate regions where observations were incomplete due to nearby gas clouds in the Milky Way.*

After this is done, the largest signal that remains is due to the radio- and millimeter-wave emission from the Milky Way. This emission is much stronger at centimeter wavelengths than at millimeter wavelengths, so if a fraction of a centimeter-wavelength map is subtracted from a millimeter-wavelength map, then almost all of the emission from the Milky Way can be canceled. A similar treatment removes the infrared and millimeter point-sources, such as ultraluminous infrared galaxies. This leaves the residual CMB signal.

Anisotropies

Anisotropy in this signal was at first undetectable. Isotropy would, however, break down at some level. The COBE spacecraft carried an instrument to search for spectral distortions and an instrument to search for anisotropy. The announcement in 1990 that the CMB spectrum measured by COBE showed no deviations from a black body at a level of 1 part in a 1000 led to a standing ovation at an AMERICAN ASTRONOMICAL SOCIETY meeting. The detection of patchy anisotropy by COBE at a level of 40 parts in a million (which is much more uniform than the most perfect white paper), was described as the 'discovery of the century, if not of all time' by Stephen HAWKING in 1992.

These anisotropies are related to the large-scale structure of the universe. The structure developed through gravitational forces acting on the extremely small density fluctuations caused by quantum mechanical processes during the first picosecond after the Big Bang, and developed into the anisotropies revealed by the CMB. Following this they developed into the major structures in the universe, such as clusters of galaxies (*see* GALAXY CLUSTER AND GROUP).

The anisotropies in the CMB show as a collection of bright and dark patches of various angular scales on the sky. This is displayed in a statistical way by plotting the average power of the fluctuations on the various scales. At small angular scales the density fluctuations have a certain scale, in part because the quantum mechanical density fluctuations have time to expand between the Big Bang and recombination. The initial density fluctuations expand as acoustic waves in the Big Bang material at the sound speed, which is high at about two-thirds the speed of light. This produces a natural scale in the size of the fluctuations after 300 000 years. Structure is most pronounced at an angular scale of 1.7°, and the height of the peak is primarily determined by the DENSITY of the universe. In 2001 new results from the Degree Angular Scale Interferometer (DASI) and further detailed analysis of results from the Balloon Observations of Millimetric Extragalactic Radiation and Geophysics (Boomerang) revealed CMB anisotropies on small scales in the sky, including the critical peak. The data indicate that the universe is flat, with its density at the critical value.

In 2003 NASA unveiled the first detailed full-sky map of the CMB. Scientists created the map using data collected by the WMAP satellite over a 12-month period. The PLANCK SURVEYOR satellite, which will be launched by the ESA in 2007, will provide the definitive map to all possible scales. The data from Planck (named for the physicist who determined the black-body spectrum) will enable the HUBBLE CONSTANT, the density of the universe, and other cosmological parameters to be determined to 1% accuracy, or it will prove the inadequacy of the current cosmological models.

Further reading: Borner G (1993) *The Early Universe* Springer; Kolb E W and Turner M S (1990) *The Early Universe* Addison-Wesley.

Cosmic ray

Cosmic rays are high-energy IONS and ELECTRONS that originate generally beyond the SOLAR SYSTEM. They consist of galactic cosmic rays, which come from our GALAXY and beyond, and of anomalous cosmic rays, which come from the outer region of the HELIOSPHERE. High-energy ions and electrons from the Sun are often called solar cosmic rays, or solar energetic particles. These are all 'primary cosmic rays' of extraterrestrial origin. 'Secondary cosmic rays' are the results of the interaction of primary cosmic rays with the Earth's atmosphere.

Galactic cosmic rays are an important constituent of the interstellar medium. Together with the anomalous component they penetrate the heliosphere into the orbit of the Earth. The most energetic, least charged cosmic rays also penetrate Earth's MAGNETOSPHERE and impinge on the upper atmosphere. The galactic cosmic rays represent unique samples of matter from beyond the solar system.

The figure on p. 107 shows a schematic diagram of the heliosphere. The cavity in the interstellar PLASMA and magnetic field is formed by the SOLAR WIND, which blows radially away from the Sun. Its pressure decreases and, at the heliopause, becomes equal to the pressure of the local interstellar medium. The heliopause is the boundary between the solar wind plasma and magnetic field, and the interstellar plasma and magnetic field. We should discover more about the heliopause in the next few years, when the VOYAGER space probe encounters it.

Victor Hess (1883–1964) discovered cosmic rays in 1912 (he won the Nobel prize in 1936). He found that a balloon-borne electrometer measured increasing conductivity of the air as its altitude increased. He attributed this phenomenon to IONIZATION of the air by a radiation whose origin, as 'primary cosmic rays,' was probably beyond the solar system. Subsequent measurements with Geiger counters and ionization chambers in the 1930s established that primary cosmic rays are mostly PROTONS with energies sufficient to penetrate the Earth's magnetic field to the top of the atmosphere, where they produce 'secondary cosmic rays' through interactions with the air. The development of the Wilson cloud chamber in 1933 enabled investigations of these natural high-energy nuclear interactions, which led to the discovery of the POSITRON (1933), muon (1937) and K-meson (1947).

In the 1930s Scott Forbush (1904–84) established that the intensity of cosmic rays was anticorrelated with solar activity.

In the 1950s John Simpson (1916–2000) developed the neutron monitor, which provided a sensitive measure of cosmic ray intensity by recording the NEUTRONS that they produced in the air. A global network of neutron monitors was established in 1957 for the International Geophysical Year. The detection of lower-energy galactic cosmic rays, anomalous cosmic rays, and most solar-cosmic-ray events had to await measurements from space.

Origin of cosmic rays

Galactic cosmic rays originate predominantly at shock waves produced by SUPERNOVAE. Such a shock wave propagates into the surrounding interstellar medium and sweeps up the ionized interstellar gas. The particles are accelerated by so-called Fermi acceleration, between approaching gas clouds, just as a Ping-Pong ball is accelerated between approaching paddles. A large fraction of the kinetic energy released by supernovae is transferred to very energetic cosmic rays.

The highest-energy cosmic rays may be produced by NEUTRON STARS AND PULSARS, but it is more likely that they are produced in intergalactic space, possibly at shocks formed by the motion of galaxies through the intergalactic medium. However, other sources, such as strong stellar winds, may also contribute to the acceleration of cosmic rays, just as the solar wind accelerates the anomalous cosmic rays. One puzzle is the origin of the highest-energy cosmic-ray electrons. They are attenuated rapidly in the interstellar medium, and must come from a very close but unknown source within approximately 500 l.y. of the Sun.

Once accelerated, cosmic rays roam randomly through the Galaxy and the GALACTIC HALO, partially confined by the galactic magnetic field and scattered by magnetic irregularities. The highest-energy cosmic rays cannot be bottled up by the magnetic field of the Galaxy. In any case, the decrease in cosmic-ray pressure toward the edge of the galactic halo drives a galactic wind and causes cosmic rays to leak from the Galaxy.

Solar cosmic rays are accelerated in discrete events as a by-product of solar activity. Rapid bursts of cosmic rays originate in solar flares. More gradual events originate with fast coronal mass ejections (CMEs) (*see* CORONA). Both gradual and impulsive events are more frequent during maximum solar activity, when flares and CMEs are more frequent.

Anomalous cosmic rays are interstellar atoms accelerated by the solar wind; for example, by the termination shock (see figure).

Interaction of cosmic rays with the Sun and Earth

Collisions between cosmic rays and matter within the heliosphere are negligible, forgetting spacecraft, cosmic-ray detectors and the Earth itself! But cosmic rays respond to the magnetic and electric fields of the Sun. They are partially excluded from the heliosphere by the solar wind in a process known as the solar modulation of cosmic rays. The reduction of their intensity within the heliospheric cavity is called 'solar modulation.' The effectiveness of solar modulation varies with the 11-year solar cycle.

The Earth's magnetic field also modulates the cosmic rays that reach the top of the atmosphere. Above each point on the surface of the Earth, at any one time, there is a geomagnetic cutoff. Cosmic rays above the cutoff penetrate to that point from beyond the magnetosphere. The cutoff changes as the Earth's magnetic field changes. Less energetic cosmic rays incident on the upper atmosphere lose energy because they ionize air molecules, and collide with air nuclei. These interactions produce 'showers' of secondary cosmic rays (electrons, protons, neutrons, gamma-rays and other particles) with a maximum intensity about 16 km above the surface of the Earth.

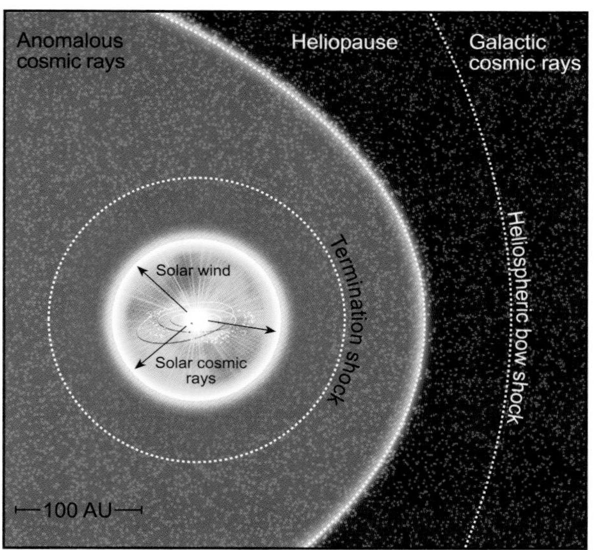

◀ ***Cosmic rays*** *Solar cosmic rays are found in the region in which the Sun's magnetic field and solar wind are dominant (the heliosphere), shown in yellow. Galactic cosmic rays, the light blue dots, come from energy phenomena in the Galaxy and are partially excluded from the heliosphere. Anomalous cosmic rays are atoms in interstellar space that have leaked into the heliosphere and been accelarated by the solar wind. They are present in the blue area within the heliopause.*

As the ionized particles move rapidly through the air, they may travel faster than the speed of light in air (although, of course, not faster than the speed of light through a vacuum). In this case they produce light, or Cerenkov radiation, which can be detected in large optical telescopes. The cosmic-ray intensity can also be measured by neutron monitors, which count the neutrons produced in parallel to the charged particles.

The interaction of cosmic rays with the atmosphere produces unstable nuclei, such as carbon-14 and beryllium-10. Diffusing through the atmosphere, these nuclei are then incorporated into the terrestrial environment, including biological material, in a timescale of 10–20 years. The abundance of carbon-14 (relative to the stable nucleus carbon-12) can be used to date wood. The consistency of this dating method is weakened because the flux of cosmic rays that is incident at the surface of the Earth modulates. This occurs partly because the flux of primary cosmic rays changes, but mostly because of changes in the solar and terrestrial magnetic fields. Conversely, material that can be accurately dated can reveal variations in the cosmic-ray flux. The abundance of carbon-14 in tree rings of known age has revealed variations in solar modulation and, by implication, variations in solar activity. There are peaks of the beryllium-10 isotope in uniformly accumulating seafloor sediments that are approximately 35 000 and 60 000 years old. The peaks may be cosmic-ray enhancements from prehistoric supernovae.

Further reading: Lee M A (1997) *Cosmic Winds and the Heliosphere* University of Arizona Press.

Cosmogony

The study of the formation of planets, stars and galaxies, but especially of the solar system. *See also* PLANETS: ORIGIN.

Cosmological constant

See DARK ENERGY AND THE COSMOLOGICAL CONSTANT

Cosmological model

An idealized picture of the overall structure of the UNIVERSE and its evolution that may be tested by comparing its predictions with observational data. For simplicity, most cosmological models ignore individual GALAXIES and CLUSTERS, and treat the matter and radiation content of the universe as if it were smeared out into an idealized smooth distribution that is sometimes called the 'substratum.' Most cosmological models incorporate the COSMOLOGICAL PRINCIPLE (that the universe is homogeneous and isotropic) and assume the universality of physical laws (that the laws of nature are everywhere the same). If the universe is homogeneous (the same everywhere), then all fundamental observers (observers who are at rest relative to the substratum in their vicinity) see the same sequence of events in the history of the universe, this sequence of events defining a universal cosmic time.

Most cosmological models are based on the general theory of relativity, in which gravitation is treated as a phenomenon arising from the curvature of space (or, strictly, four-dimensional SPACETIME) that is induced by the presence of massive bodies.

Cosmological principle

See UNIVERSE: COSMOLOGICAL THEORY

Cosmology

The study of the structure, origin and evolution of the UNIVERSE as a whole. *See also* UNIVERSE: COSMOLOGICAL THEORY.

Cosmos

A series of more than 2300 Russian/Soviet military, research and scientific satellites. Most operated for a few weeks or months. Few details of their payloads or results were released.

Côte d'Azur Observatory

This French institute was established in 1988 by a merger of the Nice Observatory (founded in 1881 by rich banker and amateur astronomer Raphael Bischoffsheim, 1823–1906) and the Research Center for Geodynamics and Astrometry (CERGA, founded in 1974 and located near Grasse). The observatory's telescopes are located on a third site on the Calern Plateau outside Grasse. The largest telescope is a 76-cm REFRACTOR, commissioned in 1887. The dome of the telescope was built by Gustave Eiffel (1832–1923).

Coude telescope

A telescope in which light is brought by a system of mirrors to a focus on the polar axis. *Coude* is French for 'elbow,' and refers to the light path. Since the focus is at a fixed point, bulky or orientation-sensitive equipment can be put there. A coude spectrograph would be a large typical example.

Crab Nebula (M1, NGC 1952)

A SUPERNOVA remnant in the constellation Taurus, produced by the supernova of AD 1054. This reached a magnitude of –6 and was visible in daytime. The nebula was discovered in 1731 by John BEVIS and independently in 1758 by Charles MESSIER, prompting him to compile the MESSIER CATALOG of objects that might be confused with comets. The nebula was named by Lord Rosse (*see* ROSSE, THIRD EARL OF) for its superficial resemblance to a crab. The Crab is 6 × 4 arcmin in extent and of magnitude 8. Its outer regions consist of twisting filaments of HYDROGEN expelled by the supernova. These appear red on photographs and travel outward at over 1000 km s^{-1}. The inner region glows with the pale yellow light of SYNCHROTRON RADIATION triggered by ELECTRONS emitted by the CRAB PULSAR at the center (the core of the star that exploded as a supernova). This inner region makes the Crab Nebula the best-known example of a plerion – a supernova remnant with a 'filled' center.

Known since 1948 as the powerful radio source Taurus A, the Crab Nebula was discovered in 1964 to be a powerful X-ray source (Taurus X-1), and was the first such to be optically identified beyond the solar system.

Crab Pulsar

The pulsar PSR 0531+21 (previously known as NP 0532) is situated at the heart of the CRAB NEBULA in the constellation Taurus. One of the first pulsars to be discovered (in 1968), it had a dramatically short period of only 33.3 ms. In 1969 optical observations of the south-westernmost of a pair of stars near the center of the nebula showed the same rapid fluctuations. This star of magnitude 16 was the first positive optical identification of a pulsar.

Subsequent observations have shown the Crab Pulsar's rotation to be decelerating by 36.4 ns per day. This loss in rotational energy arises from its conversion into SYNCHROTRON RADIATION, which is emitted into the surrounding nebula along paths constrained by a strong dipolar magnetic field and gives rise to the lighthouse-like 'flashes' as the star rotates. Rudolph MINKOWSKI's suggestion that a neutron star is the power source of the nebula has thus been triumphantly vindicated. Measurements of the expansion of the nebula have confirmed its identification with a SUPERNOVA explosion observed in AD 1054. This indicates that the Crab is one of the youngest pulsars yet discovered, hence its very rapid rotation. *See also* NEUTRON STAR AND PULSAR.

Crater

(1) A crater is a bowl-shaped cavity that was formed by an explosion. On Earth craters can be of volcanic origin, but on other planets (for example, Mars and Venus) volcanoes are not explosive, and nearly all craters there result from the impact of a METEOROID at very high speed (tens of kilometers per second). Lunar craters have been known since the time of the telescope, but the first recognized impact crater on Earth was METEOR CRATER, identified in about 1903 by Daniel Barringer (1860–1929). The MARINER flybys showed Mars and Mercury to be larger versions of the Moon. The PIONEER and VOYAGER missions to the systems of OUTER PLANETS have observed enormous impact features, such as the Valhalla basin on Callisto, Jupiter's second-largest Galilean satellite. More recently, close-range observations of ASTEROIDS have shown craters with sizes up to 30% of the diameter of the body.

(2) *See* CONSTELLATION.

Crescent

The phase of a body in the solar system when less than half of its sunlit side is visible. The only objects to show crescent phases to observers on the Earth are those that can pass between the Earth and the Sun: Mercury, Venus and the Moon. However, objects outside the Earth's orbit have been imaged in the crescent phase by spacecraft.

Crimean Astrophysical Observatory (CrAO)

The Crimean Astrophysical Observatory is one of the largest scientific centers in the Ukraine and former Soviet Union.

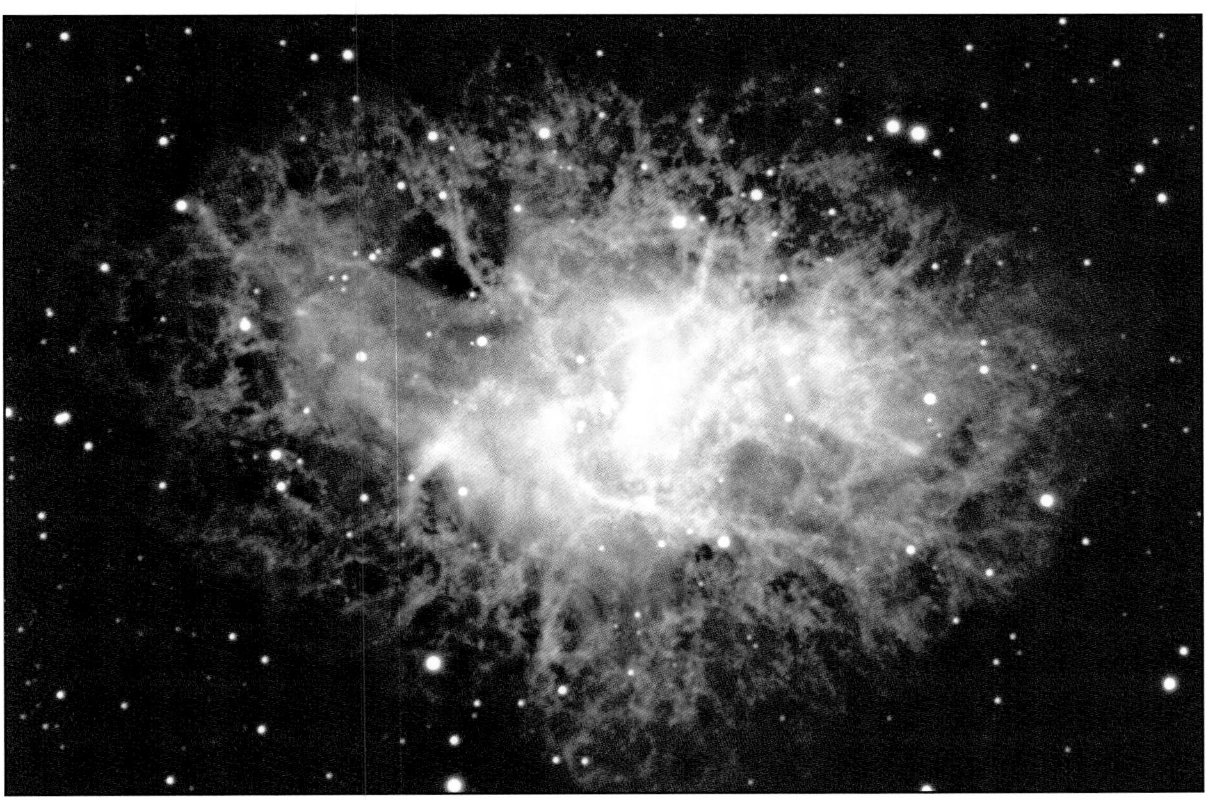

◀ ***Crab Nebula*** *(M1)The first object identified by Charles Messier in his famous catalog of astronomical objects. The nebula is an 8.5-magnitude blur in small-aperture telescopes. A large aperture is needed to see the filaments. They were first glimpsed with the 72-in telescope at Birr Castle, Ireland. David Malin made this image using archived photographs from the 200-in telescope at Mount Palomar, USA.*

The main part of the observatory is located in Nauchny, about 12 km southeast of Bakhchisarai. The main telescopes are: the 2.6-m Shajn telescope; two 1.25-m telescopes; a ground-based gamma-ray telescope (GT-48); the 1.2-m Solar Tower telescope; as well as some other smaller instruments. The Department of Radioastronomy, with its 22-m RADIO TELESCOPE (RT-22) for millimeter and centimeter radio waves, is located near Simeiz. RT-22 participates in international programs in the global VLBI network for geodynamical and astrophysical investigations. The Simeiz station, located on Mount Koshka at an altitude of 346 m, participates in the global network of satellite laser ranging stations for studies of Earth dynamics.

Critical density

In COSMOLOGICAL MODELS, the critical density is the mean density of a universe that is just capable of expanding forever. In such a universe (an Einstein–de Sitter universe) the rate of expansion is slowed by gravity, and approaches ever closer to zero. If the mean density exceeds the critical value, the expansion will cease at a finite time in the future and the universe thereafter will begin to collapse. If the mean density is less than the critical value, the universe will continue to expand at a rate that decreases toward a constant value.

The ratio of the actual mean density to the critical density (the density parameter) is denoted by the symbol Ω (omega). Ω is greater than 1 if the actual density exceeds the critical value; Ω equals 1 if the actual density is equal to the critical density; and Ω is less than 1 if the actual density is less than the critical value.

Crust

The thin outer layer of a differentiated planetary body (*see* DIFFERENTIATION). Crusts consist of rock and/or ice (rock predominates in the inner solar system, ice in the outer), and have a different composition from the underlying mantle. Primary crust is the surface layer formed when the planet first differentiates. It preserves the cratering record from the early bombardment in the solar system's history and is found, for example, in the highlands of Mercury and the Moon, and in many of the satellites in the outer solar system. The secondary crust is formed when the mantle melts and volcanism leads to the first phase of resurfacing. The floodplains of the lunar maria (*see* MARE) are secondary crust, as is the whole surface of Venus. Tertiary crust is produced on geologically active worlds by the recycling of primary and secondary crust material. On the Earth and Io (a satellite of Jupiter), continuing volcanism produces tertiary crust. On the Earth, plate tectonics is continually recycling the crust, which is also modified by processes such as sedimentation and erosion.

Neutron stars are believed to have a crust that consists predominantly of iron.

Crux *See* CONSTELLATION

Cubewano

Any of the main stream of Kuiper Belt objects, at 41–47 AU mean distance from the Sun, with an INCLINATION of orbit ranging up to 30°, and ECCENTRICITY up to 0.1. It is named for the designation of the first Kuiper Belt object, 1992 QB1, which was discovered by Dave Jewitt (1958–). *See also* OORT CLOUD AND KUIPER BELT.

Culmination

The passage of a celestial body across an observer's MERIDIAN (also known as TRANSIT). Upper culmination (upper transit) is the passage of the body through the point at which its

▲ ***Cygnus Loop*** *This image from the FUSE satellite shows a tiny portion of the immense Cygnus Loop. Different colors show emission from different parts of the optical and ultraviolet spectrum.*

ALTITUDE above the horizon is greatest (and its ZENITH distance is least). For example, the Sun reaches upper culmination at noon. Lower culmination (lower transit) is the crossing farthest from the zenith. If the body is CIRCUMPOLAR, the point of lower culmination will be above the horizon; otherwise, lower culmination occurs when the body is at its maximum angular distance below the observer's horizon.

Curtis, Heber Doust (1872–1942)

American astronomer. Curtis spent 18 years at LICK OBSERVATORY from 1902, before becoming director of the ALLEGHENY OBSERVATORY of the University of Pittsburgh. From 1930 he was director of the astronomical observatories of the University of Michigan. He surveyed nebulae with the Lick Observatory's Crossley REFLECTING TELESCOPE, and identified that some long, thin-looking nebulae were identical to spiral nebulae, but seen edge on with a band of dust like that seen centrally in the Milky Way. Curtis interpreted the zone of avoidance, in which few or no spiral nebulae were seen near the Milky Way, as being due to obscuration by the Milky Way's obscuring matter, and concluded that spiral nebulae were Milky Way galaxies outside our Galaxy. He engaged in the Great Debate on the scale of the universe in Washington in 1920 with Harlow SHAPLEY, arguing correctly against Shapley's view that spiral nebulae were minor objects within an immensely bigger Milky Way structure.

Cusp cap

A brightening at one or other of the tips (cusps) of the CRESCENT phase of Venus, as seen from the Earth. Cusp caps were first reported by the German amateur astronomer Baron Franz Paula von Gruithuisen (1774–1852) in 1813, and have been recorded by telescopic observers ever since. Early observers fancied that they were seeing glimpses of a possibly Earth-like surface through breaks in the Venusian cloud cover. On the side of a cusp cap away from the cusp itself there is often a darker band known as a cusp collar. Both caps and collars show short- and long-term variations in brightness and size. It was once thought that the cusp caps were contrast effects, but images from the MARINER 10 and PIONEER Venus spacecraft showed that the brightenings, along with other 'deformities' in the planet's LIMB and TERMINATOR as observed from Earth, are real and related to the planet's atmospheric circulation.

Cygnus *See* CONSTELLATION

Cygnus A (3C 405)

The brightest radio source in the constellation of Cygnus, and the second-brightest cosmic radio source in the sky.

Cygnus A, which lies at a distance of about 750 million l.y., has a radio output a million times more powerful than that of a conventional galaxy like the Milky Way. At radio frequencies it has a classic double-lobed structure, the main body of radio emission emanating from two elongated clouds that extend to a distance of some 200 000 l.y. on either side of the center of the galaxy. The central radio source is linked to the outer lobes by two long, narrow filaments that are composed of energetic radio-emitting ELECTRONS that have been ejected from the core of the galaxy. The galaxy itself, which is exceptionally massive (about 10^{14} solar masses), is also an X-ray source. The X-ray emission comes from a distribution of hot gas with a temperature of about 10^{8} K.

Cygnus Loop

A large SUPERNOVA remnant in the constellation Cygnus, measuring nearly 3° across. Some arcs of the Loop, known collectively as the Veil Nebula (or sometimes the Cirrus Nebula), are visible at optical wavelengths. Radio, infrared and X-ray images reveal the complete Loop. The object is so large (six times the diameter of the full Moon) that its brighter parts were given separate NGC numbers. The brightest part of the Veil is NGC 6992 and NGC 6995, to the east; the western section is NGC 6960. Other sections are NGC 6974 and NGC 6979.

Cygnus X-1

Cygnus X-1 is one of the strongest X-ray sources. It has been identified with the supergiant HDE226868, a single-line spectroscopic BINARY STAR and ellipsoidal variable with an orbital period of 5.6 days. It is the first celestial object for which there was reasonably convincing evidence of a BLACK HOLE. X-ray observations showed that the X-ray flux varies with the orbital period and another period found in the LIGHT CURVE, thereby confirming that Cygnus X-1 is a massive X-ray binary. The radial velocities (*see* ASTROMETRY) of the emission lines in the optical SPECTRA show that mass is flowing from HDE226868 toward Cygnus X-1. Because Cygnus X-1 is not an eclipsing system, we cannot determine the mass of the X-ray source with precision. However, strong limits can be placed on the mass. Analysis yields a mass of Cygnus X-1 of more than seven solar masses, which is more than twice the upper limit for the mass of a neutron star. Cygnus X-1 is therefore likely to be a black hole.

Dactyl
Small moon of asteroid (243) IDA, discovered by the Galileo probe on its way to Jupiter in 1993. Dactyl was the first natural satellite of an asteroid to be found. It measures approximately 1.6 × 1.2 km and orbits Ida at a distance of about 90 km.

Dark adaption
On moving from a well-lit area into darkness the eyes take a while to adapt and to see faint objects. Under poor illumination the retina secretes the hormone rhodopsin. This stimulates the rods in the eye, and sensitivity can increase for up to half an hour, although most improvement occurs in the first 10 min. The most sensitive part of the retina is an annular ring around the center, so observers find it easier to see a faint object by looking slightly to one side of it and using AVERTED VISION.

Dark ages
Term applied by Martin REES to the interval early in the development of the UNIVERSE when it was without light. This extended from the time when the COSMIC MICROWAVE BACKGROUND radiation cooled below 3000 K and became infrared radiation, to the time when stars, galaxies and BLACK HOLES in QUASARS lit up the universe again. The dark ages extended from about 0.5 million years after the Big Bang to about 0.5 billion years after that. This was the time of the formation of structure, when the INTERGALACTIC MEDIUM became clumpy and formed galaxies and galaxy clusters and groups. The infrared-sensitive JAMES WEBB SPACE TELESCOPE is intended to look into the dark ages.

Dark energy and the cosmological constant
Imagine taking a region of space and emptying it – removing MATTER, radiation and everything so that it is much more empty than the space between planets or stars. The result is a vacuum.

The vacuum is a physical state, and there is no reason in principle for its energy to be zero. In the absence of gravity, however, there is no way of measuring the energy of a state on an absolute scale. The best we can do is to compare energy differences, but the vacuum energy itself would be arbitrary. In the general theory of relativity, however, any form of energy has a gravitational effect, so the vacuum energy might be a crucial ingredient in the evolution of the universe. Colloquially, the vacuum energy is known as 'dark energy.'

The vacuum is approximately the same everywhere in the universe. The vacuum energy density is a universal number, which in relativity is called the 'cosmological constant,' denoted by the Greek letter Λ.

Albert EINSTEIN introduced Λ as a parameter in cosmology for mistaken reasons. The force of gravity attracts all galaxies together. If gravity were the only force, it would be impossible for the universe to be static. Not knowing at the time that the galaxies are expanding and pulling away from each other against the mutual force of their gravity, Einstein believed that GENERAL RELATIVITY should be able to describe a static universe, and he invented the cosmological constant as a repulsive effect to counter gravity. When Edwin HUBBLE found that the universe was not static but expanding, the cosmological constant became unnecessary. Einstein famously called it his 'greatest blunder.'

When scientists understood that the cosmological constant measured the energy density of the vacuum, they realized that the vacuum energy is potentially important to the dynamical history of the universe. A cosmological constant has the tendency to cause galaxies to accelerate away from us, in contrast to the tendency of ordinary forms of energy (such as matter) to slow them down.

In a universe with both matter and vacuum energy, as ours is proving to be, there is a competition between the tendency of Λ to cause acceleration and the tendency of matter to cause deceleration. The ultimate fate of the universe depends on the precise amounts of each component.

The most direct way to measure the cosmological constant is to determine the proportionality between redshifts and distances of faraway galaxies. Nearby galaxies have redshifts that are related to their distances by the HUBBLE CONSTANT. Galaxies far away (further back in time) show a different Hubble constant, according to whether they are accelerating or decelerating. Two independent groups have used type Ia SUPERNOVAE as distance indicators of faraway galaxies. They discovered many supernovae by carefully observing deep into small patches of the sky, and proved the distance of the supernovae with observations (by the HST and the Keck Telescope) of their brightness and of the redshifts of the parent galaxies. The cosmological constant is not zero and Einstein's 'greatest blunder' seems in fact to have been one of his greatest intuitions.

The evidence suggests that approximately 70% of the density of the universe is vacuum energy (dark energy), and 30% is matter (95% of this being DARK MATTER).

See also QUINTESSENCE.

Further reading: Goldsmith D (1997) *Einstein's Greatest Blunder?* Harvard University Press.

Dark matter
What makes up most of the MASS in the universe? With conventional astronomical methods we can see only the 'luminous MATTER,' notably in the form of stars. But when we compare this with the mass of galaxies (determined from their dynamical properties), there is a huge discrepancy: their mass is much more than the luminous matter seen. If the usual law of gravity is correct there are large amounts of dark matter, a term first introduced by Fritz ZWICKY in 1933 when he studied the dynamics of GALAXY CLUSTERS AND GROUPS.

Evidence for dark matter
SPIRAL GALAXIES provide perhaps the most impressive evidence. They have a thin disk that rotates around a central bulge. The mass of the galaxy can be determined from the disk's orbital velocity. In fact, the change of orbital velocity with radius can be used to map the distribution of mass in the galaxy's interior regions. Until the early 1970s, most of the rotation data for spirals came from optical observations of the luminous inner regions and seemed consistent with the distribution of luminous matter. With the construction of RADIO TELESCOPES like the WESTERBORK SYNTHESIS RADIO TELESCOPE in the Netherlands, it became possible to measure the rotation of the hydrogen gas in spiral galaxies. The hydrogen in many spirals extends far beyond the starlight, allowing us to measure the rotation of the galaxy at large radii from the center.

In general, the orbital velocity of gas around a galaxy is faster if it orbits a larger mass, but slower if it is far from the mass. The rotational velocity curve rises from the center outward, because the increased amount of mass that the gas is orbiting more than compensates for the increased distance. It reaches a value of the order of 100–200 km s^{-1}. From here it would be expected to fall, because most of the

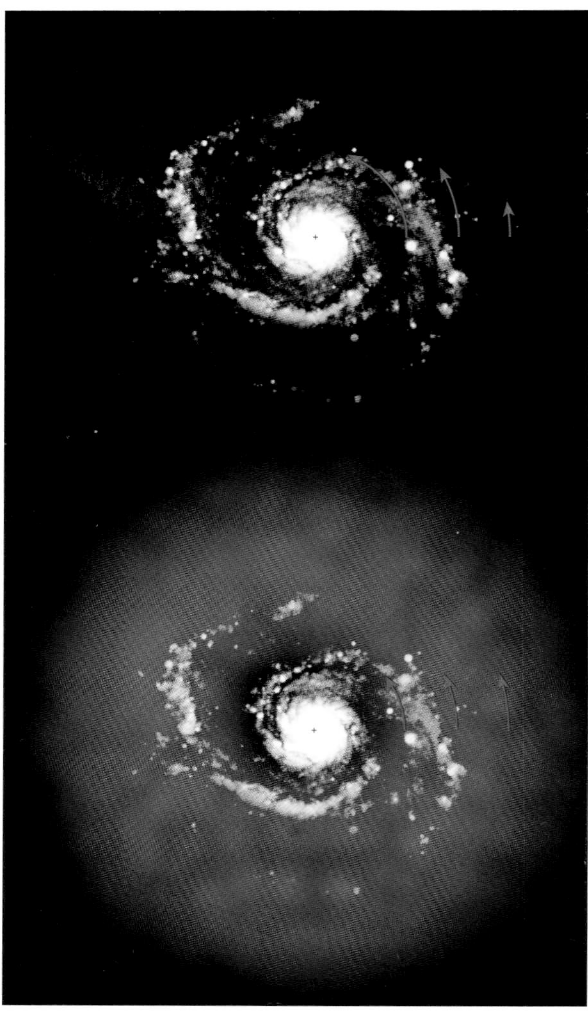

► **Dark matter** *Most of the luminous mass of a spiral galaxy lies at its center. In such a case, the stars and gas clouds in the spiral arms rotate around the galaxy like planets around the Sun, and might be expected to move more slowly at greater distances from the center (represented by shorter arrows) (top). In fact they move at the same speed right out beyond the visible spiral arms. This can happen if the galaxy is surrounded and immersed in a halo of dark matter (bottom).*

mass of a galaxy (the stars) is concentrated near its center, and gas outside it would orbit according to the expectation that the orbital velocity decreases with distance. But the figure shows that the orbital velocity stays constant out to the largest radius where there is any gas, far beyond the stars. The discrepancy is ascribed to the gravitational effect of dark matter. The resulting picture of spiral galaxies is of the bulge and disk, immersed in a huge dark-matter halo.

Galaxy clusters are the largest gravitationally bound systems in the universe. The velocities of the member galaxies are so large that huge amounts of dark matter are needed in galaxies to keep them in orbit around each other. Recently it has become possible to measure the amount of dark matter in a galaxy cluster from the distortion of background galaxies caused by a cluster's gravitational lensing effect (*see* GRAVITATION, GRAVITATIONAL LENSING AND GRAVITATIONAL WAVES).

The contribution of a given matter component to the overall density of the universe is usually expressed in terms of the omega (Ω) parameter (*see* CRITICAL DENSITY). The critical value for omega is the value of the density that fits a certain model of the universe (one in which the spatial geometry of the universe is flat). All the results point to the same conclusion: stars provide about 2% of the critical density, but galaxies as a whole, including their dark-matter halos, provide about 30%.

What is dark matter?

The first thought is that dark matter is made of the same matter as everything else (hydrogen, helium and the heavier elements), but invisible. Ordinary matter whose mass consists of the BARYONS, namely, electrons, protons and neutrons, is called 'baryonic matter.'

Could dark matter be baryonic matter in some non-luminous form? Perhaps stellar remnants, such as neutron stars, that are difficult to observe? The overall baryon density is severely constrained by calculations from the BIG BANG. The relative amounts of the light elements made in the Big Bang depend on the cosmic baryon density. Deuterium is the element that is most sensitive to the baryon density. It can be measured in intergalactic hydrogen clouds by observing quasar absorption lines, and implies that the baryonic density is about 2%. Luminous stars account for all of this, leaving no scope for dark baryonic matter.

If dark matter is not baryonic, it might consist of ELEMENTARY PARTICLES made in the Big Bang. Ramanath Cowsik and J McClelland speculated in 1973 that NEUTRINOS could play this role. However, while neutrinos do have some mass, not all dark matter can be neutrinos. Could it instead consist of some hitherto unknown elementary particles? The problem of dark matter is a key subject of ASTROPARTICLE PHYSICS.

The COSMIC MICROWAVE BACKGROUND (CMB) hints at the nature of the dark-matter particles. The granulations, or anisotropies, that have been observed in the CMB are small. Galaxies and clusters of galaxies do not form from such small lumps if the Big Bang medium consists only of baryons and radiation, because the bath of PHOTONS stops the baryons from gathering together. If there are particles that interact weakly they have a less repulsive effect than photons, and if they have mass they help pull the matter together into galaxies. This is the motivation for suggesting that weakly interacting massive particles (WIMPS) are the source of dark matter.

Another feature that prevents galaxies and clusters of galaxies forming from the anisotropies in the CMB is heat, so it helps if the dark matter is cold. There is now almost universal consensus that some variant of cold dark matter was involved in the formation of our universe, as opposed to baryons and low-mass, fast-moving (hot) particles such as massive neutrinos.

The search for dark matter

If the dark matter consists of some form of WIMP, then the Milky Way galaxy and our laboratories should be filled with a gas of these particles. Numerous groups have made more and more sensitive detectors to search for WIMPs. Usually they try to measure the tiny energy deposited by a rare collision of a WIMP with a nucleus in a crystal. The extremely small signal rate depends on the assumed WIMP properties and the target material, but a typical number is below one event per kilogram of crystal per day. To reduce natural radioactive contamination, researchers must use extremely pure substances for the detectors and shield them from COSMIC RAYS by conducting the experiments underground (for example, in deep mines at Boulby in Yorkshire, UK). Axions are WIMPs, a kind postulated by the theory of the strong interaction among QUARKS. Two experiments to find them are now in operation, one in Livermore, California (US Axion Search), the other in Kyoto, Japan (CARRACK).

Other indirect search methods do not depend on the properties of the WIMPs. For example, WIMPs will occasionally collide with a nucleus in the Sun and lose

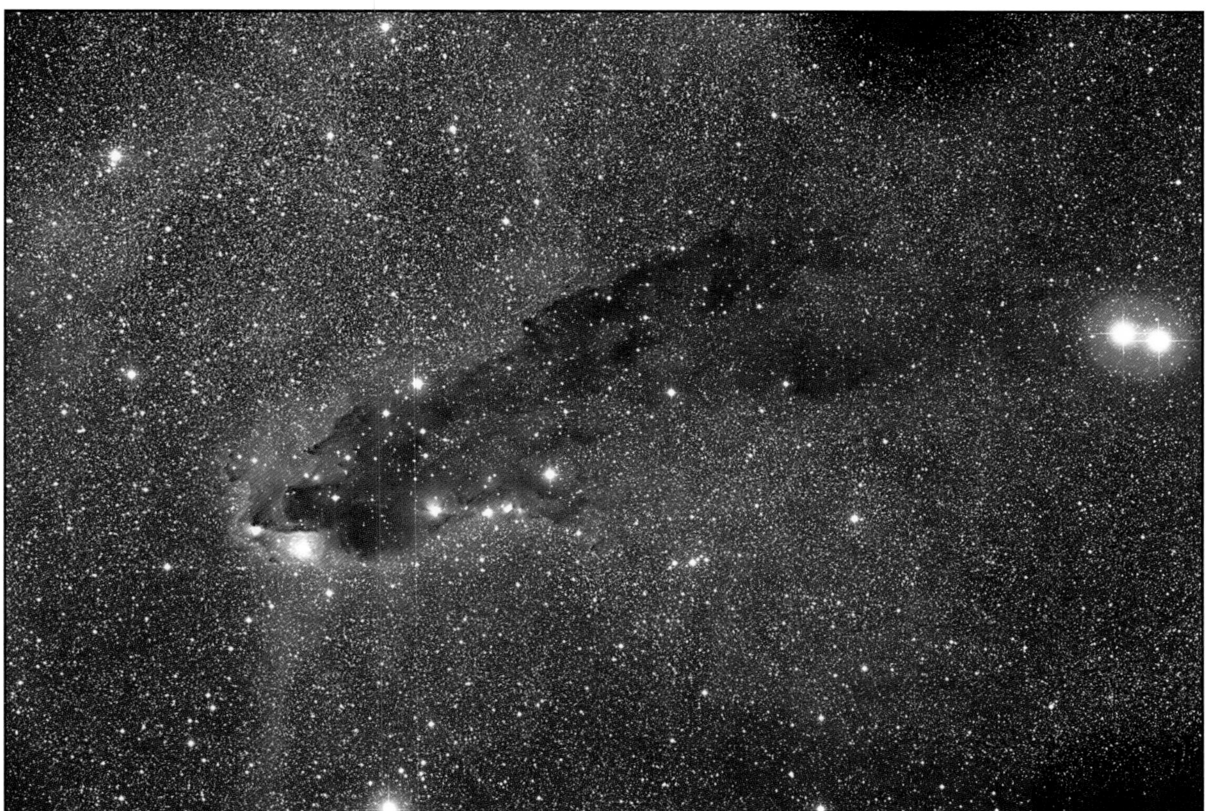

◀ ***Dark nebula*** *This unnamed dark nebula in the constellation Scorpius has a swept-back shape due to radiation from very luminous stars some distance away. The bright red rims at the end of the nebula are exposed to radiation from a group of very hot stars about 5000 l.y. away from the Earth, known as the Scorpius OB association.*

enough energy to get trapped. Ultimately a trapped WIMP annihilates with others, producing high-energy neutrinos. These neutrinos may show up in neutrino telescopes (*see* NEUTRINO ASTRONOMY).

As long as particle dark matter has not been found we can continue to speculate that there is something wrong with our theories of gravity, that the arguments against baryons are wrong and that galactic halos consist of ordinary matter in some hard-to-see form. Such dim stars are termed massive compact halo objects (MACHOS), to contrast with WIMPs. MACHOs could be stars that are too small to shine brightly (brown dwarfs or M dwarfs), or burnt-out stellar remnants (white dwarfs, neutron stars, black holes).

If such stellar remnants exist, a special scenario would have to be invented to have made them. One possibility is that black holes formed in the early universe, although no plausible production mechanism is known. Another possibility is that there are large numbers of white dwarfs in galaxies, the remnants of a very early and very massive burst of star formation.

In an attempt to look for MACHOs, the Large and Small Magellanic Clouds have been monitored since the early 1990s by the MACHO and the French EROS collaborations. If a MACHO happens to pass near the line of sight to a background star, it produces a gravitational deflection of the starlight and a temporary lensing effect that causes the star to brighten (gravitational microlensing). The MACHO experiment has detected numerous lensing events. The masses of the lensing stars are about half the solar mass, which is typical of white dwarfs rather than black holes. MACHOs could provide some dark matter, but not most of it.

Without a direct and positive identification of the dark-matter particles or objects, the nature of dark matter remains a problem whose solution is complicated and may not yet have been thought up.

Further reading: Börner G (1993) *The Early Universe* Springer; Kolb E W and Turner M S (1990) *The Early Universe* Addison-Wesley; Persic M and Salucci P (eds) (1997) *Dark and Visible Matter in Galaxies* (ASP Conference Proceedings Series 117) Astronomical Society of the Pacific.

Dark nebula

An interstellar cloud of gas and dust that absorbs light from nearby sources. Also known as an absorption nebula. The light is re-emitted as infrared radiation or scattered, making the nebula appear dark. At higher galactic latitudes, away from the Milky Way, dark nebulae appear merely as star-poor regions of the sky; but near or against the Milky Way they contrast strongly with the bright starfields around them. Prominent dark nebulae seen in silhouette against the Milky Way include the broad, dark band in the northern hemisphere known as the Cygnus Rift, and the COALSACK in the southern hemisphere. The famous HORSEHEAD NEBULA in Orion is a small but distinctive dark nebula. The smallest dark nebulae are known as GLOBULES. Dark nebulae consist predominantly of molecular hydrogen and are believed to be sites of star formation. The true nature of dark nebulae, that they are not simply voids, was first recognized by Edward BARNARD.

Dark skies and good outdoor lighting

See practical astronomy feature overleaf.

Darwin, George Howard (1845–1912)

English mathematician and astronomer. The son of Charles Darwin (1809–82), George Darwin became Plumian professor of astronomy and experimental philosophy at Cambridge

Dark skies and good outdoor lighting

Light pollution is a relatively new topic, both for scientists and for the general public. It is a concern not just for astronomers, but for the general population, but it rarely receives the attention it deserves. It is an issue that must be addressed in order for future generations to be able to engage in the kind of Earth-based observing we enjoy today.

What is it?

Most define light pollution as any of the many adverse effects of poor nighttime lighting. These include artificial sky glow, light trespass (obtrusive light), glare, energy waste and impacts on human health and the ecosystem. Almost everyone is affected in some way by these problems.

Why does it happen?

There is not yet much public knowledge of the difference in value between good nighttime lighting and poor lighting. There are many valid reasons for good night lighting; most relate to improving vision at night and to having a more effective and more comfortable nighttime environment.

It makes sense to light only where and when needed, and to light only to lighting levels actually required for the situation. These levels can and do change a lot depending on the specific task and the general locale where lighting is needed at night (city center versus countryside, for example). Wasting light is clearly also light pollution.

The effects of light pollution

One of the key results of light pollution is a loss of the view of the beauty of the night sky, with stars, aurorae and other natural phenomena no longer being as readily visible. This is especially a problem for professional and amateur astronomers, but it is also a huge loss for the public. The table shows the loss in value of a major telescope with varying degrees of light pollution. These increases in light pollution exist; increases by even as much as a factor of four in or near major cities are common.

Light pollution encompasses a very wide range of concerns. These include:

- viewing the universe we live in under a dark sky;
- preserving sea turtles, birds and other wildlife;
- preserving our circadian rhythms and thus maintaining human health and sleep patterns;
- saving energy by not wasting light at night;
- improving visibility at night (and, hence, our ability to move around and to enjoy ourselves at night safely and securely).

The overall nighttime ambiance is improved with good night lighting, not compromised.

Loss of value in a 4-m telescope due to sky glow

Fractional increase in sky glow	Telescope aperture in meters	Percent of original value
1.00	4.00	100
1.10	3.81	88
1.20	3.65	78
1.25	3.58	74
1.50	3.27	58
2.00	2.83	39
3.00	2.31	23
4.00	1.79	11

(A 10% increase in sky glow is a factor of 1.10 above, and so forth.)

What can be done?

We can educate ourselves and then others about the value of dark skies and of good nighttime lighting. We must then use only such lighting and push to change the existing poor lighting. Education is the key to finding solutions, followed by action.

To be more explicit about solutions: shine the light down, not up where it only brightens the sky, nor sideways to cause glare. Don't over light. Use only the amount needed. 'The more light the better' is a myth. Wipe out glare (that is, any blinding light). Use energy-efficient sources in an efficiently designed lighting installation. Support outdoor lighting-control ordinances, which help improve community lighting for all.

▶ ***Sky glow*** *above a typical US city. The lack of any visible stars is very visible evidence of the way wasted light has an adverse effect on the night environment and of our view of the universe.*

▶▶ ***Same part of the sky,*** *photographed from a location well away from the city.*

Where to find more information

The International Dark-Sky Association (IDA) is a non-profit organization, incorporated in 1988, whose goals are to preserve and restore the pristine dark skies that most of our ancestors had, while at the same time maximizing the quality and efficiency of nighttime outdoor lighting. IDA is a membership-based organization, and these members educate their neighbors and their communities about these goals; many reference materials are produced and provided by IDA for these purposes.

As of the end of 2002, IDA had over 9800 members, from all of the states in the United States and from 70 other countries. These include astronomers, both amateur and professional; lighting engineers and designers; environmentalists; ecologists; utility companies; concerned members of the public; and organizations involved with astronomy, the lighting industry, civil engineering, city planning and so forth.

Lots of useful information is available on the IDA website. It also lists other IDA resources, including a regular newsletter, many information sheets, images, slides, CDs and several videos.

IDA can supply sample outdoor lighting control ordinances and the website lists communities that have adopted such lighting controls. There are formal IDA Sections and Affiliates in many locations, groups that act as local centers of information, resources and activities.

David L Crawford is the Executive Director of the International Dark-Sky Association and a Fellow of the Illuminating Engineering Society of North America.

◀ ***Los Angeles in 1908*** *as seen from Mt Wilson.*

◀ ***Los Angeles in 1988*** *as seen from Mt Wilson. An increase in population has brought an increase in light pollution. Notice the direct light from 'nearby' fixtures.*

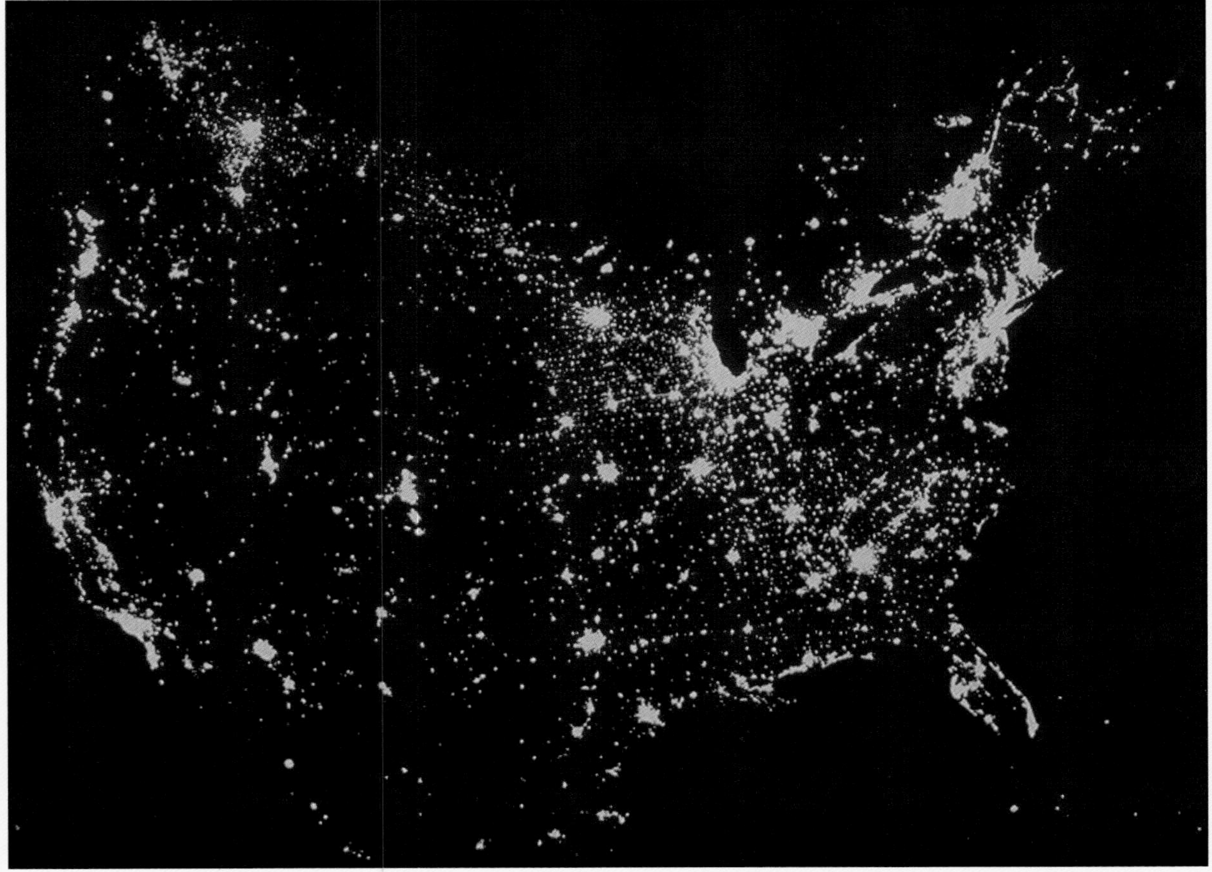

◀ ***Satellite image of the USA*** *at night, as seen from space. The large amount of interesting detail that can be seen is indicative of all too much wasted light. All images courtesy of the International Dark-Sky Association.*

▶ ***David Dunlap Observatory*** *The dome of the 1.88-m reflecting telescope.*

University, UK. He applied mathematical theory to the dynamics of the Sun-Earth-Moon system, and added physical effects to gravitational theory, including the frictional effects of tidal action and the shapes that the Earth and Moon take up as they rotate. Darwin proposed a theory of the origin of the Moon, namely that it was pulled from a molten Earth early in its history by tidal action of the Sun. This was the first time that the origin of a planet had been explained from its current dynamical state.

Databases, data processing

See INFORMATION HANDLING IN ASTRONOMY

David Dunlap Observatory

The David Dunlap Observatory is operated as a facility of the Department of Astronomy at the University of Toronto, Canada. It is located 25 km north of the center of Toronto, in the town of Richmond Hill. The first director, Clarence Chant, received a generous donation for the observatory from the Dunlap family in 1928. The observatory is the site of a 1.88-m REFLECTING TELESCOPE, the largest in Canada. The telescope, which saw first light in May 1935, has been upgraded with computer controls and modern instrumentation and is dedicated to spectroscopic studies of mainly galactic objects. Two smaller telescopes and support facilities are also available. The observatory offers public programs, courses and school visits.

Dawes, William Rutter (1799–1868)

English clergyman and amateur astronomer. Dawes observed DOUBLE STARS, and discovered Saturn's Crêpe Ring independently of William BOND. He gave a useful empirical formula for the RESOLVING POWER of a telescope, known as DAWES' LIMIT.

Dawes' limit

An empirical measure of the RESOLVING POWER of a telescope devised by William DAWES, a keen-eyed observer of DOUBLE STARS. According to the Dawes criterion $R = 0.115/D$, where R denotes resolving power expressed in arcsec and D denotes the aperture of the telescope in meters. For example, Dawes' limit for a telescope of 0.1-m (100-mm) aperture would be 0.115/0.1 = 1.15 arcsec. Dawes' limit gives resolving powers that are some 20% better than the theoretical values given by the RAYLEIGH LIMIT. This reflects the fact that skilled observers, under ideal conditions, may be able to resolve double stars that are marginally closer together than the theoretical limit. Dawes' limit is a convenient, but to some extent arbitrary, characterization of telescopes.

Day

Traditionally, the period when the Sun is above the horizon (contrasting with night). Astronomically, it is the time taken for the Earth to rotate once on its axis. In the accurate measurement of TIME a day is defined as 86 400 s, where the second is defined as a certain (large!) number of the oscillations of a cesium atom. There are several different definitions of the astronomical day.

The sidereal day is the time interval between two successive upper transits of the vernal EQUINOX; it is, in effect, the time taken for the Earth to rotate on its axis through an angle of 360° with respect to the background stars.

The apparent solar day is the time interval between two successive noons (upper TRANSITS of the Sun across an observer's MERIDIAN). It is the day measured by a SUNDIAL. Because the Earth moves around the Sun in an elliptical orbit at a variable rate, and the apparent annual path of the Sun relative to the stars (the ECLIPTIC) is tilted at an angle to the CELESTIAL EQUATOR, the right ascension of the Sun changes at a non-uniform rate and the time interval between successive noons (and hence the duration of the apparent solar day) is not precisely constant.

The mean solar day is the time interval between two successive upper transits of a hypothetical object called the mean Sun, which moves along the celestial equator, relative to the background stars, at a uniform rate (equal to the average angular rate at which the real Sun appears to move along the ecliptic). It is the day measured by a clock running at a constant rate, which agrees on average with a sundial. On the assumption that the Earth rotates on its axis at a uniform rate (which is not strictly accurate), mean solar days are of equal duration.

Because the Earth revolves around the Sun at an average angular rate of just under 1° per day, the Earth has to rotate through an angle of nearly 361° to position successive upper transits of the Sun across an observer's meridian. Consequently, the mean solar day is just under 4 min longer than the sidereal day, and the sidereal day is equal to 23 hours, 56 min, 4 s of mean solar time.

DBV pulsating star

Member of the small group of pulsating WHITE DWARF stars with almost pure HELIUM atmospheres that exist in a narrow temperature range near 25 000 K. Collectively they go by two names: V777 Her stars, named after the VARIABLE STAR designation of the first known star in the class and, more informatively, DBV stars. The name DBV follows the standard convention for white dwarf stars and tells us that they are of spectral type DB (indicating a nearly pure helium atmosphere), and that they are variables of that spectroscopic class. *See also* ASTROSEISMOLOGY.

Deceleration parameter

A quantity, denoted by the symbol q, that describes the rate at which the expansion of the UNIVERSE is slowing down. The value of the deceleration parameter at the present epoch in the history of the universe is denoted by q_0.

Different values or ranges of values of q_0 correspond to different cosmological models. For example, if q_0 is greater than 0.5 the universe is closed and will eventually collapse; whereas if q_0 is less than 0.5 the universe is open and will expand forever at a finite rate. The flat, or Einstein–de Sitter, universe corresponds to $q_0 = 0.5$. The case $q_0 = 0$ corresponds to a universe that expands at a constant rate. A negative value of q_0 corresponds to a universe in which the expansion is accelerating.

In principle, it should be possible to determine the value of q_0 observationally. For example, for a population of uniformly luminous sources (such as a set of identical SUPERNOVAE within remote galaxies) the relationship between apparent brightness and redshift is dependent on the value of the deceleration parameter. Although measurements of this kind are notoriously difficult to make and to interpret, recent observations tend to favor q_0 values of less than 0.5, and hence favor open-universe models.

de Chéseaux, Jean Philippe Loys (1718–51)

A Swiss astronomer, de Chéseaux observed several clusters and 'nebulous stars,' and compiled a catalog of their positions. De Chéseaux was the first to formulate OLBERS' PARADOX,which he applied to the Milky Way.

Declination

The angular distance of a celestial body north or south of the CELESTIAL EQUATOR. In other words, the angle between the celestial equator and a star, measured in a direction perpendicular to the celestial equator. Declination may take values between 0° and 90°, and is positive or negative for objects north or south of the celestial equator, respectively. Declination is often abbreviated to dec., or denoted by δ. The CELESTIAL COORDINATES of a star are normally expressed in terms of right ascension and declination.

Deep Impact

A NASA Discovery mission scheduled for launch in 2004. It is intended to send a 500-kg copper projectile into Comet Tempel 1 in July 2005. A camera and infrared spectrometer on the spacecraft will study the resulting icy debris and pristine interior material.

Deep sky

A term used by amateur astronomers to describe observations of objects beyond the solar system.

Deep Space

A series of NASA space technology missions under its New Millennium Program. Deep Space 1 (launched in October 1998) successfully tested 12 advanced space technologies, including autonomous navigation and ion propulsion, and it captured excellent images of COMET BORELLY. The mission came to an end on December 18, 2001. Deep Space 2, launched in January 1999 aboard the failed Mars Polar Lander mission, was to have slammed two miniature probes into the Martian soil. Other missions in the program are Earth Observing 1 (launched 2000) and 3 (launch scheduled for 2005 or 2006), and Space Technology 5 (launch 2004), all of which aim to develop and test advanced technology in space flight.

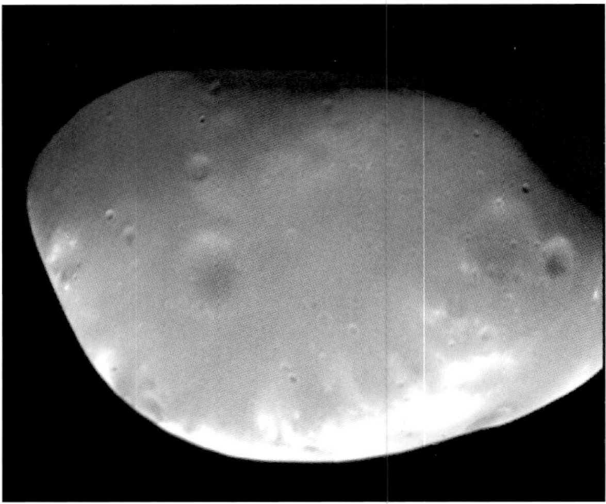

◀ ***Deimos*** *This picture of the smallest moon of Mars is a mosaic of images taken by the Viking 2 spacecraft during one of its close approaches in 1977. Dusty regolith makes its craters indistinct, and its surface markings streak down-slope, despite Deimos's low gravity.*

Degenerate matter

Highly compressed MATTER in which the normal atomic structure has broken down and which, because of quantum-mechanical effects, exerts a pressure that is independent of temperature.

At the very high temperatures and pressures that exist inside stars, matter is almost completely ionized and forms a gas of nuclei and ELECTRONS (a PLASMA). When a star has consumed all of its nuclear fuel, it shrinks under the action of its own gravity to form, in most cases, a compact WHITE DWARF. As the star shrinks and its constituent electrons are forced closer together, the volume of space available to each electron rapidly decreases. The PAULI EXCLUSION PRINCIPLE implies that no more than two electrons (electrons with opposite spin) can have the same position and momentum. Under these conditions the pressure generated depends only on the density of the gas, not on its temperature. Matter in this state is said to be electron degenerate, and the pressure exerted by the electrons is called electron-degeneracy pressure. At much higher densities a similar phenomenon, called BARYON degeneracy, occurs for PROTONS or NEUTRONS.

If the mass of a WHITE DWARF exceeds about 1.4 solar masses (the Chandrasekhar limit), gravity will overwhelm electron degeneracy and further collapse will ensue. During the ongoing collapse, electrons combine with protons to form neutrons. Because neutrons, too, are subject to the Pauli exclusion principle, at high enough densities they can form a neutron-degenerate gas that prevents further collapse, so producing a neutron star. The maximum mass that can be supported by baryon-degeneracy pressure is thought to be in the region of two to three solar masses. This gives an upper limit for the mass of a neutron star. If a neutron star exceeds this mass (for example, by merging with another one) it will form a BLACK HOLE.

Deimos *See* PHOBOS AND DEIMOS

Delisle, Joseph-Nicolas (1688–1768)

French astronomer and member of the Academy of Sciences, he hired CHARLES MESSIER as a draftsman and as a recorder of astronomical observations. Delisle visited Isaac NEWTON in London in 1724. He calculated the return path for COMET HALLEY in 1758–9 and published a map that guided the recovery search, showing the predicted path of return calculated at 10-day intervals.

Delphinus *See* CONSTELLATION

Delta Aquarids

A METEOR SHOWER that takes place in late July and early August. There are two radiants in the constellation Aquarius. One, at declination –17°, comes to maximum on July 29; the other, slightly weaker, at declination –10°, comes to maximum around August 6. The parent body of this shower is unknown.

Delta Cephei

A VARIABLE STAR discovered in 1784 by John Goodricke (1764–86), a young English astronomer who two years previously had explained the variability of ALGOL. The star is the prototype of an important class of pulsating variable stars now known as CEPHEID VARIABLES. Cepheids are characterized by the regularity of their periods and the form of their LIGHT CURVES, and provide a means of measuring distances through the period-luminosity law. This is especially valuable in the case of Cepheids found in other galaxies.

The apparent magnitude of Delta Cephei varies between 3.48 and 4.37 in a period of 5.36634 days. *See also* VARIABLE STAR: PULSATING.

Delta Scorpii

Central star in the head of Scorpio, the star is normally magnitude 2.3. It is 400 l.y., spectral type B0 (*see* STELLAR SPECTRUM: CLASSIFICATION). It has a faint companion in an eccentric 10.6-year orbit. In July 2000, at the time of the close approach of the companion, Delta Scorpii brightened to about magnitude 1.5 and ejected a gas disk, becoming a BE STAR.

Delta Scuti star

Relatively young, pulsating variable star. Delta Scuti stars are of spectral type A0–F5 III–V, have light amplitudes of several hundredths of a magnitude, and periods from 0.01 to 0.2 days. The shapes of the LIGHT CURVES, periods and amplitudes can vary greatly. *See also* VARIABLE STAR: PULSATING.

de Mairan's Nebula (M43) *See* MESSIER CATALOG

Deneb

The star Alpha Cygni, the 'Tail of the Swan.' One of the three bright stars that form the ASTERISM of the Summer Triangle (*see* ALTAIR; VEGA). It is a white supergiant of spectral type A2Ia, the type-star of the ACYG class of pulsating supergiant variable stars with EMISSION SPECTRA, varying erratically over a range of ±0.04 magnitudes. *See also* VARIABLE STAR: PULSATING.

Denning, William Frederick (1848–1931)

An English amateur astronomer. In 1877 he demonstrated a steady night-by-night movement in the PERSEID meteor RADIANT, which proved that METEORS came from showers of dust distributed along the path of a COMET. He was cited in H G Wells's *The War of the Worlds* (1898) as 'Denning, our greatest authority on meteorites....' He published a catalog of meteor radiants, and also discovered several comets and Nova Cygni 1920.

Density

The amount of mass contained within a unit volume of material. The mean density of an object is equal to its total mass divided by its volume. In practice the density of an astronomical body increases toward the center. Thus, the density of the surface rocks on the Earth is about half the mean density, while the central density is about two-and-a-half times the mean value. Astronomical bodies exhibit a wide range of densities. Examples are given in the table.

Descartes, René (1596–1650)

French mathematician and philosopher who settled in Holland. His work, *La Géométrie* (Geometry), formulated geometry in terms of algebra, from which came the concept of Cartesian coordinates. He wrote *Principia Philosophiae* (Principles of Philosophy) and attempted to put the whole universe on a mathematical basis. Since he did not believe in action at a distance, he assumed that the universe was filled with MATTER that constituted a system of vortices. These carried the Sun, the stars, the planets and COMETS in their paths, which for unexplained reasons were ellipses. This idea was championed in France for nearly 100 years.

Descending node *See* ORBIT

de Sitter, Willem (1872–1934)

Dutch cosmologist who started with a traditional career in astronomy, observing with David GILL at the Cape of Good Hope and working on dynamical problems of stars and solar system objects. He became director of the LEIDEN OBSERVATORY in the Netherlands, and turned to a range of problems in the theory of GENERAL RELATIVITY. He proved observationally the fundamental tenet that the velocity of light, *c*, does not depend on the velocity of the source. His mathematical theories inspired Arthur EDDINGTON's expedition to measure the gravitational deflection of light rays passing near the Sun during the 1919 solar eclipse. De Sitter created a general relativistic theory of the universe and found solutions that described an expanding universe, a theoretical result that was spectacularly confirmed by Edwin HUBBLE's observations of the recession of galaxies. He worked with Albert EINSTEIN on the Einstein–de Sitter model of the universe, which is the basis for current, more complex BIG BANG models.

Deuterium

A heavy, stable ISOTOPE of HYDROGEN. Its nucleus contains one PROTON and one NEUTRON. Deuterium is also known as 'heavy hydrogen.'

Diamond ring

An effect observed at the onset and end of the totality phase in a total eclipse of the Sun. Just as the extreme edge of the Sun's disk is about to disappear behind (or emerge from behind) the Moon's disk, a bright arc of sunlight is seen. As the last (or first) of BAILY'S BEADS of sunlight shines through the irregularities on the lunar LIMB, the arc has the appearance of a diamond ring.

Dichotomy

The phase of a body in the solar system when exactly half of its sunlit side is visible. The term is used in particular for the half-phases of the INFERIOR PLANETS, MERCURY and VENUS. When referring to the Moon, the terms first quarter and last quarter are preferred. Only spacecraft can image other bodies in the solar system at dichotomy.

Density of astronomical bodies

Object	Mean density (water = 1)	Mean density (kg m^{-3})
Neutron star (typical)	4×10^{14}	4×10^{17}
White dwarf (typical)	10^{6}	10^{9}
Earth	5.5	5.5×10^{3}
Jupiter	1.3	1.3×10^{3}
Sun	1.4	1.4×10^{3}
Red giant (typical)	10^{-4}	10^{-1}
Red supergiant (typical)	10^{-7}	10^{-4}
Interstellar gas cloud	10^{-22}	10^{-19}

Dicke, Robert Henry (1916–97)

American physicist and professor at Princeton University (1946–84). He did not believe Albert EINSTEIN's theory of GENERAL RELATIVITY and developed numerous experiments to test the physics of gravity. He attempted to determine whether the Sun was oblate (flattened at the poles) and could thus alter the orbit of Mercury. To a good degree of approximation, Mercury orbits in an ellipse as calculated by Newton's theory of gravitation, but there are small changes. The changes in orbit produced by solar oblateness could mimic changes produced by general relativity, as proposed by Einstein, so it was important to exclude other possibilities. Dicke also developed a theory of gravity as an alternative to Einstein's. To look for effects that would provide evidence of his own theory, Dicke remeasured the gravitational deflection of starlight by the Sun, measured the position of the Moon by laser beams reflected back to Earth, and determined the age of the oldest stars. But this was all in vain and his challenge to Einstein was unsuccessful. Dicke did, however, believe that the universe began with a BIG BANG, and he identified the possibility that there would be a microwave remnant of that event. He was building a RADIO TELESCOPE to look for this microwave remnant when Robert WILSON and Arno PENZIAS found it while engaged in other studies. *See also* COSMIC MICROWAVE BACKGROUND.

Differential rotation

The variation with latitude of the angular ROTATION rate of the Sun or other non-solid bodies. It was first observed on the Sun by English astronomer RICHARD CARRINGTON from studies of the motion of sunspots across the solar disk, carried out between 1853 and 1861. Recent measurements, obtained by measuring Doppler shifts in the wavelengths of spectral lines at different points on the Sun (*see* DOPPLER EFFECT), indicate that the sidereal rotation period of the photosphere is 24.8 days at the solar equator, 26.1 days at latitude 30°, about 31 days at latitude 60°, and approximately 35 days at the poles. HELIOSEISMOLOGY indicates that differential rotation extends down to the base of the convective zone, located at a depth equal to 29% of the Sun's radius. Below this level the solar interior has a near-uniform rotation period of just under 27 days.

Differential rotation is also observed on Jupiter's clouds.

Differentiation

The process by which a planetary body (a solid planet or major satellite) acquires a layered structure, with dense materials in a core at its center and less dense materials in an overlying mantle and crust. A body that has formed by ACCRETION has a homogeneous composition, with materials of different densities distributed uniformly throughout its volume. The interior of the body must melt for its constituent materials to begin separating out. The heat required for melting comes from several sources. Once a body is at least partially molten, the differentiation process, driven by gravity, can begin. Gravity also plays a part in the melting that makes differentiation possible, so the process can take place only in bodies that have grown sufficiently large (typically, more than 200 km in diameter).

Diffraction

The bending and spreading out of waves (for example, light waves or water waves) that occurs when they pass by the edge of an opaque object or through a narrow slit or aperture.

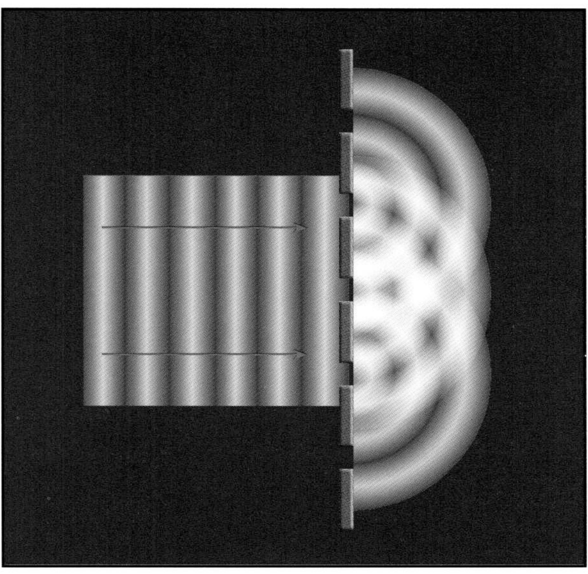

◀ ***Diffraction pattern*** *The crests and troughs of light waves are represented as white and blue. When the waves passing through the slits interfere constructively (add together) as the crests and troughs overlap, they are seen as bright bands. When they interfere destructively (cancel out) they are seen as dark bands.*

Light waves from a distant, point-like source advance as a series of wavefronts (crests and troughs) that lie at right angles to the direction in which the light is propagating, like ocean waves rolling up onto a beach. When the wavefronts from the source are interrupted, secondary wavefronts spread out from the edges of the obstruction causing some of the light to pass into the geometric shadow of the object. This effect, called diffraction, causes the edges of shadows cast by opaque objects to be fuzzy rather than perfectly sharp. When light passes through two or more slits, the secondary wavefronts interfere to form a diffraction pattern of light and dark bands. These may be seen if projected on a screen.

When light passes through a circular aperture, the resulting diffraction pattern consists of a bright central spot (called an AIRY DISK) surrounded by a series of concentric light and dark rings (or fringes). The magnitude of any diffraction effects depends on the wavelength of the light and the size of the aperture. The longer the wavelength, or the smaller the aperture, the larger the diffraction pattern.

Diffraction grating

A plate on which a large number of parallel grooves or slits have been cut, which spreads a collimated beam of light into its constituent wavelengths (*see* COLLIMATION), so producing a SPECTRUM. A DIFFRACTION grating may be a transmission or reflection grating depending on whether it transmits light or reflects light from the surfaces of the grooves. Most astronomical spectrographs employ reflection gratings. The RESOLVING POWER of a grating (its ability to reveal fine detail in a spectrum) depends on the number of grooves that it contains. Astronomical gratings typically have between 100 and 1000 grooves per millimeter, and contain between 1000 and 50 000 grooves in all. A simple grating spreads the incoming light into a large number of spectra. In order to concentrate most of the light (up to about 90%) into one of these spectra, the reflecting surfaces of the grooves are oriented at a particular angle (the grating is said to be 'blazed'). *See also* ECHELLE GRATING.

Dione

A mid-sized, icy satellite of SATURN, discovered by Giovanni Cassini in 1684 (*see* CASSINI DYNASTY). Dione's high density indicates that it is composed of rock and ice. There are two basic terrain types: on one hemisphere, bright streaks overlay a lightly cratered terrain, while the other hemisphere is more heavily cratered. The largest crater is Amata, with a diameter of 231 km. The longest trough is the 394-km Palatine Chasma, which is up to 8 km wide. A small satellite, Helene, is co-orbital with Dione.

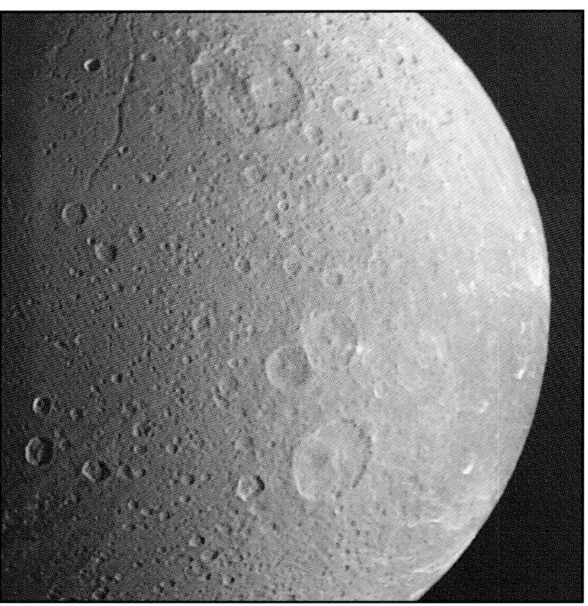

▶ ***Dione***
In this mosaic from the Voyager mission, Dione shows its heavily cratered leading surface. The largest crater (top) is Aeneas (1500 km in diameter).

Direct motion

Angular motion in the prevailing direction, also known as prograde motion (as opposed to RETROGRADE MOTION). The motion of a body in the solar system, either axial ROTATION or orbital REVOLUTION, is direct if it is in the same direction as that of the Sun's rotation: counter-clockwise as viewed from the Sun's north pole. The majority of solar system objects show direct motion, which is a relic of the motion of the rotating disk of material from which the Sun and the planets formed, the PROTOSOLAR NEBULA. The term is also used for the regular movement of solar system bodies from west to east on the CELESTIAL SPHERE.

Disconnection event

A discontinuity in the GAS TAIL of a COMET, in which the tail appears to break off from the COMA and a new tail begins to grow. A disconnection event occurs when the comet crosses a sector boundary of the SOLAR WIND (the two-dimensional surface that separates regions of north and south polarity in the spiraling magnetic-field lines). A comet can undergo several disconnection events in a single apparition if it makes several crossings of sector boundaries. For example, COMET HALLEY underwent 19 disconnection events during its return in 1986.

Disk galaxy

A GALAXY with a flat, circular disk of stars, dust and gas, in contrast to elliptical or irregular galaxies. Includes SPIRAL GALAXIES, BARRED SPIRAL GALAXIES and LENTICULAR GALAXIES.

Dispersion

In optics, the spreading of a beam of light into its constituent wavelengths (colors) through refraction or DIFFRACTION. The dispersion of a spectrograph is a measure of the extent to which the various wavelengths are spread out at the focal plane of the instrument. The more spread out the spectrum, the more the detail that can be seen (higher resolution), but of course more light is required to do this. Dispersion (more correctly, 'reciprocal dispersion') is usually expressed for a particular wavelength in nanometers (nm) or ångstroms (Å) of wavelength per millimeter (mm) of length along the spectrum. The dispersions of astronomical spectrographs usually lie in the range 200 to 0.01 nm mm^{-1} (2000 to 0.1 Å mm^{-1}). The lower resolutions are generally used to survey large numbers of faint stars or galaxies quickly.

Distance

Like an extending ladder, there is, in astronomy, a series of distance scales that are calibrated one against the other in the overlap regions. Each scale in the series progressively probes into more distant regions.

The Earth

The distance scale starts on the Earth, at locations that are mapped by surveyors and geographers. The first successful attempt to measure the Earth on a global scale was by ERATOSTHENES OF CYRENE. His value of the circumference of the Earth was apparently within 8% of the modern value. The shape of the whole Earth is now measured to centimeters by Earth-observation satellites. The location of specific geodetic markers has been measured to millimeters by ranging from them to reflective satellites with lasers, like radar, and by measuring the time it takes for radio signals to reach the markers from the GPS navigational satellites.

The solar system

From terrestrial baselines astronomers extend the distance scale into the solar system. If the distance to one planet is determined (say the Sun-Earth distance, known as the ASTRONOMICAL UNIT, AU), the distance to the others follows from theoretical considerations. Johannes KEPLER himself identified the possibility of using his third law this way (*see* CELESTIAL MECHANICS). Setting a precedent for projects of international cooperation in astronomy, the AU was determined to an accuracy of 10% by timing the transits of Venus in 1761 and 1769 as seen from different locations across the Earth. In effect, the time of entry and egress of Venus onto and off of the Sun's disk, as seen from various places, measures the angle subtended by the Earth from the LIMB of the Sun, and hence the AU. Simon NEWCOMB improved the accuracy to about 1% in his analysis of the measurements of the transits of Venus in 1874 and 1882.

It is difficult to triangulate on a large planet seen against the bright Sun. Johann GALLE suggested that MINOR PLANETS might be better, as they showed no visible disk. In 1900 a newly discovered minor planet, (433) EROS, came to OPPOSITION from Earth at a distance of only 0.27 AU, and direct triangulation on it then and at its subsequent close opposition in 1930 set the AU near 140 million km. Measurements by radar of the distances to nearby asteroids and planets now provide the accurate scale of the solar system in terms of light seconds: 499.00578 light seconds for the AU, which is equivalent to 149 597 870 km.

The stars

Given the scale of the solar system, the motion of the Earth around the Sun can be used as a baseline to triangulate on the nearest stars and determine a star's PARALLAX, the angle that the AU subtends at a star (*see* ASTROMETRY). For convenience, the astronomer's unit of distance, the parsec, is defined as the distance at which a star's parallax would be one second of arc (arcsec), the unit in which parallaxes are traditionally measured. One parsec equals 3.2616 l.y. (light years), where the light year is the distance that light travels in a year.

The nearest known star, PROXIMA CENTAURI (α Cen C), has a parallax of 0.77 arcsec. A typical naked-eye star's parallax is about 0.01 arcsec (10 milliarcsec). In the 1990s, astrometry was revolutionized by ESA's HIPPARCOS satellite, which measured parallaxes for 118 000 stars to 1 milliarcsec. Scanning the sky for just over three years, Hipparcos observed 15 times more stars than had been observed in the previous 100 years. Some 66 000 of the fainter stars were selected for astrophysical or astrometric interest. Additionally, the satellite's star mapper measured parallaxes of modest precision for 1.06 million stars.

Trigonometric parallaxes calibrate the brightnesses, or luminosities, of different types of nearby stars, which makes it possible to extend the distance scale from nearby stars to the rest of the Galaxy and to nearby galaxies.

The sequence for doing this is:

- identify intrinsically bright stars of a certain class (for example VARIABLE STARS, or stars with a distinctive SPECTRUM);
- find some nearby examples and measure their parallax;
- measure how bright the stars appear and use the distance measurements to calibrate their absolute brightness;
- correct for interstellar absorption;
- if the identified stars are found to be STANDARD CANDLES (all stars of a given type having the same brightness), use similar stars in more distant places (in star clusters or galaxies) to determine their distance.

In a textbook example of exploiting data from the Hipparcos satellite, the parallaxes of nearby field stars were used to calibrate the brightnesses of metal-poor stars, which were compared with similar stars in galactic GLOBULAR CLUSTERS to derive their distances.

Hipparcos strained to measure the distances to some stars that are used as important standard candles (such as RR LYRAE STARS and the Cepheids), but unfortunately these stars are rare in the solar neighborhood. For example, RR Lyrae is the only star of that type in the *Hipparcos Catalogue* that has a measurable parallax. Hipparcos observed more than 200 CEPHEID VARIABLE STARS, but at low accuracy. Many of the measurements are individually useless, but they have statistical value. For a set of stars at virtually the same distance, such as in a star cluster, the parallaxes can simply be averaged to obtain the cluster's parallax. Hipparcos observed stars in more than 200 OPEN CLUSTERS. The distance (d) of the Hyades star cluster was determined to better than 1% precision ($d = 46.3 \pm 0.3$ parsec), with 18 other clusters well measured. If a cluster (or a galaxy) contains examples of two sorts of standard candle it provides a cross-check of the consistency.

Future space missions (Gaia and the SPACE INTERFEROMETRY MISSION) are expected to greatly improve on Hipparcos's accuracy and scope.

Cepheid period–luminosity law

The distances to individual Cepheid variable stars, to star clusters containing Cepheids, and to the LARGE and SMALL MAGELLANIC CLOUDS, determined the absolute brightness of the Cepheid variables. Cepheid variable stars are not immediately standard candles, but there is a correlation (discovered by Henrietta LEAVITT) between the luminosity of a Cepheid variable and its period. Cepheids are good standard candles because:

- they are bright and can be seen from far away;
- they are readily identifiable because of their characteristic variability;
- their brightness can be readily deduced from their period.

The Cepheid distance scale links our Galaxy and the Magellanic Clouds with all the galaxies in which Cepheids can be identified.

The HST was, in part, designed to identify Cepheids in order to measure the distances of galaxies as far out as the Virgo cluster of galaxies. This extends the Cepheid distance scale by some 10 or 20 times. The HST found extragalactic Cepheids, determined their magnitudes and periods, and measured the distances to more than 24 galaxies, out to the Virgo and Fornax clusters. Galaxies in which Type Ia SUPERNOVAE had occurred were calibrated as 'standard bombs' (*see* Supernovae on p. 122).

The Tully–Fisher relation

The Cepheid variable stars are crucial for establishing the distances of galaxies within 20 megaparsecs. Beyond this, however, Cepheids are too faint, even for the HST. Astronomers now use the TULLY–FISHER RELATION, which relates the rotational velocity of a SPIRAL GALAXY to its luminosity. Spiral galaxies contain both Cepheid variables and hydrogen gas (used to measure the velocity). This makes it relatively easy to calibrate the Tully–Fisher relation using nearby systems.

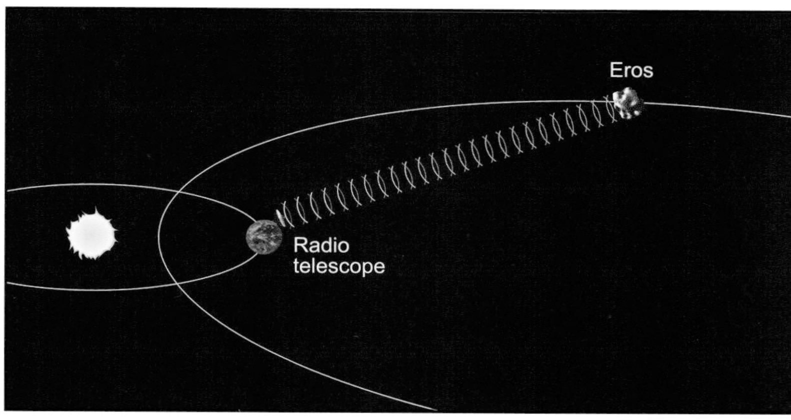

▲ ***Radio ranging*** *Radio telescopes are used as radar transmitters and receivers to measure the distances to asteroids that pass close to Earth. Measuring the orbits of the asteroids gives the scale of the solar system.*

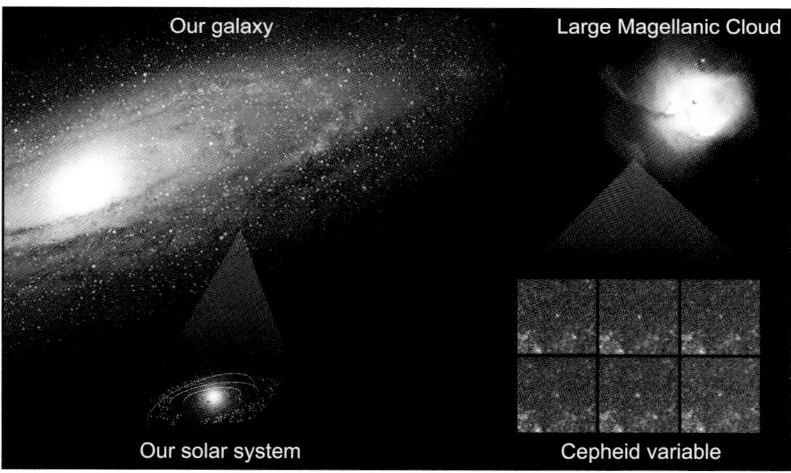

▼ ***Cepheid variable stars*** *From the solar system, the distance scale can be extended into the Galaxy by triangulation on stars. Cepheid variable stars (an example followed through its light cycle, lower right, at the center of each square picture) provide the link between our Galaxy and others nearby.*

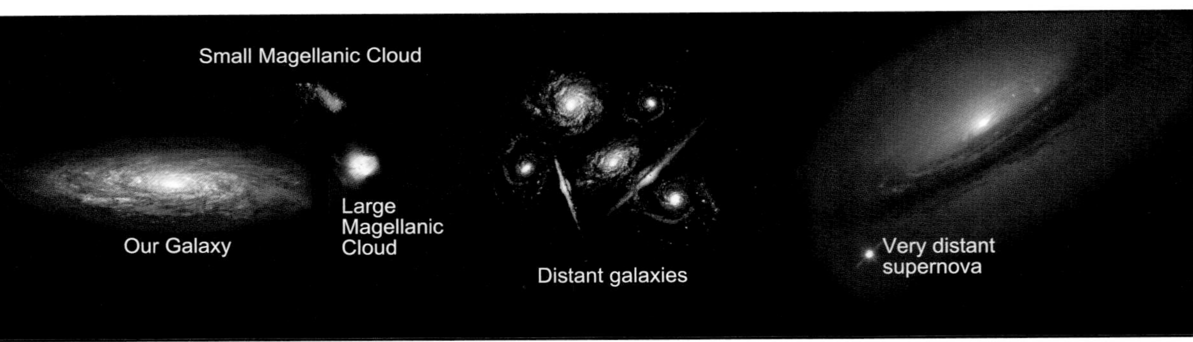

▶ ***Tully–Fisher relation*** *This extends the cosmic-distance scale into the universe of galaxies. Supernovae provide a check on the Tully–Fisher relation and provide the distances of the furthest galaxies.*

The Tully–Fisher relation is used like this:
- measure the speed of rotation of the hydrogen gas in a galaxy with a radio telescope;
- use the calibration to determine the galaxy's luminosity;
- measure the galaxy's apparent brightness;
- correct somehow for interstellar absorption both in our Galaxy and in the galaxy being measured;
- compare the apparent brightness of the galaxy with its luminosity, and deduce its distance by using the inverse square law.

Supernovae

Supernovae extend the range of the Cepheid period-luminosity law and the Tully–Fisher relation to the most distant parts of the universe. Supernovae of Type Ia are explosions of WHITE DWARFS that are probably accreting in BINARY STAR systems. Type Ia supernovae are more or less standard candles, or 'standard bombs,' presumably because white dwarfs are stars that are built, more or less, to a standard pattern. There are some significant variations in maximum brightness from supernova to supernova, but they correlate with the rapidity of the decline of the LIGHT CURVE from maximum, and can be taken into account. Supernovae are very bright, so they have made it possible to probe the distant universe and measure its acceleration, the density of DARK ENERGY AND THE COSMOLOGICAL CONSTANT.

Over the universe at large, the astronomical distance scale has been applied to determine the HUBBLE CONSTANT of the rate of expansion of the universe to 10% or so. Conversely, if the recession speed of a galaxy can be measured, its distance follows by application of the Hubble constant. The labor and painstaking detail that has gone into establishing a consistent distance scale, focused through the operation of the HST, is a heroic achievement of science.

Further reading: Binney J and Merrifield M (1998) *Galactic Astronomy* Princeton University Press; Giovanelli R (1997) *The Extragalactic Distance Scale* Cambridge University Press; Van Helden A (1985) *Measuring the Universe: Cosmic Dimensions from Aristarchus to Halley* University of Chicago Press.

Distance modulus

A star's absolute magnitude, M, may be calculated from its apparent magnitude, m, and distance, d, in parsecs as follows: $M = m + 5 - 5 \log d$.

The quantity $m - M$ is therefore a measure of distance and is called the distance modulus. It is used, for example, to measure the distances of stars or galaxies. If there is interstellar absorption, this dims a star, increasing its apparent magnitude. For the distance modulus to be effective in measuring distance, the magnitude of the star must be corrected by the value of interstellar absorption, A: $m - A - M = -5 + 5 \log d$.

Diurnal motion

The apparent daily (diurnal) motion of a celestial body across the sky from east to west, caused by the axial rotation of the Earth. Diurnal motion causes stars to appear to trace out circles centered on the celestial poles. For an observer at the north or south pole of the Earth, the diurnal motion of stars is along circles parallel to the horizon, while, viewed from the equator, stars trace out paths that intersect the horizon at right angles. In general, the arcs along which stars appear to move intersect the horizon at an angle equal to 90° minus the observer's latitude.

Dobsonian telescope

A simple, inexpensive and portable form of NEWTONIAN TELESCOPE that has an ALTAZIMUTH MOUNTING. It was invented by Californian amateur astronomer John Dobson (1915–) in 1956, and is also called a sidewalk telescope (since it is used to show the wonders of the universe to passersby). A well-balanced Dobsonian telescope can be moved quickly and smoothly by hand to point toward any part of the sky. It is attached to a fork mounting that rotates in ALTITUDE (elevation) and AZIMUTH (parallel to the horizon), and has Teflon strips as the bearing surfaces. Because of its simple mode of construction, aperture for aperture the Dobsonian is significantly cheaper to produce than other designs. Consequently, amateur astronomers can acquire instruments of relatively large aperture and light-gathering power for a given budget.

Dog Star *See* SIRIUS

Dollond, John (1706–61)

An English instrument maker, he was the first to manufacture achromatic lenses commercially, starting in 1758 (*see* CHROMATIC ABERRATION). The achromat consisted of two lenses made of different types of glass. One type, crown glass, was relatively common, but the other, flint, was produced in small quantities. English glass makers gave Dollond first choice, and with the protection of a patent he exploited his position and gave British astronomers access to excellent instruments.

Dome

(1) A circular, shallow-sided hill on the surface of a planetary body. Lunar domes are found in the maria (*see* MARE), and typically measure 10–15 km across, with gradients of no more than a few degrees. They are volcanic in origin, and some have a small pit at their center. They may mark the

tops of the volcanic vents through which the lava that formed the maria emerged.

(2) Popular name for the housing of a telescope which may be dome-shaped, but not necessarily so. A movable aperture in the roof allows access to the open sky.

Dominion Astrophysical Observatory

The Dominion Astrophysical Observatory is located on Little Saanich Mountain near Victoria, British Columbia, Canada. It began operating its 1.8-m telescope in 1918, added a 1.2-m telescope in 1962 and joined the National Research Council of Canada in 1970. The site now contains the headquarters of the Herzberg Institute of Astrophysics, the institute's administration, groups supporting optical astronomy and submillimeter astronomy, instrumentation and the Canadian Astronomy Data Center.

Dominion Radio Astrophysical Observatory

The observatory began operating in 1959 and joined the National Research Council of Canada in 1970. It became part of the Herzberg Institute of Astrophysics in 1975. The site, near Penticton, British Columbia, Canada, has a 26-m RADIO TELESCOPE; a seven-antenna synthesis telescope on a 600-m baseline (*see* APERTURE SYNTHESIS); and two telescopes that monitor the solar radio flux at 10.7 cm. This part of the institute also supports the development of correlators and the design of future radio telescopes.

Doppler, Johann Christian Andreas (1803–53)

Born in Austria, Doppler studied and taught mathematics in Vienna and later Prague. Despite a heavy teaching schedule, he was able to carry out some research of his own. In 1842 he read a paper to the Royal Bohemian Society 'On the colored light of the double stars and certain other stars of the heavens' in which he articulated the DOPPLER EFFECT. This relates the frequency (or wavelength) of a source to its velocity relative to an observer. He foresaw that the effect would 'in the not too distant future offer astronomers a welcome means to determine the movements and distances of distant stars.'

Doppler effect

The change in the observed frequency and wavelength of a source of radiation, resulting from its motion toward, or away from, the observer. If a source is approaching, the frequency of the radiation is increased (more wave crests per second reach the observer than would be the case if the source were stationary relative to that observer) and its observed wavelength decreased. If the source is receding, the frequency is decreased and the wavelength is increased.

Johann DOPPLER suggested the principle in 1842. In 1848 the French physicist Hippolyte Fizeau (1819–96) showed that the wavelengths of spectral lines would be affected in the same way. Consequently, lines in the SPECTRUM of a receding source would be displaced toward the longwave (red) end of the spectrum, and those in the spectrum of an approaching source would be displaced towards the shortwave (blue) end. The term REDSHIFT is applied to the former case and blueshift to the latter.

The observed wavelength, λ', differs from the rest-wavelength (the wavelength measured in a frame of reference in which the source is stationary), λ, by an amount $\Delta\lambda$, which depends upon the radial velocity (speed directly toward or

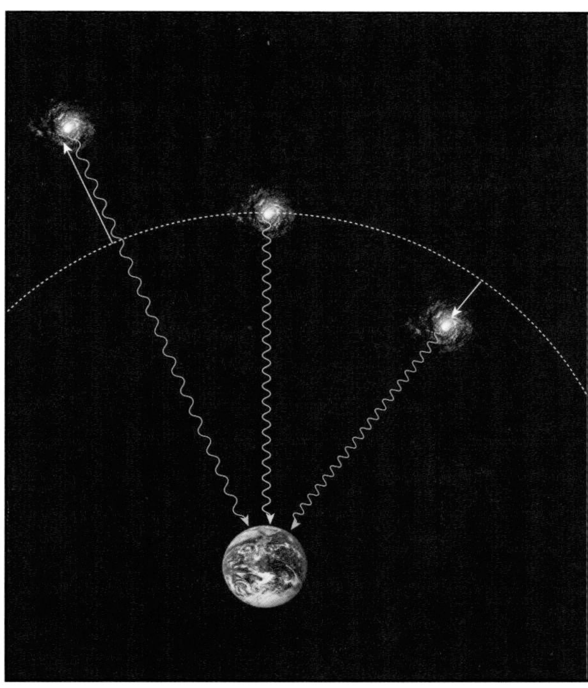

◀ ***Doppler effect*** *The wavelength of a galaxy moving away from the Earth increases and its frequency decreases (left), compared to a stationary galaxy (center). The wavelength of a galaxy moving toward us decreases and its frequency increases (right). Essentially, a given number of waves emitted in a given time are fitted into a greater or lesser distance, respectively.*

away from the observer) of the source. For velocities, v, that are small compared with the speed of light, c, the wavelength shift is given by: $\Delta\lambda/\lambda = (\lambda' - \lambda)/\lambda = v/c$. If the velocity of the source relative to the observer is an appreciable fraction of the speed of light, the following relativistic expression must be used: $\Delta\lambda/\lambda = ((c + v)/(c + v))^{1/2} - 1$.

The velocity of a light source along the line of sight (the so-called 'radial velocity') may be determined by comparing the observed wavelengths of its spectral lines with the wavelengths that those lines would have if the source were stationary. The measured velocity, v, is negative if the source is approaching the observer, or positive if it is receding. The Doppler effect, therefore, provides the means by which the radial and rotational velocities of stars and galaxies may be determined.

Dorado *See* CONSTELLATION

Double Cluster

A close pair of clusters in the constellation Perseus. They are known as h and Chi Persei (NGC 869 and NGC 884). The clusters were observed in ancient times and cataloged by HIPPARCHUS. The distance of both is about 7500 l.y.

Double Double *See* EPSILON LYRAE

Double quasar

The double QUASAR Q0957+561 is in fact two images of the same distant object, produced by an intervening gravitational lens. The images are separated by 6 arcsec and are observed in the optical, radio and X-ray bands; their SPECTRA are almost identical. Identical variability of the sources confirms that the two images are of the same object. A difference in light travel time produces a time delay of about 1.5 years between changes observed in the two images. Q0957+561 provided the first confirmed demonstration of gravitational lensing in action. *See also* GRAVITATION, GRAVITATIONAL LENSING AND GRAVITATIONAL WAVES.

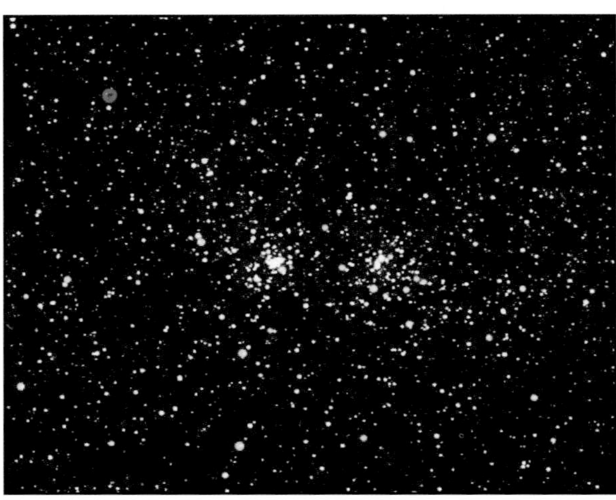

▶ ***Double cluster*** *These clusters are separated by a few hundred light years, and lie at a distance of over 7000 l.y. They are both quite young: h (left) is 5.6 million years old, and Chi is 3.2 million years old. h is approaching the Earth at 22 km s^{-1} and Chi at 1 km s^{-1}.*

Double stars

Wide BINARY STARS that may be seen individually are called 'double stars.' Optical double stars are stars at different distances from us, seen in the same direction but unconnected. Intrinsic double stars are pairs of stars that orbit periodically around their common center of mass. They interact by gravitation. (Close binaries also interact non-gravitationally. *See* VARIABLE STAR: CLOSE BINARY STAR.)

Double stars resolvable with the naked eye, like MIZAR (the middle star in the handle of the Big Dipper ASTERISM) and Alcor, were known to the ancients. In 1767 John MICHELL suggested that on statistical grounds many closer double stars must be physically connected, because there were too many to be chance alignments. This was confirmed in 1804 by William Herschel (*see* HERSCHEL FAMILY), who found orbital motion in the 5-arcsec pair, CASTOR A and B.

Multiple stars are systems that contain more than two stars (*see* Multiple stellar systems, on this page).

Double and multiple stars are designated with their discoverers' codes; for example, Alpha Ursae Majoris (Dubhe) is named Bu 1077, after its discovery as a double by Sherburne BURNHAM. These codes are unambiguous and memorable, but cumbersome.

The well-used *Aitken Double Star Catalog* (known as the *ADS Catalog*, published by Robert Aitken in 1932) contains double stars that were discovered before 1927 in the sky north of –30° DECLINATION. The master catalog of double stars, which is maintained at the US NAVAL OBSERVATORY and is being revised, has 100 000 entries. About 1000 have had their orbits fully observed, and several hundred have sufficiently accurate orbital elements to determine masses.

Observing techniques

- The FILAR MICROMETER contains two thin wires, one movable, with a precision screw to measure their separation. It is mounted on a rotatable frame at the focus of a high-power eyepiece. The micrometer supplies the position of one star relative to the other in polar coordinates: the position angle (direction in the sky) is counted from N over E, S, W from 0° to 360° by orientation of the frame and the separation.
- The double-image micrometer is a modification of this. It uses a birefringent prism, made of a material that splits the light of the double star into two polarized images that are separated in angle. The four star images produced from a double star are brought into a measurable configuration by rotating and shifting the prism.
- Extensive data has been obtained by photographing wider pairs, typically over 2 arcsec. Such work can now be done with digitizing cameras (CCDs).
- INTERFEROMETRY is used rarely, owing to high instrumental demands. A successful variant is the eyepiece interferometer, constructed and widely used by William Finsen (1905–79).
- Speckle interferometry results in superior precision by correlating many short-exposure images (speckles). These are recorded by a CCD camera with an image intensifier as detector, and fed to computers for fast processing.
- The HIPPARCOS satellite measured numerous double stars and added some 3000 new pairs, mostly with large differences in brightness whose detection from the ground was impeded.

The apparent orbit of a visual binary in the sky is a projection of the true orbit ELLIPSE. Seven constants, the so-called ORBITAL ELEMENTS, describe the motion. Three of them are angles specifying the projection. The other elements fix the motion within the orbit, as with planets. It is necessary to know the distance to get the masses.

Multiple stellar systems

Multiple stars include triple, quadruple and larger stellar systems, all the way up to star clusters. There are two categories: hierarchical and non-hierarchical. Hierarchical stellar systems consist of physically separate single or binary stars that orbit each other in approximately Keplerian fashion. The ALPHA CENTAURI system is a hierarchical triple-star system comprising a close binary pair (Alpha Centauri A and B) and a much more distant, and less massive, companion (Alpha Centauri C, or PROXIMA CENTAURI). Non-hierarchical systems, such as star clusters, are best described by statistical quantities rather than by orbital parameters.

About 30% of all binary systems are triple: systems with more component stars are much rarer. Only a few systems with up to six components are known; for example, CASTOR and the TRAPEZIUM.

Formation and evolution

Usually, binary and multiple stars form together out of a single gas cloud (like the Trapezium). Occasionally, however, binary- and multiple-star systems originate in star clusters. Previously single stars may pair off, a star may exchange between a pair and a single star or between two pairs, or two pairs may join together to make a foursome. Other stars close by in the cluster may give a nudge or absorb some of the pairing momentum. Subsequent interactions may exchange some member stars with more massive ones.

The hierarchical triple system containing the pulsar PSR B1620–26 in the GLOBULAR CLUSTER M4 was probably formed by a binary-binary interaction. Continual encounters with stars in M4 are likely eventually to disrupt the system.

Very wide double stars have only a weak gravitational bond that may be disrupted by single encounters with passing stars. The limit of SEPARATIONS above which a pair is expected to become unbound at some stage is 10 000–20 000 AU, or 0.05–0.1 parsecs, corresponding to orbital periods of millions of years.

Wide, non-hierarchical stellar systems like the Trapezium are unstable and therefore young and rare. They reassemble into stable hierarchical configurations, often ejecting some lower-mass stars in the process.

Further reading: Batten H (1973) *Binary and Multiple Systems of Stars* Pergamon; Eggleton P P and Pringle J E (eds) (1983) *Interacting Binaries* Kluwer; Kopal Z (1959) *Close*

Binary Systems Wiley; Sahade J, McCluskey G and Kondo Y (eds) (1993) *The Realm of Interacting Binaries* Kluwer.

Double stars, observing

See practical astronomy feature overleaf.

DQ Herculis

The old NOVA DQ Herculis (Nova Herculis 1934) is the prototype of a subclass of cataclysmic VARIABLE STARS, containing an accreting, magnetic, rapidly rotating WHITE DWARF. DQ Herculis (DQ Her) stars are characterized by strong X-ray emission, high excitation spectra, and very stable optical and X-ray pulsations in their LIGHT CURVES. They are similar to the AM HERCULIS stars, although DQ Her stars are also characterized by spin-orbit synchronism and the presence of strong circular POLARIZATION. At the time of writing there are about 20 DQ Her stars, also known as intermediate polars (having intermediate polarization).

DQ Her itself is an eclipsing binary system (*see* BINARY STAR) with a period of orbit of 4.56 days. In the late 1950s DQ Her held the record for the shortest photometric period (the 71-s white dwarf spin period) in an astronomical source. The variability has an observed amplitude of roughly 1%. What remains remarkable today is the system's high INCLINATION of approximately 89.5°. The white dwarf primary is eclipsed by an edge-on ACCRETION disk, and only a thin sliver of the back face of the accretion disk contributes to the pulsed radiation received from DQ Her. Variations in the phase of the 71-s signal during eclipse ingress and egress, first detected in 1972, led to the rotating magnetic accretor model for DQ Her and related systems.

Draco *See* CONSTELLATION

Drake, Frank Donald (1930–)

American astronomer whose principal activities have been to search for life elsewhere in the universe (*see* EXOBIOLOGY AND SETI). In 1958 at the newly founded NATIONAL RADIO ASTRONOMY OBSERVATORY in Green Bank, West Virginia, he attempted the first search for extraterrestrial broadcasts, with a two-week observation of the stars Tau Ceti and Epsilon Eridani. He invented the Drake equation – a way to estimate how many intelligent, communicating civilizations there are in our Galaxy (N_i). The Drake equation multiplies the number of stars in the Milky Way (N) by the fraction that have orbiting planets (f_p), then by the number of planets per star that are capable of sustaining life (N_p), then by the fraction of these planets where life actually evolves (f_L), then by the fraction of these where the life is intelligent and communicates (f_i), and then by the fraction of the planet's life during which the communicating civilizations live (f_t). $N_i = N.f_p.N_p.f_L.f_i.f_t$. The real value of the equation is not the answer itself, but the questions that are prompted when attempting to come up with an answer.

Draper, Henry (1837–82)

American pioneer of astronomical photography, born in Prince Edward County, Virginia. He made the first photograph of a stellar SPECTRUM (of VEGA). His assistant, Williamina FLEMING, analyzed the objective prism photographs of stellar spectra and formed the basis for the classification of stars. The data was eventually published by Annie CANNON as the *Henry Draper Catalogue of Stellar Spectra. See also* HENRY DRAPER CATALOG; STELLAR SPECTRUM: CLASSIFICATION.

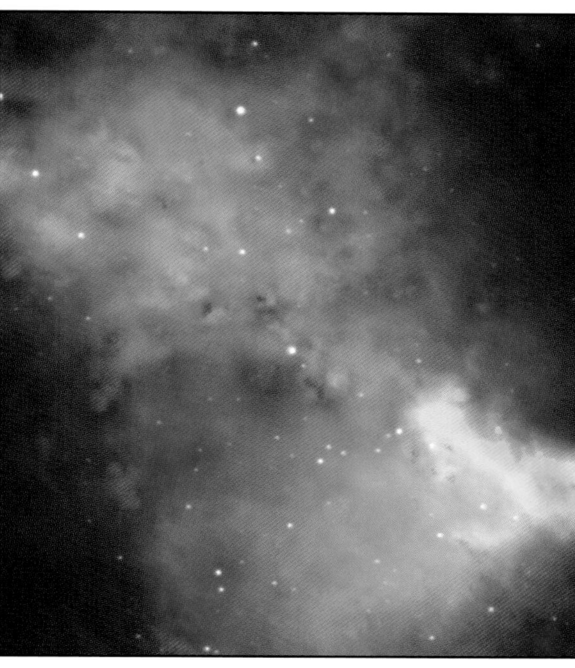

◀ **Dumbbell Nebula** *(M27) So called because of the two lobes of gas ejected from the hot central star.*

Dreyer, John Louis Emil (1852–1926)

Danish astronomer who, after working at Birr Castle in Ireland, became director of the ARMAGH OBSERVATORY. There he studied the many nebulae he had previously observed at Birr. At a time when it was not clear whether spiral nebulae were inside our own Galaxy (the Milky Way) or were, as was suspected, island universes outside, he established through careful measurements that they had no proper motion and were likely to be distant (*see* ASTROMETRY). Dreyer compiled the NEW GENERAL CATALOGUE OF NEBULAE AND CLUSTERS OF STARS.

Dudley Observatory

The Dudley Observatory, in Schenectady, New York, USA, is a private foundation that supports research and education in astronomy, astrophysics and the history of astronomy. Chartered in 1852, it is the oldest organization in the USA, outside academia and government, that is dedicated to the support of astronomical research. For more than a century it was a world leader in ASTROMETRY, with such achievements as the publication in 1937 of the *General Catalogue of 33,342 Stars for the Epoch 1950.*

Dumbbell Nebula (M27, NGC 6853)

The first PLANETARY NEBULA to be discovered and entered into the MESSIER CATALOG by Charles Messier in 1764. It is located in the constellation Vulpecula, measures 8 × 4 arcmin, and is of magnitude 8. The bipolar expulsion of material from the magnitude 13 central star gives it its twin-lobed appearance. It was named by Thomas Webb, though it resembles a bow tie more than a dumbbell.

Dunsink Observatory

Dunsink Observatory (1783) was designed by Henry Ussher (1741–90), the first astronomy professor of Trinity College, Dublin, Ireland. He incorporated modern designs, such as ventilation in the observation room and a free-standing telescope support column. Among the past directors were the mathematician William Hamilton, from 1827 to 1865,

Observing double stars

Measurements of the POSITION ANGLE (PA) and SEPARATION with time of visual DOUBLE STARS allow the apparent orbit of the two stars to be calculated. With this information, the total mass of the two stars can be derived. It is still one of the best ways of calculating stellar masses, but the long period of many visual binary systems (typically a few hundred years) means that the number of reliable orbits is still very small (currently about 50).

Amateur observers have a long and distinguished record in the annals of double-star observing. Sherburne BURNHAM started out with a small REFRACTING TELESCOPE and later made it his life's work, holding down a day job as a clerk in a Chicago court and on weekends observing at YERKES OBSERVATORY. He later became a professional. In England, Thomas Espin (1858–1934) discovered several thousand faint and wide pairs with his REFLECTING TELESCOPES and significant contributions were also made by William DAWES and Theodore Phillips (1868–1942) (all three, incidentally, clergymen). The greatest amateur was the Frenchman Paul Baize (1901–95), by day a pediatrician, who made 24 000 measurements and computed nearly 500 orbits. He is tenth on the all-time list of observers.

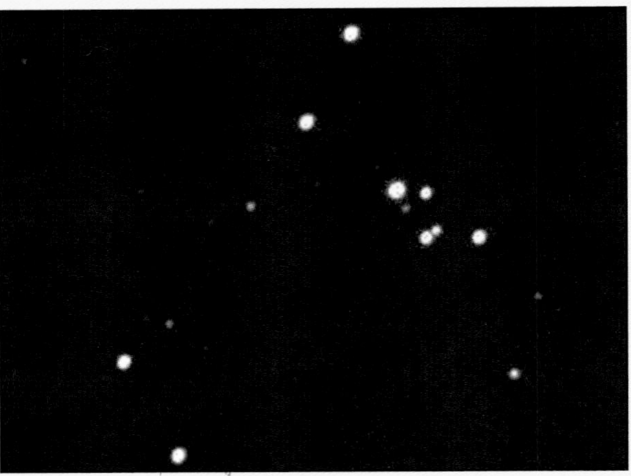

▶ ***Multiple star HJ 684*** *The bright close pair center right is CD - magnitudes 10.3 and 12.5, PA is 308.72° and separation 10.26 arcsec. The field size is 10 × 6 arcmin. In this image, north is upward and east is to the left.*

Equipment

The FILAR MICROMETER, invented in the 1640s and still in use today, is one of the chosen instruments for those who want to push their telescopes to the limit of resolution. The human eye is still the best detector for pairs that are close or unequal, and a good 35-cm telescope will allow pairs as close as 0.25 arcsec to be measured. Another effective instrument is the double-image (or Lyot) micrometer. The double-image is better than the filar for small separations, but cannot match the filar in the range of separations that it can handle.

While many bright pairs have orbits that require little further adjustment, there are a large number of fainter pairs that require attention. In addition, the HIPPARCOS satellite has discovered thousands of new pairs, most of which have not yet been followed from the ground; amateurs can contribute by helping to fill in these gaps in our knowledge.

While the filar micrometer has been the mainstay of this work for centuries, the advent of CCD cameras means that there is an opportunity for amateurs who are equipped with small to medium apertures to make substantial contributions. In particular, there are many thousands of faint pairs on the various Schmidt sky surveys that have never been recorded and measured. Using a 12-in SCHMIDT–CASSEGRAIN TELESCOPE and an SBIG CCD for instance, Martin Nicholson, working near Northampton, UK, has found many new pairs using the *Guide Star Catalogue*. He routinely images pairs as close as 5 arcsec and as faint as magnitude 15 with his 12-in Meade LX200 and SBIG ST7-E CCD. Reduction of the CCD frames

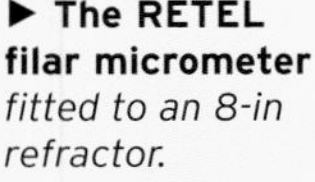

▶ **The RETEL filar micrometer** *fitted to an 8-in refractor.*

is made using the IRAF (Image Reduction and Analysis Facility) suite of image-processing routines.

Another advantage of the CCD camera is that, used with the appropriate filters, it can be used to determine the colors of each component of resolved double stars. Many pairs in the *Washington Double Star (WDS) Catalog* have not been observed photometrically, and much remains to be done here.

Contributing to science

In the field of double-star observing, there is much that modern amateurs can do. They can continue to monitor the longer-period pairs and lay down measures for future orbital determination. The United States Naval Observatory (USNO) website points to all the data needed. The WDS catalog is as inclusive as its maintainers can make it (it currently contains over 99 000 entries) and the *Sixth Orbital Catalog* lists all the orbits deemed worthy of printing. The quality of each orbit has been determined – grade 1 is definitive and grade 5 is indeterminate. Any pairs with orbits of grade 2 or lower could usefully be observed.

Amateurs with a large enough aperture (30 cm or more) can examine and measure the new double stars in the HIPPARCOS or TYCHO STAR CATALOGS. All are listed in the WDS catalog, available both online and as a CD-ROM which can be requested free of charge from USNO.

In addition, the power of the modern Schmidt surveys can be utilized to measure and catalog many faint pairs that have been ignored until now.

Finally, a CCD could be used to great effect on a photometric program of double stars. Many pairs have the magnitudes assigned to them by their discoverers – usually eye estimates. Little is known about the colors of these stars or how many are VARIABLE STARS. With a double star, there is a built-in comparison for PHOTOMETRY.

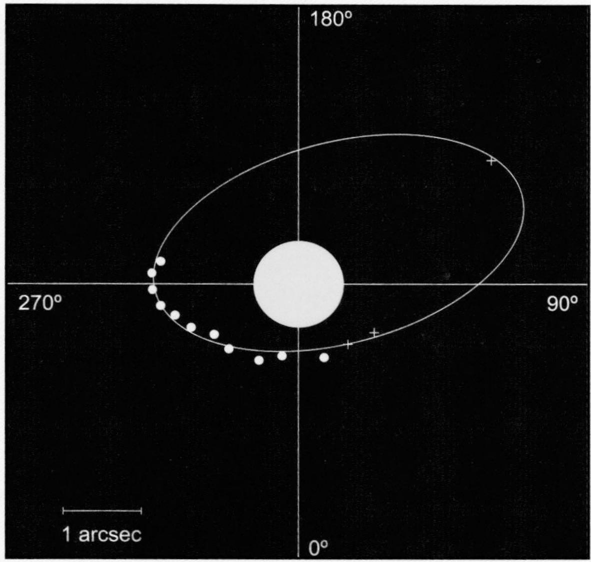

◀ ***Xi UMa,*** *observed by Bob Argyle between 1970 and 2002. The solid line shows the current orbit for this pair; the motion is clockwise. The solid dots represent means of 5 or more nights, the crosses 1-5 nights and the solitary single cross is an observation made in 1970. The radius of the central circle shows the angular resolution of the telescope - in this case 0.56 arcsec.*

Those intending to make a serious contribution must consider how they wish to publish their measurements. Their work can be submitted to professional journals such as *Astronomy & Astrophysics*, or one can choose to publish through an amateur group. The WEBB SOCIETY encourages observation and publishes measures on a regular basis. The Iberoamerican League for Astronomy (LIADA) and the Double Star Section of the long-established French group, the Société Astronomique de France, do the same.

Bob Argyle works at the Institute of Astronomy, Cambridge, UK, and uses the 8-in refractor there for regular double-star measurements.

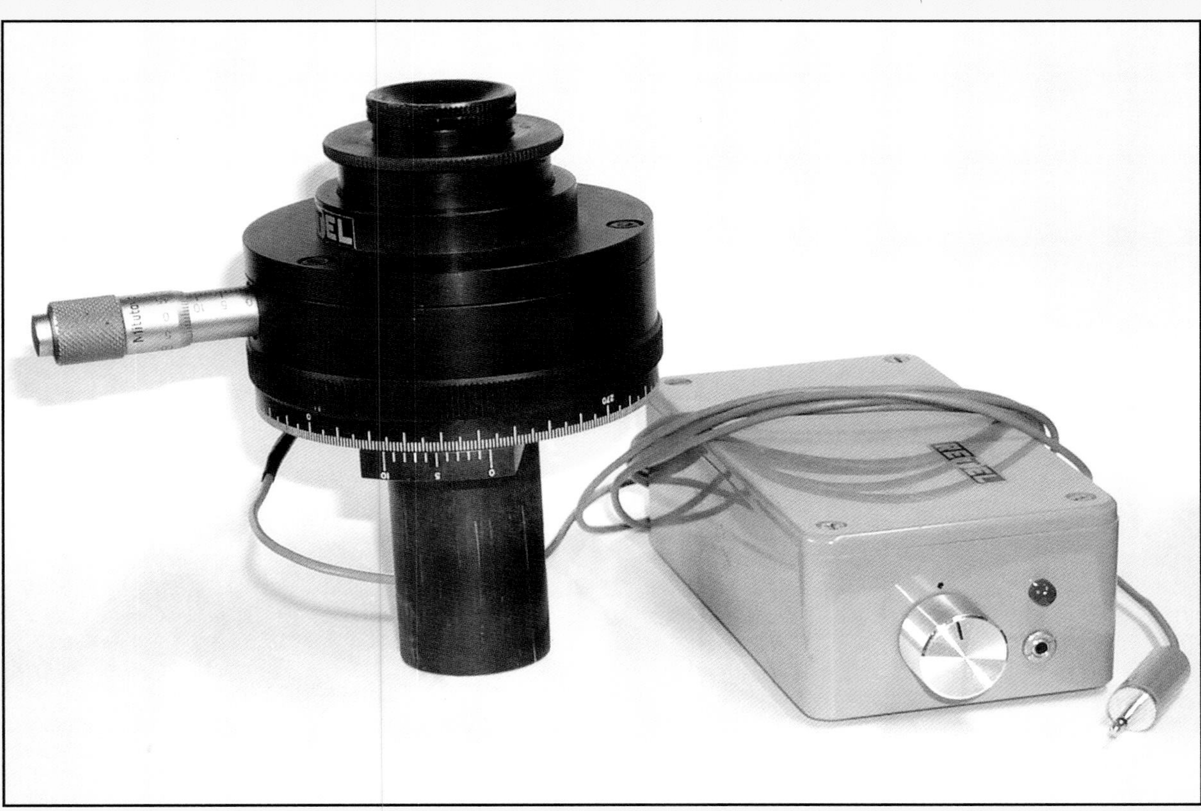

◀ ***The RETEL filar micrometer*** *uses parallel wires (one movable and one fixed) to measure separation, and a third wire at right angles to measure position angle. The whole field is illuminated by means of a variable-intensity red LED.*

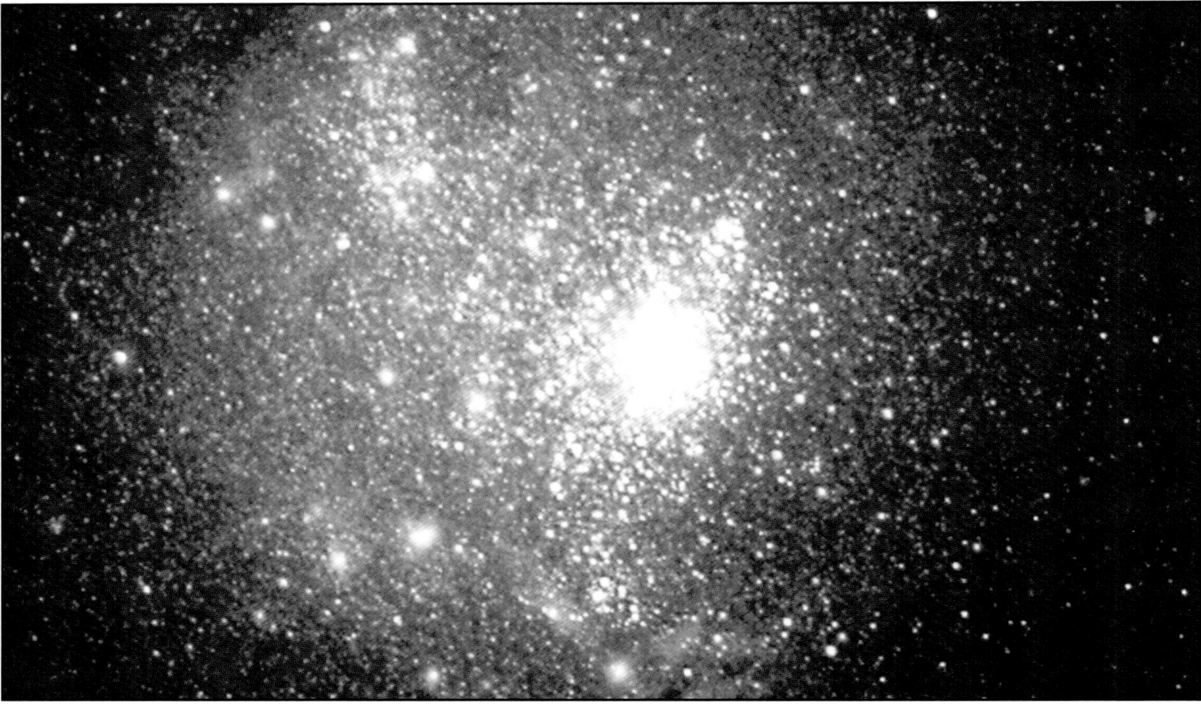

▶ ***Dwarf galaxy*** *The central region of the dwarf irregular galaxy NGC 1705 glimmers with the light of 'only' a few million stars.*

and stellar dynamicist Henry Plummer, from 1912 to 1921. The principal remaining instrument is a Grubb 12-inch REFRACTOR. The observatory has been part of the Dublin Institute for Advanced Studies since 1947, and now concentrates on high-energy studies of extragalactic objects.

Dust tail

The part of a COMET that consists of dust grains carried away from the nucleus by the outflow of gas from the COMA during a close approach to the Sun, and propelled away from the comet's head by radiation pressure. Not all comets develop a dust or GAS TAIL, but both types of tail always point away from the Sun. The combined effects of radiation pressure and the Sun's gravity acting on the dust particles make dust tails curved (the POYNTING–ROBERTSON EFFECT), in contrast to gas tails, which are straight. Dust tails shine only by reflected sunlight, so their color is yellow. Their length is typically 10^6–10^7 km, and like gas tails they begin to form at around 2 AU from the Sun. Most of the particles in a dust tail are less than 1 μm across. Larger particles remain in the comet's orbit and form METEOR STREAMS.

Dwarf galaxy

A small GALAXY with very low mass, LUMINOSITY and surface BRIGHTNESS. There are several classes of dwarf galaxy, but the differences are often a question of semantics and have not been precisely defined. Dwarf galaxies may be elliptical or irregular in shape, and both types are usually sparsely populated with stars. Because of their low surface brightness and low overall luminosity they are hard to detect. It is usually possible to look straight through a dwarf galaxy.

Dwarf ellipticals tend to be populated by old stars and contain little, if any, gas, whereas dwarf irregulars contain younger stars and substantial quantities of gas. Dwarf elliptical galaxies typically contain from a few hundred thousand to a few million stars, have diameters of a few thousand light years, and luminosities ranging from about 10^5 to about 10^7 solar luminosities. They are probably the most abundant species of galaxy in the universe. Dwarf elliptical galaxies are denoted by 'dE' in the Hubble Classification Scheme but do not fit onto the HUBBLE TUNING-FORK DIAGRAM. Some authors include dwarf spheroidal galaxies (dSph) as a subclass of dwarf ellipticals.

Dwarf irregulars (dIrr), which appear to be less abundant than dwarf ellipticals, contain from about 10^6 to 10^8 (or, exceptionally, up to about 10^9) stars. They have one or more compact star-formation regions. There are also a small number of transition objects that have extremely low rates of star formation but evidence of more active star formation in the recent past.

The existence of transition objects leads to the hypothesis that dwarf galaxies may be the same type of objects viewed at different evolutionary phases. Dwarf galaxies all appear to have undergone several discrete episodes of star formation. The most luminous dwarf galaxies are also on average the most metal rich, whether elliptical, spheroidal or irregular; conversely the lowest luminosity galaxies tend to be metal poor, thus representing a sample of galaxies that is still composed of primordial matter. In astronomy, 'metals' include all elements with an atomic number greater than two, that is, heavier than hydrogen or helium. Among the 41 galaxies in the LOCAL GROUP OF GALAXIES there are 29 dwarf galaxies of varying types.

Dwarf nova

Also known as a U GEM STAR, after the first to be discovered. These are a class of cataclysmic variable stars (*see* CATACLYSMIC BINARY STAR; NOVA) in which the primary is a WHITE DWARF and the secondary is a cooler MAIN SEQUENCE star of spectral type G or K. The components are of similar mass (about 0.7 to 1.2 solar masses). The secondary star undergoes irregular expansion, filling its ROCHE LOBE. Hydrogen-rich gas streams from the secondary into an ACCRETION disk around the white dwarf. The gas then spirals down onto the surface of the

◀ Dwingeloo Observatory *The official opening in 1956 of the 25-m Dwingeloo Telescope by Queen Juliana of the Netherlands. Dutch astronomer Jan Oort is by her side.*

white dwarf, causing irregular explosions. This results in an increase in brightness of 2–5 magnitudes. *See also* VARIABLE STAR: CLOSE BINARY STAR.

Dwarf star

In general a small star (like a WHITE DWARF), but more particularly a star on the MAIN SEQUENCE in the HERTZSPRUNG–RUSSELL DIAGRAM and therefore in the hydrogen-burning phase of its evolution.

Dwingeloo Observatory

The 25-m telescope at Dwingeloo, the Netherlands, was for a time the largest telescope in the world. Inaugurated in 1956, for many years the Dwingeloo Telescope was the bread-and-butter instrument of the Dutch radio astronomy community. Its main operating frequency band around 1400 MHz allowed studies of the distribution of neutral HYDROGEN in our Galaxy and other nearby galaxies. The all-sky survey of the 21-cm line radiation of neutral hydrogen, and the discovery of an unknown galaxy behind the plane of the Milky Way, show that the Dwingeloo Telescope was a viable instrument until the end of its active life. The telescope was closed for observations in 1998.

Dyer Observatory

The Arthur J Dyer Observatory is the principal astronomical facility of Vanderbilt University, USA. It is located about 16 km south of Nashville near Brentwood, Tennessee, at an elevation of 345 m above sea level. Built in 1953, the observatory houses the Seyfert 60-cm telescope and the DeWitt 30-cm CASSEGRAIN TELESCOPE, which is equipped for photoelectric photometry and long-focus photography.

Dynamics Explorer

Two NASA satellites, DE-1 and DE-2, both launched in August 1981 in the EXPLORER series to investigate the processes coupling the hot, tenuous PLASMAS of the Earth's MAGNETOSPHERE, and the cooler, denser plasmas and gases of the IONOSPHERE and upper atmosphere. The two satellites were placed in polar orbits, one under the other, permitting simultaneous measurements to be made at high and low altitudes. The lower satellite, DE-2, lasted 18 months, and the higher, DE-1, lasted 10 years. DE-1 is also known as Explorer 62, and DE-2 as Explorer 63.

Dynamo, solar and stellar

Dynamo theory covers the processes that led to the existence of magnetic fields in the Sun and other stars. The simplest hypothesis for the origin of these fields is that they are simply the remnants of magnetic fields carried in by accreting gas during the formation phase of the stars. However, several observational facts make this explanation untenable for stars such as the Sun.

The observed fields are globally relatively weak, but locally very strong (much stronger than would be expected of decaying remnant fields dating back to the formation of the Sun). The global magnetic fields are observed to vary systematically in a roughly periodic fashion (in the Sun's case, this periodicity is about 22 years). These observations argue for the presence of an oscillating dynamo process that generates and maintains the overall magnetic fields of stars like the Sun. The strengths of the magnetic fields are in the range 1–10 Gauss, not much different from the surface magnetic field of the Earth.

Magnetic activity is observed in single stars all the way along the MAIN SEQUENCE, from F-stars to the coolest M-stars, and long-period cycles ranging from 2.5 years to 25 years have been observed in about 20% of those surveyed. However, surveys of low-mass brown dwarfs suggest that some of them may be inactive (*see* EXOPLANET AND BROWN DWARF). Magnetic activity is found exclusively in stars that have envelopes in which there is CONVECTION, while the strength of the activity also depends on the rate of rotation of the star. The most active single stars are young, relatively rapidly rotating stars that have recently arrived on the zero-age MAIN SEQUENCE. Thereafter the star's rotation slows down because of magnetic braking, and the magnetic activity declines. The rotation of the star and circulation of material inside constitute the movement of the 'dynamo.'

A subclass of late B-A main-sequence stars shows the presence of extremely strong surface magnetic fields, with examples reported in excess of 3 T.

The more active stars of spectral type K and M can display substantial variations in luminosity, which have been attributed to the presence of large 'starspots' on their surface (*see* RED DWARF AND FLARE STAR). While sunspots typically occupy areas of 0.001% of the visible solar disk, these starspots can occupy 30% or more of the visible surface of the active star. Theories that explain small-scale sunspots cannot be extended to explain these enormous magnetic features.

Dyson, Frank Watson (1868–1939)

English astronomer, he became ASTRONOMER ROYAL for Scotland and afterwards for England. With Arthur EDDINGTON he directed the 1919 eclipse expedition that confirmed the deflection of starlight by the Sun's gravity, as predicted by Albert Einstein's theory of GENERAL RELATIVITY. Dyson directed measurements of terrestrial magnetism, latitude and time at the ROYAL OBSERVATORY, GREENWICH, and initiated the radio broadcast of TIME.

Eagle Nebula
An EMISSION NEBULA (IC 4703) in the constellation Serpens. It measures about 30 arcmin across and surrounds the OPEN CLUSTER M16 (NGC 6611); it is often given the cluster's designation in the MESSIER CATALOG. The nebula is the site of star formation, especially in huge, twisting columns of dark nebulosity that are nicknamed 'elephant trunks' or the 'Pillars of Creation.'

Early-type star
One of the hotter stars, with spectral type O or B (*see* STELLAR SPECTRUM: CLASSIFICATION).

Earth
The Earth, third planet from the Sun, is the archetype of the TERRESTRIAL PLANETS (Mercury, Venus, Earth and its Moon, and Mars). However, of all the planets in our solar system, the Earth is the only one that has oceans of water and an oxygen-rich atmosphere.

The planet formed about 4.6 billion years ago and is still evolving, as proved by earthquakes and volcanoes. In its first 500 million years, warmth from the Sun and heat from planetary ACCRETION, including the impact of the Mars-sized PLANETESIMAL that formed the Moon, melted most, perhaps all, of the Earth. Dense elements, such as iron, settled down to form the core, while lighter elements rose to the surface. This led to the chemical DIFFERENTIATION of our planet into a series of concentrically nested spheres, the main ones being, from the outside in, the crust, mantle and core.

Crust
The surface of the protoplanet cooled and the crust formed. Continued impact bombardment fractured the crust, however, allowing parts of the still-molten interior to flow to the surface. An atmosphere formed. With time, rainfall led to the accumulation of liquid water on the surface. Tectonic activity and volcanism formed mountains and valleys. Running water, glaciers and weathering eroded the high areas, producing sediments that filled in the low areas. However, much of the Earth's surface results from internal processes, such as CONVECTION within the mantle. Weathering of the crust obliterated most of the CRATERS that were formed early in the planet's history.

There are two main divisions to the Earth's crust. The oceanic crust is young, 8 km thick and basaltic, made of silicate rocks rich in iron and magnesium. The continental crust is old, up to 70 km thick and granitic, composed of silicate rocks rich in aluminum, silicon and calcium.

The crust is the upper part of the lithosphere. The lithosphere consists of seven major plates and many smaller plates that slide about on an underlying plastic layer (the asthenosphere). The energy responsible for the plate motion comes from the heat generated by the decay of radioactive elements, plus residual heat resulting from planetary accretion. Because of plate motion, 80% of the Earth's history (before 600 million years ago) has been lost through crustal recycling. The Mohorovicic discontinuity marks the transition from the bottom of the crust to the top of the mantle.

Mantle
The mantle that surrounds the iron-nickel core is 3000 km thick and composed mostly of hot plastic silicate materials. It is subdivided into the lower part of the lithosphere (a cool, rigid layer about 100 km thick), the asthenosphere (about 250 km thick), and the mesosphere (the more solid lower mantle below 700 km, not to be confused with the mesosphere of the atmosphere, namely the layer on top of the stratosphere). The lithosphere and asthenosphere are divided by the low-velocity zone, 50–100 km thick, where seismic waves move unusually slowly.

Core
From an analysis early in the twentieth century of the propagation of seismic waves, seismologists Beno Gutenberg and Emil Weichert identified a major discontinuity about 2900 km down, at the boundary between the mantle and the core. The core is much denser than the crust, as the density of the entire globe is about 5.5 g cm^{-3} while crustal material like granite has a density of about 2.8. The core is in fact mainly iron. Solids transmit both shear waves (motions that jiggle from side to side) and sound waves (motions that push to and fro). The core does not transmit shear waves, and that proves it is a liquid (because side-to-side motions in a liquid just continue, and dissipate without becoming waves). However, the inner part of the core, below a depth of about 5000 km, is solid; presumably an iron sphere roughly the size of the Moon.

The Earth's magnetic field is the result of convection currents in the liquid outer core, by a dynamo mechanism.

◄◄ ***Eagle Nebula*** *The finger-like 'Pillars of Creation' (near center) in M16 are the result of the dust nebula's local erosion by radiation pressure from nearby young ultraviolet-emitting stars. The radiation that sculpted these intricate forms should wear them away completely within the next million years, leaving behind only the stars that formed within.*

Atmosphere
Earth's original atmosphere contained mainly carbon dioxide and water vapor, with possibly traces of methane and other hydrogen-rich gases, and was lost to space. The atmosphere which evolved into that of the present day originated from the out-gassing of light gases from the interior during the first 100 million years of the Earth's life, and from the impact of COMETS and METEORITES over a few hundred million years.

The average atmospheric pressure at ground level is 1013 mbar, with a column density of 1 kg cm^2. The average temperature at the surface of the Earth is 288 K, or 15 °C. The atmosphere is mainly nitrogen and oxygen (see table). As much water is present in the atmosphere, either as vapor or in the form of water droplets or ice particles (clouds), as would cover the surface of the Earth to a uniform depth of a few centimeters of liquid. The proportion of water vapor by volume is highly variable, but is everywhere less than 1%.

Because the Earth's atmosphere contains large amounts of molecular oxygen and therefore ozone, it screens out solar ultraviolet (UV) radiation. Greenhouse gases such as

Earth	
Mean distance from Sun	150 × 10^6 km (1 AU)
Revolution period	1 year
Orbital eccentricity	0.017
Inclination to ecliptic	0°
Rotation	1 day
Diameter	12 756 km
Mass (Earth masses)	1
Density	5.5 g cm^{-3}
Escape velocity	11 km s^{-1}
Surface (main materials)	basalt, granite, water
Atmosphere	nitrogen (78%)
	oxygen (21%)
	argon (0.9%)
	carbon dioxide (0.03%)

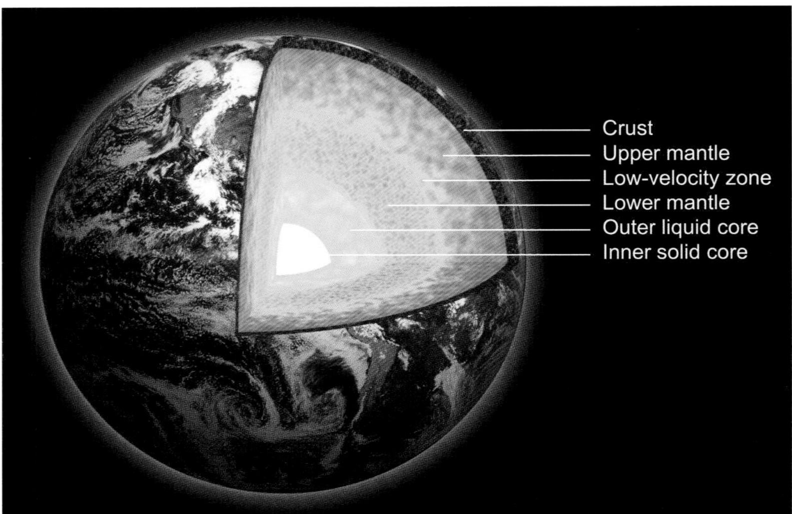

▲ ***Earth*** *from core to crust.*

water vapor and carbon dioxide (despite its small relative abundance of 0.04%) maintain mild climatic conditions. Thus the atmosphere preserves an environment that is suitable for life at the Earth's surface.

The oxygen that is now in our atmosphere is the result of photosynthesis in the biosphere. It progressively accumulated during the Precambrian era and reached its present level 300 million years ago, when the first land animals appeared. It is regulated by marine activity, which removes carbon dioxide from the atmosphere. A small part of this precipitates downward through the ocean into sea-floor sediments and is trapped, typically for several hundred million years (the timescale for sea-floor expansion). Oxygen atoms are released by the reduced carbon into the atmosphere. Ultimately, carbon is ejected by volcanoes or resurfaced metamorphic rocks, and is oxidized by the atmosphere, closing the cycle.

The edge of the atmosphere

The ionized atmosphere, or IONOSPHERE, has a layered structure, because of the way different energetic radiations (such as ultraviolet, X-rays and cosmic rays) interact with different gases. The main ionospheric layers are denoted by D, E and F. They extend from an altitude of 60 km (the base of the D layer) up to 400 km. Major ions are oxygen and nitrogen monoxide. The ionosphere reflects radio waves.

The upper boundary of the Earth's atmosphere is visible in an AURORA. Aurorae are extended sources of light of different forms and colors that are observable at high latitudes and sometimes at midlatitudes. They may reach the brightness of full moonlight. Most aurorae can be observed at the auroral oval around each magnetic pole. POLAR CAP aurorae (inside the oval) and daytime aurorae are usually weak and diffuse, while at the oval they are more intense and highly variable. Aurorae are generated by particles of the SOLAR WIND, which collide and chemically interact with the atoms and molecules of the Earth's upper atmosphere. Their spectra are correspondingly complex. Aurorae are usually between 80 and 300 km high. They follow solar activity, showing, for example, the 11-year solar cycle.

Oceans

Beneath the atmosphere, 70% of the solid surface is covered by an ocean 4 km deep. The oceans are held in basins formed between moving tectonic plates; thus they are temporary. The Atlantic Ocean, for example, is a geographic feature of only the last 2% of the age of the Earth. Large-scale currents in the oceans are driven by the wind, subject to the Coriolis force, and constrained by the boundaries of the oceanic basins. They strongly affect climate. The Moon raises tides by an amplitude of about 1 m in the free ocean, and by a much greater amplitude in narrow sea channels.

Rotation

The Earth rotates once per day, but there are fluctuations in the ROTATION speed of several dozen parts per billion (corresponding to a variation of a few milliseconds (ms) in the length of a day). These fluctuations are mostly seasonal (every year or every six months), but they show oscillations at longer and shorter periods. Observations of the solar eclipse by Babylonians, Greeks, Arabs and Chinese show that the length of the day increases by about 2 ms per century. These changes are due to torques acting on the Earth, especially the tidal forces of the Moon, and to changes in the distribution of mass within the Earth, which alter its moment of inertia.

Further reading: Brown G C, Hawkesworth C J and Wilson R C L (eds) (1992) *Understanding the Earth* Cambridge University Press.

Earthgrazer

A COMET or ASTEROID whose orbit brings it near to the Earth. A NEAR-EARTH ASTEROID is an Earthgrazer that comes within the orbit of the Moon.

Earthshine

A faint illumination of the Moon's night side by sunlight reflected from the Earth. When the Moon is a thin crescent, it is visible as a pale-gray ghostly glow over the remainder of the disk. Earthshine is traditionally known as 'the old Moon in the new Moon's arms'; alternative names are earthlight and ASHEN LIGHT (also used for VENUS). Earthshine is difficult to see less than a day after new Moon because the Moon is then very low in a twilit sky. It is most noticeable between one and four days before or after new Moon. It is best seen at about two days, as the Moon moves away from the Sun into the darker night sky, and before the area of brightly lit Moon overpowers the earthshine, and the lit face of the Earth toward the Moon diminishes. The intensity of earthshine varies as a result of changing levels of cloud cover over the Earth, and is an indicator of climate change.

Eccentricity

A parameter used to describe the shape of any curve from the family of CONIC SECTIONS (circles, ellipses, parabolas and hyperbolas). The eccentricity of an ellipse is its degree of flattening; more precisely the ratio of the distance from a focus of the ellipse to its center divided by its semi-major axis. Eccentricity, e, lies between 0 and 1, with large eccentricity indicating a long, thin ellipse. A circle has zero flattening and perfect symmetry, so e equals 0. The eccentricity of a parabola is 1, and for a hyperbola it is greater than 1. In CELESTIAL MECHANICS, where orbits under gravity are closely conic sections, e is one of the key ORBITAL ELEMENTS.

Echelle grating

The essential part of an echelle spectrograph. It is a DIFFRACTION GRATING in which the angles within the grooves are right angles (the shape of a staircase). Light is shone directly onto one face of the grooves. An echelle grating produces an interference pattern of many overlapping SPECTRA with high dispersion. In an astronomical echelle spectrograph, the overlapping spectra are separated by a cross-dispersing prism or a second diffraction grating, and stacked one above the other in the focal plane of the spectrograph's camera. This enables the astronomer to record a large range of wavelengths at high resolution on a small CCD detector (*see* CCDS AND OTHER DETECTORS and the figure accompanying ABSORPTION SPECTRUM).

Eclipse and occultation

An occultation is the phenomenon observed when one celestial body moves in front of another. The occultation of the Sun by the Moon is called a 'solar eclipse.' The word 'occultation' is usually reserved for a star being occulted by a planet, a satellite or an asteroid.

Lunar occultation

The abrupt disappearance or reappearance of a bright star behind the Moon can be observed with the naked eye. Nicolaus COPERNICUS saw an occultation of ALDEBARAN in 1497. These observations prove that the Moon is nearer to the Earth than the stars, that the stars are nearly point sources, and that the Moon has little or no atmosphere. Sometimes a BINARY STAR can be seen to disappear behind the Moon in two steps.

High-speed light recorders have revealed rapid fluctuations of starlight at the LIMB of the Moon due to optical interference, and the structures known as Fresnel fringes. Sometimes a progressive fade is observed if the star is large and close.

Solar system occultation

Occultations of stars by planets in our solar system are rare and difficult to predict accurately, except that mutual occultations of the satellites of Jupiter and Saturn occur each time the Earth passes through the orbital planes of the satellites, every six years in the case of Jupiter. The phenomena are full of interest because:

- the exact position of the occulting object in its orbit can be determined;
- if several observatories view the disappearance and reappearance of the star, the LIGHT CURVES show sections of the occulting object and therefore its shape (*see* figure in the practical astronomy feature Observing occultations, p. 136). Sometimes satellites can be detected (as with the minor planets Herculina, Melpomene and Lucina);
- the layered structure of a planet's atmosphere can be revealed by the extinction of a star's light behind it. Pluto's atmosphere was discovered by this method.

Observations of occultations led to the discovery of the ring systems of Uranus (in 1977) and Neptune (in 1984) (*see* PLANETARY RING).

Exoplanetary occultation

Searching for the transits of exoplanets (*see* EXOPLANET AND BROWN DWARF) in front of their sun (a faint occultation) is a promising way to investigate planets outside our solar system.

Solar eclipse

A solar eclipse occurs when a new Moon is especially well aligned. Solar eclipses can be total (the Sun is completely covered by the Moon), annular (the central part of the Sun is covered, with a ring of its surface showing around the lunar limb), or partial (the Sun is partly covered). Total and annular solar eclipses occur when the sizes of the Sun and Moon as seen from Earth are about the same, although in absolute diameter they differ by a factor of 400. The mean semidiameters of the Sun and the Moon are 960 arcsec and 931 arcsec, respectively. The ellipticities of 0.055 and 0.017 of the orbits of the Moon around the Earth and the Earth around the Sun, respectively, provide enough range in apparent sizes to produce annular eclipses (with the Moon appearing as much as 10% smaller than the Sun) and total eclipses (the Moon appearing up to 8% larger than the Sun).

Types of solar eclipse

At a total eclipse, the Earth's surface intersects the central cone of complete lunar shadow, forming an elliptical shadow (UMBRA) within which the Sun is entirely blocked. The shadow traces a path across the Earth that, over a period of hours, is thousands of kilometers long but only up to about 300 km wide. Observers in a much larger area around the umbra are in the PENUMBRA, and see a partial eclipse. For an observer in the umbral path, totality can last from just an instant up to about 7 min. During totality the Sun's CORONA is revealed. Since the PHOTOSPHERE is approximately one million times brighter than the full Moon, an annular eclipse allows so much sunlight to reach the Earth's atmosphere that the sky remains blue and the corona is not visible.

Total and annular eclipses are known as central eclipses. Between two and five solar eclipses occur in any given year, most of them partial with the umbra never passing across the Earth. About every 18 months a total eclipse is visible from somewhere on Earth. An observer waiting at one location on Earth would see a total solar eclipse on average once every 375 years, a rate that depends on latitude (solar eclipses are more frequent at the equator). The umbral shadow moves across the Earth at about 3400 km h^{-1}. The Earth's rotation diminishes the velocity relative to a given point on the surface by as much as half. The supersonic aircraft Concorde kept up with the umbral shadow of the 1973 eclipse for 74 min.

Predicting eclipses

Eclipses are predicted using a method descended from that worked out by Freidrich BESSEL in 1824. For each total eclipse Fred Espenak and Jay Anderson produce a NASA reference publication, posting tables and maps on the World Wide Web. Computer programs that accurately predict eclipse paths are also available. It is possible to predict an eclipse without using orbital calculations from a coincidence in the cycles of the Sun and the Moon: eclipses with approximately the same characteristics are repeated every 18 years and 11.3 days, a period that is called SAROS.

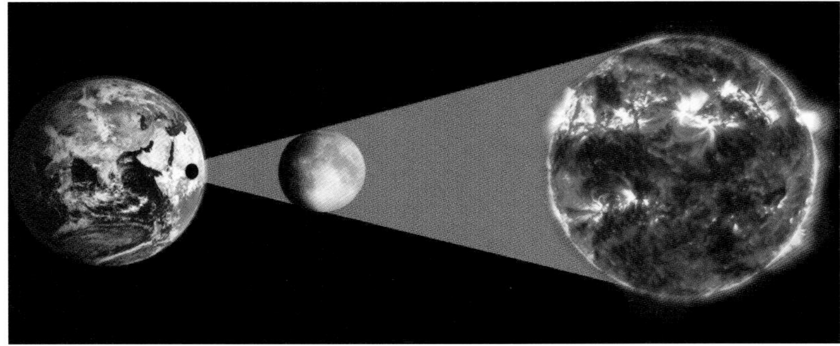

▼ ***Solar eclipse*** *From the small region of totality on Earth, the Sun is completely covered by the Moon.*

Observing eclipses

Eclipse calculations specify the times of the contacts. First contact is the first meeting of the disks of the Moon and Sun; the ingress of the Moon is noticeable to an observer looking through a filter with unaided eye a few seconds afterwards. Second contact is the moment when the Moon first completely covers the solar PHOTOSPHERE, the beginning of totality. Third contact is the first appearance of the solar photosphere at the end of totality. Fourth contact marks the departure of the lunar disk from the photosphere. The interval between first and fourth contact is commonly three hours, and can be as long as four hours.

During the partial PHASES (between first and second contacts, and between third and fourth) the crescent shape of the Sun is projected by fortuitous pinhole cameras in the canopy of trees onto the ground. For the last 15 min or so before second contact, shadows grow strangely sharper, since they are now being cast by a crescent instead of a sphere. The sky noticeably changes color. During the last couple of minutes before totality, shadow bands, low-contrast ripples of light and dark, are often seen to move rapidly across the landscape. They are caused by phenomena in the Earth's upper atmosphere.

Just before totality a phenomenon known as BAILY'S BEADS results from the last bits of solar photosphere shining through the deepest valleys on the lunar LIMB. The last Baily's bead glows so brightly compared with the other solar phenomena visible – the CHROMOSPHERE and/or the innermost CORONA – that it is known as the DIAMOND RING effect.

As the diamond ring diminishes, a pinkish narrow rim on the Sun becomes visible. It got the name 'chromosphere' from its colorful appearance at eclipses. Depending on how much larger in angular size the Moon is than the Sun, you may be able to see the full chromosphere at one time or you may see only the leading edge at second contact and then the trailing edge at third contact. The pink color comes largely from hydrogen-alpha radiation, and has an admixture of other chromospheric emission lines.

As the diamond ring and chromosphere disappear, you have totality. The shape of the corona becomes visible, with coronal streamers obvious. The shape of the corona varies over the solar-activity cycle (SOLAR CYCLE). At solar minimum, most of the streamers visible are equatorial, and polar tufts show near the poles. Contour-like lines through areas of equal brightness (called isophotes) of the corona appear elongated. At solar maximum, streamers are visible at more latitudes and isophotes of the corona appear nearly round. PROMINENCES have become visible during the diamond-ring effect and chromosphere period, and if sufficiently large may remain visible throughout totality. All too soon, totality is over and the eclipse phenomena recur in reverse order.

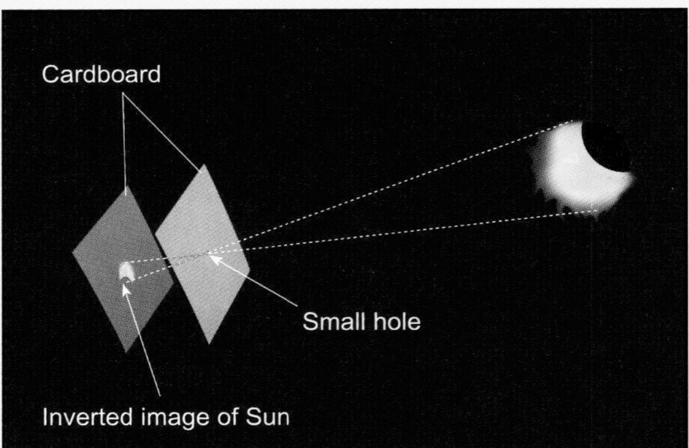

► ***A pinhole camera*** *is the most simple way to view an eclipse. The solar image is inverted.*

Eclipse safety

The brightness of the solar corona during totality is approximately that of the full Moon, and the corona during totality is equally safe to watch. However, staring at the crescent or other partial phases of the eclipse can be hazardous to eyesight, so warnings should be given about how to watch an eclipse safely. In general, the partial phases are hazardous, but totality is both glorious and safe. In fact, you will not experience the full glory of totality unless you view it directly.

On a normal day, one is not tempted to stare at the Sun, and the eye-blink reflex normally prevents one from doing so. However, during the partial phase of an eclipse, people's eyes are much more likely to be directed toward the Sun.

► ***The corona*** *at solar minimum (left) and solar maximum (right). Coronal streamers appear at more latitudes at solar maximum. Images by Douglas Arnold.*

◀ **Points of contact** *from left to right: first, second, third and fourth contact.*

During the crescent phases, the total intensity of sunlight may be inadequate to activate the eye-blink reflex even though the specific intensity of regions of the solar photosphere that are still visible remains high. Thus, you should observe the Sun only through special filters. These filters cut out all but about 1/100 000 of the photospheric light, reducing the photosphere to a safe level for observation.

The key factor in safe solar filters, aside from their absorbing or reflecting sufficient levels of solar intensity, is that they do so evenly across the spectrum; that is, that they are of neutral density (ND). Filters that pass only 1/100 000 the light are ND5. Solar filters made of aluminized Mylar, a coated plastic, are very popular and inexpensive. As long as these Mylar filters are undamaged, without creases or pinholes, they are safe to look through.

For telescopic use, filters made of chromium deposited on glass pass the requisite amount of photospheric light to photograph and view the partial phases while also giving images of excellent definition. A very slight increase in transmission at the red end of the spectrum often gives a pleasing tint to the solar image. Welders' glass of #14 or #15 grade is also safe for visual or photographic viewing, although the color is usually less pleasing. Photographic 'ND filters' (like Wratten filters, for example) are not safe for use at eclipses since they are transparent in the infrared rather than being ND across the spectrum. Similarly, crossed polarizers are also unsafe.

A very safe way to view the partial phases is with a pinhole camera. Such a device is no more than a small hole, perhaps 2–5 mm across, in a piece of cardboard or aluminum foil (perhaps taped in place over a larger hole in cardboard). This small hole is held 0.5–1 m or so above a piece of paper or cardboard; observers look down at this second surface with the Sun behind them. A small hand-mirror a few millimeters across will likewise project the partial phases of the Sun onto a distant screen, such as a white wall.

Though the eye is not damaged by occasional unaided glances at the Sun, such as those that often occur accidentally on normal days, it can be immediately and permanently damaged by the focused image of the Sun through a telescope or through binoculars. So, if either are used in a direct-view mode, a solar filter must be taped or otherwise secured very carefully to the front of the optical device so that it cannot blow or fall off. Filters that go on the eyepiece end are always liable to crack or be otherwise damaged under the concentrated solar radiation, so are much less safe than filters placed above the objective lens. Filters such as smoked glass, photographic negatives or CDs are not safe enough.

Jay Pasachoff is director of Hopkins Observatory, Williams College.

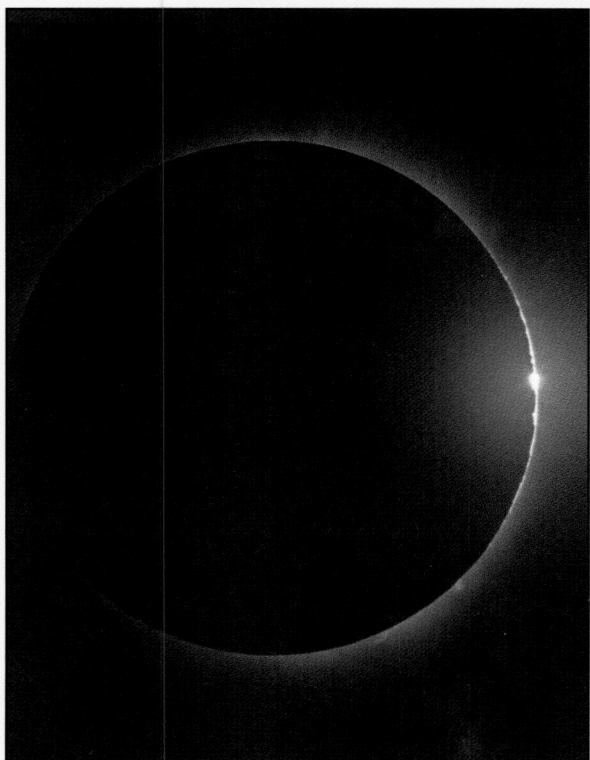

◀◀ ***Baily's beads*** *photographed from the Atacama Desert, Chile, during the eclipse of November 1994, by Douglas Arnold.*

◀ ***The diamond ring effect*** *photographed from Baja, California, USA, in July 1994, by Douglas Arnold.*

Observing occultations

An occultation can be an exciting event for an observer; there is something magical about watching a planet slowly disappear behind the face of the Moon. It can be interesting too; in particular, a grazing occultation can give a graphic demonstration of the rugged nature of the Moon's surface (and provide the interesting logistical challenge of having to ensure that one is exactly on the narrow track). Observing occultations can add variety to a night's observing and provide a much-needed incentive to go out and take another look at well-known objects.

An occultation occurs when one body passes in front of another. The most spectacular of all occultations is a solar eclipse, when the Moon passes in front of the Sun. (A lunar eclipse, by contrast, is a true eclipse with the Moon moving into the shadow of the Earth, but this won't be dealt with here.) Similarly, a TRANSIT of Mercury or Venus across the face of the Sun is really an occultation. Transits of Mercury are common, and provide no more than an observational curiosity; their past importance for measuring the scale of the solar system has been superseded. Transits of Venus are rarer and occur in pairs almost a century apart. One interesting effect is the BLACK DROP; as Venus begins to transit the Sun's disk, it seems to elongate. The cause of this effect is still debated, but it seems to be an optical illusion.

The vast majority of occultations involve the Moon. Its large angular diameter and rapid movement around the sky make the likelihood of a bright star being intercepted by its path far greater than that for any of the planets. Tables of predicted occultations can be easily obtained via an Internet search, and such events, especially when a really bright star is involved, can make good targets for photography through the telescope.

Of particular interest are grazing occultations, which occur when the path of the Moon causes a star to pass along the LIMB of the satellite, blinking as it disappears behind each mountain and appearing in the valleys. Obviously, these events will only be observed along a narrow track, and it is essential to obtain accurate predictions. Setting up equipment in the middle of a country lane, having allowed insufficient time to locate a suitable observing site, is not entirely unknown! Things move rapidly, and it is difficult to take still photographs quickly enough. Video is an ideal medium, but it is absolutely essential that accurate timings (to at least the tenth of a second) are made; these days the easiest way to obtain such timings is via GPS equipment. One cannot rely upon the camera's clock.

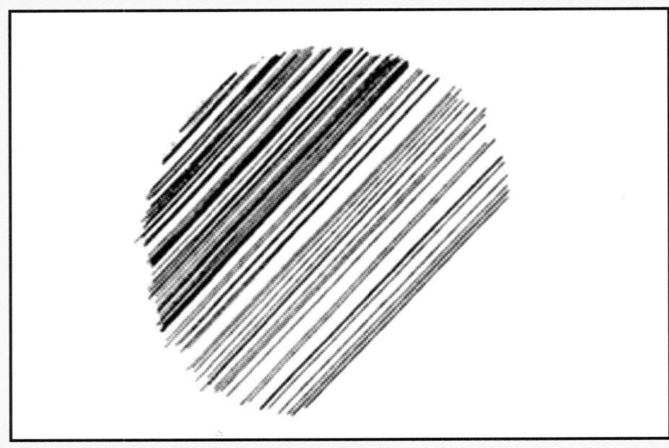

▶ ***Limb profile*** *Profile of the asteroid (2) Pallas deduced from timings of the disappearance of the star 1 Vulpeculae from various locations during the May 29, 1983 occultation.*

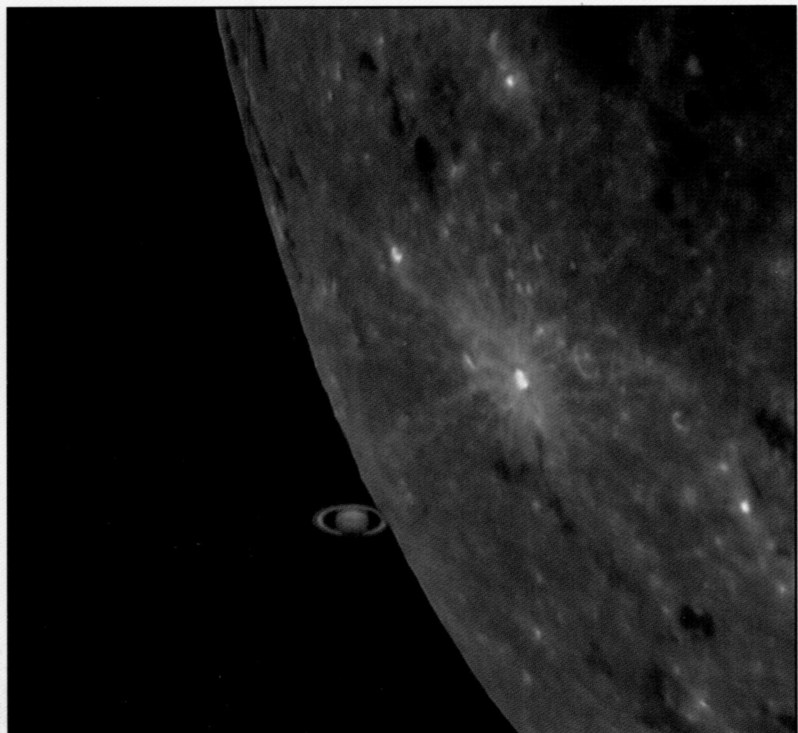

▼ ***Occultation of Saturn by the Moon,*** *imaged on December 1, 2001 from Sant Llorenc de la Muga, Girona, Spain. Image composite made from several sequences taken with a Philips webcam ToUCam Pro coupled to a 6-in reflector, f/18. Image by Juan Casado.*

Of greater scientific import are occultations of stars by planets and asteroids. Detailed photometric studies have successfully provided information about the atmospheres of planets, in particular Uranus, Neptune and Titan, Saturn's largest moon. A temperature profile can be developed and the possibility of haze can be determined. A sudden spike of brightness may be seen at the center of the occultation due to the focusing of the star's light by deep atmospheric layers. Occultations are also the source of important information about ring systems around the giant planets (*see* PLANETARY RING). Once more, the key to the successful observing of these objects is accuracy, both in timing and in the recording of the star's position. It is also undoubtedly true that occultations of giant planets provide fabulous photographic targets for producing interesting images.

The most useful of all occultation observations are those of asteroids; for the huge majority of these, we have no idea of their shape or structure. The method employed is simple. When an occultation is predicted to occur, bulletins are issued and are available on the World Wide Web. An individual observer records his or her exact position, and precise timings for the beginning and end of the event. These results can then be combined with those of other observers to produce a limb profile for the asteroid. It should be noted that a negative result (no occultation seen) is also important in constraining the measured profile, so observers close to, but outside of, the predicted track should also aim to be observing at the predicted time. Such studies have produced excellent results that, when combined with the known mass of many asteroids, gives a DENSITY. These densities, then, give information about their possible compositions. Occasionally, secondary minima are seen, and observers should be aware of the possibility that they may need to record these; these minima are the result of double asteroids such as (243) Ida and Dactyl (discovered by the Galilean satellite).

Chris Lintott recently graduated from Magdalene College, Cambridge and is now pursuing a PhD at University College, London.

Historical eclipses

Records of eclipses can be found on cuneiform tablets from ancient Babylonia, and in records from China and classical Greece and in medieval Islamic writings. In 1997 F. Richard Stephenson analyzed ancient records of eclipses to determine the rotation rate of the Earth, since whether an eclipse is total or not at a given point depends on the rotational position of the Earth under the umbral shadow. In 500 BC the day was approximately 42 ms shorter than it is now, and the Earth 20 000 s of rotation shifted (almost 90°). Much of the change in the rotation rate of the Earth derives from lunar and solar tides.

Eclipse science

Eclipse expeditions began in the eighteenth century. The first photograph of the corona was taken in 1851 in Königsberg (now Kaliningrad, Russia). During the expeditions of 1860–71, PROMINENCES and the corona were discovered to be intrinsic to the Sun rather than artifacts. At the 1868 eclipse in India a spectrograph revealed the brightest emission lines of the CHROMOSPHERE for the first time, including a new bright-yellow line near to the characteristic yellow emissions of sodium. The element discovered was called HELIUM, from the Greek Sun-god Helios.

Starting with the total eclipse of 1869, additional emission lines were discovered that came from the corona and were said to come from 'coronium,' later identified as 13-times-ionized iron (Fe XIV) and nine-times-ionized iron (Fe X). This proved that the corona is very hot, containing gas at millions of kelvins.

At the 1919 eclipse expeditions, Arthur EDDINGTON discovered that stars very near the Sun were deflected slightly, as predicted by Albert EINSTEIN's theory of GENERAL RELATIVITY.

Solar space observatories like SOHO view the corona every day, which has diminished the value of coronal eclipse research. But eclipse observations can extend the regions accessible to study within SOHO's capability, which shows the corona only beyond 1.8 solar radii.

Lunar eclipse

Sometimes, when the Sun, Earth and Moon are exactly lined up, the full Moon moves into the shadow cast by the Earth. A lunar eclipse is a solar eclipse as seen from the Moon, so there are umbral and penumbral regions to the cone of shadow. As the shadow passes across the surface of the Moon the supply of sunlight and heat is reduced. If the whole of the Moon passes into the umbra, the lunar eclipse is called total, but even in a total lunar eclipse the Moon is not completely dark. Sunlight reddens and refracts in the Earth's atmosphere onto the Moon, which often turns a coppery color. Lunar eclipses occur less often than solar eclipses (0–3 times every year), but a lunar eclipse can be seen from any point on the Moon-facing side of the Earth, rather than only from the thin solar eclipse totality path. Thus, total lunar eclipses are seen much more often than total solar eclipses.

Further reading: Stephenson F R (1997) *Historical Eclipses and Earth's Rotation* Cambridge University Press; Zirker J (1995) *Total Eclipses of the Sun* Princeton University Press.

Eclipsing binary star

One of a pair of orbiting stars that are oriented such that each periodically passes in front of the other (with the orbital plane nearly edge-on), thereby dimming light from the system. Bright eclipsing stars such as ALGOL and EPSILON AURIGAE excite the interest of even casual observers, and many such stars can be seen through binoculars. Eclipses of only one star are possible, but in most examples both stars are eclipsed (although usually one of the alternating eclipses is much more prominent than the other). Initial mental images may be of simple spherical stars, but in fact a host of phenomena occur, including tides, winds, impacting streams, and magnetic starspots. The eclipsing binary concept has grown to include eclipses of and by circumstellar disks, eclipses of fluorescent circumstellar gas by one or both stars, and eclipses of the stars by attenuating clouds of circumstellar material. *See* VARIABLE STAR: CLOSE BINARY STAR.

A complete eclipse is one that is either total or annular, while a partial eclipse is neither total nor annular (*see* ECLIPSE AND OCCULTATION). The deepest eclipses occur when the brighter star (in a given photometric bandpass) is totally eclipsed, as in RW Tauri. Annular eclipses, partial eclipses and eclipses of stars with low relative luminosity are typically much less deep and may even be undetectable. Total eclipses that are detectable only in X-rays can occur when a normal star has a hot neutron star companion.

Among the phenomena that make LIGHT CURVES informative are tidal and rotational distortions, mutual heating, gravity brightening, limb darkening, magnetic starspots, and ACCRETION hot spots. Brightness variations due to tides and heating are collectively known as proximity effects. They tend to be prominent in eclipsing light curves because the same characteristic that makes eclipses likely (closeness) also enhances tides and heating.

There are many contenders for the title of strangest binary, but BETA LYRAE and Epsilon Aurigae have continued as leading candidates for a century. *See also* BINARY STAR.

Ecliptic

A great circle on the CELESTIAL SPHERE that represents the apparent annual path of the Sun in its motion relative to the background stars. It is so called because eclipses can occur when the Moon crosses it. The ecliptic, in fact, represents the intersection of the orbital plane of the Earth with the celestial sphere. Because the Earth's equator is inclined by an angle of approximately 23.5° to the orbital plane, it follows that the ecliptic is inclined to the celestial equator by the same amount (known as the 'obliquity of the ecliptic'). The ecliptic intersects the celestial equator at two points, the vernal EQUINOX and the autumnal equinox. *See also* CELESTIAL COORDINATES.

Ecosphere

The shell-shaped region around a star within which the temperature at the surface of a planet might support life. The region depends on the type of star, in particular on its

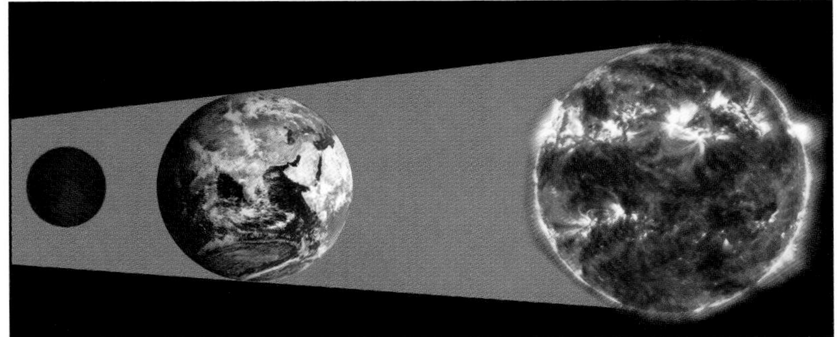

▼ ***Lunar eclipse*** *The Sun always casts a long, thin shadow behind the Earth. When the Moon enters the shadow zone, a lunar eclipse takes place.*

▶ ***Effelsberg Radio Telescope*** *A masterwork of large-scale precision engineering, the Effelsberg 100-m-diameter radio telescope dwarfs the adjacent control building.*

surface temperature. In the solar system, the ecosphere extends from just outside the orbit of Venus to just within the orbit of Mars, making Earth the only suitable planet for life, by this definition. The Moon is in the Sun's ecosphere, but it does not possess other suitable conditions for life. In general, hotter stars have a larger and wider ecosphere; cooler stars a smaller and narrower one. A DWARF STAR would have a much smaller and narrower ecosphere than a giant star with the same surface temperature.

This definition of an ecosphere applies to surface-dwelling life-forms (on land or in the ocean) similar to those found on Earth. Micro-organisms known as extremophiles have been found, which, as the name implies, thrive in extreme environments, in particular at temperatures much higher or lower than other organisms can tolerate. In addition, deep-ocean and subsurface organisms exist that do not rely on photosynthesis, and that can withstand temperatures of 1100 °C and 1200 °C respectively, at depths of 5 km. The definition of ecosphere needs to be rewritten.

Eddington, Arthur Stanley (1882–1944)

Astrophysicist, born in Kendal, Westmoreland, England, he became Plumian professor of astronomy and director of the Cambridge Observatory. Eddington's work on the theory of relativity was described by Albert EINSTEIN as 'the finest presentation of the subject in any language.' From Greenwich he led one of the two 1919 solar eclipse expeditions, which confirmed the predicted deflection of starlight by gravity. His lifetime's work concerned the internal structure of stars. He discovered the mass-luminosity relationship, calculated the abundance of hydrogen, and explained the pulsation of CEPHEID VARIABLE STARS and the very high densities of WHITE DWARFS. He was one of the first to identify nuclear reactions as the source of power in stars. He studied the pressure of the transport of energy through a star by radiation, and showed that if a star was above a certain brightness (the 'Eddington limit') it would be disrupted as radiation pressure overcame the force of gravity. In his later years he pointed out some surprising numerical coincidences between atomic and cosmological quantities (*see* ANTHROPIC PRINCIPLE).

Edgeworth, Kenneth Essex (1880–1972)

Amateur astronomer, born in Streete, County Westmeathe, Ireland. His 1943 paper 'The evolution of our planetary system' is the first reference to a reservoir of COMETS beyond the planets. This theory foreshadowed the concept of the Kuiper Belt, often referred to in Europe as the Edgeworth–Kuiper belt (*see* OORT CLOUD AND KUIPER BELT).

Effective temperature

The effective temperature of a star is defined to be the temperature of a black-body radiator with the same radius and the same total energy output (luminosity) as the star (*see* BLACK-BODY RADIATION). The effective temperature corresponds quite closely to the temperature of the visible surface of a star, and is related to the star's COLOR INDEX or spectral type.

Effelsberg Radio Telescope

The 100-m-diameter radio telescope near Effelsberg, Germany, has been operated by the Max Planck Institut für Radioastronomie since 1972. The telescope is fully steerable and its mechanical design was pioneering, in that it minimizes deformation of the figure of the dish under gravity as the telescope is tilted to different elevations. This makes it possible to observe routinely to wavelengths as short as 6 mm, even though the flexure is much bigger than this. The telescope frequently participates in VLBI observations.

Egg Nebula

A PLANETARY NEBULA (CRL 2688) in the constellation Cygnus. It is young, possibly having left its RED GIANT phase only a few hundred years ago. The central star, which is obscured by dust, is emitting powerful JETS and has shed several shells of material.

Einstein, Albert (1879–1955)

Physicist born in Ulm, Württemberg, Germany. Einstein described the photoelectric effect (for which he received the Nobel prize in 1921), and in 1905 created the theory of SPECIAL RELATIVITY. He developed this in his spare time, while an employee of the Swiss patent office. The theory of relativity was based on two hypotheses: that the laws of physics had to have the same form in any frame of reference; and that the speed of light remained constant in all frames of reference. He deduced that mass and energy were equivalent. He sought to extend the special theory of relativity to phenomena involving acceleration by using the principle of equivalence (*see* GENERAL RELATIVITY; GRAVITATION, GRAVITATIONAL LENSING AND GRAVITATIONAL WAVES). While a professor in Prague he predicted the gravitational deflection of light; for example, how light from a distant star, passing near the Sun, would appear to be bent slightly, shifting the star's apparent position away from the Sun. This was verified by the observations of Arthur EDDINGTON and Frank DYSON at the solar eclipse of 1919.

This success began to establish Einstein as an international scientific icon. By contrast, his work in Berlin was disrupted by demonstrations with an anti-Jewish taint, and eventually Einstein took up a post at Princeton University, USA, as the Nazis took power. At Princeton he began his work to unify the laws of physics – an ambitious, possibly overambitious task, which physicists continue. He became active in the peace movement and took on the role of a wise and independent scientific counselor to the world. The EINSTEIN OBSERVATORY was named for him.

Einstein Observatory

The second in the series of HIGH ENERGY ASTROPHYSICAL OBSERVATORIES (HEAO) was launched by an Atlas-Centaur rocket on November 13, 1978. Soon after its insertion into a 470-km circular orbit inclined at 23.5° to the equator, HEAO-2 was named the Einstein Observatory, in celebration of the centenary of Albert EINSTEIN's birth.

This groundbreaking NASA mission revolutionized X-RAY ASTRONOMY. As the first satellite to be equipped with focusing X-ray mirrors, Einstein observed sources 10 000 times fainter than did its predecessor, UHURU, and it was the first observatory to image X-ray sources with a telescope. The field of view varied from instrument to instrument within the satellite, but was typically about one square degree, with a spatial resolution as fine as 2 arcsec. The leader of the project and the principal investigator for all five instruments was Riccardo GIACCONI. The Einstein Observatory's ability to detect faint sources meant that it could study objects such as supernova remnants, X-ray binary stars, galaxy clusters, active galaxies and quasars. Its high spatial resolution meant that many of these could be linked to optical counterparts for the first time.

The mission ended in April 1981, when the satellite ran out of attitude-control gas. It re-entered the atmosphere on March 25, 1982.

Ejecta

Material thrown out, as in the case of a SUPERNOVA or formation of an impact CRATER. Crater ejecta appear in several forms: rays that extend radially from the crater; an ejecta blanket,which is a continuous deposit of ejecta surrounding the crater; and secondary craters, caused by larger fragments of ejected debris. Crater ejecta, which consist of freshly exposed subsurface material, are frequently highly reflective.

Electromagnetic radiation

Radiation consisting of an electric and a magnetic disturbance, which travels in a vacuum at a characteristic speed known as the velocity of light (299 792.458 km s^{-1}). In other words, a light wave consists of periodically varying changes in electric and magnetic fields. Most of our information about the universe has been obtained by measurements of electromagnetic radiation of one kind or another reaching us from space.

In James Clerk MAXWELL's theory of electromagnetism, electromagnetic radiation is thought of as a wave motion, the distance between successive crests being the wavelength. Visible light has a wavelength of a few hundred nanometers, while a radio wave has a wavelength of the order of meters. The number of wave crests that pass a fixed point in one second is the frequency. This equals the velocity of light divided by wavelength. Thus, short wavelength corresponds to high frequency. The full range of electromagnetic radiation is called the ELECTROMAGNETIC SPECTRUM.

In QUANTUM MECHANICS, electromagnetic radiation is described in terms of quanta or PHOTONS ('packets' or 'particles' of energy). The energy associated with electromagnetic radiation is directly proportional to frequency. Thus, the shorter the wavelength, the more energetic the photon. Electromagnetic forces are conveyed (or mediated) between charged and magnetic particles by photons.

Electromagnetic spectrum

The complete range of ELECTROMAGNETIC RADIATION, from the shortest to the longest wavelength. By convention, the electromagnetic spectrum is divided into different wavebands. The principal divisions are: gamma-rays (wavelengths shorter than 0.01 nm), X-rays (0.01–10 nm), ultraviolet (10–390 nm), visible (390–700 nm), infrared (700 nm – 1 mm), and radio (1 mm upward).

The human eye responds to different wavelengths of light by recognizing different colors (red light has a longer wavelength than blue light). Although the human eye is sensitive only to wavelengths in the range from about 390 to 700 nm, the 'optical' or 'optical–infrared' (optIR) region of the spectrum is generally taken to encompass the wider waveband from 310 to 1100 nm (310 nm – 1.1 μm). (*See* OPTICAL SPECTRUM; OPTIR SPECTRUM.) Further subdivisions in common astronomical use are: hard X-ray (0.01–0.1 nm), soft X-ray (0.1–10 nm), extreme ultraviolet (EUV, or XUV, 10 –91 nm), near-infrared (1–4 μm), mid-infrared (4–40 μm), far-infrared (40–350 μm, that is 40 μm–0.35 mm), submillimeter (0.35–1 mm), millimeter-wave (1–10 mm), and microwave (1 mm–0.3 m).

Most incoming radiation is absorbed by the ATMOSPHERE or reflected back into space, but radiation in the wavelength range 310–1100 nm, and from about 2 cm to about 30 m, can penetrate to ground level. These bands are known, respectively, as the optical and radio 'windows.' Several narrow bands of radiation in the near- and mid-infrared regions of the spectrum, and radiation in the submillimeter- and millimeter-wave regions, can be studied from high-altitude observing sites.

Electron

A stable elementary particle with a negative unit charge and a mass of 9.1091×10^{-31} kg (equivalent, in energy terms, to 0.511 MeV). The electron has a spin of 1/2 (in units of the Planck constant) and is thus a LEPTON. POSITRONS are related to electrons as anti-particles.

Electrons can exist either as free particles or as the negatively charged components of atoms. A flow of electrons (for example, along a wire) constitutes an electric current. In the history of radioactive decay and nuclear physics, electrons were referred to as beta (β) particles, and a stream of electrons as beta radiation. An electron is thus denoted not only by

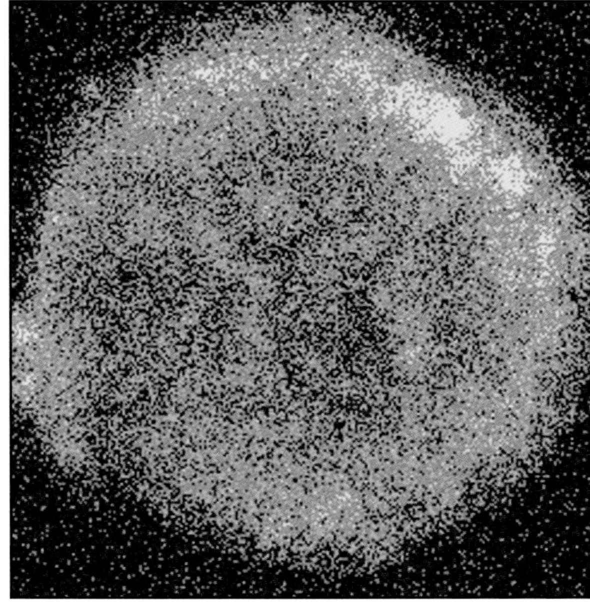

◄ ***Einstein Observatory*** *This supernova remnant, an X-ray-emitting shockwave of gas and dust, was imaged by the sensitive X-ray telescopes of the Einstein Observatory. The Danish astronomer Tycho Brahe watched the supernova that generated this nebula in 1572.*

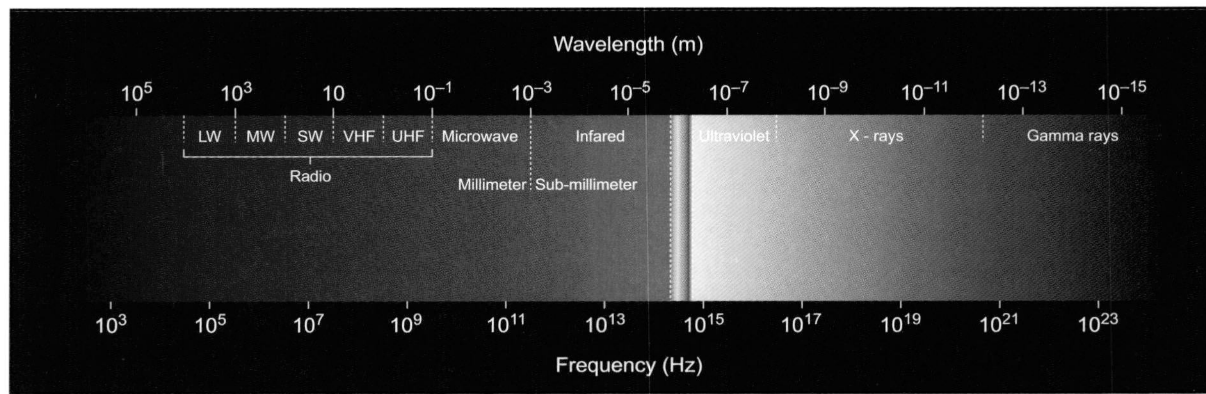

▶ ***Electromagnetic spectrum*** *Astronomy was originally confined to the tiny portion of the spectrum that is visible to the human eye, the optical spectrum, seen here as the central 'rainbow' band. Modern astronomy looks at the sky at wavelengths from radio to gamma-rays.*

the symbol 'e' but also β, or 'e⁻' and β^- if it is necessary to emphasize the distinction with a positron.

The magnitude of the charge on the electron (the 'electronic charge'), which is denoted by the symbol '*e*,' is used as the unit for describing the charges on subatomic particles and atomic nuclei.

Electron-volt

A unit of energy that is used to describe the energies of subatomic particles and PHOTONS, or the energy levels of ATOMS. One electron-volt (symbol eV), which is the kinetic energy gained by an ELECTRON when it is accelerated through a potential difference of one volt, is equivalent to 1.602×10^{-19} J. Commonly used multiples are keV (1000 eV), MeV (1 million eV), and GeV (1 billion eV).

Element, chemical

A substance composed of ATOMS, each of which has the same atomic number (that is, contains the same number of PROTONS in its nucleus), and which cannot be broken down by chemical processes into two or more other substances. The atomic number determines the number and arrangement of the ELECTRONS that orbit the atom's nucleus. The chemical properties of the element are determined by the arrangement of the electrons in the shells that surround the nucleus.

Each chemical element is denoted by a chemical symbol. For example, hydrogen is denoted by H, helium by He, carbon by C and iron by Fe. The basic symbol can be amplified by adding the number of protons and neutrons (known as the mass number, *A*) as a superscript, and, if necessary for clarity, the atomic number (*Z*) as a subscript. Thus, element *X* would be denoted, symbolically, by $^{A}_{Z}X$. For example, hydrogen (with one proton and no neutrons in its nucleus) is represented by $^{1}_{1}\mathrm{H}$; helium (with two protons and two neutrons, so that $A = 4$ and $Z = 2$) by $^{4}_{2}\mathrm{He}$, iron (26 protons and 30 neutrons), by $^{56}_{30}\mathrm{Fe}$, and so on. Of the 92 'naturally occurring' elements, the heaviest of which is uranium, 90 have been found on Earth. The remaining two (technetium and promethium) have been synthesized experimentally. In addition, at least 15 transuranic elements (elements with atomic numbers greater than 92, that of uranium) have been synthesized by nuclear processes.

Elementary particle

One of the many subatomic (smaller than the atom) particles; for example electrons, protons, mesons, neutrinos and quarks, some of which have even smaller constituents. The only true FUNDAMENTAL PARTICLES are the leptons, quarks and their antiparticles and the force-carrier particles.

Elements, formation of

In 1835, in a famously inaccurate forecast, the French philosopher Auguste Comte (1798–1857) wrote of stars: 'We understand the possibility of determining their shapes, their distances, their sizes and their movements; whereas we would never know how to study by any means their chemical composition.' At the close of the twentieth century, analyzing the visible surface layers of stars has become a relatively straightforward astrophysical procedure. Such measurements give the abundance of the approximately 100 chemical elements, which are each distinguished as ATOMS by the number of ELECTRONS orbiting each nucleus. Changes in the arrangement of the electrons produce light, which is what can be measured spectroscopically.

Each nucleus in an atom has a number of PROTONS, exactly balancing the number of electrons. It also has an approximately equal number of NEUTRONS, but the number can vary from atom to atom in a given element. Each different nucleus is an ISOTOPE of the given element, identified by the name of the atom and the number of protons and neutrons, such as helium-4 (two protons and two neutrons), also written $^{4}\mathrm{He}$. The process that produces the different sorts of nuclei, and thus the elements in the universe, is called nucleosynthesis.

The abundance of the isotopes of the elements is measured mostly through laboratory analysis of meteorite samples (*see* COSMIC ABUNDANCE OF ELEMENTS). The abundance of the chemical elements is a thread that ties together much of what we know about the evolution of stars, galaxies and the universe as a whole. The origin of the universe in the BIG BANG yielded the most abundant of the elements, hydrogen ($^{1}\mathrm{H}$) and helium-4. Most of the remaining elements and isotopes found in nature have been produced by the nuclear reactions that power the stars during the normal course of their evolution, or by rapid nucleosynthesis during explosions of certain stars. The isotopes of several elements have been created by the interaction of heavier nuclei with COSMIC RAY particles in the INTERSTELLAR MEDIUM.

Some of the products of nuclear reactions that occur within a star are injected back into the interstellar medium as the star casts off most of its mass during the brief final stages of its evolution. There they provide the raw material for the formation of later generations of stars, planets and life itself.

Nucleosynthesis in the early universe

The early universe was hot and dense, behaving as a cosmic nuclear reactor during the first 20 min of its evolution. It was, however, a 'defective' nuclear reactor, expanding and cooling very rapidly. As a result, only a handful of the lightest

NUCLIDES were synthesized before the density and temperature dropped too low. After hydrogen (^{1}H – protons) the next most abundant element to emerge from the Big Bang was helium-4 (ALPHA PARTICLES). Isotopes of these nuclides, deuterium (^{2}H) and helium-3, are the next most abundant. After helium, there is a large gap to the much lower abundance of lithium-7.

Reconciling the measured cosmic abundances, the theory of the Big Bang, and the physics has been interesting. There were major uncertainties in the physical data until experiments at CERN in the late 1980s proved that there are three neutrino species, and the half-life of the neutron was measured accurately. The largest uncertainties are now related to the astronomical determination of the primordial abundance of deuterium, helium-3, helium-4 and lithium-7.

Cosmologists have agreed on a reference theory of the Big Bang, called the 'standard hot Big Bang cosmological model.' It is built on assumptions that are both simple and simplistic (for example, it assumes that the universe is more or less the same in every direction and everywhere, and it assumes our theories of gravity and particle physics are accurate). Nevertheless, the standard Big Bang theory is remarkably successful, and primordial nucleosynthesis (the only probe of the physical universe during its very early evolution) is one of its main pillars. In particular, astronomers are striving to become more certain about the abundances of deuterium and helium, because these observational parameters help to determine the conditions in the Big Bang.

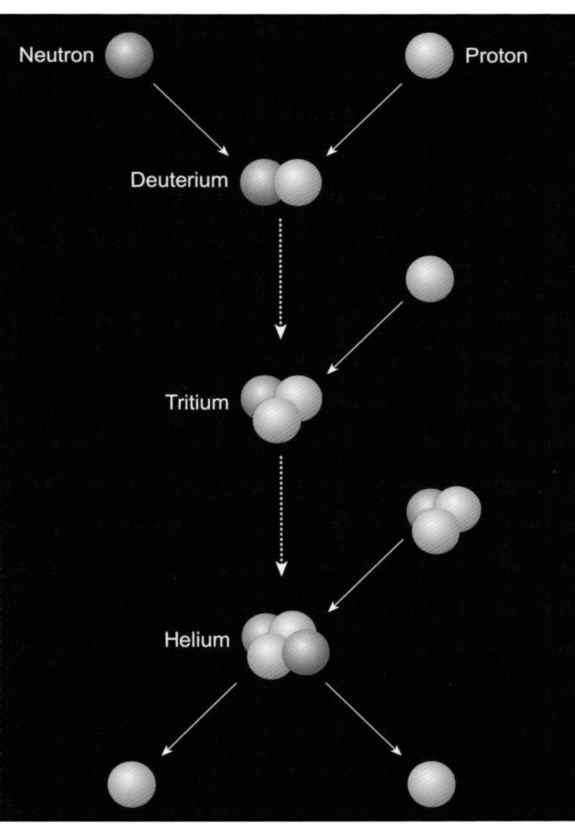

◀ ***Nuclear burning*** *Hydrogen nuclei (protons) fuse together to create helium in the pp (proton-proton) chain reaction. Two protons combine to form deuterium (an isotope of hydrogen); another proton joins to make the isotope tritium. Finally, two tritium isotopes combine to form helium-4, and two protons are emitted.*

Nuclear burning

After the Big Bang, galaxies formed, and stars, in which further nucleosynthesis occurred. Stars process lighter nuclei to heavier nuclei during successive nuclear burning processes in an onion-skin structure of concentric shells around a central core:

- hydrogen burning converts hydrogen to helium-4 by means of pp (proton-proton) chains or the CNO (carbon-nitrogen-oxygen) cycles;
- helium burning converts helium to carbon;
- carbon nuclei receive a helium nucleus to form oxygen;
- oxygen and subsequent burning produce neon and silicon, up to iron.

During helium burning, neutrons are liberated. These are responsible for the slow-neutron-capture nucleosynthesis process (s-process). Neutrons are added successively to the heavier nuclei produced by nuclear burning, but slowly, so that radioactive nuclei at intermediate stages have long enough to decay by the emission of electrons (beta decay). Starting with existing heavy nuclei around iron, the process produces elements up to lead and bismuth.

The r-process

The r-process is nucleosynthesis by rapid neutron capture, in which neutrons are added to existing nuclei by an explosive process. This occurs so quickly that the intermediate radioactive nuclei in the chains of reaction do not have time to decay. No one knows exactly where in the universe the r-process occurs.

Release of the elements into the interstellar medium

Stars of low and intermediate mass (less than six times the solar mass) contribute to the interstellar medium via stellar winds and/or the ejection of PLANETARY NEBULAE. They then form WHITE DWARFS made of carbon and oxygen. Heavier stars (with a mass more than eight times that of the Sun) develop an onion-like structure and then collapse at the end of their evolution, as SUPERNOVAE. This produces an abundance of elements and nuclei close to iron. Nickel-56 produced by the core collapse is radioactive and decays to cobalt-56 and then to iron-56. The gamma-rays that followed these stages of nuclear decay were observed in Supernova 1987A (*see* GAMMA-RAY ASTRONOMY). The energy released by this radioactivity powers the supernova's LIGHT CURVE. The ejecta of supernovae themselves are observed to be rich in oxygen, titanium and iron. Observations of supernova remnants show abundant elements like carbon, oxygen, silicon, chlorine, argon, cobalt and nickel.

An interesting residual question is at what mass a progenitor produces a black hole instead of a neutron star and a supernova explosion. Hypernovae are failed supernovae that produce black holes and are probably powered by black hole ACCRETION and astrophysical JETS. Astronomers would expect hypernovae to be active in producing some of the elements, but their nucleosynthesis products have yet to be analyzed.

There are strong observational and theoretical indications that Type Ia supernovae are thermonuclear explosions of accreting white dwarfs. They too are sites of nucleosynthesis. Others include events in BINARY SYSTEMS such as NOVAE (small hydrogen accretion rates onto a white dwarf and explosive ejection of the accreted layer), and X-ray bursts (hydrogen accretion onto a neutron star and explosive processing of the accreted layer).

Further reading: Arnett W D (1996) *Nucleosynthesis and Supernovae* Princeton University Press; Cox P A (1989) *The Elements: Their Origin, Abundance and Distribution* Oxford University Press.

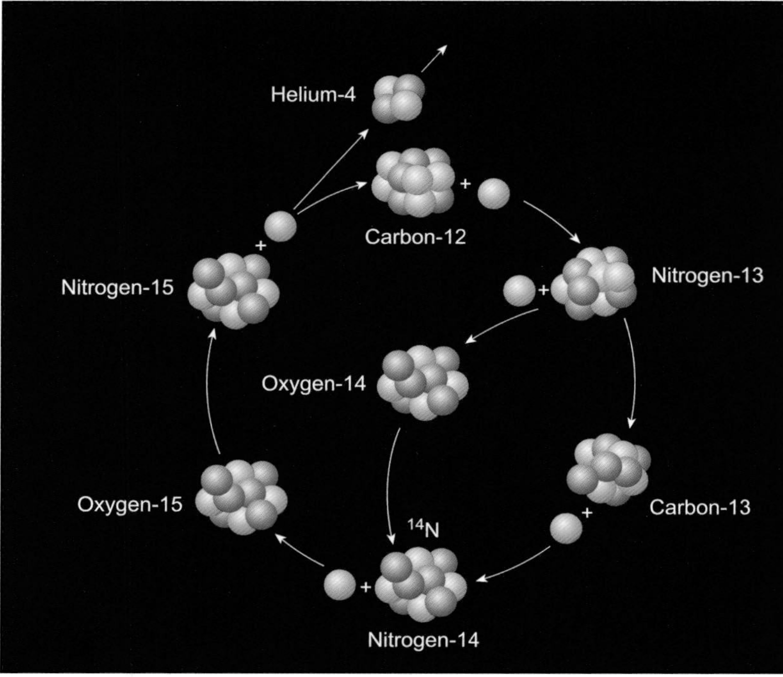

▲ ***CNO (carbon-nitrogen-oxygen) cycle*** *In an endless nuclear circuit inside a typical star, four protons (hydrogen nuclei) combine to produce one helium nucleus. Carbon acts as a catalyst, and is transformed into nitrogen and oxygen, before returning to carbon at the beginning of a new cycle.*

Ellipse

The CONIC SECTION obtained by cutting a cone by a plane at an angle so as to produce a closed curve. An ellipse is an oval shape. An alternative geometric definition is that an ellipse is the locus of the points such that the sum of the distances from two points called foci (singular, *focus*) is a constant. It is possible to draw an ellipse using this definition by putting pins in a piece of paper at the position of the foci and looping a loose thread around the pins. A pencil is then used to pull the thread taut and to draw the shape of the ellipse as it is moved around the pins. The greatest diameter of an ellipse is known as the major axis (half of this is the semi-major axis), and the smallest diameter is the minor axis. The ellipse is symmetrical about both its axes. When rotated about either axis, the curve forms the surface called an ellipsoid of revolution, or a spheroid.

The foci are located on the major axis, one on either side of the center of the ellipse; the greater the separation of the foci, the more flattened the ellipse. A measure of the separation of the foci is given by the ECCENTRICITY, e, which can take values between 0 (a circle) and almost 1 (a PARABOLA). For an ellipse of semi-major axis a and eccentricity e, the distance from the center of the ellipse to a focus is ae.

As discovered by Johannes KEPLER and proved by Isaac NEWTON, the path of a heavenly body moving around another in a closed orbit under the influence of gravity is an ellipse. In the solar system one focus of such a path around the Sun is the Sun itself. Strictly, this result is accurate only if both bodies are point masses (or, at least, spherical and solid), and there are no other disturbing bodies or non-Newtonian forces (such as light pressure or GENERAL RELATIVITY).

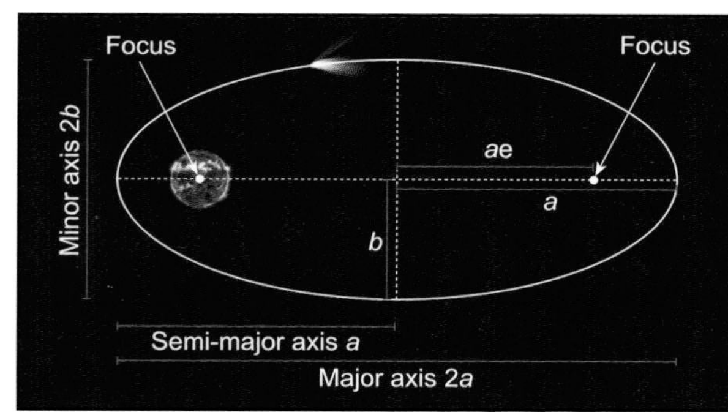

► ***Ellipse*** *The two foci are offset from the center of the ellipse. The larger the offset, the larger the eccentricity,* e.

Elliptical galaxy

An elliptical galaxy is an elliptical-appearing, smooth, quiescent star-pile that is devoid of any of the spectacular structures found in a SPIRAL GALAXY, such as spiral arms, a disk or large amounts of gas and dust. There are no obvious signs of star formation, such as nebulae or star clusters. Elliptical galaxies are old single-stellar populations. This makes them the simplest galactic systems. In the HUBBLE TUNING-FORK DIAGRAM elliptical galaxies are characterized by a single number: the ellipticity = $10(1 - b/a)$, where b and a are the projected angular extent of the short and long axis of the galaxy on the sky. E5 galaxies are the flattest ellipticals, with a long axis twice the length of the short axis. It is possible to determine the three-dimensional structure of elliptical galaxies from their appearance in the sky through statistical analysis. Most are triaxial, with the three axes of the solid galaxy being of different lengths. This means that they are not ellipsoids of revolution (*see* ELLIPSE).

When scrutinized, elliptical galaxies show departures from uniformity, the most common being a systematic increase in flattening with radius (that is, the outer, fainter contours are flatter than the inner, brighter contours). Some ellipticals show spectacular shells in their otherwise smooth structure. And the centers of ellipticals often contain stellar disks, and patches or disks of dust. The centers sometimes host powerful radio sources (*see* ACTIVE GALAXY).

Early theories suggested that elliptical galaxies formed following the collapse of a massive gas cloud. This collapse resulted in a huge starburst, which, about 10 billion years ago, simultaneously turned most of the gas in a HALO into stars. There is emerging evidence for the alternative theory that elliptical galaxies formed that long ago through a sequence of galactic mergers (*see* GALAXIES, COLLIDING).

Elongation

The angular separation between the Sun and a planet or other body orbiting the Sun, or between the Sun and the Moon (the angle Sun-Earth-object). The practical definition, which takes account of the fact that the orbits of bodies orbiting the Sun are inclined to the plane of the ECLIPTIC, is the difference between a planet's CELESTIAL LONGITUDE and the Sun's. Elongation is measured in degrees east or west of the Sun. If a body's elongation is 0° it is at CONJUNCTION, if its elongation is 90° it is at QUADRATURE, and if its elongation is 180° it is at OPPOSITION.

The elongations of the inferior planets, Mercury and Venus, are restricted between certain maximum limits: greatest elongation east, when the planet sets at the latest time after the Sun; and greatest elongation west, when it rises at the earliest time before the Sun. Because Mercury's orbit is so eccentric, its greatest elongation varies between 18° at PERIHELION and 28° at APHELION. For Venus, greatest elongation varies only between 45° and 47°.

Emersion

The reappearance of a star or other body at the end of an eclipse or occultation (*see* ECLIPSE AND OCCULTATION).

Emission nebula

A GASEOUS NEBULA that is self-luminous because it is made of hot ionized gas that, on recombining, emits light. Emission nebulae comprise PLANETARY NEBULAE, the remnants of SUPERNOVAE and H II regions (also called 'diffuse nebulae'). *See also* NEBULAE AND INTERSTELLAR MATTER.

Emission spectrum

A SPECTRUM that consists of a number of bright emission lines, each line corresponding to the emission of light at a certain wavelength. An ATOM or ION will emit a PHOTON of a particular energy (and hence of a particular wavelength) when one of its constituent ELECTRONS makes a 'downward transition,' spontaneously dropping from one of its higher ('excited') permitted energy levels to a lower energy level. The various permitted transitions correspond to a series of lines of different energies and therefore wavelengths. MOLECULES can undergo electronic transitions in the same way as atoms and ions, and they can also emit RADIATION at characteristic wavelengths through downward transitions in their vibrational and rotational states.

To emit radiation through downward transitions, atoms, ions and molecules must first be excited to higher energy levels by absorbing energy. This may occur when an atom, ion or molecule absorbs an incoming photon or collides with another, or captures an electron ('recombination').

An emission spectrum is characteristic of a high-temperature, low-density gas. Typical sources of emission spectra include fluorescent lights, the chromospheres and coronae of stars, luminous nebulae, AGN and quasars.

Enceladus

A mid-sized icy satellite of SATURN, discovered in 1789 by William Herschel (*see* HERSCHEL FAMILY). Its diameter is 500 km and it orbits Saturn at a distance of 238 000 km. It has two distinct types of terrain: areas with a medium density of cratering, and almost crater-free plains with extensive fault features (straight and curved grooves and ridges, such as Daryabar Fossa and Samarkand Sulci). The surface of Enceladus is very fresh. So, not only has the satellite been geologically active in the past, it may still be so today. The energy for the activity could be supplied by tidal heating, produced by its slightly elliptical orbit. It orbits at the center of Saturn's tenuous E ring. Meteoric debris from the satellite is believed to be the source of the ring's small particles.

Encke, Johann Franz (1791–1865)

German astronomer, born in Hamburg. He became professor and director of the observatory at Berlin. In 1805 Encke discovered a comet, which he later found to be identical to a comet observed in 1818. COMET ENCKE has a period of around 3.3 years, the shortest of any comet, and Encke predicted that it would return in 1822 (this was seen only in the southern hemisphere) and in 1825 (Encke observed this return from the Seeberg Observatory near Gotha). At the Berlin

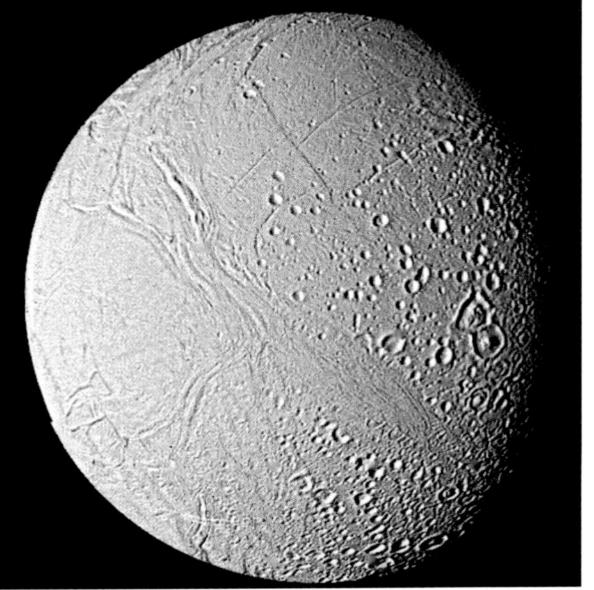

◀ ***Enceladus*** *Voyager 2 passed Saturn in 1981 and took this picture of Saturn's moon, Enceladus. Though the spacecraft was 119 000 km away, structures down to 2 km are visible, including not only craters (Aladdin and Ali Baba are the large craters to the right) but also the grooves and ridges of Samarkand and Sarandib in the left half.*

Observatory he presided over work to calculate the orbits of asteroids, and the discovery of Neptune by Johann GALLE (who used star charts that Encke had edited). In 1838 Encke discovered the gap between the A and F rings around Saturn. This is now known as the Encke Division.

Energy

A physical quantity traditionally defined to be the ability to do work, where 'work' is the application of a force over a distance. The energy that a body possesses by virtue of its motion is called kinetic energy. To change the velocity of a body, a force must be applied; the resultant change in the kinetic energy of the body is equal to the work done by the force. The energy possessed by a body by virtue of its location (for example, in a gravitational field) is called potential energy.

Energy exists in other forms. For example, heat is essentially the kinetic energy contained in the motions of atoms, molecules or ions. ELECTROMAGNETIC RADIATION is a form of energy, the energy of a PHOTON being inversely proportional to the wavelength of the radiation. According to the special theory of relativity, energy (E) and mass (m) are equivalent and interchangeable, the relationship between the two being given by $E = mc^2$, where c denotes the speed of light. Because c is a large number, a small mass is equivalent to a large quantity of energy. The conversion of mass to energy through nuclear reactions powers the stars.

The law of conservation of mass-energy states that the total amount of mass-energy in an isolated system is constant (one form can change into the other, but the total sum of mass and energy remains the same).

The SI unit of energy is the joule (J). A unit commonly used in particle physics is the electron-volt (eV).

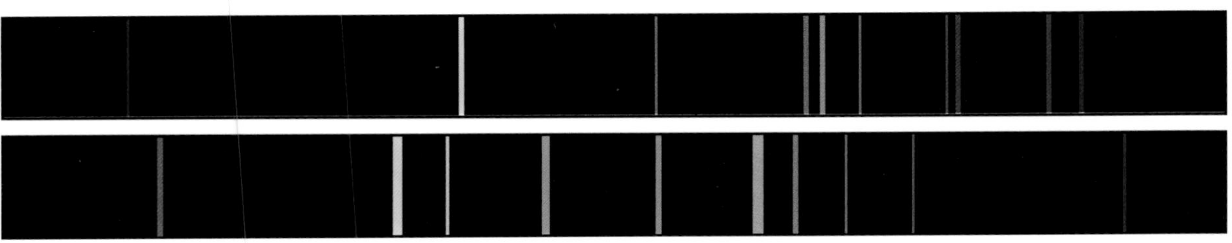

◀ **Emission spectrum** *The emission spectra of a gas of helium (top) and carbon (bottom) in the visible range show that each element has a distinct emission pattern. Spectroscopic analysis enables astronomers to compare the electromagnetic radiation from a radiating nebula to such spectral fingerprints, revealing its elemental composition.*

▶ Equatorial mounting *The rotation of an equatorial telescope parallels exactly the rotation of the Earth.*

Eötvös, Baron Loránd (Roland) von (1848–1919)

Hungarian physicist, born in Pest (now part of Budapest), where he later became professor of experimental physics. He worked on many physical problems, including gravitation, and invented the Eötvös balance (a torsion balance). With this he tested the equivalence principle, in what became known as the Eötvös experiment. *See also* GRAVITATION, GRAVITATIONAL LENSING AND GRAVITATIONAL WAVES.

Ephemeris

The Greek root of the word 'ephemeris' (plural: ephemerides) indicates something that lasts a day. By extension, the astronomical ephemeris is a table that gives the positions of a celestial body at given times. Ephemerides have been published since the fourteenth century. The first ephemerides intended for use in planetary dynamics appeared in the French *Connaissance des temps (Knowledge of Time)*, Paris, 1679. In 1767 the *Nautical Almanac and Astronomical Ephemeris* was published in London; this was designed to improve astronomy, geography and navigation. In 1855 the US Nautical Almanac Office published the first *American Ephemeris and Nautical Almanac*. From 1960 the same tables were published in both the UK and the USA; in 1981 these merged to become *The Astronomical Almanac*. The tabular presentations of the ephemerides were also represented by empirical formulas called Chebyshev polynomials. These were much more compact and well adapted to the then newly developed microcomputers.

The main ephemerides published today are:

- *Apparent Places of Fundamental Stars*, by the Astronomisches Rechen Institut, Heidelberg, Germany;
- *Astronomical Almanac, Nautical Almanac*, by HM Nautical Almanac Office of the former Royal Greenwich Observatory, now at the Rutherford Appleton Laboratory, UK, and by the Nautical Almanac Office of the US Naval Observatory;
- *Connaissance des temps, annuaire du bureau des longitudes, ephémérides nautiques*, by the Institut de Mécanique Céleste/Bureau des Longitudes, France;
- *Ephemerides of Minor Planets*, by the Institute of Applied Astronomy, St Petersburg, Russia.

It is also possible to calculate positions using special servers on the Internet. Some of these can be found at the websites of the following organizations: the Jet Propulsion Laboratory, Pasadena, California; the US Naval Observatory, Washington, DC; the Minor Planet Center, Cambridge, Massachusetts; the Institut de Mécanique Céleste/Bureau des Longitudes, France; and HM Nautical Almanac Office, UK.

Epoch

A particular instant of time used for reference purposes. In the context of measuring stellar positions, the epoch of observation is the date on which the observational data were obtained. A catalog 'for the epoch 2000.0,' for example, would list positions valid for that date. To obtain positions at some other epoch, the effects of proper motion must be calculated (*see* ASTROMETRY). Other effects, such as NUTATION and ABERRATION, have also to be taken into account to obtain a precise position for an object on a particular date.

Note that the epoch of observations of star positions does not have to be the same as the EQUINOX to which the coordinate system is referred (*see* CELESTIAL COORDINATES). The positions of a star with high proper motion at epochs 1972.3, 1984.5 and 1999.1 could be plotted in the coordinate frame of the equinox 1950.0, for example.

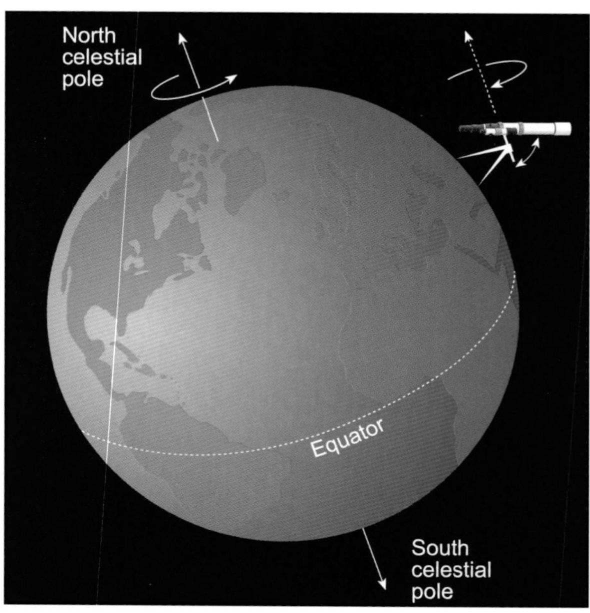

In the context of the motion of planets or binary stars, the term 'epoch' may be used to describe the time of PERIHELION passage, this being one of the elements of the ORBIT from which positions at subsequent times may be determined.

Epsilon Aurigae

This star, sometimes known as Almaaz or Al Anz, is an ECLIPSING BINARY of the same type as ALGOL, but of exceptionally long period. The eclipses last for around a year and occur at intervals of 27.1 years. The next eclipse is due in 2010.

Epsilon Lyrae

One of the best-known star systems among amateur astronomers. The system, situated 1.6° northeast of VEGA at a distance of 160 l.y., consists of a pair of DOUBLE STARS and is commonly known as the Double Double. On a clear night it is just possible to see the wide pair with the naked eye. With binoculars the double stars are easily seen, being 3.5 arcmin apart at a position angle of 173°. A telescope with an aperture larger than 75 mm reveals all four stars. The first double, 1 Lyr, which has a combined apparent magnitude of about 4.7, consists of two white stars. A is magnitude 5.06 and spectral type A3V; B is magnitude 6.19, spectral type F1V. These are separated by 2.5 arcsec at position angle 353°, with an orbital period of about 1165 years. The second system, 2 Lyr, has a combined magnitude of about 4.6 and is similar to the first pair. It also comprises two white stars. C is magnitude 5.30 and spectral type A5V; D is magnitude 5.51, spectral type F0Vn. These stars are separated by 2.3° and have an orbital period of 585 years.

Equation of time

The function that gives the difference between APPARENT SOLAR TIME (sundial time) and mean time. A positive value means that apparent time is ahead of mean time. A negative value means that apparent time is lagging behind mean time. The difference arises for two reasons. First, because the Earth moves around the Sun in an elliptical orbit, when the Earth is near PERIHELION it moves faster than when it is near APHELION; consequently, the apparent motion of the

Sun relative to the background stars is more rapid near perihelion than near aphelion. This variation in the Sun's apparent motion affects the length of the apparent solar day. Second, although the Sun's real apparent motion is along the ECLIPTIC, the variation in the length of the apparent solar day depends upon the daily change in the RIGHT ASCENSION of the Sun. As a result, apparent time is ahead of mean time between April 16 and June 15, and again between about September 2 and December 26, the difference reaching maxima of about 3.75 min and 16.3 min on May 15 and November 4, respectively. Apparent time lags behind mean time during the rest of the year, reaching 14.3 min and 6.3 min on February 12 and July 27, respectively.

Equatorial mounting

A telescope mounting in which one axis of rotation (the polar axis) is aligned parallel to the Earth's axis, and the other (the DECLINATION axis) is aligned at right angles to this axis, in the plane of the CELESTIAL EQUATOR. Rotation about the polar axis allows the telescope to be pointed toward celestial bodies that have different RIGHT ASCENSIONS or hour angles, while rotation about the declination axis enables the telescope to be swiveled perpendicular to the celestial equator in the direction of increasing or decreasing declination.

The advantage of the equatorial mounting is that the apparent DIURNAL MOTION of a celestial body can be followed by driving the telescope in the opposite direction to the Earth's rotation (driving from east to west) at a rate of one revolution per sidereal day around the polar axis only. The disadvantages of the equatorial mounting are that it is more complex and expensive to construct than the ALTAZIMUTH MOUNTING; and that gravitational loadings change in a complex way as the telescope moves, making it difficult to compensate for structural flexure. Nevertheless, until recently most large telescopes were set up on equatorial mountings, mostly of the following types:

- German mounting – the declination axis sits across the top end of a short polar axis, with the telescope at one end and a counterbalancing weight at the other;
- cross-axis mounting – a long polar axis is supported at top and bottom by two pillars, or piers, and the telescope is attached to a declination axis that crosses the polar axis roughly midway between its top and bottom ends. Once again, the telescope is counterbalanced by a weight placed at the opposite end of the declination axis;
- fork mounting – the telescope swings in declination around two pivots attached at opposite sides of the upper end of an open U-shaped fork, the bottom of which is attached to a short polar axis;
- horseshoe mounting – the telescope swings in declination around two pivots attached across the ends of a tilted circular horseshoe, the outline of which rotates on supporting pads around the polar axis through the center of the horseshoe.

Equator-S

German satellite to investigate Earth's equatorial magnetosphere. Operations began in February 1999 but ceased after three months due to a system malfunction. It was the first satellite to receive positional information from GPS satellites while beyond geostationary orbit.

Equinox

An instant at which the Sun crosses the CELESTIAL EQUATOR; the Sun is then vertically overhead at the equator, and day and night have equal duration at every point on the Earth's surface. The apparent annual path of the Sun on the CELESTIAL SPHERE is inclined to the celestial equator and intersects it at two points. The terms vernal equinox and autumnal equinox are applied to these points. Star positions are measured in CELESTIAL COORDINATES based on a particular vernal equinox; for example, the equinox of the year 2000. Thus, the word 'equinox' applied in the context of the position of a star defines the coordinate system being used. *See also* EPOCH.

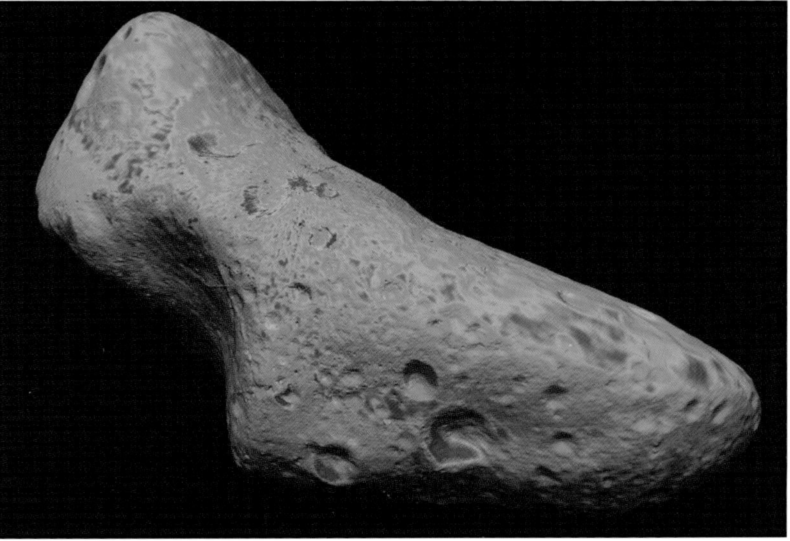

▲ ***Eros*** *The irregular shape of (433) Eros and the fact that it is spinning can conspire to create peculiar patterns of force on its surface. This image in false color shows the effective local 'slope' of the surface, with blue representing a flat 0° and red representing an incline of 35°.*

Equuleus *See* CONSTELLATION

Eratosthenes of Cyrene (*c.*276 – *c.*196 BC)

Astronomer, philosopher, geographer and geometrist. He was born in Cyrene (now Shahhat, Libya) and became a librarian in Alexandria, Egypt. Eratosthenes made a measurement of the size of the Earth. Having heard that at the summer SOLSTICE the Sun illuminated the bottom of a well at Syene (now Aswan, south of Alexandria), and was therefore overhead, he observed the angle of the Sun off the ZENITH on the same day at Alexandria. He measured the distance between the two cities, allegedly by counting the rotations of the wheel of a carriage driven between them, and determined the circumference of the Earth as 250 000 stadia. Depending on exactly which unit he used, this seems to be equivalent to 47 000 km, which is surprisingly close to the true value of 40 000 km. He also measured the tilt of the Earth's axis with great accuracy.

Erfle, Heinrich Valentin (1884–1923)

Optician, born in Duerkheim, Germany. He became head of the telescope department for optics manufacturer Carl Zeiss in Jena. The Erfle widefield EYEPIECE is named for him.

Eridanus *See* CONSTELLATION

Eros

The first Amor ASTEROID to be discovered, independently by Gustav Witt and Auguste Charlois in 1898. Its designation is (433) Eros. It orbits the Sun at a mean distance of 1.46 AU (perihelion 1.13 AU, aphelion 1.78 AU), in a period of 1.76 years. Inclination is 11°, eccentricity 0.22, and rotation period is 5.3 hours. In the 1930s Eros figured in the campaign

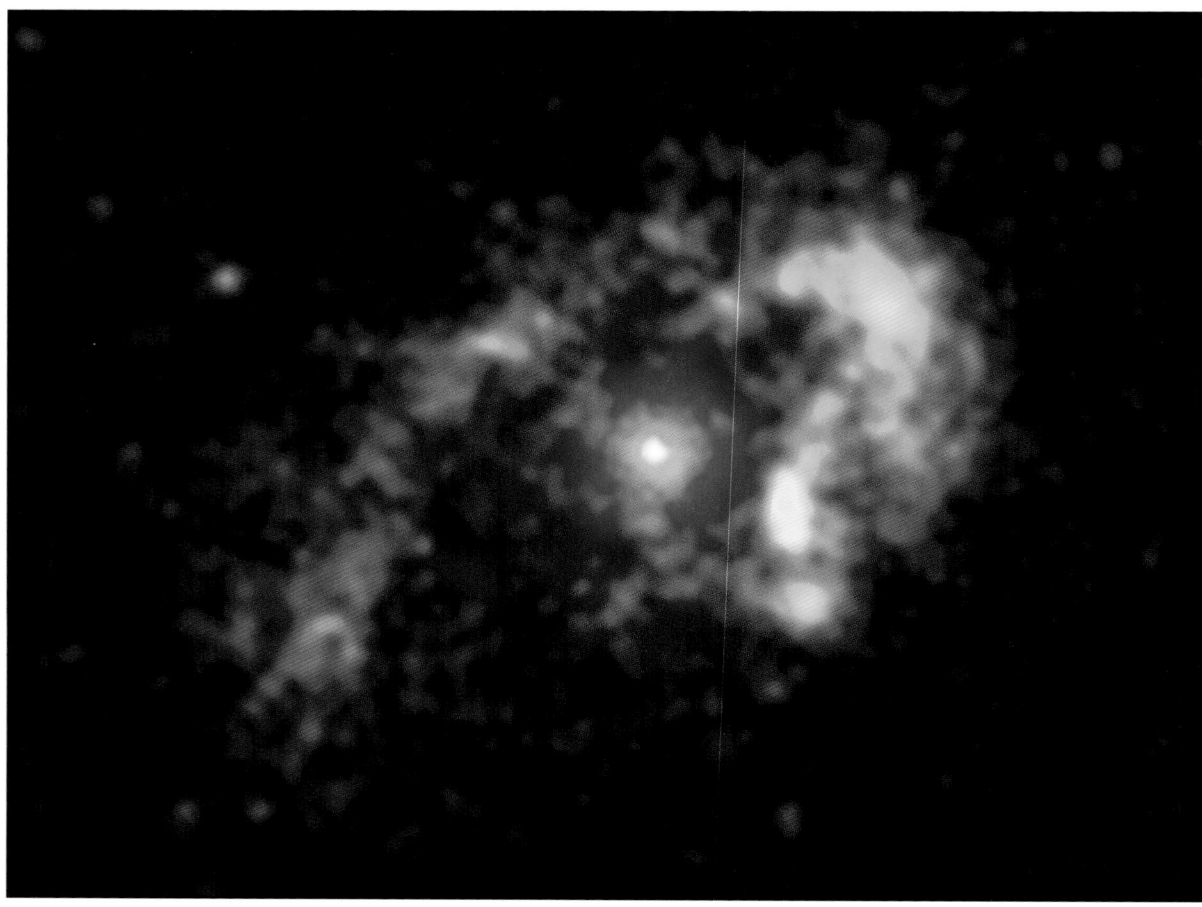

► ***Eta Carinae*** *(NGC 3373) This X-ray image of Eta Carinae, taken by the Chandra Observatory, shows a horseshoe-shaped cloud 2 l.y. in diameter that is invisible in the optical spectrum. The structure was presumably caused by high-speed collisions of ejected matter.*

to measure the scale of the solar system (*see* DISTANCE).

Eros is an S-type (stony) asteroid, with a reflection spectrum similar to that of ordinary CHONDRITES and STONY-IRON METEORITES. It was the prime target of the NEAR Shoemaker spacecraft, which entered orbit around Eros on February 14, 2000. Images returned showed an irregular, elongated object measuring 33 × 13 km, with several craters. Some of these are square, indicating that the asteroid material has inherent strength (it is not a loose accumulation of independent rocks – the 'rubble pile' model). Its surface is covered with dust and blocks of rock, and some craters have 'ponded' material in their bottoms. These flat accumulations of dust cannot have been drained into the craters by water, and their mobility has been identified only as due to some form of SPACE WEATHER. The slight deviation to the probe's trajectory revealed that Eros has a bulk density of about 2.5 g cm^{-3}. On February 12, 2001 the NEAR Shoemaker probe landed on the surface of Eros, the first spacecraft ever to land on an asteroid.

Escape velocity

The minimum free-flying velocity required for an object to leave the surface of, or depart from any point within the gravitational field of, a massive body and recede to an indefinitely large distance. In principle, if atmospheric resistance is ignored, a projectile fired vertically from the surface of a massive body at its escape velocity will continue to move away with a speed that decreases with increasing distance, but that does not decline to zero until it has receded to an infinite distance. The escape velocity from the surface of the Earth is about 11.2 km s^{-1}; from the less massive Moon it is about 2.4 km s^{-1}. BLACK HOLES have an escape velocity that exceeds the velocity of light. A planet's atmosphere will leak into space if the planet's escape velocity is near the average velocity of its gas molecules.

Eskimo Nebula *See* CLOWN FACE NEBULA

Eta Aquarids

A METEOR SHOWER that occurs in late April and May. The RADIANT lies in the constellation Aquarius. The Eta Aquarids occur when the Earth intersects the descending NODE of the METEOR STREAM from COMET HALLEY; the ORIONIDS in October are produced by the Earth's passage through the ascending node. Because Halley's orbit is retrograde, Eta Aquarid meteoroids impact the Earth at a high relative velocity and produce fast meteors. Records of this shower date back to 74 BC.

Eta Carinae

An S DORADUS-type eruptive LUMINOUS BLUE VARIABLE star, situated at the heart of the bright, diffuse nebula NGC 3373 in the constellation Carina. Eta Carinae is obscured by the HOMUNCULUS NEBULA, a bright cloud of gas and dust ejected by the star during a major outburst in 1843. At that time it reached an apparent magnitude of –0.8, becoming temporarily brighter than all other stars except Sirius. Its apparent magnitude is now 6.2, but this varies erratically over decades and has been recorded as faint as 7.9. Eta Carinae is a powerful source of infrared radiation, and is believed to be a highly luminous blue star much of whose visible radiation is absorbed by the nebula and re-emitted in the infrared. Because of the obscuration its spectral type, parallax and absolute magnitude cannot be determined accurately. It has been suggested that Eta Carinae is a very young star in its pre–MAIN SEQUENCE stage, or an extremely old star in the closing stages of its evolution. The latter is now thought more likely. The last phase of its existence could result in a spectacular SUPERNOVA explosion.

Eta Carinae Nebula (NGC 3373)

A bright, diffuse nebula surrounding the star ETA CARINAE. It measures about 2° across and is divided by a dark, V-shaped obscuring dust lane. It contains several interesting stars and star clusters, and also two major nebulae: the bright HOMUNCULUS NEBULA and the dark, absorbing KEYHOLE NEBULA. It is estimated to be about 8000 l.y. away.

Ether

A transparent, weightless medium that was believed by nineteenth-century physicists to fill all space, and to provide the medium through which electromagnetic waves could propagate. Experiments were devised to try to measure the motion of the Earth through the ether (*see* MICHELSON–MORLEY EXPERIMENT), but they failed to detect any such motion. This result was explained by Albert EINSTEIN's theory of relativity, first published in 1905. Because the ether could not be detected, it became an unnecessary hypothesis, and the concept was abandoned.

Eudoxus of Cnidus (*c.*400 – *c.*347 BC)

Greek astronomer and mathematician, born in Cnidus (on the Resadiye peninsula, now Turkey). Eudoxus was the first astronomer to produce a detailed description of the stars and CONSTELLATIONS. His work is now lost but known through Aratus of Soli's poem *Phaenomena*, which was written in about 275 BC. Eudoxus proposed a geocentric system for the solar system in which the Sun, Moon and planets moved in spheres centered on the Earth. The model failed to account for variations in the observed diameter of the Moon or changes in the brightness of planets, which were correctly interpreted to indicate that their distances were changing. When Eudoxus observed some of these discrepancies he tried to adjust the model by postulating that each sphere had its poles set at an angle to the next sphere. His model contained no mechanical explanation and was only a mathematical description, but the spheres were later regarded as having a physical reality. Some of his mathematics became a foundation for parts of Euclid's treatise on mathematics, *The Elements*, and indeed may have been due to him.

Europa

The smallest of the four large Galilean satellites of JUPITER.

Europa Orbiter

Proposed mission to investigate Jupiter's moon Europa. It is part of NASA's Outer Planets/Solar Probe Project and is scheduled for launch in March 2008. The spacecraft is planned to carry a radar sounder to measure the thickness of the surface ice and detect any underlying liquid ocean.

European Southern Observatory (ESO)

The ESO is an intergovernmental European organization with its headquarters and scientific and technical divisions in the German town of Garching. The ESO was founded in 1962 to establish and operate an astronomical observatory in the southern hemisphere, and to promote and organize cooperation in European astronomical research. Its member states are Belgium, Denmark, France, Germany, Italy, the Netherlands, Portugal, Sweden, Switzerland and the UK.

▲ ***European Southern Observatory*** *This view shows the VLT telescopes on the summit of Mount Paranal, Chile at sunset. The laboratory where the telescopes' light beams are interferometrically merged is mostly underground, and lies between the two rearmost telescopes. The grid patterns on the left are the rails upon which the 1.8-m auxiliary telescopes will run.*

The original observatory is located at La Silla, 600 km north of Santiago, in the Chilean Atacama Desert. Since the land was acquired in 1964, 14 optical telescopes have been built on La Silla mountain, 2400 m above sea level. The most recent addition, in 1989, was the 3.5-m New Technology Telescope. The largest telescope at La Silla measures 3.6 m. Other facilities include the MPG (Max Planck Gesellschaft (Institute)) /ESO 2.2-m widefield instrument, the 1.52-m telescope with its new FEROS (Fiber-fed Extended Range Optical Spectrograph) spectrograph, and the Danish 1.54-m telescope. The newly installed Swiss 1.2-m Leonard Euler telescope with the CORALIE echelle spectrograph is successfully 'hunting' for planets around other stars. The 15-m Swedish-ESO submillimeter telescope became operational in 1987.

ESO operates a second observing site on Mount Paranal, 130 km south of Antofagasta in Chile, where it has erected the Very Large Telescope (VLT) 2635 m above sea level. The VLT comprises four 8.2-m reflecting telescopes, and will eventually have several moving 1.8-m auxiliary telescopes. The VLT is the largest optical telescope in the world as measured by the combined collecting area of all four large telescopes working together. Each telescope can be operated independently or in conjunction with the others, when their light beams are sent through underground tunnels and combined at a single focus in the VLT INTERFEROMETER. Two of the VLT telescopes have been operated successfully in this way, providing an interferometric measurement of the southern star ACHERNAR (Alpha Eridani) of 1.9 milliarcsec. At a distance of 145 l.y., this corresponds to a size of 13 million km. The observation is equivalent to measuring a 4-meter-long car on the surface of the Moon.

European Space Agency (ESA)

An international organization whose task is 'to provide for and to promote, for exclusively peaceful purposes, cooperation among European states in space research and technology and their space applications.' ESA has 15 member states: Austria, Belgium, Denmark, Finland, France, Germany, Italy, Ireland, the Netherlands, Norway, Portugal, Spain, Sweden, Switzerland and the UK. Canada participates in some projects under a cooperation agreement.

The agency was born in 1975 from the merging of two existing organizations: the European Space Research Organization (ESRO), which developed satellites; and the European Launcher Development Organization (ELDO), which was responsible for developing a European launch vehicle. ESA's activities cover space science; Earth observation; telecommunications; space segment technologies that include orbital stations and platforms, ground infrastructures and space transportation systems; and basic research in microgravity.

ESA's headquarters are in Paris. Other ESA establishments include: the European Space Research and Technology Centre (ESTEC) in the Netherlands; the European Space Operations Centre (ESOC) in Germany; the European Space Research Institute (ESRIN) in Italy; the European Astronaut Centre (EAC) in Germany; and the European spaceport, the launch center at Kourou in French Guiana.

The science program, in which all member states participate, is one of the agency's mandatory activities, and has consisted so far of about 20 scientific missions (including the HST). It is based on a long-term plan, known originally as Horizon 2000. The most expensive and ambitious missions, called Cornerstones, include SOHO and CLUSTER II (combined as the Solar-Terrestrial Science Program), the XMM-NEWTON mission, the ROSETTA comet mission, and the HERSCHEL SPACE OBSERVATORY. ESA has also launched several small- and medium-class missions, including HIPPARCOS, ULYSSES, the INFRARED SPACE OBSERVATORY, the CASSINI/HUYGENS MISSION and the INTERNATIONAL GAMMA-RAY ASTROPHYSICS LABORATORY.

Evening star

A name given to VENUS when it is visible in the west after sunset. Ancient astronomers believed that morning and evening apparitions of Venus were of two different planets; the evening apparition was called Hesperus. The name 'evening star' is sometimes also given to Mercury. *See also* MORNING STAR.

Event horizon

The boundary of a BLACK HOLE from which light rays cannot escape. Because of this, events occurring within the event horizon cannot be seen from outside. The event horizon of a spherical mass has a radius called the Schwarzschild radius.

Evershed, John (1864–1956)

English astronomer. He discovered the radial motion of the gases in sunspots, now known as the Evershed effect (also called the Evershed–Abetti effect because it was discovered independently by Giorgio ABETTI).

Excitation and ionization

Cold material, such as gas that is isolated in interstellar or intergalactic space, settles into the lowest energy level (known as 'the ground state'). If the material is energized, its constituent atoms and molecules become excited into more energetic states and move between them, emitting and absorbing radiation during these transitions. If the excitation breaks electrons off of the atoms, the material becomes ionized and other electronic transitions may occur; for example, during recombination of the ion and its free electron.

Exit pupil

The image of a telescope's OBJECTIVE LENS or MIRROR formed by its EYEPIECE. All of the rays collected by the objective pass through the exit pupil, which is where the pencil of rays emerging from the eyepiece has its smallest diameter and the illumination of the image is greatest. The exit pupil, therefore, is the best place at which to place the pupil of the eye (or camera lens) when observing with a telescope.

Exobiology and SETI

The word 'exobiology' was coined by Nobel-prizewinner Joshua Lederberg (1925–) to describe the study of the origins, evolution and distribution of life in the universe. The terms 'bioastronomy' and 'astrobiology' are also used. SETI, the Search for Extraterrestrial Intelligence, assumes that life has evolved, as on Earth, to produce technologically advanced civilizations that transmit detectable signals.

The modern discussion of life on other worlds began with the heliocentric theory of Nicolaus COPERNICUS, in 1543, which envisaged each planet as a world, more or less like the Earth. In 1584 the Dominican monk Giordano BRUNO argued that there was an infinite number of planetary systems. His burning at the stake in 1600 by the Roman Inquisition, principally for his other 'heresies,' had a chilling effect on exobiology. Even so, GALILEO GALILEI noted from his telescopic observations in 1610 that the surface of the Moon was 'not unlike the face of the Earth,' Johannes KEPLER conjectured that one large and particularly circular lunar crater might be an artificial city, and in 1638 Bishop John Wilkins (1614–72) made a broader case for lunar inhabitants.

The French philosopher René DESCARTES envisaged a universe full of atoms that formed vortices around every star. This theory implied that inhabited solar systems were ubiquitous (according to Bernard de Fontenelle (1657–1757) in 1686). With Isaac NEWTON's laws of gravitation, the formation of solar systems proved to be complex. The modern study of exobiology is connected to the idea of cosmic evolution, which is extended from Charles Darwin's theory of evolution in biology.

The twentieth-century debate about life on other worlds has returned repeatedly to the planet Mars, searching for intelligence, vegetation, organic molecules and fossils. Percival LOWELL claimed in 1895 that he had seen linear markings on Mars, which he interpreted as artificial CANALS built by Martians to channel water on their dying planet. In 1909 the French astronomer Eugenios Antoniadi (1870–1944) resolved the 'canals' into dark patches, few of which had a physical basis. From 1924 to the 1960s, changes in the dark patches, supported by spectroscopic evidence in the late 1950s, were interpreted as seasonal changes in vegetation. The two Viking lander spacecraft detected no organic molecules when they landed on Mars in 1976. However, investigations by a series of spacecraft and by the MARS PATHFINDER lander discovered geological evidence for water flowing in the past.

In parallel with the search for life in our solar system, astronomers have searched for and found other solar systems (*see* EXOPLANET AND BROWN DWARF). Planets must have the right conditions for life to originate, but we do not know how this happens. During the twentieth century three hypotheses were proposed:

- following George Darwin's (1809–1964) suggestion of life originating in a 'warm little pond,' in the 1920s Alexandr Oparin (1894–1980), John Haldane (1892–1964) and others proposed that life began in a soup of dilute chemicals;
- the PANSPERMIA hypothesis, championed by Svante ARRHENIUS in 1908, claimed that life came from outer space;
- at the end of the century the discovery of bacteria existing under extreme conditions of temperature, pressure and acidity originated the idea that life began underground in a 'deep biosphere' and gradually emerged to the surface.

The APOLLO space program in the 1960s aimed to send men to the Moon to bring back lunar samples. The expedition carried two risks: it was essential to avoid contaminating the Moon with terrestrial organisms; more importantly, any returned lunar organisms should not contaminate the Earth.

The nature of life

The only example of life we have – the terrestrial one – suggests that its characteristics are:

- autoreproduction;
- mutation with a high level of information transfer;
- autoregulation against the constraints induced by the environment;
- evolution.

The last common ancestor, a primitive cell, evolved into a fantastic biological diversity of terrestrial species of fundamental unity. If there is life elsewhere it will have evolved differently from life on Earth, but it may be similar at molecular level, especially if the panspermia hypothesis is correct. If so, extraterrestrial life is based on carbon chemistry and liquid water, even though details in the biochemistry may differ.

All living systems are made of structural units called cells, from one (unicellular organisms) to billions (pluricellular). All cells have similar macromolecules, which are isolated from the outside by a membrane made of lipids. Through complex chemistry, the membrane selects molecules to build, develop and replicate the cell.

Chromosomes made of nucleic acids program the replication of cells. These molecules include the genetic information that makes one living system different from another. The message is stored in deoxyribonucleic acid (DNA). The genetic code synthesizes proteins, which act as powerful catalysts (enzymes), as a transport and storage molecule (for example, hemoglobin, which transports oxygen), or as a mechanical support (such as collagen).

Nucleic acids consist of many units, called nucleotides. The smallest DNA molecule has several thousand, the largest several billion. However, all DNA uses only four different nucleotides. From this alphabet of only four letters nature makes a very large number of words, some of which represent viable life forms.

Proteins are made of relatively simple building blocks, called amino acids. Of the large number of different amino acids, proteins use only 20. These molecules have an asymmetric carbon atom. Consequently they exist in two different configurations, called enantiomers, which are symmetric but not superimposable, like a left and a right hand. (This property is called 'chirality,' and the molecules are said to be 'chiral.') Only one enantiomer is used by living organisms.

The origin of life

Stanley Miller (1930–) provided the first chemical evidence of the origin of life in 1953 with an experiment in 'prebiotic chemistry.' Into a reactor of gases representing the primordial atmosphere of the Earth and water, he injected energy (simulating lightning, volcanoes and so on). Miller obtained amino acids. Since then, hundreds of similar experiments have been carried out using various conditions (such as different energy sources and gas mixtures). If the starting atmosphere has been hydrogen-rich, they have all produced biologically active molecules, including amino acids and nucleotides. Some ingredients of the starting atmosphere are particularly important because of their chemistry in liquid water. These include formaldehyde (HCHO), hydrogen cyanide (HCN), cyanoacetylene (HC_3N) and other nitriles (organic cyanides), which are among the 110 different molecules detected in the interstellar medium, and which were in the presolar nebula that accreted onto the Earth.

The conditions in the primitive atmosphere are not known for certain (*see* EARTH). The first atmosphere may have been hydrogen-rich, but some recent models suggest that it was the reverse, namely oxidizing. The Miller experiments, which seemed to demonstrate that life began from a soup of chemicals, may thus be unrealistic. Life could also have started in deep-sea hydrothermal vents. Another possibility is that biotic chemistry fell from space on METEORITES and COMETS.

Whatever its start, life was widespread on the Earth 3.5 billion years ago, the age of the oldest stromatolites (small geological structures resulting from the interaction of microorganisms with their environments). By this time plant life was already complex and capable of photosynthesis. The evolution of life produced and still maintains oxygen in the Earth's atmosphere.

The search for extraterrestrial life

Direct evidence about extraterrestrial life is sparse. Until 1969 the only extraterrestrial samples available were meteorites, in particular CARBONACEOUS CHONDRITES. There are no extraterrestrial microorganisms in those objects, not even in the famous (or notorious) martian meteorite ALH 84001, as was claimed in 1996. However, carbonaceous chondrites are rich in organic compounds, including amino acids and nucleotide bases, and some contain an excess of the biologically active amino acid enantiomers.

The two most promising planets of the solar system on

▼ ***Exobiology***
The Viking landers are the only two spacecraft which specifically looked for life on Mars. The surface was assessed to have been sterilized by ultraviolet light from the Sun, so they tested the soil beneath for biological activity. Several tests were negative and one was ambiguous. The landers investigated arid areas of Mars and there is still hope that life may have evolved in wetter areas.

which to search for life are Mars and EUROPA (one of the satellites of Jupiter). And TITAN (the largest satellite of Saturn) shows many similarities with primitive Earth.

Since 1996 more than 100 exoplanets have been discovered around other stars, all of them about the size of Jupiter. In the future we may detect more potentially biologically active planets or satellites in these systems. We may even be able to detect signatures of life on them. For example, ozone in a planet's infrared spectrum indicates that it has an oxygen atmosphere and suggests it has life.

SETI

The Search for Extraterrestrial Intelligence has attempted to leapfrog all of these arguments about how life evolves and whether it could evolve in other places in the universe, by detecting an artificial signal. As proposed by Guiseppe Cocconi (1914–) and Philip Morrison (1915–) in 1959, radio waves are the most appropriate channels for interstellar communication. There are 'magic' wavelengths to listen to, such as the radio line of hydrogen at 21 cm. Frank DRAKE carried out the first SETI experiment in 1960 (called Project Ozma), listening for several weeks to Epsilon Eridani and Tau Ceti. The SETI INSTITUTE's Project Phoenix, the University of California's Project SERENDIP (Search for Extraterrestrial Radio Emissions from Nearby Developed Intelligent Populations), and Harvard's BETA (Billion-channel Extraterrestrial Assay) Project are three of the most important SETI searches now under way. No clearly identified signal of extraterrestrial intelligence has been detected. Most searches have reached only 100 l.y. and have used only a few of the billions of channels possible.

Although we have not received any clear message from an extraterrestrial civilization, we have sent some symbolic messages. The Pioneer 10 and 11 spacecraft carry a plate engraved with information about the solar system, the Earth and terrestrial life; and Voyager 1 and 2 carry a similar videodisk. A message has also been sent from the Arecibo radio telescope toward the globular cluster M13.

Further reading: Dick S J (1996) *The Biological Universe: The Twentieth Century Extraterrestrial Life Debate and the Limits of Science* Cambridge University Press; Dick S J (1998) *Life on Other Worlds* Cambridge University Press; Jakosky B (1998) *The Search for Life on Other Planets* Cambridge University Press.

Exoplanet and brown dwarf

Stars, brown dwarfs and exoplanets (planets outside the solar system, also called extrasolar planets) are bodies in space that hold together under their own force of gravity, supported by an internal pressure.

Astronomers distinguish between them as follows. Stars have (or had earlier in their lives) such a large weight to support that their internal pressures and temperatures sustain (or once sustained) hydrogen FUSION reactions. They have a mass above 0.08 the solar mass (about 100 Jupiter masses). A brown dwarf is a body incapable of sustained hydrogen fusion, but massive enough briefly to burn DEUTERIUM. They have a mass between about 0.01 solar masses (10 Jupiter masses) and the lower stellar mass limit (100 Jupiter masses).

Planets are less massive than stars and brown dwarfs. In addition to the criterion based on nuclear burning, a planet can also be defined as a body formed in association with a star through the condensation of PLANETESIMALS in a protostellar disk, whereas brown dwarfs and stars may be formed in isolation. Calculations about circumstellar PROTOPLANETARY DISKS suggest that the boundary between planets and brown dwarfs on this criterion is a mass of about 10 Jupiters.

The discovery of free-floating objects with masses that appear to be below 15 Jupiters, and of an object in a planetary system with a mass of 17 Jupiters makes the distinction between planets and brown dwarfs arguable. So too is the distinction between brown dwarfs and stars. In 2001, two small, cool stars were discovered that resemble brown dwarfs but are actually the remnants of stars that have been whittled down to cool, Jupiter-mass bodies by losing material onto a companion star. In theory, the estimated masses of LL Andromedae and EF Eridani are 40 Jupiters.

Discovery of exoplanets and brown dwarfs

The possibility that there exist planets other than our own has been discussed since classical times (*see* EXOBIOLOGY AND SETI), but the first convincing detections of both exoplanets and brown dwarfs were very recent. The first exoplanetary system was found in surprising circumstances: the host is the 6.2-ms pulsar PSR 1257+12, a billion-year-old neutron star formed from a SUPERNOVA. From precise timing of its radio pulses, Alexander Wolszcan and Dale Frail in 1992 inferred the orbits of three terrestrial-mass companions (this figure has since increased to four or even five). It is thought that such planets form when the neutron star, post-supernova, captures a small amount of material from a companion star into a protoplanetary disk. Evidently planets can form under a wider variety of conditions than was at first conceived.

Apart from this exception, the first Jupiter-sized exoplanet was discovered orbiting the Sun-like G-type star 51 Peg by Michel Mayor and Didier Queloz of the University of Geneva in 1995. In the same year, a group of scientists from Johns Hopkins University and the California Institute of Technology detected the first proven brown dwarf, a resolved, distant companion to the nearby M dwarf star, Gliese 229.

Direct detection techniques

GL 229B was identified by its radiation, which led to estimates of its temperature and brightness, and from there to its size, surface gravity and mass. It is difficult to identify planets this way because the reflected light from a Jupiter-sized planet is typically one billionth that of its star. Infrared emission is one ten-thousandth. A planet or brown dwarf near to a star can only be imaged if scattered light from the atmosphere and the camera is rigorously controlled. In the future, adaptive optics systems with coronographs or space INTERFEROMETERS may detect planets like this. The present capability is close: ESO's ADONIS (Adaptive Optics Near Infrared System) instrument has imaged a 65-AU-radius dust disk around Iota Horologii, but not its Jupiter-like planet at 1 AU.

Astrometric detection

The astrometric signature of an orbiting planet is a periodic wobble in the parent star's proper motion (*see* ASTROMETRY). The system's center of mass (the BARYCENTER) proceeds in a straight line, but the center of light (the main star) oscillates as the unseen companion swings the star from side to side around the barycenter. A Jupiter-sized planet at 4 AU from a solar-type star 10 parsecs away would produce an astrometric amplitude of only 0.5 milliarcsec. In the future, as precision is refined to microarcseconds (from space, by the Gaia satellite, for example), this technique might become powerful, but it has not detected any exoplanets to date.

Photometric detection

Planets and brown dwarfs may partly occult (*see* ECLIPSE AND OCCULTATION) their parent star if their orbits are accurately edge-on to the Earth . Seen from a distant place in the ECLIPTIC plane, Jupiter would dim the Sun by 1% (the ratio of surface areas), and the Earth by only 0.01%. One such exoplanetary system (HD209458, 150 l.y. from the Earth) has been detected, leading to the first determination of the radius of an exoplanet. The planet has a mass 60% that of Jupiter and its radius is 30% larger than Jupiter's. The density is thus consistent with a GAS GIANT planet and no other plausible body. HST detected the spectroscopic presence of sodium in light filtered through the planet's atmosphere when it transited its star. Thus, from transits we learn about a planet's constitution.

Another photometric detection method is to monitor dense star fields for occasions when a star and its planets successively pass across the line of sight to a more distant star. The gravitational microlensing (*see* GRAVITATION, GRAVITATIONAL LENSING AND GRAVITATIONAL WAVES) associated with each body may cause the brightness of the background star to increase temporarily in a series of pulses.

The COROT MISSION, an ESA-sponsored space mission to be launched in 2006 to detect planetary transits, should identify tens of earth-like planets.

Detection by radial velocity variations

More than 100 exoplanets and many brown dwarfs have been detected by monitoring stars for small variations in radial velocity, which are caused in the same way as astrometric wobbles. Precision spectrographs have been designed for this purpose. The Sun orbits the barycenter of the solar system with a typical velocity of only 13 m s^{-1}, principally under the attraction of Jupiter and Saturn. Several groups have measured the radial velocity of brighter stars to an accuracy of 3 m s^{-1}, which is accurate enough to detect a Jupiter in a nearby clone of the solar system if it was monitored for over a decade (the period of Jupiter is 12 years). The technique is limited by the 'noise' due to motions in the stellar atmospheres, and by the star's rotational velocity, which correlates with the stellar activity.

The mass of the companion brown dwarf or planet can be determined by this technique, but only as a lower limit. This is because the INCLINATION of its orbit is unknown (unless the planet also occults its star), which limits the conclusions that may be drawn.

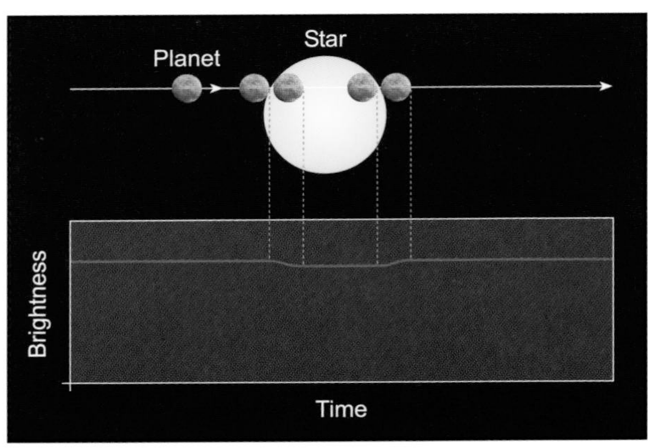

◄ ***Exoplanet occultation*** *As a planet transits across its parent star (assuming we are in the plane of the planet's orbit), the star dims slightly and briefly, this wink hinting at the planet's existence.*

Observing exoplanets

It is really surprising that astronomers, and even amateur astronomers, can detect planets circling faraway stars, but it is possible; about 100 exoplanets have already been found. The professional astronomers are using the world's largest telescopes with high-end spectrographs to detect the minute telltale radial speed variations caused by planets (*see* RADIAL VELOCITY). However, PHOTOMETRY, another method of observing exoplanets, is available to amateurs and requires only quite modest equipment compared with what is used in large observatories.

Only one special case of exoplanets can be observed photometrically: exoplanets that are transiting their parent stars (*see* TRANSIT). These are planets that travel over the star's face, thereby making the star appear a tiny fraction dimmer than usual. This dimming can be measured quite easily by means of CCD photometry. If the planet is already known from spectroscopic observations, then it is known when the transit is likely to happen, and when it is best to observe. The first observation of a transiting exoplanet was made by the author in September 2000. The planet HD209458-b has been observed repeatedly since then by amateurs with modest equipment.

A photometric exoplanet observation is accomplished by taking several CCD images before, during and after the transit and measuring very carefully the changes in star brightness. The dimming caused by a planet is very small, a few hundredths of a magnitude at best, so the accuracy in measuring must be quite good. Even more important, the accuracy must remain just as good during the several hours of the transit.

Measuring parent stars is not usually difficult, because they are often bright nearby stars. Therefore, when performing differential photometry, one needs to measure the comparison star or stars with the same accuracy – and therein lies the challenge. Quite often the parent star is the only bright star on the CCD image, and when one adjusts to measure the faint background stars, the parent star may easily saturate the CCD. A trick one can use to avoid saturation is to unfocus the telescope a bit so that the light spreads over more pixels. The same technique helps if using an under-sampled system, that is a short focal length telescope with a small-pixel CCD.

As the observing runs tend to be quite long, the target star is evidently observed through different air masses at different altitudes. This can cause photometric errors larger than the magnitude drop from the transit, making detection very difficult. One way to avoid these problems is to use a photometric filter (*see* CAMERA AND FILTER) to make the stars more or less the same color. Using photometric filters also makes comparing your results with those of other observers easier. If the photometry is not good enough on single frames, the observer can either co-add several consequent frames to increase signal-to-noise ratio or use statistical tools; for example, averaging the results of several measurements. The statistical approach is usually easier; moreover, as an extra bonus, it will give true error analysis as well.

Arto Oksanen is a computer and system administrator and uses the 16-in telescope of the Nyrölä Observatory in Finland.

Properties of exoplanets

• About 5% of MAIN SEQUENCE stars have a giant planet at a distance less than 2 AU. It is of course not known if there is the same proportion of small, solid, Earth-like planets, since they are not yet detectable.

• It had been expected that Jupiter-sized planets would be found at a distance of several AU from solar-type stars. This is the distance at which the temperature allows for the formation of icy cores of GIANT PLANETS. Many exoplanets lie as close as 0.05 AU from their parent star, where the temperatures are as high as thousands of degrees. It is thought that when a giant planet has formed at that distance in a protoplanetary disk of gas and dust, its gravity creates a local hole in the disk, which shears into a circular gap. Tidal forces then occur between the planet and the inner and outer edge of the gap. These tidal forces pull the planet toward the star. This is known as orbital migration.

• Half of the planets have orbits with unexpectedly large ECCENTRICITIES. One such planet orbits the star 16 Cygni B, which is one member of a binary system; the planet's orbital eccentricity comes perhaps from perturbation by the other member, 16 Cygni A. In other cases, the eccentricity of the orbits might be the result of orbital migration, during which one planet expels another from the system and sees its own orbit perturbed.

Free-floating brown dwarfs

Dozens of brown dwarfs have been found in photometric surveys of several young clusters of stars, such as the Pleiades. Dusty disks surround some brown dwarfs in the ORION NEBULA, confirming that they are more star-like than planet-like, with the potential to form planetary systems.

There are more than 100 brown dwarfs among the 'field' infrared point sources detected in DENIS (Deep Near Infrared Survey) and 2MASS (Two Micron All Sky Survey). In the solar neighborhood there are as many brown dwarfs as other stars, though they contribute only modestly to the mass of the Galaxy.

Further reading: Boss A (1998) *Looking for Earths* Wiley; Clark A (1998) *Extrasolar Planets: The Search for New Worlds* Wiley-Praxis.

Exosat (European X-ray Observatory Satellite)

ESA X-ray observatory launched in 1983. It was designed to locate and identify cosmic X-ray sources. The orbit was highly elliptical (356 × 191 581 km), with the satellite spending most of its time moving slowly at the APOGEE region. This allowed continuous observations of X-ray sources for up to 72 hours without interference from radiation or obstruction, as occurs at low altitudes. Exosat's payload consisted of three instruments to study the X-ray SPECTRUM between 0.04 and 50 keV. The satellite suffered several equipment failures but made important studies of binary stars and galaxies, including the discovery of several new transient X-ray sources. The satellite's operations ended in April 1986.

Expanding universe *See* HUBBLE LAW

Explorer

A series of small NASA experimental, scientific and research satellites, numbering nearly 80 missions. Explorer 1, launched in 1959, discovered the Earth's VAN ALLEN BELTS. Past satellites included the Interplanetary Monitoring Platforms (IMP, 10 satellites which investigated the Earth-Moon magnetic environment over a complete solar cycle), the Radio Astronomy Explorers (Explorers 38 and 49), and the SMALL ASTRONOMY SATELLITES (Explorer 42, named UHURU; 48 , known as SAS-B; and 53, called SAS-C). Recent noteworthy missions, some of which are still operating, include Explorers 50 (IMP 8), 57 (IUE), 66–74 (COBE, EUVE, SAMPEX, RXTE, FAST, ACE, SNOE, TRACE and SWAS), 77 (FUSE) and 78 (IMAGE).

Extragalactic background light

The integrated light from all sources in the universe, including those that are not individually detected. Because we do not have a privileged position in the universe, it appears homogeneous and the same in every direction, so the extragalactic background light (EBL) is more or less uniform everywhere in the sky. The cosmological significance of the EBL was first recognized as early as the 1700s, when Heinrich OLBERS formulated what has become known as OLBERS' PARADOX.

Extragalactic sources contribute less than 1% of the sky brightness from the ultraviolet (UV) to the infrared (IR). Terrestrial, zodiacal and Galactic sources produce foreground emission, which makes the EBL difficult to measure. Most of the EBL at UV to IR wavelengths is produced by stars in galaxies at all distances up to the distance corresponding to the time at which the first stars formed, after the Big Bang. The first galaxies appear now at redshifts of about 10. The infrared background 'light' is from the COSMIC MICROWAVE BACKGROUND (the cooling fireball of the Big Bang explosion), and from unresolved individual galaxies (both active galaxies and galaxies that contain large amounts of warm dust, heated by newly born stars).

Extrasolar planet

A planet outside the solar system. *See also* EXOPLANET AND BROWN DWARF.

Extreme Ultraviolet Explorer (EUVE)

NASA spacecraft for ultraviolet astronomy launched in 1992 to carry out a full-sky survey in the extreme ultraviolet (EUV) range of the spectrum. It carried three EUV telescopes, each sensitive to a different wavelength band. A fourth telescope performed a high-sensitivity search of a limited sample of the sky in a single EUV band. Scientific operations ended in December 2000, and the EUVE re-entered the Earth's atmosphere on January 31, 2001. EUVE observed more than 1000 sources.

Eyepiece

A magnifying lens used to view and enlarge the image produced at the FOCUS of a telescope. For any given telescope, the MAGNIFICATION achieved is inversely proportional to the FOCAL LENGTH of the eyepiece. A telescope used for visual work will normally have a range of eyepieces of different focal lengths. The eyepiece is normally mounted in a short tube (the drawtube) that can be slid in and out to enable the sharpest focus to be achieved. For visual use, the eyepiece is normally placed beyond the focus of the telescope, at a distance such that the rays emerging from the eyepiece from each part of the image enter the eye along parallel paths. The eye is then relaxed (as if looking at an infinitely distant object) and less likely to become tired. For an eyepiece consisting of a single lens, this condition is achieved when the distance between the focal plane of the telescope and the eyepiece is equal to the focal length of the eyepiece.

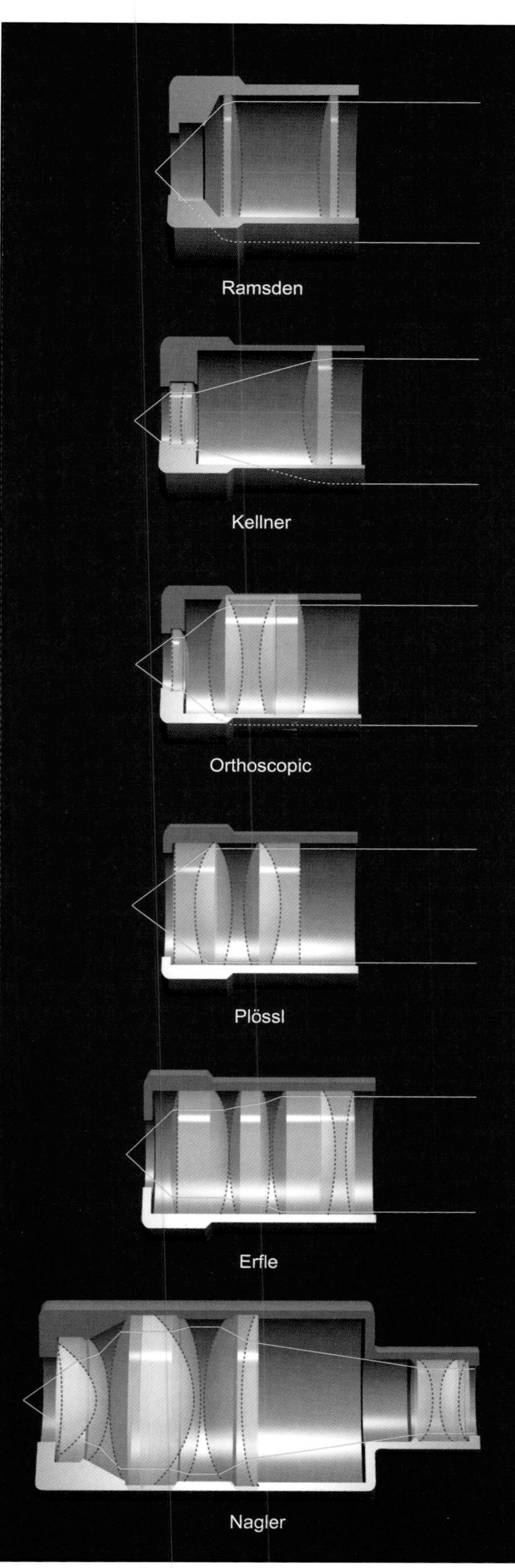

◄ ***Eyepiece***
A range of typical eyepieces.

The main parameters of an eyepiece are focal length and apparent angular field of view. Focal length determines the magnification. The apparent angular field of view, or acceptance angle, is the angular diameter of the circle of vision that can be seen when the eye is placed at the exit pupil. Apparent fields of view for eyepieces range from about 30° to as much as 85°, depending on the design. The observed field of view (the angular diameter of the region of sky that is visible through the telescope) is equal to the field of view of the eyepiece divided by the magnification. For example, if an eyepiece with an acceptance angle of 50° produces a magnification of ×100, the observed field of view will be 0.5°. In general, the more accurately an eyepiece is to perform the more component lenses it needs. One lens (a singlet) is good, two lenses (a doublet) are better, and a triplet is better still.

Types of eyepiece include:

- a single biconvex lens. This is the simplest design, but it suffers from various ABERRATIONS, in particular CHROMATIC ABERRATION and SPHERICAL ABERRATION. In practice, eyepieces consist of at least two lenses: a field lens, which accepts rays of light from the telescope objective or mirror; and an eye lens, from which light emerges into the eye;
- Huygenian and Ramsden eyepieces. These consist of two plano-convex lenses. (With the Huygenian eyepiece, the field lens is placed inside the focal plane of the telescope and the image is formed, inconveniently, between the two lenses.) Although cheap to produce, the eyepieces suffer from chromatic aberration and have modest fields of view (35–45°).
- the Kellner. This is similar to the Ramsden, but uses as its eye lens a doublet in which each lens is made of different sorts of glass chosen so as greatly to reduce chromatic aberration;
- the Orthoscopic eyepiece (in which the field lens is a triplet and the eye lens is plano-convex), and the Plössl (which consists of two identical achromatic doublets). These are more complex and provide better correction for aberrations, good fields of view (about 50°), and good EYE RELIEF;
- the Erfle and Nagler eyepieces. These are widefield and suitable for observing star clusters or nebulae of large size. Eyepieces of this kind, which consist of many elements (typically six or more), offer angular fields of view of up to 85° but are expensive to produce.

Eye relief

The distance between the EYEPIECE and the EXIT PUPIL. In the case of a compound eyepiece consisting of two or more lenses, the eye relief is the distance from the final lens to the exit pupil. For comfortable visual observing, the eye relief should be about 6–10 mm.

Faber–Jackson relationship

The correlation between the luminosity of an ELLIPTICAL GALAXY and the velocity dispersion (the spread in velocities) of the stars in its central regions, established in 1976 by Sandra Faber and Robert Jackson. The light that emerges from the central regions of an elliptical galaxy is the combined light of large numbers of stars, some of which (measured relative to the center of the galaxy) are approaching the observer, and others receding. Consequently, the DOPPLER EFFECT broadens the lines in the galaxy's spectrum by an amount that depends on the spread of velocities among its constituent stars. This spread is found to depend on the mass of the galaxy. As a general rule, the higher the galaxy's mass, the more stars it contains. Consequently, it is logical that there should be a link between velocity dispersion and luminosity. This is analogous to the TULLY–FISHER RELATION for spiral galaxies.

Fabricius, David (1564–1617) and Fabricius, Johannes (1587–1616)

David Fabricius, a Lutheran pastor and astronomer from northwest Germany, discovered in 1596 the first known VARIABLE STAR, Mira Stella ('wonderful star'), now called MIRA (Omicron Ceti). Fabricius initially thought it was a NOVA, but eventually Mira's long periodic cycle was recognized. In 1611 David and his son Johannes discovered spots on the disk of the rising Sun, noting how with time the spots rotated behind the Sun and back into view. Johannes wrote a tract on sunspots that was overshadowed by Christoph Scheiner's (1575–1650) similar discoveries a year later. *See also* SUNSPOT, FLARE AND ACTIVE REGION.

Facula

A patch of enhanced brightness in the solar PHOTOSPHERE. Faculae are often seen in white-light images of the SUN. They can normally be observed when close to the LIMB, where the background brightness of the photosphere is less intense than at the center of the disk. Faculae form inside ACTIVE REGIONS where localized magnetic fields heat their material to temperatures several hundred kelvin higher than the temperatures of their surroundings. Although they are often associated with sunspot groups, they may also be seen before spots emerge and after they disappear. *See also* SUNSPOT, FLARE AND ACTIVE REGION.

False color image

An image in which bright and dim parts are represented as colors (for example, red = bright, blue = dim, and other colors between). Contrast with 'true color image,' in which three images, observed in different spectral bands, are combined as a color picture, even if the three bands are not visible to the eye and the colors are thus not those that can be distinguished by a human being.

False Cross

A cross-shaped ASTERISM in the southern hemisphere formed by the four stars Delta and Kappa Velorum, and Epsilon and Iota Carinae, all of magnitude 2. The asterism is so named because it is sometimes mistaken for the CONSTELLATION Crux (the Southern Cross), which is more compact and comprises brighter stars.

◄◄ False Cross *The False Cross (top right) and the Southern Cross (bottom left) lie in the Milky Way, either side of the constellation Carina. The image has been rotated anticlockwise through 90°.*

Far-Infrared and Submillimetre Telescope (FIRST)

See HERSCHEL SPACE OBSERVATORY

Far-Ultraviolet Spectroscopic Explorer (FUSE)

Funded by NASA as part of its Origins Program, FUSE is the first large-scale space mission to be fully planned and operated by a US academic establishment, Johns Hopkins University in Maryland. It began as a NASA-managed project in the mid 1980s, but was later restructured to reduce costs and development time. FUSE was designed to conduct high-resolution spectroscopic studies at far-ultraviolet wavelengths. Launched in June 1999, it is surveying the cosmic abundance of DEUTERIUM (heavy hydrogen), which must have been formed in the Big Bang. FUSE is also studying the hot gas content of the Milky Way and nearby galaxies, providing new information about galaxy evolution and STAR FORMATION. *See also* COSMIC ABUNDANCE OF ELEMENTS; EXPLORER.

FAST (Fast Auroral Snapshot Explorer/ Explorer 70)

NASA EXPLORER mission to observe and measure rapidly changing electric and magnetic fields, and particle flows in the acceleration region above Earth's AURORA. Launched in August 1996.

Faulkes Telescopes

Two 2-m robotic telescopes that are being built on the island of Maui in Hawaii and at Siding Spring near Coonabarabran in Australia. They are funded by the Dill Faulkes Educational Trust. The aim is to give schools and colleges access to world-class telescopes, enabling students to participate in real research programs mentored by professional astronomers. The telescopes will be operated remotely from control centers in the UK, Hawaii and Australia. They will be available to users in the UK during their daytime, and to local users in their evening or early morning.

FG Sagittae

FG Sge is a peculiar star undergoing extremely rapid evolutionary changes. It has evolved at a remarkable pace over the last century, changing among other properties its appearance from a hot (O-type) star to a luminous, cool (K-type) giant. FG Sge is now experiencing its second stage as an asymptotic giant branch (AGB) star. The best evidence that FG Sge is indeed a 'born-again giant' is its surrounding

▼ Faulkes Telescope *Two robotic telescopes are being built with funding from the Dill Faulkes Educational Trust. The picture shows one of the instruments, in Maui, Hawaii. The other is at Siding Spring, Australia.*

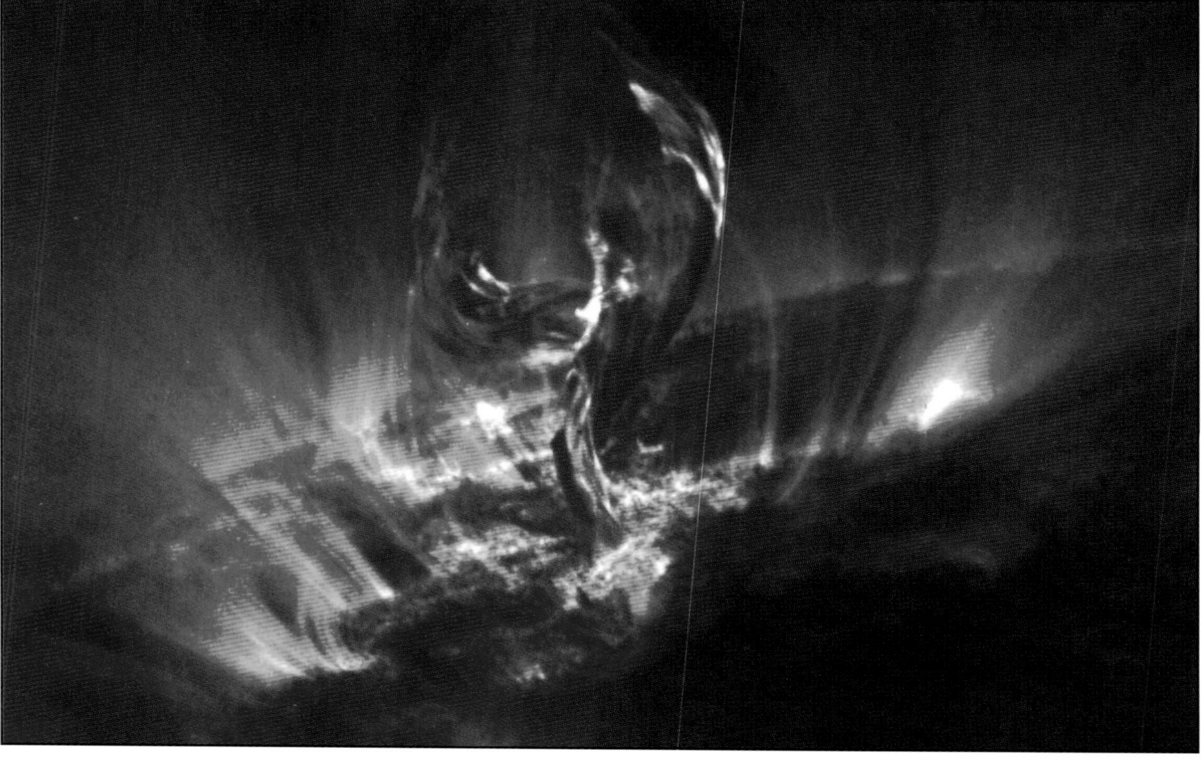

▶ ***Filament***
The TRACE satellite pictured this filament lifting off from the Sun's surface on July 19, 2000. The filament is 100 000 km long.

nebula, Henize 1-5, which implies that FG Sge must very recently have been the nucleus of a hot PLANETARY NEBULA. Theoretical estimates suggest that in about 10% of stars going through the planetary nebula phase, the pre-WHITE DWARF will experience one more helium-shell flash, causing an additional loop in the HR diagram. Owing to the heating of the interior during the gravitational collapse of the core, the temperature may rise sufficiently to again trigger helium burning. The newly liberated nuclear energy has ballooned the star back to giant dimensions and simultaneously cooled the stellar surface. As one of just a few such known stars (other examples are V605 Aql and SAKURAI'S OBJECT, V4334 Sgr), FG Sge provides astronomers with an unusual opportunity to study stellar evolution and NUCLEOSYNTHESIS in real time.

Field of view

The angular diameter of the area of sky that is visible through a telescope or binoculars. When a telescope is used visually, this equates to the area of sky that can be seen when the eye is placed at the EXIT PUPIL of the EYEPIECE. For most large telescopes, the field of view is less than 1°, a few tenths of a degree being typical. SCHMIDT TELESCOPES, however, have photographic fields of view of up to 10°.

Filament

A long, dark absorption feature seen against the solar disk in monochromatic light (the light of one particular wavelength). Filaments are clouds of gas that are suspended above the CHROMOSPHERE by magnetic fields. They have temperatures in the range 5000–10 000 K, and are tens of thousands to hundreds of thousands of kilometers long. They are normally located along the boundary between regions of opposite magnetic polarity on the solar surface (the 'neutral line').

Filaments appear dark because they absorb light from the background solar disk. However, they also emit light, and when solar rotation carries a filament across the edge of the solar disk it appears as a bright feature (a PROMINENCE) against a dark background. Most filaments have lifetimes of about a month, but larger ones survive for several months. If the rising filament crosses the solar limb (the edge of the visible disk) it is seen as an eruptive prominence.

Filar micrometer

A device that is used for the accurate visual measurement of the angular SEPARATION and relative orientation of two neighboring astronomical objects, such as the two component stars of a visual binary (*see* BINARY STAR). It consists of a fixed 'horizontal' wire and two 'vertical' wires, one of which is fixed and the other movable. The movable wire is adjusted by turning a finely threaded screw. The wires are positioned at the focal plane of the telescope and viewed through an EYEPIECE in which the wires and the astronomical objects are both in focus. The measured separation can be converted to the angular separation of the stars on the sky if the FOCAL LENGTH of the telescope is known.

Filter *See* CAMERA AND FILTER

Finder

A small telescope of low magnification and relatively wide FIELD OF VIEW that is attached to the main telescope and optically aligned with it. The finder's wide field of view enables the observer to locate an object of interest (or a bright star nearby). The controls of the telescope are then adjusted to bring it to the center of the finder, which is usually marked by crosswires; the object should then be readily visible within the field of view of the main telescope. Modern computer-controlled telescopes can position themselves on a celestial object to arcsecond accuracy, and have little need for a finder telescope. *Contrast* GUIDE TELESCOPE.

Fireball

An exceptionally bright METEOR, traditionally defined as one that exceeds Venus in brightness, that is, a meteor with a magnitude of –5 or greater. About one meteor in every thousand recorded is a fireball. Some METEOR SHOWERS, for example the GEMINIDS, produce a lot of fireballs. Among SPORADIC METEORS, fireballs are most likely to occur in the spring of either hemisphere because that is the time when, seen from one hemisphere or the other, the meteors penetrate to the lowest regions of the atmosphere (and are bright) and are at a high elevation (and readily visible). Fireballs are produced by larger-than-average meteoroids and usually originate from ASTEROIDS rather than COMETS, so are more dense. A sufficiently large object will not only produce a fireball but will be only partially consumed by its passage through the Earth's atmosphere, and will fall to the surface as a METEORITE. A fireball that generates a sonic boom is known as a BOLIDE.

Fireball observations

If a fireball is recorded from two or more stations, the true path of the METEROID through the atmosphere can be calculated, as well as its orbit in space before the encounter with Earth. The orbit will also point to an impact area where METEORITES may be found. Most of the observing stations are maintained by amateur astronomers.

The American Meteor Society collects and coordinates fireball observations in North America. The international Meteorological Society does the same with view to collecting fallen meteorites. The Meteorite and Impacts Advisory Committee of the Canadian Space Agency collects fireball sightings in Canada with a view to recovery. The international Meteor Organization coordinates METEOR work by amateurs. The program is coordinated by the ONDREJOV OBSERVATORY.

The European Fireball Network of observing stations across Europe is equipped with wide-angle cameras to photograph bright meteors. The cameras contain a rotating shutter allowing the fireballs' apparent velocity to be measured and a clock records the time of the observation.

First point of Aries

The position on the CELESTIAL SPHERE where the ECLIPTIC crosses the CELESTIAL EQUATOR as the Sun progresses north in the northern spring. This defines the origin of RIGHT ASCENSION at a particular EPOCH. It is an alternative term for vernal EQUINOX. At one time (some 2000 years ago) the vernal equinox lay in the constellation of Aries but, because of PRECESSION, this is no longer the case and it has moved into Pisces, and will move next into Aquarius. When the transition is made, depending on the boundary of the constellations, it will be the 'dawning of the Age of Aquarius.'

First point of Libra

An alternative term for autumnal EQUINOX. Owing to PRECESSION this point no longer lies in the constellation of Libra but, instead, is located in Virgo. *See also* FIRST POINT OF ARIES.

Fission

The breaking apart of a body into smaller fragments. In the context of nuclear physics, the term 'nuclear fission' refers to the splitting of a heavy atomic nucleus such as uranium, thorium or plutonium into two or more lighter nuclei, with the release of energy. Prior to fission the mass of the nucleus is greater than the combined masses of the fragments, the difference in mass, Δm, being released as a quantity of energy, ΔE ($\Delta E = \Delta mc^2$, where c denotes the speed of light). *See also* FUSION.

Five College Radio Astronomy Observatory

The Five College Radio Astronomy Observatory (FCRAO) is a research organization within the University of Massachusetts, USA. FCRAO operates the 14-m telescope located in New Salem, Massachusetts, for observations within the 3-mm wavelength band.

Flagstaff Field Center

The US Geological Survey's Flagstaff Field Center was established in 1963 to provide geologic information about the Moon and to help train astronauts. After APOLLO, research expanded to support robotic exploration of the planets, and the study of climate and human impacts on the natural environment.

Flame Nebula (NGC 2024)

A bright nebula in the constellation Orion. About 0.5° across, its illuminating source is hidden behind a patch of dark nebulosity that crosses its center.

Flammarion, Nicolas Camille (1842–1925)

French astronomer and geophysicist. He observed DOUBLE STARS and planets from his observatory at Juvisy, south of Paris. A well-known popularizer of astronomy, he produced many lavishly illustrated books (such as *L'Astronomie populaire)* and the journal *L'Astronomie*, which is still published. It was Flammarion who, as shown by Arthur Beer and Bruno Weber, drew, sometime before 1888, a much reproduced 'German medieval woodcut' that showed the Ptolemaic universe with a pilgrim looking through the CELESTIAL SPHERE to the mechanism beyond. *See also* ASTROLOGY for an illustration.

Flamsteed, John (1646–1719)

English astronomer. Flamsteed was the first director of the ROYAL OBSERVATORY, GREENWICH, and the first ASTRONOMER ROYAL. He had originally proposed the need for an observatory to King Charles II, and when it was completed, in 1675, the King appointed Flamsteed as director at a salary from which he had to provide his own instruments.

Flamsteed was a skilled observer, and he made the most accurate measurements of the positions of the stars and Moon then known. These were used for navigation, in particular for determining longitude. Flamsteed was a perfectionist and he clashed with Isaac NEWTON when the latter, sitting on the board of governors of the observatory, pressed Flamsteed to release preliminary data on the Moon, to test the law of gravitation. Edmond HALLEY intervened on behalf of Newton and edited the observations for immediate publication, over Flamsteed's objections. Flamsteed bought and burned 300 of the 400 published copies. His definitive observations were published by his widow as the *Historia Coelestis Britannica* ('British Catalog of the Heavens'). Flamsteed introduced a system of cataloging stars that is still in use. Known as Flamsteed numbers, stars are numbered in order of RIGHT ASCENSION within a CONSTELLATION (for example, 61 Cygni).

▶ *Flare* *This ultraviolet image of the Sun was captured by SOHO in September 2001. The brightest spot on the Sun is so bright that it has caused a streaky defect in the CCD image. It is a powerful solar flare, classified as an M9 event by the NOAA Space Environment Center, erupting near a sunspot. The flare emitted X-ray and extreme ultraviolet radiation, causing radio blackouts on Earth.*

Flare

A localized, explosive release of energy on the Sun, usually from a site located within a complex ACTIVE REGION. A typical flare reaches peak brightness within the first 5 min of the event and declines more slowly over the next 20 min. Some flares can last for up to 3 hours. Flares emit radiation over practically the entire ELECTROMAGNETIC SPECTRUM, from hard (short-wavelength) X-rays, or even gamma-rays, to radio waves. The bulk of the energy is emitted at X-ray and extreme-ultraviolet wavelengths. Flares eject streams of high-energy subatomic particles, including electrons (which in some cases are accelerated to speeds in excess of half the velocity of light), protons and small numbers of heavier nuclei, and expel bulk clouds of PLASMA through the CORONA into interplanetary space. As the high-speed particle streams and plasma clouds plow through the corona, they stimulate the emission of microwave and radio radiation across a wide range of frequencies. These dramatic events also send shock waves across the surface of the Sun and into its interior, triggering PROMINENCE phenomena. Flares are believed to be caused by a process called magnetic reconnection, in which magnetic energy that had previously been stored in the twisted magnetic fields of complex sunspot groups or active regions is suddenly released. Flares affect the Earth as a component of SPACE WEATHER.

Flare star

An intrinsically faint red dwarf star with emission lines in its spectrum that show sudden short-duration flares in brightness. The flares may last for 10 to 60 min, with output peaking in the near-ultraviolet region of the spectrum. These VARIABLE STARS are also known as UV Ceti (or UV) variables, for the prototype of the same name.

Flash spectrum

The emission-line SPECTRUM of the solar CHROMOSPHERE. The name derives from the fact that when the Sun is observed with the aid of a spectroscope during a total solar eclipse, the faint chromospheric emission lines suddenly flash into view when the brilliant light of the PHOTOSPHERE is blotted out by the disk of the Moon. American astronomer Charles Young (1834–1908) first observed the flash spectrum in 1870. During a total solar eclipse, the appearance of the flash spectrum changes rapidly as the Moon's disk advances across the thin chromospheric layer at the solar limb.

Fleming, Williamina Paton Stevens (1857–1911)

Born in Scotland, she emigrated to Boston, USA, with her parents and worked as a maid in the home of EDWARD PICKERING, the director of the Harvard Observatory. Noticing her potential, Pickering hired her to do clerical work and mathematical calculations. Fleming devised a system of classifying stars according to their SPECTRA, obtained by Pickering from an objective prism spectrograph. This resulted in a catalog of the spectral types of 10 000 stars, published in 1890. This in turn formed the basis of the several volumes of the definitive *Henry Draper Catalogue of Stellar Spectra* drawn up by Annie CANNON (*see also* Henry DRAPER). Fleming discovered 222 VARIABLE STARS.

Flux density

A measure of the radiation arriving from a source at a particular frequency. The flux of radiant energy is the quantity of

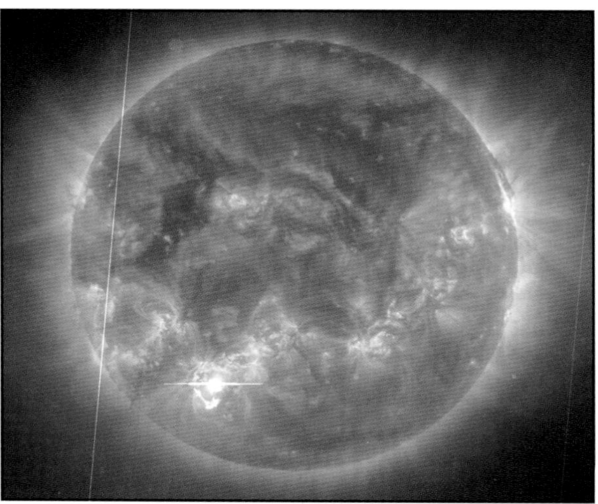

energy per second (measured in watts, or W) passing through unit surface area (1 m^2) perpendicular to the direction of the source, and is expressed as W m^{-2}. The flux density is the flux of radiant energy within a unit interval of frequency (a frequency band with a width of 1 hertz), and is expressed as W m^{-2} Hz^{-1}. The flux densities for most cosmic radio sources are extremely low. A unit of flux density used by radio astronomers is the jansky (Jy), where 1 Jy = 10^{-26} W m^{-2} Hz^{-1}.

Focal length

The distance between the center of a lens, or the reflective surface of a mirror, and its focal point, or FOCUS.

Focal ratio

The ratio of the FOCAL LENGTH of a lens or mirror to its aperture. For a lens or mirror of focal length F and aperture D, the focal ratio, f, is given by $f = F/D$. For example, a lens with a focal length of 10 m and an aperture of 2 m would have a focal ratio of 10/2 = 5; this would be written as 'f/5'.

The focal ratio, or f-number, of a camera lens may be altered by means of an adjustable diaphragm that changes its aperture. For a lens or mirror of a given focal length, increasing the aperture (and thereby reducing the focal ratio) increases the illumination of the image. Because, in general photography, this decreases the exposure time that is needed to record a particular image, the term 'focal ratio' is synonymous with the 'speed' of an optical system (for example, an f/2 system is faster than an f/16 system).

Focus

(1) In optics, the point at which parallel light rays from a distant point source are brought together by a lens or mirror to form an image. Light rays traveling parallel to the optical axis (the line that passes perpendicularly through the center of a lens or mirror) of an ideal convex lens or concave mirror converge to intersect at the focus, or focal point, of the lens or mirror. Rays arriving at an angle to the optical axis, from point sources or extended objects (sources of finite angular size), form images at the focal plane, a plane that passes through the focal point perpendicular to the optical axis. In practice, the focal plane may be slightly curved.

In a conventional REFRACTOR, the distance between the OBJECTIVE LENS and the focus is the FOCAL LENGTH. In a REFLECTOR, the focal point of light rays reflected from the primary mirror is called the prime focus. Most reflecting telescopes use one or more additional mirrors to reflect the converging cone of light, and the focal point, to a different location.

(2) In an ELLIPSE, a focus (plural: foci) is one of two points on the major axis such that the sum of the distances from each of the two foci to any point on the ellipse is a constant.

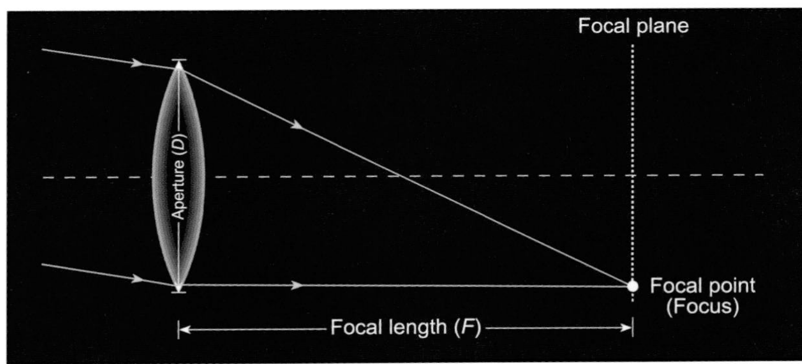

▲ ***Focus*** *The basic ray diagram for a convex lens shows how distant light rays are brought into focus.*

Forbidden line

Emission lines seen in the SPECTRA of GASEOUS NEBULAE and other objects of very low density and high mass. They are 'forbidden' in the sense that the probability of seeing them under normal laboratory conditions is low. These lines are normally indicated by square brackets. A good example of forbidden lines are those of doubly ionized oxygen (O III). These consist of a strong pair of lines at 500.7 nm and 495.9 nm, and a weaker line at 436.3 nm. The strong lines are responsible for the green colour often seen in PLANETARY NEBULAE and AURORAE.

Forbidden lines, also observed in EMISSION NEBULAE and in ACTIVE GALAXIES, are produced in the following way. The gas is ionized by radiation from a hot central source. Collisions between IONS and energetic ELECTRONS (freed in the ionization process) knock an ion's outer electron into a higher energy level that is 'metastable.' This means that the electron can spend an unusually long time in this excited state before it returns to the lower level by radiative decay, thereby emitting a PHOTON of energy at the appropriate frequency. Under normal conditions the atom would undergo many collisions in the time required, and the electron would have returned to the ground state or moved to a higher energy level. Radiative decay is possible in rarefied conditions with gas densities of about 1000 atoms cm^{-3} (similar to the best vacuums on Earth but high compared with the interstellar value of about 1 atom cm^{-3}). The presence of forbidden lines gives valuable information about the density, pressure and chemical composition of the medium in which the atoms exist.

Fornax *See* CONSTELLATION

Foucault, Jean Bernard Léon (1819–68)

French physicist, he became a professor at the PARIS OBSERVATORY and a member of the Longitude Institute. Foucault collaborated with Hippolyte Fizeau (1819–96) in the measurement of the velocity of light using the Fizeau wheel. Foucault was the first scientist, with Fizeau, to photograph the Sun (using an early photographic process called daguerreotype). He developed the FOUCAULT PENDULUM. Appointed to the staff of the Paris Observatory, he set about improving its telescopes. Here he invented the knife-edge test for the figure of a MIRROR, the process of silvering glass mirrors (replacing the speculum metal mirror), and pneumatic mirror-support systems.

Foucault pendulum

A freely swinging pendulum mounted in such a way as to reduce its physical contact with the Earth to an absolute minimum. An ideal pendulum of this type would, if set swinging in a particular plane at one of the Earth's poles, continue to swing in that same plane relative to the 'fixed stars', and the Earth would rotate underneath it. To the observer on the Earth's surface, the pendulum would be seen to rotate its plane of swing through 360° each day. The period of rotation of the plane of the pendulum increases with decreasing latitude on the Earth's surface, from a minimum of 23.93 hours at the poles. No rotation is observed for a pendulum at the equator. Observation of the behavior of such a pendulum is direct evidence that the Earth rotates on its axis. This type of pendulum is named for French physicist, Léon FOUCAULT.

Fowler, William Alfred (1911–95)

American nuclear physicist, he worked at the California Institute of Technology in Pasadena and won the Nobel prize in 1983 with Subrahmanyan CHANDRASEHKAR. Fowler studied the nuclear reactions of PROTONS with the ISOTOPES of carbon and nitrogen, the very reactions in the carbon–nitrogen cycle identified at that time by Hans Bethe (1906–) as occurring in the stars. Fowler established the science of nuclear ASTROPHYSICS. He found that there was a gap in the sequence of stable nuclei at mass 8, as at mass 5. These mass gaps spelled doom for George GAMOW's idea (the alpha-beta-gamma theory) that all nuclei heavier than HELIUM (mass 4) could be built by neutron addition, one mass unit at a time, in the BIG BANG. Fowler measured the rates of most nuclear reactions of astrophysical interest, and co-authored with Margaret Burbidge (1919–), Geoffrey Burbidge (1925–) and Fred HOYLE the famous 1957 paper, 'Synthesis of the elements in stars,' known as B^2FH after the authors' initials. They showed how the cosmic abundances of essentially all the nuclides from carbon to uranium could be explained as the result of nuclear reactions in stars, using hydrogen and helium produced in the Big Bang (*see* COSMIC ABUNDANCE OF ELEMENTS; ELEMENTS, FORMATION OF). After 1964 Fowler worked on theoretical problems involving SUPERNOVAE, gravitational collapse and nucleo-cosmo chronology.

Franklin-Adams, John (1843–1912)

English businessman and amateur astronomer, he compiled the *Franklin-Adams Photographic Atlas of Star Positions* (published in 1913). The Franklin-Adams camera was a 25-cm aperture, 10° field telescope, used first in England and later re-erected in Johannesburg, South Africa.

Fraunhofer, Joseph von (1787–1826)

German instrument-maker. Fraunhofer worked first as a wood turner then as a glazier, and educated himself in optics, joining the instrument workshop of Georg Reichenbach (1771–1826) and Joseph Utzschneider (1763–1840). He calculated, designed and tested every instrument made in the workshop and oversaw their installation, including the big REFRACTOR for the Russian observatory in Dorpat, which was used by Friedrich STRUVE to measure 3000 DOUBLE STARS. This was the first telescope mounted equatorially and driven about the polar axis with a clock. In about 1813 Fraunhofer began to research the chromatic properties of different sorts of glass, and rediscovered the dark lines in the Sun's SPECTRUM. These were first seen by William Wollaston (1766–1828) but are now known as FRAUNHOFER LINES.

Fraunhofer line

A dark line in the SPECTRUM of the Sun, first studied by Joseph von FRAUNHOFER in 1814. He labeled the most prominent of these by the letters A to K, ranging from the red end of the spectrum to the violet. Most are absorption lines, caused by the absorption of light by atoms in the outer regions of the Sun (*see* ABSORPTION SPECTRUM); for example, a prominent pair of close dark lines in the yellow part of the solar spectrum are the 'D' lines of the element sodium. A few Fraunhofer lines originate in molecules in the Earth's atmosphere.

Free-free (or bremsstrahlung) radiation

ELECTROMAGNETIC RADIATION may be emitted or absorbed in a variety of ways. For example, if an ELECTRON moving freely through a gas makes a close encounter with an ION, and is not captured by that ion, its kinetic energy will be changed by a finite amount, the change in energy corresponding to the emission or absorption of a PHOTON (a 'packet' of electromagnetic energy). Radiation emitted in this way is called free-free radiation (the electron moves freely before and after the encounter). In a PLASMA in which a large number of electrons and ions are involved, their individual free-free emissions add together to produce a CONTINUOUS SPECTRUM of radiation. Because the temperature of the plasma determines the energies of the electrons, radiation produced in this way is often called thermal emission. *See also* BLACK-BODY RADIATION.

Freja

Swedish-German satellite, launched in October 1992 to study the Earth's aurora and associated wave and particle phenomena. It ceased operating in October 1996.

Frequency

The rate at which specific events occur. In the context of ELECTROMAGNETIC RADIATION the term is taken to mean the number of wave crests per second passing a particular point. For light of wavelength λ, since the speed of light equals c, the frequency, f, equals c/λ.

The SI unit of frequency measurement is the hertz (Hz). This measures the number of waves passing a fixed point per second. Thus, one wave per second would correspond to a frequency of 1 Hz. Common multiples are kilohertz (kHz, 1000 Hz), megahertz (MHz, 1000 000 Hz), and gigahertz (GHz, 1000 000 000 Hz). An older unit of frequency is cycles per second (c/s); 1 Hz = 1 c/s.

Friedmann, Aleksandr Aleksandrovich (1888–1925)

Russian mathematician. After exciting wartime and revolutionary experiences he returned to Petrograd (later Leningrad, now St Petersburg), where he became interested in Albert EINSTEIN's general theory of relativity, and showed in a classic paper, 'On the curvature of Space,' that the radius of curvature of the universe can be either increasing or a periodic function of time. Friedmann's solution foreshadowed the concept of the expanding universe.

Fundamental catalog

A catalog listing stars whose positions and proper motions have been accurately determined over many years by positional measurements relative to telescopes, rather than to other stars. A number of early catalogs are still of value because the determination of proper motions, in which old observations are combined with the latest measurements, is an ongoing exercise; a notable example is the *General Catalog of 33 342 Stars*, compiled by the American astronomer Benjamin Boss (1880–1970) and published in 1936–7.

The most important fundamental catalogs are the FK series, begun in Berlin in 1879. The latest edition, FK5, was published in 1988 by the Astronomisches Rechen-Institut, Heidelberg, on behalf of the IAU, using observations of stars made at observatories around the world. The catalog positions are calculated for a standard EPOCH, so that they provide a fundamental reference frame to which observations of the positions of all other bodies may be referred.

Fundamental force

A force that governs the various kinds of interaction between particles. The four fundamental forces, or interactions, are: gravitation, the electromagnetic force, the strong nuclear force and the weak nuclear force.

Fundamental particle

A subatomic particle that cannot be divided into smaller components. The fundamental particles are the leptons (ELECTRONS, muons, tau particles and associated NEUTRINOS), and the six 'flavors' of QUARKS and their antiparticles and the force-carrier particles (photons, gluons, W and Z bosons and gravitons).

Fundamental star

A star whose position and proper motion have been accurately determined over many years. A definitive list of 1535 fundamental stars was adopted by the IAU in 1935. Positional data, which are updated regularly by high-precision observations, are published in the FK series of FUNDAMENTAL CATALOGS, providing a reference-frame for position- and motion-determining observations of all other bodies.

Fusion

The process in which two lighter atomic nuclei are combined to form a heavier atomic nucleus. Very high temperatures are normally required for atomic nuclei to collide with sufficient energy to overcome their mutual electrostatic repulsions (each atomic nucleus has a positive charge, the magnitude of which depends on the number of PROTONS it contains). Fusion that occurs at high temperature is called 'thermonuclear fusion'. Fusion reactions involving light elements release large amounts of energy. The mass of the resulting nucleus is less than the combined masses of the two original nuclei, the difference in mass, Δm (known as the mass deficit or deficit), being released as energy (E) in accordance with Albert EINSTEIN's relationship $E = \Delta mc^2$, where c denotes the velocity of light. The fusion of elements up to iron (atomic mass 56) results in the release of energy. The fusion of elements heavier than iron requires an input of energy. Thermonuclear fusion provides the energy that powers stars. In MAIN-SEQUENCE stars, such as the Sun, fusion reactions (the proton-proton reaction or the carbon-nitrogen-oxygen cycle) convert hydrogen to helium. In RED GIANT STARS helium is converted to carbon by a process called the triple-alpha reaction and, in highly evolved high-mass stars, fusion reactions synthesize a succession of elements up to iron. *See also* ELEMENTS, FORMATION OF; FISSION.

G

Galactic bulge

Bulges are the centrally visible, spheroidal star systems in DISK GALAXIES. Some galaxies have prominent bulges, like the Sombrero Galaxy, NGC 4594. In others, like our own Milky Way Galaxy, the bulge is much less prominent relative to the disk. The stars in a galactic bulge are old, so it will have been formed early in a galaxy's lifetime, in the first billion years of cosmic history – probably by the sudden collapse of a cloud of protogalactic gas condensed from the BIG BANG. At the center of a galactic bulge is a GALACTIC NUCLEUS.

Galactic center

The very center of our Galaxy, around which it rotates. It is assumed to be identical to the AGN at the center of our Galaxy, the radio source SAGITTARIUS A.

Galactic cluster *See* OPEN CLUSTER

Galactic disk

Our MILKY WAY GALAXY's spiral arms lie in a flat disk, termed the Galactic disk. The disk has two components, a thin disk embedded in a thick disk. The thin disk has more stars, and is the most visible. The disk is seen edge-on and is manifest as the Milky Way, except for the GALACTIC BULGE in the direction of the center of our Galaxy. The thin disk is made up of stars of a large range of ages (but younger than the GLOBULAR CLUSTERS) including the galactic star clusters, and a thin sheet of neutral hydrogen, rotating around the galactic center. These components of our Galaxy are called POPULATION I.

The galactic thick disk is intermediate in its properties between those of the GALACTIC HALO and the galactic thin disk. In a well-studied example, NGC 891, the thin disk has a height of 400–650 parsecs and is not unlike the situation one finds in our Galaxy. The thick disk has a height of 1500–2500 parsecs. In our Galaxy, some 2% of the stars are in the thick disk while the rest are in the thin disk, which explains why the thick disk was only discovered in the 1980s.

The thick disk is old and contains the younger globular clusters, including 47 Tucanae, which are 12 billion years old. The thick disk may have been caused by a thickening of the distribution of some of the stars of the thin disk, a remnant of a merger with another galaxy or a remnant of the formation process of our own Galaxy.

Galactic halo

A spherical or spheroidal distribution of GLOBULAR CLUSTERS and other old stars that surrounds a SPIRAL GALAXY. The diameter of a galactic halo is comparable to, or larger than, the overall diameter of the disk within which the spiral arms are located (about 100 000 l.y. in the case of the Milky Way).

Old stars are believed to have been formed at an early stage in the evolution of galaxies from clouds of hydrogen and helium that contained very little in the way of heavier elements. In the Milky Way Galaxy these POPULATION II stars circulate around the galactic center in orbits that are tilted at random angles to the GALACTIC PLANE. Because these stars do not share in the orderly motion of stars in the GALACTIC DISK, their velocities relative to the Sun are much higher than those of stars in the disk population. Consequently, they are known as HIGH-VELOCITY STARS.

Studies of the motions of halo-population objects show that the total mass of the Milky Way is about 10^{12} solar masses, about 10 times greater than the combined mass of all the stars, gas and dust that can be detected directly, and that most of this mass is contained in the halo. These observations, and similar results for other spiral galaxies, imply that up to 90% of the mass of a typical spiral galaxy is made up of DARK MATTER. While part of this total may be provided by objects with extremely low luminosities, such as brown dwarfs or black holes, much of it may exist in the form of exotic elementary particles.

Galactic latitude and longitude

Coordinate system that is useful for displaying galactic phenomena as seen from the position of the Earth. The galactic equator is the centerline of the Milky Way defined by radio observations of galactic neutral hydrogen and the GALACTIC CENTER. Galactic latitude is the angular distance between the galactic equator and a celestial body, measured perpendicular to the galactic equator, and taking values between –90° and 90°. Galactic longitude is the angular distance between the direction of the galactic center and the point on the galactic equator perpendicularly below (or above) a celestial body; it is measured in an anticlockwise direction, and takes values between 0° and 360°. The galactic longitude and latitude are denoted by (l^{II}, b^{II}), where the superfix relates to a definition adopted in 1959 by the IAU. They were earlier denoted (l^{I}, b^{I}), or simply (l, b).

Galactic nucleus

The core of a galaxy. In ELLIPTICAL GALAXIES and the central GALACTIC BULGES of SPIRAL GALAXIES, the nucleus corresponds to the location where the optical brightness reaches a maximum, and to the region where stars are most densely concentrated. IRREGULAR GALAXIES do not contain well-defined nuclei. ACTIVE GALAXIES, which are only a few percent of all galaxies, contain intensely bright, compact nuclei called active galactic nuclei (AGN). Some galaxies have two nuclei, one of which may be the core of another galaxy that has been absorbed following a galactic collision.

Galactic plane

The plane of the GALACTIC DISK. It is the central line of the distribution of neutral hydrogen in our Galaxy as determined by radioastronomers, and follows a line running more or less through the center of the MILKY WAY GALAXY. It is inclined to the celestial equator by an angle of about 62°.

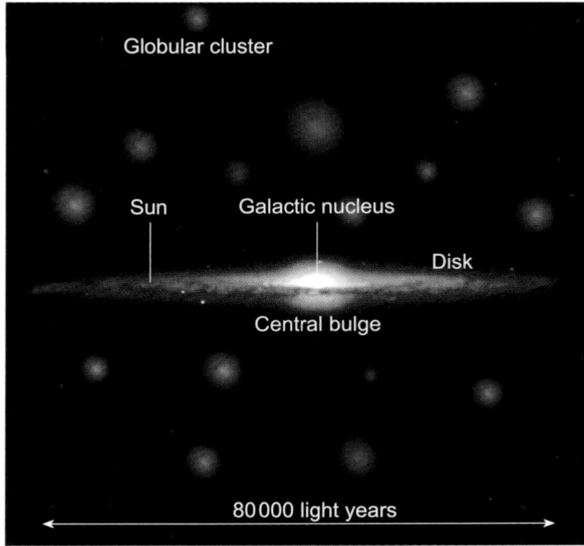

◀ ***The Milky Way Galaxy*** *How the Galaxy might look from far away in the galactic plane. Globular clusters mark the galactic halo, which is itself surrounded by a halo of invisible dark matter.*

Galactic year

A term that is sometimes used to describe the orbital period of the Sun around the GALACTIC CENTER, approximately 225 million years.

Galaxies, colliding

For much of the twentieth century, galaxies were pictured as 'island universes,' evolving from their initial state by internal processes with no external influence. As more galaxies were surveyed, however, many examples of paired galaxies were found, as well as many peculiar galaxies with long, luminous 'plumes' and 'tails' emanating from their bodies. The galaxies were given evocative names such as the Mice, the Antennae and the Tadpole. These galaxies have anomalous blue colors, indicating that they contain many recently formed massive stars and implying that their star-forming activity is different from normal.

At first many astronomers believed that the long streamers of stars and gas were shaped by magnetic fields or nuclear jets; the thinness and linearity of these features made a gravitational origin seem unlikely. However, computer models of interacting galaxies in the early 1970s by Alar and Juri Toomre showed that these streamers were the consequence of gravitational tides between rotating disk galaxies.

Computer simulations of the interactions employ a technique known as *N*-body modeling: two galaxies are represented by *N* discrete particles whose initial positions and velocities represent a sample of the stars in each galaxy. Current simulations employ up to 10 million particles, so that each particle represents up to 100 000 stars. The computer model is advanced forward in time by calculating the gravitational acceleration acting on each particle from all the other particles, and then advancing each particle forward in time from its position, given its velocity and acceleration. For the interstellar gas, hydrodynamic forces must also be considered.

The simulations showed how tidal forces act to stretch the galaxies radially, just as our own Moon raises two bulges of water on the surface of the oceans, which produce diametrically opposite oceanic tides. The radial stretching, combined with the galaxies' rotation, causes the stars and gas in the outskirts of each galaxy to 'shear off' from their parent galaxies. Material on the far side of each disk (away from the companion galaxy) is ejected into long, thin 'tidal tails,' while material on the near side is drawn toward the companion, possibly even forming a physical 'bridge' between the galaxies, along which material may flow from one galaxy to another.

Certain types of collision lead to very special results. For example, if a small galaxy plunges directly through the center of a larger disk-shaped galaxy, a collisional ring galaxy can result (*see* CARTWHEEL GALAXY). The ring in this case represents a shock wave moving outward, much like the water wave that results when a pebble is dropped into a pond. Another ring shape results when a small companion galaxy, usually a gas-rich irregular, passes close over the pole of a gas-poor S0 disk galaxy. In this circumstance, the interaction can disrupt the companion and capture it into a polar orbit. Eventually, the material that made up the companion spreads along the whole orbit, producing a polar ring galaxy.

An alternative picture of galaxy evolution began to grow. Galaxies are found in clusters and groups (*see* GALAXY CLUSTER AND GROUP). Perhaps most galaxies have interacted strongly with another one during their lifetime. Rather than being exceptional, galaxy interactions may be the dominant process shaping the evolution of galaxies, and galaxies that have remained isolated may be the rare ones.

Mergers

One frequent outcome of the models of colliding galaxies is that, after interacting and separating once, the galaxies fall back on one another, rapidly merging to form a single ellipsoidal galaxy. Effectively, the orbital energy of the two galaxies is transferred to internal motions of the stars and the dark matter in each galaxy. It was this energy transfer that earlier provided the energy necessary to launch the tidal tails. The tidal tails represent only a few percent of the lost orbital energy, however, as most of it is soaked up by the dark-matter halos of the galaxies.

The dark-matter halos of each galaxy extend to many tens or even hundreds of kiloparsecs. Encounters where a galaxy seems to be passing near another may in fact be intimate encounters, with each galaxy penetrating deep into the other's dark halo. Each sets up a trailing wake in the other's halo. This causes a gravitational drag on the galaxies, and the merger. Without halos, mergers would be rare indeed.

The disturbance to the gas in interacting galaxies causes a burst of star formation, hence the blue colors of the distorted galaxies. Even more extreme was the discovery by IRAS, launched in 1983, of ULTRA-LUMINOUS INFRARED GALAXIES, or ULIRGS. This emission is believed to come from dust that is re-radiating energy from a central source. Nearly all ULIRGs show distorted shapes or strong tidal features, as if they had recently collided. The collision of two disk galaxies drives gas inward from the galaxies' disks into their nuclei. Once the gas reaches the inner kiloparsec, it can fragment and form stars (at a rate of approximately 100 solar masses per year, compared to 1 solar mass per year in a normal galaxy). The gas can also flow further inward and fuel the central BLACK HOLE, creating ACTIVE GALAXIES.

Once a merger is complete, and the ULIRG phase (if any) is over, what is left behind? In 1977 Alar Toomre suggested that the remnants of the mergers of disk galaxies could be the elliptical galaxies. The merging process would effectively 'scramble' the stellar disks, giving the remnant the smooth look characteristic of elliptical galaxies. The peculiar galaxy NGC 7252 is a missing link. This galaxy possesses two

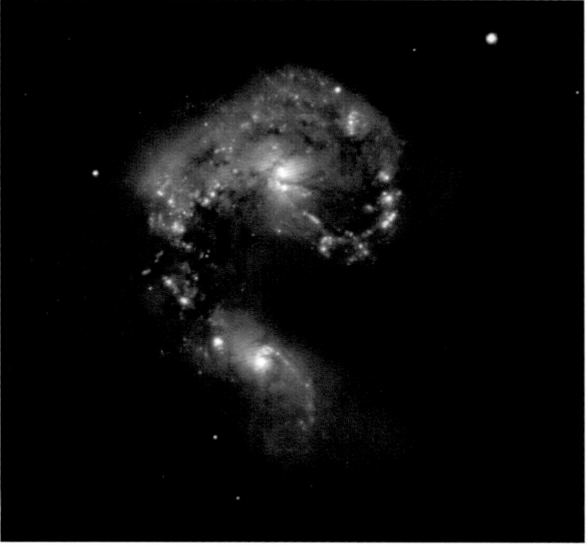

► ***Colliding galaxies*** *Face-on spiral galaxy NGC 4038 (top) is shown in collision with tilted spiral galaxy NGC 4039, and many star clusters have formed from the gas between them. The pair are known as the Antennae.*

◀ ***Galaxies***
(Left) M65 is an 'early' spiral galaxy of type Sa, seen inclined to the line of sight. Its spiral arms are tightly wound and their dust shows on the nearer rim. (Center) NGC 2997 is a 'late' spiral galaxy of type Sc with a smaller bulge and looser arms. (Right) M87 is an elliptical galaxy. Its almost circular shape makes it type E0 or E1.

gas-rich tidal tails, indicating that it is the merger of two spirals, but it has the surface-brightness profile expected for an elliptical galaxy. François Schweizer argued that NGC 7252 is the birth of such a galaxy. However, the question of whether most ellipticals were formed through major merger events remains open.

Minor mergers

Major mergers are the mergers of two large, roughly equal-mass disk galaxies. These are the most spectacular collisions, but minor mergers between a galaxy and a small satellite companion are much more common. Even these minor mergers can have a dramatic impact on the evolution of galaxies.

Our own MILKY WAY GALAXY has undergone, is undergoing and will undergo minor mergers. It is surrounded by 14 known satellite galaxies, the largest being the Large and Small Magellanic Clouds. Tidal forces from the Milky Way have torn a long stream of gas (the Magellanic Stream) from these companions.

In 1978 Leonard Searle and Robert Zinn proposed that the halo of the Milky Way formed through continual ACCRETION from intergalactic space. Certainly part of the halos of SPIRAL GALAXIES formed through satellite mergers. The Milky Way's halo has substructure, co-moving groups of stars, and groups of GLOBULAR CLUSTERS of different ages. These are the fossil traces of the mergers of satellite galaxies with the Milky Way in the past. Recently, our Galaxy has acquired some of its younger globular clusters from the Magellanic Clouds. The discovery of the SAGITTARIUS DWARF GALAXY plunging into the far side of our Galaxy has given us a close view of a full merger in progress.

Further reading: Kennicutt R C, Schweizer F and Barnes J E (1998) *Galaxies: Interactions and Induced Star Formation* Springer; Toomre A (1977) *The Evolution of Galaxies and Stellar Populations* Yale Observatory Press.

Galaxy

A galaxy, like our own MILKY WAY GALAXY, is a large assemblage of stars, gas and dust held together by mutual gravitational interaction. Galaxies hold between a few million and 10 trillion stars, together with different proportions of interstellar material (gas and dust). Their diameters mostly range from a few thousand to a few hundred thousand light years.

Normal galaxies have many different shapes. Classification provides order to the daunting variety, and a logical framework for further studies. We should, however, keep in mind that galaxies are 90% DARK MATTER, which is invisible, so the dominant shape of a galaxy is unknown in any detail.

Factors influencing galaxy shapes

- Most galaxies have some symmetry, but are oriented randomly to the line of sight, making their appearance very dependent on viewing geometry.
- Galaxies are spread over great distances, and the larger they are, the harder it is to see their shape.
- Galaxies have different mixtures of hot stars (which emit blue light) and cool stars (which emit red light), so their appearance depends on the color of the radiation with which they are recorded.
- In highly flattened galaxies, interstellar dust collects in a thin plane at the mid-section. This has a wavelength-dependent effect on the appearance of a galaxy, especially when recorded with blue light.
- At longer (radio) wavelengths, emission from the stars can be very weak, and the radio appearance of a galaxy can be completely different from optical images.
- The cosmological REDSHIFT shifts and stretches galaxies' spectral energy distribution. This greatly alters the appearance of distant galaxies.
- Because the speed of light is finite, very distant galaxies are seen as they were when the universe was much younger, and their appearance is influenced by their young age and the different conditions then.

The Hubble classification system

From 1781 to 1847, William Herschel and his son John (*see* HERSCHEL FAMILY) were able visually to detect different degrees of central concentration, apparent flattening and mottling in some 'nebulae.' These were later recognized as what were then called 'extra-galactic nebulae,' eventually shortened to 'galaxies.' William Parsons recognized the 'spiral' structure of some galaxies in 1845 (*see* ROSSE, THIRD EARL OF AND FOURTH EARL OF). When photography became important in astronomy, Edwin HUBBLE developed a classification system for galaxies that is still in use today.

The HUBBLE TUNING-FORK DIAGRAM system identifies several types of galaxies:

Elliptical galaxies

The surface brightness of an ELLIPTICAL GALAXY, symbolized by the letter E, is smooth and largely featureless. They have

Observing galaxies

Amateur astronomers have observed galaxies from the early days of the telescope. Originally they were classed as nebulae and many were discovered by amateurs, although the nature of these elusive faint smudges did not become clear until the 1920s. Important discoveries were made by astronomers who were strictly speaking amateurs, albeit the Grand Amateurs of the Nineteenth Century. Such discoveries include the visual observation of spiral structure in M51 by Lord Rosse (*see* ROSSE, THIRD EARL OF AND FOURTH EARL OF), and the photographic structural details of many others by Isaac Roberts (1829–1904). Modern amateurs have a love–hate relationship with galaxies, the definitive 'faint fuzzies' to the visual observer becoming detailed wonders chiseled from stars, gas and dust to the CCD enthusiast.

Of all classes of deep-sky object, galaxies are the least impressive to the new visual observer. The majority of galaxies, with a couple of dozen exceptions, are small, faint, diffuse smudges of light that do not contrast well against the bright sky glow of the modern world. While the EYEPIECE view of most galaxies is unimpressive, when coupled with the knowledge of the ASTROPHYSICS and some imagination of what is on display, the rewards of tracking galaxies down can be great.

Equipment and techniques for observing

Galaxies do require larger apertures for rewarding views, and the big Dobsonian reflectors transported to darker sites helped revive the interest in this field in the 1980s, with North American observers taking the lead (*see* DOBSONIAN TELESCOPE). Light-pollution reduction (LPR) filters are of limited benefit with galaxies, and narrow-band filters are of no use at all, but some techniques can help track down the quarry. Dark adaptation and the exclusion of extraneous light by the use of screens and even a dark cloth draped over the observer's head can help. The AVERTED VISION technique can be very helpful, as it allows the observer to make use of the most light-sensitive area on the retina of the eye. Observing on Moon-free nights of good transparency is best, and once the galaxy is found with low power, the use of higher magnification to increase contrast and tease out detail is recommended.

Photography certainly helped record more detail in the larger and nearby galaxies, but it was the advent of the CCD in the early 1990s that made the imaging of galaxies so popular, and extended the range of amateur involvement in their study.

The majority of observations of galaxies are now made using CCDs and most observers record these building blocks of the universe for pleasure and satisfaction. The best amateur CCD images can rival professional photographs taken with the world's largest telescopes of only two decades ago, and so many galaxies are now gaining their own identities in a way unimagined ten years ago. Visual observers still enjoy the chase and the challenge of identification and of the teasing out of details of these subdued beauties, but the CCD will reign supreme for the foreseeable future.

Amateur contributions to science

While observing for recreation is the most popular reason for observing galaxies, scientifically valuable projects are now possible with amateur equipment. A small number of dedicated amateurs around the world, notably from the UK, Italy, Japan, Finland and the USA, are making serious contributions to science by patrolling for extragalactic

► ***NGC 253*** *Imaged by Nik Szymanek with a Meade 10-in LX200 at f/10, an AO-7 Adaptive Optics unit and an SBIG ST-8E CCD camera.*

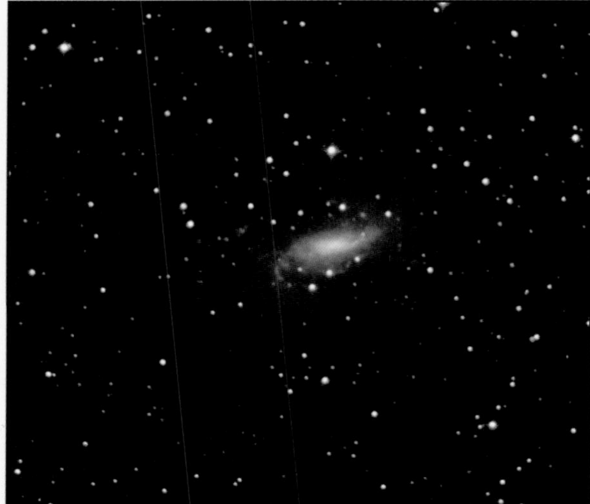

◀ ***M101***
A nearby Sc spiral in Ursa Major seen face-on. Amateur astronomer Adrian Catterall has captured its delicate spiral arms, and HII regions in this CCD image.

◀◀ ***NGC 925***
LRGB color composite of NGC 925 in Triangulum. Made from images taken by Al Kelly with a 12-in Takahashi Mewlon and FLI 'Dream Machine' CCD camera. The scope is an automated, remotely-operable instrument located near Cloudcroft, New Mexico. Twenty-eight subexposures were processed in AIP4WIN.

SUPERNOVAE and NOVAE, imaging GRBs, and monitoring AGN.

Supernova explosions can briefly equal the energy of the host galaxy, and can be seen across many millions of light years. They are important to the professional astronomers for many reasons, especially as cosmic yardsticks, the distance to the host galaxy being measured more accurately if a Type 1a supernova erupts within it. Although professional supernova search programs have now been developed, amateurs can still compete, and remain remarkably successful in discovering supernovae. There have been some 70 discovered between 1996 and 2002 from the UK alone.

To be successful, the patroller must image as many galaxies as possible as often as conditions allow; it takes true dedication, but the rewards are great. Most observers have developed sophisticated automated patrols, often with more than one telescope and CCD. However, the visual feats of Australian Bob Evans, who discovered over 30 supernovae with a 400-mm telescope in the 1980s and 1990s, deserve a special mention.

Novae in Local Group galaxies can be detected by amateurs armed with moderate-aperture telescopes and CCDs. Classical novae will peak at around magnitude 15, and can be followed down to magnitude 19, so have become interesting targets. A handful of amateurs worldwide monitor Messier 31 and 33, and the antipodeans can patrol the MAGELLANIC CLOUDS.

GRBs are likely to be extreme supernovae, and amateurs can occasionally detect their afterglows. The host galaxies cannot be seen, so great are their distances. They are very transient phenomena, so rapid alert systems, such as developed by the *Astronomer* organization in the UK, are needed to coordinate the effort. This is likely to become an important Pro-Am collaboration for the next few decades.

Some types of galaxy show variability in their nuclei, due to the presence of massive black holes at their centers. These galaxies with AGN include the Seyfert galaxies, BL Lacertae objects and QUASARS. Some show quite extraordinary changes of brightness within the nucleus on surprisingly short time-scales that betray the compact nature of the phenomena. Monitoring such changes in a select few AGN can be undertaken visually, photographically and with CCDs, and the results made available to professionals.

There are many organizations around the world that pool the results of amateur efforts. Most countries have a society that will accept results from amateurs. The AAVSO, the Deep Sky Section and Variable Star Section of the BAA and *The Astronomer* are long-standing and highly regarded organizations that serve this purpose, analyzing and archiving the data, then making it available to all. More recently, VSNET has developed as a World Wide Web-based organization (in Japan) that makes variable star measures and nova, supernova and AGN results rapidly available on the Internet.

However, most amateurs will continue to observe these most beautiful and pure works of nature to wonder at them, to record them for pleasure, and to illustrate magazines, journals, lectures and talks at local and national societies.

Nick Hewitt is director of the Deep Sky Section of the BAA. He earns his living as a doctor in Northamptonshire, UK.

◀ ***NGC 2903***
This LRGB color composite of dusty galaxy NGC 2903 was made by Al Kelly from filtered images taken by Arne Henden of the US Naval Observatory in Flagstaff, Arizona, using the 3.5-m WIYN telescope and a 2048 x 2048 CCD array.

a round or elliptical appearance, and Hubble added a numerical index of ellipticity, n, such that E0 galaxies look round while E6 galaxies are highly elongated. In many elliptical galaxies, the light declines smoothly and gradually with increasing distance from the center, but some 40% of ellipticals have dust, scattered in patches or collected in planes of symmetry. An alternative classification of elliptical galaxies is based on the deviations of the isophotes of ellipticals from a perfect elliptical shape, from 'boxy' to 'pointy' or 'disky.'

Dwarf elliptical galaxies

The most common galaxies are dwarf elliptical (or dwarf spheroidal) galaxies, such as MAFFEI 1, with much lower masses and surface brightnesses than the giant ellipticals. *See* DWARF GALAXY.

Spiral galaxies

The symbol for spiral galaxies in the Hubble classification system is S. They are highly flattened disk-shaped systems with spiral 'arms' in the disk and a smooth, relatively spherical bulge of old stars, as if there is an elliptical galaxy inside every spiral. Spirals differ according to the presence or absence of a central bar (suffix B), from the ends of which the spiral arms start. They also differ in the prominence of the bulge, whether the arms are open or tightly wrapped, and whether the arms are lined by nebulae and star clusters. The last three aspects seem to correlate, in such a way that tightly wound spiral arms with few star clusters are wrapped around spiral galaxies with large central bulges (type Sa), while loosely wound spiral arms, small bulges and prominent star clusters go together in type Sc; Sb lies between.

Galaxies of types Sa and SBa are frequently referred to as 'early-type' spirals, and types Sc and SBc are frequently referred to as 'late-type' spirals, as if there is a process that transforms one into the next (although there is not). Harlow SHAPLEY and Gérard de Vaucouleurs (1918–95) extended the spiral sequence to types Sd and SBd, in which the bulge is very small or absent and the structure is open and highly resolved. The MAGELLANIC CLOUDS, earlier thought of as irregular galaxies, have a rudimentary spiral structure and no bulge component, and are types Sm and SBm.

Some spiral galaxies show a 'grand-design' spiral structure, consisting of two well-defined symmetric and long spiral arms. If not, spiral patterns are called 'flocculent.' Radio observations show that spirals contain a larger proportion (typically about 10%) of gas and dust than ellipticals. The hydrogen of spiral galaxies extends far outside the starlight. The gas content of irregulars varies considerably, but is typically about 20%.

The galaxy to which the Sun belongs is a spiral, known as our Galaxy (with a capital 'G'), or the Milky Way Galaxy.

Irregular galaxies

Irregular galaxies were originally characterized as 'nondescript' and lacking rotational symmetry. It is now known that many irregulars do rotate and have bar-like structures in their disks. Irregular galaxies resembling the Magellanic Clouds but without any spiral structure are called Im or IBm types.

Lenticular galaxies

In 1936, Hubble revised his classification system to include a fourth major galaxy class, LENTICULAR GALAXIES. They are armless disk galaxies representing the transition from ellipticals

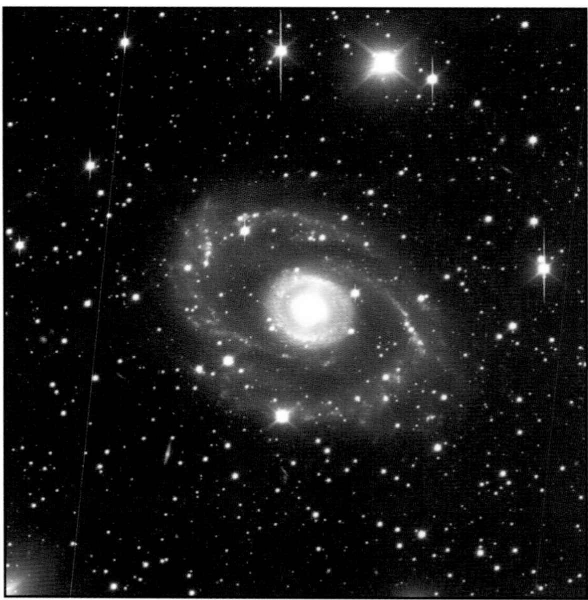

► ***Galaxies*** *ESO 269-57 is a spectacular spiral galaxy with an inner ring and two outer rings, of type Sa(r).*

to fully developed spirals and are symbolized by S0. S0s might be spirals that were stripped of their residual gas by interactions in clusters, or they may indeed be true transition forms between spirals and ellipticals. Some S0s have bars.

Ringed galaxies

A ringed galaxy is a normal disk galaxy (S or S0), with a ring pattern as part of the light distribution. The classification symbols 'nr,' 'r' and 'R' are used for nuclear, inner and outer rings, respectively, and all three types may coexist. The symbols are added in brackets to the type of the main disk galaxy, like S(nr). Of the many features of galaxies, rings are among the best understood. They are generated by a bar, which can redistribute gas in a galactic disk into ringed-shaped patterns that eventually condense into stars. Other ring galaxies are the results of collisions of galaxies (*see* GALAXIES, COLLIDING).

Peculiar galaxies

Peculiar galaxies do not look like any of these shapes and may be disturbed by interactions with other galaxies or be ACTIVE GALAXIES.

Why are galaxies of different types?

In 1960, Sidney van den Bergh (1929–) demonstrated that galaxy shape is influenced by its total luminosity, and he assigned luminosity classes to galaxies, like stellar luminosity classes (the most luminous are luminosity class I, the least luminous class V). In general the luminosity and surface-brightness distribution of a galaxy depend on the kind of stars that are radiating and their number. In turn, the number of stars determines the mass of the galaxy and this causes the galaxy to rotate faster or slower.

These underlying connections are the reason for the TULLY–FISHER RELATION, which is a tight correlation in general between the luminosity of galaxies and their rotational velocity. This is a powerful relationship for two reasons. Firstly, it is one way that astronomers can estimate the distances of distant galaxies; and secondly, it shows a way to connect the almost bewildering variety of the shapes of galaxies to physical quantities, such as mass and angular momentum, that are important in their formation and evolution. The masses of galaxies range from a few million solar masses, in the case of dwarf ellipticals, to a few trillion, in the case of the most massive ellipticals. Their luminosities range from about 200 000 to a trillion times that of the Sun.

Bulges and disks of galaxies differ because of the composition of their stellar populations and interstellar medium, and because of their dynamics. Galaxies are large concentrations of mass, and would collapse under the effect of their own self-gravity unless some opposing force kept them in equilibrium. Disks are kept in equilibrium by rotation, which provides the centrifugal force to oppose gravity. Bulges have modest or no rotation, and are sustained against gravity by the 'velocity dispersion' of their stars, which is similar to the motions of molecules in a hot gas.

Massive galaxies tend to be much more structured, better ordered, more luminous and of higher average surface brightness than low-mass galaxies, such that catalogs over-represent high-mass galaxies and under-represent low-mass ones. Because of this selection effect, the galaxy-classification systems in use today apply mainly to massive galaxies, and most galaxies listed in catalogs are high-luminosity spirals.

William MORGAN identified an extremely important rare type of galaxy known as a cD galaxy. These are very large elliptical-like systems found at the centers of rich galaxy clusters and result from the mergers of other galaxies (*see* GALAXY CLUSTER AND GROUP). Like cD ellipticals, different types of galaxies are found in different regions of the universe. Elliptical and S0 galaxies are common in rich galaxy clusters, but spirals dominate looser clusters and the general field. In fact Alan Dressler discovered that the shape of a galaxy is most determined by the density of other galaxies nearby ('the galaxy morphology–local density relation'). The shapes of present-day galaxies are the outcome of different evolutionary histories, including mergers (see GALAXY FORMATION AND EVOLUTION).

Further reading: Sandage A (1961) *The Hubble Atlas of Galaxies* Carnegie Institution of Washington; Sandage A and Bedke J (1994) *The Carnegie Atlas of Galaxies* Carnegie Institution of Washington; van den Bergh S (1998) *Galaxy Morphology and Classification* Cambridge University Press.

Galaxy cluster and group

Most galaxies are found in groups of no more than a few dozen members. Groups range from the satellite systems of giant galaxies (like the MILKY WAY GALAXY and its satellites the MAGELLANIC CLOUDS) to loose associations, to compact cores of rich galaxy clusters. The common thread linking these examples is that the galaxies interact more with each other than they do with the rest of the universe.

Groups are important for the evolution of galaxies and large-scale structures. Galaxy formation is a drawn-out process, involving the collapse of primordial perturbations, ACCRETION of gas and dark matter, outright merging of distinct objects, and outflows of gas enriched by SUPERNOVAE. Most galaxies do this in a group. However, while the galaxies in a group are forming, the group itself may be separating out from the cosmic expansion, collapsing under the influence of gravity, accreting new members, and finally merging with other groups to build clusters and superclusters.

Group evolution begins when several galaxies form together; the Local Group is now at this stage, with the Milky Way and M31 approaching each other for the first time. One of our nearest neighbors, the M81 group, is at a more advanced stage; three galaxies are linked by a complex structure of hydrogen, indicating that they have already undergone at least one passage. STEPHAN'S QUINTET is a compact group, containing an extended, possibly tidal feature in neutral hydrogen, and a central cloud of hot gas visible in X-rays; this gas may have been heated by a fast, interpenetrating encounter between two galaxies. Finally, V Zw 311 is actually the core of a cluster of galaxies, Abell 407.

Clusters of galaxies

Groups are not the only associations between galaxies. Clusters of galaxies contain many thousands of members, although they contain just 1% of all the galaxies and are rare (one cluster per million cubic megaparsecs).

Clusters of galaxies are the largest stable structures in the universe. They were first tentatively recognized in the sky surveys of nebulae during the nineteenth century. The most prominent in the northern sky is the VIRGO CLUSTER OF GALAXIES, which is the approximate center of the local supercluster at a distance of about 20 megaparsecs from the Milky Way, and can be recognized in the MESSIER CATALOG because of the large number of nebulae that Messier listed in this area. The Virgo cluster lies near the north galactic pole and provides us with a convenient local sample of a relatively low-richness cluster. The COMA CLUSTER is a spectacularly 'rich' cluster at a distance of about 100 megaparsecs, also located in the north galactic cap.

The Palomar Sky Survey of the 1950s provided the first nearly uniform optical coverage of the northern celestial sphere, and provided the source material for the systematic identification of clusters of galaxies by George Abell (1927–83) in 1958 (*see* ABELL CLUSTER). Abell devised a set of approximately redshift-independent criteria for identifying clusters of galaxies, classifying them according to richness (or number of galaxies). His list of about 3000 clusters, culled from approximately 10 000 square degrees of sky, has been exceptionally durable. Abell's sample entered the twenty-first century as the only really large one.

The first imaging X-ray telescopes in space established

▼ Galaxy cluster *The richest part of the Virgo cluster is centered on the giant elliptical galaxies M84 (the most prominent galaxy near the center) and M86 (right).*

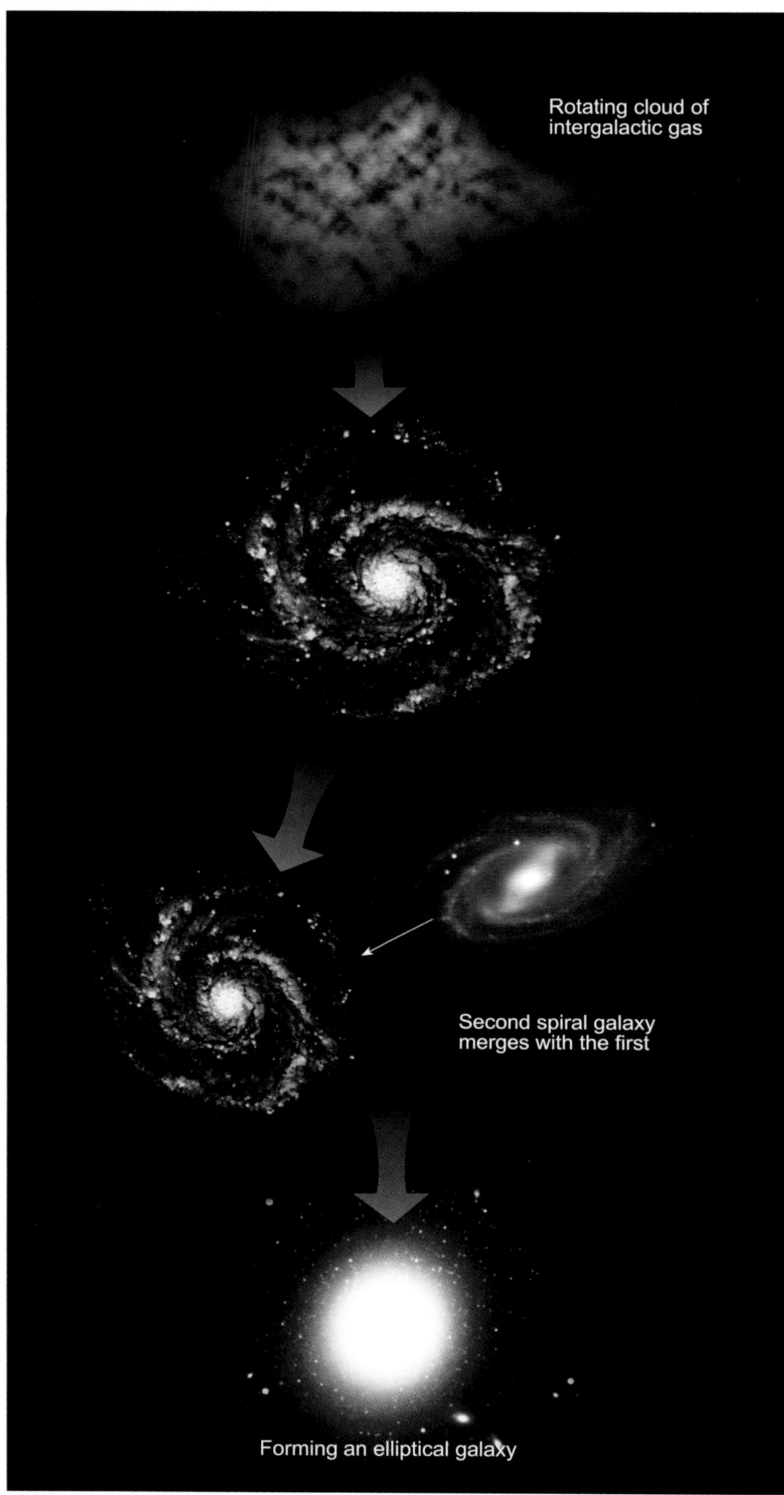

▲ ***Galaxy creation*** *An elliptical galaxy is formed when two spirals merge.*

that large clusters of galaxies emit X-rays. X-ray-selected samples have found objects slightly beyond redshift 1, but there are only about 12 beyond even redshift 0.5. However, carefully defined samples of galaxy clusters to redshifts well beyond 1 are becoming more available, thanks to the CHANDRA X-RAY OBSERVATORY and the XMM-NEWTON MISSION. These samples are beginning to open up a very large new range of astrophysical investigations.

The nature of clusters of galaxies

When galaxy clusters were first noticed, the size and structure of our own Galaxy were not known, nor was even the true nature of galaxies in general. The first step in the physical investigation of galaxy clusters came in 1933 when Fritz ZWICKY measured the velocities of individual galaxies in the Coma cluster. Then, making the bold assumption that the cluster, as a physical entity, was in a stable equilibrium, he derived its total mass. The ratio of the mass to the total luminosity of the Coma cluster galaxies is an astonishing value of about 200 solar masses per solar luminosity (adjusted to the current distance scale). The Sun, as a typical star, has a mass-to-light ratio of unity, and mixes of stars in an average galaxy have mass/luminosity values ranging from about 1 to about 10, depending on the galaxy's type. Zwicky concluded that most of the mass was 'missing' from the census of visible light; that there is much more mass in a cluster of galaxies than shines as stars.

This conclusion was controversial until the 1970s, when X-ray telescopes showed that large clusters of galaxies contain hot gas. The internal velocities in clusters of galaxies result in any gas in the cluster being heated to a temperature of between 10 million and 100 million K. Gas at these temperatures radiates primarily in the X-ray part of the spectrum. The X-ray-emitting gas ultimately derives most of its energy from its gravitational energy release as it falls into the cluster. This confirmed the mass of clusters of galaxies.

In the 1980s gravitational lensing was discovered in galaxy clusters, more or less as predicted by Zwicky (*see* GRAVITATION, GRAVITATIONAL LENSING AND GRAVITATIONAL WAVES). This elegant technique is both simple and powerful. The data consist of a picture of a cluster at a moderate redshift, say about 0.5. The cluster's gravitational field acts like a lens, which causes the background galaxies to be slightly elongated on the average, more than expected by projection effects of disk-like galaxies, and preferentially in the direction perpendicular to the radius from the center of the cluster. This directly measures the density of the cluster, again confirming the existence of DARK MATTER and its distribution.

A famous paper by Rashid Sunyaev (1943–) and Yakov ZELDOVICH in 1969 predicted that as COSMIC MICROWAVE BACKGROUND (CMB) radiation passed through a cluster of galaxies its radiation would be affected. By comparing the CMB radiation near the cluster with that seen coming through it, the SUNYAEV–ZELDOVICH EFFECT can be used to map the distribution of hot gas in clusters and for cosmological investigations.

Clusters contain more ELLIPTICAL GALAXIES and fewer SPIRAL GALAXIES than the rest of the universe. The cD galaxy type, a supergiant elliptical with an extended envelope, is unique to clusters. Cluster galaxies are older than field galaxies and a larger fraction of their stars are spheroidal as opposed to a flat disk. One possible inference is that galaxy merging is more important in clusters than in the field, with cD galaxies being the results of many mergers (*see* GALAXIES, COLLIDING). The cD galaxy is also fed by the inward fall of the X-ray-emitting gas. The gas cools as it flows inward, and within the central 10 kiloparsecs of many 'cooling flows' there are clear signs of massive star formation. In extreme cases this may amount to many tens of solar masses of new stars per year. The central galaxy appears blue toward its core, contrary to the red appearance of a normal old elliptical galaxy.

Further reading: Kolb E W and Turner M S (1990) *The Early Universe (Frontiers in Physics)* Addison-Wesley. The

Abell catalog of clusters of galaxies is in Abell G O (1958) *Astrophysical Journal Supplement* 3 211.

Galaxy formation and evolution

The GALAXIES that we observe in the universe at the present time are remarkably varied. Like stars, they come in a wide range of masses, with many more smaller galaxies than larger ones. Unlike stars, they also exhibit a great variety of shapes. There are two basic types of galaxies: galaxies with disks and galaxies that are spheroidal. But across the two basic types and the internal structures (including bars and spiral arms) there is a continuum, with most disk galaxies containing small spheroidal components, and most spheroidal galaxies containing small disks. So how did this variety come about?

Formation of galaxies

Galaxies began as clouds of primordial gas, hydrogen and helium, and large amounts of DARK MATTER. Before they condensed into distinct galaxies, infinitesimal density fluctuations were present in the expanding universe of almost homogeneous dark matter. These originated in the period of inflation at the start of the universe. They accumulated into larger lumps, each drawing in surrounding matter by gravity. Eventually, however, they stopped growing, and separated from each other. The dark-matter lumps then collapsed under their self-gravity to form galaxies' dark-matter halos.

The gas in the lumps cooled and flowed onto the center of the halos. Eventually, the gas fragmented into stars. Since we can't see the dark matter but can see the stars, we have to look into this process to understand galaxies' shapes.

As structures grew and collapsed in the early universe, they exerted tidal torques on each other, and this gave each collapsing mass some angular momentum. This momentum determined the properties of the galaxies that formed. If the collapsing structure had enough angular momentum (in other words, enough rotation), it pancaked and eventually settled into a disk, where the centrifugal force balances gravity. Without sufficient angular momentum, there was no centrifugal force to oppose gravity, and the collapse continued until it was halted by the stars' velocity dispersion. The equilibrium configuration in this case was that of a spheroid. Different galaxies had different angular momentum, and this accounts for the continuum in the proportions of disk structure to spheroidal structure in disk galaxies today.

The history of the collapse affected star formation. The violent collapse that originated a spheroid drove a violent burst of star formation throughout the structure that quickly converted all the available gas into stars. When the burst ends because of gas exhaustion, no more significant star formation is possible, unless the galaxy accretes additional gas from other galaxies. The stars that formed in the collapse evolved into today's old red giants, with little gas mixed with them. This accounts for the red colors of the stars in the spheroidal components of spiral galaxies and why they are essentially gas-free.

By contrast, in the disk of spiral galaxies star formation proceeds at a relatively slow but continuous pace. The luminosity and colors of disk galaxies, therefore, largely reflect those of the stellar population that continues to form. A reservoir of gas in the outer regions of the galaxy continuously feeds the disk and enables stars to be born.

The collapse of a completely non-rotating galaxy may be how elliptical galaxies are formed. They may also be formed by the mergers of spiral galaxies. There are numerous examples seen today of disk galaxies merging to form elliptical galaxies, and it is quite conceivable that mergers were more frequent in the past. Nearly half of elliptical galaxies possess features such as shells, indicative that mergers have occurred.

When two or more galaxies merge, the outcome depends on their relative masses and their types. If the two merging systems are spiral galaxies, their disks will be destroyed. If the merging systems have very different masses, and consist of one massive galaxy and at least one that is sub-massive, the event is less traumatic. Although the small galaxy is destroyed and loses its identity, it only causes a minor perturbation to the structure of the massive galaxy, which tends to retain its properties and morphology.

Look-back times (LBT)

Distance (Mpc)	Redshift (*z*)	Age (billion years)	LBT	Fractional age (%)
0	0.0	13.48	0.0	100
460	0.1	12.2	1.3	90
1550	0.3	10.1	3.4	75
2800	0.5	8.4	5.0	63
5000	0.8	6.6	6.8	49
6600	1.0	5.8	7.7	43
15 500	2.0	3.2	10.2	24
19 400	2.4	2.7	10.8	20
25 400	3.0	2.1	11.4	16
35 900	4.0	1.5	12.0	11
46 700	5.0	1.2	12.3	9
103 800	10.0	0.5	13.0	3.5
1 272 500	100.0	0.02	13.46	0.15
1 785 000	1300.0	0.0004	13.47	0.003

Seeing evolution from past to present

The differences that appear in present-day galaxies are evidence of different evolutionary histories. The stellar populations of elliptical galaxies as well as the bulges of spirals are very old, so star formation must have ended early during cosmic evolution. The age of the stars dates their formation to when the universe was about 20% of its current age, with most ellipticals and bulges already assembled into stars by the time the universe was around 3 billion years old.

The finite speed of light means that representative galaxies at earlier cosmic epochs can be studied simply by looking far away, which is equivalent to looking back in time. For very distant galaxies, the look-back time is a large fraction of the entire age of the universe. The table shows the relationships between distance, redshift, age and look-back time for a universe that is 13.5 billion years old (Hubble constant H_0 of 70 km s^{-1} Mpc^{-1}). The epoch when the universe was half its present age corresponds to a redshift of about 0.8 or a distance of about 5000 Mpc. The bulk (say 80%) of cosmic evolution, (that is, since the time the universe was 20% of its present age) has taken place from redshift 2.4 to the present time (redshift 0).

Galaxies at redshifts up to 1 have been studied fairly thoroughly. At the high end of this range there were many more faint-blue galaxies than now, identified by the HST as irregular galaxies undergoing a robust burst of star formation. Today's small galaxies seem to have undergone most of their evolution since then. Although we do not know for certain the present-day counterparts of most of the faint-blue galaxies, there is mounting evidence that these systems have merged with larger galaxies and disappeared.

But spirals and ellipticals were fully formed by the same time. This implies that they completed most of their formation at earlier epochs, and that to understand their origin one must search even further back. However, from redshift z~1 up to 2.5 there is a gap in the observations due to the difficulty, with current technology, of measuring galaxies with redshifts in this range. This means that it is not possible to make the link between 3 and 6 billion years after the Big Bang. This will be a major research activity when more sensitive instruments for large ground-based telescopes, and the Next Generation Space Telescope, are ready.

At redshifts higher than 2.5 and up to 5, some galaxies have been discovered in large numbers, thanks to a special observing strategy called the 'Lyman-break technique,' fine-tuned to be sensitive in that redshift range. These galaxies are observed during intense star formation, which created between a quarter and a half of all the stars that we count in the present-day universe. It appears that the Lyman-break galaxies are the direct progenitors of present-day spheroids, observed when they were forming the bulk of their stars.

Further reading: Liddle A (2003) *Introduction to Modern Cosmology* (2nd ed.) Wiley; van den Bergh S (1990) *Dynamics and Interactions of Galaxies* Springer.

GALEX

The Galaxy Evolution Explorer, a NASA small Explorer-class mission, launched in April 2003 on a 29-month mission to survey galaxies for ultraviolet astronomy. Unusually, the launch vehicle was a Pegasus XL rocket carried by an L-1011 Stargazer aircraft to 12 km over the Atlantic Ocean, and air-launched to a near-circular orbit at 700 km. The 280-kg satellite carries a 50-cm, wide-angle Ritchey–Chrétien telescope sensitive to wavelengths between 135 and 280 nm.

Galilean satellite

See CALLISTO; EUROPA; GANYMEDE; IO; JUPITER AND ITS SATELLITES.

Galilean telescope

A REFRACTING TELESCOPE that uses a long-focal-length converging lens and a short-focus diverging lens to produce a magnified image of a distant object. Named for the Italian scientist GALILEO GALILEI, who designed and built a telescope of this type in 1609, the Galilean refractor produces an upright image but has a very small field of view. Opera glasses have this design, because it is compact and cheap to produce.

Galileo Galilei (1564–1642)

Galileo is one of the most significant figures in the long history of astronomy. Within just two years (1609–11) he made telescopes with which he saw details of the Sun, Moon and planets that ended the ancient era of astronomy and set off the new.

Born near Pisa in Italy, Galileo was appointed in 1592 to the chair of mathematics at the University of Padua. In 1604, a new star was observed in Sagittarius (KEPLER'S SUPERNOVA), challenging the Aristotelian doctrine of the unchangeability of the heavenly bodies. Already critical of Aristotelian views on motion, Galileo offered three public lectures on the supernova, his first venture in astronomy. He concentrated on explaining how the star's lack of PARALLAX implies its great distance from Earth and hence the changeability of the non-terrestrial region.

In 1609 a chance event diverted him from his research in mechanics for 25 years. He heard about a spyglass invented in Holland that made distant objects seem near, and immediately grasped its potential. Between 1609 and 1611 he made all the major astronomical discoveries that were within the reach of a telescope of this type (*see* GALILEAN TELESCOPE). By January 1610 he had discovered Earth-like irregularities on the Moon, four satellites orbiting Jupiter, and stars beyond number that were invisible to the naked eye. Later that year came moving sunspots, the phases of Venus and the oddly changing shape of Saturn. Galileo rushed his first discoveries into print, dedicating the *Sidereus Nuncius* (Sidereal Messenger) to his former pupil, now Cosimo II (1590–1621), Grand Duke of Tuscany, and dubbing Jupiter's moons the 'Medicean' planets. Other works quickly followed with new astronomical discoveries, notably the periods of Jupiter's satellites and properties of sunspots, mocking the views of the Jesuit astronomer, Christopher Scheiner (1575–1650).

Galileo refuted the Aristotelian claim that the Earth is the immobile center of rotation of the planets, but this was contrary to passages in Scripture that describe the Sun as in motion. In February 1616 the Roman Congregation of the Index banned Nicolaus COPERNICUS's *De Revolutionibus Orbium Coelestium* (On the Revolutions of Celestial Spheres) 'until it be corrected,' declaring it 'contrary to Scripture.' Cardinal Robert Bellarmine (1542–1621), the leading theologian of the Congregation of the Holy Office, was instructed to inform Galileo officially that the Copernican doctrine was not to be held or defended; if he proved resistant, he was to be given a personal injunction not even to discuss the forbidden view. Whether and how this injunction was delivered has been a matter of debate.

However, forced to abandon his plans for a Copernican treatise, Galileo turned in 1618 to the problem of the nature of comets. Three bright comets appeared in that year, setting off a flurry of speculation. Tycho BRAHE had carefully tracked the comet of 1577 and had concluded that it lay far beyond the Moon, that it was orbiting the Sun not the Earth, and that its motion was either non-uniform and circular or oval. Galileo published the *Discorso delle Comete* (Discourse on the Comets) in 1619 under the name of his friend, Mario Guiducci (1585–1646), claiming that Tycho's parallax arguments would work only if the comets were physical bodies, and not (as Galileo maintained) mere reflections of sunlight from vapors of earthly origin lying beyond the Moon.

In 1623 Cardinal Maffeo Barberini (1568–1644), an avowed admirer of Galileo's work, was elected Pope, as Urban VIII. Galileo, encouraged, asked permission to resume work on his Copernican treatise. Urban set a condition that the treatment be hypothetical. Galileo evidently took it to mean that he could make the best case possible for the Copernican hypothesis, but Urban wanted to ensure that the explanations did not challenge divine omnipotence.

The treatise gradually took shape, delayed by ill health. There were lengthy negotiations with the censors in Rome and Florence, but eventually the *Dialogo sopra i due massimi sistemi del mondo* (Dialog on Two Chief World Systems) appeared in 1632. Roman theologians noted that the *Dialogo* clearly did defend the proscribed view. Urban's reaction was personal; in his eyes Galileo had betrayed his trust. Galileo was summoned to Rome and put on trial in early

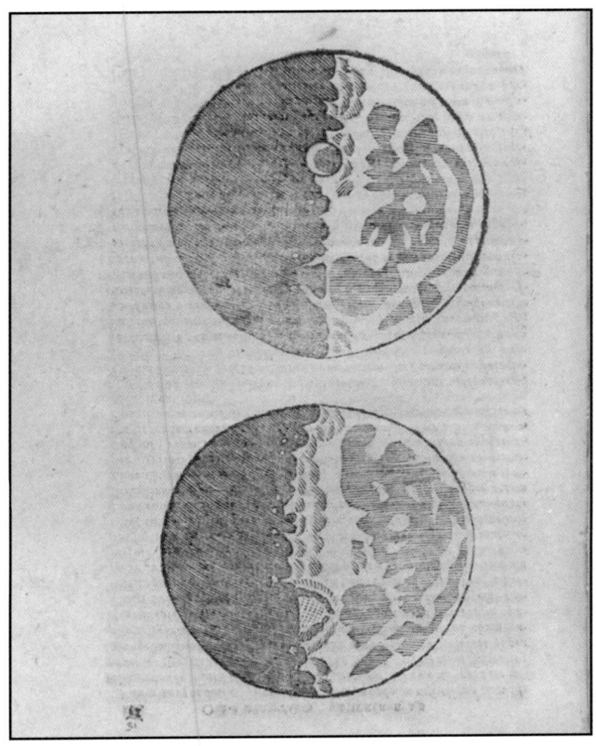

◀ ***Galileo*** *In 1610 Galileo made the first topographic drawing of the Moon, using his modest reflecting telescope. He also worked out the height of the lunar mountains via geometry and the length of their shadows.*

1632. A plea-bargain was struck; Galileo agreed to abjure publicly the prohibited claims of the Earth's motion and the Sun's rest and was sentenced to prison, the sentence promptly commuted to house arrest.

Back finally at home, he returned to the long-interrupted manuscript on mechanics. In 1638 the *Two New Sciences* was published in Holland. According to his secretary, Vincenzio Viviani (1622–1703), he climbed the Leaning Tower of Pisa and dropped cannon balls of different masses, showing they fall together (the equivalence principle; *see* GENERAL RELATIVITY). This story is believed to be untrue but Galileo did describe a 'thought experiment' like it. He died, blind, four years later.

Galileo mission to Jupiter

The Galileo mission to Jupiter was the first spacecraft to sample an outer-planet atmosphere. It made the first extended study of an outer planet, its satellites and magnetosphere. Moreover it took the first images of an asteroid, and confirmed that asteroids can have gravitationally bound moonlets.

Galileo was launched on October 18, 1989, by the Space Shuttle Atlantis from Kennedy Space Center, USA. Its trajectory included a close flyby of Venus (February 1990) and two of Earth (December 1990 and December 1992) for gravity assists to gain sufficient energy to reach Jupiter. By luck this trajectory allowed the first flyby of two asteroids (GASPRA and IDA, discovering its moon, DACTYL) and a direct view of the impacts with Jupiter of COMET SHOEMAKER–LEVY 9.

The spacecraft had an atmospheric-entry probe and an orbiter. The probe was separated from the orbiter on July 13, 1995, and entered the atmosphere of Jupiter on December 7 that same year. The probe returned data for just over 1 hour, until it reached a pressure level in the atmosphere of 22 bar. The orbiter was placed in orbit around Jupiter. Its original two-year mission was extended three times, particularly to reconnoiter the jovian moon EUROPA. In the later stages of the mission, controllers risked ever more daring flybys close to the surface of all the Galilean moons. In November 2002, its final flyby, it dipped within about 500 km of the irregularly shaped moon Amalthea, and determined its density to be about the same as water ice. This implies that Amalthea is a loose assembly of rocks. Galileo's final orbit took an elongated loop away from Jupiter, returning in September 2003 for a direct impact, burning up in Jupiter's atmosphere.

The Galileo probe made direct measurements of Jupiter's atmosphere as it penetrated its cloud layers, including its temperature, composition and wind speeds. The orbiter's observations of the Galilean satellites are of high scientific interest. It found that the three inner satellites, Io, Europa and Ganymede, are strongly differentiated (*see* DIFFERENTIATION). The best fit is a three-layer model consisting of a metallic core, rocky envelope and (except for Io) water-ice outer shell. CALLISTO, the outermost Galilean satellite, is partially differentiated with an ice–rock mixture, with a greater proportion of rock toward the core. Spectra of the satellite surfaces imply that, besides water ice, the surfaces of Ganymede and Callisto contain CH, CN (molecules containing carbon and hydrogen or nitrogen), carbon dioxide and sulphur dioxide. Certain impact scars on Europa seem to contain saline hydrates mixed with the dominant constituent, water ice, which floats on a water ocean. The surface of Io does not contain water ice because of the active heat release by volcanism, but, besides silicate materials, sulphur dioxide ice grains are almost everywhere.

Galle, Johann Gottfried (1812–1910)

German astronomer. In 1846, at the Berlin Observatory, he discovered the planet Neptune, whose position had been calculated by Urbain LEVERRIER. Leverrier had written to Galle asking him to search for the 'new planet' at a predicted location. Galle, together with his assistant Heinrich d'Arrest, began a search on the same night that they received the letter. At d'Arrest's suggestion, Galle used the latest star chart that had only just been produced by Johann ENCKE. In 30 minutes they had located a star not on the map, confirming that it was the new planet on the following night by its motion relative to the other stars.

Gamma-ray astronomy

Gamma-rays (also written as 'γ-rays') are the most energetic form of ELECTROMAGNETIC RADIATION. Gamma-rays have wavelengths shorter than 0.01 nm and photon energies greater than about 30 000 eV, the boundary between gamma-rays and X-rays. Despite their high energies and penetrating power, gamma-rays from cosmic sources are absorbed by the atmosphere, and do not reach below about 10 km altitude. Gamma-ray emission from space can be detected either by the effect of the gamma-rays on the atmosphere or by nuclear instrumentation carried in stratospheric balloons, rockets and satellites.

Gamma-ray astronomy faces several serious handicaps:

- conventional mirrors are transparent to gamma-rays, and so they cannot be focused. The detectors themselves are not very directional and must be very large;
- gamma-ray detectors are bombarded by energetic cosmic-ray particles, which greatly confuse the signal;
- although astrophysical sources emit great energies, each gamma-ray photon is itself very energetic. Thus only small numbers of gamma-ray photons are emitted, and astronomers detect few of them.

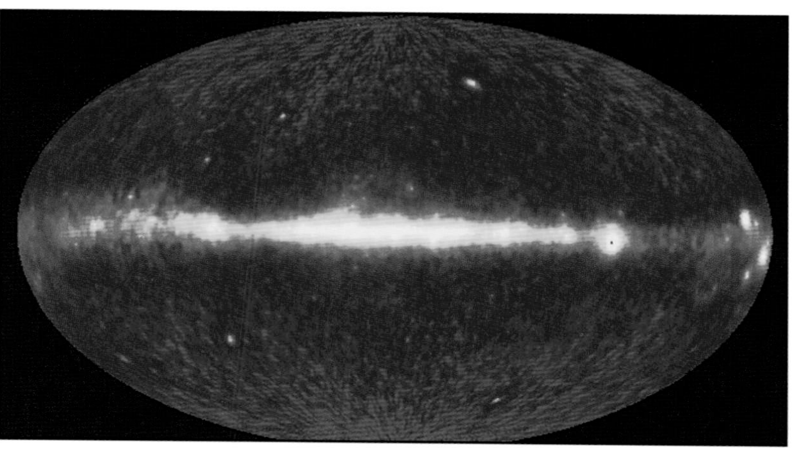

▲ ***Gamma-ray astronomy*** *In this map by the EGRET satellite of the brightness of the sky as seen in gamma-rays, the galactic center is at the middle. The rays are brightest along the galactic plane, stretching either side. They come from the regions with energetic stars (like supernovae) and with magnetic fields. The bright patches at the far right of the picture are the Crab and Geminga pulsars, and the brightest source to the right of center in the galactic plane is the Vela supernova remnant. The quasar 3C273 is the isolated patch at the top.*

These difficulties explain why gamma-ray astronomy is the last window of the electromagnetic spectrum to be opened onto the universe.

Gamma-ray astronomy from space

The first extraterrestrial gamma-rays were detected during the solar flare of March 1958. Within 10 years, gamma-ray emitters outside our solar system were detected by OSO-3 (1961), which recorded a few hundred gamma-rays from the inner Galaxy. In 1972 a US team released secret data on 'gamma-ray bursters' from the military Vela satellites (which were monitoring tests of nuclear weapons). The SAS-2 and Cos-B satellites were the first to detect extragalactic gamma-ray sources, and in 1987 gamma-rays were seen from the supernova SN1987A by the SMM satellite. Present knowledge has been stimulated by the SIGMA mission on the MIR space station (*see* GRANAT) and CGRO. INTEGRAL was launched in 2002 and the Gamma-Ray Large Area Space Telescope (GLAST) is scheduled to be launched in 2006.

Ground-based telescopes

When a high-energy gamma-ray strikes the atmosphere, it produces an electron and a positron, triggering a cascade of energetic photons and particles along the trajectory of the original gamma-ray. The secondary particles all move at nearly the speed of light and the shower front arrives at the ground as a disk, only 1 m thick but 0.1 sq. km in area, which can sweep across an array of detectors. The timing of the arrival of the disk at each detector shows the arrival direction and hence the source of the radiation. This is one of the few astronomical techniques where the Earth's atmosphere plays an essential positive role. The HEGRA EXPERIMENT in the Canary Islands at an elevation of 2.4 km and a densely packed array of scintillation detectors in Tibet at 4.3 km are gamma-ray detectors of this type.

Additionally, when the shower of particles traverses the air, they radiate Cerenkov light, which can be detected by a simple light detector (a mirror, plus a phototube, plus fast electronics). Early telescopes were built around ex-World War II searchlight mirrors. Some groups use large solar arrays (STACEE, the Solar Tower Atmospheric Cerenkov Effect Experiment, in the USA, and CELESTE, the Cerenkov Low Energy Sampling and Timing Experiment, in France). Although their optics are crude, solar arrays have very large mirror areas (>1000 m^2). Specially built Cerenkov telescopes have large tesselated mirrors at a dark mountain-top observatory. The largest include the Whipple Observatory (Mount Hopkins in Arizona), MILAGRO (Multiple-Institution Los Alamos Gamma-Ray Observatory), located near Los Alamos, and CANGAROO (the Collaboration of Australian and Nippon for a Gamma-Ray Observatory in the Outback) in Woomera.

Future larger arrays include VERITAS (Very Energetic Radiation Imaging Telescope Array System), six 10-m aperture Cerenkov telescopes in Arizona; and HESS (High Energy Stereoscopic System), four 12-m telescopes in Namibia.

Emission processes of gamma-ray photons

Nature creates gamma-ray photons in a variety of ways:

- high temperatures (>100 million K) produce gamma-rays. The close neighborhood of a BLACK HOLE accreting matter from a companion star reaches such high temperatures;
- high-energy electrons radiate gamma-rays by interacting electromagnetically with nuclei, photons or intense magnetic fields. These processes are called bremsstrahlung, inverse Compton emission and synchrotron emission, respectively;
- nuclear interactions of protons produce unstable particles, pions and mesons, which decay to gamma-rays;
- just as optical photons can be produced by electron transitions in atoms, gamma-rays result from the much more energetic transitions between energy levels in atomic nuclei. Spectroscopy can identify the exact isotopes concerned;
- the annihilation of an electron and a positron gives rise to gamma-rays at an energy of 0.511 MeV.

The Sun

The Sun emits gamma-rays continuously, but particularly during brief periods of intense activity at the maximum of the 11-year cycle of solar flares. Energy stored in unstable magnetic configurations of the solar-surface layers is suddenly liberated, then solar particles are accelerated to high energies and produce gamma-rays. Additionally, part of the Sun's emission is due to transitions in atomic nuclei.

Stellar explosions and remnants

Radioactive nuclei are produced by thermonuclear reactions in supernovae (and novae) and can be identified by gamma-ray spectroscopy. The gamma-ray intensity shows how much of the isotope has been synthesized. The shape of the lines may give information on the structure (velocity and density profiles) of the supernova. Cobalt-56 was observed by SMM in Supernova 1987a. The long-lived radioactive nucleus titanium-44 (with a half-life of 60 years) was detected by CGRO in the 300-year-old supernova remnant, CASSIOPEIA A. Supernova remnants aged tens of thousands of years radiate gamma-rays produced by accelerated electrons and protons interacting with the ambient interstellar medium. The Crab Nebula, and five other sources detected by CGRO, belong to this class.

Gamma-ray pulsars

The CRAB PULSAR, VELA PULSAR and five others emit gamma-rays periodically. One of them, GEMINGA, is the only known pulsar with no detectable radio emission.

Accreting compact objects

According to theory, the inner accretion disks around neutron stars reach temperatures less than 100 million K (X-ray emission), whereas disks around black holes may become 1 billion K (gamma-rays). Some black-hole candidates in our Galaxy show steady emission, while others are variable (presumably as a result of variation in the accretion disk). Typical transient sources (or gamma-ray novae) were GRS 1124-68, alias Nova Muscae (detected in January 1991), and 1E1740.7-2942 (October 1990).

Diffuse gamma-ray emission in the galaxy

Some products of stellar activity, including cosmic rays, long-lived radioactive nuclei and positrons, produce gamma-rays. Cos B and CGRO have mapped diffuse gamma-ray emission from our Galaxy and the LARGE MAGELLANIC CLOUD.

Gamma-ray bursts (GRBs)

Cosmic GRBs are randomly occurring flashes of high-energy radiation which briefly dominate the gamma-ray sky (from a few milliseconds to a few seconds) then fade away. The BATSE instrument aboard CGRO detected about one burst per day for several years and established that their distribution in the sky is random and isotropic, indicating that they are at cosmological distances (in the order of 1 billion l.y.). There is about one burst per galaxy per million years, making them much rarer than supernovae.

The satellite BEPPOSAX quickly and accurately located several GRBs, so that some long-lived X-ray, optical and radio 'afterglows' were identified. The GRBs were indeed located in distant galaxies. GRBs radiate huge amounts of energy, more than supernovae. They originate in some kind of supernovae, which must be nearly 'naked' (there must be little material surrounding them, or the gamma-rays would be absorbed). A 'fireball' of nearby material expanding at relativistic energies produces the afterglow.

Active galactic nuclei (AGN)

Cos-B detected the first known extragalactic gamma-ray source, the bright quasar 3C 273. Several dozen other gamma-emitting AGN have since been found (*see* ACTIVE GALAXY). The gamma-rays are emitted from the hot inner regions of the accretion disk surrounding the black hole and by relativistic particles moving in opposite jets. A uniform, diffuse, isotropic gamma-ray emission in the sky comes from numerous individually unresolved extragalactic sources, including Seyfert galaxies, blazars and other kinds of AGN.

Further reading: Dermer C (1995) *The Gamma-Ray Sky with CGRO and SIGMA* Kluwer; Fichtel C E and Trombka J I (1997) *Gamma-Ray Astrophysics: New Insight into the Universe* NASA Reference Publication 1386.

Gamma-ray burst

See GAMMA-RAY ASTRONOMY

Gamma-ray bursts, observing

See PRACTICAL ASTRONOMY FEATURE OVERLEAF.

Gamow, George (1904–68)

American physicist born in Odessa, now in Ukraine. Gamow worked with Edward Teller (1908–2003) on beta decay, and devised the theory of stellar nuclear reactions in terms of the tunneling of a colliding nucleus through another nucleus's potential barrier. Even though there is only a small probability that a nucleon in the Sun will tunnel through the electrostatic repulsion barrier of another (say once every 10 000 million million million collisions), collisions in the solar core are so frequent that enough nuclei do get close enough to fuse and release solar energy. Gamow developed a BIG BANG cosmology in which a primordial matter (ylem) existed at the origin of the universe. He supported Georges LEMAÎTRE's Big Bang model, predicting that its consequence was a residual cosmic fireball of background radiation with a temperature of 10 K (Arno PENZIAS and Robert WILSON later discovered the COSMIC MICROWAVE BACKGROUND radiation at 3 K, surprisingly close to Gamow's prediction).

◀ ***Ganymede*** *Ganymede has dark and bright terrains, and a number of relatively young craters with bright rays of ejecta. The large dark area is Galileo Regio.*

Ganymede

One of Jupiter's icy moons and one of the Galilean satellites, Ganymede is the largest planetary satellite in the solar system. This moon, which is larger than the planet Mercury (Ganymede's radius is 2638 km, Mercury's is 2439 km), generates its own magnetic field and has large polar caps composed of thin frosts which reach an average latitude of 40°. On a global scale the surface of Ganymede can be divided into two basic terrains: dark terrain and bright terrain.

Dark terrains are older, more heavily cratered regions that are generally crosscut by areas of the younger bright terrain. The low ALBEDO of the dark terrain is thought to be due to mixtures of rocky material with water ice on the surface. The dark material is believed to have resulted in part from the non-ice components of the many meteors and comets that have hit Ganymede over geologic time. The higher-albedo regions within dark terrain have been brightened by both tectonic and impact processes. Dark material also appears to have moved down slopes, exposing bright icy walls, and collecting in valleys and depressions. Dark terrain, which is almost as old as the solar system itself, has been fractured by tectonic activity as well as by large impact craters. Concentric furrow systems, similar to multi-ring structures on neighboring CALLISTO, were formed when large asteroids or meteors impacted Ganymede early in its history. The bright terrain is covered with troughs and grooves created by tectonic activity.

Ganymede is an icy satellite. Its relatively low mean density (1926.8 kg m^{-3}) means that it is 60% rock and 40% water ice. Ganymede is differentiated into a metallic core surrounded by a silicate mantle. The deep metal-rock interior of Ganymede is covered by a thick (800 km) ice-liquid water shell. Ganymede thus has a three-layer internal structure consisting of an inner metal core, a middle rock shell and an outer water (ice-liquid) shell.

Ganymede's magnetic field is large enough to carve out its own MAGNETOSPHERE within the jovian magnetosphere. It comes from dynamo action within a still liquid part of

Observing gamma-ray bursts (GRBs)

GRBs are some of the most enigmatic objects studied by modern astrophysics. Detected initially by satellites operating for GAMMA-RAY ASTRONOMY, it is only in the last two decades that any optical signature has been found.

It is perhaps surprising, therefore, that amateurs have contributed to classical optical afterglow PHOTOMETRY. The unusual burster GRB980425/SN1998bw was observed in 1998 by amateurs working at MOUNT STROMLO OBSERVATORY (the name indicates that it was discovered on April 25, 1998). The first GRB observation with an amateur telescope was by Warren Offutt for 990123; he used a 0.6-m telescope, Bessell R-band filter (*see* CAMERA AND FILTER) and CCD to provide photometry about a day after the burst. Subsequent amateur observations were made by the Buffalo Astronomical Association (GRB000301C), with a 0.35-m telescope, and Nyrölä Observatory, Finland using a 0.4-m telescope. To date, about a dozen amateur observatories have observed GRB afterglows and provided photometry to the professional community. In August 2003, South African amateur astronomer Berto Monard discovered such an afterglow before the professionals (many of whom were attending an IAU conference away from their telescopes).

To observe a GRB afterglow, you need a telescope that can point accurately, a CCD camera, and preferably a standard photometric filter. Unfiltered discovery observations are accepted, but for accurate light curves, professionals prefer that the observation is taken through a filter. The Cousins RC bandpass is most common since it covers the range of wavelengths at which a typical CCD is most efficient and provides some contrast improvement under moonlit skies.

Positions provided from the HETE-2 and Integral satellites (observing in gamma-rays) have errors in the arcmin-to-tens-of-arcmin range; therefore the observer should have CCD fields of view in the 10–30 arcmin range. Start observing as quickly as possible after notification, which is usually obtained via the Internet. Use short exposures if you are observing within an hour or two after the burst, but increase your exposure time for later observations. Most amateur observations will be completed within the 24 hours after the burst since accurate photometry after that point will require multiple-hour exposures and so is better performed by larger telescopes.

The planned Swift satellite promises even better localizations; for these afterglows, R_c or I_c (Cousins red or infrared) filters will complement the onboard photometric capabilities.

If you are observing a GRB field where an afterglow has not yet been announced, you can use the Digital Sky Survey (DSS) to provide a finding chart. Comparing the DSS with your CCD image will highlight any new object in a similar way to supernova observations. Continued observations will indicate whether this object is fading as rapidly as expected for an afterglow candidate. If an afterglow has been announced, then try to set your exposure times so that the candidate has a photometric error of at most a few percent. Once secondary standards are made available for the field, you can perform differential photometry of your images with respect to published comparison stars.

Data can be posted to the AAVSO International GRB Network. Joining the AAVSO network is recommended, as they provide finding charts and other observing aids.

In addition, efforts are being made to search for so-called prompt emission; the afterglows seen cannot be associated directly with the gamma-rays as they arrive too slowly afterwards. In fact, they are believed to result from interaction with the surrounding material. The only prompt optical emission detection of a GRB was for GRB990123 by the ROTSE-I experiment. This emission peaked around a V-band magnitude of 9, but decayed to 14 about 10 minutes after the burst. ROTSE-I used 10-cm telephoto lenses plus a CCD camera, so almost any amateur set-up could make the same detection, provided rapid positions were available and observations could be made quickly.

Dr Arne Henden is a senior research scientist forthe Universities Space Research Association (USRA) at the US Naval Observatory.

► ***Gamma-ray burst*** *Color composite of the sky around the position of GRB 000131, detected in January 2000. Based on images with the VLT at Paranal, Chile. The object is indicated with an arrow. The bright foreground star, over a million times brighter than the faintest objects in the photo, is causing some strange imaging effects.*

Ganymede's metallic core, similar to the origin of the Earth's magnetic field.

Convection through the ice layer transfers Ganymede's internal heat to its surface and maintains its liquid core. The heat is mainly from radioactivity in the silicate component of the satellite, but some still comes from the cooling of Ganymede from its early hot state, when it was being pounded by asteroids and settling. Unlike Io, tidal heating is not important.

Gaseous nebula

Three main kinds of EMISSION NEBULAE – those that radiate emission lines – inhabit our Galaxy. The ORION NEBULA and its kind are known as 'gaseous nebulae,' as 'diffuse nebulae' from their fuzzy appearance, and as 'H II REGIONS' because they are largely made of ionized hydrogen. *See also* NEBULAE AND INTERSTELLAR MATTER.

Gas giant

An alternative name for the GIANT PLANETS.

Gaspra

An ASTEROID discovered by Grigorii Neujmin (1885–1946) in 1916, designated (951) Gaspra. In October 1991 it became the first asteroid to be imaged from nearby when the Galileo probe passed it at a distance of 16 000 km on its way to Jupiter (*see* GALILEO MISSION TO JUPITER). It is an angular body measuring 18 × 11 × 9 km, pitted with small craters. It has few large craters, so either Gaspra has a strong, metal-rich surface or it is a young object. Galileo also detected a significant interaction with the local interplanetary magnetic field, suggesting that there is remnant magnetism. Gaspra orbits the Sun at a mean distance of 2.21 AU, near the inner edge of the main asteroid belt, in a period of 3.28 years; the orbital inclination is 4° and the eccentricity 0.17. It is an S-type asteroid, with a composition similar to that of ordinary CHONDRITES and STONY-IRON METEORITES.

Gas tail

The part of a COMET consisting of ions and electrons formed from gas molecules expelled from the comet during a close approach to the Sun; also known as an ion tail or plasma tail. The molecules are ionized by ultraviolet light from the Sun and drawn away from the head of the comet by the solar wind. Not all comets develop a gas tail. A gas tail always points away from the Sun, its direction aligned with the interplanetary magnetic field at its position. Gas tails are straight, but may appear kinked or discontinuous as a result of a disconnection between two regions of magnetic field trapped in the solar wind. As in COMET HALE–BOPP, gas tails are predominantly blue, the characteristic color of emission by ionized carbon monoxide at 420 nm. Gas tails begin to form typically at 2 AU from the Sun and can extend to 100 million km.

Gauge boson

A particle that transmits the FUNDAMENTAL FORCES. If two ice-skaters throw a heavy ball to one another they will drift apart. This resembles how particle physicists think of the forces between particles, caused by the exchange of gauge bosons. Photons convey the electromagnetic force between charged particles, intermediate vector bosons (the W and Z particles) transmit the weak nuclear force (which governs the radioactive decay of atomic nuclei), and gluons carry the strong nuclear interaction (or 'color force') between the quarks that comprise hadrons and mesons. The existence of an equivalent force carrier for gravitation, the graviton, remains unproven.

Gauss, Carl Friedrich (1777–1855)

Scientist, born in Brunswick (now in Germany), who worked in mathematics and physics, including number theory, analysis, differential geometry, geodesy, magnetism, astronomy and optics. A child prodigy, at an early age Gauss independently discovered BODE'S LAW, and various advanced mathematical techniques such as the binomial theorem and how to construct a regular 17-sided figure by ruler and compasses. In 1801 (1) Ceres, a minor planet discovered by Giuseppe PIAZZI, was lost behind the Sun. Several predictions of its position on reappearance were made from the few discovery observations, including a prediction by Gauss which differed greatly from the others. When Ceres was recovered by Franz Xaver von Zach (1754–1832), on its reappearance it was almost exactly where Gauss had predicted it would be.

Although he did not disclose his reasoning at the time, Gauss had used his 'least-squares method' to determine its orbit. This is a statistical method to determine a curve from observations that have observational errors, so that the mathematical curve does not pass exactly though all the points. The technique adjusts the equation of the curve and picks out the one that best fits the data, so that the sum of the square of the discrepancies (or 'residuals') is minimized. A 'Gaussian distribution' (also 'normal distribution') is a term still used for the name of the theoretical distribution of residuals in statistical data, and makes the least-squares method possible.

Gegenschein *See* ZODIACAL LIGHT

Geminga

A gamma-ray source in the constellation of Gemini, its name deriving from 'Gemini gamma-ray source,' or alternatively from the Milanese dialect word for 'nothing there' (no optical counterpart on the charts available). In 1992, the X-ray satellite ROSAT discovered pulsed emission from Geminga. Located at a distance of 300 l.y., Geminga is a pulsar with a period of 0.237 s and, as such, is one of the nearest neutron stars (*see* NEUTRON STAR AND PULSAR). It does have an optical counterpart, a very faint star that has a high proper motion because of its nearness and its high speed through space (presumably the result of its creation in a supernova explosion).

Gemini *See* CONSTELLATION

Geminids

One of the year's most prolific METEOR SHOWERS, its main appearance being between December 6 and 19. The RADIANT lies in the constellation Gemini. Uniquely, the parent body of the Geminids is not a comet but the asteroid 3200 Phaethon, discovered in 1983. The PERIHELION distance of Phaethon is a mere 0.14 AU. It may be that the Geminid METEOR STREAM consists of particles that have crumbled away from the asteroid's surface under the intense heat of its perihelion passage. An asteroidal source is consistent with the Geminid meteoroids' high density of around 2 g cm^{-3}, about an order of magnitude higher than other METEOROIDS, which are cometary dust particles. These dense particles tend to produce quite slow, bright METEORS of long duration.

Gemini Observatory

The Gemini Observatory is an international collaboration between the US National Science Foundation, the UK, Canada, Chile, Argentina, Brazil and Australia. It consists of twin 8-m telescopes in each hemisphere, one on Hawaii's Mauna Kea (Gemini North) and the other on central Chile's Cerro Pachón (Gemini South).

General relativity

The general theory of relativity, created by Albert EINSTEIN between 1907 and 1915, is a theory of both gravitation and SPACETIME structure (*see* GRAVITATION, GRAVITATIONAL LENSING AND GRAVITATIONAL WAVES). The general theory extended Einstein's previous work on SPECIAL RELATIVITY.

General relativity embodies the concept that the three dimensions of space (length, breadth and height) and the dimension of time are linked together into a four-dimensional spacetime. Gravitation is regarded as a geometric property of spacetime rather than a force acting directly between individual massive bodies. The effect of a distribution of mass (or energy) is to induce curvature into (to bend) spacetime in its vicinity. Conversely, particles and rays of light follow paths that are determined by the curvature of spacetime in their locality. In the words of physicist John Wheeler (1911–), known for inventing the phrase 'black hole,' 'matter tells spacetime how to curve, and spacetime tells matter how to move.'

Equivalence principle

Both Isaac NEWTON and GALILEO GALILEI knew that the gravitational force on a body is directly proportional to the mass (the 'gravitational mass'), and the acceleration is inversely proportional to the mass (the 'inertial mass'). If every object had a 'gravitational mass' exactly proportional to its 'inertial mass,' then these two values for the mass would cancel out from the equations of motion (the 'principle of equivalence'). All objects moving in a given gravitational field have the same acceleration, and, starting together, move together. Galileo and Newton experimented on falling masses with different composition using dropped objects and pendula. Loránd von EÖTVÖS and others carried out several laboratory experiments with exquisite precision in the late 1800s without the equivalence breaking down. Galileo's legendary experiment at Pisa, with two falling weights, was dramatically repeated on the Moon by the Apollo astronauts with a weight and a feather, but without the complicating effects of air resistance. Most precisely, 30 years of laser ranging to several passive reflectors placed on the lunar surface by the Apollo missions confirmed that the Earth and Moon fall at equal rates toward the Sun with a precision of 1 part in 10 million million. The Earth has a substantial iron core while the Moon is silicate materials, so this confirms the composition independence.

What the principle of equivalence implies is that an observer in a small closed box (say, an elevator or a windowless spacecraft) cannot tell whether that box is subject to uniform acceleration or is at rest in a uniform gravitational field. Both situations would give the observer a feeling of 'weight.' Conversely, if a test particle floats freely inside a closed box, an observer cannot tell whether the box is in the depths of space, far from a gravitating body, or falling freely in the gravitational field of a massive body. For example, an observer inside an elevator falling freely down its shaft would have no sensation of weight.

Einstein incorporated the equivalence principle into his general theory of relativity. From it, he concluded that clock rates should be reduced near gravitating matter when compared with rates of more distant clocks. In 1977 a hydrogen maser clock was put on a suborbital rocket flight, and its rate was compared with a clock on Earth, confirming the predicted shift with a precision of almost 1 part in 10 000.

Celestial mechanics

Although Newtonian theory is perfectly satisfactory for calculating most orbits of celestial objects, general relativity can step in where Newtonian theory is inadequate.

In particular, general relativity has an effect on the elliptical orbit of the planet Mercury. The orbit does not exactly repeat, but twists around in such a way that the perihelion position of the planet moves in space. It advances in the same sense as the planet revolves in orbit, by an amount that exceeds the Newtonian prediction by 43 arcsec per century. A discrepancy of this size had already been established by the end of the nineteenth century, and its explanation was the first and immediate triumph of general relativity in 1916. Today, radar ranging from Earth to Mercury's surface, a less precise procedure than the ranging to specific transponder sites on Mars, nevertheless confirms general relativity to a precision of about 1 part in 1000.

Even more stringent tests are provided by two pulsars. Pulsars are very stable clocks when isolated. A pulsar in a binary system is thus a moving clock, a good tool to test a relativistic theory. Many orbital data can be extracted from analyzing the pulse arrival times and comparing them with the theory's predictions. More precisely, one can fit the pulse arrival times to a general formula taking into account any possible timing effect. The formula contains several parameters. Five of them describe the approximately Keplerian orbit of the pulsar. The relativistic corrections to this Keplerian orbit are described by eight 'post-Keplerian' parameters. The Hulse–Taylor binary pulsar, PSR B1913+16, has been continuously observed since its discovery in 1974. This is a rapid (8-hour period) and very eccentric (0.6) system. The binary pulsar PSR B1534+12 is also a rapid (10 hour) and eccentric (0.3) system. Observed only since 1991, it is much closer to the Earth than PSR B1913+16, and its pulses are stronger and narrower. In both, it has been possible to measure the post-Keplerian parameters with good precision. General relativity passes these tests with flying colors.

Gravitational waves

In general relativity, the gravitational field propagates like a wave. Its value in the vicinity of an event A can influence a later event B only because the field propagates at the speed of light. Any change in a gravitational field such as the rotation of one star around another results in the emission of gravitational waves. The pulsar PSR B1913+16 shows a decrease in its orbital period, due to the gravitational waves that it is radiating, in agreement with general relativity to a precision of 0.35%. This is held to confirm the existence of gravitational radiation, which has not been seen directly.

Horizons and black holes

In the same way that an event today cannot be seen by an observer yesterday, not all parts of spacetime can be seen

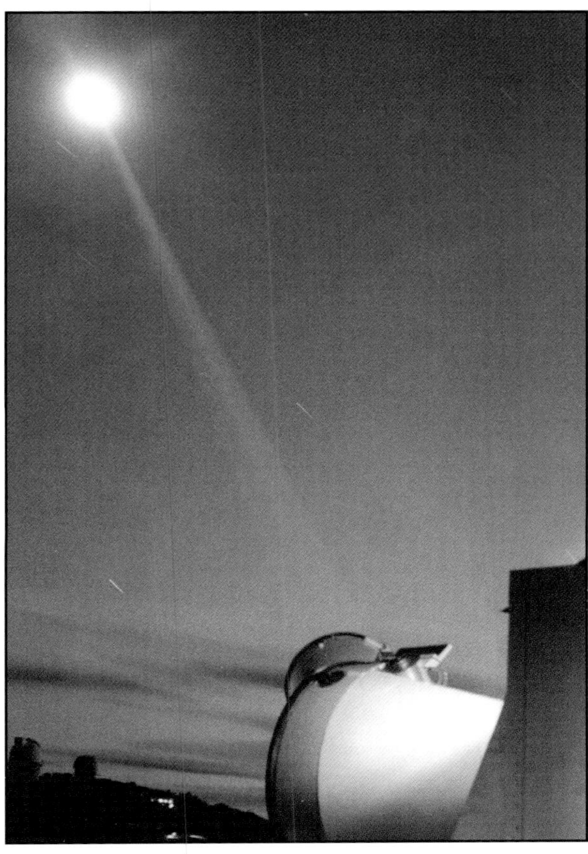

◀ ***Laser ranging*** *A green laser beam takes more than 2 s to travel to the Moon and back. The McDonald Observatory laser ranger shown here measures the distance of the Moon to 1 cm.*

by all observers. If a spherical star collapses so much that the ratio of mass:radius of the star reaches the 'Schwarzschild limit,' events inside will never reach an outside observer. This 'event horizon' separates events that can be seen from outside from those that are hidden from sight. The part of spacetime thus hidden is called a BLACK HOLE; the horizon is its theoretical surface.

Cosmology

Of the four basic interactions which act between all kinds of matter (the strong and the weak forces in nuclei, the electromagnetic force and the gravitational force), gravitation dominates the behavior of matter on large scales in spite of its extreme weakness. Therefore, general relativity provides the basis for cosmology. The exact form of general relativity that describes the universe is under discussion. Einstein seems to have made a good guess for reasons now known to be shaky. He tried to make a theory about a static universe before it was known to be expanding. In 1917 he supplemented his equations of 1915 by a 'cosmological term,' parameterized by the cosmological constant. This contribution was later interpreted as the energy of the vacuum, or dark energy, for which there is now some good (but not yet convincing) evidence (*see* DARK ENERGY AND THE COSMOLOGICAL CONSTANT).

The status of general relativity

After almost 100 years of the theory of general relativity, we are left with a wide variety of experimental evidence that it is the most accurate description of the long-range interaction between bodies.

Further reading: Hawking S W and Israel W (eds) (1987) *Three Hundred Years of Gravitation* Cambridge University Press; Hawking S W and Israel W (eds) (1997) *General Relativity: An Einstein Centenary Survey* Cambridge University Press.

Genesis

NASA Discovery mission to collect samples of charged particles in the solar wind. On November 16, 2001, the Genesis spacecraft took up its orbital station around the Lagrangian point L_1, a million miles from the Earth toward the Sun, where it travels alongside the planet in its annual orbit. It collects solar-wind particles, monitors the speed, density, temperature and approximate composition of the solar-wind electrons and ions, and separates and collects elements like oxygen and nitrogen. Isotopes of oxygen, nitrogen, the noble gases and other elements of the solar wind will be returned to Earth in a capsule (planned for September 2004).

Geocentric coordinate system

System of coordinates, such as latitude and longitude, centered on the Earth.

Geodesy

Measurement of the shape and size of the Earth.

Geodynamo

The large magnetic field in the Earth is powered by dynamo action, caused by electric currents driven from motions in the planet's liquid-iron core. The field is predominantly dipolar, but its orientation changes, as has been known since 1634. The position of the pole migrates over the centuries. This is known as the secular variation. In addition studies of remnant magnetism of rocks and ocean sediments show that the main dipole field has reversed direction many times, irregularly but with a typical interval of 100 000–1 million years.

Geomagnetic storm

A fluctuation in the Earth's magnetic field caused by the arrival of a disturbance in the solar wind; also known as a magnetic storm, and one of the phenomena called 'space weather.' The sequence of events in a typical major storm is as follows:

- sudden storm commencement (lasting a few minutes): a cloud of PLASMA from the Sun, a result of a coronal mass ejection, compresses the Earth's magnetosphere and increases the geomagnetic field at ground level;
- initial phase (a few hours): the plasma cloud flows past the Earth and the geomagnetic field strength remains higher than normal;
- main phase (a few hours to a day): charged particles flow through the inner magnetosphere, creating an electrical current around the Earth (the 'ring current'). The ring current generates a magnetic field opposite to that of the Earth and causes a pronounced drop in the geomagnetic field;
- recovery phase (a few days): the ring current declines and the field strength returns to normal.

Geomagnetic storms are often accompanied by displays of the AURORA, and surges (perhaps damaging) of electric current in telephone lines, power cables and metal pipelines.

Geotail

Japanese-US mission, part of the International Solar-Terrestrial Physics program, launched in July 1992 to study the dynamics

▶ ***Ghost of Jupiter*** *(NGC 3242) This HST picture of the Ghost of Jupiter shows the planetary nebula and its central star, as well as the mysterious red 'fliers' (fast low-ionization emission regions) flung out in opposite directions.*

of the Earth's magnetotail over distances ranging out to 200 Earth radii initially. The magnetotail is the part of the Earth's MAGNETOSPHERE downstream of the SOLAR WIND. The spacecraft was brought closer to Earth in 1995.

Ghost of Jupiter (NGC 3242)

A PLANETARY NEBULA in the constellation Hydra, right ascension 10 h 24.8 m, declination –18° 38'. It is of magnitude 8 and has a magnitude 12 central star. It is bluish, and with its size of 16 arcsec it can appear as a faint version of Jupiter.

Giacconi, Riccardo (1931–)

Astrophysicist, born in Genoa, Italy. An American astronomer and international project director, Giacconi won a Nobel prize in 2002 for his work in founding X-RAY ASTRONOMY. With a rocket-borne X-ray telescope intended to study the composition of the Moon, he discovered SCORPIUS X-1, the first known X-ray source outside the solar system. Giacconi built the UHURU X-ray satellite that made the first surveys of the X-ray sky. He also led the construction and successful operation of the EINSTEIN OBSERVATORY, the powerful X-ray satellite, and later the completion of the VLT.

Giacobinids

A periodic METEOR SHOWER, also known as the Draconids, associated with Comet Giacobini–Zinner. The shower takes place between October 10 and 27, and has its radiant in the CONSTELLATION Draco. Zenithal hourly rates were high in 1933 (up to 450), 1946, 1985 and 1998. Radar observations in 1946 detected a peak rate of 170 per minute and represented the first identification of a meteor shower by this technique. *See also* COMET.

Giant impactor theory

The theory that the twin Earth-Moon system was formed when a planetoid about the size of Mars collided with the newly formed Earth. *See also* MOON.

Giant planet

A term for the major planets Jupiter, Saturn, Uranus and Neptune, the four largest planets in the solar system. They are characterized by their size (of the order of 100 000 km in diameter), their density (close to liquid water), their largely gaseous nature, with no solid surface, and their great distance from the Sun (5–30 AU). Most exoplanets are more massive than Jupiter, and may also be classed as giant planets. *See also* EXOPLANET AND BROWN DWARF.

Gibbous

The appearance or phase of a body in the solar system when more than half of its sunlit side is visible, but not all of it. From the Earth, the INFERIOR PLANETS (Mercury and Venus) and, as is readily seen, the Moon show pronounced gibbous phases, the Moon when aged 7 to 14 and 21 to 28 days. Of the SUPERIOR PLANETS, only Mars can appear markedly gibbous (but always with at least 84% of its disk illuminated). Of the OUTER PLANETS, Jupiter can show a slight gibbous phase, but the planets beyond are too distant.

Gill, David (1843–1914)

Astronomer, born in Aberdeen, Scotland, who became the ASTRONOMER ROYAL for Scotland. Inspired by the unexpected discovery in 1882 of stellar images on a photograph of a comet, he pioneered astronomical photography.

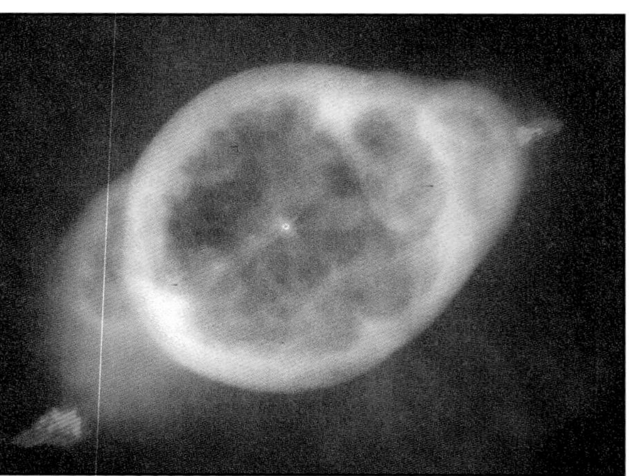

Ginga (Astro-C)

Japanese X-ray satellite, launched February 1987. Ginga carried three instruments: a large-area proportional counter array, an all-sky monitor and a gamma-ray-burst detector. The satellite observed supernova 1987A soon after its appearance, and discovered several pulsars and other transient sources. Ginga re-entered the atmosphere in November 1991. Its name means 'galaxy.' *See also* ASTRO.

Giotto

The Giotto mission investigated the nucleus of COMET HALLEY at close quarters during its 1986 PERIHELION passage. It was ESA's first deep-space mission and was named after the Italian artist, Giotto di Bondone (*see* ART AND ASTRONOMY). Five space probes – two Soviet, two Japanese and one European – were launched toward the comet. The Japanese spacecraft (Sakigake and Suisei) made long-distance measurements, and the Soviet VEGA probes, launched from the two VEGA spacecraft when they were investigating Venus, acted as pathfinders for Giotto by accurately locating the comet's nucleus.

Giotto was launched on July 2, 1985. It first detected the comet on March 12, 1986, recording hydrogen ions 7.8 million km from the nucleus. Twenty-two hours later, Giotto crossed the bow shock between the SOLAR WIND and the ions from the comet and entered the COMA. The first of 12 000 dust impacts was recorded 62 min before closest approach (290 000 km out), but the rate of impacts rose sharply as the spacecraft passed through a jet of material from the nucleus. Only 7.6 s before closest approach, the spacecraft was sent spinning by an impact from a 'large' (about 1 g) particle. Contact with the Earth was lost for half an hour while the thrusters stabilized its motion. By then Giotto had passed the nucleus of the comet, getting within 596 km on the sunward side. The last impact was detected 49 min after closest approach. A unique series of images revealed the comet nucleus to be a dark body (ALBEDO 4–5%, like coal). It was peanut-shaped, about 15 km long and 7–10 km wide. Seven jets, covering about 10% of the warmer sunlit side, threw out 3 tonne s^{-1} of material, which was 80% water. There were also substantial amounts of carbon monoxide (10%) and carbon dioxide (2.5%), with traces of other hydrocarbons, iron and sodium. Most of the dust was small particles, one type made of the light 'CHON' elements (carbon, hydrogen, oxygen and nitrogen) while the other was rich in sodium, magnesium, silicon, iron and calcium.

Despite the damage sustained during the flyby, Giotto

survived. It was hibernated and the mission was extended to encounter Comet P/Grigg-Skjellerup. Giotto was reawakened on May 4, 1992, and swept past the comet at a distance of only 100–200 km just after it had passed perihelion.

Global Oscillation Network Group (GONG)

GONG is an international project, operated by the NATIONAL SOLAR OBSERVATORY for the US National Science Foundation, to conduct a detailed study of the internal structure and dynamics of the Sun over an 11-year solar cycle using HELIOSEISMOLOGY. Measurements are obtained by a six-station network located at BIG BEAR SOLAR OBSERVATORY (California), Mauna Loa Observatory (Hawaii), Learmonth Solar Observatory (Western Australia), Udaipur Solar Observatory (India), TEIDE OBSERVATORY (Canary Islands) and the CERRO TOLOLO INTER-AMERICAN OBSERVATORY (Chile), providing coverage of the Sun that is 90% continuous.

Globular cluster

A globular cluster is a spherical star system of approximately 10 000–1 million gravitationally bound stars, highly concentrated to the center, spread over a volume that is a few dozen to more than 300 l.y. in diameter. Globular clusters resemble shining old islands orbiting a galaxy.

The density of stars in the cluster's center is so high (up to a few thousand stars per cubic light year) that it is generally impossible to separate the individual stars from ground-based observations. Only recently has the HST allowed astronomers to dig into the very central regions of many Galactic globular clusters, where stars (sometimes peculiar or even exotic) move randomly like molecules of gas, interacting according to the basic laws of gravity.

The Galaxy hosts about 200 globular clusters (147 are cataloged, but some are hidden behind the Milky Way). They form a GALACTIC HALO of roughly spherical shape, which is highly concentrated around the Galactic center. The most distant Galactic globular clusters (such as NGC 2419) are located far beyond the edge of the Galactic disk, at distances out to 300 000 l.y. Almost all galaxies have a system of globular clusters, some of which (such as M87), number several thousands.

Ages of globular clusters and their evolution

Most of the globular clusters are orbiting our Galaxy in highly eccentric elliptical orbits, with orbital periods of about 100 million years or even longer. During these orbits the clusters are perturbed by tidal forces from the parent Galaxy, passage through the Galactic plane, stars escaping after a close encounter with others, and so on. This implies that the existing globular clusters are the survivors of a much larger population, which was spread out throughout the Galactic halo and far beyond. Within the next 10 billion years or so, most of the present Galactic globular clusters could disappear. On the other hand, our Galaxy has recently acquired four clusters in Sagittarius (M54 in particular), which were members of the SAGITTARIUS DWARF GALAXY, currently merging into the Milky Way.

The stars of the globular cluster in our Galaxy are typically metal-poor and old. This suggest that they were born during the early stages of our Galaxy's formation and are archeo-astronomical sites where the universe in its youth can be studied. In fact, since globular clusters contain the oldest stars known, they provide a lower limit on the age of the universe. The latest estimates, based on calibrations provided by the Hipparcos satellite, yield an age of 12–14 billion years for most of the globular clusters. The age of the oldest globular cluster, M92, is 13–14 billion years. The youngest Galactic globular clusters are Ruprecht 106, Palomar 12, Terzan 7 and Arp 2. They all, but especially the first two, appear to be around 3 billion years younger than the others. Perhaps they originally belonged to a satellite galaxy tidally disrupted and captured by the Milky Way.

While all the globular clusters in our Galaxy and in M31 are old (aged about 10 billion years, at least), there are galaxies, such as the two Magellanic Clouds and M33 (the Triangulum Galaxy), hosting much younger globular clusters (aged a few billion years, or less). HST images of colliding galaxies show large numbers of bright, blue compact clusters of stars, whose color, magnitude and size are consistent with those expected for globular clusters in our own Galaxy when they were young. They make up an appreciable fraction of the recent star formation in some starbursts and mergers. In the most energetic starbursts observed, 20% of the light of the galaxy is blue and ultraviolet light from bright, compact young star clusters. The formation of stars in this way must be very efficient. However, given that old globular clusters make up only about 1% of the light in old stellar populations such as the Galactic halo and in giant ellipticals, most of these clusters do not survive for very long.

Globular cluster stars

Galactic globular clusters are old, metal-poor and most belong to POPULATION II, although clusters almost as metal-rich as the Sun (and perhaps even more) have been recently discovered near the Galactic center. The stars in a given globular cluster are remarkably homogeneous, unlike a typical galaxy where there is a spread of compositions and ages, suggesting that star formation probably occurred in a different way, in a single event at a single location.

As in Galactic OPEN CLUSTERS, members of each globular cluster (usually) share a common history and differ only in their initial mass. Globular clusters are ideal laboratories for testing the theories of stellar structure and evolution, through the calculation of isochrones in the HERTZSPRUNG–RUSSELL DIAGRAM. The detailed shapes of HR diagrams of globular clusters (particularly the shape of the horizontal branch) differ because the stars have different compositions from cluster to cluster (the 'first parameter'). However, other differences are unaccountable. This is the long-standing 'second parameter problem': another parameter besides composition must be at work, perhaps age or helium abundance.

Thanks to the very large number of stars, almost every evolutionary stage (even those with very short lifetimes, down to a few ten thousand years) is represented in globular clusters. They contain strong and weak X-ray sources, neutron stars and millisecond pulsars, white dwarfs, cataclysmic variables, close binary stars, blue stragglers, planetary nebulae, and so on. Moreover, they contain large numbers of variable RR LYRAE STARS. According to the periods of their RR Lyraes, Galactic globular clusters can be separated into the two Oosterhoff groups. The groups differ in their metal abundance, with variables in metal-poor clusters having longer mean periods than those in the metal-rich ones. The large numbers of globular clusters and their RR Lyrae stars provide material for a good calibration of such effects, and these stars are ideal STANDARD CANDLES to measure distances to other galaxies.

Twelve of the 147 globular clusters associated with our

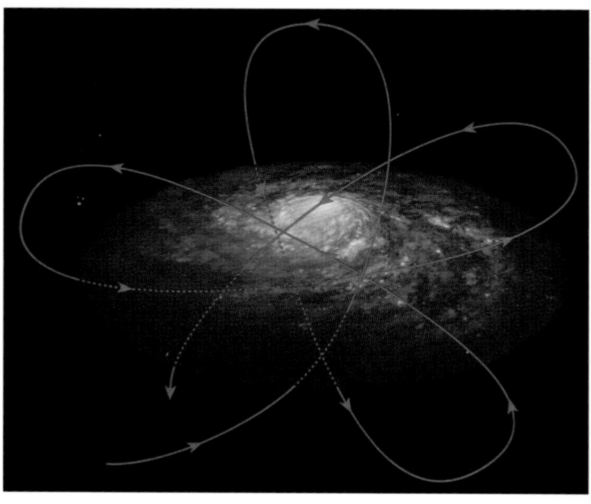

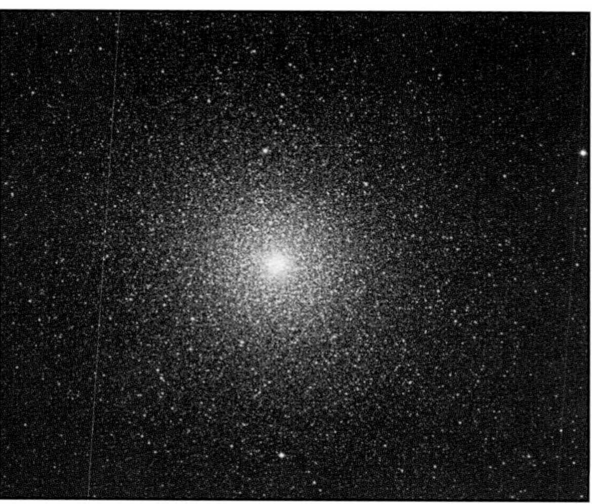

► ***Globular clusters*** (Left) *The orbits of globular clusters plunge into the galaxy, round the galactic center and out again, in a rosette pattern. (Right) 47 Tucanae is the second-brightest globular cluster, after Omega Centauri.*

Galaxy contain a bright X-ray source; many more contain dim X-ray sources, of which more than 40 are now known. X-ray sources have also been detected in globular clusters in M31. The bright X-ray sources are thought to be binaries in which a neutron star accretes matter from a low-mass companion, but the dim X-ray sources are more diverse. A single cluster can contain several dim sources. Soft X-ray transients, cataclysmic variables, recycled radio pulsars, and RS CANUM VENATICORUM binaries can all have X-ray luminosities in the range observed for the dim sources, and are known to exist in globular clusters. Twelve bright cluster sources correspond to about 10% of the bright X-ray sources of our Galaxy, a surprisingly large fraction if one considers that the globular-cluster system holds less than 0.1% of the stars in our Galaxy. The efficient formation of X-ray binaries in globular clusters is ascribed to two processes: tidal capture and exchange encounters. Tidal capture occurs when a neutron star transfers some of its kinetic energy to tides in another star during a close passage, and enough tidal energy is dissipated to bind the neutron star in orbit around its captor. An exchange encounter occurs when a neutron star ejects one of the stars in a binary in a close encounter, and takes its place. These processes are especially efficient in globular clusters with dense cores, where the stars are close.

Further reading: Ashman K M and Zepf S E (1998) *Globular Cluster Systems* (Astrophysics Series, no. 30) Cambridge University Press.

Globular clusters, observing

See practical astronomy feature, p. 84.

Globule

A small, round, DARK NEBULA thought to represent a stage in the collapse of a concentration in a molecular cloud toward a protostar. Globules range in mass from about 1 to more than 1000 solar masses and in size from about 10 000 AU to 3 l.y., and are visible in silhouette against bright nebulae such as the ROSETTE NEBULA, or against bright Milky Way starfields. BART BOK was the first to suggest that they might be the precursors of protostars; they are sometimes called Bok globules.

Goddard Space Flight Center

The Goddard Space Flight Center (GSFC), in Greenbelt, Maryland, USA, was established in 1959 as NASA's space science center. It was named for Robert Goddard (1882–1945), a pioneer in rocket research.

Gold, Thomas (1920–)

Astronomer born in Vienna, Austria, who emigrated to Britain and the USA. Noted for provocative, imaginative and sometimes wholly incorrect theories, Gold proposed the now discredited STEADY-STATE THEORY of cosmology (with Hermann Bondi (1919–) and Fred HOYLE). At Cornell University he directed the Center for Radio Physics and Space Research, where he named the MAGNETOSPHERE and studied pulsars, proposing correctly that they were rotating magnetic neutron stars (*see* NEUTRON STAR AND PULSAR). Gold caused some concern for the safety of US astronauts when he incorrectly theorized that the APOLLO lunar landers would sink into meters of lunar dust. Recently he proposed that terrestrial carbon-fuel deposits (coal and oil) are not of biological origin but are cosmic carbon welling up from the Earth's central regions.

Goodricke, John (1764–86)

Born of English parents in the Netherlands, he discovered in 1782 that ALGOL, identified as a variable star by Geminiano Montanari (1633–87)in 1669, was an ECLIPSING BINARY STAR and estimated its period. He also discovered the variability of DELTA CEPHEI.

Go To telescope

See REFLECTING TELESCOPE; SOFTWARE IN ASTRONOMY

Gould, Benjamin Apthorp (1824–96)

Astronomer, born in Boston, Massachusetts, USA, whose early work in Germany was on the observation and motion of comets and asteroids. Gould's greatest work was his mapping of the stars of the southern skies.

Gould's Belt

Following work in the southern hemisphere by John Herschel in 1847 (*see* HERSCHEL FAMILY), Benjamin GOULD identified in 1874 a zone of bright stars lying off the track of the MILKY WAY, and consisting of the constellations Orion, Canis Major, Carina, Puppis, Vela, Crux, Centaurus, Lupus, Scorpius, Taurus, Perseus, Cassiopeia, Cepheus, Cygnus and Lyra. The stars define a GREAT CIRCLE with an INCLINATION of 20° to the galactic equator; dust and gas in the solar neighborhood follow the same distribution. This feature is a young (60 million years old), flattened structure, 700 parsecs in size, with the Sun located in its interior, which is nowadays known as 'Gould's Belt.' Seen from outside our Galaxy, it would appear as a tilted spur on the inner edge of a spiral arm.

Granat
Soviet GAMMA-RAY ASTRONOMY mission, launched December 1989. Granat carried the French SIGMA telescope designed for high-resolution imaging, together with instruments for X-ray imaging and spectroscopy, and X-ray/gamma-ray-burst detectors. Granat made detailed studies of the GALACTIC CENTER, and operated for nine years, ceasing transmission in November 1998.

Grand unified theory
Theories that attempt to show that the strong nuclear force, the weak nuclear force and the electromagnetic force (*see* FUNDAMENTAL FORCE) are different manifestations of a single, unified, fundamental force. Many physicists believe that a more comprehensive theory (a 'theory of everything') that unifies all four forces, including gravity, will eventually be found. The first step towards unification was achieved by the Weinberg–Salam–Glashow theory (recognized by the Nobel Prize for Physics 1979), which unified the electromagnetic and weak nuclear forces (confirmed in 1983 by the discovery of the predicted W and Z bosons).

In the present phase of the universe, which is relatively cool (3 K), the four forces are separate and distinct. At much higher energies, beyond what can be achieved in terrestrial particle accelerators, they behave as a single, unified 'superforce.' Such energies would, according to the BIG BANG model, have been present during the first 10^{-35} s of the universe's existence, when the temperature everywhere was in excess of 10^{28} K.

Granulation
The mottled structure exhibited by the visible surface of the Sun (the photosphere), when it is observed under good conditions. Because of CONVECTION, hot gas is rising and cooler gas descending in the outer layers of the Sun, and this leads to the granules, typically 1200 km in diameter, which are about 100 K hotter than the duller regions which separate them. Granulation was discovered in 1860 by James NASMYTH, who noticed a 'willow-leaf pattern.'

Gravitation, gravitational lensing and gravitational waves
There are four basic forces: the strong and the weak forces in atomic nuclei, the electromagnetic force between electric charges and magnetic poles, and the gravitational force between masses (*see* FUNDAMENTAL FORCE). The weakness of gravity is obvious. If a child picks up a book, he defeats the cumulative gravitational pull of the entire Earth on the object. The strength to do this comes from the chemical forces in the muscles, which come from electromagnetic forces.

In spite of its weakness, gravitation always attracts, cannot be shielded, and its range is limitless. It therefore dominates the behavior of matter on large scales and plays a significant role in astronomy. The attempts by early philosophers, like ARISTOTLE OF STAGIRA, and astronomers like Nicolaus COPERNICUS, Tycho BRAHE and Johannes KEPLER, to describe the motions of the planets, led to the identification of the force of gravity. GALILEO GALILEI's investigations of dynamics and gravitation inspired Isaac NEWTON to apply calculus to problems of CELESTIAL MECHANICS, such as the motion of planets and comets. The first deficiencies in Newton's laws of gravitation emerged in 1845 with observations of the elliptical orbit of Mercury. Albert EINSTEIN developed the theory of GENERAL RELATIVITY, which immediately, in 1916, solved the problem of Mercury's orbit, and has become established as the state-of-the-art theory of gravitation.

Lensing
Einstein predicted that matter would deflect light (and radio waves). More than a century earlier, John MICHELL and others had reached the same conclusion based on a particular theory of light, but Einstein's theory of 1916 predicts double the rate of deflection. This was first confirmed by Arthur EDDINGTON, who observed the positions of stars seen beyond the Sun during the eclipse of 1919. Modern repeats of this experiment include measuring changes in the positions of thousands of distant radio sources using VLBI, and of brighter stars by the HIPPARCOS satellite. These show that the general relativistic prediction is correct to 1 part in 1000. Solar-system experiments measuring the time-of-flight delays of radio signals to spacecraft orbiting Mars and to radar transponders placed by the Viking landers on its surface give similar results.

This deflection produces a general distortion of the astronomical universe as we look out past one mass towards stars and galaxies behind. The effect on the images of distant objects is more like a piece of bathroom glass than a well-focused lens, but the term 'gravitational lensing' or simply 'lensing' has stuck. 'Weak lensing' changes the shape of the image, stretching it along one direction; very weak lensing might produce stretching that is almost imperceptible but statistically measurable from images of many sources close together in the sky, as all the background galaxies systematically 'shear.' 'Strong lensing' does more than stretch the image: it separates the source into multiple images, one or more being brighter than the unlensed source would have been. This sometimes makes it possible to examine the background source, using the gravitational lens as a light-gathering telescope.

The bending of starlight by the Sun can be measured directly, but all other known examples of lensing involve very distant objects (lensing is easier to find for more distant objects). Lensing of stars in the Milky Way or nearby galaxies by foreground stars is now routinely observed (*see* DARK MATTER). Lensing by foreground galaxies of quasars is also common. Gravitational lensing by clusters of galaxies is ubiquitous – a long-exposure image of the galaxies behind any rich cluster of galaxies will show obvious weak lensing and possibly also strong lensing.

The lensing effect depends only on gravity and geometrical factors such as source and lens positions, not other (possibly poorly understood) parameters. Thus astrophysical inferences made from lensing are particularly robust. This makes lensing very important as a probe of dark matter. Both strong lensing and the 'shear' distortion of thousands of distant galaxies can be used to map the dark matter in a galaxy or cluster of galaxies, to produce an 'image' of its distribution.

Gravitational waves
According to the theory of relativity, gravitational forces are transmitted at the speed of light. So when the gravitational field of an object changes, the changes ripple outward and

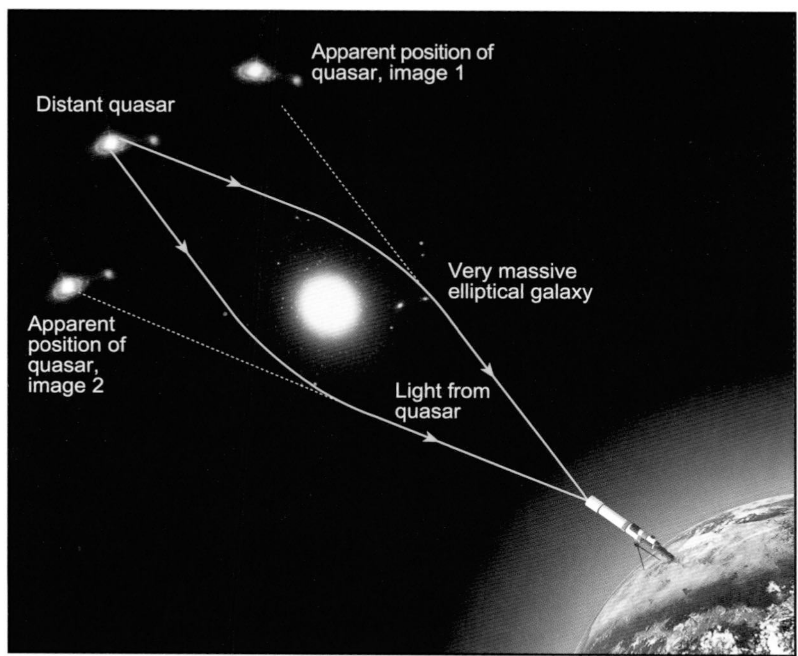

▲ ***Gravitational lensing*** *A massive elliptical galaxy forms a gravitational lens that produces two images of a distant quasar.*

take a finite time to reach other places. These ripples are called gravitational radiation or gravitational waves ('gravity waves' are the wave motion of a floating object, like a wave on the surface of water or some waves in the atmosphere of the Sun). There is strong indirect evidence from the observation of pulsars that gravitational waves follow the predictions of general relativity.

Gravitational waves are weak. The human eye readily sees the planet Jupiter, but several times a week a gravitational wave, from a very distant galaxy, carries a similar amount of energy into the eye, and we do not notice it. Virtually all the energy in the light from Jupiter that enters the eye is absorbed, but the gravitational wave passes right through. Almost none of its energy is left behind, because gravity is so weakly interacting.

Gravitational radiation is one of the last unopened windows into the universe, but we can expect a novel view. The earliest detectors were deployed by Joseph Weber (1919–2000) of the University of Maryland about 30 years ago. Development of his type of detectors has continued at the Universities of Rome and Padua in Italy, Louisiana in the USA, and Perth in western Australia, without successful detections. Newer, broader-band detectors are being developed, which are simple in principle although difficult to make in practice. To detect the weak effect of the gravitational waves all sources of interference (earth tremors, thermal noise and so on) must be minimized.

The principle is to monitor the distance between two nearby free-falling particles, such as, on Earth, freely suspended pendula or, in space, masses floating in spacecraft. If the test masses are genuinely free, the passage of a gravitational wave causes them to bob like corks on the sea, and their separations change. The bigger the separation of the particles, the bigger the change in their separation.

Gravitational-wave detectors are as big as finances and practicality allow. Laser radiation is sent between mirrors glued to the test masses and the beams combined by interferometry in order to sense the motion produced by gravitational waves. It is practical on Earth to make such interferometers up to a few kilometers in separation, so ground-based detectors can observe sources whose radiation is emitted at frequencies above a few hertz; but space-borne interferometers are being designed with arms 1 million km long and can detect lower frequencies. The following are among the gravitational-wave detectors of long baseline being designed or already being built:

- LIGO consists of two detectors in Hanford, Washington, and Livingston, Louisiana, USA, with arms 4 km long;
- the VIRGO GRAVITATIONAL-WAVE INTERFEROMETER is a detector with arms 3 km long being built at Cascina near Pisa in Italy;
- GEO 600 in Germany is being built with arms 600 m long;
- TAMA 300 has arms of length 300 m, under construction at the astronomical observatory in Tokyo, Japan;
- a space-borne detector, LISA, proposed by a collaboration of European and US research groups, has been adopted by ESA as a future Cornerstone Mission in collaboration with NASA. It consists of an array of three drag-free spacecraft (unaffected by the solar wind) at the vertices of an equilateral triangle with sides of 5 million km. This trio is placed in an Earth-like orbit. Free-floating masses inside the spacecraft (two in each) form the end-points of three separate but not independent interferometers, with which to detect the gravitational waves. LISA (planned launch date 2011) will be the largest man-made construction ever.

Where do gravitational waves come from?

There is a fundamental theorem, proved by Newton, that the gravitational field outside a spherical body is not only spherical but the same as that of a point mass located at the origin of the body. This is true even if the star pulsates in a spherical manner. This theorem is essentially the same as a theorem in general relativity known as Birkhoff's theorem (for the US mathematician George Birkhoff, 1884–1944). However, if the pulsation is non-spherical, the outside field will change. In general relativity, the changes generally propagate as a wave. So gravitational waves will be emitted by non-spherical motions.

The likely sources of detectable gravitational radiation are phenomena in which bulk matter (solar masses, more or less) changes its configuration both non-spherically and quickly (in minutes, seconds or faster), namely:

- Spinning neutron stars (*see* NEUTRON STAR AND PULSAR). If a rapidly spinning neutron star has an irregularity, a bump, on one side, the bump will emit gravitational radiation. The radiated energy would cause a spin-down of the star. Neutron-star crusts are not strong enough to support large asymmetries, but neutron stars in X-ray binary stars are accreting, and may have bumps.

- Close BINARY STARS. Close binaries can radiate more energy in gravitational waves than in light. The radiation of energy causes the orbit to shrink and the binary period to increase, making any observed gravitational waves progressively increase in strength and frequency. This is called a chirp. The measurable characteristics of the chirp provide a calibration of its intensity, so gravitational-wave chirps from binary stars will be 'STANDARD CANDLES,' or sources whose luminosity is so well known that they provide accurate distance calibration for the galaxies in which they are found.

Orbital shrinking by gravitational radiation has already been observed in the binary radio pulsar PSR1913+1 and LISA will observe all similar systems in our Galaxy. Any similar binary system with gravitational radiation that is strong enough to be observed from the ground will coalesce into a single object (a large neutron star or BLACK HOLE) within about one year. This exciting event will be well heralded so we will know where to look, but happens less than once every 100 000 years in our Galaxy. Ground-based detectors must be able to register these events in a volume of space containing at least 1 million galaxies to have a hope of seeing one. When detectors reach this sensitivity (sometime in the first decade of the twenty-first century), astronomers will be able to use LISA observations of chirping binaries to measure distance scales in the universe.

• SUPERNOVAE. Neutron stars are formed by gravitational collapse in a supernova. If the collapse is non-spherical, gravitational waves will be emitted. But we have little evidence about this point. The effect could be large, in which a good fraction of the energy released by the collapse is radiated in gravitational waves, or it could be negligibly small.

• Black holes. Black holes are surrounded by an event horizon. This boundary is a dynamical surface. If any mass-energy falls into the hole, the horizon will generally wobble. These wobbles settle down quickly, emitting gravitational waves, and leaving a smooth and slightly larger horizon afterwards. Black holes are formed from supernovae (solar masses) or are super-massive at the centers of galaxies (millions to billions of solar masses). Both kinds can radiate gravitational waves. Stellar black-hole gravitational radiation will be in the ground-based frequency range, while supermassive black holes are detectable only from space.

The biggest bursts of gravitational radiation from black holes would come from the merger of a binary black-hole pair. The large mass of the black hole makes the system visible from a great distance, and the merger of supermassive black holes would be detectable no matter where in the universe they occured. The event rate for such mergers is hard to predict: it could be zero, but it may be large, with supermassive black-hole binaries forming because of colliding galaxies. This would give an event rate of near-weekly.

• BIG BANG. Gravitational waves can come from extraordinarily early in the history of the universe. Just as the COSMIC MICROWAVE BACKGROUND carries a picture of the universe at it was about 400 000 years after the Big Bang, gravitational waves would picture the universe when it was much less than a second old. The gravitational waves come from the shifting initial density perturbations that evolve into galaxies as the universe expands. Today these perturbations should show as a random background of gravitational radiation.

Further reading: Kenyon I R (1990) *General Relativity* Oxford University Press; Pais A (1982) *Subtle Is the Lord: The Science and Life of Albert Einstein* Oxford University Press.

Gravitational constant

According to Isaac NEWTON's law of gravitation, the force between two bodies is proportional to their masses (m and M) and inversely proportional to the square of the distance between them (r). The constant of proportionality that relates the force, masses and distance in appropriate units is called the gravitational constant, usually denoted G. Newton's law is thus written $F = G\,m\,M/r^2$. G is measured by observing the attraction between two massive balls placed in a balance so they oscillate under their gravitational force. Nevil MASKELYNE also measured G by measuring the deflection of a pendulum by a massive mountain.

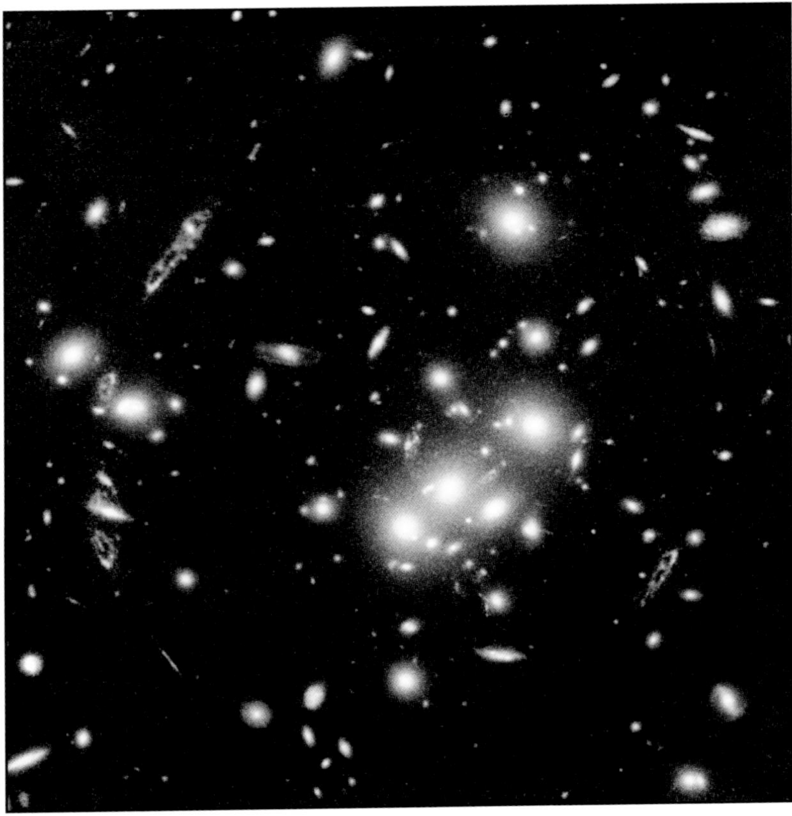

▲ ***Lensing in action*** *The cluster of galaxies 0024+16 acts as a gravitational lens to produce distorted images of a distant irregular galaxy.*

Gravitational redshift

An effect of GENERAL RELATIVITY by which a photon loses energy ('redshifts') when it travels away from a region of strong gravity. The effect is seen in WHITE DWARF stars in binary star systems, where the white dwarf, while clearly in orbit around its companion, also seems to be receding from it, by showing a redshift. The effect is also seen in the light from gas orbiting BLACK HOLES in AGN.

Graviton

The graviton is the force-carrying particle (GAUGE BOSON) hypothesized as part of the quantum-mechanical theory of gravitation. *See also* GRAVITATION, GRAVITATIONAL LENSING AND GRAVITATIONAL WAVES.

Gravity Probe B

A relativity gyroscope experiment developed by NASA and Stanford University, USA, to test two unverified predictions of Albert EINSTEIN's general theory of relativity. Scheduled for launch into polar orbit in 2004, it will precisely monitor tiny changes in the direction of spin of four gyroscopes on the satellite. Free from disturbance, the gyroscopes will measure how much the Earth warps space and time. *See also* GENERAL RELATIVITY.

Great Attractor

Name given to the mass concentration, located at about 60 megaparsecs distance, which disturbs the uniform outflow of the galaxies caused by the expansion of the universe. *See also* TULLY–FISHER RELATION.

Great circle

The circle obtained on the surface of a sphere by the intersection with the surface of a plane that passes through its center. Any plane that intersects the sphere but does not pass through the center meets the sphere in a 'small circle.'

Great Observatories Program

A NASA program consisting of a family of four orbiting observatories, each observing the universe in a different wavelength region. The missions in this program are the HST, CGRO, the CHANDRA X-RAY OBSERVATORY and SIRTF.

Great Red Spot

A huge, permanent anticyclone in JUPITER's southern hemisphere, visible as a reddish oval at a jovian latitude of 20°S. The earliest unequivocal observation was by Heinrich SCHWABE in 1831. Sightings by Robert Hooke (1635–1703) in 1664 and by Giovanni Cassini (*see* CASSINI DYNASTY) in 1655 may have been of similar but different spots. The Great Red Spot was a striking feature around 1880 and in the early 1970s, when it was deep red, but at other times the color has been less pronounced. It fades on occasion to a pale pink and sometimes disappears completely, leaving 'the Red Spot hollow.' Its color variations may be caused by variations in the concentration of phosphine (PH_3). Its size has varied from about 10 000 to 14 000 km (north-south) by 24 000 to 40 000 km (east-west). The spot drifts back and forth in longitude. Over the course of the twentieth century it completed about three circuits around the planet. There is also a 90-day oscillation in latitude that takes it nearly 2000 km either side of its mean position. It rotates counter-clockwise with a period of about seven days. It is not clear what drives its rotation – perhaps energy is released as latent heat from the condensation of gases below.

Great Rift

A prominent division of the Milky Way, consisting of a string of dark molecular clouds. It stretches from the constellations of Cygnus (a section known as the Cygnus Rift), down through Aquila, to Sagittarius.

Great Wall

A vast, sheet-like aggregation of galaxies that is about 800 million l.y. long, 280 million l.y. high, but only about 15–20 million l.y. thick. One of the largest known structures in the universe, it lies at a distance of about 300 million l.y. The COMA CLUSTER, one of the nearest great clusters of galaxies, forms part of the 'wall.'

Green Bank (National Radio Astronomical Observatory)

Located in Green Bank, West Virginia, USA, and the site of the world's largest fully steerable radio telescope (*see* RADIO TELESCOPES AND THEIR INSTRUMENTS). The 100-m telescope, completed in 2000 to replace a 91-m telescope, has actual surface dimensions of 100 × 110 m. Its structure allows the telescope to view the entire sky above 5° elevation. Of the other instruments at Green Bank, the 42-m telescope closed on July 19, 1999, while the 91-m telescope collapsed on November 15, 1988, owing to the sudden failure of a key structural element in the box girder support structure. The Green Bank Interferometer is also operated by the observatory. It includes three radio telescopes of 26-m diameter engaged in monitoring variable radio sources, X-ray binary stars and AGN. The VLBI tracking station in Green Bank is one of four such NASA facilities dedicated to the support of very-long-baseline interferometry satellites.

Greenhouse effect *See* ATMOSPHERE

Gregorian telescope

A type of REFLECTING TELESCOPE designed in 1663 by the Scottish mathematician James Gregory (1638–75), and first implemented by Robert Hooke (1635–1703) and John Hadley (1682–1744). It utilizes a concave paraboloidal primary mirror and a concave ellipsoidal secondary mirror that is located outside the focus of the primary. Light reflected from the secondary travels back down the telescope tube, through a central hole in the primary, to the eyepiece. The Gregorian optical system produces an upright image but has a very small field of view. Gregorian instruments, constructed by instrument-makers such as James Short (1710–69), became popular during the eighteenth century largely because their concave secondaries were easier to make than the convex secondaries of the similar CASSEGRAIN TELESCOPE. Several modern solar telescopes are based on the Gregorian design, since it produces a real image of the Sun, which can be occulted by a disk (to reduce scattered light or extract heat).

Group of galaxies *See* GALAXY CLUSTER AND GROUP

Grus *See* CONSTELLATION

Guide telescope

Like a FINDER, a second telescope attached to, and aligned parallel to, the main telescope, which is used to monitor the position of the object of interest. Because of changes in atmospheric refraction, and imperfections in the telescope mounting, the flexure of the main telescope, its mounting or its drive mechanism, the object that is being observed will wander during a long exposure. To compensate for this, the telescope must be adjusted continually or at regular intervals to ensure that the object remains stationary in its field of view. Traditionally, this is done by the observer adjusting the telescope controls, but an AUTOGUIDER is an electronic device that senses when the object drifts from its correct position and automatically generates a compensation signal. To avoid any problems due to motion between the guide telescope and the main telescope, it is more secure to guide on a star seen in the main telescope but off to one side of the object of interest.

Gum Nebula

A very large, near-circular EMISSION NEBULA, approximately 36° in diameter, in the constellations Puppis and Vela. The largest known nebula in the sky, it was discovered by the Australian astronomer Colin Gum (1924–60) in 1955, and is believed to be an ancient SUPERNOVA remnant more than 1 million years old. It is a convoluted mass of nebular wisps and loops, many of them very faint, but there are also numerous brighter parts. Its distance has been estimated at 1300 l.y., indicating that the nebulosity is 840 l.y. across. The much more recent VELA PULSAR and VELA SUPERNOVA REMNANT also lie within the Gum Nebula.

◀ ***H II region***
The 'Pillars of Creation' are dark clouds of dust that intrude like oil into vinegar into the H II region known as the Eagle Nebula (M16).

H I region

A cloud of interstellar matter consisting predominantly of neutral hydrogen atoms. The name H I (pronounced 'H one') arises from the convention of using the Roman numeral I to denote a neutral (not ionized) atom. In this case each neutral hydrogen atom consists of a proton and its lone electron. Higher Roman numerals denote one more than the number of electrons lost by ionization (*see* EXCITATION AND IONIZATION). For example, in H II REGIONS ('H two'), the hydrogen is ionized, each atom having lost its only electron.

In mass and volume, neutral hydrogen accounts for about half of all interstellar matter. A typical H I region is 15–20 l.y. across, and contains 50 solar masses of hydrogen at a temperature of about 100 K. H I regions emit no visible light, but are detectable by their radio emission at a wavelength of 21 cm (frequency about 1420 MHz). The detection of this so-called 21-cm radiation allowed the form of our Galaxy to be mapped in the mid-twentieth century: neutral hydrogen is present in the disk of the Galaxy, concentrated in the SPIRAL ARMS, and 21-cm radiation is little affected by intervening interstellar matter. H I regions are also typical constituents of other spiral galaxies and irregular galaxies, though they are largely absent from elliptical galaxies.

H II region

An interstellar gas cloud consisting mainly of ionized hydrogen. H II ('H two') regions are normally associated with regions of active star formation. Stars form in the densest parts of molecular cloud complexes, and the radiation fields and stellar winds of the newly formed stars excite and stir the surrounding gas. While stars of low and intermediate mass affect only a small volume of the cloud, a few massive stars can clear out the dense gas from the star-forming regions, exposing the recently formed group of stars. The ultraviolet flux from these stars ionizes the surrounding gas, creating the beautiful and complex nebulae known as H II regions.

In external gaseous (spiral and irregular) galaxies, the population of H II regions traces the locations where vigorous star-formation activity, or a 'starburst,' is taking place. In SPIRAL GALAXIES, H II regions may be located either along the

SPIRAL ARMS or in ring-like distributions surrounding the dynamical centers of the host galaxies. A beautiful example of a spiral arm distribution is found in the grand-design spiral galaxy M51. H II regions are divided into categories by their average sizes. An understanding of the processes involved enables giant H II regions to be used as distance indicators for their host galaxies.

H II regions show complex structure on large and small scales. The 'Pillars of Creation' in M16, the EAGLE NEBULA, are a spectacular example. Many H II regions also display dark clouds of neutral material silhouetted against the nebula. The ORION NEBULA is also just one of several star-forming regions in the Orion molecular cloud complex. Here both low- and high-mass star formation is taking place.

Hadron

A subatomic particle that is composed of QUARKS and that is acted on by the strong nuclear force (*see* FUNDAMENTAL FORCE).

HALCA (Highly Advanced Laboratory for Communications and Astronomy)

Japanese-US radio astronomy observatory. HALCA was the first VLBI satellite and forms part of VSOP (the VLBI Space Observatory Program). It was launched in February 1997 into a highly elliptical Earth orbit. HALCA, which is equipped with an 8-m antenna operating at frequencies near 1.6, 4.8 and 22 GHz, has returned high-resolution images of QUASARS and distant ACTIVE GALAXIES. Formerly called Muses-B, the satellite was renamed HALCA after launch. An alternative name is Haruka, the Japanese pronunciation of the HALCA acronym. 'Haruka' is Japanese for 'far away.'

Hale, George Ellery (1868–1938)

American astrophysicist. As an undergraduate at the Massachusetts Institute of Technology, he invented the SPECTROHELIOGRAPH, which made it possible to photograph the Sun's PROMINENCES in daylight. At the University of Chicago he used a 40-in lens as the basis for the telescope of the YERKES OBSERVATORY, which was completed in 1897. Hale founded the MOUNT WILSON OBSERVATORY, where he discovered the magnetic fields in sunspots. He planned and completed the observatory's 60-in and 100-in telescopes, and the 200-in telescope on Mount Palomar was named for him after his death. Hale was the first astronomer to be officially called an astrophysicist, and he started the *Astrophysical Journal*.

Halley, Edmond (1656–1742)

An English astronomer and scientist, he became professor at Oxford University and second ASTRONOMER ROYAL at the ROYAL OBSERVATORY, GREENWICH. From the island of St Helena he cataloged the positions of about 350 southern-hemisphere stars, including his own discovery of the globular cluster OMEGA CENTAURI; he also discovered the globular cluster M13. He dedicated a PLANISPHERE of the southern-hemisphere stars to King Charles II, which flatteringly included a now disused constellation (in Carina) that Halley named 'Robur Carolinum' – the oak tree in which Charles hid after his defeat by Oliver Cromwell following the battle of Worcester. Halley observed a TRANSIT of Mercury and invented the idea of using transits of Mercury and Venus to determine the distance of the Sun. Using PTOLEMY's catalog, he deduced that the stars had moved relative to each other and detected this 'proper motion' in three stars (*see* ASTROMETRY). Halley was a friend of Isaac NEWTON. Using Newton's theory of cometary orbits, Halley calculated that the comet of 1682 was periodic, and correctly predicted that what we now know as Halley's Comet would return in 76 years (*see* COMET HALLEY). Halley is considered the founder of geophysics, especially for his work on trade winds, the tides and the magnetism of the Earth. His *Breslau Table of Mortality* also laid the actuarial foundations for life insurance and annuities.

Halo

(1) An atmospheric phenomenon that produces an apparent ring of light around a celestial body, most commonly seen around the Sun or Moon. The effect is often produced by ice crystals at an altitude of 10–15 km. The ice crystals may become preferentially orientated in the atmosphere, resulting in halos of radius 22° (most common) or 46°. Halos often show some color, and the sky inside the ring is noticeably darker than outside. Halos are also produced by other atmospheric aerosols such as water droplets or dust.

(2) Any similar phenomenon around other celestial objects, such as a REFLECTION NEBULA around a star.

(3) The GALACTIC HALO.

Halo orbit *See* LAGRANGIAN POINTS

Halo population

Stars found in the GALACTIC HALO of the MILKY WAY or other SPIRAL GALAXY. These are the oldest, most metal-poor stars in the galaxy, POPULATION II.

Haro, Guillermo (1913–88)

Mexican astronomer, director of the Mexican Institute of Astronomy and the Tonantzintla Observatory. Independently of George HERBIG, he identified small EMISSION NEBULAE in the ORION NEBULA, associated with but differing from T TAURI STARS. These objects, now known as Herbig–Haro objects, are thought to be shock-excited impact regions, with their energy derived from outflowing jets from young stars.

Harriot, Thomas (c.1560–1621)

English mathematics teacher who traveled with Sir Walter Raleigh to Virginia as cartographer and navigator. Harriot introduced simplified notation for algebra, which is still in use today. He corresponded with Johannes KEPLER on optics, and discovered Snell's law of refraction before the person after whom it was named, Willebrord Snell van Royen (1580–1626). He observed COMET HALLEY in 1607 and a second comet in 1618. He made numerous telescopic astronomical observations, including the moons of Jupiter, unaware of GALILEO's discovery. Harriot's drawing of the Moon, the first recorded, preceded Galileo's by several months, and he was also first to record sunspots (*see* SUNSPOT, FLARE AND ACTIVE REGION). He made many observations of the Sun over three years and deduced the Sun's rotation period. Harriot published none of his discoveries but when he died he left a large number of manuscripts, and his scientific genius emerged. He did not influence the development of optics or astronomy, although his influence as a teacher on the development of mathematics in England was profound.

Harrison, John (1693–1776)

English clockmaker. In 1713 the British government offered a prize for the invention of a method to accurately determine longitude. Harrison developed a series of clocks and in 1726 invented the bimetallic pendulum, which compensated for thermal expansion caused by the variations of climate experienced on a long sea voyage. He made three spring-pendulum chronometers (called H1, H2 and H3) that worked regularly even during the rolling of a ship. They were finally surpassed in accuracy by a watch-like marine chronometer (H4), which, during a voyage to Jamaica in 1761–2 determined longitude with an accuracy of two geographical miles. Harrison was treated shamefully by the Board of Longitude, which invented conditions to delay awarding him the prize, and was finally awarded it in his old age on the intervention of King George III. Harrison's restored chronometers are displayed in the ROYAL OBSERVATORY, GREENWICH.

Hartebeesthoek Radio Astronomy Observatory

The Hartebeesthoek Radio Astronomy Observatory (HartRAO) is located 50 km northwest of Johannesburg, South Africa, at a former NASA satellite-tracking station. As the only operational radio telescope in Africa, it plays a cardinal role in several VLBI networks.

Hartmann, Johannes Franz (1865–1936)

German astronomer, a professor in Potsdam. Disappointed with the performance of a new 80-cm REFRACTING TELESCOPE, Hartmann developed the method of testing optics that is now named for him. The Hartmann test consists of partly obscuring the telescope aperture with a Hartmann shutter, and revealing differences of optical focus in the various zones.

Haruka *See* HALCA

Harvard classification

Classification system for stellar spectra developed at the Harvard College Observatory by Edward PICKERING and co-workers toward the end of the nineteenth century. This scheme formed the basis for the current system of spectral classification. *See also* STELLAR SPECTRUM: CLASSIFICATION.

Harvard–Smithsonian Center for Astrophysics

The Center for Astrophysics (CfA) is a joint facility of the Smithsonian Institution and Harvard University, USA. Observational facilities include the multipurpose Fred Lawrence Whipple Observatory on Mount Hopkins in Arizona and the Oak Ridge Observatory in Massachusetts, as well as a 1.2-m radio telescope at the CfA headquarters in Cambridge, Massachusetts (*see* RADIO TELESCOPES AND THEIR INSTRUMENTS). The major instrument on Mount Hopkins, the Multiple Mirror Telescope (MMT), is operated jointly with the University of Arizona. Also located there is a 10-m REFLECTING TELESCOPE to detect gamma-rays, and several smaller telescopes which are used for optical and infrared work.

Special laboratories are maintained at Cambridge to study METEORITES and lunar samples, for SPECTROSCOPY of atoms and molecules, and to develop instrumentation. Major projects include the development of a submillimeter telescope array in Hawaii, conversion of the MMT to a single-mirror telescope 6.5 m in diameter, and participation in the Magellan Project to build two 6.5-m telescopes at Las Campanas, Chile.

Numerous facilities serving the general astronomical community are located at the CfA, including the IAU's Central Bureau for Astronomical Telegrams and Minor Planet Center, the US gateway for SIMBAD (an international astronomical computer database), and Harvard's extensive collection of astronomical photographic plates. The Control Center of the Astrophysics Data System, a vast and freely bibliographical archive operated on behalf of NASA, is also based there.

▲ ***Solar halo***
This is a fine sight for many people but not for optical astronomers, since it signifies the presence of atmospheric aerosols, which both scatter sunlight and make the sky murky.

Harvest moon

The full Moon nearest to the time of the autumnal EQUINOX (around September 23) in the northern hemisphere or the vernal equinox (around March 21) in the southern hemisphere. At this time the INCLINATION of the Moon's path to the horizon is very shallow, drawing attention to the full Moon, which appears to hang near the horizon for a period at a similar time over several days. Its name, appropriate for the northern hemisphere, signifies how its light in the evening can extend the hours of harvest.

Hat Creek Radio Observatory

Located 400 km north of Berkeley, California. It is the site of the BIMA (Berkeley Illinois Maryland Association) Millimeter Array, a 10-antenna aperture synthesis telescope that operates at wavelengths of 3 mm (70–116 GHz) and 1 mm (210–270 GHz). The telescopes are 6.1 m in diameter. The antennae may be located at various stations along an approximately T-shaped track to allow antenna separations ranging from 7 m to 2 km. The four standard configurations provide angular resolutions of roughly 0.4, 2, 6 or 14 arcsec at 100 GHz.

Haute Provence, Observatory of

The observatory is a French national facility, established in 1937 some 100 km north of Marseilles. It has a 76-in telescope that has been operating since 1958, a 60-in built in 1967 and smaller telescopes. *See also* MARSEILLE-PROVENCE OBSERVATORY.

Hawaii Institute for Astronomy

During the last 30 years, Hawaii has become the most sought-after location in the world for the construction of large ground-based telescopes. The focal points for this construction are Mauna Kea (4200 m) on the island of Hawaii and Haleakala (3000 m) on Maui. The air above these isolated high-altitude sites is remarkably clear, dry and still.

The mountain Haleakala (Sacred House of the Sun) is home to the Mees Solar Observatory, the Lunar Ranging Experiment (LURE Observatory), the University of Tokyo's 2-m MAGNUM telescope and the US Air Force's 3.7-m Advanced Electro-Optical Systems (AEOS) Telescope. Hawaiian students will also share the use of the 2-m robotic FAULKES TELESCOPE with students in the UK.

On Mauna Kea the UH (University of Hawaii) 2.2-m telescope was commissioned in 1970. By the end of the 1970s, there were three new 4-m class telescopes – the IRTF, UKIRT and the Canada–France–Hawaii Telescope (CFHT). The IRTF and UKIRT were specifically designed to collect infrared radiation (*see* INFRARED ASTRONOMY); the dryness of the atmosphere above Mauna Kea is particularly advantageous at these wavelengths.

Submillimeter telescopes include the Caltech Submillimeter Observatory (CSO) and the JCMT. More recently, the Smithsonian Institution has built an array of eight 6-m submillimeter antennae, which are designed to work together as a single telescope. The 1990s saw the construction of a new series of giant optical/infrared telescopes on Mauna Kea. The twin 10-m telescopes of the W M KECK OBSERVATORY are the largest optical/infrared telescopes in the world. In 1999 both the Japanese 8.3-m SUBARU TELESCOPE and the 8.1-m Gemini North telescope came into operation. At a lower altitude, on the southern flank of Mauna Kea is the Hawaii antenna of the Very Long Baseline Array, which is part of an 8000-km wide system of ten 25-m radio dishes that work together as the world's largest dedicated full-time astronomical instrument.

Hawking, Stephen William (1942–)

English cosmologist and theoretical astrophysicist who studied at Oxford University before moving to Cambridge to do research in GENERAL RELATIVITY and COSMOLOGY. Hawking worked in the mathematics of 'singularities' in the central regions in BLACK HOLES, where conventional mathematics breaks down. Uniting QUANTUM THEORY and general relativity, he showed that black holes can emit energy; this is now known as Hawking radiation. He investigated the conditions in the energy released in the BIG BANG at the creation of the universe, and predicted that many mini-black holes would be created, perhaps weighing up to 10 tonnes but only the size of a PROTON. These small-mass black holes would lose energy by Hawking radiation, and would have evaporated in the lifetime of the universe. They therefore had an effect on the formation of structure early in the universe's history but have now disappeared.

Another remarkable proposal by Hawking is his 'no-boundary proposal,' explained by him as meaning '...that both time and space are finite in extent, but they don't have any boundary or edge...There would be no singularities, and the laws of science would hold everywhere, including at the beginning of the universe.' Hawking suffers from motor neuron disease and, confined to a wheelchair, he speaks through a voice synthesizer. Nevertheless, he has taken up an arduous and highly successful career as a popularizer of science through lectures, television and other public appearances, and has published the best-selling popular books on cosmology, *A Brief History of Time* and *The Universe in a Nutshell*.

Hayabusa *See* MUSES-C

Haystack Observatory

A multidisciplinary research center of the Massachusetts Institute of Technology located at Westford, Massachusetts, USA. Today its emphasis is on VLBI observations at 3 mm wavelength aimed at high-resolution imaging of QUASARS, and on the application of VLBI geodetic techniques to aspects of Earth science. Haystack's facilities include 37-m- and 18-m-diameter radio telescopes (*see* RADIO TELESCOPES AND THEIR INSTRUMENTS), and a multistation VLBI correlator. The radio astronomy program is guided by the Northeast Radio Observatory Corporation, a consortium of 12 educational institutions.

Heavy element

A chemical element of relatively high atomic weight. Because less than 2% of the mass of a solar-type star is made up of elements heavier than helium, the term 'heavy element,' in an astrophysical situation, is applied to all elements that have mass numbers greater than that of helium (4). The terms 'heavy elements' and 'metals' are synonymous in astrophysics.

HEGRA experiment

The cosmic-ray observatory HEGRA (High-Energy Gamma-Ray Astronomy) is an experiment located at the ROQUE DE LOS MUCHACHOS OBSERVATORY on the Canary island of La Palma, and operated by institutes in Germany, Spain and Armenia. The main scientific goal is GAMMA-RAY ASTRONOMY in the energy range from 500 GeV to 100 TeV, by detecting showers of ELECTRONS produced in the atmosphere by the impact of a gamma-ray in that energy range ('air showers'). A unique feature of the HEGRA experiment is the use of five Cerenkov telescopes to simultaneously measure the same air shower by stereoscopic observation. This technique has given great precision, for example, in observations during flaring outbursts of the BLAZAR Markarian 501.

Hektor

An ASTEROID discovered in 1907, designated (624) Hektor. It belongs to the largest group of TROJANS, orbiting ahead of Jupiter around the L_4 LAGRANGIAN POINT. Hektor orbits the Sun at a mean distance of 5.17 AU in a period of 11.76 years. It is the largest and brightest of the Trojans. From its LIGHT CURVE, its rotation period is 6.92 hours. It appears to be elongated, measuring about 300 × 150 km, but alternatively it could be shaped like a dumbbell, or a contact or close binary asteroid. Like the majority of Trojans, Hektor is D-type, with a reddish reflectance spectrum indicating a carbon-rich surface.

Heliacal rising

The term is commonly used to refer to the date at which an object first becomes visible in the dawn sky, rising just before the Sun. The heliacal rising of SIRIUS was used by ancient Egyptians to predict the annual flooding of the River Nile, which was an important event in the farming calendar.

Heliocentric

Heliocentric means 'centered on the Sun.' A heliocentric theory of the solar system, in which the planets moved in circular orbits around the Sun, was proposed by ARISTARCHUS OF SAMOS in the third century BC. The theory was abandoned when it became clear that it did not fit the observations in favor of PTOLEMY's geocentric model. Nicolaus COPERNICUS returned to the heliocentric model, which he published in 1543, maintaining that the planets do move around the Sun and that the apparent movements of celestial bodies across the sky are explained by the rotation of the Earth.

Heliocentric coordinate system

Coordinates specifying the position of an object as seen from the center of the Sun. Heliocentric coordinates are based on the plane of the ECLIPTIC. Thus the position of a planet in its orbit at any instant is often specified in terms of heliocentric ecliptic coordinates, its angles around, or above or below, the ecliptic.

Heliopause

See SUN; *see also* COSMIC RAY for illustration.

Helios

Two German–US satellites designed to study the SOLAR WIND, interplanetary dust and Galactic COSMIC RAYS in the inner solar system. Helios 1, launched in December 1974, passed within 48 million km of the Sun. Helios 2, launched in January 1976, approached to within 45 million km of the Sun.

Helioseismology

The study of the internal structure and dynamics of the SUN through the analysis of solar oscillations. The Sun vibrates like a gong with periods ranging from minutes to hours, but principally in the range 3–20 min. These oscillations can be detected as small, periodic Doppler shifts in the wavelengths of spectral lines emitted by localized regions of the solar surface as they rise toward, and fall away from, the observer (*see* DOPPLER EFFECT).

The oscillations are produced by sound waves (pressure waves) that propagate through the solar globe. The speed of sound depends on various factors, including temperature and density, both of which increase with increasing depth below the solar surface. Consequently, a wave moving downward from a point on the surface is refracted and eventually curves back to meet the surface at another point. The sharp change in density at the surface then reflects the wave back down into the solar interior, enabling it to bounce repeatedly around the Sun and thereby produce standing waves that cause different parts of the solar surface to vibrate up and down in a systematic fashion. The deeper the wave penetrates, the fewer the points at which it meets the surface.

By analyzing the millions of different modes of oscillation, and separating out those that penetrate to different depths, solar physicists can study the structure of the solar interior in the same sort of way as geophysicists use seismic waves to study the interior of the Earth. Furthermore, by comparing the speeds at which waves travel in the same direction as, and in the opposite direction to, the rotation of the Sun itself, it is possible to determine how the rotation rate of the solar interior varies with depth and latitude.

Helioseismology provides information on the variation with depth of temperature, density, pressure, chemical composition and rotational velocity. Observations made by ground-based projects such as GONG and by instruments carried on spacecraft such as SOHO, have shown, for example, that the boundary between the radiative and convective zones occurs 71.3% of the way from the center of the Sun to the surface, and that the DIFFERENTIAL ROTATION that is observed in the photosphere extends down to the base of the convective zone.

Heliosphere

The region of space surrounding the Sun that defines the extent of the Sun's influence. The heliosphere extends well beyond the planets to a distance of about 100 AU.

Heliostat *See* SIDEROSTAT

Helium

The second lightest element which, after HYDROGEN, is the second most abundant in the universe. Helium (He) was first detected, in 1868 by Norman LOCKYER and Jules JANSSEN, in the spectrum of the solar chromosphere during an eclipse – hence its name derives from Helios, Greek god of the Sun. The most abundant ISOTOPE of the neutral helium atom consists of a nucleus made up of two PROTONS and two NEUTRONS around which revolve two ELECTRONS. Thus, the atomic mass is 4, and the atomic number (the charge on the nucleus) is 2. The nucleus of such a helium atom is also known as an alpha particle. Helium nuclei with only one neutron also exist but are more rare.

In terms of the number of atoms, nearly 8% of all the directly observed matter in the universe is helium. In terms of mass, helium accounts for about 25%.

It is generally believed that the heavier elements are built up from the lighter ones (starting with hydrogen) in a chain of nuclear reactions (*see* ELEMENTS, FORMATION OF). Helium is produced from hydrogen by means of nuclear reactions inside stars, but this does not adequately account for the amount of helium actually observed. If the BIG BANG theory of the origin of the universe is correct, then the observed quantity of helium could have been produced in the first few minutes of the existence of the universe, while the temperature lay in the region of 10^9 K. The observed abundance of helium is one of many factors that favor the Big Bang theory.

Helium flash

The helium flash is the rapid onset of helium burning in a medium mass star (such as the Sun) which has reached the tip of the red giant branch, moving it rapidly onto the horizontal branch (HB) (*see* HORIZONTAL-BRANCH STAR). This star has left the MAIN SEQUENCE in the HERTZSPRUNG–RUSSELL DIAGRAM, having converted all the hydrogen in the center of the star to helium. Hydrogen continues to burn in a shell outside the core, and the star starts moving up and to the right in the diagram, mapping out the red giant branch (RGB). At this stage the helium is at too low a temperature to burn and the core contracts, raising the temperature in the central regions of the star. At the tip of the RGB, the helium in the core of the star ignites, in the so-called helium flash. If the mass of the star is small, it lands on the left extension of the HB; if the mass of the star is larger, it will fall farther to the right on the HB. The locus of points that is defined by stars arriving on the HB, as a function of mass, is called the zero-age horizontal branch (ZAHB). During this stage of its life the star is burning helium in the core and hydrogen in a shell surrounding the core. Stars may then evolve back up to the tip of the RGB along the so-called 'asymptotic giant branch' in which case there can be a second (or even a third) helium flash with helium burning now taking place in a thin shell (*see* SAKURAI'S OBJECT).

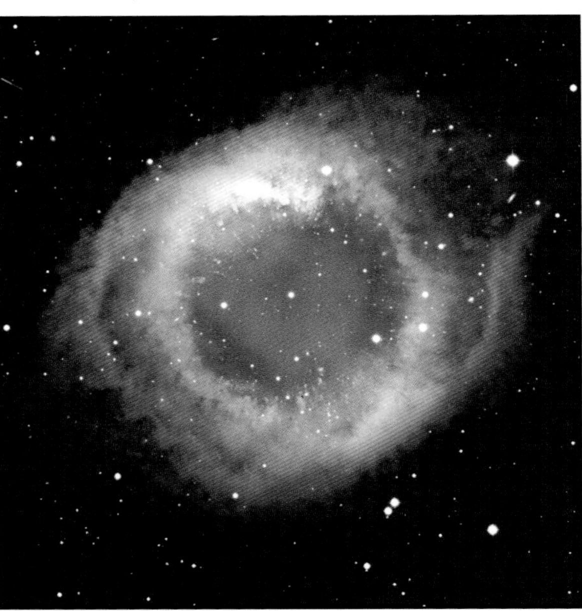

▶ ***Helix Nebula** (NGC 7293) This picture shows the brighter parts of the nebula, revealing various ionization levels within the shell of matter ejected from the central star. The greenish middle portion is evidence of excited oxygen atoms, while the outer red is mainly light from nitrogen and hydrogen.*

Helix Nebula (NGC 7293)

A PLANETARY NEBULA in the constellation Aquarius. At a distance of about 450 l.y. it is the nearest planetary nebula to Earth, and it has the largest angular size (about 16 arcmin diameter). Its magnitude of 6.5 makes it the brightest, though like most extended objects it has a low surface brightness. It is illuminated by a hot magnitude-13 central star. The Helix gets its name from its appearance of a spiral viewed from above. An alternative name of the Sunflower Nebula comes from the appearance of the streaks inside the ring. They are caused by the nebula expansion overtaking fragments left in space after the evolution of the progenitor star.

Hellas Planitia

The official name for the Hellas Basin, the largest impact feature on MARS and indeed the largest in the solar system. The inner part of Hellas Planitia is about 2100 km across and is surrounded by a rim 2 km high. At its deepest it is 9 km below Mars's average surface level (the greatest depth on the planet). The outer part of the basin extends for about 4000 km. The impact that formed Hellas was probably the last big impact event on Mars, and may have created the uplift of the Tharsis Bulge on the opposite side of the planet – the highly volcanic region containing THARSIS MONTES.

Helmholtz, Hermann Ludwig Ferdinand von (1821–94)

Born in Germany. Helmholtz had a varied career in medicine and physics. He studied the source of energy of the Sun, at the same time as but independently of Lord Kelvin (Sir William Thompson, 1824–1907). He suggested that instead of radiating by consuming chemical energy, which would be used up in a lifetime of a few thousand years, the Sun was gradually settling under gravity so the source of its energy was gravitational potential energy. There was enough potential energy in the Sun to supply its radiation for several million years. Even this was shorter than the age of the Earth as deduced from geological evidence, and the problem was not solved until the discovery of solar nuclear energy. However, Helmholtz's conjecture proved to be applicable when stars first settle from interstellar clouds. The cloud's gravitational potential energy heats the protostar and ignites its nuclear reactions. The duration of the settling process is called the Kelvin–Helmholtz timescale.

Henderson, Thomas (1798–1844)

Scottish astronomer who began his career as a lawyer's clerk. He was appointed as His Majesty's Astronomer at the Cape of Good Hope, South Africa, but he soon left because of poor health. Henderson returned to Scotland as the first Scottish ASTRONOMER ROYAL. From observations made in South Africa, he was the first person to measure the distance of a star, namely ALPHA CENTAURI, a member of the nearest star system to the Sun. Friedrich BESSEL's independent measurements of the star 61 Cygni were, however, published three months earlier.

Henry, Paul (1848–1905) and Henry, Prosper (1849–1903)

French brothers (known as the Frères Henry) who worked together at the PARIS OBSERVATORY and built telescopes to search for minor planets. With the encouragement of the then director of the observatory, Admiral Ernest Mouchez (1821–92), they built, in 1885, a 13-in photographic REFRACTOR. This was the foundation of the Carte du Ciel project, led by France, to photograph the sky in its entirety. The brothers made half of the telescopes used in the international project. A close pair of craters on the Moon, called the Frères Henry, were named for the brothers.

Henry Draper (HD) Catalog

Catalog of the position and spectral type of 225 000 stars brighter than magnitude 10, compiled by Annie CANNON at the Harvard College Observatory and published in 1924. The catalog is named for Henry DRAPER, an early pioneer in astronomical spectroscopy, and was funded by his wife, Anna, in his memory. The extension (HDE), published in 1936, contains a further 47 000 stars, extending the catalog to magnitude 11. Stars are still widely known by their HD or HDE numbers.

Herbig, George Howard (1920–)

American astrophysicist. Herbig worked in California, mostly at the LICK OBSERVATORY, and in Hawaii. He is known for his spectroscopic studies of recently formed stars and the INTERSTELLAR MEDIUM, from which he studied the diffuse interstellar lines. He and Guillermo HARO discovered independently the Herbig–Haro objects, gas clouds now thought to be caused by ejected material from unstable young stars. Herbig showed that the abundance of lithium is correlated with age in young stars, and he investigated the rotation speeds of stars of different spectral types.

Hercules *See* CONSTELLATION

Hercules Globular Cluster (M13)

See MESSIER CATALOG

Hercules X-1

Her X-1 was detected with the first X-ray satellite, UHURU, in 1971. It is an accreting X-ray BINARY STAR, consisting of a neutron star that rotates every 1.24 s while orbiting a 'normal' stellar companion of about twice the mass of the Sun every 1.7 days. The 1.7-day orbital motion shows up clearly in Doppler shifts of the 1.24 s period (*see* DOPPLER EFFECT); the

X-ray source shows distinct eclipses every 1.7 days; and a previously known optical variable star HZ Herculis (HZ Her) at the same position is found to have an identical period. The fact that we see total eclipses implies that our line of sight is within a few degrees of the orbital plane. The X-ray source shows sharp eclipses, attesting to its compact size. Variation in optical light by 1.8 magnitudes over the 1.7-day orbital period is attributed to X-ray heating on the side of the companion facing the active neutron star.

The system also has an approximately 35-day 'superorbital' period in which the X-ray emission displays so-called main and short 'on' states, which are separated by relative X-ray quiescence. One plausible theory suggests that this is due to PRECESSION of a tilted, co-rotating ACCRETION disk. Her X-1 is one of only about six known X-ray binaries that show a superorbital period.

Her X-1, which is bright over a broad range of energies, provides an excellent opportunity to study characteristics of matter and radiation under the intense gravitational and magnetic fields of the compact object.

Hermes

An ASTEROID discovered in 1937, provisionally designated 1937 UB. Hermes is the only asteroid to have been given a name without first having received a permanent number and having had its orbit accurately calculated. On October 30, 1937, it passed the Earth at a distance of 733 000 km, a close-approach record that stood for over 50 years. The parameters of its orbit are only approximate: mean distance from the Sun 1.64 AU, perihelion 0.62 AU, aphelion 2.66 AU, period 2.11 years. These dimensions put Hermes in the class of 'Apollo asteroids,' which cross the Earth's orbit but with average orbital diameters greater than that of the Earth. A NEAR-EARTH ASTEROID 2002 SY50, detected by several observatories in September 2002, has an orbit like Hermes and may be the same asteroid, a hypothesis that awaits further observations during future approaches. Their orbits are similar to the orbit of the Cetid METEOR SHOWER. 2002 SY50 has been seen by radar to consist of two components. With an estimated diameter of 900 m and a potential closest-approach distance of 0.003 AU, Hermes remains near the top of the list of potentially hazardous asteroids.

Herrick, Edward Claudius (1811–62)

American bookseller and librarian, self-educated by association with Yale University academics. In 1839 he confirmed that the METEOR SHOWER he had first observed in August 1837 was an annual shower with its RADIANT in Perseus, the PERSEIDS. He also identified the LYRIDS and the BIELIDS as annual showers. These discoveries followed the earlier identification of the LEONIDS as an annual event.

Herschel family

William Herschel (1738–1822), his sister Caroline (1750–1848) and his son John Frederick William (1792–1871) freed astronomy from its preoccupation with the solar system and helped to make the study of the stars, the nebulae and the cosmos part of mainstream science.

William and Caroline were the children of a humble bandsman in the Hanoverian Guards. William joined his father's military band but in 1757, after the defeat of the Hanoverian army by the French, he fled to England, where later (in 1766) he was appointed organist to a fashionable chapel in the spa resort of Bath. Despite his busy life as a performer, conductor and teacher of music, William's financial security now allowed him to develop other interests, including astronomy.

In 1772 Caroline joined her brother in England with the intention of pursuing her singing career, but she was soon to share her brother's love of astronomy.

William was self-taught and from the start set himself the goal of understanding 'the construction of the heavens.' He realized that he would need telescopes as large as possible. His first success was a NEWTONIAN TELESCOPE of 20-ft focal length and with a 12-in mirror, but it was simply slung from a pole and used by an observer perched precariously on a ladder. On the first page of his first observing book in 1774 William sketched the ORION NEBULA. In 1778 he finished polishing a mirror of 7 ft focal length to great precision. He began to collect stars that were double, hoping that they might prove useful for detecting annual PARALLAX. On March 13, 1781, William noticed a 'star' with unusual appearance in the constellation Taurus. Re-examining it four days later, he found that it had moved and was therefore not a star. William pointed out the object to Nevil MASKELYNE, the ASTRONOMER ROYAL. Maskelyne confirmed the object's motion, and thus the first planet was discovered, later named URANUS.

William became famous overnight, and he was invited to take his 7-ft reflector to Greenwich to be compared with Maskelyne's instruments. He then went to the royal court in London, and afterwards to Windsor Castle, the residence of King George. Suitably impressed, the king awarded William a modest but adequate pension, for which he agreed to settle near Windsor and show the heavens to the royal family when asked. Toward the end of 1783 William had built a new and larger 20-ft reflector with a stable platform. With this he began to 'sweep' the sky for nebulae, in a campaign that was to take two decades and increase the number of known nebulae from about 100 to 2500. Caroline shared the night watches, recording their observations and later arranging their results for publication. On her spare nights she searched for comets and was to discover no fewer than eight, the first known to have been discovered by a woman.

William and Caroline made many other contributions to astronomy, including the discovery of satellites of Saturn and Uranus. William also came across the PLANETARY NEBULA now known as NGC 1514 without understanding its nature. Re-examining DOUBLE STARS after a long interval, he found that in some the two stars had moved in orbit around each other, though he did not have enough information to demonstrate that the force binding them together was gravitational. He carried out star counts to investigate the shape of the galaxy, and from a study of proper motions

◀ ***Herschel family*** *(Left) Portrait of William Herschel as president of the Royal Astronomical Society, 1814. (Right) Caroline Herschel, painted in 1847.*

▶ ***William Herschel's*** *40-ft telescope, which was completed in 1789. Its main mirror was of 48-in diameter, the largest in the world at the time.*

determined the direction of motion of the solar system. His experiments with sunlight led him in 1800 to the discovery of infrared radiation. William also made telescopes for sale, and many of the crowned heads of Europe were among his clients. For himself he completed in 1789 a 40-ft reflector with 4-ft mirrors. But it was cumbersome and the mirrors tarnished easily.

In 1788 William married, and in 1792 his son, John, was born. Where William had been self-taught in astronomy, John was to enjoy a Cambridge education, and by 1815 was embarked on a career as a Cambridge don. In 1816, however, his father, now in failing health, persuaded John to return home and become his astronomical apprentice.

Caroline later recataloged the 2500 nebulae, for which she was awarded the medal of the Royal Astronomical Society (RAS). John had played a leading part in the founding of the RAS, and William had been its nominal first president.

John undertook a mammoth task to revise and extend his father's and aunt's surveys. First he re-examined the known nebulae in the northern hemisphere, resulting in his 1833 catalog of 2306 nebulae. He then took two telescopes to the Cape of Good Hope, South Africa, where from 1834 to 1838 he surveyed the southern sky. His results were published in a sumptuous volume in 1847, 'being a completion of a telescopic survey of the whole surface of the visible heavens,' for John had examined the entire heavenly sphere with his 20-ft telescope.

Two of John's sons, Alexander Stewart Herschel (1836–1907) and Col. John Herschel (1837–1921), made their own contributions to astronomy, the former particularly in cometary and meteoric studies. Later members of the family carefully preserved the manuscripts of their forebears, which are now largely in permanent collections, notably the library of the RAS.

Herschel Space Observatory

An ESA 'Cornerstone' mission, originally named the Far-Infrared and Submillimetre Telescope (FIRST), which is due to be launched in 2007. It will carry a 3.5-m-diameter mirror, the largest ever used for an infrared space telescope, and will orbit at the L_2 Lagrangian point between the Earth and the Sun, 1.5 million km from the Earth. Its three instruments (two cameras and a high-resolution spectrometer) will observe far-infrared and submillimeter emissions from gas and dust heated by young stars, and should detect galaxies forming in the very early universe. The instruments will be cooled close to absolute zero (–273 °C), using liquid helium cryostat technology developed for the ISO mission.

Hertzsprung, Ejnar (1873–1967)

Danish astronomer, formerly a chemist, who worked in Göttingen and Potsdam with Karl SCHWARZSCHILD, and then at the Leiden Observatory in the Netherlands, where he became director. Hertzsprung determined the positions of BINARY STARS, and used photography to measure the proper motions as well as the brightnesses and colors of stars. In 1911 he plotted these quantities in what became known as a color-magnitude diagram for the Hyades star cluster, and discovered the distinction between giant stars and MAIN SEQUENCE stars. Two years later Henry RUSSELL plotted an equivalent diagram for nearby stars. The two diagrams have the same topography, and collectively are known as the HERTZSPRUNG–RUSSELL DIAGRAM. Hertzsprung also determined the distance to the SMALL MAGELLANIC CLOUD by use of the statistical PARALLAX of a group of CEPHEID VARIABLE STARS. Although he made a mistake, putting the stars a factor of 10 too close, it is interesting that the distance was so large that a factor of 10 too small did not immediately stand out.

Hertzsprung gap

The region to the right of the MAIN SEQUENCE in the HERTZSPRUNG–RUSSELL DIAGRAM, where very few stars are seen. This represents a stage where stars are evolving rapidly. The core hydrogen has been used up, and hydrogen is burning in concentric shells before core helium burning begins.

Hertzsprung–Russell (HR) diagram

A star's color is related to its temperature, and its brightness relative to others in a cluster all at the same distance is related to its luminosity. In 1911 Ejnar HERTZSPRUNG plotted these quantities for stars in the HYADES star cluster, in what became known as a color-magnitude diagram. He discovered the distinction between giant stars and MAIN SEQUENCE stars. Two years later Henry RUSSELL plotted an equivalent diagram for stars whose spectral types were known and which, although not in a cluster, were so near that their distances and thus their luminosities could be determined. The two diagrams show the same distribution of stars. An apparently independent parallel study of the 'relation between brightness and spectral type in the Pleiades' was undertaken by Hans Rosenberg (1879–1940) of Göttingen. His paper, published in 1910, contained the first published color–magnitude diagram, but it was overlooked. The plot of the stars' luminosity against their temperature is known as the Hertzsprung–Russell (HR) diagram.

The HR diagram is a powerful tool for studying the evolution of stars. Every star at every instant of its life has a given luminosity and surface temperature. These are dictated by the mass and structure of the star, including its overall radius. The internal structure is dictated by the star's composition and the nuclear FUSION inside. The star evolves because energy leaks from the surface, nuclear fuel is consumed, and its composition changes. As a result, the star's luminosity and temperature alter and it moves along a track in the HR diagram.

HR diagrams of cluster and field stars

Different STELLAR POPULATIONS are of different compositions and ages, and have different HR diagrams. In the HR diagrams of a GLOBULAR CLUSTER like 47 Tucanae:

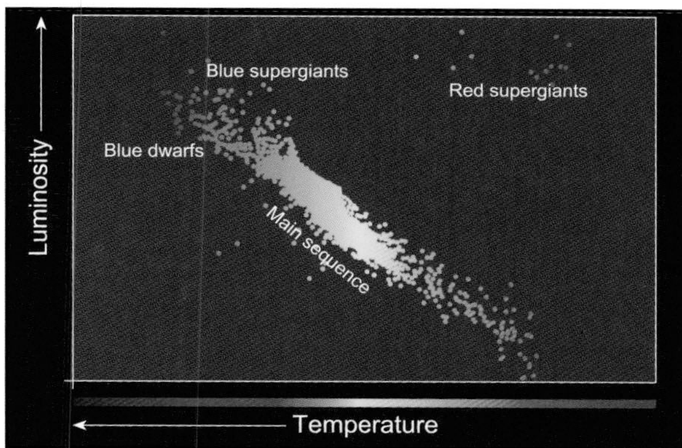

◀ ***HR diagram of young stars*** *This diagram of Population I stars in an open cluster consists of a main sequence with a few blue and red supergiants (as a result of the rapid evolution of some of the most massive stars).*

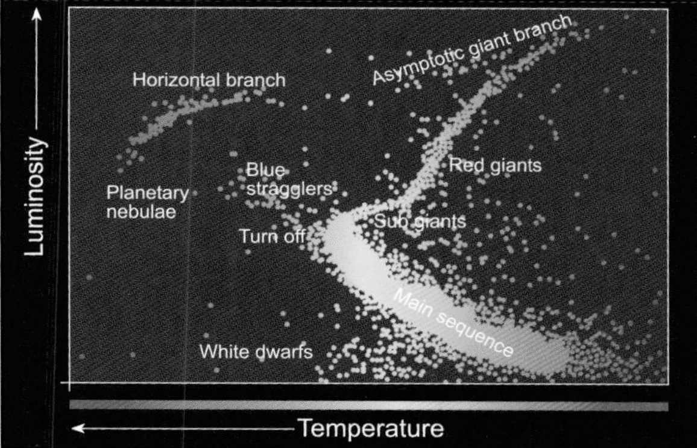

◀ ***HR diagram of old stars*** *These Population II stars in a globular cluster have a shorter main sequence with many of the brighter, more massive stars transformed to black holes and white dwarfs. There are many red giants, and some of these have progressed to the horizontal branch and asymptotic giant branch on their way to becoming white dwarfs. The red giants 'turn off' the main sequence at the place marked, but some stars straggle to the blue end of the main sequence.*

- Most stars lie on the main sequence, running from low-luminosity red stars to high-luminosity blue stars. The bending portion of the top of the main sequence defines the so-called turn-off point. The main sequence is populated by stars that are burning hydrogen in their cores.
- The subgiant branch and the red-giant branch are so named because the stars there are large to a greater or lesser degree. The hydrogen in the cores of the subgiant and red-giant branch stars is spent and has turned to helium, but has not yet ignited in helium burning. Their energy comes from a shell of hydrogen burning.
- The stars that are slightly bluer than the red-giant branch and that crowd into a sort of blue tail belong to the horizontal branch – called horizontal because, in many clusters, this feature lies at constant luminosity. HORIZONTAL-BRANCH STARS are burning helium in the core.
- Just above the red-giant branch and in a region converging with it are the few asymptotic giant branch stars. They have completed helium burning in their cores and have two shells of nuclear fusion, burning helium and hydrogen (the 'double shell' stage).
- White dwarf stars, the dying embers of globular cluster stars, lie in a sequence that is nearly parallel to the main sequence, but considerably fainter.

By contrast, the HR diagram of an OPEN CLUSTER of stars is almost only a main sequence. Some of the most massive stars of the main sequence become red supergiants.

What drives the evolution of stars and produces such different HR diagrams? An HR diagram is a snapshot of the state of the stars in a cluster as they are at this moment, and the locus that the stars map in the HR diagram is called an isochrone (meaning 'at the same time'). Any individual star reached its present position by following an evolutionary track. In general, the more massive stars move faster, so massive stars move off the main sequence before less massive ones. The turnoff point – the point of the main sequence above which there are no stars because they have turned off it – is the most sensitive indicator of the age of the population of stars in an HR diagram.

A star moves in fits and starts from one area of the HR diagram to the next as its internal structure changes, according to the nuclear fusion process in its interior. It spends most time in the areas where its structure is relatively stable and only slowly changing – for example, while burning hydrogen on the main sequence. According to University of California astronomer John Faulkner, 'motion in the HR diagram is the outward and visible sign of an inner nuclear turmoil.' By calculating the tracks of stars of different masses and plotting the positions of these stars at a given time it is possible to calculate theoretical isochrones to match onto a given HR diagram.

Evolution of solar-mass stars

The stars we see today in old globular clusters and old open clusters are about the same mass as our Sun. Solar-mass stars are of primary interest in interpreting HR diagrams, and are also of parochial interest. Stars between 0.5 and 2.2 solar masses evolve as follows:

- Within some millions of years after its formation the star settles down on the main sequence, where it quietly burns hydrogen. The hydrogen content in the center decreases to zero after nearly 10 billion years. A core of helium builds up, and the hydrogen-burning zone occupies a shell surrounding the core. The star leaves the main sequence.
- The star moves into the subgiant branch in the HR diagram. It expands and becomes a red giant. The helium core grows and the star begins to burn helium. It loses mass by a stellar wind. Stars more massive than about 1 solar mass lose little mass in a stellar wind, but stars of about 0.8 solar mass lose 20% of their mass this way. The duration of central helium burning is a few hundred million years.
- The star builds up a carbon-oxygen core. When helium is exhausted, the core cannot ignite because solar-mass stars are too small. However, the star continues to evolve for a few million years thanks to the helium- and hydrogen-burning shells. The helium shell goes through a series of paroxysms alternated with quiescence, which causes the star's envelope to expand and contract. Strong stellar winds strip off the whole envelope, revealing the hot inner regions with temperatures in excess of about 100 000 K. The expelled gas is illuminated by the very hot central stars. This is a PLANETARY NEBULA, which lasts about 10 000 years as it disperses into space. While this happens, the star turns into a WHITE DWARF. After an initial phase of rapid cooling, the white dwarf takes a long time to radiate away its energy.

What about other stars?

- Stars between 0.08 and 0.5 solar mass are very long lived, lasting several hundred billions of years, so the universe is not old enough for any of them to have left the main sequence. If nothing else intervenes they will become helium white dwarfs.
- As stars between 2.2 and 8 solar masses move off the main sequence they skip the red-giant branch and settle on the core helium-burning band, nearly parallel to but brighter and cooler than the main sequence. Stars of up to about 6 solar masses will become planetary nebulae and white dwarfs. More massive stars ignite carbon while in the

asymptotic giant branch. A thermonuclear runaway disrupts the whole star in a Type II SUPERNOVA. No stellar remnant is left.

• Stars over 8 solar masses proceed through the whole series of nuclear-burning stages up to the formation of an iron core. The evolution of their cores is relatively straightforward, but their path in the HR diagram is uncertain because they are heavily affected by mass loss. In brief, stars less massive than about 30 to 40 solar masses evolve like stars of seven solar masses, exploding as Type II supernovae while shining as red supergiants. The situation is uncertain for more massive objects. They probably explode as another type of supernova, leaving either a neutron star or a BLACK HOLE, depending on how the core collapses.

Further reading: Clayton D D (1983) *Principles of Stellar Evolution and Nucleosynthesis* University of Chicago Press; Cox J P and Giuli R T (1968) *Principles of Stellar Structure* Gordon and Breach.

Herzberg, Gerhard (1904–99)

German-born Canadian molecular spectroscopist, winner of the Nobel Prize for Chemistry in 1971 'for his contributions to the knowledge of electronic structure and geometry of molecules, particularly free radicals.' This included methylene, since identified in nebulae and other interstellar objects. In 1934 Herzberg escaped the Nazi regime and emigrated to Saskatoon in Canada. After a brief period in the USA at the University of Chicago and the Yerkes Observatory, he returned to Canada to establish a laboratory for fundamental research in spectroscopy, now known as the HERZBERG INSTITUTE OF ASTROPHYSICS. He identified the water in COMET KOHOUTEK, and in 1980 discovered triatomic hydrogen.

Herzberg Institute of Astrophysics

The National Research Council of Canada's Herzberg Institute of Astrophysics (NRC-HIA), whose headquarters are in Victoria, British Columbia, was named in honor of Nobel prizewinner Gerhard HERZBERG. It manages Canada's involvement in major astronomical observatories in Hawaii and Chile. NRC-HIA also operates two of Canada's premier astronomical facilities, the DOMINION ASTROPHYSICAL OBSERVATORY and the DOMINION RADIO ASTROPHYSICAL OBSERVATORY.

Hevelius, Johannes (1611–87)

Astronomer and instrument-maker, born in Danzig, now Gdansk, Poland. Influenced by Tycho BRAHE in subject and instrumentation, Hevelius erected in Danzig what must have been then the world's finest observatory, sadly destroyed by fire in 1679. He made several large telescopes and some of the last large, open-sighted instruments, to which he introduced the vernier scale for reading the scales and determining star positions. He wrote a book on instruments, *Machina Coelestis* (Instruments of the Skies), and his widow, Elisabeth, posthumously published his catalog of star positions in 1687, and his celestial atlas in 1690. The atlas introduced several constellations, most of which are still in use. His atlas of the Moon dubbed the gray, flat regions 'maria' (*see* MARE), but his crater names were superseded by those of Giambattista Riccioli (1598–1671). Hevelius discovered four comets, and was one of the first to observe the transit of Mercury.

Hewish, Antony (1924–)

English radio astronomer, he won the Nobel Prize for Physics in 1974 for pioneering research in radio astrophysics and

► ***Johannes Hevelius*** *and his second wife, Elisabeth, are seen using Hevelius's 6-ft sextant in this engraving from Hevelius's book on instruments.*

the discovery of pulsars (*see* NEUTRON STAR AND PULSAR). Hewish worked in war-time radar development and later became professor of radio astronomy at Cambridge University, leading the Mullard Radio Astronomy Observatory from 1982 to 1988. He first worked on the SCINTILLATION, or twinkling, of radio 'stars', which he used to investigate the properties of the PLASMA clouds in the ionosphere and the solar wind. He built a RADIO TELESCOPE to use this technique to obtain very high angular resolution observations of radio sources. The observational requirements were precisely those needed to detect pulsars, and it was with this telescope that Hewish's graduate student, Jocelyn BELL, noticed them.

Hidalgo

An ASTEROID discovered by Wilhelm BAADE in 1920, designated (944) Hidalgo. It was the first known Jupiter-crossing asteroid, having an orbit that takes it from 2.01 AU, near the inner edge of the main belt, to 9.69 AU, just beyond the mean distance of Saturn, in a period of 14.15 years. Hidalgo's mean distance from the Sun is 5.85 AU; its diameter is uncertain but is about 40 km. Its highly inclined (42°) and eccentric (0.66) orbit suggests that it may be a large extinct cometary nucleus. It is a D-type object, indicating a dark surface.

High Altitude Observatory (HAO)

The observatory, in Boulder, Colorado, USA, was established in 1940. It is dedicated to research in solar and solar-terrestrial physics, with an emphasis on solar variability and its impact on the Earth. HAO is a division of the US National Center for Atmospheric Research, and is sponsored by the National Science Foundation. HAO conducts its research in co-operation with the university community, NASA and NOAA, and with research centers worldwide.

High Energy Astrophysical Observatory (HEAO)

Series of three NASA orbital observatories. HEAO-1, launched in August 1977, successfully completed the most accurate all-sky survey of X-ray sources up to that time. It discovered the Cygnus Superbubble, created by a series of SUPERNOVAE. HEAO-2 (later known as the EINSTEIN OBSERVATORY), launched in 1978, used and demonstrated the effectiveness of the grazing incidence technique for X-RAY ASTRONOMY. HEAO-3, launched in September 1979, carried a gamma-ray spectrometer and two cosmic-ray detectors, and operated in the range 50 keV–10 MeV. It carried out an all-sky survey for narrow gamma-ray line emission.

High Energy Transient Experiment (HETE)-2

A small satellite for GAMMA-RAY ASTRONOMY launched in October 2000 whose prime objective is to determine the origin and nature of GRBs. The program is an international collaboration led by the Center for Space Research at the Massachussetts Institute of Technology. HETE-2 carries instruments sensitive to X-ray and gamma-ray radiation and can locate the source of a burst to within tens of arcseconds. The first HETE was lost due to the launch failure of its Pegasus rocket in 1996.

Because of the short-lived nature of bursts it is vital to alert other telescopes as soon as possible after a burst is detected. The optical counterpart of gamma-ray burst GRB021211 (whose name signifies that it was discovered on December 11, 2002) was observed 65 s after the detection by the ground-based RAPTOR (Rapid Telescopes for Optical Response) automatic telescope at the Los Alamos National Laboratory, New Mexico. On September 21, 2001, HETE detected a rare optical afterglow of a gamma-ray burst in the constellation Lacerta. The source was relatively close, only about 5 billion l.y. from the Earth. The opportunity to see the afterglow in optical light provides crucial information about what is triggering these mysterious bursts, which scientists speculate to be the explosion of massive stars, the merging of neutron stars (*see* NEUTRON STAR AND PULSAR) and BLACK HOLES, or possibly both.

High-velocity cloud

A cloud of neutral HYDROGEN whose velocity exceeds, typically by 100 to 200 km s^{-1}, the rotational speed of the Galaxy (which is 220 km s^{-1}). Some high-velocity clouds (HVCs) are thought to have been produced by gravitational interactions (*see* GALAXIES, COLLIDING). For example, and most notably, the MAGELLANIC STREAM, a huge bridge of material between the Galaxy and the MAGELLANIC CLOUDS, is presumed to have been drawn out during a close passage in the past. Other HVCs may be clouds of intergalactic hydrogen falling into the Galaxy, or material expelled out of the GALACTIC PLANE as SUPERNOVA remnants and now falling back.

High-velocity star

A star that is moving with very high velocity relative to the Sun (usually 65 km s^{-1} or more), faster than the average velocity of stars in the solar neighborhood. High-velocity stars are usually part of the GALACTIC HALO, and they move in very elliptical orbits that are highly inclined to the GALACTIC PLANE. Hence they do not share the revolution of the Sun and other stars in the spiral arms of the Galaxy around the galactic center. They may be remnants of a very early stage in the evolution of the Galaxy, so high-velocity stars are RUNAWAY STARS.

Hind, John (1823–95)

English astronomer who discovered in 1852 a small nebula in the constellation Taurus. In 1861 this was found to have disappeared but it later reappeared. Hind's Variable Nebula, as it came to be called, demonstrated that at least some nebulae were small, as nothing larger than a light year in dimension can disappear in a year. The nebula (NGC 1555) is a REFLECTION NEBULA that varies with the changing illumination of the nearby variable star T TAURI. Hind discovered 11 ASTEROIDS, NOVA Ophiuchus 1848, and became superintendent of the Nautical Almanac Office.

Hinotori *See* ASTRO

Hipparchus (2nd century BC)

Hipparchus was the most highly regarded Greek astronomer in antiquity before PTOLEMY. It is likely that he lived from 190 to 120 BC. In a series of writings, only one of which has survived, he investigated a range of problems in observational and mathematical astronomy. In his most important works he studied the periodicities of the motions of the heavenly bodies, and used observations to test and quantify geometrical models for those motions.

Hipparchus was born at Nicaea in Bithynia (now Iznik, Turkey). While living in Bithynia, he recorded observations of weather correlated with the solar year and the risings and settings of fixed stars. He is believed to have compiled the first catalog of star positions.

This work is known to us only because it is cited in Ptolemy's ALMAGEST. The earliest dated observations that Hipparchus appears to have made are determinations of the date of the autumnal EQUINOX in 162–158 BC. Ptolemy informs us that Hipparchus was on the island of Rhodes when he made observations in 141 and 128–127 BC, and it is thought that he resided there most of his working life. Hipparchus was in contact with astronomers in Alexandria, and very likely also with those in Babylon. Hipparchus incorporated ancient Babylonian data with his own careful observations of the length of the year, which led to his most famous discovery of the PRECESSION of the equinoxes. This is due to the slow rotation on the plane of the sky of the Earth's axis of rotation. Hipparchus also made a careful study of the motion of the Moon, and developed a theoretical model of the motion of the Moon based on epicycles. He used observations of an eclipse of the Moon in 190 BC to calculate the distance of the Moon.

Hipparcos

The Hipparcos space ASTROMETRY mission was dedicated to accurately measuring star positions, and their distances and motions in space. It was launched by ESA in August 1989 and operated until mid 1993, with two resulting star catalogs being completed and published in 1997. The *Hipparcos Catalogue* is a high-accuracy compilation of nearly 120 000 star positions, distances and space motions; the TYCHO STAR CATALOGUE (named for the Danish astronomer Tycho BRAHE) is a lower-accuracy catalog of slightly more than a million stars. Together they represent astronomers' best understanding of space distribution, space motion and physical properties of stars in the solar neighborhood. The work was carried out within a collaboration of nearly 200 mainly European scientists.

The name of the mission, as well as being an acronym for High Precision Parallax Collecting Satellite, remembers the contribution of the ancient Greek astronomer HIPPARCHUS.

Hirayama family

Groups or families of ASTEROIDS each of whose members share similar ORBITAL ELEMENTS, in particular a closely similar semi-major axis and orbital INCLINATION. These similarities, first pointed out by Kiyotsugu Hirayama (1874–1943) in 1918, are taken to indicate that the family members share a common origin in the collisional break-up of a large parent body. Several dozen Hirayama families have been identified with a fair degree of certainty. Spectral similarities within some families support the idea of a common origin, and this is true of the most populous families, for example the Themis family. However, other families (such as the Alexandra family) contain a variety of types, making it unlikely that they originated from a single object.

Hiten (Muses-A)

First Japanese Moon mission, launched in January 1990. It was named Hiten after a Buddhist angel who plays music in heaven. The mission verified the swingby technique by using lunar gravity. It returned engineering data, detected cosmic dust, and released a 12-kg orbiter called Haroromo.

Hoba meteorite

The largest known single METEORITE mass (also known as Hoba West meteorite), weighing an estimated 60 tonnes. It still lies where it was discovered in 1920, at Hoba Farm near Grootfontein in Namibia. It produced no impact crater but sits partially embedded in a hollow in the ground. The Hoba meteorite is an ataxite, a variety of iron meteorite, and measures about 2.7 m in its longest dimension. It is considerably weathered, and the surrounding surface is covered by a rusty deposit, the amount of which suggests that when the meteorite landed it may have weighed 75 tonnes.

Hobby Eberley Telescope

A large optical telescope of unusual design at the MCDONALD OBSERVATORY, Texas, USA. The mirror consists of 91 identical segments that form a hexagon measuring 10 × 11 m. The working aperture is 9.2 m. The telescope, which began operation in 1997, cost only a fraction of a conventional 8-m telescope. The Hobby Eberley Telescope is fixed at an angle of 55° to the horizontal but can rotate through 360° of AZIMUTH.

▼ ***Hoba meteorite*** *The largest known single meteorite in the world.*

Hoffleit, Dorrit (1907–)

American astronomer and compiler of the revised *Yale Bright Star Catalog*, which contains basic data for about 9000 stars. She also discovered more than 1000 VARIABLE STARS.

Hoffmeister, Cuno (1892–1968)

German astronomer and founder of the Sonnenberg Observatory. He discovered thousands of VARIABLE STARS through repeated photography of the sky and his technique of 'fly-spanking,' in which he compared the size of the stellar images to identify changes in brightness.

Hogg, Helen (née Battles) (1905–93)

Canadian astronomer, also known as Helen Sawyer Hogg. She became 'tied...to astronomy for life' after a total eclipse of the Sun in 1925. She worked at Harvard University with Harlow SHAPLEY on star clusters, then moved to Victoria, British Columbia, Canada, where she started an observing program with the 72-in telescope there to study VARIABLE STARS in GLOBULAR CLUSTERS, accompanied by her astronomer husband Frank Hogg (*c.*1901–51). The work continued at Toronto's DAVID DUNLAP OBSERVATORY. Hogg compiled her work and that of others into the *Catalogs of Variable Stars in Globular Clusters*. Throughout her distinguished career she received numerous awards and honors, including the naming for her of an observatory (in Ottawa, Canada) and a telescope (in Chile).

Holmberg, Erik Bertil (1908–2000)

Swedish astronomer who investigated the photometric properties of galaxies. He identified a characteristic size for a galaxy in terms of its brightness (the Holmberg radius); and by comparing elliptical galaxies tilted at different angles to the line of sight found that galaxies do not differ significantly in brightness, concluding that they were relatively free of dust obscuration.

Homunculus Nebula

A cloud of dust surrounding the unstable star ETA CARINAE. The cloud has been shed from the star during outbursts that began in 1843, and now measures about 17 × 12 arcsec. It is part of NGC 3372, the Carina Nebula. The name derives from its shape, which resembles a small human figure.

Horizon

A plane perpendicular to the line from an observer to the ZENITH. The astronomical horizon is the great circle formed by the intersection of the plane of the horizon with the CELESTIAL SPHERE. The term is also used to separate the boundary beyond which objects or events cannot be seen, as in the EVENT HORIZON of a BLACK HOLE.

Horizontal-branch star

A star found on the horizontal part of the HERTZSPRUNG–RUSSELL DIAGRAM of a GLOBULAR CLUSTER. This is a thin strip of stars to the right of the MAIN SEQUENCE and to the left of the red-giant branch. The small number of stars here is seen as evidence that evolution across this region is rapid. According to theories of stellar evolution, these stars are burning helium in the core and hydrogen in a shell surrounding the core.

◀ ***Horsehead Nebula*** *A dark, dusty cloud (lower half of picture) abuts an emission nebula, IC 434, illuminated by the star Zeta Orionis (the brightest star in the photo). The Horsehead Nebula (center, right) pushes from the cloud into the nebula. The yellowish nebula close to Zeta is NGC 2024, the Flame Nebula. This image is usually shown in this orientation with north to the left of the picture to highlight the horsehead appearance.*

Low-mass stars start helium-core burning with a helium flash, and the star moves rapidly to the left of the horizontal branch. More massive stars are found to the right of the horizontal branch. RR LYRAE STARS occur where the horizontal branch crosses the instability strip in the HR diagram. There is considerable variation between the horizontal branches of different globular clusters, which is attributed by different astronomers to differences such as age or metal abundance.

Horizontal coordinates

Coordinates that use the HORIZON as a plane of reference. Also known as altitude–azimuth coordinates.

Horologium *See* CONSTELLATION

Horrocks, Jeremiah (1619–41)

English astronomer. Horrocks was admitted to Cambridge University as a sizar (a scholar without money) at the age of 13 and left three years later without taking a degree. He served as a parish assistant for two years and may have been a teacher. He became friendly with William Crabtree (1610–44?) and corresponded with him on the orbits of the Moon and planets.

Horrocks studied the predications of PTOLEMY and Johannes KEPLER and, finding fault with both, developed his own theories of planetary dynamics, which Isaac NEWTON and John FLAMSTEED were later to acknowledge. Kepler had predicted the transit of Venus in 1631 but it was not visible from Europe. Horrocks calculated that transits occurred in pairs eight years apart, and he and Crabtree separately observed the transit of 1639 by a telescopic projection method. From the observation Horrocks calculated the solar PARALLAX, finding a much greater distance of the Sun and scale of the solar system than anyone before him – a fact that some of his contemporaries found incredible. Horrocks and Crabtree observed the occultation of the Pleiades by the Moon in 1637, and as the light of each star was snuffed out instantaneously this demonstrated that stars are point sources of light.

Horsehead Nebula (B33)

A DARK NEBULA in the constellation Orion. Although small (6 × 4 arcmin), it has the distinctive shape of a chess knight. It is one of the best known of all astronomical images, projected in silhouette against the bright, red EMISSION NEBULA IC 434.

Hour angle (HA)

The angle between an observer's MERIDIAN and the HOUR CIRCLE of a celestial body measured westward along the CELESTIAL EQUATOR. It is a measure of the time elapsed since the star last crossed the meridian and is usually expressed in hours, minutes and seconds of SIDEREAL TIME (one hour is equivalent to 15° at the celestial equator). The hour angle is zero for a star crossing the meridian (at upper transit); six hours later the celestial sphere has rotated through 90°, and the value of hour angle is 6 h. After 24 hours the star returns to the meridian.

Hour circle

A great circle on the CELESTIAL SPHERE passing through any object and the north and south CELESTIAL POLES. These are lines of constant RIGHT ASCENSION. DECLINATION is measured from the CELESTIAL EQUATOR along an hour circle.

Hourglass Nebula (MyCn 18)

There is some confusion over the name 'Hourglass,' as this shape is common among PLANETARY NEBULAE. However, the HST, which first imaged well the planetary nebula MyCn 18 in the constellation Sagittarius, named it the Hourglass and this is now the identification commonly used. MyCn 18 is only 4 arcsec in size. The twin lobes of ejected matter from the central star form a distinctive hourglass shape, while recently expelled gas at the center strongly resembles the human eye. The Hourglass Nebula, which is also known as the Etched Hourglass Nebula, lies at the heart of the Lagoon Nebula (M8) in Sagittarius (*see* MESSIER CATALOG).

Hoyle, Fred (1915–2001)

An English astrophysicist, controversialist and science fiction writer, Hoyle founded the Institute of Theoretical Astronomy, which later merged with the Cambridge Observatory as the Institute of Astronomy. He joined Hermann Bondi (1919–) and Thomas GOLD in developing steady-state cosmology. Hoyle coined the term BIG BANG for the explosive event at the start of the universe, intending it to be derisive of evolutionary theories. The STEADY-STATE THEORY is generally regarded as disproved (by the existence of the COSMIC MICROWAVE BACKGROUND), although it survives in a version in which a large steady-state universe experiences a continuous sequence of randomly occurring Big Bangs. Hoyle, in collaboration with Margaret Burbidge (1919–), Geoffréy Burbidge (1925–) and William FOWLER, produced a landmark paper on the synthesis of the elements, known from its authors' initials as B²FH. With Chandra Wickramasinghe (1939–), Hoyle was also a modern proponent of Svante ARRHENIUS'S PANSPERMIA hypothesis, that life on Earth comes from space, in particular in the form of viruses delivered by comets. Hoyle published 14 novels, a play and numerous works to popularize science.

Hubble, Edwin Powell (1889–1953)

American astronomer who trained as a lawyer with a Rhodes scholarship to Oxford University, England. Hubble switched to astronomy at the University of Chicago, and became a staff astronomer at MOUNT WILSON OBSERVATORY, where he had access to the best telescope in the world at that time – the 100-in telescope.

Between 1922 and 1936 he made four discoveries that have changed our view of the universe and have endured. First he proposed classification systems for nebulae (*see* NEBULAE AND INTERSTELLAR MATTER). Within the Milky Way he recognized both EMISSION NEBULAE and REFLECTION NEBULAE. He divided extragalactic nebulae using the now-standard Hubble classification system, producing the HUBBLE TUNING-FORK DIAGRAM. Second, with his discovery of CEPHEID VARIABLE STARS in NGC 6822, M33 and M31, he demonstrated that extragalactic nebulae were galaxies comparable to our own Milky Way at great distances. This settled a question that had been argued between Heber CURTIS, Knut Lundmark (1889–1958), Harlow SHAPLEY and Ernst ÖPIK, to name only four. He then showed that the distribution of galaxies is uniform, and that they occupy a space that is cosmological. Finally, and most importantly, working with Milton HUMASON, Hubble showed in 1929 that galaxies are moving away from us with a speed proportional to their distance (HUBBLE'S LAW).

The explanation for this led to the discovery of the expanding universe. The slope of the HUBBLE DIAGRAM of the velocities of galaxies plotted against their distances is known as the HUBBLE CONSTANT, and its reciprocal, Hubble time, is an estimate of the age of the universe. The HST, which is named for him, continues his investigations.

Hubble classification

See HUBBLE TUNING-FORK DIAGRAM

Hubble constant

The constant of proportionality, denoted by H, that relates the velocities at which galaxies are receding to their distances. According to the HUBBLE LAW, velocity of recession (V) is directly proportional to distance (D), a relationship that may be expressed as $V = H_0 D$, where H_0 is the value of the Hubble constant at the present epoch in the history of the universe.

The Hubble constant, or Hubble parameter, is a measure of the rate of expansion of the universe and is, in effect, a measure of the fractional increase in the scale of the universe in unit time. Its value is usually expressed in units of velocity (expressed in kilometers per second), divided by distance (expressed in megaparsecs), that is in km s^{-1} Mpc^{-1}. There is still a significant degree of uncertainty about the precise value of H_0, although there is wide general agreement that it lies between 50 and 100 km s^{-1} Mpc^{-1}. The uncertainty arises primarily because it is hard to obtain reliable measurements of the distances of galaxies. Many recent research programs using the apparent brightnesses of CEPHEID VARIABLE STARS or Type Ia SUPERNOVAE as distance indicators have yielded values in the range 57–80 km s^{-1} Mpc^{-1}, with a concentration of values in the range 60–70 km s^{-1} Mpc^{-1}. According to the astronomers who interpreted the data from the WILKINSON MICROWAVE ANISOTROPY PROBE, the best value is 71 kms^{-1} Mpc^{-1} with a 5% uncertainty.

The inverse of the Hubble constant has the dimensions of time and is called the Hubble time. If the universe had always expanded at a constant rate then $1/H_0$ is the time that the universe would have taken to expand to its present size, and is thus a measure of the age of the universe. If H_0 lies in the range 50–100 km s^{-1} Mpc^{-1}, values of the Hubble time lie between 20 and 10 billion years, respectively. If, for example, H_0 = 65 km s^{-1} Mpc^{-1}, then the value of the Hubble time is 15 billion years. The Hubble time is related to the age of the universe through a model of its expansion (*see* UNIVERSE: COSMOLOGICAL THEORY). In the standard model, the age of the universe is two-thirds the Hubble time.

Hubble Deep Field (HDF)

The Hubble Deep Fields are two small areas of the sky that were selected for deep observations by the HST. They represent the deepest optical observations to date and reveal galaxies as faint as magnitude 30, 4 billion times fainter than can be seen with the unaided eye.

Ten-day observing campaigns were carried out by the HST in December 1995 on a field in the northern hemisphere (HDF-N), and in October 1998 on a field in the southern hemisphere (HDF-S). The fields were chosen to avoid bright stars, dust or nearby galaxies interrupting the view of more distant galaxies. To make the most efficient use of the HST, the fields were also required to be in the 'continuous viewing zone,' at about ±60° declination. Other than that, the fields are essentially random 'core samples' of the universe. The observations were released to the public within weeks of being obtained, and have been the subject of extensive research and follow-up observations with other telescopes. The HDFs represent the deepest observations not only of the HST but also of some of the largest ground-based telescopes and other space observatories.

Both fields are roughly 3 arcmin square and were imaged by the HST's Wide-Field Planetary Camera-2 through four broadband filters. Each field showed about 3000 galaxies brighter than magnitude 29.

Hubble diagram

A plot of the apparent magnitude of galaxies versus the REDSHIFT of their spectral lines. The resulting straight-line relationship forms the basis for the HUBBLE LAW.

Hubble law

The relationship that states that the recessional velocities of distant galaxies are directly proportional to their distances; the constant of proportionality is known as the HUBBLE CONSTANT, or Hubble parameter. It is named for the American astronomer Edwin HUBBLE who, in 1929, published his discovery that the REDSHIFTS in the SPECTRA of galaxies are proportional to their distances. This was one of the key discoveries in cosmology, for it showed that the galaxies are receding and the universe as a whole is expanding.

The law is consistent with a universe undergoing uniform expansion whereby each galaxy (or cluster of galaxies) is receding from every other one with a speed proportional to the separation between them. In such a universe, observers in any galaxy would see all other galaxies receding with speeds proportional to their distances, and no one galaxy is the unique center of this expansion. The smooth, idealized recession of the galaxies from each other (ignoring local variations caused, for example, by the gravitational influence of clusters and superclusters of galaxies in accordance with the Hubble law is called the Hubble flow.

Hubble Space Telescope (HST)

After the HST was deployed from the Space Shuttle Discovery on April 25, 1990, its commissioning tests caused concern. Each time the HST passed from or into sunlight the observatory's huge solar arrays deformed. The resultant shake could not be counteracted by the onboard attitude control system. Even worse, the telescope's images showed SPHERICAL ABERRATION. The primary mirror was wrongly shaped – it was 2 μm too flat near its edge. While the spatial resolution of the core of the images of around 0.1 arcsec was close to specification, the telescope's sensitivity was seriously compromised since it was focusing only 15% of incoming light. Some image properties could be restored by computer processing, but the only long-term solution was to compensate for the faulty optics with a Corrective Optics Space Telescope Axial Replacement (COSTAR) system.

COSTAR's 10 coin-sized, adjustable mirrors were inserted into the optical train of three of HST's instruments during the first servicing mission in December 1993. With the HST captured by the remote manipulator arm and located upright in the Shuttle's payload bay, the astronauts also replaced the warped solar arrays and installed a modified Wide-Field Planetary Camera (WFPC). The 'refurbished' HST went on to revolutionize observational and theoretical astronomy with a stream of remarkable observations and beautiful images that have stunned astronomers and public alike.

The telescope

The HST is named in honor of the American astronomer EDWIN HUBBLE, commemorating the breakthroughs in cosmology that he made during the 1920s. The idea for a space telescope above the turbulence of the atmosphere was first proposed in 1946 by Lyman SPITZER, who also noted the importance of ultraviolet (UV) sensitivity.

Although it began and was led as a NASA project, the ESA partnered HST, receiving approximately 15% of the available observing time in return for providing the Faint Object Camera and two solar arrays. At the time of its launch, HST was the largest non-military optical telescope ever placed into orbit. The observatory weighs approximately 11.5 tonnes, and its main body is 13.3 m long and 4.3 m in diameter.

The HST has a RITCHEY–CHRÉTIEN design, which uses a 2.4-m aluminum-coated primary mirror made of ultra-low-expansion glass. Light is reflected from the primary mirror to a 33-cm secondary located 4.6 m in front of it. This reflects the light back through a 66-cm hole in the primary mirror to the scientific instruments behind it in the aft shroud section.

The observatory was designed with in-orbit maintenance in mind. To aid astronauts during extravehicular activity, the HST has 68 m of handrails and 31 foot restraints on its surface. It contains 70 modules that are designed to be replaced in space.

The HST was originally equipped with two cameras, two spectrographs and a photometer:

- the WFPC, developed by the California Institute of Technology and the Jet Propulsion Laboratory;
- the ESA's Faint Object Camera;
- the Goddard High Resolution Spectrograph for UV observations, developed at NASA's Goddard Space Flight Center;
- the Faint Object Spectrograph, developed at the University of California, San Diego;
- the High Speed Photometer, developed at the University of Wisconsin.

The High Speed Photometer was replaced during the December 1993 servicing mission to make room for COSTAR. At the same time, the WFPC was replaced by a version with modified optics (WFPC-2) to offset the HST's spherical aberration. In February 1997, the High Resolution and the Faint Object Spectrographs were replaced by the Space Telescope Imaging Spectrograph, covering a wider wavelength range (1050–11 000 Å) than the previous instruments combined. In addition, the Near-Infrared Camera and Multi-Object Spectrometer (NICMOS) gave the observatory near-infrared (0.8–2.5 μm) observing capability. The Advanced Camera for Surveys replaced the Faint Object Camera in the summer of 2002.

Control and communication

The HST slews very slowly (90° in 18 min). At all times the HST must be oriented away from the Sun so that its sensitive optics are not damaged. It can maintain position on a target while occulted by the Earth, picking it up when next in sight, but it is efficient to slew between close targets while they are hidden. HST is therefore maneuvered by an optimized preplanned control process. For attitude control it has three rate sensor units, each containing two gyroscopes. These gyros have failed frequently, particularly in November 1999, when the observatory had to be shut down for two months after four of them stopped functioning (the HST needs at least three operational gyros).

Fine-guidance sensors are used to pinpoint the particular target by fixing on pairs of guide stars in the telescope's field of view. After locking on to the guide stars, these sensors

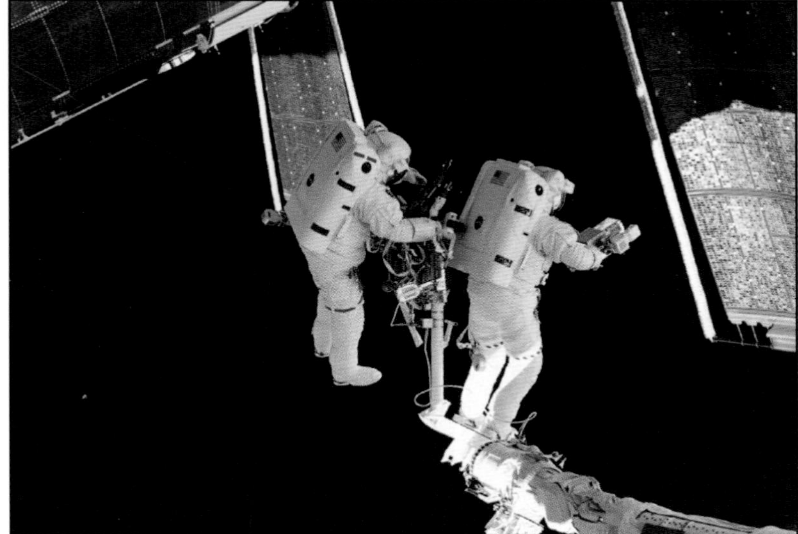

▲ ***Hubble Space Telescope*** *(Top) Astronauts working on the HST during the second servicing mission. (Bottom) HST fixed by the grappling arm to the Space Shuttle for servicing.*

aim the telescope with an accuracy of 0.01 arcsec for up to 24 hours. The *Guide Star Catalog* used by the HST to lock onto celestial targets includes the precise locations of 15 million stars down to magnitude 14.5, and constitutes a resource used for the same purpose by ground-based observatories.

A revolution in astronomy

The HST has observed seven of the nine planets and many of their larger satellites, studied COMETS such as Hyakutake and Hale–Bopp, imaged numerous ASTEROIDS, and searched for Kuiper Belt objects (*see* OORT CLOUD AND KUIPER BELT). It made the first maps of Pluto, its moon Charon, and the large asteroid Vesta, and detailed views of the fragmented COMET SHOEMAKER–LEVY 9 as it collided with Jupiter.

Within the MILKY WAY GALAXY the HST found evidence for BROWN DWARFS; studied the changing environment around SUPERNOVA 1987A as its encircling rings of glowing gas expand outward and interact; discovered circumstellar disks in which planets may be forming; found stars incubating inside tall pillars of gas and dust in the EAGLE NEBULA; and determined the ages of GLOBULAR CLUSTERS.

The HST identified evidence for massive BLACK HOLES in the cores of many galaxies. The epoch-making image of the HUBBLE DEEP FIELD in the northern hemisphere was assembled from 342 separate exposures by the WFPC-2 over 10 consecutive days in December 1995. A similar deep-field view was obtained for the southern hemisphere in October 1998. Both images contained about 2500 galaxies with apparent brightness down to magnitude 30. These galaxies, at look-back times of more than 90% of the age of the universe, show how galaxies looked soon after being formed.

Astronomers have also been able to use the HST to map DARK MATTER by means of gravitational lenses, and have observed supernovae in distant galaxies to find evidence for DARK ENERGY. Other cosmic puzzles, such as the origin of GRBs, the age of the universe and the value of the HUBBLE CONSTANT, have also been attacked using the space telescope. Studies of CEPHEID VARIABLE STARS in galaxies up to 65 million l.y. away have pinned down the Hubble constant to 70 km s^{-1} Mpc^{-1}, with an uncertainty of 10%.

Although the HST was intended originally to have an operational lifetime of 15 years, the successful servicing mission 3A in December 1999 extended the telescope's life expectancy until about 2008. Further servicing missions have been ruled out as a result of the safety concerns following the Columbia Shuttle disaster. By that time the JAMES WEBB SPACE TELESCOPE may be in orbit.

Hubble time *See* HUBBLE CONSTANT

Hubble tuning-fork diagram

A classification scheme for galaxies, devised in its original form in 1925 by Edwin HUBBLE and still in use today. The Hubble classification recognizes four principal types of galaxy – elliptical, spiral, barred spiral and irregular – and arranges these in a sequence that is called the tuning-fork diagram. *See also* GALAXY.

Hubble's Variable Nebula (NGC 2261)

A REFLECTION NEBULA in the constellation Monoceros. It is small (2 × 1 arcmin) but of quite high surface brightness. The nebula's average magnitude is 10, but, as Edwin HUBBLE discovered in 1916, it varies in brightness, mirroring the variability of its illuminating star, R Monocerotis.

Huggins, William (1824–1910) and Huggins, Margaret Lindsay (née Murray) (1848–1915)

William was a wealthy amateur English astronomer who invented the stellar SPECTROSCOPE in his own observatory. This followed discoveries by Gustav KIRCHHOFF and Robert Bunsen (1811–99) about terrestrial and solar spectral emission and absorption lines. At first William examined the spectra visually, but in 1875, after many experiments, he recorded the spectrum of VEGA on a dry gelatine plate. That same year William married the talented Margaret Lindsay Murray, who had also been working on astronomical spectroscopy. Together they produced some of the earliest spectra of astronomical objects, particularly the ORION NEBULA. William published the papers but acknowledged his wife's assistance. He showed that some nebulae, including the Orion Nebula, have pure emission-line spectra and thus must be truly gaseous, while others, such as the ANDROMEDA GALAXY, have continuous spectra that are characteristic of stars. He imaged solar PROMINENCES in the red hydrogen-alpha emission line (*see* HYDROGEN SPECTRUM). With chemist William Miller (1817–70) in 1868 he made the first measurement of the Doppler shift in a star (*see* DOPPLER EFFECT).

Hulse, Russell (1950–)

American radio astronomer and PLASMA physicist. In 1993 he won the Nobel Prize for Physics with Joseph TAYLOR for the discovery of the first known binary pulsar (*see* NEUTRON STAR AND PULSAR), the Hulse–Taylor pulsar (PSR 1913+16), which opened up avenues for the study of gravitation (*see* GRAVITATION, GRAVITATIONAL LENSING AND GRAVITATIONAL WAVES). They made the discovery using the 300-m radio telescope at the ARECIBO OBSERVATORY in Puerto Rico. The pulsar is in orbit around another neutron star, each being the mass of the Sun but only about 10 km in size and separated by only several times the Moon's distance from the Earth. Since their mutual gravitational pull is large, their orbit shows deviations from Isaac NEWTON's gravitational laws. In particular, it was found that the period of the orbit is declining at the rate of about 75 millionths of a second per year, because the pulsar system loses energy by emitting gravitational waves. The rate is fully consistent with the predictions of GENERAL RELATIVITY. This was the first, albeit indirect, proof of the existence of gravitational waves.

Humason, Milton La Salle (1891–1972)

American astronomer who began his career as a muleteer and janitor at the MOUNT WILSON OBSERVATORY, and became a night assistant. He learned photography and became Edwin HUBBLE's research assistant, and then an astronomer in his own right, measuring the speeds of faint galaxies and the properties of the SUPERNOVAE in them.

Huygens, Christiaan (1629–95)

Dutch scientist who first studied law and mathematics at Leiden. He made microscopes and telescopes, and in 1655 discovered Titan, the first moon of Saturn. In 1659 Huygens explained the apparent changes in the shape of Saturn's rings. His telescopes, of enormous FOCAL LENGTH, with the objective lenses supported on masts, were far in advance of the rest of the field. In 1656 Huygens patented the first pendulum clock. He also derived the law of centrifugal force for uniform circular motion, which, combined with Johannes KEPLER's third law, led Christopher Wren (1632–1723), Edmond HALLEY and others to the inverse square law of gravitation and proved by experiment that momentum is conserved in a collision between two bodies. Huygens argued in favor of a wave theory of light, and in the final years of his life he wrote a book, published posthumously, containing a discussion of extraterrestrial life. The Huygens probe that will parachute to the surface of Titan in 2004 is named for him (*see* CASSINI/HUYGENS MISSION).

Hyades

The Hyades star cluster in the constellation Taurus is centered near to the bright star ALDEBARAN (which, however, is not a member). The cluster's 40 or so brightest stars are visible to the naked eye, the brightest being Theta-2 Tau at magnitude 3.4. It is the nearest moderately rich star cluster, so near that Charles MESSIER omitted to catalog it. According to the HIPPARCOS satellite, it is at a distance of 46.3 parsecs (151 l.y.), accurate to better than 1%. Its stars move in paths that are parallel and directed toward a point slightly eastward of Betelgeuse. It has more than 300 members, some of them in the direction opposite to the direction from the Sun to the Hyades. So the Sun, although not a member, lies within the confines of the cluster, passing through its further reaches. Stars beyond about 10 parsecs from the cluster center are liable to be lost in time through the tidal pull of the Galaxy.

Because the Hyades is so close and well studied it is the typical POPULATION I open cluster. Its stars define the ZERO-AGE MAIN SEQUENCE. Its members form the basis of the distance scale to the OB and similar stars (*see* OB ASSOCIATIONS), and therefore determine the map of the spiral arms of the Galaxy. About half its brighter stars are BINARY STARS. Its total mass is about 1200 times the mass of the Sun, a value that comes from dynamical simulations of the cluster as it passes through the Galaxy and accounts for unseen WHITE DWARFS and other stars, embedded gaseous material, and DARK MATTER. The age of the cluster is 625 million years. Its lowest-mass star is LH0418+13, which is 0.083 times the mass of the Sun, so it is a star rather than a brown dwarf (*see* EXOPLANET AND BROWN DWARF). The Hyades contains 10 known white dwarfs.

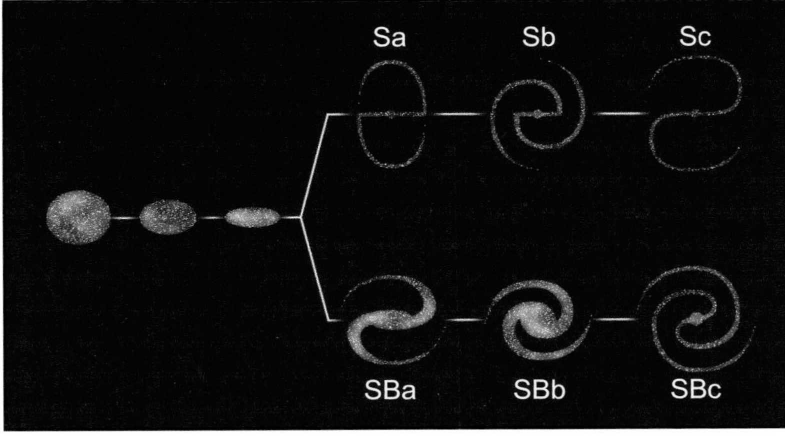

▲ ***Hubble classification*** *Hubble's tuning-fork diagram shows elliptical galaxies becoming more flattened, and progressing into spiral galaxies in two parallel sequences, one with bars and one without. This is a way to systemize galaxies, not a progression of evolution.*

Hydra *See* CONSTELLATION

Hydrogen

The lightest and most common element in the universe. Of all the directly observed matter in the universe, hydrogen (H) accounts for about 92% by number of atoms, and about 75% by mass. The most common form of hydrogen atom consists of one positively charged particle, the PROTON, and one negatively charged particle, the ELECTRON, which orbits the proton. The atomic mass is 1 and the atomic number (charge on the nucleus) is also 1. Hydrogen nuclei also exist with one or two neutrons in addition to the one proton. These forms of 'heavy hydrogen' are called DEUTERIUM and tritium, respectively. Regions of neutral hydrogen are referred to as H I REGIONS. Regions of ionized hydrogen are H II REGIONS.

Hydrogen spectrum

The series of absorption or emission lines that are characteristic of the HYDROGEN atom. According to the Bohr theory of the ATOM, devised by Danish physicist Neils Bohr (1885–1962)

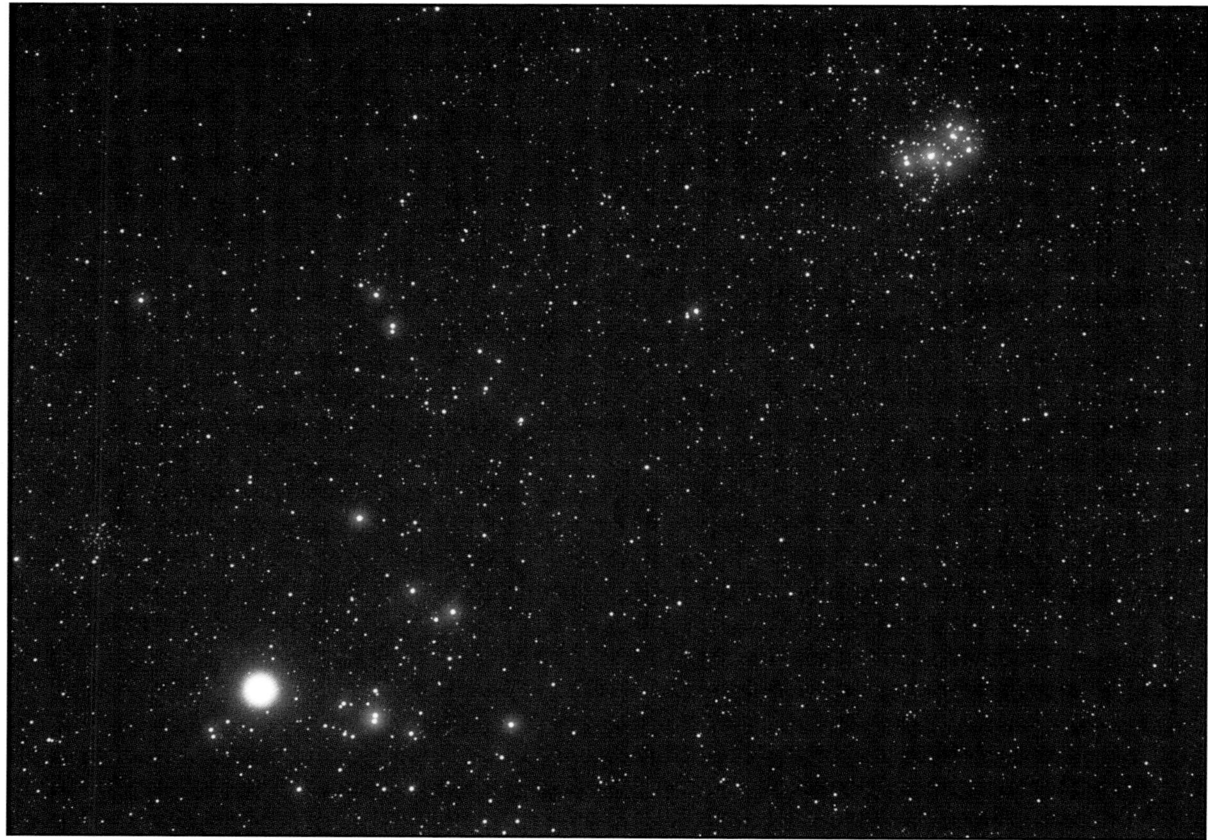

► ***Hyades***
The Hyades star cluster occupies the lower left quarter of this image. The dominant star Aldebaran is not part of the cluster, but a foreground star. The Pleiades cluster is at top right.

in 1913, the hydrogen atom can be envisaged as consisting of a central nucleus (a PROTON) around which a single ELECTRON revolves. The electron is located in one of a number of possible permitted orbits, each of which corresponds to a different energy level. If an electron drops down from a higher to a lower level (makes a 'downward transition'), a PHOTON of energy equal to the difference in energy between the two levels is emitted, and this corresponds to the emission of light of a particular wavelength. Conversely, if an electron absorbs a photon of the correct energy, it can make an upward transition from a lower to a higher level with the corresponding absorption of particular wavelengths of light. This mechanism is responsible for the production of the various lines in the hydrogen spectrum.

In the hydrogen atom, the various energy levels are denoted by a number (a quantum number), n, with $n = 1$ corresponding to the lowest level, $n = 2$ to the second level (the first 'excited' level) and so on. The line due to the transition between $n = 2$ and $n = 3$ lies toward the red end of the visible spectrum at wavelength 656.3 nm and is called the hydrogen-alpha ($H\alpha$) line. This is the first in a series of lines, known as the Balmer series, which is produced by the various possible transitions between the $n = 2$ level and higher levels. The second line, hydrogen-beta ($H\beta$), which occurs at a wavelength of 486.2 nm, corresponds to transitions between the second and fourth levels ($n = 2$ and $n = 4$); the third, hydrogen-gamma ($H\gamma$), at 434.1 nm, and so on. The series comes to an end at 364.6 nm ($n = 2$, $n = \infty$), which corresponds to the electron being completely removed from the atom.

Transitions between the ground level ($n = 1$) and higher levels involve larger energy gaps and, because the energies of photons are inversely proportional to their wavelengths, give rise to lines of shorter wavelengths than those of the Balmer series. The resulting series of lines, which occur in the ultraviolet region of the spectrum, is called the LYMAN SERIES. The series begins with Lyman alpha, at 121.5 nm, and ends at the series limit of 91.16 nm. Transitions between the third level ($n = 3$) and higher levels correspond to small energy gaps and give rise to the Paschen series, in the infrared region of the spectrum.

Hydrus *See* CONSTELLATION

Hyperbola

The CONIC SECTION obtained when a right circular cone is cut by a plane that makes an angle with the base greater than that made by the side of the cone. It is an open curve; that is, it does not close on itself like an ELLIPSE.

Hyperion

A mid-sized icy satellite of SATURN, discovered in 1848 by William BOND and, independently, by William LASSELL. Hyperion orbits at a distance of 1 481 000 km, and is an irregular object, measuring 330 × 260 × 220 km. Hyperion was probably once a larger object that suffered a number of near-catastrophic collisions. Most of the impact debris would have been swept up by TITAN, but some would have peppered RHEA, giving the latter its heavily cratered surface. Hyperion has chaotic rotation: it tumbles along in its orbit instead of rotating about a single axis. This adds credence to the collision hypothesis. Its surface is heavily cratered and has a scarp more than 200 km long named Bond–Lassell Dorsum in honor of the satellite's discoverers.

Iapetus

A mid-sized icy satellite of SATURN, discovered by Giovanni Cassini in 1671 (*see* CASSINI DYNASTY). It is 1440 km in diameter, and orbits at a distance of 3 561 000 km. Iapetus is unique in the solar system because of its harlequin appearance: the leading hemisphere (facing the direction of orbital motion) is dark, with an ALBEDO of around 0.05, while the trailing hemisphere is bright, with an albedo of 0.5. Its magnitude thus changes during its orbit, as Cassini himself noticed.

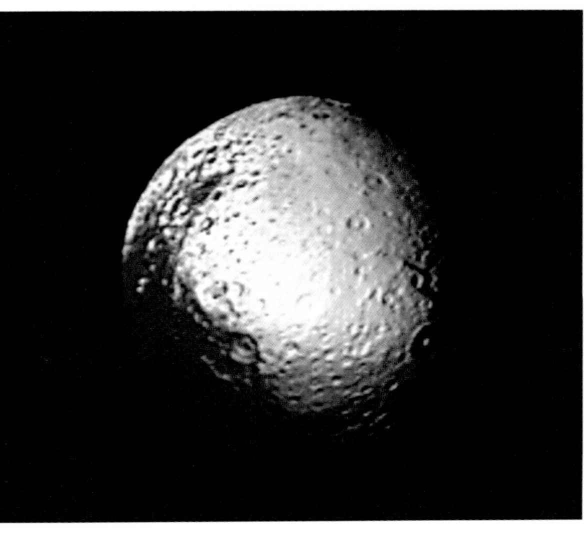

◀ ***Iapetus*** *Its leading surface is stained dark, perhaps by impacting material in the Saturn system, but it is not known why the boundary between light and dark is so sharp.*

Icarus

An Apollo asteroid discovered by Wilhelm BAADE in 1949, designated 1566 Icarus. When Icarus passed just 0.04 AU from the Earth in 1968, it was observed by radar and found to be about 1 km in diameter and to rotate in 2.27 hours. Icarus is one of only a handful of asteroids with a PERIHELION distance (0.19 AU) within the orbit of Mercury; APHELION is at 1.97 AU, and its mean distance from the Sun is 1.08 AU. Icarus is a NEAR-EARTH ASTEROID classed as potentially hazardous.

Ida

An asteroid discovered by Johann Palisa (1848–1925) in 1884, designated (243) Ida. In August 1993 it became the second asteroid to be imaged in close-up by a spacecraft when the Galileo probe passed within 2400 km (*see* GALILEO MISSION TO JUPITER). Ida was found to be a heavily cratered object measuring 56 × 24 × 21 km, possessing a small companion about 1.5 km in diameter later named Dactyl – the first confirmed asteroidal satellite. From Dactyl's orbital motion, Ida's density was found to be consistent with a composition similar to that of ordinary CHONDRITES, as suggested by its reflectance spectrum. It orbits the Sun in the main asteroid belt at a mean distance of 2.86 AU in a period of 4.84 years.

Imbrium Basin

A large, multi-ringed impact structure on the nearside of the MOON, measuring 1123 km at its largest diameter. It is visible as the lava floodplain known as the Mare Imbrium (Sea of Showers). Dating techniques show the Imbrium Basin to be the second youngest of the Moon's large impact features, from near the end of the LATE HEAVY BOMBARDMENT era; if the impact had been any greater it would probably have shattered the Moon completely. EJECTA from the impact covers much of the Moon's nearside. Lava flooding took place over the next 600 million years.

Inclination

The angle at which one plane is tilted to another. Taking three examples, first, the inclination of a planetary orbit is normally taken to mean the angle between the plane of the orbit and the plane of the Earth's orbit (the ECLIPTIC plane). The closest and most distant planets have the greatest orbital inclinations: Mercury's orbit is inclined at 7° and Pluto's at 17°. Second, the equatorial inclination of a planet is the angle between the plane of its equator and its orbit plane. Third, the inclination of the orbit of an artificial Earth satellite is the angle at which its orbit plane intersects the plane of the equator.

Index Catalogue (IC)

A catalog of nebulae and clusters prepared by Danish astronomer John DREYER as a sequel to the NEW GENERAL CATALOGUE OF NEBULAE AND STAR CLUSTERS, which had been published in 1888. The *Index Catalogue* was published in two parts, in 1895 and 1908, and the second of these volumes contained notes and corrections on the earlier work. Objects in this catalog are denoted by the letters 'IC' followed by a number, for example 'IC 2220.'

Indus *See* CONSTELLATION

Inertia

The resistance of a massive body to any change in its velocity. Any body will continue in a state of rest or of uniform straight-line motion unless acted upon by a force (the first of NEWTON'S LAWS OF MOTION). Thus a spacecraft, when its rocket motors are switched off, follows a ballistic trajectory because of its inertia. In Newton's second law of motion, the relationship between the acceleration, a, of a body and the applied force, F, is $F = ma$, where m is the inertial mass. As far as can be determined, the inertial mass is exactly equal to a body's gravitational mass, which determines its gravitational attraction. This is called the equivalence principle (*see* GENERAL RELATIVITY). There does not seem to be any *a priori* reason why these two masses are exactly equal, and there is still considerable debate on the nature of inertia.

Inertial reference frame

A frame of reference in which NEWTON'S LAWS OF MOTION hold.

Inferior conjunction *See* CONJUNCTION

Inferior planet

A term for the major planets Mercury and Venus whose orbits lie within that of the Earth, as opposed to the SUPERIOR PLANETS (major planets whose orbits lie outside). *Compare* INNER PLANET.

Inflationary universe

The inflationary hypothesis was introduced in 1981 by American mathematical physicist Alan Guth (1947–), to tackle a number of outstanding issues in the standard BIG BANG model of COSMOLOGY. *See also* UNIVERSE: COSMOLOGICAL THEORY.

Information handling in astronomy

Information handling encompasses the collection, analysis and dissemination of data, as well as the way astronomers

▲ ***Information handling in astronomy*** *The encyclopedia's editor in chief, Paul Murdin, in the library of the Institute of Astronomy, Cambridge, UK. The stacks of the* Astrophysical Journal *in front of him show the dramatic increase in astronomy publishing from 1895 to 2002.*

publish, interact, and communicate with other communities, amateur astronomers and the public. Information handling in astronomy thus reflects the way astronomers work.

Collecting data

Astronomical observations depend on access to ground-based and space-borne instruments. Candidates have to submit proposals saying what they aim to achieve and identifying the equipment configuration and scientific targets. The proposal needs to build on the available literature and take account of past observations. Even if access to a telescope is automatic, writing a proposal is a good way to get an overview of a planned project.

Nomenclature of celestial objects

Solar-system objects and their surface features are named according to the recommendations of professional committees confirmed by the IAU. The Central Bureau of Astronomical Telegrams (CBAT) and the Minor Planet Center (MPC) handle the discoveries of new bodies. They assign provisional designations until ratification by the IAU. COMETS are now the only bodies routinely receiving the names (up to three) of their discoverers, together with an alphanumerical identification.

The IAU is the sole internationally recognized authority for naming celestial bodies. Those names are not sold.

Catalogs, surveys and archives

Catalogs are organized according to object types and/or the type of data they offer. For example, the *Hubble Guide Star Catalog* (*see* HUBBLE SPACE TELESCOPE) catalogs accurate positions of stars and galaxies. The data centers (see below) are essentially compilations of catalogs.

Surveys are the systematic coverage of the sky with specific instruments, such as the famous PALOMAR OBSERVATORY Sky Survey (POSS), a two-color photographic atlas of the sky by the Palomar Schmidt telescope.

Astronomers often need to refer to past observations, making the archiving of data a critical issue. Astronomers are now creating 'virtual observatories' of multiwavelength data. The greatest challenge is technical, as the average lifetime of storage media and user interfaces is about four years. Thus any long-term archiving policy has to make provision for regularly transferring data to new media.

Databases and information hubs

Beyond catalogs and collections of catalogs, extremely powerful databases are now available through data centers. The Strasbourg Astronomical Data Center (Centre de données astronomiques de Strasbourg, or CDS) is recognized as the world leader for astronomy and includes the following:

- The SIMBAD database holds more than 2 200 000 objects under 5 500 000 identifiers, together with more than 105 000 bibliographical references including 3 000 000 object citations. Anyone seriously studying an astronomical object must visit SIMBAD first.
- VizieR is a search-and-shop individual-catalog service, with access to large tables published in professional journals.
- Aladin is an interactive digitized sky atlas.
- The CDS bibliographical service provides access to abstracts from several major journals.
- The dictionary of nomenclature gives details on more than 4000 different catalog acronyms.
- AstroGlu is a discovery tool that helps to locate database servers providing relevant information.

StarPages are 'yellow-page' resources comprising:

- StarWorlds: 6300 entries and more than 5500 Web links on astronomical organizations;
- StarHeads: more than 5200 home pages on individual astronomers and related scientists;
- StarBits: 140 000 abbreviations and acronyms.

The NASA/IPAC Extragalactic Database (NED) is a master list of extragalactic objects containing cross-identifications of names, accurate positions, redshifts, basic data, and bibliographic references (IPAC is NASA's Infrared Processing and Analysis Center).

The Astrophysics Data System (ADS) is a freely available NASA-funded project whose main resource is an abstract service (about 500 000 abstracts for astronomy and astrophysics), together with links to scans of more than 40 000 journal articles. ADS also provides access to astronomical data catalogs and data archives (particularly from NASA space missions) and StarPages.

The National Space Science Data Center (NSSDC) provides access to a wide variety of data from and complementing NASA missions.

The Canadian Astronomy Data Centre (CADC) and the Astronomical Data Analysis Center (ADAC) at the National Astronomical Observatory of Japan (NAOJ) are national astronomical resources.

Data processing

Advanced astronomical experiments (telescopes or ground stations for spacecraft) have their own specific image-processing software. General-purpose software systems for data reduction include ESO's Munich Image Data Analysis System (MIDAS) and NOAO's Image Reduction and Analysis Facility (IRAF). There are a number of specific software packages and libraries. The Starlink network helps UK-based astronomers with a range of services.

The Flexible Image Transport System (FITS) is a machine-independent way to encode definitions of astronomical data and the data themselves. The FITS format has become the standard for exchanging image data between observatories, and has been adopted in other disciplines.

Publishing and information sharing

Publishing is motivated by the noble aims of educating and sharing information. It is also a means to gain recognition, which in turn leads to positions (amassing the human resources for astronomy), acceptance of proposals (gaining access to resources) and achieving funding (allowing the materialization of ideas).

The most important general professional journals include the *Astrophysical Journal* and the *Astronomical Journal* published by the AMERICAN ASTRONOMICAL SOCIETY, the *Monthly Notices of the Royal Astronomical Society*, and *Astronomy and Astrophysics*, which resulted from the merger in 1969 of a number of professional European journals.

Astronomers also communicate via a spectrum of publications ranging from informal newsletters to books, including collections of review papers by specialists on specific topics. Conferences, colloquiums, workshops and meetings provide ways of exposing oneself to current work.

Astronomers also contribute to public education. Many countries have their own national astronomical journal, but *Sky & Telescope* magazine has the largest audience worldwide.

Information is increasingly exchanged electronically, both dynamically by e-mail and passively through websites. However, the 'electronization' of the handling of astronomical information presents some new challenges.

Further reading: Boroson T, Davies J and Robson I (eds) (1996) *New Observing Modes for the Next Century* Astronomical Society of the Pacific; Heck A (ed) (1997) *Electronic Publishing for Physics and Astronomy* Kluwer.

Infrared astronomy

Infrared (IR) radiation is like light, but has less energy and is emitted by colder objects. Astronomical sources observed in the INFRARED SPECTRUM have temperatures of a few kelvin to a few thousand kelvin. The most prominent sources are planets, cool stars like red dwarfs and red giants, nebulae, ULIRGs, the cosmic microwave background and some quasars and AGN.

The long wavelength of IR waves means that dust affects them less than optical waves: IR waves that encounter micron-sized dust grains pass around them, like waves of water pass around the legs of a pier. By contrast, light waves are absorbed by dust grains, which therefore hide what lies beyond. Viewing with IR lets astronomers penetrate dust-obscured regions of space.

Distant objects in the universe undergo REDSHIFT, so the visible to ultraviolet spectra of very distant galaxies and quasars are shifted into the IR.

History

In 1800 William Herschel (*see* HERSCHEL FAMILY) dispersed solar light through a glass prism and projected the colors onto a row of thermometers. He discovered the warming effect of 'infrared' in the dark space 'beyond the red.' Thomas Edison observed a total solar eclipse in 1878 in the IR. In the early 1900s the Moon, planets and bright stars were observed in the IR by Seth Nicholson (1891–1963) and Edison Pettit (1889–1962). Frank Low fully opened the field to astronomy in the early 1960s by inventing a sensitive gallium-doped germanium BOLOMETER.

CCDs do not detect much IR radiation. IR array detectors have been developed (up to 2048 × 2048 pixels) based on a mixture of the elements indium antimonide (InSb, pronounced 'inns-bee') and mercury-cadmium-tellurium, pronounced 'mercatel' (HgCdTe). This was a revolutionary development – astronomers could now take IR images instead of constructing pictures point by point by scanning a single detector across the sky.

▲ ***Infrared astronomy*** *Pictured in 1998 by the HST's infrared-sensitive NICMOS instrument, Uranus shows bright spring-time clouds that bubbled up from the lower clearer atmosphere. The planetary ring system is prominently visible in infrared.*

The submillimeter domain remains behind the IR as far as detector development is concerned. The detectors at the millimeter end are partly based on radio techniques (heterodyne receivers). At the shorter-wavelength end of the submillimeter range, there are sensitive bolometers, and arrays of bolometers make a submillimeter camera. SCUBA (the Submillimetre Common User Bolometer Array) on the JAMES CLERK MAXWELL TELESCOPE is the largest currently in operation, with up to 91 pixels. Now that the technology has been proved, larger arrays are being developed.

Telescopes and techniques

In the far-IR (FIR) and most of the mid-IR (MIR) domains the atmosphere is opaque. In the other IR domains there are some partially transparent atmospheric windows. Even in the short-wavelength near-IR (NIR) domain, observations are contaminated by variable emission from atmospheric gases, particularly water vapor. Over most of the IR region, faint astronomical sources have to be detected in spite of copious local emission. Everything at a temperature of about 300 K (the ambient temperature) radiates in the IR. This includes the air, the telescope and the instrument. The most sensitive measurements are only possible from cold, high mountains, and from balloons, airplanes and satellites. Instruments and detectors have to be cooled, and in space this is practical even for telescopes. The telescope design excludes support structures from the beam (so they do not radiate into the detector), or keeps them as small as possible.

Because the background radiation is so large, IR measurements are made by comparing a place in the sky where there is a source with a place without one. 'Nodding' means making measurements at alternating positions by moving the telescope back and forth, and 'chopping' by moving a mirror in the beam at about 1 Hz. At a certain level, of course, there are no places in the sky without faint sources – you may measure the emission of a faint source by comparing it with an almost empty area of sky that in fact contains an equal or slightly brighter source. Obviously the

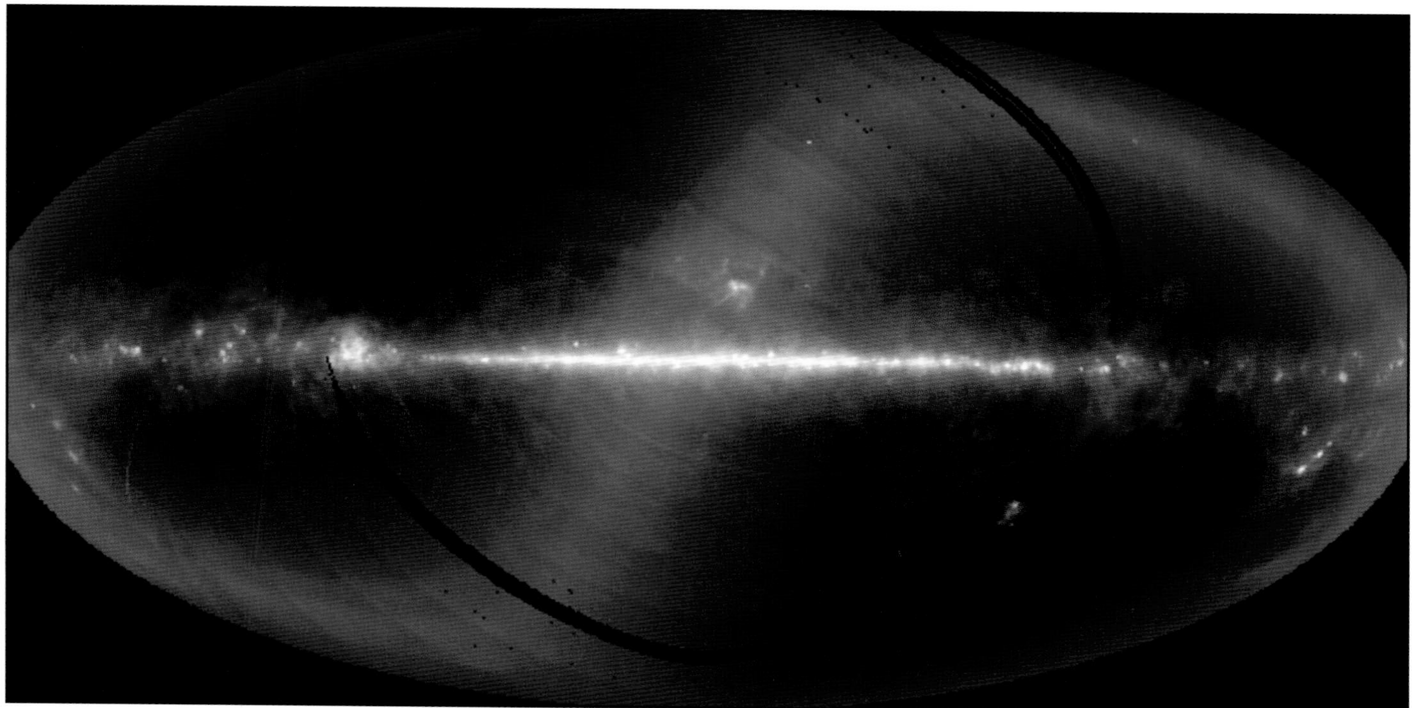

▲ ***Infrared astronomy*** *IRAS surveyed the whole sky, missing out only a few strips. This map shows the plane of the Milky Way (across the picture) and infrared cirrus (everywhere) as well as individual IR sources.*

measurement will be spurious; this is called 'confusion.'

A rare instrumental advantage for IR astronomy is that ADAPTIVE OPTICS is more effective with IR telescopes. It is possible to obtain images from several ground-based telescopes that rival the HST.

IR astronomy was developed with telescopes in the 2–4 m class, such as UKIRT near the summit of Mauna Kea in Hawaii, the IRTF and the New Technology Telescope of the ESO. It is coming to flower with the large-aperture telescopes that are now on stream with dedicated NIR and MIR facilities, like the 10-m telescope of the W M KECK OBSERVATORY in Hawaii, the VLT on Paranal in Chile (four 8-m telescopes), the multinational GEMINI OBSERVATORY twin 8.1-m telescopes in Chile and Hawaii, and the 8.3-m diameter SUBARU TELESCOPE in Hawaii.

MIR and FIR observations have been carried out from above the atmosphere using the Kuiper Airborne Observatory, which operated between 1974 and 1995 and is being followed in about 2004 by SOFIA, a telescope on a Boeing 747. In the submillimeter domain, large single-dish antennae collect the weak radiation, like the NOBEYAMA RADIO OBSERVATORY 40-m dish in Japan, the 30-m telescope of IRAM in Europe and the 15-m James Clerk Maxwell telescope in Hawaii. FIR dishes have to maintain their profile to within tens of microns and if they are big they tend to sag out of shape. The 15–40-m dishes are reaching the limits of what can be done. The near future will see the IR-optimized JAMES WEBB SPACE TELESCOPE (JWST) and interferometer arrays with a very large collecting area, like ALMA in northern Chile, as well as improved focal-plane array receivers for single-dish telescopes.

Space missions

In space, a satellite is heated by the Sun, just as the Earth is. Like the night-time side of the planet, if the satellite is in shadow, it cools. On the other hand, it may still be seeing the Earth or the Moon, which are sources of radiation and heat it to some extent. The satellite, especially the electronics, has to be kept warm (near to room temperature). The temperature control of an IR telescope and its equipment in a satellite is thus no trivial matter, and its thermal design is complicated.

The first IR satellites took coolant into space with them, and operated until it was exhausted (by inevitable escape from the recycling system). IRAS was the first IR mission, and carried liquid helium to cool the telescope to about 3 K. It made the first scans of significant areas of the sky at long wavelengths, and was principally a survey mission, covering 96% of the CELESTIAL SPHERE and detecting 350 000 IR sources.

The ISO, a follow up to IRAS, carried more helium and lasted three times as long. It pointed to specific sources and examined them in detail, contributing to the astrophysics that is described elsewhere in this encyclopedia. SIRTF will operate in the same way and for even longer.

The COSMIC BACKGROUND EXPLORER satellite had helium-cooled IR detectors for a very specific purpose, namely to measure the COSMIC MICROWAVE BACKGROUND radiation. Much effort went into calibrating the instruments to provide extraordinary precision of 0.005%.

The HST was fitted in February 1997 with the Near Infrared Camera and Multi-Object Spectrometer (NICMOS). NICMOS became inactive after two years when its coolant was depleted prematurely by a short circuit that overheated the instrument. A new cooling system was installed in mid-2002, and NICMOS was reactivated.

The James Webb Space Telescope is intended to be an IR successor to the HST. It will detect NIR radiation and be placed in orbit alongside the Earth at a quasi-stable position directly opposite the Sun, beyond the Moon. It will not rely on coolant. Instead its thermal design will include a large sunshade to shield the telescope and instruments from sunlight, moonlight and earthlight, all coming from the same direction. The telescope will look into the DARK AGES.

To help all these new IR space and ground-based telescopes choose interesting targets for study, NASA's Infrared

Processing and Analysis Center (IPAC) at the California Institute of Technology has created the Two Micron All Sky Survey (2MASS) containing more than 300 million faint IR stars, galaxies and other objects.

Further reading: McLean I S (1997) *Electronic Imaging in Astronomy: Detectors and Instrumentation* Praxis; Watt G D and Webster A S (1990) *Submillimetre Astronomy* Kluwer.

Infrared Astronomy Satellite (IRAS)

A very successful US-Dutch-UK INFRARED ASTRONOMY satellite. Launched on January 25, 1983, IRAS made the first full-sky infrared (IR) survey in four wavebands (12, 25, 60 and 100 μm) and determined the positions of galactic and extragalactic IR sources to an accuracy of 0.5 arcmin. IRAS also contained a low-resolution spectrometer. Its 0.6-m telescope and detectors were cooled by liquid helium to only 4 K. IRAS detected 350 000 objects including six new comets, a strange asteroid named (3200) PHAETHON which may be an extinct comet, and disks of dust around stars such as VEGA and BETA PICTORIS. It also discovered about 75 000 starburst galaxies, bright in the IR owing to their intense star formation, and pinpointed thousands of young stars and protostars. IRAS was also the first to find IR cirrus: warm dust associated with diffuse clouds mostly at high galactic latitudes. Operations ceased when its coolant ran out on November 21, 1983.

Infrared Imaging Surveyor

The Infrared Imaging Surveyor (IRIS), also known as ASTRO-F, is Japan's second infrared (IR) astronomy mission. IRIS is scheduled for launch in February 2004 and will be placed into a sun-synchronous polar orbit at an altitude of 750 km (*see* ARTIFICIAL SATELLITE). Working in collaboration with universities and institutes in Japan and elsewhere, IRIS will carry out an IR sky survey with much greater sensitivity than IRAS, and is expected to add significant information on many astrophysical problems. It will also provide a database for closer study by future observatory-type missions. IRIS has a 0.7-m telescope cooled to 6 K with liquid helium. It will carry two focal-plane instruments: a far-IR surveyor, which will survey the entire sky in the 50–200 μm wavelength range with angular resolutions of 30–50 arcsec; and an IR camera that will take deep images of selected sky regions in the near- and mid-IR range. The in-orbit lifetime of the coolant is expected to be 500 days, but near-IR observations can be continued once the helium runs out.

Infrared Space Observatory (ISO)

The true successor to IRAS was ESA's ISO, though its mission was to focus on specific targets rather than carry out an all-sky survey. ISO was launched into a highly elliptical orbit on November 17, 1995. It carried a 0.6-m telescope, four scientific instruments and enough helium coolant to be able to operate for 28 months. The camera, photometer and two spectrometers between them covered the full range of IR wavelengths from 2 to 200 μm. Significant early findings included the discovery that water is much more widespread throughout our Galaxy than expected. For example, it is found in the atmospheres of Saturn's moon Titan and the outer planets, in planetary nebulae and in giant molecular clouds. The satellite was also well placed to study the prodigious dust and gas emissions from COMET HALE–BOPP. ISO also revealed the existence of complex organic molecules around young stars.

Infrared spectrum

The infrared (IR) part of the ELECTROMAGNETIC SPECTRUM covers the wavelength range from about 1 mm to about 1 μm. The IR band is divided as follows:

- submillimeter (1 mm–0.3 mm);
- far-infrared, or FIR (0.3 mm–50 μm);
- mid-infrared, or MIR (50–10 μm);
- near-infrared, or NIR (10–1 μm).

At the long-wavelength end the IR connects to the radio domain and at the short end to the optical. IR for human detectors (the retina) includes radiation with wavelengths longer than 0.7 μm. *See also* OPTIR SPECTRUM.

Infrared Telescope Facility (IRTF)

A 3.0-m telescope, optimized for infrared observations, located at the summit of Mauna Kea, Hawaii. It was established in 1979 primarily to provide infrared observations to support NASA programs. The observatory is operated and managed for NASA by the University of Hawaii Institute for Astronomy, located in Honolulu. Observing time is open to the entire astronomical community, with half of the time reserved for studies of solar-system objects. Since August 2002 it has been possible to observe with the IRTF remotely.

Infrared Telescope in Space (IRTS)

The first Japanese Infrared Telescope in Space consisted of a cryogenically cooled 15-cm telescope, launched on March 17, 1995, aboard a multi-purpose space platform. It operated for a few weeks until its supply of helium was exhausted, then remained dormant until the other onboard experiments were completed and the satellite could be recovered by the Space Shuttle in January 1996.

Initial mass function (IMF)

A mathematical expression to describe the distribution of stellar masses that is obtained when stars form from clouds of gas and dust. First studied by Edwin SALPETER in 1955 it is also known as the Salpeter mass function. The IMF provides clues to the processes of stellar formation and is also important when the light observed from distant clusters or galaxies is dominated by a few of the brightest stars. Knowing the IMF can help work out the size of the iceberg when only the 'tip' can be seen.

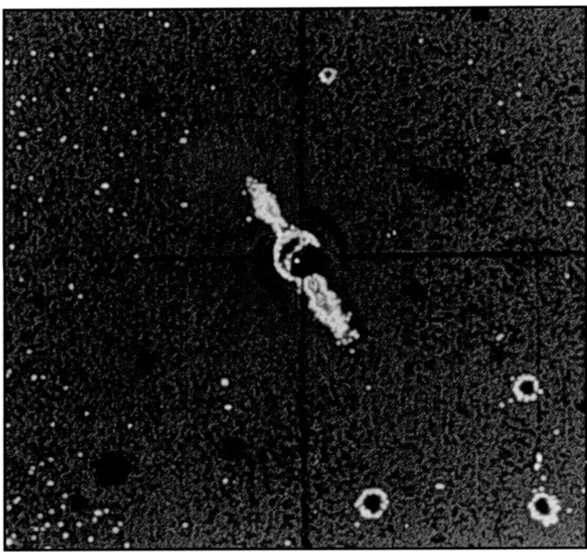

◄ ***IRAS*** *In 1983 IRAS discovered excess infrared emission from Beta Pictoris, which astronomers inferred was from dust surrounding the star. This was spectacularly confirmed in 1984 by Brad Smith and Richard Terrile, who imaged the star with an obstruction to block off the bright light and revealed a protoplanetary dust disk.*

▶ ***Interferometer***
(a) A parallel beam of radio waves from a radio galaxy is focused by a huge radio-telescope dish to a receiver. (b) Radio astronomers need the resolution of a huge telescope but cannot afford one. They can tolerate a waste of radio waves (from a smaller collecting area) and need only two small dishes separated by the diameter of the huge one. Their first design concept is not practical. (c) The radio astronomers make a radio telescope from two identical small dishes, each with its own receiver. They feed their output into a 'correlator' that combines the signals into one. But one telescope picks up the signal ahead of the other by an amount that changes as the source tracks across the sky. This causes the source to fade and brighten as one telescope interferes with the other. (d, e) The astronomers realize that the telescopes are mapping out the diameter of the huge disk as the source tracks across the sky seeing a foreshortened gap between the two telescopes. (f, g) They add more telescopes in a line with the other two, and some to form a cross, and allow the motion of the source to synthesize most of the aperture of the huge dish over time. They have built an 'aperture synthesis radio interferometer.'

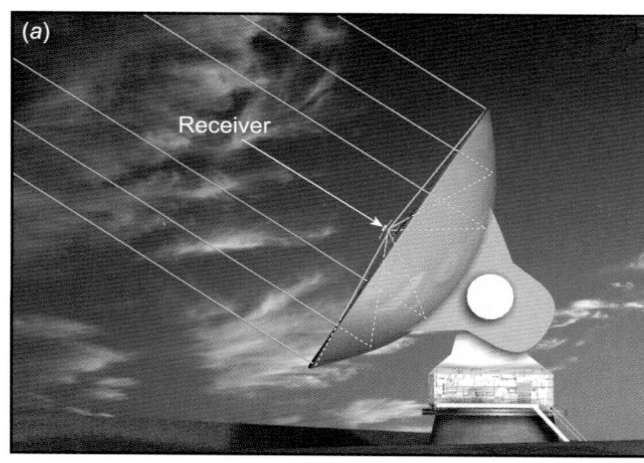

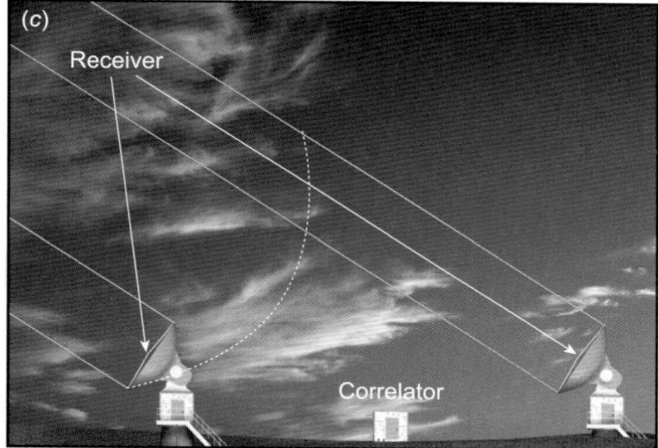

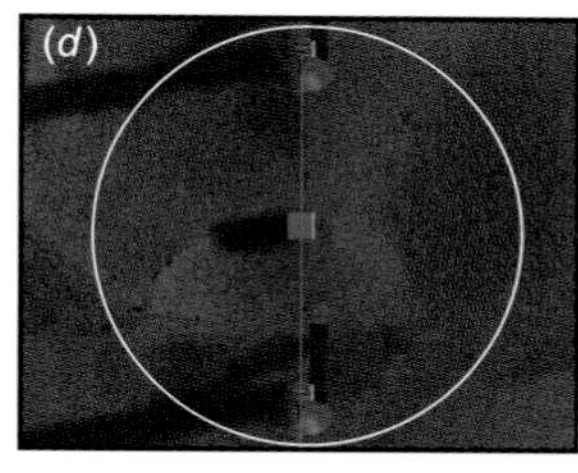

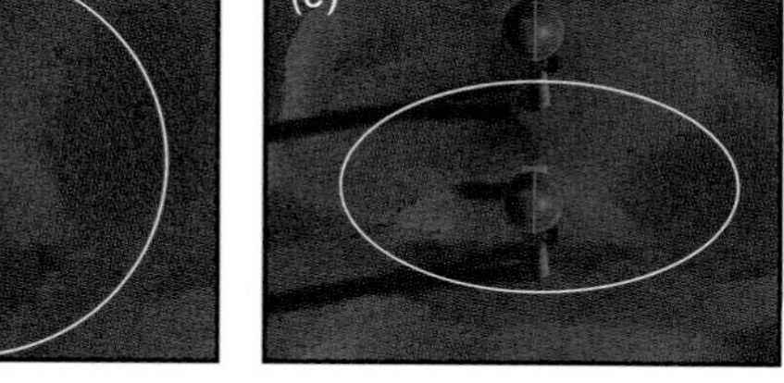

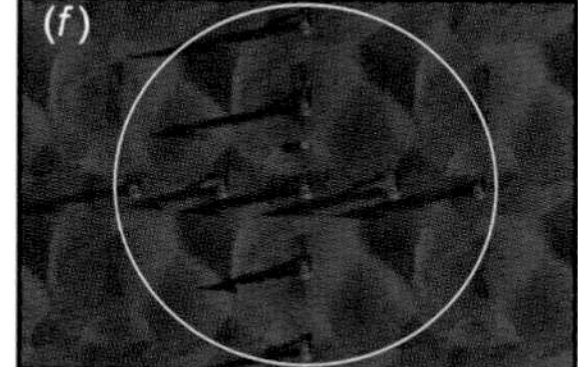

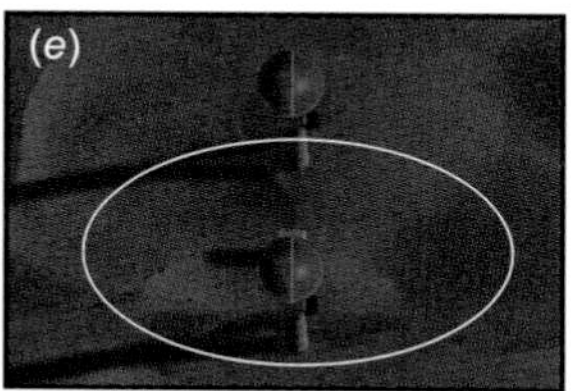

Inner planet

A collective term for the MAJOR PLANETS Mercury, Venus, Earth and Mars: those whose orbits lie closest to the Sun, inside the main ASTEROID belt, as opposed to the OUTER PLANETS, which orbit beyond it. *Compare* INFERIOR PLANET.

Instability strip

The region in the HERTZSPRUNG–RUSSELL DIAGRAM where most of the pulsating variable stars are found (*see* VARIABLE STAR: PULSATING). The strip extends upwards from the MAIN SEQUENCE at spectral type A-F through the horizontal branch and includes the location populated by the RR LYRAE and CEPHEID VARIABLE STARS. Most stars cross the strip at least once during their lifetime with the location of the crossing depending on their mass.

Institute of Millimeter-Wave Radio Astronomy (IRAM)

The headquarters of the Institute of Millimeter-Wave Radio Astronomy (Institut de Radio Astronomie Millimétrique) is in Grenoble, France. It is a German-French-Spanish collaboration founded in 1979 to study the sky at millimeter wavelengths. In this spectral range thousands of molecular emission lines allow the study of physics and chemistry in the INTERSTELLAR MEDIUM, and the coldest parts of the universe can be observed through the continuum emission of dust. The institute operates two observatories. One is a radio INTERFEROMETER, which will eventually have fifteen 15-m diameter antennae on the Plateau de Bure in the French Alps. The other is a 30-m-diameter telescope, located at an altitude of 2.9 km on the Pico Veleta in southern Spain, with a support base in Granada.

Interball

Solar-terrestrial study involving two Russian Prognoz spacecraft and two small subsatellites made in the Czech Republic. Interball was designed to study plasma processes in the Earth's magnetosphere by using two pairs of spacecraft, one above the polar aurora and one in the magnetospheric tail. The Tail Probe and its subsatellite S2-X were launched in August 1995. The Auroral Probe and S2-A were launched in August 1996.

Interference

The pattern of light and dark obtained when two beams of light combine or interfere with each other. If two waves of the same wavelength and amplitude are in phase then the crests will enhance each other to produce a bright 'fringe.' If they are exactly out of phase, the crests of one wave and the troughs of the other will cancel each other out, producing a dark fringe. These patterns are also referred to as interference fringes. Such patterns may be formed with any type of electromagnetic radiation (or with other types of waves, including water waves), and the interpretation of these interference patterns forms the basis of interferometry. See also DIFFRACTION.

Interferometer

A device that utilizes INTERFERENCE effects between two or more waves to attain higher angular resolutions than a single telescope or other device can provide. The pattern of fringes, or interference pattern, that is produced can be analyzed to provide information about the angular size of a source, the angular separation between different parts of the same source, the angular separation between two separate point sources, and the location of the source in the sky.

The interferometer was invented by Albert MICHELSON in 1881 to measure differences in the path length of light beams, and has since had a wide variety of applications. A subsequent version, the Michelson stellar interferometer, consisted of two apertures, or slits, placed in front of a large telescope. In principle, when light from a single star, or a double star, passed through the slits and recombined at the FOCUS, analyzing the resulting interference pattern allowed the angular diameter of a large star or the angular separation between the components of the binary to be measured. The RESOLVING POWER of an interferometer of this kind depends on the separation between the apertures or slits; the larger the separation the higher the resolution ($R = \lambda/2D$, where R = angular resolution in radians, λ = wavelength and D = the separation of the apertures). Later versions extended the separation between the apertures by means of mirrors placed on a beam located across, and extending beyond, the aperture of the telescope itself.

Interferometry has been used in radio astronomy for many years and the analysis of the signals received by two or more radio dishes is called APERTURE SYNTHESIS. The longest baseline so far achieved (*see* VERY LONG BASELINE INTERFEROMETRY) is between California and Australia, some 10 600 km, with which a resolving power of 0.001 arcsec has been achieved at a wavelength of 13 cm.

Because light waves are many orders of magnitude shorter than radio waves, the tolerances necessary for optical interferometry are much, much finer than those that radio astronomers have to achieve. Consequently, although optical fringes produced by two separate telescopes were first observed by French astronomer Antoine Labeyrie in 1975, it is only in recent years that optical interferometry has become a realistic research tool. The COAST system achieved optical resolutions of around 0.01 arcsec in 1995. The CHARA array of six telescopes on Mount Wilson, California, USA, began to work in 1999 and can achieve resolutions of 200 microarcsecs (the angular size of a nickel coin seen across the diameter of the Earth). The twin 10-m Keck telescopes (on Mauna Kea, Hawaii) and the four 8-m mirrors of the VLT (on Mount Paranal, Chile) achieved milliarcsecond resolutions when they became fully operational as interferometers in 2001.

Intergalactic medium

As the universe entered the DARK AGES, when the ever-fading COSMIC MICROWAVE BACKGROUND radiation cooled and became infrared and then radio wavelengths, the ionized helium and hydrogen that filled the universe recombined and became neutral atoms. The dark ages persisted until the stars and the quasars in the first galaxies had formed and generated ultraviolet (UV) radiation that reheated and re-ionized the remaining material. This material still exists in intergalactic space as the intergalactic medium. The intergalactic medium contains most of the ordinary baryonic MATTER left over from the Big Bang – the more attention-grabbing, light-emitting galaxies and quasars are considerably less important in this respect. The intergalactic medium has developed since those early times; it has become clumpy and the stars have polluted it with heavier elements. But even now, the intergalactic medium is the gas that rains onto galaxies, cooling and, to a certain extent, replenishing them (*see* GALAXY FORMATION AND EVOLUTION).

The intergalactic medium reveals itself in the spectra of QUASARS. Astronomers detect the spectra of high REDSHIFT quasars in optical light, but when this light set out on its journey it was in the quasar's UV spectrum. Quasars therefore often show in their optical spectrum the UV Lyman-alpha emission line of hydrogen, which is normally at a wavelength of 121.6 nm. It has been known since the late 1960s that the spectra of quasars also exhibit large numbers of discrete, isolated, narrow absorption lines, at first unidentified.

It was very quickly noted that there are many more absorption lines at wavelengths shorter than the quasar's Lyman-alpha emission lines than at longer wavelengths. Almost all are Lyman-alpha absorption lines produced by gaseous material that by chance intervenes along the lines of sight to a bright, distant background quasar, at redshifts less than it. The absorption arises because on its way to us the quasar's light passes through gas: the intergalactic medium. Absorption by the gas modifies the spectra of the background objects and imprints a record of the gas clouds' physical and chemical states on the spectrum of the background quasar. The whole arrangement resembles a giant cosmic slide projector, where a quasar plays the role of the light bulb and the intervening gas clouds are the slides, changing the colors of the light source by absorbing parts of the (white) spectrum.

Collectively, the absorption lines shortward of Lyman-alpha are known as the LYMAN-ALPHA FOREST, because, in a spectrum, they give the impression of a thick forest of tree trunks. These absorption lines are the only direct observational evidence we have for the intergalactic medium.

Most of the Lyman-alpha forest of absorption lines are from low-density, diffuse hydrogen clouds. Some come from denser, more compact clouds, and some of these show absorption lines from heavier elements (for example, carbon, oxygen, magnesium, silicon and iron, plus rarer elements such as titanium, chromium, nickel and zinc). This indicates that the clouds once contained stars, some of which have exploded as supernovae and caused this metal pollution. Thus, some, many or most of the denser Lyman-alpha absorption systems probably arise in the central parts of galaxies. Some of these galaxies have been detected but others have not, even though astronomers know exactly where to look. This hints at the existence of galaxies of kinds yet to be identified.

Even in some of the less dense clouds there are absorption lines from heavy elements. Rather surprisingly, there is no pronounced trend for heavy elements to become more abundant as redshift decreases. Astronomers expect the heavy elements to become more abundant with time, owing to the conversion of gas into stars. It rather looks as if most of the heavy elements were formed quite early in the history of the universe, and that star formation has been petering out since early times, affecting heavy-element abundances less than we would have thought.

The least-dense clouds of intergalactic gas have never formed into conventional galaxies. Pure Lyman-alpha forest absorption in a quasar spectrum was predicted and first detected by Jim Gunn and Bruce Peterson (1965). Their basic idea was as follows. Going back in time, an increasing fraction of the universe's total matter must be in the form of gas (before it became stars). Even if only a small fraction of the total mass of the universe was in the form of hydrogen, its Lyman-alpha lines should completely absorb a part of the spectrum of any background light source. Gunn and

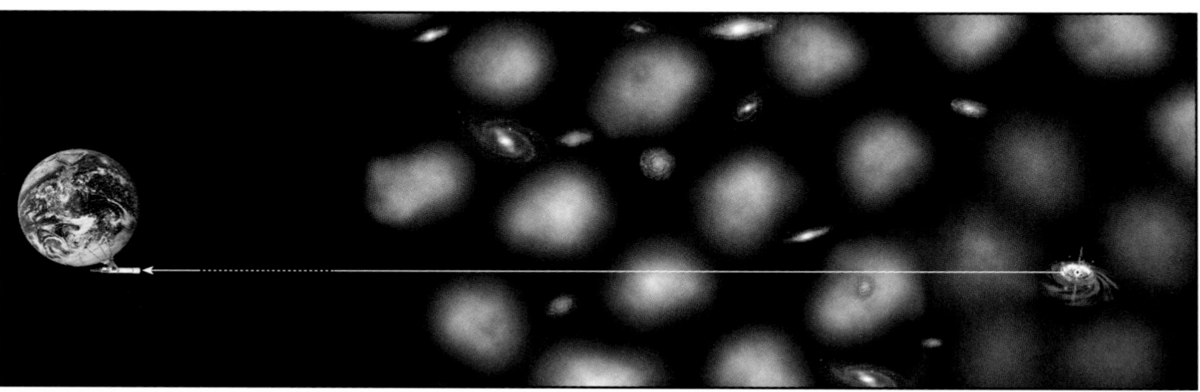

► ***Intergalactic medium*** *Light from a distant quasar penetrates intergalactic clouds (some of them centered on individual galaxies) which leave their traces on the light as the 'Lyman-alpha forest.'*

Peterson predicted not a Lyman-alpha forest but a Lyman-alpha 'wall.' This particular absorption pattern is referred to as 'the Gunn–Peterson effect.' Gunn and Peterson did detect the absorption they were looking for but it was much less pronounced than they expected. The relative weakness of the absorption could have meant that there is not much intergalactic medium. In fact, it means there is a lot of intergalactic medium, but most of the hydrogen is not in neutral form where it produces Lyman-alpha absorption. The surprise was that it is fully ionized.

This led to the discovery that it is the UV radiation from stars and, particularly, quasars that keeps most of the intergalactic medium highly ionized. Indeed, in any quasar spectrum there is a relative lack of Lyman-alpha tree trunks in the parts of the forest close to the quasar. There is a 'proximity effect' – UV radiation produced by the quasar itself dramatically reduces the neutral hydrogen close by.

The gas that gives rise to Lyman-alpha absorption systems came to be referred to as 'Lyman-alpha clouds' or 'intergalactic clouds,' to distinguish it from the gas associated with galaxies. The gas is arranged in what has been called the 'cosmic web.' It consists of relatively small sheet-like structures or pancakes of gas, that are connected by more filamentary structures to large, round clouds, which line up on the sight line to a quasar like pearls on a string. This web closely mimics the structures calculated by cosmologists modeling the development of structure in the DARK MATTER of the universe. The conclusion is that the intergalactic clouds seen in the spectra of quasars follow the same distribution in the universe as dark matter.

International Astronomical Union (IAU)

The IAU was founded in 1919, and plays a key role in promoting and coordinating worldwide cooperation in astronomy. The personal involvement of more than 8300 members from 67 countries is emphasized. The tasks of the IAU range from the definition of fundamental astronomical and physical constants and unambiguous nomenclature, the rapid dissemination of discoveries, the organization of international observing campaigns, and the promotion of educational activities in astronomy through to early informal discussions of possible international large-scale facilities. The IAU is also the sole internationally recognized authority for giving designations and names to celestial bodies and their surface features.

International Atomic Time

The international reference scale of TIME based on the continuous intercomparison of numerous atomic clocks in laboratories and institutions around the world. Since 1971 international atomic time, denoted by the abbreviation TAI, has been provided by the Bureau International de l'Heure (International Time Bureau) in Paris, France.

International Cometary Explorer (ICE)

See INTERNATIONAL SUN–EARTH EXPLORER.

International Dark-Sky Association

See practical astronomy feature, p. 114.

International Gamma-Ray Astrophysics Laboratory (INTEGRAL)

INTEGRAL, launched in October 2002, is an ESA GAMMA-RAY ASTRONOMY mission. There are four scientific instruments: a gamma-ray imager, gamma-ray spectrometer (20 keV to 8 MeV), X-ray monitor (3–35 keV) and optical camera, all of which observe the same region of sky simultaneously. These locate gamma-ray sources to within a few arcminutes and measure their radiation energy with unprecedented accuracy. During its first months of operation, INTEGRAL detected gamma-ray bursts at a rate of nearly one per day, with about one per month occurring near the centre of its field of view. The four instruments help to locate these short-lived events and enable follow-up observations to be made by other observers. The INTEGRAL Science Data Centre (ISDC) is located in Versoix, Switzerland, at the Geneva Observatory.

International Space Station (ISS)

The construction of the International Space Station began with the launch of the first element, the Russian Zarya Control Module, in November 1998. Construction will require 43 space flights and completion is scheduled for around 2010, provided difficulties caused by the Columbia Space Shuttle disaster can be overcome. When complete it will be larger than a five-bedroom house and will measure 100 m from end to end. It orbits at an average altitude of 354 km at an inclination of 51.6° to the equator. The ISS draws on the expertise of 16 cooperating nations. In 1988 US president Ronald Reagan gave the station its original name, Freedom, but it was renamed following financial difficulties and the involvement of other nations. The plans to use it for scientific research, including astronomy, are increasingly peripheral.

International Sun-Earth Explorer (ISEE)

A series of three US satellites, all members of NASA's EXPLORER series, designed to study the SOLAR WIND and its interaction with the Earth's MAGNETOSPHERE. The coordinated spacecraft allowed a three-dimensional interpretation of this interaction. ISEE-1 (Explorer-56) and ISEE-2 were launched together in October 1977 and placed into highly elliptical Earth orbits. ISEE-3 (Explorer-59) was launched in August 1978 and placed in a halo orbit at the L_1 LAGRANGIAN POINT between the Sun and the Earth. It gave advance warning of solar storms heading toward Earth. After completing its original mission, ISEE-3 was reactivated and diverted to pass through the tail of Comet Giacobini–Zinner on September 11, 1985, becoming the first spacecraft to encounter a comet. For this part of its mission it was renamed the International Cometary Explorer (ICE). ICE also observed COMET HALLEY from a distance of 28 million km in March 1986.

International Ultraviolet Explorer (IUE)

NASA-ESA-UK ultraviolet (UV) space observatory (NASA EXPLORER-57), launched in January 1978. IUE operated for 24 hours a day from geosynchronous orbit. Short- and long-wavelength spectrographic cameras covered UV wavelengths from about 1200 to 3400 Å. IUE observed planets, comets, novae, supernovae and galaxies. In July 1994 it observed the collision of Comet Shoemaker–Levy 9 with Jupiter. IUE confirmed the nature of the precursor star that exploded as supernova 1987a, and revealed the chemical elements present in the star Nova Cygni 1978. It provided a Final Data Archive containing more than 100 000 spectra. IUE ceased to operate in September 1996.

Interplanetary dust

For more than two decades NASA has collected samples of interplanetary dust in the Earth's stratosphere using a modified U-2 aircraft, the ER-2. These tiny particles include samples of stardust and molecular-cloud material suggesting that they have remained essentially unchanged since the PROTOSOLAR NEBULA formed 4.5 billion years ago. Some of this dust is the debris that survived the formation of the larger solar-system objects. *See also* MICROMETEORITE.

Interstellar dust

A component of the INTERSTELLAR MEDIUM in the region between a galaxy's stars. It consists of small solid grains (0.01–0.1 μm in size) believed to be of carbon, silicates and iron with mantles of water, ammonia and/or carbon dioxide ices. The dust, thought to be about 1% of the mass of the gas in the MILKY WAY GALAXY, causes the dimming, reddening and POLARIZATION of starlight.

Interstellar medium

The material in the region between the stars within the MILKY WAY GALAXY. It constitutes about 10% of the mass of the Galaxy. It is mostly concentrated in a thin layer in the galactic plane and tends to be concentrated in the spiral arms. Its constituents include clouds of ionized hydrogen (H II REGIONS), clouds of neutral hydrogen (H I REGIONS) and regions of hot gas inferred by the presence of a diffuse background of X-ray emission and ultraviolet absorption lines. There are also very cool (10 K), dense clouds of molecular hydrogen and other molecules. It also contains INTERSTELLAR DUST.

Io

Io is the innermost of the four large Galilean satellites of JUPITER discovered by GALILEO GALILEI in 1610. Io's mean radius (1821 km) and bulk density (3.53 g cm^{-3}) are comparable to the Moon's. However, long before the VOYAGER spacecraft encounters, it was apparent from Earth-based observations that Io is very different from all other solar-system bodies: it has an unusual color and anomalous thermal properties, and is surrounded by immense clouds of ions and neutral atoms. During its fly-by of Io in 1979, the Voyager 1 spacecraft discovered active volcanoes which have been proved to be high-temperature eruptions of silicate lava.

Ion

An atom or molecule that has a net electrical charge because it has lost, or gained, one or more electrons relative to the normal complement. An ordinary, neutral atom normally has as many electrons as it has positively charged protons in its nucleus. A positive ion has fewer electrons (a net positive charge) than a neutral atom and a negative ion has more (a net negative charge).

Ionization

The process whereby atoms lose one or more electrons to become positively charged ions. Ionization occurs if an atom absorbs a sufficiently energetic photon or if it suffers a sufficiently violent collision.

The minimum energy needed to remove an electron to infinity from the ground state of a given atom is called the ionization potential, or ionization energy. For a HYDROGEN atom, with its single electron in the ground state (the lowest energy level of the atom), the ionization potential is 13.6 eV. If the electron is at a higher (excited) level, less energy is needed. An atom that has lost one electron is identified by the Roman numeral II (the neutral atom, by I) or by the superscript +. For example, neutral hydrogen is denoted by HI and ionized hydrogen by HII or H^+. Doubly ionized helium (helium that has lost both electrons) is denoted by He III or He^{++}. In many astrophysical situations very high ionization stages are encountered, and the Roman convention is normally used; for example, Fe XVI represents an iron atom that has been stripped of 15 out of its total complement

▼ ***ISS*** *The space station will be built in 43 space flights and is scheduled to be completed in about 2010.*

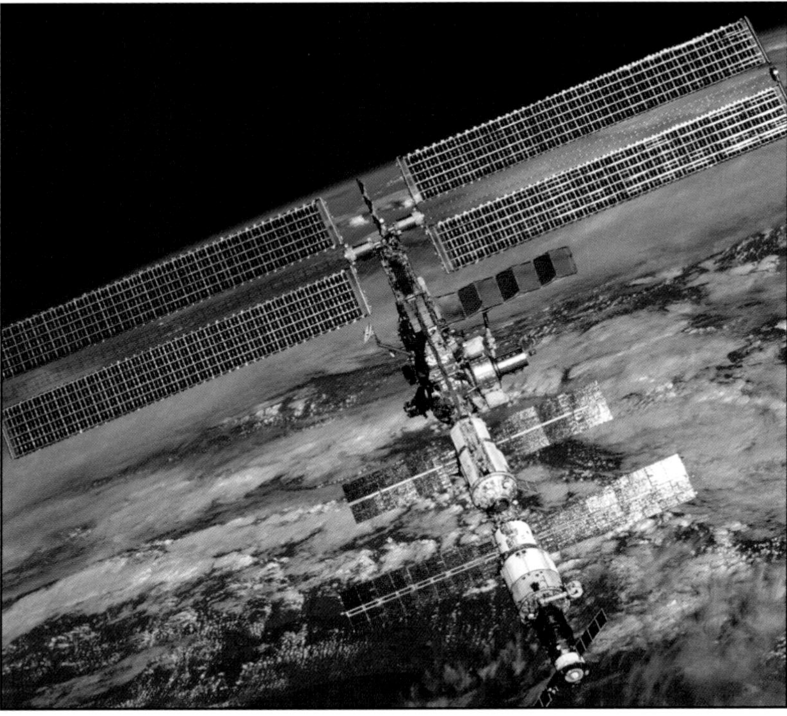

▲ ***Isaac Newton Group of Telescopes*** *The three telescopes show on the skyline of La Palma, Canary Islands, with the 'beehives' housing the HEGRA gamma-ray detectors in the foreground.*

of 26 electrons. Ions that retain at least one bound electron can absorb or emit radiation, so producing spectral lines (emission and absorption lines) that differ in wavelength from those produced by neutral atoms (*see* ABSORPTION SPECTRUM; EMISSION SPECTRUM). Photons are also emitted when an ion captures, or recaptures, an electron ('recombination').

Ionized gases are encountered in a wide variety of astrophysical situations, including the interiors and atmospheres of stars, luminous nebulae and high-temperature intergalactic gas.

Ionosphere

Most planets and many satellites in our solar system are surrounded by ATMOSPHERES. The interaction of radiation and charged particles from space with these gases produces weak IONIZATION that creates a layer of PLASMA embedded within the denser atmosphere. This is an ionosphere. In the inner solar system, the sources of energy are the SOLAR WIND plus X-rays and ultraviolet (UV) light from the Sun. Additional sources of ionization that are relatively more important in the outer solar system include the resonance scattering of solar radiation by interplanetary atomic hydrogen, COSMIC RAYS and UV starlight from the southern Milky Way and the Orion region.

The Earth's ionosphere was discovered using reflected radio waves by Edward Appleton (1892–1965), who later won the Nobel Prize for Physics in 1947. Appleton labeled it the 'E-layer' (height 95–135 km), and named additional layers the D-layer (60–95 km) and the F-layer (135–200 km or more).

Ion tail *See* COMET; GAS TAIL

Iron meteorite

A METEORITE composed mainly of nickel-iron, with traces of other metals; formerly known as a SIDERITE. Iron meteorites account for more than 6% of all meteorite specimens. They are the easiest type to identify, being heavy, magnetic and rust-colored; their metallic sheen tarnishes quickly on the Earth's surface, but otherwise irons resist weathering more than other meteorites. Iron meteorites are believed to have originated from the cores of asteroidal parent bodies that differentiated before being shattered by impact (*see* ASTEROID; DIFFERENTIATION). They are divided into three main classes according to the crystal structure and nickel content of the nickel–iron: hexahedrites (4–6% nickel), octahedrites (6–12%) and ataxites (more than 12%). The largest known meteorites are irons. The most massive are the HOBA METEORITE (Namibia, ataxite, 60 tonnes) and CAPE YORK METEORITE (Greenland, octahedrite, fragments total 58 tonnes).

Irregular galaxy

A galaxy with an irregular appearance and no well-defined shape or structure. Irregular galaxies can be divided into two broad classes. Irr I galaxies contain many OB ASSOCIATIONS and H II REGIONS and significant quantities of gas (typically about 20% of the total mass). Some contain hints of incipient spiral structure. Irr II galaxies are amorphous and often contain substantial amounts of gas and dust. Some have distorted shapes that appear to have been caused by collisions, or close encounters, with other galaxies or by violent internal activity. *See also* GALAXY.

Isaac Newton Group of Telescopes (ING)

The largest optical telescope for astronomy in western Europe stands on the rim of an extinct volcano in the ROQUE DE LOS MUCHACHOS OBSERVATORY at an altitude of around 2350 m on the island of La Palma, Canary Islands. The ING consists of the 4.2-m William Herschel Telescope, the 2.5-m Isaac Newton Telescope and the 1.0-m Jacobus Kapteyn Telescope. Its first scheduled use by astronomers was in May 1984, and thousands of research projects have been carried out. Many ground-breaking achievements have been made, such as the first detection of the optical afterglow of GRBs, the discovery of BLACK HOLES in binary star systems, and the study of supernova explosions in distant galaxies. The ING also operates an instrument-development program, which is vitally important for world-class astronomical research.

Isotope

Any of two or more forms of one chemical element whose atomic nuclei have the same number of protons (and therefore have the same atomic number) but different numbers of neutrons (and therefore have different mass numbers). For example, oxygen has three stable isotopes, oxygen-16, oxygen-17 and oxygen-18, denoted symbolically by ${}^{16}_{8}O$, ${}^{17}_{8}O$, and ${}^{18}_{8}O$ (the subscript denotes atomic number, and the superscript mass number). Here each nucleus contains eight protons, together with eight, nine and ten neutrons.

Isotropic, isotropy

The same in all directions.

◀ **James Clerk Maxwell Telescope** *The JCMT is shown at dusk, illuminated from within by high-intensity sodium lamps. The telescope itself is hidden in its building behind a large, translucent sheet of material, which serves both as a windbreak and as a parasol.*

James Clerk Maxwell Telescope (JCMT)

The 15-m-diameter James Clerk Maxwell Telescope is the largest facility in the world designed specifically to operate in the submillimeter region of the spectrum. It became operational in 1987, and is situated close to the summit of Mauna Kea, Hawaii, at an altitude of 4.092 km. On July 11, 1991, it became the largest telescope ever to observe a total solar eclipse. Its performance has been enhanced by the addition of SCUBA (the Submillimeter Common-User Bolometer Array), a camera and a photometer for MILLIMETER AND SUBMILLIMETER ASTRONOMY. SCUBA, cooled to 0.1 K, has two arrays of bolometric detectors: one operates in the 750 and 850 μm atmospheric transmission windows, while the other is used for 350 and 450 μm. SCUBA, which was designed and constructed by the ROYAL OBSERVATORY, EDINBURGH in collaboration with Queen Mary and Westfield College, London, UK, began observations in May 1997. Its discoveries include a ring of dust particles around Epsilon Eridani; a distant galaxy hidden in dust; and possible new solar systems in formation around VEGA and Fomalhaut.

James Webb Space Telescope (JWST)

The James Webb Space Telescope, formerly the Next Generation Space Telescope, will succeed the HST as a general purpose space observatory. It will be a 6.5-m infrared-sensitive telescope, optimized for the waveband from 0.6–28 μm, operating for 10 years near the Earth-Sun second LAGRANGIAN POINT (L_2), 1.5 million km away from the Sun. It will be operated by the SPACE TELESCOPE SCIENCE INSTITUTE for NASA, ESA and the Canadian Space Agency. Its segmented mirror will unfold from its stowed position within an Ariane 5 launch vehicle. The telesope will be cooled by radiating the heat that it gains from the Sun, Earth and Moon into space, without the consumption of coolant . It is currently under design, with a planned launch of 2011.

JWST will look into the DARK AGES to observe the first generations of stars and galaxies, including supernovae out to REDSHIFTS of 5–20.

James Webb (1906–92) was the administrator of NASA from 1961 to 1968, during the Apollo program.

Further reading: Stockman, Hervey S (ed) (1997) *Visiting a Time When Galaxies Were Young*, AURA.

Jansky, Karl Guthe (1905–50)

American radio engineer and the founder of RADIO ASTRONOMY. Jansky was employed by Bell Telephone Laboratories to reduce noise ('static') in shortwave transatlantic radio-telephony. Using a crudely directional, rotating shortwave receiver array operating at 20.5 MHz and 45 kHz, he noticed a continuous but varying interference. Originally it was thought this came from the Sun, but the time of the peak shifted over the course of the year, and a partial solar eclipse in 1932 did not reduce the signal. Jansky realized that the signal was associated with the Milky Way, particularly the direction of Sagittarius, where the Milky Way was brightest. The discovery was widely publicized, appearing in the *New York Times* of May 5, 1933. Because this static was outside the control of radio engineering, he moved on to other work. In 1973 the radio-astronomical unit of FLUX DENSITY (10^{-26} W m^{-2} Hz^{-1}) was named the 'jansky' (Jy) in his honor.

Janssen, Pierre Jules César (1824–1907)

French solar astronomer. Janssen founded and became head of the Astrophysical Observatory at Meudon, and observed the bright line spectrum of a solar prominence in the total eclipse of 1868, discovering an unknown spectral emission

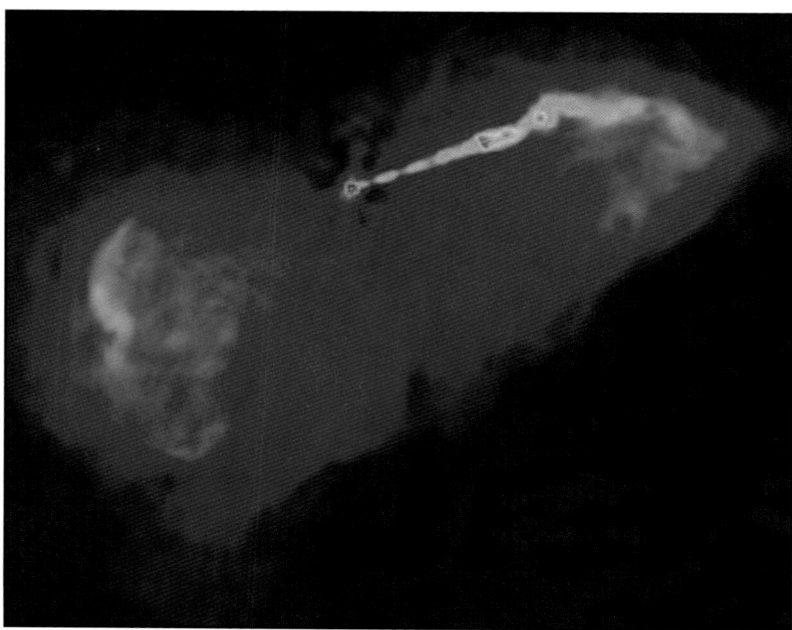

▲ ***Jets*** *This radio image of M87 shows giant bubble-like structures where radio emission is thought to be powered by jets of subatomic particles from the galaxy's massive central object.*

proved by William Ramsay (1852–1916) to be the new element helium. The brightness of the emission lines suggested to him that they should be visible even outside an eclipse, and so they were (*see* Norman LOCKYER). Janssen also photographed the Sun (publishing in 1904 his *Atlas de photographies solaires* – Atlas of Solar Photographs), and discovered solar GRANULATION (*see* James NASMYTH), as well as establishing a solar observatory on Mont Blanc.

Jeans, James Hopwood (1877–1946)

English astrophysicist. Jeans worked at Cambridge University, as well as in the USA at Princeton, New Jersey, and MOUNT WILSON OBSERVATORY in California. He retired early to devote himself to research. Jeans worked on physical topics such as thermodynamics, applying the physics to astronomy, about which he wrote lucid accounts in a number of books. He was a brilliant broadcaster and writer of popular astronomy, including the best-selling *The Universe Around Us* (1929), published when the scale of the expanding universe was becoming clear. He proposed a tidal theory for the creation of the solar system, based on the hypothesis of a close approach of a star to the Sun, which drew out a cigar of material from which the planets condensed. The 'Jeans length' is the characteristic size of a cloud of gas that will collapse under its own gravitational attraction into, say, a star or a cluster of stars.

Jeffreys, Harold (1891–1989)

English geophysicist and mathematician. He became Plumian professor at Cambridge University, worked in many areas of science including geophysics, where he studied earthquakes, and was the first to conclude from the waves they initiated that the Earth has a liquid core. In astronomy he studied the outer planets, proposing models for their structures. He also studied the origin of the solar system.

Jet

There are many circumstances in astronomy where gas circulates around a central gravitating mass. This happens in BINARY STARS, where one component is a compact object (a white dwarf, neutron star or black hole) and mass is transferred from the normal stellar companion. It occurs in AGN, where the gravitating object is a massive black hole, and also happens during the earliest phases in the evolution of a PROTOSTAR.

The orbiting gas flows are called ACCRETION disks. The gas flows in toward the central object, which cannot swallow gas as fast as it is being delivered. It rejects some of the flow, so inflowing accretion disks are frequently accompanied by rapid outflows, launched in antiparallel directions, along the rotation axes of the disks. The outflows are called 'jets.'

Heber CURTIS discovered the first example of an astrophysical jet in 1918 using the LICK OBSERVATORY in California, USA. He observed the galaxy M87 in the Virgo cluster and found a curious straight ray, apparently connected with the nucleus by a thin line of matter.

What we observe directly in M87 is a fairly straight feature, some 6000 l.y. in length. It can be seen from long radio wavelengths to high-energy X-rays and exhibits about eight regions of high intensity ('knots') along its length. The jet can be traced as close as 0.01 l.y. to the central source, less than 100 times the radius of the 3 billion solar mass black hole that has been found to lie in the galaxy's nucleus.

Gas is flowing away from the black hole in M87. VLBI provides images of the knots as they move out. Some parts of the jet appear to be moving outward as much as six times faster than the speed of light. This phenomenon is called SUPERLUMINAL MOTION. However, there is no violation of the theory of SPECIAL RELATIVITY (which stipulates that all material motion occurs more slowly than the speed of light). Superluminal motion is only an illusion, and the gas is moving at less than the speed of light, but toward us. Unless the material moved at near-light speeds it would not exceed the escape velocity from the zone just outside the event horizon around the central black hole. Because they move at nearly the speed of light, jets like M87's are often styled 'relativistic.'

There is only one jet seen in M87, but its antiparallel counter-jet has not been detected. This is quite common. A given knot is an isotropic source of radiation (it radiates in all directions equally) but when it moves at near-light speed, effects of relativity beam its emission along its direction of motion. Consequently, if the jet is approaching us, it will appear to be very bright, and vice versa. This 'relativistic beaming effect' explains why we detect only one jet in M87 and why brighter jets exhibit superluminal motion – they are the jets that are beamed toward us.

Some accreting stellar-sized black holes show jets. The star SS433 has two antiparallel jets moving at just over one-quarter the speed of light. Some stellar sources are small versions of extragalactic radio sources. They exhibit giant outbursts, superluminal expansion and beaming, and are sometimes called 'microquasars' as a result.

Jets are also associated with some new stars. Such stars have an orbiting disk made of the gas that has not yet settled onto the star (or condensed into planets). Their jets move at several hundred kilometers per second, which is comparable to the escape velocity from the central protostar. The jets may collide with surrounding gas clouds to produce bow shocks. Fragments of the excited shocked material are known as Herbig–Haro objects (*see* Guillermo HARO; George HERBIG).

Jet Propulsion Laboratory (JPL)

Located in Pasadena, California, USA, JPL is a research, development and flight center operated for NASA by the

California Institute of Technology (Caltech). Its primary role is the exploration of the solar system using robotic scientific spacecraft. JPL is also responsible to NASA for supporting research and advanced development related to flight projects.

JPL was founded in the 1930s when Caltech professor Theodore von Karman (1881–1963) conducted pioneering research into rocket propulsion. From 1944 it performed research for the US Army, and in November 1957 was chosen to design and build the first American satellite. Data from the 14-kg EXPLORER 1 led to the discovery of the VAN ALLEN BELTS.

On December 3, 1958, JPL was transferred to NASA. It has since participated in most of the agency's planetary missions, including the Ranger and Surveyor lunar projects, the Mariner series, the Viking missions to Mars, the Voyager missions to the outer solar system, the Magellan Venus radar mapper, Galileo and Cassini. Current projects include the series of Mars Surveyor missions and proposals to send spacecraft to Europa and Pluto. JPL was also the US manager of the IRAS project to map the infrared sky. In addition, JPL conducts studies of the Earth and its environment through satellites such as the Solar Mesosphere Explorer and Topex-Poseidon. A number of JPL experiments have been flown in the Space Shuttle's payload bay, including several imaging radar missions. The laboratory also designs and operates NASA's Deep Space Network of antennas, which is used to communicate with lunar and interplanetary spacecraft.

Jewel Box

The brilliant OPEN CLUSTER NGC 4755 in the southern constellation Crux.

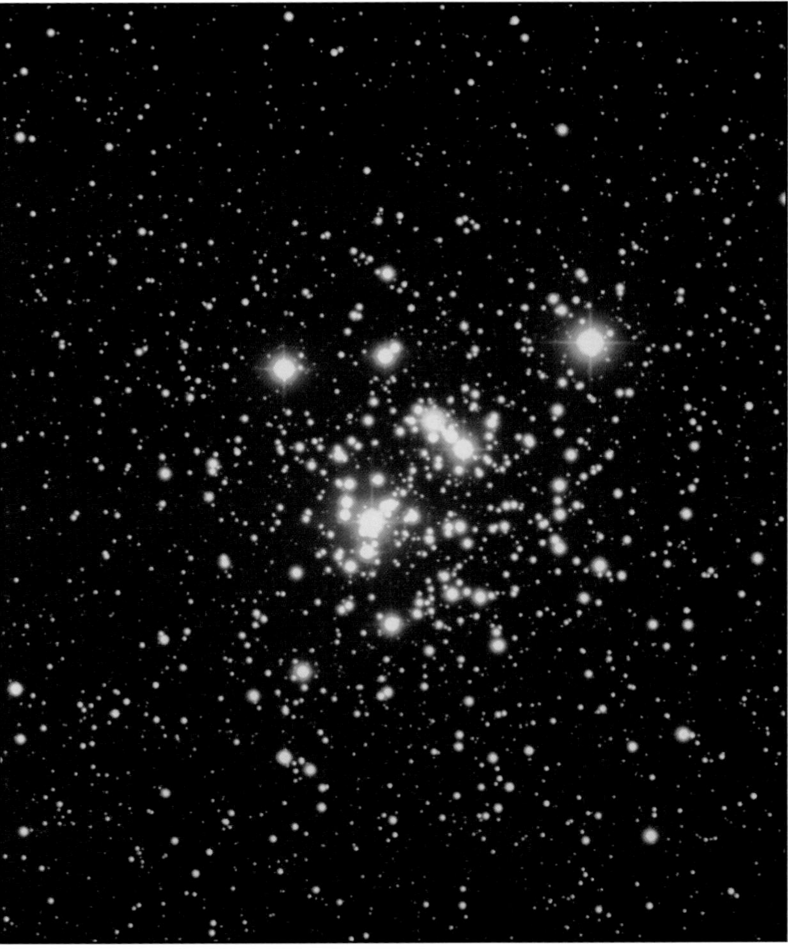

▲ ***Jewel Box***
NGC 4755 is an open cluster of stars likened by John Herschel to brilliantly colored jewelry on black velvet.

Jodrell Bank Observatory

The Jodrell Bank Observatory is part of the University of Manchester, England, and was founded by Alfred LOVELL in December 1945. Its prime instrument, the 76-m MK1 RADIO TELESCOPE, was completed in 1957. It was given a major upgrade in 1971 and is now known as the Lovell Telescope. In its early years it pioneered the technique of VLBI, which led to the discovery of QUASARS. A major use now is in the study of pulsars (*see* NEUTRON STAR AND PULSAR). In 1964 a second 26–32 m telescope, the MK2, was completed. Given a new surface in 1987, it is now used as one element of the MERLIN/VLBI NATIONAL FACILITY. Observations by the Lovell and MK2 telescopes led to the discovery of the first gravitational lens, the double quasar (0957+561), and their studies continue to form a major part of the observatory's research efforts. It also operates a number of telescopes, including the Very Small Array, on Mount Teide, Tenerife in Spain, to study fluctuations in the COSMIC MICROWAVE BACKGROUND, and is building receivers for the PLANCK SURVEYOR spacecraft to extend this work in the future. A further upgrade finished in 2003 saw the complete replacement of the reflecting surface of the Lovell telescope, and the installation of a high-precision drive. Together they allow it to operate over a range of frequencies that is four times greater.

Johnson, Harold Lester (1921–80)

American astronomical photometrist. Johnson had a peripatetic career at observatories such as Lick, Lowell, Washburn, Yerkes, McDonald and Tucson in the US, and at Tonanzintla and the National Observatory in Mexico. At Lick he used the knowledge of electronics he acquired in wartime service to develop a photomultiplier, and established that the brightnesses of the standard stars then used to determine the distances of galaxies gave results that were out by a factor of two. He worked with William MORGAN to establish photoelectrically measured standards, now known as the Johnson–Morgan, or the U, B, V system (*see* MAGNITUDE AND PHOTOMETRY). This became the international standard of stellar (and later galactic) photometry, which is used to construct color-magnitude diagrams of clusters and galaxies (*see* HERTZSPRUNG–RUSSELL DIAGRAM). Johnson demonstrated how star clusters of different ages have different color-magnitude diagrams, and, in particular, how the turnoff point (where stars begin to depart from the ZERO-AGE MAIN SEQUENCE) can be used to estimate fairly precisely the clusters' ages. Johnson's design for the stellar photometer became the benchmark and it was replicated worldwide. In Texas, Johnson worked with Frank Low to extend the U, B, V system to the longer wavelengths of the near-infrared, the R, I, J, K, L and N bands. He thus produced absolute energy curves for all sorts of important categories of stars.

Joint Institute for VLBI in Europe (JIVE)

Established in 1993 and located in Dwingeloo, the Netherlands, JIVE has played a leading role in the construction and development of a new (Mark IV) generation of VLBI correlator and operates a purpose-built data processor supporting the data analysis of the European VLBI Network (EVN).

Jovian

Pertaining to the planet JUPITER.

Joy, Alfred Harrison (1882–1973)

Astrophysicist who was a staff member of MOUNT WILSON OBSERVATORY in California, USA, for nearly 60 years, starting in 1915. Joy worked on spectroscopy and determining the radial velocities of stars. He used the results to find the absolute magnitudes of CEPHEID VARIABLE STARS (as a contribution to the determination of the intergalactic distance scale), the distance of the GALACTIC CENTER, and the Sun's rate of revolution about it. During the observational study of dark clouds, he identified a population of irregular variable stars with chromospheric emission lines; these were named T TAURI STARS after the bright prototype.

Julian date

Also known as the Julian period, the number of days, including fractions, which have elapsed since noon on January 1, 4713 BC. Thus 1800 hours UT on March 1, 1980, corresponds to JD 2 4444 3000.25. This is the system of dating that is most convenient when analyzing the frequency with which events occur over long periods of time, for example. The Julian period was proposed by Joseph Justus Scaliger (1540–1609) in 1583 and named by him after his father, Julius Caesar Scaliger (1484–1558). *See also* CALENDAR; UNIVERSAL TIME.

Juno

The third ASTEROID to be discovered, by Karl Harding in 1804, and so designated (3) Juno. Its diameter is 268 km, and it orbits the Sun in the main asteroid belt at a mean distance of 2.67 AU in a period of 4.36 years. Juno rotates in 7.21 hours. It is an S-type asteroid, with a reflectance spectrum indicating compositional similarities to STONY-IRON METEORITES and ordinary CHONDRITES; the ALBEDO is quite high at 0.24.

Jupiter and its satellites

At 5.2 AU from the Sun, Jupiter is the closest and the most massive (318 terrestrial masses) of the GIANT PLANETS. It has a ring system (*see* PLANETARY RING) and 48 satellites (confirmed, out of an estimated 100 or more). This is the most of any planet, and includes the four main ones known as the Galilean satellites, discovered by GALILEO GALILEI in 1610.

Jupiter	
Semimajor axis of the orbit	5.202 AU
Orbital eccentricity	0.048
Inclination to ecliptic	1° 18′ 28″
Sidereal period	11 years 314.84 days
Mean orbital velocity	13.06 km s^{-1}
Equatorial diameter	142 796 km
Equatorial diameter relative to the Earth	11.19
Polar diameter	133 540 km
Flattening	0.062
Mass relative to the Earth	317.9
Mean density	1.31 g cm^{-3}
Surface gravity (pressure = 1 bar)	24.8 m s^{-2}
Surface gravity relative to the Earth	2.64
Escape velocity	59.64 km s^{-1}
Rotation period	9 hours 55 min 30 s
Inclination of equator versus orbital plane	3° 4 arcmin
Albedo	0.45
Atmosphere	hydrogen (89%) helium (11%) methane (0.2%)

Jupiter has a low density, and is principally made of hydrogen and helium. It rotates quickly (its 'day' is under 10 hours). This rapid rotation causes its atmospheric system of dark 'belts' and white 'zones,' parallel to the equator. Their features have been monitored continually for over 300 years. The first atmospheric gases identified were methane and ammonia (discovered by Rupert Wildt in 1932), but the most abundant gas, hydrogen, was only suspected theoretically until its identification in 1960, and helium remained undetected until the space era.

Jupiter, with its great mass, is second only to the Sun in controlling the passage of bodies in the solar system. Jupiter captured and broke up COMET SHOEMAKER–LEVY 9, which in July 1994 collided with the planet, proving by example that cometary material is still being added to planets. Jupiter both brings incoming comets in a rain down from the outer reaches of the solar system (*see* OORT CLOUD AND KUIPER BELT) and intercepts some of them, sheltering the Earth.

Jupiter is a strong radio source. Radio emission at decimetric and decametric wavelengths (10 cm to 10 m) arises in the planet's strong magnetic field, causing AURORAE.

The PIONEER MISSIONS in 1973 and 1974 and the two VOYAGER spacecraft in 1979 were the first space investigations of Jupiter. The GALILEO MISSION TO JUPITER, launched in 1989, dropped a probe into Jupiter's atmosphere and, as an orbiter, explored the Jupiter system until 2003. The CASSINI/HUYGENS MISSION observed the planet from October 2000 to March 2001 while on its way to Saturn.

Atmosphere

The Galileo probe determined the properties of Jupiter's atmosphere below the cloud tops, at pressure levels up to 20 bar. The temperature of Jupiter's atmosphere decreases because sunlight is progressively absorbed. It reaches a minimum of 110 K at the tropopause, where the pressure is about 100 mbar. The temperature then rises toward the interior. Some of the atmospheric gases condense in the upper troposphere: ammonia condenses at around 0.5 bar and is responsible for the white color of Jupiter's clouds. Calculations suggest that at the center of the planet the pressure is 50 million–100 million bar and the temperature is 23 000 K. The heat comes from progressive gravitational settling of the core. Jupiter's interior consists of a solid nucleus of about 10–15 terrestrial masses, probably made of rocks and metals at the center, surrounded by a convective ocean of metallic hydrogen, where the planet's magnetic field is generated. Above the metallic-hydrogen ocean lies an envelope of molecular hydrogen.

Jupiter radiates into space twice the heat that it receives from the Sun. Decades of painstaking telescopic observations by amateur and professional astronomers gave a first general description of Jupiter's wind system. Large-zonal winds are permanently present and roughly correlate with the banded structure of the clouds. Jets alternate in prograde and retrograde direction, producing the planet's apparently differential rotation. We view Jupiter from outside and it has been practical to express the longitude of features with respect to the average rotation of the cloud tops, not with respect to the body of Jupiter itself. The cloud tops at the equator make a full rotation in 9 hours 50 min 30 s (defining system I of longitude) while at the mid-latitudes they rotate in 9 hours 55 min 40 s (system II).

The modern system III of latitudes is linked to the periodicity of the radio emission of Jupiter, which is itself related to the planet's magnetic field, and is more likely to express latitudes realistically. In system III, the sidereal period of rotation is 9 hours 55 min 30 s, and the winds, measured with reference to this system of longitude, reach about 120 m s^{-1} at the cloud tops at the equator (prograde).

Spacecraft observations have shown that the zones are colder and cloudier than the belts. Zones are rising gases, while the belts, which are drier, are descending. North and south of each zone, CORIOLIS FORCES create winds of opposite direction, generating eddies. The biggest structure (20 000 km wide) is the GREAT RED SPOT, which is a giant anticyclone (a rotating column of ascending motion). Although the white color of the zones can be reasonably interpreted as ammonia ice, the composition of the clouds and/or gases present in the eddies is still a puzzle, especially in the case of the Great Red Spot. Its red color suggests the presence of unidentified phosphorus or sulfur compounds.

The clouds inside the Great Red Spot, observed by the Galileo orbiter, spiral upward to the center. The center rotates slower than the boundary, or even counter-rotates. The theory of the Great Red Spot is incomplete because although large vortices often form in turbulent rotating flows, the reason why is one of the most difficult problems in physics and has not been worked out for the Great Red Spot.

The Great Red Spot is not the only interesting, long-lasting atmospheric feature. The anticyclonic White Oval Spots were first observed in 1938 in the South Temperate Belt. In 1998 and again in 2000, two of them merged into one as their separation was at a minimum, observed by the Galileo orbiter. The Galileo probe descended by chance into one of the North Equatorial Belt Hot Spots. These are regions of low cloud opacity, which radiate warm thermal emission from below.

Aurorae

On Earth, an aurora is observed when charged particles from the solar wind precipitate into the atmosphere, exciting the atmospheric molecules which then emit light. Aurorae were first detected at Jupiter's poles by the IUE satellite in 1979, followed by the Voyager observations of Jupiter's auroral ring in 1979 and 1980. The differences from the Earth's aurorae are mainly due to the difference in atmospheric composition, with hydrogen dominating Jupiter's atmosphere instead of oxygen and nitrogen.

The Cassini satellite made its closest approach to Jupiter on December 30, 2000, at a distance of about 9.7 million km. It discovered an auroral zone on Jupiter's atmosphere at the location of a magnetic footprint below IO. It appears as a spot of auroral emission that remains fixed underneath Io as Jupiter rotates. The planet's magnetic field channels the plasma emitted from the surface of the satellite. The plasma forms an electric current of around 1 million amperes down through Jupiter's ionosphere where it impacts on the atmosphere and causes the aurorae.

Satellites

The Galilean satellites are large enough to be seen easily with inexpensive binoculars, devices that are far superior to Galileo's telescope. A few decades after Galileo, Ole RØMER of Denmark realized that the irregularities in the times of the eclipses and mutual occultations of the Galilean satellites (used by mariners to determine longitude) were due to the changing distance between Jupiter and the Earth. Rømer used this to make the first determination of the speed of light. The eclipses and occultations are still of interest.

▲ ***Jupiter*** *This processed color image of Jupiter was produced in 1990 by the US Geological Survey from a picture captured by Voyager 2 in 1979.*

Between 1610 and 1974, nine more Jovian satellites were detected by ground-based telescopic observations, and three were discovered by Voyager 1 and 2 in 1979. The number of satellites tripled in 1999–2003, and might yet increase after surveys by a deep widefield CCD camera on the Canada-France-Hawaii and Subaru telescopes. The satellites of Jupiter fall into four categories:

- four small (16–170 km in diameter), irregularly shaped inner satellites in regular orbits (circular, prograde, low inclination); they lie embedded within Jupiter's tenuous rings and are the likely sources of the rings' dust-sized particles;
- four big (3100–5300 km), spherical Galilean satellites, also in regular orbits;
- eight small, irregularly shaped satellites (4–180 km), in two groups, that have prograde but irregular orbits (with high inclinations and eccentricity);
- more than 50 irregularly shaped satellites (2–60 km), which have orbits that are both irregular and retrograde.

The Galilean satellites

Io is covered with sulfur and/or sulfur compounds, which make it reddish. It is volcanically active, thanks to internal heating from the frictional dissipation of tidal energy generated by its orbital resonance with EUROPA. Io's density is high (3.5 g cm^{-3}), and its interior is mostly silicates, like the Moon. It has a very tenuous atmosphere of sulfur dioxide. In addition, material removed from Io's surface (atoms of hydrogen, oxygen, sulfur, sodium and potassium) forms a torus of plasma surrounding Io's orbit.

Europa, the smallest Galilean satellite, has a very smooth white surface of pure water ice, with few topographic features. The paucity of craters shows that the surface is relatively young. A net of linear structures covers Europa's surface.

▶ ***Jupiter*** *can be easily viewed and imaged with modest amateur equipment. This image of the the Great Red Spot, by Jamie Cooper, was captured through a 14-in Dobsonian telescope. The camera, an Olympus Comedia 30-30 digital camera, was attached to the telescope using an 11-mm Plössl eyepiece.*

Observing Jupiter and its satellites

Jupiter is often referred to as the amateurs' planet. Its size and large apparent diameter make Jupiter easy to observe with modest amateur equipment. Due to the efforts of amateurs and amateur organizations such as the Association of Lunar and Planetary Observers, the BAA and others around the world, there exists a continuous visual record of Jupiter going back over 170 years. Even in this advanced space age, one must acknowledge that continuous spacecraft observations are relatively short-lived. Without the contributions of amateurs, there would be serious gaps in the observational record.

Equipment used

A 4-in REFRACTING TELESCOPE or a 6-in REFLECTING TELESCOPE is considered to be the minimum aperture necessary for serious study of Jupiter. The best telescope design is one that produces high image contrast. Therefore, large secondary mirrors are to be avoided. However, the author owns and has made many useful observations with a SCHMIDT–CASSEGRAIN TELESCOPE. Regardless of the telescope used, it should be in good working order with perfectly collimated optics (*see* COLLIMATION). The contrast of Jupiter's features is subtle. Misalignment of the optics will smear the fine detail that would otherwise be seen.

The author prefers high-quality orthoscopic or Plössl EYEPIECES with good optical coatings. Both types of eyepiece provide good performance at high magnification. However, magnification should be limited to ×40 per inch of aperture as Jupiter goes soft quickly.

The serious observer of Jupiter should possess a collection of color filters of known wavelength transmission (*see* CAMERA AND FILTER). The use of color filters can increase the contrast of subtle features on the planet and help in their identification. Blue filters will enhance the contrast of reddish features. Red filters can enhance blue and gray features. Green filters can increase the contrast of blue and red features. Yellow filters can be used to increase the overall contrast of features, especially that of bright ovals against the grays of the south temperate belt (`) and polar regions. In fact, the author prefers the light-yellow Wratten No. 8 filter as the best general-purpose filter, as it gently increases contrast over the planet without obscuring subtle detail. One should experiment and determine which filters perform best for one's own eyes.

Typical observations made by amateurs

A full disk drawing seeks to capture the appearance of the entire planet. Belts, zones, and other features are sketched on a preprinted form depicting the correct ELLIPSE of the planet. Since the planet rotates completely in less than 10 hours, it is necessary to complete a sketch in 20 min or less.

A strip sketch allows the observer to concentrate on a narrow band of latitude as the planet rotates, and to record features continuously and in more detail than time would allow in a full disk drawing. Again, a standard form is used.

The observer may also record an estimate of the intensity of the belts, zones, and other features seen. These intensity estimates may be recorded on the same form used to make drawings. ECLIPSES AND OCCULTATIONS of Jupiter's moons may also be observed.

The most valuable observation the amateur astronomer can make of Jupiter is the central meridian (CM) TRANSIT timing. The dark and bright features seen in the cloud tops of Jupiter are not stationary, but change position in longitude, drifting and moving in the various wind currents and jet streams. Tracking the behavior of these features can provides clues to help characterize and understand the atmospheric conditions on the planet. Transit data collected in a careful, systematic manner provide this invaluable record.

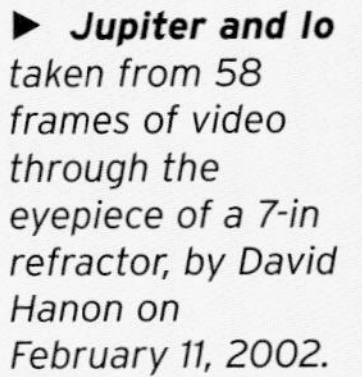

▶ ***Jupiter and Io*** *taken from 58 frames of video through the eyepiece of a 7-in refractor, by David Hanon on February 11, 2002.*

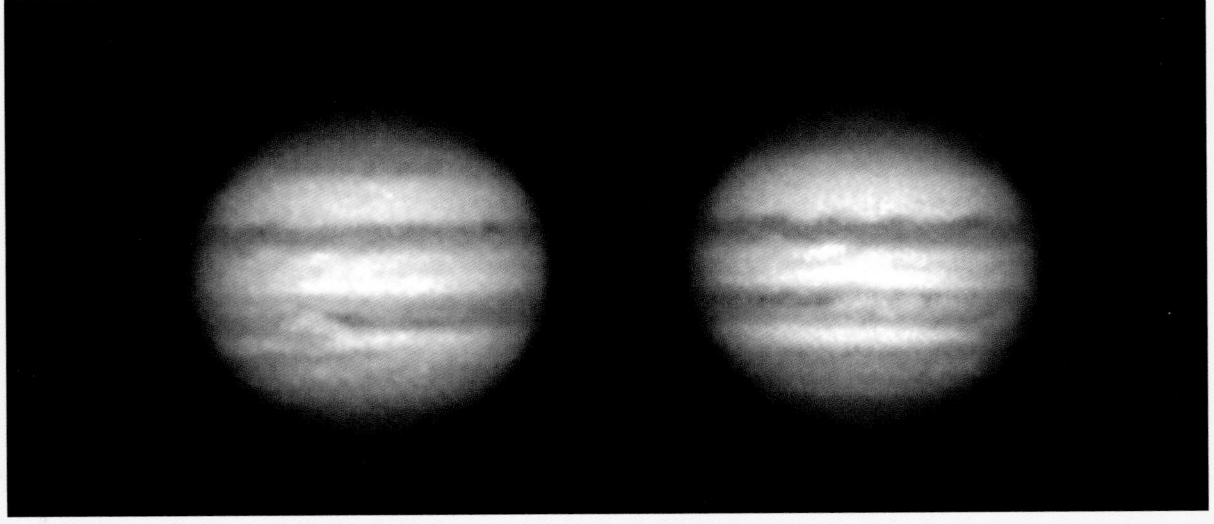

▲ ***Jupiter and its satellites*** *Europa, Io and, far right, Ganymede. Imaged by Jamie Cooper on October 28, 2002.*

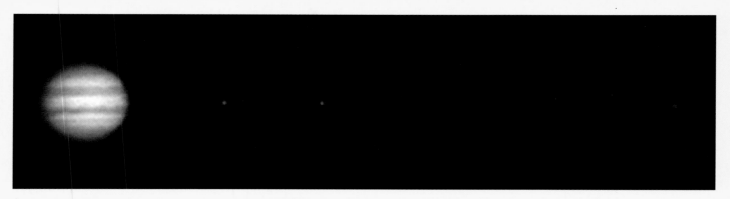

The CM is an imaginary line that passes through the planet's north and south pole, bisecting the planet. As the planet rotates, features appear to move from the following limb to the preceding limb, crossing the CM. Using an ephemeris, one can determine the longitude of the CM at any given time. If one notes the time, to the nearest 30 s, that a feature is seen on Jupiter's CM, and one determines the longitude of Jupiter's CM at that time, then one can also determine the longitude of the feature that was seen on the CM. If one observes the transit of the same feature over time, then a determination of drift rate can be made by plotting the feature's longitude versus time on a drift rate chart. Thus, over time the behavior of the feature may be determined.

All of the research formerly done with visual observations and sketches can now also be completed by analyzing and scaling CCD images. Some amateurs are now imaging Jupiter in methane, something unheard of just a few years ago.

Some recent amateur contributions

During June 1998, amateurs discovered a small, but extremely intense dark spot in Jupiter's STB. Amateurs brought this new feature to the attention of the professional community. With keen interest, professional astronomers began monitoring the spot with the assistance of amateurs around the world. These observations ultimately determined that the feature was not a spot at all, but a small opening in Jupiter's cloud tops. Thus, the value of amateur and professional cooperation was demonstrated.

During the 1997–1998 and 1999–2000 apparitions, bright ovals in the STB merged, ending a long chapter in Jupiter's history. Amateurs and professionals alike kept watch. Due in great part to amateur observations, there is a detailed record of the behavior of these ovals leading up to the merger. Amateurs have faithfully observed the remaining oval since that time.

Amateur observations may be reported to the Association of Lunar and Planetary Observers or the BAA, both of which also publish observing aids and instructional materials.

Since 1997, John McAnally has held the position of Assisant Coordinator for Transit Timings in the Jupiter Section of the Association of Lunar and Planetary Observers. He has been an avid amateur astronomer since 1967 and published his first paper in 1974.

◄ ***Jupiter's scars*** *from the impact of Comet Shoemaker–Levy on 28 July, 1994 (top) and 3 August (bottom). The row of impact sites appears dark, due to deep material sprayed up onto the clouds at the foot of the picture. By David Hanon.*

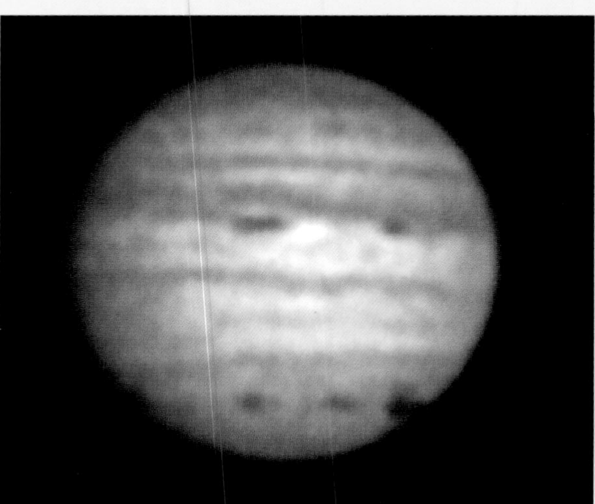

▼ ***Comet Shoemaker–Levy*** *broken up by Jupiter's gravity. By David Hanon.*

Jupiter's satellites

No.	Name	a	i	e	Period	Mag.	Diameter	Year	Discoverer(s)
Small inner regular satellites									
XVI	Metis	0.128	0.02	0.001	0.30	17.5	44	1979	Voyager
XV	Adrastea	0.129	0.03	0.002	0.30	18.7	16	1979	Voyager
V	Amalthea	0.181	0.39	0.003	0.50	14.1	168	1892	E E Barnard
XIV	Thebe	0.222	1.07	0.018	0.68	16.0	98	1979	Voyager
Galilean satellites									
I	Io	0.422	0.04	0.000	1.77	5.0	3643	1610	Galileo
II	Europa	0.671	0.47	0.000	3.55	5.3	3122	1610	Galileo
III	Ganymede	1.070	0.17	0.001	4.16	4.6	5262	1610	Galileo
IV	Callisto	1.883	0.31	0.007	16.69	5.7	4821	1610	Galileo
Themisto prograde irregular group									
XVIII	Themisto	7.507	43.08	0.242	130.0	21.0	9	2000	S Sheppard et al.
Himalia prograde irregular group									
S/2003 J_6		10.973	22.41	0.759	233.8	22.6	4	2003	S Sheppard et al.
S/2003 J_1		11.029	35.17	0.792	237.0	22.6	4	2003	S Sheppard et al.
XIII	Leda	11.165	27.46	0.164	240.9	20.2	18	1974	C Kowal
VI	Himalia	11.461	27.50	0.162	250.6	14.8	184	1904	C Perrine
X	Lysithea	11.717	28.30	0.112	259.2	18.2	38	1938	S Nicholson
VII	Elara	11.741	26.63	0.217	259.6	16.6	78	1905	C Perrine
S/2000 J_{11}		12.555	28.30	0.248	287.0	22.4	4	2000	S Sheppard et al.
Retrograde irregular group									
S/2003 J_3		18.340	143.7	0.241	504.0	23.4	2	2003	S Sheppard et al.
S/2001 J_{10}		19.394	145.8	0.143	553.1	23.1	2	2001	S Sheppard et al.
S/2001 J_7		21.017	148.9	0.230	620.0	22.8	3	2001	S Sheppard et al.
XXII	Harpalyke	21.105	148.6	0.226	623.3	22.2	4	2000	S Sheppard et al.
XXVII	Praxidike	21.147	149.0	0.230	625.3	21.2	7	2000	S Sheppard et al.
S/2001 J_9		21.168	146.0	0.281	623.0	23.1	2	2001	S Sheppard et al.
S/2001 J_3		21.252	150.7	0.212	631.9	22.1	4	2001	S Sheppard et al.
XXIV	Iocaste	21.269	149.4	0.216	631.5	21.8	5	2000	S Sheppard et al.
XII	Ananke	21.276	148.9	0.244	610.5	18.9	28	1951	S Nicholson
S/2001 J_2		21.312	148.5	0.228	632.4	22.3	4	2001	S Sheppard et al.
S/2001 J_6		23.029	165.1	0.267	716.3	23.2	2	2001	S Sheppard et al.
S/2002 J_1		23.064	163.1	0.244	715.6	22.8	3	2002	S Sheppard et al.
S/2001 J_8		23.124	165.0	0.267	720.9	23.0	2	2001	S Sheppard et al.
XXI	Chaldene	23.179	165.2	0.251	723.8	22.5	4	2000	S Sheppard et al.
XXVI	Isonoe	23.217	165.2	0.246	725.5	22.5	4	2000	S Sheppard et al.
S/2001 J_4		23.219	150.4	0.278	720.8	22.7	3	2001	S Sheppard et al.
S/2003 J_4		23.258	144.9	0.204	723.2	23.0	2	2003	S Sheppard et al.
XXV	Erinome	23.279	164.9	0.266	728.3	22.8	3	2000	S Sheppard et al.
XX	Taygete	23.36	165.2	0.252	732.2	21.9	5	2000	S Sheppard et al.
XI	Carme	23.404	164.9	0.253	702.3	17.9	46	1938	S Nicholson
S/2001 J_{11}		23.547	165.2	0.264	741.0	22.7	3	2001	S Sheppard et al.
XXIII	Kalyke	23.583	165.2	0.245	743.0	21.8	5	2000	S Sheppard et al.
VIII	Pasiphae	23.624	151.4	0.409	708.0	16.9	58	1908	P Melotte
XIX	Megaclite	23.806	152.8	0.421	752.8	21.7	6	2000	S Sheppard et al.
S/2003 J_7		23.808	159.4	0.405	748.8	22.5	4	2003	S Sheppard et al.
S/2001 J_5		23.808	151.0	0.312	749.1	23.0	2	2001	S Sheppard et al.
IX	Sinope	23.939	158.1	0.250	724.5	18.3	38	1914	S Nicholson
S/2003 J_5		24.084	165.0	0.210	759.7	22.4	4	2003	S Sheppard et al.
XVII	Callirrhoe	24.102	147.1	0.283	758.8	20.8	7	1999	J V Scotti et al.
S/2001 J_1		24.122	152.4	0.319	765.1	22.0	4	2001	S Sheppard et al.
S/2003 J_8		24.514	152.6	0.264	781.6	22.8	3	2003	S Sheppard et al.
S/2003 J_2		28.570	151.8	0.380	982.5	23.2	2	2003	S Sheppard et al.

In 2003, 21 further members of the group were discovered, for which only preliminary data are available.
a = semimajor axis in millions of kilometers; i = inclination in degrees; e = eccentricity; period = time of one revolution around Jupiter in days; Mag. = optical magnitude; diameter = diameter in kilometers; year = year of discovery

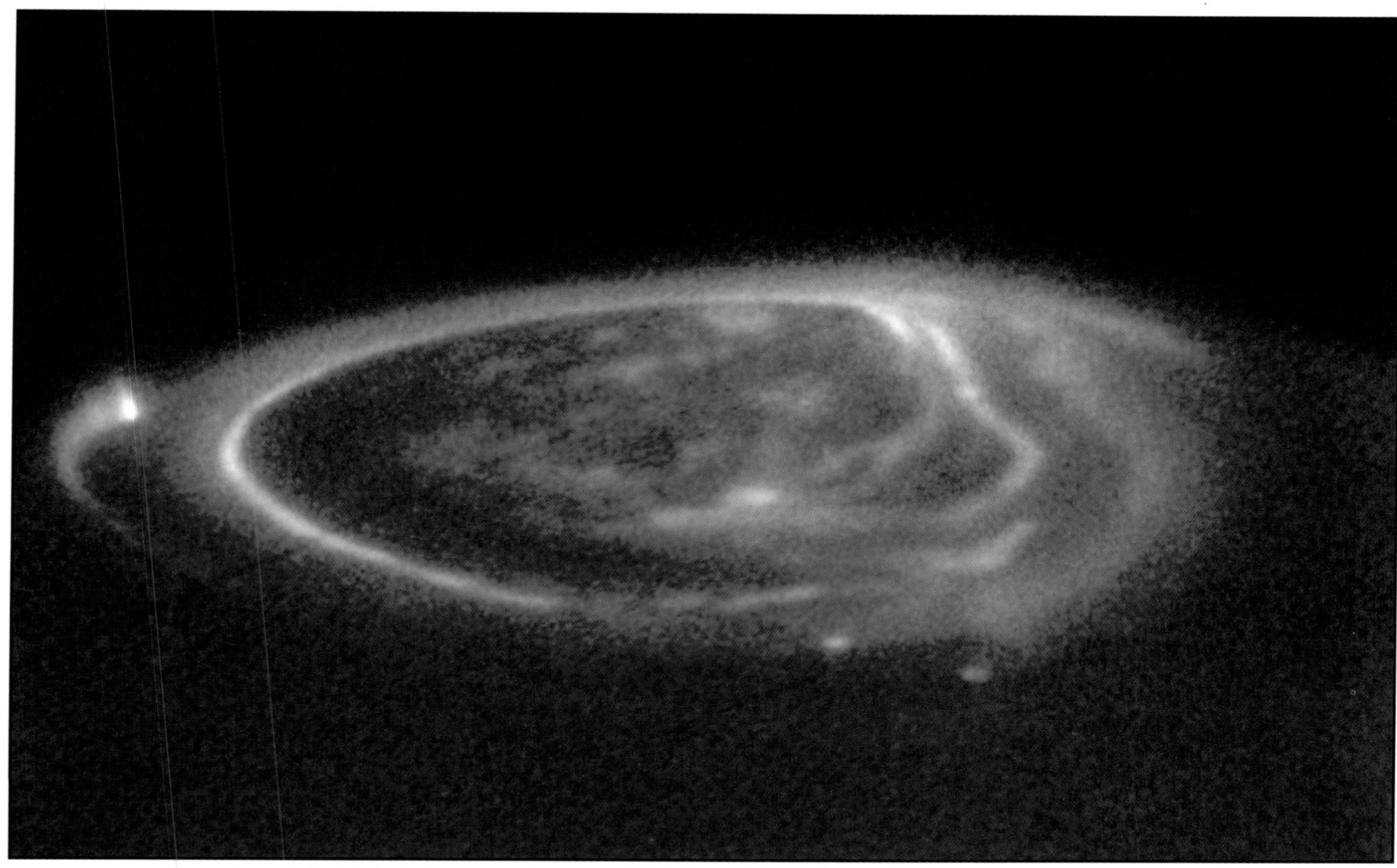

High-resolution images by the Galileo orbiter suggest that these structures are cracks in the ice as it floats and shifts on liquid water below, pushed from side to side by tidal forces. Ice 'rafts' the size of cities have broken off the ice sheet. The cracks show characteristic scallop shapes induced by the periodic tidal forces, while frozen 'puddles' of ice smooth over older cracks, and warmer material bubbles up from below the surface. Possessing more water than the total amount found on Earth, Europa has a salty ocean beneath its icy surface, and evaporative salts tint the white ice. Indeed, since it is the only other planet or satellite in the solar system that has an ocean of liquid water, Europa is of great interest for exobiology (*see* EXOBIOLOGY AND SETI). Below this ocean, the interior is likely to contain silicates. The satellite has a thin oxygen atmosphere and an ionosphere.

GANYMEDE is the largest satellite in the solar system; it is even larger than Mercury and Pluto. The surface of Ganymede is quite dark, which implies, together with its density (1.94 g cm^{-3}), that it is composed of roughly equal amounts of water ice and silicates. A major result from Galileo was the discovery of a magnetic field around Ganymede; its origin is unclear.

Finally, CALLISTO, Jupiter's outermost Galilean satellite, shows a heavily cratered surface, much older than the others. Being farther from Jupiter, its tides are weak and it has no internal energy source. Like our own Moon, its surface has therefore changed little since its formation.

Formation of Jupiter and its satellites

At great heliocentric distances, the more abundant molecules like carbon dioxide, water, methane and ammonia were ices, which built up the cores of the giant planets. The cores accreted the surrounding GASEOUS NEBULA, mostly formed of hydrogen (75%) and helium (25%), as is interstellar material in general. Jupiter's present composition is dominated by these elements.

Jupiter somewhat resembles the smallest stars, but is only 0.1% of the solar mass. Jupiter could be considered a low-mass BROWN DWARF.

Jupiter's gravitational perturbations disturbed the formation of a planet from the material immediately inside its orbit. Some of the several VESTA- and PALLAS-sized minor planets that did succeed in forming, all at the same orbital radius from the Sun, were doomed to collide. Jupiter is thus responsible for the main belt of asteroids.

The formation of Jupiter and its prograde inner satellites (including the Galilean satellites) was analogous to that of the solar system. In contrast, the retrograde outer satellites are most likely captured comets or asteroids. They are, however, in such profusion and such similar orbits that their capture must be generic rather than by chance, an unexplained feature of the formation of Jupiter in its first few hundred million years.

Further reading: Gehrels T (ed) (1976) *Jupiter* University of Arizona Press; Morrison D (ed) (1988) *Satellites of Jupiter* University of Arizona Press.

▲ ***Jovian aurora***
This ultraviolet image of Jupiter's north polar region was taken with the HST Imaging Spectrograph. The bright emissions are auroral lights, similar to those seen at the Earth's poles.

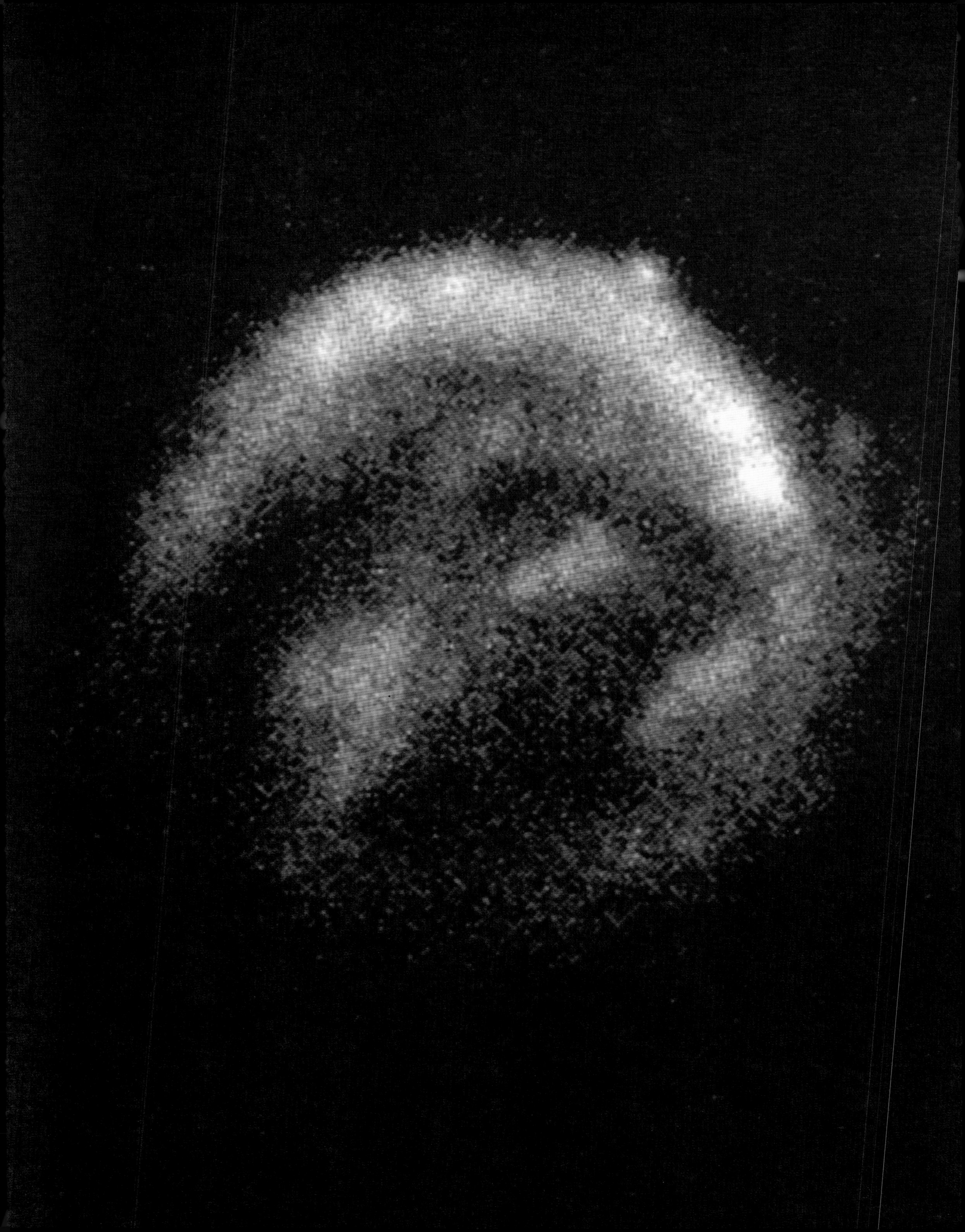

◀◀ ***Kepler's supernova***
In 1604 Johannes Kepler observed the explosion of a star in our Milky Way. Today, in its place a supernova remnant shines brightly in X-rays. The X-ray emission stems from star fragments and interstellar gas heated up to several million degrees.

Kant, Immanuel (1724–1804)

German philosopher who published his view of the universe in *General History of Nature and Theory of the Heavens* (1755). His nebular hypothesis of solar-system formation was much like the present theory that the Sun and planets formed from the condensation of a rotating disk of interstellar material. Kant identified the Milky Way as a lens-shaped collection of stars in orbit around its center like the rings of Saturn, and as one of many 'island universes' (he coined the term). He suggested that the tides raised by the Moon were the reason for its always presenting the same face toward us, and that tidal friction was slowing the rotation of the Earth.

Kapteyn, Jacobus Cornelius (1851–1922)

Dutch astronomer who, at the University of Groningen, and with the aid of convicts from a prison nearby, measured photographic plates taken by David GILL at the Cape of Good Hope. He compiled a catalog of a half-million southern stars, from which he derived values for the density of stars as a function of distance, brightness and spectral class, extending John Herschel's work (*see* HERSCHEL FAMILY). He confirmed that the MILKY WAY GALAXY was lens-shaped, but by not realizing the significance of interstellar absorption concluded that the Sun was near the center. Kapteyn identified a selection of typical areas of the sky located at intervals of galactic latitude and longitude so as to form a collection that would be representative of the Milky Way. He proposed that astronomers worldwide should sample the faint stars in these 'Selected Areas' and thus map the Galaxy. The project was undermined by World War I, but stimulated many individual discoveries. He measured the velocities of stars and discovered 'star streaming,' which proved to be a manifestation of the rotation of the Galaxy. He discovered KAPTEYN'S STAR and successfully explained the expanding halo around Nova Persei as a LIGHT ECHO, reflecting the nova flash from surrounding dust clouds. The Jacobus Kapteyn Telescope on La Palma in the Canary Islands is named for him.

Kapteyn's Star

A faint RED DWARF situated in the southern constellation of Pictor, about 12 arcsec northwest of CANOPUS. Discovered by the Dutch astronomer Jacobus KAPTEYN in 1897, it is notable for having the second-largest proper motion known: its motion of 8.67 arcsec per annum is exceeded only by that of BARNARD'S STAR. It is also one of the closest stars to the Sun: at 12.8 l.y., it has a PARALLAX of 0.255 arcsec and its absolute magnitude is 10.9.

Keck Observatory *See* W M KECK OBSERVATORY

Keeler, James Edward (1857–1900)

American astronomer. In 1881 Keeler went with a pack train of mules, loaded with mirrors, telescopes and bolometers, to the 4.4-km summit of Mount Whitney, California (the highest elevation in the contiguous USA). His aim was to measure, undisturbed by the atmosphere, the infrared spectrum of the Sun and its flux, the 'solar constant.' Keeler became director of the ALLEGHENY and LICK OBSERVATORIES, and established that Saturn's rings were not solid by showing that they rotated with motions that conformed to KEPLER'S LAWS (as James MAXWELL had predicted), implying that they consisted of meteoritic particles in individual orbits. He founded the *Astrophysical Journal* in 1895 with George HALE; its first volume contained Keeler's paper on Saturn's rings – published only a month after he had taken the spectra.

Kellner eyepiece *See* EYEPIECE

Kepler, Johannes (1571–1630)

Kepler was arguably the most innovative astronomical theorist in the period between PTOLEMY'S *Almagest* (*c.*AD 150) and Isaac NEWTON'S *Principia* (1687). Before Kepler, planetary and lunar theory consisted of combining circular motions, as the planets were taken to be perfect beings for whom the only thinkable motion was circular. Kepler changed all this by envisaging planets and satellites as bodies moving in orbits under the action of forces issuing from the central body. Since the eighteenth century these rules have been called 'laws.'

Growing up in Weil (today in Baden-Württemberg, Germany) in a disunited family, Kepler was a sickly child. His father became a mercenary and did not take care of them; his mother narrowly missed being burnt at the stake for witchcraft. As he early proved himself to be a prodigious child, Kepler benefited from the performance-based educational system founded by the Dukes of Württemberg. As Lutherans, they had promoted such a system to turn out the erudite clergy and the efficient administration they needed. From 1589 to 1594 Kepler attended the University of Tübingen, aiming to become a Lutheran pastor. The Tübingen theological faculty prohibited the teaching of Nicolaus COPERNICUS'S heliocentric astronomy as contrary to Holy Writ, but his tutor Michael Mästlin (1550–1631) privately helped a few exceptional students like Kepler to understand the kinematic economy of the heliocentric arrangement.

In 1594 Kepler became mathematics teacher at a Protestant school in Graz, Austria, a post he accepted only on the proviso that he might later return to complete his theological studies. However, in 1598 the school was closed down and most Protestants were exiled. Kepler himself was exiled in 1600 for refusing to convert to Catholicism. He moved to Prague to assist Tycho BRAHE in preparing his work for publication, but it was not until September 1601 that he was granted a stipend by Emperor Rudolph II to collaborate with Brahe on assembling new planetary tables. On October 24 Brahe died, and two days later Kepler was appointed imperial mathematician in his place. The next 10 years in Prague would be among the most fruitful of his life.

Kepler's *The Optical Part of Astronomy* (1604) contained major new results for both optics and astronomy: the geometrical explanation of the formation of images through small apertures, as in the camera obscura (used in the observation of solar eclipses); the most accurate table of astronomical refractions yet to appear, based on a law of refraction closely approximating the sine law; the first enunciation of the inverse-square law for the intensity of illumination from a point source; and the discovery of the formation of inverted images on the retina of the eye. In 1604, Kepler was also fortunate to observe a galactic supernova, which now bears the name KEPLER'S SUPERNOVA. In planetary theory the revolutionary work was Kepler's *New Astronomy* (1609), a study of the motions of Mars, which contained what became his first and second laws. His third law was published in 1618 in *Harmony of the Universe*. In 1627, three years before his death, Kepler published the *Rudolphine Tables* in honor of his protector which presented the planetary EPHEMERIS based on his three laws.

►► ***Keyhole Nebula*** *(NGC 3324) The dark dusty nebula (left of center) gets its name from its unusual shape. The Keyhole Nebula is a smaller region superimposed on the larger Eta Carina Nebula. These nebulae were created by the luminous blue variable star Eta Carinae.*

Kepler's laws *See* CELESTIAL MECHANICS

Kepler's Supernova

This is the most recently observed confirmed SUPERNOVA in the Milky Way Galaxy. It was discovered on October 9, 1604, by several people including John Brunowski, a court official and amateur astronomer in Prague, who notified Johannes KEPLER. Kepler first saw it on October 17, and, inspired by Tycho BRAHE's work on the supernova of 1572, started a systematic study of the phenomenon which he described in his book *De stella nova* (On the New Star). The supernova reached a maximum apparent magnitude of –2.5 and remained visible to the naked eye for about a year. The records indicate that it was a type Ia supernova. The supernova remnant, consisting of faint filaments and knotted fragments of nebulosity about 40 arcsec in extent, was discovered in 1941, approximately 2.2° east of Xi Ophiuchi. It is a strong source at radio wavelengths, with the identification 3C 358.

Keyhole Nebula (NGC 3324)

A DARK NEBULA in the constellation Carina, right ascension 10 h 44.3 m, declination –59° 53'. It is seen in silhouette against the bright ETA CARINAE NEBULA.

Kinetic theory of gases

The theory, developed in the nineteenth century, notably by Rudolf Clausius (1811–99) and James MAXWELL, that the temperature, pressure and other properties of a gas could be described in terms of the motions (and kinetic energy) of its molecules. The theory has wide implications in astrophysics. In particular, the perfect gas law, which relates to the pressure, volume, temperature, and number of molecules in a gas, and which is fundamental to theoretical models of the interior of stars, is consistent with this theory.

Kirchhoff, Gustav Robert (1824–87)

Born in Königsberg, Prussia (now Kaliningrad, Russia), Kirchhoff became professor of physics at Heidelberg University, where he collaborated with Robert Bunsen (1811–99), and later a professor at Berlin University. With his laws of electricity he extended the work of Georg Ohm (1789–1854). He worked on black-body radiation and spectrum analysis. With his laws of radiation he explained the dark lines in the Sun's spectrum as being caused by the absorption of particular wavelengths as light from a hot source passes through a cooler gas (*see* ABSORPTION SPECTRUM). He showed that a given element produced the same characteristic pattern of emission and absorption lines (*see* EMISSION SPECTRUM). This began the topic of stellar ASTROPHYSICS, by which the composition, density, temperature and other physical conditions of a star's atmosphere can be found.

Kirkwood, Daniel (1814–95)

American astronomer and mathematician who became professor of mathematics at the universities of Delaware and Indiana. He explained the distribution of ASTEROIDS in the asteroid belt as a result of the influence of Jupiter. An asteroid resonates with Jupiter when its orbital period is a simple fraction (1/2, 2/3…) of the planet's. The asteroid's orbit is disturbed whenever it has a close approach, so that it is eventually ejected into an eccentric, Earth-crossing orbit. The resulting lack of asteroids with these periods in the asteroid belt became known as 'Kirkwood gaps.'

Kitt Peak National Observatory

Kitt Peak was selected in 1958 as the site for a national observatory after a three-year survey of more than 150 mountain ranges across the USA. In 1982 the observatory became part of the new organization, the NOAO. Kitt Peak, located in the Quinlan Mountains of the Sonoran Desert, southwest of Tuscon, Arizona, comprises 200 acres of the nearly 3-million-acre Tohono O'odham Nation. This land is leased by NOAO from the Tohono O'odham under an agreement that is valid for as long as scientific-research facilities are maintained at the site.

Kleinmann–Low Nebula (KL Nebula)

An extended, dusty region of star formation behind the ORION NEBULA, discovered in 1967 by Douglas Kleinmann and Frank Low. There are several discrete sources of infrared radiation, including the BECKLIN–NEUGEBAUER OBJECT, thought to be young stars or protostars.

Konkoly Observatory

Konkoly Observatory was founded in 1874 in Ogyalla, Hungary, as the private observatory of Miklos Konkoly-Thege (1842–1916). Since 1929 the observatory's main area of work has been the study of variable stars.

Kraus, John (1910–)

American radio astronomer who became professor of electrical engineering and astronomy at Ohio State University, and director of the Ohio State–Ohio Wesleyan Radio Observatory. Kraus is the author of widely used textbooks on radio techniques, and the inventor of many types of radio antenna. He has surveyed the radio sky at centimeter-wavelengths, and mapped and cataloged 20 000 radio sources, including many quasars. Kraus is also the instigator of an extensive search for artificial radio signals from extraterrestrial sources.

KREEP

A most unusual type of rock brought back from the MOON by astronauts on the Apollo missions. It is rich in potassium (K), rare earth elements (REE) and phosphorus (P). KREEP is believed to have formed early in the history of the Moon during the solidification of its molten stage.

Kreutz sungrazing comets *See* COMET

Kuiper, Gerard Peter (1905–73)

Dutch astronomer, educated at Leiden, who moved to the USA, where he became director of YERKES OBSERVATORY. He also founded in 1960 and became director of the Lunar and Planetary Laboratory of the University of Arizona. Kuiper spectroscopically detected the methane atmosphere on Saturn's satellite Titan and the carbon dioxide atmosphere of Mars. He discovered Miranda (the fifth moon of Uranus), and Nereid (the second moon of Neptune). Kuiper also predicted the existence of the belt of comet-like debris at the edge of the solar system now known as the Kuiper Belt (*see* OORT CLOUD AND KUIPER BELT). He was a key advisor in the selection of the landing sites on the Moon for the RANGER probes and APOLLO landings. The Kuiper Airborne Observatory for infrared astronomy (used extensively for remote sensing of the planets) was named for him.

Kuiper Belt *See* OORT CLOUD AND KUIPER BELT

Lacaille, Abbé Nicholas Louis de (1713–62)

French astronomer. From 1750 to 1754 Lacaille led an expedition to the Cape of Good Hope, where he was the first to measure the curvature of the Earth in South Africa, and, with a small telescope, compiled a still-used catalog of nearly 10 000 southern stars. In the course of his survey he discovered 50 nebulous objects (eight are now regarded as non-existent), which he classified in three categories: nebulae, nebulous star clusters and nebulous stars, foreshadowing the modern astrophysical interpretation. In his *Atlas*, he named 15 new southern CONSTELLATIONS: Antlia, Caelum, Circinus, Fornax, Horologium, Mensa, Microscopium, Norma, Octans, Pictor, Pyxis, Reticulum, Sculptor and Telescopium, and renamed the constellation Musca. *See also* Johann BAYER.

Lacerta *See* CONSTELLATION

Lagoon Nebula (M8, NGC 6523)

See MESSIER CATALOG

Lagrange, Joseph-Louis (1736–1813)

Italian mathematician of exceptional ability. As a young man, he published work on the calculus of variations, developed a theory of dynamics based on the principle of least action, and wrote about fluid mechanics (where he introduced the Lagrangian function – *see* LAGRANGIAN POINTS) and the orbits of the Moon, Jupiter and its moons, and Saturn. He was also interested in number theory. A protégé of Leonhard Euler, he succeeded him as director of mathematics at the Berlin Academy of Science, and worked with him on a number of mathematical problems. At the age of 51 he moved to Paris, where he was acclaimed as a French mathematician through the publication of *Mécanique analytique* (Analytical Mechanics). This summarized all the work in mechanics since the time of Isaac NEWTON, using differential equations to transform mechanics into the branch of mathematical analysis that it is today.

Lagrangian points

Five neutral points in the combined gravitational field of two massive bodies (mass M_1 and M_2) which are orbiting their center of mass. Here an object of much smaller mass can exist in equilibrium, because it experiences no net gravitational force. They are so named because Joseph-Louis LAGRANGE was the first to find them in 1772 as solutions to a restricted case of the three-body problem (*see* N-BODY PROBLEM). Three of the points lie on a line joining the centers of the two massive bodies. The inner Lagrangian point, L_1, lies between M_1 and M_2, while L_2 and L_3, the outer Lagrangian points, lie on either side. L_4 and L_5 are in the orbit of M_2 about M_1, respectively 60° ahead of and behind M_2 in its orbit. L_4 and L_5 therefore form equilateral triangles with M_1 and M_2. A body at L_1, L_2 or L_3 is in unstable equilibrium, very prone to gravitational perturbations; but a body at L_4 or L_5 is less prone to perturbations and has long-term orbital stability.

In practice, the situation is complicated by the ellipticity of real orbits and the presence of gravitational perturbations. In the solar system the best example of bodies in stable orbits at Lagrangian points are the TROJAN ASTEROIDS at the L_4 and L_5 points of Jupiter's orbit around the Sun. A 'halo orbit' is an orbit around a Lagrangian point. SOHO was placed in a halo orbit around the Earth's inner Lagrangian point, L_1, between the Earth and the Sun.

Lalande, Joseph-Jérôme Lefrançais de (1732–1807)

French astronomer. In one of the first internationally coordinated scientific campaigns (to determine the Moon's PARALLAX) he was sent in 1752 to Berlin, Germany by the French Academy of Sciences. Other astronomers took up positions at six other sites located on more or less the same meridian from Stockholm to the Cape (where Lalande's colleague Nicholas LACAILLE was located). They were able to use triangulation to determine the Moon's distance. Lalande edited *La Connaissance de temps* (The Knowledge of Time), the French almanac, and became professor of astronomy at the Collège de France, and later director of the Paris Observatory. His chief work is *Traité d'Astronomie* (Treatise on Astronomy) from 1764, and he also produced a comprehensive star catalog, recording the planet Neptune on two occasions before its discovery without recognizing that it was not a star.

Lambert, Johann Heinrich (1728–77)

This French mathematical physicist had the 'lambert,' the unit of light intensity, named for him. In orbital dynamics 'Lambert's problem' is: given a planet A at a given time and planet B at a later time, with what velocity must you launch a projectile to get from A to B by orbiting under the influence of the Sun? This is the fundamental question of interplanetary space travel.

La Palma

See HEGRA EXPERIMENT; ISAAC NEWTON GROUP OF TELESCOPES; NORDIC OPTICAL TELESCOPE; ROQUE DE LOS MUCHACHOS OBSERVATORY

Laplace, Pierre-Simon (1749–1827)

French celestial mechanician, who became professor of mathematics at the Ecole Militaire in Paris, examining the cadet Napoleon Bonaparte. He later became Napoleon's minister of the interior. He researched probability theory, calculus, and celestial mechanics. His work was interrupted by the Reign of Terror. Even though he wisely exiled himself from Paris, Laplace, as a member of the metrication committee of the Academy of Sciences, was consulted about rationalization of the calendar and metrication of angles. As more stable times returned, Laplace became head of the Paris Observatory. He presented the 'Laplace nebular hypothesis' in *Exposition du système du monde* (1796) (Explanation of the System of the World), which viewed the solar system as originating from a large, flattened, slowly rotating cloud of hot gas. He also expressed another remarkably modern view, of the impact of comets on the Earth: 'The small probability of collision of the Earth and a comet can become very great in adding over a long sequence of centuries.'

In 1799 Laplace published the first volume of his most important, five-volume work, *Traité du mécanique céleste* (Treatise on Celestial Mechanics). This work set up the differential equations of dynamics, solving them to describe the motion (including the orbits) of the planets, the rotation and shape of the Earth, and the tides. The book contains what is now known as 'Laplace's equation' (which was, in fact, known earlier). In Laplace's later work *Théorie analytique des probabilités* (Analytical Theory of Probability), he applied probability theory to errors in observations, the determination of the masses of Jupiter, Saturn and Uranus, surveying and GEODESY.

◄◄ ***Lagoon Nebula*** *(M8) This picture reveals a pair of half light-year long 'twisters' – funnels and twisted rope structures – in the heart of the nebula (at upper left). In the same region, the star Herschel 36 is the primary source of ionizing radiation.*

La Plata Observatory

The La Plata Observatory, officially founded on November 22, 1883, owes its existence to a French-Argentine expedition to Bragado, a province of Buenos Aires, Argentina to observe the transit of Venus on December 6, 1882. The expedition was equipped with a 21.6-cm telescope, incorporated soon afterwards into the observatory. A 2.15-m telescope like the one installed in Kitt Peak was set up in Calingasta Valley (in the province of San Juan), seeing first light in 1986 and becoming a national observing facility.

Large Binocular Telescope (LBT)

The Large Binocular Telescope is a collaboration to build a twin interferometric telescope on Mount Graham in Arizona, USA, shared between institutions in Arizona, Ohio, Germany and Italy. The two 8.4-m borosilicate honeycomb primary mirrors for the LBT were cast at the STEWARD OBSERVATORY's Mirror Laboratory. The telescope structure was fabricated in Italy, and the telescope is expected to see first light in 2004.

Large Magellanic Cloud

The larger of two nearby companions of the MILKY WAY GALAXY that can be seen with the naked eye in the southern hemisphere sky. Both MAGELLANIC CLOUDS are named for the Portuguese navigator, Ferdinand Magellan, whose sailors observed them in 1519 during his circumnavigation of the world. The Large Magellanic Cloud was originally classified as an irregular galaxy (type Irr I in the Hubble classification; *see* GALAXY). Because it has a conspicuous central bar and shows hints of what may be an incipient spiral arm, it is now considered to be an irregular barred spiral, the prototype of a class designated Sm. It contains the TARANTULA NEBULA, a huge H II region, with a diameter of about 900 l.y., which surrounds a vigorous star-forming region, and is one of the largest H II regions known.

Las Campanas Observatory

One of the large observatories in Chile, Las Campanas Observatory is located at the southern edge of the Atacama Desert. The local conditions make Las Campanas an ideal site for optical and infrared observations. Las Campanas is the main observational facility of the CARNEGIE OBSERVATORIES and has been active since 1971. Currently functioning telescopes include the 2.5-m du Pont and the 1-m Swope telescopes, and the two 6.5-m telescopes of the Magellan Project which commenced operations in 2000 and 2002 respectively. The Magellan Project is a joint enterprise with Harvard University, the Massachusetts Institute of Technology, and the universities of Michigan and Arizona. The universities of Warsaw (Poland), Nagoya (Japan) and Birmingham (UK) operate other telescopes at Las Campanas. Observations at Las Campanas have provided insight into many fundamental problems including the GREAT ATTRACTOR, the universe at high redshift, the age and distance scale of the universe, and the structure and chemical composition of our Galaxy. The supernova 1987A was discovered on a photographic plate obtained at Las Campanas.

La Silla *See* EUROPEAN SOUTHERN OBSERVATORY

Lassell, William (1799–1880)

A wealthy English businessman (a Liverpool brewer), Lassell built progressively larger reflecting telescopes of innovative design. He moved away from the tradition of wood and altazimuth mounting to telescopes with massive, strong iron frames, equatorially mounted on high-quality bearings. He commissioned engineer/astronomer James NASMYTH to make his 24-in telescope, with a fine, machine-figured 168-kg speculum mirror. According to family legend, Lassell failed to discover Neptune because a letter communicating John Couch ADAMS's predicted position was destroyed by a too-zealous maid. However, later in 1846 Lassell observed that Neptune revealed a distinct disk and had a satellite, Triton. An apparent ring system turned out to be a feature of a maladjusted mirror support system. In 1851 Lassell discovered two moons around Uranus (Ariel and Umbriel) and in 1858 co-discovered with William BOND the eighth moon of Saturn (Hyperion). He moved his 24-in telescope to the clear skies of Malta and built his 48-in telescope there, making a particularly fine study of the Orion nebula. The 24-in telescope has been reconstructed in Liverpool.

Late heavy bombardment

In tracing the geological record of the TERRESTRIAL PLANETS, it is clear that there was a period of intense cratering during their final stages of formation around 3.8–4.0 billion years ago which dropped off rapidly thereafter. Evidence for this bombardment is seen in the distribution of craters on the Moon, Mercury and Mars. It seems certain that the Earth underwent the same bombardment, but much of the evidence has disappeared owing to erosion.

Late-type stars

Relatively cool stars of spectral types K, M and 'later' types. It was originally thought (wrongly) that this represented a later stage of evolution than EARLY-TYPE STARS. The term is applied equally to giants and dwarfs.

Leap second *See* TIME

Leap year *See* CALENDAR

Leavitt, Henrietta Swan (1868–1921)

American astronomer who became a volunteer then a staff member at the Harvard College Observatory in Massachusetts. Leavitt was directed by Edward PICKERING to work on VARIABLE STARS in the MAGELLANIC CLOUDS, measuring their brightness on repeated exposures of photographic plates. She set up standards of photographic magnitudes (brightnesses) of stars that were accepted worldwide and could be used to measure the magnitudes of other stars. Leavitt also discovered the relation between the period and luminosity of classical CEPHEID VARIABLE STARS. This period-luminosity (P-L) relation made possible measurements of the distances of clusters and galaxies and thus the scale of the universe. Leavitt's P-L relation was recalibrated and used by the HST to determine the HUBBLE CONSTANT. Leavitt is at the center of the photograph.

Legendre, Adrien-Marie (1752–1833)

French mathematician who carried out research into celestial mechanics and the gravitational attraction of ellipsoidal masses (by which could be modeled rotating planets or stars), during which he introduced what are now known as 'Legendre functions.' This work brought him membership of the Academy of Sciences. He worked on a triangulation survey between the Paris and Greenwich observatories to measure the Earth, and was a member of the committee at the academy to standardize weights and measures. He lost his wealth in the Revolution in 1793, but was able to continue scientific work with the support of his wife. He published a book describing how to determine the orbit of a comet, under simplifying assumptions, and giving the least-squares method of fitting a curve to the data (*see* Carl Friedrich GAUSS).

Leiden Observatory

The Leiden Observatory was founded in 1633 and is the oldest university astronomy department in the world. It is part of the Faculty of Mathematics and Natural Sciences of Leiden University, the Netherlands.

Leiden Observatory hosts the Sackler Laboratory for Astrophysics and the NOVA-ESO Expertise Center for VLTI (a center for optical interferometry). During the twentieth century Leiden Observatory was associated with and produced many outstanding astronomers, including Willem DE SITTER, Ejnar HERTZSPRUNG, Jan OORT and Hendrik VAN DE HULST. The Dwingeloo and Westerbork radio telescopes were designed and developed there.

Lemaître, Abbé Georges Edouard (1894–1966)

Belgian priest, civil engineer and astrophysicist who became professor of the theory of relativity at Louvain University, and researched cosmic rays and the three-body problem. In 1927 he proposed, independently of Aleksandr FRIEDMANN, an evolving cosmological model in GENERAL RELATIVITY. This mathematical solution to Albert EINSTEIN's equations indicated that the universe had begun in a BIG BANG, and originated from a dense mass concentration which Lemaître envisaged as a primeval atom. Edwin HUBBLE's discovery of the expansion of the galaxies immediately gave credibility to this theory of the origin of the universe.

Lenticular galaxy

A lens-shaped galaxy that has a central bulge and disk but no spiral arms. Lenticular galaxies, which are denoted by 'S0' or 'SB0' in the Hubble classification scheme, appear to be intermediate in type between elliptical and spiral galaxies. They are flatter than the flattest ellipticals ('E7' ellipticals) but show no signs of spiral structure in their disks. Possibly they are spiral galaxies that have been stripped of their gas and dust.

Leo *See* CONSTELLATION

Leo Minor *See* CONSTELLATION

Leonids

A periodic METEOR SHOWER seen in mid-November with RADIANT in the constellation Leo. In most years meteor activity is low as debris from its parent, COMET TEMPEL–TUTTLE, has spread around the ORBIT. But the greatest concentration of METEOROIDS remains near the comet, so when the comet returns to perihelion every 33 years a storm is possible. In 1966, for example, some observers estimated that rates were as high as 200 000 per hour in one 40-min period. However, not every return of Tempel–Tuttle produces storm activity. Following the great storms of 1799, 1833 and 1866, there was little more than average activity in 1900 and 1933. In 1999 the shower peaked at around 1500 per hour over a two-hour period. The main Leonid swarm (sometimes called the 'ortho-Leonids') is thought to consist of a number of dense ribbons of meteoroids, and a storm occurs only if the Earth intersects one of these ribbons. Tempel–Tuttle's retrograde orbit gives the incoming meteoroids a high relative velocity, producing very fast meteors. There are records of Leonid storms going back more than 1000 years.

◄ ***Leonid meteor*** *The meteor streaks across startrails in this time exposure. It brightens as it plunges deeper into the atmosphere but is slowed, cools and fades again.*

Lepton

Subatomic particle that is not acted on by the strong nuclear force and is not built of QUARKS. Leptons are FUNDAMENTAL PARTICLES, having no internal structure and unable to be broken down into component particles. The leptons are the ELECTRON, the muon, the tau (or tauon), and the three types of NEUTRINO, together with their antiparticles.

Lepus *See* CONSTELLATION

Leverrier [Le Verrier], Urbain Jean Joseph (1811–77)

French celestial mechanician who worked at the Paris Observatory under Dominique ARAGO. He later became director, but his drive for efficiency and for total control created unrest among the staff, and Leverrier was removed from the post, although he was later reinstated under a ruling council. Leverrier had found fame as a young man after Arago had suggested that he calculate the position of Neptune from irregularities in Uranus's orbit (*see* John Couch ADAMS). Neptune was found at the predicted place by the German astronomer Johann GALLE. Leverrier attempted to repeat this success by examining a discrepancy in the motion in the PERIHELION of Mercury, and attributed this to an intra-mercurial

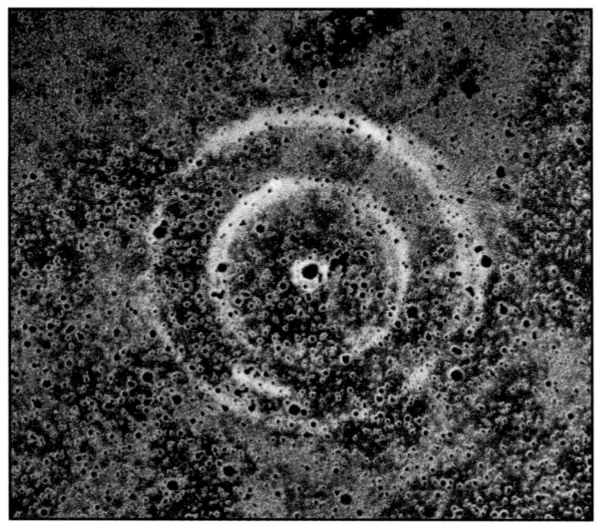

▶ ***Light echo*** *The bright flash of Supernova 1987A was deflected by two sheets of dust (and dust around the supernova and in localized patches). The echo of light rippled outwards in growing rings. In this photo, stars have been removed and show as black circles.*

planet, which he called VULCAN (the discrepancy was due to an effect of GENERAL RELATIVITY).

Libra *See* CONSTELLATION

Libration

An oscillation of a celestial object around some mean position. The term is mostly used in connection with the MOON, to describe effects that cause the hemisphere presented to an observer on the Earth to vary slightly as the Moon proceeds around its orbit. The Moon appears to rock slightly from side to side, and to nod slightly up and down. Although the Moon has SYNCHRONOUS ROTATION, rotating on its axis in the time it takes to orbit the Earth, the different types of libration make it possible for 59% of the Moon's surface to be visible over the course of 30 years or so.

Lick Observatory

Lick Observatory, located on Mount Hamilton in northern California, at an elevation of 1.28 km, serves astronomers throughout the University of California system. Its headquarters are on the Santa Cruz campus which also serves as the University of California liaison with the W M KECK OBSERVATORY. UCO/Lick provides the technical facilities and staff to design and fabricate state-of-the-art instrumentation. Lick has seven major research telescopes the largest of which is the 3-m Shane reflector. The original observatory building houses the 0.9-m refractor, the second largest in the world, as well as a 1-m reflector. Also on the mountain is a 0.5-m twin astrograph and a 0.75-m robotic telescope. The research undertaken at Lick spans the diverse interests of the astronomical community. In 1892, using the 0.9-m refractor, Edward BARNARD discovered Jupiter's fifth moon, the first found since Galileo. Continuing that tradition, the Shane reflector, coupled to the Hamilton high-dispersion echelle spectrograph, has become the world's premier planet finder (for planets of other stars). Other ongoing programs include SUPERNOVA searches, ADAPTIVE OPTICS development and research into the inertial proper motions of stars.

Light

Electromagnetic radiation of those wavelengths to which the human eye responds, that is from just below 400 to just over 700 nm. Different wavelengths within this range correspond to different colors as perceived by the eye. In order of decreasing wavelength they are red, orange, yellow, green, blue, indigo and violet. Wavelengths shorter than violet are referred to as ultraviolet, while wavelengths longer than red are infrared. Light may also be regarded as particles or photons, the longer-wavelength radiation corresponding to lower-energy photons. In common with other forms of electromagnetic radiation, light travels in a vacuum at a speed of 300 000 km s^{-1}.

Light curve

A graph of the change in brightness of a variable object with time. For periodic variables the curve may be folded over a number of repeating cycles. The term may also be used to describe variations in other wavelength bands, such as an X-RAY light curve.

Light echo

The pulse of light from a NOVA or SUPERNOVA may reflect from dust in the vicinity and form a temporary REFLECTION NEBULA. The temporary reflection is known as a light echo. The phenomenon was first identified in Nova Persei 1901, and was seen again surrounding supernova 1987a, and after the 2002 nova-like outburst of V838 Monocerotis.

Light-gathering power (light grasp)

A measure of the amount of light collected and brought to a focus by a telescope. The light collected is proportional to the surface area of the collector (objective or primary mirror), and therefore proportional to the square of the APERTURE. For example, a telescope with an aperture of 2 m will collect four times as much light as one with an aperture of only 1 m.

Light year (l.y.)

A unit of measurement equal to the distance traveled by a ray of light (or other form of electromagnetic radiation) in a vacuum in one year. Since light travels at 299 792 458 m s^{-1}, this is equivalent to 9.46×10^{12} km (9.46 million million km) or 63 240 AU. The unit is commonly used to express large distances in the universe. A parsec is approximately 3.262 l.y.

LIGO (Laser Interferometer Gravitational Wave Observatory)

LIGO is a facility dedicated to the detection of gravitational waves (*see* GRAVITATION, GRAVITATIONAL LENSING AND GRAVITATIONAL WAVES). It consists of two installations 3218 km apart in Louisiana and Washington within the USA. The detectors each consist of a 1.2-m-diameter vacuum pipe arranged in the shape of an L with 4-km arms. Since gravitational waves penetrate the Earth unimpeded, these installations need not be exposed to the sky and are entirely shaded under a concrete cover. At the vertex of the L and at the end of each of its arms hang test masses with mirror surfaces. Laser beams traversing the vacuum pipes measure the effect of gravitational waves on these masses, and the two observing stations eliminate the effects of any local disturbances. The first data were taken in September 2002.

Limb

The edge of a celestial body which appears as a disk, either to the naked eye or telescopically. As the body moves across the sky, the 'leading' part of the limb is termed 'the preceding limb,' and the opposite, 'lagging,' part 'the following limb.'

Limb darkening

The decrease in brightness toward the edge of the Sun's disk as seen at visible wavelengths. Near the center of the Suns' disk, light is radiated directly toward us. But light reaching us from the LIMB of the Sun has to pass obliquely through a greater thickness of the cooler upper layers of the solar atmosphere. Limb darkening has also been observed in other stars, principally in the light curves of ECLIPSING BINARY STARS.

LINER galaxy *See* ACTIVE GALAXY

Lipperhey, Hans (c.1570–1619)

German spectacle-maker who settled in Zeeland, the Netherlands, and was apparently the inventor of the telescope. Lipperhey applied for a patent and was supported by a letter from the government of Zeeland, dated September 25, 1608. The patent application, the first record of a telescope (although not the first of the idea), was turned down because the device could not be kept secret. Other claimants to the invention appeared immediately afterwards in the Netherlands.

LISA (Laser Interferometer Space Antenna)

The main objective of this NASA/ESA mission is to observe gravitational waves such as may be produced in galactic and extra-galactic binary systems or near very massive BLACK HOLES (*see* GRAVITATION, GRAVITATIONAL LENSING AND GRAVITATIONAL WAVES). LISA consists of three spacecraft flying 5 million km apart in an equilateral triangle. The center of the triangle will be in the ecliptic plane 1 AU from the Sun and 20° behind the Earth close to the L_5 LAGRANGIAN POINT. The spacecraft will act as a giant Michelson INTERFEROMETER, measuring the distortion of space caused by passing gravitational waves. Each spacecraft will contain two free-floating 'proof masses.' These will define optical paths 5 million km long, with a 60° angle between them. Lasers in each spacecraft will measure changes in the optical-path lengths to a precision of 20 picometers (1 picometer = 10^{-12} m). If approved, the project will start in 2004 with a planned launch in 2011.

Literature and astronomy

Just as artists have used astronomical themes as a backdrop to human affairs, so too have writers (*see* ART AND ASTRONOMY). Astronomical references abound in the work of William Shakespeare. His plays contain numerous speeches for and against ASTROLOGY, views against being expressed mostly by the more pushy, rebellious characters. Shakespeare was born after Nicolaus COPERNICUS published his theory of the solar system in 1543, and was a contemporary of enthusiastic English proponents of the new view of the universe, such as Thomas Digges (1545/6–95), William Gilbert (1544–1603) and John Dee (1527–1609). Nevertheless, no trace of this controversy is found in Shakespeare's works, save in disputes about the validity of astrology.

Thomas Hardy was a poet and prolific novelist of the English countryside. He described his novel *Two on a Tower* as a 'slightly-built romance' that 'was the outcome of a wish to set the emotional history of two infinitesimal lives against the stupendous background of the stellar universe, and to impart to readers the sentiment that of these contrasting magnitudes the smaller might be the greater to them as men' (from the preface to the 1895 edition). Peter Ackroyd, an English novelist known for his handling of the relativity of time in his work, also set his novel *First Light* (1989) in a west-country landscape, which shows a palimpsest of time in its tumuli, folk memories and modern observatory.

It is possible in literature to make explicit comment on quite abstract astronomical ideas. The size of the universe is one of astronomy's imaginative pulls, and the very word 'astronomical' has gathered a subsidiary meaning of 'immense' because the quantities in astronomy are so vast. Sometimes the reaction to the distances of stars is awe and fear, as in *Pensées ('Thoughts')* (1670) by French philosopher Blaise Pascal: 'The eternal silence of those infinite spaces strikes me with terror.'

Because it takes so long for light from a star to reach us, the image of the sky that we perceive today originated long ago and may show the memory of a star that no longer exists. 'The Ode to Charles Sumner' (1893) by American poet Henry Wadsworth Longfellow illustrates this idea:

Were a star quenched on high,
For ages would its light,
Still travelling downward from the sky,
Shine on our mortal sight.

So when a great man dies,
For years beyond our ken
The light he leaves behind him lies
Upon the paths of men.

American poet Robert Frost had a lifelong interest in astronomy, starting from the age of 15 when he installed in the upper room of his parents' house a small telescope, bought with money from selling magazine subscriptions. In a speech two months before his death in 1963, he gave his philosophical reaction to astronomy: 'How stirring it is, the Sun and everything. Take a telescope and look as far as you will. How much of the universe was wasted just to produce puny us. It's wonderful... fine.' In 'A Star in a Stone Boat' (1921) he described how celestial material was part of the fabric of the Earth and its contents:

Never tell me that not one star of all
That slip from heaven at night and softly fall
Has been picked up with stones to build a wall.

Primo Levi was an Italian-Jewish poet and writer. Trained as an industrial chemist, he had a dramatic life as a Piedmont partisan, and was a witness, participant and survivor from the concentration camp at Auschwitz. He committed suicide, apparently in despair at the human condition. His poem 'In the Beginning' (1970) describes the same thought as Frost, in relation to the BIG BANG:

From that one spasm everything was born:
The same abyss that enfolds and challenges us,
The same time that spawns and defeats us,
Everything anyone has ever thought,
The eyes of a woman we have loved,
Suns by the thousands
And this hand that writes.

However, again just as in fine art, not all references in literature to astronomy are favorable, and Isaac NEWTON is in the firing line. Four literati attended a drunken dinner on December 28, 1817 – Benjamin Haydon, William Wordsworth, Charles Lamb and John Keats. Lamb abused Haydon for putting a bust of mathematical scientist Newton into a picture that he had just painted. Lamb derided Newton as 'a fellow who believed in nothing unless it was as clear as three sides of a triangle.' Lamb proposed a toast to 'Newton's health and confusion to mathematics.' Keats agreed, and, with Wordsworth, lamented the destruction of beauty by science. This derision of science appears in Keats's and Wordsworth's poetry. No scientist would agree with the sentiment expressed by Keats in lines 229–238 of Part II of *Lamia* (1819):

...Do not all charms fly
At the mere touch of cold philosophy?
There was an awful rainbow once in heaven:
We know her woof, her texture; she is given
In the dull catalogue of common things.
Philosophy will clip an Angel's wings,
Conquer all mysteries by rule and line,
Empty the haunted air, and gnomed mine –
Unweave a rainbow...

Science fiction

Astronomy is associated with a distinct literary genre, namely science fiction. 'Science fiction is that class of prose narrative treating of a situation that could not arise in the world we know, but which is hypothesised on the basis of some innovation in science or technology, or pseudo-technology, whether human or extra-terrestrial in origin' (Kingsley Amis, *New Maps of Hell,* 1960). Science fiction has a long and honorable history. It goes back to the second century AD; the first true novel about an expedition to the Moon was written at that time by a Greek satirist, Lucian of Samosata. He called it the *True History*, because it was made up of nothing but lies from beginning to end. It describes things that he had 'neither seen nor suffered nor learned from another, things which are not and never could have been; therefore my readers should by no means believe them.' Lucian combined a cool brain with a gift for fluent, easy writing, and he possessed a strong sense of humor.

His travelers are sailors, who were caught up in a water spout as they passed through the Pillars of Hercules (our Straits of Gibraltar), and were hurled upward so violently that after seven days and seven nights they landed on the Moon. They were arrested by lunar warriors and imprisoned by the King of the Moon. Associating the Aristotelian perception of celestial bodies as more perfect than the Earth with the Moon, Lucian described the Moon-men as far more advanced than the people of our world. Anything unclean or impure was abhorrent. Sex was either unknown or ignored, and when a Moon-man died he merely dissolved into smoke, so that no remains should be left for burial. The conviction that celestial beings are more pure than we are survives to this day in opinions held by some believers in extraterrestrial visitors and unidentified flying objects.

Johannes KEPLER's *Somnium* (Dream) was very different. In 1593, as a 22-year-old student at the University of Tübingen, Germany, Kepler wrote a dissertation, from a Copernican standpoint, on how the heavens would appear to an observer on the surface of the Moon. Debate on this radical thesis was suppressed. *Somnium*, written by 1609 and published in 1634, expands the thesis into a work of imagination. It was, and was meant to be, educational as well as entertaining (there were 223 notes explaining its references). In its emphasis on what to Kepler was science rather than on the adventure story, the work was the first science-fiction story (hard core, in which the science is the most important feature, as opposed to science fantasy). Kepler's hero, an Icelander named Duracotus, is carried to the Moon by a demon. Kepler knew that the Earth's atmosphere does not extend all the way to the Moon, and that there must be a neutral point where the gravitational pulls of the Earth and Moon balance. Kepler also knew that the Moon always keeps the same face turned toward the Earth, and explains that conditions on the two hemispheres are quite different. He describes a lunar eclipse as seen from the Moon (that is, an eclipse of the Sun by the Earth) and what a solar eclipse looks like to a moon-dweller (that is, the transit of the umbral shadow across the face of the Earth).

French novelist Jules Verne's *From the Earth to the Moon* appeared in 1865 and merited a sequel, *Round the Moon*. Verne made a conscious effort to keep to the facts as he knew them, and on the whole he succeeded remarkably well, even though his basic method of travel was wrong. His travelers were fired to the Moon from the barrel of a huge cannon, and set off at a speed of 7 miles s^{-1}. Numerically Verne was correct; 7 miles s^{-1} is the Earth's escape velocity, and a projectile launched at this speed would never return. Unfortunately, the projectile would at once be destroyed by friction against the atmosphere, and in any case the shock of departure would certainly reduce any travelers to jelly – but at least Verne made it all sound plausible. His description of the Moon was based on the best information available, and his plot is ingenious: the projectile encounters a minor satellite of the Earth, and its path is changed so that instead of reaching the Moon the travelers are boomeranged back home.

H G Wells's *The War of the Worlds* (1898) describes how Earth is invaded by grotesque monsters from Mars, fleeing the desiccation of their planet. They cause devastation until they are destroyed by terrestrial bacteria, against which they have no immunity. In 1938 a radio dramatization of the novel, produced by Orson Welles, caused widespread panic in parts of the USA, after many listeners mistook it for a real news bulletin.

A novel of a very different kind was written by Konstantin Eduardovich Tsiolkovsky, born in 1857 in Ijevsk, a remote village in Russia. He realized that the only practicable method for space travel is the rocket, which depends on the principle of reaction and can work in the vacuum of space. His only novel, *Beyond the Planet Earth*, was probably complete by 1895, although it did not appear in print until 1920 and its English translation was delayed until 1960. As a story, and as a literary effort, it can only be described as atrocious, but as a scientific forecast it was years ahead of its time. In his novel, the travelers used a liquid-fuel rocket motor instead of one fuelled by solids such as gunpowder; he described the 'step' principle of mounting one rocket on top of another (what we would call a multistage rocket). He also gave a perfectly accurate description of the causes and effects of weightlessness, or zero gravity. He was equally aware of the many problems involved in long space journeys, and he proposed taking along various types of green plants to remove excess carbon dioxide from the atmosphere inside the spacecraft and replace it with free oxygen.

During the 1930s large numbers of mass-market science-fiction (and detective-story) magazines appeared. These generated the name 'pulp fiction,' now a synonym for storytelling without sophistication. Nevertheless, some established writers started in this market. Among authors

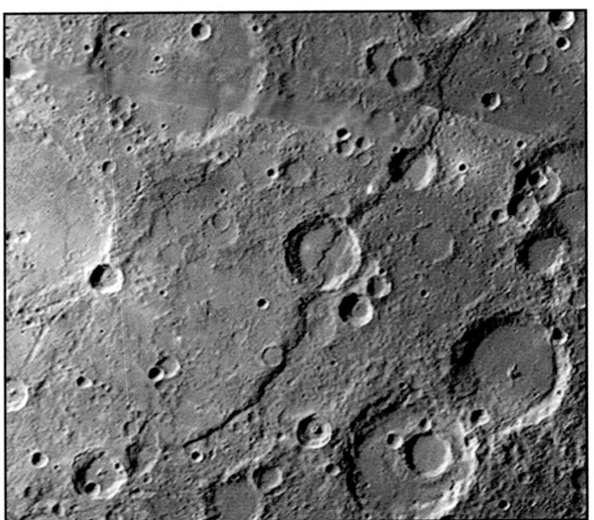

◀ ***Lobate scarp*** *The Discovery Rupes scarp is 550 km long and up to 1500 m high. This thrust fault on Mercury has deformed the floors and walls of two craters, and detoured around the walls of others.*

whose books contain a great deal of sound science (in contrast with the bulk of pulp fiction) are the space-travel visionary Arthur C Clarke, author of *2001: A Space Odyssey* (1968), filmed by Stanley Kubrick. The astronomer Fred HOYLE was also an established writer, whose first and best-known work of science fiction, *The Black Cloud*, includes mathematical formulae (1957).

Lithosphere

The stiff upper layer of a planetary body including the crust and part of the upper mantle. The Earth's lithosphere is broken into rigid 'plates' that can slide over the area beneath, as explained by the theory of plate tectonics.

Little Dumbbell Nebula (M76, NGC 650-51)

See MESSIER CATALOG

Lobate scarp

A long, sinuous, cliff-like feature consisting of a series of connected lobes. They are found chiefly on Mercury, where they are interpreted as thrust faults caused by compressive forces. They vary from 20 to 500 km in length and from a few hundred meters to 2 km in height. Their planet-wide distribution is taken as evidence of a global shrinking early in Mercury's history when the mantle and core cooled and contracted, and the crust 'puckered.' One of the longest is Discovery Rupes ('rupes' is Latin for cliff), which crosses craters; others are found in the Caloris Basin. The global contraction must therefore have happened after the era of major cratering (*see* LATE HEAVY BOMBARDMENT), which ended about 3.9 billion years ago.

Local Group of galaxies

Not long after Edwin HUBBLE established that galaxies are 'island universes' similar to our own Milky Way, he realized that 11 are considerably closer to us than any others. He coined the term 'Local Group' in *The Realm of the Nebulae* (1936) to identify them. There are 43 galaxies identified now as the members of the Local Group. In them we see close up normal galaxies and galaxies interacting with one another in a relatively small volume of space.

The brightest members of the Local Group (the first members of the table) are so close that they can be seen with the naked eye. But the faintest are so faint that more than half have been discovered in only the last 30 years and there may be more to come. They are DWARF GALAXIES, whose low brightness is so spread out that their surface brightness is lower than that of the night sky, making them very difficult to detect. Additionally, searches for galaxies near the Milky Way are severely hindered by the massed stars and by the clouds of gas and dust within the plane of our Galaxy.

Some of the galaxies listed in the table may actually only be 'passing through the neighborhood.' Moreover, the distances to some are unreliable. Thus we can expect not only additions to the table but also subtractions.

The individual galaxies of the Local Group span a large range of basic properties – luminosities, sizes and types. The ANDROMEDA GALAXY (M31), the Milky Way and M33 are SPIRAL GALAXIES, each a slightly different class.

All the remaining members of the Local Group are dwarf galaxies. The most massive is the LARGE MAGELLANIC CLOUD (LMC). Indeed, the LMC is so large that it has incipient spiral structure and some astronomers classify it as a spiral galaxy. The LMC's companion, the SMALL MAGELLANIC CLOUD, is a true irregular with little sign of large-scale structure. The remaining irregular galaxies are substantially smaller and less luminous than the LMC. Some of them are among the most metal-poor galaxies known. They are forming stars in large quantities for the first time in their lives, in the case of IC 10 at an unsustainably rapid rate. If its starburst were to continue, IC 10 would exhaust its raw materials in only a few million years. Unless it started out with an astoundingly large reservoir of gas, IC 10 has probably been caught during a particularly active (but short-lived) phase in its star-formation history.

The remaining dwarf galaxies of the Local Group are gas-poor. M32 is a dwarf elliptical galaxy and harbors a massive black hole in its extremely bright nucleus. It is distorted by a strong gravitational interaction with M31. The remaining dim, unassuming dwarf spheroidal galaxies ('dSph galaxies') are the most common type in the entire universe.

Between the galaxies of the Local Group are isolated clouds of gas, usually neutral hydrogen. This may be primordial matter that has yet to collapse into small galaxies or it may have been expelled from the galaxies, like the MAGELLANIC STREAM. Some remote globular star clusters are 'free-floating' members of the Local Group that were ejected from their parent galaxies during past interactions.

The motions of galaxies within the Local Group

The Local Group is not uniformly filled with galaxies. There are three or four clumps. The 13 satellites of the Milky Way are in orbit about our Galaxy. M31 has a similar stable of small companions. These situations are analogous to the moons that orbit the planets of the solar system: although Jupiter and Saturn orbit the Sun, both planets possess their own large families of bound satellites. The close companions of M31 (NGC 205 and M32) both show distortions due to strong tidal effects. The Magellanic Clouds are similarly distorted by our Galaxy, and have lost the gas of the Magellanic Stream, pulled from the Clouds as they passed close to the Milky Way. In the next few billion years the Clouds will fall into and merge with the Milky Way. The SAGITTARIUS DWARF GALAXY, only discovered in 1994, is already at this stage.

The clumps of the Local Group interact with each other. This makes some galaxies appear to violate the HUBBLE LAW that the universe is expanding. The most striking example is M31, which is approaching our Galaxy at nearly 50 km s^{-1}.

Local Group

Galaxy	Other name	Year	RA		Dec.		Hubble type	Sub.	Dist.	Lumin.	Mass
			h	m	°	′					
M31	NGC 224	-	00	42.7	+41	16	SbI-II	M31	2.5	25 000	700 000
Milky Way	-	-	17	45.7	-29	01	Sbc	MW	0.03	8300	350 000
M33	NGC 598	-	01	33.9	+30	40	ScII-III	M31	2.7	3000	30 000
LMC	-	-	05	23.6	-69	45	IrrIII-IV	MW	0.16	2100	20 000
SMC	NGC 292	-	00	52.7	-72	50	IrrIV-V	MW	0.19	580	1000
WLM	DDO 221	1923	00	02.0	-15	28	IrrIV-V	LGC	3.0	500	150
M32	NGC 221	1749	00	42.7	+40	52	E2	M31	2.6	380	2120
NGC 205	M110	1864	00	40.4	+41	41	E5p/dSph-N	M31	2.6	370	740
NGC 3109	DDO 236	1864	10	03.1	-26	10	IrrIV-V	N3109	4.1	160	6550
IC 10	UGC 192	1895	00	20.4	+59	18	dIrr	M31	2.7	160	1580
NGC 185	UGC 396	1864	00	39.0	+48	20	dSph/dE3p	M31	2.0	130	130
NGC 147	DDO 3	1864	00	33.2	+48	31	dSph/dE5	M31	2.3	130	110
NGC 6822	DDO 209	1864	19	44.9	-14	48	IrrIV-V	LGC	1.6	94	1640
IC 5152	-	1895	22	02.7	-51	18	dIrr	LGC	5.2	70	400
IC 1613	DDO 8	1906	01	04.9	+02	08	IrrV	M31	2.3	64	795
Sextans A	DDO 75	1942	10	11.1	-04	43	dIrr	N3109	4.7	56	395
Sextans B	DDO 70	1955	10	00.0	+05	20	dIrr	N3109	4.4	41	885
Sagittarius	-	1994	18	55.1	-30	29	dSph-N	MW	0.08	18	-
Fornax	-	1938	02	40.0	-34	27	dSph	MW	0.45	16	68
Pegasus	DDO 216	1958	23	28.6	+14	45	dIrr/dSph	LGC	3.1	12	58
EGB0427+63	UGCA 92	1984	04	32.0	+63	36	dIrr	M31	4.2	9.1	-
Sag DIG	UKS1927-177	1977	19	30.0	-17	41	dIrr	LGC	3.4	6.9	9.6
And VII	Cassiopeia	1998	23	26.5	+50	42	dSph	M31	2.5	5.7	-
UKS2323-326	UGCA 438	1978	23	26.5	-32	23	dIrr	LGC	4.3	5.3	-
Leo I	DDO 74	1955	10	08.5	+12	19	dSph	MW	0.81	4.8	22
And I	-	1972	00	45.7	+38	00	dSph	M31	2.6	4.7	-
GR 8	DDO 155	1956	12	58.7	+14	13	dIrr	GR8	4.9	3.4	7.6
Leo A	DDO 69	1942	09	59.4	+30	45	dIrr	MW	2.2	3.0	11
And II	-	1972	01	16.5	+33	26	dSph	M31	1.7	2.4	-
Sculptor	-	1938	01	00.2	-33	43	dSph	MW	0.26	2.2	6.4
Antlia	-	1985	10	04.1	-27	20	dIrr/dSph	N3109	4.1	1.7	12
And VI	Peg dSph	1998	23	51.7	+24	36	dSph	M31	2.7	1.4	-
LGS 3	Pisces	1978	01	03.9	+21	53	dIrr/dSph	M31	2.6	1.3	13
And III	-	1972	00	35.3	+36	31	dSph	M31	2.5	1.1	-
And V	-	1998	01	10.3	+47	38	dSph	M31	2.6	1.0	-
Phoenix	-	1976	01	51.1	-44	27	dIrr/dSph	MW	1.4	0.9	33
DDO 210	Aquarius	1959	20	46.8	-12	51	dIrr/dSph	LGC	2.6	0.8	5.4
Tucana	-	1985	22	41.8	-64	25	dSph	LGC	2.9	0.6	-
Leo II	DDO 93	1950	11	13.5	+22	09	dSph	MW	0.66	0.6	9.7
Sextans	-	1990	10	13.1	-01	37	dSph	MW	0.28	0.5	19
Carina	-	1977	06	41.6	-50	58	dSph	MW	0.33	0.4	13
Ursa Minor	DDO 199	1955	15	09.2	+67	13	dSph	MW	0.21	0.3	23
Draco	DDO 208	1955	17	20.3	+57	55	dSph	MW	0.27	0.3	22

WLM = Wolf-Lundmark-Melotte galaxy in Cetus; LGS-3, also called the Pisces Dwarf, was the only one of five dwarf galaxies found in 1978 that proved to be in the Local Group (LGS = Local Group (Suspected)); DDO = entry in the David Dunlap Observatory Catalog of Galaxies by Sydney van den Bergh (1959); LGC = Local Group Cloud; Year = year of discovery; RA = right ascension in 2000; Dec. = Declination in 2000; Sub. = Hubble subgroup; Dist. = distance in millions of light years; Lumin. = luminosity in millions of Suns; Mass = mass in millions of Suns

If it is on a collision course, the two galaxies will meet in about 8–10 billion years. More likely, M31 and the Milky Way make up a system like a BINARY STAR in which the two galaxies orbit their common center of gravity. As with binary stars, their orbit can be used to estimate their mass. Much as a projectile thrown in the air arcs up then falls toward the Earth, they were thrown apart by the Big Bang and are now falling toward each other on their first approaching orbit. Their orbital period can be estimated as the age of the universe. The resulting mass of M31 and the Milky Way is about 1000 billion times that of the Sun. Virtually all the matter of the Local Group is located in these two galaxies, so this is also our best estimate of its total mass. As they continue to orbit, each will raises tides on the other and this will cause them to merge in the distant future.

As massive as the Local Group is, the total light from all its galaxies is 'only' 30 billion times the luminosity of the Sun. This is 50 times less than astronomers would normally expect from its mass, suggesting that the group is dominated by DARK MATTER. The Local Group galaxies show the same statistic individually. Thus, the dark matter within the Local Group is not spread out uniformly, but is concentrated on

◀ ***Antlia dwarf galaxy*** *The individual stars of this member of the Local Group can be seen, along with many background galaxies.*

the individual galaxies, in dark halos. The Local Group may even contain 'dark galaxies' of dark matter with little luminous material, although no such systems have been detected.

The Local Group is itself interacting with other nearby groups of galaxies. It has stretched the Sculptor Group to such an extent that there is no clear boundary between them. The same is true to a lesser extent with the M81-Maffei 1 Group on the opposite side of the sky. On top of the interactions with neighboring groups, the Local Group is falling into the nearby VIRGO CLUSTER OF GALAXIES.

Further reading: Hodge P, Skelton B P and Ashizawa J (2003) *An Atlas of Local Group Galaxies* Kluwer; van den Bergh S (2000) *The Galaxies of the Local Group* Cambridge University Press.

Local time *See* TIME

Lockyer, Joseph Norman (1836–1920)

English astronomer, who started his career as a civil servant. Observing an annular solar eclipse in 1858 stimulated his interest and he erected an observatory at his home in Hampstead, London. In time, he rose to become director of the Solar Physics Observatory, Kensington (which in 1911 was moved to Cambridge University). Lockyer then retired to Devon, UK, where he built himself a private observatory. Lockyer showed spectroscopically that phenomena seen

during a total eclipse could be seen in daytime on the SUN and were not lunar phenomena. He and Pierre JANSSEN, reaching the same conclusions independently, reported them to the same meeting of the Academy of Sciences and both received a medal. Lockyer identified all the emissions in the solar spectrum except for one, which he suggested was due to a new element. He named it HELIUM after 'helios,' the Greek word for Sun. Helium was isolated on the Earth 25 years later by William Ramsay (1895), and Lockyer was knighted.

Long-period comet

A comet whose PERIOD exceeds 200 years. Comets with lesser periods are 'periodic' or SHORT-PERIOD COMETS. Long-period comets may also be 'periodic,' but one can only be incontrovertibly classified as periodic when it has been observed on at least two returns. The choice of 200 years as a cutoff is traditional, but arbitrary. No comet with a calculated period of more than 200 years has yet been observed to return. The first to do so will be Comet Peters (C/1857 O1), expected to return in 2092. About one-third of all long-period comets are found to have periods close to 1 million years, corresponding to APHELIA of rather less than 50 000 AU, while the periods of the remainder are fairly evenly spread between 1 million and 200 years. The million-year peak points to the origin of these comets in the Oort Cloud (*see* OORT CLOUD AND KUIPER BELT).

Lovell, Alfred Charles Bernard (1913–)

English physicist and astronomer who worked on cosmic rays at Manchester University, researched radar in World War II, and afterwards with James Hey procured ex-army equipment and attempted to detect cosmic-ray showers. They moved to the university's botanical research station at Jodrell Bank (now the Nuffield Radio Astronomy Laboratories) to avoid electrical interference in Manchester, and constructed the 76-m-diameter Mark 1 Jodrell Bank Telescope, completed in 1957. The Mark 1 telescope was thrust into public view that year when it received signals from Sputnik and American space vehicles. Lovell was knighted in 1961. He has written many popular books on astronomy, and on radio astronomy in particular.

Lowell, Percival (1855–1916)

American businessman, traveler, diplomat and, eventually, astronomer in a distinguished family (his younger brother was a president of Harvard University, and his sister was a writer). At the age of 37 he traveled with William PICKERING to observe the 1894 Mars opposition from the clear skies of Arizona. They founded the LOWELL OBSERVATORY at Flagstaff to investigate the possibility of life on other worlds in the solar system. Lowell had already been convinced by Giovanni SCHIAPARELLI'S identification of a network of canals on Mars. He studied the planet for 15 years, noting the seasonal changes and drawing Mars in exquisite but imaginary detail, covering its map with a network of hundreds of straight 'canals' intersecting in 'oases.' However, the editors of *Astrophysical Journal* refused to publish any of Lowell's submissions. In his later years Lowell conducted a search for a trans-Neptunian planet, having analyzed the orbit of Uranus and found discrepancies even after allowing for Neptune. He was unable to find the planet, but his search was continued after his death by Clyde TOMBAUGH at Flagstaff, and Pluto was discovered in 1930. There is great resistance by astronomers to attempts to name planets after living people, but this tradition was sidestepped by making the first two letters of Pluto, and the symbol for the planet, Percival Lowell's initials.

Lowell Observatory

Founded in 1894 by Percival LOWELL and endowed by him, Lowell Observatory is one of the largest independent, privately managed research observatories in the world. The large REDSHIFTS of galaxies were discovered by Lowell astronomer Vesto SLIPHER, and here in 1930 Clyde TOMBAUGH discovered Pluto. Today Lowell Observatory has a staff of 50, including 20 astronomers. The observatory operates eight telescopes: four at its original campus on Mars Hill in Flagstaff, Arizona and four at a dark-sky site on Anderson Mesa, southeast of Flagstaff. Lowell is a partner with the US NAVAL OBSERVATORY and the Naval Research Laboratory in the Navy Prototype Optical Interferometer (NPOI) at Anderson Mesa, and with Boston University in the shared use and development of Lowell's 1.8-m Perkins Telescope. The observatory is engaged in research spanning many areas of modern astronomy and astrophysics, while maintaining its traditional emphasis on the study of the solar system.

Low-surface-brightness galaxy

A low-surface-brightness (LSB) galaxy emits much less light per area than a normal galaxy. Because of their lower contrast with the night sky LSB galaxies are hard to find, so their contribution to the general galaxy population has long been underestimated. LSB galaxies are contrasted with high-surface-brightness (HSB) galaxies, which for all practical purposes are normal SPIRAL GALAXIES. Investigations have found that LSB galaxies are more gas-rich than HSB galaxies, but the gas surface densities are well below the critical threshold for star formation, leading to very low star-formation rates. LSB galaxies are also bluer than normal late-type galaxies, suggesting that they are slowly evolving with sporadic periods of star formation. A small amount of recently observed star formation is already enough to make the colors significantly bluer.

The rotation curves of LSB galaxies show that they are dominated by DARK MATTER almost all the way into their centers (*see* ROTATION). The curves rise less steeply than those of HSB galaxies, so the LSB halos are probably less dense. LSB galaxies can be said to be trapped in their current evolutionary state. Even external influences, like interactions, might not be enough to accelerate their evolution significantly.

The above results apply to blue LSB galaxies that have been discovered in photographic plates. Recent CCD surveys turned up a class of red LSB galaxies, showing that even the deep photographic plates suffered from selection effects. However, not enough is known about these red LSB galaxies to determine whether they are a cosmologically significant component of faded galaxies. The most important reason for studying LSB galaxies is to expand our understanding of the full range of nearby galaxies. If a large population of hitherto unseen objects exists, the true extent of galaxy properties may be much larger than assumed. Furthermore, investigations of individual LSB galaxies show that they form an alternative track of galaxy evolution, free from the instabilities and interactions that have shaped the Hubble sequence. They therefore give us the opportunity to study unevolved galaxies in great detail.

Luminosity

Brightness of a star or galaxy measured in some standard way; that is, not depending on distance. In a star, it is measured by its absolute magnitude (*see* MAGNITUDE AND PHOTOMETRY). In a star's spectrum, luminosity is characterized by its 'luminosity class' (I for a supergiant, III for a giant and V for a dwarf, and numbers between for intermediate cases, such as III–IV). In physics units, luminosity is measured in watts (W), or their equivalent.

Luminosity function

The relative numbers of stars (or other objects) of a particular luminosity in a given volume of space. For stars the volume considered is usually a cubic parsec, for galaxies a cubic megaparsec. The luminosity may be measured in the optical, radio or any other waveband. The luminosity function for stars in the solar neighborhood peaks at around absolute magnitude +15, showing that the numbers of intrinsically faint stars dominate our part of the Milky Way galaxy.

Luminous blue variable

Luminous blue variable stars (LBVs) are among the most luminous hot stars, and suffer very irregular and unpredictable changes in brightness. The visual brightness of the star can vary by about a magnitude, although some have suffered very large eruptions. LBVs are very rare: there are only five confirmed LBVs in our Galaxy (ETA CARINAE, AG Carinae, HR Carinae, P Cygni and HD 160529), and a few tens in other galaxies, including the bright star S Doradus in the LARGE MAGELLANIC CLOUD. This is because they represent a relatively short phase in the life of the most massive stars. Yet this phase is critical to their evolution: the large mass-loss prevents them from becoming red supergiants, and they evolve directly into very hot, helium-rich stars (WOLF–RAYET STARS).

LBVs are recognized by their strange variability on many timescales. Some LBVs are changing all the time, the light curves meandering up and down from year to year; others go through phases of roughly constant brightness that may last 10 years. Satellite observations have shown that when these stars are faint in visual light, they are bright in ultraviolet radiation and vice versa, demonstrating that the variations are due to changes in their radius and surface temperature. During the visually faint phases, the stars are relatively hot and small. When they are visually bright their surface temperature drops but their radius increases by a factor of 2–10. The reason for these changes is still unknown. Several LBVs have been observed to go through large eruptions, when they brighten by as much as a factor of 100. Such eruptions are rare and may occur only once every few thousand years.

During an eruption the LBV ejects a large amount of mass, roughly the total mass of the Sun. The most famous LBV is Eta Carinae, which had an eruption in late 1830s when it suddenly became one of the brightest stars in the southern sky. The star is now surrounded by a magnificant bipolar nebula consisting of the gas ejected during the eruption.

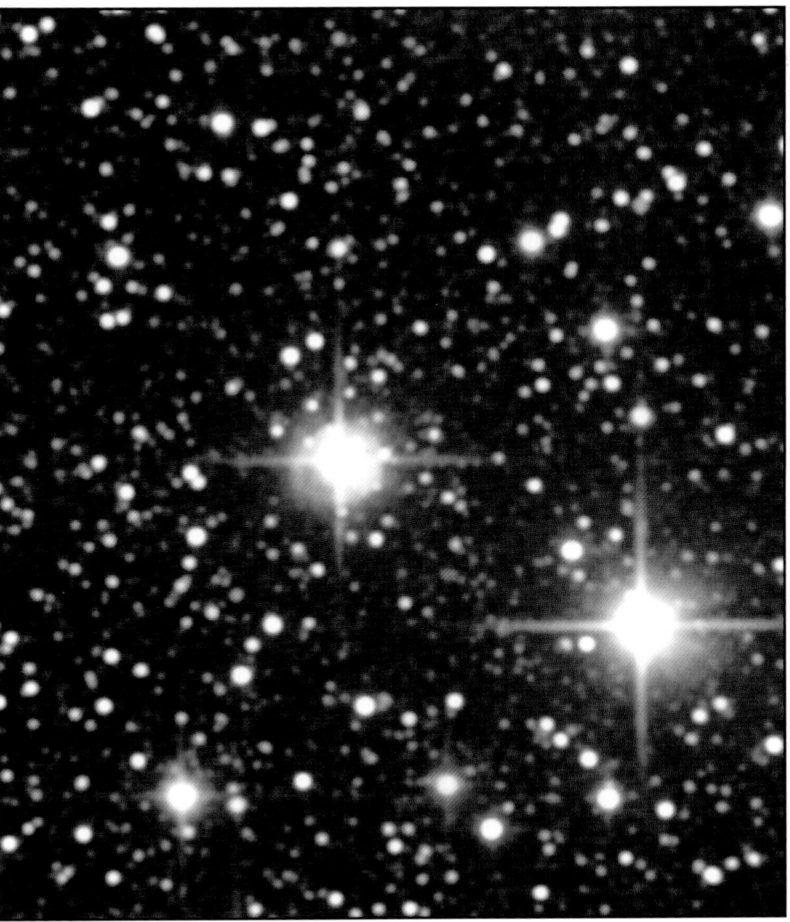

▲ ***Luminous blue variable*** *The central star in this image is AG Car, a prototypical LBV. One million times as luminous as the Sun, AG Car's luminosity has varied greatly over the last two decades. A dusty circumstellar nebula is visible as a blue ring around the star, probably the result of an earlier, brief red supergiant phase.*

Luna

Series of Soviet automated Moon missions launched in 1959–76. Lunas 16, 20 and 24 were successful sample-return missions. Lunas 17 and 21 carried the first automated Moon rovers, known as LUNOKHODS 1 and 2. Luna 3 returned the first pictures of the lunar far side, while Luna 9 sent back the first pictures from the Moon's surface.

Lunar-A

Japanese lunar orbiter, scheduled for launch in 2004. The orbiter will fire two penetrators 1–3 m into the lunar surface to detect moonquakes and measure soil temperature ('penetrators' burrow into the surface, in contrast with 'probes' which land on it, and 'rovers' which are mobile on it).

Lunar Orbiter

Series of five NASA moon-orbiting spacecraft, launched 1966–67, and designed to map potential APOLLO landing sites. Perturbations in the orbit of Lunar Orbiter 1 provided the first data on lunar mass concentrations (mascons) and associated gravity anomalies. This information was used to create the first detailed lunar atlas, with spatial resolution down to 1 m.

Lunar Prospector

NASA mission in the Discovery class, intended to contrast with the largest missions by being 'faster, better, cheaper,' in the words of the slogan. Launched in January 1998, and designed to spend one year in orbit around the MOON. Discovery carried a neutron spectrometer which gave strong indications of water ice in shadowed craters at both lunar poles. The instrument also gave high-resolution gravity data and returned information on the composition of the lunar crust. Discovery was lowered into a 30-km orbit in January 1999 on completion of its primary mission. The extended mission ended in July 1999 when the spacecraft was directed to crash into a crater near the south pole. This was part of an experiment to confirm the existence of water ice on the Moon, but it produced no observable signature.

Lunar transient phenomenon

A purported localized and short-lived change in the appearance of a feature on the surface of the MOON. Lunar transient phenomena (LTPs) tend to be reported by amateur observers . They take various forms, including temporary colorations (usually red), bright flashes (visible especially in shadows or on the night side), extended cloudy patches and obscurations of normally visible features. Permanent changes to lunar features reported before the days of close-range photography from spacecraft are now discounted, being ascribed to deficiencies in mapping, observational error, or wishful thinking. Temporary changes reported more recently may be genuine, but they remain controversial. Unequivocal sightings of LTPs could be evidence that the Moon is not the geologically inert body it is generally held to be. There is a connection between LTPs and moonquakes

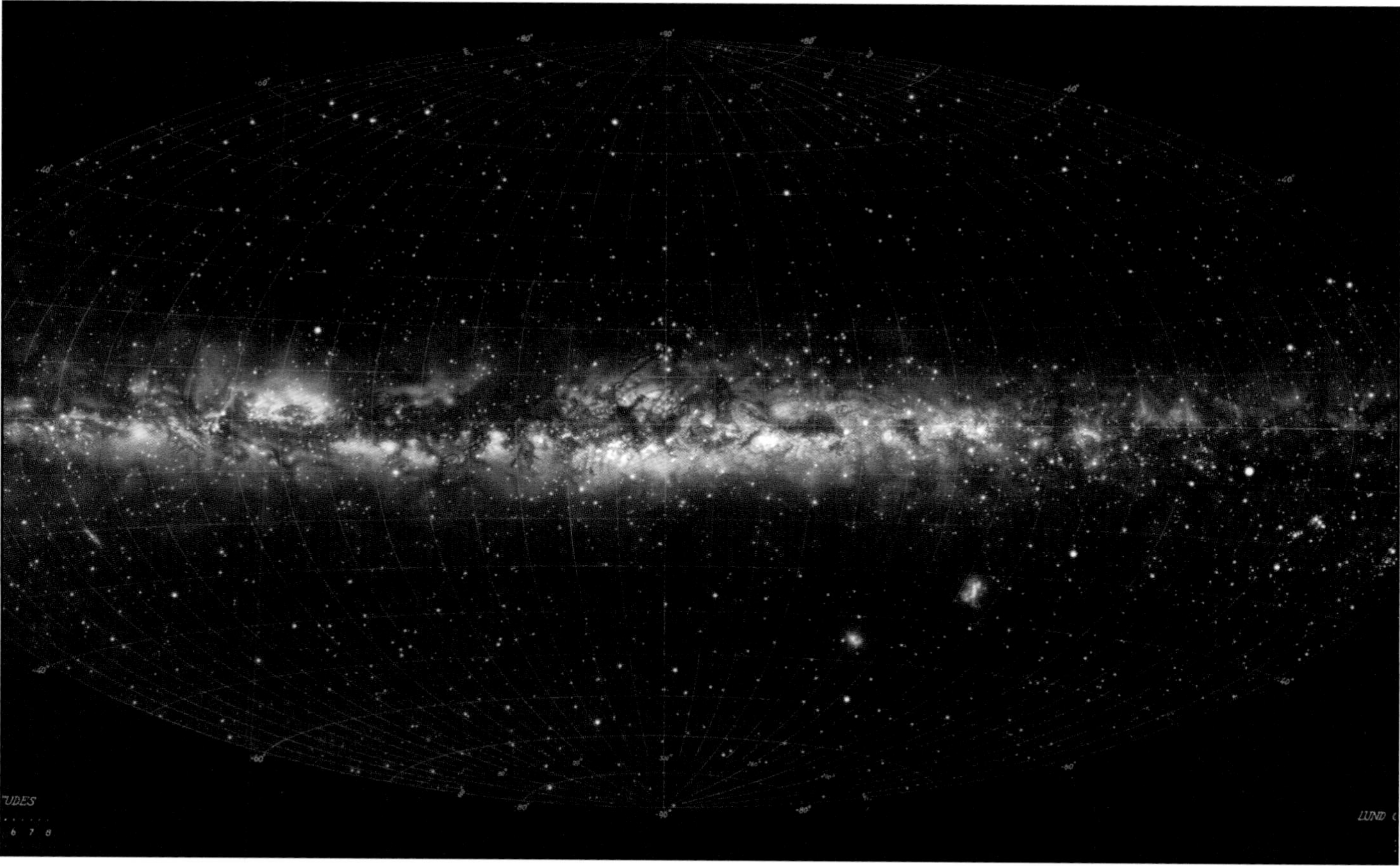

▲ ***Milky Way*** *Knut Lundmark of the Lund Observatory supervised this drawing of the Milky Way and 7000 individual stars, painted by two engineers over the years up to 1955. It remains the best all-sky representation of the naked-eye Milky Way.*

– seismic tremors detected by instruments left on the Moon during the APOLLO missions that originate near the crust-mantle boundary. Both phenomena are more common when the Moon is at PERIGEE. It has been suggested that tidal flexing is triggering the release of gas or dust. It is possible that brief LTPs are caused by the impact of small METEORITES. There was initial excitement when a fresh-looking crater observed by CLEMENTINE in 1994 appeared to coincide with the position of a bright flare recorded on camera by an amateur astronomer in Oklahoma vin 1953, but this identification is now doubtful.

Lunation

The time taken for a complete cycle of the phases of the Moon, such as from one full Moon to the next. A lunation lasts for 29.53 days and is the same as the SYNODIC PERIOD.

Lund Observatory

Lund Observatory is part of Lund University, Lund, Sweden, active in astronomy since 1668. The past century featured stellar statistics (Carl Charlier) and (extra-)galactic studies (Knut Lundmark, who produced a Milky Way panorama in the 1950s). Current projects include stellar astrophysics, space astrometry, the design of large optical and radio telescopes, and involvement in various international observatories.

Lunokhod

Two Soviet Moon rovers, the first automated roving vehicles to operate on another world. They were operated from Earth by remote control, and carried stereo cameras, a laser reflector, a magnetometer, an X-ray spectrometer and a cosmic-ray detector. Lunokhod 1 was launched onboard LUNA 17 in November 1970. The rover landed on Mare Imbrium and operated for 322 days, covering 10.5 km. Lunokhod 2 was launched onboard Luna 21 in January 1973. It landed to the east of Mare Serenitatis and operated for four months, covering 37 km.

Lupus

See CONSTELLATION

Luyten, Willem Jacob (1899–1994)

American astronomer who worked at the University of Minnesota, determining the proper motions of more than 120 000 stars as a means of finding nearby or high-velocity stars and determining their distances. He repeated the PALOMAR OBSERVATORY Sky Survey, building an automated, computerized plate-measuring machine to compare it with the earlier survey, and thus determined the proper motions of 400 000 more. With these data he discovered the great majority of known WHITE DWARFS.

Lyman-alpha forest

An absorption phenomenon which may be seen in the spectra of high-redshift QUASARS and GALAXIES. *See also* INTERGALACTIC MEDIUM.

Lyman-break technique

A technique for identifying high-redshift objects based on their overall energy distribution. Star-forming galaxies are very luminous at ultraviolet wavelengths and have a characteristic SPECTRUM. In a diagram that plots the light intensity as a function of wavelength, the spectrum of a star-

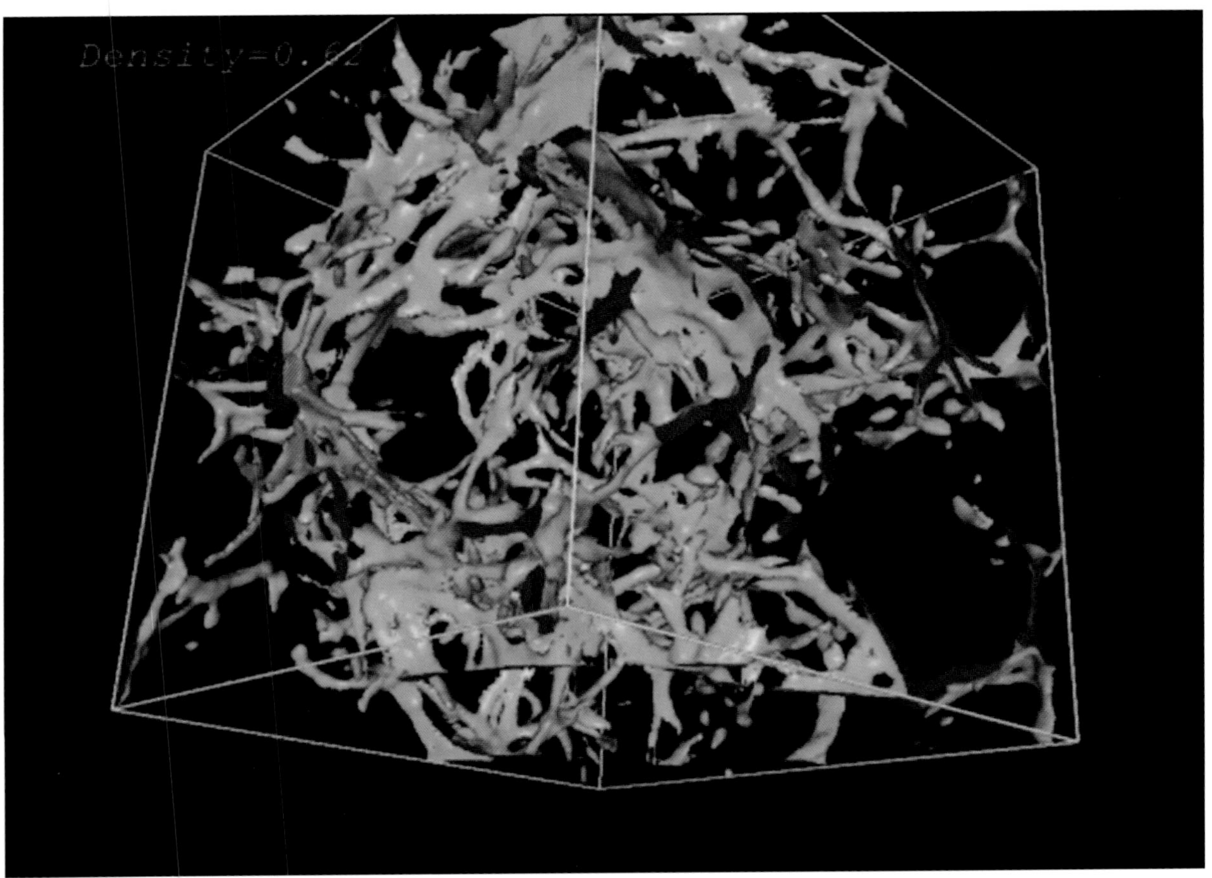

◀ ***Lyman-alpha forest*** *One possible simulation of the distribution of intergalactic clouds as they may have been 11 billion years ago. Each side of the box is 30 million l.y. across. The sight line to a distant quasar cuts through the clouds to produce Lyman-alpha absorption – one 'tree' of the forest for each crossing of a cloud.*

forming galaxy is approximately horizontal. However, the Lyman break at a rest wavelength of 91.2 nm interrupts the 'flatness' of these galaxies' spectra, and provides the telltale clue to identify them. For example, at redshifts around $z \sim 3$ the Lyman break is shifted from 91.2 nm in the rest frame to 370.0 nm in the observer's frame, well into the optical band and observable from the ground. When observed through a set of filters that straddle the Lyman break for a particular redshift, the galaxy images are relatively bright in filters longward of the break, and extremely faint (or not visible at all) in filters that probe shortward of the break. Thus it is possible to find high-redshift galaxy candidates from the much more abundant galaxies of similar apparent LUMINOSITY at modest distances. These candidates are followed up with spectroscopic measurements to confirm their redshifts. In practice, the Lyman-break technique is most efficiently used with telescopes of middle size (for example, 4–5 m), while the spectroscopic measurements to confirm the redshifts require telescopes of the 8-m class or larger.

Lyman series

Emission and/or absorption lines in the spectrum of neutral hydrogen (H I) produced by transitions to and from the lowest energy level (the ground state) – *see* ABSORPTION SPECTRUM; EMISSION SPECTRUM. For example, the Lyman-alpha emission line is produced by the emission of a photon of energy as the electron drops from the first excited state to the ground state (*see* LYMAN-ALPHA FOREST). All the lines in the series are in the far ultraviolet with Lyman-alpha at a wavelength of 121.6 nm and the series converging at the Lyman limit at 91.2 nm.

Lynx *See* CONSTELLATION

Lyot, Bernard Ferdinand (1897–1952)

French astronomer who worked at the Meudon Observatory and invented the coronagraph, a device that creates an artificial eclipse in a telescope and allows the Sun's CORONA to be observed at any time. With it he recorded the SPECTRUM of the corona in new detail, and the first time-lapse pictures of solar prominences. He also invented the Lyot filter, a birefringent INTERFERENCE filter with alternating layers of Polaroid and calcite plates. He also pioneered the study of the POLARIZATION of light reflected from the surface of the Moon and of the planets, finding that the lunar surface behaves like volcanic dust and that Mars has sandstorms.

Lyra *See* CONSTELLATION

Lyrids

A meteor shower that takes place in April, sometimes known as the 'April Lyrids.' The RADIANT lies in the constellation Lyra, close to the border with Hercules, near the star Vega. There are occasional outbursts, most recently in 1982, where the usual peak ZENITHAL HOURLY RATE of 10–15 increases to 100 or more. The parent COMET, C/1861 G1 Thatcher, has the longest period (415 years) of any comet known to be associated with a meteor shower. Its high orbital inclination of 80° means that the meteor stream is little affected by planetary perturbations, and Lyrid activity therefore shows a long-term constancy.

Mach, Ernst (1838–1916)

Czech physicist, scientist and philosopher. The basis of his natural philosophy was that all knowledge is a matter of experiments, indeed sensations, so that 'laws of nature' are summaries of experience provided by fallible senses. Albert EINSTEIN, a former student of Mach, expressed much the same thing in his theory of relativity by incorporating the speed of light into concepts of simultaneity. 'Mach's principle' is a philosophical statement which says that the properties of a body (such as inertia) are influenced by the properties of all the other bodies in the universe. This principle is not easy to interpret scientifically, but it also influenced Einstein, in his concept of spacetime and general relativity. Willem DE SITTER, however, found solutions to Einstein's field equations in the absence of matter – that is, a body to which the equations are being applied has inertia even in an empty universe. In GENERAL RELATIVITY, therefore, the inertia of a body does not arise because of other bodies, and Mach's principle is wrong in this theory. Mach's work on acoustics is remembered in the term 'Mach number' – the ratio of a body's speed to the speed of sound.

MACHO

(1) A MACHO is a Massive Compact Halo Object, that is, hypothetically one of the black holes, brown dwarfs or planets that make up the dark matter in the halo of the Milky Way Galaxy. *Contrast* WIMP.

(2) The MACHO project is an experiment to determine the nature and amount of dark matter in the halo of the Milky Way. If a MACHO passes in front of a distant extragalactic star, for example, one in the Magellanic Clouds, the light of the star will be amplified by the gravitational lens effect. *See also* DARK MATTER; GRAVITATION, GRAVITATIONAL LENSING AND GRAVITATIONAL WAVES.

Maffei 1 and 2

Two relatively nearby galaxies that lie so close to the plane of the Milky Way that they are almost completely obscured by interstellar dust. They were discovered in 1968 as a result of infrared observations conducted by the Italian astronomer Paulo Maffei. Maffei 1 appears to be an ELLIPTICAL GALAXY, located at a distance of about 330 000 l.y. Maffei 2 is a SPIRAL GALAXY that lies at a distance of 15 million l.y.

Magellan

Highly successful NASA Venus orbiter, launched from the Space Shuttle Atlantis in May 1989. It arrived at VENUS in August 1990. Magellan used synthetic aperture radar to map 98% of the surface of Venus at a resolution of 120–300 m, and provided global altimetry data. It revealed more than 1000 impact craters and 1100 volcanic features. The orbiter also discovered sinuous valleys, and unique geological structures known as coronae and arachnoids. Precision radio tracking of the spacecraft measured Venus's gravitational field. Magellan entered the planet's atmosphere in October 1994.

Magellanic Clouds

The LARGE MAGELLANIC CLOUD and SMALL MAGELLANIC CLOUD are two naked-eye IRREGULAR GALAXIES (or in some classifications, BARRED SPIRAL GALAXIES) in the southern sky. Extending over approximately 7° × 7°, the Large Magellanic Cloud has a prominent stellar bar superposed on a disk, which is tilted by about 40° with respect to the sky. The Small Magellanic Cloud, about 4° × 3°, has a less distinct bar-like structure, with its long axis stretched along the line of sight. The Large and Small Magellanic Clouds are at distances of 50 kiloparsecs and 60 kiloparsecs, respectively, and are believed to be satellite galaxies of our Milky Way and members of the LOCAL GROUP OF GALAXIES. Their masses, 20 billion and 2 billion solar masses, are much smaller than the mass of the Milky Way, which is a few hundred billion solar masses.

The Magellanic Clouds are a prototypical young stellar population. There were three epochs in their history of star formation: an initial burst, middle age and a recent burst. The recent burst began approximately 3 billion years ago, possibly triggered by a close approach to our Galaxy.

The Magellanic Clouds are close enough to each other and to the Milky Way for tidal interactions to have altered significantly the distribution of their masses. Tidal interactions are responsible for the formation of the stellar bar in the Large Magellanic Cloud as well as the elongated geometry of the Small Magellanic Cloud. Large-scale observations of neutral hydrogen in the Magellanic Clouds show that tidal interactions have produced the Magellanic Bridge between the two galaxies, the MAGELLANIC STREAM and a leading arm in the opposite direction. The Magellanic Clouds are not independent galaxies; they form a Magellanic system and are a laboratory for the study of colliding galaxies (*see* GALAXIES, COLLIDING).

Further reading: Westerlund B E (1997) *The Magellanic Clouds* Cambridge University Press.

◄◄ ***Magellanic Clouds*** *The Large Magellanic Cloud (pictured) and its sister galaxy, the Small Magellanic Cloud, are close enough for their nebulae, star clusters and even individual stars to be seen with a 10-in telescope.*

Magellanic Stream

The Magellanic Stream, which covers an arc of more than 90° across the sky, is a thin band of neutral HYDROGEN. One end is connected with the MAGELLANIC CLOUDS, and the other end trails away from them. While it seems most likely that the Magellanic Stream is the result of gravitational interactions there is no completely satisfactory theory.

Magnetars *See* NEUTRON STAR AND PULSAR

Magnetic fields in stars

Magnetism (the force that deflects the needle of a compass) and magnetic fields have been found in hundreds of stars during the past 50 years. Magnetic fields have been detected in T Tauri stars and other pre–main sequence stars, several types of main sequence stars, white dwarfs and neutron stars.

George HALE detected the first stellar magnetic fields in sunspots in 1908. Thirty-nine years later Horace BABCOCK found the first magnetic field in a star other than the Sun, namely 78 Virginis. This is a 'chemically peculiar' main-sequence star, and one of 200 or so with spectral classification Ap or Bp (*see* STELLAR SPECTRUM: CLASSIFICATION). These stars have unusual surface chemical compositions, arising from nuclear reactions on their surfaces and chemical stratification of their atmospheres caused by the magnetic fields. The typical field strength of such stars is about 0.1 T, some 3000 times greater than the strength of the Earth's magnetic field.

The discovery of pulsars in 1967 by Susan BELL-BURNELL and Antony HEWISH was soon recognized to be the discovery of neutron stars with magnetic fields. They are born with a field strength of about 100 million to 1 billion T, which decays in strength by a factor of about 100. Three years later James Kemp, John Swedlund, John Landstreet and Roger Angel detected the first magnetic field in a white dwarf. These fields range from about 10 to 100 000 T in strength.

Magnetic fields in stars are often variable. The magnetic field is thought to have the form of a dipole, typically inclined (oblique) to the rotation axis of the star by a large angle. As the star rotates, an observer usually sees one pole of the bipolar distribution, and then the other. This model is known as the 'oblique rotator model.'

Astronomers generally accept two origins for observed stellar fields. In the first, a field may be generated by the interplay between CONVECTION in the outer layers of a star and the overall ROTATION of the star, which act as a dynamo. This mechanism seems to be the cause of the magnetic field observed in the Sun. In the second, the field is the result of a star or huge gas cloud collapsing to form a tiny star, which traps a small fraction of the aboriginal magnetic field in the electrically conducting gas. As the field lines of the aboriginal magnetic field are squeezed together, the strength of the entrained field is amplified.

Magnetohydrodynamics

The study of the interaction between a magnetic field and a plasma. Most of the universe is not a normal gas but is instead a plasma. We are all familiar on Earth with the three states of matter (solid, liquid and gas). Matter changes from one state to another according to the temperature (for example, heat turns ice to water), and if you raise the temperature of gas sufficiently it changes to the fourth state of matter, namely plasma. In the plasma state the atoms have split into positive ions and negative electrons, which can flow around freely, so the gas becomes electrically conducting and a current can flow. In a normal gas, such as the air we breathe, there is virtually no interaction with a magnetic field. But in a plasma the close coupling with the magnetic field means that whatever the plasma is doing intimately affects the magnetic field and vice versa. Indeed, on Earth we are in an extremely unusual part of the cosmos, a separate small island of solid, liquid and gas. But as soon as we go up to the IONOSPHERE, the plasma universe begins. This includes the region between the Earth and the Sun, the Sun itself, as well as the interstellar and intergalactic media and the stars and galaxies contained in them.

The magnetic field has several physical effects:

- It exerts a force, which may accelerate plasma or create structure, such as JETS;
- It stores energy, which may later be released as, for example, a solar flare;
- It acts as a thermal blanket, which may protect it from any surrounding hot gas, like the phenomenon on the Sun called a CORONAL CAVITY;
- It channels fast particles and plasma, as in the SOLAR WIND and comet tails;
- It drives instabilities, like coronal mass ejections, and supports waves.

Magnetopause

The location where the internal PLASMA and magnetic field of a MAGNETOSPHERE balance the external plasma and magnetic field of the outside universe. The magnetopause is usually considered the boundary of the magnetosphere. The Earth's magnetosphere is immersed in the SOLAR WIND, and typically the distance from the center of the Earth to its magnetopause is about 10 times the radius of the Earth. The Sun's magnetosphere is called the HELIOSPHERE and is immersed in the INTERSTELLAR MEDIUM. The Pioneer 10 satellite (*see* PIONEER MISSIONS) is on its way to the Sun's magnetopause (or heliopause) but, when, after a journey of 30 years, it was last contactable (in February 2003), it had still not encountered the heliopause even up to a distance of 82 AU. The heliopause must be some 100 AU in radius.

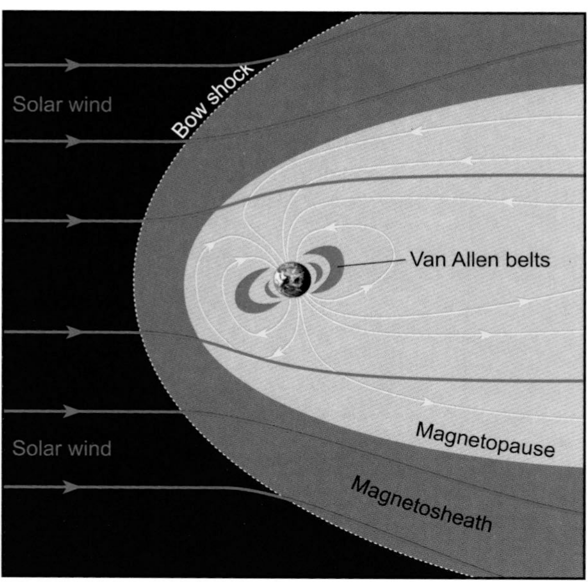

▶ Magnetosphere *The Earth's magnetosphere, including the Van Allen belts of radiation trapped within, is enclosed in a sheath formed by the solar wind, and swept back in the downwind direction.*

Magnetosphere

The region of effect of a planet's magnetic field (*see* PLANETARY MAGNETOSPHERE), the similar region of a star (*see* MAGNETIC FIELDS IN STARS) or of the Sun. In the latter case it is called the HELIOSPHERE.

Magnification

When a telescope is used visually, its magnification is the ratio of the angular diameter of the image as seen by the eye to the angle subtended by the object when viewed directly without the aid of the telescope. The magnification, or magnifying power, of a telescope is given by the ratio of the FOCAL LENGTH of the objective or mirror to the focal length of the EYEPIECE. The magnification of a telescope is also equivalent to the ratio of the aperture of the collector (objective or primary mirror) to the diameter of the EXIT PUPIL (the image of the collector produced by the eyepiece). The maximum practicable magnification is limited by factors such as image contrast (the higher the magnification, the fainter the image) and by 'seeing' conditions (turbulence in the Earth's atmosphere causes sources to twinkle, shimmer and shake; this affects the quality of the image). For small and moderate-sized telescopes an approximate guide to the highest practicable magnification that is likely to give usable images is about 20 per centimeter of aperture (for example, about 200 for a telescope of 10 cm aperture). In practice it is seldom worthwhile to use magnifications greater than 200–300.

Magnitude and photometry

More than any other aspect of astronomy, the subject of the brightness of the stars (PHOTOMETRY) is encumbered by history. In standard scientific units the intensity, *I*, of light from a star is expressed as the flux of energy passing through a square meter per second; but the brightness of stars is usually expressed in magnitudes, whose definition is based on Greek science. The first known catalog of stars was made by HIPPARCHUS in about 120 BC and contained 850 stars.

Hipparchus rated their brightness on a scale of 1 to 6, the brightest being 1.

In the nineteenth century, astronomers began using instruments to measure the brightness of stars accurately. Norman Pogson (1829–91) found in 1856 that each magnitude (as defined by Hipparchus) is about 2.5 times brighter than the next greater magnitude. Around the same time Gustav Fechner (1801–87) and Wilhelm Weber (1813–94) were investigating the response of the eye to light, and they proposed a general psychophysical law that related perception of a stimulus to its logarithm. This and Pogson's relation led to the fundamental definition of stellar magnitude.

Magnitude

The magnitude, m, of a star relative to another of magnitude m_0 is related to its intensity, I, by $m - m_0 = s \log (I/I_0)$, where m is a perceived brightness and s is a constant. Pogson chose s such that a difference in magnitudes of five units (from magnitude 1 to magnitude 6, for example) corresponds to a change in brightness of exactly 100 times. This means that a difference of one magnitude corresponds to a brightness ratio equal to the fifth root of 100, so Pogson's ratio $s = 2.51188643150958$, which is usually rounded to 2.512. One star has to be defined as of an arbitrary magnitude (called the zero point of the magnitude scale), and other stars have magnitudes expressed relative to it. Pogson's scale was originally fixed by assigning POLARIS a magnitude of exactly 2. Astronomers discovered that Polaris is slightly variable, so other stars have displaced Polaris from this role, but the principle remains the same.

Apparent magnitude

The apparent magnitude of a star, often designated by m, is a measure of the intensity of the star's radiation within a particular wavelength interval (bandpass). If the wavelength interval corresponds to the sensitivity of the eye, the magnitude is referred to as a visual magnitude, m_{vis}. If the wavelength interval is defined by particular yellow-colored glass (*see* CAMERA AND FILTER), with a bandpass approximating to what the eye sees, the magnitude is called V (for visual). Other wavelength intervals are defined by filters, and the corresponding magnitudes are designated with the capital letters U (ultraviolet, centered at 350 nm), B (blue, 435 nm), V (yellow, 550 nm), R (red, 640 nm), I (infrared, 800 nm), and J, H, K, L, M and N (bandpasses defined by interference filters deeper into the infrared at approximately 1.22, 1.63, 2.19, 3.45, 4.75 and 10.4 μm). The ratio of light in two different bandpasses, expressed in magnitudes, is called the COLOR INDEX of the star.

Extinction

Stars nearer the horizon appear dimmer than those overhead. This is due to atmospheric extinction (gas molecules, dust and aerosols reduce the intensity of light passing through them). The measured magnitude of a star is corrected to the value that it would have had overhead by the 'extinction correction.' The usual assumption is that the atmosphere is plane stratified over horizontal ground, in which case the correction is of the form A secant (z), where z is the zenith distance of the star (angle from the ZENITH) and A depends on location and on the bandpass used (blue bandpasses absorb more). For the U, B and V bandpasses, A is approximately 0.65, 0.34 and 0.2 magnitudes, respectively. *See also* AIR MASS.

Photometric systems

A photometric system consists of a series of measurements of so-called 'standard stars' distributed all over the sky that are made with a particular light detector, a telescope, filters and a method to correct for atmospheric extinction. Using as close a match to the standard equipment as possible, astronomers make a measurement of a star they are interested in (such as a VARIABLE STAR) relative to the standard stars. They may go through an intermediate step of measuring 'secondary standard stars' near the star of interest, and subsequently measuring its brightness relative to those (for example, in a single photograph or CCD picture).

All photometric systems enable the measurement of relative fluxes in the different wavebands. These relative flux measurements are called a color index. Astronomers have put a great deal of effort into accurately measuring and calibrating colors in terms of temperatures and other stellar parameters.

In principle it is possible to transform a measurement in one system to another: for example, to use V measurements to determine m_{vis} for stars near to a variable star, so that amateurs can estimate its brightness by eye observations. To a first approximation $m_{vis} = V$. To a better approximation $m_{vis} = V + \alpha\,(B\text{-}V)$, where α is a number that has to be determined for a given person (with varying color-sensitive sight) in given circumstances (for example, with a given telescope) and (B-V) is the color index. This is called the 'color equation.'

Absolute magnitude

Absolute magnitude is the magnitude that the object would have were it situated at a distance of 10 parsecs (about 32.6 l.y.) from the Sun. The relation between apparent magnitude, m, and absolute magnitude, M, is $m - M = 5 \log d - 5$, where the distance, d, is in parsecs.

Bolometric magnitude

The total energy, integrated over all wavelengths, received at the Earth from an object is expressed as the bolometric magnitude. The difference between the bolometric magnitude, M_{bol}, and the magnitude A in bandpass A is called the bolometric correction BC_A. BC without a qualifier normally refers to the correction to the visual magnitude, BC_V.

The zero point of the bolometric magnitude scale is usually set by adopting for the Sun $M_{bol} = 4.75$, in which case the Sun's bolometric correction is BC = –0.07.

Maidanak Observatory

Located on Mount Maidanak in Uzbekistan. The most important instrument is a 1.5-m telescope designed for high-resolution imaging. Until 1991 institutes from several republics of the former Soviet Union had their observatories at Maidanak, including the Moscow Sternberg Astronomical Institute and the Astronomical Observatory of Kharkov University. After the fall of the Soviet Union the facility became the property of the Ulugh Beg Astronomical Institute in Tashkent. Subsequent economic difficulties have meant that the observatory underexploits its good atmospheric properties.

Main sequence

The location in the HERTZSPRUNG–RUSSELL DIAGRAM of the DWARF STARS (stars burning hydrogen to helium). Most of the stars in the universe are in this state.

Major planet

A term for the nine largest planetary bodies in the solar system (as opposed to asteroids, minor planets and their natural satellites). The major planets are Mercury, Venus, Earth, Mars, Jupiter, Saturn, Uranus, Neptune and Pluto. Pluto, included in this category since its discovery in 1930, is smaller and less massive than seven of the solar system's planetary satellites, and only twice the size of the largest asteroid, CERES. Its orbit, eccentric and with an inclination of 17°, seems to put it in the Kuiper Belt among the TRANS-NEPTUNIAN OBJECTS (*see* OORT CLOUD AND KUIPER BELT). Calls for Pluto's demotion from major planet status led to several attempts to define 'major planet' more closely. Arbitrary size limits or mass limits (set so as to exclude Ceres but include Pluto) have been proposed. A possible structural definition is that a major planet is large enough for its own gravity to have overcome the rigidity of its material, so it is spherical (this would include Pluto and Ceres, but also other large asteroids).

Maksutov, Dmitri Dmitrievich (1896–1964)

Russian optician and telescope maker. After fighting in the Russian revolution and World War I, he worked on astronomical optics at Odessa, Moscow and Pulkovo, and invented the MAKSUTOV TELESCOPE.

Maksutov telescope

A modification of the SCHMIDT TELESCOPE, devised in the early 1940s by Dmitri MAKSUTOV in Moscow and, independently, by Albert Bouwers in Holland. It uses a thin concave meniscus lens, located at the front of the telescope tube, to compensate for the SPHERICAL ABERRATION of its concave spherical primary mirror. The convex side of the meniscus lens, which itself has a spherical curve, faces toward the primary mirror. Because its corrector lens can be placed closer to the FOCUS than that of a conventional Schmidt telescope, the Maksutov is shorter and more compact. It has a curved FOCAL PLANE, which lies inside the instrument. It can be adapted for visual observation with a small secondary mirror, reflecting light to the side of the tube (Maksutov–Newtonian) or to a focus at the rear of the primary mirror (Maksutov–Cassegrain). In the Maksutov–Cassegrain design, like the Schmidt–Cassegrain, the secondary mirror is a reflective surface on the rear of the lens.

Malmquist, Gunnar (1893–1982)

Swedish astronomer who worked on statistical astronomy; for example, the populations of stars in our Galaxy. The Malmquist bias is a statistical effect by which the fainter members of a population are more represented in a brightness-limited sample than they should be. This is because there are more of them to cross above the cutoff line by accidental measuring error than there are above the limit to fall below.

Mantle

The Earth's mantle is composed of a thick layer of solid rock extending from the core, 2891 km below the surface, to the so-called Mohorovicic discontinuity (or Moho), a few kilometers below the surface. *See also* EARTH.

Mare

A dark lunar plain. The name 'mare' (plural: maria), which is Latin for 'sea', was first used in the seventeenth century, when astronomers believed the Moon's light and dark areas were land (sometimes called 'terrae') and water. In fact the maria are composed of BASALT – solidified lava which erupted some 3 billion years ago after the Moon had suffered large impacts that weakened or, in some cases, penetrated its crust. The maria range in size from huge lava floodplains such as Oceanus Procellarum (2568 km across) and the large impact features Mare Orientale and Mare Imbrium, down to the 150-km Mare Anguis. Maria are much more prevalent on the Moon's nearside; on the farside the crust is thicker and withstood impacts better. The term 'mare' was also formerly used for dark regions on Mars (for example, Mare Tyrrhenum, named after the Tyrrhenian Sea between Italy and Sicily), which in the nineteenth century were also assumed to be seas. However, in general they do not correspond to topographic features, and are now used only on albedo maps (showing areas that appear light and dark) of Mars.

Mariner missions

The name Mariner was given to the earliest US space missions to explore the planets, and to the spacecraft developed at the JPL to carry them out.

Mariner 4

This spacecraft took 21 photographs of the martian surface, revealing a region that it is covered with impact craters. The mission also revealed that the atmosphere on Mars is mainly carbon dioxide at a pressure of less than 7 mbar.

Mariner 5

Mariner 5 determined that the magnetic field of Venus is much weaker than the Earth's, and too weak to hold off the SOLAR WIND and produce an Earth-like MAGNETOSPHERE. However, when the solar wind reaches the top of Venus's atmosphere, it is deflected by the IONOSPHERE and flows around the planet without touching the surface. The mission determined the properties of the venusian atmosphere, and measured the surface temperature at 748 K and the ATMOSPHERIC PRESSURE at 90 atm.

Mariners 6 and 7

The two spacecraft acquired 143 images of Mars before encounter, which showed the entire visible disk or a considerable fraction of it, and 59 images near encounter, which covered small areas (about half of these had resolutions of 0.2 km or better). All the close-up pictures were of the southern hemisphere and, although they covered only about 10% of the surface, they revealed new surface features and laid to rest the myth of the martian CANALS. Later missions, however, showed that most of the interesting features of

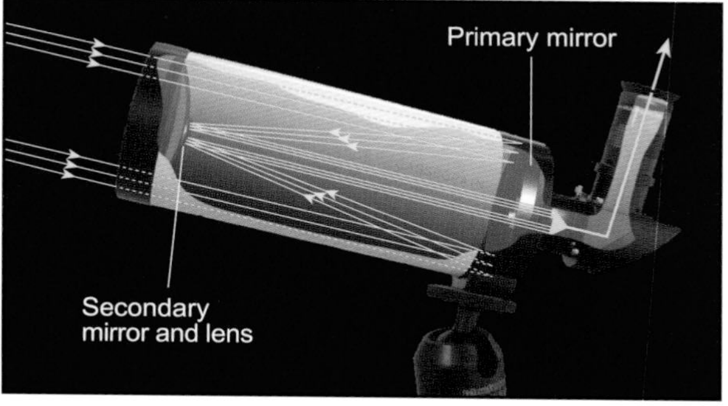

▶ ***Maksutov telescope*** *The Maksutov–Cassegrain design is compact as well as convenient. It is always used with an elbow/diagonal eyepiece, as here.*

Mars had been missed or misinterpreted. The missions measured a temperature of 150 K on the south polar cap, indicating it to be carbon dioxide ice and not water ice.

Mariner 9

When Mariner 9 arrived at Mars the planet was shrouded by the most intense global dust storm that had ever been observed. Settling into its intended orbit the spacecraft waited out the storm, and in 349 days of operation it returned 7329 photographs. Mariner 9 discovered many volcanoes, including the 27-km-high OLYMPUS MONS; an enormous system of deep canyons, dubbed VALLES MARINERIS, that stretch about one-quarter of the way around the planet; a plethora of channels of five different types, many of them appearing to be ancient river beds; numerous evidences of eolian (wind) erosion and deposition; and a variety of meteorological phenomena.

Mariner 10

To reach Mercury, Mariner 10 was put on a course to Venus in the first use of the 'gravity-assist' technique of interplanetary navigation. The spacecraft was deflected into an orbit that took it around the Sun and looped twice in a flyby past Mercury.

Its TV camera returned more than 1000 images of the Earth and the Moon, 3500 of Venus and 3700 of Mercury, a small number of which had a resolution as small as 134 m. They provided detailed coverage of the clouds of Venus (made possible by the ultraviolet sensitivity of the cameras) and the surface of Mercury. Mercury was found to look much like the Moon, but with some distinctly nonlunar features, including large scarps or cliffs nearly 3 km high and up to 500 km long, probably indicative of crustal shrinkage. The density of small craters was surprisingly similar to that on the Moon and Mars, implying that all of these planets received similar intensities of bombardment by meteorites, contrary to earlier assumptions. A major and unexpected discovery was an intrinsic magnetic field, with a MAGNETOSPHERE that was a miniature copy of the Earth's.

The exploration of the solar system by the Mariner series was continued by the VIKING MISSIONS and VOYAGER MISSIONS.

Successful Mariner missions

Spacecraft	Target	Date	Mission
2	Venus	1962–3	Flyby
4	Mars	1964–5	Flyby
5	Venus	1967	Flyby
6, 7	Mars	1969	Flybys
9	Mars	1971–2	Orbiter
10	Venus/Mercury	1973–4	Flyby of Venus, three flybys of Mercury

Mars and its satellites

Mars is the fourth planet from the Sun. The 'red planet' is the most Earth-like planet of the solar system, but is smaller, colder, drier and has a much thinner atmosphere. Life may have evolved on Mars and may still survive there in niche environments. The planet has been the target of more than 30 spacecraft, and will probably be the first that humans visit. Mars has two moons (PHOBOS AND DEIMOS), which orbit close to the planet in its equatorial plane. They are irregular, very dark and very low density, indicating that they are probably captured ASTEROIDS whose orbits have circularized, but the capture mechanism is not known.

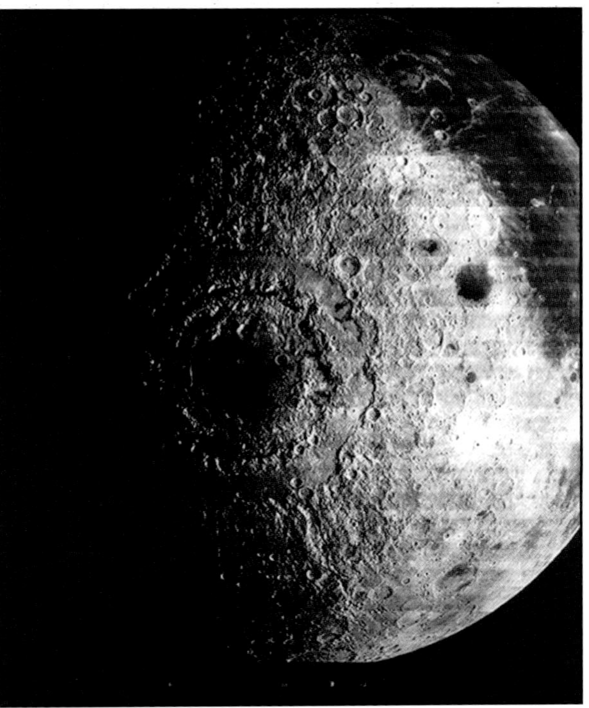

◀ ***Mare*** *With a total diameter of 930 km, Mare Orientale (left of center and bisected by the lunar terminator) is one of the largest impact basins in the solar system. It was formed about 3 billion years ago when an asteroid shattered the Moon's crust, causing magma to flood out and fill the resulting basin. Such was the power of the impact that the ripples in the crust formed three concentric rings of mountain ranges.*

Surface features

The spherical shape of Mars, its polar caps, its variable markings and its rotation are readily visible in small telescopes (*see* CASSINI DYNASTY (Gian Cassini); GALILEO GALILEI; Christiaan HUYGENS). In 1840 Wilhelm Beer (1797–1850) and Johann von Madler (1794–1874) made the first maps of the planet. The dark areas were generally thought to be seas, following William Herschel (*see* HERSCHEL FAMILY), but Emmanuel Liais (1826–1900) suggested in 1860 that they could be large patches of vegetation, thus explaining the changes in color and brightness. Giovanni SCHIAPARELLI's map, based on precise measurements made during the close opposition of 1877, included old 'continents' resolved as 'islands' and 'bays,' and numerous CANALS. Percival LOWELL interpreted the 'canals' as artificial constructions. Eugenios Antoniadi (1870–1944) subsequently recognized that the canals were subtle surface structures seen through the constantly moving terrestrial atmosphere.

The first close-up images, by Mariner 4 in 1965 (*see* MARINER MISSIONS), showed that the surface of Mars was more Moon-like than Earth-like, with prominent craters. This is thought to be because the internal temperature of Mars never reached that of the Earth, owing to its smaller size. The crust therefore stabilized early in the planet's history, and plate tectonics, the most effective resurfacing process on Earth, never developed. Plate tectonics has erased most craters from the Earth, but the absence of plate tectonics on Mars has preserved very old terrains, formed 3.8 billion years ago when impacts were numerous.

There are younger terrains. As Mariner 9 discovered, volcanic activity has formed 60% of the surface. Most of Mars's plains were formed by successive lava flows. Because the sides of the volcanoes show few craters they must have been formed after the period of LATE HEAVY BOMBARDMENT.

Chronology of events affecting the martian surface

Epoch	Age range	Event	Surface (%)
Noachian			
Early	4.6–3.92	Formation of the planet	4
		Outgassing of early atmosphere	
Middle	3.92–3.85	Early development of Tharsis	24
		Highland formation and large basins	
Late	3.85–3.5	Hot and humid climate?	
		Dissipation of early atmosphere	12
		Highland valleys, fluvial resurfacing	
		Highland volcanism	
Hesperian			
Early	3.5–3.1	Main development of Tharsis and Elysium	16
		Earliest volcanism	
		First volcanic plains	
		Development of Valles Marineris	
Late	3.1–1.8	Giant volcanoes	19
		Relatively hot climate?	
		Liquid water may be stable	
		Layered terrains in Valles Marineris	
		Outflow channels and chaotic terrains	
Amazonian			
Early	1.8–0.7	Late volcanic plains	11
		Erosion of layered terrains in Valles Marineris	
Middle	0.7–0.25	Polar layered deposits	8
		Outflow channels are still active	
Late	0.25–0	Eolian resurfacing	7
		Final activity of Tharsis volcanoes	
		Final evolution of Valles Marineris	

Age range = billions of years ago (approx.); surface = percentage of present surface

The most recent volcanic eruptions occurred a few billion years ago.

There is an asymmetry between the two hemispheres, often referred to as the 'martian dichotomy.' Old, densely cratered highlands are concentrated in the south, and younger, low volcanic plains in the north. The difference in altitude is about 3 km on average.

The overall height difference between the highest mountain and the deepest valley is 37 km, twice the range of relief on Earth. The enormous size of OLYMPUS MONS, THARSIS MONTES and other martian volcanoes is a consequence of the absence of plate tectonics. All the material accumulates on the same spot over a magma source instead of building a series of volcanic constructs as the plate shifts, as has happened in the Hawaiian Islands.

Other geological processes were identified by Mariner, VIKING and other spacecraft, and studied in exquisite detail from the MARS GLOBAL SURVEYOR in 2001. A huge canyon system, VALLES MARINERIS, extends eastward from Tharsis along the equator. It is 4000 km long, and reaches 600 km wide and 7 km deep. The canyons are related to the faulting that followed the formation of the Tharsis Shield.

Catastrophic flooding has produced channels that are tens of kilometers wide and hundreds long. Most of them originate in the chaotic terrains east of Valles Marineris, cross the old terrains, and merge with the low plains at 45°N. Within some craters there are distinct, thick, sedimentary layers of rock, suggesting that they contained lakes or shallow seas. Many martian craters are surrounded by lobate blankets (petal-shaped, like sunflowers). These craters were formed by meteor impacts in soils with high water content. There are also glacial features on Mars, including thermokarst, patterned grounds, pingos, eskers and ridges similar to glacial moraines.

These forms suggest that water and ice have played an important role in the evolution of Mars, when the climate was different. More speculatively, Mars may have had an ocean in the distant past: images from Viking suggest that there is an ancient shoreline at the contact between highlands and plains in the northern hemisphere. Radar observations from the Mars Global Surveyor show that the 'shoreline' is indeed close to an equipotential surface (as expected from a sea level), and that the terrains are smooth below this level.

The ice and dry ice deposits at the poles are 2–3 km thick with thin layering (less than 20 m), cut by spiraling valleys. In the north, a vast, uncratered dune field surrounds them.

Geological ages on Mars

The ages of planetary surfaces are estimated by comparing the number of craters with those on lunar terrains (whose ages are known from the APOLLO samples). Applying this dating to Mars is uncertain, but the relative age of terrains can be estimated. The time scale is divided into three main systems (Noachian, Hesperian and Amazonian), and eight subsystems (*see* table).

Surface characteristics

The SNC METEORITES and the Viking and MARS PATHFINDER landers have provided detailed knowledge of small portions of the surface of Mars. The two Viking landing sites were chosen to be relatively safe for the landing, free of hills or mountains. The landing sites are deserts, with rocks standing in a layer of fine sand. The rocks are angular and range from centimeters to meters in size; they are EJECTA thrown to the sites from nearby meteor impacts. The dust particles are wind eroded from fractured rocks. They are less than 10 μm in size but often aggregate in larger clods. In contrast to the Viking sites, the Mars Pathfinder's Sojourner landing site was selected for its geological interest: it is located in an outflow channel east of Valles Marineris, and hence it contained materials coming from different source regions. Although the landscape is similar to the previous two sites, it shows evidence of flooding (such as water-rounded rocks). But wind has been the major cause of erosion for a long time.

All three landers analyzed the surface materials. The soils are very similar at the three landing sites: they are the weathered products of basalts, with high degrees of oxidation providing the planet's red color.

The SNC meteorites provide the most detailed mineralogical information. These meteorites are believed to be igneous martian rocks ejected by meteor impacts, which have found their way to the Earth. The SNC meteorites form three families: the youngest one, composed of basaltic rocks dominated by two pyroxene mixtures (typified by the Shergotty and Zagami meteorites formed 180 million years ago); an older one dominated by calcium-rich pyroxenes and olivine, with very low concentrations of aluminum (typified by the Chassigny and Nakhla meteorites formed 1.3 billion years ago); ALH 84001 is much older (4.5 billion years ago). The three groups were ejected from three separate, unidentified places during three different impact events.

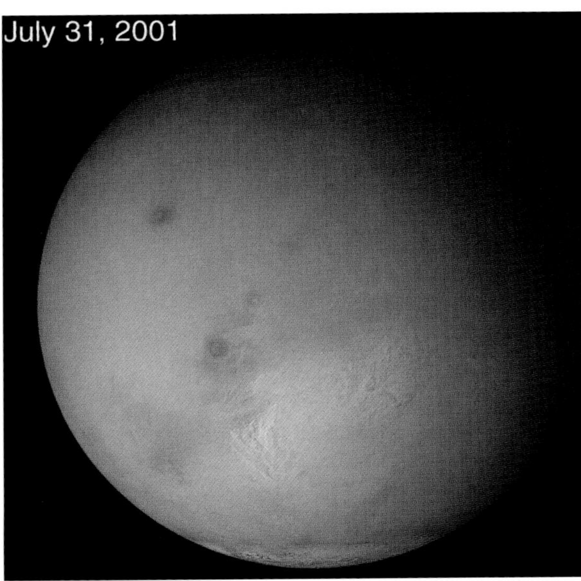

◀ ***Mars*** *The Mars Orbiter Camera aboard the Mars Global Surveyor recorded Mars on a relatively calm and clear day (left). Some white clouds can be seen condensing over the peaks of large volcanoes, and the southern carbon dioxide frost cap shows clear. Three weeks later (right), dust storms hid the surface detail.*

Mars	
Mean distance from Sun	228 × 10^6 km (1.52 AU)
Revolution period	687 days
Orbital eccentricity	0.0934
Inclination to ecliptic	1° 51′
Rotation	24.6629 hours
Diameter	6779.84 km
Mass relative to the Earth	0.107
Density	3.9335 g cm^{-3}
Escape velocity	5 km s^{-1}
Surface (main materials)	basalts, pyroxene, hematite
Atmosphere	carbon dioxide (95%) nitrogen (2.7%) argon (1.6%)
Atmospheric pressure	5.6 mbar
Surface temperature	−63 °C, 210 K

Climate and atmosphere

The atmosphere of Mars is much thinner than Earth's. However, even the simplest visual telescopic observations show atmospheric events such as the seasonal exchange of material between the polar caps, the passing appearance of clouds, and changes of visibility of dark regions on the disk of the planet, sometimes due to dust storms.

The large orbital eccentricity of Mars causes a marked difference in the length of the seasons in the two hemispheres, with a longer, colder winter in the south. Accordingly the polar caps are different, the northern ice cap being water ice, and the colder southern ice cap being dry ice (over water ice).

As with the other terrestrial planets, changes in the orbit of Mars cause dramatic changes in its climate (*see* Milutin MILANKOVITCH). The seasonal differences in the north and south reverse every 25 000 years. On a time scale of 10 million years, the obliquity of Mars (the tilt of its rotational axis relative to its orbital plane) changes from 15° to 35°. This changes the magnitude of seasonal effects. On longer time scales the variation of the obliquity is larger and chaotic. This is unique among terrestrial planets because Mars is not stabilized by a massive satellite, like the Earth, nor by tidal dissipation, like Mercury and Venus.

The most abundant gaseous components of the martian atmosphere are carbon dioxide, nitrogen and argon. Other identified gases are so-called minor or trace constituents: oxygen, carbon monoxide, water vapor, ozone, helium, neon, krypton and xenon.

The average water concentration in the atmosphere is equivalent to an approximately 10-μm layer of precipitated water covering the whole surface of the planet. The atmosphere is extremely dry. Almost all water is hidden in permafrost, polar caps and hydrated minerals. According to neutron radiation measurements by MARS ODYSSEY released in 2003, the equivalent liquid water depth of subsurface water is some 13 cm.

During the northern winter and southern summer (Mars near PERIHELION), great dust storms may cover low and moderate latitudes – most of the planet. There may be one, two or even three dust storms lasting up to 150 days. Sometimes local dust storms are observed.

Magnetic field

In 1999 the Mars Global Surveyor did not detect a global magnetic field, but it did detect strong magnetism in the southern highlands. This is the remnant of a time when the planet had a magnetic field. The remnant field occurs in a regular pattern of stripes with alternating polarity. The stripes are 1000 km long and 200 km wide, oriented east-west. They are similar to magnetic patterns observed on the terrestrial seafloors. The stripes are remnants of successive reversals of the polarity of the early magnetic field, trapped in surface materials drifting over the mantle and the convective core of the planet (*see* GEODYNAMO). At some point the core stopped convecting and the martian dynamo died away. The atmosphere was exposed to the solar wind (because it had no protective MAGNETOSPHERE) and thinned to its current state.

Biology and the search for life

As summarized above, Mars was once a very active planet, with water on the surface possibly forming lakes or even oceans. In these conditions it is possible that life originated there (*see* EXOBIOLOGY AND SETI). If so, it is possible that it has

Observing Mars

Mars is arguably the planet that has most captured man's imagination. The strong red-orange color reminded pre-telescopic stargazers of a blood-red warmonger. It comes to opposition at the longest interval (780 days) of the naked-eye planets. Coupled with this, Mars' eccentric orbit leads to it being closer to Earth and brighter if opposition coincides with aphelion than if it occurs at perihelion when it may be easily overlooked. Mars comes closest to Earth (opposition) at 2-year intervals, but the eccentricity of its orbit means that its size and brightness are greatly variable. The disk subtends some 24 arcsec at perihelic opposition, 13.5 arcsec at aphelic opposition and around 3.5 arcsec when Mars passes behind the Sun.

Mars repays close study either by eye or by the CCD-equipped observer. Prior to the visit of space missions to the planet, observers concentrated on the general mapping of its surface features. Accounts of amateur work published prior to 1940 gave great credence to the maps derived from drawings by the observers. Time-sensitive observation allowed the shrinking of the polar cap (whichever was presented towards the Earth at that opposition) to be followed. Dust storms occur often, the size and frequency of which depend on the point in the Martian year. There are also diurnal clouds and other atmospheric activity which are worth observing.

We now know that the light markings are due to dust and that the dark markings are areas where dust has either never collected or been blown away. The linear markings often reported in the past are either valleys or (more frequently) illusions. The light Hellas basin is the lowest point on Mars and it is subject to early morning hazes. It is also one of the areas prone to dust storms. Indeed, it can be the starting point of the occasional global dust storms. The most impressive feature, olympus mons, had been observed long before we knew that it is a giant volcano; prior to this discovery it was known as Nix Olympica (the Olympic Snow), as observers misinterpreted the clouds which often form above it.

► ***The Sinus Sabaes feature** at Mars's central meridian and the breakup of the southern polar cap (seen at the top of the image) are visible in this image made with a home-made, 16-in Newtonian. Using eyepiece projection into a minidv camcorder, 2 min of video were obtained. He then stacked the 200 best images using image-processing software. Image courtesy of David Hanon.*

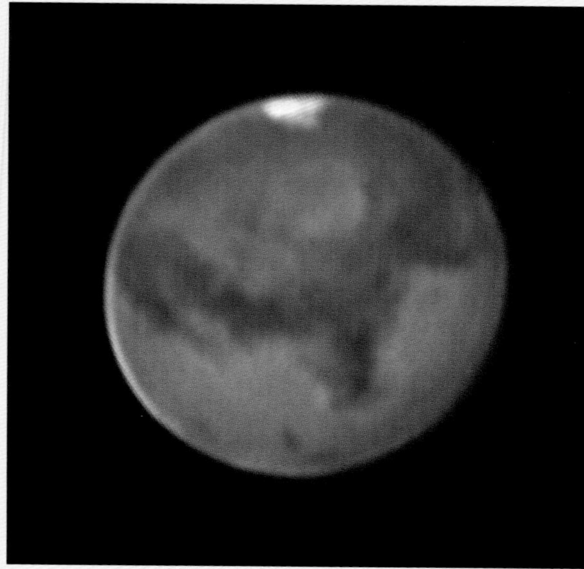

► ***Syrtis major,** a feature of Mars that somewhat resembles the continent of Africa, can be seen in David Hanon's September 9, 2003 image of Mars. Again, the southern polar ice cap detail can be seen at the top of the image.*

We have been following Mars since the first telescopes were turned towards it in the 1600s, but modern observers have a huge advantage over their predecessors: forearmed with knowledge from numerous spacecraft orbiting and mapping Mars, we are freed from previous misinterpretations of canals, vegetation and the inhabitants of Mars. Today, we can relate our observations to the detailed data from spacecraft imagery that describes the planet's surface. The historical record can also benefit from the application of the topographical information obtained from space missions.

As with all planetary work, the visual observer continues to make drawings of what he sees while the CCD imager strives to increase the clarity and detail of his work by taking and stacking multiple images and then processing them to reveal detail. As would be expected, the larger the apparent disk of Mars, the easier it is to detect fine detail; imaging when the disk subtends 7 arcsec has been carried out, but does not allow the detection of fine detail.

Amateur planetary work has now moved away from general mapping, as most features are now well established and mapping is, in any case, better done through space missions. The argument is no longer between the images gathered by space probes and those drawn by observers, but rather between the images gathered by space probes and those generated by CCDs. Some of the images taken from Earth using 12-in scopes in the hands of experienced operators are magnificent. However, there are many other challenges for observers. In particular, a useful study that can be undertaken is looking for changes on the planet's surface. The first requirement for the amateur observer is knowing what a feature normally looks like, a fairly fundamental necessity and one that can best be satisfied from regular observation or, at least, from consulting reference material. *The Planet Mars* by E M Antoniadi is some 70 years old, but nonetheless still relevant for an Earth-based amateur.

Objectives of contemporary observation

There are several main objectives that the amateur observer might keep in mind when observing Mars. These objectives include the following:

- checking for the appearance of new areas of light or dark markings, which represent the removal or depositing of dust due to the activity of wind. The appearance of the dark Pandora Fretum, just south of the Sinus Meridiani, is a frequently occurring example.

◀ ***Mars** seen from the Picos de Europa in northern Spain. The photograph, by Jamie Cooper, was taken with a tripod-mounted Nikon Coolpix 5700 digital camera with a 40-s exposure.*

• looking for signs of dust storm activity, which almost always appears as bright yellow clouds. The areas affected tend to be bigger than areas involved in depositing, although they change as the storm increases or shrinks. Sometimes the storm grows so that it envelops the entire visible surface of the planet. The first sign of such a storm subsiding is the detection of the tops of the volcanoes on the Tharsis Ridge; these appear as dark patches to the observer.

• looking for evidence of cloud activity. A blue filter can help in this endeavor by suppressing the surface of the planet. Clouds can be early-morning or late-evening hazes that will be close to the limbs of the planet. There are also topographic clouds that are associated with mountains such as Olympus Mons or the Tharsis volcanoes.

• following the retreat of the polar cap. The caps grow when the pole is pointed away from Earth and hence cannot be detected. During the period of an opposition, the visible cap will melt and shrink; drawings can be used to follow this retreat. There may be temporary halts or changes during the course of the melting. The writer observed a case in which the cap changed from a whitish to a pinkish shade; it is believed that this was the result of a layer of dust that had been deposited in the growing cap after a dust storm.

Observations from Earth also allow global conditions to be monitored, which can be difficult to achieve from space. They therefore provide a useful piece in the jigsaw of our understanding of the planet. In addition, comparison with the approximately 400 years of historical observations is easiest when observers are working in the same way: drawing and recording observations at the eyepiece. Ultimately, however, amateur observing, for the love of it, is the best security we have of a continuing knowledge base. If something unusual happens on the planet and there is no space mission looking in the right place, the only way we will know is if alert amateurs are watching.

Andrew Hollis is a chartered structural engineer, and director of the BAA's Asteroids and Remote Planets Section.

▼ ***Sequence of five images** of Mars taken between August 10 and September 4, 2003 by Jamie Cooper.*

▲ ***Mars Express*** *Mars Express left Earth on June 2, 2003, when the positions of the two planets made for the shortest possible route. This condition occurs once every 26 months, but the 2003 occurrence was particularly favorable to send the largest possible payload to Mars.*

survived the worsening conditions.

Three experiments on the Viking landers tried to detect biological activity. All results are consistent with simple oxidation reactions in inorganic materials, although some results were considered marginally positive at first and are not completely understood today.

In 1996 David McKay and his group announced that it had found fossil life in one of the SNC meteorites, ALH 84001. The possible fossils are primitive bacteria from about 3.6 billion years ago. The group identified three types of peculiar objects that look like terrestrial bacteria, and small mineral grains and organic compounds as produced by bacteria on Earth. All of these structures could have a different origin not related to life, but taken together they were considered as evidence for life on Mars. After much scientific activity, the issue is unproven.

Apart from its intrinsic interest, the study of early martian life would shed light on how life developed on Earth. The search for extant life is still a priority for future missions to the planet.

Further reading: Carr M (1996) *Water on Mars* Oxford University Press; Sheehan W (1996) *The Planet Mars: A History of Observation and Discovery* University of Arizona Press.

Marseille-Provence Observatory

The Observatoire de Marseille-Provence (OAMP) was formed in January 2000 by a merger of the two institutions, the Laboratoire d'Astrophysique de Marseille (which includes the Marseille Observatory) and the Observatoire de Haute-Provence (OHP). It is an institute of the French Centre National de la Recherche Scientifique (CNRS).

The Marseille Observatory was founded by the Jesuits in 1702 and taken over by the state in 1763. A century later it was transferred to its present site on the Longchamp Plateau in Marseille, southern France.

The observatory owns an 80-cm telescope, built in 1867 by Jean FOUCAULT, which was the first large telescope with a silvered mirror. It was used for observations until 1960 and designated an historical monument in 1993.

The OHP, founded in 1937 as a national facility for French astronomers, is situated close to the village of Saint Michel at an altitude of 650 m. The observatory has several telescopes, the largest of which is a 1.93 m which has been operating since 1958.

Mars Exploration Rovers

NASA's twin robot geologists, the Mars Exploration Rovers named Spirit and Opportunity, were launched in June and July 2003, and reached their destination in January 2004. From the landing sites on opposite sides of the planet each rover, a sort of mechanical geologist walking the surface of Mars, will travel about a kilometer performing investigations with the particular hope of finding evidence of past water on Mars.

Mars Express

ESA's Mars orbiter was launched in June 2003 and reached Mars in December 2003. Mars Express will carry out a scientific survey of the planet. It carried a small lander called Beagle 2, which disappeared on approach to the planet.

Mars Global Surveyor

NASA Mars orbiter that was launched in November 1996 and arrived at Mars in September 1997. It carries an advanced camera system to map the planet at high resolution, a thermal-emission spectrometer and two magnetometers. The Mars Global Surveyor has discovered regional magnetic fields and mapped global topography. Its successful prime mapping mission, delayed when a faulty solar array prevented aerobraking, began in March 1999 and provided close-up images of geological features on Mars, many of which were formed by water. The mission has been extended twice.

Mars Odyssey

NASA's 2001 Mars Odyssey was launched in April 2001 and reached Mars in October that year. Odyssey's primary science mission is to map the amount and distribution of chemical elements and minerals that make up the martian surface. The spacecraft has looked for hydrogen, in the form of water ice, in the shallow subsurface of Mars, and within the top few feet has detected the equivalent of 13 cm uniform depth of liquid water across the planet. Mars Odyssey also records the radiation environment in low Mars orbit to determine the radiation-related risk to any future human explorers. During and after its science mission, the Odyssey orbiter will provide the communications relay for US and international landers such as the MARS EXPLORATION ROVERS. The name '2001 Mars Odyssey' was selected as a tribute to science-fiction author Arthur C Clarke.

Mars Pathfinder

The first of NASA's Discovery missions. Launched in December 1996, it arrived at Mars in July 1997. Mars Pathfinder was intended mainly as a technology demonstration mission, but it was also outstandingly successful in science and in public education. It used airbags to cushion the landing on Mars. The Carl Sagan Memorial station returned images of an ancient flood plain in Ares Vallis. The 10 kg Sojourner rover used an X-ray spectrometer to study the composition of rocks and traveled about 100 m. Operations ceased after 83 days, in September 1997.

Mars Probes

Between 1960 and 1974 the USSR attempted a number of missions to Mars under this name, with only moderate success. Those that successfully left Earth orbit were named Mars 1, 2 and so on, up to 7. As well as technical problems, the Soviets had some bad luck. Mars 2 and 3 arrived during the 1971 global dust storm that obscured the planet's surface. Some images were returned by Mars 4 and 5 orbiters.

Atmospheric data were returned during the descent of the Mars 6 lander, but contact was lost before touchdown.

Maskelyne, Nevil (1732–1811)

ASTRONOMER ROYAL, born in London, UK. Maskelyne's interest in astronomy began when he saw the solar eclipse of 1748. After studying mathematics at Cambridge University, he was sent by the Royal Society in 1761 to the South Atlantic island of St Helena to observe the transit of Venus. In 1764 he traveled to Barbados, testing John HARRISON's chronometer, and was appointed Astronomer Royal on his return. Maskelyne initiated publication of *The Nautical Almanac* in 1766. Following Pierre Bouguer (1698–1758) and Charles-Marie La Condamine (1701–74), he proposed an experiment for determining the Earth's density with the use of a plumb line, whose plumb-bob was attracted sideways by the gravitational pull of a mountain. He carried out the experiment in 1774 on Schiehallion, a mountain in Perthshire, Scotland (chosen because it was a regular conical shape and its volume could be calculated accurately), and found the Earth's density to be approximately 4.5 times that of water.

Mass

A measure of the amount of MATTER in a body or of its inertia. The mass of a body is one of the factors that determines the strength of its gravitational field. Mass, in this context, is called 'gravitational mass.' Mass is also the extent to which a body resists acceleration when a force is applied to it (Isaac NEWTON's third law of motion, the law of inertia). Mass in this context is called 'inertial mass.' Inertial and gravitational mass are identical (the equivalence principle, tested by the Eötvös experiment – *see* GENERAL RELATIVITY). The mass of a stationary body is its 'rest-mass.' According to the special theory of relativity, the mass of a moving object is greater than its rest-mass. Mass and energy are equivalent and interchangeable, the relationship between mass (m) and energy (E) being $E = mc^2$, where c denotes the speed of light.

The unit of mass in the SI system is the kilogram (kg). Particle physicists use the term 'mass' to describe the energy that is equivalent to the mass of a subatomic particle, and quote 'masses' in energy units (electron-volts). For example, the 'mass' of a proton is 938.3 MeV.

Mass-luminosity relation

The relationship between mass (M) and luminosity (L) for MAIN SEQUENCE stars is expressed as $L \propto M^n$. The exponent n varies according to mass.

Mathilde

Located in the outer part of the asteroid belt, asteroid (253) Mathilde was discovered in 1885 and is believed to be named for the wife of astronomer Maurice Loewy (1833–1907), then vice-director of the Paris Observatory.

Mathilde is a C-type asteroid and is one of the darkest objects in the solar system, reflecting only 4% of the light falling on it. It has a very slow rotation rate (17.4 days); only two asteroids, (288) Glauke and (1220) Clocus, have longer rotation periods.

It was imaged by the NEAR EARTH ASTEROID RENDEZVOUS MISSION (NEAR–Shoemaker) in June 1997. On the 50% of the surface imaged by the spacecraft there are five craters with diameters between 19 and 33 km. A total of 91 craters have been identified, and 70 have diameters larger than 0.6 km. There is evidence of downslope movement on crater walls, indicating that the surface of Mathilde is covered with regolith. There has been some retention of EJECTA produced during impacts. Mathilde's ALBEDO is between 0.035 and 0.05. NEAR determined the asteroid's mass to an accuracy of about 5%. Its size approximates an ellipsoid 66 × 48 × 46 km, and the asteroid has a low density, of 1.1 to 1.5 times that of water, indicating a porous structure. It has no satellite that could explain its long rotation period, but if it was made to rotate quickly enough, for example, by being struck, its porous structure could mean that it would disintegrate.

Matter

A physical substance that occupies finite volumes of space and that has MASS. Bulk quantities of matter can exist in any of the following states: solid, liquid, gas or plasma. On the microscopic scale, matter exists as molecules, atoms, ions, subatomic particles or elementary particles. An atom is composed of a nucleus of protons and neutrons, surrounded by a cloud of much lighter electrons. Because protons and neutrons belong to a class of particles called BARYONS, normal matter is often referred to as 'baryonic matter.' In addition to luminous matter (the stuff of which stars are composed), which is detectable by its radiation, the universe appears to contain DARK MATTER that emits no discernible radiation and has been detected only through its gravitational influence. It may be 'non-baryonic' (composed of hypothesized particles that are not baryons).

Mauna Kea *See* HAWAII INSTITUTE FOR ASTRONOMY

Maunder, Edward Walter (1851–1928) and Maunder, Annie Scott Dill (1868–1947)

Edward and Annie Maunder were solar astronomers. Edward became assistant for spectroscopic and solar observations at the ROYAL OBSERVATORY, GREENWICH under George AIRY, aided by Annie, his second wife. In 1890 Edward identified the MAUNDER MINIMUM. In 1904 he invented the BUTTERFLY DIAGRAM of sunspots, showing the solar latitude at which sunspots appear as a function of time (*see* SUNSPOT CYCLE).

Maunder minimum

In 1890, while studying the numbers of sunspots over a 300-year time span, EDWARD MAUNDER noticed the scarcity of spots during the period 1645–1715. This so-called Maunder minimum was confirmed by Jack Eddy in 1976 to be a real effect rather than simply a scarcity of observations: old tree rings showed a reduction of carbon-14 during the same period (carbon-14 production in the Earth's atmosphere is modulated by solar activity). The Maunder minimum is due to an unexplained reduction of the solar magnetic field for about 50 years.

Maxwell, James Clerk (1831–79)

Born in Edinburgh, Scotland, he worked at Cambridge University, England on electric and magnetic fields and their interrelation. Maxwell entered a prize-winning essay for a competition on the subject 'The Motion of Saturn's Rings,' and showed that the rings could be stable only if they consisted of numerous small solid particles in orbit. Maxwell invented the three-color process used today by David Malin, for example, in astrophotography. Maxwell worked on the kinetic theory of gases, showing with a Maxwellian velocity distribution that the temperature of a gas was connected to

▶ ***Maxwell Montes*** *How the mountains might look to a pilot approaching from the northwest at an altitude of 7 km. Clouds and haze add to the perception of depth and distance.*

the motion of its molecules. He wrote a *Treatise on Electricity and Magnetism*. This unified what appeared to be two separate forces into one set of electromagnetic equations, known as 'Maxwell's equations,' and showed that light is a form of electromagnetism. The JAMES CLERK MAXWELL TELESCOPE in Hawaii and MAXWELL MONTES on Venus are named for him.

Maxwell Montes

The highest mountain range on Venus, situated in the upland region Ishtar Terra. It extends for 797 km and contains the highest point on the planet, nearly 12 km above Venus's average surface level (comparable to Earth's Mount Everest). It is the only venusian feature to bear a male name, honoring James MAXWELL.

McDonald Observatory

McDonald Observatory, near Fort Davis in western Texas, USA, is the astronomical observatory of the University of Texas at Austin. Discoveries at McDonald Observatory include water vapor on Mars, the abundance of rare-earth chemical elements in stars, the discovery of planets circling around nearby stars, and the use of the measurements of rapid oscillations in the brightness of WHITE DWARF stars to deduce their ages. Telescopes at McDonald Observatory include the 9.2-m-diameter HOBBY EBERLEY TELESCOPE, which came into operation in 1999 and specializes in astronomical spectroscopy. Other telescopes include the 2.7-m Harlan Smith Telescope, the 2.1-m Otto Struve Telescope, the 0.7-m widefield survey telescope, and a 0.7-m laser-ranging telescope that measures the distance to reflectors on the Moon.

Mean solar time

The TIME system used for most civil and many astronomical purposes. It is based on the motion of a hypothetical object called the mean Sun, the right ascension of which increases from day to day at a uniform rate, keeping pace on average with the real Sun. The local mean solar time is the local hour angle of the mean Sun plus 12 hours. Greenwich mean time (GMT) is taken as the standard for reference; the modern term UNIVERSAL TIME (UT) is synonymous.

Measuring machine

Surveys of the sky at optical wavelengths expand our knowledge of the universe, mapping the number and distribution of stars and galaxies, and identifying rare and therefore transitory objects. Up to the end of the twentieth century, all of the major all-sky optical surveys were carried out using photographic plates taken on wide-angle SCHMIDT TELESCOPES (*see* SLOAN DIGITAL SKY SURVEY for the beginning of the future). Notable surveys of the northern hemisphere were made by the Palomar Schmidt telescope, and of the southern hemisphere by the ESO Schmidt telescope in Chile and the UK Schmidt telescope in Australia.

Four gigabytes of information can be stored on a Schmidt plate. Around 2000 such plates are needed to cover the whole sky in one passband. Add another passband or two to provide color information, add possibly different epoch observations to open up proper motion and variability studies, not to mention special plates for specific studies, and the total volume of data recorded is more than a staggering 50 Tbytes (50×10^{12} bytes).

Formerly, this material could only be studied visually with low-power binocular microscopes. Even large teams could mine only a minute fraction of the available wealth of information available. Objective quantitative studies were difficult in many cases and impossible in most. However, in the late 1970s and early 1980s fast optical micro-densitometers, or 'measuring machines,' were developed to extract automatically the enormous amount of information contained in these survey plates. The measuring machines produce the sort of information astronomers are really interested in, such as whether the black spot on the photograph is a star or a galaxy, its position, magnitude, color, and so on. Most of the area on a survey plate belongs to the sky background. Faint images and all stellar images are reducible to simple descriptors such as those above. It is only the handful of bright galaxies where more detailed mapping is required. If, apart from the bright galaxies, the images are recorded by the simple descriptors, then the original data volume is reduced by factors of typically ×100 with little or no loss of pertinent information. Nowadays it is common practice among astronomers to use the Internet to access a virtual observatory to see catalogs of objects and two-dimensional pictures.

The initial breakthrough came in the 1950s, when Peter Fellgett proposed a machine, called GALAXY, that could find and measure the black spots on photographs automatically. GALAXY used a 'flying spot' produced by a cathode-ray tube (CRT). When the image of a star on the plate blocked off the light from the CRT the machine registered an object, centered on it, and measured its position and size. GALAXY, completed in 1968 at the Royal Observatory, Edinburgh in Scotland, by Vincent Reddish, represented a significant advance over previous efforts. It could process 1000 stars per hour but, as each survey-grade Schmidt plate contains more than 250 000 stars, measuring a complete plate still took weeks. GALAXY's successor, COSMOS, took about a day. COSMOS was superseded in 1992 by SuperCOSMOS, a multichannel system based on a CCD detector. A complete Schmidt plate has a turn-round time of some three hours.

Meanwhile in Cambridge, England, also in the 1970s, Ed Kibblewhite, an electronics graduate, started work on a different system: the APM machine. Kibblewhite's break-through was to use a laser beam to scan the plate, and a series of high-speed special-purpose computers to analyze the data in real time. An acousto-optic deflector generates the 'flying spot' necessary for fast scanning, by using a sound wave of varying frequency to diffract parts of the incident laser beam. The plate is held on a massive X-Y table that moves under the laser beam, and the scanner samples the plate 250 000 times per second with a positional accuracy of a few tenths of a micron.

In the USA, Willem LUYTEN had pioneered the use of a manual BLINK COMPARATOR for locating stars of high proper motion. He and the Control Data Corporation implemented in the late 1960s a fully automated, computer-controlled blink machine based on laser optics. This machine played a major role in the production of Luyten's Palomar proper motion catalogs. After a major electronic upgrade in the 1980s under the direction of Roberta Humphries, the machine was named the APS. The APS is based on a laser spot deflected with a rotating prism and can scan two plates simultaneously in a few hours. In 1966 astrometric measurements at LICK OBSERVATORY for the Lick Proper Motion program were automated with the introduction of the Lick Automatic Measuring Engine (LAME) by Stanislavs Vasilevskis.

One of the main drivers for the latter-day expansion of measuring-machine effort in the USA was the requirement that acquisition and guidance stars be used to point and stabilize the HST. The catalog had to cover the whole sky as faint as magnitude 15. Barry Lasker at the Space Telescope Science Institute in Baltimore initiated a round-the-clock program using two commercially made and trademarked PDS microdensitometers to complete the task, prior to the launch of the HST. This catalog has also become the mainstay for ground-based telescope acquisition and guidance systems. All the data are available on the Internet for public access.

Mensa *See* CONSTELLATION

Mercury

Mercury is the second smallest planet in the solar system (Pluto is smaller). It has no known satellites and almost no atmosphere. It is the planet closest to the Sun and is never seen from the Earth in a dark sky so it is difficult to view telescopically, but, as an INNER PLANET it shows pronounced phases. To date, Mariner 10 is the only spacecraft that has explored the planet, and it remains the best source of data (*see* MARINER MISSIONS). It imaged only about 45% of the surface at a resolution comparable to telescopic Earth-based coverage of the MOON before space flight. Recent ground-based observations have imaged parts of its unmapped surface and discovered new constituents in its atmosphere. Because Mercury is so close to the Sun, has no insulating atmosphere and has such a long day, it experiences the greatest range in surface temperature of any planet or satellite in the solar system: 635 K. Because of the heat of the nearby Sun, space missions to Mercury are technically difficult, even more so than to the outer solar system. However, new missions to Mercury are in construction, including NASA's MESSENGER (launch in 2004) and ESA's BEPICOLOMBO (launch due 2011).

Rotation and orbit

Mercury's rotation period and period of revolution (*see* table) are in the ratio of 3:2 – it turns exactly three times on its axis for every two orbits around the Sun. This resonance was acquired over time through tidal friction. Because of it, a solar day (sunrise to sunrise) lasts two mercurian years or 176 Earth days. The obliquity of Mercury (tilt of its axis) is close to 0° and, therefore, it does not experience seasons. Consequently, the polar regions never receive direct rays of sunlight and are always frigid compared to the very hot, sunlit equatorial regions.

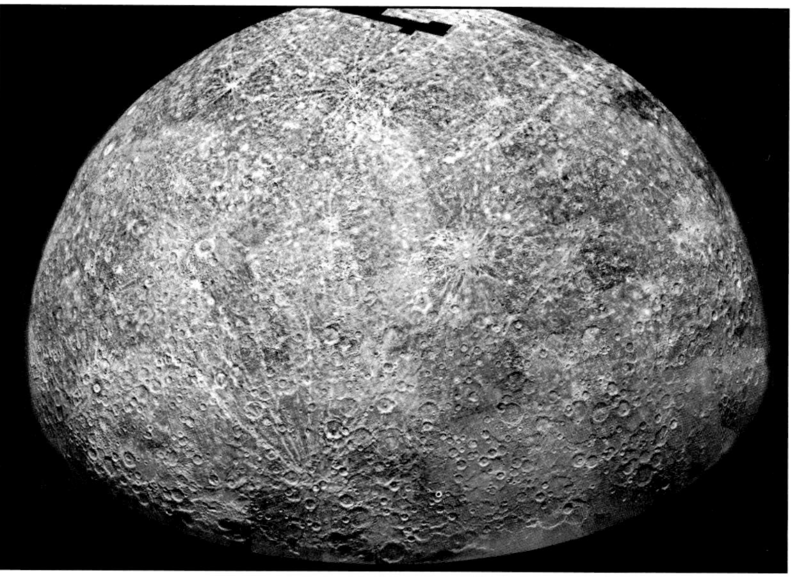

▲ Mercury *Mariner 10 took this photo-mosaic of the southern hemisphere. A barren world unable to maintain a protective atmosphere, Mercury is very vulnerable to meteor impacts. The south pole is near Chao Meng-Fu, the large shadowed crater at the bottom center. Several rayed craters are prominent in the hemisphere to the north of it.*

One effect of the 3:2 resonance and the eccentric orbit is that an observer on Mercury will at PERIHELION typically witness (depending on location) a double sunrise or a double sunset, or the Sun will backtrack in the sky at noon.

Atmosphere

Mercury is so hot that its atmosphere is extremely tenuous, with a surface pressure a trillion times smaller than the surface pressure of Earth. It is, therefore, an atmosphere where atoms rarely collide; their interaction is primarily with the surface. Mariner 10 identified hydrogen, helium and oxygen. The hydrogen and helium are probably largely from the SOLAR WIND, although a portion of the helium may be from radioactive decay of Mercury's rocks. There is some oxygen, sodium and potassium in the atmosphere, derived from the planet's surface.

Internal structure

Mercury's internal structure is unique in the solar system. Its mean density is only slightly less than Earth's. Earth has large internal pressures, however, and its equivalent uncompressed density is smaller than Mercury's. This means that Mercury contains a much larger fraction of iron than any other planet or satellite in the solar system. The diameter of its iron core is about 75% of the planet diameter, and its silicate mantle and crust is only about 600 km thick. Earth's iron core is only 54% of its diameter.

Mercury has a significant magnetic field, like the Earth's. It has sufficient strength to hold off the solar wind most of the time, but the solar wind actually reaches the surface at times of highest solar activity. Mercury's magnetic field suggests that it has a fluid outer iron core surrounding a solid inner core (*see* GEODYNAMO).

Three hypotheses have been put forward to explain the enormous iron core:

- It came from enrichment processes as Mercury assembled by ACCRETION in the innermost part of the solar system;
- Intense bombardment by solar radiation from the newly-

Observing Mercury and Venus

Mercury presents a solid surface at low resolution, while Venus offers for inspection only a visually opaque but dynamic upper atmospheric layer. Past amateur study is largely the story of visual techniques applied with moderate instrumentation in order to build up a pictorial and descriptive record. Now, however, amateurs use sophisticated techniques to monitor a broader spectral range and there is scope for comparing visual observations and CCD results.

► ***Mercury*** *Imaged by Frank Melillo, courtesy of the BAA.*

As recorded in *The Interior Planets* (Oliver and Boyd, 1968), Axel Firsoff used observers' drawings to sum up the history of telescopic observation of Mercury: 'Most show a brighter disk with a few dark areas, but others appear to see it in the negative, as it were, with bright markings in the minority. Yet most experienced students of the planet seem to concur in the delineation of the main features.'

Today, electronic imaging is also capable of recording gross ALBEDO features on Mercury. Visual users of 20-cm telescopes can also hope to record disk features. Mariner 10 (*see* MARINER MISSIONS) imaged only one hemisphere and so the Mercury 'jigsaw' is incomplete. Thus, telescopic observers can continue to delineate the albedo variations that stimulated various mapping attempts since Giovanni SCHIAPARELLI's chart of 1889. Note that even the best amateur map does not cover latitudes greater than 60° north and south. Drawings should be made on a circular disk of 5 cm.

There are bright areas and faintly contrasted dark markings. Overall, Mercury's disk exhibits low-contrast features, but discrete bright spots should be watched for and should be accurately positioned on sketches. These are potentially significant, because a connection has been made between them and areas of bright or rayed craters for some regions imaged by Mariner.

Asymmetries and irregularities of the cusps are sometimes reported, chiefly blunting of the southern cusp. Slight surface albedo variations may be responsible, but very unusual effects, such as cusp extensions are most difficult to explain on an airless world! Irregularities that are seen at the TERMINATOR may also possibly be due to slight albedo variations on the disk.

W80A light blue or W58 green color filters are suitable for darker areas (*see* CAMERA AND FILTER). W22 orange or W8 yellow and W13 green stacked generally give distinct views and stronger contrast with bright features. One should make intensity estimates of these features on the normal planetary scales. Unfortunately, the BAA scale of 0 = bright white to 10 = black is the reverse of the Association of Lunar and Planetary Observers (ALPO) scale where 0 = black, so it is necessary to specify which scale is used.

Observing Venus presents different challenges. There are extremely faint albedo patterns on the disk and, occasionally, bright or darkish spots appear. Records of features that are suspected in one fleeting moment are not reliable: visual observers should record only those impressions that are repeatedly glimpsed. Again, drawings should be made on a scale of 5 cm. There is much discussion of 'UV sensitivity' among visual observers, but while that debate continues, it is undeniable that concordant work is produced by practiced observers using visual filters that effectively

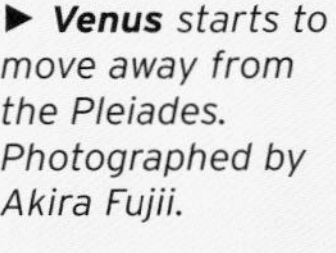

► ***Venus*** *starts to move away from the Pleiades. Photographed by Akira Fujii.*

◀ ***Venus** in the near-UV. CCD images by David Moore, courtesy of the BAA.*

block UV; and also with achromatic REFRACTORS that cannot possibly focus near-UV features at the normal visual focus of such telescopes. Therefore, it is most likely that there is a faint corollary of the strong near-UV features that extends into the ordinary OPTICAL SPECTRUM. There is no best telescope to use. What is necessary is steady SEEING. It is also vital to dim the dazzling brightness of the image in order to reveal faint contrasts. The English planetary observer David Gray, who observes Venus with a grossly dimmed image and takes precautions to exclude extraneous light from the EYEPIECE, likens the process of revealing the faint disk patterns to glimpsing deep-sky objects.

The standard method of brightness reduction is to combine eyepiece filters with a magnification as high as both the telescope and the observing conditions will permit. Standard filters are W15 yellow and W25 red filters, used singly or stacked together. Apodising screens may be employed with apertures of 20 cm or larger. Apart from the need to use a standard filter for phase estimates, the choice of visual filters for general disk examination seems largely subject to instinctive personal visual comfort. Some recommend deep blue filters such as W47, whereas a few observers stack W15 yellow and W58 green filters. A W80A light-blue filter enhances extensive bright areas on the disk.

The terminator shows contour anomalies and darker spots. Red filters, such as W25, enhance terminator features. It is normal to allocate intensity estimates to the terminator shading and also to any disk features, according to the standard BAA or ALPO scales. The BAA Venus scale is specifically tailored to suit the narrow range of intensities displayed there:

0 = extremely bright (visible white spots, for example);
1 = bright areas;
2 = general hue of disk;
3 = shading near limit of visibility;
4 =shading well seen;
5 = unusually dark shading (for Venus).

On the BAA Venus scale 5 is equivalent to 3 on the BAA general planetary scale (the usual intensity of Saturns' A ring).

The bright CUSP CAPS of Venus are variable together with the presence of dusky borders or collars. Careful comparison may reveal anomalous cusp features, such as blunting, sharper appearances, odd shapes to one horn, or detached points or patches of illumination. Some find that a W58 green filter enhances cusp caps.

The bright LIMB band varies in breadth and in extent. Short-term phenomena appear occasionally at the limb, such as bright spots and apparent projections. The bright limb band appears broader in a blue filter.

The ASHEN LIGHT is the most controversial planetary phenomenon. Isolated light spots are occasionally reported at the dark limb and, near inferior CONJUNCTION, mottling of the dark side between the horns of the ultra-thin crescent as well as a dull rusty color. Thus, a red filter has been advocated for ashen light. Ashen-light searches are conducted at dusk or dawn. For visual searches, an occulting device in the eyepiece is necessary to block out the planet's bright crescent. There are doubts as to the objective reality of the ashen light. A common daylight impression is that of the unilluminated hemisphere appearing darker than the daytime sky in the crescent phase.

It is standard practice to measure the phase from a drawing prepared at the eyepiece using a W15 yellow filter. Another way is to make a direct estimate at the eyepiece (with W15) while referring to a sheet of prepared outlines, such as that produced by the Union of Italian Amateur Astronomers (UAI) planetary group.

When Venus is a narrow crescent a bright equatorial band is sometimes prominent. For a few days around inferior conjunction, cusp extensions may glimmer in the eyepiece and, indeed, the whole disk may be encircled by a woolly thread of diffused light. The extended cusps may not be symmetrical, unbroken, or of the same brightness and color. Atmospheric clarity is important and high magnification is not necessary at this time, since the planet's apparent diameter is almost one arcmin. If the sky is deep and clear blue up to a distance of an outstretched hand's width from the Sun, conditions are ideal, but extensions are often reported in much less than ideal conditions. Various filters may be tried to enhance the effect. Unusual effects, such as mottling of the dark side between the horns of the ultra-thin crescent should also be watched for. Care must be taken at this stage when the planet is typically within 8° of the Sun. Studies close to inferior and superior conjunction are for experienced planetary observers, and not the newcomer.

Superior conjunction presents the only opportunity to observe the whole disk for any faint markings that might be present.

Robert Steele is director of the Mercury and Venus section of the British Astronomical Association.

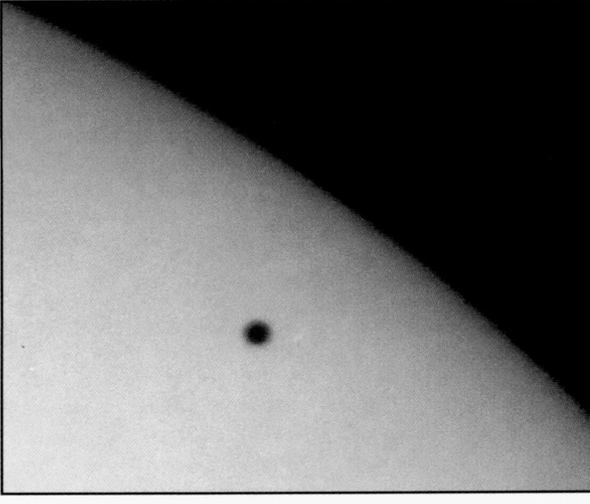

◀ ***Transit of Mercury** in front of the Sun, as imaged on May 7, 2003 by Jamie Cooper, using a Nikon Coolpix and a Meade ETX-90EC fitted with a 'white-light' solar filter.*

born Sun vaporized much of the silicate material of Mercury as it formed;
- A planet-sized object impacted Mercury and blasted away much of its silicate mantle, leaving the core largely intact.

Surface

Mercury is superficially like the Moon. It is covered with a soil consisting of fragmented ejecta from meteor impacts over billions of years. Its surface is heavily cratered with smooth planes filling and surrounding large impact basins. Long LOBATE SCARPS traverse the surface for hundreds of kilometers, and large expanses of plains fill regions between craters in the highlands.

In general, the surface of Mercury can be divided into four major terrains:
- heavily cratered regions;
- intercrater plains;
- smooth plains;
- hilly and lineated terrain.

As on Mars, the heavily cratered uplands record the period of LATE HEAVY BOMBARDMENT that ended about 3.8 billion years ago. The crater population superimposed on the smooth plains within and surrounding the CALORIS BASIN shows a much lower crater density. This suggests that they formed near the end of the heavy bombardment.

Fresh impact craters on Mercury are like others in the solar system. The freshest have extensive ray systems, some of which extend over 1000 km. The largest relatively fresh impact feature seen by Mariner 10 is the 1300-km-diameter Caloris Basin. Its floor consists of closely spaced ridges and troughs arranged in both a concentric and radial pattern. Directly opposite the Caloris Basin (the antipodal point) is an unusual hilly and lineated terrain that cuts across pre-existing landforms, like crater rims. The hilly and lineated terrain is thought to be the result of seismic waves generated by the Caloris impact and focused at the antipodal region.

Mercury's older plains are volcanic plains erupted through a fractured crust. They are probably about 4 billion to 4.2 billion years old. The younger smooth plains are primarily associated with large impact basins. They also fill smaller basins and large craters, like the lunar maria (*see* MARE), and are, therefore, lava flows that erupted later. They have an age of about 3.8 billion years, older than the lunar maria.

Three large radar-bright features have been identified on the unimaged side of Mercury. One is a fresh impact crater. One has a radar signature unlike any other in the solar system. The third has a more or less structureless radar-bright halo (500 km diameter) and a radar-dark center (70 km diameter). This is a fresh, large shield volcano (formed by layers of lava flows), as large as OLYMPUS MONS on Mars.

Mercury has a tectonic framework of compressive thrust faults called lobate scarps. Individual scarps vary in length from about 20 km to over 500 km, and have heights from a few 100 m to about 3 km. The scarps were probably caused by the shrinking of Mercury by 2 km as its crust cooled.

Very little is known of the composition of the surface of Mercury. However, high-resolution radar images of Mercury taken in 1991 show very high reflectivities and polarization in the areas centered on the poles, typical of water ice. Mariner 10 images of Mercury's poles show cratered surfaces where ice could be concentrated in permanently shadowed portions of the craters. Because the planet has no seasons the temperature in the shaded polar regions is always less than 112 K, and water ice does not evaporate, even over billions of years. The ice could originate from comet or water-rich asteroid impacts that released the water, forming a temporary steamy atmosphere, some of which froze in the permanently shadowed craters (as may have happened on the Moon).

Further reading: Vilas, F (ed) (1988) *Mercury* University of Arizona Press.

Mercury	
Mean distance from the Sun	57.9×10^6 km (0.3871 AU)
Period of revolution	87.969 days
Orbital eccentricity	0.206
Inclination to ecliptic	58.646 days
Diameter	4878 km
Mass relative to the Earth	0.0554
Density	5.44 g cm^{-3}
Escape velocity	4.2 km s^{-1}
Surface (main materials)	Little known
Atmosphere (main components)	Largely the solar wind (hydrogen and helium)
Atmospheric density	100 000 atoms cm^{-3}
Surface temperature	725 K at perihelion at noon on the equator, 90 K at night, and similarly cold at the poles

Meridian

(1) A great circle on the Earth's surface, running perpendicular to the equator and passing through the north and south poles. To an observer located at a particular position on the Earth, the meridian represents his north-south line. The Greenwich or prime meridian (the meridian passing through the ROYAL OBSERVATORY, GREENWICH) is taken as the zero of longitude measurement; thus the longitude of a point on the Earth's surface is the angle (measured parallel to the equator) between the Greenwich meridian and the meridian passing through that point. The term 'central meridian' refers to the line joining the north and south poles of a planet that crosses the center of the visible disk of that planet.

(2) For purposes of astronomical position and time measurement, the meridian is taken to be a circle on the CELESTIAL SPHERE passing through the north and south celestial poles and the observer's ZENITH. For an observer on the Earth's surface, the celestial meridian is the projection of his terrestrial meridian onto the celestial sphere.

MERLIN/VLBI National Facility

The Multi-Element Radio Linked Interferometer Network (MERLIN) is a unique synthesis RADIO TELESCOPE originally developed by the University of Manchester, England, in the late 1970s. The telescopes are spread across the UK. MERLIN operates in discrete wavelength bands within the range 2 m to 1 cm, and has corresponding angular resolutions ranging from 1.4 arcsec to 8 milliarcsec. At its prime operating wavelength of 6 cm it has a resolution that matches that of the HST.

Merrill, Paul Willard (1887–1961)

American astronomer, staff member of the MOUNT WILSON OBSERVATORY, spectroscopist who studied variable stars. He identified the element technetium in an S-type star R Andromedae. The atoms of this element have no stable nuclear isotopes, so the element must have been made within the last million years, although the star was much older. The discovery proved that the element must have been made recently in the star. The process by which it was made was identified by B²FH (*see* William FOWLER) as the 's-process' of NUCLEOSYNTHESIS.

Meson

A particle that is composed of a QUARK and an antiquark. Mesons are short-lived particles that decay rapidly into LEPTONS or PHOTONS, either directly or by first decaying into other types of meson. *See also* FUNDAMENTAL FORCE.

MESSENGER (Mercury Surface, Space Environment, Geochemistry and Ranging mission)

NASA Discovery mission scheduled for launch in 2004. It is intended to fly past Mercury twice in 2008, then enter orbit around the planet in September 2009. MESSENGER will map the entire planet and study the magnetosphere and composition of its thin atmosphere.

Messier, Charles (1730–1817)

Comet seeker, nebula avoider, born in Badonvillier, Lorraine, France. He went to the Paris Observatory, where he was employed by Joseph-Nicolas Delisle (1688–1768) to search for COMET HALLEY on its return in 1757. While doing so, he discovered a comet-like patch in the constellation of Taurus on the comet's expected path. The patch did not move, and he later identified it as a nebula (now known as the CRAB NEBULA), found earlier by John BEVIS. Messier decided to make a list of nebulous objects that simulate comets, and Bevis's nebula became the first entry, M1. *See* MESSIER CATALOG.

While discovering two comets in 1763–4, Messier recorded two more nebulae: M2 (previously discovered by Jean-Dominique Maraldi, 1709–1788) and M3 (the to him unresolved GLOBULAR CLUSTER, his first original nebular discovery). He undertook systematic searches for comets and nebulae in the sky and the literature. This eventually brought his catalog to 110, of which 42 objects were discovered by himself.

Messier Catalog

In the eighteenth century the French astronomer Charles MESSIER drew up a catalog of 103 of the brighter nebulae, star clusters and galaxies. The published catalog was expanded by Messier himself in manuscript form, and by modern historians using Messier's manuscript archives. In its modern form it contains 110 objects, some spurious. Objects in this catalog are denoted by the letter M followed by a number; for example, M31 is the ANDROMEDA GALAXY. Messier published a preliminary list of 45 nebulous objects in 1771, and compiled the bulk of his catalog 10 years later. The catalog is still a valuable guide for amateur astronomers, although it has been superseded by the more comprehensive NEW GENERAL CATALOGUE OF NEBULAE AND CLUSTERS OF STARS and by modern astrophysically-based catalogs. *See* table overleaf.

Meteor

A transient luminous phenomenon visible in the night sky, popularly called a 'shooting star.' Small, solid cosmic bodies called METEOROIDS cause meteors by friction with air. Most meteoroids are disintegrated COMETS and ASTEROIDS, but a small percentage may be of interstellar origin and a few are manmade space debris. Meteoroids producing typical meteors visible to the naked eye measure only millimeters across.

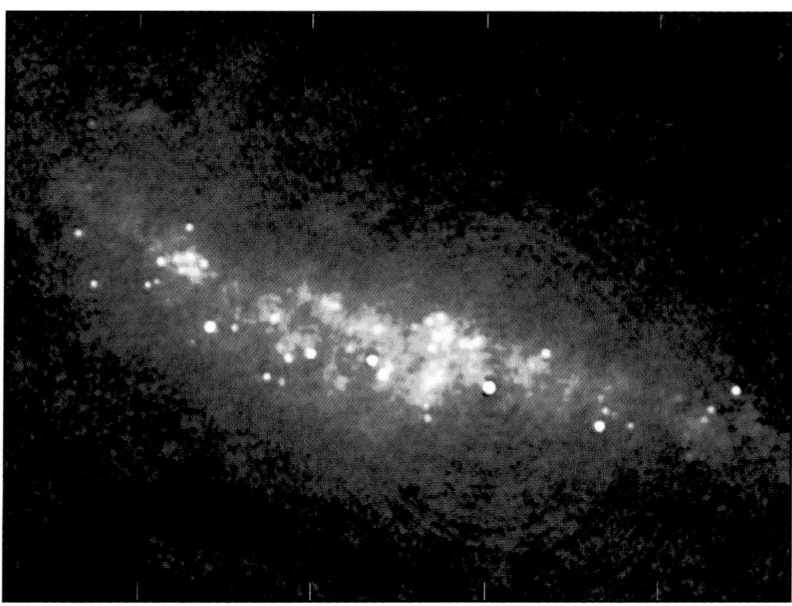

▲ ***MERLIN*** *This rendition of the nearest starburst in M82 was made by the MERLIN radio telescope. The areas of varying emission are color-coded, with reds and yellows denoting the strongest signals.*

Meteors brighter than the planet Venus are called FIREBALLS or BOLIDES. Some of them are brighter than the full Moon and, rarely, some are visible during daylight. It is only generally true that the brighter the fireball, the larger the meteorite. Fragments of sufficiently large and/or sufficiently robust meteoroids can survive their fall and reach the ground as METEORITES.

Most meteors come from random directions. These are called SPORADIC METEORS. At certain times of year, however, meteors come from the same direction on parallel trajectories and with the same speed. This is a METEOR SHOWER. The meteors of a shower have a common origin in one parent body, in most cases a comet. Meteoroids released from a comet spread along the comet trajectory and form a meteoroid stream or, somewhat inconsistently, a METEOR STREAM. The direction from which the meteors come is called the RADIANT. Extremely rich meteor showers are called METEOR STORMS. These occur a few times each century and last only a few hours.

Bright fireballs and meteor storms belong to the most impressive astronomical phenomena observable by the naked eye.

History

Meteor astronomy as a science started with the observation from North America on November 13, 1833, of a meteor storm, when the radiant was obvious. The parent meteor shower was named the LEONIDS. The radiant was then recognized for less impressive annual meteor showers such as the PERSEIDS, which was related in 1867 by Giovanni SCHIAPARELLI to COMET SWIFT–TUTTLE.

Ernst Chladni (1756–1827) first recognized in 1894 the connection between meteors and meteorites. His idea was accepted only at the beginning of the nineteenth century, with a report by Jean-Baptiste BIOT on the meteorite falls at L'Aigle, France, for the French Academy of Sciences.

Meteor observations can be made by eye, by photography including spectroscopy, by radar, and by video camera. In photography and television techniques, widefield cameras

Messier Catalog

M	Right ascension		Declination		Name(s)	Type	NGC	Constellation
	h	m	°	'				
1	05	34.5	+22	01	CRAB NEBULA	SNR	1952	Tau
2	21	33.5	-00	49		GC	7089	Aqr
3	13	42.2	+28	23		GC	5272	CVn
4	16	23.6	26	32		GC	6121	Sco
5	15	18.6	+02	05		GC	5904	Ser
6	17	40.1	-32	13	Butterfly Cluster	OC	6405	Sco
7	17	53.9	-34	49	Ptolemy's Cluster	OC	6475	Sco
8	18	03.8	-24	23	Lagoon nebula (contains the HOURGLASS NEBULA)	DN	6523	Sgr
9	17	19.2	-18	31		GC	6333	Oph
10	16	57.1	-04	06		GC	6254	Oph
11	18	51.1	-06	16	Wild Duck Cluster	OC	6705	Sct
12	16	47.2	-01	57		GC	6218	Oph
13	16	41.7	+36	28	Hercules globular cluster	GC	6205	Her
14	17	37.6	-03	15		GC	6402	Oph
15	21	30.0	+12	10		GC	7078	Peg
16	18	18.8	-13	47	EAGLE NEBULA (or Star Queen Nebula, IC 4703)	OC	6611	Ser
17	18	20.8	-16	11	Omega, Swan, Horseshoe, or Lobster Nebula	DN	6618	Sgr
18	18	19.9	-17	08		OC	6613	Sgr
19	17	02.6	-26	16		GC	6273	Oph
20	18	02.6	-23	02	TRIFID NEBULA	DN	6514	Sgr
21	18	04.6	-22	30		OC	6531	Sgr
22	18	36.4	-23	54		GC	6656	Sgr
23	17	56.8	19	01		OC	6494	Sgr
24	18	16.9	-18	29	Sagittarius Star Cloud, Milky Way patch	Other	incl 6603	Sgr
25	18	31.6	-19	15		OC	IC 4725	Sgr
26	18	45.2	-09	24		OC	6694	Sct
27	19	59.6	+22	43	DUMBBELL NEBULA	PN	6853	Vul
28	18	24.5	-24	52		GC	6626	Sgr
29	20	23.9	+38	32		OC	6913	Cyg
30	21	40.4	-23	11		GC	7099	Cap
31	00	42.7	+41	16	ANDROMEDA GALAXY	SG	224	And
32	00	42.7	+40	52	Satellite of ANDROMEDA GALAXY	EG	221	And
33	01	33.9	+30	39	TRIANGULUM GALAXY	SG	598	Tri
34	02	42.0	+42	47		OC	1039	Per
35	06	08.9	+24	20		OC	2168	Gem
36	05	36.1	+34	08		OC	1960	Aur
37	05	52.4	+32	33		OC	2099	Aur
38	05	28.4	+35	50		OC	1912	Aur
39	21	32.2	+48	26		OC	7092	Cyg
40	12	22.4	+58	05	Double Star Winnecke 4 (WNC4)	Other	-	Uma
41	06	46.0	-20	44		OC	2287	Cma
42	05	35.4	-05	27	ORION NEBULA	DN	1976	Ori
43	05	35.6	-05	16	de Mairan's Nebula (part of ORION NEBULA)	DN	1982	Ori
44	08	40.1	+19	59	Praesepe, the Beehive Cluster	OC	2632	Cnc
45	03	47.0	+24	07	The PLEIADES (Seven Sisters), Subaru	OC	-	Tau
46	07	41.8	-14	49		OC	2437	Pup
47	07	36.6	-14	30		OC	2422	Pup
48	08	13.8	-05	48		OC	2548	Hya
49	12	29.8	+08	00		EG	4472	Vir
50	07	03.2	-08	20		OC	2323	Mon
51	13	29.9	+47	12	WHIRLPOOL GALAXY, Lord Rosse's 'Question Mark'	SG	5194	CVn
52	23	24.2	+61	35		OC	7654	Cas
53	13	12.9	+18	10		GC	5024	Com
54	18	55.1	-30	29		GC	6715	Sgr
55	19	40.0	-30	58		GC	6809	Sgr
56	19	16.6	+30	11		GC	6779	Lyr
57	18	53.6	+33	02	RING NEBULA	PN	6720	Lyr

Messier Catalog

M	Right ascension h	m	Declination °	′	Name(s)	Type	NGC	Constellation
58	12	37.7	+11	49		SG	4579	Vir
59	12	42.0	+11	39		EG	4621	Vir
60	12	43.7	+11	33		EG	4649	Vir
61	12	21.9	+04	28		SG	4303	Vir
62	17	01.2	-30	07		GC	6266	Oph
63	13	15.8	+42	02	Sunflower Galaxy	SG	5055	CVn
64	12	56.7	+21	41	Black Eye Galaxy, Sleeping Beauty	SG	4826	Com
65	11	18.9	+13	05	Leo Triplet (with M66 & NGC 3628)	SG	3623	Leo
66	11	20.2	+12	59		SG	3627	Leo
67	08	50.4	+11	49		OC	2682	Cnc
68	12	39.5	-26	45		GC	4590	Hya
69	18	31.4	-32	21		GC	6637	Sgr
70	18	43.2	-32	18		GC	6681	Sgr
71	19	53.8	+18	47		GC	6838	Sge
72	20	53.5	-12	32		GC	6981	Aqr
73	20	58.9	-12	38	Asterism of 4 stars	Other	6994	Aqr
74	01	36.7	+15	47		SG	628	Psc
75	20	06.1	-21	55		GC	6864	Sgr
76	01	42.4	+51	34	Barbell, Little Dumbbell, Cork or Butterfly Nebula	PN	650-651	Per
77	02	42.7	-00	01	Cetus A (a Seyfert 2 galaxy)	SG	1068	Cet
78	05	46.7	+00	03		DN	2068	Ori
79	05	24.5	-24	33		GC	1904	Lep
80	16	17.0	-22	59		GC	6093	Sco
81	09	55.6	+69	04	Bode's Galaxy, a pair with M82	SG	3031	UMa
82	09	55.8	+69	41	Cigar Galaxy	IG	3034	UMa
83	13	37.0	-29	52	Southern Pinwheel Galaxy	SG	5236	Hya
84	12	25.1	+12	53	Member of the VIRGO CLUSTER OF GALAXIES	LG	4374	Vir
85	12	25.4	+18	11		LG	4382	Com
86	12	26.2	+12	57	Member of the VIRGO CLUSTER OF GALAXIES	LG	4406	Vir
87	12	30.8	+12	24	VIRGO A	EG	4486	Vir
88	12	32.0	+14	25		SG	4501	Com
89	12	35.7	+12	33	Member of the VIRGO CLUSTER OF GALAXIES	EG	4552	Vir
90	12	36.8	+13	10	Member of the VIRGO CLUSTER OF GALAXIES	SG	4569	Vir
91	12	35.4	+14	30		SG	4548	Com
92	17	17.1	+43	08		GC	6341	Her
93	07	44.6	-23	52		OC	2447	Pup
94	12	50.9	+41	07		SG	4736	CVn
95	10	44.0	+11	42		SG	3351	Leo
96	10	46.8	+11	49		SG	3368	Leo
97	11	14.8	+55	01	Owl Nebula	PN	3587	UMa
98	12	13.8	+14	54		SG	4192	Com
99	12	18.8	+14	25	Coma Pinwheel	SG	4254	Com
100	12	22.9	+15	49		SG	4321	Com
101	14	03.2	+54	21	TRIANGULUM GALAXY, Pinwheel Galaxy	SG	5457	UMa
102*	15	06.5	+55	46	Spindle Galaxy?	LG	(5866?)	Dra
103	01	33.2	+60	42		OC	581	Cas
104	12	40.0	-11	37	Sombrero Galaxy	SG	4594	Vir
105	10	47.8	+12	35		EG	3379	Leo
106	12	19.0	+47	18		SG	4258	CVn
107	16	32.5	-13	03		GC	6171	Oph
108	11	11.5	+55	40		SG	3556	UMa
109	11	57.6	+53	23		SG	3992	UMa
110	00	40.4	+41	41	Satellite of M31	EG	205	And

SNR = supernova remnant, GC = globular cluster, DN = dark nebula, OC = open cluster, PN = planetary nebula, SG = spiral galaxy, EG = elliptical galaxy, LG = lenticular (SO) galaxy, IG = irregular galaxy

Observing meteors

METEOR observing must be one of the most accessible and most enduring forms of amateur observing. As with the observation of many other celestial objects, amateurs with the necessary equipment can produce amazing results. However, meteor observing is unique in that the results of observations obtained with the naked eye can still be scientifically useful.

A dark site is essential for all but the most active METEOR SHOWERS, and observers should aim to look at regions of the sky separated from the RADIANT by at least 60° (*see* DARK SKIES AND GOOD OUTDOOR LIGHTING). When observing in a group, the whole sky can be monitored by assigning regions to each observer. Although the most obvious aim is to produce a count as an indication of meteoric activity, other information may also be of use and should be recorded. Examples include any color that may be seen (usually on the brightest meteors), the presence and duration of any train (which remains glowing after the 'head' of the meteor has been extinguished) and the exact time the meteor was seen (useful for triangulation when compared with observations from other sites). Tracing the trail backwards will give an indication of whether the meteor belongs to a particular meteor shower, and it is often instructive to plot the positions of the meteors on a chart for later analysis. To achieve this, the low-tech aid of a piece of string can be used; held up against the rapidly-moving meteor, it allows leisurely observation of the path taken. Finally, an indication of the magnitude of the meteor should be recorded; often this is an approximate guess, or the estimate can be made more quantitative by recording the relative magnitude of nearby stars.

This wealth of information should serve to keep the meteor-watcher entertained, but the recording of all these details can become difficult to maintain at the peak of especially-active displays. Some observers record observations onto tape for later transcription so that their eyes never leave the sky, but this may be time consuming. If one is lucky enough to witness a METEOR STORM such as those provided by the LEONIDS in 1966 and 2001, then perhaps the best advice that can be given is try to keep a rough count while enjoying the spectacle!

The final question is when to observe; this is easily answered by consulting a table of METEOR SHOWERS. Best conditions for observing are to be found when the Moon is out of the way (with the Moon in the sky, the large number of faint meteors which make up the majority of the crop are lost), and in the early hours of the morning; rates for both shower and sporadic meteors usually peak after midnight. Otherwise, detailed knowledge of each shower should be used to find the best times to observe; some have sharp peaks lasting only an hour or two, whereas others will maintain a constant rate of activity for a few nights.

Observations obtained make much more sense and provide a more reliable picture when combined with those of other observers. This is usually completed by a central organization: the International Meteor Organization in the USA and the BAA in the UK.

▼ ***Meteor*** *from the Leonid meteor shower in 1995. Photographer Graham Boots had to keep trying for 15 years before he caught such a bright meteor in the field of view of his camera.*

Meteor photography

Meteor photography can be compared to night-time fishing. A lot of effort is put in, and perhaps you catch one, or perhaps you don't. But photography is a marvelous way to record the wonders of the heavens, especially if you are not prepared to purchase the high-tech equipment needed for electronic imaging. The equipment can be simple and inexpensive; you need no more than what is required to photograph star trails. All that's necessary is a static single-lens reflex camera (SLR) on a tripod with a cable release, loaded with an ISO 400-speed film and pointed at the sky with the shutter open for around five minutes at a time. A second-hand 50-mm SLR camera can be purchased for less than $150. However, you will get a much better photograph if the equipment used can track the stars.

You should use a lens that embraces a large area of sky, such as one with a 50-mm or 28-mm FOCAL LENGTH set at infinity. Setting the aperture wide open at a focal ratio of *f*/2.8 or *f*/2.0 improves your chances of capturing a meteor. Suitable films are Ilford HP5 or Kodak Tri X Pan for black and white photography, and Kodak Ektachrome for color slides. Although color may seem more flashy at first, many prefer black and white as it gives a clearer photograph.

As already stated, a 5-min exposure time is recommended but a shorter exposure might be preferred if there is a bright moon, fog or an artificially light-polluted sky. The problem with long exposures is that sky glow builds up on the film and, as a result, contrast is lost. Periodically the lens should be checked to see if it has accumulated dew and, if so, it should be wiped very gently with a soft linen cloth or a clean paper towel. If the rate of meteor activity is less than 500 per hour (which it normally is), the camera should be positioned to photograph the sky some 40°–50° away from the radiant. (The reason for this is that the meteor needs to travel for some time and distance before it heats up enough to become bright and visible to the observer. Meteors are not normally seen at the radiant unless they are travelling in the observer's general direction, which only applies to a small percentage.)

If a meteor goes through or near your chosen FIELD OF VIEW, the time should be noted and the camera shutter should be closed to stop the build up of sky glow. The start and end times of each exposure should be recorded, as well as the area of sky being photographed (which may well be the same throughout a photographic session).

If a driven camera system is used to successfully photograph a meteor, the time should be noted to the minute; this result can be used with other results to work out the

position in the sky of the radiant. If a meteor is photographed from two or more locations between 40–80 km apart, triangulation can be used to work out the meteor's ALTITUDE.

Observing meteors by video and image intensifier

In the past 20 years, image intensifiers have become more available; video cameras, particularly those using CCDs, have become vastly more sensitive and affordable; and computers can now be used to capture and process video. These developments, alone and in conjunction with other techniques, have made it possible to record automatically and analyze the dynamics of meteors and other astronomical events. Modern video cameras, including camcorders and even webcams, are sensitive enough to record the output from image intensifiers, and these form the basis of intensified video systems. Such systems allow meteors to be video-recorded down to naked-eye level with only slightly less accuracy than photography, enabling much more meteor data to be gathered than was previously possible. The latest CCD video cameras are not yet as sensitive as intensifiers but, used alone, can still record meteors with an accuracy similar to that of photography. Using such systems, groups of astronomers have also developed fireball networks to help pinpoint meteorite falls, and the first successful intensified meteor SPECTRA have been obtained.

Intensified meteor cameras are generally purpose-built to specifications that depend on the type of intensifier and camera available, and the constructor's expertise. Various adapters are required to couple the intensifier, camera and lenses together correctly. An intensifier can be coupled to a miniature low-light security camera. A 16-mm semi-fisheye lens produces an image of the sky on the intensifier's 25-mm photocathode. The field of view is about 90° in diameter, but interchangeable lenses can give different fields if required. The intensifier produces an 'amplified' image of the sky on an output screen that resembles a miniature television. A 12-mm video lens, adjusted for close focusing, relays the image to the camera's CCD chip. The output from the camera is a standard composite video signal that can be input directly into most television monitors and video cassette recorders for focusing and recording. A time-inserter module imprints the current date and time, both of which are visible on-screen for timing purposes. A sufficiently sensitive camcorder may be used in place of the camera, time inserter and VCR, making the equipment even more portable.

Recording of meteors with such systems is absolutely straightforward. During the maxima of major meteor showers, over 100 meteors per night can be recorded easily (several hundred during recent Leonid 'storms'). The observer then has a permanent recording that can be scrutinized later to extract the details of each meteor. Used in conjunction with multi-station photography, precise astrometric measurements of the meteor images from each site can be combined with the precise timings from the video recording, enabling the meteor's atmospheric trajectory and heliocentric ORBIT to be determined very accurately.

Manual examination of video tapes is very laborious, and automated meteor detection software is now being developed for use with PC video-capture hardware. Foremost in this field is German amateur Sirko Molau, who has developed 'Metrec,' software which can process the video signal in real time on fast PCs. Metrec efficiently detects meteors passing through the video field, then accurately records their positions and velocities, assigns them to known showers and saves an image of each trail to disk. The accurate position of meteor-shower radiants can be determined from subsequent processing of the data. Although Metrec was developed for only one specific capture card (aptly named the Matrox 'Meteor'), more general software solutions are now being investigated.

◄ ***Single-lens reflex camera*** *supported by a driven camera platform, and powered by eight AA batteries or a 12-volt dry-cell. It is counterweighted at the opposite end of the declination axis from the camera, and the polar axis is pointed to the Pole Star with the aid of a finder scope.*

The equipment described here is still relatively expensive, and the work has been confined to a small but growing number of professional astronomers and dedicated amateurs. The cost of new intensifier tubes is prohibitive and government restrictions in some countries make them difficult or impossible to obtain. However, amateurs may be able to obtain ex-military or second-hand units where available. An Internet search will also identify suppliers, although import restrictions may apply. Intensifiers are manufactured in a range of sizes, but 18-mm tubes are probably the smallest that should be considered for serious work. Caution should be exercised when choosing an intensifier, as some low-cost 'consumer' units have small photocathodes with a resolution that is too poor for serious astronomical use. Sensitive and affordable CCD video cameras are widely available, with a wide range of suppliers and prices. Manually-set time inserters are available from similar outlets, and made-to-order units which automatically synchronize to radio time signals or GPS are sometimes available.

The use of intensified video to record meteors is a new technique which is still under development, but it is already clear that video techniques will lead to a huge increase in the existing body of data about meteors which should enable continuing improvements in our understanding of them.

Graham Boots is the Curator of the Observatory for the Worthing Astronomical Society, UK; Andrew Elliott is the Occultation Coordinator of the BAA's Lunar Section.

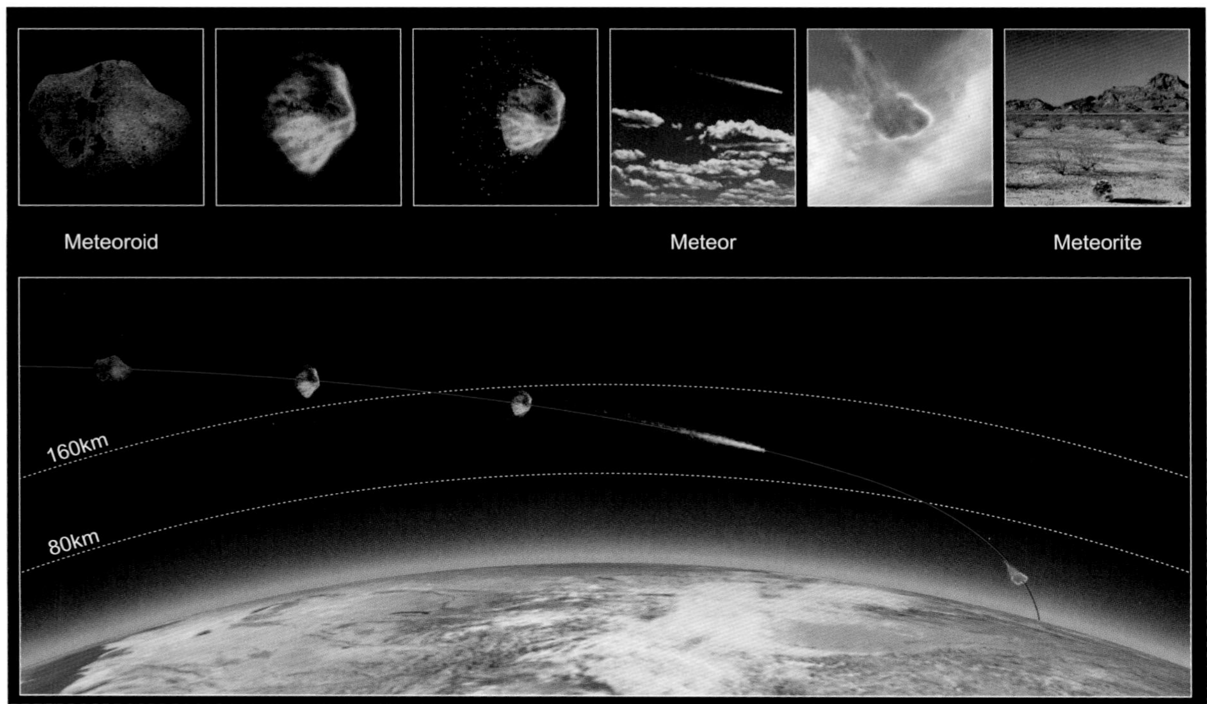

► ***Meteor*** *A meteoroid that enters Earth's atmosphere brakes, slows, shows as a meteor and eventually falls almost vertically to the ground as a meteorite.*

can be opened to wait for a sufficiently bright meteor. Records from two stations of the same meteor as seen against the background of stars triangulate on the meteor's train and provide its trajectory.

Reflections from meteor trails were first observed by military radar in World War II, but kept secret until peacetime.

Passage through the atmosphere

When a meteoroid of typical size enters the Earth's atmosphere at heights of 300–100 km, its surface is warmed by friction with air. This 'preheating' phase lasts from several seconds to tens of seconds. In meteoroids over a millimeter in size the interior remains cool, but at about 130–80 km the surface temperature rises to 2500 K and 'ablation' starts (the surface material falls off because of melting, evaporation and fragmentation). The meteoroid gets hotter and the ionized gases (ablated material and air) produce the meteor phenomenon, lasting typically 0.1–10 s.

As the larger meteoroids penetrate deep into the atmosphere they are decelerated. When their speed drops below 3 km s^{-1} there is not enough friction to heat the body and ablation stops. If the meteoroid has not completely disintegrated, it falls for several minutes without ablation or radiation. This phase is called dark flight, starting usually at a height of about 30 km and ending at the Earth's surface as a meteorite. Some meteorite falls are accompanied by sonic booms; these are heard during dark flight, but generally before the meteorite lands. Rumbles follow, with those produced in the lowest part of the trajectory arriving first.

Typical meteorites of mass 0.01–10 kg impact the surface with velocities of 10–100 m s^{-1}. The impact of a dark-flight meteor creates a pit comparable in size with the meteorite. Large meteoroids, which are still luminous when they reach the ground, explode on impact, and a much larger crater is formed.

Small meteoroids (measuring less than 0.01 mm, *see* INTERPLANETARY DUST) produce no meteors. They are decelerated during the preheating phase and do not get hot enough to ablate. The particles settle gently to the Earth's surface.

Meteor brightness is usually given in magnitudes as determined by comparison with stars (*see* MAGNITUDE AND PHOTOMETRY). About 700 000 meteors of magnitude 0 appear over the entire Earth every day. Typically, 5–10 sporadic meteors are seen from one location each hour; the rate is more frequent in the morning than in the evening.

The absolute magnitude of a meteor is the magnitude as seen from a distance of 100 km. The light curve of a meteor (the meteor brightness displayed as a function of time or height) may show flares (sudden increases of brightness) lasting between 0.01 and 0.1 s. A meteor wake is a 'tail' of radiation, up to several kilometers long, behind the meteor 'head.' A meteor train is an afterglow, left in the meteor path for up to an hour after the meteor's disappearance.

Meteor radar observes the reflection from ionized meteor trails. Radar is reflected in exceptional cases also from a moving source near the meteoroid. This phenomenon, called 'head echo,' is still unexplained.

Meteor trajectories and orbits

The trajectory of a meteor is a straight line, unless it is unusually long lasting. A meteor's velocity through the atmosphere is influenced by the Earth's rotation and orbital velocity, and by the Earth's gravitational attraction, and its speed decreases in flight owing to atmospheric drag. The meteoroid's original heliocentric velocity can be obtained after correcting for these effects. This gives the meteoroid's orbit around the Sun. If the orbit matches that of a comet (or asteroid), this indicates that the meteoroid and the comet are related, particularly if there are several meteoroids whose orbits coincide and form a meteor stream.

If the speed of the meteoroid is above a certain value (74 km s^{-1}), it cannot have been in a bound orbit round the Sun but must have entered the solar system from outside.

No interstellar meteoroid has been confirmed among bright photographic meteors, but about 1% of faint television meteors (produced by meteoroids with mass of about 0.1 mg) are interstellar, and the percentage is higher among radar meteors.

Even sporadic meteors have diffuse radiants and define diffuse meteor streams. They are called the Helion, Antihelion, Apex, Northern Toroidal and Southern Toroidal streams. The first three lie near the ecliptic, at about 20°, 160° and 90° west from the Sun, respectively. The toroidal streams lie at +60° and –60° from the ecliptic, above and below the apex. The Helion and Antihelion meteors originate from asteroids and short-period comets. The Apex meteors originate from long-period comets. The origin of the toroidal sources is not known.

The precise trajectories and orbits of fireballs that have produced recovered meteorites are of particular interest. There are only five such cases: three meteorites that fell in North America, at Lost City (Oklahoma), Innisfree (Alberta) and Peekskill (New York); and the others at Příbram (Czech Republic) and Neuschwanstein (Bavaria, Germany). All five meteorites are ordinary CHONDRITES with orbits like asteroids. Curiously, the Neuschwanstein and Příbram meteorites had the same orbits, although they are of different composition.

Meteor spectra

These belong to the hot vapors of the meteoroid and the atmosphere. The strongest lines are usually those of sodium, magnesium, calcium, iron, manganese, chromium, silicon and atmospheric nitrogen and oxygen. There are two SPECTRA present, one from material with a temperature of 3500–5000 K and one of about 10 000 K. The high-temperature component is much stronger in faster meteors, but little is known about the reason for this. Although the composition of the vapor is not the same as the composition of the meteoroid (because the refractory elements are not always vaporized), meteor spectra relate to the main compositional classes of meteoroids.

Composition of meteors

Meteors differ in their physical aspects, depending whether the meteoroid derives from an asteroid (and if so what sort), or a comet. The height at which the meteor trail begins and ends reveals some differences, which relate to the properties of the meteoroid's material (heat conductivity, heat capacity, density, radiation emissivity, and so on). Four main types can be recognized (*see* table overleaf). The five recovered falls of meteors confirm that Type I meteoroids are stones (ordinary chondrites). ACHONDRITES may also belong to Type I. Members of typical cometary meteor showers belong to types IIIA and IIIB. Iron meteoroids lie between types II and IIIA, and are best recognized from meteor spectra. Achondrites can also be distinguished by their spectra, but chondrites and cometary meteoroids have similar spectra because they have a similar chemical composition despite their different structure.

Further reading: Bronshten V A (1983) *Physics of Meteoric Phenomena* Reidel.

Meteor Crater

A terrestrial impact crater 55 km east of Flagstaff in the Arizona desert, USA, discovered in 1891. It is known by several other names, including the Barringer Crater after Daniel Barringer (1860–1929) who, starting in 1902, was the first to investigate it and the first to suggest its extraterrestrial origin. Meteor Crater has an average diameter of 1.2 km, the floor is 180 m below ground level, and the rim is 30–45 m high. It is estimated to have been produced 50 000 years ago by the impact of an iron METEORITE about 40 m across with a mass of some 250 000 tonnes. The energy of the blast produced by the impact, which excavated 175 000 tonnes of limestone, would have been equivalent to 20 megatonnes or more of TNT. Thousands of meteoritic nickel-iron fragments have been found at distances up to 7 km from the crater, suggesting that the incoming body fragmented shortly before impact. Excavations have revealed no masses of nickel-iron below the crater floor, and the main crater-forming mass of the meteorite is assumed to have vaporized.

Meteorite

A METEOROID that has fallen onto the surface of the Earth. As a METEOR it traversed our atmosphere at high speed. This modified its external aspect, but, if the meteoroid was large enough, left the interior intact. Hence the scientific interest in meteorites: they enable us to study in the laboratory parts of the solar system that humans have not yet explored, such as Mars, the far side of the Moon and asteroids. There is an even more important reason. Because asteroids did not, in general, experience the geological processes that have reprocessed all the materials of the large planets like Earth, Mars, or even the Moon, most meteorites are samples of 'primitive matter.' They remember the processes that gave birth to the solar system. Meteorites let us study the materials from which the planets were made, and establish the age of the solar system at 4.5 billion years.

Meteor falls

The fall of a meteorite is rare but impressive. If the meteorite fragmented in flight, the larger fragments are less efficiently braked and travel farther, so that the fragments are found distributed on an elliptical strewn field with the largest fragments at the most distant end.

Because of their fiery arrival, the surface of meteorites is black and smooth, with rounded angles. The fusion crust is not usually thicker than a fraction of a millimeter. A few millimeters below the fusion crust, the meteoritic materials are unaffected by the fall.

The largest known meteorite, the HOBA METEORITE, found in Namibia, has a mass of about 60 tonnes. The CAPE YORK METEORITE is marginally smaller. Stony meteorites in the size range 10–100 m (mass about 10 000 tonnes) explode in the atmosphere. Such events occur about eight times each year, without being noticed from the ground. This is probably what happened on June 30, 1908, above the region of the River Tunguska in Siberia (*see* TUNGUSKA EVENT), but in this

◀ ***Meteor Crater*** *The scale of the 1.2-km-wide Arizona crater in this image can be appreciated from the size of the Visitors' Center on the raised rear crater rim to the right.*

Finding and collecting meteorites

▶ Appley Bridge *A stone from the Appley Bridge chondrite, seen to fall in Appley Bridge, Lancashire, England, in October 1914.*

METEORITES fall at random over the Earth's surface, at a rate of around 1000 per year for samples weighing 10–100 kg, and at about 10 000 per year for samples weighing between 10–100 g. Many of these specimens fall in the ocean, and are lost. Those that fall unobserved in temperate or wet regions rapidly weather and break down over a timescale of about 100–200 years. In contrast, meteorites that fall in desert regions are preserved for up to 1–2 million years; in such environments, meteorite numbers build up over time, and deserts (hot and cold) are now areas of active meteorite collection, visited by both government-sponsored and privately funded expeditions. The lack of vegetation ensures that the meteorites are easy to spot, and many thousands of meteorites have been found. To put numbers into perspective: the *Catalogue of Meteorites* lists some 1030 meteorites observed to fall, over 3000 found in hot deserts (particularly in North Africa and the Nullarbor region of Australia), 19 900 found in Antarctica and around 2300 found elsewhere.

Naming desert meteorites

Meteorites are traditionally named after a local geographic feature or center of population. However, where large numbers of meteorites are recovered from one area, this convention is not possible to follow. In such cases, meteorites are numbered. Antarctic specimens collected by government-led expeditions are given a year and number combination with a prefix recording the icefield from which they were retrieved (for example, Allan Hills 84001). Alternatively, meteorites collected in hot deserts are simply numbered incrementally by region (for example, Dar al Gani 262). The Nomenclature Committee of the International Meteoritical Society assigns names to meteorites, and keeps track of the total number of reported specimens.

Antarctica

In 1969, a team of Japanese glaciologists recovered nine specimens, representing at least six different meteorites, from a restricted area of the Yamato Mountains in Antarctica. Since then, Antarctica has become the region from which the highest number of meteorites has been retrieved. This is not because meteorites fall over Antarctica any more frequently than they do elsewhere over the globe, but is a result of the unique mechanism that acts to preserve and concentrate meteorites in specific areas of the continent.

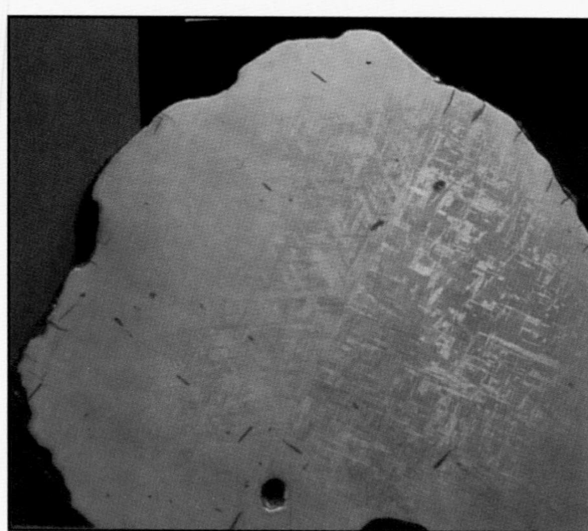

▶ Gibeon *A polished slice of the Gibeon iron meteorite found in Namibia, measuring 20 cm across.*

As meteorites fall on the ice, they are transported with the ice as it flows off the polar plateau to the sea. Gradually, the meteorites become buried in the ice. In some places, the passage of ice to the sea is impeded, either by convergence with a faster-flowing glacier, or against a mountain chain. Ice cannot move and builds up, bringing a constant supply of meteorites to the barrier. The surface of the ice is constantly stripped away by the katabatic (gravity) winds that stream off the plateau. The combined effect of the wind and ice pressure gives the ice a characteristic blue color, and blue 'ice regions' are rich harvest grounds for meteorites.

The same effect is not observed in the Arctic, because there is no major land mass with areas where ice accumulates against land barriers as it does in the Antarctic. Meteorites recovered from Antarctica are covered by the Antarctic Treaty, and should only be collected and removed by recognized expeditions. Almost all Antarctic meteorite-collecting trips, therefore, are government funded, and the specimens are returned to national collections for curation, classification and distribution for research. Antarctic meteorites very rarely reach the international market.

Hot deserts

In contrast to Antarctica, there is no specific meteorite concentration mechanism in hot deserts, other than time. The reduced rate of both physical and chemical weathering in the dry environment serves to allow meteorite numbers to build up. The recovery of meteorites from hot deserts has become more significant in terms of meteorite numbers over the past decade or so, mainly because the deserts are more accessible than Antarctica.

Meteorites recovered from the Nullarbor region of Australia are the property of the Australian government, and may not be removed from the country without an official permit.

Ownership of meteorites removed from the Sahara Desert is less clear, and subject to local rules. Many of the teams exploring the deserts of northern Africa are private, rather than government led, and often the meteorites they recover are bought, sold and exchanged widely in an international

◀ ***Ice fields*** *Searching for meteorites on an icefield in Antarctica.*

market. Because the precise find location of many African desert meteorites is unknown, they have been given the noncommittal name of Northwest Africa, abbreviated NWA, as in NWA 1234.

Why continue?

So many meteorites have been recovered from hot and cold deserts that the question of why we should continue to explore such localities has often been raised. The answer is simple: each expedition has the potential to return with a new, unusual and different type of meteorite, derived from an asteroidal (or planetary) parent that has not been sampled before, and is thus not represented in collections. Meteorites have been found in deserts that extend the range and variety of martian rocks available for study and lunar meteorites have been found only in deserts. The recovery of meteorites from both hot and cold deserts is a valuable method for the acquisition of rare and unusual material for research and display.

Monica Grady is the current Head of the Petrology and Meteoritics Division of London's Natural History Museum, where she serves as curator of the Meteorite Collection and editor of the *Catalogue of Meteorites*.

◀ ***Hot plains*** *Its very dry atmosphere helps make the Nullarbor Plain in Australia one of the richest meteorite fields on Earth.*

Classification of meteor trails

Type	Beginning (km)	Trail height end (km)	Density (g cm^{-3})	Material
I	80	30	3.7	stony
II	90	45	2.0	carbonaceous
IIIA	100	60	0.8	cometary
IIIB	110	70	0.3	soft cometary

case the explosion occurred low enough to destroy approximately 2000 km^2 of forest. Still larger stony meteorites, and iron meteorites over 10 m (greater than 5000 tonnes), reach the ground with almost their original velocity. They are entirely vaporized upon impact. Sporadic meteors are the major source of cosmic material on the Earth. On average, 150 000 tonnes of cosmic material hits the Earth each year. Half arrives very infrequently (one large asteroid impact every 10 million years makes half the average annual rate). Half comes from meteoroids in the mass range 100–100 000 tonnes (mostly cometary bodies approximately 10 m in diameter) with several impacts each year. A few hundred tonnes comes from visual meteors, and several thousand tonnes comes from INTERPLANETARY DUST.

Composition

Meteorites are of different sorts. The simplest and traditional classification is STONY METEORITES, IRON METEORITES and STONY-IRON meteorites. There is, however, a major distinction between differentiated and non-differentiated meteorites, and this new classification helps direct us to the origins of meteorites. DIFFERENTIATION is the process by which melted rocks separate into rocks of different compositions. Non-differentiated meteorites (also called CHONDRITES) have not been melted, and remember the birth of the solar system. Differentiated meteorites have been melted in their parent planets. Melting modified the chemical composition and structure of the materials from which their parent planets were built. But although differentiated meteorites have forgotten the formation of the solar system, they remember subsequent planetary processes.

The various meteorite classes and groups, which are designated by names or symbols, are summarized in the table on the facing page.

Non-differentiated meteorites

Over 80% of meteorites are chondrites and non-differentiated. Their chemical composition is similar to that of the Sun, except for the most volatile elements. Subtle differences have led us to sort them into various groups, as shown in the table. Each group probably corresponds to different parent asteroids.

Spherical structures, called chondrules, occur in the primitive chondrites. These chondrules have diameters in the range 50 μm to several millimeters. The space between chondrules is filled by a matrix made of very fine grains (measuring less than 1 μm) of similar minerals. The CI carbonaceous chondrites are the most primitive of all chondrites, and their composition most closely resembles that of the solar nebula. The high proportion of chondrules in chondritic matter strongly suggests that the chondrule formation process, which is not identified, was common and an effective process in the early solar system. No asteroids have spectra that look like the surfaces of ordinary chondrites. The Galileo spacecraft's flyby of the asteroids (951) GASPRA and (243) IDA suggests why: space weathering has changed the surfaces of most asteroids. Younger terrains look more like ordinary chondrites than older ones.

Carbon is present in all chondrites, and is nearly 5% of the mass of those that are richest in carbon. Where hydrogen is combined in organic molecules (containing carbon), it is strongly enriched in its heavy isotope, DEUTERIUM. This was synthesized in the INTERSTELLAR MEDIUM before the birth of the Sun. Some of the organic molecules are complex, and include sugars and amino acids. Sugars, and closely related compounds called polyols, are critical to life and are components of the nucleic acids RNA and DNA. Dihydroxyacetone and several other sugar-like substances, including glycerol, have been identified in the Murchison and Murray meteorites that fell at Murchison near Melbourne, Australia, in 1969, and near Lake Murray, Oklahoma, USA, respectively, about 110 million years ago (the latter was recognized as a meteorite in 1952). The Murchison meteorite also contains amino acids, which build proteins. In fact, the meteorite contains a small excess of the form of several amino acids associated with life. If these discoveries are not caused by terrestrial contamination, they support the hypothesis that extraterrestrial organic matter is the origin of life on Earth (*see* EXOBIOLOGY AND SETI).

Differentiated meteorites

Unlike chondrites, differentiated meteorites have compositions widely different from that of the Sun. The extreme case is iron meteorites, which are made of only two major elements: iron and nickel. They are sorted into 13 groups corresponding to different parent asteroids, but some 13% of them do not fit into any group. Iron meteorites are believed to be fragments of the metallic cores of larger, shattered asteroids.

Eucrites, diogenites and howardites are magmatic rocks ejected from the crust of a differentiated asteroid, probably (4) VESTA. There is a striking similarity between the spectra of basaltic achondrites and Vesta, and the 20 smaller, similar asteroids (measuring less than 10 km in diameter). According to images from the HST, Vesta has a huge impact crater (460 km in diameter). This crater may well be the site of origin of the small asteroids and the basaltic achondrites.

Pallasites probably originate from the interface between the metallic core and the rocky mantle of differentiated asteroids. Mesosiderites result from a collision involving a metallic and a silicate asteroid. Angrites are like eucrites but are probably from a different parent body. The origin of ureilites might be differentiation of a carbonaceous asteroid and/or an asteroid collision.

The lunar origin of 13 achondrites has been established beyond any doubt by comparing them with the lunar samples returned by the APOLLO and LUNA missions. These meteorites are from the lunar highlands and from the lunar maria (*see* MARE). The SNC METEORITES are from Mars.

Age of the solar system

Two isotopes present in rock can sometimes be used like a clock to determine its age. One of the isotopes (the daughter) is produced by the radioactive decay of the other (the parent); for instance, rubidium-87 decays to strontium-87, potassium-40 decays to argon-40, and uranium-238 to lead-206. The decay occurs at a known rate, and the clock started at the time when no more of either isotope was added to or

Classification of meteorites

Type (subgroups in brackets)	Fall frequency (%)
Non-differentiated meteorites	
Carbonaceous chondrites (CI, CM, CO, CV, CK, CR, CH)	4
Ordinary chondrites (LL, L, H)	79
Enstatite chondrites (EL, EH)	1.5
The chondrite groups are arranged in decreasing order of oxidation from top to bottom, and from left to right in the subgroups. The differences probably reflect different local conditions in the solar nebula at the time of formation.	
Differentiated meteorites	
Irons (IAB, IC, IIAB, IIC, IID, IIE, IIF, IIIAB, IIICD, IIIE, IIIF, IVA, IVB)	5
Pallasites	0.4
Mesosiderites	0.7
Basaltic achondrites (eucrites, diogenites and howardites)	6
Angrites	<1
Ureilites	0.5
Enstatite achondrites (aubrites)	1
Lunar meteorites (anorthosites, basalts)	<1
Martian meteorites (shergottites, nakhlites, chassigny)	0.5
Primitive achondrites (acapulcoites, iodranites, brachinites)	<1

removed from the sample studied, except for that due to radioactive decay. This happened when the temperature of the sample became low enough to crystallize the rock.

The temperature of primitive meteorites has always remained low. Their age is thus the time elapsed since the end of the high-temperature event in which they formed. The uranium-lead clock shows that chondrites have an age of up to 4.566 billion years, with an uncertainty of only 2 million years. No older material has ever been found, and this age of 4.566 billion years is taken as the age of the solar system. Terrestrial rocks crystallized long after the formation of the Earth – the oldest ones are only 3.8 billion years old. The ages of the differentiated meteorites vary between 4.56 and 4.45 billion years, which shows that the oldest asteroids formed very early in the history of the solar system. The youngest SNC meteorites from Mars (Shergotty and Zagami) are very young, around 180 million years, and the older ones (Chassigny and Nakhla) are 1.3 billion years old. Since no asteroid was still hot at such a recent time, these meteorites come from a large planet with recent magmatic activity. The SNC meteorite is much older (about 4.5 billion years).

When a meteoroid is extracted from its parent body it is exposed to irradiation by cosmic rays, from which it was protected before by overlying rocks. Energetic cosmic-ray particles induce nuclear reactions in the meteoroid within the first few meters below the surface, breaking atomic nuclei and creating new ones. This irradiation stops when the meteorite reaches the surface of the Earth, whose atmosphere protects it against cosmic rays. The duration of the exposure (the time it took the meteorite to travel from its parent body to the Earth) is typically 1 million–100 million years for chondrites, while those of irons extend to more than 2 billion years.

Further reading: Hewins R H, Jones R H and Scott E R D (eds) (1996) *Chondrules and the Protoplanetary Disk* Cambridge University Press; Norton O R (2002) *The Cambridge Encyclopedia of Meteorites* Cambridge University Press.

Meteoroid

A piece of debris from a larger parent body, pursuing its own orbit around the Sun, which has the potential to strike the Earth and produce a METEOR or METEORITE. There is no agreed dividing line between a meteoroid and a small ASTEROID, but the word 'meteoroid' is normally used to describe a sub-meter-sized particle.

Meteor shower

The appearance in the sky, at a particular time of the year, of higher-than-average numbers of meteors, all appearing to emanate from a specific point on the CELESTIAL SPHERE known as the RADIANT. Most showers are named after the constellation in which the radiant lies, by adding the suffix '-id': for example, the Lyrid shower (or Lyrids) have their radiant in the constellation Lyra. The Quandrantid shower is named after the obsolete constellation Quadrans Muralis, now part of Boötes. Where there is more than one radiant in a constellation the showers are distinguished by adding the Bayer letter of a bright star near the radiant, as for example with the Delta Aquarids and Eta Aquarids (*see* Johann BAYER). A few showers are named after their parent comets – the comets that produced the meteor streams. For example, the Giacobinids are named after COMET GIACOBINI–ZINNER.

A shower occurs when the Earth intersects a METEOR STREAM. The activity of a shower, which is measured by the ZENITHAL HOURLY RATE (ZHR), lasts for as long as the Earth is passing through the stream, and peaks where the stream is at its most concentrated. Young, tight meteor streams tend to produce showers of short duration which build quickly to a high peak ZHR and then tail off sharply. Older, more evolved streams which have spread either side of the primary orbit are of longer overall duration.

The 10 or so most active showers, with a ZHR of about 10 or more, are known as major showers. There are dozens of less active, minor showers. The table gives details of the most active major showers. In most cases the parent body of a meteor shower is a comet. A rare exception is the Geminid shower, which is associated with the asteroid (3200) PHAETHON.

Meteors that belong to showers are called shower meteors, as distinct from SPORADIC METEORS (produced by meteoroids that do not come from identifiable meteor streams). Annual

Principal annual and periodic meteor showers

Shower	Position of radiant at maximum: Right ascension h	m	Declination °	Duration	Date of maximum declination	Maximum zenithal hourly rate	Parent body
Quadrantids	15	28	50	1–6 January	3–4 January	120	96P/Machholz 1*
Lyrids	18	00	32	19–25 April	21 April	15	C/1861 G1 Thatcher
Eta Aquarids	22	20	−1	24 Apr–20 May	5 May	50	1P/Halley
Arietids	02	58	23	22 May–2 July	8 June	50	96P/Machholz 1*
Delta Aquarids	22	36	−17	15 Jul–20 August	29 July	25	96P/Machholz 1*
Perseids	03	04	58	25 Jul–20 August	12–13 August	80	109P/Swift–Tuttle
Giacobinids	17	23	57	7–10 October	8 October	**	21P/Giacobini–Zinner
Orionids	06	24	15	15 Oct–2 November	20–22 October	30	1P/Halley
Taurids	03	44	14	15 Oct–25 November	3 November	10	2P/Encke
Leonids	10	08	22	15–20 November	17 November	***	55P/Tempel–Tuttle
Geminids	07	26	32	7–15 December	13 December	120	3200 Phaethon

** Association is uncertain.*
*** Outbursts possible every 12 or 13 years. Up to 450 per hour in 1933.*
**** Possibility of meteor storm every 33 years. May have reached 200 000 per hour in 1833.*

showers are those whose ZHRs normally vary little from one year to the next, being produced by meteor streams in which the meteoroids are evenly spread around the orbit. Meteor streams in which the meteoroids are concentrated in a swarm in one part of the orbit give rise to periodic showers, which are highly active only in years when the Earth intersects the swarm, and show little or no activity in other years. Details of two periodic showers, the LEONIDS and the GIACOBINIDS, are given in the table. The Arietids are a daytime shower observable only by radar.

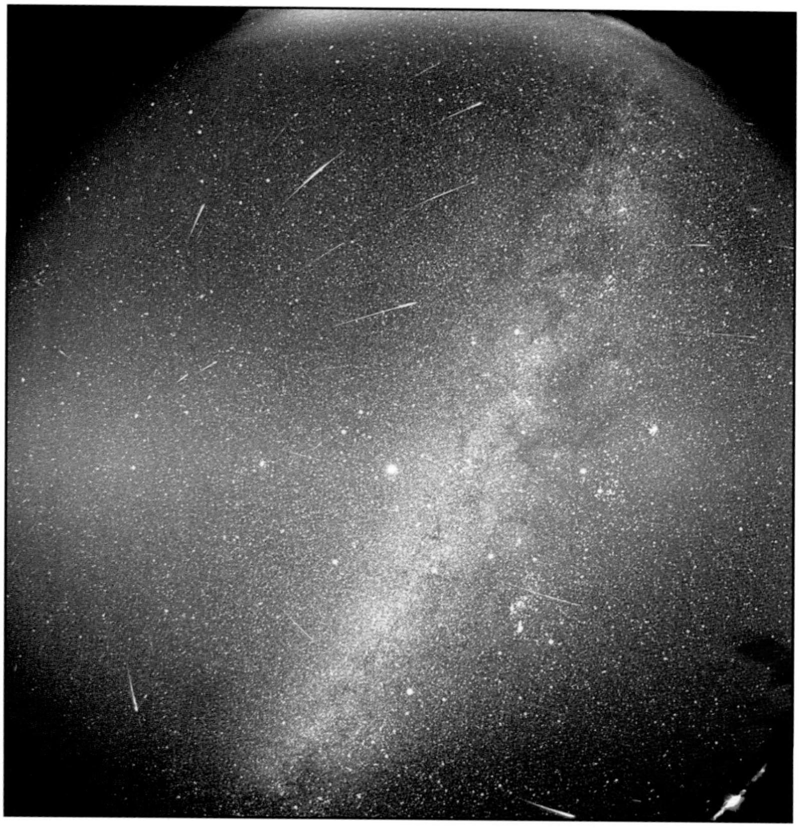

▼ ***Meteor storm*** *All-sky camera view of the Milky Way and a storm of meteors, streaking across the sky from their radiant.*

Meteor storm

An intense burst of METEOR activity of short duration (typically about an hour) associated with a periodic METEOR SHOWER. There is no generally agreed quantitative definition of what constitutes storm-level activity, but an hourly rate of 1000 is a reasonable lower limit. Meteor storms are rare events. The showers that produced meteor storms in the last 200 years were LEONIDS (1799, 1833, 1866, 1966, 1999), Andromedids (1872, 1885), and Draconids (1933, 1946). The Leonids are the best-known meteor storm. The Leonid meteor storm of 1833 led the Sioux native American Indians to name the year 'stars all falling down year.' According to some reports,the 1966 Leonids reached a zenithal hourly rate of 144 000.

Some annual meteor showers occasionally show sharp rises above their normal hourly rates; such outbursts are sometimes called substorms.

The year-to-year observation of the activity of meteor showers provides information about the distribution of METEOROIDS along the cometary orbits and, consequently, about the ejection processes from COMETS. The storms are probably related to the cometary dust trails observed in infrared light.

Meteor stream

A continuous ribbon of METEOROIDS in a closed orbit around the Sun. (The more correct term 'meteoroid stream' is used less often.) A meteor stream consists of dust particles shed by a COMET, which have gradually spread around the comet's orbit. When the Earth passes through a meteor stream, there is a METEOR SHOWER. Most of the debris from young comets remains bunched together over a small arc of the orbit in what is known as a meteor swarm. When the Earth passes through a swarm the result is a METEOR STORM. As a meteor stream ages, and the meteoroids spread around the orbit, the stream assumes the shape of an elliptical toroid. Within the stream there may be filaments containing a greater concentration of meteoroids shed by the parent comet on its most recent circuits of the Sun. The evolution of individual filaments and the stream as a whole is affected by planetary perturbations of both the meteoroids and the parent comet, and the structure of a meteor stream can grow increasingly complex. Eventually, under repeated perturbations, a meteor stream will disperse to the point where the concentration of meteoroids is reduced to that of the interplanetary dust.

Metonic cycle

An interval of 19 years, after which the phases of the Moon recur on the same days of the year. This occurs because 235 lunar months is almost exactly equal to 19 tropical years (235 lunar months is equivalent to 6939.689 days, whereas 19 tropical years is equal to 6939.602 days, a discrepancy of about two hours). This cycle appears to have been discovered about 432 BC by the Greek natural philosopher Meton, and about 50 years later by the Babylonians. Because the Earth, the Moon and the Sun are independent bodies with quite arbitrary periods, attempts to fit together days (Earth's rotation period), months (the lunar orbital period) and the year (the period of the solar orbit) never exactly work, so that, for example, the year is not a whole number of days. But the Metonic cycle almost fits, and it thus provided the basis of the rules for determining the date of Easter (which is set by solar/lunar calendrical considerations). *See also* CALENDAR.

Meudon Observatory

The Meudon Observatory was founded on an old royal estate south of Paris in 1876. It was built with public funds and placed at the disposal of astronomer Jules JANSSEN to allow him to develop his research on solar SPECTRA far from urban pollution. Several instruments were installed by 1893, the largest of which is a 1-m reflector, which was restored in 1969. Later additions included a solar tower, which was used for spectroscopic studies of the Sun. Meudon merged with the PARIS OBSERVATORY in 1926. The section is now devoted to theoretical astrophysics, with departments dedicated to studies of stars, the interstellar medium, galaxies and cosmology.

Mexico National Astronomical Observatory

Located in the mountains of the Sierra San Pedro Martir in Baja, the Observatorio Astronómico Nacional operates three telescopes, a 2.1 m, 1.5 m and 0.84 m.

Michell, John (1724–93)

Cleric, geologist and astronomer, born apparently in Nottingham, England. He invented the idea of BLACK HOLES in a speculative lecture to the Royal Society in 1783 about the gravitational attraction on light in the vicinity of the Sun. He thought that a body more massive than the Sun would attract light, such that it could not escape from the Sun's surface without traveling faster than the speed of light (according to Rømer's value; *see* Ole RØMER). Michell also showed by statistical methods (which he initiated) that many of the DOUBLE STARS cataloged, for example, by William Herschel (*see* HERSCHEL FAMILY) were truly associated and gravitationally bound together. He also proposed the experiment with a torsion balance carried out by Henry CAVENDISH to weigh the Earth by determining the gravitational constant *G*.

Michelson, Albert (1852–1931)

German-American physicist. He performed the MICHELSON–MORLEY EXPERIMENT with Edward Morley (1838–1923) to detect the motion of the Earth through the ether. Michelson went on to apply interferometry to astronomy by mounting an interferometer on a steel beam on the 100-in Mount Wilson telescope, measuring the diameters of six stars.

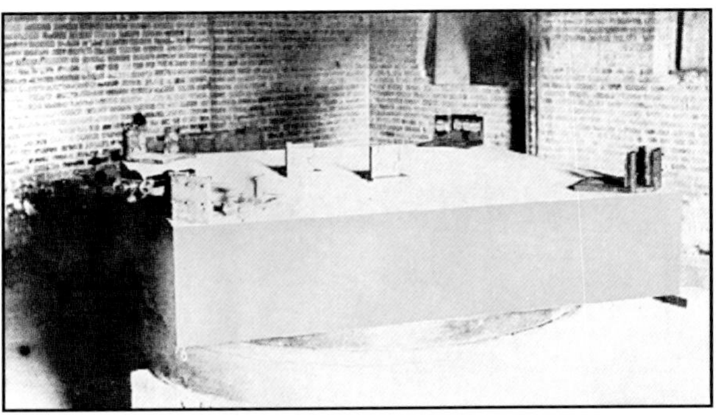

◄ ***Michelson–Morley Experiment*** *The mirrors of the interferometer were laid out on a heavy masonry table to minimize vibration.*

Michelson–Morley experiment

A celebrated experiment carried out by Albert MICHELSON in 1881 and by Michelson and Edward Morley (1838–1923) in 1887, which attempted and failed to detect the motion of the Earth through the ether. The basis of the experiment was that if a beam of light traveled a known distance in the direction in which the Earth was supposed to be moving through the ether, and another beam traveled the same distance at right angles to this direction, then, if light moved at a constant velocity through the ether, the two beams should take different times to cover the distance; they didn't. This result was explained by the Fitzgerald–Lorentz contraction, and was a foundation for Albert EINSTEIN's theory of SPECIAL RELATIVITY.

Michigan Radio Astronomy Observatory

Located in Ann Arbor, close to the University of Michigan campus. The primary instrument at the US observatory is a 26-m-diameter parabolic reflector with an 11-m focal length, constructed in 1958. It is used to study total FLUX DENSITY and linear POLARIZATION from active extragalactic objects. The discovery that such objects vary in brightness with timescales of weeks to a few years was made using the Michigan instrument in 1964–5, and the discovery of variability in polarization followed in 1966.

Microlensing

Refers to the special case of gravitational lensing where the multiple images produced are too close together on the sky to be observed as separate images. The multiple images appear as a single object of increased apparent brightness. This can be detected as lensing if the lens moves across the Earth-source line, as with the PROPER MOTION of a star. Typically the source will appear to brighten, then fade symmetrically back to normal over the course of weeks or months; this is a 'microlensing event,' predicted by Albert EINSTEIN in 1936, and first detected in QUASARS in the 1980s. Nearly 1000 such stellar events have been detected since 1993. *See also* DARK MATTER; GRAVITATION, GRAVITATIONAL LENSING AND GRAVITATIONAL WAVES; MACHO.

Micrometeorite

A small INTERPLANETARY DUST particle (or a particle ablated from a meteoroid) that has not vaporized during its passage through the atmosphere as a METEOR. Micrometeorites are recoverable from cores of sediments drilled from the ocean floor and from the Antarctic ice, and from rainwater. They can also be collected before they reach the ground by aircraft. Interplanetary dust particles recovered from the atmosphere are sometimes called Brownlee particles for Donald Brownlee, the first person to investigate them systematically.

MICROSCOPE mission

MICROSCOPE ('Microsatellite à traînée Compensée pour l'Observation du Principe d'Equivalence') is a CNES microsatellite project designed to test the Principle of Equivalence postulated by Einstein that all bodies, regardless of their mass and internal composition, will acquire the same acceleration in a gravitational field (assuming identical initial conditions). This principle is one of the assumptions of GENERAL RELATIVITY. The accuracy is expected to be to 1 part in 10^{15}, three orders of magnitude better than ground-based measurements. MICROSCOPE will also test drag-free satellite technology in orbit. It is to be launched in 2006.

Microscopium *See* CONSTELLATION

Milankovitch, Milutin (1879–1958)

Yugoslav mathematician and astronomer who, starting in 1912, showed how the radiation received by the Earth varies cyclically as the planet's orbit varies. There are three significant Milankovitch cycles: the ECCENTRICITY of the Earth's orbit varies from more elliptical to more circular with a period of about 100 000 years; the obliquity of the Earth's axis oscillates between two extremes with a period of 41 000 years; and the spin axis precesses with a 26 000-year cycle (*see* PRECESSION). Paleoclimatic oscillations with these periods have been discovered in data from deep-sea cores dating back 6 million years. The data measured global ice volume (indicated by oxygen isotopes in fossil organisms) and fluctuating productivity in surface waters (indicated by calcium carbonate sedimentation), and thus correlate with the ice ages. Although the cycles are named after Milankovitch, his ideas were predated by those of Joseph Adhemar (1797–1862) in 1842, and James Croll (1821–90) in 1875.

Milky Way Galaxy

The GALAXY (the assemblage of stars, gas and dust) of which the Sun is a member. The Milky Way Galaxy is the second largest but the most massive member of the LOCAL GROUP OF GALAXIES. The Milky Way Galaxy (also 'the Galaxy') is most likely a BARRED SPIRAL GALAXY of Hubble type SBc (*see* HUBBLE TUNING-FORK DIAGRAM). In brief, the main structural components of the Galaxy are the GALACTIC DISK, including the barred GALACTIC BULGE with a central cusp; SPIRAL ARMS; a GALACTIC HALO; and a more extended CORONA. The total mass of the Galaxy is about 2×10^{11} solar masses.

We are situated within the outer regions of its spiral disk, only about 14 l.y. above the equatorial symmetry plane but about 8 kiloparsecs or 26 400 l.y. from the GALACTIC CENTER. The Milky Way forms a luminous band of stars spanning all around the sky. This symmetry plane is also called the GALACTIC EQUATOR. In addition to the luminous stars there is also a dark band caused by the obscuration of stellar light via gas and dust within the GALACTIC PLANE. The center of the Galaxy, the radio source and black hole SAGITTARIUS A, lies in the direction of the constellation Sagittarius, close to the border of both neighboring constellations Scorpius and Ophiuchus.

Because the Sun is located within the Milky Way disk, it is difficult to determine its large-scale structure and dynamics. The reasons for this are mainly distance ambiguities and the location of the Sun within the Galactic obscuring dust layer.

Stars in the Galaxy are divided into two principal categories, or populations: POPULATION II stars (old stars that formed early in the history of the Galaxy), and POPULATION I stars (second, or later, generation stars that formed from gas clouds which had been seeded with heavier elements generated in, and expelled from, earlier generations of stars). The molecular gas and dust as well as the youngest, brightest stars of population I are located in a flat disk structure. The older population I stars form a less flattened distribution. This part of the Galaxy is surrounded by a probably ellipsoidally-shaped halo of population II stars out to a radius of about 20 kiloparsecs. Population II stars are also found in the barred bulge of the Galaxy. The mass of this system is comparable with the mass in the inner section of the Milky Way. The corona extends out to approximately 100 kiloparsecs and exceeds by far the mass contained in the spiral disk and bulge region.

The entire solar system is orbiting the Galactic Center at a distance of about 8 kiloparsecs, on a nearly circular orbit. It is moving at about 250 km s^{-1} toward RIGHT ASCENSION 21 h 12 m, DECLINATION +48° 19'. At this velocity it takes about 220 million years to complete one orbit. The Sun has orbited the Galactic Center approximately 20–21 times since its formation some 4.6 billion years ago.

The spiral arms can be traced by the distribution of atomic and molecular gas as well as the distribution of young, bright stars and the EMISSION NEBULAE. The corresponding directions and distances with respect to the Sun can then be used to map out the local spiral structure. The spiral arms manifest themselves via an increase in projected density of brightness of the corresponding tracers, especially when the line of sight is tangential to the spiral arm. Spiral arms are named after the constellations in which they are most prominent. The results indicate that the Sun lies near the inner edge of the so-called ORION ARM. The Perseus arm is located about 2 kiloparsecs beyond the Sun, and the Aquila and Carina arm sections are located approximately 2 kiloparsecs from the Sun toward the Galactic Center. From the study of H II REGIONS, molecular clouds and the Galactic magnetic field, it appears that the Milky Way may have four main spiral arms. The spiral arms have a number of shorter segments, one of which – the Orion 'spur' – contains the Sun and the Orion star-forming region. GOULD'S BELT is a manifestation of this structure.

Further reading: Hoskin M (1982) *Stellar Astronomy: Historical Studies* Science History Publications; Paul E R (1993) *The Milky Way Galaxy and Statistical Cosmology 1890–1924* Cambridge University Press.

Millimeter and submillimeter astronomy

The millimeter wavelength band extends from 4 mm (a frequency of 75 GHz) to 1 mm (300 GHz), while the submillimeter band extends from 1 mm to 100 μm (3 THz). The submillimeter band is relatively unexplored because of the technological challenges and because of severe atmospheric absorption by the Earth's atmosphere, particularly water vapor. To a lesser degree these challenges also affect millimeter astronomy. Studies in the millimeter and submillimeter bands cover a range of topics:

- the physics and chemistry of the INTERSTELLAR MEDIUM, especially the structure and content of the interstellar molecular clouds;

◀ ***Milky Way Galaxy*** *The massed star clouds of the center of the Galaxy are split by lanes of dust lying nearer to us and seen in silhouette.*

- star formation;
- the interaction of stars with molecular clouds, especially during the very early and very late phases of stellar evolution;
- the distribution of molecular clouds in external galaxies and high-redshift objects;
- solar system objects;
- the 2.73 K COSMIC MICROWAVE BACKGROUND.

The first millimeter and submillimeter observations were of emissions from the Sun, planets and a few quasi-stellar objects. Molecular spectroscopy provided a great impetus for millimeter and submillimeter astronomy. Radio spectroscopy of molecules began with the detection of the oxygen-hydrogen radical (OH) at a wavelength of 18 cm in 1963. In 1968 the detection of ammonia and water vapor, both at 1.3 cm, was made with the Berkeley 6-m telescope. In 1969 formaldehyde was discovered with the NRAO's 140-ft telescope at 6 cm. The most widespread polar molecule, carbon monoxide, was detected at 2.6 mm in 1970 with the NRAO's 36-ft telescope. After the discovery of carbon monoxide, there was a rapid increase in the number of interstellar molecules. As of 2003, 123 molecules have been found in the interstellar medium; the most complex molecule, with 13 atoms, is the cyano-polyyene $HC^{11}N$.

The largest millimeter telescopes are the 30 m, operated by IRAM, and the 45 m in Nobeyama, Japan; the 50-m Large Millimeter Telescope is being constructed near Mexico City for completion in about 2004. The 10.4-m-diameter telescope of the CALTECH SUBMILLIMETER OBSERVATORY (CSO) and the 15-m JAMES CLERK MAXWELL TELESCOPE (JCMT) are submillimeter telescopes on Mauna Kea, Hawaii, both put into operation in the late 1980s. Both are constructed of aluminum and are housed in astronomical dome-like shelters.

Submillimeter measurements are also made from aircraft. The Kuiper Airborne Observatory provided such an opportunity with a 91-cm-diameter paraboloid. The follow-up facility, SOFIA, which begins operation in 2004, will have a 2.5-m-diameter mirror. At the normal flight altitude of 14 km the water vapor content is extremely low. Thus SOFIA can be used for measurements at wavelengths at which the atmosphere is opaque from ground-based sites. SWAS was launched late in 1998, SIRTF was launched in mid-2003 and the HERSCHEL SPACE OBSERVATORY, formerly FIRST, is due for launch in 2007.

Millimeter-wave INTERFEROMETERS consist of a number of parabolic antennae. The receiver outputs are combined to produce a response corresponding to the overall size of the array. Interferometers provide higher angular resolutions and positional accuracies than single-dish telescopes. The first millimeter-wave interferometer, at the Berkeley Radio Astronomy Laboratory, consisted of two 6-m antennae. There are four systems in the millimeter range: the BIMA array near Hat Creek, California, USA (*see* HAT CREEK RADIO OBSERVATORY); the Caltech Owens Valley Interferometer near Bishop, California (*see* OWENS VALLEY RADIO OBSERVATORY); the Nobeyama Millimeter Array in Japan; and the IRAM array on the Plateau de Bure, France. In addition, the 10.4-m CSO and 15-m JCMT telescopes are occasionally joined into a submillimeter interferometer. A dedicated submillimeter array consisting of eight 6-m dishes has been constructed by the Smithsonian Astrophysical Observatory and the Academica Sinica Institute for Astronomy and Astrophysics, Taiwan, on Mauna Kea, Hawaii. The ALMA project will be built in Chile at 5 km elevation by the US National Science Foundation and ESO, with 64 elements.

For wavelengths over 0.3 mm the most efficient receivers are heterodyne systems, much like domestic radio receivers, that scan through and record the profiles of spectral emissions. Bolometers measure the total incident radiated energy, like CCD arrays; these are preferred for single-dish measurements of, for example, thermal radiation, and, in contrast to heterodyne systems, can be made in arrays to give pictures. The largest is SCUBA, a 91-pixel system on the JCMT in Hawaii, but larger arrays with thousands of pixels are being built.

Millimeter and especially submillimeter observing sites are selected to minimize the effect of weather on the observations. Signals are absorbed by oxygen molecules and water vapor. The scale height of water vapor is approximately 2 km, while that of oxygen is about 8 km, so the influence of water in particular is reduced at high altitudes. Because the signal levels are small compared with atmospheric or receiver noise, all measurements must be made by taking the difference of two signals. For spectral line measurements, the receiver compares data taken at two different positions or frequencies. For continuum measurements there is a rapid switching by movements of the telescope subreflector or by wobbling a flat mirror between two regions of the sky, one of which is presumed to be empty.

The most well-studied Galactic millimeter and submillimeter sources are:

- IRC+10216, the circumstellar envelope of the RED GIANT STAR CW Leo, about 150 parsecs from the Sun;
- the Kleinmann–Low nebula, at 500 parsecs, near the Trapezium cluster in Orion;
- the dark cloud center TMC-1 in Taurus;
- Sgr B2, a complex of dense molecular hydrogen clouds and nebulae at 8.2 kiloparsecs, near the center of our Galaxy;
- Sgr A, the GALACTIC CENTER itself.

The following external galaxies have also been studied well: Arp 220, IC342, M82, M51 and NGC 253.

Interstellar chemistry can be studied by looking at millimeter and submillimeter spectral transitions – the chemical reactions do not occur on Earth. Millimeter and submillimeter radiation penetrates dust and in particular reveals what happens inside dusty star-formation regions and the dust cocoons of stars losing mass. These observations have shown what happens in young stellar objects (YSOs). Material that is excess to the formation of the YSO is expelled in the form of a bipolar high-velocity stream of material, which is mainly molecular. Such outflowing material has been traced in millimeter wavelength lines of carbon monoxide and other molecules. A dense disk of material, perpendicular to the direction of the bipolar outflow, usually surrounds the YSO.

A completely different type of outflow occurs at the end of the life of stars of roughly a few solar masses, typically red giants that lose mass, whose features are revealed with carbon monoxide lines. The outflow velocities are about 30 km s^{-1}. The mass-loss mechanism is thought to be radiation pressure on dust grains, which drags the gas along with it.

Submillimeter spectroscopy of galaxies reveals warmer regions of exceptional star-forming activity. The most studied galaxy in the millimeter and submillimeter is M82, which has an abundance of warm molecular gas. Other interesting objects include the galaxy CENTAURUS A and the face-on spiral IC342, which is thought to be an analog of our own MILKY WAY GALAXY.

Mills, Bernard Yarnton (1920–)

Australian engineer and astronomer who worked at the Radiophysics Division of the Australian Council for Scientific and Industrial Research. In 1954 he constructed the Mills Cross radio interferometer (which operated at 85 MHz and had 457-m arms) with which he mapped the Magellanic Clouds and conducted source surveys. Mills led the group that constructed the Molonglo Cross, an instrument with 1.6-km-long arms operating at 408 MHz. Built in 1967, it was converted in 1981 to a synthesis telescope (*see* MOLONGLO OBSERVATORY SYNTHESIS TELESCOPE). Mills discovered the VELA PULSAR at the center of the VELA SUPERNOVA REMNANT.

Mimas

The innermost of Saturn's mid-sized icy satellites, discovered in 1789 by William Herschel (*see* HERSCHEL FAMILY). It orbits the planet at a distance of 186 000 km, within the tenuous E ring and near its inner boundary. Mimas is a heavily cratered world, with many craters overlapping. With one exception they are less than 30 km across. The exception is Herschel (named to honor the satellite's discoverer), whose diameter of 130 km is nearly a third of the satellite's diameter, 400 km, and must have come close to disrupting Mimas. The satellite's density, 1200 kg m^{-3}, indicates that it is composed largely of water ice. Mimas's proximity to Saturn's ring system gives it a controlling influence: it keeps the Cassini Division clear, for example, and is responsible for 'corrugations' in the A ring.

Minkowski, Rudolph Leo Bernhard (1895–1976)

German astronomer born in Strassburg. Minkowski worked on atomic spectroscopy at Hamburg. After fleeing the Nazi persecution, he joined Wilhelm BAADE at the MOUNT WILSON OBSERVATORY in the USA. He investigated nebulae, including remnants of supernovae (especially the CRAB NEBULA), and classified supernovae into types I and II. This led to their identification as two similar implosions in different kinds of stars. He headed the National Geographic Society–Palomar Observatory Sky Survey, which photographed the entire northern sky with the Palomar 48-in SCHMIDT TELESCOPE. With Baade he optically identified many of the early radio sources, including CYGNUS A. In 1960, on his very last observing night with the 200-in Palomar Telescope, he found a galaxy (3C 295) with the then highest REDSHIFT, a record which stood for 15 years. After he made the discovery, the night was declared overcast and the occasion toasted with whisky. Minkowski also discovered Comet 1950b, and a small, faint, provocatively shaped proto-planetary nebula in Cygnus is named Minkowski's Footprint.

Minor planet *See* ASTEROID

Mira

The star Omicron Ceti, a red giant (*see* RED GIANT AND FLARE STAR), the first VARIABLE STAR ever identified as such. It is the type-star of the most common class of long-period pulsating variables, which are usually described as Mira-type stars. Its variability was first recorded by the German theologian David FABRICIUS in 1596. It is visible to the unaided eye for only about four months around each maximum, and is a telescopic object for the next seven months – hence Fabricius initially believed it to be a NOVA. It was for many years the only known case of a star varying in brightness. This behavior

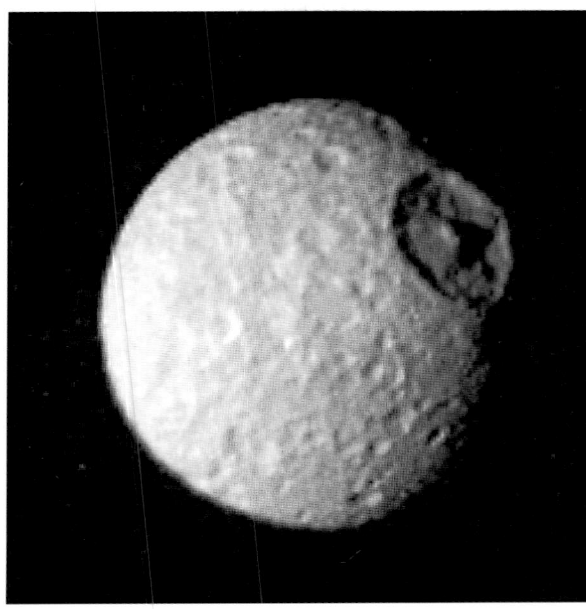

◀ ***Mimas*** *The satellite is heavily cratered, with its largest crater, Herschel, visible on the terminator at upper right.*

gave rise to the name Mira ('Wonderful'), which is attributed to Johann HEVELIUS. Despite its long absences it was recorded by Johann BAYER, who showed it as a star of magnitude 4 in his *Uranometria* star atlas of 1603.

Mira's variable period averages 331.96 days. The spectral type varies during the light cycle, between M5e and M9e. Its apparent magnitude at maximum is typically 3, but can be anywhere in the range 2–5; minima usually lie between 8.5 and 9.5, but can be as low as 10.1. Mira has a close binary companion, the WHITE DWARF VZ Ceti, which is believed to accrete material from Mira's stellar wind. The existence of this faint (magnitude 13) companion was predicted in 1920 by Alfred JOY of the MOUNT WILSON OBSERVATORY, from periodic variations in Mira's spectrum. VZ Ceti is visible only at times when Mira itself is very faint; it was first detected optically with the 36-in refractor at the LICK OBSERVATORY in 1923 by the great DOUBLE STAR observer Robert Aitken (1864–1951). Mira is probably the most thoroughly observed of all variable stars; every maximum since 1638 has been recorded. More than 4000 Mira-type variables are now known. *See also* VARIABLE STAR: PULSATING.

Miranda

Miranda is the smallest of URANUS's five major satellites, and one of the most unusual objects in the solar system. It was discovered in 1948 by Gerard KUIPER.

Miranda is small, only about 470 km in diameter. Very little was known about Miranda until the Voyager 2 spacecraft flew through the uranian system in 1986 (*see* VOYAGER MISSIONS). Miranda was the great surprise of the encounter. Its surface is incredibly diverse: the expected heavily cratered terrain is present, but the surface is also criss-crossed by a network of rift-like canyons that is more characteristic of a much larger object.

The unusual and unexpected collection of features on Miranda has prompted a number of competing explanations for their origin. One early suggestion was that Miranda had been catastrophically shattered into many large pieces after it had partially melted into layers rich in dark rock and light ice. The shattered pieces then fell back together in random order, with darker fragments formerly in the deep interior forming the coronae now visible on the surface. The problem is in understanding how temperatures high enough to cause such extensive geologic activity can occur in as small a body as Miranda.

Mira variable

Class of RED GIANT variable star named for MIRA. *See also* VARIABLE STAR: PULSATING.

Mirror

An optical component from which light is reflected. Because light is reflected from the front surface of an astronomical mirror, it does not have to travel through the material of which the mirror is made. An astronomical mirror usually consists of a suitably shaped substrate (the body of the mirror) on which a thin layer of highly reflective metal, such as silver or aluminum, is deposited. The substrate is normally a material that expands and contracts as little as possible when its temperature changes. Typical materials are Pyrex (a low-expansion glass), quartz and ceramics.

Telescopes other than optical telescopes (for X-rays or radio, for example) use mirrors (or REFLECTORS) in analogous ways; however, the materials of the substrate and reflecting surface are adapted to the properties of the radiation.

Mitchell, Maria (1818–89)

First acknowledged female astronomer in the USA, born in Nantucket, Massachussets (and one of the most famous American scientists of her day). She worked at its Atheneum Library and helped her father in his amateur observatory. In 1847 she discovered a comet (since named Miss Mitchell's Comet), and became a 'computer' at the US Nautical Almanac Office, calculating the motion of Venus. After completing a tour of Europe, she was presented with a telescope bought with money collected by American women. She became professor of astronomy and director of the observatory at the newly opened Vassar College in Poughkeepsie, New York, where she photographed Jupiter, Saturn and the stars, and furthered women's education.

Mizar and Alcor

Mizar is the name given to the center star in the handle of the BIG DIPPER by Joseph Justus Scaliger (1540–1609). It is Arabic for 'the Girdle.' The star, in Ursa Major, was formerly known as Merak or Mirak, from the Arabic 'al-Marakk' ('the Loin' of the bear), but this duplicated the name given to another star, Beta Ursae Majoris.

With Alcor ('the Rider,' from its position at a time when the Dipper was imagined as a cart pulled by a train of horses), Mizar forms a hierarchical MULTIPLE-STAR system. Mizar is believed to be the first BINARY STAR ever discovered; it was found in 1650 by Giovanni Riccioli (1598–1671). In 1889 its brighter component also became the first spectroscopic binary to be announced, by Antonia Maury (1866–1952) of the Harvard College Observatory, with a period of 20.5 days. The spectroscopic binary, separation about 0.01 arcsec, was first resolved telescopically in 1925 by Francis Pease (1881–1938), using a 20-ft beam INTERFEROMETER attached to the 100-in REFLECTOR at the MOUNT WILSON OBSERVATORY. This was an epoch-making instrument devised by Albert MICHELSON, with which the first stellar diameters were also measured. The secondary star of the main binary system is itself also a spectroscopic binary, with a period of 175.5 days. Because of its multiplicity, the distance of Mizar is not certain. Alcor (80 UMa) is a main-sequence DWARF STAR, of spectral type A5V and magnitude 3.99. It is a little more distant than Mizar at 81 l.y., but may be a distant member of the same gravitational system.

Molecule

The smallest particle of a chemical compound that can exist by itself and which has all the chemical properties of that compound. A molecule consists of two or more ATOMS, of the same elements or of different elements, linked together by chemical bonds. The molecular mass (often referred to as 'molecular weight') of a molecule is its mass expressed in atomic mass units, and is equivalent to the sum of the atomic masses of all its constituent atoms. Molecules range in size and mass from simple molecules with two atoms (diatomic), such as hydrogen, to macromolecules with molecular masses of more than 10 000. A molecule has a physical size and shape that depends on the separation and relative orientation of its constituent atomic nuclei. Consequently, it exhibits three types of discrete (quantized) energy levels and transitions (changes between levels):

- electronic transitions, which involve changes between energy states in the shared cloud of electrons that surrounds its constituent atomic nuclei;
- vibrational transitions between its various permitted states of vibration (these involve the separations between nuclei);
- rotational transitions, which involve discrete changes in the way in which the molecule rotates around various axes.

Each of these kinds of transition is accompanied by the emission or absorption of radiation of a specific wavelength, often in the infrared, submillimeter or millimeter wavebands.

Molonglo Observatory Synthesis Telescope

Radio telescope located near Canberra, Australia, and operated by the University of Sydney (*see* RADIO TELESCOPES AND THEIR INSTRUMENTS). A large radio interferometer was built there along the model of the Mills Cross (*see* Bernard MILLS), one arm of which has been converted into the Molonglo Observatory Synthesis Telescope (MOST). MOST consists of two cylindrical 778 × 12 m paraboloids, 15 m apart and aligned east-west. The telescope is steered by mechanical rotation of the cylindrical paraboloids about their long axis, and by phasing the feed elements along the arms.

Monoceros *See* CONSTELLATION

Month

A unit of time based on the motion of the Moon around the Earth. The synodic (lunar) month is the mean time interval between two successive new moons (that is, the time taken for the Moon to pass through its cycle of phases) and is equal to 29.53059 mean solar days. The synodic month is the basis of the lunar calendar. CALENDAR months are synodic months rounded to a whole number of days, and the number of days in each calendar month is varied between 28 and 31 so that 12 successive months fit into a year.

The sidereal month (duration 27.32166 days) is the time taken for the Moon to move through an angle of 360° relative to the background ('fixed') stars. The tropical month (27.32158 days) is the time interval between two successive conjunctions of the Moon with the vernal EQUINOX.

Moon

The Moon, Earth's satellite, is our closest neighbor. The Earth-Moon system is unique in the solar system with its ratio of less than 4 between the sizes of the planet and its satellite, and can be considered as a double planet. Because of the large mass of the Moon and its proximity to the Earth, lunar tides are nearly three times larger than solar tides. The eccentric and high-inclination orbit of the Moon is quite different from that of other large satellites (with the exception of Triton, around Neptune), which are nearly circular and lie in the equatorial plane of their primaries.

The Moon is a prime target of naked-eye astronomy, and has been since the stone age. The phases of the moon (new Moon, first quarter, full Moon, last quarter) are linked to the angle between the directions of the Moon and the Sun, which repeats every 29.53 days (the lunar MONTH). With the naked eye one can readily see the bright zones (the highlands) and the dark zones (the MARE), the latter covering less than 35% of the visible regions. In central and eastern regions the dark zones appear nearly circular, with diameters of up to 1100 km (Mare Imbrium). On the western side they merge into the vast Oceanus Procellarum. It is obvious that the Moon always turns the same side toward the Earth, which means that both its rotation period and its orbital period are tidally locked at 27.3 days. The variations of the orbital velocity along the slightly elliptical lunar orbit lead to an apparent back and forth rotation of the Moon as seen from the Earth, which is called the LIBRATION. Almost three-fifths of the lunar surface can be observed from the Earth.

In 1609 GALILEO GALILEI identified with his telescope countless round features: the lunar CRATERS. Forty years later, the major features of the near side had been mapped, with the names that are still in use today. Ray craters such as TYCHO had been observed, with bright lines extending radially over thousands of kilometers. The nearly instantaneous disappearance of occulted stars demonstrated that the Moon has no atmosphere.

In the eighteenth and nineteenth centuries the orbit of the Moon was determined with great accuracy (pushed by the desire to predict the orbit for navigational purposes), and this led to the discovery of the Moon's evolution. Tidal dissipation within the Earth's mantle and oceans is leading to a long-term evolution of the orbit of the Moon, which is slowly drifting away from the Earth.

Space exploration (phase I)

The Moon was the obvious target for spaceflight beyond low-Earth orbit. Indeed, only two years elapsed from the first Sputnik (1957) to the first Soviet lunar probe. The first lunar flyby by Luna 3, in 1959, revealed a far side almost entirely covered by highlands. In 1961 American President John F Kennedy announced the APOLLO program to land a man on the Moon in less than 10 years. The surface of the Moon was mapped from orbit, with high-resolution images of the potential landing sites. Remote sensing showed that the Moon had a very small or no core (at most a few hundred kilometers in radius), a thick mantle, then a crust which is thinner on the near side (60 km) than on the far side (100 km). The Moon has about half as much iron as the Earth.

The Moon	
Semimajor axis	384 400 km
Eccentricity	0.0549
Inclination to the ecliptic	5.15°
Radius	1740 km
Density	3.34
Mass ratio (Moon/Earth)	1/81

The Apollo program returned nearly 400 kg of lunar material (BRECCIA and BASALT). Apart from sampling, the astronauts performed a range of scientific investigations, using, for example, a SOLAR WIND collector, seismometers, and reflectors for laser ranging. The set of instruments left on the surface remained operational years after the departure of the astronauts.

History of the Moon

The determination of lunar geologic history by radioactive decay methods was the most important result from analysis of the Apollo samples. This demonstrated that the Moon formed at the same time as the Earth, 4.55 billion years ago. After at most 200 million years, the DIFFERENTIATION between a core, a mantle and a crust was complete. Over nearly 500 million years, the Moon (and hence also the Earth) was submitted to an intense bombardment from the left-over planetesimals in the inner solar system. This period resulted in the formation of giant impact basins, measuring hundreds of kilometers in diameter and tens of kilometers in depth, with fractures in the underlying crust, while the surface was thoroughly pockmarked by overlapping craters. All the 'mountains' on the Moon (up to 8 km high) result from the accumulation of debris, or the rebound of the lunar crust associated with the formation of basins. The far side, still dominated by highlands, gives a good idea of what the whole lunar surface looked like 4 billion years ago.

In the meantime, heat from radioactive decay accumulated slowly in the lunar interior. From 4 billion to 3 billion years ago melted mantle material escaped through the fractures of the crust. Gigantic lava flows filled out the basins and form the lunar maria. This was much easier on the near side than on the far side, where the crust was twice as thick, which explains the near absence of maria on the far side. The cooling of lava flows led to the formation of specific features, such as faults, domes and valleys.

Another important result of high-precision isotopic analyses was the discovery that the Earth and Moon were formed from a single isotopic reservoir, different from that of the parent bodies of METEORITES. The minerals constituting the Moon are also similar to those of corresponding units on the Earth.

Later impacts resulted from a small steady-state influx of impactors from the ASTEROID belt and the Kuiper Belt (*see* OORT CLOUD AND KUIPER BELT). The main outcome of this secondary bombardment was the formation of sparse fresh craters (the most recent large one is Tycho, a few hundred million years old), and of the lunar regolith from the slow, grinding effect of small micrometeoroids. The relative age of surface units can be determined from crater counts, calibrating on the age of the nine sampled regions. The resulting evolution is extremely slow: the footprints left by the Apollo astronauts may well outlast humanity, with a lifetime of a few million years.

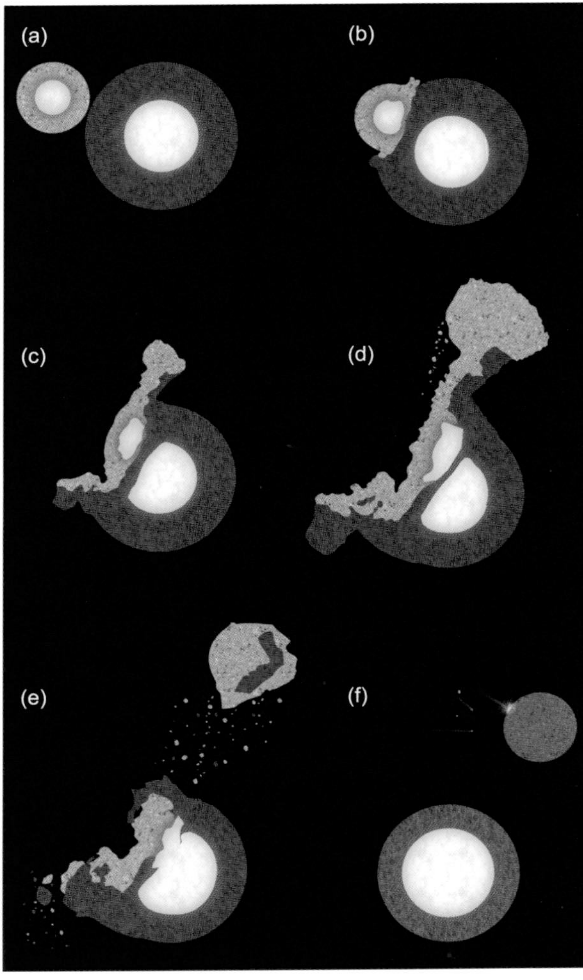

◀ ***Moon*** *The Moon formed through the collision at (a) of two planets, each with a mantle and a core. The cores coalesced into the larger planet, Earth, at (d) and (e), and the mantle debris became the Moon at (e) and (f).*

Laser echos using the corner reflectors left behind on Apollo allowed the recession rate of the Moon to be measured with precision (4 cm each year). Isotopic analysis techniques were also used to follow the growth pattern of fossilized corals over time. This yielded the number of days and lunar months per year in the distant past. Integrating this evolution backward in time led to a highly elliptic and inclined orbit.

The major breakthrough in identifying the origin of the Moon occurred in the early 1980s when it was shown that the orbits of planetary bodies can be stable for millions of years and then evolve chaotically as a result of perturbations by Jupiter and Saturn. This generates jaywalking planets and asteroids that produce giant collisions in the early solar system. Simulations showed that an off-center impact of a Mars-sized body with the Earth caused the ejection of part of the mantles of the impactor and the proto-Earth into orbit (making the Moon), while the iron-rich cores fused (forming the Earth). This giant impact model leads naturally to a highly inclined and eccentric early orbit, which evolves into the present one. It is consistent with the geochemical evidence.

The investigation of the dynamical behavior of the Earth-Moon system also showed that the INCLINATION of the rotation axis of the Earth could increase to 60° or more if the Moon was not there to stabilize it. This would produce vast climate variations, each pole alternately facing the Sun (contrast this with MARS). The consistent evolution of life on our planet happened because we have a large satellite.

Observing the Moon

▶ ***The crater Plato** is at top center in this image. The high mountain Mons Pico is near the center and the flat plane is Mare Imbrium. Imaged by Jamie Cooper using an Intes Micro MN-78 telescope and a Philips Tollcam Pro webcam. The telescope is a Maksutov Newtonian with a 7-in mirror.*

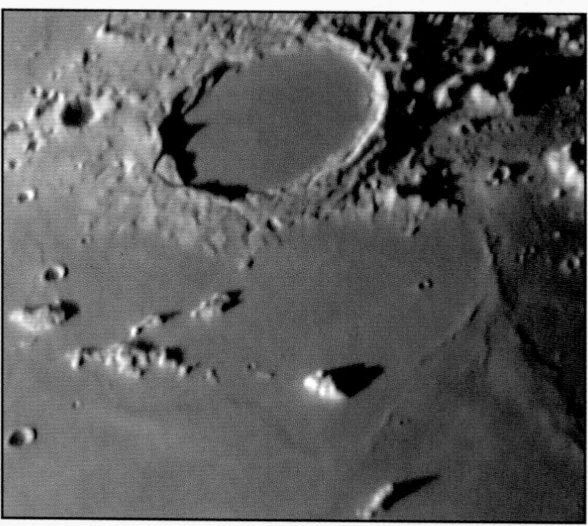

There was a time, not many decades ago, when professional astronomers as a class took very little interest in the surface of the Moon. Amateur work was of great importance, and most of the best lunar maps were of amateur construction. Of course, the situation today is very different; the entire surface has been mapped in great detail, not only from EARTH but also from spacecraft. Therefore, it may be claimed that observing the Moon with small or moderate-sized TELESCOPES has become pointless.

In fact, this is not so. Quite apart from the fact that lunar observation is immensely enjoyable, there is still a certain amount of really useful research to be undertaken, and a telescope of, say, 16-in APERTURE is adequate.

In any case, the first essential is to gain real experience. The lunar surface may be virtually changeless, but the apparent changes due to the shifting solar illumination are obvious over periods of even a few minutes, and a CRATER which is spectacular when seen at or near the TERMINATOR (the boundary between the sunlit and dark areas) may be very obscure when looked at when the Sun is higher above it, and the shadows are smaller. Except for some specialized branches of observation, full Moon is the very worst time to begin regular research. There are almost no shadows, and the whole scene is dominated by the rays from some of the craters, notably TYCHO in the southern uplands and the COPERNICUS CRATER in the Mare Nubium (*see* MARE). It must be remembered that a typical lunar crater does not resemble a deep, steep-sided mineshaft; in profile, it is more like a saucer, with a depressed floor and a rim which rises to only a modest height above the outer surface. Near full PHASE, even a large crater is easy to find only if it has a very dark floor (as with Plato and Grimaldi) or exceptionally bright walls (as with Aristarchus or Proclus).

One excellent, though laborious method of becoming familiar with the main surface features is to take an outline map and, using a small telescope – a 7.5-cm REFRACTOR will do very well – make several sketches of each major formation, spread over a period of several weeks. It is important not to draw too large an area at a single observation; a scale of 2.5 cm per 100 km is reasonable – so that, for example, the majestic crater Theophilus (diameter 110 km) will be drawn to a diameter of about 2.6 cm. Great care should be taken in positioning the shadows, and the time of the observation should be noted because the shadows change so quickly.

The outline map will also show the most conspicuous of the minor features: the isolated peaks and hills, the system of RILLES (also known as rills or clefts), the valleys, and the low swellings known as DOMES, many of which are crowned by summit craters. Systematic observation over a few lunations will be sufficient for the lunar enthusiast to begin some useful work.

LIBRATION effects are very important; for example, there are times when the dark-floored Grimaldi seems to be almost touching the LIMB, while at others it is well on the disk, and considerable detail can be seen beyond it. There are areas which are carried in and out of view according to libration, and these are all so very foreshortened that they are difficult to map; before the space age, our charts of them were rough and incomplete. Note, too, that there has been an official change in nomenclature. Classically, Mare Crisium was near the western limb with Grimaldi in the east; this was reversed in the 1960s by official edict of the IAU when the definitions of east and west were reversed. (The Mare Orientale, or Eastern Sea, was so named because at extreme libration a very small part of it can be seen over the limb, some way from Grimaldi; the IAU ruling now means that the Eastern Sea appears over the western limb – though, of course, almost the whole of it lies on the far side of the Moon and can never be seen from Earth.)

Structural changes on the Moon belong to the remote past; if the dinosaurs had used telescopes, they should have seen the Moon just as it is today. Claims of definite change must be discounted. The most famous changes involve Linné, a small formation on the Mare Serenitatis, and the Messier twins in Mare Fœcunditatis. It was said that between 1838 and 1866 Linné changed from a deep craterlet into a shallow pit surrounded by a white nimbus, while Messier and Messier A, now dissimilar, were once exactly alike. Neither claim stands up to scrutiny. Messier and Messier A can sometimes look identical, though under most illuminations they are seen to differ in shape as well as in size. The apparent changes are due to nothing more significant than lighting effects.

During a lunar eclipse, the surface temperature drops abruptly, and it has been suggested that there could be detectable effects in some formations, but this now seems to be highly unlikely.

Flashes on the surfaces have been reported from time to time, and it is not impossible that these were due to the impacts of METEORITES, but no case of a newly formed craterlet has been established, even though there have been several false alarms. It was also claimed that during the Leonid meteor storms of 2000–2002 (*see* LEONIDS; METEOR STORM), surface flashes indicated impacts, but this idea can quickly be ruled out, because a METEOR of sand-grain size could not possibly produce a flash bright enough to be seen with an Earth-based telescope; such a flash would indicate a meteorite – and meteorites are not associated with meteor showers. Note also that the lunar atmosphere is too tenuous to produce shooting star trails.

On the other hand, the situation is very different for the local glows and obscurations known as transient lunar phenomena (TLP); *see* LUNAR TRANSIENT PHENOMENA. They have been seen by many long-term observers, and there is some photographic evidence. It is here that the observer equipped with a moderate telescope comes into his own.

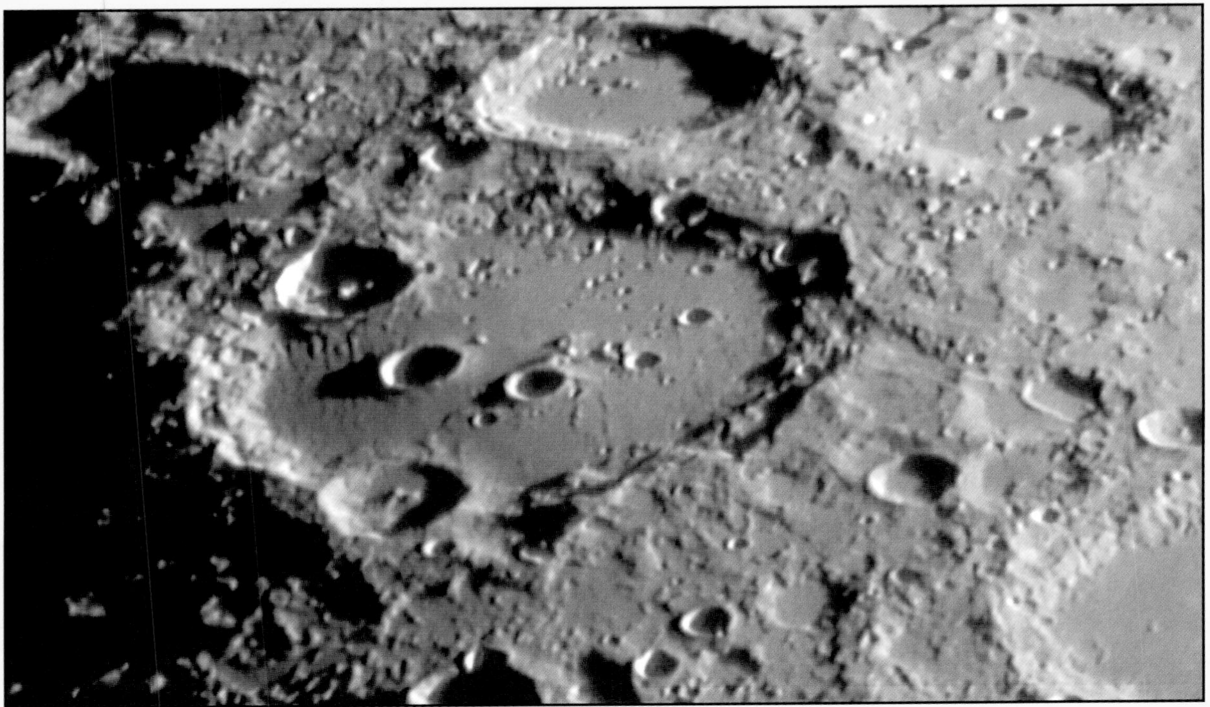

◀ ***Clavius*** *is the large, old crater in this picture, with its floor pitted with smaller, fresher, more recent craters. Taken using the same equipment detailed opposite, Jamie Cooper used 6 images of an AVI video clip, processed using Adobe Photoshop.*

TLP are localized, and do not usually last for long. They are most frequent in areas around the peripheries of the regular 'seas' and in areas which are rich in rilles, such as the floor of the crater Gassendi, but the most event-prone area is that of Aristarchus on the Oceanus Procellarum, the brightest feature on the Moon; it is 40 km in diameter, with walls and central peak. (The strange dark bands running down the walls were once attributed to vegetation!) The TLP question became prominent in 1958, when the Russian astronomer Nicolai Kozyrev (1908–83) observed a red glow in the large-walled plain Alphonsus, and obtained a SPECTRUM. This was followed in 1963 by reports of red patches near Aristarchus made by astronomers at the LOWELL OBSERVATORY in Flagstaff, Arizona, USA. It was then thought that TLP must be generally red, and observers of the Lunar Section of the BAA introduced a 'Moon-blink' device. This consists of a drum-shaped chamber in which colored filters can be mounted and which rotates by means of simple, hand-operated gearing. The suspect area is then examined in quick succession with red and blue filters; a red glow will be suppressed through the red filter and enhanced with the blue filter, so that it will appear to 'blink.' The device is easy to make and has given good results, although it now seems that only certain types of TLP are red.

TLP are far from common, and the greatest care must be taken in searching for them, because many times reported TLPs turn out to be mistakes caused by lighting effects. If, for example, a suspect area shows a 'blink,' checks should be made on other areas too. If they show blinks, no TLP can be involved. It is unwise to place too much credence upon a TLP report unless confirmed by someone observing from a different location. Obviously, an adequate telescope is needed, and probably 12 in is the minimum really useful aperture (for a REFLECTOR).

What are TLP? For many years their reality was questioned, mainly because most of the reports (not all) were recorded by amateurs, but then, in 1992, the eminent French astronomer Audouin Dollfus (1924–), using the 83-cm Meldon refractor, wrote that on December 31, 'glows have been recorded in the lunar surface, on the floor of the crater Langenus. They were not present the day before. Their shape and brightness were considerably modified three days later ... they are apparently due to dust grain levitations above the lunar surface, under the effect of gas escaping from the soil. The Moon appears as a celestial body which is not totally dead.' No doubt this is the correct explanation, but our knowledge of TLP is still very incomplete, and more observations will be of great value.

Some features of the Moon also seem to vary in brightness, and it is just possible that effects other than changing solar illumination play a part. A CED (crater extinction device) again consists of rotating filters, suitably mounted as with the blink device, but of varying degrees of intensity. For example, compare the brightness of the walls of Aristarchus with that of Proclus, near the Mare Crisium; which can remain visible with increasingly darker filters? There may be some useful results to be gained from their use, but there are all sorts of associated problems with CEDs, and there is bound to be a good deal of uncertainty involved with results obtained through thier use.

Finally, remember that there is always a chance, albeit a slight one, of making a spectacular discovery; if a new impact crater is formed, for example, or if (less probably) a violent TLP appears, it is likely that it will be an amateur who first reports it. And in any case, lunar observation is as inspiring as it is enjoyable. The magic of the Moon will never fade.

Sir Patrick Moore has presented the BBC's *Sky at Night* non-stop since 1957. He is the author of over 100 books, the composer of three operettas and a Fellow of the Royal Society. He was one of the pre-Apollo Moon mappers and was also developer of the BAA's lunar section.

Main lunar exploration missions (1959–76)

Mission	Date	Mission
Luna 3 (USSR)	Oct. 1959	First images of the far side
Luna 9 (USSR)	Jan. 1966	First soft lander
Surveyor 1, 5, 6 (USA)	1966–8	Soft landers
Lunar Orbiter 4 (USA)	May 1967	Photocartography from a polar orbit
Lunokhod 1, 2 (USSR)	1970, 1973	Remotely controlled rovers (10–40 km)
Luna 16 (USSR)	Sept. 1970	Automated sample return (Mare Fecunditatis)
Luna 20 (USSR)	Feb. 1972	Automated sample return (Apollonius highlands)
Luna 24 (USSR)	Aug. 1976	Automated sample return (Mare Crisium); 2-m-deep drill core

The Apollo program (USA)

Mission	Date	Landing site and mission
Apollo 11	Jul. 1969	First manned landing (Mare Tranquilitatis)
Apollo 12	Nov. 1969	Oceanus Procellarum
Apollo 14	Jan. 1971	Fra Mauro formation (southeast of Procellarum)
Apollo 15	Jul. 1971	Hadley (mare-highland contact zone); first deep drill core; first manned rover (20 km)
Apollo 16	Apr. 1972	Descartes (highlands); deep drill core, rover
Apollo 17	Dec. 1972	Taurus-Littrow (mare-highland contact zone); field geology (by Harrison Schmitt), 35 km explored, deepest drill core (3.2 m)

Space exploration (phase II)

These theories renewed scientific interest in the Moon and restarted the lunar exploration program in 1994, with the CLEMENTINE mission. Clementine discovered the largest lunar basin, Aitken, which covers most of the southern far side. The spacecraft provided the first multicolor database for the whole lunar surface. It also observed polar regions that are always in shadow, providing possible cold traps for volatile species such as water.

A second small polar satellite, LUNAR PROSPECTOR, was launched in early 1998. It observed hydrogen trapped near the lunar poles, which may be linked to deposits of ice crystals in the permanently shadowed polar craters, although this interpretation is still controversial. Such water deposits were probably provided by cometary impacts on the Moon.

Further missions to the Moon are planned. A small ESA mission, SMART-1, was launched in September 2003 to map the surface composition. LUNAR-A is a Japanese mission with two penetrators equipped with seismometers to investigate the lunar core (possible launch in 2004); followed by an ambitious mission, SELENE.

Further reading: Murray C and Cox C B (1989) *Apollo: the Race to the Moon* Simon and Shuster; Whittaker E A (1999) *Naming and Mapping the Moon* Cambridge University Press; Wilhelms D E (1993) *To a Rocky Moon: a Geologist's History of Lunar Exploration* University of Arizona Press.

Moore, Patrick Alfred Caldwell (1923–)

Colorful British amateur astronomer, prolific writer, ubiquitous broadcaster and enthusiastic musician. Since 1957 he has presented the monthly BBC television program *The Sky at Night*, the longest-running program with the exception of the news. *See also* CALDWELL CATALOGUE.

Morgan, William Wilson (1906–94)

American astronomer, staff member of the YERKES OBSERVATORY. With Philip Keenan (1908–2000) he introduced stellar luminosity classes and the Morgan–Keenan (MK) classification scheme of stars, based on the appearance of their SPECTRA. He discovered, with his students Donald Osterbrock (1924–) and Stewart Sharpless (1926–), the two nearest SPIRAL ARMS in our Galaxy by mapping the distribution of bright O- and B-type stars whose distances were derived from their spectral classification. In 1951 Morgan presented this finding in Cleveland to tumultuous applause, unusual in a scientific conference.

Morning star

A name given to Venus when it is visible in the east in the pre-dawn sky. Ancient astronomers believed that morning and evening apparitions of Venus were of two different planets; the morning planet was given the name Phosphor, or Phosphorus. As the morning star, Venus is moving from inferior CONJUNCTION (when it lies between the Earth and the Sun) to superior conjunction (behind the Sun), and is visible for longest when it reaches the position known as greatest ELONGATION west, when its angular separation from the Sun is greatest. The name 'morning star' is sometimes given to morning apparitions of Mercury.

Mount Graham International Observatory

The observatory is located near Safford, Arizona, USA, at an elevation of 3200 m. It specializes in advanced-technology telescopic facilities that can benefit from the low water vapor and sharp images of the site, and its easy access. The 10-m-diameter Heinrich Hertz submillimeter telescope is a joint project of Arizona and the Max-Planck Institute for Radio Astronomy, Germany. The accuracy of adjustment of the carbon fiber telescope surface has surpassed its goal of 15 μm. The Vatican Observatory/Arizona Lennon telescope has a 1.8-m *f*/1 primary figured to 17 nm. The third and largest telescope is the 2 × 8.4 m (*f*/1.14) LBT (LARGE BINOCULAR TELESCOPE), under construction and expected to begin operation in 2004.

Mount Stromlo and Siding Spring Observatories

The astronomical observatories of the Australian National University (ANU). The facilities at Mount Stromlo Observatory (MSO) at Woden near Canberra were originally established as the Commonwealth Solar Observatory, which began operation in 1924. Expansion after World War II saw the MSO move into stellar astrophysics, and in 1957 it joined the ANU's Research School of Physical Sciences. The development of the city of Canberra in both size and light pollution led to the establishment of a dark-sky observatory at Siding Spring, west of Coonabarabran in New South Wales, in 1964. Ten years later the Siding Spring Observatory (SSO) became the host of the ANGLO-AUSTRALIAN OBSERVATORY with its 4-m and Schmidt telescopes.

The MSSSO's major facility is the ANU 2.3-m telescope, which opened in 1984. The telescope is a Cassegrain and Nasmyth altazimuth under full software control. It is capable of optical and infrared imaging and spectroscopy. A 1-m telescope is used for imaging, and there are smaller facilities too. The SSO also hosts the University of New South Wales patrol telescope. The Uppsala telescope is carrying out a near-Earth object survey for NASA and the University of Arizona. The 2-m FAULKES TELESCOPE is under construction. At the MSO the ANU had a 1.9-m telescope and the 1.3-m Great Melbourne Telescope, as well as its research headquarters, library and workshops. The telescopes and workshops at the site were destroyed by a bush fire in 2003.

Mount Wilson Observatory

The observatory, located in the San Gabriel Mountains near Pasadena, California, USA, was founded in 1904 by George HALE with financial support from Andrew Carnegie. In the 1920s and 1930s, working at the 2.5-m Hooker telescope, Edwin HUBBLE made two of the most important discoveries in the history of astronomy: first, that 'extragalactic nebulae' are actually 'island universes' – galaxies, each with billions of stars; second, that these galaxies are moving away from us in all directions, resulting in an expanding universe. This second discovery became the basis for the BIG BANG theory of the origin of the universe.

Mount Wilson's calm atmosphere results in the best 'seeing' (natural sharpness and quality of its telescope images) in North America. The observatory facilities include the 2.5-m Hooker telescope; the 1.5-m telescope; the Georgia State University CHARA interferometer array, consisting of six 1-m telescopes arranged in a 350-m-diameter configuration; the UC Berkeley Infrared Spatial Interferometer; the 45-m and 18-m solar tower telescopes; the Snow horizontal solar telescope; and the 60-cm and 35-cm remote-controlled telescopes, used by the Telescopes in Education program for education and student research.

Moving group

A set of stars, less densely packed than a cluster, that have similar velocities and a common origin.

Mullard Radio Astronomy Observatory

The observatory is operated by the University of Cambridge, England. It is famed for the pioneering sky surveys by Martin RYLE, who invented the technique of APERTURE SYNTHESIS, and for the discovery of PULSARS by Antony HEWISH and his student Jocelyn BELL-BURNELL in 1967. These contributions were recognized by the award of the 1974 Nobel Prize for Physics to Ryle and Hewish.

Instruments in operation include the Ryle Telescope – an array of eight parabolic dishes on a 5-km baseline, originally built in 1971 for high-resolution imaging of radio galaxies and quasars. It has been upgraded to map faint structures in the COSMIC MICROWAVE BACKGROUND caused by the SUNYAEV–ZELDOVICH EFFECT in nearby galaxy clusters and protoclusters at high REDSHIFT. Mapping the microwave background to detect primordial density fluctuations of cosmological significance has been achieved with a prototype Cosmic Anisotropy Telescope and this has led to the design of a more advanced instrument, the Very Small Array located on Tenerife, one of the Canary Islands. Milliarcsecond resolution at optical wavelengths for imaging stellar disks has been achieved with the Cambridge Optical Aperture Synthesis Telescope.

Multiple Mirror Telescope Observatory

At the time of its dedication in 1979, the 4.5-m Multiple Mirror Telescope (MMT) was the third largest optical telescope in the world. It featured so many ambitious design innovations that its completion heralded the beginning of the current generation of telescope design. The MMT was decommissioned in March 1998, and was replaced in March 1999 with a 6.5-m single-primary telescope.

The 4.5-m MMT was the first large optical telescope to incorporate multiple lightweight primary mirrors in a common mount, a rotating building instead of a dome, active secondary mirror control during all observations, a high-performance ALTAZIMUTH MOUNT, and detailed finite element analysis of the telescope's structure during the design process.

Multiple star

Any star system that comprises three or more components that are physically linked (that have a common gravitational field). *See also* DOUBLE STARS.

Musca *See* CONSTELLATION

Muses-A *See* HITEN

Muses-B *See* HALCA

Muses-C

Japanese-US sample-return mission to asteroid (25143) Itokawa/1998 SF36. It was launched in May 2003. A US-built nano-rover will image the surface and collect three samples for return to Earth in 2007. Other technologies to be tested include a solar-powered electrical propulsion system, and autonomous navigation and guidance. The spacecraft will spend three months on the asteroid.

◄◄ *NASA The launch by NASA of a Saturn V rocket from Cape Canaveral, Florida, on July 16, 1969, resulted in the first successful manned mission to the Moon, Apollo 11.*

Nadir
The point on the CELESTIAL SPHERE directly below the observer, diametrically opposite the ZENITH.

Nagler eyepiece *See* EYEPIECE

Names of celestial objects
See INFORMATION HANDLING IN ASTRONOMY; STELLAR NOMENCLATURE

Nançay Radio Observatory
The Nançay Radio Observatory has the largest radio telescopes in France. Founded in 1952, it is located 200 km south of Paris and is operated by the PARIS OBSERVATORY.

Nasmyth, James (1808–90)
Scottish engineer who invented the steam hammer and many other devices. A childhood enthusiasm for astronomy meant that he maintained an interest in telescopes, making them for William LASSELL, Warren de la Rue (1815–89) and himself. His largest telescope, made in 1845, had a novel design now known as a NASMYTH TELESCOPE, which, instead of needing the observer to climb a ladder to an eyepiece, 'brought the stars down to him.' He made observations of the Moon, constructing models of the lunar surface from plaster of paris, and of the Sun, discovering what he described as a 'willow leaf' pattern, namely solar GRANULATION.

Nasmyth telescope
An altazimuth reflecting telescope with relatively stable platforms for mounting heavy, large, delicate or developmental equipment that cannot be, or has not been, engineered to cope with attitude changes during the tracking of a star. The optical configuration is of the Cassegrain type with a primary and secondary mirror, but there is an additional flat mirror mounted at the intersection of the ALTITUDE and AZIMUTH axes (*see* CASSEGRAIN TELESCOPE). The third mirror reflects light along the altitude axis, through one of the altitude bearings, to a FOCUS at the side of the telescope. Analysis equipment (or the observer) is supported on a platform mounted on the attitude bearing. The telescope was invented in his later, less agile years by James NASMYTH, a mechanical engineer and gentleman astronomer.

National Aeronautics and Space Administration (NASA)
In 1915, US Congress created an organization that would 'supervise and direct the scientific study of the problems of flight, with a view to their practical solutions.' That organization, the National Advisory Committee for Aeronautics (NACA), evolved into NASA in 1958. American spacecraft have explored more than 60 worlds in our solar system, while methodically peering back in space and time to reveal many of the secrets of the universe. In fact, NASA's activities have had a major impact on the astronomy described in this encyclopedia. NASA's principal spacecraft, and their results, are described in separate entries.

National Centre for Radio Astrophysics
India's National Centre for Radio Astrophysics (NCRA), located on the Pune University Campus, is part of the Tata Institute of Fundamental Research. At Khodad, 80 km from Pune, NCRA has set up the Giant Metrewave Radio Telescope (GMRT), the world's largest telescope operating at meter wavelengths. GMRT consists of 30 fully steerable dishes 45 m in diameter, spread over a 25-km area. Another meter-wavelength facility operated by NCRA is the Ooty Radio Telescope (ORT), a unique 530 × 30-m steerable parabolic cylinder located along a hill slope in southern India.

National Optical Astronomy Observatory (NOAO)
The observatory was formed in 1982 to consolidate all the ground-based astronomical observatories managed by the USA's Association of Universities for Research in Astronomy (AURA). NOAO has its headquarters in Tucson, Arizona, and consists of KITT PEAK NATIONAL OBSERVATORY (KPNO), CERRO TOLOLO INTER-AMERICAN OBSERVATORY (CTIO), and the NATIONAL SOLAR OBSERVATORY (NSO); it also represents the American astronomical community in the GEMINI Telescopes Project. KPNO operates the Mayall 4-m, the 3.5-m WIYN (Wisconsin, Indiana, Yale and NOAO) Telescope, the 2.1-m and Coude Feed, and the 0.9-m telescopes on Kitt Peak Mountain, about 88 km southwest of Tucson. The CTIO is located in northern Chile, where it operates the 4-m, 1.5-m, 0.9-m, and Curtis Schmidt telescopes. The NSO has its telescopes on Kitt Peak, including the McMath–Pierce Solar Telescope Facility containing the world's three largest solar telescopes (1.6-m main and two 0.9-m auxiliaries), along with the Vacuum Telescope and the Razdow small solar patrol telescope, and on Sacramento Peak, New Mexico, that include the Vacuum Tower Telescope, the Evans Solar Facility, and the Hilltop Dome Facility. The NOAO Gemini Science Center (NGSC) is in Tucson.

National Radio Astronomy Observatory (NRAO)
The NRAO designs, builds and operates a number of advanced radio telescopes in the USA and overseas. Its headquarters are in Charlottesville, Virginia. A laboratory at Tucson is involved in the development of the international ATACAMA LARGE MILLIMETER ARRAY (ALMA) project. At GREEN BANK the NRAO operates the recently completed 100-m Robert C Byrd telescope, while one of the 26-m telescopes of the Green Bank interferometer has been monitoring pulsars daily since 1989. At Socorro, New Mexico, the Expanded Very Large Array (EVLA) will enhance the performance of the VERY LARGE ARRAY (VLA) with the addition of up to eight new stations as distant as 250 km from the current array. The Very Long Baseline Array (VLBA), also operated from Socorro, is a series of ten radio antennae spread across the USA and its territories from Hawaii to the Virgin Islands (*see* VERY LONG BASELINE INTERFEROMETRY).

National Solar Observatory
The National Solar Observatory (NSO) operates several solar telescopes on Sacramento Peak in Sunspot, New Mexico; on Kitt Peak near Tucson, Arizona; and in a network of six sites around the world. The world's first and best-instrumented vacuum-tower telescope, the 76-cm Richard B Dunn Solar Telescope, the Evans Coronal Facility (a 40-cm coronagraph), and Hilltop flare patrol are located at Sacramento Peak. The McMath–Pierce, located on Kitt Peak, is the world's largest (1.5-m) solar telescope and it is the premier facility for infrared studies of the Sun. The Kitt Peak Vacuum Tower Telescope makes daily magnetic maps of the entire visible solar disk. The six GONG telescopes continuously measure solar oscillations. Research at the NSO concentrates on the nature of solar convection and

magnetism, the origins of solar activity and variability, the structure and heating of the solar corona, and the interior structure of the Sun. The NSO was founded in 1983 when the solar programs at Sacramento Peak Observatory merged with those at Kitt Peak.

Navigation

Navigation is the knowledge required to sail a ship (or fly an airplane) by the shortest, fastest, or safest route. Oceanic navigation, out of sight of land, is possible by the observation of celestial bodies, whether stars, planets or artificial satellites. Navigation has historically been linked with astronomical science and technology.

Navigation by observation of the Sun and stars

Babylonian, Phoenician, Greek, and Arab seafarers sometimes ventured, in the relatively safe Mediterranean Sea, beyond the sight of land. During his imaginary voyage, Odysseus, the first celestial navigator, had to keep the stars of the Great Bear, as they rose (or set) each night, on his left hand: the 'method of steering by horizon stars.' To find their destination port, seafarers memorized which stars passed over it, observed the stars' positions near the ZENITH during their voyage, and adjusted their course accordingly: the 'method of zenith stars.'

Compass points were determined by observing the Sun at noon, sunrise and sunset. In northern latitudes the Pole Star indicated north. Around AD 650, Arabs used a compass rose whose points were named after the rising and setting of fixed stars.

The dawn of astronomical navigation

In the fifteenth century, the Portuguese exploration of the west coast of Africa was navigated in latitude by measuring, with an ASTROLABE, the Pole Star's altitude at night, and the Sun's at noon. The astronomer Abraham Zacuto (1452–1528) of Salamanca and his pupil José Vizinho compiled the first day-to-day declination tables of the Sun, the *Regimento do astrolabio e do quadrante* (Manual of the Astrolabe and of the Quadrant). This work was published in 1509, but Christopher Columbus sailed to America in 1492 with a manuscript version.

East-west voyages across the Atlantic Ocean (after 1492) and the Indian Ocean (1497) made it imperative to find longitude. This meant determining local time compared with a reference location, such as the home port. Before accurate traveling clocks, time there might be determined from a predicted celestial event. In 1494 Columbus used a lunar eclipse as a reference clock, and Ferdinand Magellan's astrologer Andrés de San Martín in 1519 used a conjunction of the Moon and Jupiter. These attempts failed because of inaccurate predictions. The Nuremberg astronomer Johann Werner (1468–1522), in 1514, was probably the first to suggest that time could be determined by measuring the orbit of the Moon relative to stars – the 'method of lunar distances.'

Astronomy in navigation

Between 1598 and 1714, prizes were established in Spain, Holland, France, Venice and Britain for the determination of longitude. GALILEO GALILEI received a gold chain from the Dutch-States General in 1636 for proposing the eclipses of Jupiter's satellites as clocks, but observing them was impractical from the deck of a moving vessel.

To implement Werner's suggestion, the PARIS OBSERVATORY was founded in 1667 and the ROYAL OBSERVATORY, GREENWICH in 1675. The Paris Observatory published tables of astronomical data for navigation in the *Connaissance des temps* (The Knowledge of Time) from 1678 and Greenwich first published the *Nautical Almanac* in 1765, which Captain Cook used in New Zealand in 1769.

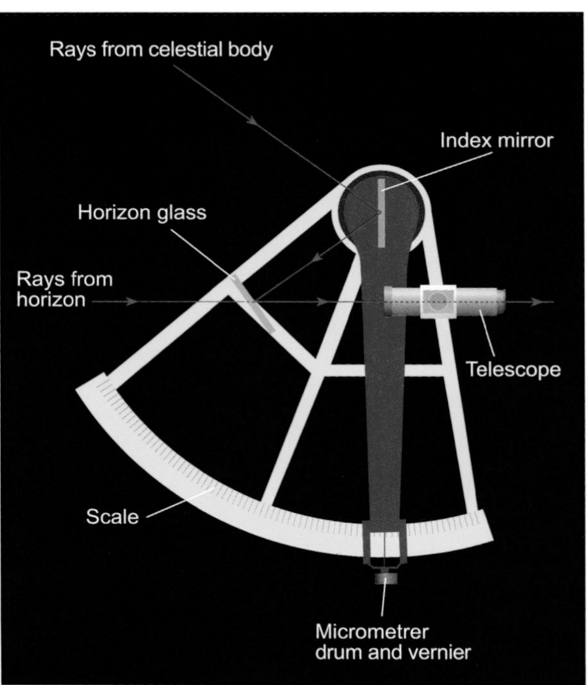

▶ Navigation *A sextant measures the angle of a celestial body (a star or planet) above the horizon. The body and horizon are viewed at the same time through a telescope, with the help of a semi-transparent mirror. The arm on which the index mirror is mounted is adjusted so that the image of the body is on the horizon and its altitude is read on the scale.*

Eventually, however, John HARRISON invented a more accurate, immeasurably simpler method of keeping time, an accurate clock called a chronometer. Nevertheless, lunar distances remained in use throughout the nineteenth century, and until 1907 the necessary tables were published annually in the *Nautical Almanac*.

Radio time signals were introduced around 1904 and satellite navigation in the late 1970s. A position could be obtained instantly, and far more accurately than by celestial navigation. In the 1980s global positioning satellite (GPS) systems started to replace sextants. In 1998 the United States Naval Institute ceased teaching celestial navigation to sailors, although of course operators of global positioning systems use established astronomical principles to provide navigation.

Further reading: Taylor E G R (1956) *The Haven-Finding Art: A History of Navigation from Odysseus to Captain Cook* Hollis and Carter; Williams J E D (1992) *From Sails to Satellites. The Origin and Development of Navigational Science* Oxford University Press.

N-body problem

Calculations involving the gravitational interaction of an arbitrary number (*N*, greater than two) of masses. The motion of two bodies is easily analyzed but not the motion of three or more. Solutions do exist for particular cases such as the 'restricted three-body problem' in which the attraction between two of the three bodies is much higher than the attractions of the third, such as Jupiter's perturbations of comets in their orbit round the Sun. Further examples requiring the solution of the N-body problem include the motion of a space probe in the solar system and the orbits of stars within a cluster.

NEAR (Near Earth Asteroid Rendezvous)

The Near Earth Asteroid Rendezvous mission (also known as NEAR–Shoemaker, for Eugene SHOEMAKER) was the first to orbit an asteroid. A spacecraft of NASA's new Discovery program, NEAR was launched on February 17, 1996, on a looping four-year trajectory to (433) EROS. On its way the NEAR spacecraft passed within 1225 km of asteroid (253) MATHILDE on June 27, 1997. It performed the first flyby of Eros on December 23, 1998, rendezvoused again on February 14, 2000, and landed on it on February 12, 2001, after orbiting it for a year.

Near-Earth asteroid

Near-Earth asteroids (NEAs) are ASTEROIDS whose orbits bring them into the inner solar system, by definition those with PERIHELION distances of less than 1.3 AU. They are also known as Earth-approaching asteroids; the term 'Near-Earth object' (NEO) is sometimes used in recognition that some of these objects are cometary in origin. There are three classes of NEA:

- Amor asteroids have a perihelion distance between 1.017 AU (the Earth's APHELION distance) and 1.3 AU, and are named for (1221) Amor. As of April 2003, 1039 were known. The largest are (1036) Ganymede, at 40 km, and (433) EROS.
- Apollo asteroids have a perihelion distance less than 1.017 AU and a semi-major axis greater than 1 AU. They are named for (1862) Apollo, and 1087 are known; the largest is the 8-km diameter (1866) Sisyphus.
- Aten asteroids have a semi-major axis measuring less than 1 AU, and are named for the 1-km diameter (2062) Aten. Only 182 are known, so few because they lie inside the Earth's orbit and remain within 90° of the Sun, often hidden in its twilight glare.

The terms Mars-, Earth- and Venus-crossing are employed to refer to asteroids that intersect the orbit of one of these three planets.

Three-quarters of NEAs are thought to be asteroids perturbed inward from the main asteroid belt. The mechanism is believed to be based on chaotic dynamics near the KIRKWOOD GAP corresponding to a 3:1 resonance with Jupiter. Others are thought to be extinct comet nuclei. The lifetime of NEAs is probably not much more than 10 million years, their likely fate being collision with one of the terrestrial planets or gravitational ejection from the inner solar system via a close encounter.

In 1980 the Spaceguard Program was established to track down what have become known as potentially hazardous asteroids (PHAs): the closest-approaching NEAs large enough to create a hazardous impact. Objects of around 10 m would burn up during their passage through the atmosphere. Objects of 100 m or less would cause local damage, as in the TUNGUSKA EVENT of 1908. A major event from a kilometer-sized asteroid might occur once in several hundred thousand years – in other words, the odds of such an occurrence in a given century are several thousand to one.

The statistics of the population of NEAs is under close scrutiny; there may be 2000 bodies of 1 km or more, and 100 000 larger than 100 m. Three objects are known to have approached the Earth to within 30% of the distance of the Moon: 1994 XM1 on December 9, 1994, 2002 MN on June 14, 2002, and 2002 XV90 on December 11, 2002.

Near-Earth object *See* NEAR-EARTH ASTEROID

Nebulae and interstellar matter

The enormous volume of space between the stars in the Milky Way and other galaxies is filled with interstellar matter – dust and gas. Stars form from the interstellar matter. Mass-loss from stars enriches the interstellar material in heavy elements. The energy produced by stars heats, ionizes, and produces structures in the interstellar matter – 'nebulae.'

Nebulae have been known since the discovery of the ORION NEBULA by Nicolas de Peiresc (1580–1637). The existence of unseen matter between the stars was first proposed by Friedrich STRUVE in 1847, based on an analysis of star counts that suggested that the number of stars per unit volume decreases with distance from the Sun. Struve proposed that the starlight was experiencing absorption proportional to distance. In 1909 Jacobus KAPTEYN realized the full significance of this interstellar extinction, caused by dust scattered between the stars. In 1945, Hendrik van de Hulst (1918–2000) predicted that it would be possible to observe neutral hydrogen in the galaxy through its radio emission at 21 cm. This prediction was confirmed in 1951.

Composition of interstellar matter

The interstellar matter in the solar vicinity has a composition by number of atoms of 91% hydrogen (H), 9% helium (He), and a trace abundance of the heavier elements. Approximately 1% of the mass of the interstellar matter is in the form of INTERSTELLAR DUST, mostly solid silicate grains; there are also graphite (carbon) grains. The grains range from very small up to several microns in diameter. In the denser interstellar regions the grains accrete atoms which form icy mantles on the grain cores; the most abundant molecule in interstellar space, H_2, forms on the grains, but does not stick. The dust also includes very small grains, called polycyclic aromatic hydrocarbons (PAHs), which are planar molecules of benzene rings with attached H atoms. These particles have nanometer sizes and are on the dividing line between large molecules and small particles.

The interstellar matter contains gas at different temperatures – cold (around 10 K), cool (100 K), warm (10^4 K) and hot (10^6 K). Most of the mass (around 80%) is contained in the neutral hydrogen gas. Cold molecular clouds, containing H_2, carbon monoxide (CO) and other molecules, comprise a small fraction of the volume of the interstellar matter, but around half its mass. They are bathed in a cool neutral medium that contains neutral hydrogen atoms, H. The line through the concentration of neutral hydrogen along the Milky Way is used to define the GALACTIC PLANE (*see* GALACTIC LATITUDE AND LONGITUDE). With a temperature of around 80 K and a density of some 40 atoms cm^{-3} it occupies only around 3% of the volume of the interstellar matter in the GALACTIC DISK. The warm neutral medium also produces H I 21-cm emission, but the molecules are separated into their atoms. It occupies a substantial fraction (some 35%) of the volume of the interstellar matter in the Galactic Disk at heights up to several hundred parsecs above the plane.

The warm ionized medium can be studied through the optical emission lines it produces in its densest parts, namely the gaseous nebulae near hot stars. Observations of radio pulsars show the overall properties of the warm ionized medium. The power required to maintain the ionization of the warm ionized medium is comparable to the total power input of all galactic SUPERNOVAE, or around 15% of the ionizing radiation emitted by galactic O-type stars.

The hot ionized medium contains low-density gas with

Observing nebulae

The search for galactic nebulae has always been one of the most challenging types of visual observation. The discovery of galactic nebulae had to wait until the invention of the telescope, and even then, only 11 galactic nebulae were contained in Charles MESSIER's famous catalog of 110 non-stellar objects. As astronomers continued to search the skies, the discovery of most deep-sky objects, including galactic nebulae, was initially the province of amateurs as the professionals were more involved in stellar and solar system astronomy.

Roughly a century ago, the rise of financially well-off amateur astronomers with large telescopes searching first visually, and then applying the new art of photography, rapidly increased the number of galactic nebulae known. The advent of professional all-sky photographic surveys culminating in the National Geographic–Palomar and ESO/SERC (Science and Engineering Research Council, UK) surveys provided a resource whereby most prominent nebulae were discovered. Further surveys in the red and infrared regions have picked up most of the remaining nebulae.

Galactic nebulae come in five main types, and each type may require its own observation techniques. The main types are diffuse nebulae, PLANETARY NEBULAE, REFLECTION NEBULAE, SUPERNOVA remnants and DARK NEBULAE. With the exception of the planetary nebulae and supernovae remnants, the others can occur in combinations; for example, M20, or the Trifid nebula, is a combination of reflection and diffuse nebulae. Many of the dark, diffuse and reflection nebulae are also associated with stars and star clusters.

With the exception of dark nebulae, most of the objects under consideration here are best seen with optical aid: in most cases a telescope. Binoculars can show some of the brighter ones and, given very dark skies, some of the larger ones can be seen with the naked eye assisted by a nebula filter (*see* CAMERA AND FILTER).

The basic observing techniques are the same as for any deep-sky object. Use a decent star chart to locate the field and then, starting with a low-power EYEPIECE, try and locate the target. As most of these objects are very faint, observers should let their eyes get well adapted to the dark, and try and shield both eyes and the telescope optics from any stray light. Try observing the object when it is on the MERIDIAN and, obviously, when the Moon is out of the sky.

The eye is better at spotting moving, rather than stationary, objects, so if the object is still not visible, try tapping the telescope slightly. It is also best to move the suspected object out of the center of the field and use the technique of averted vision, an approach which relies on the fact that the rods in the retina of the eye are more sensitive to faint objects. Once the object is located, increase the power and try to make out as much detail as possible. Increase in contrast can often make the object easier to see.

Many of the nebulae may be quite small and some of the planetary nebulae in particular benefit from using high power. They can also show color as their high brightness and small size can give a surface brightness high enough to stimulate the color detectors in the eye. Very few other astronomical objects show color.

▼ ***Eagle Nebula***, *imaged by Robert Dalby and Nik Szymanek. This LRGB CCD image was based on 4 × 5 min luminance exposures, using a Meade 10-in LX200 af f/10; an SBIG ST-8 CCD camera and CFW-8 filter wheel; and an SBIG AO-7 adaptive optics unit.*

Observers of diffuse nebulae, planetary nebulae and supernova remnants also have benefited from the invention of the nebula filter. The nebula filter is an INTERFERENCE filter, which passes only certain wavelengths of light. The success of this technology relies on the fact that most emission nebulae emit their light in only certain wavelengths, and the most prominent of these wavelengths in the visual spectrum are the lines from doubly-ionized oxygen (OII) at 495.9 and 500.7 nm, along with the hydrogen-beta line at 486.1 nm. The design of filters that pass only these wavelengths opened up new techniques that amateurs could use to observe, and thus also opened up more fields which they could study.

Later, filters were designed which passed only the lines of OII; these made a huge difference to the observing of faint planetary nebulae and supernovae remnants. Filters designed to pass only the light from the hydrogen-beta line opened up the possibility of observing a number of very faint diffuse nebulae.

There are three main techniques to learn when using these filters. The first is just to hold the filter in front of the eye. This will allow one to observe large objects such as the CALIFORNIA NEBULA and the NORTH AMERICA NEBULA. The second method is known as 'blinking' as it involves holding the filter in front of the eyepiece and moving it rapidly in and out of the field so that one can spot small bright objects. The third and most common method is screwing the filter into the eyepiece and then searching for the object. The invention of nebula filters is part of the reason for the rise in popularity of observing planetary and, to a lesser extent, diffuse nebulae.

Reflection and dark nebulae do not benefit as much from the nebula filter revolution as they effectively shine from the reflected light of stars, and narrow-band filters will not help much here. The best method to use when observing these objects is to observe on the darkest and most transparent nights, making sure all optics are as clean as possible; the first clue as to the existence of these objects is often what appears to be a haze around a star, and it is nice to know that this is not due to grease on the eyepiece!

Most visual observations are recorded by making either notes or pencil drawings. These will have limited scientific use, but a great deal of pleasure can be obtained from sharing one's observations with others.

Most of the forgoing has dealt with objects apart from dark nebulae. In general, dark nebulae are quite large and are best seen with binoculars and rich field telescopes. As they do not shine at all, they are best seen in front of stars, in regions like the Milky Way. Many of them were discovered by Edward BARNARD using widefield photography, and many of the best still bear his catalog numbers.

The size of nebulae makes them good targets for both photography and CCD imaging, and the techniques used here can be applied to the observation of other deep-sky objects as well.

There is very little science that can be done today by amateurs on these types of objects. One area in which they might contribute is in the long-term monitoring of variable nebulae. These are objects associated with star formation and can be interesting to observe, as both their shape and brightness change with time.

There are several organizations dealing with deep-sky observing of this type, including the National Deep Sky Observers Society in the USA, and the WEBB SOCIETY and the BAA's Deep Sky Section in the UK. Most other national societies will have sections devoted to this kind of observing.

The popularity of this nebulae observing has led to a vast amount of information now being made available on the Web, along with specialist news groups devoted to the discussion of the various types of deep sky objects. Good places to begin looking for online resources include the Webb Society, BAA, National Deep Sky Observers Society, Adventures in Deep Space, and the NGC/IC Project.

Owen Brazell has been interested in astronomy from an early age. He works in oil exploration, and when time permits he enjoys hunting down faint nebulae.

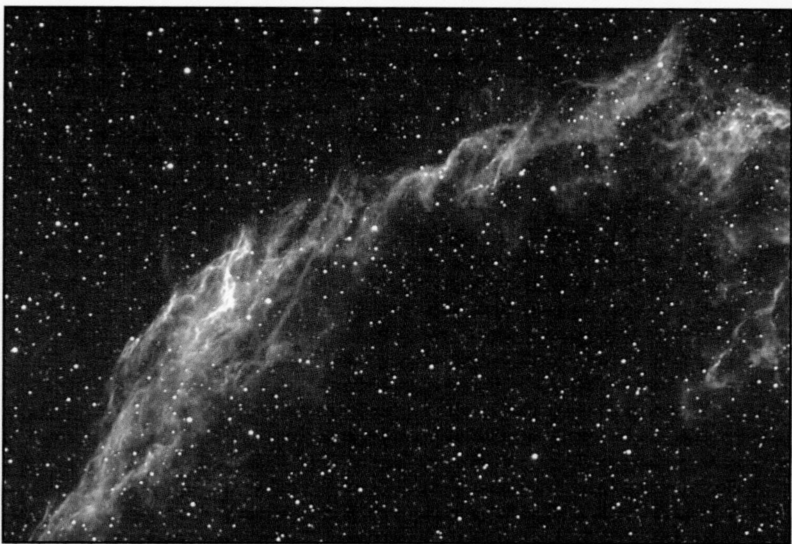

▲ ***Veil Nebula*** *in Cygnus. Image by Nik Szymanek. The LRGB CCD image was based on 9 x 5 min luminance exposures, using a Pentax SDHF 75-mm f/6.6 apochromatic refractor; Starlight Xpress SXV-H9 CCD camera; Astronomik 10-nm Hydrogen Alpha filter; True-Technology RGB filters and filter wheel; and an SBIG ST-4 autoguider.*

◀ ***North America Nebula*** *(NGC 7000) in Cygnus, imaged by Nik Szymanek. This LRGB image was based on 14 x 5 min luminance exposures, using a Pentax SDHF 75-mm f/6.6 apochromatic refractor; Starlight Xpress SXV-H9 CCD camera; Astronomik 10-nm Hydrogen Alpha filter; and an SBIG ST-4 autoguider.*

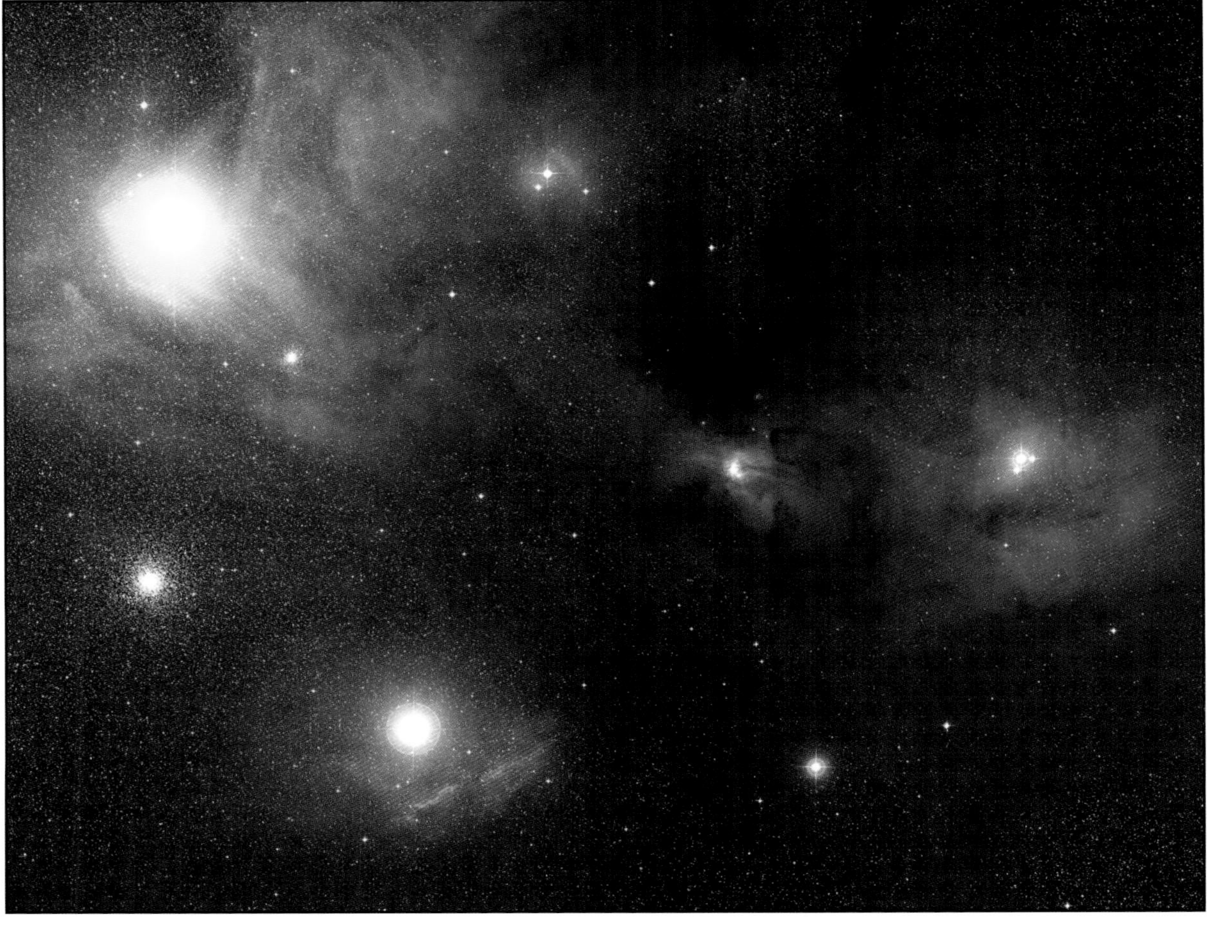

▶ ***Reflection nebula*** *The Rho Ophiuchi dark nebula contains many stars whose light reflects from the dust. The variety of colors comes from the different colors of the stars and the light-scattering properties of the dust.*

a temperature of 1 million degrees. It is dispersed up to a few kiloparsecs above the galactic plane and was detected through its soft X-ray emission. Somewhat cooler hot gas associated with the hot ionized medium was detected by the COPERNICUS (OAO-3) ultraviolet-sensitive satellite.

Nebulae

The Orion Nebula is one of a huge number of unresolved nebulae that belong to the warm ionized interstellar medium, known since the invention of the telescope. As telescopes improved, some so-called nebulae, like those cataloged by Charles MESSIER (*see* MESSIER CATALOG), were resolved into individual stars in clusters. In the nineteenth century others were resolved into galaxies. But a residual hard core could not be resolved because they were not made of stars. Turning his primitive spectroscope in 1864 to the Cat's Eye 'planetary nebula' (NGC 6543) discovered by William Herschel (*see* HERSHEL FAMILY), William HUGGINS found that it radiated emission lines. Some astronomers claimed that they had seen the Orion Nebula break into myriad stars, but Huggins found that it too displayed three characteristic emission lines: the H-beta line of ionized hydrogen at 486.1 nm and the mysterious green 'nebulium' lines at 495.9 and 500.7 nm that Ira BOWEN identified in 1928 as 'forbidden lines' of doubly ionized oxygen. If Huggins could have detected the red region of the spectrum, he would have found the even more characteristic red H-alpha line at 656.3 nm. The Orion Nebula and others like it were luminous because they were made of hot gas.

Reflection nebulae

A new type of nebula was recognized in 1913. Wilhelm Tempel (1821–89) discovered that the star Merope in the Pleiades was surrounded by a nebula. Indeed the whole star cluster was embedded in nebulosity. Vesto SLIPHER photographed the spectrum of the nebula in 1912 and discovered that it was reflecting the spectrum of the brighter stars of the Pleiades – they were not hot enough to ionize the nebula.

Reflection nebulae are made of gas mixed with dust particles that reflect the light of one or more stars too cool to ionize the gas. Turn up the stellar heat and reflection nebulae glow like Orion. There are two types:

- In accidental reflection nebulae, the dust comes from the general interstellar medium, and the illuminating star has encountered the dust by chance in its orbit round the Galaxy (like the Pleiades reflection nebula).
- In intrinsic reflection nebulae, the dust is either left over from the formation of illuminating star(s), or has been made by the star that is itself illuminating the reflection nebula, so the stars and dust are generically related. Examples are the reflection nebulae in the region of the Rho Ophiuchi dark cloud, and the Toby Jug Nebula, respectively.

It seems in pictures that reflection nebulae surround their star, and 'reflection nebula' suggests that the dust is a screen behind the star. However, interstellar dust grains scatter light mainly in the forward direction, so in most reflection nebulae the dust lies between the star and us, like the halo around the lights of an oncoming car in fog.

Emission nebulae

Emission nebulae are luminous because they are made of hot ionized gas that, on recombining, emits light. There are three types:

- PLANETARY NEBULAE, found everywhere from the Galactic Disk to the halo, are small shells of gas around, and illuminated by, hot, dying intermediate-mass stars;
- supernova remnants are the heated remains of high-mass exploding stars;
- H II REGIONS (also called 'diffuse nebulae') are largely made of ionized interstellar hydrogen, heated by nearby or embedded stars (like the Orion Nebula).

H II regions are related to very new stars and to star formation. Vastly larger than planetary nebulae, they are closely confined to the plane of the Galaxy and line up along the center line of the Milky Way. Though the brighter nebulae, like Orion, look rather greenish to the eye (as a result of the forbidden oxygen lines), photography reveals most as red, the color caused by the bright red H-alpha emission line. Reflection nebulae are also in the Milky Way and are readily identified by their usually blue colors. Just as in the sky, blue starlight is scattered much more efficiently than red.

H II regions are inside or on the outer surface of giant molecular clouds (GMCs), and related to young T Tauri stars and to the outflowing JETS that make Herbig–Haro objects (*see* George HERBIG). The nebulae are local portions of the molecular clouds that have been ionized by hot, recently born massive stars. Many have NGC catalog entries and others are included in the Sharpless catalog of 1959, which contains 313 objects identified on the red plates of the Palomar Sky Survey (which highlight H-alpha radiation). Because infrared and radio emissions are unimpeded by dust, they allow the discovery of nebulae that are hidden behind and within the thick dust of galactic GMCs. The Becklin–Neugebauer/Kleinman–Low (BN/KL) infrared nebula in Orion was discovered in this way.

The smallest H II regions like the BN/KL complex are called 'ultracompact H II regions,' under 0.1 parsecs, which are found around the newest stars embedded in GMCs. At the next scale are 'compact H II regions' with dimensions in the realms of tenths of a parsec, like the Orion Nebula. Above these are 'large' structures like the NORTH AMERICA NEBULA, and at the top the 'extragalactic giant H II regions' like 30 Doradus in the Magellanic Clouds and NGC 604 in M33, and others which define the spiral arms of open spiral galaxies such as M101.

The structure of H II regions

It takes a UV photon with wavelength shorter than 91.2 nm to ionize hydrogen. For a star to radiate at these wavelengths it must be hotter than about 25 000 K, which corresponds to spectral class B1. Thus the diffuse nebulae are the realm of the O-type stars, with masses exceeding 12 times that of the Sun. Orion's ionizing TRAPEZIUM quartet includes four stars like this. O stars do not live very long, about 10 million years. Most O stars are thus still attached to their H II regions.

The structure of diffuse nebulae was discovered by Bengt STRÖMGREN in 1939. A 'Strömgren sphere' is a bubble formed within a neutral cloud (a GMC) by a hot, ionizing star, the gas fully ionized out to a sharp edge. Each UV photon from the central star causes one hydrogen ionization, which results in one recapture of an electron by a proton, which in turn results in one Balmer photon. Similar ionization and recombination processes happen in helium atoms, and for a variety of others like oxygen, nitrogen, carbon, and neon.

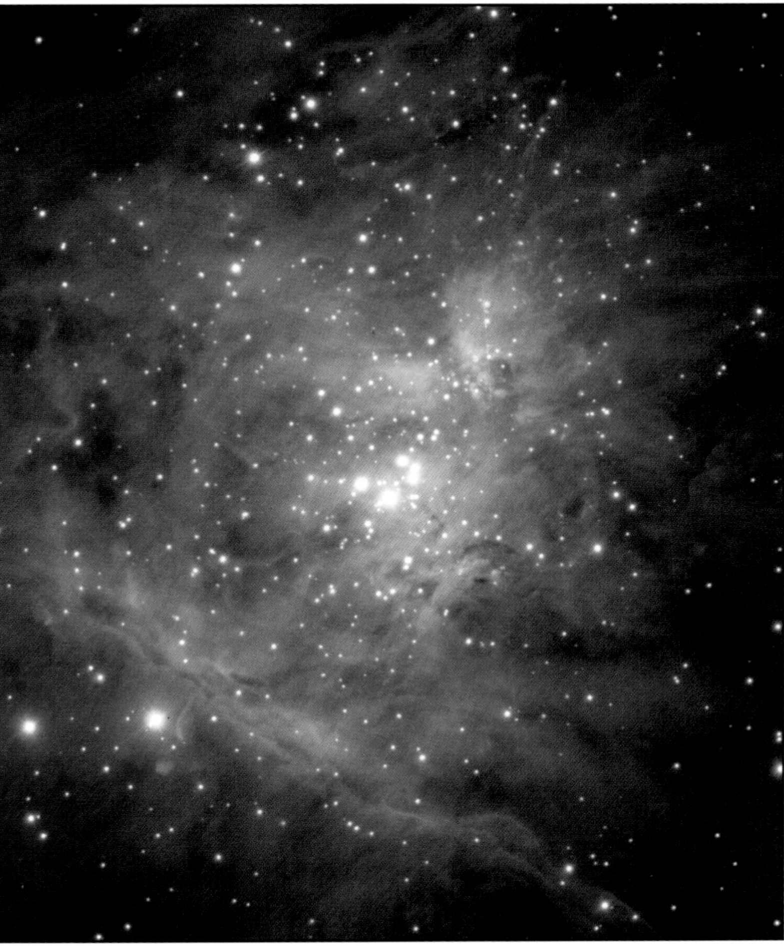

▲ ***Emission nebula*** *The Orion Nebula, M42, is centered on four stars, the Trapezium, the brightest of the stars of a central cluster. Their ultraviolet light ionizes the gas nearby, and creates an emission nebula. This image is captured in infrared.*

It is an ideal model. Strömgren spheres, indeed diffuse nebulae in general, are dynamic. The heated gas in the Strömgren sphere is at higher pressure than the cold gas outside, and the boundary – the 'ionization front' – expands into the surrounding GMC. 'Elephant trunks' of dark, dusty gas may break into the expanding ionization front, as in the ROSETTE NEBULA and the HORSEHEAD NEBULA. If the nebula becomes large enough, it can break through the outer boundary of the GMC, the high-pressure gas erupting in an ionized fountain. If O stars form near the edge of the GMC, they can also erode the molecular layers separating them from open space. Their nebula appears as a blister on the molecular cloud's surface – the stars in front of it, the nebula behind. The ORION NEBULA is such a blister, with its illuminating stars (particularly Theta[1] Orionis C) eating their way into the dense dark cloud at the same time that the BN/KL star inside is eating its way out.

Further reading: Osterbrock D E (1989) *Astrophysics of Gaseous Nebulae and Active Galactic Nuclei* University Science Books; Neckel Th. and Vehrenberg H (1985) *Atlas of Galactic Nebulae* Treugesell.

Neptune and its satellites

Neptune is the outermost of the four giant planets, and its orbit is nearly circular. It forms a pair with Uranus, and is distinct from Jupiter and Saturn.

▶ ***Neptune*** *This photograph was taken by Voyager 2 on its flyby on August 25, 1989. Neptune's thick atmosphere consists mainly of hydrogen, helium, water and methane. The methane gives the planet its blue color because it absorbs light at red wavelengths. An enormous hurricane called the Great Dark Spot is visible.*

Neptune rotates quickly, slightly more rapidly than Uranus. This is responsible for the slight flattening (or oblateness) of the planet. The equatorial plane of Neptune is highly tilted to the orbit, causing significant seasonal variations. It emits 2.6 times more energy than it absorbs from the Sun and this excess raises its temperature from 47 to 59.3 K. The excess emission derives from cooling of Neptune's hot interior.

Discovery and exploration

With a visual magnitude of 7.8, Neptune is too faint to be seen by the naked eye. Through a telescope, it is a bluish disk with an average angular diameter of 2.3 arcsec. Neptune was identified in September 1846 by Johann GALLE and Heinrich d'Arrest (1822–75) after calculations by two astronomers, John Couch ADAMS and Urbain LEVERRIER of the effect of dark matter (a new planet) on the motion of Uranus. Its largest satellite, Triton, was discovered a few weeks later.

Its atmosphere shows large changes on timescales from hours to years. There are discrete clouds rotating in the prograde direction, that is, in the same sense as the orbital revolution, which makes its rotation seem faster than it really is. In 1984, rings were discovered around Neptune (*see* PLANETARY RING).

But most of what we know about Neptune and its satellites comes from the VOYAGER 2 flyby in August 1989, including the discovery of an Earth-sized 'Great Dark Spot' (GDS). Six moons were discovered with diameters ranging between 50 and 400 km.

Atmosphere

The atmosphere of Neptune is 3500 km deep and overlies a denser fluid comprising mainly water, methane and ammonia. The mass of this outer layer, between 0.5 and 1 Earth mass, is a small fraction of that of the whole planet, in contrast with gas-rich Jupiter and Saturn. The temperature of Neptune's atmosphere is similar to that of Uranus. The reason for this is that the internal heat flow from Neptune compensates for the reduction in solar heating by its remoteness.

Neptune's atmosphere is primarily molecular hydrogen (H_2) and helium (He), at an abundance compatible with the presolar nebula. Beyond these two light elements, Neptune's observable atmosphere contains methane (CH_4), ethane (C_2H_6), acetylene (C_2H_2), water vapor (H_2O) and carbon dioxide (CO_2). The CH_4 abundance is about 2%. Absorption of sunlight at red wavelengths by methane is at least partly responsible for the bluish appearance of the planet.

Clouds and winds

In the upper levels of Neptune's atmosphere are thick hydrogen sulphide (H_2S) clouds, with additional thin CH_4 clouds. But Neptune's atmosphere is not homogeneous and shows a variety of light and dark features. Some prominent, bright cloud features, like 'the Scooter,' are likely made of CH_4 particles near the 1-bar level. Dark ovals are holes in the H_2S cloud cover.

Neptune has some of the fastest winds measured on any planet, despite the small amount of energy received from the Sun. Fast retrograde (westward) winds blow at the equator, and there are prograde (eastward) winds poleward of around 50°. Wind speeds measured at the cloud tops vary from 400 m s^{-1} at the equator to 250 m s^{-1} at 70° latitude.

The most remarkable feature seen by Voyager was the GDS, a dark oval spot in the southern hemisphere. Like the GREAT RED SPOT on Jupiter it was an anticyclone – a high-pressure center. The size of the GDS oscillated with an eight-day period between 12 000 and 18 000 km in longitude and between 5200 and 7400 km in latitude. Observations with the HST in 1994 showed that the GDS had disappeared or faded away, but in 1994 a new spot (GDS-94) of similar size appeared.

Neptune is a surprisingly active planet considering the paucity of solar energy available to power its wind system and meteorology, and the atmosphere must have low frictional losses.

Interior and formation

Beneath the hydrogen-rich envelope lies the denser interior of Neptune, composed of the abundant 'ices': water (H_2O), methane (CH_4) and possibly ammonia (NH_3). Although these molecules are liquid, they are referred to as 'ices' because they were ices in the outer solar nebula, unlike hydrogen

Neptune	
Mean distance from Sun	30.0611 AU
Orbital period	162.79 years
Orbital eccentricity	0.0097
Orbital inclination	1.774°
Mass	1.024×10^{26} kg
Mass relative to Earth's	17.14
Equatorial radius at 1 bar	24 760 km
Equatorial radius relative to Earth's	3.88
Density	1.64 g cm^{-3}
Equatorial gravity at 1 bar	10.9 m s^{-2}
Equatorial gravity relative to Earth's	1.11
Escape velocity	23.3 km s^{-1}
Oblateness	0.017
Rotation period	16.11 hours
Obliquity of rotation axis	29.6°
Effective temperature	59.3 K
Bolometric albedo	0.29

and helium. Neptune's interior includes a significant fraction of 'rocks' (about 25% by mass), mostly silicates and iron. The temperature is 8000 K at the center, where the pressure is around 8 Mbar. There may be a small rocky core (at most 1 Earth mass).

Unlike Jupiter and Saturn, most of the mass in Neptune (80–90%) is in the form of 'ice' and 'rock' rather than hydrogen and helium. In the outskirts of the solar nebula, Neptune took longer to accrete a massive core than the other giant planets, and did not have the time to trap a large envelope of gaseous hydrogen and helium before the nebula was dispersed.

Magnetic field and aurorae

Neptune has a substantial magnetic dipole field tilted from the axis of rotation by 47°. Neutral atoms escape from Triton's thin atmosphere and are ionized in the magnetosphere, causing weak AURORAE. Weak radio emissions, like those on Uranus, are controlled by the rotation of the magnetic field, which is linked to the interior of the planet and reveal its true rotation period (16 hours 6 min 30 s).

Triton

Triton is the largest satellite of Neptune but its properties remained unknown until the Voyager 2 flyby. Its orbit is nearly circular, retrograde and with a high inclination (157°). Like the Moon and the Galilean satellites, Triton's orbit is synchronous, so the satellite always turns the same hemisphere toward Neptune. The diameter of Triton is 2700 km, its density 2.08 g cm^{-3}, its surface temperature 38 K, its surface pressure 14 μbar and its atmospheric chemical composition molecular nitrogen with traces of methane (0.01%). Triton has a very high ALBEDO (0.7), making it among the most reflective of solar-system objects.

The Voyager images were mostly of the southern hemisphere. They showed a bright polar cap. North of here, a rugged terrain, reminiscent of the skin of a cantaloupe melon, is cross-cut by a pattern of intercepting ridges, possibly the remnants of a past tectonic event. Triton's surface shows comparatively few impact craters, and it is relatively young. A more-cratered region, near the equator, shows an impact density comparable to the 3 billion-year-old lunar maria. Triton's surface is made of N_2, H_2O, CO_2, CO and CH_4 ices.

Seasonal effects are very strong on Triton, probably the strongest of any solar-system object. They are due to its high inclination. As with Mars, seasonal effects induce large-scale wind motions. Voyager observed plumes (dark trails), rising vertically up to an altitude of 8 km then trailing to the west over more than 100 km. The plumes may be the result of active volcanism or solar-driven geysers.

Apart from Triton's density, we have little information about its interior. But theory suggests it has a core of metals and rocks, surrounded by a mantle of water and other ices.

Triton's curious orbit and its striking similarities to Pluto (in terms of size, density, surface chemical composition and atmospheric composition) suggest they have a common origin (see PLUTO AND CHARON). Pluto is the biggest of the Kuiper Belt objects (*see* OORT CLOUD AND KUIPER BELT). Perhaps Triton originated in the Kuiper Belt and was captured by Neptune's gravity. Such a capture, however, needs the presence of a third body to carry off the excess energy and this is quite improbable.

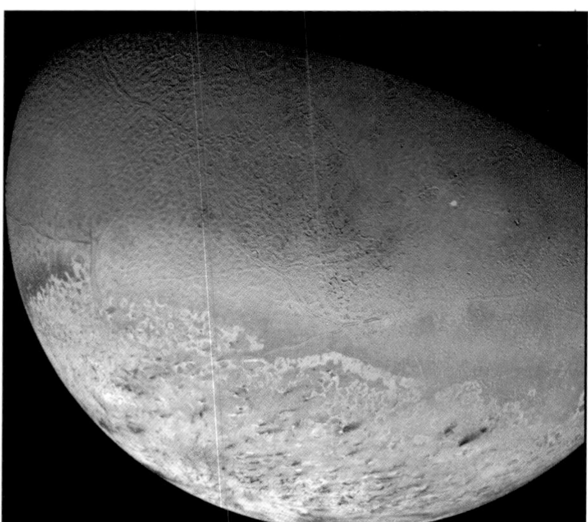

◀ ***Triton*** *This image of the southern hemisphere of Neptune's largest moon is a mosaic of images taken by Voyager 2 in 1989. It has one of the coldest surface temperatures in the solar system, 34.5 K, and the icecaps are visible at the bottom of the image, where ice volcanoes sometimes erupt. The dark streaks are from geysers of icy dust.*

Other satellites

Neptune has 13 natural satellites, the outer five of which, discovered in 2002–3, are small and distant from the planet. All but one of Neptune's eight other minor satellites orbit within or just outside its ring system; the exception is the distant object Nereid. The inner four, with approximate diameters, are Naiad (60 km), Thalassa (80 km), Despina (150 km) and Galatea (160 km). The first three lie within the Leverrier ring, to which Despina appears to act as an inner SHEPHERD MOON. All four were discovered during the Voyager 2 flyby in 1989. Diameters of the other satellites are rather more certain. The next, Larissa (195 km), was discovered in 1981 by Harold Reitsema when it occulted a star; it acts as an inner shepherd to the Adams ring. Proteus, orbiting outside the ring system at 118 000 km, is Neptune's second-largest satellite, after Triton. It is rather blocky, measuring about 435 × 400 km, and is a dark, heavily cratered body; the largest crater, Pharos, is some 250 km across. The outermost satellite, Nereid, orbiting at an average distance of 5.5 million km, has the most eccentric orbit of any known planetary satellite, at 0.753. It was discovered in 1948 by Gerard KUIPER, and has a diameter of 340 km. It may be a captured object, or it may have been perturbed into this orbit when Triton was captured (if indeed it was). *See also* SATURN AND THE OUTER GAS PLANETS, OBSERVING.

Further reading: Benner L A M (1997) *Encyclopedia of Planetary Sciences* Chapman and Hall; Cruikshank D P (ed) (1995) *Neptune and Triton* University of Arizona Press.

Neutral point

A point in space where an object experiences no net force which would cause it to move (relative to some, possibly moving, reference frame). The simplest example would be between two motionless masses, and it would lie on a line joining their centers of mass, closer to the larger one. For the simplest case of two massive bodies in circular orbits around their center of mass, there are five neutral points, known as the LAGRANGIAN POINTS, stationary relative to the orbiting pair.

Neutrino

Neutrinos are electrically neutral elementary particles that experience only the weak nuclear force and gravity (*see* ASTROPARTICLE PHYSICS and FUNDAMENTAL FORCE). Wolfgang Pauli (1900–58) hypothesized their existence in 1930 to explain features of radioactive beta decay (*see* BETA PARTICLE). Their mass is very small – theoretically zero – and they very weakly interact with matter. Hans Bethe (1906–) and Rudolf Peierls (1907–95) first calculated that a typical neutrino would travel thousands of light years in lead without interacting. In the opinion of physicists in the 1930s, neutrinos were therefore undetectable – ghosts. Nevertheless, neutrinos are produced in great numbers during nuclear reactions and laboratory experiments, and detectors have been made that trap them.

In the standard theory, neutrinos are associated with families of elementary particles. Three neutrino species have been identified in nature, called the electron neutrino (ν_e), muon neutrino (ν_μ), and tau neutrino (ν_τ). A very sensitive measurement of the number of neutrino families was performed at the Large Electron-Positron (LEP) collider at CERN in Geneva, and gave the answer 2.998 ± 0.029, so it is clear that there are exactly three families.

Neutrino oscillations

Although neutrinos theoretically have zero mass, the standard theory is wrong. Outside the standard theory, there is a phenomenon called neutrino oscillation, postulated by Bruno Pontecorvo (1914–93), by which one kind of neutrino can change into another. It only happens if neutrinos have a residual mass, even if it is small. The first convincing observation of this phenomenon was reported in 1998 by a Japanese group led by Masatoshi Koshiba (1926–, Nobel Prize 2002), using a large neutrino detector (45 000 tonnes of water) in an underground mine in Japan (Kamioka). Muon neutrinos produced by cosmic rays in the atmosphere on the other side of the Earth propagated through it to Japan, and many became tau neutrinos over a distance of about 4000 km and disappeared. The appearance of a corresponding number of tau neutrinos has yet to be observed because the Kamioka detector cannot identify them. A new experiment has to be built to make this observation. It will use a muon-neutrino beam made at CERN which will be directed into and detected at the Gran Sasso underground laboratory near Rome, 730 km away. It is planned to start in the year 2005. The Kamioka result was confirmed by measurements at the Sudbury Neutrino Observatory in Canada.

The existence of neutrino oscillations proves that neutrinos have mass. The mass is not known exactly but it is small.

Neutrino astronomy

The Sun generates energy deep in its core by nuclear-fusion processes. A SUPERNOVA (of type II) explodes and blows off the entire star by gravitational energy liberated when the central core collapses. These interesting phenomena of energy generation can never be observed with visible light or any other electromagnetic waves. They occur in places that are buried deep in dense material, and relevant information is lost as the radiation travels through the material.

Nevertheless, astronomers obviously want to probe the deep interior of the Sun or the gravitational collapse of the supernova core. NEUTRINOS are produced in these places, are electrically neutral, do not interact much and travel straight so they can be traced back toward their source. They bring out the desired information.

Solar neutrinos

Neutrinos from the Sun were first observed by Raymond Davis Jr (1914–, Nobel Prize 2002) in the Homestake gold mine, South Dakota, USA, in the late 1960s. He constructed a completely new detector, a gigantic tank filled with 600 tonnes of dry-cleaning fluid. Over 30 years he captured 2000 neutrinos from the Sun and was thus able to prove that fusion provided the energy from the Sun, but, at Earth, provides only 30–60% of what solar theory predicts.

This was the 'solar-neutrino problem.' The reason for the deficit is not that the solar engine is only 30–60% efficient, but that solar neutrinos are electron neutrinos that are converted to muon neutrinos on the way to Earth. Muon neutrinos are difficult if not impossible to observe, hence the apparent deficit of solar neutrinos.

Neutrinos from supernovae

The birth of a neutron star (*see* NEUTRON STAR AND PULSAR) was first observed with neutrinos on February 23, 1987. Kamiokande and IMB, two large neutrino detectors, observed 11 and 8 neutrino events from supernova 1987A, respectively. The Japanese researchers were led by Masatoshi Koshiba (1926–, Nobel Prize 2002).

Cosmic neutrinos

Stars and supernovae are not the only places in astronomy where neutrinos are made, but neutrinos from these sources are the only ones to have been successfully observed. About 1 s after the BIG BANG, the universe was as hot as the core of a supernova, and light particles such as electrons and positrons existed in equal numbers. The universe cooled down as it expanded. Electrons and positrons were then annihilated; they disappeared and created photons and neutrinos. The photons became the COSMIC MICROWAVE BACKGROUND. Likewise, there is also a background of cosmic neutrinos, which should still exist in the present universe, and hold important information about its early history. At one time it was suspected that, even though the mass of a neutrino is very small, so many were created in the Big Bang that their total mass might be significant and neutrinos might be a form of DARK MATTER (*see* ASTROPARTICLE PHYSICS). The measurement of neutrino oscillations shows that this is not so, and neutrinos do not contribute significantly to the evolution of the universe.

Cosmic-ray neutrinos

COSMIC RAYS are accelerated somewhere in our Galaxy or, if they are very energetic, perhaps in distant active galaxies. Extremely energetic cosmic rays might even be produced by the decay of exotic particles created in the early universe and still floating around now; their existence is conjectured in theories of elementary particles. Cosmic rays bend in galactic or intergalactic magnetic fields and cannot be traced back to their birthplaces. Even to this day, we do not know where cosmic rays are actually born. But they may interact with whatever surrounds their source and produce MESONS, which soon decay and produce neutrinos. These neutrinos may reach Earth with fruitful information.

Neutrino observatories

- Fréjus (a nucleon-decay experiment in the Fréjus tunnel in southeast France, now terminated);
- National Laboratory of Gran Sasso, National Institute of Nuclear Physics (near Rome, Italy);
- Homestake Mine (a solar-neutrino experiment);
- IMB (Irvine–Michigan–Brookhaven Experiment, Ohio, USA, intended to see proton decay);

• SAGE (Soviet American Gallium Experiment, Baksan Mountains, Russia);
• SOUDAN-2 (Tower-Soudan Iron Mine, Minnesota, USA);
• SNO (Sudbury Neutrino Observatory, Canada, a heavy-water detector);
• Kamioka Observatory (Kamiokande, Kamiokande II (a 1986 upgrade) and SUPER-KAMIOKANDE, Japan);
• UK Dark Matter Collaboration, Boulby Mine, UK.

High-energy cosmic-ray neutrinos will be detected in large under-ice and under-water detectors that are being built. We have absolutely no idea how to detect relic neutrinos born in the early universe, because they have no significant interaction with matter.

Further reading: Bahcall J N (1989) *Neutrino Astrophysics* Cambridge University Press; Winter K (ed) (1991) *Neutrino Physics* Cambridge University Press.

Neutron

An uncharged subatomic particle that is composed of three QUARKS. A neutron is one of the two basic constituents of all atomic nuclei (apart from the normal form of hydrogen, which consists of a single proton). Neutrons and protons have nearly the same mass, and the total number of protons and neutrons in an atomic nucleus defines its mass number. An isolated neutron decays into a proton, an electron and an antineutrino after about 15 minutes on average. This process is called 'beta decay.'

Neutron star and pulsar

Neutron stars are small, compact stars with densities comparable to those inside nuclei, and radii of 10–15 km. They consist predominantly of NEUTRONS and a few percent of PROTONS and ELECTRONS. In a sense they are huge, neutron-rich, atomic nuclei, bound by gravitation. Above a maximum (Chandrasekhar) mass of between two and three times that of the Sun, neutron stars are unstable and collapse to become BLACK HOLES (*see* Subrahmanyan CHANDRASEKHAR). Pulsars are rapidly spinning neutron stars whose strong magnetic fields produce cone-shaped beams of electromagnetic radiation that sweep past the Earth with each rotation of the star, creating the eponymous pulses that are observed primarily at radio wavelengths.

In 1933, only a year after James Chadwick (1891–1974) discovered the neutron, Wilhelm BAADE and Fritz ZWICKY hypothesized that neutron stars are formed in SUPERNOVA explosions. The Crab supernova was recorded by the Chinese in AD 1054. The discovery in 1967 of a rotating neutron star (a radio pulsar) in the CRAB NEBULA supernova remnant confirmed that neutron stars are formed in type II or Ib supernova explosions when massive stars run out of nuclear fuel after millions of years.

Radio pulsars were discovered by Jocelyn BELL-BURNELL and Antony HEWISH in 1967. There are now 1200 known pulsars. The radio pulses are extremely regular with periods in the range 1.5 ms to 8.5 s and the pulsars are spinning down slowly. The age of a pulsar is determined from its spin-down rate. Most pulsars are old (around 10 million years) and slowly rotating with relatively small changes in period. A few young pulsars have short, rapidly changing periods.

Binary pulsars were discovered by Russell HULSE and Joseph TAYLOR in 1973. We know of about 50, and they are mainly millisecond radio pulsars with periods between 1.56 and 100 ms. They are believed to be old recycled pulsars, which have been spun up by mass ACCRETION (the repeated impact of matter obliquely on the surface of the neutron star has accelerated its rotation). About 30 (more than half of the known millisecond pulsars) are found in BINARY STARS where the companion is a WHITE DWARF or a neutron star. Six double neutron stars are known so far.

Some binary pulsars are of particular interest:

• The Hulse–Taylor pulsar is in a high-eccentricity, short-period orbit around a second neutron star. It is an extraordinary laboratory for studying GENERAL RELATIVITY.
• In 1988, a pulsar (B1957+20) was discovered with a 0.025 solar mass companion. The pulsar is eclipsed by its companion for about 10% of each orbit, much longer than expected. The eclipse is by the wind of material streaming off the companion.
• PSR B1257+12, a pulsar with a 6-ms spin period, has two planetary companions of roughly terrestrial mass, 3.4 and 2.8 times the Earth's mass, orbiting with periods of 66.5 and 98.2 days. A third body, of roughly lunar mass, orbits in a one-month orbit. This was the first planetary system detected outside the solar system, and these remain the lightest known extrasolar planets.
• The pulsar PSR B1620-26, in the globular cluster M4, has a white-dwarf companion of around 0.3 solar mass in a low-eccentricity, half-year orbit. The binary orbits around its center of mass with a second, more distant companion, a planet or a BROWN DWARF.

By measuring the inward spiraling or orbital decay, one can determine many parameters in the binary systems such as the neutron star and companion masses. The orbital parameters provide a test of general relativity to an unprecedented accuracy. Binary neutron stars all have masses in a narrow range very close to the Chandrasekhar mass of the iron core, 1.44 solar masses.

X-ray pulsars and X-ray bursters are accretion-powered neutron stars. Almost 200 have been discovered within the last three decades by satellite-borne X-ray detectors. About 60 have had their orbital periods determined. X-ray pulsars and bursters are believed to be neutron stars accreting from high- and low-mass companions respectively. X-ray pulses are most probably due to strong accretion on the magnetic poles, emitting X-rays (as northern lights) with orbital frequency. X-ray bursts are caused by slow accretion spreading all over the surface of a neutron star before igniting in a thermonuclear flash, like a NOVA.

A half-dozen X-ray pulsars have been discovered that are slowly rotating with a period of around 10 s, but are rapidly slowing down. They have huge magnetic fields and are appropriately named 'magnetars.' They reside inside the remnants of SUPERNOVAE, as do 10 radio pulsars.

X-ray bursts are thermonuclear explosions of accreted matter on the surface of neutron stars. After accumulating hydrogen on the surface for hours, the pressure and temperature become sufficient to trigger a runaway thermonuclear explosion seen as an X-ray burst.

A typical 1.4 solar mass neutron star has a thick crust of approximately 1 km and a nuclear liquid in the interior. The outer crust is made of neutron-rich iron and cobalt nuclei. The inner crust consists of neutron-rich nuclei in intricate rod-, plate- and bubble-like structures. The properties of the core of neutron stars are not well understood, but it consists of a liquid of neutrons and protons, with a neutralizing background of negative charge from electrons, muons and other particles, such as hyperons, pions, kaons and quarks.

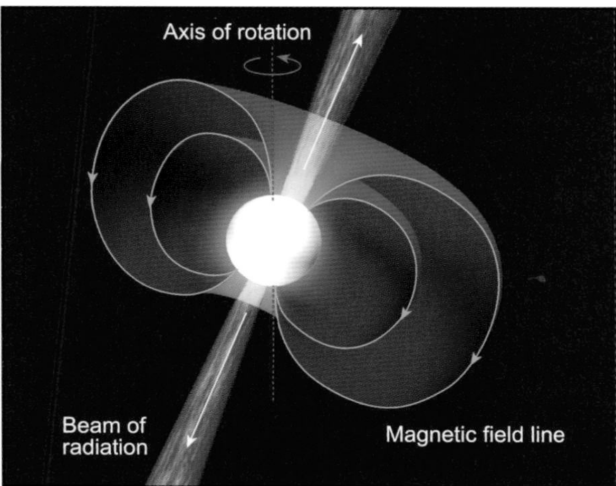

► ***Pulsar***
The beam of radiation rotates around the axis of rotation. If Earth lies in the path of the beam, then we can detect the pulsar's signature rhythmic pulses.

Timing 'glitches' are observed in some young pulsars. During a glitch event, the pulsar's period decreases suddenly. Smaller glitches are caused by a sudden cracking of the crust of the neutron star and its relaxation to a more spherical shape; larger ones are caused by sudden slipping between the superfluid interior of the star and its independently rotating crust.

Neutron stars are born with interior temperatures of the order of 1000 billion K, but cool rapidly via NEUTRINO emission to temperatures of the order of 10 billion K within minutes and 1 million K in 100 000 years. Information on neutron-star temperatures comes from X-ray or ultraviolet astronomy. Non-rotating and non-accreting neutron stars are virtually undetectable but the HST has observed one, RX J185635-3754. It is at a distance of only 200 l.y. It was born in a supernova explosion 1 million years ago.

Gamma-ray bursters or GRBs (*see* GAMMA-RAY ASTRONOMY) have enormous energy outputs and are of two kinds, of long or short duration. The longer bursts may be due to the merger of two neutron stars in a binary system to a black hole. They emit neutrinos and gravitational waves (*see* GRAVITATION, GRAVITATIONAL LENSING AND GRAVITATIONAL WAVES). Nineteen neutrinos were detected from the formation of the neutron star in SN 1987A in the Large Magellanic Cloud.

Pulsars have high space-velocities, which show as large PROPER MOTIONS corresponding to typical transverse speeds of around 300 km s^{-1}. Most pulsars are born in the explosions of massive stars found near the GALACTIC PLANE, and their high velocities carry them high above (and below) the plane.

There are two reasons for these high speeds. Firstly, many pulsars are born in binaries, whose disruption during a supernova explosion can leave the individual stars with large velocities. Secondly, the supernova birth-event itself is slightly asymmetric and kicks the neutron star to speeds of 1000 km s^{-1} or faster.

More than 40 pulsars are now known in the Galactic GLOBULAR CLUSTERS, 20 in 47 Tucanae alone. Most have short spin periods and were formed in binary stars – about half still are in binary systems. Globular clusters are old and now contain no supernovae. This implies that some neutron stars are not created by supernovae, and may be the result of the collapse of a white dwarf, pushed over the Chandrasekhar mass limit by accretion from a binary companion.

Because of their great rotational inertia, pulsars have extremely stable rotation, as precise as the best terrestrial atomic time standards. This makes pulsars very useful astrophysical probes. For example, a sharp radio pulse emitted by a pulsar is delayed and broadened during its propagation through the INTERSTELLAR MEDIUM, and pulsars have been used to map the ionized material and the magnetic field of the Galaxy.

Further reading: Lyne A G and Smith F G (1998) *Pulsar Astronomy* Cambridge University Press; Shapiro S L and Teukolsky S A (1983) *Black Holes, White Dwarfs and Neutron Stars* Wiley.

Newcomb, Simon (1835–1909)

Canadian-born astronomer who moved to the USA. He worked at the US Nautical Almanac Office (then in Cambridge, Massachusetts), studied at Harvard, and was appointed to the US Naval Observatory at Washington, DC, becoming director. In his own words, because of the 'confusion which pervaded the whole system of exact astronomy, arising from the diversity of the fundamental data made use of by the astronomers of foreign countries and various institutions in their work,' he started 'a systematic determination of the constants of astronomy from the best existing data, a reinvestigation of the theories of the celestial motions, and the preparation of tables, formulae, and precepts for the construction of ephemerides, and for other applications of the same results.' He thus used careful measurements of stellar and planetary positions to compute formulae for their future motions, with an analysis of the constants in the formulae (such as the masses of the Sun, Moon and planets), and the speed of light. This set in train a century of astronomical activity.

New General Catalogue of Nebulae and Clusters of Stars (NGC)

A basic reference list of star clusters, nebulae and galaxies compiled in 1888 by Danish astronomer John DREYER. His work is a compilation of lists made by the HERSCHEL FAMILY and others. Dreyer included 7840 celestial objects. He later extended the list by 5386 objects with his first and second INDEX CATALOGUES (IC), published in 1895 and 1908 respectively. With these supplements the NGC covers the entire sky, although many objects visible with modern instruments are not listed. A revised edition of the NGC and the two *Index Catalogues*, edited by Roger Sinnott of *Sky & Telescope* and updated to EPOCH 2000.0, was published as a single volume in 1988.

New Horizons

Proposed NASA mission to fly past Pluto and its moon Charon, and study Kuiper Belt objects. The mission has a possible launch date of 2006, with arrival in 2015. New Horizons was formerly the Pluto Kuiper Express, a mission cancelled for budgetary reasons.

Newton, Isaac (1642–1727)

Isaac Newton is known for discoveries in mathematics (calculus), optics (the white-light spectrum) and mechanics (laws of motion and gravitation). Some questions of priority are still disputed. The foundational character of Newton's chief works is absolutely clear: his *Principia* (first edition in 1687) and *Opticks* (1704).

Newton was born at Woolsthorpe near Grantham, Lincolnshire, England. Put to managing his parents' estate, he was a disaster, since he was interested only in bookish learning. So he was sent to Trinity College, Cambridge, in

July 1661. In 1669 Isaac Barrow (1630–77) saw to it that Newton became his successor as Lucasian Professor of Mathematics in Cambridge, where he remained until 1696 when he moved to London to become Warden of the Mint in late 1699 and work on alchemy; his scientifically most fruitful years were over.

Newton's most intense study of the new mathematics began in 1664, and lasted until November 1665, by which time he had made his principal discoveries, including the method of 'fluxions,' or calculus. Newton's work in optics began in early 1666 when he obtained a prism 'to try the celebrated Phaenomena of Colours.' Newton showed that white light was composed of individual colors of the spectrum. He designed and constructed a reflecting telescope (*see* NEWTONIAN TELESCOPE). It was completed in early 1669, and shown to the Royal Society in 1671, where it caused a sensation. Flattered, Newton agreed to provide an account of his theory and experiments. His essay 'New Theory about Light and Colors' appeared in the *Philosophical Transactions* of the Royal Society in February 1672 and proved controversial. By 1678 Newton had had enough of politics and dispute. Now immersed in alchemical experiments, he wanted no part in further controversy. Only after his death, he said, would his further writings be published.

Newton's study of mechanics also began in the mid-1660s. From René DESCARTES he accepted the law of INERTIA. In November 1679, Robert Hooke (1653–1703), as a secretary of the Royal Society, asked Newton what he thought of his idea of 'compounding the celestial motions of the planets of a direct motion by the tangent & an attractive motion towards the central body.' In perhaps August 1684 Edmond HALLEY visited Newton in Cambridge to ask what path a planet would follow if subject to an inverse-square force toward the Sun. By now the inverse-square variation of the solar force was accepted by Christopher Wren (1632–1723), Hooke and others as deducible from Christiaan HUYGENS' formula for centrifugal force (published in 1673). Newton reportedly answered 'An ellipse!' and promised to produce a demonstration. Given the proportionality of time to area, he derived a formula for the planet's instantaneous radial acceleration. Applying this to an elliptical orbit with center of force in a focus, he showed that the force would be inverse-square. 'Therefore the major planets revolve in ellipses having a focus in the centre of the Sun.' In the next two and a half years this small treatise expanded into the 510 pages of the *Principia*. Newton's procedure for fitting orbits to comets enabled Halley to predict the return of COMET HALLEY in 1759.

Newton failed to generate a following to test his theory. Wrongheadedly and fiercely, Newton and Gottfried Leibniz (1646–1716) wasted energy charging each other with plagiarism in the invention of the calculus. His science of gravitational orbits, known today under Pierre-Simon LAPLACE's name of CELESTIAL MECHANICS, led to new mathematics and new understanding.

Further reading: Westfall R S (1980) *Never at Rest* Cambridge University Press.

Newtonian telescope

A REFLECTING TELESCOPE in which the converging cone of light from a concave parabolic primary mirror is reflected to the side of the telescope tube by a small flat mirror set at 45° to the optical axis. The light is brought to a focus at the side of the tube, at the opposite end from the primary mirror, and this is where the eyepiece is located. Originally devised by Isaac NEWTON, this system is still widely used in small reflecting telescopes, since the observer's head does not block incoming light. The design is less practical when the telescope has a large focal length, say, larger than a person's height.

Newton's first reflector had an aperture of 1 in and a length of 6 in. As a reflector it did not suffer from chromatic aberration. Its mirrors were made from speculum metal (an alloy of various metals, predominantly copper and tin) which, even when freshly polished, reflected no more than 60% of the incident light. With two such mirrors, a tiny aperture, and a magnification of about 25 times, it produced very faint images. Furthermore, because the curve of its primary mirror was spherical rather than parabolic, it suffered from an optical defect called SPHERICAL ABERRATION.

Newton's laws of motion

Laws governing the motion of all bodies which were set out by Isaac NEWTON in 1687. They form the basis of Newtonian mechanics. The laws are as follows:

- *First law:* every body continues in a state of rest or uniform motion in a straight line unless acted upon by a force.
- *Second law:* if a body is acted upon by an external force, it accelerates, the acceleration being directly proportional to the force and inversely proportional to the mass of the body; the acceleration takes place in the direction of the force.
- *Third law:* to every action there is an equal and opposite reaction; in other words, if a force acts on one body an equal and opposite force must act on another body.
- *Law of gravitation:* every body in the universe attracts every other with a force proportional to the product of their masses and inversely proportional to the square of the distance between them.

The essence of some of these laws was appreciated by others prior to and contemporary with Newton, such as GALILEO GALILEI, Robert Hooke (1653–1703) and René DESCARTES. Newton, however, was the first to appreciate them fully and to formalize them.

Next Generation Space Telescope

See JAMES WEBB SPACE TELESCOPE

Nicolaus Copernicus Astronomical Center

The largest astronomical institution in Poland, located in Warsaw and founded in 1956. Its main research fields are theory and observations of binary stars; stellar structure and pulsations; stellar atmospheres and circumstellar matter; and cosmology and the large-scale structure of the universe.

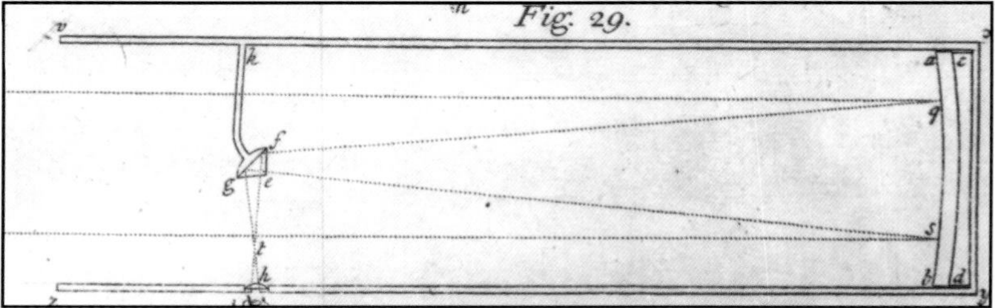

▼ Newtonian telescope *This schematic is taken from Isaac Newton's own Opticks of 1704.*

Nitrogen

The element that is the principal constituent of the Earth's atmosphere (making up, at ground level, 78% by volume). By number of atoms, nitrogen (N) is the fifth most abundant element in the universe (by mass, it is seventh).

Nobeyama Radio Observatory

Nobeyama Radio Observatory has telescopes at millimeter and submillimeter wavelengths. It was established in 1982 as an observatory of Tokyo Astronomical Observatory (National Astronomical Observatory, Japan since 1987), and operates the 45-m telescope, Nobeyama Millimeter Array, and Radioheliograph. High-resolution images of star-forming regions and molecular clouds have revealed many aspects of the first stages of stellar evolution, the chemical evolution of molecular clouds, and galaxy evolution. Activities of VLBI made it possible to realize the first space VLBI mission VSOP (also known as HALCA).

Node

Either of the points at which an orbit intersects a reference plane. On the celestial sphere, a node is either of the points at which the great circle representing the orbital plane intersects the great circle corresponding to the reference plane (usually the ecliptic or the celestial equator). In the context of planetary motion, the reference plane is the ecliptic. The point at which the orbiting body crosses the reference plane from south to north is called the ascending node, and the point where it crosses from north to south, the descending node. The line joining the two nodes, which is the line of intersection between the orbital plane and the reference plane, is called the line of nodes. In the case of the Moon, the gravitational influence of the Sun causes the line of nodes to rotate slowly around the Earth in a westerly direction, this motion being called the regression of the line of nodes.

Nomenclature of celestial objects

See INFORMATION HANDLING IN ASTRONOMY; STELLAR NOMENCLATURE

Non-gravitational force

A force which is not gravitational in origin that acts on a celestial body to alter its orbit. The term is often used in connection with COMETS, in which the out-gassing of volatile material through vents in a comet's surface can give rise to jets. Strong jets behave like rockets and alter the comet's orbit. Radiation pressure of sunlight and the pressure of the solar wind on cometary material are also non-gravitational forces. Such forces changed the orbit of COMET SWIFT–TUTTLE, which was calculated on its observation in 1862 to have a period of 120 years, but did not return until 1992 (*see* POYNTING–ROBERTSON EFFECT). A spacecraft that is controlled to nullify non-gravitational forces (as in the space interferometer LISA) is called 'drag-free.'

Non-radial pulsations

Non-radial pulsations of stars are complex, asymmetric pulsations in which part of the stellar surface moves outward (expands) while other parts of the stellar surface move inward (contract). Non-radial pulsations form sound waves moving horizontally as well as radially, traveling around the star. Pressure and gravity provide the restoring force for these displacements, known as p-modes and g-modes respectively. The g-modes involve significant movement of the gas in the deep interior of the star, whereas the p-modes have their greatest motion near the stellar surface. The use of these modes as probes of the interior and surface layers of the Sun and stars underlies HELIOSEISMOLOGY and stellar seismology.

Nordic Optical Telescope

The Nordic Optical Telescope (NOT) is a 2.56-m, (Super) RITCHEY–CHRÉTIEN, altitude-azimuth telescope with an actively controlled primary mirror sited at 2.4 km on the Roque de Los Muchachos, La Palma, Canary Islands, Spain. It was built by Denmark, Finland, Norway and Sweden, achieving first light in 1988. Iceland joined the association in 1997. Core instruments on the Cassegrain-only telescope are a high-resolution imager (HiRAC) and a faint-object spectrographic camera (ALFOSC), a photopolarimeter (TurPol), a high-resolution spectrograph (SOFIN) and an infrared spectrographic camera (NOTCam).

Norma *See* CONSTELLATION

North American Indian astronomy

As archaeoastronomer Anthony Aveni (Colgate University, New York State, USA) observed more than 20 years ago, North American Indian astronomy holds a particular fascination for North Americans, 'who now occupy the land once held by a race of people so long regarded as distant noble savages.' This interest has been sustained since about 1980, particularly in the case of those indigenous people who, in historic times (AD 1600–1900), occupied the very heartland of North America – the Plains Indians. This is a land of sun, wind and vast grasslands, stretching from north to south more than 3000 km from the Saskatchewan River in Canada almost to the Rio Grande in Mexico. The western and eastern boundaries are approximately those of the foothills of the Rocky Mountains and the Mississippi-Missouri valleys. In the northern part of the region (now Alberta, Saskatchewan and Montana) lived such tribes as the Blackfeet and Cree; to their south (present-day Wyoming, southern Montana and the Dakotas) were the Sioux, Cheyenne, Crow, and Mandan; the Southern Plains (present-day Texas, Nebraska and Oklahoma) were occupied by the Comanche, Kiowa, Wichita and Pawnee. Most of these tribes had migrated into the region after about 1650, bringing together peoples of great diversity of background and history – not least the way their ancestors incorporated astronomical knowledge into their culture.

There is considerable evidence suggesting that many earlier tribal groups, such as the Anasazi Indians (Ancestral Pueblo people, AD 600–1300), led largely sedentary lifestyles, and observed regular celestial events in order to time planting and harvesting, as do the Hopi Indians today, in northeast Arizona. The progression of the seasons was marked from observations of the position of sunrise or sunset along the features of the mountainous horizon. Given the short growing season on the Colorado Plateau (perhaps only 120 frost-free days in the year) relative to the growing time for native corn (perhaps 80–90 days), there is not much leeway in the planting time. Hunters and gatherers, however, including the Plains Indians, put more emphasis on use of the changing sky to order ritual and religious activities, with phenomena such as comets, solar and lunar eclipses, and meteor showers frequently incorporated in religious and ceremonial activities. Symbols of such phenomena also adorned various artifacts.

Annual renewal ceremonials were common throughout the Americas. Many were complex, lasting several days, even weeks. There is considerable evidence that astronomical phenomena played an important role in many of them. For example, the Mahicans (these are probably the Mohicans of author James Fenimore Cooper, 1789–1851; they were an Algonquian-speaking group who, in ancient times, occupied the Hudson River Valley of present-day Vermont) traditionally enacted a complex world-renewal ceremonial which commenced at signals from the sky. The changing sky patterns associated with the constellation Ursa Major were viewed as an annual celestial bear hunt. Four of the stars defined the body of the bear while three stars that form the handle of the Dipper represented waiting hunters, the star Alcor being their dog. This 'cosmic bear' was observed to revolve around the North Star and then, in spring, seen to leave his den – the Corona Borealis. It was then followed by seven stars representing hunters, the original three being joined by Arcturus and stars from Boötes. Throughout the summer this bear was trailed and he was finally slain in the fall. The seasonal reddening of the forest foliage was attributed to this sky hunt, the red tint being the blood of the slain bear, and the mantle of early winter snow being some of the bear's grease, dried out by the star hunters.

The ceremonials were held at night within a special structure – the Big House, Xwate'k'an, which itself was replete in sky symbolism. The constellation Ursa Major was represented on the floor of the house. The furnishings and positions of officials corresponded to the positions of the stars in the constellation, while the movements and acts of the performers paralleled those of the stars. In turn, the Xwate'k'an was said to represent the universe. This renewal ceremonial commenced again the following spring when the bear once more emerged from his celestial den.

Further west, in present-day Nebraska, the various Pawnee bands of the branch of the tribe known as the Skidi assembled for annual ceremonials, arranging themselves according to the place of their stars in the sky. The whole 'Spring Awakening' could not be undertaken without reference to the position of the stars to guide the time relating to the commencement of the appropriate ceremonials. The earth lodge itself acted as a type of astronomical observatory, the priests viewing the sky through both the smoke hole and the oriented entranceway.

Astronomical and cyclic patterns were documented and understood by the tribal intellectuals and woven into mythology, religion and ceremonials. Thus, Morning Star, Opirikus – believed to be Mars – was viewed as the leader of men who traveled with his brother, the Sun, to the land of the western stars in an attempt to overcome the power of the Moon. Credited with helping the creation of the universe and the fathering of the human race, Morning Star utilizes the power of sacred bundles (housed within the ceremonial lodges) to overcome the obstacles that Moon puts in his way and is thus able to enter the woman-star village. This aspect of Pawnee mythology was clearly based on accurate and detailed observations of the sky.

Such mythology and religious concepts, however, led to one sinister ritual. Because of his favors to mankind, and particularly to the Pawnee themselves, Opirikus demanded human sacrifice – the Morning Star Ceremonial. Tribal priests stated that the sacrifice not only ensured continued abundant crops and game but also, perhaps more than anything, was an acknowledgment of a celestial heritage and essential to the well-being of the tribe – although not all agreed.

▲ ***North America Nebula*** *NGC 7000 is a large emission nebula in the constellation Cygnus, which in outline resembles North America, delineated by surrounding dark nebulae. Its main source of illumination is believed to be the hot blue star HR 8023.*

Further reading: Williamson R A (1984) *Living the Sky: the Cosmos of the American Indian* Houghton Mifflin.

North America Nebula (NGC 7000)

A large emission nebula in the constellation Cygnus, which in outline bears a strong resemblance to the North American continent, delineated by surrounding dark nebulae, including one known appropriately as the Gulf of Mexico. NGC 7000's main source of illumination is believed to be the hot blue star HR 8023. The adjacent PELICAN NEBULA is part of the same nebulosity.

Nova

A nova (plural 'novas' or 'novae'), from the Latin 'nova stella' (new star), is a star that brightens suddenly several hundred- to a million-fold, remains bright for a few days to several months, then returns to its former, low LUMINOSITY. The word 'new' refers to the sudden appearance of a bright star where none was readily visible before. In fact, a nova is an explosion on an existing old star, and the eruption is recurrent. For most novae, the time elapsed between outbursts is thousands to tens of thousands of years. If only one outburst has been recorded, a nova is referred to as a 'classical nova.' By contrast some novae erupt at intervals of tens of years and are termed 'recurrent novae.' *See* CATACLYSMIC BINARY STAR.

Nova outbursts are explosions and novae eject mass. The nova forms a nova shell, which slowly disperses into the

Discovering novae

▶ Light curve *for a typical nova. These stars are often only visible for a few days or weeks; their visual magnitude sharply peaks, then quickly fades over time.*

Novae can rapidly rise from minimum to maximum brightness, even, on occasions, within 24 hours. In some cases, they become bright naked-eye objects, and disturb the shape of their constellation, especially if they are brighter than magnitude 3. This was the case of Nova V1500 Cygni, which reached magnitude 2 after a steep rise to maximum in 1975. Such cases are often discovered by amateur astronomers rather than professionals. TYCHO'S SUPERNOVA was discovered in 1572, not by Tycho BRAHE, but by peasants who pointed out to him the bright star disturbing the readily recognizable shape of the constellation of Cassiopeia. Dozens of amateur astronomers independently discovered Nova Cygni 1975. The reasons are that amateurs spend more time under the night sky than professionals, who are stuck in a telescope control room, and that many amateurs know the constellations better than many professionals and can spot unfamiliar stars.

Binoculars (say 8 × 30 or 10 × 50 size) bring many more stars into view and raise the chances of discovering a nova. Comets look like comets (or nebulae) but novae do not have any flag on them. Nothing says, 'I am a nova' – a nova looks like any other star, and to know one when you see it you have to know the star patterns and recognize the intruder. English amateur George Alcock (1912–2000) was legendary for his memory of every star that he could see in his binoculars. He discovered five novae. On December 1, 1999 Alfredo Pereira, an amateur nova hunter in Cabo da Roca, Portugal, discovered with binoculars a nova of magnitude 6 (Nova Aquilae 1999 No. 2) near to Delta Aquilae. Other naked-eye discoveries of novae have been made in modern times by amateurs such as Bernard Dawson (Nova Puppis 1942), Manfred Durkëfalden (Nova Cygni 1975), Warren Morrison (Nova Cygni 1978), Reverend Beckmann (Nova Cygni 1978, Nova Aquilae 1982, and Nova Vulpeculae 1987) and Gary Nowak (Nova Aquilae 1999 No. 2).

In 1992, Peter Collins of Boulder, Colorado, discovered NOVA V1974 CYGNI. Collins, America's premier binocular nova finder, has four novae to his credit (Nova Vulpeculae 1984 No. 2, Nova Cygni 1992, Nova Cygni 1978 and Nova Vulpeculae 1987). He has memorized the Milky Way area (where most novae are found) by dividing the star patterns into ASTERISMS through which he tracks, while talking himself through the names he has given them – 'down the broom handle to the mini-Dipper with the extra star in the handle,' that sort of thing. Even so, most people will need to check the patterns with a good reference star atlas such as *Norton's Star Atlas*, Hans Vehrenberg's *Handbook of the Constellations*, or *Sky Atlas 2000* by Wil Tirion.

A systematic search maximizes the chances of discovering novae. In the era of the dynasties, Chinese astrologers systematically scanned the skies for celestial portents. They discovered and recorded many novae as 'guest stars,' stars temporarily in the sky. In the modern era, the Nova Search Committee of the AAVSO was established in the early 1930s to discover novae in the Milky Way. The Milky Way region was divided into more than 100 areas, 10° in declination by about 1 h in right ascension. An observer who is interested in searching for novae within the program is assigned specific areas (some in the winter sky, some in the summer, and distributed in order to give the most complete coverage). But of course the observer is free as well to search other areas, and make a 'dome search' of the whole visible sky for naked-eye novae. The UK Nova Patrol, launched by John Hosty and Guy Hurst in 1976, works along the same principles. By January 7, 1977, its first nova was detected when Hosty, searching very close to the horizon in Sagitta from his observing base in Huddersfield, England, found what is now cataloged as HS Sagittae.

Photographic searches

For photographic searches, it is possible to patrol effectively with a camera with a 50-mm lens at a focal ratio of typically about *f*/2. A 30 second exposure on a film with a speed of ISO 400 record stars to magnitude 8, more if the Moon is absent from the sky, and light pollution is not too severe (*see* DARK SKIES AND GOOD OUTDOOR LIGHTING). No sidereal drive is necessary and the equipment can be mounted on a standard photographic tripod. The resulting star trails are minimal. Some patrollers take the equipment to a dark country site and run off an entire film.

A special, crucial requirement for photography is always to take two consecutive exposures of the same area so that film flaws can be eliminated, as many of these flaws look remarkably like genuine stars. The best approach is to compare the latest pictures with earlier masters obtained with similar equipment. BLINK COMPARATORS have been constructed by amateurs to make the task easier. They work by viewing the two pictures through a microscope and switching with a mirror from one to the other so that the nova 'blinks' on and off. Potential discoveries can be checked on the AAVSO *Variable Star Atlas* and the *Millennium Star Atlas* based on the over 1 million stars of the TYCHO STAR CATALOGS. However, many stars brighter than the quoted limiting magnitude of star atlases are missing from charts. Photographs, taken earlier, provide a better basis for checking queries. Another trap is that a passing asteroid or a variable star, usually faint but caught near maximum, can look like a nova. Variable stars can be excluded, if previously known, by checking reference works (like the AAVSO charts and the *General Catalog of Variable Stars*). Asteroids move, so

▶ 10 x 50 monocular *The instrument used by John Hosty when he discovered HS Sagittae in 1977.*

◀ ***Nova Cygni 1975*** *before (arrowed, left) and after its outburst.*

they can be eliminated by a second picture taken a half hour afterwards.

It is vital that each exposure is actually checked. Too often, the prospect of setting up temporary darkrooms and the use of the chemicals needed for film development have delayed discoveries. A sky patrol started by *The Astronomer* in 1973 recorded Nova V400 Persei two months before its discovery by a professional, but it was not spotted as the exposures were not checked. Such pre-discovery data is particularly valuable in the analysis of a light curve, but this episode was also a spur for regular checking.

Several novae have been discovered by photography by members of organized groups such as AAVSO and the UK Nova Patrol. Bill Liller in Chile has found many novae, especially in the southern sky, and similar results have been achieved in the USA by the late Ben Mayer, in Australia by Rob McNaught, and in Japan by Katsumi Haseda (who photographically discovered four novae in a single year in 2002). Despite regular battles in England with long cloudy spells, Mike Collins has been successful using a 135-mm telephoto lens.

The amateur astronomer networks not only organize the observing program, they serve another function: a member who finds a potential nova can get confidence by calling on other members to confirm it before taking the step to contact the IAU's Central Bureau at Harvard's Center for Astrophysics to announce the discovery and trigger professional follow-up.

CCD searches

The future of amateur nova discovery probably belongs to the use of a CCD with a simple lens system to record as wide a field as possible. Five-second exposures through a 50-mm lens onto an unfiltered CCD record stars as faint as magnitude 12, again with a stationary tripod. The great advantage of a CCD image is that it is immediately available for checking. CCD images can also be 'blinked' on a computer monitor screen with software programs, to compare the latest image with an earlier master shot.

Once discovered, a nova can be followed in decline, to study how it behaves as it returns to the pre-outburst stage. Many fluctuate in brightness after maximum, but remain within reach of amateur telescopes. One example is Q Cygni, which was found back in 1876. Observations can be entered into a spreadsheet, which automatically generates a raw light curve. No two novae follow exactly the same pattern.

The chances of discovering a nova in outburst are increased by examining stars which have had outbursts before. This is particularly so with DWARF NOVAE, which may outburst several times per year. In September 2002, amateur astronomers worldwide teamed up with NASA's CHANDRA X-RAY OBSERVATORY and the EXTREME ULTRAVIOLET EXPLORER (EUVE) to observe SS Cygni, in order to alert satellite controllers to an outburst. Janet Mattei, Director of AAVSO, organized the sky watch. On Saturday September 9, observing from his backyard in Espoo, Finland, Tino Kinnunen saw SS Cygni brighten through the magnitude-11 barrier that signifies the start of an outburst of this star. An hour and a half later and a continent away, Tom Burrows caught SS Cygni at magnitude 10.4 with his 6-in telescope at his home in Petaluma, California. Night moved onto Hawaii, where Mike Linnolt checked the star from the roof of his apartment in Honolulu with his 8-in telescope before going to bed. Linnolt caught SS Cygni at magnitude 9.7. On Sunday morning, Janet Mattei analyzed the data from the previous night's observations, concluded that the outburst was really on and triggered the observing program by Chandra and the EUVE.

Guy Hurst was formerly President of the BAA, and is a retired bank manager. His special astronomical interests are novae and star clusters.

▼ ***HS Sagittae*** *Photographic confirmation of the UK Nova Patrol's first discovery on Jan 9, 1977 at 17.54 UT. Peter Birtwhistle used a 300-mm telephoto lens to capture the nova discovered by John Hosty.*

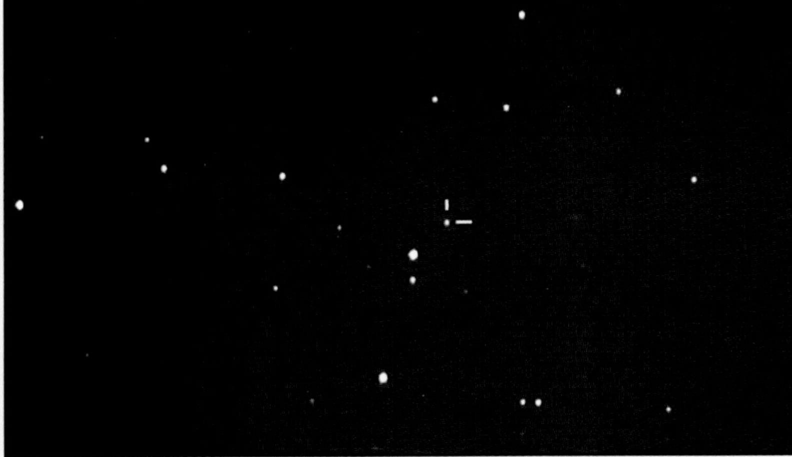

INTERSTELLAR MEDIUM. 'Nova' refers to the star that explodes but also to the explosion. Thus terms such as pre-nova, post-nova, and nova progenitor are quite common.

There are two populations of novae: disk novae, observed in the solar neighborhood, which are brighter and evenly distributed, and bulge novae, concentrated toward the Galactic center with a larger scale height, which are fainter and have a generally slower outburst development. It is estimated that the Galactic nova rate is about 40 per year, of which we actually see around 10%.

Early studies of novae revealed that they varied considerably in the rate of evolution after maximum. The time of decline by three magnitudes from maximum light characterizes the pace: from very fast (310 days) to slow (3100 days). The LIGHT CURVE has several stages:

- In the pre-nova stage the star varies a little and has the SPECTRUM of a hot star.
- The light curve rises steeply to maximum. The spectrum of the nova indicates the approaching side of an explosion.
- The optical maximum is defined by the peak in the optical light curve. The emission lines of the spectrum have complex profiles suggesting that material has been ejected in blobs, cloudlets or rings at speeds between a few hundred and a thousand kilometers per second.
- The transition stage, during which the brightness declines, is characterized by the spectrum of a nebula. The decline is attributed to the formation of dust in the nebula; the light loss is due to absorption.
- During the nebular stage the nova has become fainter by several magnitudes and a thin nebula has formed. Its mass is small, approximately 10^{-5}–10^{-4} solar masses.
- In the post-nova or quiescence stage, the nova returns to its pre-outburst appearance.

By the early 1960s enough observational evidence had accumulated, mainly through the work of Robert Kraft, indicating that novae were invariably members of close binary systems. The nova companion was found to be a low-mass MAIN SEQUENCE star. Observations of novae after (and in a few cases before) eruption showed them to be hot, compact stars. Mass estimates, albeit scarce and uncertain, suggested that the erupting stars were WHITE DWARFS. This led to the hypothesis that the red dwarf companion (*see* RED DWARF AND FLARE STAR) is extended enough (that is, it fills its ROCHE LOBE) to allow mass transfer to the hotter star (through the inner LAGRANGIAN POINT of the binary system's gravitational field). Indeed, in some cases, a rapidly rotating region, identified as an ACCRETION disk, could be detected around the hot star.

Nova outbursts happen on the surface of the white dwarf when hydrogen-rich material from its red-dwarf companion accretes on the white dwarf's surface. The material gradually accumulates and becomes compressed, and the temperature in the bottom layers rises. When it reaches 20 million K, hydrogen is ignited in a thin shell by the carbon-nitrogen-oxygen (CNO) cycle of nuclear reactions (*see* FUSION). The energy that is released raises the temperature exponentially. The nuclear reaction runs away; the shell expands rapidly, and is driven out by radiation pressure. When most of the envelope has been ejected, mass loss comes to an end. Hydrogen burning declines and the white dwarf returns to its pre-outburst state. The decline takes roughly one to several years and the white dwarf remains almost unaffected by the outburst that has taken place. Accretion resumes toward the next outburst and a new nova cycle begins.

The large variation in the features of novae is attributed by the theory to differences in the three basic parameters that characterize the binary system: the mass of the accreting white dwarf, its internal temperature (which is a measure of the system's age), and the mass-transfer rate, which is determined by the mass of the companion star and by the binary separation (orbital period).

Further reading: Payne-Gaposhkin C (1957) *The Galactic Novae* North-Holland; Warner B (1995) *Cataclysmic Variable Stars* Cambridge University Press.

Nova V1974 Cygni

On the night of February 18/19, 1992, through a moonlit sky and polluted conditions due to a recent volcanic eruption, Peter Collins of Boulder, Colorado, USA, was searching for NOVAE when he spotted with binoculars a suspect object shining at magnitude 7.2 at a location about 6° north of the bright star Deneb. His atlas listed a star near that position, so he decided that perhaps it was nothing new, but would check additional references later to be sure. Before making the comparison, Collins observed the field again and found that within six hours the suspect had brightened by an additional magnitude! After checking various catalogs to no avail, Collins contacted AAVSO headquarters and various other authorities so that the astronomical community could be alerted about his finding, which was designated Nova Cygni or V1974 Cygni. *See* VARIABLE STAR.

The nova rapidly climbed to a peak brightness of about fourth magnitude. Because of its brightness, it was studied by more telescopes at more wavelengths than any before or since. The first spectra were obtained in the ultraviolet with the INTERNATIONAL ULTRAVIOLET EXPLORER (IUE) satellite. Analysis of these spectra showed that the nova had been caught in the 'fireball' phase when the hot, dense material ejected by the explosion was in its first cooling phase. It is the gas cooling that forces the radiation from the nova to emerge at optical and infrared wavelengths then become bright in the optical. Luckily, V1974 Cyg was caught before maximum brightness in the optical.

By the time V1974 Cyg reached maximum brightness in the optical, its temperature had declined to below 15 000 K and it was emitting most of its energy in the optical and infrared, not the ultraviolet. It declined further in brightness as the ejected gases expanded and the outermost layers gradually became transparent. The increasing transparency allows us to see deeper into the gas, where the material is hotter, and thus the decline in optical light goes hand in hand with an increase in ultraviolet light. Evidence that the heat from the underlying layers was getting through the ejected material came from observations in April 1992 with ROSAT, the X-ray satellite. The first observations showed that the nova was faint in X-rays, but in a series of observations performed over the next 12 months it steadily brightened. It was later realized that ROSAT had detected the underlying hot object which was gradually appearing through the expanding shell of gas ejected by the explosion.

V1974 Cyg was also observed a number of times with various instruments on the HST. Analyses of the first HST observations showed that the nova had not ejected material in a smooth uniform sphere; rather the expanding gas had formed many dense knots inside a less-dense, faster-expanding, non-spherical hollow shell of gas. As the gaseous debris from the explosion expanded, it thinned and continued to clear so that it became possible to see completely through

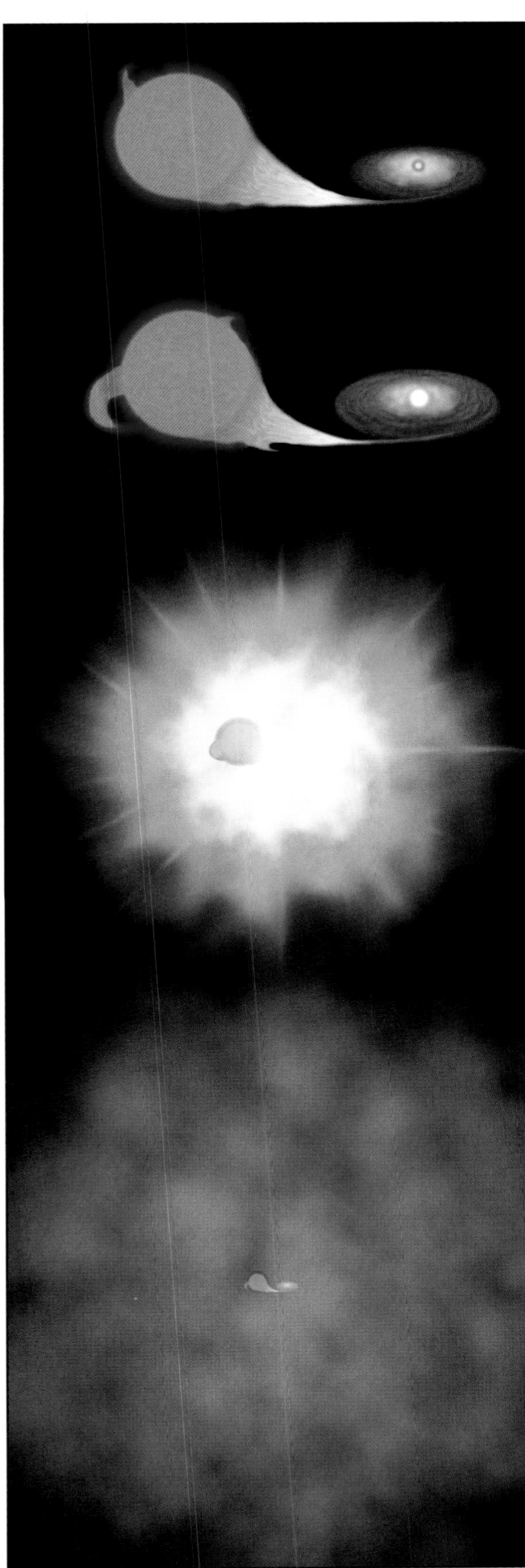

◀ ***Nova*** *A surge, or long build-up, of material flowing from a red-giant star on to a white-dwarf companion detonates to be seen as a nova. As the nebula expands, the red giant and white dwarf remain visible.*

it to the underlying object on which the explosion took place. It then became possible to study the structure of this object and determine the cause of the explosion.

Studies of V1974 Cyg after the explosion showed that it has the same structure as all other novae and the entire class of stars designated 'cataclysmic variables.' It is a BINARY STAR with one component a WHITE DWARF and the other component a larger, cooler star. The two stars had an orbital period of two hours.

This nova has provided the most important and complete datasets ever obtained for a nova in outburst, and it is sufficiently bright that it is still being observed by astronomers using observatories in space and on the Earth.

Nucleosynthesis

Nucleosynthesis is the science related to all astrophysical processes that are responsible for the abundances of the elements and their isotopes in the universe. The astrophysical sites are the BIG BANG and stellar objects, either during their evolution and wind ejection or during explosions like NOVAE and SUPERNOVAE, or possibly other events, where binary stellar systems are involved. *See also* ELEMENTS, FORMATION OF; STAR.

Nuclide

The nuclei of atoms of a given element all contain the same number of protons, but the number of neutrons may vary. Each combination of protons and neutrons is called a nuclide.

Nuffield Radio Astronomy Laboratories

See JODRELL BANK OBSERVATORY

Nutation

Short-period oscillations in the long-term motion of the pole of rotation of a body that is being acted on by external perturbing forces. The gravitational attractions of the Sun and Moon on the Earth's equatorial bulge cause the direction of its axis, and hence the positions of the celestial poles, to revolve slowly around the pole of the ECLIPTIC (the points on the celestial sphere 90° away from the ecliptic), this phenomenon being called PRECESSION. Because the orbital plane of the Moon is inclined to the plane of the ecliptic by about 5°, and revolves round the Earth in a period of 18.6 years, an additional small oscillation in the position of the celestial poles is superimposed on precession. This causes the Earth's poles (and hence the celestial poles) to tilt, or 'nod,' periodically toward and away from the poles of the ecliptic, the term 'nutation' deriving from this nodding motion, of size about 9 arcsec, and period 18.6 years.

◄◄ *Cerro Tololo*
A complex of astronomical telescopes and instruments situated 80 km east of La Serena in Chile at an altitude of 2200 m.

OB associations

Large collections of gas and young stars (of spectral type O and B) located primarily in the disk of SPIRAL GALAXIES. The number of stars varies from fewer than 10 up to tens of thousands. The latter case is sometimes referred to as an OB superassociation (NGC 206 in the ANDROMEDA GALAXY, for example).

OB associations are among the least dense star clusters. In some cases, the overpopulation of O- and B-type stars is the only factor that allows the association to be distinguished from the remaining stars in its vicinity, and because of this lack of structure they fall within the category of OPEN CLUSTERS. OB associations, like other open clusters, are highly concentrated toward the GALACTIC PLANE, meaning that the more distant ones in our Galaxy may be obscured by dust. As a result, studying the distribution of OB associations in nearby galaxies allows us to learn more about their probable distribution in the Milky Way. OB associations act as tracers for both current and recent massive star formation. This is important for understanding the environments in which massive stars are most likely to form.

Oberon

A mid-sized satellite of Uranus, discovered by William Herschel in 1787 (*see* HERSCHEL FAMILY). Its diameter is 1520 km and it orbits at a distance of 191 000 km. The images of Oberon obtained by VOYAGER 2 were not clear, but showed an extensively cratered surface, some CRATERS surrounded by bright EJECTA or rays, and some with dark floors, possibly a result of the eruption of water ice mixed with organic material. The preponderance of craters suggests that this is an ancient surface, largely undisturbed by geological activity. One hemisphere is redder than the other, perhaps as a result of dust accumulated from the two small outer retrograde satellites of Uranus.

Objective grating or prism

A diffraction grating or narrow-angled prism placed in front of the aperture of a telescope to produce a low-resolution SPECTRUM of every object in the field. This allows the principal features of the spectra and the spectral types of many stars to be studied easily and rapidly.

Objective lens

The principal lens of a REFRACTING TELESCOPE, which collects light from a distant object and forms an image of the object at its FOCUS, or focal plane. The objective lens, or just 'objective,' is also known as the object glass (OG) because, in a telescope, it is the lens nearest to the object that is being imaged. A simple objective lens suffers from a number of optical defects, or aberrations, notably SPHERICAL ABERRATION and CHROMATIC ABERRATION. Both of these aberrations are considerably reduced by the use of an achromatic lens, a compound lens consisting of two components, each made of different kinds of glass with different optical properties. The combination of two lenses is called an achromatic doublet. Adding a third lens further reduces the spread of focal positions. A three-element objective lens is called an apochromat.

As the residual chromatic aberration decreases with increasing focal ratio, achromatic doublets usually have focal lengths that are at least 10 times their apertures (focal ratio *f*/10 or more). Objective lenses with large focal ratios have small fields of view. Where a wider field of view and shorter focal ratio are required, an apochromat is used.

Oblateness

The degree of flattening of an oblate spheroid, the solid body obtained by rotating an ellipse about its minor axis. The Earth is approximately an oblate spheroid, as the equatorial diameter is slightly greater than the polar diameter. Jupiter and Saturn are considerably more oblate than the Earth. If a hypothetical planet had a polar radius r_p and an equatorial radius r_e, the oblateness would be $(r_e - r_p)/r_e$. By measuring the flattening at the poles of a planet as compared with the speed of rotation, it is possible to infer the density distribution inside it. If two planets had the same mass and bulk density, the planet with most of its mass concentrated close to the center would be more flattened by rotation. For example, from the relatively small oblateness of Uranus combined with the planet's relatively rapid rotation, it would appear that its constituents, ice and gas, are well mixed and a rocky core is small or non-existent. *See also* ACHERNAR.

Obliquity of the ecliptic

The angle between the planes of the ECLIPTIC and the equator. On the CELESTIAL SPHERE, the angle at which the ecliptic intersects the CELESTIAL EQUATOR. The year-2000 value of the obliquity (symbol ε) was 23° 26' 21". Its value varies by ±9 arcsec over a period of 18.6 years because of NUTATION. Over a much longer period (about 40 000 years) the perturbing influence of the planets causes its value to vary between 21° 55' and 28° 18'; currently ε is decreasing at an average rate of about 0.5 arcsec per year.

Observatory

An astronomical observatory is a building or institution dedicated to the observation of celestial objects to understand them, or to reckon time and maintain the calendar.

The first observatory was built during the reign of the caliph al-Ma'mn (AD 813–833), as part of a scientific academy in Baghdad. The two most successful Islamic observatories were those at Maragha (in Azerbaijan), built under the direction of Nasir al-Din al-Tusi (1201–1274), and at Samarkand by ULUGH BEG. In seventeenth-century China, Louis Lecomte recorded vigilant activity by astronomers at the Imperial Observatory in Beijing, where, as for centuries beforehand, 'five mathematicians spend every night on the tower in watching what passes overhead…nothing of what happens in the four corners of the world may escape their diligent observation.' From 1673 to 1676 the Beijing Observatory was re-equipped with a new set of Tychonic instruments, which are still in place today (*see* Tycho BRAHE).

In India the most significant observatory was built by Jai Singh (1686–1743), a Hindu prince. He had constructed large instruments of masonry at Delhi, Jaipur, Ujjain, Benares, and Mathura.

The first European observatory worthy of the name was built by Bernard Walther (1430–1504), a wealthy private citizen of Nuremberg, Germany, a pupil and a patron of the astronomer Johann Müller Regiomontanus (1436–76). An innovation at Walther's observatory was determining the times of observations by mechanical clocks instead of astronomical sightings.

The most significant observatory prior to the invention of the telescope was Tycho Brahe's, situated on the island of Hven, between Copenhagen and Elsinore in Denmark. There Tycho built Uraniborg, the 'castle of the heavens.' Uraniborg was the first scientific research institute in Renaissance Europe.

National observatories

Uraniborg was almost a national observatory, but the Paris Observatory (established 1667) certainly was. The observatory, along with those at Greenwich (1675), Berlin (1701), and St Petersburg (1725), was dedicated to practical matters of national importance, especially the determination of longitude at sea (*see* NAVIGATION).

The nineteenth century was characterized by offshoots from existing national observatories, such as Royal Observatory Cape, South Africa, 1820; by newer observatories such as the United States Naval Observatory (1839), Pulkovo in Russia (1839), the Chilean National Observatory (1852), the Argentine National Observatory (1870), the USA's Smithsonian Astrophysical Observatory (1891), and Canada's Dominion Observatory (1903); and by the rise of astrophysical observatories, such as Potsdam in Prussia (1874), and Canada's DOMINION ASTROPHYSICAL OBSERVATORY (1918).

The present era is characterized by national or international consortia and large budgets. Significant observatories include the NRAO (USA, 1956), KITT PEAK NATIONAL OBSERVATORY (USA, 1957), National Radio Astronomy Observatory (Australia, 1959), CERRO TOLOLO INTER-AMERICAN OBSERVATORY (Chile, 1963), ESO (Chile, 1964), ANGLO-AUSTRALIAN OBSERVATORY (Australia, 1967), the Kuiper Airborne Observatory (USA, 1975–95) and the SPACE TELESCOPE SCIENCE INSTITUTE (USA, 1981).

Private observatories

The most accomplished private astronomer of all time was the German-English astronomer William Herschel (*see* HERSCHEL FAMILY). He pioneered the production of large reflecting telescopes. His discovery, during a systematic all-sky survey, of the planet Uranus in 1781 won him a royal pension that enabled him to establish his private observatory to carry out deeper surveys. Herschel's son John took a large reflector to South Africa where from 1834 to 1838 he completed his father's work of cataloging DOUBLE STARS and nebulae (*see* NEBULAE AND INTERSTELLAR MATTER).

The eighteenth-century German astronomer Johann Hieronymus Schröter (1745–1816) quit his administrative post in Hanover to devote himself to astronomy. He built in Lilienthal what was at the time the largest observatory in Europe. The title of largest telescope in the world was then claimed by Ireland after William Parsons, the THIRD EARL OF ROSSE, built at Parsonstown in 1845 the 'Leviathan of Parsonstown' with a mirror 6 ft in diameter.

In the USA, George HALE established a private solar observatory in his parents' backyard in Kenwood, then a suburb of Chicago, later part of the University of Chicago. Hale became the greatest observatory entrepreneur of all time, setting up the YERKES OBSERVATORY (1897) and the MOUNT WILSON OBSERVATORY (1904). He was also the driving force behind the establishment of PALOMAR OBSERVATORY (1948).

Remote observatories

Although astronomers always recognized the importance of good skies for their observations, they had to accept the weather where they were, in the days when transportation was not easy. But urban expansion and increasing pollution from smoke and light put observatories near cities at a disadvantage. The VATICAN OBSERVATORY, founded by Pope Leo XIII in 1891, was originally located in the Vatican gardens. As light pollution increased, it was moved out of Rome to the Papal summer residence at Castel Gandolfo. By the 1970s, light pollution from the nearby city had become so bad that it was decided to relocate the observatory to Arizona. A similar process occurred with the Royal Observatory at Greenwich, relocated in the 1950s to the Sussex countryside at Herstmonceux, and split in 1990 between La Palma in the Canary Islands and Cambridge.

In a famous passage in *Opticks* (1704), Isaac NEWTON wrote that telescopes 'cannot be so formed as to take away that confusion of rays which arises from the tremors of the atmosphere. The only remedy is a most serene and quiet air, such as may perhaps be found on the tops of the highest mountains above the grosser clouds.' Modern technology in the form of ADAPTIVE OPTICS has, in recent times, proved Newton wrong in the first statement but his second statement was proved correct in 1856, when Charles SMYTH mounted an expedition to the mountain of El Teide on the island of Tenerife in the Canary Islands. He found that he could see stars that were four magnitudes fainter and distinguish closer double stars than in Scotland.

The first large observatory built on a mountain summit according to Newton's conjecture and Smyth's experiments was LICK OBSERVATORY, opened on Mount Hamilton in northern California in 1888 at an altitude of 1.28 km. Over the following decades, many leading mountain observatories sprang up in the dry, clear air of the American west, including the world's largest (at the time) reflecting telescopes. The 100-in Hooker Reflector was built in 1917 at Mount Wilson Observatory near Pasadena and the 200-in Hale Reflector in 1948 at Palomar Observatory near San Diego.

What observatories need

All observatories need road access to a level site. The astronomers need water and telecommunication facilities. The equipment needs reliable power. However, neither telecommunications nor power should degrade the electromagnetic environment of the observatory – small induced currents can interfere with electronic measurements.

Special requirements for an optical observatory are good atmospheric transparency and SEEING characteristics. For an infrared or millimeter-wave observatory, transparency is linked to a lack of atmospheric water-vapor content. Remoteness from population centers minimizes artificial sky background, industrial smoke, dust and heat plumes, and background radio noise. The less seasonal the cloud cover, the more even the access to the whole sky. If winters are severe, ice or snow may also affect building costs and access. A period of systematic site-testing is needed in order to establish the weather characteristics.

Low latitudes are generally favored. The natural sky glow decreases away from the polar regions, the nights are of more uniform length near the equator, and more of the sky passes above the horizon. The southern-hemisphere sky contains unique astronomical phenomena, such as the GALACTIC CENTER and the MAGELLANIC CLOUDS, as well as half the astronomical sources observed by satellite telescopes.

Today, most of the world's largest, most powerful optical-infrared telescopes are found in a handful of high-altitude sites. There are concentrations in the western half of the continental USA, in Hawaii, in the arid Andean mountains of Chile, and in the Canary Islands. The remoteness, effects of altitude on people and machines, earth tremors, bush fires, environmental and ecological considerations, and concerns of local people are the difficulties to be overcome.

In contrast with optical observations, radio astronomy has a relatively short history. However, it follows a similar

◀ ***Kitt Peak Observatory*** *This American observatory supports one of the most diverse collection of astronomical telescopes on Earth.*

pattern. From the earliest experiments, radio interference from terrestrial sources has been an obstacle to studies of the universe. Modern radio-astronomy observatories, such as the Goldstone and OWENS VALLEY RADIO OBSERVATORIES in California, and the Australia Telescope, are built in radio-quiet areas, far from large population centers and often shielded by hills, sometimes with the benefit of local legislation that prevents interference (such as the 150 km wide National Radio Quiet Zone in West Virginia).

The future

Three major threats to future research have been recognized by the astronomical community:

- Growing levels of radio pollution and radio interference from telecommunications satellites interfere with radio astronomy.
- Space debris threatens scientific satellites and interferes with ground-based observations as satellites or space junk leave streaks across sky images.
- Outdoor lighting, increased urbanization, and projects to launch highly luminous objects into space, such as the unsuccessful 1999 Znamya space-mirror project, also threaten optical astronomy.

The last excellent undeveloped astronomical site on Earth is Antarctica. Stratospheric air is pulled down onto the south pole by convection upward at latitudes closer to the equator, and the cold, dry and stable conditions on the high-level Antarctic Plateau allow small telescopes at the south pole to outperform much larger telescopes at temperate sites. But there are difficult operational obstacles.

Even more difficult is access to the Moon, but there is no cloud cover, atmospheric turbulence or atmospheric absorption, and in a one-sixth gravity environment, much larger, lighter, deformation-free mirrors could be installed. The thermal cycle and dusty environment are disadvantages. ESA concluded that the advantages of an optical-infrared or millimeter-wave observatory on the Moon were outweighed by the disadvantages, so that free-flying observatories like the HST and the JAMES WEBB SPACE TELESCOPE were preferred. The lunar far side would, however, be uniquely favorable for radio astronomy. Shielded by the body of the Moon from terrestrial signals, the telescopes would have an interference-free view. This might be where artificial signals from other civilizations could most readily be identified.

Further reading: Krisciunas K (1988) *Astronomical Centers of the World* Cambridge University Press; Müller P (1992) *Sternwarten in Bildern: Architektur und Geschichte der Sternwarten von den Anfängen bis ca. 1950* Springer.

Observing program

Whether an astronomer uses an optical telescope, a radio telescope or a space satellite, the stages in an observing program are the same.

The first stage is to decide what to do. Astronomers get ideas about what to observe by listening to lectures, reading books and articles, and talking to each other. They try to decide on observations that will check astronomical theories, or confirm observations that other people have made, perhaps by complementing observations made at one wavelength by making observations at another. Even if this thinking process does not lead to an observing program, it produces scientific ideas.

Usually, the program is a series of observations of a series of astronomical objects. The candidate list of objects comes from a catalog of some sort. Ideally there will be many, so that the selection of key examples is easier. Sometimes there is no catalog, and there has to be a preliminary observing program in which a catalog is compiled – a 'survey.' William Herschel's systematic optical survey for double stars and nebulae is a historical example (*see* HERSCHEL FAMILY); a more modern one is the IRAS infrared satellite survey.

Feasibility

Of course, an observing program must be feasible:

- Are the objects to be observed sufficiently above the horizon at the latitude of the observatory? At an appropriate season? It is hard, even with a radio telescope or a satellite, to observe other astronomical objects when the Sun is in the sky nearby, and with a satellite the Earth may be in the way too. There are planetarium programs for personal computers that will check on accessibility.
- If there are many objects in the program, there may be examples available to a telescope at more or less any time, and one can select the group to go for on any given night. If the program is for a single unique object (for example, a comet), or for an object predicted to do something interesting at a particular moment like a periodic variable star, then this will drive the observer to the telescope at a specific time, even if it is not convenient.
- Are the objects bright enough to be within range of the available telescopes and the available equipment?
- Some observations, for example observations of fainter nebulae and galaxies, are hard to make in moonlight, so is the phase of the Moon appropriate? Again, planetarium programs have a part to play in answering this question. Diaries often have lunar phases. Optical astronomers divide the month according to lunar phases: 'bright time' is the 10–12 nights around full Moon; 'dark time' is the 10–12 nights around new Moon; and 'gray time' is the two or three days on the boundaries between the two, when the Moon is quite bright but below the horizon for a significant part of the night.

The sky is so bright in the infrared all the time that moonlight is not a significant issue for infrared observations.

Having decided what to do, a professional astronomer may have to write an application to use a telescope or satellite, if it is owned by someone else. The application outlines the scientific motivation to make the program of observations, and demonstrates feasibility. Access to the telescope may be highly competitive – factors of 10 over-subscription for available observing opportunities are not rare. Competition for optical telescopes is most severe during 'dark time,' because this is when galaxies and nebulae can be studied. A panel of scientists will judge the best programs and award access to the telescope ('telescope time').

However, astronomers may have their own telescope, or a share of a telescope owned by a university group or a society. In the latter case, no formal competition is necessary, just networking and negotiation.

Observing

Usually, the telescope has a manager who assigns its use night by night, or week by week, to individual observers or observing groups. Conventionally the observers travel to the telescope themselves, arriving a day or so early to prepare. This may mean enviable foreign travel to a romantic mountain-top observatory, or a less romantic radio-telescope control room or satellite-tracking station. On the day before the assigned nights, they ready the telescope and equipment, and make last-minute checks. Optical astronomers begin to observe the brighter objects during the twilight hours, infrared observers even earlier, and radio astronomers can observe all day. As the observing program progresses, the weather or other environmental conditions may change, and the astronomers will adapt the program accordingly.

If the telescope is complex or valuable, or if there are questions of efficiency or safety, the observatory will provide someone to operate the telescope, and perhaps to keep notes. This will always be the case for a satellite telescope because it could be told to do something that caused irreperable damage. The telescope operator is sometimes called a night assistant, a traditional term that somehow belittles an important job.

Astronomers increasingly pass the entire responsibility for making the observations to full-time telescope experts. This is usually mandatory for telescopes in space, but also for some ground-based telescopes, like the ESO's VLT. This is called 'queue scheduling,' with approved observations joining a queue to be carried out.

Astronomers choose the order in which observations are carried out to maximize the telescope's scientific output. Scientific priority is one urgent criterion, with lower-priority observations executed if there is time (that is, if it does not get cloudy, or if the satellite orbital and telecommunication circumstances are favorable).

Efficiency

The condition of the sky is important for optical astronomy. In general the observing conditions are better in the region of sky overhead, because the length of atmosphere along the sight-line to the ZENITH is minimized. ATMOSPHERIC EXTINCTION is least and SEEING is best for a star at that point. For a given star viewed from a given observatory, the sight-line through the atmosphere (or AIR MASS) is a minimum at the time that the star is at CULMINATION; so, other things being equal, it is best to observe a star at that time.

Weather conditions are also an issue in scheduling optical observations within a given night – it may be difficult to determine a star's brightness if there is light cirrus cloud, but it may be possible to obtain a perfectly good SPECTRUM or to check its brightness relative to another nearby star. Astronomers factor the weather forecast, the time of moonrise or moonset, and the target star's culmination into their assessment of the order of the program that they wish to carry out.

An operational requirement is to minimize time wasted as the telescope slews across the sky from one object to the next. Human observers do something similar. For example, an amateur astronomer making repeated visual checks on a series of galaxies to see if one has a supernova may work from one to the next by a quick and convenient route sign-posted by recognizable star patterns.

Even when the astronomers do not actually operate the telescope themselves, they may be present in the control room. The astronomers may wish to adjust their program according to what they find out during the observations – to continue to observe a variable star that is doing something unusually interesting, for example. Often they are reluctant to give the responsibility for this scientific judgment to someone else. Some space telescopes can be operated in this way – the IUE satellite was the first, with observers

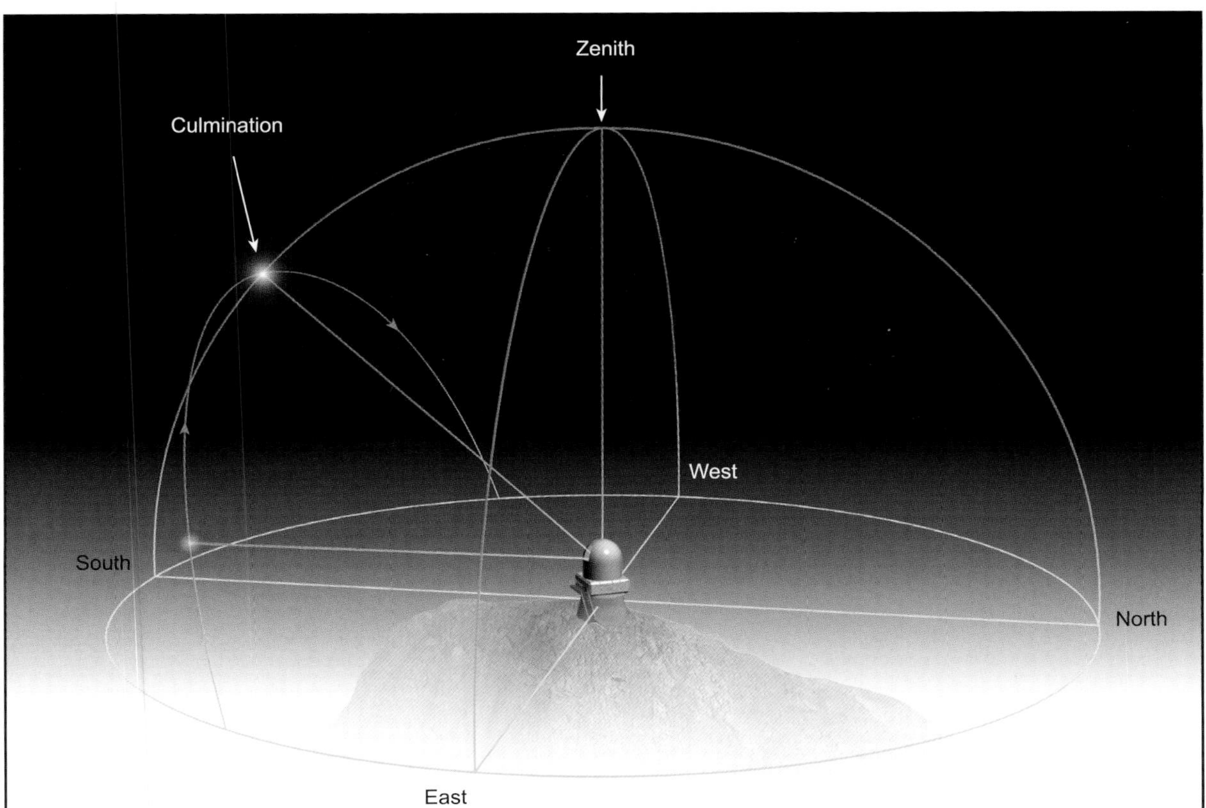

◀ ***Observatory*** *Optical telescopes are best placed on mountains above the atmospheric absorption, and at intermediate latitudes so that stars are on average high in the sky at their highest point (culmination).*

traveling to the control centers at either the NASA GODDARD SPACE FLIGHT CENTER in Maryland, USA, or ESA's Villafranca Satellite Tracking Station (VILSPA) near Madrid, Spain, for the time of their observing turn.

One potentially big time-waster is deciding on the next object to be observed, finding it and configuring the telescope equipment; for example, finding a guide star suitable for the AUTOGUIDER. Even with well-set-up computer-controlled telescopes, it may be necessary to adjust the position of the telescope after its slew, to find one star in a cluster, for example. Even on well-managed telescopes, up to half of the telescope's time may be wasted in this process. Efficient astronomers prepare for this stage in the program with well-made charts showing the stars nearby, with orientation and scale marked. The *HST Guide Star Catalog* was prepared for these very purposes, and is used to make charts by computer plots. The operators of the HST at the SPACE TELESCOPE SCIENCE INSTITUTE have taken a lead in developing an efficient 'end-to-end' observing process and have improved the telescope's efficiency with time.

Data that no one sees have no scientific point. Astronomers increasingly concentrate on developing efficient ways to complete an observing program to its ultimate potential by turning raw observations into scientifically meaningful data, and publishing them in journals, at conferences, and so on (*see* SOFTWARE IN ASTRONOMY). Paradoxically, although one might believe the busy life of an astronomer reduces time to think, their brains are stirred by the adrenaline and competition of an observing program, and the observations themselves, to produce ideas like raisins rising to the top of a pudding stirred with a spoon.

Occultation *See* ECLIPSE AND OCCULTATION

Ocean and sea

Oceans and seas are large bodies of liquid water, found on PLANETARY SURFACES. Of the nine planets of the solar system, Earth's orbit alone occupies the narrow corridor where surface liquid water is stable by virtue of the heat of the SUN, and surface oceans are possible. However, no terrestrial ocean or sea is permanent. Continental drift on Earth proceeds at a typical rate of about 2 cm per year, or more than two trips around the Earth over the age of the solar system. The Atlantic Ocean, for example, has existed for 2% of the age of the Earth. Liquid water has, however, been present since the atmosphere formed.

Sunlight is not the only heat source in the solar system. There is also geothermal energy, and with it, the possibility of underground oceans. Jupiter's satellite EUROPA is covered by an ice coat, but its temperature increases inwards at the rate of 5 K per kilometer because of tidal heating from Jupiter. There is an ocean of melted ice situated only 20–30 km below the surface.

Octans *See* CONSTELLATION

Ohio State University Radio Observatory

The Ohio State University Radio Observatory (OSURO), Columbus, Ohio, USA, started in 1951. Its director, John Kraus, designed, built and operated radio telescopes including a 96-helix array and the 'Big Ear,' a radio telescope of equivalent 52-m aperture, which ceased operation in 1998. Achievements include the Ohio Sky Survey (which measured more than 19 000 sources, over half previously undetected), and the Ohio Specials (sources with unusual radio spectra, which led to the discovery of the two most distant objects known at the time). The origin of the 'Wow!' signal, the strongest narrowband signal detected during a program to search for extraterrestrial intelligence, is still unknown (*see* EXOBIOLOGY AND SETI).

Olbers, Heinrich Wilhelm (1758–1840)

German doctor and astronomer who discovered several comets. In 1800 he joined the team of 24 'celestial police,' looking for the planet predicted by BODE'S LAW to lie between Mars and Jupiter, which led to his discovery of the ASTEROIDS (or minor planets) (2) Pallas and (4) Vesta. In contemplation of more distant problems, he formulated OLBERS' PARADOX, which asks the question 'Why is the night sky not as uniformly bright as the surface of the Sun?'.

Olbers' paradox

If the universe is infinite, static and uniformly populated with stars, any line of sight should end up on the surface of a star, and the sky should be as bright as the Sun. The fact that the sky is dark at night is the so-called Olbers' paradox. Heinrich OLBERS discussed it in 1826, but it can be traced as far back as Johannes KEPLER (1610), Edmond HALLEY and Philippe de Chéseaux (1718–51). The explanation is that the universe is expanding. Beyond a certain range, the intensity of radiation reaching us would be so greatly reduced by REDSHIFT that stars could not be detected (even if the universe were infinite). Likewise, if the universe has a finite age (as the BIG BANG theory implies), there has not been sufficient time for light to reach us from galaxies beyond a certain range. The paradox is also resolved if the universe is finite in extent.

Oljato

An Apollo ASTEROID discovered by Henry Giclas (1910–) in 1947, designated (2201) Oljato. About 2 km in diameter, it follows an orbit that shifts chaotically as a result of frequent close approaches to Earth and Venus. Its present orbit takes it from within the orbit of Venus at PERIHELION (0.67 AU) to the outer reaches of the main asteroid belt at APHELION (3.72 AU) with a period of 3.2 years. Oljato was lost after its discovery and not recovered until 1979. It is very high on the list of potentially hazardous asteroids, with a potential closest approach of less than 75 000 km (*see* NEAR-EARTH ASTEROIDS). Its reflection spectrum is unlike that of any other asteroid, meteorite or comet so far obtained. It may well be an extinct comet nucleus.

Olympus Mons

The largest volcano on Mars, and the largest in the solar system. It is visible from Earth and is also known as Nix Olympica, the 'Snows of Olympus,' after the mountain in Greece that is the legendary home of the gods. Olympus Mons rises to a height of 27 km above Mars's mean surface level, and is more than 600 km wide. The cliff walls of the 90-km central caldera tower up to 6 km above the surrounding plateau. It contains 100 times the mass of the Earth's largest volcano, Hawaii's Mauna Loa. On the Earth, plate tectonics operates to carry a volcano away from the underlying source of magma. With no such movement on Mars, volcanoes could continue to grow for as long as magma was available.

Omega Centauri (NGC 5139)

The largest GLOBULAR CLUSTER in the Milky Way Galaxy. Its total mass is about 5 million solar masses, 10 times the mass of other globular clusters and of similar mass to the smallest whole galaxies. With an apparent magnitude of 3.7, it is visible to the naked eye as a fuzzy 'star' in the southern constellation of Centaurus. Its brighter stars number several hundred thousand and they are noted for the great variation in their metal content, suggesting that the stars formed not all at the same time but in several starburst peaks, or by the merger of several different-aged globular clusters.

Omega Nebula (M17) *See* MESSIER CATALOG

Ondrejov Observatory

The Ondrejov Observatory is located 32 km from Prague in the village of Ondrejov. It was established in 1898 as a private observatory and donated to the state of Czechoslovakia in 1928. Since 1953 it has been part of the Astronomical Institute of the Academy of Sciences of the Czech Republic. The observatory's instruments include a 2-m stellar telescope, a photometric telescope, a multichannel solar-flare spectrograph, a solar magnetograph, a solar telescope, solar radio spectrographs, meteor radar and a photographic zenith tube. The observatory is part of the European Fireball Network for FIREBALL OBSERVATIONS.

Onsala Space Observatory

The Onsala Space Observatory (OSO), the Swedish National Facility for Radio Astronomy, operates the 20-m diameter millimeter-wave telescope and the 25-m decimeter-wave telescope at Onsala, and the Swedish-ESO Submillimeter Telescope, SEST, in Chile. OSO was the first European observatory to participate in high-resolution observations of radio sources using the VLBI technique, used in astronomy and in geodesy when transatlantic distances are measured with high precision. OSO is a pioneer of VLBI at millimeter wavelength, providing images of AGN with resolution of 50 microarcsec.

Oort Cloud and Kuiper Belt

The Oort Cloud

The Oort Cloud is a huge, spherical swarm of a million million COMETS surrounding the solar system and extending halfway to the nearest stars. The cloud is named for the Dutch astronomer Jan OORT who first suggested its existence in 1950 in order to explain the orbits of the observed LONG-PERIOD COMETS. Comets in the Oort Cloud are so distant (typically 30 000–60 000 AU) that they are perturbed by random passing stars, giant molecular clouds and the distorted gravitational field of the Galaxy that creates 'tides' in the cloud. These perturbations occasionally send comets back into our solar system, where we can see them.

When the comets make their first pass through the planetary system, random planetary perturbations, primarily by Jupiter, eject roughly half of the 'new' comets out into interstellar space. The other half are captured to less eccentric orbits, which nevertheless reach out well beyond the edge of the solar system, if not so far as they were before. The captured comets make only a few passes through the planets before they are evaporated or disrupted, or collide with a planet. They survive only 1 million years or so.

Interestingly, Edmond HALLEY speculated in *A Synopsis of the History of Comets* in 1705 that there was a distant comet cloud. Halley found that the 24 comet orbits that he calculated were parabolic, but he argued that the orbits would prove, with more accurate calculations, to be highly elliptic. He wrote: 'For so their Number will be determinate and, perhaps,

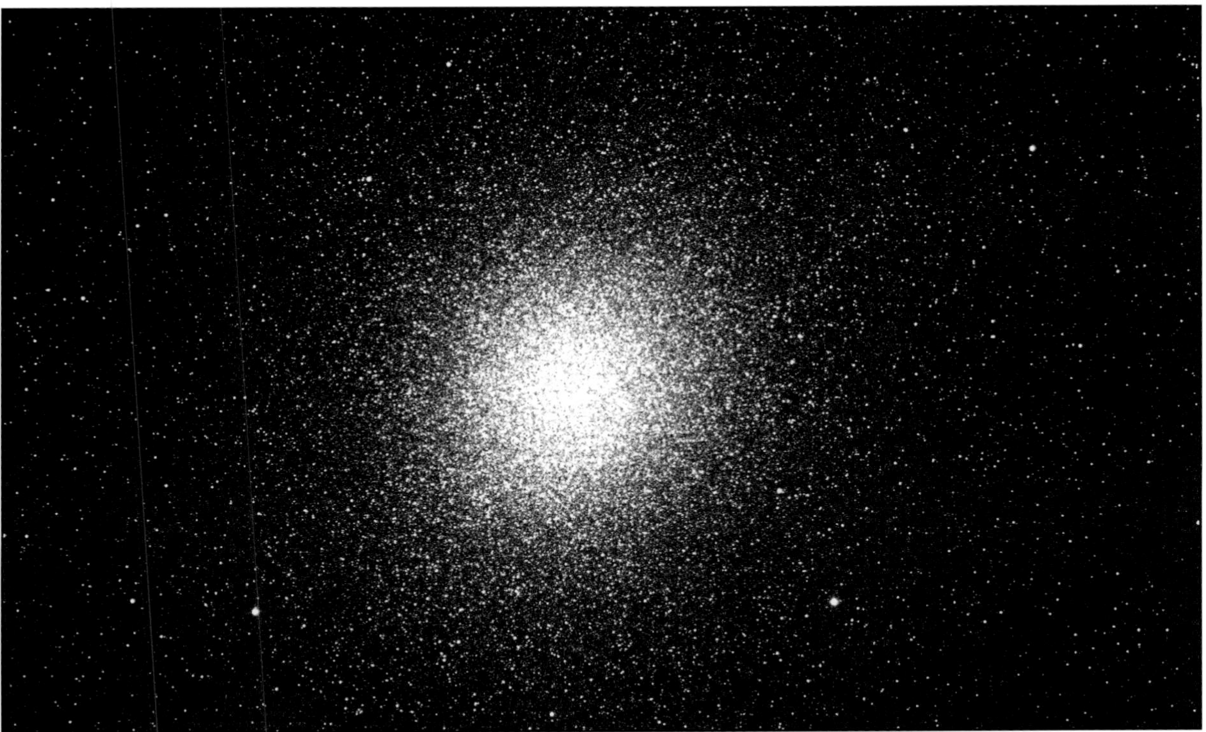

◀ ***Omega Centauri*** *NGC 5139 is the largest globular cluster in our Galaxy.*

not so very great. Besides, the Space between the Sun and the fix'd Stars is so immense that there is Room enough for a Comet to revolve, tho' the Period of its Revolution be vastly long.'

The comets in the Oort Cloud probably did not form there. Rather, the comets formed as icy planetesimals near the GIANT PLANETS and were ejected to distant orbits by gravitational encounters with proto-Jupiter and the other proto-gas giants.

The Kuiper Belt

The existence of the Oort Cloud is a conjecture, but there certainly does exist a trans-Neptunian belt of comets that was indeed left over from the formation of the solar system. Irish amateur astronomer Kenneth EDGEWORTH first mentioned it in 1949, and Gerard KUIPER elaborated the idea in 1951. Their basic argument was that there is no reason why the solar system should end abruptly at Neptune or Pluto. Perhaps large planets could not form beyond Neptune, but smaller bodies might have been able to. Edgeworth and Kuiper envisaged a population of small residual planetesimals, which, because it is cold in those distant regions, would be ice-rich, like comets.

The Oort Cloud and the Kuiper Belt are long-lived storage reservoirs for comets in the outer solar system. The Oort Cloud is the source of the long-period comets, with orbital periods exceeding 200 years. At first it was thought that the short-period comets (those with orbital periods shorter than 200 years) were derived from long-period comets, perturbed by Jupiter. In 1988, Canadian astronomers Martin Duncan, Tom Quinn and Scott Tremaine confirmed with computer simulations a hypothesis by Paul Joss (1970) that this happens too seldom. They also found that a spherical cloud like the Oort Cloud produces comets that have long, thin orbits pointing in random directions, as is the case for the long-period comets. Short-period comets, by contrast, nearly all orbit in the plane of the solar system and must have originated from a flat disk. This comet belt is what is now called the Edgeworth–Kuiper, or just Kuiper, Belt. Neptune defines the inner edge of the Kuiper Belt at 30 AU. It extends at least as far as 130 AU.

Kuiper Belt objects

In August 1992, after a five-year search, Dave Jewitt and Jane Luu proved that the Kupier Belt exists by detecting the first Kuiper Belt object (KBO), 1992 QB1 (*see* CUBEWANO). More than 700 KBOs have been discovered as of 2003, but it is estimated that there are at least 70 000 of them with diameters larger than 100 km. They are divided into distinct categories. Most KBOs have low-eccentricity, low-inclination orbits. They formed directly from the solar nebula and have preserved their primordial orbits. Some have orbits that have been influenced by other bodies, in two ways.

A number are in resonance with Neptune. An object in an *m:n* resonance would make *n* orbits around the Sun for every *m* orbits completed by Neptune. Pluto (at 39 AU) is in the 3:2 resonance, meaning that it completes two orbits for every three orbits completed by Neptune. This resonance is densely populated with KBOs called PLUTINOS.

In view of the abundance of plutinos, Pluto is better understood as the largest known KBO, rather than as a deviant planet with high-inclination and high-eccentricity orbit. Pluto has a diameter of 2320 km. The next largest KBOs are Pluto's moon Charon (1270 km) and the asteroids (number, designation and/or name), 2001 KX76 (1200 km), (50 000) Quaoar (1200 km), (28 978) Ixion (1065 km), (20 000) Varuna (900 km) and 2002 AW 197 (890 km). Neptune's satellite Triton (2700 km) possibly also originated in the Kuiper Belt, but somehow became captured by Neptune.

The third dynamical class in the Kuiper Belt is made up of the scattered KBOs. Only a half-dozen are known, with large, eccentric orbits. The object 1996 TL66 exemplifies the class. Most KBOs have circular orbits and their orbits group

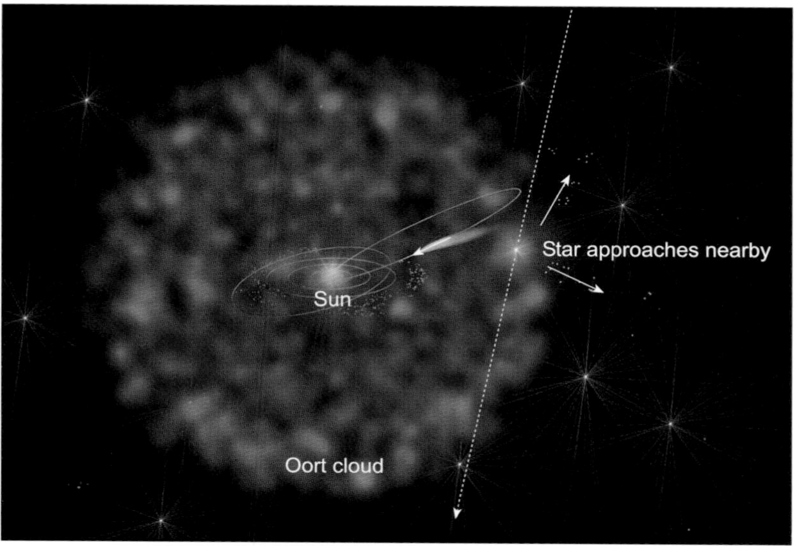

▲ ***Oort Cloud*** *Dormant comets in the Oort Cloud may be disturbed by a passing star. Some are ejected into interstellar space but some fall in elliptical orbits toward the Sun and are activated, developing tails.*

together at 50 AU, but 1996 TL66 has an eccentricity of 0.59, and reaches as far as 130 AU at aphelion. The origin of the scattered KBOs is unknown.

The mass of the Kuiper Belt between 30 and 50 AU is estimated at one-fifth the mass of the Earth. If a massive belt existed at great distances, it would have affected the motion of spacecraft such as Pioneer 10 (currently at 85 AU), and the VOYAGERS (65 and 51 AU) as well as COMET HALLEY. It didn't, and must be less than five Earth masses.

It takes about 10 million years for a KBO to go from a Kuiper Belt orbit to an inner-solar-system orbit. Thus, if the Kuiper Belt does supply short-period comets, there must be ex-KBOs making their way into the inner solar system. The missing links are the CENTAURS, whose orbits cross those of the giant planets. The brightest Centaur, cataloged as the asteroid (2060) CHIRON, is indeed a comet.

The properties of KBOs are of intense interest since these bodies have changed relatively little since they formed in the solar nebula. High-energy particle irradiation of surface ices may have altered the surface chemically and physically. The surface may have become a dark crust, rich in carbon compounds. The crust may survive the KBO's first entry into the inner solar system.

Ideally, spectra are needed in order to assess KBOs' surface composition, but they are very faint and it is too difficult. Astronomers have had to rely on COLOR INDEX information. Surprisingly, KBOs and Centaurs exhibit a wide range of colors, from nearly neutral (that is, all wavelengths are reflected equally) to very red (red wavelengths are preferentially reflected). In fact, KBOs and Centaurs are the reddest of all small solar-system bodies. The range of colors suggests that the surfaces of KBOs are diverse. This is against expectation, but KBOs may be as different in composition as the asteroids. Alternatively, if KBOs collide, fresh interior material might be revealed, and fresh impact debris would shower nearby regions. The overall color of a particular KBO would then depend on how much of its surface is covered by recent craters.

Extrasolar Kuiper Belts and Oort Clouds

Do Kuiper Belts exist around other stars? Dust disks are common around MAIN SEQUENCE stars, the most famous in this respect being BETA PICTORIS. Its 0.003 Earth-mass of dust is continuously replenished, perhaps by grinding between colliding planetesimals in a Kuiper Belt. Other possible Kuiper Belt systems include the main sequence stars HD 141569 and HR 4796A.

Likewise, Oort Clouds presumably exist around some other stars, but they have not been detected. The formation of Oort Clouds presumably depends on the presence of giant planets in order to eject enough comets to Oort Cloud distances.

Although many comets have been ejected to interstellar space in forming the Oort Cloud and over its history, no comet incoming from interstellar space on a clearly interstellar trajectory has ever been observed passing through our planetary system. This sets an upper limit on the density in space of interstellar comets. They constitute at most 100–300 times the density of other material in the solar neighborhood. So interstellar comets cannot contribute significantly to the DARK MATTER in the Galaxy.

Further reading: Davies J (2001) *Beyond Pluto* Cambridge University Press; Weissman P R (1996) *Completing the Inventory of the Solar System* (ASP Conference Series 107).

Oort, Jan Hendrik (1900–92)

Dutch astronomer who became director of the LEIDEN OBSERVATORY. He confirmed Bertil Lindblad's (1895–1965) hypothesis of galactic rotation by analyzing the motions of distant stars. He derived 'Oort constants': measurements of the Galaxy's rotation, from which he calculated the distance of the Sun from the center of the Galaxy and the period of its orbit. He inspired Hendrik van de Hulst (1918–2000) to calculate the 21-cm radio spectral line from hydrogen, and led the Dutch group that co-discovered and used the line to map hydrogen gas in the Galaxy, discovering spiral structure and the galactic center. He suggested the existence of a sphere of primordial cometary material surrounding the solar system (*see* OORT CLOUD AND KUIPER BELT). Oort also showed that light from the CRAB NEBULA is polarized, confirming Iosif Shklovskii's (1916–85) suggestion that it was SYNCHROTRON RADIATION.

Open cluster

Stars frequently form in clusters, either in rich clusters or in smaller groups. Galactic open clusters of stars are stellar systems containing from a few hundred to several thousand stars. Their central parts are readily distinguished by a density of stars much larger than that in the surrounding field. About 1200 galactic clusters have been cataloged, and about half investigated at least once. Typical clusters include the ORION NEBULA cluster, the PLEIADES, HYADES and M67.

Open clusters differ from GLOBULAR CLUSTERS. They are found mainly near the GALACTIC PLANE and have ages ranging from a few million to 10 billion years, while globular clusters populate the Galaxy's halo and are as old as the universe. Open clusters may be used to describe the history of our Galaxy after the formation of its disk.

The HST intensely observed two galactic clusters: NGC 3603 and the Orion Nebula cluster. NGC 3603 is a rich, very young cluster, the densest concentration of bright, massive stars in our Galaxy, at a distance of about 25 000 l.y. It is in the core of the most massive visible giant H II REGION. It is a starburst, comparable to the R136 cluster in the TARANTULA NEBULA in the LARGE MAGELLANIC CLOUD. HST's most impressive result on the Orion Nebula cluster, also known as the Trapezium cluster, was the direct observation of

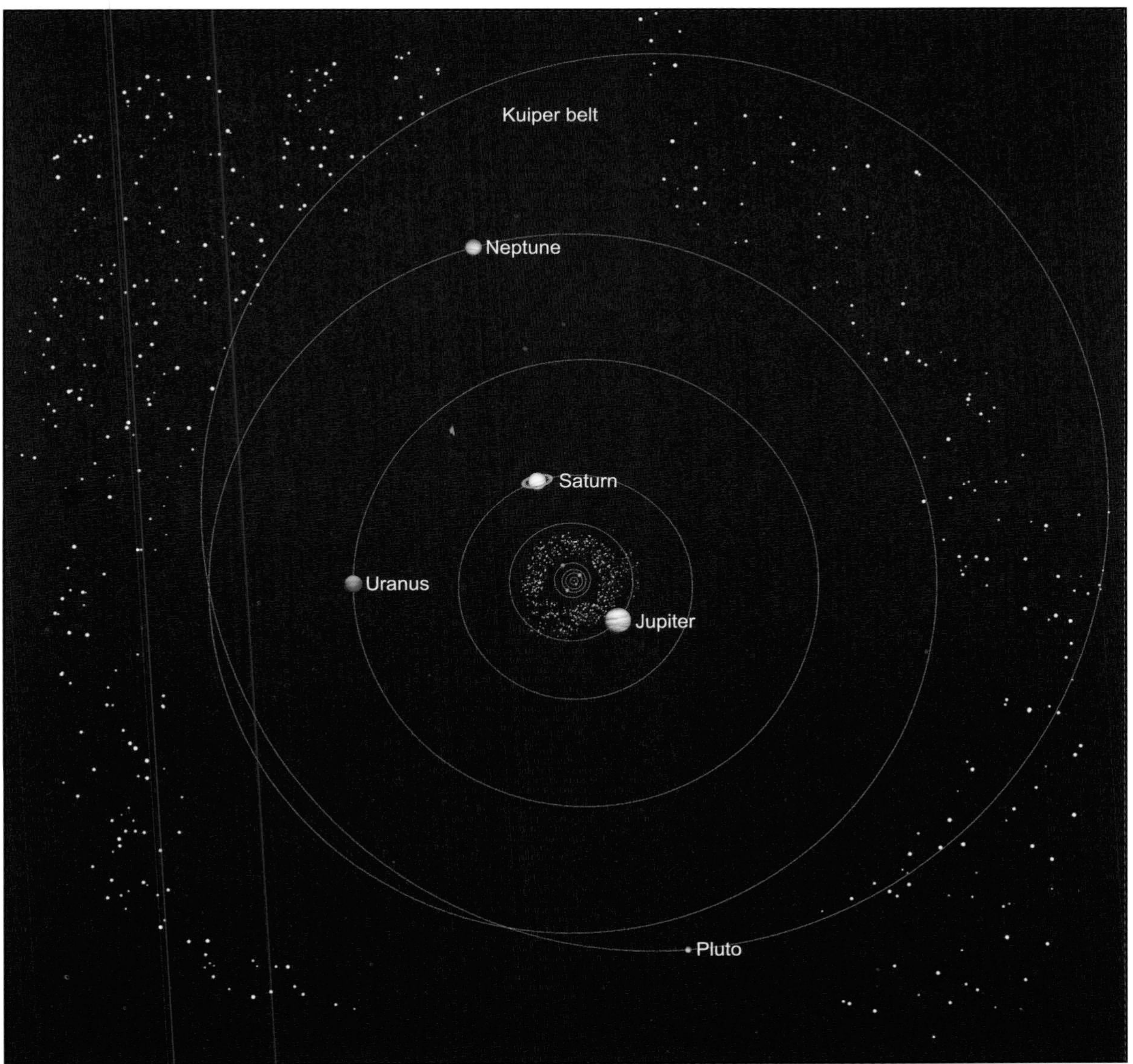

◀ ***Kuiper Belt*** *Icy asteroids or comets orbit beyond Neptune (like Pluto) or in more distant regions, known as the Kuiper Belt (or Edgeworth–Kuiper Belt).*

protoplanetary disks around its stars.

The strong interest in open clusters results from the unique properties of their constituent stars:

• They are all at the same distance from the Sun (the depth of an open cluster is negligible for open clusters at distances larger than about 300 parsecs);

• They have the same chemical composition (the scatter in the chemical composition is much smaller than we can determine);

• They all have the same age (the spread amounts only to a few million years and is important only for very young clusters that are less than 10 million years old);

• They all share the same spatial motion (if this were not the case, the group would rapidly dissolve).

The stars, however, have different masses. The most massive stars in the youngest open clusters are about 80 solar masses. The least massive are BROWN DWARFS of less than 0.08 solar masses. Consequently, open clusters offer opportunities to investigate how stellar phenomena depend on the masses of the stars or on the ages of the stars (by selecting stars of a given mass in several clusters).

Cluster members

Like other distant objects, open clusters suffer from the confusing effects of absorption of light by interstellar matter. The membership of the stars in the cluster is the second important problem to solve. On photographs or CCD frames, cluster members and field stars (which are closer or more distant than the cluster) are confused. To separate them, astronomers traditionally have measured their proper motions (*see* ASTROMETRY). The radial velocity now also provides a powerful criterion to identify the cluster members, using specialized instruments that determine the radial velocities of many stars at one hit. The cluster stars travel together. One interesting by-product of the X-ray satellite ROSAT was the discovery of some faint member stars in two nearby southern open clusters, IC 2391 and IC 2602, which are easily detected by their X-ray emission.

Evolution of stars in clusters

Because cluster stars have the same age, each cluster defines an isochrone by its HERTZSPRUNG–RUSSELL DIAGRAM (HR diagram). Thousands of stars are measured on CCD images and the data are concentrated to obtain the few parameters that describe each open cluster: the amount of interstellar absorption, distance, chemical composition and age.

Very young open clusters, with ages less than a few million years, may still be associated with the cloud from which they are born. Some clusters are even embedded in clouds that are completely dark at visible wavelengths, their stars revealed by near-infrared imaging.

The HR diagrams of young open clusters, from tens of millions to 1000 million years old, have a strong MAIN SEQUENCE. The most characteristic property of the HR diagrams of intermediate-age and old open clusters (older than 1000 million years), is that, as well as a main sequence, they each have a spectacular red-giant branch. The best-observed old cluster of this type is M67 (NGC 2682). At any age, the more massive stars of a cluster are more evolved than low-mass stars. This explains why the upper part of the main sequence has evolved away from the main sequence and is a characteristic shape for each age of cluster.

The HIPPARCOS satellite recently revealed limitations in our understanding of the main sequence. Its best observed cluster was the Hyades. Hipparcos determined the distance of the cluster center with an unprecedented accuracy: 46.34 ± 0.27 parsecs, that is 151 l.y. The PARALLAXES of individual stars enabled a three-dimensional view of the cluster, the only cluster where this is possible. Hipparcos determined good distances for nine further open clusters out to 300 parsecs, including the Pleiades. The distance of the Pleiades obtained by Hipparcos is 10% less than that determined from the ground by earlier techniques (the main sequence is more than 0.1 magnitude fainter than it should be – *see* MAGNITUDE AND PHOTOMETRY). The reason for this discrepancy has not been identified.

If all stars were single and the same age, the main sequence of an open cluster would be very thin. However, with high-precision photometric data (for example, data obtained with the HST), the main sequence presents a finite thickness. HST's color-magnitude diagram of the Orion open cluster shows that star formation was remarkably coeval: 80% of the stars have ages less than 1 million years. But 15% are aged between 1 and 4 million years. The age spread is therefore a few million years.

Most scatter in the main sequence of an open cluster is produced by BINARY STARS. Obviously, an unresolved double star is brighter than a single star of similar temperature. The addition of the fluxes of two stars makes the binary system brighter than the brighter star by 0.75 magnitude, if the stars are identical (less if one is fainter than the other). Stellar ROTATION also introduces scatter in the main sequence, depending on the rotation rate and the angle at which the star is viewed.

Clusters show how the stars' rotation changes down the main sequence. The more massive stars rotate quickly, but less massive stars, like the Sun, rotate slowly. Rotation is also linked to a star's X-ray emission, through sunspot-like activity (*see* SUNSPOT, FLARE AND ACTIVE REGION). This is connected to the star's magnetic field, which is connected to its convective interior. The ROSAT X-ray satellite showed how the intensity of X-ray emission is lower in the stars of older clusters. Older stars have been quickly braked by the interaction of their stellar winds and magnetic fields.

Pulsating variables, like the CEPHEID VARIABLE STARS, have been found in open clusters, which permits us to relate their properties to their ages and chemical composition. Very young clusters contain flare stars (*see* RED DWARF AND FLARE STAR). A kind of variable star was discovered in open clusters in the beginning of the 1980s, in which variability is caused by surface spots. Because the spots are darker than the surface, the spotty hemisphere of such a star is a little less luminous, by a few hundredths of a magnitude. The star's rotation produces a cyclic variation, at the rotation period of the star. In clusters for which observations have been obtained over a long time, variability in X-rays has also been observed, but has not yet been linked to stellar cycles like the 11-year solar cycle.

BLUE STRAGGLERS are stars that remain in the main sequence area of the HR diagrams of cluster stars after other stars have left, owing to their evolution. Blue stragglers may have formed later than usual in the cluster, they may have been pushed at a later date into being a star of high mass (by receiving mass from another star in a binary system), or they may have been formed from the coalescence of two stars.

Formation of star clusters

The stars in open clusters are from the fragmentation of the parent cloud. The numbers of new stars of different masses form what is known as the IMF (INITIAL MASS FUNCTION). What is observed is, in fact, the LUMINOSITY FUNCTION, namely the numbers of stars of different brightnesses. Brightness is converted to mass from a mass-luminosity relation. As a rule, most clusters have an IMF of the form known as the Salpeter law (from its formulator Edwin SALPETER), although its universality as a law has not yet been fully demonstrated.

In most clusters the massive (brighter) stars are found close to the center, while less massive stars are more uniformly distributed, and often have only recently been recognized to be part of the cluster. Mass segregation is caused by near collisions between stars, which make the massive stars fall toward the cluster center. On the other hand, open clusters lose stars by an evaporation process, because during collisions (especially of a single star with a binary star) a star may acquire enough speed to leave the cluster. As a result, the less massive open clusters have lives of at most a few hundred million years. Field stars result from evaporation from many clusters.

Further reading: Janes K A (ed.) (1991) *The Formation and Evolution of Star Clusters* (Astronomical Society of the Pacific Conference Series, vol. 13); Payne-Gaposchkin C (1979) *Stars and Clusters* Harvard University Press.

Open clusters, observing

See practical astronomy feature, p. 84.

Ophiuchus *See* CONSTELLATION

Öpik, Ernst Julius (1893–1985)

Born in Estonia, Öpik studied at Moscow University, and helped establish Turkestan University in Tashkent, becoming the astronomer (director) at Tartu Observatory in Estonia. He fled the Red Army during World War II and went to ARMAGH OBSERVATORY, Northern Ireland, in 1948. His wide-ranging interests are reflected in his discoveries and theories. These included the discovery of degenerate stars (WHITE DWARFS) in his calculation of the density of omicron-2 40 Eridani (1915). He calculated the distance of M31 as 450 000 parsecs from the Sun (1922). He computed by hand evolutionary models of MAIN SEQUENCE stars into giants (1938) more than a decade earlier than the computer computations

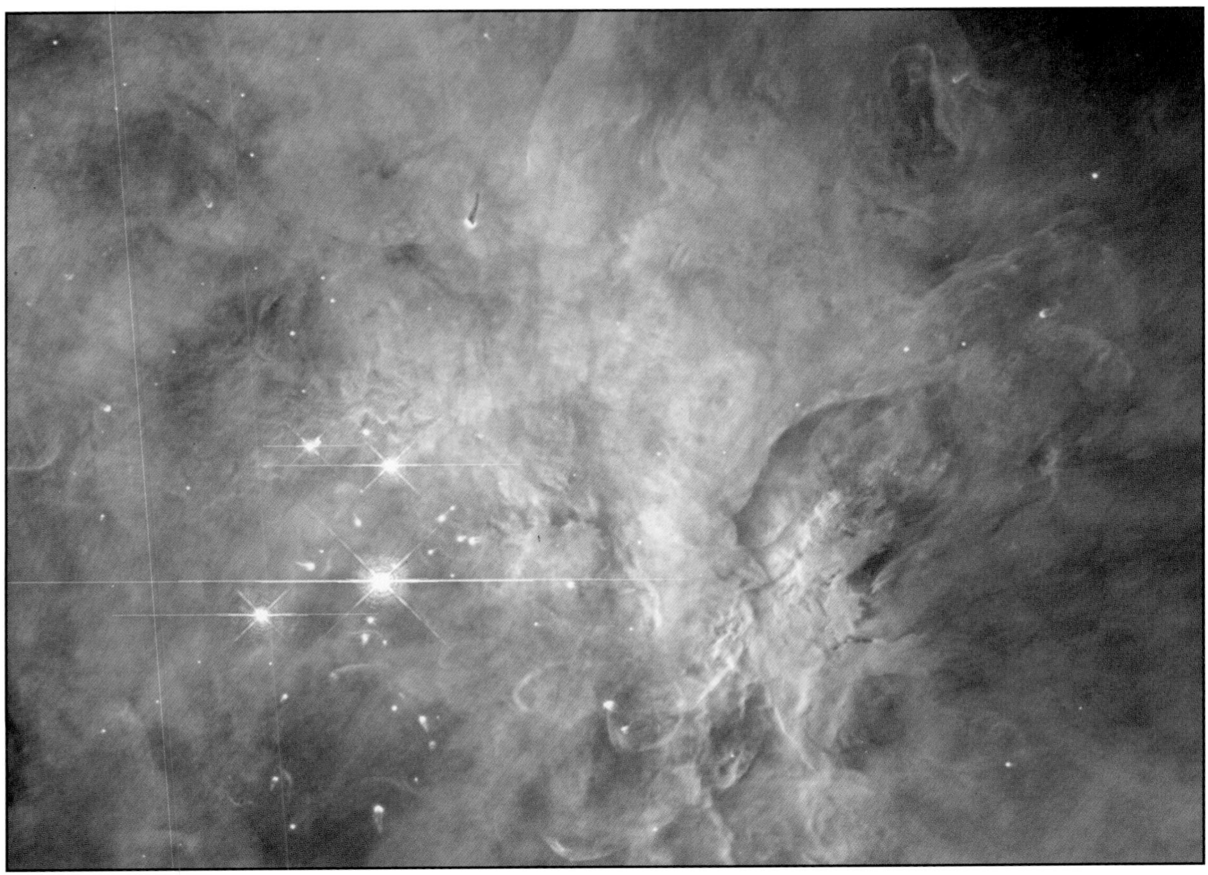

◀ ***Trapezium***
An open cluster around four massive stars at the center of the Orion Nebula.

of Fred HOYLE and Martin SCHWARZSCHILD. He predicted the density of craters on the surface of Mars, which planetary probes confirmed 15 years later.

Opposition

The position of a planet in its orbit when it is directly opposite the Sun in the sky. It then has an elongation of 180° and is on the meridian at midnight. The term is also used for the time at which this alignment occurs. Only superior planets (those outside the Earth's orbit) can come to opposition. Opposition is the best time to observe a planet as it is at its closest to Earth and is visible throughout the night. This applies in particular to Mars, whose orbit is markedly eccentric, Its distance from the Earth at opposition varies from 101 million km at aphelic opposition (furthest from the Sun), to only 56 million km at perihelic opposition (closest to the Sun). *Contrast* CONJUNCTION, and see this article for a diagram.

Optical astronomy

Astronomers increasingly concentrate on their subject matter, rather than on the observational techniques that they use. To gather all the information that is available about the cosmic experiments in which they are interested, they use all possible wavelengths to study them. Nevertheless, the different techniques of astronomy are individually complex, and in practice astronomers specialize in observational techniques too. They collaborate with others when they venture outside their specialty technique to make the necessary observations of their specialty subject matter.

Optical astronomy comprises the most common observational techniques used by astronomers to access a wide range of celestial objects in the OPTIR SPECTRUM. Radio astronomers, X-ray astronomers, and infrared astronomers turn to optical telescopes to locate the source of radio waves, X-rays, and infrared radiation that they have discovered, and to provide basic knowledge of the circumstances in which their wavelengths are radiated. Because the human eye has evolved in a location near the Sun, a typical star, it is no coincidence that many stars are of a surface temperature such that they emit significant radiation in this waveband. Because the human eye is made of a molecular fluid, and photochemical in the nature of its detector, and operates under an atmosphere, it is also no coincidence that many of the weaker electronic transitions in atoms and some in molecules involve photons in the optical range.

Optical astronomy therefore consists mainly of studying:

- thermal emission from stars of temperature above, say, 1000 K;
- thermal emission from collections of stars such as clusters and galaxies;
- spectral lines from atoms and ions in stellar atmospheres, nebulae and rarefied material in the vicinity of stars or black holes;
- light reflected by solid matter in the vicinity of a source of light, like planets and dust.

Because SYNCHROTRON RADIATION has a broadband nature, synchrotron sources like the CRAB NEBULA are also seen through optical astronomy, even if relatively weak.

The OptIR band brings into the subject matter for study:

- thermal emission from cooler stars and the dust material that surrounds them, at temperatures down to about 150 K;
- spectral lines of a number of molecules.

Techniques

The broad range of techniques that lie in the science of optics have been applied to OptIR astronomy. Glass is transparent to the OptIR spectral range. Ionic impurities can limit the spectral range of transmission, a limitation to the wider spectral range of glass or an exploitable asset to filter the light.

Mirrors reflect light and infrared. At first mirrors were of polished metal (speculum) and later of glass or ceramic material coated with aluminum, silver or gold (depending on the spectral region within or near the OptIR range that is most important). Lenses and mirrors can be used to focus and filter light and form optical images for detection and analysis. These components make telescopes and instruments like SPECTROSCOPES, cameras and filters.

The DISPERSION properties of transparent materials produce a means to form spectra for analysis. The wavelength of light is longer than the separation of atoms and substructure in glass masses, so parallel scratches can create DIFFRACTION GRATINGS, also producing spectra. The wavelength is not so small that the scratch separation or the controlled separation of parallel planes of glass is outside technological reach. The wavelength is small enough that a large density of scratches is feasible, making high spectral resolution possible in the dispersed light. These considerations facilitate spectroscopy, spectral-line analysis and the measurement of radial velocities at the levels common in the motions of stars and nebulae.

The mechanical rigidity of glass materials and their relatively low thermal-expansion coefficient make large telescope lenses and mirrors possible. This has enabled optical telescopes to double in size every 40 years, from GALILEO GALILEI's approximately 1-cm telescope through 10 generations to the 10-m telescopes of the year 2000. Optical telescopes reach low flux limits and draw into consideration a substantial fraction of the universe of stars and galaxies, with optical telescopes like the HST able to penetrate to within 1 billion years of the BIG BANG. The ability to cope with mechanical and thermal stability sets limits on the size of the individual ground-based telescopes that can be made, even with the active control systems now possible. But we have not yet reached the limits of technology, and a study by the ESO has identified no reason why optical telescopes 100 m in diameter could not be made. This has motivated the OWL project, an optical and near-infrared adaptive telescope 100 m in diameter. Because of the flux gradients from celestial objects such as the Sun and the Moon, thermal stability limits the performance of OptIR telescopes, such as the JAMES WEBB SPACE TELESCOPE, even in space.

Optical depth

A measure of the absorption of radiation of a particular wavelength as it passes through a semi-transparent medium. It is usually denoted by the symbol τ. The optical depth depends on both the opacity of the medium and on the distance. If $\tau = 0$ the medium is transparent and the radiation passes through unattenuated. If the optical depth is much less than 1 we say the medium is optically thin. If it is much greater than 1 then the medium is optically thick. In the Sun we define the base of the PHOTOSPHERE as the depth at which $\tau = 1$.

Optical double

An observed double star that consists of two quite unrelated stars that happen to lie in almost the same line of sight, and therefore appear close together in the sky. They may in fact be at very different distances from us and each other. For most pairs, only measurements of their separation and position angle repeated over many decades will confirm that they have no orbital motion and do not constitute a true binary system. *See also* DOUBLE STARS.

Optical spectrum

The part of the ELECTROMAGNETIC SPECTRUM visible to the human eye, from about 0.4 to 0.7 μm. The optical spectrum includes blue (about 0.45 μm), green (0.5 μm), yellow (0.55 μm), orange (0.6 μm), and red (0.65 μm). At the short-wavelength end, the optical spectrum is bounded by ultraviolet radiation, and at the long end by infrared.

OptIR spectrum

The OptIR (pronounced 'opteer' or 'opt-eye-are') spectrum is the part of the ELECTROMAGNETIC SPECTRUM accessible in astronomy to CCDs, running from the atmospheric cutoff in the ultraviolet at 0.32 μm to the limit of detection of silicon detectors in the near-infrared, 1.1 μm.

Orbit

The path pursued by a body moving in a field of force. In most astronomical contexts we consider motion in a gravitational field (*see* NEWTON'S LAWS OF MOTION), but the term applies equally well to motion under the action of other forces, for example to charged particles moving in a magnetic field. Bodies moving freely in the gravitational field of a massive body (such as the planets moving around the Sun, or an artificial satellite moving around the Earth) follow conic orbits, which are elliptical, circular, parabolic or hyperbolic, depending on the strength of the gravitational field and the velocity of the body. A planetary orbit is completely described by six geometric properties called the ORBITAL ELEMENTS.

Orbital element

The parameters required to define the shape and orientation in space of an ORBIT. and to fix the position in that orbit at any time of the planet or satellite moving within it. For a planet moving around the Sun, six elements are required:

- a: the semi-major axis of the ellipse;
- e: the eccentricity of the ellipse;
- i: the inclination of the orbit to the plane of the ecliptic;
- Ω: the longitude of the ascending node (the angle measured anti-clockwise in the ecliptic from the vernal equinox to the point of intersection of the orbit with the ECLIPTIC at which the planet passes from south to north of this plane);
- ω: the longitude of PERIHELION (sometimes called the 'argument of perihelion') – the angle, measured anti-clockwise in the orbit plane, from the ascending node to perihelion;
- T: the time when the planet passed perihelion.

The Sun occupies one of the two foci of the ellipse of a planet's orbit. A line drawn through the point of the planet's perihelion and APHELION passes through the Sun and is called the line of apsides or major axis of the orbit. One-half of this line's length is the semi-major axis, equivalent to the planet's mean distance from the Sun. The eccentricity of an elliptical orbit is a measure of the amount by which it deviates from a circle; it is found by dividing the distance between the focal points of the ellipse by the length of the major axis. The position of a planet at any time can be predicted if it is

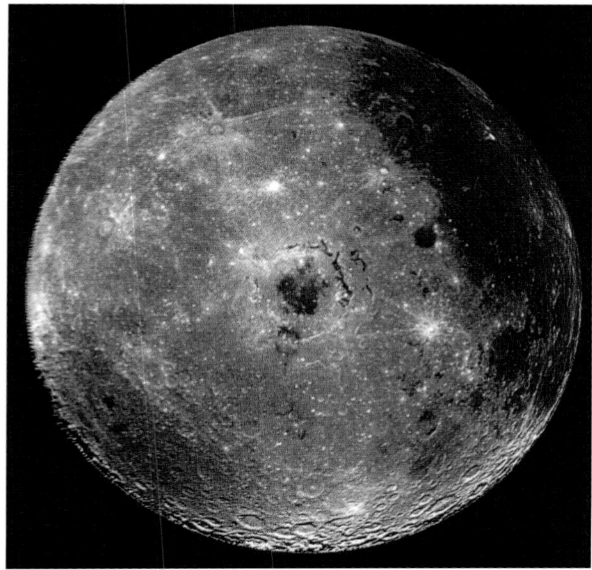

◀ ***Orientale Basin*** *The Moon seen from the Galileo probe. The dark region to the right is Oceanus Procellarum, and the dark region to the left is the South Pole-Aitken basin.*

known when it passed through any definite position; for example, its time of perihelion passage.

For a satellite in orbit around the Earth, the ascending node and inclination would be referred to the plane of the celestial equator.

In the case of a binary system of unknown mass, a seventh element is necessary: the orbital period P, or the mean motion $n (= 2\pi/P)$. In the case of the solar system this additional element, although sometimes listed, is related to the value of the semi-major axis (a) by the third of Kepler's laws (*see* Johannes KEPLER). In the case of a binary system, T relates to the time of PERIASTRON passage (the instant at which the two stars make their closest approach to each other).

Orbiting Astronomical Observatory (OAO)

Series of four NASA satellites. No data were returned from OAO-1 or OAO-B. The second satellite (OAO-2) operated for four years. It carried two experiments which surveyed the sky in the ULTRAVIOLET SPECTRUM. The most successful of the series was COPERNICUS (OAO-3), which was launched in August 1972 and operated for nine years.

Orbiting Solar Observatory

Series of eight NASA satellites launched between 1962 and 1975, and designed to study the Sun at ultraviolet and X-ray wavelengths. OSO-1 and 3 also carried gamma-ray instruments for extrasolar studies. The most successful of the series was OSO-8 which operated for more than three years and studied the diffuse cosmic X-ray background.

Orientale Basin

A large, multi-ringed structure on the MOON, about 930 km across in its largest dimension. It is only partly visible from Earth when LIBRATION brings it into view, but spacecraft like the Galileo mission have revealed it to us as if we were right above it. It has been partially filled by basaltic lava flooding, to form the Mare Orientale. Orientale is the youngest of the Moon's large-impact features and the surrounding ring structure is the best preserved of any of them. EJECTA from the impact traveled up to 1000 km, and the basin is surrounded by chains of secondary craters and other radial features, including lunar valleys.

Origins Program

Long-term NASA program stretching to 2020 and beyond to answer the basic questions, such as 'Where did we come from?' and 'Are we alone?'. The central principle of the Origins mission architecture is that each major mission builds on the scientific and technological legacy of previous ones, while providing new capabilities for the future. In this way, the complex challenges of the theme can be achieved with reasonable cost and acceptable risk. Ground-based observatories in the program include the Keck Interferometer (*see* W M KECK OBSERVATORY) and LARGE BINOCULAR TELESCOPE (LBT). Space missions in operation or proposed are HST, FUSE, SIRTF, SOFIA, Kepler, SPACE INTERFEROMETRY MISSION (SIM), JAMES WEBB SPACE TELESCOPE (JWST), Terrestrial Planet Finder (TPF), Single Aperture Far-Infrared Observatory (SAFIR), Large UV/Optical Telescope, Life Finder (LF) and Planet Imager.

Orion *See* CONSTELLATION

Orion Arm

The local SPIRAL ARM of the MILKY WAY GALAXY within which the Sun and the star-forming region of Orion are embedded. The stars and clouds of gas and dust that constitute the Orion arm curve outward (relative to the galactic center) from the general direction of the constellation of Cygnus to the constellation of Vela. Sometimes referred to as the 'Orion spur,' it is believed to be an extended arm segment rather than a major arm that originates directly in the central bulge of the Galaxy. The Orion region is the most prominent star-forming complex in the local arm.

Orion association

A large OB ASSOCIATION that contains more than 1000 stars and lies in the region of the Orion nebula.

Orionids

A METEOR SHOWER that takes place in late October. The RADIANT lies in the constellation Orion, near the border with Gemini. The Orionids occur when the Earth intersects the ascending NODE of the METEOR STREAM from COMET HALLEY; the ETA AQUARIDS in May are produced by the Earth's passage through the descending node. Because Halley's orbit is retrograde (*see* RETROGRADE MOTION), Orionid meteoroids impact the Earth at a high relative velocity and produce very fast meteors. But it is a less active shower than the Eta Aquarids, because in October the Earth encounters the meteor stream off center.

Orion Nebula (M42, NGC 1976)

A huge (four times the area of the full Moon), bright (magnitude 4) EMISSION NEBULA, the most prominent in the sky. The quadruple star theta-1 Orionis, known as the TRAPEZIUM, lies in the foreground of a dish-shaped hole that it has blown in the front surface of the nebula, and its ultraviolet light energizes the gas of the dish behind. The northwestern part of the nebula is M43 (NGC 1982). The designations of the MESSIER CATALOG recognize the separation of M43 from M42 by a dark lane of dust, named the Fish Mouth for its shape. The Orion Nebula abuts onto OMC-1, the largest of many molecular clouds that occupy much of the constellation Orion. This is a region of active star formation, the Trapezium stars being some of many young, hot stars associated with the nebulosity. The BECKLIN–NEUGEBAUER OBJECT and the KLEINMANN–LOW NEBULA

▶ ***Orrery*** *The Reverend William Pearson (1767–1847) demonstrates an orrery to wife and daughter in this portrait which hangs in the Royal Astronomical Society, London.*

lie behind the Orion Nebula. It is surrounded by BARNARD'S LOOP. The HORSEHEAD NEBULA and the reflection nebula around M78 are part of the complex.

The nebula is the closest star-formation region to Earth. Behind the Trapezium, within the nebula, lies the Trapezium open cluster. Its optical light is greatly absorbed by dust but its infrared radiation reveals its newly born stars and PROTOPLANETARY DISKS. *See also* NEBULAE AND INTERSTELLAR MATTER.

Orrery

A mechanical model that illustrates the relative positions and motions of bodies in the solar system. The name derives from the fourth Earl of Orrery, Charles Boyle (1676–1731), for whom such a device was constructed, probably in about 1725 by George Graham (1674?–1751).

Orthoscopic eyepiece *See* EYEPIECE

Oscillating universe

A universe that expands and contracts in a cyclic fashion. If the expansion ceases at some time in the future, the universe will contract, slowly at first, then ever more rapidly until it collapses back into a hot, highly compressed state: the 'Big Crunch' (*see* DECELERATION PARAMETER). The oscillating-universe theory suggests that the collapse of the universe is followed by a new BIG BANG and a new cycle of expansion and contraction. Such cycles could be repeated indefinitely. There is no known physical process that could cause a collapsing universe to 'rebound' in this way.

Outer planet

A collective term (also SUPERIOR PLANETS) for the MAJOR PLANETS whose orbits lie farthest from the Sun, beyond the main asteroid belt, as opposed to the INNER PLANETS, which orbit inside it.

OVV quasar *See* ACTIVE GALAXY

Owens Valley Radio Observatory

The Owens Valley Radio Observatory (OVRO) is in a deep dry valley 400 km north of Pasadena, the location of the California Institute of Technology, which operates the observatory. It was founded in the mid-1950s and the first major instrument was an INTERFEROMETER composed of two 26-m antennae. These dishes moved on east-west and north-south tracks, so the pair provided excellent resolution in two dimensions. At the time it was the largest such system in the world, and made many high-resolution studies especially of extragalactic radio sources. In the late 1960s a 40-m paraboloid was constructed. For many years it was the major west-coast station of the Very-Long Baseline Interferometer. More recently it has been used to study pulsars and the cosmic microwave background radiation. At present the major instrument at OVRO is an interferometer comprising six 10-m telescopes, operating at millimeter wavelengths. These dishes are on tracks and the system can synthesize a 0.5-arcsec beam at 1-mm wavelength. Extensive studies are being undertaken in star formation, molecular studies in the Milky Way and other galaxies, and in studies of galaxies at high REDSHIFT.

Owl Nebula (M97) *See* MESSIER CATALOG

Oxygen

A gas that makes up 20.95% by volume of the Earth's atmosphere at ground level, 89% by weight in seawater and

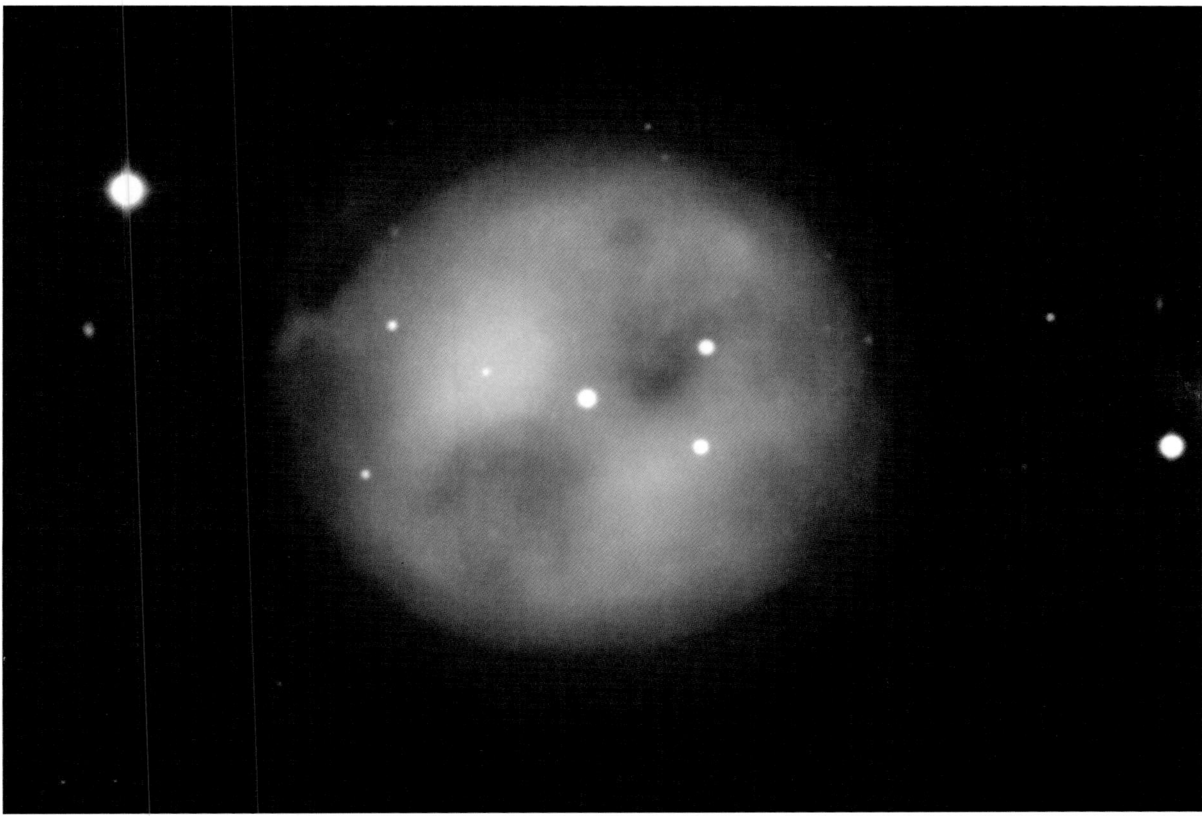

◀ ***Owl Nebula*** *The name of this planetary nebula was coined by the Third Earl of Rosse, who considered the two dark circles in the center to resemble the eyes of an owl.*

46.6% in the Earth's crust. It is the third most abundant element in the universe (after HYDROGEN and HELIUM), but its abundance in terms of number of atoms is only about 1/1500 that of hydrogen (*see* COSMIC ABUNDANCE OF THE ELEMENTS). The chemical symbol is O, and it normally occurs in the atmosphere in molecular form (two atoms linked), O_2. Triatomic oxygen, O_3, called ozone, and monatomic oxygen, O, are more predominant in the upper atmosphere, where ozone shields the Earth from the Sun's ultraviolet radiation. Its atomic mass is 15.99 and its atomic number 8. Oxygen molecules are essential to the sustenance of human and animal life on Earth. Oxygen combines with most other elements and, when this reaction occurs rapidly with the release of heat and light, we describe the reaction as combustion, or burning. Oxygen liquefies under normal pressure at about −183 °C and it becomes solid at about −218 °C. Liquid oxygen (LOX) is commonly used as the oxidant in the propellant of chemical rockets. Free molecular oxygen is almost entirely absent from the atmospheres of Venus and Mars.

Ozone

Ozone is O_3 – three OXYGEN atoms. Ozone is produced from molecular oxygen by photolysis (chemical reaction caused by light). The oxygen and ozone of the Earth's atmosphere are highly reactive and are present only because of the constant generation of oxygen by photosynthesis in plants. A terrestrial planet with life will have ozone as the unique spectral signature of a non-equilibrium atmosphere. By contrast, a biologically dead world will have an atmosphere of mainly carbon dioxide. Ozone produces a spectral absorption line at about 9.5 μm wavelength, the region planned to be used by NASA's Terrestrial Planet Finder and ESA's Darwin space missions in the search for extraterrestrial life (*see* ORIGINS PROGRAM).

Ozone layer and the ozone hole

On Earth, ozone occurs mainly in an atmospheric layer located in the stratosphere at about 25-km altitude, where there is a concentration of about six ozone molecules for every million air molecules (equivalent to a layer of pure ozone 3 mm thick at sea level). The balance between the rates of ozone production and destruction governs the amount of ozone. Ozone is destroyed in chemical processes that involve numerous constituents, including man-made chemicals released into the air, such as chlorine and nitrogen compounds and chlorofluorocarbons, as shown by Paul Crutzen (1933–) in 1970 and Mario Molina (1943–) and F Sherwood Rowland (1927–) in 1974 respectively (all of them shared the Nobel Prize for Chemistry, 1995). Despite its small abundance, ozone plays a vital role in the survival of life on Earth: stratospheric ozone absorbs most of the biologically damaging ultraviolet sunlight.

The ozone hole is a recent seasonal phenomenon, namely a significant reduction in the total amount of ozone in the Earth's atmosphere during springtime over Antarctica. The ozone hole is a sudden and near total loss of ozone over this region but is part of the general phenomenon of ozone depletion at a slower rate (a few percent per year).

P

Padova/Asiago observatories

The main complex of optical telescopes in Italy is located in two sites in the Asiago Highlands, about 90 km northwest of Padova, at an elevation of about 1000 m. The telescopes belong to two independent institutions based in Padova: the Astronomical Observatory of Padova (OAP) and the astronomy department of the local university. The OAP, a research institute, was founded in 1767 by the Venetian Republic and located in a tower built in 1242 by the tyrant Ezzelino da Romano. The construction by Padova University of a 122-cm REFLECTOR (then the largest telescope in Europe) was completed in 1942. This fostered the development of modern astrophysical research in Padova/Asiago. Three more telescopes were subsequently built by OAP: a 50/40-cm Schmidt (in 1958), a Schmidt 92/67 cm (in 1965) and the 182-cm Copernicus reflector (in 1973). (In Schmidt telescopes the first dimension is the size of the mirror and the second the size of the corrector lens.) At present they are all located at the top of Cima Ekar (elevation 1350 m).

Palimpsest

A bright, circular feature with little vertical relief on an icy surface of a planet. The term is used to describe features on the outer three Galilean satellites of Jupiter. Its original meaning is a piece of parchment prepared for reuse by erasing what had earlier been written. Palimpsests are interpreted as impact features, typically 100 km in diameter, whose original crater shape has been smoothed out. The impacts occurred where the planet's crust was thin; subsurface 'slush' filled the crater, or the crater walls slid back because the underlying material was not strong enough to be able to support them.

Pallas

The second ASTEROID to be discovered, by Wilhelm OLBERS in 1802, and so designated (2) Pallas. It is also the second largest asteroid in the main belt, irregular in shape with an average diameter of 525 km. Pallas rotates in 7.81 hours. It is a B-type asteroid, with a reflectance spectrum similar to that of CARBONACEOUS CHONDRITES, and an ALBEDO of 0.16.

Palomar Observatory

The observatory is located on Mount Palomar, north San Diego County, California, USA. It is owned and operated by the California Institute of Technology, a privately endowed educational and research institution located in Pasadena. The principal instrument is the 200-in Hale Telescope, which from 1948 to 1976 was the largest REFLECTING TELESCOPE in the world. Palomar also houses the 1.2-m Oschin Telescope, the 46-cm SCHMIDT TELESCOPE, and a 1.5-m REFLECTING TELESCOPE.

The project to build the Hale telescope began in 1928. The 200-in, 20-tonne Pyrex glass disk was cast on December 2, 1934, and after a cooling period of eight months was transported by rail to Pasadena for the long process of grinding and polishing. Construction of the building (including the 1000-tonne rotating dome) and the telescope structure (the moving parts weigh about 530 tonnes) began in the mid 1930s and was nearly complete by 1941. World War II delayed polishing of the mirror, and it was not until November 18, 1947, that the finished mirror, now weighing only 14.5 tonnes, began its two-day trip to Mount Palomar, where it was installed in the telescope. Scientific research at Palomar Observatory has been undertaken since 1948.

The 200-in telescope has been upgraded. It is now computer controlled and fitted with CCD cameras, a new adaptive optics system, and an infrared camera – the Palomar High Angular Resolution Observer (PHARO).

The 1.2-m Oschin Telescope, which is designed for widefield viewing, still uses glass photographic plates to record images. A SCHMIDT TELESCOPE, it carried out the famous National Geographic–Palomar Sky Survey in the early 1950s, producing an atlas of the entire sky north of DECLINATION –33°. Each 6° square segment of the sky was photographed on both blue- and red-sensitive plates, giving a total of 879 pairs of plates. A second sky survey is under way to obtain complete multiwaveband photographic coverage of the northern sky at a second EPOCH.

◄◄ ***Palomar Observatory*** *Painting of the Hale Telescope by J D Cremi.*

Panspermia

A theory of the origin of life that assumes that living spores are present everywhere in the universe and can travel in the cosmos, pushed by the pressure of radiation from stars, for example. When cosmic spores reached the solar system they seeded the Earth. The theory, originated in the first decade of the 1900s by Svante ARRHENIUS, was rekindled in the 1970s by Fred HOYLE and Chandra Wickramasinghe, who controversially identified spectral features in starlight as caused by bacteria in the intervening dust. They also proposed that comets carry bacteria and viruses, and deliver them to planets. A limited form of the theory suggests that life, having originated by the evolution of primitive organic chemistry on one planet or another (*see* EXOBIOLOGY), could be transported to another planet on rocky ejecta sent into space by meteor impacts, like the SNC METEORITES.

Parabola

A CONIC SECTION obtained by cutting a right circular cone by a plane parallel to the side of the cone. Such a curve is open (it does not form a closed shape such as an ellipse or circle). It has an ECCENTRICITY of 1, and stretches to an infinite distance with the two arms tending to become parallel.

Parallax

The trigonometric parallax of a star is the angle subtended by one astronomical unit (AU) at the star's distance from the Sun, and is a measure of the star's DISTANCE.

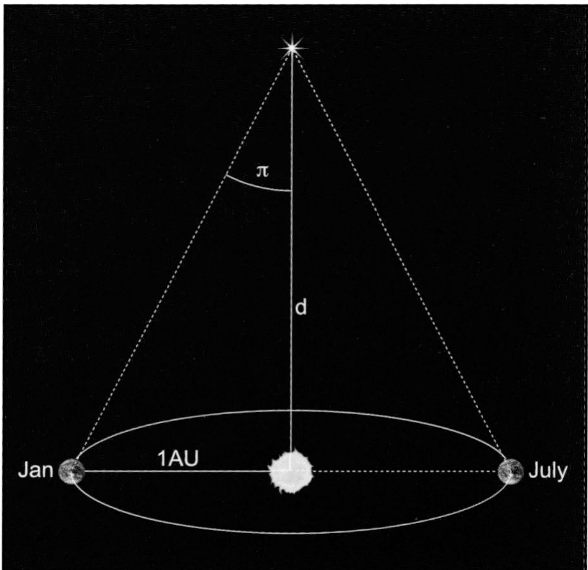

◄ ***Parallax*** *The angle π is the parallax of the star at distance d. The angle 2π can be observed from the Earth at times of the year six months apart.*

Paris Observatory

The Paris Observatory is the largest French institution dedicated to astronomy. Created in 1667 by Louis XIV to define the Paris MERIDIAN, and then devoted to metrology, CELESTIAL MECHANICS and positional astrometry, it further developed many branches of astrophysics and is now established on three sites: the initial Paris Observatory; the MEUDON OBSERVATORY (southwestern suburb of Paris), added in 1926; and the radio astronomy facility of Nançay (central France), added in 1953. The observatory provides the astronomical community with services related to its research activities (for example, time and frequency metrology, solar monitoring, and the International Earth Rotation Service).

Parkes Observatory

Located in rural New South Wales, 400 km west of Sydney, Australia. It is home to the Parkes Radio Telescope, a 64-m fully steerable paraboloid. Commissioned in 1961, the telescope is operated as part of the AUSTRALIA TELESCOPE NATIONAL FACILITY, and much of its time is allocated to international pulsar, galaxy and VLBI programs, with occasional support for NASA DEEP SPACE missions (it was famously used in 1970 to communicate with the stricken Apollo 13 spacecraft). The H I Parkes All-Sky Survey for galaxies is a major project.

Particle physics

The physics of ELEMENTARY PARTICLES. *See also* ASTROPARTICLE PHYSICS.

Pauli exclusion principle

A principle of QUANTUM THEORY, devised in 1925 by Wolfgang Pauli (1900–58), which states that no two fermions (particles with spin 1/2, like electrons and neutrons) may occupy exactly the same quantum state. For example, the two electrons in the lowest energy level of a helium atom have opposite values of spin. The principle limits the extent to which electrons or neutrons can be squeezed together. This gives rise to a pressure, called degeneracy pressure, which supports compact stars such as white dwarfs and neutron stars, and the cores of ordinary stars.

Pavo *See* CONSTELLATION

Payne-Gaposchkin, Cecilia Helena (née Payne) (1900–80)

Astronomer, born in England. She was the first woman to become a full professor at Harvard University, USA. She worked on stellar atmospheres, and in her 1925 dissertation suggested correctly that the great range in strength, from star to star, of absorption lines in stellar SPECTRA was due to differing amounts of IONIZATION (differing temperatures), not differing chemical composition. She suggested that HYDROGEN was the most abundant element in stars. This conclusion was at first resisted but accepted by 1929, in particular by Henry RUSSELL. In 1934 she married the astronomer Sergei Gaposchkin (1898–1984).

Peculiar galaxy

A galaxy that cannot be easily classified as being spiral, elliptical or irregular, based on its optical shape. Peculiar galaxies constitute between 5% and 10% of the known galaxy population, and the vast majority can be attributed to strong gravitational tides generated in the close passage of two galaxies, to the extent that the terms 'peculiar galaxy' 'colliding galaxy' and 'interacting galaxy' are virtually synonymous.

Pegasus *See* CONSTELLATION

Pelican Nebula (IC 5070)

An EMISSION NEBULA in the constellation Cygnus, at RIGHT ASCENSION 20 h 50.8 m, DECLINATION +44° 21'. Its eastern border, the 'pelican' profile, is delineated by dark nebulosity which separates it from the NORTH AMERICA NEBULA (NGC 7000).

Penumbra

In an eclipse, the region of partial shadow. From points on the Moon in the penumbra in a lunar eclipse, the Sun may be seen partially covered by the Earth. From points on the Earth in the penumbra of a solar eclipse, the Sun may be seen partially covered by the Moon, as a partial solar eclipse. *See also* SUNSPOT, FLARE AND ACTIVE REGION; *contrast* UMBRA.

Penzias, Arno Allan (1933–)

Radio scientist born in Munich, Germany. He won the Nobel prize in 1978 for the discovery of COSMIC MICROWAVE BACKGROUND radiation. A refugee from Germany at the age of six, he found his way to America. Penzias joined Bell Laboratories in Holmdel, New Jersey, and searched for and investigated line emission from the interstellar oxygen–hydrogen (OH) molecule. He gained the use of a large radio telescope (*see* RADIO TELESCOPES AND THEIR INSTRUMENTS), the Holmdel horn, and a new ultra-low-noise maser receiver, and with ROBERT WILSON began a series of radio astronomical observations intended to make the best use of the extreme sensitivity of the system, including a measurement of the radiation intensity from the Galaxy at high latitudes. They discovered the cosmic microwave background radiation. They also made a millimeter-wave receiver and discovered a number of interstellar molecular species.

Periapsis

The point in an elliptical orbit at which the orbiting body is closest to the body it is orbiting. The prefix 'peri-' may be attached to various words or roots depending on the body being orbited. *Contrast* APOAPSIS, the furthest such point.

Periastron

The PERIAPSIS of one component of a BINARY STAR to the other (the closest approach). The term is also used for the closest position of a planet orbiting a star other than the Sun. *Contrast* APHELION, the furthest such point.

Perigee

The PERIAPSIS of the Moon or an orbiting spacecraft to the Earth. *Contrast* APOGEE, the furthest such point.

Perihelion

The PERIAPSIS of a planet or other object to the Sun. The Earth reaches perihelion on January 3–5, when it is about 147.5 million km from the Sun. *Contrast* APHELION, reached on July 4–6 (152.6 million km).

Perseids

The best-known METEOR SHOWER and one of the most prolific, occurring in late July and August. In medieval times the shower was known as the Tears of Saint Lawrence, who was martyred on August 10, AD 258, close to the time of maximum Perseid activity. The RADIANT begins in the constellation Cassiopeia but moves most of the time through northern Perseus.

Perseus *See* CONSTELLATION

PG 1159 stars

There are about 30 PG 1159 stars, which are named after the prototype PG 1159-035 (GW Virginis). The PG prefix comes from its discovery in the Palomar–Green catalog of blue stars. PG 1159 stars are very hot, X-ray emitting stars. They are located in the HERTZSPRUNG–RUSSELL DIAGRAM between the central stars of PLANETARY NEBULAE and the hottest WHITE DWARFS. They are thought to be stars making the transition between the two.

Phaethon

An Apollo asteroid (*see* NEAR-EARTH ASTEROID) discovered by Simon Green and John Davies in 1983 from data returned by the IRAS satellite; it is designated (3200) Phaethon. The asteroid follows a highly elliptical orbit. At PERIHELION it is just 0.14 AU from the Sun, well within the orbit of Mercury, and at APHELION 2.40 AU. Its orbital period is 1.43 years, INCLINATION 22°, ECCENTRICITY 0.89. It rotates with a period of about 4 hours. The discovery of Phaethon solved the mystery of the GEMINID meteor shower's 'missing' parent comet, for the orbital elements of the asteroid match those of the Geminid METEOR STREAM. Phaethon may therefore be an extinct comet.

Phase

In the different phases of its motion a celestial body of the solar system, as seen from the Earth, shows different apparent shapes. This comes from the relative positions of the Earth (the location of the observer), of the body (that is observed) and of the Sun (which is illuminating the body). The phase angle is the Sun-object-observer angle, for example, the angle between the Sun and an observer on the Earth as seen from the Moon. The different successive shapes of the illuminated body are the named phases.

Moon phases

One complete cycle of the Moon's phases is termed a lunation, and is completed in just over 29.5 days (the Moon's synodic period). The new Moon at opposition is, strictly, invisible, but the name 'new Moon' is informally applied to the thin crescent visible a day or two afterwards (at the earliest about 18 hours afterwards). The phases of the Moon are:

- new Moon or conjunction, phase angle is 180°, age 0 days;
- crescent Moon, age 1–7 days;
- first quarter or quadrature, phase angle is 90°, age about 7 days, magnitude –10.1;
- waxing gibbous Moon, age 8–14 days;
- full Moon or opposition, phase angle is 0°, age about 15 days, magnitude –12.7;
- waning gibbous Moon, age 16–21 days;
- last quarter or quadrature, phase angle is 90°, age about 22 days, magnitude –10.1;
- last crescent, age 23–29 days;

Maximum phase angles of outer planets

Planet	Maximum phase angle (°)
Mars	20
Jupiter	6
Saturn	3
Uranus	2
Neptune	1

- new Moon or conjunction, phase angle is 180°, age about 29 days.

The numerical phase of the Moon is the illuminated area expressed as a fraction of the disk, that is, 0.5 [1 + cos (phase angle)], varying from 0 at New Moon to 1.0 at Full Moon.

Phases of the planets and their satellites

All the bodies of the solar system, and especially the planets, show phases. For the inner planets (Mercury and Venus), the phases are just like the Moon's. The phases of Venus were first seen by GALILEO GALILEI in the seventeenth century, and proved the heliocentric theory.

In the case of the outer planets and their satellites (Mars, Jupiter, Saturn, Uranus and Neptune), the phase angle is always much less than 90° since these planets are never between the Earth and the Sun (*see* table).

The distribution of light over the illuminated surface of a planet at different phases provides information about the nature of the planet. The 'flat' aspect of the Moon, independent of the phase, shows that the Moon has no atmosphere. In contrast, the dimming of the LIMB of the planet Jupiter shows that it does have one. Whether the line marking the terminator between the illuminated side and the dark side of a planet is sharp or not is an indicator of the duration of twilight due to an atmosphere. The nature of the surface of the planet affects the light distribution, including geometrical and diffusing effects (Lambert's law) and the granularity of the

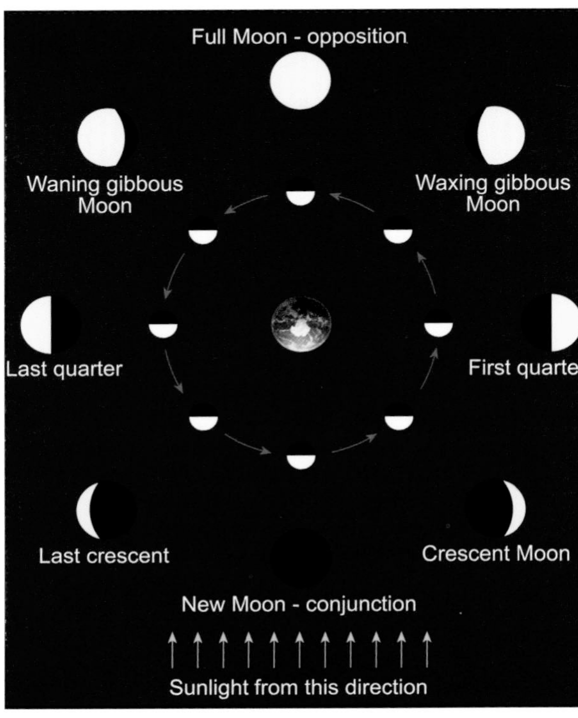

◀ ***Phases of the Moon** through its 29.5-day cycle.*

Phobos and Deimos		
	Phobos	**Deimos**
Semi-major axis	9878.5 km (2.76 Mars radii)	23 459 km (6.92 Mars radii)
Period	7 h 39 min	30 h 18 min
Eccentricity	0.0152	0.0002
Inclination	1.02°	1.82°
Mass	10.8×10^{15} kg	1.8×10^{15} kg
Axes (radii)	13.4 × 11.1 × 9.3 km	7.5 × 6.2 × 5.4 km
Density	1.9 g cm^{-3}	1.7 g cm^{-3}
Escape velocity	12 m s^{-1}	6 m s^{-1}
Albedo	0.071	0.068

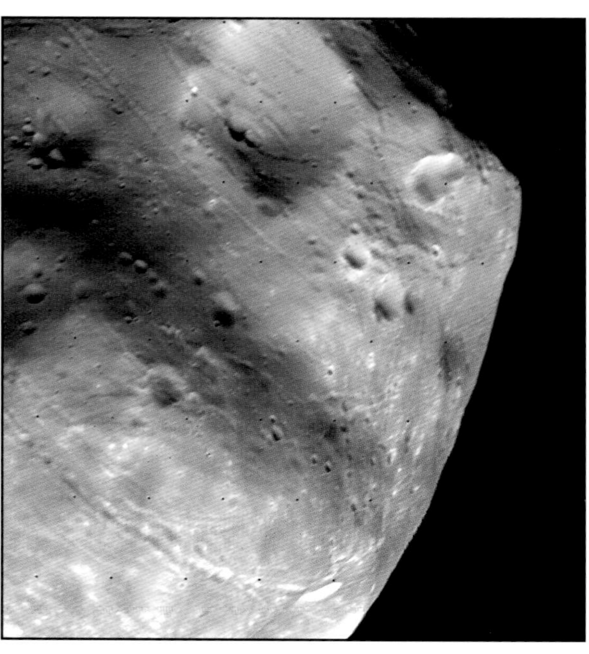

► ***Phobos*** *Viking 1 orbiter view of Phobos showing the south pole and Hall crater. The crater is at the top center, and the south pole is just inside the rim of the crater. The image spans about 21 km from top to bottom.*

dust and surface features (Hapke's law). At opposition, the light from an irregular surface increases suddenly owing to the disappearance of shadows.

Phobos and Deimos

Asaph Hall (1829–1907) discovered Phobos and Deimos, the two moons of Mars, in August 1877. They already had a literary history, however, because two satellites of Mars are mentioned by Jonathan Swift in *Gulliver's Travels* (1726) and by Voltaire in *Micromégas* (1750). These were presumably guesses based on Johannes KEPLER's view that Mars should have two satellites since Mercury has none, Earth has one, and Jupiter four.

Both have a direct circular motion in the equatorial plane of Mars, at very short distances from the planet (9400 and 23 500 km). Phobos is so close that it is located well inside the synchronous orbit of Mars, so it rises in the west and sets in the east twice a day as seen from Mars. This also means that Phobos is slowed down by tidal forces and is spiraling toward Mars. It will crash to the surface in 40 million years time. Like most satellites in the solar system they are in SYNCHRONOUS ROTATION, with their longest axis turned toward the planet and the shortest one perpendicular to the orbital plane.

They are very dark and small (with diameters of approximately 27 and 15 km) and are difficult to observe from the ground, particularly Phobos, the closer to Mars. The satellites were imaged by Mariner 7 (in 1969), Mariner 9 (1971), Viking (1977) and Phobos-2 (1989). They are both very irregular. Phobos has three large impact craters: Stickney (10 km diameter), Hall and Roche (5 km each). On Deimos the largest identified crater is only 3 km across. Deimos is smoother than Phobos. Phobos has a REGOLITH up to 200 m thick, a consequence of the impacts. On Deimos the regolith is much thinner, but it fills most of the craters.

The two satellites are ASTEROIDS formed well beyond the orbit of Mars, captured after the ACCRETION of the planet, and brought to a circular and equatorial orbit around it. The capture and evolution mechanism is unknown, however. The orbit of Deimos in particular is stable on very long time scales and cannot result from a more eccentric, high-inclination orbit under the single influence of tidal forces. Theories put forward include chaotic tumbling, collisions, gas drag on a dissipating nebula, and the splitting apart of a single object in synchronous orbit by a collision.

Further reading: Sheehan W (1996) *The Planet Mars: A History of Observation and Discovery* University of Arizona Press.

Phoenix *See* CONSTELLATION

Photometry

The science of the measurement of light. *See also* MAGNITUDE AND PHOTOMETRY.

Photon

An ELEMENTARY PARTICLE of, for example, light, X- or gamma-rays, that is a discrete packet, or 'quantum', of electromagnetic energy. As expressed by wave-particle duality, photons also behave like waves of ELECTROMAGNETIC RADIATION. The photon is the force-carrying particle (gauge boson) that conveys the electromagnetic force between charged particles. It has zero mass, zero charge, and a spin of 1.

Photosphere

The region of a star's atmosphere from which most of its light comes. It is the region that we see as the 'surface' of the Sun, the region of the atmosphere of a star characterized by its continuous spectrum, with absorption and emission spectral lines coming from the thinner regions above.

Piazzi, Giuseppe (1746–1826)

Born in Ponte di Valtellina, Italy, he became a monk, professor of theology and professor of mathematics in Palermo, where he set up an observatory in 1789. He discovered the first minor planet, CERES, on January 1, 1801, but made only three observations before it was lost behind the Sun. Fortunately, Carl Friedrich GAUSS had developed mathematical techniques with which he could calculate the orbit and Ceres was recovered. The thousandth ASTEROID discovered was named Piazzia in his honor, as was the British astronomer Charles Piazzi SMYTH.

Pic-du-Midi Observatory

The Pic-du-Midi Observatory is located in the Pyrenees, 2876 m above sea level; it is the highest observatory in France, and is part of a multidisciplinary research unit called the Observatoire Midi-Pyrénées (OMP) which also encompasses space research, geophysics and so on. Construction of the observatory began on the site of a meteorological station in 1878. The observatory was affiliated

◀ ***Pipe Nebula*** *The dark nebulae (B59, B65–7, B78) are silhouetted against the massed star clouds of the Milky Way, stretching out toward Antares. The smaller S-shaped Snake Nebula (B72) can be seen in the 'smoke' coming out of the pipe.*

to the Toulouse University Observatory in 1903. The main instrument is the 2-m Bernard Lyot Memorial Telescope, which began operation in 1980. The observatory is being developed as a cultural center for the Bigorre region in the Hautes-Pyrénées.

Pickering, Edward Charles (1846–1919) and Pickering, William Henry (1858–1938)

Brothers, born in Boston, Massachusetts, USA. Edward became director of the Harvard College Observatory and observed the brightnesses of 45 000 stars. He hired a number of women, including Williamina FLEMING, Annie CANNON, Antonia Maury (1866–1952) and Henrietta LEAVITT, and produced the HENRY DRAPER CATALOG, classifying the spectra of hundreds of thousands of stars. He and Hermann Vogel (1841–1907) discovered independently the first spectroscopic BINARY STARS.

Edward's brother, William, worked in Harvard's astronomy department with the new astronomical detector, dry-plate photography. He was the first person to discover a satellite (Phoebe, Saturn's ninth moon) by photography, in 1899. William established an outpost observatory for Harvard at Arequipa, Peru, and used its telescopes to publish sensationalist accounts of martian canals, rather than for the stellar spectroscopy for which they had been established. For this and William's inability to stay within budget his brother dismissed him from his post at Arequipa. William formed an alliance with Percival LOWELL and the two of them, astronomer and businessman, founded the LOWELL OBSERVATORY to observe Mars. The careers of William and Lowell overlapped later when William, independently of Lowell, predicted the position of a trans-Neptunian planet using the discrepancies in both the orbits of Uranus and Neptune as data, without it being found.

Pictor *See* CONSTELLATION

Pinwheel Galaxy (M101) *See* MESSIER CATALOG

Pioneer missions

Development of the first US Pioneer space missions began with the advent of the space age in 1957. The series included the first probes of space outside the Earth's influence, the first passage through the asteroid belt, the first flyby of Jupiter, the first flyby of Saturn, and the first to escape the solar system. The final Pioneer mission was the USA's first intense study of Venus.

Pioneers 1–5 (1958–60) were the USA's first probes into space beyond altitudes of the early satellites, to measure radiation and detect micrometeoroids.

Pioneers 6–9 (1965–9) orbited the Sun independently from the Earth, not far inside and outside Earth's orbital path, to measure the SOLAR WIND.

Pioneers 10 and 11 (1972 and 1973) explored beyond Mars (1.6 AU), through the asteroid belt (2.5–4 AU), near Jupiter (5 AU) and beyond. Pioneer 11 later made a close reconnaissance with Saturn (10 AU). As the first man-made craft to escape the solar system, Pioneer 10 carried a plaque to identify where in our Galaxy it originated and to introduce ourselves to its discoverers. Pioneer 11 carried a duplicate of the plaque.

The Pioneer Venus Orbiter and Multiprobe spacecraft (1978) mapped Venus's cloud-shrouded surface by radar, examined the planet's upper atmosphere, and explored the interaction of the solar wind with Venus's atmosphere over a 14-year period. The Orbiter also measured the rate of water loss from COMET HALLEY. The Multiprobe deployed four probes into Venus's atmosphere, describing the atmosphere down to the surface where lead and zinc would melt.

Pipe Nebula (B59, B65–7, B78)

A large DARK NEBULA in the constellation Scorpius that resembles a tobacco pipe. Edward BARNARD assigned five separate catalog numbers to sections of the nebula. B78 is the 'bowl,' about 3.5° by 2.5°, while B59 and B65–7 comprise the 'stem.'

Pisces, Piscis Austrinus *See* CONSTELLATION

Plage

A region of intensified emission in the solar CHROMOSPHERE. The name derives from the French word for 'beach,' an allusion to the fact that these features stand out like bright sandy beaches against the fainter background of the chromosphere as a whole. Plages are regions of enhanced temperature and density that float in the chromosphere, heated by magnetic fields.

Planck Surveyor

ESA mission to measure temperature fluctuations in the COSMIC MICROWAVE BACKGROUND with ultimate precision and angular resolution, including measurements of POLARIZATION. The spacecraft, named after Nobel prizewinning physicist Max Planck (1858–1947), is planned for launch with FIRST in 2007 (*see* HERSCHEL SPACE OBSERVATORY). It will carry a 1.5-m telescope and two instruments to cover frequencies between 30 and 857 GHz.

Planet

The definition of a planet has evolved as astronomers have gained greater understanding of the structure of the solar and other planetary systems. Originally a 'wandering' body in the sky, the word 'planet' was at first the name for the Moon, the Sun, Mercury, Venus, Mars, Jupiter and Saturn – that is, the celestial bodies that are not 'fixed' stars. With GALILEO GALILEI's discoveries, the nature of the Sun as a star became clearer, together with the identification of the Earth as a planet like Mercury, Venus, Mars, Jupiter and Saturn, and the Moon as a satellite like the moons of Jupiter. The Sun is self-luminous, whereas a planet's (and a satellite's) light is reflected sunlight. A planet came to be defined as a body both in an independent orbit around the Sun (or star) and not self-luminous. The word 'independent' is meant to distinguish planets from satellites, which accompany a planet.

As the nature and history of solar system objects emerged, the vocabulary failed to keep up with the understanding. In the discussion of some topics, like planetary formation and evolution, solid telluric planets like the Earth would be distinguished from gas-giant planets like Jupiter. The larger satellites could then be identified as planets in such a discussion, the Galilean satellites having formed around Jupiter and evolved analogously to the telluric planets revolving and evolving around the Sun. On the other hand, although COMETS are in independent orbits and have no light of their own, they are not planets, on grounds of their different composition and behavior (for example, they are ice, not solid or gas, and develop tails). When discovered, ASTEROIDS were classified as planets (they orbit the Sun, reflecting its light), but denoted MINOR PLANETS, to distinguish them from the MAJOR PLANETS, with a boundary at about 1500 km diameter (which separates Pluto at 2300 km from Ceres at 950 km).

Some modern definitions of planet, based on size, considered that worlds with a diameter greater than about 400 km would have a strong enough gravity to settle to a spherical shape, and that this feature might be considered to be part of the definition of a planet. This definition would make several of the larger asteroids 'planets.' In any case, recognizing their distinctive history, astronomers came to prefer the word 'asteroid' to 'minor planet.' METEOROIDS also behave like planets but are considered too small to be planets or even minor planets, so there is a lower size limit for a planet (presumably about 1 m). The newly discovered Kuiper Belt objects (the largest of them about 800 km in diameter) are asteroid- and/or comet-like (*see* OORT CLOUD AND KUIPER BELT). With the growing realization that Pluto has been a Kuiper Belt object, its status as a planet has been called into question. But a poll of astronomers in 1999 showed that they were reluctant to downgrade Pluto's status. Like asteroids, at least some Kuiper Belt objects could be called small planets.

With the discovery of EXTRASOLAR PLANETS, the definition of a planet has become clear again in principle, mainly because virtually the only property of an extrasolar planet that can be estimated is its mass. An object is discovered, orbiting another star and with a mass somewhere between Jupiter and the Sun. Is it a star or a planet? Although there is no agreed lower size limit that determines what constitutes a planet, there is a consensus about the upper limit, based on the consideration that stars are self-luminous and planets are not. This criterion needs to be carefully drawn to take account of the ability of planets to radiate infrared radiation derived from radioactive decay or gravitational settling. This points to a definition in which a planet has no internal nuclear fusion processes that lead to it radiating energy from its surface. A planet does not burn hydrogen (as MAIN SEQUENCE stars do), nor even DEUTERIUM (like BROWN DWARFS). This leads to an upper limit to the mass of a planet of between 10 and 80 Jupiter masses. If a planet is defined in this way there is no reference to its environment or formation process, so on this definition there is no assumption that the planet has been formed in association with a star, or is in orbit around one. Isolated planets, not orbiting a star and perhaps formed independently in space like one, are a possibility, although their properties might be different from planets formed in planetary systems. *See also* PLANETS: ORIGIN.

Planetarium

A domed building housing a projector that is used to simulate the night sky; or the associated institution devoted to popular education in astronomy. The term 'planetarium' was originally used to describe teaching devices, known as ORRERIES, designed to portray the orbital motions of the planets and their satellites. The first planetarium was opened at the Deutsches Museum in Munich, Germany, in 1923. The larger planetariums have an extensive exhibition area, museum collections, and can seat more than 600 people. Since the 1940s the most common opto-mechanical projectors have been manufactured in Germany, Japan and the USA. With the aid of gear-driven rotating mounts, they can place the planets, Sun and Moon in their correct locations among the stars for thousands of years in the past and the future. These devices are now being superseded by computer-controlled, optically-projected bright screens. The screens are highly versatile (for example, they can project simulations of a supernova, star evolution or the Big Bang) but cannot yet match the opto-mechanical projectors for the brightness range and spatial resolution that we see in the night sky.

Planetary interior

To understand the formation and history of a planet requires knowledge of its composition and how the constituents are distributed. Many features of the planet's surface have their roots inside, such as volcanoes, magnetic fields or heat flow. The pressure inside a planet is high because of the weight of overlying material. The internal pressure of the Earth's Moon is typically around 50 000 bars (50 kilobars). For the Earth itself, the pressure is 2 million bars (2 megabars). For Jupiter, the largest planet, the estimated 'typical' pressure is about 10 megabars. These numbers are of interest because it takes about one electron-volt per atom to compress material to a megabar pressure, and the electron-volt is the unit that measures the natural energy of an atom. In other words, the pressures within planets are sufficient to change the structure of materials and modify greatly their behavior.

Moreover, the material of a planet forms an insulating blanket, so planets have high internal temperatures, irrespective of whether they are hot or cold externally. The central temperature of the Earth is about 6000 K, and temperatures deep within Jupiter are 10 000 K or more. Even the Moon has a central temperature of around 1600 K. Planets are hot enough to be partly, or, in the case of GIANT PLANETS, mostly molten. In planets larger than Mars, the high internal temperatures are assured just by the large gravitational energy of formation, much of which is unavoidably converted into heat. These high temperatures cause gravitational DIFFERENTIATION (the separation of a dense iron core from a less dense silicate mantle), particularly during the planet's early history.

Composition

Planets are made principally of 'gases,' 'ices' and 'rock.' These labels refer to composition and are listed in order of decreasing volatility (their tendency to evaporate easily or resist condensation). Quotation marks are needed to remind us that these labels do not refer to the state of the material (for example, 'gas' can be a metal, as explained below). 'Gases' refers primarily to HYDROGEN and HELIUM, which do not condense (to form a solid or a liquid) under conditions encountered during the formation of planets. Jupiter and Saturn are the only planets in our solar system that are predominantly gas. 'Ices' refers to the dominant compounds involving oxygen, carbon and nitrogen, and especially to water. The ices partially condense under the cold conditions encountered beyond the asteroid belt and are thus important in all objects from Jupiter outward. 'Rock' refers to the least volatile materials: the silicates, oxides and metallic iron that dominate all bodies interior to Jupiter. The behavior of all these materials can be very different from our everyday experience, because of the high pressures and high temperatures. Thus, 'gas' becomes a liquid metal or semi-metal at the extreme pressures inside Jupiter and Saturn.

The giant planets, Jupiter, Saturn, Uranus and Neptune, are ices and gases. At similar densities but much smaller radii there are bodies that have comparable amounts of ice and rock (Ganymede, Callisto, Titan, Pluto and Triton). The bodies that are primarily rocky include Mercury, Venus, the Earth, the Moon, Mars, Io and perhaps Europa (though the latter has a significant component of water).

Structure

The internal structure of a planet can in principle be determined from seismology. Seismology of the Earth has enabled scientists to determine the thickness of the crust, to identify major mineral phase transitions in the mantle, and to establish that the outer core is fluid and the inner core is solid. However, seismology requires seismometers in place on the surface: we have only limited seismic data for our Moon, little yet of value for Mars, and none for any other planet.

Planets are not spherically symmetrical and, as a consequence, the external gravity field has a rich structure that influences the trajectory of natural satellites and spacecraft. This structure is our main way of estimating the internal structure of all giant planets and also the Galilean satellites.

Internal heating

There are two major sources of heat in planets: radioactivity and gravitational energy release. In the terrestrial planets (including the Moon), the dominant source of heating is energy released by long-lived radioactive elements (primarily uranium-238, thorium-232 and potassium-40). In Jupiter, Saturn, Uranus and Neptune, gravitational energy dominates, partly because these bodies are so massive but also because only a small fraction of the mass is in the rocky component that carries radioactive elements.

Gravitational energy was converted into heat as the planet formed and perhaps differentiated, and may be an ongoing process (the gradual settling of heavier material). Tidal flexing is an additional source of heat. Some satellites, notably Io and Europa, are in eccentric, close orbits and are flexed by the tides. Heat is dissipated by the flexing. Io's heat flux is roughly 30 times greater than the Earth's, and as a consequence Io is the most volcanically active body in the solar system.

The mantles of terrestrial planets consist of a mixture of minerals and so have a range of melting temperatures rather than a single melting point. The melting is expressed at the surface as volcanism and the formation of crust. The melt freezes and forms basalt, which is also a mixture of minerals (those that melt most readily). Basaltic volcanism has dominated the volcanic history of all but the earliest periods for all terrestrial bodies.

Magnetism

Magnetism can arise from the permanent magnetism of iron or magnetite, but also from large-scale currents in an electrical conductor. At most locations on Earth, permanent magnetization accounts for only one part in 1000 or less of the magnetic field. In planets, large dipole fields (for example, surface fields of about 1 microTesla) are caused by macroscopic electrical currents sustained by structures like the GEODYNAMO. Earth is the only terrestrial planet that possesses a large field, but Mars shows evidence of an ancient large field (expressed in unusually high permanent magnetism of ancient rocks). Mercury has a weak global field and Venus has none. The Moon may once have had a large field.

Planetary magnetosphere

The region surrounding a planet that has its own magnetic field. MAGNETOSPHERES are primarily a topic of the space age, starting with the discovery of the VAN ALLEN BELTS around the Earth in 1957.

A magnetosphere is a magnetic cavity in the SOLAR WIND. The term magnetosphere does not imply a spherical shape but is used in a looser sense, as in the phrase 'sphere of influence.' A planet's satellites may be embedded within its magnetosphere, but if a moon has sufficiently strong magnetization it may carve out its own magnetosphere within the magnetosphere of the parent planet.

Planetary magnetospheres

Planet	Rotation period (days)	Dipole moment (Earth = 1)	Field at equator (Gauss)	Tilt of dipole	Source of plasma	Typical distance of magnetopause (million km)
Mercury	59	0.0007	0.003	+14°	W	0.004
Earth	1	1	0.305	+10.8°	A, W	0.07
Jupiter	0.41	20 000	4.2	-9.6°	S, A, W	6.0
Saturn	0.44	600	0.20	<-1°	S, A, W	1.2
Uranus	0.72 (retro)	50	0.23	-59°	A, W	0.5
Neptune	0.74	25	0.14	-47°	S, A, W	0.6

W = solar wind; A = atmosphere; S = satellites or rings
Note: The Gauss is a convenient, but now non-standard, unit to measure weak magnetic fields such as those of planets. 1 G = 100 microTesla

An illustration of a 'generic' magnetosphere accompanies the magnetosphere entry. A surface called the MAGNETOPAUSE divides the magnetosphere from the solar wind. Upstream of the magnetosphere, in the solar wind, there is a standing shock wave called the bow shock. After passing through the bow shock, the solar wind mostly flows around the sides of the magnetic barrier formed by the planet's magnetic field but some solar wind plasma leaks inside. The size of the magnetosphere is the distance, on the sunward side, of the magnetopause from the center of the planet, where the internal pressure of the planet's magnetic field balances the external pressure of the solar wind. However, the magnetic field accompanying the solar-wind plasma merges with that of the planet and stretches the magnetic field out to produce a long, turbulent magnetotail, or wake, on the downwind side of the planet, which can extend up to 100 times farther down stream than the subsolar magnetopause distance. Thus magnetospheres are far from spherical, having an aspect ratio (1:100) that is similar to that of a COMET (or a long pencil).

Magnetospheres contain considerable amounts of PLASMA (electrically charged particles with equal proportions of positive charge on ions and negative charge on electrons) from various sources. The Sun is the main source of plasma in the solar system. However, a small fraction of a planet's IONOSPHERE may have sufficient energy to escape up magnetic field lines and into the magnetosphere. Moreover, the interaction of magnetospheric plasma with any natural satellites or ring particles that orbit within the magnetosphere can generate significant quantities of plasma, such as sulfur and oxygen ions in the case of Jupiter's satellite IO, and water-product ions from the rings and icy satellites (DIONE and TETHYS) of Saturn.

Where do these energetic particles go? Most diffuse inward toward the planet, streaming into the upper atmosphere where they can excite aurorae and deposit large amounts of energy.

The planets in the solar system have retained their heat of formation but lost their original magnetic fields. If a planet has a magnetic field it is because there is an internal dynamo: the planet has a large interior region that is fluid, electrically conducting and convective. The planets and many larger satellites contain electrically conducting fluids: the TERRESTRIAL PLANETS and the larger satellites have liquid iron cores, while the high pressures in the interiors of the giant planets Jupiter and Saturn make hydrogen behave like a liquid metal. In Uranus and Neptune a water-ammonia-methane mixture forms a deep conducting 'ocean.' The fact that some planets and satellites do not have dynamos tells us that their interiors are stable – stratified and not convecting.

Magnetospheres of the individual planets

Six of the eight planets that have been explored by space craft (all of the major planets except Pluto) have magnetic fields. Most are shaped approximately like a dipole, with a north and a south magnetic pole near to the poles of rotation, connected by magnetic field lines which flow around the planet past its equator. The magnetic fields of Uranus and Neptune are the least like a dipole, and the magnetic axes of these two planets are strongly tilted to their rotation axes. There is something different in the way their magnetic fields are generated.

Venus has no magnetic field and does not have an active dynamo. Mars has no large-scale magnetic field, but does have residual traces of a past magnetic field frozen into the lava of its surface, as discovered by the Mars Global Surveyor in September 1997. Mercury has a magnetosphere but little atmosphere. This results in a unique situation where the magnetosphere interacts directly with the outer layer of the planetary crust. Dense gusts of the solar wind break through the magnetosphere and directly strike the planet's surface. At all of the other planets the topmost regions of their atmospheres become ionized by solar radiation to form ionospheres.

The magnetospheres of Jupiter and Saturn are gigantic, dwarfing all the other planetary magnetospheres. The strong planetary magnetic fields are linked to the planet's rotation, which accelerates particles to high energies. The magnetospheric plasma generates powerful radio emissions and stimulates strong auroral emissions (radio and optical) in the planets' polar atmospheres. In fact the discovery of radio emission from Jupiter in 1955 predated the space-age identification of the Earth's radiation belts and the solar wind.

The orientation of a planet's magnetic field is described by two angles: the tilt of the magnetic dipole field with respect to the planet's spin axis; and the angle between the planet's spin axis and the solar wind direction, which is generally radially outward from the Sun. The direction of the spin axis with respect to the solar wind direction varies over a planetary year. Earth, Jupiter and Saturn have both small dipole tilts and small obliquities, so their magnetospheres vary little and slowly.

In contrast, the large dipole tilt angles of Uranus and Neptune mean that the orientation of their magnetic fields with respect to the interplanetary flow direction varies considerably over a planetary rotation period, resulting in complicated, variable magnetospheres. Furthermore, Uranus's large obliquity means that the configuration of the planet's magnetosphere will have strong seasonal changes over its 84-year orbit.

Magnetospheres of natural satellites

Ganymede and Io are the only moons known to have substantial magnetic fields. Earth's Moon has a negligibly small planet-scale magnetic field, although localized regions of the surface are highly magnetized.

Neptune's large satellite TRITON is comparable in size with Ganymede, which raises the question of whether it too could have an internal magnetic field. The large moon orbits at 14.6 Neptune radii, well inside the magnetosphere. Voyager 2 detected protons and nitrogen ions in Neptune's magnetosphere, probably from Triton.

Further reading: Cravens T E (1997) *Physics of Solar System Plasmas* Cambridge University Press; Gombosi T I (1998) *Physics of the Space Environment* Cambridge University Press.

Planetary nebula

A shell of illuminated gas surrounding an old star that is small but hot. A planetary nebula forms from the outer envelope of a once-giant star, and the central star from its core. The central star, which is on its way to becoming a WHITE DWARF, illuminates the envelope, which appears as a planetary nebula. Planetary nebulae were identified by William Herschel, who in 1785 referred to the 'planetary' or 'disk-like' appearance of the object we now call NGC 7009 (*see* HERSCHEL FAMILY). He later discovered that NGC 6543 had a central star with which it was clearly associated. Some 2000 planetary nebulae are known today, their sizes ranging from less than 1 arcsec to 0.25°, one of the most prominent being the Ring Nebula in the constellation Lyra.

William HUGGINS examined NGC 6543 in 1864 with a visual SPECTROSCOPE and found three emission lines, demonstrating that the nebulae are made of low-density gas. The mechanism producing the emissions is the same as for GASEOUS NEBULAE or H II REGIONS. The blue central stars are very hot, above 25 000 K, and emit copiously in the ultraviolet (UV). The energetic UV radiation ionizes the material of the nebula. As the material recombines, it emits the spectral lines.

The relative strengths of the spectral lines yield the conditions in the nebula: temperatures, densities and chemical compositions. Planetary nebulae have broadly normal compositions but with strong enrichments of helium, nitrogen and carbon. These elements were created in the progenitor when it was a giant star and were dredged to the surface of the star as it made the nebula.

The size of a planetary nebula follows from measurements of its angular diameter and distance; distances of planetary nebulae, however, are poorly known. Some planetary nebulae are associated with companions, clusters or, as in the MAGELLANIC CLOUDS, galaxies, and that is enough of a guide. The volume of a planetary nebula and the density of its gas as found from the spectral lines produce the nebula's mass, which is typically a few tenths that of the Sun. But invisible un-ionized gas often surrounds a planetary nebula, so this is a lower limit.

The DOPPLER EFFECT on the emission lines shows the expansion of the shells at typical velocities of a few tens of kilometers per second. The ages of planetary nebulae, from their sizes and expansion velocities, range from a few hundred to tens of thousands of years.

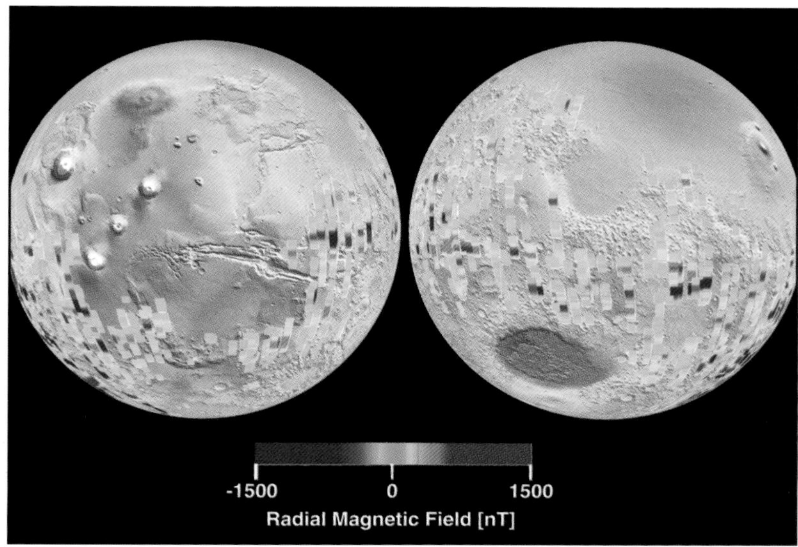

▲ Mars's magnetism *The southern highlands of Mars are magnetized in stripes, evidence of its once-strong magnetic field. Mars now has no living magnetosphere or active geodynamo.*

The central stars

Excepting neutron stars (*see* NEUTRON STAR AND PULSAR), the central stars of planetary nebulae are the hottest known and are exceedingly luminous. They are faint only because most of their energy is radiated in invisible UV radiation. Their temperatures range from about 25 000 K to nearly 300 000 K. Their luminosities range from a few solar luminosities to tens of thousands.

Temperatures and luminosities allow the central stars to be placed on the HERTZSPRUNG–RUSSELL DIAGRAM (HR diagram). The stars enter the planetary nebula area of the HR diagram at high luminosity and low temperature, moving to the left, increasing in temperature to between 100 000 K and 200 000 K. The stars then decrease in luminosity while cooling to temperatures of 70 000 K. After this the stars cannot be tracked because the nebulae have dissipated and the stars cannot be identified. But the track leads directly toward the white dwarf region.

The central stars of planetary nebulae are therefore inferred to be the spent carbon-oxygen nuclear cores of giant stars that are turning into white dwarfs. The central stars range from about 0.6 solar masses for what started as a solar mass star, to about 1.4 solar masses for what started as a 10 solar mass star. In the latter case the mass lost from the progenitor star would be several solar masses, most of it cocooning the planetary nebula.

The chemical enrichments seen in the planetary nebulae have taken place in the progenitors. While the progenitor is on the red-giant branch of the HR diagram, CONVECTION in its envelope penetrates inward to the hydrogen-burning shell and dredges up its products. In higher-mass stars (over 4 solar masses) further dredge-ups take place after helium burning. This brings to the surface helium, nitrogen and carbon. Most of the carbon and much of the nitrogen in the INTERSTELLAR MEDIUM come from planetary nebulae.

Shapes

Some planetary nebulae look circular or elliptical and are therefore presumably spherical or ellipsoidal in shape. Most nebulae have double or even multiple shells, and sometimes vast, extended outer halos. Many are also 'point symmetric,' in which each feature on one side of the nebula corresponds to a similar feature reflected onto the opposite side.

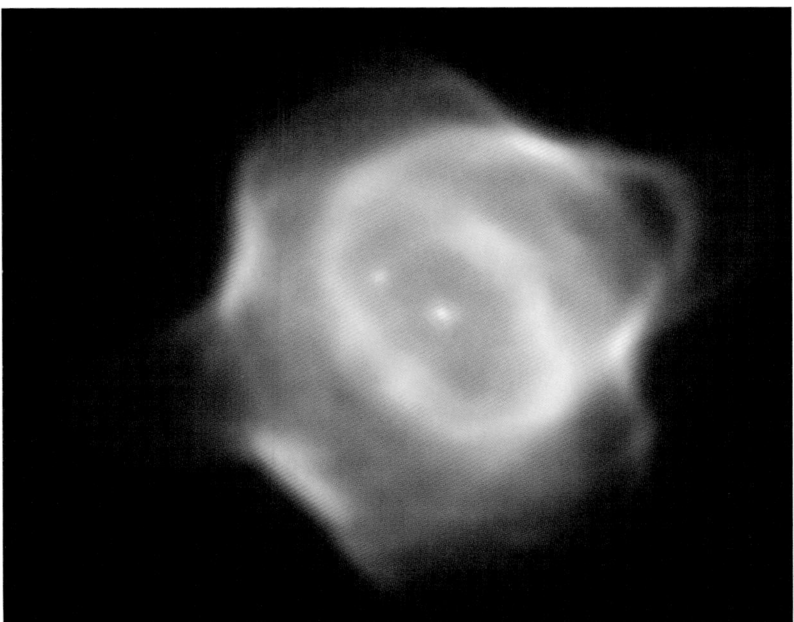

▲ ***Stingray Nebula*** *The bright central star is in the middle of the green ring. There is a companion star diagonally above it, which may have played a part in creating the 'stingray' shape. The red curves represent bright gas heated by a 'shock' caused when the central star's wind hit bubbles of gas. The nebula has appeared in the last 20 years, the central star rapidly heating enough to make the nebula glow.*

Some have complex shapes. Some have two lobes ('bipolar nebulae'). Some circular planetary nebulae, like the OWL NEBULA, have double features (like the owl's eyes) that suggest they are bipolar nebulae seen end on. The more extreme bipolar nebulae have a tight 'waistband' that encircles the middle between two extended poles, giving an hourglass shape. The bipolar nebulae have distinctive common features. They tend to reside close to the plane of the Milky Way, suggesting that they are the progeny of the higher-mass stars. The inference is that high-mass stars eject their envelopes in a different way from stars of lower mass. The bipolar nebulae have small, unexplained 'microstructures.'

Some planetary nebulae have rod-like features that extend along the major axis. They can extend past the main nebula to form jug-handle 'ansae.' In some cases little pieces of the nebula move rapidly, and are called 'FLIERs' (fast low ionization emission regions). The HELIX NEBULA, the closest planetary nebula, contains thousands of comet-like knots that point away from the central star. The knots are filled with neutral gas, molecules and dust, their surfaces heated by starlight. These features all somehow emerged from the progenitor giant stars. The key to the process lies in protoplanetary nebulae.

Protoplanetary nebulae are planetary nebulae in their developmental state, like the EGG NEBULA in Cygnus. It is a bipolar REFLECTION NEBULA whose bright lobes reflect the light of a class F supergiant buried within its thick, dark waistband. Shining along the poles of the torus (donut) of dust are twin beams of starlight that illuminate dozens of shells of bright reflective dust (cooled carbon grains from the progenitor star). The shells were ejected from the star, one every couple of hundred years. As the hot core of the progenitor star in a protoplanetary nebula is uncovered, the old, cool, dusty slow wind of the progenitor star is overtaken by a fast hot wind and shoveled aside, as if by a snowplow. Presumably this is the origin of the microstructures and FLIERs.

If mass is lost symmetrically, the resulting planetary nebula will be round. If the wind is thicker at the equator than at the poles (as in our Sun, *see* CORONAL HOLE), the hot wind will blow more vigorously through the poles, and the nebula will take on an elliptical shape. If the slow wind is yet more concentrated by a waistband, then the resulting nebula becomes bipolar.

It is not known why some planetary nebulae are so asymmetric. Perhaps the progenitor interacts with a binary companion. There is a BINARY STAR explanation for the point-symmetric nebulae, involving PRECESSION of the progenitor caused by an orbiting companion (if not a star then a planet). Indeed, it has been suggested that new planets might form within the dusty waistbands of protoplanetary nebulae.

The final stages

As time passes, the planetary nebulae become so large and tenuous that they and their motions through space are slowed by the interstellar medium. They gradually merge with it. The central stars leave their nebulae behind, the star of one old object actually leaving a wake. They become dim, cooling white dwarfs that orbit the Galaxy forever.

Yet there may be a last surprise. As the central star cools along the track that leads to the white dwarf zone on the HR diagram, the old helium-burning shell can reactivate itself in a final thermal pulse. The new energy source re-expands the star to giant proportions. New thermonuclear reactions dramatically change the chemical composition of the surface. FG Sagittae, which has brightened and cooled over the past several decades, and whose spectrum has become increasingly bizarre, seems to be one of these stars 'raging against the dying of the light.'

Further reading: Gurzadyan G A (1997) *The Physics and Dynamics of Planetary Nebulae* Springer; Pottasch S R (1984) *Planetary Nebulae* Reidel.

Planetary ring

There are planetary rings around all the GIANT PLANETS. The rings are disks of particles that follow nearly circular orbits near the equatorial plane of the central planet. Saturn's rings extend over hundreds of thousands of kilometers and have been known for centuries, but the rings of all of the other planets were discovered in the last decades of the twentieth century.

Saturn's rings

In 1610 GALILEO GALILEI's telescope was not good enough to see the rings of Saturn for what they were. In 1659 Christiaan HUYGENS's telescope revealed them: '[Saturn] is surrounded by a thin flat ring, which does not touch him anywhere and is inclined to the ecliptic.' Soon their structure was revealed, when in 1675 Gian Domenico Cassini observed a gap in what Huygens saw as a single ring (*see* CASSINI DYNASTY). This gap is now known as the Cassini Division. It is a ring less dense than and separating the A ring and the B ring.

The first theories of Saturn's rings, by Pierre-Simon LAPLACE in 1787, suggested that the rings were narrow, solid structures. But in 1849 Edouard Roche (1820–83) calculated at what distance an extended solid in orbit, like a satellite, will break up under the tidal effects exerted by a planet (this distance is known as the ROCHE LIMIT). His answer of 2.44 times the radius of the planet was 'a little farther than the external radius of Saturn's rings,' so solid rings could not survive. Then in 1857 James MAXWELL showed that Saturn's rings could only be an indefinite number of very small particles. This was confirmed in 1895 when James KEELER measured the Doppler shift of the rings. He found that they were orbiting with the inner parts circling the planet quicker than the outer parts (*see* DOPPLER EFFECT). Solid rings cannot do this, but an assembly of individual satellites can (according to Johannes KEPLER's third law).

Astronomers discovered further substructures. In 1850 a ring was discovered inside the B ring independently by amateur astronomer William DAWES, and William and George BOND. The faint, dusky ring (labelled the C ring) extended inward from the B ring almost halfway to the apparent surface of Saturn. William LASSELL christened it the Crêpe Ring. Johann ENCKE discovered the Encke Division in the A ring in 1895. Further rings were discovered photographically and by the PIONEER 11 and VOYAGER spacecraft.

The main rings (C ring, B ring and A ring) are the most opaque. Their total mass is approximately that of the satellite MIMAS. Their particles are primarily water ice. Most are between 1 cm and 5 m in size, but some are as big as a small satellite (10 km). By contrast, the G, D and E rings are thin and consist of micron-sized particles.

Voyager 1 discovered dark, nearly radial features in Saturn's B ring; these finger-like markings are called spokes. They are about 8000 km long and 2000 km wide, develop in several minutes, follow the rotation of the rings, and disappear several hours later. The spokes are dark in reflected light and bright in scattered light, indicating that they are made of micron-sized particles, which are easily moved by electric and magnetic forces. They remain unexplained.

The F ring is intermediate in properties between the main rings and tenuous rings. It is a narrow, inclined, multistranded ring (A, B and C) lying the 3400 km beyond the edge of the main ring system and shows remarkable clumps, kinks and braided structures.

Uranus's rings

Uranus has 10 narrow rings, one diffuse ring, and some tenuous material between the narrow rings. The confusingly varied names of the 10 narrow rings (in order of increasing distance from the planet) are 6, 5, 4, Alpha, Beta, Eta, Nu, Delta, Lambda and Epsilon; they are less than 12 km in extent. The name of the diffuse ring is 1986U1R.

Although these rings can now be imaged from the Earth, they were discovered through stellar occultations (*see* ECLIPSE AND OCCULTATION). In 1977 airborne and ground-based high-speed photometers were used to observe a star that was being occulted by Uranus, with the intention to study the fading of the starlight by the planet's atmosphere. The starlight was unexpectedly blocked by the rings. The rings were imaged in 1986 by the Voyager 2 spacecraft.

Uranus's rings are elliptic and slightly inclined. This is a result of PRECESSION, caused by the planet's ellipsoidal shape.

◄ ***Rings of Saturn*** *Seen by Voyager 2 from a distance of 4 million km. Radial 'spoke' features may be seen at the top left. The spokes are, apparently, radial regions in the rings where the dust density is reduced. Perhaps the dust particles are electrostatically charged and this accounts for the spoke structures.*

Jupiter's rings

These were discovered by the Voyager spacecraft in 1979, and further studied by the GALILEO MISSION. These faint rings are about a million times less dense than those of Saturn, and they are close to the bright planet.

Jupiter's rings are very thin. The main ring is a relatively bright, narrow ring about 7000 km wide. A toroidal halo lies inside the main ring, extending up to 10 000 km above and below the plane. The Gossamer Rings are very tenuous and lie outside the main ring. Seen as a single structure in the Voyager image, Galileo has shown that they are in fact two faint rings bounded by the orbits of the small ring-moons, Amalthea and Thebe.

Orbital properties of Uranus's rings

Ring	Semimajor axis (km)	Radial width (km)	Eccentricity
6	41 837	1.0-2.5	0.001 013
5	42 235	1-7	0.001 899
4	42 571	1-7	0.001 059
Alpha	44 718	4.5-10.5	0.000 761
Beta	45 661	5.5-12	0.000 442
Eta	47 176	55	~0.004
Nu	47 627	1-8	0.109
Delta	48 300	2-8	0.004
Lambda	50 024	1.3-2.5	~0.0
Epsilon	51 149	20-96	0.007 936

Saturn's rings and inner satellites

	Radial distance (km)	Period (hours)	Particle or satellite size	Mass (g)
Saturn's cloud tops	60 330	10.66	-	-
D ring	66 000-74 500	4.9-5.6	10-100 μm	-
C ring	74 500-92 000	5.6-7.9	1-500 cm	1×10^{21}
B ring	92 000-117 580	7.9-11.4	1-500 cm	3×10^{22}
Cassini Division	117 580-122 170	11.4-12.1	1-750 cm	0.5×10^{21}
A ring	122 170-136 780	12.1-14.3	1-500 cm	6×10^{21}
Pan (and Encke Division)	133 580	13.8	10 km	4×10^{18}
Atlas	137 670	14.4	15 km	9×10^{18}
Prometheus	139 353	14.7	45 km	3×10^{20}
F ring	141 220	14.9	0.01 μm-10 cm	$1 \times 10^{17\pm1}$
Pandora	141 700	15.1	40 km	2×10^{20}
Epimetheus	151 420	16.7	60 km	5.5×10^{20}
Janus	151 470	16.7	90 km	1.98×10^{21}
G ring	166 000-173 000	19.9	<0.03 μm	-
Mimas	185 540	22.6	390 km	4.5×10^{22}
E ring	181 000-483 000	~33	1 μm	7×10^{11}
Enceladus	238 040	33	500 km	8.5×10^{22}

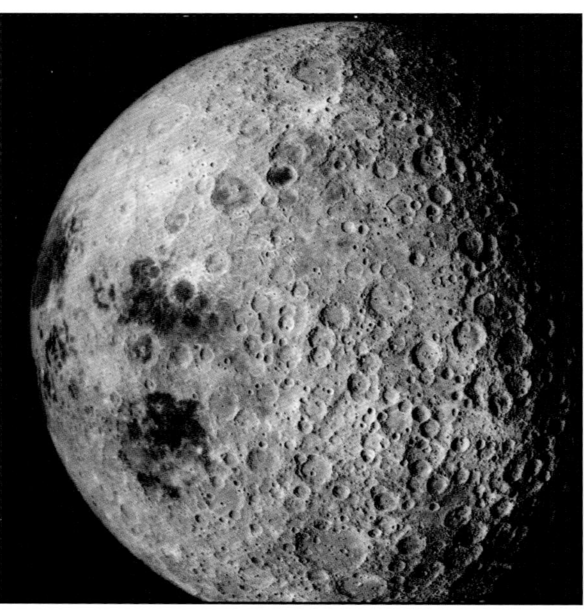

▶ ***Far side of the Moon*** *The Moon's surface is a record of the history of the Moon from its birth about 500 million years ago, through the time of heavy bombardment by meteorites about 3900 million years ago, up to the present day. Apollo 16 astronauts took this picture. The three large dark 'seas' on the left are (from the top) Mare Crisium, Mare Marginis and Mare Smythii.*

Neptune's rings

Like Uranus's rings, Neptune's were found by observations of stellar occultations, in July 1984. A dip in the starlight was observed on one side of Neptune, indicating that it had a ring, but also that the ring was incomplete. This discovery led NASA to reprogram observations of Neptune by the Voyager spacecraft, both to learn more about its ring system and also to avoid colliding with it.

Neptune has four dusty narrow rings named for astronomers involved in the discovery of Neptune. The Adams ring is the outermost ring at a radius of 63 000 km from Neptune (*see* John Couch ADAMS). The Leverrier ring is very narrow at 53 200 km radius (*see* Urbain LEVERRIER). Between the two is Lassell , a broad ring 4000 km wide (*see* William LASSELL). At an orbital radius of 42 000 km lies the broad ring Galle, which is 2000 km wide (*see* Johann GALLE). Finally, a very faint and narrow feature orbits Neptune at the same distance as the satellite Galatea, 1000 km inside the Adams ring.

The Adams ring has three main arcs (named Liberté, Egalité and Fraternité) that span a region of 40° in longitude. There are at least two more arcs, and Egalité itself is separated into two distinct arcs. They are clumps of material that is denser than average.

The nature of the ring particles is uncertain, but there are indications that they are composed of 'dirty ice,' perhaps with silicates or some carbon-bearing material.

Nature and origin

All the rings of the giant planets are located within the Roche limit. The lifetime of the particles in their orbits is much shorter than the age of the solar system. Any primordial rings should have disappeared already. To replenish Saturn's rings would take the equivalent mass of a 200-km-sized satellite every 500 million years. The progenitor of Saturn's denser rings, an icy satellite or a captured comet, drifted within the Roche limit and was disrupted by tidal effects. Interparticle collisions in the rings disperse the particles radially and longitudinally. The gaps between rings have been long known, but the most remarkable discovery in the past 30 years has been the unexpected diversity of structures in planetary rings, such as divisions, sharp edges, spiral waves, narrow ringlets, or longitudinal asymmetries (such as arcs and clumps in narrow rings, or spokes in Saturn's rings).

The gaps between rings are caused by nearby satellites, which 'shepherd' the particles (*see* SHEPHERD MOON), taking them away from resonant orbits. The edge of Jupiter's main ring is maintained by the close satellite Adrastea. The outer edge of the B ring is located at the 2:1 resonance with Mimas, and corresponds to the Huygens gap in the Cassini Division. The outer edge of Saturn's A ring is located at a 7:6 resonance with the satellite Janus. The Encke gap is carved by the moonlet Pan.

Two shepherding moons on each side of a ring confine the ring radially. Saturn's F ring is shepherded by Prometheus and Pandora, and Uranus's ring by Cordelia and Ophelia.

The discovery of Neptune's incomplete arcs set the new question of what confines a ring in longitude. Shepherding theories have predicted satellites, but none have been found. The origin and evolution of the arcs are not understood.

The particles in the diffuse rings are typically micron sized. Mutual collisions between particles in the rarefied rings are uncommon, and their evolution is dominated by electromagnetic effects and radiation drag forces, rather than gravity. The small particles are short lived in the high-radiation environment.

They are replenished with the ejecta produced by meteoroid impacts on the close satellites. The peak density of Saturn's E ring is correlated with the orbit of the icy satellite Enceladus. Jupiter's Gossamer Rings are made of dust that is kicked off of Amalthea and Thebe when they are struck by meteoroids at speeds greatly magnified by Jupiter's huge gravitational field.

Further reading: Elliot J and Kerr R (1984) *Rings: Discoveries from Galileo to Voyager* MIT Press; Greenberg R and Brahic A (eds) (1984) *Planetary Rings* University of Arizona Press.

Planetary rings, observing

See SATURN AND THE OUTER GAS GIANTS, OBSERVING

Planetary Society

Carl SAGAN, Bruce Murray and Louis Friedman founded the non-profit Planetary Society in 1979 to advance the exploration of the solar system and to continue the search for extra-terrestrial life. The society has its headquarters in Pasadena, California, USA, but is international in scope, with 100 000 members worldwide; this makes it the largest space interest group in the world. The society funds a variety of projects and programs, including Red Rover Goes to Mars (for students), the Mars Microphone on the Mars Polar Lander mission, SETI@home, and the Gene Shoemaker Near Earth Object Grants.

Planetary surface

The surface features of planets and their moons result from the complex interaction between internal processes. This leads to tectonic and volcanic activity, atmospheric phenomena, and external processes (for example, irradiation by solar particles and bombardment by meteorites). All of these processes leave scars of history on planetary surfaces, from seasonal to geological time scales.

Heavily cratered terrains

Impact CRATERS, as on the Moon and Mercury, are the result of the very intense bombardment that occurred early in the

◀ ***Planetary surface*** *Volcanoes, tectonic motions and geysers are some of the forces which shape the surfaces of telluric planets. Atmospheric erosion, lightning and running water may be important on the planets with atmospheres.*

history of the solar system, and of sporadic later impacts from residual METEOROIDS and ASTEROIDS. If pictures of the inner planets had been taken 4 billion years ago, immediately after the bombardment, they would all look the same. On the inner planets now, records of this era remain in the form of impact craters of all sizes. Bodies that have not evolved much still show their impacted surface.

Asteroids, planetary satellites and planet areas are labeled 'old' when their surfaces appear heavily cratered, which means that they have not been further remodeled by global activity. As a rule, the smaller the objects are, the less they have evolved after the early bombardment. Thus, by studying the Moon and Mercury we have access to the first billion years of solar system evolution. By contrast, the surfaces of Venus and Europa are very young and show what happened yesterday, in cosmogenic terms.

On the largest of the inner planets, the Earth, geology has erased the record of that early history, but some craters exist from impacts in more recent times. Mars is an intermediate case: about half of its surface has been transformed by more recent activity, while heavily cratered highlands on its southern hemisphere are preserved.

Ice on the Moon, the Earth and Europa (and possibly elsewhere) is in part deposited from cometary impacts, and, whether moving or melted, is a significant externally induced erosion agent. Solid material (meteoroids) is deposited on Earth at the rate of some 60 000 tonnes each year, although this is insignificant compared with the amount of dust mobilized by atmospheric erosion processes.

Tectonic and volcanic features

TECTONICS and volcanism produce features on the surfaces of planets if they have interior activity that penetrates through the crust. On the Moon, the only surface features associated with his internal activity are the maria (*see* MARE). These resulted from lava flows filling the largest, deeper-impact-produced basins.

On larger bodies, internal activity has lifted great areas, such as THARSIS MONTES on Mars or Ishtar Terra on Venus. The rise of these features is sometimes associated with great faults or canyons, like VALLES MARINERIS on Mars. On Earth, internal activity floated and mobilized large tectonic plates (continental drift) that collide and, in folding, produce mountain ranges that are rare or unique in the solar system. In a few cases, tidal effects initiate volcanic activity. The most striking example is IO, the closest of Jupiter's four Galilean satellites, whose entire surface is remodeled by volcanic activity on time scales of only thousands of years.

Features induced by interaction with the atmosphere

Surface structures can also be developed by the atmospheres of planets. Polar caps and frosts occur on Mars, as on Earth; the results of liquid, ice and aeolian (wind) erosion are everywhere. Some erosion processes on Mars happened billions of years ago, such as valley networks; while dunes are formed by dust transportation processes that occur seasonally. Giant dust storms are driven by the condensation of carbon dioxide at the poles, which induces drastic changes in atmospheric pressure. On Earth, the seasonal changes

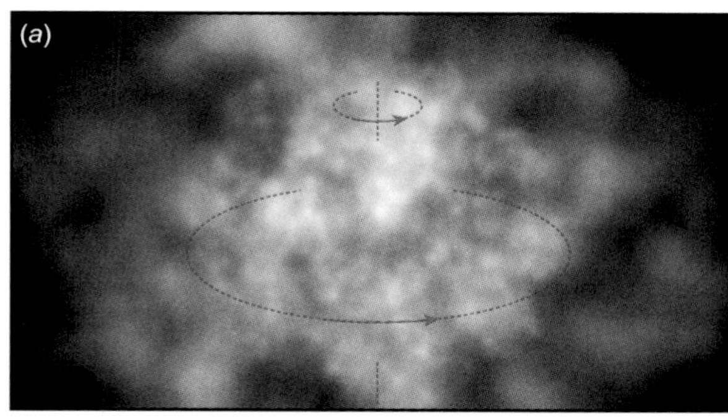

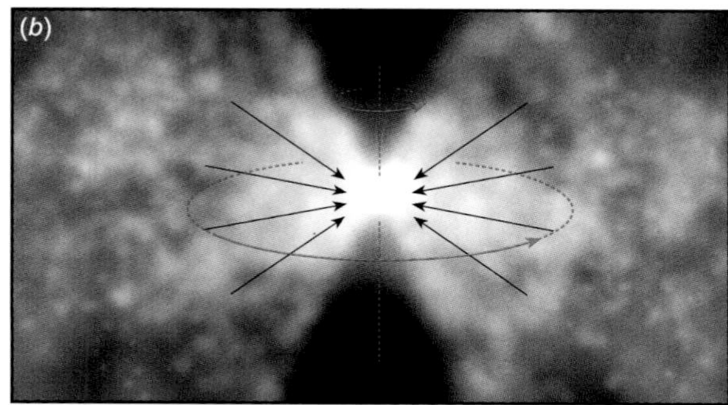

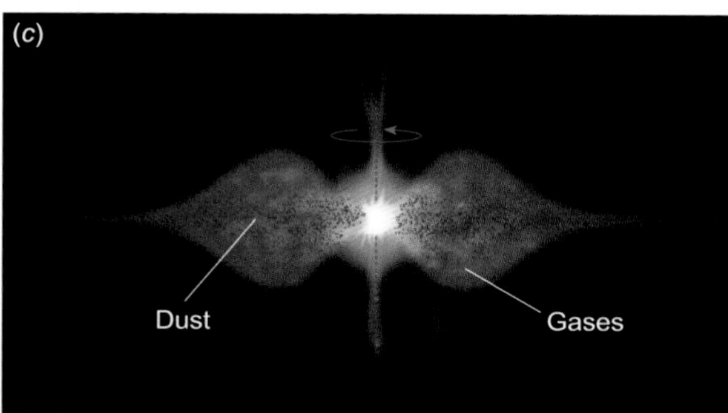

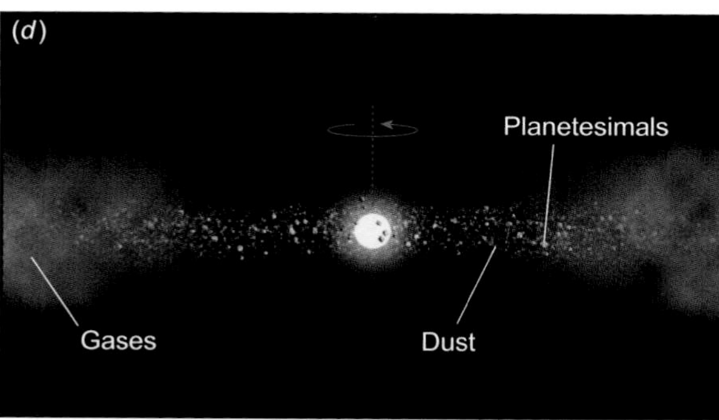

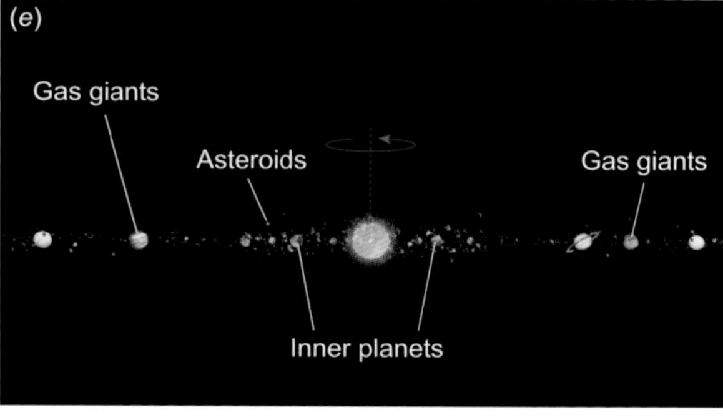

▶ ***Origin of the planets*** *A slowly rotating interstellar cloud (a) begins to collapse (b), more quickly at the poles than at the equator, where it is supported by centrifugal force. (c) A star forms at the center of a disk, which ejects material from its poles in anti-parallel jets. (d) Dust in the disk material sticks together in planetesimals, and the gases and ices in the disk near to the star are expelled. Farther from the star it is cold enough for the gases and ices to collect into gas giant planets.*

are driven by changes in the orientation of the planet's rotation axis to the Sun (on time scales of a year, the PRECESSION period and the Milankovitch cycles; *see* Milutin MILANKOVITCH).

Planetesimal

In cosmogony, a planetesimal is one of numerous solid bodies of subplanetary size orbiting the Sun (or another star), from which planets may accumulate (*see* PLANETS: ORIGIN). The term was coined near the beginning of the twentieth century by Thomas Chamberlin (1843–1928) and Forest Moulton (1872–1952), who developed the 'planetesimal hypothesis' for the origin of the solar system.

Planetology

Planetology is the identification and study of the processes that shape the surfaces and atmospheres of the bodies of our solar system: PLANETS, SATELLITES, COMETS and ASTEROIDS. *See also* ATMOSPHERE; CRATER; IONOSPHERE; POLAR CAPS; SOLAR SYSTEM; TECTONICS.

Planets: origin

COSMOGONY tells us about the origin of our solar system. The planets formed as a by-product of the birth of the Sun in the solar nebula.

The solar nebula

Astronomers think that the solar nebula was formed by the collapse of part of an interstellar molecular cloud – a very dense, very cold cloud of dust and gas. The molecular cloud was mostly hydrogen and helium. These elements are not only much more abundant than the other elements, they are also gaseous down to temperatures near 0 K. This means that almost 98% of the solar nebula remained gaseous.

Other elements behave differently. They assembled into molecules that condense at temperatures of 10–100 K. These molecules are called ices. The common molecules were made of hydrogen, carbon, nitrogen and oxygen, the most abundant elements (excluding helium, which does not make molecules at all). Example molecules included carbon monoxide, which, although very volatile, condenses at around 25 to 30 K, and water, which condenses at more than 110 K. These ices were common in the cold parts of the nebula (the regions away from the warming Sun), but not on the periphery of the cloud, which was illuminated by external stars.

Other elements react with oxygen to form oxides. The most important reactions were the combination of silicon dioxide (silica) with magnesium, sulfur, iron, aluminum or nickel to form silicates. Silicates condense at high temperatures, up to 1000 °C. With metals, which condense into crystals, they are called refractories. They existed as solids even in the inner, warm parts of the nebula.

Telluric (Earth-like) planets are made of refractories. The Earth formed in the solar nebula from an Earth's mass of refractories. However, refractories account for only 0.3% of the mass of the nebula (about 1/300), so we need 300 earth masses of the nebula to form the Earth. The same estimates can be done for Mercury (15 earth masses), Venus (300), Mars (30) and the ASTEROIDS (0.15). Gas-giant planets contain essentially hydrogen and helium, with a small proportion of ices and refractories (probably more in the case of Uranus and Neptune than for Jupiter and Saturn). The nebula mass from which Jupiter formed was approximately 1000 earth masses, and 500 each for Saturn, Uranus and Neptune. When we add all these values we obtain around 3000 earth masses, which is 1% of the mass of the Sun. Even if the nebula did not end sharply at Neptune's orbit, only a few per cent of the solar mass was left over to form the planets.

How long did it take to form the planets?

The newborn Sun and the solar nebula drifted clear of the molecular cloud. The HST has observed PROTOPLANETARY DISKS (proplyds), for example in the great molecular cloud in the constellation Orion. Their gas is concentrated in a disk rotating around a central mass, just as we visualize the solar nebula. Moreover, the rotation of the disk obeys Kepler's laws, so that we can estimate the value of the central mass at about the mass of the Sun (*see* CELESTIAL MECHANICS).

A T TAURI STAR is a kind of VARIABLE STAR whose SPECTRUM indicates that it is a new star surrounded by a nebula of gas. It is the next stage in the formation of the Sun. The star itself ejects a stellar wind in every direction, which progressively blows the disk away. There is a subclass of T Tauri stars that are at this stage; these are called 'naked' T Tauri stars.

The time between emerging from the cloud to blowing away the disk (the lifetime of a solar nebula) is around 5 million years. This is only one-thousandth of the age of our solar system. The formation of the planets takes place in a short time from a nebula that is not very massive.

From dust to planetesimals

The dust grains in the solar nebula were born in the atmospheres of giant red stars by the condensation of heavier elements (carbon, metals and silicates). They progressively acquired a three-layer structure: a nucleus of refractories, a mantle of carbonaceous materials, and a covering of ices. The ice covering is mainly amorphous water (H_2O). This ice is different from the crystals common on Earth, existing only at very low temperatures. The hydrogen and helium in the cloud move randomly and collide with each other and with the grains. The grains are knocked slightly and gain some thermal velocity. They therefore collide with each other from time to time. The amorphous ice surface is somewhat sticky, so the grains aggregate into larger particles.

Their mass is more than the masses of individual grains, so their thermal velocity decreases as they grow. But the solar nebula is a very turbulent medium. Micron-sized particles are sufficiently small that they remain caught up in the turbulent flow of the gas that surrounds them. When two vortices collide, the particles can come into contact and stick. They form larger and larger aggregates. They grow up to several centimeters.

As they become heavier these aggregated particles progressively decouple from the motion of the gas, blowing about only in the more powerful vortices. When massive particles collide at high speeds they tend to break apart instead of sticking together. At this point, turbulent aggregation is no longer efficient. Another mechanism takes over. Centimeter-sized flakes, decoupled from the turbulence of the gas, fall down to the equatorial plane of the nebula. They orbit the Sun. But the surrounding gas moves more slowly than the solids because internal pressure in the gas helps support it. To remain an orbiting nebula, the gas does not have to revolve so fast and with the same centrifugal force. The aggregated flakes have to move through a slower wind. They lose velocity and spiral toward the center. This spiral motion is the last step where the biggest flakes catch and 'eat' the others. The resulting size of the agglomerates is estimated to be around several tens of meters.

The disk fragments further, according to models proposed by V Safronov, Peter Goldreich and W Ward. The lumps formed in this way are 5 to 10 km in size; these are PLANETESIMALS.

The age of the chondrites

Astronomers and geologists have searched the oldest rocks for evidence of the timescale of the formation of the solar system. Rocks that have not been modified by geological processes are hard to find.

On Earth, the most ancient rocks are from the Precambrian Eon, which is subdivided into the Hadean Eon (from which no rocks have yet been discovered), the Archaeon Eon (4.03–2.5 billion years ago), and the Proterozoic Eon. The oldest rocks are Acasta gneiss: they are pods of metamorphosed granitic rock (now gneisses) separated by greenstone belts. Found in 1999 in the Northwest Territories, Canada, they are 4.03 billion years old. The Isua Formation of ancient sedimentary rocks in Greenland contains the oldest evidence of life (3.8 billion years). Metasediments, which contain the oldest known terrestrial minerals (zircons dated at 4.2–4.3 billion years), have been found in Jack Hills, Western Australia.

All these rocks are altered from their primitive states and have lost their memory of the matter from which our planet formed. Some of the lunar rocks brought back from the APOLLO missions date from 4.4 billion years, but they are transformed too.

Until samples are returned from COMETS and ASTEROIDS, the most primitive material known is found in meteoroids called CHONDRITES. Chemical and isotopic analysis of chondrites has led to two conclusions: they are almost 4.6 billion years old; and, except for the volatile ices, their composition is almost the same as the composition of interstellar matter and the solar nebula.

Some chondrites have very refractory inclusions of calcium, aluminum (Al) and alumina (Al_2O_3). The refractory inclusions in the ALLENDE METEORITE are the oldest material ever found in the solar system. They formed at the beginning of our history and are 4.56 billion years old.

The age of chondrites without refractory inclusions is around 4.555 billion years, that is, they were formed 5 million years after the formation of the refractory inclusions. Five million years is precisely the lifetime of the solar nebula. The inclusions formed during the initial process of the fall of the cloud downward into the nebula, while the chondrites themselves formed at the end of the life of the nebula.

The final step to planets

The compositions of the planetesimals varied with their distance from the Sun. In the internal regions, where the terrestrial planets formed, planetesimals consisted mainly of refractory rocks and metals, but farther out, where we now find the giant planets, they contained much ice and volatile gases.

The sizes of the planetesimals were some tens of kilometers and their masses therefore about 1000 billion tonnes. Tens of billions of them would be needed to build the Earth. The disk of planetesimals consisted of several tens of billions of bodies in the region of the terrestrial planets, orbiting the proto-Sun, almost in the same plane. They moved on almost concentric orbits with almost the same velocity but, given their enormous number, collisions frequently occurred, with

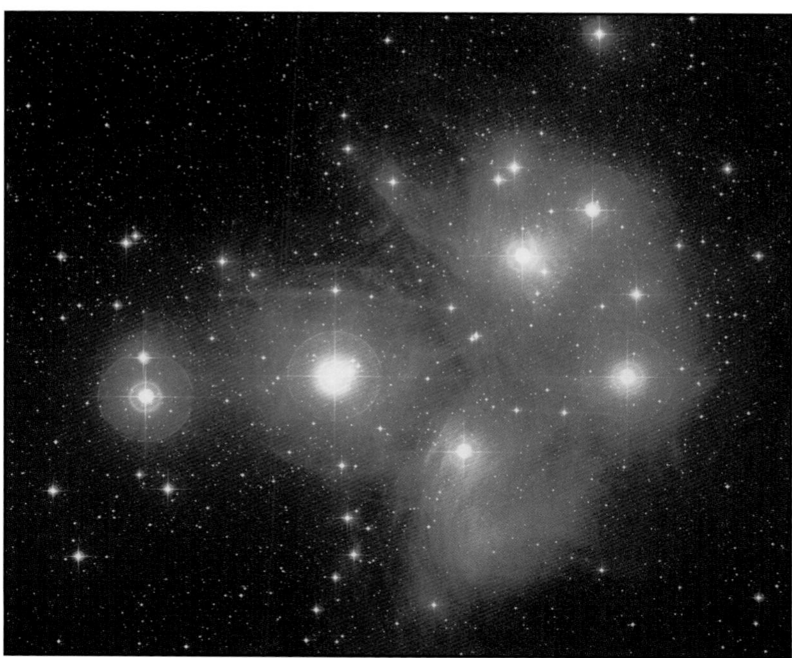

▲ ***Pleiades*** *(M45)* *The most famous star cluster in the sky, surrounded by a blue reflection nebula.*

a velocity of several meters per second. During such low-velocity collisions a kind of 'gluing' occurred, which led to the formation of greater and greater planetesimals.

At a certain size the gravity of the planetesimals became sufficiently high to attract others. A planetesimal that initially became large was more able to attract others. The more it grew, the more its gravity increased, and the more it grew, and so on. This process led to runaway ACCRETION, where the largest body in a given part of the disk enlarged at the expense of the others. When the largest bodies had attained the size of the Moon or Mercury, they had emptied their immediate neighborhood and their growth stopped.

At this stage, in the region of the terrestrial planets, 100 or so small protoplanets, with sizes of hundreds to thousands of kilometers, revolved around the proto-Sun on orbits much more widely spaced than the orbits of the previous planetesimals. Jupiter grew the fastest and accreted much of the gas of the nebula, so that its gravity perturbed the orbits of these terrestrial protoplanets. Some were moved into noncircular orbits, and giant impacts occurred between them. These impacts aggregated some of these protoplanets and shattered many others (forming asteroids). One impact neither shattered the protoplanets into small pieces nor fused the colliding bodies into a single planet: it gave birth to a 'twin planet,' the Earth and the Moon. The numerous pieces rained in a heavy bombardment on the surfaces of many of the growing planets, forming, for example, large craters on Mercury and the Moon (*see* MARE). At this stage, the solar system had taken its present form. *See also* PLANET.

Planet X

The name given to a hypothetical tenth MAJOR PLANET once believed to exist in the outer SOLAR SYSTEM, beyond the orbit of Neptune. The 'X,' which stood for 'unknown,' is also the Roman numeral for 10. The label 'Planet X' was originated by Percival LOWELL. From the late nineteenth century, he and others, including William PICKERING, worked out orbits for a large tenth planet that they believed was responsible for gravitational perturbations of Uranus; even since the discovery of Neptune in 1846, Uranus did not appear to be following its calculated orbit. In 1930 Clyde TOMBAUGH discovered Pluto. Though Pluto was only 5° from Lowell's predicted position, and in a similar orbit, it soon became clear that the new planet was too small to be Planet X. Tombaugh continued searching, but did not find a Planet X.

From the 1940s further predictions were made for a tenth planet whose perturbations would account for the deflection of comets from an outer reservoir into orbits that would bring them into the inner solar system. Charles Kowal's (1940–) search from 1977 to 1984 turned up the 180-km CHIRON, orbiting between Saturn and Uranus, which was hailed briefly in the media as the tenth planet. Others, including Robert Harrington (1942–93) and Thomas Van Flandern, returned to the problem of discrepancies in the orbits of Uranus and Neptune, and issued a variety of predictions for a tenth planet – but searches revealed nothing. In the 1990s the need to invoke a Planet X disappeared. VOYAGER 2 in 1989 had established more accurately the mass of Neptune, and Myles Standish and others demonstrated that the discrepancies in the orbital motions of Uranus and Neptune could be explained away by this and by observational errors. The accurately calculated and observed orbit of PIONEERS 10 and 11 showed that the outer solar system contained very little mass. The discovery from 1992 onwards of the Kuiper Belt objects showed that the Trans-Neptunian region had been well surveyed, without revealing a large planet (*see* OORT CLOUD AND KUIPER BELT). This seems finally to have brought the Planet X chapter to a close.

Planisphere

A simple analog computer that displays the stars visible from a particular place at a particular time and date. It consists of a disk-shaped chart of the stars that can be seen from a certain latitude, centered on the CELESTIAL POLE, around which pivots a disk-shaped mask in which an aperture has been cut to represent the hemisphere above the horizon. Using graduated scales of time and date around the perimeters of the two disks, the mask may be set so that the aperture reveals the stars that are visible at a particular time. The principle of the planisphere is broadly similar to the principle of the ASTROLABE.

Plasma

An almost completely ionized gas that contains equal numbers of negative ELECTRONS and positive IONS moving freely and independently of each other. Although a plasma is electrically neutral (because it contains equal numbers of positively and negatively charged particles), it is highly conductive. Plasmas, which normally have very high temperatures, are found in a variety of astrophysical contexts: for example, deep inside stars, where matter is highly ionized, or in the high-temperature outer atmospheres of stars (CORONAE). The solar corona and the SOLAR WIND are both examples of plasmas that consist predominantly of electrons and PROTONS.

Plato (428–347 BC)

Greek philosopher. He was born in Athens and taught by Socrates. His real name is said to have been Aristocles; 'Plato' (meaning 'broad') was a nickname derived from the breadth of his knowledge, shoulders, or forehead. In Athens he founded, in 387 BC, a school of learning called the Academy. This was devoted to research and instruction in philosophy and the sciences, particularly to the development of

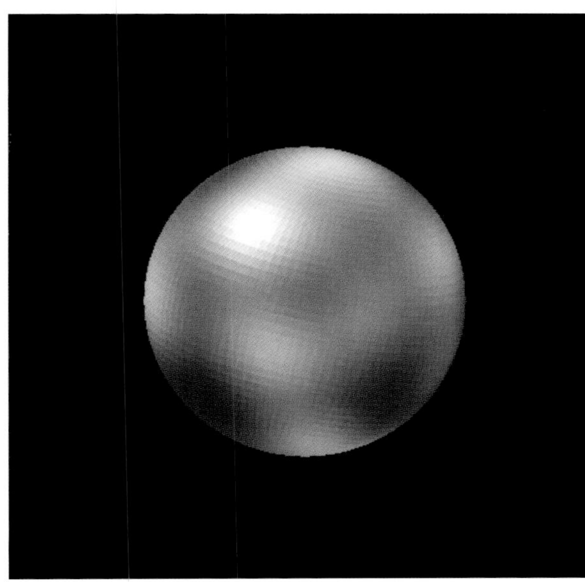

◀ ***Pluto*** *Viewing surface detail of Pluto from Earth is like trying to read the printing on a golf ball 53 km away. The HST was able to see light and dark surface features, which alter in seasonal effects.*

mathematical models of natural phenomena. Plato envisaged the universe as a series of shells of crystalline spheres on which the stars, planets, Sun and Moon moved around the Earth. He believed that the Moon shone by reflected sunlight and that astronomers were the wisest people. Plato's Academy was deemed a pagan establishment and closed in AD 529 by the Christian Emperor Justinian, but at 900 years of age it was the longest surviving university.

Pleiades (M45)

An open cluster in Taurus, also known as the Seven Sisters. Known from ancient times, the Pleiades has six stars readily visible to the naked eye, and more can be seen under good conditions by those with good eyesight. The Pleiades has at least 500 stars, spread over a 2° field.

Long-exposure photographs (and rich-field telescopes and binoculars) show that the Pleiades are embedded in a REFLECTION NEBULA, which the cluster encountered by chance as its stars moved together around the Galaxy. The age of the Pleiades star cluster is about 100 million years and it will survive as a cluster for not much longer than this time again, after which it will have 'evaporated,' its stars spread along the cluster's orbital path like meteoritic dust behind a comet. The distance of the Pleiades cluster as measured by HIPPARCOS is 380 l.y. As the Pleiades are situated within 4° of the ecliptic, occultations of the stars of the cluster by the Moon occur frequently.

Plössl, Georg Simon (1794–1868)

Optical-instrument maker born in Wieden near Vienna. The Plössl EYEPIECE is a four-element design for a telescope eyepiece used by amateur astronomers.

Plough

British-English name for a well-known ASTERISM, the most readily recognizable part of the disjointed constellation of Ursa Major. In North American-English it is called the BIG DIPPER.

Plutino

Any of the Kuiper Belt Objects in or near the 3:2 mean orbital resonance with Neptune, like Pluto itself. Plutino means little Pluto, and Pluto is considered to be the largest (by far) of the group. A third of the known TRANS-NEPTUNIAN OBJECTS are plutinos. It is estimated that there are 25 000 larger than 100-km diameter.

Pluto and Charon

Pluto is the smallest and the most distant from the Sun of all nine MAJOR PLANETS. Its orbit is unusually eccentric and highly inclined to the ECLIPTIC. It has a satellite, Charon, whose radius is half of Pluto's. Pluto's surface is among the coldest and, except for IAPETUS, is the most contrasty of all PLANETARY SURFACES. Pluto is the only planet never to have been visited by a space probe (*see* NEW HORIZONS).

Although Pluto was discovered by Clyde TOMBAUGH in 1930, prediscovery photographs establish positions back to 1914. Pluto has thus been observed for only one-third of its 250-year orbital period. Pluto's orbit is 'Neptune-crossing' in the sense that for approximately 20 years of its orbit it can get closer to the Sun than Neptune, whose PERIHELION distance is 29.8 AU. However, due to the high-inclination of its orbit, Pluto never gets closer than 17 AU to Neptune. The reason for this is that there is a 3:2 resonance between Pluto and Neptune. Over 495 years, Pluto orbits the Sun twice while Neptune orbits three times. This stabilizes Pluto's orbit.

Charon, discovered in 1978, is the only known satellite of Pluto, with a magnitude of 17.25 at opposition. The volume of space around Pluto in which satellites could exist in stable orbits extends to about 100 Charon orbit radii, but none have been detected. The orbital period of Pluto and Charon about each other is the same as the rotational period of Pluto, and Charon must be tidally locked too. Each body keeps the same hemisphere turned toward the other. As seen from the surface of one, the other is either never visible or always in the same place in the sky. Mutual rotational synchronism is the stable end of tidal evolution; Pluto and Charon are the only known case. The best determination of Charon's orbit is from HST imaging.

Every 124 years (every half-orbit of Pluto), for a period of about five years, Charon's orbit is seen edge on from the Earth (and the Sun), allowing observers to view Pluto and Charon passing directly in front of each other. This happened most recently in 1985–90. The 'mutual events' are scientifically productive, providing accurate orbital characteristics, sizes, scans that resolve the two bodies, and maps.

Charon's orbital INCLINATION and thus Pluto's spin axis are inclined by 119.6° to Pluto's orbital plane. This means that the latitude of Pluto's subsolar point varies between –60° and +60° over a Pluto year. This is the largest range experienced by any planet except Uranus, and must produce large seasonal changes.

The total mass of the Pluto-Charon system is 1/4000 of the mass deduced by Percival LOWELL in 1915 to account for perturbations seen in the orbit of Uranus. Therefore although Pluto was indeed discovered somewhere near the position predicted by Lowell, this appears to be a mere coincidence (*see* PLANET X).

Composition

Pluto's surface has bright and dark regions (its average ALBEDO is high at 0.58). It has a large bright southern POLAR CAP, a broad dark equatorial band, and a large region of intermediate brightness in the northern hemisphere. The equatorial and midlatitude regions are very patchy. Brightness contrast ratios over the planet can reach values of 6:1, nearly as high as on Iapetus. The brightness variations result from the SUBLIMATION of volatiles that move and condense somewhere else. Charon's surface is not so well known, but it does not change as much. Its albedo is 0.38, substantially less than Pluto's, suggesting that it is not as frosty.

Pluto's surface is covered by ices, mostly nitrogen, with traces of carbon monoxide and methane ices. Water ice has been identified on Charon, and possibly on Pluto. The surface temperatures of Pluto and Charon remain uncertain. At 30 AU from the Sun, the maximum the temperature could be is 72 K. The surface temperature on the visible disk (on the day side) averages 50 K, but the areas of nitrogen ice on Pluto are colder (40 K).

The slow fading of a star observed in a stellar occultation in June 1988 showed that Pluto has an ATMOSPHERE. Methane is the only gas to have been detected. However, nitrogen must dominate because it is the most volatile of the ices on Pluto and the most abundant. Nitrogen probably accounts for 99% of the atmosphere, the rest being methane and carbon monoxide. The occultation of another star observed in 2002 showed that the structure of Pluto's atmosphere had changed since 1998, due to the seasonal changes as Pluto passed its midwinter.

The density of Pluto is about 2 g cm^{-3} and of Charon 1.65 g cm^{-3}. These suggest that they are both mixtures of ice and rock. Pluto's interior is probably differentiated, with a 300-km-thick water ice layer and an 850-km layer of rock, but Charon's is mixed (*see* DIFFERENTIATION).

Formation

The most popular early hypothesis of the origin of Pluto and Charon, based on the fact that Pluto's orbit is Neptune-crossing, was that Pluto was formerly a satellite of Neptune. Pluto was believed to have been ejected to its present orbit by a collision with TRITON (reversing Triton's orbit). In fact, Pluto's 3:2 orbit resonance with Neptune, which prevents future close approaches between the two planets, implies that they were never close.

Like the Earth and Moon, Pluto and Charon form a twin planet (with a satellite-to-planet mass ratio of about 1:9). Like the Moon, Charon may have been formed from a collision. The high inclination of the system also suggests that the system originated in a collision. Pluto and its impactor (the Charon progenitor) must have been members of a large population of small planets to make a collision likely. These planets were presumably the forerunners of the TRANS-NEPTUNIAN OBJECTS (TNOs) of the Kuiper Belt (*see* OORT CLOUD AND KUIPER BELT). Pluto and the Charon progenitor formed independently in a near-circular, low-inclination heliocentric orbit, beyond Neptune, with other TNOs. Due to perturbations by Neptune, these bodies migrated and were captured in a succession of stable resonances, ultimately the 3:2 resonance. The migration amplified their ECCENTRICITY and inclination. At some point, probably when the two bodies already had significant eccentricities, Pluto and the Charon progenitor collided and formed the binary system. Charon's orbit then quickly (in a few hundred million years) evolved to synchronism. Neptune perturbed the 35–50 AU zone; cleared the Kuiper Belt of most of its mass, leaving the present low population of TNOs; and prevented any further growth of Pluto.

Further reading: Stern S A (1997) *Pluto and Charon* University of Arizona; Stern S A and Mitton J S (1999) *Pluto and Charon – Ice World on the Ragged Edge of the Solar System* Wiley.

Pluto	
Orbital period	248 years
Mean distance from Sun	39.54 AU
Eccentricity	0.249
Perihelion distance	29.7 AU
Aphelion distance (AU)	49.85 AU
Orbital inclination	17.1°
Rotation period	6.38726 days
Diameter	2320 km
Mass relative to the Earth	0.002 143

Charon	
Mean distance from Pluto	19 636 km
Mean distance from Pluto	17 Pluto radii
Eccentricity	0.0076
Orbital period	6.38722 days
Diameter	1200 km
Mass relative to Pluto	0.119

Pluto-Kuiper Express

Formerly proposed NASA mission to Pluto, superseded by the mission called NEW HORIZONS.

Poincaré, Jules Henri (1854–1912)

Born in Nancy, France, Poincaré became professor of mathematics at the Sorbonne university in Paris. He contributed to all areas of mathematics, including the three-body problem of CELESTIAL MECHANICS and studies of the stability of the solar system. He was the first to consider the possibility of CHAOS in planetary orbits. His work began as the answer to a question in a contest to show rigorously that the solar system, as a multibody modeled by Isaac NEWTON's equations, is dynamically stable. Poincaré realized that the prediction of the outcome was sensitive to the initial conditions: 'If we knew exactly the laws of nature and the situation of the universe at the initial moment, we could predict exactly the situation of that same universe at a succeeding moment. But even if it were the case that the natural laws had no longer any secret for us, we could still only know the initial situation approximately…it may happen that small differences in the initial conditions produce very great ones in the final phenomena. A small error in the former will produce an enormous error in the latter. Prediction becomes impossible…' Poincaré also determined the shape of a rotating fluid that was subject only to gravity. He wrote *Les Méthodes nouvelles de la mécanique celeste* (New Methods of Celestial Mechanics) (1892–9) and *Leçons de la mécanique celeste* (Lessons of Celestial Mechanics) (1905). He wrote on the philosophy of science, worked on equations of mathematical physics, and, with Albert EINSTEIN and Hendrick Lorentz, developed the special theory of relativity.

Pointers

Name for the ASTERISM consisting of Dubhe and Merak (the two leading stars of the BIG DIPPER) that are frequently used to locate POLARIS, and hence the north CELESTIAL POLE.

Polar

Space mission, part of NASA's Global Geospace Science program and the International Solar Terrestrial Physics program. Polar was launched in February 1996 in order to study the interaction between the Earth's MAGNETOSPHERE and the SOLAR WIND at high latitudes. It carried imaging instruments to measure visible, ultraviolet and X-ray SPECTRA of the polar regions.

Polar caps

Bright, icy surface deposits in the polar regions of a planet or satellite. In the solar system, the Earth, Mars, Pluto and Neptune's major satellite Triton have polar caps. The Earth's polar caps are permanent (but progressively thinning and shrinking due to global warming) and consist of water ice. The northern cap lies over Greenland and floats in a thin sheet (a few meters thick on average) on the Arctic Ocean; the southern cap has an average thickness of 2.3 km (up to 4 km) and covers the continent of Antarctica. Both caps show seasonal variations, with different limits of sea ice in winter and summer. Mars's polar caps show a marked seasonal variation. At their minimum extent (about 600 km across for the northern cap, 400 km for the southern) they have a distinctive swirled pattern, the ice filling valleys that curve outward from the poles. In the winter the southern cap extends to around 60°S (50°S in the Argyre Basin), the northern to about 65°N, as carbon dioxide sublimes out of the atmosphere to form extensive deposits of frost. While the permanent southern polar cap consists of carbon dioxide ice, the northern one is of water ice. Triton has a pink southern polar cap of frozen nitrogen and frost. There may be seasons, and the southern polar cap is expected to migrate to the north pole as the angle of solar illumination changes over the course of Neptune's 165-year orbit. Pluto's polar caps are frozen nitrogen.

Polarimetry

Our knowledge of the universe comes almost entirely through observing the ELECTROMAGNETIC RADIATION emitted from everything within it. The intensity of that radiation gives only a fraction of the information that is potentially available. The POLARIZATION of light (and other radiation) reveals some of the processes in which it originated, or by which it has propagated. The techniques for measuring the polarization of light are called polarimetry.

Polaris

The star Alpha Ursae Minoris, the (north) Pole Star. Polaris is at a distance of 430 l.y. and is a CEPHEID VARIABLE with a period of 3.97 days, magnitude range 1.9–2.1. Polaris is also a BINARY STAR, having a magnitude-9 companion at a separation of 18.4 arcsec.

Polarization

The extent to which the vibrations of an electromagnetic wave display a non-random orientation. Light, and other forms of ELECTROMAGNETIC RADIATION, consists of a periodically varying electric and magnetic disturbance that travels through space as a transverse wave motion (a wave that vibrates perpendicular to the direction in which it is propagating, like a wave on water). A beam of light is said to be unpolarized if it contains waves vibrating with equal amplitudes (heights) in all directions perpendicular to the direction in which the beam is propagating. By contrast, a polarized beam contains waves vibrating in one plane only. If, as a wave advances, the direction of vibration rotates (clockwise or anticlockwise, with the same frequency as the wave itself) but the amplitude remains the same, it is said to be circularly polarized. The state of polarization of a beam of radiation provides information about the mechanism responsible for producing the radiation at its source, and/or on what has happened to that radiation between leaving the source and arriving at the observer.

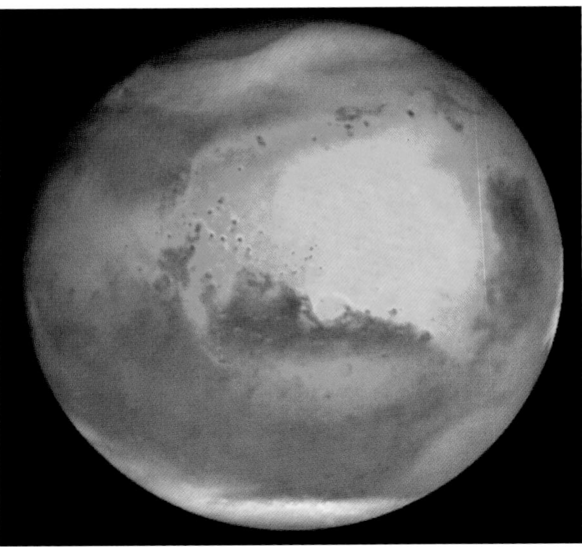

◀ ***Polar caps*** *Mars in June 2001 has a frosty southern polar cap. Frosty water ice clouds swirl around both the north and south caps, and orange dust storms hide the surface markings in some intermediate latitudes. The impact basin Schiaparelli is prominent at the center of the HST picture.*

Thermal radiation and BLACK-BODY RADIATION are unpolarized, whereas SYNCHROTRON RADIATION (radiation emitted by charged particles moving at high speeds in magnetic fields) is highly polarized. Unpolarized light is polarized when it scatters off electrons (Thomson scattering) or dust grains (Rayleigh scattering). Scattered light allows regions to be observed that are normally obscured from direct view. Examples are young stars and AGN (*see* ACTIVE GALAXY), both of which are surrounded by thick, dusty disks. These disks often obscure the central source, but light can escape along the poles of the disk and be scattered to us by particles there. The scattered (polarized) flux shows how the source and dust are arranged and allows us to look at the central source almost as clearly as by reflection in a mirror. Starlight is usually unpolarized, but it becomes partially polarized when it passes through clouds of interstellar dust grains that are lined up in the galactic magnetic field. Radio radiation from pulsars (*see* NEUTRON STAR AND PULSAR) and SUPERNOVA remnants, which is produced largely by the synchrotron process, is strongly polarized.

Polar motion

The movement of the axis about which the Earth rotates. It consists of two motions. The largest is a mostly circular path that repeats after 435 days and is called the Chandler motion after Seth Chandler (1846–1913). It is caused by the misalignment of the Earth's rotation axis with respect to its axis of symmetry. The second motion is one that follows an elliptical path and repeats annually. This motion is caused by the redistribution of atmospheric mass during the year; for example, as water migrates toward the winter polar region. These two motions cause a spiral motion of the pole on the Earth's surface of approximately 6 m radius. Polar motion is associated with periodic changes in the length of the day, of up to 5 ms. The motion of the pole is determined by measuring the position of the Earth's surface relative to objects located in space. Modern accurate methods include the use of lasers to range to the Moon; the use of the Global Positioning System to determine the positions of satellites; and the use of radio telescopes to measure the positions of quasars. The International Earth Rotation Service, which combines all the measurements, is in Frankfurt, Germany, and at the US Naval Observatory, Washington DC.

Polar ring galaxy

This type of galaxy has a galactic disk and a ring of stars orbiting in a plane that is almost perpendicular to the disk (almost polar). The prototype is NGC 4650A. Polar ring galaxies originate in the merger between two unequal DISK GALAXIES. When a small galaxy companion is tidally destroyed and swallowed by a larger galaxy the stars are spread all around, since stars cannot collide and dissipate energy. The gas, on the contrary, does dissipate its energy and settles into a disk perpendicular to the average spin of the two galaxies, as in the case when SPIRAL GALAXIES are first formed. Usually, the gas becomes aligned with the larger galaxy. In rare cases, however, the gas can be trapped in a stable polar orbit. If the gas disk is massive enough it may be stable, stars can form there, and a polar ring galaxy is born.

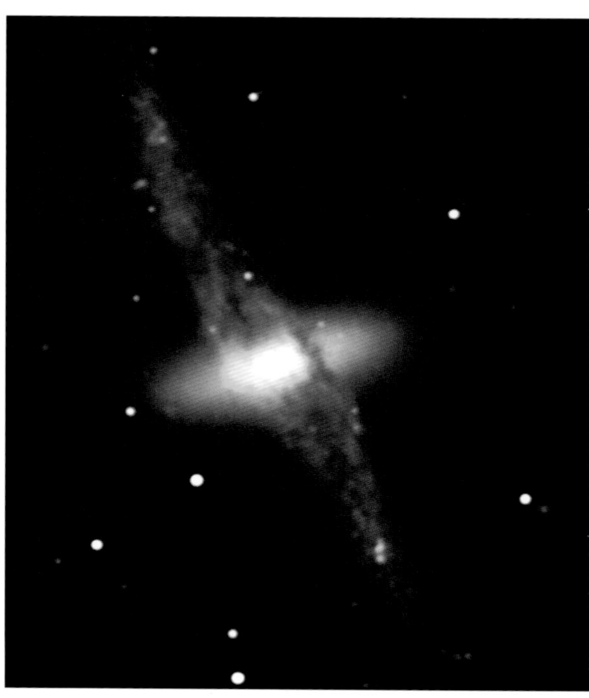

▶ Polar ring galaxy *NGC 4650A is the prototype for galaxies with a flattened disk of stars with a bright core and a sharply tilted outer ring of gas, dust and stars, orbiting over the poles of the inner disk.*

Polar satellite

An artificial satellite that orbits over the poles (for example, to survey high latitudes or investigate the MAGNETOSPHERE).

Pole star *See* POLARIS

Pond, John (1767–1836)

As an amateur astronomer Pond had used the mural quadrant at the ROYAL OBSERVATORY, GREENWICH and analyzed its inaccuracies. He succeeded NEVIL MASKELYNE as ASTRONOMER ROYAL and concentrated on upgrading the instrumentation at the observatory. Its measurements of star positions and time were raised to an accuracy never before achieved. During his administration the observatory was taken over by the Royal Navy, for which the observatory regulated chronometers; and in 1833 Pond instituted the public dissemination of time through the dropping of a time ball at 1 p.m. each day (this still happens). Owing to the pressure of all this new work the *Nautical Almanac* suffered. It fell into disrepute and was transferred to a separate office, called the Nautical Almanac Office.

Population I

The relatively young, metal-rich stars that are found in the disks and SPIRAL ARMS of SPIRAL GALAXIES, with the associated interstellar clouds. Population I was differentiated from POPULATION II by Wilhelm BAADE in 1944. Population I stars are youngish stars, with circular orbits about the GALACTIC CENTER, that lie in the Milky Way plane. These stars are strongly concentrated toward the galactic plane. The Sun belongs to this population, together with the stars in older OPEN CLUSTERS, many A stars, giants, and emission-line (Me) dwarfs. Extreme Population I stars have an extremely uneven distribution within galaxies' spiral arms. They include PRE–MAIN SEQUENCE STARS, T TAURI STARS, OB ASSOCIATIONS, supergiants, classical Cepheids (*see* VARIABLE STAR: PULSATING), and young OPEN CLUSTERS.

Population II

This consists of the old red stars that occur throughout ELLIPTICAL GALAXIES, and in the halos, GLOBULAR CLUSTERS and central bulges of SPIRAL GALAXIES. They differ from POPULATION I stars in having a very low metal content, and those heavier elements that they do contain are thought to have been produced by massive, short-lived earlier stars which form a hypothetical POPULATION III. The oldest stars with the lowest metal content form a spherical distribution around the Galaxy, and are known as HALO POPULATION II stars. This group includes the stars in globular clusters, long-period RR Lyrae stars (period greater than 0.4 days), and the W Virginis stars, also known as Type II Cepheid variables (*see* VARIABLE STAR: PULSATING). The stars of Intermediate Population II include most HIGH-VELOCITY STARS and many MIRA-type variables, and have a flattened spheroidal distribution. The youngest Population II stars are those of the galactic bulge, where they intermingle with the very oldest Population I stars and form part of the Disk Population. They include short-period RR Lyrae stars (period less than 0.4 days).

Population III

A hypothetical population of stars that may have existed in the very early universe. An early population of massive stars, long extinct, that would have exploded as SUPERNOVAE, and thus account for the metal content of the POPULATION II stars in GLOBULAR CLUSTERS. The remnants of this early population would be neutron stars or black holes, and could thus account for some of the DARK MATTER postulated to explain the missing mass in GALACTIC HALOS.

Position angle

A measure of the orientation of one object relative to another on the CELESTIAL SPHERE (for example, the direction of the fainter component of a double star from its brighter companion). Together with the separation, this defines the form of the system. (*See* DOUBLE STARS, OBSERVING.) Position angle (PA) is measured eastward (anticlockwise as seen on the sky), from the northerly direction. PA also records the location of a feature on the disk of a planet or the Sun, or the direction of an axis of rotation.

Positron

The antiparticle of the ELECTRON. It has the same mass, spin and other properties as the electron except that its charge, the same size as the electron's, is positive. Electron-positron pairs are produced in high-energy events (from gamma-rays, for example). Conversely, when electrons and positrons collide they liberate large amounts of energy (GAMMA-RAYS, HADRONS and QUARKS).

Poynting–Robertson effect

When a PHOTON strikes a dust particle in circular orbit around the Sun the photon may be reflected and gives a kick to the particle, pushing it outward. If the photon is absorbed, on the sunward side, it also pushes the particle outward. The photon then warms the particle, which reradiates the energy. The energy is radiated isotropically (in all directions). This radiation leaving the particle produces no net impetus. The overall result of a photon arriving and leaving the dust particle is an outward push, called 'radiation pressure.' However, the photon impinges not directly 'side on' but slightly on the particle's leading side (the 'forward' side in terms of its orbital motion), because of the compounding of the motion of the particle and the photon. This gives a slight impetus to the dust particle against the direction of motion. As a result the particle loses orbital speed, so it tends to spiral inward, toward the Sun. John Poynting (1852–1914) was the first to describe this effect, in 1903; Howard Robertson (1903–61) derived it from relativity theory in 1937.

Dust particles have different shapes and spins; a ring of dust particles of the same mean radius but with a distribution of shapes would gradually broaden out inward and outward, like planetary ring systems.

Praesepe (M44) *See* MESSIER CATALOG

Precession

Discovered by HIPPARCHUS, precession (or 'precession of the equinoxes') is the slow periodic change in the orientation of the Earth's axis of rotation. It is caused by gravitational attractions of the Sun and Moon on the non-spherical globe (geoid) of the Earth. The term is also applied to the periodic change in the rotation axis of any spinning body (for example, a top or artificial satellite). The axis of the Earth is inclined to the perpendicular to the ECLIPTIC by an angle of 23.44°, so the CELESTIAL POLE traces out a circle of this radius around the pole of the ecliptic in 25 800 years. The north celestial pole is now near POLARIS, but was close to Thuban in the constellation Draco 4500 years ago, and will be near Vega in 12 000 years. The DECLINATION of any given star will thus change with time, even if the star is 'fixed.' Its RIGHT ASCENSION also changes, because, as the celestial pole moves, so the CELESTIAL EQUATOR moves relative to the ecliptic. The point where these two circles intersect (the vernal equinox) moves clockwise along the ecliptic at a rate of 360° in 25 800 years. It follows that the right ascension of a given star must increase due to precession.

Because of precession, star catalogs and charts must be drawn up for a particular date of equinox (for example, 12 am on January 1 for the years 1900, 1950, 2000 and so on).

Prehistoric astronomy

The earliest claim for astronomical observations relates to the Upper Paleolithic period, when fine depictions of animals first appeared in western European caves. Part of an eagle's wing from a cave at Abri Blanchard in the Dordogne valley, France, contains a series of notched marks in a serpentine pattern. In the early 1970s Alexander Marshack studied the microscopic structure and overall arrangement of the marks, and concluded controversially that they were nightly tallies of lunar phases.

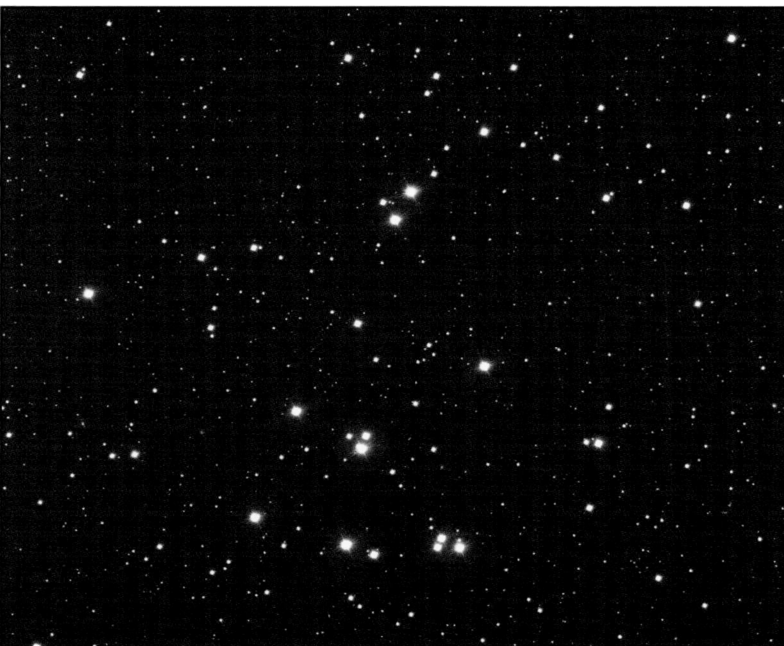

▲ **Praesepe** *(M44) (Latin for 'manger'), also known as the Beehive Cluster, is an easy naked-eye object more than twice the apparent size of the full Moon, and has thus been noted since prehistoric times. It was one of the first objects on which Galileo trained his newly acquired telescope. It is one of the nearest open clusters, close to 600 l.y. away.*

In the subsequent Neolithic period, early farmers around the Atlantic fringes of Europe constructed communal tombs and longhouses that cluster in orientation. Each region has a distinctive 'signature,' many related to the motions of the Sun: for example, spreads between northeast and south are quite common, and may reflect a need for entrances to face the rising Sun. This also holds for tombs and other forms of stone monument such as the megalithic rings and rows built in remote parts of Scotland and Ireland during the Neolithic period and well into the Bronze Age, perhaps as late as 1000 BC. The orientation of some of the later groups, such as the Scottish 'recumbent stone circles' and short stone rows in southwest Ireland, suggest that the Moon rather than the Sun was of primary importance, and perhaps regulated the timing of rituals or ceremonials.

In the 1960s and 1970s Alexander Thom (1894–1985) propounded the idea that many standing stone monuments incorporated intentional high-precision alignments on the Sun and Moon, reflecting their use as high-precision 'observatories.' These ideas did not stand up to close scrutiny, the main problem being the fair selection of data. Nonetheless, certain Neolithic chambered tombs do seem to have incorporated more specific solar alignments, a famous example being Newgrange in Ireland, where a specially constructed roof-box above the entrance permitted sunlight to enter the tomb for a few minutes after dawn on days around the winter SOLSTICE, shining down the entire length of the 19-m-long passage to light up the bones of the dead in the central chambers. Similarly, Stonehenge in Wiltshire, England – which for some 500 years had existed as an unexceptional ditch-and-bank enclosure around various timber structures – was reoriented on the solstices *c.*2500 BC, when the familiar stone edifice was built.

▶ ***Stonehenge*** *The reconstructed view from the center of the sarsen stone horseshoe of megaliths, with their lintels in place, toward the two Heel Stones, at mid-summer sunrise. The second (western) Heel Stone has now disappeared and many of the lintels in this drawing have fallen.*

In other parts of the world astronomy developed in distinctive ways. Polynesian navigators, for example, developed a practical astronomy based on 36 linear star-to-star 'constellations' that formed a three-dimensional 'star compass' enabling navigators to locate distant islands. Central to Inca cosmology was the system of ceques – conceptually straight lines radiating outward from the Coricancha temple in the Incaic capital Cuzco – which reflected Inca knowledge of the sky, dictated the layout of the city and organized social groups within it, and functioned as a calendar.

The Mayans developed a quantitative, predictive astronomy capable of forecasting eclipses, the appearances of Venus and Mars, and the passage of the Moon through the zodiac. Their calendar was a series of integer multiples of different cycles: thus 73 cycles of a basic 260-day count equals 52 years of 365 days; and 46 cycles equals 405 lunations (11 960 days). The calendar was developed not in a spirit of abstract enquiry but because there was a cosmic rule that needed to keep ritual activity in tune with various celestial cycles.

Many investigations of prehistoric astronomies now take place within the framework of archeoastronomy, which arose in the 1970s from the need to resolve the disputes raised by Thom. In northwest Europe archeoastronomy was mainly concerned with statistical appraisals of alignments among stone monuments, but in the Americas it quickly began to assess such alignments in the context of cultural evidence about the nature of astronomy in pre-conquest times. This broader approach highlights particular themes such as concepts of space and time, calendar development, and the place of celestial signs in schemes of 'sacred geography' reflecting cosmology in the physical landscape. The sky is a symbolic resource, not only because many of its cycles are convenient and reliable for regulating human activity, but also because some individuals convince others that they wield control over celestial events. *See also* ASTROLOGY; NORTH AMERICAN INDIAN ASTRONOMY.

Further reading: Aveni A F (ed) (1989) *World Archaeoastronomy* Cambridge University Press.

Pre–main sequence star

A pre–main sequence (PMS) star is one that has not yet fully contracted to the MAIN SEQUENCE and is not yet fusing HYDROGEN at its core. Such a star is recently formed, normally less than 10 million years old. PMS stars are often found still associated with the parent molecular clouds out of which they formed. They are also often surrounded by protostellar (and, likely, protoplanetary) disks.

PMS stars are contracting, shine by reradiating gravitational potential energy, and are relatively bright. Several thousand examples are known, even though their lifetime is only about 0.1% of a main sequence star. Louis Henyey constructed the first detailed, numerical models of young stars in the 1950s. These were improved by Chushiro Hayashi in the early 1960s. The path of a PMS star in the HERTZSPRUNG–RUSSELL DIAGRAM is known as a Hayashi track. T TAURI STARS and Herbig stars of spectral types Ae and Be are PMS stars.

Primary mirror

The main mirror in a REFLECTING TELESCOPE.

Primeval fireball

A term coined by James Peebles (1935–) to describe the early phase of the BIG BANG universe during which space was filled with an opaque, high-temperature mixture of

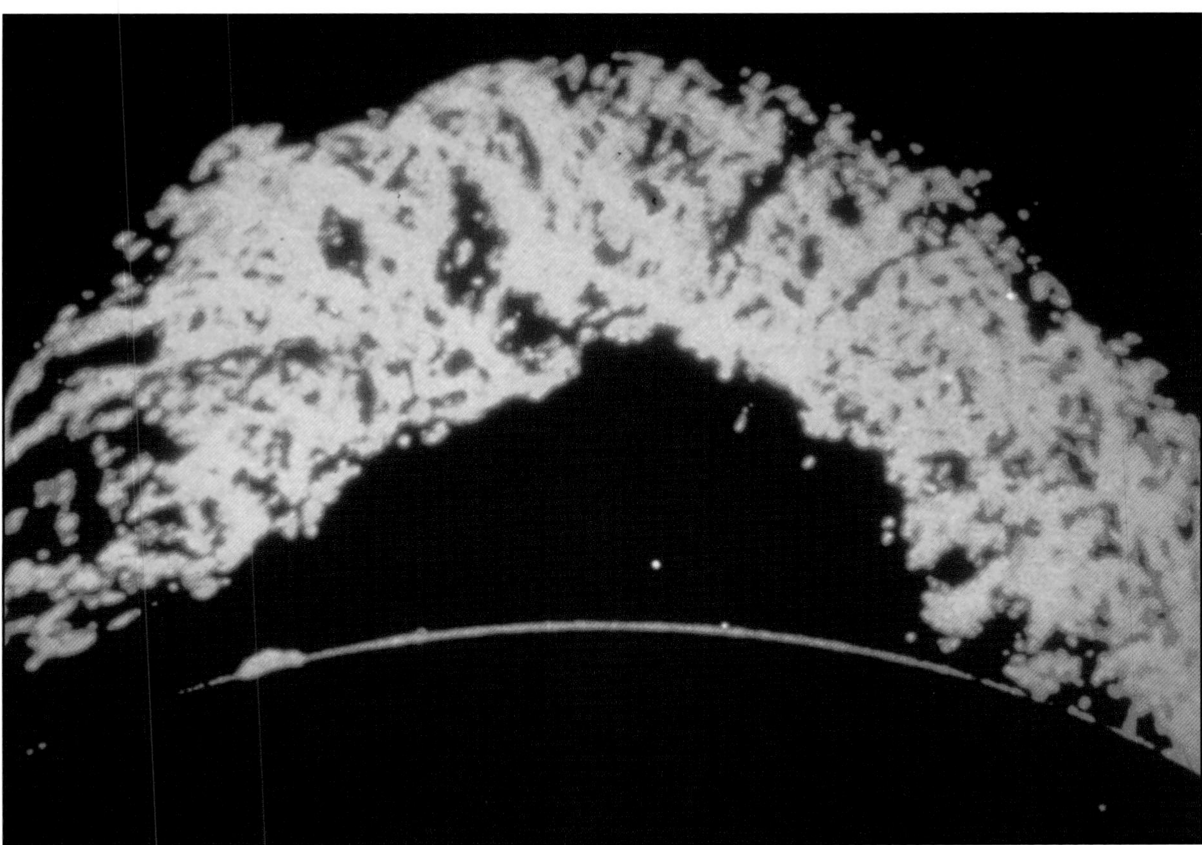

◀ ***Prominence*** *This solar prominence is looping high above the top of the chromosphere.*

radiation (PHOTONS) and particles (protons, neutrons, electrons and others). Also known as the 'primordial fireball.'

Prognoz

A series of 10 Soviet spacecraft launched between 1971 and 1985 to investigate the SOLAR WIND and its interaction with the Earth's MAGNETOSPHERE. Prognoz-9 (launched July 1983) carried experiments to study the Sun, X-ray and GRBs, as well as the Relict-1 radio astronomy experiment. Prognoz 10 (launched April 1985) carried instruments to study the PLASMA, the magnetic field, electric and magnetic components of waves, solar flare X-ray bursts and kilometer radiation. *See also* INTERBALL; RELICT.

Prominence

A plume of luminous gas in the solar atmosphere visible beyond the solar limb (the edge of the visible disk). The visible light consists mainly of emission lines of hydrogen and ionized calcium and, except during total eclipses of the Sun (when they may be seen directly by eye), they are normally studied in light corresponding to particular emission lines, such as the hydrogen-alpha line in the red part of the SPECTRUM.

There are two principal types of prominence: quiescent and active. Quiescent prominences hang like clouds in the CORONA for weeks or months with little overall change, whereas active prominences undergo rapid changes. Quiescent prominences have lengths ranging from several tens of thousands of kilometers to several hundred thousand kilometers, and heights of around 30 000 km. They are suspended above the CHROMOSPHERE by magnetic fields. Active prominences (which are associated with ACTIVE REGIONS on the Sun, and are often associated with, or triggered by, solar flares) are short-lived phenomena, most of which survive no more than a few hours. Among the most active are loop prominences, structures that connect regions of opposite magnetic polarity and within which coronal material condenses and flows down into the chromosphere. Giant eruptive prominences (often loop-shaped) surge upward to heights of 500 000 km or more.

Proper motion

The change of position of an object on the CELESTIAL SPHERE. *See also* ASTROMETRY.

Protogalaxy

A protogalaxy, also called a 'primeval galaxy,' is the progenitor of a present-day (normal) GALAXY. It is a galaxy in the early stages of formation, and is usually seen at high REDSHIFT (that is, it existed early in the history of the universe). In practice this might mean:

- a galaxy showing its first major burst of star formation, for example ULIRGS;
- a large but gaseous galaxy-mass body before any star formation has taken place, like clouds of the INTERGALACTIC MEDIUM;
- an assembly of dark halos able to produce a galaxy;
- a denser, galaxy-mass region of DARK MATTER in the very early universe, destined to become gravitationally bound and to collapse.

Proton

(1) A positively charged ELEMENTARY PARTICLE that is composed of three QUARKS. It has a charge of +1, equal in magnitude

to the charge on the electron, and a mass 1836 times greater than the mass of the electron. The proton is a nucleon, one of the two basic constituents of an atomic nucleus (the other being the marginally heavier neutron). The number of protons in an atomic nucleus (its atomic number) determines its net charge and, therefore, the chemical element of which that ATOM is an example. For example, a hydrogen nucleus contains one proton, a helium nucleus contains two protons, and a uranium nucleus contains 92 protons. The proton has a lifetime of at least 10^{32} years and may be completely stable (*see* ASTROPARTICLE PHYSICS).

(2) A series of four large Soviet magnetospheric research satellites launched from 1965 to 1968. They carried experiments to investigate ultra-high-energy cosmic particles, such as COSMIC RAYS.

Protoplanet
An object lying somewhere between a PLANETESIMAL and a PLANET as an intermediate stage in the process of planet formation.

Protoplanetary disk
A disk of dust and/or gas that orbits a very young star. *See also* PLANETS: ORIGIN.

Protoplanetary (proplyd) nebula
A star and its nebula that are in transition towards being a planetary system.

Protosolar nebula
The nebula from which the Sun and solar system formed 4.6 billion years ago; a PROTOPLANETARY NEBULA. *See also* PLANETS: ORIGIN.

Protostar
A star in the act of formation. *See also* PROTOSOLAR NEBULA; T TAURI STAR.

Proxima Centauri
The star Alpha Centauri C, a red dwarf (M5Ve) member of the Alpha Centauri multiple-star system, magnitude 11.01. At a distance of 4.22 l.y., Proxima is the closest known star to the solar system (hence its name, which means 'the nearest' star). The separation of Proxima from the other stars of the system (0.17 l.y. in radial distance and 2.2° in angle) is so great that no orbital motion has been detected.

Ptolemaic system
The geocentric view of the universe as expounded by PTOLEMY. According to this view the Earth lies at the center of the universe, and around it, in order of distance, move the Moon, the planets Mercury and Venus, the Sun, and the planets Mars, Jupiter and Saturn. Beyond the outermost planet lies the sphere of the stars. To account for the retrograde and variable motion of the planets like Mercury and Mars, Ptolemy used an epicycle, a small-scale circular motion whose center moves on an eccentric orbit. It did not work quite well enough to describe the motions of the planets, so Ptolemy introduced the 'equant point,' around which the revolution of the epicycle is uniform, not around its center. In contrast to the closely reasoned treatment of the planetary models so far as they concern motion along the ECLIPTIC, Ptolemy's complicated system of additional wobbles was presented as a *fait accompli*, unsupported by specific observations. The Ptolemaic system was superseded by the COPERNICAN SYSTEM. This had many similar mathematical features but, having the Sun at the center of the solar system, was considerably simpler. The mathematical features were abandoned with the realization by Johannes KEPLER that planetary orbits were ellipses.

Ptolemy
Claudius Ptolemaeus (Ptolemy) worked at or near Alexandria in Egypt during the middle decades of the 2nd century AD. He began making astronomical observations at least as early as AD 127, and by 147 he had completed a preliminary version of his PTOLEMAIC SYSTEM of the planets. He presented this in full in a large treatise in Greek whose title translates as *Mathematical Composition*, which much later came to be called the ALMAGEST (from the Arabic for 'the greatest'). Ptolemy wrote several other works on physical science, most of them after the *Almagest*.

The work whose title translates as *Planetary Hypotheses* represented Ptolemy's last word on the celestial motions. In this work he treated the models of the *Almagest* in a more physical way, as a system of rotating, nested spheres. This idea influenced medieval astronomy profoundly.

Ptolemy's Cluster (M7) *See* MESSIER CATALOG

Pulsar *See* NEUTRON STAR AND PULSAR

Puppis *See* CONSTELLATION

Purcell, Edward Mills (1912–97)
American physicist, born in Taylorville, Illinois. He won the Nobel prize (with Felix Bloch, 1905–83) in 1952 for his studies of the nuclear magnetic moment of HELIUM. With graduate student Harold Ewen in 1951, he formed one of the three groups that independently detected cosmic 21-cm radio radiation from the spin transition of the electron in neutral hydrogen, predicted by Hendrik VAN DE HULST. The detection was soon confirmed in the Netherlands by C Alex Muller and Jan OORT, and in Australia by Wilbur Christiansen and Jim Hindman.

Purple Mountain Observatory
The observatory, built in 1934, is situated in the eastern suburb of Nanjing, China. The observatory has three observing stations outside Nanjing. A 13.7-m millimeter-wave telescope was set up at the Qinghai station.

Pyxis *See* CONSTELLATION

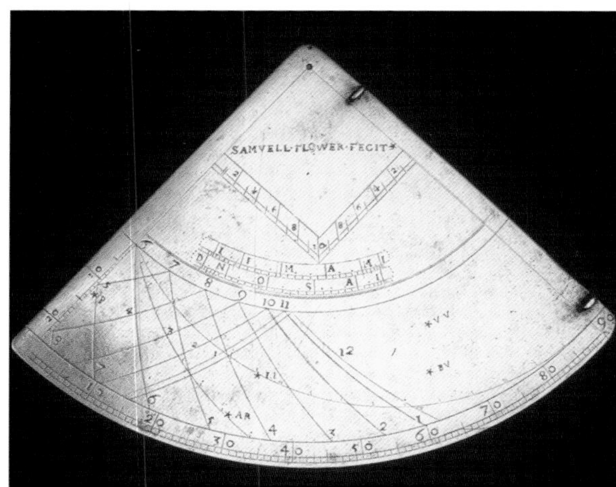

◀ ***Quadrant*** *This Moorish quadrant was made in 1804. It is held with the corner upwards, the astronomer looking to a star through its sights on the upper edge. The star's altitude is read by the fall of the pendulum (now lost) over the scale. The mathematical functions graphed on the face serve the same purpose as an astrolabe, without being as intuitive.*

Quadrant

An instrument dating back to antiquity for measuring the altitudes and angular separations of stars. It consisted of a graduated arc of 90°, usually of brass. In a mural quadrant, such as used by Tycho BRAHE at Uraniborg, the graduated arc was attached to a wall oriented so as to lie in the observer's MERIDIAN. Tycho's great mural quadrant measured more than 1.8 m in radius.

Quadrantids

A METEOR SHOWER with a peak of activity around January 3–4. The number of meteors seen varies from year to year, but it can be one of the most prolific showers with a peak zenithal hourly rate of 120 in a short sharp maximum of about 12 hours. The brightest meteors show a blue or yellow-green tinge. The RADIANT lies in the constellation Boötes, near its border with Hercules. The shower is named for the obsolete constellation Quadrans Muralis (the Mural Quadrant) which used to occupy this region. Perturbations of the stream by Jupiter cause it to oscillate up and down through the ECLIPTIC.

Quadrature

The position of a planet or the Moon when it is at right angles to the Sun as seen from the Earth. At these times its ELONGATION is 90° or 270°. The Moon is at quadrature when it is at first or last quarter.

Quantum mechanics

A development of QUANTUM THEORY that was initiated in the 1920s by Werner Heisenberg (1901–76) and Erwin Schrödinger (1887–1961). The theory drew on a proposal that particles have wavelike properties (the wave-particle duality) and that an ELECTRON, for example, could in some respects be regarded as a wave with a wavelength that depended on its momentum. In this new formulation, the permitted energy levels in the Bohr theory of the hydrogen atom corresponded to electron orbits into which integral numbers of electron wavelengths could fit (*see* HYDROGEN SPECTRUM). The same principles apply to atoms of other elements, and form the underlying basis for astronomical spectroscopy, making it possible to identify the composition, physical conditions, motions and so on, of celestial bodies.

Schrödinger developed a formulation of quantum mechanics known as wave mechanics. An orbiting electron is treated as a standing wave, represented by a 'wave function,' which is described by the Schrödinger wave equation (an equivalent formulation, based on matrices, was developed by Heisenberg). In accordance with the Heisenberg uncertainty principle (which implies that it is not possible to determine simultaneously the position and velocity of an electron), the wave function gives only the probability that an electron is in a particular state or at a particular point at a particular time. Therefore, Bohr's precise orbits are replaced by a set of fuzzy 'orbitals' within which electrons of a particular energy will lie.

The quantum state of an electron (or other subatomic particle) is defined by a set of numbers (quantum numbers) that specify quantities such as energy, angular momentum and spin, and which are associated with solutions to the wave equation. Each energy level within an atom (or other system of ELEMENTARY PARTICLES) corresponds to a unique set of quantum numbers, and various selection rules determine which transitions between levels, and hence which spectral lines, are permitted. For electrons and neutrons, or any other particles with spin 1/2, each particle in the system must have a different defining set of quantum numbers. Put another way, if such a particle has a certain set of quantum numbers, it excludes any other particle from having the same set. This is the PAULI EXCLUSION PRINCIPLE discovered by Wolfgang Pauli (1900–1958), winner of the Nobel prize in 1948. It is the underlying reason for the existence of WHITE DWARFS and neutron stars (*see* NEUTRON STAR AND PULSAR).

Quantum theory

A theory based on the premise that, on the microscopic scale, physical quantities have discrete, rather than a continuous range of, values. The theory was devised in the early part of the twentieth century to account for certain phenomena that could not be explained by classical physics. In 1900, the German physicist Max Planck (1858–1947) was able to describe precisely the previously unexplained distribution of energy radiated by a black body by proposing that light, and other forms of ELECTROMAGNETIC RADIATION, exists in the form of discrete 'packets' of energy called quanta (singular: 'quantum'). The energy (E) of a quantum of electromagnetic radiation is directly proportional to the frequency (f) of the radiation and inversely proportional to its wavelength (λ); thus, $E = hf = hc/\lambda$, where h is a constant of proportionality called Planck's constant (or the Planck constant), and c denotes the speed of light. The value of the Planck constant is 6.63×10^{-34} J s.

Quark

A FUNDAMENTAL PARTICLE that joins with others to form HADRONS (BARYONS and MESONS). There are six varieties, or 'flavors,' of quark, each with an equivalent antiparticle. The six flavors, in order of increasing mass, are called up, down, strange, charm, bottom, and top. Quarks possess fractional charge (they carry an electrical charge that is a simple fraction of the charge on the electron) and have half-integer values of spin. For example, an up quark (symbol 'u') carries a charge of +2/3 and an anti-up ($\overline{u}$) –2/3; a down quark (d) carries a charge of –1/3 and an anti-down ($\overline{d}$) +1/3. Quarks join together in clusters of three to form baryons, such as protons and neutrons, and as quark–antiquark pairs to form mesons. For example, a proton consists of two ups and a down (net charge: 2/3 +2/3 –1/3 = 1) and an antiproton of two anti-ups and an anti-down (net charge: –2/3 –2/3 +1/3 = –1). Of the six flavors, only the up and down quarks are needed to make the nucleons (the building blocks of atomic nuclei)

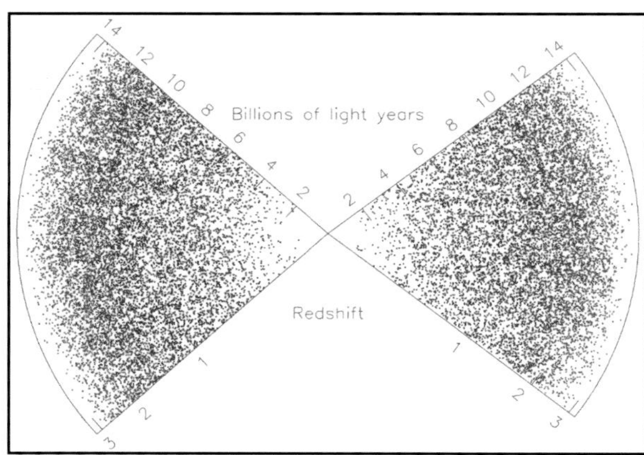

▶ Quasars *The density of quasars increases outwards from the Earth. In this plot (which has empty quadrants where quasars are hard to find among the stars of the Milky Way) the Earth is at the center, in a nearly blank area. The density of quasars decreases after about 11 billion l.y., into the 'dark ages' before they had formed.*

and pions, the other varieties being required to construct some of the many short-lived particles that are found among cosmic rays and created in high-energy particle accelerators.

In addition to electrical charge, quarks possess another property, called 'color' or 'color-charge'; this is the charge associated with strong interactions. There are three types of color (called red, blue, and yellow by analogy with the primary colors), each of which can be positive or negative. In a hadron (a particle that consists of quarks), the net color has to be neutral (or 'white'). This can be achieved by combining three different colored quarks to make a baryon, or by linking a quark with an antiquark (in which case the opposite color charges neutralize each other) to form a meson.

The interiors of some particularly small neutron stars may be so dense as to be made of quark material (*see* NEUTRON STAR AND PULSAR). The CHANDRA X-RAY OBSERVATORY discovered such a possible 'quark star,' RX J185635-375, in 2003.

Quasar

A quasar is an ACTIVE GALAXY with a powerful AGN that outshines the rest of it, so that the galaxy has a very small angular size and, except under the closest scrutiny, looks point-like. The first quasars discovered were radio sources that coincided in position with what looked like stars. They were called 'quasi-stellar radio sources,' which was then abbreviated to quasar.

In 1960 Tom Mathews and Allan Sandage (1926–) identified the radio source 3C 48 (the 48th object in the *3rd Cambridge Catalog of Radio Sources*) with what looked like a star of magnitude 16. Its optical spectrum showed unusual emission lines that could not be recognized. In 1962, Cyril Hazard, Brian Mackey and John Shimmins established a very accurate radio position for another source, the bright quasar 3C 273, fixing its position as it disappeared and reappeared while it was being occulted by the Moon. Maarten Schmidt (1929–) identified 3C 273 with a magnitude-13 star, with a similarly puzzling spectrum. But he suddenly realized that the emission lines were from hydrogen and oxygen redshifted by 15.8% of their rest wavelength. A redshift of 0.158 was unprecedented for such a bright object. The natural explanation was that it was cosmological; that is, it arose from the expansion of the universe. At the inferred distance, 3C 273 had a luminosity 100 times that of an entire galaxy.

Following Schmidt's breakthrough, it was immediately realized that the redshift of 3C 48 was 0.37. By 1965, 3C 9 was found to have a redshift of 2.01. According to the modern scale of the universe (*see* GALAXY FORMATION AND EVOLUTION) its light has taken 75% of the age of the universe, or more than 10 billion years, to reach Earth.

3C 273 had been shown to be of enormous luminosity. Its luminosity is not only large, it changes from month to month. The scientific jargon for this is that '3C 273 varies on a timescale of months.' For all the parts of the quasar to vary together within this time, the characteristic size of the emission region had to be light-months – solar-system sized. How could such a small object produce more light than an entire galaxy?

The main observed properties of quasars are:

- They are basically galaxies. Indeed, with the HST faint emission close to the quasar can often be detected from the host galaxy. The galaxies can be elliptical or spiral, but are often, perhaps always, interacting or colliding (*see* GALAXIES, COLLIDING).
- The nuclei of the galaxies are starlike.
- They have spectra showing broad emission lines, indicating high orbital motions of gas in the quasar.
- Radio images of quasars show JETS and other structures, which range in angular extent from milliarcsec to tens of arcsec, and are often observed to be expanding outward, showing that quasars have an engine that can accelerate bulk material to high speeds.
- They radiate power at energies ranging from gamma-rays to radio frequencies.
- They have redshifts ranging from 0.1 to 6.
- They vary in brightness within a week, so they are small. Indeed, some quasars, called BLAZARS, vary as quickly as one day. Variability of quasars has been observed at different wavelengths in intensive campaigns and programs extending for years, including work by professional and amateur astronomers, using satellites and ground-based telescopes of all sizes.
- Even though quasars are small, they have luminosities as high as 100 000 billion suns, or say, 1000 galaxies.

All these properties are brought together in the so-called unified theory of AGNs, which are thought to be massive black holes, orbited by gaseous material which falls through an accretion disk and releases prodigious amounts of energy. Quasars are AGNs viewed from near the pole of the accretion disk, from which the power and luminosity readily escape.

Because quasars are so bright, their light probes the INTERGALACTIC MEDIUM. This revealed intergalactic clouds, the main component of the baryonic content of the universe.

The chemical abundances in quasars are similar to those observed in present-day stars and nebulae, even in quasars at the highest redshifts. This is a striking result when one considers that the most distant quasars correspond to a time when the universe was less than 10% of its present age. The apparently normal abundance of the elements implies that quasars were preceded by many stars in the early universe, the so-called POPULATION III.

Originally quasars were identified through their radio emissions. However, most of the tens of thousands of quasars now known have been found by mass surveys aimed at finding objects whose optical spectra are different from stars and galaxies. At redshifts greater than 2.2, quasars radiate more near-ultraviolet light than common stars, because the ultraviolet emission lines Lyman-alpha and others are redshifted into the near-UV and optical spectrum. Other techniques to identify quasars include looking for variability of what otherwise look like stars, and looking for objects that do not move in the sky (that is, with zero PROPER MOTION) and must therefore be distant.

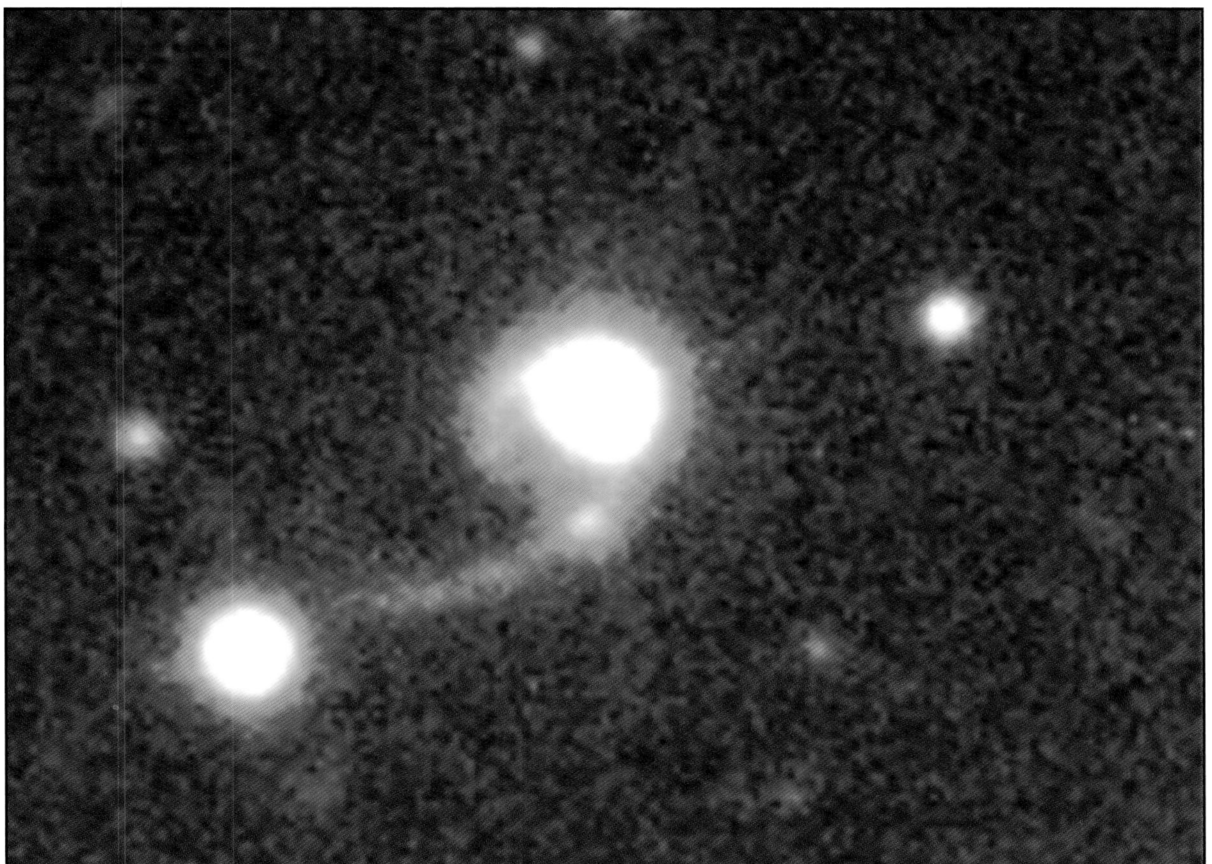

◀ ***Quasar*** *The quasar HE 1013-2136 is the bright image at the center, in the center of a galaxy (surrounding fuzz). A second galaxy lies below the quasar. The interaction between these two near-colliding galaxies has ripped stars from the galaxies in tidal waves, one of which points by coincidence to a star in the same direction. The quasar has been triggered into outburst by the interaction.*

Two major modern surveys are the 2dF Quasar survey and the SLOAN DIGITAL SKY SURVEY (SDSS). In Australia, a multi-object fiber spectrograph, which surveys a field of 2° at a time, recently completed the so-called 2dF (2° field) Quasar survey, with the discovery of 23 424 quasars. The SDSS is imaging a quarter of the sky in five spectral regions, to identify 1 million galaxies. The brighter quasar candidates are confirmed by mass spectroscopy. The SDSS has progressively pushed out the distance record for quasars. The first quasar with a redshift of more than 5 was found in 1998. In June 2001, two quasars with redshifts of 6.0 and 6.2 were found. In April 2003, the distance record was 6.4. Light from this quasar set out when the universe was less than 1 billion years old.

One of the most striking properties of quasars that has come from such surveys is their distribution with distance. As Schmidt first noticed, there are too many quasars at high redshift. The density of quasars is more than a thousand times greater at a redshift of 2 than in the local universe. Beyond a redshift of 3, their density declines steeply. A logical interpretation is that we are seeing back beyond the epoch of peak quasar activity.

It evidently takes about 1 billion years for supermassive black holes to grow to the mass needed to make a quasar. This is enough time for a few generations of massive stars to form and evolve. The overall rate of star formation and quasars in the universe grew as galaxies continued to form, interact and assemble, and reached a peak when the universe was aged 3.5 billion years. After that, the numbers of quasars and newborn stars have declined for 10 billion years, because galaxies became fully assembled and separated as a result of the expansion of the universe.

Further reading: Peterson B M (1997) *An Introduction to Active Galactic Nuclei* Cambridge University Press.

Quasi-stellar object *See* QUASAR

Quintessence

Quintessence is a mysterious fluid of matter and radiation that has been postulated to pervade the entire UNIVERSE. Remembering that mass and energy are equivalent through Albert EINSTEIN's famous equation $E = mc^2$, it is a form of dark energy distinct from any normal forms of matter, radiation and DARK MATTER (*see* DARK ENERGY AND THE COSMOLOGICAL CONSTANT). It could drive the accelerating expansion of the universe. If so, it makes up about 70% of the density of the universe. As the universe expands its density decreases less rapidly than the density of ordinary matter, but it does change with time.

The term 'quintessence' has a historical precedent, being, in Aristotlean philosophy, the fifth element constituting the heavens after everything in the 'sublunary sphere' consisting of earth, water, air, and fire (*see* ARISTOTLE OF STAGIRA). In modern cosmology it is the fifth element after baryons, radiation, neutrinos and dark matter.

In 1998 two independent teams reported that distant Type Ia SUPERNOVAE were dimmer than they ought to be from the REDSHIFTS of their parent galaxies. This was taken to mean that the universe expanded less quickly in the distant past than at present. In other words, the expansion is accelerating. This implies an extra cosmic force which operates as a cosmic repulsion that overwhelms gravity, and can be interpreted as evidence for the existence of quintessence.

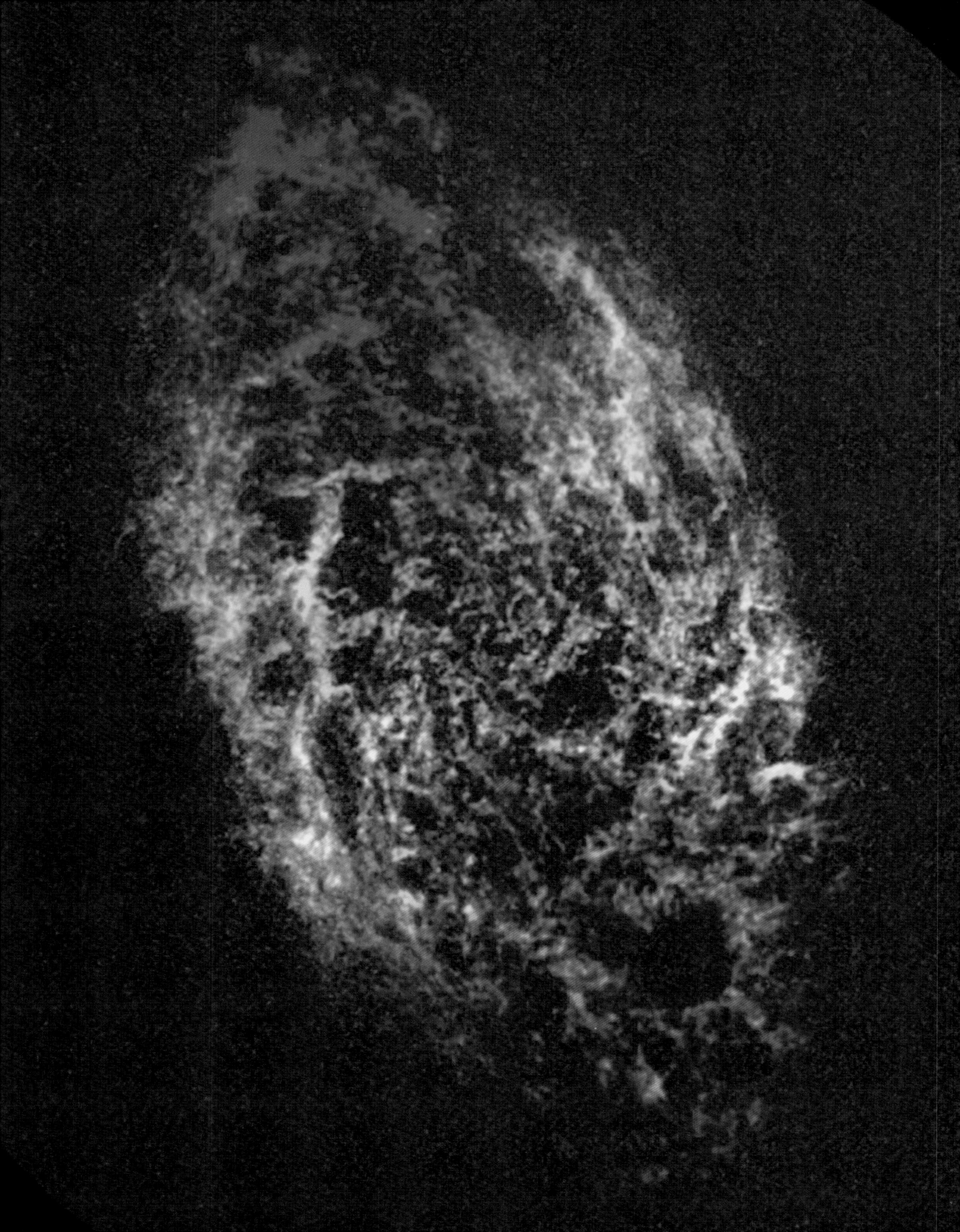

Radar astronomy

The branch of astronomy that is concerned with the investigation of solar-system bodies by radar methods. Radar (an acronym for 'radio direction and ranging') involves transmitting a beam of microwave radiation from a radio telescope and analyzing the faint echo that returns from the target's surface. Radar techniques can yield information on the distance to the body, the nature of its surface (including its roughness and composition), and its rotation.

The distance to a planet (Venus) was first measured in 1960. From radar ranging of Venus and asteroids, astronomers measure the scale of the entire solar system.

If the target object is rotating, part of the reflected pulse will be Doppler-shifted by the approaching side and part by the receding side (*see* DOPPLER EFFECT). This alters the frequency characteristics of the reflected pulse. This technique enabled the retrograde rotation of the cloud-covered surface of Venus to be determined in 1962 and the rotation of MERCURY to be measured in 1965. Recent developments make it possible not only to determine the rotation PERIOD on the assumption that the target is spherical, but to determine the shape of the target itself.

Radar astronomy has been used to determine the shape of asteroids (for example, (4179) Toutatis and the unusual dog-bone shape of asteroid (216) Kleopatra, 217 km long and 94 km wide). It has also been used to identify satellites of asteroids (such as 1999 KW4, which has a small satellite orbiting with a period of 16 hours, and the binary asteroid 2000 DP 107, two asteroids 800 and 300 m in diameter, separated by 2.5 km). The technique works only for NEAR-EARTH ASTEROIDS, because the range of radar techniques is limited by the faintness of the echo signal. Nevertheless, more than 200 asteroids have been detected using the technique.

Radar astronomy has probed further out to larger targets, namely the Galilean satellites and the rings of Saturn.

The characteristics of the reflected pulse of radar (for example, polarization) can be used to show the nature of the reflecting surface. Ice on the shadowed floors of craters on Mercury was identified from the backscatter characteristics of radar pulses from this planet in 1994.

The world's primary facilities used for radar astronomy are the National Astronomy and Ionosphere Center's Arecibo Observatory in Puerto Rico and NASA's Goldstone Solar System Radar (part of the Deep Space Network) in California. Because of its size, Arecibo has twice the range of Goldstone; but Goldstone, which can steer to view the whole sky, can provide twice the sky coverage as well as much longer tracking times.

Radial velocity

The component of the motion of a star or galaxy along the line of sight to the observer. Radial-velocity measurements reveal the systematic motions of stars around the Galaxy, around each other in BINARY STARS, and around their exoplanets (*see* EXOPLANET AND BROWN DWARF), and the motion of galaxies around each other and in the expansion of the universe. *See* ASTROMETRY for a diagram.

Radiant

The point on the celestial sphere from which the METEORS of a particular METEOR SHOWER all appear to diverge. Some showers have more than one radiant, representing different strands of meteoroids within the meteor stream. The radiant is in the direction of the sum of the velocities of the meteor stream and the Earth. The Earth's orbital motion curves, and as a result the radiant of a meteor stream moves eastward at a rate of 4 min of right ascension per day.

Radiation pressure *See* POYNTING–ROBERTSON EFFECT

Radioactivity

Radioactivity, or radioactive decay, is the spontaneous decay of certain unstable atomic nuclei through the emission of alpha particles (helium nuclei), beta particles (electrons) or gamma radiation. This process changes ('transmutes') the original, 'parent' atomic nucleus into a 'daughter' nucleus of another element. If the resulting nucleus is itself unstable, further decay will ensue until the original nucleus has been transmuted into a stable nucleus.

The decay of any nucleus in a sample is a matter of chance, and the overall trend for the numbers that do tails off with time in a curve known as 'exponential decay.' The time taken for half the nuclei in a sample of radioactive material to decay is known as its half-life, each radioactive NUCLIDE (a nucleus of a particular isotope) having its own characteristic half-life. The half-lives of the various radioactive nuclides range from fractions of a microsecond to billions of years. For example, ^{238}U, by far the most abundant of the 14 different isotopes of the element uranium, undergoes alpha decay with a half-life of 4.5 billion years.

Radioactive decay is used in planetary science to estimate the time that has elapsed since a rock has solidified. The ages of the Earth, of meteorites and of samples of lunar rock have been calculated by this technique. Younger samples of once-living material can be dated by means of the shorter-lived radioactive decay of carbon-14 (^{14}C). The radioactive decay of cobalt to nickel and then to iron was observed in SUPERNOVA 1987A. Similarly, spectroscopic identification in RED GIANT STARS of technetium, the longest-lived isotope of which has a half-life of much less than the age of the universe, shows that the synthesis of the elements is happening still (*see* ELEMENTS, FORMATION OF).

Radioastron (Spectrum-R)

Long-delayed, still-proposed Russian–US radio-astronomy satellite, designed as part of a very-long-baseline RADIO INTERFEROMETER. The satellite will carry a 10-m antenna.

Radio astronomy

The spectral range over which ground-based radio astronomy can be carried out extends from a wavelength of about 150 m (as low a frequency as 2 MHz) to a wavelength of 0.35 mm (1 THz at the high-frequency end). Karl JANSKY made the first radio-astronomy observations in the early 1930s. Jansky was working for Bell Telephone Laboratories, investigating the sources of interference that might affect transatlantic phone calls. He built a rotatable antenna sensitive to emissions at a wavelength of 15 m and found sources of 'static' coming from local thunderstorms, distant thunderstorms in the tropics and an unknown steady source. The peak of the steady source drifted in position at the rate stars drift across the sky from night to night (about 4 min per day), indicating an extraterrestrial origin for the emissions. Jansky had discovered synchrotron emission associated with energetic electrons accelerated in the magnetic field of the Milky Way.

The astronomical community did not fully appreciate the significance of Jansky's discovery for decades. One of the few who followed up his results was Grote REBER. Reber

◀◀ ***Radio astronomy*** *This true-color radio image of the Triangulum Galaxy shows gas moving toward us in blue, and gas that is moving away from us in red, with other colors depicting gas at intermediate velocities. The motions are inferred from observations of the atomic hydrogen gas in the disk of this galaxy which is emitted at one frequency, but this frequency becomes Doppler-shifted, depending on whether the gas is moving toward us or away from us.*

Radio astronomy for amateurs

Karl JANSKY is considered the father of RADIO ASTRONOMY. During the 1930s, Jansky worked for the Bell Telephone Laboratories studying the origin of static noise from thunderstorms. During the course of this work he discovered that some signals had an extraterrestrial origin. However, it was Grote REBER, a professional radio engineer and radio amateur, who carried out further investigations. In 1937, he built a 9.5-m diameter radio dish in his back yard in a suburb of Chicago, Illinois. He mapped the Milky Way at a frequency of 160 MHz and detected discrete celestial sources in the constellations of Cassiopeia, Cygnus and Sagittarius. His results, published in engineering and astronomical journals, brought this new field of research to the notice of professional astronomers and led to the establishment of major radio observatories during the postwar years.

A radio telescope consists of three basic components: an antenna, a receiver and a recording device. Specific requirements are dictated by the particular field of investigation and the depth to which it is to be studied. Ideally, one would use a large antenna with a preamplifier, connected by high-quality cable to a high-gain, low-noise receiver, all situated in an area with little radio interference. However, useful work can be done with less sophisticated equipment.

Solar system astronomy

The intense X-ray and ultraviolet radiation released during a solar flare will affect the Earth's IONOSPHERE. The disturbance, known as a sudden ionospheric disturbance (SID), enhances sub-100 kHz radio signals. Various time-signal and maritime stations operate in this very low frequency (VLF) band. When the Earth's MAGNETOSPHERE is shocked by the arrival of a stream of particles originating from a flare, the normally steady signal strength will undergo a marked change. Stronger disturbances may alert the observer to the possibility of auroral activity (*see* AURORA) arising 24–48 hours later when the slower-speed particles from the SUN arrive at the Earth. A simple antenna, consisting of several hundred turns of wire looped around a square frame of about 50 cm per side, can be connected to a frequency converter which is, in turn, attached to a communications receiver. Alternatively, an effective receiver can be built very easily at a cost of no more than a few dollars. The AAVSO publishes designs of receivers suitable for home building. Anyone with a little experience of soldering electronic components could construct one in just a few hours.

Meteors may be detected by the technique of forward-scatter. This is similar to bouncing radar signals from the ionized trails left by meteors as they burn up in the atmosphere, except that a transmitter operating in the very high frequency (VHF) broadcast band provides the 'radar' signal. Reflections are reasonably strong, allowing commercial receivers and small Yagi antennas to be used. A scanning receiver will work, but a better choice is a communications receiver costing around a few hundred dollars or more. Cheaper second-hand receivers are available from suppliers of amateur radio equipment. A suitable Yagi antenna will cost about $30. Typically transmitters 1000–2000 km away from the observer are monitored as these cannot be heard under normal propagation conditions. The best frequencies are around 50 MHz (where there are plenty of European television transmitters operating), and between 60–70 MHz. Signals in the latter band, used in Eastern Europe for local radio, are diminishing as the Western European 90–108 MHz band is becoming more widely adopted.

The Sun is a very strong radio source and may be observed over a range of frequencies from VHF to microwave. As signals are strong, commercial receivers can form the basis of a station. Yagi antennas for VHF and ultra high frequency (UHF), and dishes of about 1-m diameter for the higher frequencies work well. For continuous monitoring, the antenna must be driven to track the Sun across the sky. Commercial dish mountings are designed either to hold the dish in one position, or rapidly slew between satellites. A home-brew solution is needed to move the dish at the solar rate. At 12 GHz frequencies, an old Ku-band analog satellite television receiver and dish can be used. Offset-feed dishes of up to 1 m in diameter can be easily set up in a garden. A DC voltage, of the order of tens of millivolts, can be taken from across the tuning meter of the satellite television receiver

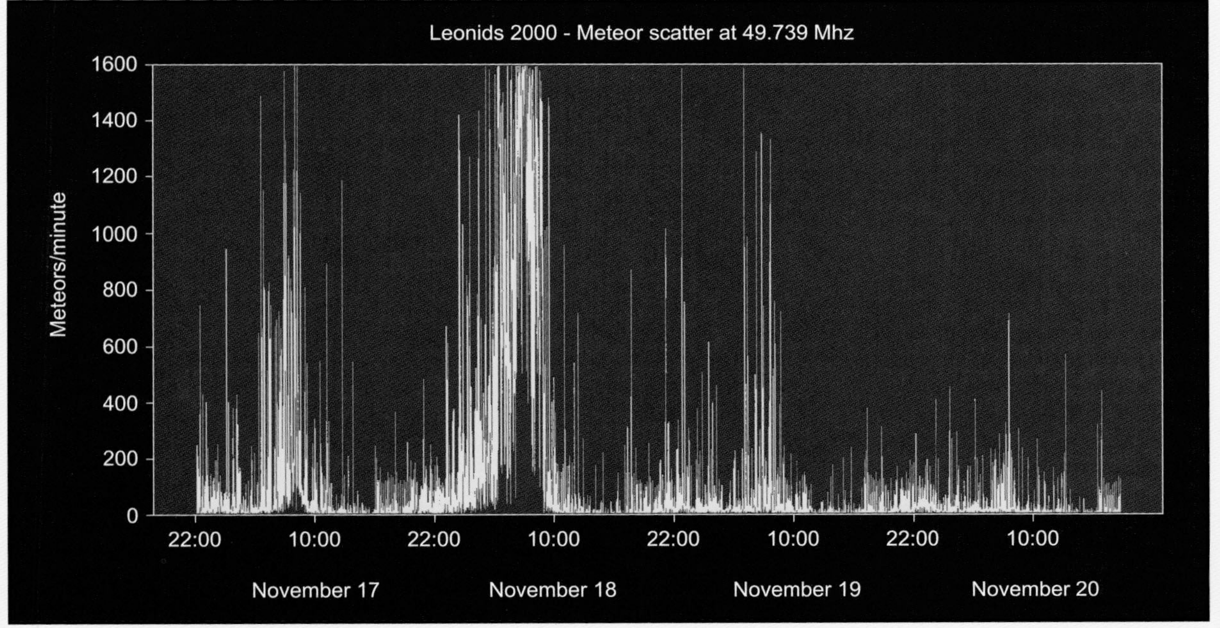

► ***Meteor scatter signals*** *received from the Leonid meteor shower over consecutive days during November 2000.*

◀◀ ***70-MHz Yagi*** *This antenna is used by the author for meteor scatter work.*

◀ ***Small radio telescope*** *This radio telescope, available in kit form, is capable of continuum and spectral line observations in the L-band (1.42 GHz). The SRT is a standard 2-m-diameter satellite television dish mounted on top of a fully motorized Az-El mount. This unique mounting arrangement allows the observer to perform total power measurements and contour mapping. Software is provided for controlling the antenna and selection of sources.*

and fed into the logging device. On most receivers, this is available via a connection to which an external meter was originally attached, allowing positioning of the dish during installation.

Jupiters's noise storms are detectable at decameter wavelengths, typically around frequencies of 18–40 MHz. A half-wave dipole, possibly with reflector and director elements, is a suitable antenna. The higher the gain, the more important it is to keep the antenna pointing towards the planet. Commercially produced or home-built receivers of moderate sensitivity can be used. A frequency not being used by transmitting stations (commercial, utility, military, amateur, and so on) must be found. At times of sunspot minimum, the higher short-wave frequencies are less effective for communication, and the detection of jovian signals becomes easier. Noise from Jupiter shows a correlation with IO as it orbits the planet. Several websites give predictions regarding when these events might occur.

Beyond the solar system

Detecting radio sources outside the solar system requires large dishes (of several meters in diameter) feeding sensitive receivers. Specially built receivers (or total power radiometers) operating at 1420 MHz, the frequency of neutral hydrogen, are available from around a thousand dollars. Alternatively, home-built equipment can reduce the overall cost, although such projects do require the constructor to have experience in building complicated electronic circuits. Wide bandwidths and variable integration times (milliseconds to tens of seconds) are used to extract the extremely faint signals from the noise. The strongest radio sources detectable by the amateur are SUPERNOVA remnants (Cassiopeia A and the Crab Nebula), RADIO GALAXIES (Cygnus A and Virgo A) and NEBULAE (Orion A and M16, M17 and M20). QUASARS are an order of magnitude fainter still, but 3C273 has been detected by amateur equipment.

To capture the data for later analysis, paper chart recorders may be used, but personal computers or stand-alone data loggers are more effective. As computer equipment is upgraded, an otherwise redundant machine can be pressed into service. Data-logging software is available at low or even zero cost.

The areas open to investigation that are listed above are mentioned in order of increasing complexity. SID and meteor detection systems can be set up and left unattended to collect data for many days. Solar and jovian work requires a bit more involvement from the observer who is required to switch equipment on and off each day and to check the tracking. Radio observatories designed to log more distant radio sources may require the observer to be present throughout the run.

Equipment is becoming ever more sensitive, but the radio environment, like the visible spectrum, is becoming ever more polluted by devices such as cellphones, pagers and artificial satellites. If the amateur can find a quiet location, then digital signal processing (DSP) techniques might reveal pulsars, high-energy pulses from the GALACTIC CENTER, GRBs or even ET.

Nick Quinn has been interested in space and astronomy since he was inspired by the Apollo Moon landings. He works as a computer systems developer for a multinational company.

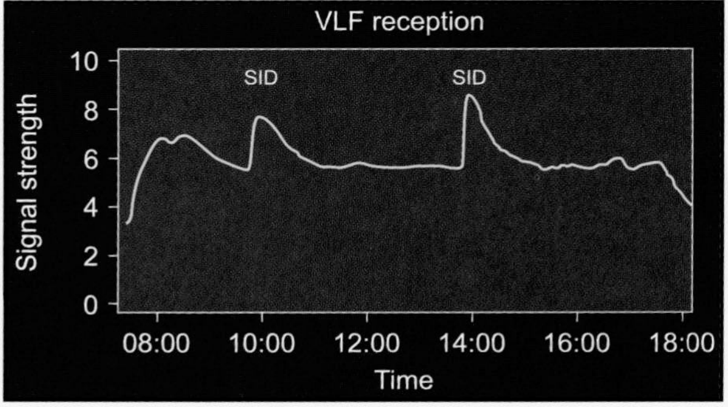

◀ ***Sudden ionospheric disturbance*** *caused by radiation that was emitted during a solar flare disturbing the Earth's ionosphere.*

▶ ***Jansky's antenna***
Built at Bell Laboratories to research a transatlantic radio telephone service at short wavelengths. Jansky recorded signals coming from the center of our galaxy. His antenna, called the Merry-go-round, was effectively the first radio telescope.

built a 9-m diameter parabolic reflector to investigate the strength of the radio emission from the Milky Way at meter and centimeter wavelengths. His antenna-receiver system was not sensitive enough to detect the Milky Way at centimeter wavelengths, but he did detect and map it at 1.9-m wavelength.

Both Jansky and Reber detected broadband radiation, analogous to the continuous static in radio reception found between man-made radio channels. Dutch astronomers sought a spectral feature, radiation at a particular frequency analogous to the narrow-band emission from a radio station. Spectral features, or lines, in the radio band come from atoms or molecules, similar to those seen optically. With a spectral line, one can measure a frequency (or Doppler) shift indicative of the relative motion between the source and the observer (*see* DOPPLER EFFECT). A transition of atomic hydrogen at a wavelength of 21 cm provided a candidate, and some years later this line was detected at Harvard University. Observations of atomic hydrogen have been used to study the structure of the Milky Way as well as to measure the mass of other galaxies. A major advantage of observations at radio waves, compared with visible light, is that one can see through interstellar dust grains, which hide most of the Milky Way from view at optical wavelengths.

Radio astronomy grew rapidly following World War II, aided by the significant wartime investment in radio equipment for use in radar systems. The Sun was shown to be a strong radio source, along with some other isolated sources. The strongest source in the constellation of Taurus, called Taurus A, was identified with the CRAB NEBULA, the remnant of the AD 1054 supernova. Another radio source, Cygnus A, became the archetypal member of the important class of objects called RADIO GALAXIES.

Large portions of sky were observed at 1-m wavelength, such as by the third Cambridge (or 3C) survey, and astronomers sought optical identifications for the newly discovered sources. While many radio sources could be associated with remnants of SUPERNOVAE in the Milky Way or with radio galaxies, one class appeared as bluish, star-like objects, which came to be called QUASARS. Quasars and radio galaxies are types of AGN.

Some quasars had a small number of 'blobs,' which appeared to expand outwards at speeds exceeding that of light. This surprising phenomenon, called SUPERLUMINAL MOTION, is explained quite simply in terms of the constant and finite speed of radio waves.

In the mid-1960s, Antony HEWISH in England led the construction of a telescope capable of studying variable radio emissions. Some radio sources were known to be variable, owing to a scintillation effect in the interplanetary medium. In 1967, Jocelyn BELL-BURNELL unexpectedly found pulsars with this telescope, which proved to be neutron stars formed by supernova explosions. Supernova explosions also create shock waves, which expand into the interstellar environment, creating shell-like radio-emitting regions called supernova remnants.

In the early 1970s, an intense radio source was discovered in the Milky Way. This radio source, called SAGITTARIUS A, has within it a small source, Sgr A*, which is the center of our rotating galaxy.

One of the predictions of Albert EINSTEIN's theory of GENERAL RELATIVITY is that gravitational mass bends light, and indeed radio waves, and therefore acts as a lens. In 1979, 0956+571, a source discovered in a large radio survey, was found to

◀ ***Radio galaxy*** *The radio emission from galaxy 0313-192 is mapped in red by the VLA. HST's view in visible light shows that a spiral galaxy, edge on, is producing the radio emission. Usually elliptical galaxies are radio galaxies, not spirals.*

consist of two compact objects separated by only about 6 arcsec. They were identified as optical quasars with nearly identical spectra. It seemed unlikely that two identical quasars could form so closely. Recently, fluctuations in the two quasars have been shown to be nearly identical, 'except' for a time difference of roughly one year. The two images are a single quasar whose light reaches us along two different paths, one longer by a light year. This provides conclusive evidence of gravitational lensing (*see* GRAVITATION, GRAVITATIONAL LENSING AND GRAVITATIONAL WAVES).

In the 1970s, radio astronomers started to discover many molecules, including water, ammonia and simple organic molecules, in clouds of interstellar gas. At centimeter wavelengths, detectable spectral lines are rare. However, at millimeter and especially submillimeter wavelengths, the spectrum is nearly completely covered with molecular lines. More than 100 different molecular species have been identified in molecular clouds.

One of the early surprises of molecular spectroscopy in the radio band was the discovery of extremely strong emission lines from the hydroxyl (OH) radical. These lines were not only thousands of times stronger than expected, they were also extremely narrow and highly polarized. Originally dubbed 'mysterium' because of these remarkable properties, the emission comes from a natural amplification process, called a maser, analogous to an (optical) laser.

As early as the 1940s, the Sun was found to be a very strong and complex source of radio emission. Intense, sporadic, low-frequency emissions were detected by radar equipment during World War II, but these discoveries were kept classified until the end of the war. A variety of solar flares were seen and found to be associated with sunspots and other active regions (*see* SUNSPOT, FLARE AND ACTIVE REGION).

Among the planets, early radio measurements of the brightness of Venus indicated very high temperatures, due to a strong 'greenhouse effect.' Jupiter, surprisingly, was found to have strong emission at about 10-cm wavelength from its MAGNETOSPHERE.

The far future for radio astronomy will probably include radio telescopes in space and perhaps on the Moon. The weightless environment of space should allow large, low-mass telescopes to be constructed, and INTERFEROMETERS with extremely large separations. In space too the exposure to man-made interference will be reduced, particularly in regions behind the Moon.

Further reading: Verschuur G L and Kellermann K I (eds) (1988) *Galactic and Extragalactic Radio Astronomy* Springer.

Radio galaxy

A galaxy that is an intense source of radio-frequency emission. One group consists of the star-forming galaxies, including SPIRAL GALAXIES and starburst galaxies. Their radio emission is associated with the byproducts from energetic stars, such as particles accelerated by SUPERNOVAE. The radio image of such a galaxy is not unlike the optical picture showing the star-forming regions. In fact, there is an even stronger correlation between the radio emission from such galaxies and their far-infrared emission, presumably because young massive stars are the common energy source (for example, they make supernovae and they warm interstellar dust).

In the second, much more powerful group of radio galaxies, most of the radio emission comes from two large 'lobes.' These clouds of radio-emitting material are well outside, and often on either side of, the visible galaxy, which is usually elliptical. The lobes of the giant radio galaxy 3C 236 extend nearly 20 million l.y. High-resolution images of many radio galaxies reveal a compact central radio source from which there emerges a JET, or a pair of oppositely directed jets, of radio-emitting material, pointing outward toward the distant lobes. The lobes are clouds of electrically charged particles (predominantly electrons) that have been expelled from a central 'powerhouse' in the core of the galaxy, a supermassive black hole. The jets are streams of highly energetic electrons that have been accelerated to a substantial fraction of the speed of light, also generated by the 'powerhouse.' Where the outflowing jets of particles plow into the distant lobes they make intense spots of radio emission. The first radio galaxy of this type to be identified was CYGNUS A.

Soon after radio galaxies were discovered, censuses showed that they were different in the past, because there were many more fainter, more distant galaxies than expected from an extrapolation from the number of strong ones nearby. This effectively ruled out the then-popular STEADY-STATE THEORY of the universe and showed that the universe had a history. But ironically, because much of this evolution was due not to the expansion of the universe but to the change in radio galaxies from the distant past to now, radio galaxies did not turn out to be the powerful new tool for cosmology for which astronomers had hoped.

Radio interferometer

A device that uses interference effects between radio waves to obtain higher angular resolutions than individual radio telescopes achieve. The simplest form of radio interferometer consists of two antennae or dishes that point in the same direction but are separated by a distance (the 'baseline') that is large compared with the wavelength of the radio radiation. Depending on the angle between the direction of the source and the direction of the baseline, wavefronts arriving at the two dishes from a source may be in phase (a wavecrest arrives simultaneously at each dish), out of phase, or somewhere in between. As a point source moves across the sky, and its direction relative to the baseline changes, the telescope separations (as seen from the source) alter because of the Earth's rotation. The signals will shift in and out of phase with each other, producing an interference pattern. From the multiple interference patterns of neighboring point sources (or an extended source) their distribution can be determined. A radio interferometer has the angular resolution, but not the light-collecting capacity, of a giant telescope the overall size of the array of telescopes. Martin RYLE pioneered this technique, called APERTURE SYNTHESIS.

Large-scale interferometers involving many individual telescopes have been built in the Netherlands (Westerbork) and in the USA (the VLA) and have achieved angular resolutions of better than 0.1 arcsec. In the mid-1960s some sources of radio emission (namely RADIO GALAXIES and molecular masers) were found to contain structures much smaller than could be resolved with those interferometers. This discovery motivated radio astronomers to seek even higher resolution by combining signals from telescopes in an array across the Earth. This was achieved by tape-recording the signal received at each telescope with a precise timing signal from atomic clocks. The tapes were shipped to a correlation facility where the data streams were played back, carefully synchronized, and cross-multiplied, bringing together the signals just as if they had been combined at the time of observation by wires. This technique, called VLBI (VERY LONG BASELINE INTERFEROMETRY), achieved the astounding angular resolution of 0.001 arcsec and revealed remarkable properties of bright sources.

Recently, VLBI techniques have been extended to link a radio telescope in space (the Japanese HALCA spacecraft with the 8-m VSOP telescope) with arrays of ground telescopes, yielding interferometer baselines of about three times the Earth's diameter. *See also* RADIOASTRON.

Radio telescopes and their instruments

A radio telescope collects radio waves from cosmic sources. In essence, a radio telescope has one or more collectors of radiation, each with a detector, an amplifier (to enhance the weak cosmic signals) and an instrument (to analyze, store, monitor and display the output from the system). The archetypal collector is a parabolic dish that, like the primary mirror of an optical REFLECTING TELESCOPE, reflects incoming radiation to a focus. The largest steerable dish is the 100-m instrument at Effelsberg, near Bonn, Germany. The largest fixed dish is the 300-m instrument that is built into a natural hollow at Arecibo, Puerto Rico.

Because the wavelength of radio waves is so much longer than that of visible light, radio dishes do not have to be shaped as precisely as the surfaces of optical mirrors. For example, an instrument that is studying radio emissions from clouds of hydrogen, at 21 cm, can tolerate irregularities in the reflecting surface of 1 cm (deviations of less than one-twentieth of the wavelength). At shorter wavelengths (centimeter or millimeter waves), the tolerances are proportionately finer. At longer radio wavelengths (above, say, 20 cm), a wire mesh tacked onto a skeleton frame is a good enough reflecting surface. This lightens the dish considerably, and reduces its distortion under the effects of gravity and the wind.

The commonest type of radio-astronomy receiver is a heterodyne receiver, in which the incoming signal is successively amplified and mixed with local oscillators to produce a signal at low frequency. This can then be processed in 'back-end' instrumentation.

In the early days of radio astronomy, the radio spectrum of a source was measured with a set of filters, each tuned to a slightly different frequency. The filters were inflexible and unstable. In the 1970s, Fourier transform or autocorrelation spectrometers ('correlators') were developed. In these instruments, the signal was successively delayed electronically. The delayed samples were then cross-correlated with the original signal. The resulting 'autocorrelation function' was then analyzed to produce a spectrum of the signal. The advantages of this technique were much higher stability and flexibility. The SETI project (*see* EXOBIOLOGY AND SETI) uses correlators with up to 1 million channels.

Early single-dish radio telescopes typically used just one detector at their focus. The entire telescope moved back and forth in a raster pattern, and built up an image of the source line by line. Increasingly, there is pressure to mount an array of detectors there. The first multibeam array to make a major impact was the seven-beam receiver developed by NRAO and successfully deployed first on the GREEN BANK 70-m telescope in the USA and then on the PARKES OBSERVATORY 64-m telescope in Australia.

Virtually all the types of instrumentation fitted to a single-dish radio telescope can also be fitted to a RADIO INTERFEROMETER, but have to be duplicated many times over, one for each telescope in the array. Because of this, interferometers have had modest spectroscopic capability. But the increasing speed and power of general-purpose computers means that some functions that once had to be performed in hardware can now be done in software, increasing the versatility of interferometers. As a bonus, finished results are available in real time.

Another change is the increasing contamination of radio-astronomy signals by radio interference, for example, by out-of-band emissions from low Earth-orbiting satellites. Techniques are being developed to cancel this interference. Though these are in their infancy, they will become increasingly important as the Square Kilometer Array, the next-generation radio telescope, is developed.

Ramsden eyepiece *See* EYEPIECE

Ranger

Series of NASA missions launched 1961–65, designed to send back detailed images of the Moon before crash landing on the lunar surface. Only the last three, Rangers 7, 8 and 9, were successful.

Ray

A bright linear feature on a planetary body formed by EJECTA from a recent crater-forming impact. Surface material darkens with age through exposure to radiation, so recently produced rays are bright by contrast. In addition, some impacts have sufficient force to vaporize surface material, which can resolidify in a reflective, glassy form. Rays extend from their associated crater, often in a radial pattern known as a ray system. The best-known ray systems are lunar, and the most prominent surrounds the crater TYCHO. This is very prominent at full moon, individual rays extending for 1000 km or more. Other solar-system bodies with rays are Mercury and Ganymede.

Rayleigh, Lord *See* John STRUTT

Rayleigh limit

The theoretical RESOLVING POWER of a telescope according to a criterion devised by Lord Rayleigh (*see* John STRUTT). Because of DIFFRACTION, the image of a point source of light (such as a star) produced even by a perfect optical instrument consists of a central bright spot (the AIRY DISK) surrounded by concentric dark and light rings. If two point sources are very close together, the resulting image will consist of overlapping diffraction patterns, and analysis of this situation forms the Rayleigh limit. The empirical form of the Rayleigh limit is DAWES' LIMIT.

R Coronae Borealis star

R Coronae Borealis (R CrB) stars fade suddenly from time to time. They dim dramatically and unpredictably for a few weeks, then gradually recover their original brightness. R CrB stars are evolved stars, which are rich in carbon (and helium) and pulsate. The pulsations cause the carbon to condense into sooty clouds, which obscure the star's surface. The star's brightness returns as the cloud dissipates. Edward Pigott, an English amateur astronomer, discovered the behavior of R CrB itself in the spring of 1795.

The origin of R CrB stars is unclear. One view is that they form from normal RED GIANT STARS. Usually such a star would become a PLANETARY NEBULA then a WHITE DWARF. It may happen that sufficient unprocessed helium remains on the surface of the white dwarf that nuclear reactions can be reignited, and the star becomes a helium-burning red giant for a second time. Convection mixes the outer layers of this star to give the helium and carbon seen in R CrBs.

In another model, it is supposed that a BINARY STAR of white dwarfs, one of helium and a more massive one of carbon, merges. The helium ignites in nuclear reactions and, like the previous model, the merged star becomes a helium-burning giant.

Three stars have been seen to change through the R CrB stage. V605 Aql is now the central star of a planetary nebula (Abell 58), having been in 1919 a slow nova; its spectrum at one time resembled an R CrB star. FG Sge was a faint blue star in 1900. Since then, it has become progressively redder and brighter; in 1992 it became an R CrB star. In 1996, an unremarkable faint blue star became an R CrB star, now known as SAKURAI'S OBJECT or V4334 Sgr.

▲ ***Parkes Radio Telescope*** *The Parkes Telescope in New South Wales, Australia, has a 64-m steerable parabolic dish. It was built in 1961.*

Reber, Grote (1911–2002)

American radio engineer. After reading Karl JANSKY'S articles on radio astronomy he built (1937) the world's first radio telescope (a 9-m tiltable paraboloid) in his backyard in Wheaton, Illinois. With it he detected radio emission ('cosmic static') at 1.9-m wavelength from the Milky Way. With an improved receiver in 1941 he detected the Sun and a strong source in Cassiopeia (Cas A). As a self-employed researcher in Hawaii and Tasmania he mapped the background radiation at 1–2 MHz. He produced the first radio maps of the sky.

Recurrent nova *See* NOVA

Red dwarf and flare star

Red dwarf stars are stars of spectral class M that are MAIN SEQUENCE stars (dM stars). Their mass range is 0.51–0.08 solar masses. The upper mass limit corresponds to a spectral type of M0 V and the lower limit is the smallest mass of a star that ignites hydrogen burning (below this mass limit lie BROWN DWARFS and PLANETS). Red dwarfs are the most common stars, and number at least 80% of the stars in the Galaxy. The radii of the red dwarfs span from 0.60 to 0.18 solar radii, while their surface temperatures are in the range 2500–4000 K.

Their luminosities are low relative to the Sun, ranging from less than a few tenths of a percent to about 8% of the solar luminosity. Like the Sun, a red dwarf has a large convection zone inside, indeed for most red dwarfs the whole interior is convective. This property is key to understanding those red dwarfs that are flare stars.

Flare stars are red dwarfs that brighten on timescales of seconds to minutes, like solar flares. They have a spectral type of dMe, namely a red dwarf with emission lines from an extended atmosphere.

The light from red dwarfs can vary periodically. This is due to large, cool spots, analogous to sunspots (*see* SUNSPOT, FLARE AND ACTIVE REGION), on the surfaces of these stars, which are carried by rotation across the near side. Rapidly rotating red dwarfs are the ones that show as dMe stars. They are also X-ray sources, with hot CORONAE, and radio sources.

Underlying the activity of the dMe stars are magnetic fields. The magnetic fields and the star-spots are caused by internal dynamo action, generated by the rapid rotation and the convection inside the dMe stars.

See also VARIABLE STAR: ERUPTIVE.

Red giant star

Although DWARF STARS derive their energy from hydrogen burning, giant stars have already exhausted the hydrogen in their cores (although they may still burn it in a surrounding shell). Instead, they burn helium or even heavier elements in their cores. Their atmospheres are rich in oxygen- or carbon-bearing molecules such as titanium oxide (TiO), vanadium oxide (VO), C_2, CH and CN, with other elements like ytterbium, zirconium and lanthanum dredged up from the nuclear-burning zones below. Conclusive proof that this material stems from an ongoing process in the star itself is provided by the presence of technetium, an element with a radioactive half-life of only 200 000 years, so that no significant quantity would remain from the time when the star was formed.

Giant stars of spectral type M are luminous and large. Many red giants have extensive dust shells, some so thick that the star itself is invisible at optical wavelengths. The very low surface gravity and high luminosity of a red giant causes mass loss from its atmosphere, which occurs in a succession of episodes. The ejected material appears for a while as a PLANETARY NEBULA, when the core of the star has been exposed and is on its way to becoming a WHITE DWARF.

There are also red giants of unusual surface composition, known as CH STARS AND BARIUM STARS.

The coolest and most luminous red giants are VARIABLE STARS, with periods of a few hundred days in the case of the stars of larger amplitude. These are the long-period variables, divided into semiregular and MIRA VARIABLES.

Redshift

The redshift (or blueshift) of an object is the displacement of its spectral features to longer (or shorter) wavelengths due to a combination of the GRAVITATIONAL REDSHIFT, its RADIAL VELOCITY and the general expansion of the universe, as discovered by Edwin HUBBLE. Gravitational redshifts have been measured in the Sun and a small number of WHITE DWARF stars.

Rees, Martin John (1942–)

British cosmologist. Rees became Plumian professor of Astronomy at Cambridge University, and, in 1995, ASTRONOMER ROYAL. His forte is modeling any astrophysical phenomenon with physical insight, and he has successfully applied this technique to quasars, X-ray sources, gamma-ray bursts, galaxy formation, galaxy clustering and the cosmic background radiation. Rees is also a successful lobbier for and popularizer of science.

Reflecting telescope (reflector)

The first optical telescopes (1608 onwards) were REFRACTING TELESCOPES, using lenses to form an image. The first practical optical reflecting telescope was made by Isaac NEWTON around 1668. In a Newtonian reflector, the starlight falls upon a concave mirror. The light is then directed via a 45° flat mirror through the side of the tube to focus outside the incoming light beam – the observer's head or instrument does not obstruct the incoming beam.

The Newtonian pattern is still widely used by amateur reflectors of the 15–50 cm class. The 1.9-m Radcliffe Telescope, originally erected in 1948 at Pretoria in South Africa, is largest Newtonian telescope constructed. Its Newtonian focus is accessed from a cage attached to the dome shutter mechanism – this facility (intended for photography) is almost unused nowadays.

For apertures larger than about 50 cm, the CASSEGRAIN TELESCOPE, invented in 1673, is more convenient. The combined action of the concave primary and convex secondary mirrors makes the telescope more compact and reduces the off-axis aberrations. The short length directly reduces the size of its building, saving money. The focus of a Cassegrain telescope is conveniently close to the observing floor.

William Herschel's largest reflecting telescope, of altazimuth design, had a mirror 1.5 m across (*see* HERSCHEL FAMILY). This was not exceeded until 1845, when the THIRD EARL OF ROSSE, in Ireland, produced a Newtonian telescope with a 1.9-m mirror. Despite the limitations of its extraordinary mounting (it swung on ropes and chains between two massive stone walls on which the observers stood), the Rosse telescope discovered SPIRAL GALAXIES. The telescope was restored to working order (although not with its original mirror) in 1998.

The early reflectors had 'speculum' metal mirrors. Glass mirrors are coated with a highly reflective substance such as silver or aluminum, coated to optimize the wavelength bands used by the telescope; for example, the infrared.

As a telescope tracks a star, or points to different stars in the sky, its attitude changes relative to the direction of gravity. This produces mechanical flexure not only of the telescope structure but also of the glass mirrors. Supporting mechanisms or mirror mounts spread the support evenly and at all attitudes of the mirror.

Off the axis of a reflecting telescope the images are aberrated by COMA. A large reflecting telescope thus has a small field of view, perhaps only an arc-minute or two, unless a correcting lens is used just ahead of the prime focus. Fields of 1° are possible. To make an even larger field of view, an entirely new optical system was developed in 1930 by Bernhard SCHMIDT, with another invented in 1941 by Dmitri MAKSUTOV. These reflecting telescopes have a spherical main mirror and a glass corrector plate. The detector (a photographic plate or a mosaic of CCDs) is situated at the prime focus, bent to conform to the curved focal plane. Sharp star images can be obtained over very wide fields, often several degrees across in such a telescope, like the Oschin Schmidt telescope at Mount Palomar. SCHMIDT TELESCOPES have a collecting area set by the aperture of the

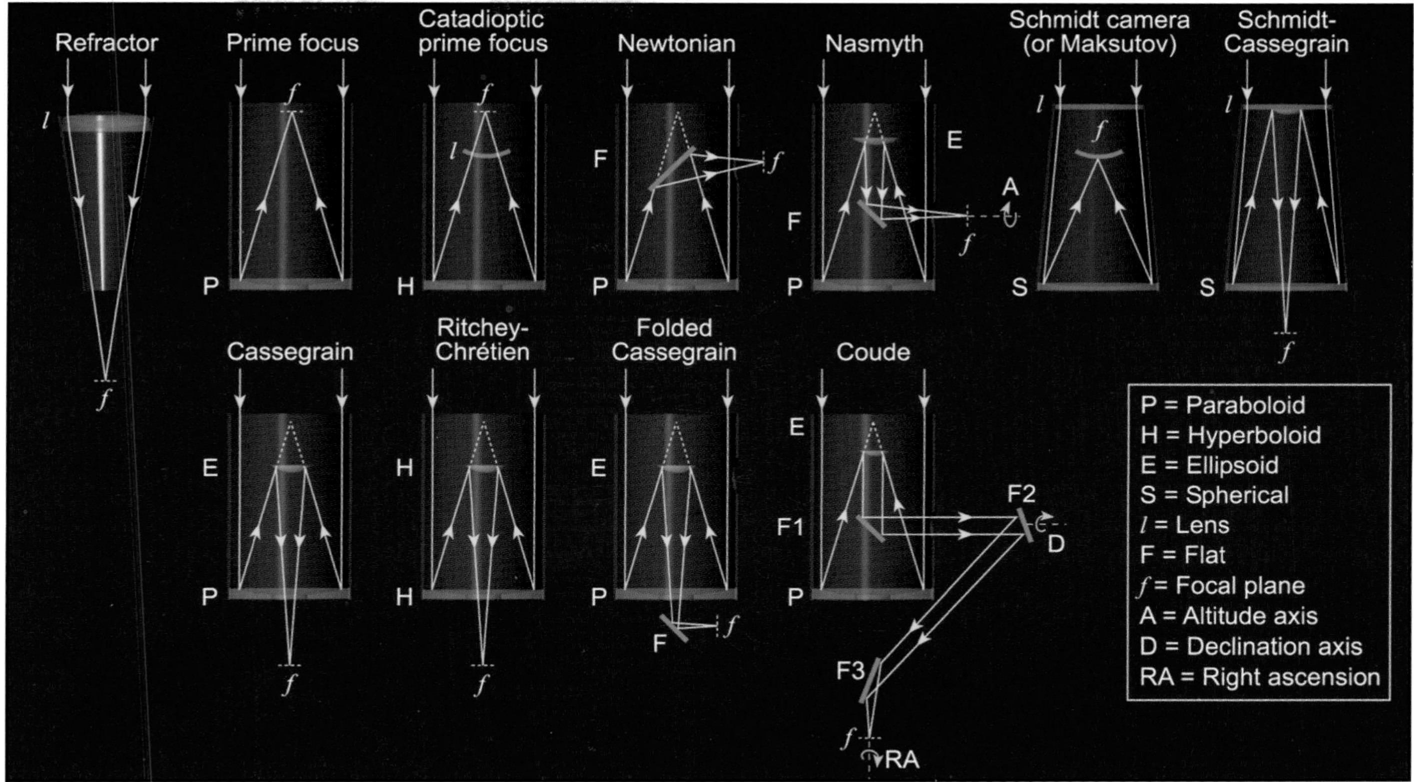

▲ ***Reflector and refractor telescopes***
Optical layout of different types.

correcting plate, but the main, primary mirror is larger than this. The largest Schmidt telescope is at Tautenberg in Germany and has a 1.35-m diameter corrector and a 2.0-m mirror. Sometimes both figures are given in the description of such a telescope, as for example 'the 135/200-cm Tautenberg Schmidt telescope.' A 400/500-cm meridian-mounted Schmidt telescope, the Large Sky Area Multi-Object Fiber Spectroscopic Telescope (LAMOST), is under construction near Beijing, China. The hybrid Cassegrain–Schmidt design can be mass-produced for amateur use.

In 1908 George HALE masterminded the construction of a 1.5-m reflector on Mount Wilson, California. The tube was a structure of rigid, skeletal girderwork, rather than a solid walled tube. This reduced weight, and improved air flow and seeing, but still held the optical components correctly. Reflecting telescopes often have open support structures, which nevertheless are called telescope 'tubes.' Another design driver, very important for INFRARED ASTRONOMY, is to reduce structure intruding into the infrared beam, to minimize infrared background in the telescope.

Hale set up the 2.5-m Hooker telescope on Mount Wilson in 1948 and planned the reflector with a 5-m mirror on Mount Palomar named in his honor. It contained innovations such as the 'Serrurier Truss,' invented by Mark Serrurier of the California Institute of Technology in 1935. This structure bent as little as possible, and if it did deflect (as was inevitable to some extent), it kept the primary and secondary mirrors in the correct relative positions.

An ideal equatorially mounted telescope tracks by rotation about one axis, at constant speed. However, flexure and other mechanical imperfections change the direction of the telescope, and atmospheric refraction shifts the stars. These effects can be compensated by computer control, first achieved at the arc-second level in 1974 by the 3.9-m Anglo–Australian Telescope, and now available for amateur use. This paralleled the successful computer control of altazimuthally mounted optical NASMYTH TELESCOPES, like the 6-m Great Altazimuth Telescope (Bolshoi Teleskop Azimutalnyi, or BTA, in Russia).

The telescopes at the GEMINI OBSERVATORY, the VLT and the SUBARU TELESCOPE at 8 m have probably reached the practical size limit of telescopes with a monolithic mirror. In a segmented mirror, various smaller components are fitted together to make a much bigger mirror. This technique was developed in reflecting telescopes working at wavelengths longer than light (such as for radio and millimeter astronomy). The 4.5-m Multi-Mirror Telescope in Arizona, USA, was the first such large optical telescope. The Keck telescopes on Mauna Kea in Hawaii have segmented hexagonal reflecting mirrors 12 m across (10-m diameter effective circular area). As the largest telescopes in the world, the Keck telescopes are setting the pace of discovery.

The overall mirror figure of a segmented mirror is maintained through active computer control of the mirror support system, referenced to laser beams. In ACTIVE OPTICS, the shape of the main mirror is continuously reformed by computer-controlled supporting pads. In ADAPTIVE OPTICS, a relatively bright star in the field of view is monitored, either a naturally occurring star or an artificial star generated in the upper atmosphere by a laser beam. Reacting to the degradation of the image of the monitor star, a flexible optical component in the telescope beam (a small mirror) is continuously modified in shape, to remove distortions due to turbulence in the Earth's air.

An alternative approach with large reflectors to increase angular resolution is to combine beams from separate telescopes situated tens or hundreds of meters apart. The two Keck telescopes, the Large Binocular Telescope and

The largest reflecting telescopes			
Name and abbreviation	**Location**	**Size (m)**	**Date**
Very Large Telescope (combination of VLT 1–4)	Cerro Paranal, Chile	16.4	2003
Keck Telescope I	Mauna Kea, Hawaii, USA	10.0	1991
Keck Telescope II	Mauna Kea, Hawaii, USA	10.0	1996
Gran Telescopio de Canarias (GTC)	La Palma, Spain	10.0	2004
Hobby Eberley Telescope (HET)	Mount Fowlkes, Texas, USA	9.2	1999
South African Large Telescope (SALT)	Sutherland, South Africa	9.2	2004
Large Binocular Telescope (LBT)	Mount Graham, Arizona, USA	8.4	2002/04
Subaru	Mauna Kea, Hawaii, USA	8.3	1999
Antu - Very Large Telescope 1 (VLT 1)	Cerro Paranal, Chile	8.2	1998
Kueyen - Very Large Telescope 2 (VLT 2)	Cerro Paranal, Chile	8.2	1999
Melipal - Very Large Telescope 3 (VLT 3)	Cerro Paranal, Chile	8.2	2000
Yepun - Very Large Telescope 4 (VLT 4)	Cerro Paranal, Chile	8.2	2000
Gemini North	Mauna Kea, Hawaii, USA	8.1	1999
Gemini South	Cerro Pachon, Chile	8.1	2002
Multiple-Mirror Telescope (MMT)	Mount Hopkins, Arizona, USA	6.5	2000
Magellan Telescopes I and II	Las Campanas, Chile	6.5	2000
Large Zenith Telescope (LZT)	Vancouver, British Colombia, Canada	6.1	2003
Bolshoi Teleskop Azimutalnyi (BTA)	Mount Pastukhov, Russia	6.0	1975
200-in Hale Telescope	Mount Palomar, California, USA	5.0	1948
William Herschel Telescope (WHT)	La Palma, Spain	4.2	1987
Southern Observatory for Astronomical Research (SOAR)	Cerro Pachon, Chile	4.2	2001
Victor Blanco Telescope	Cerro Tololo, Chile	4.0	1976
Large Sky Area Multi-Object Fiber Spectroscopic Telescope (LAMOST)	Xinglong Station, China	4.0	2004
Anglo-Australian Telescope (AAT)	Siding Spring, Australia	3.9	1975
Mayall Reflector	Kitt Peak, Arizona, USA	3.8	1973
UK InfraRed Telescope (UKIRT)	Mauna Kea, Hawaii, USA	3.8	1978
3.6 m Telescope	La Silla, Chile	3.6	1977
Canada-France-Hawaii Telescope (CFHT)	Mauna Kea, Hawaii, USA	3.6	1979
James Webb Space Telescope (JWST)*	Lagrangian point L_2	6.5	2011
Hubble Space Telescope (HST)*	Low Earth orbit	2.4	1990
Faulkes Telescope North** (FTN)	Haleakala, Hawaii, USA	2.0	2003
Faulkes Telescope South** (FTS)	Siding Spring, Australia	2.0	2004

* = the largest reflecting telescopes in space; ** = the largest reflecting telescopes for public education; Date = date of completion

the world's greatest telescope at Paranal in Chile (the four telescopes known as the VLT) are used this way, as INTERFEROMETERS.

Seeing conditions at the mountain-top observatories where large reflectors are nowadays constructed are excellent, but 'perfect' seeing exists only outside the Earth's atmosphere. The HST, launched from the Space Shuttle in April 1990, has a 2.4-m mirror. In spite of its relatively small size, limited by the capacity of the Shuttle's cargo bay, the HST is the reflecting telescope with the most impact of modern times. The 6.5-m reflecting James Webb Space Telescope, optimized as a successor to the HST for observations in the near-infrared, will be made of segments that unfold like an umbrella.

A completely different approach is to make a reflecting telescope of rotating mercury, spinning such that centrifugal force builds up the correct parabolic dish shape (as first proposed by Newton). The 3-m NASA Orbital Debris Observatory (NODO) telescope in New Mexico and 2.7-m Liquid Mirror Telescope of the University of British Columbia, Canada, are as far as this technology has been proved. Such telescopes are restricted to staring at the ZENITH, and can be used to sample the sky for space debris, stars or galaxies.

Another innovative large reflecting telescope is the segmented11 × 10 m Hobby Eberley Telescope, its 91 hexagonal segments each 1 m across, giving it an effective aperture of 9.2 m. The telescope structure is tilted at a fixed angle of 55° above the horizon. It can be rotated to catch stars and galaxies as they rise or set across the small circle centered on the zenith at 55° altitude, and track them via a moving instrument carriage for up to two and a half hours.

It is believed that there are no theoretical limitations to the manufacture of ground-based reflecting telescopes up to the 30–100 m class. Design studies are in progress for the OWL (Overwhelmingly Large Telescope) and MaxAT (Maximum-size Astronomical Telescope); they will surely be projects with global participation.

Ever since GALILEO GALILEI had such spectacular success in 1608 when he first turned a telescope to the sky, astronomers have been seeking, in their quest for more light, ways to make telescopes larger and more accurate. Engineers, technologists, scientists, administrators and politicians have been inspired to work together internationally to address

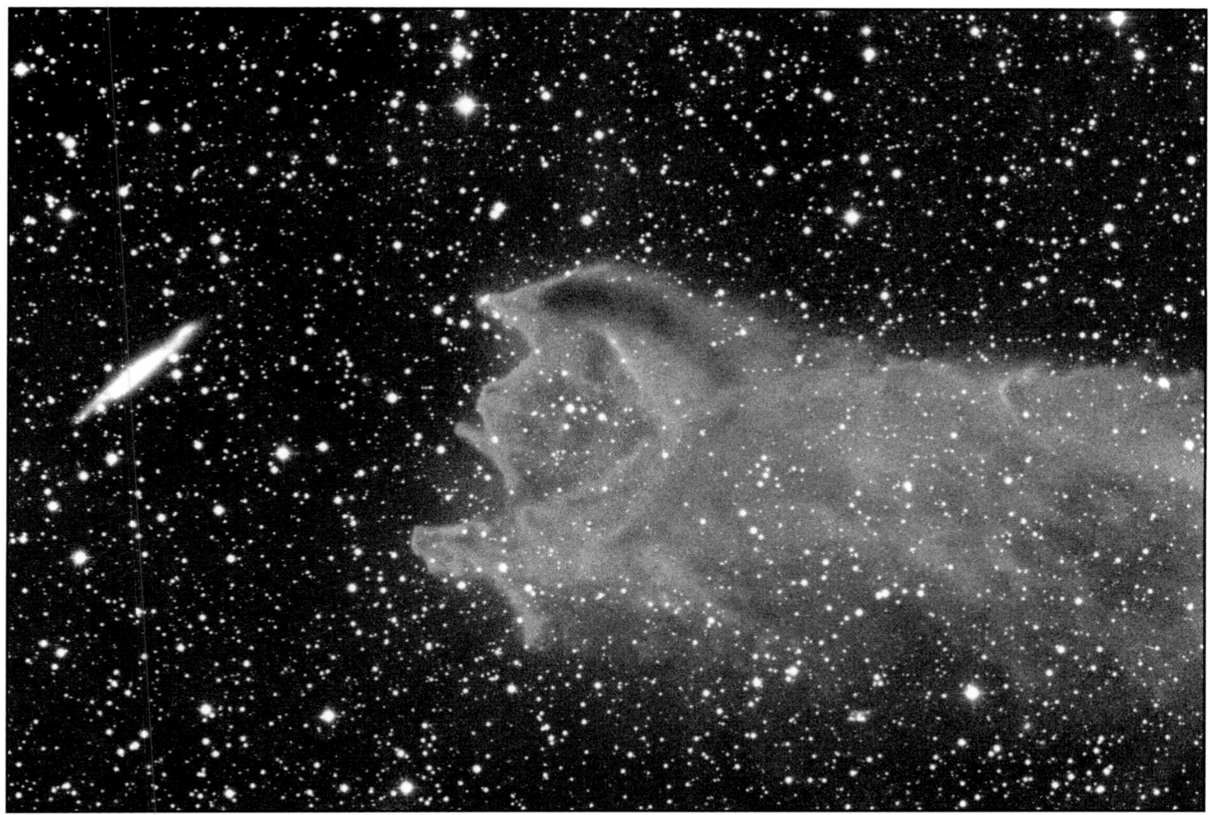

◀ ***Reflection nebula*** *The head of cometary globule CG4 (an isolated, relatively small cloud of dust and gas) is illuminated by light from hot nearby stars. Their energy is gradually destroying the head, sweeping away tiny particles that scatter the starlight as a faint bluish reflection nebula.*

questions that, in the twenty-first century, are the modern equivalent of those tackled by Galileo in the seventeenth. In the words of astronomer Fred HOYLE, put into the mouth of Britain's Prince Charles at the dedication of the Anglo-Australian Telescope in 1974, 'reflecting telescopes are examples of the sorts of things our civilization does well.'

Reflection nebula

Dust in interstellar space interacts with light by scattering. When a concentration or cloud of dust happens to be near a bright source of light such as a star or group of stars, the resultant scattered light stands out against the sky background and identifies this cloud as a reflection nebula. A defining characteristic of reflection nebulae is the star-like spectrum, matching that of the illuminating sources, but bluer in color, because the scattering efficiency of interstellar dust particles is greater at blue wavelengths.

Reflector *See* REFLECTING TELESCOPE

Refracting telescope (refractor)

Nobody is sure when the first telescope was made. It may be as far back as the early 1550s in England, when there is evidence that Leonard Digges (*c.*1520–59) constructed a telescope of some sort. The Dutch spectacle-maker Hans LIPPERHEY built in 1608 the first telescope of which we have definite proof. He is reputed to have discovered the principle when, by chance, he noticed that distant objects were seen to be enlarged when he placed one lens in front of another. GALILEO GALILEI made the first systematic telescopic series of observations in early 1610; his telescope was a small refracting telescope. Whatever he could see can be seen in modern binoculars. The Galilean refractor used a convex ('positive') objective and a concave ('negative') eyepiece that was placed in front of the focal plane of the objective. It produced erect images, but its field of view was very small and the instrument was difficult to use. Johannes KEPLER published in 1611 an improved design, which is the basis of the modern refractor. The Keplerian, or 'astronomical,' refractor uses a convex objective and a convex eyepiece that is placed behind the focal plane of the objective. Although it produces an inverted image, it has a wider field of view than the Galilean instrument. Because the eyepiece is located on the observer's side of the focal plane, it is possible to focus simultaneously on the image and on micrometer wires placed at the focal plane. In a refractor, the light from the target object is collected by a glass lens, also known as an object glass or objective. The light is passed down the telescope and brought to focus, where an image is formed and is enlarged by a second lens, termed an eyepiece or ocular. The eyepiece magnifies, and the object glass collects the light. The distance between the object glass and the focus is termed the FOCAL LENGTH of the telescope. MAGNIFICATION is given by the focal length of the telescope divided by the focal length of the eyepiece. The focal length of the telescope divided by the diameter of the object glass gives the FOCAL RATIO. Thus if a 3-in refractor has a focal length of 36 in, its focal ratio (f) is 36/3 = 12. If the 3-in refractor has an eyepiece of focal length 0.5 in, the magnification will be 36/0.5 = 72, often written as ×72.

Suppose, with our 3-in refractor, we use an eyepiece of an eighth of an inch (0.125) in focal length. The magnification will then be 36/0.125 = 288. If an image is enlarged it becomes fainter, and with this power on this telescope the image would be too faint. The maximum satisfactory magnification is ×50 per inch of aperture, so that for our 3-in telescope the highest power that can be properly used is 3 × 50 = 150. For

Some large refracting telescopes

Name	Observatory	Size (in)	Date
Yerkes 40-in	Yerkes, Williams Bay, Wisconsin, USA	40	1897
Lick 36-in	Lick, Mount Hamilton, California, USA	36	1888
Meudon Refractor	Meudon, Paris, France	33	1889
Potsdam Refractor	Potsdam, Germany	31	1899
Lunette Bischoffsheim	Nice, France	30	1886
Thaw Refractor	Allegheny, Pittsburgh, Pennsylvania, USA	30	1985
Grosser Refraktor	Archenhold, Treptow, Germany	27	1896
26-in Equatorial	US Naval Observatory, Washington, DC, USA	26	1873
Innes Telescope	Johannesburg, South Africa	26	1926
McCormick Refractor	Leander McCormick, Charlottesville, Virginia, USA	26	1880
Thompson Refractor	Herstmonceux, England	26	1897
Vienna Refractor	Vienna, Austria	26	1880
Newall Refractor	Athens, Greece	25	1862
Lowell Refractor	Lowell, Flagstaff, Arizona, USA	24	1895

Date = date of completion

a higher magnification it is necessary to have a larger telescope. With any telescope it is desirable to have several eyepieces: one to give low magnification and a wide field, suitable for observing objects such as star clusters; one with a moderate magnification, for views of the Moon and planets; and one with high magnification, for use on really good, clear nights. With our 3-in, *f*/12 refractor, suitable eyepieces might well be of focal length 1 in (36/1 = 36), 0.5 in (36/0.5 = 72) and 0.25 in (36/0.25 = 144).

The main problem with a refractor is that it introduces CHROMATIC ABERRATION. This can be reduced by an achromatic object glass, in which there are several component lenses made of different kinds of glass, chosen such that the errors tend to cancel each other out. Some false color always remains, but with a good achromatic objective it is not really serious.

An astronomical refractor will give an inverted image. In fact any refractor will do this, but in a telescope made for terrestrial use an extra lens system is put into the optical train to make the image erect. However, each time a ray of light passes through glass it is slightly weakened. This does not matter in the least when looking at birds, or ships out at sea, but it matters very much to an astronomer, who is anxious to collect every scrap of light available. Therefore, the erecting lens system is left out, although an erecting eyepiece can always be obtained.

▶ ***Refracting Telecope*** *Zeiss 30-cm refractor at the Griffith Observatory, Los Angeles, USA.*

Before the invention of the REFLECTING TELESCOPE, efforts to eliminate false color meant that refractors were made with very long focal lengths. This reduces chromatic aberration, but makes the telescopes very unwieldy. For example, Christiaan HUYGENS, probably the best observer of the early seventeenth century, constructed a telescope with an aperture of 2 in and a focal length of 10.5 ft and used it to discover Titan, the largest of the satellites of Saturn. He then built a telescope with a focal length of 23 ft, and with it discovered the true nature of Saturn's ring system. The telescope must have been incredibly awkward to use, and even more so were the refractors made and used by another pioneer observer, Johannes HEVELIUS of Danzig (now known as Gdansk). One of Hevelius's telescopes had a focal length of no less than 150 ft, but was so subject to wind disturbance that it could seldom be used to its full potential.

Next came the 'tubeless' telescope. Here, the object glass was fixed to the top of a mast, and the observer sighted it by looking along guide wires, which could be used to turn the object glass to the right position. The observer then held the eyepiece by hand. One of Huygens's 'aerial telescopes' had a focal length of 210 ft, and it is said that a refractor with a focal length of 600 ft was planned, although there is no record that it was actually built.

Obviously these long-focus telescopes could never be really satisfactory, but in 1733 a wealthy amateur astronomer, Chester Moor Hall (1704–71), constructed the first achromatic or compound objective, with one component made of flint glass and the other of crown glass. John DOLLOND took up the idea, and set up an optical company to make such lenses. In 1765 John's son, Peter, took one of his achromatic objectives to the Royal Observatory at Greenwich, and its performance was found to be far better than that of the observatory's best long-focus telescope. The old aerial refractors promptly became obsolete. Moreover, Dollond's telescopes looked attractive. From about 1783 they were made with brass tubes; when set up on well-made mahogany stands they were easy to use and were fitted with slow motions.

Greatly improved glass-making techniques, developed by the German astronomer Joseph von FRAUNHOFER, enabled larger objectives of higher quality to be produced – his discovery of wavelength standards in the solar spectrum (*see* FRAUNHOFER LINES) enabled him to control his glass-making process. By the first part of the nineteenth century really good objectives could be made. Then, in 1862, Thomas Cooke (1808–68), in England, built the first of the 'great refractors'; it is known as the Newall Telescope and had an objective of 25 in across (it is still in use, at the Athens Observatory on Mount Penteli in Greece). Between 1870 and 1900 even larger refractors were built, culminating in 1897 with the 40-in telescope which was set up at the Yerkes Observatory in Wisconsin, USA. At the time, and for some years afterwards, it was the most powerful telescope in the world, and it proved to be a great success. It is still in use on every clear night.

Is it possible to build a refractor bigger than the Yerkes 40 in? Theoretically, yes, but there are serious practical difficulties. If a lens is too large it will begin to distort under its own weight, and this will make it useless. An attempt was made at the very end of the nineteenth century, when

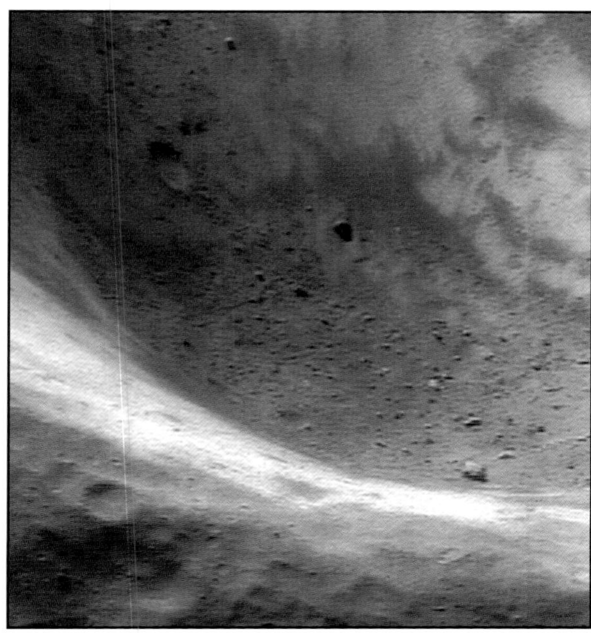

◀ ***Regolith***
A regolith visible on the surface of the asteroid Eros, seen at a distance of 50 km by the NASA spacecraft, NEAR–Shoemaker.

a 49-in objective was made in France. The 180-ft tube could not be mounted in the conventional way; instead it was left horizontal, and the light was brought to it using movable mirrors. The telescope was shown at the Paris Exposition of 1901, but it was never used for any astronomical work. Before long it was dismantled, and the fate of the object glass is unknown. The fate of a 41-in refractor that was destined for the Pulkovo Observatory in Russia, was even worse; before the optics could be completed, the mounting of the telescope rusted away.

Yet, in amateur hands, the refractor remains ideal. It is much less delicate than a reflector and will need little maintenance; if it is treated with care, it will last a lifetime and more. Although its color correction can never be as good as that of a reflector, it will give superb, clear-cut images.

Choosing a refractor for home use must be undertaken with care, because a low-quality telescope does not betray itself at a glance. Some small refractors are sold according to the power available, or so it is claimed; in one recent advertisement it was said that the telescope offered 'would magnify 750 times.' The aperture of the object glass was, however, only 3 in, so that the maximum power (magnification) which could be used to advantage was no more than 150. Aperture is all-important, and a telescope advertised without giving the diameter of the objective should be avoided. Watch too for a ring fixed inside the top of the tube, masking the outer part of the objective and thereby cutting down the aperture; this is a trick designed to conceal defective optics.

Generally, it is not really sensible to spend much money on a refractor with an object glass less than 3 in in diameter. A refractor of aperture up to 4 in is easily portable, although a larger instrument will be too heavy to move around, and will be best set up in a permanent observatory.

Refractor *See* REFRACTING TELESCOPE

Regolith

The loose material covering the surface of a rocky planetary body. Regolith consists of dust and tiny fragments of rock produced by the impact of small METEORITES and MICROMETEORITES, and is characteristic of worlds that have no atmospheres and have been geologically inactive for a few billion years. The lunar regolith is 2–8 m deep on the maria, and as much as 15 m deep in the highlands, which have had longer exposure to meteoritic bombardment. Its composition largely reflects that of the underlying rock. Mercury has a very similar ALBEDO to the Moon, and is assumed on this basis to possess a similar regolith. Boulders and dust on the surface of the ASTEROIDS (such as (243) IDA and (433) EROS) show they have a regolith.

Regulus

Alpha Leonis, a blue-white star of spectral type B7V, 78 l.y. distant. Its name, which only looks as if it dates from antiquity, means 'little king' and was given by Nicolaus COPERNICUS in the sixteenth century, marking the fact that it had been regarded as a principal star in the sky for millennia.

Relative sunspot number

An index of sunspot activity devised in 1858 by Rudolf Wolf (1816–93) of the Federal Observatory of Zurich, Switzerland, and otherwise known as the Zurich sunspot number. The relative sunspot number, denoted by the letter R, is given by the formula $R = k(f + 10g)$, where g is the number of visible sunspot groups, f is the total number of individual spots visible on the face of the Sun and k is an efficiency factor that takes account of the size of telescope used, atmospheric conditions and the performance of the individual observer. For example, if on a particular occasion there were four sunspot groups containing 3, 7, 9 and 12 spots respectively, g would be 4 and f would be 31. Assuming $k = 1$, the value of R would be $1(31 + 40) = 71$.

The relative sunspot number places greater emphasis on the number of groups than on the number of individual spots, because the former is more closely related to the number of underlying active regions. Although the relative sunspot number gives a reasonable guide to the changing level of sunspot activity throughout successive sunspot cycles, sunspot areas give a better measure of the overall level of solar activity.

Relict

Two Soviet astronomical missions. Relict-1 was launched on the PROGNOZ-9 satellite in July 1983. Two small radio telescopes were used to search for any anisotropy (unevenness) in the cosmic background radiation. Relict-2 was launched into a halo orbit at the L_2 LAGRANGIAN POINT, 1.5 million km in the anti-sunward direction, in January 1994. It returned submillimeter-wavelength astrophysical and magnetotail plasma data.

Remote sensing

Use of an ARTIFICIAL SATELLITE as a proxy human being to determine the conditions on a planet.

Resolve

To separate the components of an image; for example, to distinguish one star of a binary system from another.

Resolving power

A measure of the ability of a telescope to separate objects that are apparently close together. Resolving power, or resolution, is usually expressed as the minimum angular

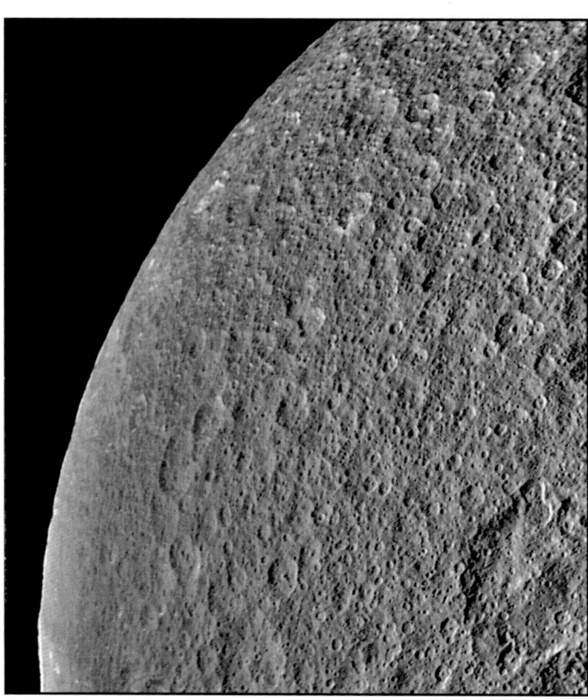

► ***Rhea***
Seen from the Voyager 1 spacecraft in 1980, a uniform surface of old craters.

separation at which two identical point sources of light (for example, two identical stars) can just be distinguished as separate points (if the sources are closer together than this minimum angle, they will appear as a single point). *See also* DAWES' LIMIT; RAYLEIGH LIMIT.

Resonance

In CELESTIAL MECHANICS a resonance occurs when the PERIODS of two or more interrelated celestial bodies have a ratio close to an integer fraction. For many satellite pairs of the outer planets a resonant relation can be found at a high level of accuracy. Within the saturnian system the pairs Mimas–Tethys and Enceladus–Dione are both in the 1:2 resonance, Titan–Hyperion 3:4 and Titan–Iapetus 1:5. The three Galilean satellites closest to Jupiter, Io, Europa and Ganymede, represent an outstanding example of a multiple-mean-motion resonance. Io–Europa and Europa–Ganymede are both found in a 1:2 resonance: thus the periods of revolution of Io and Ganymede are also in the ratio 1:4. The 'great inequality' is a 2:5 resonance that rules the motion of the two most massive planets of our solar system, Jupiter and Saturn, while the 2:3 commensurability between Neptune and Pluto provides the basic dynamical mechanism for avoiding too-close approaches between them. For the Saturn–Uranus (1:3) and Uranus–Neptune (1:2) pairs the corresponding resonant relations are satisfied less accurately.

The TROJAN ASTEROIDS, grouped around the triangular LAGRANGIAN POINTS of the Sun–Jupiter system, follow the same orbit as Jupiter and have a 1:1 resonance. The question of whether the Earth itself has Trojans, small enough to have escaped direct observation, has been repeatedly posed. The advent of orbiting telescopes has allowed us to detect the existence of a thin cloud of dust trapped in the vicinity of the 1:1 resonance (a possible explanation for the origin of the ZODIACAL LIGHT), while the recent survey of the near-Earth asteroid population has allowed us to identify (3575) Cruithne, an asteroid in an extraordinary horseshoe orbit in resonance with the Earth.

Resonances might be closely surrounded by chaotic regions. This is the dynamical mechanism responsible for the formation of the 1:3 gap in the asteroid belt: sudden chaotic jumps in the eccentricity of the orbit may have caused a near-resonant asteroid to become a Mars-crosser and eventually be removed from the main belt by close encounters with that planet.

The contradictory tendency to prefer certain resonances and avoid others can be observed not only in the asteroid belt but also in other densely populated systems, such as PLANETARY RINGS.

Reticulum *See* CONSTELLATION

Retrograde motion

Rotation in the opposite direction to the prevailing direction of rotation in the solar system. DIRECT MOTION or 'prograde' motion is rotation in the same direction as the solar system (anticlockwise as viewed from its north pole), a memory of the direction of motion of the solar nebula (*see* SOLAR SYSTEM). Hence retrograde motion indicates some event that modified a body's motion. The major planets Venus, Uranus and Pluto have retrograde rotation, possibly as a consequence of impacts. The outermost satellites of Jupiter, Saturn and Uranus, and also Neptune's largest satellite, Triton, have retrograde orbits, indicating that these are captured objects. Many long-period comets have retrograde orbits originating in planetary encounters as the comets fell in from the OORT CLOUD AND KUIPER BELT.

Retrograde motion is also used for the temporary motion from east to west on the CELESTIAL SPHERE of a superior planet, in contrast with its usual direct eastward motion against the background of stars. Retrograde motion occurs around the time of a planet's OPPOSITION and is particularly noticeable for Mars. As a planet nears opposition, the Earth is approaching it and traveling faster. The planet's rate of motion slows down, and then reverses direction. Its retrograde motion is fastest at opposition. The reverse occurs after opposition, when the Earth is moving away from the planet: the retrograde motion slows, halts and returns to direct. The two points at which the motion changes direction are called stationary points. Because the planets' orbits are inclined to the ECLIPTIC, the direct-retrograde-direct sequence causes the planet to describe what is known as a retrograde loop on the celestial sphere, or sometimes an S-shaped curve. The more distant a planet, the smaller its retrograde loop. In the PTOLEMAIC SYSTEM of the solar system, 'epicycles' were invented to describe retrograde motion. Each planet was thought to move in a circular motion, an epicycle, whose center itself moved uniformly around the Earth. Retrograde motion occurred when the planet was traveling on the epicycle inside the orbit of its center.

Reverberation mapping

The sound generated from a lightning strike may, as it spreads out, reflect from the surfaces of clouds, or from walls or buildings on the ground. It creates a roll of thunder, possibly multiple bursts of sound from the different surfaces. It is possible, measuring the time delays between echoes and their direction, to map where the structures are. This is the principle of reverberation mapping, used to map structures in AGN from the successive brightening and fading of parts of the accretion disk when illuminated by a burst of light from the central black hole.

Revolution

The motion, or orbit, of one celestial body around another; for example, the motion of a satellite around a planet or of a planet around the Sun. The Earth completes one revolution of the Sun in one year. *Contrast* ROTATION.

Rhea

The largest of the mid-sized icy satellites of SATURN, with a diameter of 1528 km, discovered by Giovanni Cassini in 1672 (*see* CASSINI DYNASTY). It orbits at a distance of 527 000 km. The surface was well imaged by Voyager 1. Unlike the other regular mid-sized satellites of Saturn, Rhea shows little diversity of terrain: the surface is saturated with craters. The 250-km diameter crater Izanagi is the exception: most of Rhea's craters measure around 20 km. It has been suggested that this even population of craters was produced by debris from the collision that left the satellite HYPERION, orbiting farther out, with its irregular shape. There are some ridges and scarps, interpreted as resulting from a global compression, as is thought to have occurred on Mercury. The lack of any orbital RESONANCES with other satellites is what has left Rhea for the most part geologically inactive.

Riccioli, Giambattista [Giovanni Battista] (1598–1671)

Italian astronomer who became a Jesuit and came into conflict with the Copernican system, refuting it while acknowledging its use as a mathematical hypothesis. Riccioli mapped the Moon and introduced some of the names still used, in a chart published in the *New Almagest* in 1651. As a follower of the PTOLEMAIC SYSTEM, he named major lunar craters for HIPPARCHUS, PTOLEMY and Tycho BRAHE (and a big crater for himself), and gave the names of Copernicus and Aristarchus to smaller ones. His observation in 1650 that Mizar was a double star was the first discovery of such a star with a telescope. He also discovered Venus's ASHEN LIGHT phenomenon.

Right ascension (RA)

The angle between the hour circles passing through the vernal EQUINOX and a celestial body, measured eastwards (anticlockwise) from the vernal equinox and expressed in time units (hours, minutes, and seconds) where 24 hours is equivalent to 360°. In other words, it is the angle between the vernal equinox and a point on the CELESTIAL EQUATOR such that the angle between the vernal equinox, this point and a star is a right angle. The position of a star is normally expressed in terms of right ascension and DECLINATION. *See also* CELESTIAL COORDINATES.

Rille

A long, narrow depression on the Moon's surface. There are several types. Linear rilles (or straight rilles), forked rilles and arcuate rilles are mostly graben (flat-bottomed strips of land that have collapsed between parallel fault lines). They may be straight or arc-shaped, and are often associated with large-impact features. Examples are Rimae Mersenius and Rimae Hippalus on opposite sides of Mare Humorum ('rima' is the name used for this type of feature).

Different in origin are sinuous rilles. These are long, steep-sided channels, sometimes discontinuous, that wind their way across the floors of the lunar maria. They are in fact collapsed lava tubes, similar to those on Earth. The surface of a narrow lava flow can solidify while the lava beneath is still flowing, so that when the flow ceases an empty tube is left. A sinuous rille is produced when all or part of the roof of the tube collapses. An example is the Hadley Rille, visited by the crew of Apollo 15. Sinuous rilles have also been identified on Venus and Mars. Some features on Uranus's satellite ARIEL strongly resemble sinuous rilles, and may have a similar origin in a 'cryovolcanic' process in which melted ices can behave like lava.

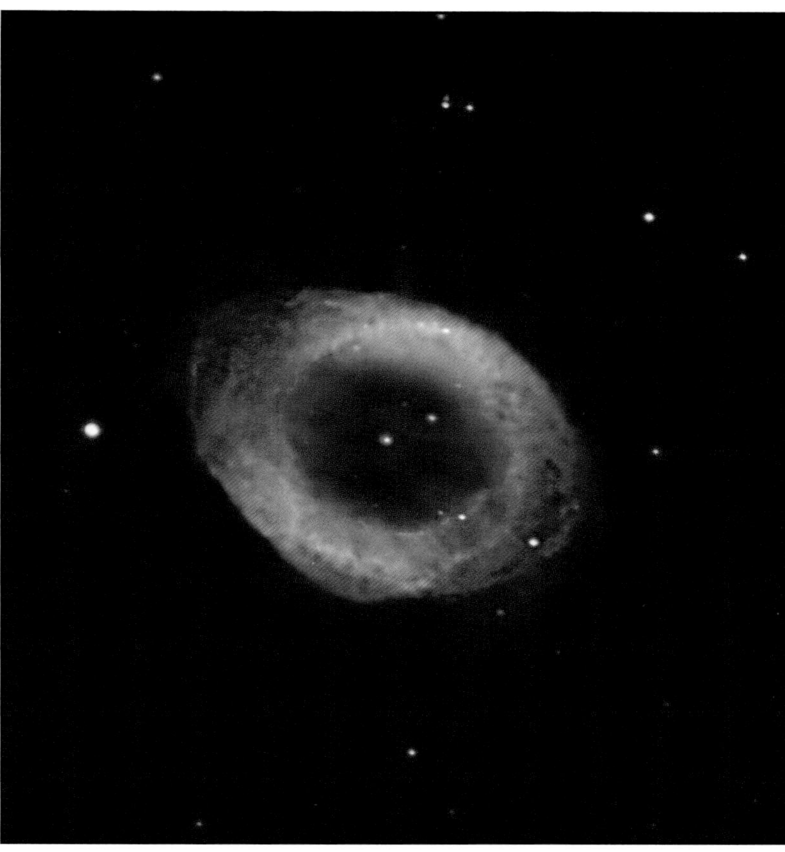

▲ ***Ring Nebula*** *(M57) Named for its resemblance to a smoke ring, the nebula is a planetary nebula, composed of material ejected from the hot white dwarf star at its center.*

Ring Nebula (M57, NGC 6720)

A PLANETARY NEBULA in the MESSIER CATALOG; the central star is only fifteenth magnitude. Its telescopic appearance resembles a smoke ring.

Ritchey, George Willis (1864–1945)

At first a furniture-maker and woodworker, Ritchey became an instrument-maker and especially an optician when he obtained part-time work at the observatory of the University of Cincinnati, Ohio, USA. Ritchey met George HALE in Chicago, Illinois, and volunteered to assist him, preparing photographic plates and learning to use the camera to photograph stars and nebulae. He became a full-time optician and supervisor of the instrument shop at YERKES OBSERVATORY and made its 40-in refracting telescope work properly. He built the 24-in Ritchey reflecting telescope with which he discovered the expanding nebulosity around Nova Persei 1901. He also built the horizontal solar telescope, and the Snow horizontal telescope that replaced it. Ritchey followed Hale to the MOUNT WILSON OBSERVATORY, and made the 60- and 100-in mirrors for the large Californian telescopes. He discovered a nova in a spiral galaxy, NGC 6946, showing the distances of galaxies as 'island universes.' However, Hale fell out with

Ritchey and fired him. Ritchey went to live in France, where he implemented the design of the RITCHEY–CHRÉTIEN TELESCOPE (like the ANGLO-AUSTRALIAN TELESCOPE). He returned to the USA, and began to build a 40-in Ritchey–Chrétien telescope for the US NAVAL OBSERVATORY in Washington, DC. But when he dropped the mirror he had been working on, the project lost confidence in him, and Ritchey retired.

Ritchey–Chrétien telescope

A variant of the CASSEGRAIN TELESCOPE developed by George RITCHEY and Henri Chrétien (1879–1956). The primary mirror is normally a hyperboloid in this optical configuration (rather than a paraboloid) and together with an appropriate hyperbolic secondary the system corrects for both SPHERICAL ABERRATION and COMA. High quality images are produced over a wider field of view than in the conventional Cassegrain system. Most large telescopes are of this design. *See also* REFLECTING TELESCOPE.

Roche limit

The minimum distance from a planet at which a satellite can resist being gravitationally disrupted by tidal forces. It is given by $2.5R(p/s)^{1/3}$, where R is the planet's radius, and p and s are the densities of the planet and the satellite. This expression (without the density term) was derived in 1849 by Edouard Roche (1820–83), and played a part in theories of the formation of Saturn's rings (*see* James MAXWELL). Material within the Roche limit would be prevented from accreting into a larger body, while the break-up of a satellite that ventured inside it would perhaps give rise to a ring system. The Roche limit as defined above applies to a 'fluid satellite' (a hypothetical body of zero tensile strength) whereas real satellites are solid objects with a degree of structural integrity, able to offer some resistance to tidal disruption. It would also apply to a loose 'sand-pile' assembly (an ASTEROID). Small solid satellites can in theory exist within the Roche limits for their bulk densities.

Roche lobe

Within the region surrounding the components of a binary system are two competing gravity fields. The Roche lobe, named for the nineteenth-century French mathematician Edouard Roche, is the volume of space within which any matter is controlled by the gravity field of a single star. It is said to be bound by an 'equipotential surface' in which the gravitational force exerted by the star is equal at all points. The Roche lobes of each component of a binary system, which in isolation would be spheroidal, are drawn out into cone-like extensions meeting at the LAGRANGIAN POINT between the two stars where the gravitational attractions of the two components are exactly equal. The position of this point along the line joining the centers of the two stars depends upon their relative masses. If one of the component stars expands to fill its Roche lobe, mass transfer will occur through the Lagrangian point toward the other star.

Rockets in astronomy

Prior to the launch of the first artificial satellite, Sputnik 1, in 1957, the only way to place scientific instruments above the atmosphere was to launch them on rockets. Suborbital sounding rockets have continued to play a role in improving our knowledge of the Earth, geospace and the universe.

Sounding rockets take their name from the nautical term 'to sound,' which means to take measurements. When the rocket motor has used its solid fuel, it separates from the vehicle and falls back to Earth. The payload continues into space for up to 30 min in a parabolic trajectory, re-enters the atmosphere and is parachuted back to Earth for retrieval and perhaps re-use.

Using rockets to probe the upper atmosphere was proposed in the 1930s by Robert Goddard (1882–1945). Between 1946 and 1951, 69 captured V-2 missiles were launched from the US Army's White Sands Missile Range in New Mexico, reaching altitudes up to 185 km. As the supply of captured missiles dwindled, new rocket designs were introduced. These included the Aerobee, which was developed under the direction of space scientist James VAN ALLEN, whose main interest was to observe COSMIC RAYS before they struck the upper atmosphere.

The Russians built and flew their own V-2 rockets from Kapustin Yar between 1947 and 1987. The rocket flights for atmospheric sounding were supervised by Soviet academician Anatoli Blagonravov (1895–1975). The most-used vehicle during the 1950s was a modified V-2, known as the R-1.

A major impetus for rocket research was the International Geophysical Year (IGY) between July 1, 1957, and December 31, 1958. This global program allowed scientists to participate in coordinated observations of geophysical phenomena. The USA launched almost 300 suborbital rockets during this period, with another 175 sent up by the Soviet Union. Other countries that participated in IGY sounding rocket programs included Australia, Canada, France, Japan and the UK. These experiments were mainly designed to investigate the properties of the upper atmosphere, particularly the IONOSPHERE.

The Earth's ozone layer absorbs most of the incoming ultraviolet radiation, so the first detection of ultraviolet radiation from the Sun had to wait until a spectrometer was carried to high altitude by a captured V-2 rocket in 1946. This was followed during the late 1950s by the first ultraviolet observations of stars from rockets carrying instruments with pointing capability. A breakthrough was achieved by Don Morton and Lyman SPITZER of Princeton University, New Jersey, who recorded ultraviolet spectra for two stars in the constellation of Scorpius.

Also in 1946, T R Burnight from the US Naval Research Laboratory, reported fogging of a piece of photographic film carried aloft by a V-2. He deduced that solar X-rays caused the fogging. These were confirmed by Herbert Friedman, with a V-2 rocket from White Sands in 1949. On June 18, 1962, Riccardo GIACCONI used three X-ray detectors on an Aerobee suborbital rocket to look for X-rays from the Moon and found, in the image of the sky beyond the Moon, the first celestial X-ray source, Scorpius X-1.

The first space investigation of infrared sources was also carried out in the course of another investigation, this time Project HISTAR, prompted by the US Air Force's desire not to confuse heat from an incoming enemy missile with the infrared emission from harmless astronomical sources. A 16.5-cm telescope, cooled by liquid helium, on a series of nine rockets, observed over 2000 cosmic infrared sources.

Sounding rockets are still widely used in the USA, Europe and Japan. NASA currently uses 15 different types and launches about 30 each year, from Wallops Island, Virginia and White Sands and Poker Flat Research Range, Alaska, as well as sites in Canada, Norway and Sweden. The Japanese space agency ISAS has also been building and launching sounding rockets for more than 30 years. Its main rockets are the small MT-

◀ ***Robert Goddard*** *American rocket pioneer, seen with his rocket 'Nel' at his aunt's farm in Auburn, Massachusetts, USA. Nel reached an altitude of 12.5 m and flew for 2.5 s.*

135, which is used primarily for middle-atmosphere research such as ozone-layer depletion; the medium-range S-310; and the powerful S-520 which is capable of launching 100 kg of payload above 300 km. These are usually launched from Kagoshima Space Center on Kyushu Island.

Andoya Rocket Range in Svalbard, Norway, is the world's northernmost permanent launch facility for sounding rockets. It is favorably located for studying the ionosphere, the dayside aurora and processes in the magnetospheric boundary layer. The range has conducted more than 700 rocket launches.

Rømer, Ole Christensen (1644–1710)

Danish astronomer. Rømer studied at the University of Copenhagen under Thomas Bartholin (1616–80) and Erasmus Bartholin (1625–98). Erasmus gave him Tycho BRAHE's manuscripts to edit and his own daughter to wed. Rømer accompanied Bartholin and Jean Picard (1620–82) to Hven to measure the position of Tycho's observatory, the better to reduce Tycho's observations. He went on to the PARIS OBSERVATORY, where he made and used instruments for measuring star positions. While in Paris, he timed the eclipses of IO, the innermost of Jupiter's moons. He discovered that the PERIOD between eclipses was changing. This was because of a Doppler shift (*see* DOPPLER EFFECT), caused by the changing velocity of Earth relative to Jupiter, and led Rømer to a value of 11 min for light's travel time from the Sun to Earth (the 'light equation'). The modern value is 499 s.

Roque de Los Muchachos Observatory

The Observatorio del Roque de los Muchachos, at a height of 2.4 km above sea level, is situated on the island of La Palma in the Canary Islands, and belongs to the Canary Islands Institute of Astronomy. It contains one of the world's most extensive collections of telescopes operated by several countries, the largest being the William Herschel Telescope (4.2 m) of the ISAAC NEWTON GROUP OF TELESCOPES. Apart from several nocturnal telescopes, the observatory houses two solar telescopes, an automatic meridian circle and an array of cosmic-ray detectors. First light for the Gran Telescopio Canarias (GTC), an optical-infrared telescope of 10.4-m diameter, is planned for 2004

ROSAT (Röntgen Satellite)

The German-US-UK X-ray observatory ROSAT was launched into a 550-km circular orbit in June 1990. It carried two imaging telescopes operating in the soft X-ray (0.1–2.4 keV) and extreme-ultraviolet (0.06–0.2 keV) ranges. ROSAT performed the first all-sky surveys with imaging X-ray and EUV telescopes leading to the discovery of 125 000 X-ray and 479 EUV sources. In addition, the diffuse galactic X-ray emission was mapped with unprecedented angular resolution (1 arcmin). Many discoveries were made using both the all-sky survey and the pointed observations of ROSAT:

- X-ray shadows cast by cool interstellar clouds on the soft X-ray background in the Galaxy and nearby galaxies;
- X-rays and EUV light from the sunlit side of the Moon;
- X-rays from BROWN DWARFS;
- supersoft X-ray sources – nuclear burning on accreting WHITE DWARFS;
- X-ray emission from the photospheres of single neutron stars (*see* NEUTRON STAR AND PULSAR);
- X-ray emission from millisecond pulsars;
- X-ray pulsations of GEMINGA and its identification as a neutron star;
- X-ray and EUV emission from COMETS, such as Comet Hyakutake and Comet Levy;
- 80% of the cosmic X-ray background resolved into discrete sources, mostly AGN;
- superluminous flares at the centers of normal galaxies indicating episodic accretion onto supermassive BLACK HOLES.

ROSAT was switched off in February 1999 after eight years of successful operation.

Rosetta

Rosetta is an ESA mission to rendezvous with a COMET, and to deploy high-resolution remote-sensing instruments to follow the comet's activity as it approaches the Sun. Part of the payload will orbit the comet's nucleus at 5–25 radii; the orbiter will study the composition and development of the volatile gases and dust material released by the nucleus. A small vehicle will land on the comet; the lander will focus on the *in situ* study of the composition and structure of the material of the nucleus.

The Rosetta mission was thrown into uncertainty by problems with the Ariane 5 rocket, as a consequence of which it missed the launch window in 2003 that would have taken it to Comet Wirtanen. Its target was re-selected to be COMET CHURYUMOV–GERASIMENKO (67P). The satellite has a launch date in February 2004 and a rendezvous with the comet in February 2014. It will stay with the comet until its perihelion passage in 2015.

Rosette Nebula (NGC 2237-39, 2246)

A complex EMISSION NEBULA in the constellation Monoceros. It surrounds the star cluster NGC 2244, the stars of which

▶ ***Rosette Nebula*** *(NGC 2237-39) Named for its resemblance to a partly opened rose. Near the center is a cluster of blue stars which formed from the surrounding gas and now illuminate it.*

energize its gas. Different parts of the nebula received separate NGC designations; the whole resembles a rosette or wreath.

Rosse, Third Earl of (1800–67) and Rosse, Fourth Earl of (1840–1908)

An Irish astronomer and landowner, William Parsons, the Third Lord Rosse, became interested in astronomy and made at the family castle in Birr a 36-in reflector with the same design as William Herschel's (*see* HERSCHEL FAMILY). Rosse mapped the Moon and observed nebulae with the intent to resolve them into stars. He developed the technology at Birr Castle to make speculum mirrors, using the estate's blacksmiths, laborers and materials (such as peat for the fuel for the furnaces). In 1842, he successfully cast a 72-in mirror for the 'Leviathan of Parsonstown,' which stood as the largest telescope in the world at the time (1845–1917), slung between two massive walls aligned on the MERIDIAN. The telescope, restored to working order in 1998, was raised in elevation, and tracked for a small angle across the meridian, by a system of ropes and pulleys. The observer had access to the eyepiece from chairs mounted on the walls and was able to observe an object for at most an hour and a half.

The potato famine in Ireland delayed the astronomical work of the telescope until 1848, from when it was used to view planets, satellites and nebulae. In more than a dozen nebulae, Rosse, his son Laurence, astronomical assistants and visiting observers (like John DREYER and Thomas Robinson, (1792–1882)) were able to resolve SPIRAL ARMS, indicating that they were more than just collections of gas-galaxies. In fact, they went so far as conclude that 'no real nebulae seem to exist…all appeared to be clusters of stars.' Rosse also realized that some elliptical and lenticular nebulae were edge-on spirals. His drawings of the nebulae were strikingly accurate, although there was a curious lapse in the telescope's first record of M1, where a fanciful drawing with many radial 'legs' inspired the enduring name CRAB NEBULA.

Laurence Parsons, who became the Fourth Earl of Rosse, fitted both the 72-in and the 36-in telescopes with spectroscopes. His observations proved that some nebulae had bright-line spectra, including all the planetary nebulae, while others, including the ANDROMEDA GALAXY, were made up of stars. He also measured the temperature of the Moon by focusing its light onto a thermocouple, watching the temperature drop during an eclipse (*see* ECLIPSE AND OCCULTATION).

Rossi X-ray Timing Explorer (RXTE/Explorer 69)

NASA mission to study variations in X-ray emissions on timescales from microseconds to years. The mission was named in honor of X-ray astronomy pioneer Bruno Rossi, who died in 1993. The explorer launched in December 1995, and carries a Proportional Counter Array and a High-Energy X-ray Timing Experiment to study compact objects with X-rays in the energy range 2–200 KeV. The explorer also has an All Sky Monitor, which scans more than 70% of the sky each orbit. Its key features are flexible operations through rapid pointing, high data rates and nearly continuous receipt of data. The mission discovered millisecond X-ray pulsars and high-frequency X-ray pulses from neutron stars in binary systems. *See also* EXPLORER.

Rotation

The motion of a body about an internal axis; for example, the Earth rotates on its polar axis in a period of 23 hours

◀ **Whirlpool Nebula** (M51) *Drawn by William Parsons, Third Earl of Rosse, using the 36-in reflector at Birr Castle in 1848.*

56 min (one sidereal day). All celestial bodies, such as planets, stars and galaxies, exhibit some degree of rotation, as a result of their accumulation from individual pieces that fell together non-radially, as well as any subsequent collisions or tidal interactions that may have accelerated or slowed the original rotation. *Contrast* REVOLUTION.

Royal Astronomical Society (RAS)

The RAS represents professional astronomers and geophysicists in the UK. It was founded in 1820 as the Astronomical Society of London by John Herschel and 13 other well-known astronomers and scientists (*see* HERSCHEL FAMILY). The society received the grant of a royal charter from King William IV in 1831. It has 3000 members and its headquarters are in Piccadilly, London.

Royal Astronomical Society of Canada

The Royal Astronomical Society of Canada, originally known as the Toronto Astronomical Club, was founded in 1868 by eight charter members. On February 25, 1890, in Toronto, it was incorporated with a constitution and bylaws. In 1903 it became known as the Royal Astronomical Society of Canada. Today, the society has 23 affiliated centers from coast to coast, and a membership of close to 4000.

Royal Observatory, Edinburgh (ROE)

The ROE comprises the UK Astronomy Technology Centre (ATC) and the University of Edinburgh's Institute for Astronomy. The Institute for Astronomy is a research and teaching group within the university's Department of Physics and Astronomy. The ATC is the UK's national center for the design and production of state-of-the-art astronomical technology. It includes an applied optics group, which designs and builds instruments for both ground-based telescopes (UKIRT, GEMINI, ING) and space telescopes (FIRST, the JAMES WEBB SPACE TELESCOPE). Other groups in the ATC deal with mechanics, electronics and software. Major technology research is undertaken into CCD detectors/cameras, infrared systems, submillimeter instrumentation and cryogenics.

The ROE manages UK telescope sites and data-archive resources. It is the home of the Plate Library for the UK Schmidt telescope, which contains more than 17 000 plates, and the SuperCOSMOS advanced photographic-plate digitizing machine.

The ROE was founded in 1811 by a group of amateur and professional astronomers. Originally known as the Astronomical Institution of Edinburgh, it was renamed the Royal Observatory in 1822 after a visit to the city by King George IV. In 1834, the University of Edinburgh took over its administration on condition that the government would provide the salary for a professor who would hold the title of Astronomer Royal for Scotland. The link continued until 1995, when this honorary title was separated from the post of director of the ROE.

Royal Observatory, Greenwich

The Royal Observatory, founded at Greenwich, London, UK, in 1675, is the location of the Airy Transit Telescope that defines the prime MERIDIAN of the world and is the home of the Harrison Chronometers (*see* John HARRISON). The observatory was founded by King Charles II with the ultimate purpose of providing an accurate star catalog and model of the Moon's motion, to enable mariners to find their longitude (*see* NAVIGATION). During the twentieth century the observatory became increasingly involved in astrophysics and the research arm, known as the Royal Greenwich Observatory, moved from the Greenwich site in the 1950s and was eventually closed in 1998. The Greenwich site is now a museum, visited by 500 000 visitors per year, containing clocks, astronomical instruments, including a 70-cm refractor, and a planetarium. It is administratively part of the nearby National Maritime Museum. The buildings, including Flamsteed House designed by Christopher Wren (1632–1723), and collectively now known as the Royal Observatory, Greenwich, have been restored.

RR Lyrae star

VARIABLE STARS whose variations in light, radius and temperature are caused by radial pulsations. They are giant A-type stars of 0.5–0.8 solar masses, and average radii of five times that of the Sun. The amplitudes of their LIGHT CURVES range between 0.2 and 1.8 magnitudes. Their PERIODS lie between 0.25 and 1.2 days. An earlier name for them was 'short-period Cepheid stars' because the shapes of their light curves are so similar to CEPHEID VARIABLE STARS. But it was soon realized that the two groups were different.

The long-period classical Cepheids are concentrated to the Galactic disk (POPULATION I STARS) while the RR Lyrae stars lie in the halo around the center of the Galaxy (POPULATION II STARS), including 2000 in GLOBULAR CLUSTERS (hence their third name of 'cluster variables'). In fact, Harvard astronomer Solon Bailey (1854–1931) first discovered RR Lyrae stars in 1893 in globular clusters. Only later was the brightest of the class discovered by Williamina FLEMING at Harvard in July 1899 – the eponymous star, RR Lyrae itself. There are now more than 7000 RR Lyraes known; but, considering all those that are still unidentified, there are over 100 000 RR Lyraes in the Galaxy.

Bailey (1902) divided the stars into three groups, based on the shape of the light curves of the RR Lyrae stars in the globular cluster Omega Centauri. Bailey types a and b have a steep rise to maximum light, while type c is more symmetrical. RR Lyrae stars vibrate like a bell, and the difference in the light curves is because type c variables vibrate in the first overtone, whereas type a and b variables vibrate in the fundamental mode.

The RR Lyrae stars occupy a unique place in the HERTZSPRUNG–RUSSELL DIAGRAM. They lie in a narrow 'instability strip' on the horizontal branch, and have only a small spread in brightness. Harlow SHAPLEY used this fact in 1918 to derive the first reliable DISTANCES to globular clusters. RR Lyrae stars have been used as standard candles to measure the structure of the Galaxy, both the disk and the halo, and distances to other galaxies. Shapley's radical new model of the Galaxy was that its distant center was in the direction of Sagittarius, unlike the previously held model by Jacobus KAPTEYN, which had the Sun at the center of the Galaxy. *See also* VARIABLE STAR: PULSATING.

RS Canum Venaticorum

RS Canum Venaticorum (RS CVn) is a close BINARY STAR. In addition to its variations because of its eclipses, the light from both stars flickers periodically. Marcello Rodono discovered in 1965 that this is due to large 'star spots,' which rotate across the face of the stars. The stars' rotation is not quite synchronized with the binary revolution, and the periodic flickering forms waves in the LIGHT CURVES that march through the eclipse cycle. RS CVn is the prototype of the class of close binaries with this behavior (*see* VARIABLE STAR: CLOSE BINARY STAR). Their spots and flares, which are occasionally observed as well, are enormously more powerful and bigger than the Sun's.

Runaway star

A star that is moving with very high velocity, typically hundreds of kilometers per second, relative to others. Such exceptionally high velocity presumably indicates an 'explosive' departure of the star from the location of its formation. One explanation is that the star was a minor component of a binary system, ejected when the primary component exploded as a SUPERNOVA; alternatively the star was ejected from a MULTIPLE STAR like the TRAPEZIUM, as it became unstable (a non-hierarchical multiple star).

Russell, Henry Norris (1877–1957)

American astronomer who spent nearly all his life working at Princeton University, New Jersey. He studied eclipsing BINARY STARS to determine the masses of their component stars. At first collaborating with Arthur Hinks at Cambridge University, he started to measure stellar PARALLAXES and, plotting the absolute magnitudes of stars whose distance he had thus measured against their spectral types, he found the correlation now known as the MAIN SEQUENCE of the HERTZSPRUNG–RUSSELL DIAGRAM. He distinguished between giant stars and dwarfs, and proposed that stars cooled along this correlation. This theory of stellar evolution was superseded, but the HR diagram remains a useful tool for testing theories. Russell applied Meghnad Saha's (1893–1956) theory of ionization to stellar atmospheres and determined the abundance of the elements in stars, including, after the foundation work by Cecilia PAYNE-GAPOSCHKIN, the great abundance of hydrogen. In the theory of stellar structure, he showed that the size, temperature and so on of a star at each stage of its evolution can be found solely from its mass, chemical composition and age (the Vogt–Russell theorem).

Russian Aviation and Space Agency (RosAvia Kosmos)

The Russian Space Agency (RKA) was created on February 25, 1992, by a decree issued by the president of the Russian Federation. It was formed after the break-up of the Soviet Union and the dissolution of the Soviet space program, and uses the technology and launch sites that belonged to the program. This includes payment to Kazakhstan for use of the Baikonur Cosmodrome. The agency now oversees a growing number of bilateral and multilateral accords with other national space agencies, such as the Shuttle–MIR program and the ISS.

In 1994, the agency acquired 38 enterprises of rocket and space industry and shares in 22 more. In 1999 it gained control of 350 companies of the aviation industry and was renamed the Russian Aviation and Space Agency, or RosAviaKosmos.

Ryle, Martin (1918–84)

British radio astronomer who won a Nobel prize in 1974 for pioneering research in radio astrophysics, in particular for inventing the APERTURE SYNTHESIS technique of radio interferometry (*see* RADIO INTERFEROMETER). During World War II Ryle worked on the development of radar and joined the Cavendish Laboratory, Cambridge, becoming professor there. In 1972 he was appointed Astronomer Royal, the first such appointment as an honorary post unconnected with the Royal Observatory. He determined accurate positions and fluxes for radio sources. This led to the identification of many sources with galaxies that could be studied in visible light, thereby revealing their nature (QUASARS, for example). He compiled catalogs that provided a body of statistical information about the numbers of sources at fainter and fainter fluxes (and therefore greater and greater distances) which proved that the universe had evolved; that is, it started from a BIG BANG. This was the first of several near-fatal blows to the credibility of the STEADY-STATE THEORY, championed by his disputative Cambridge colleague Fred HOYLE, before the theory was finished off by the discovery of microwave background radiation.

Through inventing aperture synthesis, by which an INTERFEROMETER, rotating on the Earth, could fill in the aperture of a virtual large radio telescope, Ryle created the method by which radio astronomers can produce unsurpassed angular resolution, studying phenomena close to the central BLACK HOLES in radio galaxies, for example.

Sagan, Carl Edward (1934–96)

American planetary scientist who became professor at Cornell University, Ithaca, New York, USA and director of its Laboratory for Planetary Studies. He directed programs concerned with understanding the origin of life on Earth and the possibility of finding evidence of past or present life in the solar system and elsewhere. He was very influential in NASA, playing key roles in the APOLLO program to the Moon, and the landmark expeditions to the planets. He helped identify the runaway greenhouse effect on Venus which produces its high temperature, the dust storms which cause seasonal changes on Mars, and the complex organic molecules that cause the reddish haze of Titan and are the building blocks of proteins and nucleic acids. He inspired the continuing search for extraterrestrial intelligence (*see* SETI INSTITUTE), and his novel on the subject, *Contact*, was made into a successful Hollywood film. He was a very successful popularizer of astronomy, with a book and television series called *Cosmos*.

Sagitta, Sagittarius *See* CONSTELLATION

Sagittarius A (Sgr A and Sgr A*)

Sgr A is the brightest radio source in the constellation of Sagittarius. Located at the center of the Milky Way, at a distance of some 25 000 l.y., it consists of two principal components, Sgr A East and Sgr A West. Sgr A East, which is a source of SYNCHROTRON RADIATION, appears to be a bubble of ionized gas, possibly a supernova remnant. Embedded within Sgr A West, which is a cloud of hot gas, is an intense, very compact, variable radio source called Sagittarius A* (pronounced 'A-star') which has an angular diameter of less than 0.002 arcsec. At the distance of the galactic center, this angular size corresponds to a linear diameter of less than 15 AU. This is smaller than the diameter of the orbit of the planet Saturn.

Sgr A* is believed to mark the exact center of the Galaxy. The orbital velocities of gas clouds in the vicinity imply that some 6 million solar masses must lie within a light year of the galactic center, of which stars (as we know from infrared observations) contribute about 3 million solar masses. The remaining 3 million solar masses are believed to be contained in a central BLACK HOLE. The underlying energy source for Sgr A* is an ACCRETION disk of hot gas swirling round this black hole. Sgr A* therefore appears to be similar in nature to an active galactic nucleus (*see* ACTIVE GALAXY), though it is on a smaller and less energetic scale.

Evidence for this interpretation is becoming compelling. During observations over extended periods in 2002, NASA's CHANDRA X-RAY SATELLITE detected almost daily outbursts of X-rays from the gas around the central source. Large lobes of multimillion-degree gas extending for dozens of light years on either side of the black hole and a suspected jet, 1.5 l.y. in length, due to high-energy particles ejected from the vicinity of the black hole, were also detected.

Sagittarius dwarf galaxy

The closest member of the Milky Way's entourage of satellite galaxies (*see* LOCAL GROUP OF GALAXIES). Despite being located at a mere 16 000 parsecs behind the galactic bulge and 6000 parsecs below the GALACTIC DISK, it was only discovered by chance in 1994. Its presence had previously been overlooked because it is largely hidden by the most crowded regions of our own Galaxy, with which it is merging.

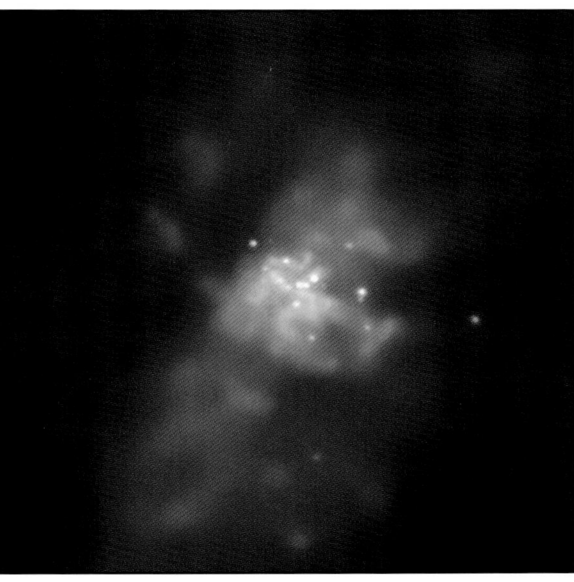

◀ ***Sagittarius A*** *The nebula filling the picture made with the Chandra X-ray satellite is Sagittarius A East, a supernova remnant. Sagittarius A* is the brightest point source near the picture's center. It is the site of our Galaxy's black hole, seen in the same line of sight as the supernova remnant, and marks the galactic center. Jets of hot X-ray emitting material shoot out of Sagittarius A* to upper right and lower left. Other less bright point X-ray sources are X-ray stars in the galactic center region.*

Sagittarius Star Cloud (M24) *See* MESSIER CATALOG

Sakigake

Japan's first deep-space mission. Launched in January 1985, it approached to within 7 million km of COMET HALLEY on March 11, 1986, to study radio and plasma waves. Sakigake means 'pioneer.'

Sakurai's Object

Sakurai's Object (also known as variable star V4334 Sgr) is the central star of a faint PLANETARY NEBULA which experienced a late HELIUM FLASH, presumably at the end of 1994 . It was discovered by amateur astronomer Yukio Sakurai in February 1996 as a rapidly brightening object, then at magnitude 11. It remained bright for two years, before dimming in mid-1998.

Sakurai's Object is the third 'final helium flash' star monitored so far. The first one was V605 Aql whose outburst and subsequent fading occurred during 1919–23. Its characteristics were very similar to those of Sakurai's Object. The second one, FG Sge, had a longer evolution timescale. It presumably underwent its final helium flash at the beginning of the nineteenth century, reached its maximum visual brightness in mid-1970, and faded in 1992.

Final helium flash stars are post-asymptotic giant branch (post-AGB) stars on their way to the WHITE DWARF stage which experience a late thermal instability in their helium burning shell. The blue compact post-AGB star evolves back to a red giant configuration on a timescale of a few years to a few decades. As a result, its effective temperature decreases from above 90 000 K to less than 7000 K, but its bolometric luminosity is expected to remain more or less constant (after a possible rapid initial increase at the onset of outburst). The rapid increase in the visual magnitude by several orders of magnitude essentially results from the surface color change (*see* HERTZSPRUNG–RUSSELL DIAGRAM).

Salpeter, Edwin Ernest (1924–)

Nuclear astrophysicist who emigrated from Austria to Australia in his teens and was educated in Sydney, and later at Birmingham University, England. He became professor at Cornell University, Ithaca, New York, USA, worked with Hans Bethe (1906–) on atomic physics, and in 1951 explained

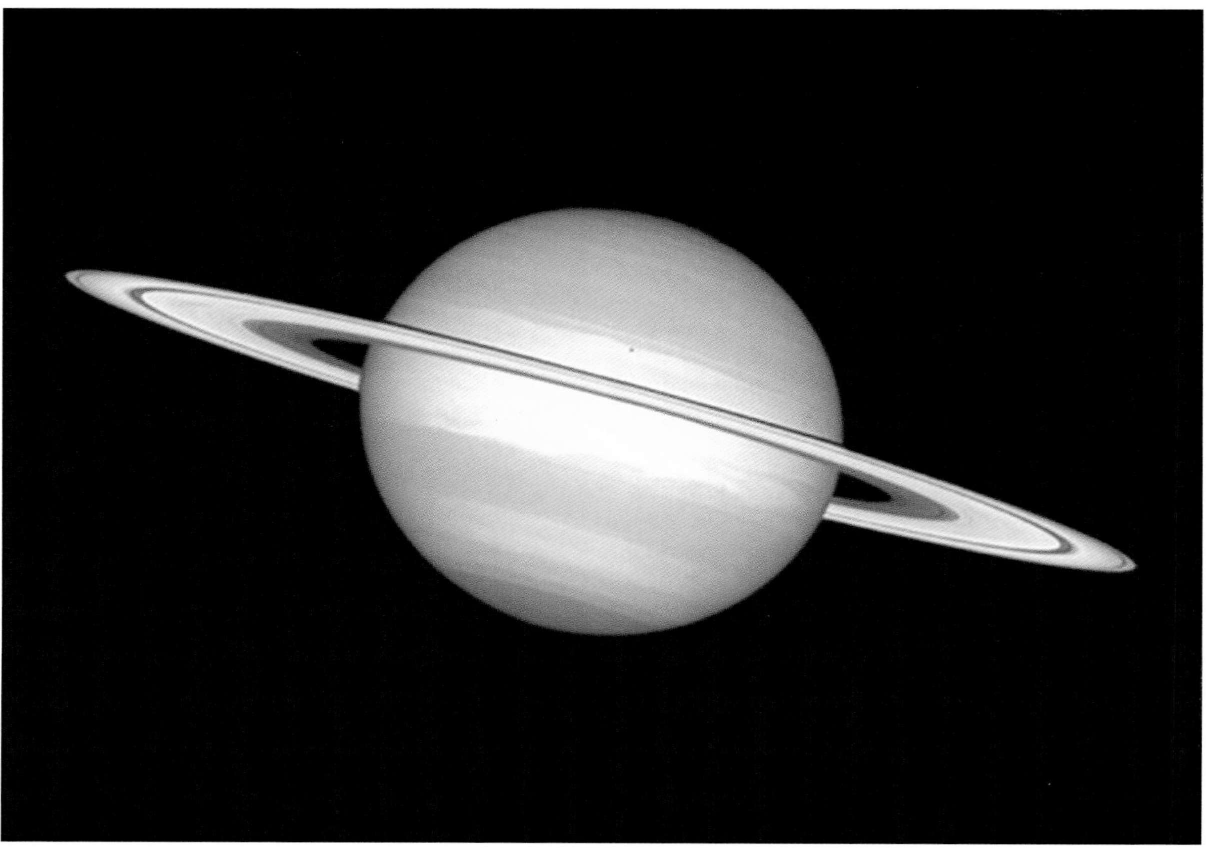

► *Saturn* is encircled by its spectacular rings and shows bands of clouds that are colored like Jupiter's but less bright.

the 'triple alpha' reaction, also called the Salpeter process, which in RED GIANT STARS makes carbon-12 from three helium nuclei. This led him to investigate stellar evolution, and, from this theory and observations of stars of different LUMINOSITIES, he worked back to the IMF (INITIAL MASS FUNCTION), the rate of formation of stars of different mass in the Galaxy, now known as the Salpeter function. This important concept identifies a target for theories of star formation to explain.

SAMPEX (Small Anomalous and Magnetospheric Particle Explorer)

First of NASA's Small EXPLORER missions, launched July 3, 1992. SAMPEX was designed to investigate cosmic rays, solar energetic particles and magnetospheric electrons.

Sandage, Allan Rex (1926–)

American astronomer who worked with Wilhelm BAADE and Edwin HUBBLE and became professor at Mount Wilson and Palomar observatories. His research has focused on the biggest problems in cosmology. He determined the ages of the oldest known stars found in GLOBULAR CLUSTERS. In 1960, with Jesse Greenstein (1909–2002), he identified the radio source 3C48 as a blue star with an unusual, unidentifiable SPECTRUM. This led to Maarten SCHMIDT's breakthrough recognition of the nature of 3C273, the first known QUASAR. Discovering many more quasars, Sandage worked on their properties. Realizing that the key question in cosmology was (and perhaps still is) the age and distance scale of the universe, he invested his career in the calibration of STANDARD CANDLES (Cepheids, supernovae, and so on). These are used to determine the distances of remote galaxies, most recently with the HST (for which he drew up an influential preparatory atlas of galaxies). This has led him to progressively more accurate values of the HUBBLE CONSTANT and the DECELERATION PARAMETER of the universe.

San Fernando Observatory

The observatory is located on the north side of the San Fernando Valley, approximately 48 km northwest of downtown Los Angeles, California, USA. It was built by the Aerospace Corporation in 1969 as ground-based support for NASA's space program, and was donated to the California State University, Northridge, in 1976. It has continued to play an important role in solar astronomy ever since.

Saros

The period of 223 synodic months (new Moon to new Moon) or 6585.32 days (about 18 years and 11 days) that elapses before a particular sequence of solar and lunar eclipses will repeat in the same order and with approximately the same duration (*see* SYNODIC PERIOD). After this interval the Earth, Sun and Moon and the nodes of the Moon's orbit return to almost the same relative positions. An eclipse repeated after one saros occurs 0.32 days later and hence is 115° west of its predecessor. The word 'saros' is of Chaldean origin and the cycle has been known since ancient times.

Satellite

A natural satellite is, in its most general sense, any celestial object in orbit around a similar larger object. Thus, for example, the MAGELLANIC CLOUDS are satellite galaxies of our own Milky Way. The term 'natural satellite' distinguishes these bodies from 'artificial satellites' that are spacecraft placed in orbit around the Earth or another celestial body.

In the solar system, all the major planets except Mercury and Venus possess one or more natural satellites (alternatively called 'moons'). A few asteroids also have satellites in orbit around them.

Saturn and its satellites

At the beginning of telescopic observations, in the seventeenth century, it was difficult to study the disk of Saturn, because of its distance and confusion with the rings. The first indications that there were atmospheric features were seen in 1676, when Gian Domenico Cassini discovered an equatorial belt (*see* CASSINI DYNASTY). The flattening of the planet was measured by William Herschel only in 1789, when the rings were edge-on (*see* HERSCHEL FAMILY). Herschel measured the period of rotation in 1790. Our knowledge of Saturn leapt forward with the space age, when the system was visited by Pioneer 11 in 1979 (*see* PIONEER MISSIONS) and the VOYAGER MISSIONS in 1980–1. The Cassini spacecraft will actually enter the saturnian system at the end of 2004 to start a close-up study (*see* CASSINI/HUYGENS MISSION).

◀ ***Prometheus*** *Saturn's satellite, which shepherds the F ring, is seen silhouetted against Saturn in this Voyager picture. The A ring and Encke gap are visible at upper left.*

Atmosphere

Apart from its PLANETARY RINGS, Saturn resembles a pale Jupiter. The composition of its atmosphere, mostly molecular hydrogen and helium, has been measured by ground-based and space observations, in particular Voyager and ISO. The less abundant chemicals control the structure of the stratosphere, and significantly affect its colors and SPECTRUM.

The structure of Saturn's atmosphere is very similar to that of Jupiter. Its density profile has been probed by observing stellar occultations (*see* ECLIPSE AND OCCULTATION). Ultraviolet photons penetrate the stratosphere, react with the methane and produce a haze of complex hydrocarbons. A recent and unexpected discovery by the ISO satellite was that dust particles from the rings and from the interplanetary medium fall into the atmosphere of Saturn at the rate of 12 000 tonnes per day. The oxygen and water in this material add further organic molecules to the haze.

Clouds appear on the GIANT PLANETS when gases condense into liquids or solids at the cold temperatures found in the upper atmosphere. Water clouds are expected in the troposphere, but are hidden by upper cloud layers. There are two overlying layers of clouds, one of them of ammonium hydrosulfide, NH_4SH, and another of ammonia ice, NH_3. Individual clouds and spots are probably the result of the lower, less uniform cloud layers. The colors shown in pictures of Saturn stem from some as yet unidentified minor atmospheric constituents.

As on Jupiter, the wind system on Saturn is banded parallel to the equator. There are strong westward winds at the equator, and eastward jet streams north and south. The equatorial wind speeds are as high as 500 m s^{-1}, four times larger than on Jupiter, and directed in the sense of rotation. An internal heat source is the most likely source of the winds.

There is no feature on Saturn like the Great Red Spot on Jupiter, but several atmospheric features were observed by Voyager, including a 'vortex street,' a succession of spiraling vortices of clouds with alternate directions of rotation. Generally, Saturn's atmosphere has fewer large storms than Jupiter's and they do not last as long. The Voyager 1 and 2 spacecraft recorded the North Polar Spot, named 'Big Bertha' by the Voyager imaging team. The first Great White Spot was observed in 1876 by Asaph Hall (1829–1907). Others were seen in 1903 (by Edward BARNARD), 1933 (by English amateur astronomer and comedian Will Hay, 1889–1949), 1960 (by South African amateur J H Botham) and 1990 (by the HST). The 30-year periodicity in the appearance of such storms is the same as the orbital period and is presumably related to the seasons of Saturn – if so, 'white spot' storms will next appear in 2020.

Seasonal effects arise from the 27° inclination of Saturn's AXIS. The north and south polar regions are alternately in shadow during part of the orbital period, and have 13-year arctic nights.

Interior

Saturn's EFFECTIVE TEMPERATURE (which measures the power it emits) is higher than its equilibrium temperature (its temperature if it were heated solely by the Sun). The Pioneer and Voyager missions measured the excess power, and showed that Saturn's internal supply of heat is comparable to the amount it absorbs from the Sun. The excess energy comes from a residual contraction of the planet and a persistently falling rain of liquid helium. Both transform gravitational energy to radiated heat.

The interior of Saturn has very large pressures, ranging between 0 and almost 50 Mbar (1 Mbar is 1 million Earth atmospheric pressures). At the center, the temperature rises

Saturn	
Revolution period	29.6 years
Orbital period	29.46 years
Mass	0.568×10^{27} kg
Mass relative to Earth's	95.7
Density	0.7 g cm^{-3}
Equatorial radius at 1 bar	60 268 km
Flattening	9.8%
Rotation period (magnetic field)	10 hours 39 min 22 s
Atmospheric helium mass fraction	0.06
Equilibrium temperature	82.4 K
Effective temperature	95.0 K
Emitted/received energy	1.78
Gravity at the equator (1 bar)	9.1 m s^{-2}
Obliquity of rotation axis	27°

Observing Saturn and the outer gas giants

Observing the distant planets is a more difficult challenge than observing Jupiter. Not only are the apparent (and actual) sizes of the planets much smaller, but, in addition, the disk itself shows much less detail. The disk of Saturn appears only 19 arcsec in diameter, but the ring system, when at its widest, covers some 46 arcsec. However, Uranus appears only 3.5 arcsec in diameter while Neptune reaches only 2.5 arcsec.

Which planets you can observe well may depend on your latitude. Uranus is moving northwards and, for northern hemisphere observers, is now reaching a good ALTITUDE. Neptune is still far to the south and is easily accessible only to southern and equatorial-based observers. Saturn has a SYNODIC PERIOD of 29 years and so is easily accessible to observers in both hemispheres for many years.

To an observer familiar with the detail shown in images from the VOYAGER and Cassini probes (*see* CASSINI/HUYGENS MISSION) and the HST, the initial view in the telescope can be a disappointment. The constant motion of the Earth's atmosphere motion blurs the image. An observer using a CCD camera can compensate by taking many images, discarding the worst and combining the best. Software processing can then correct for the blurring introduced by the atmosphere and, after very skilled processing, can produce images that may even rival those from the HST.

The visual observer has a more difficult task, but one that well repays careful study and practice. The objective for visual observing (distinguished from simple sightseeing) is to produce a drawing that accurately represents the view of the planet. The first thing to note is that you will get truly clear seeing conditions for only an occasional 0.1 to 0.5 s every few seconds. One must recognize this and be ready to translate the information gained at this time to your drawing. It is a skill that can be learned by any willing observer, and there is no shortcut to experience.

Saturn

The body of Saturn appears rather like a quieter version of Jupiter. Space probes have demonstrated that the atmosphere of the planet has the highest velocity winds in the solar system. There is, therefore, less turbulence and so features are less pronounced. Recent CCD images taken by amateurs have shown detail in the belts, but visual observers have only recently started to record this. Whether this is a feature of new improvements in equipment or a more active period for the planet is yet to be decided.

Establishing the number and visibility of the belts and the measurement of their latitudes is a project that has been underway for some years. As the belts are driven by both internal processes and by solar heating, they will be modified when passing under the shadow of the rings. As such, any drawing should accurately represent the rings visible to the observer and the positions of their north and south edges. Naturally, any spots should be recorded, other observers should be notified rapidly, and the times of central MERIDIAN passage should be accurately timed as described in the article on Jupiter (*see* Observing Jupiter and its satellites, p. 218). Diagrams can then be generated which show the drift of the spot in longitude and, therefore, the rotation period for the specific latitude of the spot.

Uranus and Neptune

Uranus orbits the Sun lying on its side; the pole is inclined at 97° to the pole of its orbital plane. Though hard to see, detail can be observed from Earth visually or using a CCD (preferably with a methane band filter or something similar). When the pole of Uranus is pointed toward the Earth, a dark cap is seen. However, an analysis by the author has shown that this is not centered on the pole, and is therefore most probably a contrast effect, not a real feature. Currently, the equatorial regions are moving into Earth's view. Here, belts have been recorded and the indications are that these are probably real.

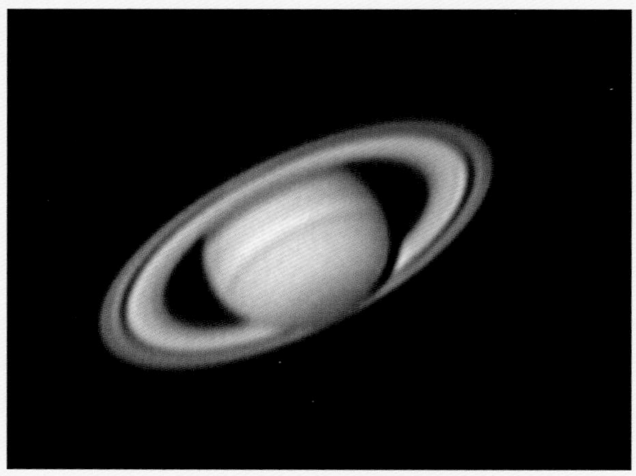

▶ ***Saturn*** *Imaged by Jamie Cooper using an Intes Micro Maksutov Newtonian with a 7-in mirror and a Philips ToUcam Pro webcam.*

Neptune is a bland, blue planet. At the time of the Voyager flyby, there was activity with a Great Dark Spot and several smaller ones together with bright spot and linear features. Such a prominent feature as the Great Dark Spot has not been shown by the HST more recently. There is no record, at least known to the author, of features on the planet being recorded by Earth-based amateur observers. Indeed, in amateur-sized telescopes, there are doubts about the reported observing of the planet's disk itself; what is seen may be the AIRY DISK caused by the unsteadiness of the atmosphere.

Uranus and Neptune are sufficiently small that they appear almost stellar to small telescopes. As such, the brightness can be estimated relative to field stars using the charts provided in the *BAA Handbook* or *Sky & Telescope*. Whole-disk estimates, as these are, can give information about conditions on the planets at a global scale. Analysis by the author has confirmed and refined the eight-year-or-more periodicity proposed for Uranus based on a few years' observation in the early 1950s. Further observations are being sought to see whether this period is constant or whether it is evolving in a cyclical or sudden manner. The observation record now stretches back for a period of over 50 years and so will soon extend to an orbital cycle for the planet (84 years). The record for the planet shows episodes when the planet fades for a period of a few days by perhaps half a magnitude; more information is required to confirm the reality of this occurrence. A CCD camera and filter can be employed to improve precision when observing.

Observing planetary rings

As far as observers on Earth are concerned, the main object to study is the Saturn ring system. Because of the polar INCLINATION of the planet and the fact that the rings are virtually in the equatorial plane, these can appear inclined to the Earth at 28° when at their widest open. They are so thin that, at the time when the Earth and Sun pass through the ring plane, they are at a zero inclination and hence difficult, if not impossible, to detect from Earth.

Attention should be concentrated on detail. The Cassini division (which is between the outer A ring and the main B ring) is usually easily seen in the Ansae (the 'handles' to each side) and, when the rings are widest open, can be followed virtually round the planet. The Encke division is in the A ring and can be seen in the Ansae under suitable conditions. The faint inner ring (C or Crepe ring) can be detected under good conditions with a suitable telescope. There is no apparent gap between the B and C rings.

Voyager has shown that the structure is much more complex than that seen from Earth, with each of the rings being composed of many thousands of thinner rings. At the inner edge of the B ring, lines of shading in the form of 'spokes' were detected which are thought to be shadows cast by electrically charged material from the main ring plane.

Now that we have knowledge of the conditions on the planet, we have a wealth of information that is very helpful in determining what we see, draw and image from Earth. The historical record has many examples of strange, unexplained observations that were discounted at the time, but are now considered probably to be real. The writer had such an experience when he saw the spokes in 1989, pre-Voyager. In a later discussion with amateur Paul Doherty, we realized that we were both observing at the same time and saw the same ring features. I included this in my report, though not on the drawing, while Paul did not. Therefore, a valuable confirmed sighting was lost. The moral is that when observing, always include everything. A confirmed observation, however unlikely it is thought to be, is very valuable and may turn out to be useful in ways that do not seem sensible at the time.

With what is known now about the system, care should be taken to record any detail that can be seen. Attention shouldn't be restricted to drawing just the Cassini and Encke divisions and the main rings. Not only can the shadings due to the spokes be seen in the Ansae, but subtle variations in the brightness of the rings will also give information about their small-scale structure. If the detail is too fine or elusive, then at least a description should be made in the observing notes describing what was seen and that it was considered too elusive or detailed to draw.

At the time of the passage through the ring plane by the Earth and Sun, the edge of the rings can be seen as a thin bright line, sometimes with a satellite or two split by it. The ring is very narrow, so conditions need to be right to see the bright line. During the ring passage that occurred in the early 1980s, there were occasions when the bright line was visible only on one side of the planet, so the variation from its normal appearance could be detected.

The other gas giants

Though Saturn's is the only direct ring-system observation that can be made, the opportunity for stellar occultations by the ring systems of all the gas giants should not be missed. The most useful and easily analyzed form of observation can be made by those who have access to a photoelectric system based on a photomultiplier and who can take continuous measurements for display and analysis, either digitally or using a chart recorder. As discussed in OCCULTATIONS, OBSERVING, it is absolutely crucial that accurate timings of each appearance and disappearance are recorded, and it is best if this is achieved automatically. If the occulted star is bright, a good trace can be achieved with a short integration time, showing great detail. This was successfully achieved in 1989 in the USA when the rings occulted 28 Sgr. Occultation methods can lead to important results such as the discovery of the rings of Uranus. The telltale signature is the symmetry of the pattern of appearances and disappearances on either side of the planet as the star passes behind both halves of the ring system. This expected pattern was broken with the occultation of Neptune, leading to the conclusion that a fragmented ring system existed. This, in turn, led to the prediction that there must be a then-unknown MOON interfering in the system. This prediction was verified by the visit of the Voyager 2 spacecraft. The advantage of occultation methods is that spatial resolution of just a few kilometers is possible, while ordinary observations are limited by DIFFRACTION. If results can be combined from a number of telescopes in different locations, or from more than one occultation, the orbits of the rings can be determined with extreme accuracy.

Andrew Hollis is a chartered structural engineer, and director of the BAA's Asteroids and Remote Planets Section.

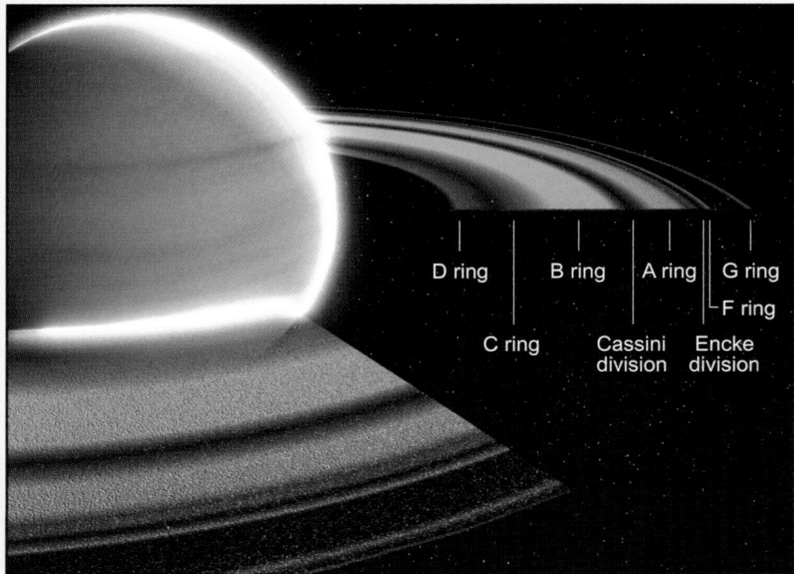

▲ ***Rings of Saturn***

▼ ***Rings of Jupiter***

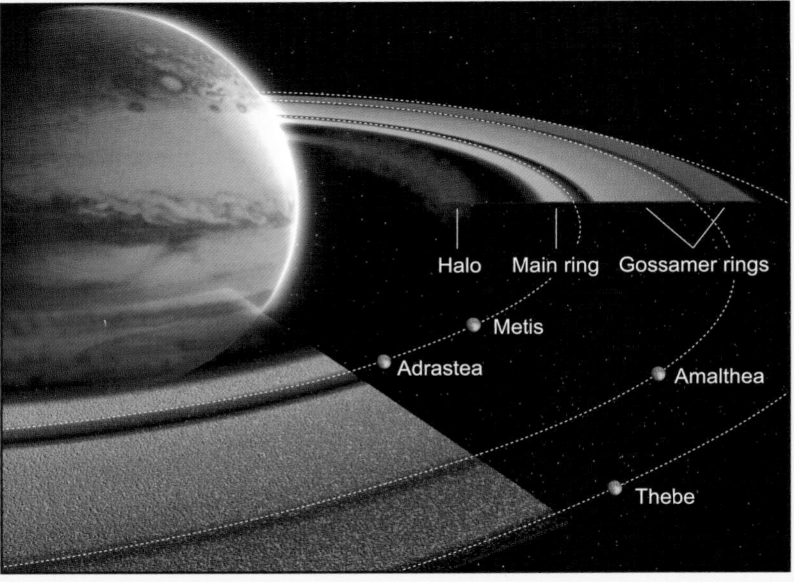

Saturn's satellites

Name and designation(s)			a	i	e	Period	Size	Density	Year	Note
Regular Satellites										
Start of innermost D ring			0.066							
XVIII	Pan	S/1981 S3	0.134	0	0	0.575	20		1981	
XV	Atlas	S/1980 S28	0.138	0	0	0.602	32		1980	A ring shepherd
XVI	Prometheus	S/1980 S27	0.139	0	0.002	0.613	100		1980	inner F ring shepherd
XVII	Pandora	S/1980 S26	0.142	0	0.004	0.629	84		1980	outer F ring shepherd
XI	Epimetheus	S/1980 S3	0.151	0.335	0.021	0.69	119	0.63	1966	co-orbital
X	Janus	S/1980 S1	0.152	0.165	0.007	0.70	178	0.65	1966	co-orbital
I	Mimas		0.186	1.566	0.021	0.94	397	1.12	1789	
II	Enceladus		0.238	0.010	0.000	1.37	499	1.00	1789	
XIII	Telesto	S/1980 S13	0.295	1.158	0.001	1.89	24		1980	co-orbital; trailing Lagrangian
III	Tethys		0.295	0.168	0.000	1.89	1060	0.98	1684	co-orbital
XIV	Calypso	S/1980 S25	0.295	1.473	0.001	1.89	19		1980	co-orbital; leading Lagrangian
IV	Dione		0.377	0.002	0.000	2.74	1118	1.49	1684	co-orbital
XII	Helene	S/1980 S6	0.377	0.212	0.000	2.74	32		1980	co-orbital
End of outermost E ring			0.480							
V	Rhea		0.527	0.327	0.001	4.518	1528	1.24	1672	
VI	Titan		1.222	1.634	0.029	15.95	5150	1.88	1655	
VII	Hyperion		1.464	0.568	0.018	21.28	266		1848	
VIII	Iapetus		3.561	7.570	0.028	79.33	1436	1.03	1671	
Irregular Groups										
		S/2000 S5	11.365	46.16	0.334	449.2	16		2000	
		S/2000 S6	11.440	46.74	0.322	451.5	12		2000	
IX	Phoebe		12.944	174.80	0.164	548.2	120		1898	retrograde orbit
		S/2000 S2	15.199	45.13	0.364	686.9	22		2000	
		S/2000 S8	15.647	152.70	0.270	728.9	8		2000	retrograde orbit
		S/2000 S11	16.404	33.98	0.478	783.5	32		2000	
		S/2000 S10	17.616	34.45	0.474	871.9	10		2000	
		S/2000 S3	18.160	45.56	0.295	893.1	40		2000	
		S/2000 S4	18.247	33.51	0.536	925.6	15		2000	
		S/2000 S9	18.709	167.50	0.208	951.4	7		2000	retrograde orbit
		S/2003 S1	18.719	134.60	0.352	956.2	7		2003	retrograde orbit
		S/2000 S12	19.463	175.80	0.114	1016.3	7		2000	retrograde orbit
		S/2000 S7	20.382	175.80	0.470	1086.9	7		2000	retrograde orbit
		S/2000 S1	23.096	173.10	0.333	1312.4	18		2000	retrograde orbit

a = semimajor axis in millions of kilometers; i = inclination (°); e = eccentricity; Period = time of one revolution around Saturn in days; Size = diameter in kilometers; Density = density in gm cm^{-3}; Year = the year of discovery or first record

to about 10 000 K. At high pressures (3–5 Mbar), hydrogen changes to a metal (this is a hypothetical state of hydrogen, never seen in the laboratory because the pressures are not achievable). By contrast, helium condenses to a liquid state, dripping down into the planet, releasing energy and depleting the upper atmosphere. In spite of the rather low abundance of helium in the atmosphere, the overall composition of Saturn must be similar to that of the solar nebula.

In a journey inwards from Saturn's external layers, we encounter first the boundary between metallic hydrogen and a molecular hydrogen envelope at about 2.8 Mbar, at about half the radius of Saturn. Is the change in density at the boundary large enough to generate giant waves, like at the surface of a monstrous ocean? Nobody knows. Deeper in the planet we find a boundary between an ice core, enriched in helium, and the hydrogen-rich envelope at about three-quarters of the way into the planet's interior. Finally, Saturn has a rock core at its center, sized at 12% of Saturn's radius, mass up to 10 Earth masses.

Magnetosphere

Although less intense than Jupiter's, Saturn's magnetic field produces a magnetosphere, including AURORAE. Observed from the HST and by infrared, the auroral emissions are located along ovals, as on Earth and Jupiter.

Satellites

In 1655 Christiaan HUYGENS discovered TITAN, the second biggest satellite after Jupiter's GANYMEDE. Cassini discovered the next four largest satellites of Saturn: RHEA, IAPETUS, DIONE and TETHYS. The eighteenth, Pan, was discovered nearly 10 years after the Voyager flybys, embedded in the A ring. The most recently discovered are from sensitive, ground-based, widefield camera images.

Altogether, Saturn has 31 known satellites, forming a diverse set of bodies. They range from the planet-like Titan to small, barren objects of irregular shape.

• Titan is the only satellite in the solar system with a significant atmosphere. Catalan astronomer Jose Comas Solà (1868–1937) claimed in 1908 to have observed LIMB DARKENING. James JEANS showed in 1925 that Titan might have an atmosphere, in spite of its small size and weak gravity, if its temperature was low enough. Titan's atmosphere was confirmed by Gerard KUIPER in 1944 when he detected methane in its spectrum. Its active organic chemistry is similar, in some respects, to that of the early Earth. This similarity is the motivation for the Huygens probe, which will fall from the Cassini spacecraft into Titan's atmosphere early in 2005.

Titan's atmosphere is thick enough to make it a mysterious world. Even from nearby, as Voyager 1 found out in 1980, the dense cloud that surrounds the small world obscures any glimpse of its surface, as it also does on Venus. The atmosphere is mainly formed of nitrogen (a property shared only by the Earth) and methane (molecular oxygen is lacking). Some 4 billion years ago, our Earth, like Titan, may have been shrouded in a thick atmosphere of nitrogen, deprived of oxygen and rich in methane.

• Half a dozen of the saturnian satellites exceed 200 km in diameter. The surfaces of the larger satellites are all covered with some type of frozen volatile, primarily water ice, but also carbon dioxide, methane products and ammonia. Their densities (from 1.0 to 1.9) are such that ice must be a major constituent of their interiors, also. The most probable structure for each of them (and Titan) is a rocky center surrounded by a thick shell of ice.

The giant planets formed their satellites in the same way that the Sun formed the solar system. After formation they were heated by the heavy bombardment of their surfaces by remaining debris. The bright regions of Iapetus (especially the north pole) are heavily cratered, like Mercury and the Moon. HYPERION is similar to Iapetus in terms of surface cratering. MIMAS and Tethys have impact craters (named Herschel and Odysseus), caused by bodies that must have been rather large. Hyperion is believed to have been disrupted in the past (as a result of an important impact?).

Nowadays the bombardment continues, but at a significantly lower rate. Impacts pulverize the surface, forming a fine dust, the REGOLITH, and exposing fresh material. Like the Moon, the satellites are synchronously rotating, locked in orbit with a leading/trailing side where most METEOROIDS fall. As a consequence, the larger satellites of Saturn have a brighter 'leading' than 'trailing' hemisphere.

The satellites interact through tides and by orbital RESONANCE with each other and Saturn. This interaction is another source of frictional energy that heats the satellites. The Voyager and Galileo spacecraft revealed that several satellites have melted surfaces, reformed from liquid ice or ice silicate mud, like terrestrial lava flows. The ridged and grooved terrain found on ENCELADUS and Tethys are the result of tectonic activity. Explosive volcanic eruptions occur on Enceladus.

• Saturn also has a number of unique small satellites under 200 km in diameter. There are three classes.

The shepherding satellites, Atlas, Pandora and Prometheus, play a key role in defining the edges of Saturn's A and F rings. Of the shepherds, Atlas lies several hundred kilometers from the outer edge of the A ring. The other two orbit on either side of the narrow F ring, constraining its width and causing its kinky appearance.

The CO-ORBITAL SATELLITES Janus and Epimetheus move in almost identical orbits at about 2.5 Saturn radii.

The Lagrangian satellites Calypso, Helene and Telesto orbit in the LAGRANGIAN POINTS of the larger satellites Dione and Tethys. No other satellites in the solar system are Lagrangians, but the TROJAN ASTEROIDS orbit in two of the Lagrangian points of Jupiter.

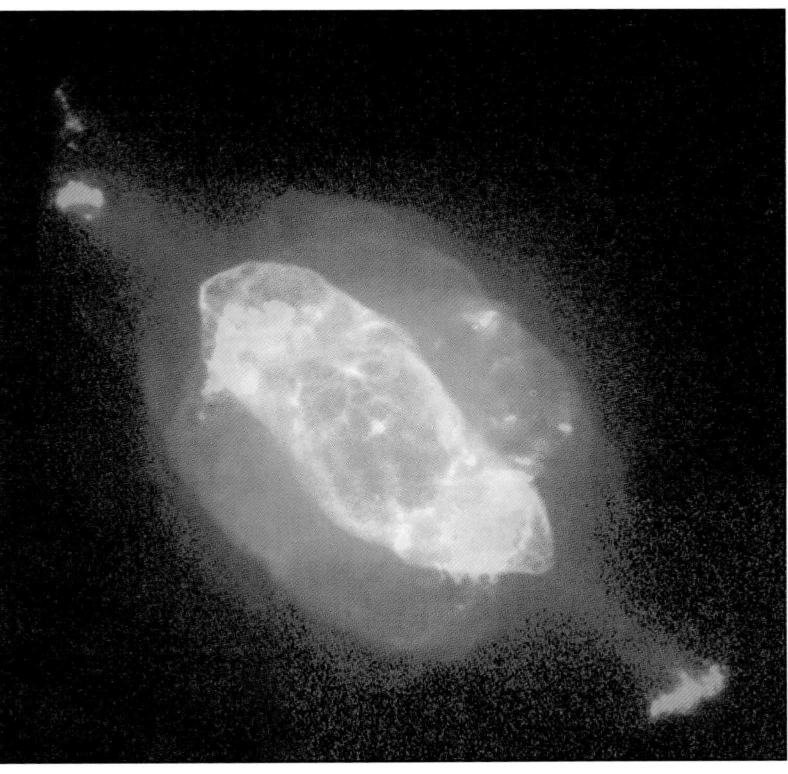

▲ ***Saturn Nebula*** *(NGC 7009) The central star expelled the green gas, which now confines stellar winds, creating the red jets.*

• Small satellites. The innermost satellite, Pan, was discovered in 1990 in Voyager 2 images dating back to 1981. It lies hidden within the A ring and helps to keep the Encke Division clear of particles (*see* JOHANN ENCKE).

The recently discovered outermost dozen satellites are moving in irregular, tilted orbits. They fall into groups, and may be the remnants of captured larger satellites that have been fragmented by collisions.

Further reading: Gehrels T (ed) 1984 *Saturn* University of Arizona Press; Beatty J K and Chaikin A (eds) 1990 *The New Solar System* Sky Publishing.

Saturn Nebula (NGC 7009)

A planetary nebula in the constellation Aquarius measuring 25 arcsec in size and shining with a greenish hue. It is of magnitude 8 and has a high surface brightness for a planetary nebula. Two small lobes on either side give it the appearance of the planet Saturn.

Schiaparelli, Giovanni Virginio (1835–1910)

Italian astronomer who studied in Germany and in Russia, and became director of Brera Observatory, Milan, Italy. He wrote authoritative accounts of the early astronomy of many cultures. He observed COMETS, inferring from their tails that there was a repulsive force from the Sun (SOLAR WIND and radiation pressure). He explained the regular METEOR SHOWERS as the result of the dissolution of comets and proved it for the PERSEIDS. From his observations, he mapped Mars, naming martian 'seas' and 'continents' and connecting items with linear features which he called channels or, in Italian, 'canali.'

The implication that these features, mistranslated 'canals,' were artificial water-distribution systems stimulated Percival LOWELL to a sensational search for life on Mars.

Schmidt, Bernhard Voldemar (1879–1935)

Estonian optical designer. In spite of an accident at the age of 15 in which he lost his right forearm, he successfully made fine optics, initially for amateur astronomers. Later he invented the SCHMIDT TELESCOPE (or Schmidt camera).

Schmidt, Maarten (1929–)

Dutch astronomer who worked on galactic dynamics and the mass distribution of the Galaxy with Jan OORT at Leiden University, went to the California Institute of Technology, and became director of the Hale Observatory. In 1963 he observed 3C273, a radio source that had been identified with a blue star by Cyril Hazard as a result of a lunar occultation. With a 5-m telescope picture, he discovered that there was a JET protruding from the 'star.' He obtained a SPECTRUM, which, like 3C48 as studied by Allan SANDAGE, showed emission lines at unusual positions. Schmidt's breakthrough was to recognize that the object exhibited ordinary hydrogen lines, but at a REDSHIFT far greater than any previously seen. This was the first completed discovery of a QUASAR. Schmidt went on to investigate the distribution of quasars, discovering that they were much more abundant when the universe was young.

Schmidt–Cassegrain Telescope

A REFLECTING TELESCOPE that combines some of the features of the Schmidt and the Cassegrain systems. The principal optical components of the Schmidt–Cassegrain are a concave primary mirror, a thin corrector lens located close to the focus of the primary, and a convex secondary (often attached directly to the inside of the corrector lens), which reflects light through a central hole in the primary to a focus. The curved secondary changes the angle at which light rays from the primary mirror are converging and, as with the conventional Cassegrain, increases the effective focal length of the instrument while keeping its overall length down. The corrector lens gives a wider field of sharp focus than that of the conventional Cassegrain. The Schmidt–Cassegrain is an example of a catadioptric system, one that uses both lenses and mirrors to collect light.

Schmidt Telescope

A REFLECTING TELESCOPE, invented by Bernhard SCHMIDT, that is used to photograph large areas of the sky. The Schmidt telescope (or Schmidt camera) uses a concave spherical mirror as its light collector and corrects for SPHERICAL ABERRATION by means of a specially shaped thin lens, or corrector plate, which is located at the front end of the telescope tube. It is smaller than the primary mirror and sets the light grasp of the telescope. An instrument such as this, which uses both lenses and mirrors to collect light, is called a catadioptric instrument. By contrast with a conventional telescope, where the field of view may be, at best, a few tens of arc minutes, the field of view may be 6–10° across. This makes it eminently suitable for surveying large areas of sky for stars, galaxies and nebulae of different kinds. A disadvantage of the system is that the surface of sharp focus (the focal plane) is curved rather than flat. This is overcome by bending the photographic plate, film, or detector to match the shape of the surface of sharp focus. The maximum size of a Schmidt telescope is limited by the practical problems associated with supporting the thin correcting lens (which can be supported only around its edges). *See also* REFLECTING TELESCOPE.

Schröter effect

A phenomenon in which the observed and predicted phases of Venus do not coincide. It applies in particular to the predicted and observed times of DICHOTOMY (half-phase), at which the terminator (the boundary between the illuminated and unilluminated portions of Venus's disk) should be a straight line. At eastern elongation, when the planet is visible in the evening sky, dichotomy usually comes a day or two earlier than it theoretically should, while at western elongation, when Venus is visible before sunrise, dichotomy tends to occur a day or two later. The effect is named for Johann Schröter (1745–1816), who first described it in 1793. There is no satisfactory explanation for the Schröter effect. It may be purely subjective, or it may have a physical cause, perhaps the scattering of twilight in Venus's thinner upper atmosphere.

Schwarzschild, Karl (1873–1916)

German mathematical physicist, who at first worked on CELESTIAL MECHANICS, including the tidal deformation of moons and the origin of the solar system. He became professor at Göttingen and Potsdam Universities. He wrote on relativity and quantum theory. He early on proposed that space was not geometrically flat but 'non-Euclidean,' giving a lower limit for the radius of curvature of space as 2500 l.y. He gave the first exact solution to Albert EINSTEIN's equations about the geometry of space near a point mass leading to the first relativistic study of BLACK HOLES. The size of the 'event horizon' within which light cannot escape from a black hole is called the Schwarzchild radius, and his mathematical formulation of the description of the geometry of space is called the Schwarzchild metric. He served in the army in World War I, contracting an illness from which he died on returning home.

Schwarzschild, Martin (1912–97)

German-born American astrophysicist, son of Karl SCHWARTZSCHILD, who became professor at Princeton University, New Jersey, USA. Working with mathematician John von Neumann (1903–57), Schwarzschild used the powers of the newly developed electronic digital computers to develop the theory of stellar structure and evolution. He improved the understanding of pulsating stars and how stars evolve off the main sequence in the HERTZSPRUNG–RUSSELL DIAGRAM to become RED GIANT STARS. He summarized his work in an influential textbook, *Structure and Evolution of the Stars*. In the 1950s and 1960s he pioneered the use of space telescopes, with the use of the balloon-borne Stratoscopes I and II for imaging the Sun, planets and stellar systems.

Scintillation

Rapid fluctuations in the RADIATION from stars due to its passage through the atmosphere. The more familiar term for the effect on visible light is 'twinkling.' The effect is most marked for a point source of light, such as a star. An extended source (covering a finite area) does not exhibit the effect so appreciably (thus a planet tends to twinkle less obviously than a star). Scintillation is caused by turbulence in layers of the atmosphere (which may have differing temperatures, refractive indices, velocities, and so on). Seen from space, stars do not appear to twinkle.

At radio frequencies it is possible to detect scintillation due to the passage of radiation through the SOLAR WIND. Analogous to the twinkling of stars, the distortion of radio emissions in the Earth's IONOSPHERE also occurs, at wavelengths longer than about 20 cm (1.5 GHz). *See also* SEEING.

Scorpius *See* CONSTELLATION

Scorpius X-1

The brightest cosmic X-RAY source in the constellation of Scorpius and the first cosmic X-ray source to be discovered. Detected by Riccardo GIACCONI and collaborators for the first time in 1962 by instrumentation carried to an altitude of 225 km by an Aerobee rocket, Scorpius X-1 is, apart from occasional transient sources, the brightest cosmic source of X-radiation in the sky. It is a low-mass X-ray binary with an orbital period of 0.787 days and an X-ray luminosity of about 2×10^{37} W (about 5000 times the optical luminosity of the Sun) and is located at a distance of about 2300 l.y. The binary is believed to consist of a low-mass star and a neutron star, the X-radiation being emitted from the surface of the neutron star and from its accretion disk (*see* NEUTRON STAR AND PULSAR).

Sculptor, Scutum *See* CONSTELLATION

S Doradus

A LUMINOUS BLUE VARIABLE star, the brightest star of the LARGE MAGELLANIC CLOUD, and the main star of the cluster NGC 1910. S Dor varies between magnitude 9 and 10 in an irregular fashion. It is an unstable, very massive star passing through a short-lived settling-down phase in its youth, and it gives its name to a rare class of variable star (*see* VARIABLE STAR: ERUPTIVE).

sdO star

Subdwarfs of spectral type O (sdO stars) are hot evolved stars with a lower LUMINOSITY than MAIN SEQUENCE O stars. The sdO stars cover a wide range in the HERTZSPRUNG-RUSSELL DIAGRAM, stretching from the extension of the extreme blue horizontal branch and the helium main sequence up to the region where the central stars of PLANETARY NEBULAE are found. Some of the most luminous ones are indeed associated with a planetary nebula. sdOs show a wide range of chemical composition with the hydrogen-to-helium ratio varying from about the solar value to an extreme helium-rich composition with no hydrogen detectable. sdO stars are predominantly found at high galactic latitudes and kinematic studies shows that they have an affiliation with the old disk population. Several sdO stars have been found by ultraviolet surveys of GLOBULAR CLUSTERS.

Search for Extraterrestrial Intelligence (SETI)

See ARECIBO OBSERVATORY; Frank DRAKE; EXOBIOLOGY AND SETI; SETI INSTITUTE

Seasons

Natural environmental and climatic changes occurring as a planet completes one orbit of the Sun. Any planet experiences seasons as long as its rotation axis is not at 90° to the plane of its orbit or its distance from the Sun changes markedly. Seasonal effects due to the inclination of rotation axis are greatest for Uranus and due to eccentricity of orbit for Pluto. But these changes are so slow, because the orbital periods are long, that they have not been observed well. Seasonal effects are very marked on Earth and Mars, especially at the poles. Also on Earth there are conventionally (especially in temperate latitudes) four seasons: winter, spring, summer and fall. Again conventionally, they begin in the northern hemisphere on the winter SOLSTICE (December 22 or 23), the vernal EQUINOX (March 20 or 21), the summer solstice (June 21 or 22), and the autumnal equinox (September 22 or 23). In the southern hemisphere, summer and winter are reversed, as are spring and autumn.

At the poles there is continuous darkness all winter and daylight or twilight all summer. In low latitudes, where the range of the annual solar radiation and temperature cycle is very small, seasonal changes are based largely on rainfall, and conventionally there are two seasons, wet and dry.

Inclined about 6° 33' to the orbital plane, the Earth's axis maintains a nearly constant orientation in space as the Earth orbits the Sun (see OBLIQUITY OF THE ECLIPTIC). During a six-month half of each orbit, each hemisphere in turn is inclined toward the Sun, resulting in more hours of daylight and more general heating as the Sun's rays travel more directly through the atmosphere than for a point in the opposite hemisphere.

The amount of solar radiation arriving at the surface of a planet is also affected by the eccentricity of its orbit. The Earth currently reaches PERIHELION in early January when it is both closest to the Sun and also (by Johannes KEPLER's second law) traveling slightly faster than at APHELION. So the input of solar radiation during the southern summer is slightly greater but for a shorter time than the input to the northern hemisphere during its summer, and vice versa. These effects are even more marked on Mars for which the orbital eccentricity is greater than that of Earth (0.0934 compared with 0.0167).

Because of the PRECESSION of the equinoxes, the times at which planets reach perihelion or aphelion migrate slowly through the seasons; for example, in about 13 000 years, the Earth will reach perihelion during the northern-hemisphere summer. *See also* Milutin MILANKOVITCH.

Secchi, Angelo (1818–78)

Italian Jesuit. After a period in exile at Georgetown Observatory, Washington, DC, he returned to Italy in 1849 as director of the Roman College (Vatican) Observatory, constructing a new observatory dome on top of the main pillars of the incomplete church of Saint Ignazio. Secchi was one of the first astronomers to concentrate on physical properties rather than positions, and is thus a founder of ASTROPHYSICS. He classified the SPECTRA of more than 4000 stars into five classes, work superseded by the more detailed Harvard system of Edward PICKERING and his co-workers. He discovered that Jupiter was gaseous. He drew the dark lines that join areas of Mars and used the word 'canali' to describe them, an idea taken up by Giovanni SCHIAPARELLI.

Secondary crater

When an impact crater is formed on the surface of the Moon or other planetary body, the ejected material (EJECTA) may result in the formation of smaller secondary impact craters. Secondary craters from recent impact events, such as those associated with the lunar crater TYCHO, tend to be easily identifiable as secondary features. They may be clustered around the primary crater, and are usually irregular, having been formed by low-angle, low-velocity impacts. Others, produced by ejecta on higher trajectories traveling faster, are found quite far away and tend to be more circular. They can occur as crater chains, as on the Moon for example (though some crater chains are volcanic in origin), and in the so-called intercrater plains of Mercury. Other bodies on which secondary craters have been identified include GANYMEDE and EUROPA.

Sedna

Discovered in 2004 and designated 2004 DW, Sedna is the most distant large member of the solar system. It has a high-inclination eccentric orbit, with semi-major axis of 39.4 AU (5.9 billion km). Named after the Inuit goddess of the sea, Sedna is a Kuiper Belt object of about 2000 km in diameter. *See also* OORT CLOUD AND KUIPER BELT.

Seeing

The twinkling of a STAR image. The sharpness and steadiness of an image are determined by the turbulence in the Earth's atmosphere. Temperature alters the refractive index of air, so bubbles of air at different temperatures, blown around by the wind across the APERTURE of a telescope, cause the instantaneous image to dance around. The result is that the image does not focus to a point (or to a near-point-like diffraction disk). The sizes of the bubbles of air relative to the aperture of the telescope are crucial. Small-aperture telescopes (or the naked eye) view the star through not many bubbles. As the wind blows, the star image dances in the focal plane and the image of the star may be displaced a lot. By contrast, a large telescope views a star image through many bubbles and the image is stationary but diffuse. Seeing is quoted in arc seconds across a star image although there is no clear convention as to how this is measured. Eye-estimates are not very reliable because the eye is a quick-response detector and picks out the brighter core of the image, minimizing the apparent size of the seeing disk. Seeing is crucial to the performance of a telescope. For this reason astronomers choose their observing locations with care and seek to eliminate all local sources of heat in the dome ('dome seeing'). *See also* OBSERVATORY.

Selene (Selenological and Engineering Explorer)

Japanese lunar exploration project conducted jointly by the National Space Development Agency of Japan (NASDA) and ISAS, with a planned launch in 2005.

Separation

The angular distance between two objects measured in arc units on the CELESTIAL SPHERE. This is most commonly used in measurements of double and multiple stars, and together with the position angle defines the observed relative positions of the system. It is also used to determine the observed distance of a satellite from its primary planet.

Serpens *See* CONSTELLATION

SETI Institute

The SETI (Search for Extraterrestrial Intelligence) Institute, founded in 1984, is a private non-profit center for research and education with a mission to explore, understand and explain the origin, nature and prevalence of life in the universe. It has been chosen as a lead team for NASA's Astrobiology Institute (NAI), the international research consortium coordinated through NAI's offices at NASA's Ames Research Center.

The SETI Institute runs many projects, the most visible of which is Project Phoenix, the world's most sensitive and comprehensive search for extraterrestrial intelligence. It is an effort to detect extraterrestrial civilizations by listening for radio signals that are deliberately beamed our way, or are inadvertently transmitted from another planet. Phoenix is the successor to the ambitious NASA SETI program that was cancelled by a budget-conscious Congress in 1993. Phoenix began observations in February, 1995, using the Parkes 64-m radio telescope in New South Wales, Australia, and has since used many other telescopes worldwide. Phoenix scrutinizes the vicinities of nearby, Sun-like stars which are most likely to host long-lived planets capable of supporting life. Stars that are known to have planets are also included. There are about 1000 stars targeted for observation by Project Phoenix. All are within 200 l.y.

Another project, SETI@home, uses data collected with the radio telescope of the ARECIBO OBSERVATORY in Puerto Rico as part of Project SERENDIP (Search for Extraterrestrial Radio Emissions from Nearby Developed Intelligent Populations), and involves thousands of Internet-connected PCs. The data can be downloaded and analyzed when the home computer is otherwise idle. Although the research is not part of the program of the SETI Institute, the institute is a major supporter of the project. The results of the analysis are sent back to the SERENDIP team, combined with the crunched data from the other SETI@home participants, and used to help in the search for extraterrestrial signals. Interesting signals must be followed up at a later date. *See also* EXOBIOLOGY AND SETI.

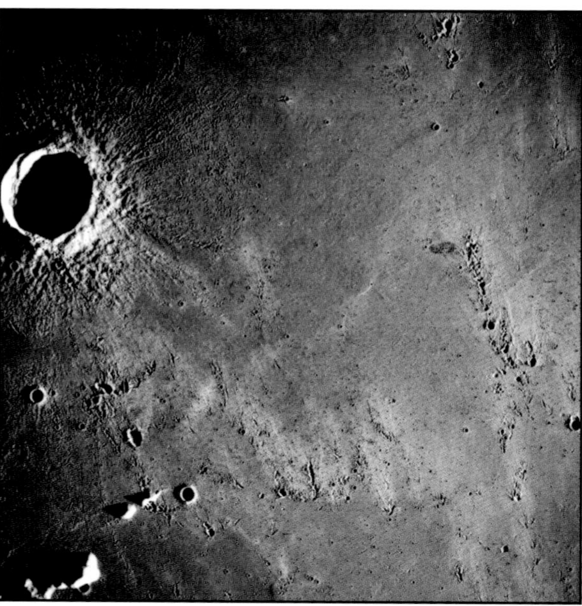

► ***Secondary craters*** *in the Moon's Mare Imbrium region are strewn east (to the right) of the large crater Euler (28 km in diameter, at upper left). The chains of secondary craters are oriented from the crater Copernicus about 380 km to the southeast – one conspicuous chain nearly 100 km long is prominent at the right hand edge. Other splashes of smaller secondary craters overlie the chains from Copernicus and clearly postdate the Copernicus impact. These secondary craters come from the Aristarchus impact, 580 km west.*

Sextans *See* CONSTELLATION

Seyfert galaxy *See* ACTIVE GALAXY

Shanghai Astronomical Observatory

The observatory is located in Shanghai, China. It was established in 1962 combining Xujiahui and Sheshan observatories, which were founded in 1872 and 1900 respectively. Its main fields are ASTROPHYSICS and space geodynamics. The major facilities are a 1.56-m telescope, 25-m radio telescope, 60-cm satellite laser ranging (SLR) system and 40-cm astrograph.

Shapley, Harlow (1885–1972)

American astronomer who studied at Princeton University, New Jersey, with Henry RUSSELL, where he analyzed Russell's observations of the LIGHT CURVES of 90 eclipsing BINARY STARS. In this work he established that CEPHEID VARIABLE STARS, whose spectral lines mimic the spectroscopic binary stars, are in reality pulsating stars. He moved to the MOUNT WILSON OBSERVATORY in California, where he began a study of GLOBULAR CLUSTERS. He discovered Cepheid variables in some and used Henrietta LEAVITT's period-luminosity relation to determine their distances. Shapley identified the asymmetric distribution of globular clusters in the sky, calculating that the center of the Milky Way Galaxy is some 50 000 l.y. in the direction of the constellation of Sagittarius. Although Shapley had underestimated interstellar absorption and thus overestimated the size of the Galaxy, his work established the shape, orientation and approximate size of our stellar system. Shapley successfully demonstrated that our Galaxy was a star system comparable to other galaxies.

Shapley Concentration

Although the universe is homogeneous on very large scales, it is clearly clumpy even on scales of a few million parsecs. Thousands of galaxies, together with gas and DARK MATTER, make up gravitationally bound clusters, and several associations of clusters, known as superclusters, have been identified. The Shapley Concentration or 'Shapley supercluster' identified by Harlow SHAPLEY in the constellation of Centaurus in the 1920s is probably the largest concentration of galaxies in our nearby universe.

Shepherd moon

A minor satellite whose gravitational influence on the particles of a planetary ring constrains the extent of the ring. The ring systems of the GIANT PLANETS are all shaped by gravitational perturbations of the particles that make up the rings. A lone shepherd moon can give a broad ring a sharply delineated outer edge, while pairs of shepherd moons can 'squeeze' the particles of narrow rings into well-defined orbits.

Shinsei

First Japanese scientific satellite, launched September 1971. Shinsei returned data on the IONOSPHERE, SOLAR WIND and COSMIC RAYS, and its name means 'new star.'

Shklovskii, Iosif Samuilovich (1916–85)

Soviet astrophysicist. He investigated the solar CORONA, showing that its temperature is around 1 million K, and made theoretical and radio studies of SUPERNOVAE and supernova remnants. In 1953 he proposed that the continuum radiation from the CRAB NEBULA is SYNCHROTRON RADIATION, because of its broad spectral distribution, and correctly predicted that the radio emission would be polarized by the magnetic fields causing the synchrotron radiation. Shklovskii's book *Intelligent Life in the Universe* was translated and expanded by CARL SAGAN.

Shoemaker, Eugene ['Gene'] Merle (1928–97)

American astrogeologist. Shoemaker became chief scientist at the US Geological Survey Field Center in Flagstaff, Arizona, and professor of geology at the California Institute of Technology. He worked with his wife Carolyn on cratering, both on the Moon and the Earth, by volcanoes and by meteor and cometary impact. He discovered, with Edward Chao, coesite, a type of silica produced in a violent impact and a signature that a terrestrial crater is meteoritic. With Eleanor Helin, he and Carolyn searched for potential Earth-impacting asteroids. As a result of this search they also discovered 32 comets that now bear the Shoemaker name, including COMET SHOEMAKER–LEVY 9, which impacted on Jupiter in 1996. Shoemaker died in a car crash in Australia while visiting meteor craters.

Shooting star *See* METEOR

Short-period comet

A COMET whose period is less than 200 years, also known as a periodic comet. Comets with longer periods are termed LONG-PERIOD COMETS. As of June 2003 there are 156 known short-period comets.

Sidereal period

A period of time of rotation or revolution measured relative to the 'fixed' stars. In normal usage the term refers to the time taken by a planet to complete one orbit of the Sun, returning to its original position relative to the position of the stars, or the time taken for a satellite to complete one orbit of its parent planet, again measured relative to the background stars. In the case of the Earth, this period of time is referred to as the sidereal year and is equal to 365.2564 mean solar days. *See also* SYNODIC PERIOD.

Sidereal time

A time system based on the rotation of the Earth measured relative to the background stars, which for this purpose are regarded as fixed in position. Relative to the stars, the Earth rotates on its axis in a period of 23 hours 56 min 4.1 s of mean solar time (ordinary civil time), and this period is called the sidereal day, which is, in turn, divided into sidereal hours. The value of the local sidereal time at any instant, also called the local hour angle, is the angle measured clockwise from the MERIDIAN of the vernal EQUINOX. Thus, when the vernal equinox is on the meridian, its hour angle is zero, and the sidereal time is zero hours.

Siderite

An obsolete term for an iron METEORITE.

Siderolite

An obsolete term for a STONY-IRON METEORITE.

Siderostat/heliostat/coelostat

A plane mirror that is driven about an axis so as to reflect light from a particular celestial object along a fixed direction. The words describe their specialist functions: siderostat for a star, heliostat for the Sun, and coelostat for the sky. In practice two plane mirrors are usually involved. One is driven around an axis parallel to that of the Earth at half the Earth's rotation rate in order to counteract the apparent rotation of the CELESTIAL SPHERE. The other mirror is oriented so as to reflect the beam of light from the first mirror into a

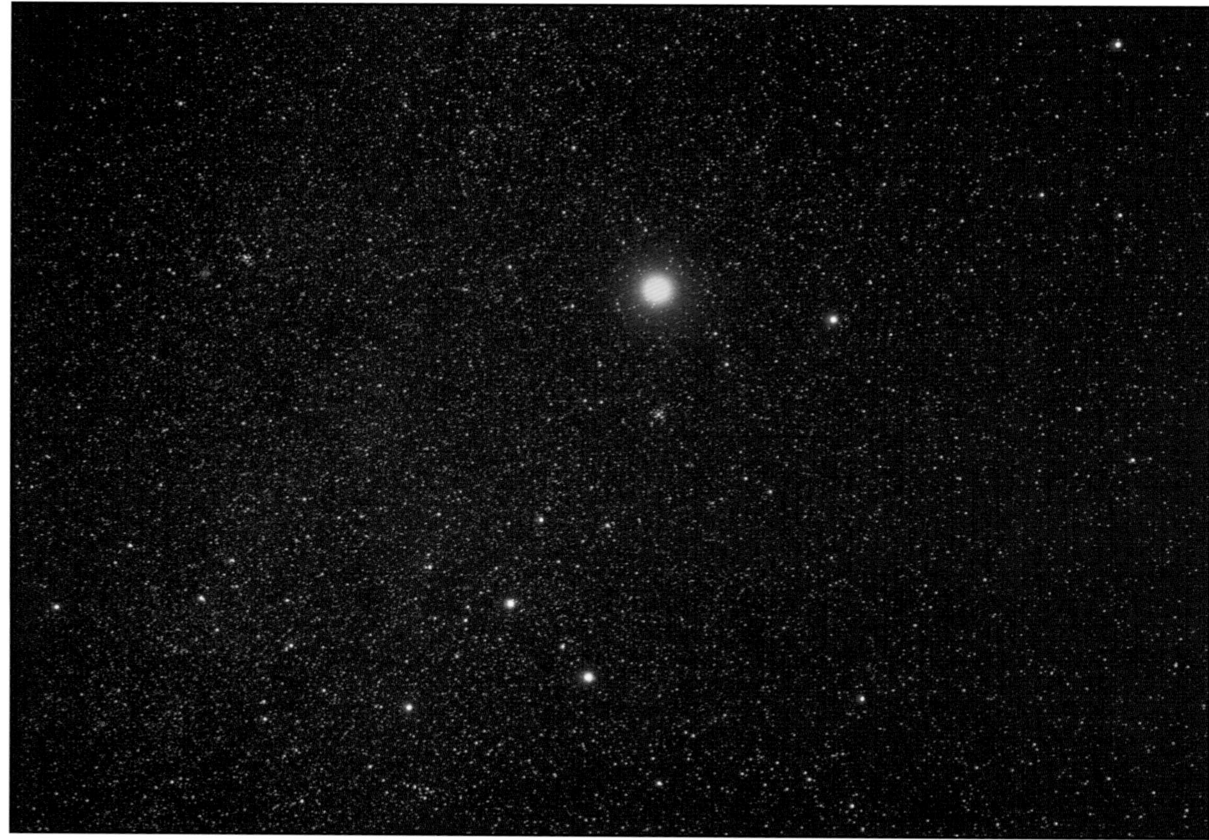

► ***Sirius***
Outstandingly bright star in the constellation of Canis Major. To its right is Mirzam, and directly below Sirius is the binocular cluster M41.

fixed telescope. Coelostats suffer from the drawback that although an object at the center of the field of view remains stationary, the rest of the field of view rotates around that central point. A siderostat may be used to feed the light of a single star into a spectrograph (*see* SPECTROSCOPE AND SPECTROGRAPH). Heliostats are often used in science centers to display a live image of the Sun to the public. They are also used in a number of specialist solar telescopes. For example, the McMath solar telescope at Kitt Peak, Arizona, uses a 2-m-diameter heliostat, mounted at the top of a 30-m tower, to reflect light down a cooled inclined tunnel, about 150 m long. A reflecting telescope system then produces a fixed image of the Sun, some 76 cm in diameter, in a laboratory.

Sikhote Alin Meteorite

An iron METEORITE that fell in the Sikhote Alin mountain range in southeast Siberia on February 12, 1947. It was preceded by a brilliant FIREBALL. The incoming body fragmented at an estimated altitude of about 5 km, and the fragments produced nearly 400 small impact craters, the largest 27 m across. The total of 23 tonnes that was collected included a fragment with a mass of 1.7 tonnes. According to some estimates there may be as much as 75 tonnes of material uncollected.

Sirius

Sirius (Alpha CMi) is the brightest star in the night sky and lies in the constellation Canis Major. It is a visual binary system consisting of a MAIN SEQUENCE star (Sirius A) of spectral type A1 V and a WHITE DWARF companion (Sirius B). Another name for this star is the Dog Star. The companion to Sirius predicted in 1844 by Friedrich BESSEL from the wobble in its position was confirmed by Alvan Clark (1832–97) in 1862 (*see* CLARK FAMILY). He was able to resolve Sirius A and B while testing a new telescope lens he had made for Dearborn Observatory, Illinois, USA. The distance to Sirius AB is 8.6 l.y. The apparent magnitudes are –1.5 (A), 8.3 (B); their separation about 4 arcsec and orbital period 50 years. They have masses 2.3 and 1.0 times solar mass, and radii 1.7 and 0.02 times solar radius. Sirius B was probably at one time the more massive star, at about 4 times solar mass. It has now passed through its red-giant phase and lost its outer envelope, collapsing to a white dwarf as its core is exhausted of nuclear fuel. The dawn (heliacal) rising of Sirius was important to ancient Egyptians because it marked the onset of the annual flooding of the Nile.

Sitterly, Charlotte Emma Moore (1898–1990)

American astrophysicist and atomic physicist. Sitterly worked with Henry RUSSELL at Princeton University, New Jersey, on BINARY STARS and their masses. She worked at MOUNT WILSON OBSERVATORY with Charles St John and Harold BABCOCK, analyzing the atomic lines in the sunspot spectrum. At the National Bureau of Standards and the Naval Research Laboratory she analyzed laboratory data on the solar spectrum and the atomic data by which spectral lines are characterized. Toward the end of her long life she was still working, extending the tables into the ultraviolet for use with data from space instruments.

Sixty-one Cygni (61 Cyg)

61 Cyg was the first star to have its PARALLAX measured, which was reported by Friedrich Bessel (1784–1846) in 1838. Its large PROPER MOTION of 5 arcsec per year identified it as likely to be one of the nearest stars and therefore a good candidate

for a measurable parallax. 61 Cyg is a binary star consisting of two late-type dwarf stars (HIP 104214 and 104217) of spectral type K5V and K7V, which orbit each other with a period of about 700 years.

Skylab

The NASA Skylab space station was launched from Cape Kennedy, Florida, on May 14, 1973, with the Saturn V rocket. It went into a 93-min low-Earth orbit with an altitude of 433 km and an inclination of 50°. Observations were conducted over three manned missions (SL-2, 3 and 4), each consisting of a crew of three astronauts who lived and worked on the station. The SL-2 mission lasted 28 days in May–June 1973; SL-3 lasted 59 days in July–September 1973; and SL-4 lasted 84 days, from November 1973 to February 1974. These missions permitted about nine months of solar observations. After the SL-4 mission, Skylab was inoperative and re-entered the atmosphere on July 11, 1979.

The Apollo telescope mount on the space station carried a suite of instruments of unprecedented spatial resolution which made significant advances in solar physics. These showed, for example, that coronal structures consisted of magnetic loops. Great strides were also made in the understanding of coronal holes and coronal mass ejections and some phenomena, such as bright points, were completely new. After the Skylab mission, it was generally concluded that the astronauts had played key roles in its success. Not only had they carried out solar observations, but they had also repaired instruments. Astronauts removed cloth threads from the front occulting disk of the S052 coronagraph and moved a jammed filter wheel into a filter position in the S054 X-ray telescope. They also pinned open experiment doors that had failed shut and replaced failed or troublesome film-transport assemblies. However, they also provided some vision to a future time when humans would no longer be needed on board spacecraft to control experiments.

Slipher, Vesto Melvin (1875–1969)

American astronomer who became director at the LOWELL OBSERVATORY in Arizona. He used SPECTROSCOPY to determine the rotation periods of several planets, and identified the constituents of their atmospheres. He recorded the first radial velocities of galaxies. The exposure times for these early photographs were as long as a week. The data were extended by Edwin HUBBLE to discover the expansion of the universe. With the data, Slipher measured the rotations of the SPIRAL GALAXIES. He also supervised the search for the ninth planet, Pluto.

Sloan Digital Sky Survey (SDSS)

The Sloan Digital Sky Survey employs a 2.5-m telescope located at the APACHE POINT OBSERVATORY in New Mexico, USA, to create a five-color digital map of 10 000 square degrees of sky, to be complete by about 2005. The images will be used to autonomously select 1 million objects for SPECTROSCOPY, via a multi-object spectrograph (*see* SPECTROSCOPE AND SPECTROGRAPH) mounted on the same telescope, yielding REDSHIFTS, and thus DISTANCES, for each galaxy. A huge number of previously unrecognized quasi-stellar objects and peculiar stars will also be revealed. Although originally motivated by problems concerning the large-scale structure of the universe, the survey will constitute a permanent public data bank applicable to a great variety of problems in astronomy, ranging from nearby ASTEROIDS to the most distant QUASARS. The survey is conducted by an international consortium of 10 universities and research laboratories.

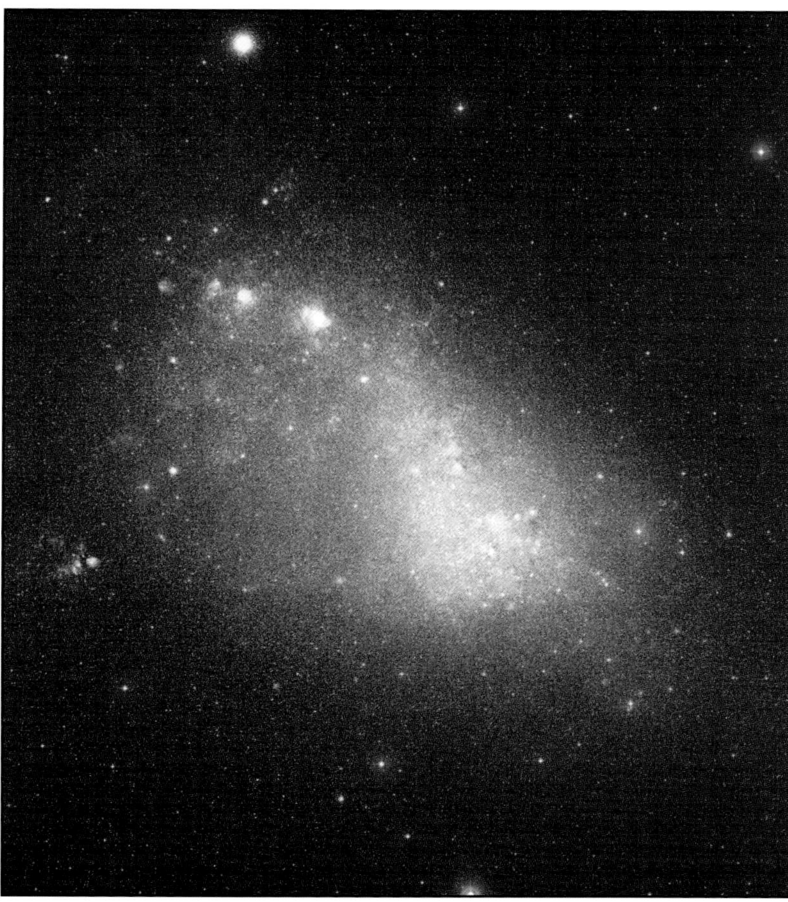

▲ ***The Small Magellanic Cloud*** *(SMC) shows pink with patchy nebulae excited by hot stars. The globular cluster NGC 362 (top) is in our Galaxy at one-sixth the distance of the SMC.*

Small Astronomy Satellite (SAS)

Series of three pioneering NASA scientific satellites, launched 1970–5. SAS-1, also known as UHURU ('freedom' in Swahili), was the first dedicated X-ray astronomy satellite. SAS-2 was the first satellite to detect gamma-rays and it detected the enigmatic GEMINGA source. SAS-3 also carried X-ray experiments.

Small Circle

See GREAT CIRCLE

Small Magellanic Cloud (SMC)

The Small Magellanic Cloud is the smaller of the two MAGELLANIC CLOUDS, nearby companions of the Milky Way Galaxy that can be seen with the naked eye in the southern-hemisphere sky and are named for the Portuguese navigator Ferdinand Magellan. Located in the constellation of Tucana, at a distance of about 190 000 l.y., the SMC is an IRREGULAR GALAXY (Hubble Class Irr I). Although the SMC contains substantially less gas and dust than the LARGE MAGELLANIC CLOUD, its content is broadly similar.

SMART-1

The ESA's Science Program encompasses small, low-cost 'flexi' missions, in addition to the ambitious Cornerstone and medium-sized missions. Named SMART (Small Missions for Advanced Research in Technology) missions, their purpose is to test new technologies that will eventually be used on bigger projects. SMART-1 is the first in this program.

Its primary objective is to flight-test solar-electric primary propulsion as the key technology for future Cornerstones in a mission representative of a deep-space one. ESA's BEPICOLOMBO mission to explore the planet Mercury could be the first to benefit from SMART-1's demonstration of electric propulsion. The SMART-1 mission, launched in September 2003, will take about 15 months to reach the Moon, which it will then orbit for a nominal period of six months. It is the first time that Europe has sent a spacecraft to the Moon. SMART-2 is due for launch in 2006 and will test key technologies for the Cornerstone missions LISA (launch 2011) and Darwin (launch 2015).

Smyth, Charles Piazzi [also Piazzi-Smyth] (1819–1900)

Son of an amateur astronomer and named for the Italian astronomer Giuseppe PIAZZI. Smyth became director of the Astronomical Institution on Calton Hill in Edinburgh, Scotland, which evolved into the ROYAL OBSERVATORY, EDINBURGH (ROE) in a new building on Blackford Hill. He estimated the amount of heat radiation received from the Moon and studied the ZODIACAL LIGHT. However, he did not distinguish himself in the role of director, being responsible for the neglect of the instruments and for commissioning a telescope that never worked. Moreover, he was obsessed with pyramidology, publishing successful popular books on this topic. When the Royal Society refused to publish his cranky papers he became the only person ever to resign his fellowship, apparently misjudging whether a threat to do so would be accepted.

He resigned the directorship of ROE and devoted the rest of his life to cloud photography in England's Lake District. Smyth made one outstanding contribution to world astronomy. In 1856, while on a camping honeymoon with his wife Jessica on Tenerife in the Canary Islands, he discovered that he could see fainter stars from the top of Mount Teide than he could with the same telescope in Edinburgh, and that he could resolve closer DOUBLE STARS. Mount Hamilton Observatory was the first mountain observatory set up on the basis of Smyth's site-testing expedition, and many others have followed (*see* OBSERVATORY).

SNC meteorite

One of a small number of ACHONDRITE METEORITES believed to have originated on Mars. SNC (pronounced 'snick') stands for the three main classes: shergottites, nakhlites and chassignites. As of July 2003, 28 SNCs had been discovered. The largest at 40 kg is the first known nakhlite, which broke up as it fell at Nakhla in Egypt on June 28, 1911. The Zagami meteorite (18 kg) which fell in Nigeria in 1962 is the largest Mars meteorite ever found.

Société française d'astronomie et d'astrophysique

The French Society of Astronomy and Astrophysics (SF2A), previously SFSA for Société française des spécialistes d'astronomie); has a membership of about 500 French astronomers and astrophysicists.

Software in astronomy

Software refers to the programmed sequences of instructions that are executed by a digital computer. Computers now influence almost all aspects of astronomy:

- planning observing time;
- scheduling observations;
- acquiring and reducing data;
- final dissemination of data across networks, particularly the Internet;
- presenting scientific results in papers or at talks.

In some cases, general-purpose software can be used, such as Microsoft's PowerPoint for presentations, or LaTeX to prepare typeset finished papers. But the actual obtaining of data and its subsequent reduction and analysis are carried out with software written especially for astronomy.

Data reduction

Most data reduction has an initial stage, which is essentially the removal of the instrumental effects from the data. The ideal result is a calibrated image. It might be a two-dimensional picture, each element of which is at a known sky coordinate and whose value is an absolute intensity. The ideal is that all instrumental effects have been removed and the result is as the target object produced it. Subsequent processing will depend on the scientific problem the data were taken to solve. Stellar photometry, for example, can be broken down into three stages:

- Pre-processing: the raw digital images from the telescope are calibrated to an intensity scale.
- Processing: each calibrated digital image is searched for the images of individual stars, whose positions and magnitudes are determined.
- Post-processing: the results of stage two from different images are combined and calibrated to fundamental magnitude scales by comparison with similar observations of other stars with known magnitudes. Measurements made with different filters can then be compared to form a COLOR INDEX and/or observations made at different times can be assembled into LIGHT CURVES for variable stars.

Optical data in the form of images taken by a camera with a CCD detector often contain spots due to the passage of COSMIC RAYS through the detectors. They must be identified by comparing one or more exposures of the same target field – the stars and galaxies will reproduce from one exposure to the next, but not the cosmic rays. Images usually require division by a 'flat field' (the image of a uniformly lit surface) to calibrate pixel-by-pixel sensitivity variations. If the detector read-out system has a bias or the sky has a significant background intensity, the background must be determined and subtracted. This can be done for a typical astronomical exposure, which is mostly empty sky with some stars and galaxies, by determining the most common value of the sky brightness and constructing a 'median sky frame.'

Radio data, particularly when produced by RADIO INTERFEROMETERS, need more complex processing even to get to a two-dimensional image of a target. Even a perfect point-source RADIO GALAXY gives a messy picture with complexities from the interferometry. An iterative algorithm invented by Swedish radio astronomer Jan Hogböm, called CLEAN, is usually used to tidy this up.

Data files are passed, on tape, disk, CD or through a network, from the data-acquisition system at the telescope to the data-reduction system to be used by the astronomer, typically off-mountain. The ability to transfer information between computer systems is fundamental, and the differences between machines can complicate this – even transmitting simple text messages by e-mail can produce surprising results. However, for astronomy a much more complicated problem is the higher-level file formats.

A 'file format' refers to the conventions adopted when writing data to a file. At the very least, to read the file a program needs to determine the dimensions of the image. For practical purposes a great deal of additional information is needed – the name of the object under study, instrumental parameters, time of observation, coordinates of the target object, and so on. Without this information carried as an intrinsic part of the file, the records get in a mess or are less easily manipulated by the computer. In the early days, *ad hoc* formats proliferated, and the only software that could read data written by a given instrument was the package that wrote it. However, FITS (Flexible Image Transport System) has been used since 1981 to define conventions based on the use of a header, which contains pairs of keywords and values describing the actual data. FITS became the *de facto* standard for exchanging astronomical data. A standard library of routines for accessing FITS data (CFITSIO) is available from GODDARD SPACE FLIGHT CENTER. Most large astronomical software packages can import and export data in FITS format.

A number of data-reduction frameworks exist. AIPS (Astronomical Image Processing System) and its development AIPS++ are the mainstay of radio-astronomical data reduction. For optical data reduction, the dominant framework is IRAF, the Image Reduction and Analysis Facility. IRAF provides a programmable command language and a large and increasing number of applications, intended originally for all instruments operated by the NOAO and the HST. Many other observatories use IRAF for processing data from their own telescopes, providing customized 'layered packages' tailored to reduce data from their instruments. Other less widely accepted reduction packages include the UK's ADAM (Astronomical Data Acquisition Monitor) and the ESO's MIDAS (Munich Image Data Analysis System).

Personal computers (both Intel-based systems running a version of Microsoft Windows and to a lesser extent Apple Macintoshes) are used extensively by astronomers, since they have to write papers and give presentations. There are a number of packages aimed at amateurs that, for example, process CCD images on PCs, and there are also some excellent 'planetarium' programs available. However, there is little professional astronomical data-reduction software available for Windows or the Mac. The emergence of Linux (a Unix-like operating system that is freely available and runs on PCs of both kinds) has made it possible for astronomers to run astronomical Unix-based packages on their home computers. This has a number of implications for the way astronomers will work in future: many find they have a more powerful machine available at home than they have available at work.

Telescope control and data acquisition

The emergence of the mini-computer in the late 1960s and early 1970s meant that it was feasible to provide computerized control systems for the large telescopes then being built. Earlier telescopes, in the main, relied on engineering devices such as the equatorial mount and AUTOGUIDERS to keep them tracking their targets. Computers made it possible to model the telescope errors introduced, for example, by mechanical flexure; allowed for atmospheric refraction; could precess coordinates; could slew the telescope in the most efficient way; could control raster scan; and could make use of feedback systems such as autoguiders and ACTIVE OPTICS. The Anglo-Australian Telescope (AAT) set the standard for such systems, being able to set with an accuracy of better than 1.5 arcsec when it came online in 1974. The AAT was developed by a team led by Patrick Wallace.

At first used to make equatorial optical telescopes more accurate, computerized control systems built on the experience of radio telescopes (*see* RADIO TELESCOPES AND THEIR INSTRUMENTS) to handle the more complex control requirements of optical telescopes with ALTAZIMUTH MOUNTINGS. Amateur 'Go-To' telescopes now have microprocessor versions of this software.

Modern data-acquisition software goes further. Not only does it control the telescope, it integrates the telescope, instruments and detectors. Telescopes need to be pointed at their targets. Detectors and instruments need to be configured and the data recorded, generally with some feedback to the astronomer about its quality. Radio telescopes and space instruments even more urgently need similar computer control. All of these aspects need to be coordinated by some higher-level system that may even allow the order of pre-planned observations to be changed in response to scientific, atmospheric and other conditions.

The data-acquisition software environments used at major observatories are now almost all based on a framework of some sort. A number of observatories, such as Gemini, have adopted the EPICS (Experimental Physics and Industrial Control System) database system. A similar system, the commercial RTAP (Real-Time Applications Platform) system from Hewlett-Packard, is used by the ESO's VLT project. The W M Keck telescopes use a system called KTL (Keck Task Library). IRAF is beginning to provide some of the communications facilities that are needed by some instrumentation projects.

There are now some common elements to astronomical computing systems. In recent years the trend in professional science has been for workstations, almost all running UNIX, to become faster, and almost all are now networked. The systems generally include workstations for overall control and online data reduction, with dedicated real-time systems based on microprocessors in direct control of the instruments and the telescopes themselves.

Observation planning and data archives

Software is becoming increasingly important both before observations are made and well after the data are reduced and published. For ESO's VLT, for example, and for space-based observatories like the HST, observations need to be planned well in advance using software that generates observation 'blocks' that the control system will execute much later at the telescope. The HST's observation-planning system has dramatically increased the efficiency of the telescope over its lifetime. Optical telescopes have not been used anything like as efficiently, and ESO made an effort to reap the benefits of HST experience when building the VLT.

The information introduced at that planning phase is associated in an end-to-end system with the data as they become part of the archive. The archive can be thought of as a virtual observatory, and its store of data can be 'mined' years later, reducing the need for observations that replicate earlier ones and providing long-term study of particular objects.

Large digital sky surveys, over a broad range of wavelengths, both from the ground and from space observatories, are major sources of astronomical data. Some examples include the SLOAN DIGITAL SKY SURVEY (SDSS) and

Data analysis software

Although obtaining astronomical data is a necessary and enjoyable first step, much more information and pleasure can be gained from subsequent processing of data to gain results. Some of the programs that do this are available on the Internet free of charge (freeware). Some are shareware that can be used for a short time before registration (and the payment of a fee) is requested. There are also some programs that must be bought before they can be used. It is unfortunate that a program must be purchased before you can find out if it will be suitable for its intended use, as 'try before you buy' seems such a sensible approach.

The general processing required for a CCD image is described in the book by Richard Berry and Jim Burnell entitled *The Handbook of Astronomical Image Processing*. This book (and the program included) covers everything from taking images through to the final processing required to obtain the final image for analysis publication, including ASTROMETRY and PHOTOMETRY. The program is under continuing development, and the current update can be downloaded from the Willman–Bell website. There are other documents, which can be downloaded from the Internet, describing the steps to take, and some websites also include downloadable processing software.

Astrometry is regularly carried out for ASTEROIDS and COMETS. The most frequently used tool is Astrometrica by Herbert Raab. This is a shareware program usable for a 100-day period before registration, when payment is required. There is also CHARON from Project Pluto (which also produces the GUIDE planetarium program) and the Italian CIRCE program. All make reference to the HIPPARCOS astrometry for reference star position so that high-precision reductions are possible. Output is in a format suitable for sending to the Minor Planet Center at the Harvard–Smithsonian. These programs can be downloaded from the Internet.

Most data-processing applications are primarily directed to the VARIABLE STAR observer, although the techniques can be applied to other bodies that have a variable light output such as asteroids or comets. The information required depends on what object has been followed. For an ECLIPSING BINARY STAR, the Heliocentric Time of minimum is required; for a pulsating Cepheid-type star, the Heliocentric Time of maximum. For an asteroid, the Asteroid Centric rotation period is to be determined; the same is true for comets. When a long timescale of data is available, period searching can be carried out using several mathematical techniques such as the discrete Fourier transform. Some suitable software is available on the Variable Star Network (VSNET) website, although a search on the Internet may well turn up some other programs.

Many software authors make their programs available for free download on the World Wide Web. The astronomical marketplace is very small, and most authors realize (or come to realize) that they will not get rich by charging and they may well get feedback to improve their programs if they make them available. The basic recommendation is to carry out an Internet search to identify possible programs suitable for your needs, and then try them out to see if they will do the intended job. With the wealth of data available there should be something to suit most requirements for data processing.

Andrew Hollis is a chartered structural engineer, and director of the BAA's Asteroids and Remote Planets Section.

Ephemeris and sky-simulation software

One of the most popular uses of computers in amateur astronomy has been software that can display the sky and locate objects of interest. These applications – often called 'planetarium software' because, like a planetarium, they simulate the night sky – allow users to display the sky at any given time from any location on Earth, and in some cases off Earth as well. Planetarium software can be much more versatile than conventional printed star charts, allowing the user to not only see the sky for a particular time and location, but also to choose what types of objects to display and learn additional information about them. In two decades, these applications have advanced from simple programs that used rudimentary graphics, including letters and numbers, to display the sky, to graphically rich tools with a great wealth of features.

Planetarium software has evolved in two directions in recent years. One approach has been to take advantage of the improving graphics capabilities of computers to generate more realistic displays. Applications like Starry Night (SPACE.com Software) and Redshift 4 (Maris Technologies) specialize in providing realistic displays of the night sky and celestial objects, and include such features as light-pollution effects and customizable horizons. These applications can also display high-resolution color images of objects. An alternative approach taken by some software developers is to emphasize advanced features. These applications, like The Sky (Software Bisque) and Voyager III (Carina Software) do not have the highly realistic graphics of their counterparts, although they do have colorful, accurate displays. Instead, these applications include many advanced features, including access to catalogs of stars, asteroids, and satellites; HERTZSPRUNG–RUSSELL DIAGRAMS and additional datasets; and other tools designed to assist observers. These programs also often include the ability to directly control telescopes, either as part of the application itself or through a companion program.

There are several criteria for choosing planetarium software, starting with how it will be used. In an educational setting, a program that includes good graphics, detailed data, and interactive features may be the best choice, while observers may prefer applications with detailed catalogs, telescope controls, and similar observing-oriented features. Software prices also vary widely. Some basic applications are available for free or for a very small charge, such as xsky, an interactive sky atlas for Linux and other Unix computers. Full-featured commercial applications can be considerably more expensive: versions of Starry Night cost \$30–\$180, while versions of The Sky cost \$129–\$249. Many popular applications are available for both Windows and MacOS, although some are available for only one operating system. Choices are more limited for Linux and other operating systems.

Planetarium software will likely become even more powerful in the future, taking advantage of the increasing power of desktop and laptop computers. Applications will also make increasing use of the Internet to allow users to obtain the latest information about astronomical objects. Software is also becoming widely available for handheld devices like Palms and PocketPCs, allowing someone literally to hold the universe – or a significant subset of it – in the palm of the hand.

Jeff Foust is a planetary scientist from MIT and author of *The Astronomer's Computer Companion* (No Starch Press, 1999).

Imaging software

Sensitive electronic detectors and computers have completely revolutionized astrophotography. Detectors, ranging from cooled scientific-grade CCD cameras to inexpensive webcams, are now routinely used in place of photographic film. The results of this transition have been dramatic.

The photographic darkroom has also been replaced with computers and sophisticated software. While standard tools like Adobe Photoshop are widely used, special-purpose programs have been created that meet the particular needs of astronomical imaging. Commonly used packages include AIP, AstroArt, CCDSoft, and MaxIm DL. Many of these both operate the camera and process the resulting images.

Capturing the image

The basic imaging setup includes an equatorially mounted telescope with a sidereal tracking drive and a CCD camera. The camera is attached in place of the eyepiece and connects to the computer via a cable. Various accessory devices are frequently used, such as electric focusers, filter wheels, autoguiders, and the telescope's own drive electronics. These are all connected to the computer and they are controlled by software.

Imaging is simple in concept but more complex in practice, because the user cannot 'look through' the camera visually. The telescope must be accurately focused, and special camera-calibration frames taken. Then the desired target must be located and centered. The drift of the telescope must be compensated for in order to avoid trailed images. Although a few cameras can directly produce color images, it is usually better to use a filter wheel and shoot separate red-, green-, and blue-filtered images. Software can help with all of these steps.

Up until recently, focusing was the most difficult part of the process. The image on the computer screen updates relatively slowly, and the image constantly fluctuates due to atmospheric seeing. New software tools, such as FocusMax, have now made this process completely automatic. FocusMax measures the 'half flux diameter' of a star image at multiple focus positions; this allows it quickly to locate the optimum focus even in poor seeing conditions.

It is often difficult to locate the target with lower-cost cameras, which can only see a very small piece of the sky. When using a Go To mount, software can help in several different ways. If the object is only slightly off target, auto-center software can quickly center it. If the target is completely off the CCD sensor, software such as PinPoint can match star patterns to a catalog and then calculate the exact position of the telescope. Another approach is to use telescope-mount modeling software such as MaxPoint or TPoint, which can correct for various mechanical pointing errors.

All telescopes slowly drift off target. This is a major problem with long exposures. One solution is to use a second, smaller CCD camera to repeatedly image a star near the target field. Software measures these images and sends commands to the telescope mount to keep the guide star stationary.

An alternative method to counter drift is to take shorter exposures. Instead of taking a single 20-minute exposure, take a series of 20 individual 1-minute exposures. Afterwards, add the 20 images together to produce almost the same image quality as the single long exposure. As an added bonus, any bad images in the series can be simply deleted. By re-aligning the images before combining them, the computer can correct for any telescope drift.

Image processing

Creating beautiful images is equal parts technology and art. Fortunately, the necessary technology is available off the shelf; you merely have to learn how to use the tools.

CCD cameras have excellent performance but are far from perfect. Fortunately a simple calibration procedure can all but eliminate these minor flaws. A special set of images is taken with the camera: dark frame (shutter closed but same exposure time), bias frame (shutter closed with zero exposure time), and flat field (image of a uniform target). The software utilizes these frames to correct for various imperfections in the camera, such as variations in sensitivity across the sensor.

To combine a series of exposures, individual images must first be aligned to each other; this is easily and automatically done by software. Once aligned, the images are simply added together to make one long exposure. More sophisticated algorithms like median or sigma clip can help by throwing out any bad pixels that might be present in a few of the images.

For color pictures, separate red-, green-, and blue-filtered images are required. Often a fourth unfiltered frame, called luminance, is also used. This LRGB technique allows you to combine long-exposure, high-resolution luminance images with shorter binned color images. First the binned images are resized to match the luminance frame. Next the color frames are aligned to the luminance frame, and then all four are combined together mathematically to build the final color image. The background level in the color images must then be equalized to remove any color cast. Since filters rarely produce the exact right color balance, the exposure times must be adjusted or the images scaled mathematically. A popular technique is to image a G2V star – the same spectral class as our Sun – and adjust the balance to make it appear white.

Two more steps are typically necessary for the best results. The first is to 'filter' the image to try and optimize its resolution. Lucy-Richardson and Maximum Entropy Deconvolution can enhance the sharpness of images, compensating for atmospheric seeing, poor focus, and optical aberrations. Simpler but highly effective Unsharp Mask, Digital Development Processing, and various types of kernel, FFT, and wavelet filters are also frequently used to sharpen or smooth images. When processing LRGB images, the luminance frame can be filtered before combining it with the RGB frames.

The final step is to 'stretch' the brightness and contrast of the image for best viewing. CCD cameras can generate images with tens of thousands of brightness levels; unfortunately computer monitors rarely even approach the 256 brightness levels they are theoretically capable of displaying. To represent both bright and faint details at the same time requires a 'non-linear' stretch. A popular technique is to use a graphical 'curves' command that allows the user to carefully fine-tune various ranges in the image.

It may require hours of work to produce the final picture, but the results can be spectacular. CCD imaging provides the unique opportunity to enjoy both the clear nights at the telescope and the cloudy nights at the computer.

Douglas George, co-discoverer of comet Skorichenko–George and also the National Past President of the Royal Astronomical Society of Canada, is now a successful supernova hunter.

the Digital Palomar Observatory Sky Survey (DPOSS) in the visible; the Two-Micron All-Sky Survey (2MASS) in the near-infrared; the NRAO VLA Sky Survey (NVSS); and the Faint Images of the Radio Sky at Twenty Centimeters (FIRST) in the radio. While most surveys are exclusively imaging, large-scale spectroscopic surveys also exist. In addition, a number of experiments with specific scientific goals, such as searches for NEAR-EARTH ASTEROIDS, are generating comparable volumes of data. Resulting datasets are in the range of tens of terabytes of digital information, including data about many millions or even billions of sources.

Making proper use of the huge data archives, particularly those from survey telescopes, is a new challenge for astronomical software. Projects such as AstroVirTel are under way to provide a common overall system to interrogate and relate information from different archives. Many websites now provide access to archival data (Centre des Données Astronomiques de Strasbourg, or Strasbourg Data Center), survey data, preprints of articles (Astro-ph), journals that are completely electronic in nature, and abstracting and bibliographic information (Sinbad, Astrophysics Data System). NASA is exemplary for its dissemination of information widely through the Internet, both to professional astronomers and to the public. *See also* INFORMATION HANDLING IN ASTRONOMY.

SOHO (Solar and Heliospheric Observatory)

A project of international cooperation between ESA and NASA to study the SUN, from its deep core (using the techniques of helioseismology) to the outer solar CORONA, and the SOLAR WIND. SOHO was launched on December 2, 1995, on top of an Atlas/Centaur combination, from Cape Canaveral Air Force Base in Florida, USA. It reached its operating orbit around the L_1 Sun-Earth LAGRANGIAN POINT in mid-February, 1996. This unique vantage position has allowed nearly uninterrupted observations of the Sun, 24 hours a day, 365 days a year. To achieve its objectives SOHO carries 12 state-of-the-art instruments, developed and furnished by 39 institutes from 15 countries. Besides watching the sun, SOHO has become the most prolific discoverer of sungrazing COMETS in astronomical history. As of May 2003, it had found more than 620.

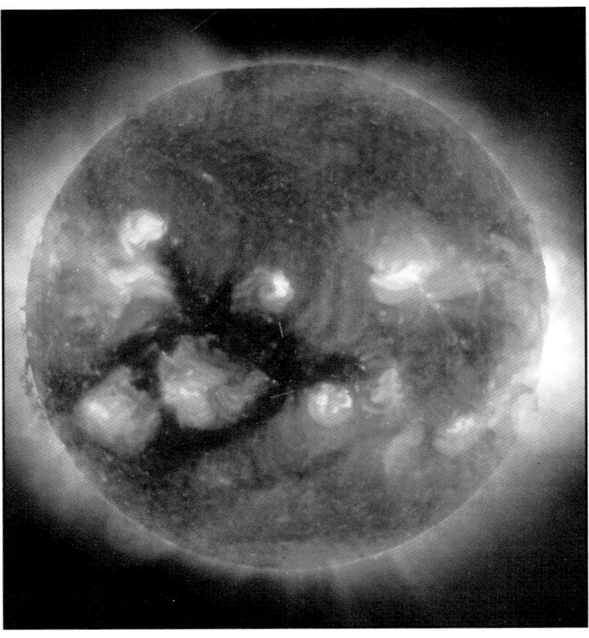

▶ ***SOHO*** *The Sun, seen by the Extreme Ultraviolet Imaging Telescope (EIT), one of SOHO's 12 instruments.*

Solar

Relating to the SUN.

Solar abundances

The first quantitative analysis of the chemical composition of the solar photosphere was made by Henry RUSSELL in 1929. Using eye estimates of the spectral line intensities (*see* SPECTRUM), he succeeded in deriving the abundances of 56 elements. He also showed that the Sun and, finally, the universe were essentially made of HYDROGEN. The Sun has always been considered as the typical star, to which the abundance analyses of all other stars are compared. Its abundance distribution has become the basic set of data that must be reproduced by nucleosynthesis theories, and it also plays a key role in modeling the chemical evolution of galaxies and the universe. Chemical-composition data can also be measured in other objects of the solar system such as the Earth, Moon, planets, comets and meteorites. These data are very important for modeling the evolution of the solar system.

Solar abundances can be derived by many different techniques and for different parts of the Sun from the interior to the outermost coronal layers, but the largest amount of available data is from the solar photosphere. A total of 65 elements, out of 83 stable ones, are present in the photospheric spectrum. The abundances are usually given by mass of hydrogen (X), helium (Y) and all the other elements together, the metallicity (Z) where $X + Y + Z = 1$. In these units the mass abundances of the Sun are $X = 0.735$, $Y = 0.248$ and $Z = 0.017$ with $Z/X = 0.023$. Although helium is the second most abundant element, it was not discovered until 1868, when it was observed in the solar spectrum during an eclipse of the Sun. One reason is that under the conditions in the solar photosphere, with a typical temperature of 5000 K, no line of helium falls within the wavelength range covered by the photosphere spectrum. *See also* COSMIC ABUNDANCE OF ELEMENTS; ELEMENTS, FORMATION OF.

Solar-B

A Japanese-led mission in partnership with the USA and UK proposed as a follow-up to the highly successful YOHKOH (Solar-A) collaboration. The mission consists of a coordinated set of optical, extreme ultraviolet (EUV) and X-ray instruments that will investigate the interaction between the Sun's magnetic field and its CORONA. It will improve our understanding of the variability of the Sun and of SPACE WEATHER AND THE SOLAR-TERRESTRIAL CONNECTION. The spacecraft will accommodate three major instruments: a 0.5-m solar optical telescope (SOT), an X-ray telescope (XRT), and an EUV Imaging Spectrometer (EIS). The spacecraft is scheduled for launch in late 2005. It will be placed in a SUN-SYNCHRONOUS POLAR ORBIT around the Earth. This will keep the instruments in nearly continuous sunlight with no day/night cycling for nine months each year.

Solar constant *See* SOLAR IRRADIANCE

Solar cycle *See* SUNSPOT CYCLE

Solar irradiance

The power from a source of radiant energy that is incident on unit area of a surface located at some distance from the source. The Sun's total irradiance is 1365 ± 2 W m^{-2}. This is the total radiant energy per unit time (that is, power) at all wavelengths that the Earth receives on unit area of its surface from the entire solar disk, when separated from the Sun by 1 AU. In the past it has been termed the solar constant but in fact it varies with solar activity. *See* SPACE WEATHER AND THE SOLAR-TERRESTRIAL CONNECTION.

Because the Earth travels round the Sun in an elliptical orbit, its distance from it ranges from 1.0167 to 0.9833 AU. Consequently, over the course of a year, the flux of solar radiation at the top of the atmosphere varies from its mean value by about ±3%. Precise measurements of the solar constant have shown that the LUMINOSITY of the Sun fluctuates on timescales of days and weeks by up to 0.2%, from changes in the numbers of dark sunspots (*see* SUNSPOT, FLARE AND ACTIVE REGION) or bright FACULAE on its surface. There is a general variation in solar luminosity that follows the solar cycle, the Sun being about 0.08% more luminous at solar maximum (when there are more active regions on its surface) than at minimum. From time to time there are brief dips in the solar irradiance as a coronal mass ejection clouds our view of the sun (*see* CORONA).

Solar Maximum Mission (SMM)

NASA satellite designed to study solar flares during the peak of the solar cycle (*see* SUNSPOT, FLARE AND ACTIVE REGION). Launched in February 1980, SMM carried seven instruments to record flares in visible, X-ray, ultraviolet and gamma-ray wavelengths. It also measured the solar constant. The satellite discovered X-rays originating from the bases of loop structures in flares. It failed after nine months, but was repaired by Space Shuttle astronauts in 1984. SMM re-entered the atmosphere in 1989.

Solar system

The objects that constitute the solar system are:

- the SUN;
- the TERRESTRIAL PLANETS (including Earth's Moon);
- Pluto;
- the gas GIANT PLANETS (including Jupiter and Saturn);
- their larger satellites;
- the ASTEROIDS, and the asteroids captured by major planets as moons;
- the METEOROIDS and INTERPLANETARY DUST;
- the COMETS and the TRANS-NEPTUNIAN OBJECTS in the OORT CLOUD AND KUIPER BELT.

Most of the mass of the solar system is contained within the Sun. Most of the rest is in Jupiter.

The Sun interacts with all the bodies of the solar system. Its gravity attracts them and holds them in orbit; its light and heat warm them; its atmosphere, the SOLAR WIND, with embedded magnetic fields, puts pressure on their MAGNETOSPHERES, atmospheres or surfaces. The extent of the solar system is defined by the extent of the solar wind. The solar system is contained within the HELIOSPHERE, and the boundary of the solar system is at the so-called heliopause. Beyond is interstellar space. *See* COSMIC RAY for an illustration.

The solar system had its origin in the solar nebula, and in the accretion of its gases and dust grains into the Sun and the planets (*see* PLANETS: ORIGIN).

The solar system remembers the rotation of the solar nebula. All the planets and most large satellites revolve in the direction that the Sun rotates (prograde), nearly in circles and nearly in the same orbital plane, the ECLIPTIC, which is close to the equatorial plane of the Sun's rotation. The Sun and most planets and large satellites rotate prograde on axes nearly perpendicular to the ecliptic. There are exceptions, however. These include the eccentricity and high inclination of Pluto's orbit, the tilted rotation of Uranus and Pluto, the high inclination of the orbits of the outer satellites of Jupiter, and the comets, revolving in elliptical orbits, some of them retrograde (*see* RETROGRADE MOTION). These cases are thought to be due to a gravitational disturbance by a passing planet or asteroid, or to a collision.

Because of the effects of the Sun's heat, there is a progressive change in the composition of the solar system with distance from the Sun. The composition progresses from the terrestrial planets (also called the inner planets), which are composed of dense elements such as silicon, iron and nickel; to the gas giant planets, which retain almost the original composition of the solar nebula; to the comets, now in the Oort Cloud and Kuiper Belt, which are nearly pristine solar-nebula material. The asteroids and meteoroids lie between the inner planets and gas giants in space and in some aspects of composition. To an extent, solar-system material has been redistributed by the flow of asteroids, comets and meteoroids among the planets and their impact on them (*see* PANSPERMIA).

The stability of the solar system

The question of the stability of the solar system was raised with the discovery of Isaac NEWTON's law of gravitation, published in 1687 and studied by Jules POINCARÉ. In the ideal case (one Sun, one planet, well separated) the solar system would last in the same state indefinitely. In fact, because there are many planets, each interacting with all the others, the motion of the solar system is continually evolving, with orbits flipping from one form to another. Over 10 million years the motion of the planets can be forecast with great accuracy. But calculations beyond about 100 million years have no hope of being correct.

The first modern calculations over this scale of time were carried out in 1986 at the Massachusetts Institute of Technology on Orrery, a computer specially designed for the task, and they are now performed on powerful workstations. These studies have shown that the solar system, particularly the system of inner planets (Mercury, Venus, Earth and Mars), is chaotic (*see* CHAOS). For example, an error of 15 m in the Earth's initial position gives rise to an error of about 150 m after 10 million years, but 150 million km (1 AU) after 100 million years.

Chaotic changes in the eccentricity and inclination of the Earth and its orbit were identified by Milutin MILANKOVITCH as the astronomical causes of climate change. The eccentricity of both Venus and the Earth varies from small to about 0.08. The change in Sun-Earth distance is modest enough to have kept the Earth in the 'habitable zone' of the solar system for the billions of years of the evolution of life, but is significant in terms of climate change. The eccentricity of Mars' orbit can reach a more serious 0.2 in a few billion years, and Mercury's exceeds 0.5. Over time, the orbits of the inner planets sweep out large areas of the disk of the ecliptic plane.

As a result of the variability of the orbits over time, the solar system is 'full.' The zone swept by the orbit of Mars to its maximum eccentricity reaches the limits of the asteroid belt, most asteroids closer than this having been ejected. There has been over the lifetime of the solar system a fair probability of a collision between Mercury and Venus,

▲ ***Solar system*** *seen from the perspective of Neptune in this artist's impression.*

although in fact it has not happened. Any planet placed among the outer planets remains there for only a few hundred million years, apart from some particular zones of stability or beyond Neptune, in the Kuiper belt. If there was another planet in the solar system, it would probably collide rapidly with one of the existing planets.

The solar system is unstable, but the catastrophic phenomena that lead to its destruction occur over billions of years and have mostly already happened. At early times, there may have been more planets than today, but collisions or ejections took place. One example was the impactor on the Earth that caused the formation of the Moon. Chaos determined the structure of the solar system, which thus organized itself toward increasing stability and regularity. This may be conjectured to have contributed to BODE'S LAW.

Solar-system exploration

As a result of space exploration, each of the planets and their satellites is now revealed as a diverse and complex world, described in this encyclopedia under individual entries. At first the missions took place one by one in an *ad hoc* way. Now the space agencies meet regularly to plan the pattern of exploration. For example, the International Mars Exploration Working Group of representatives from the space agencies meets to plan the progressive discovery of Mars.

The time to initiate a space-exploration mission must be carefully chosen. 'Planetary windows' occur when the Earth and the planet under study return to the optimum relative position. The Moon, Mars and Venus are relatively easy targets, but Mercury, the OUTER PLANETS and small bodies require a very large launcher or a particularly efficient mission scenario. A lander or entry-probe mission is much easier if there is an atmosphere to act as a free brake (as with Venus, Mars, the giant planets and TITAN), and the return of a sample from a planetary body is going to be the most demanding mission of all.

Other planets may help propel a probe toward a distant planet. As the spacecraft swings by the intermediate planet it picks up assistance from its gravity. Such gravity assists swing the spacecraft on to its destination, like Mariner 10 (*see* MARINER MISSIONS), Galileo (*see* GALILEO MISSION TO JUPITER) and Cassini (*see* CASSINI/HUYGENS MISSION). The strategies can be quite surprising: recent missions to the outer planets started off traveling inwards to Venus. The drawback is the very long mission time: eight years of travel are required to rendezvous with Saturn or a COMET. The interval of about 15 years between the definition of instruments and the scientific results is a management challenge. Ion-propulsion systems are being developed to accelerate spacecraft in orbit to reduce journey times (*see* NEAR EARTH ASTEROID RENDEZVOUS MISSION; SMART-1). Ion propulsion derives energy from solar radiation, a small flux; however, significant energy is accumulated over years in orbit.

Once a spacecraft has come close to a planet, the scientific investigations form two major categories: remote-sensing techniques and laboratory techniques.

Remote-sensing techniques take advantage of the whole ELECTROMAGNETIC SPECTRUM to obtain information from distances of hundreds or thousands of kilometers. Laboratory techniques require direct contact with the surface, the ATMOSPHERE or the MAGNETOSPHERE. All the technology operates within very stringent limits on the mass and power of the equipment.

The logical sequence of scientific missions for the exploration of a planet is as follows:

• The discovery mission (or missions). This first step is a flyby, which provides the first high-resolution remote-sensing information. Like the LUNA 3 image of the far side of the Moon in 1959 or the Voyager flybys of the giant planets, it reveals unexpected features and helps define future missions. All major planets, except Pluto, and representative asteroids and comets have been explored in this way.

• The in-depth exploration missions. The second step is a series of orbiter missions, to map the planet by remote sensing. These provide a global view of the body and enable the selection of areas of interest for *in situ* studies. The VIKING MISSION orbiter, MAGELLAN, and Galileo are examples. The Moon, Venus, Mars, Jupiter and Eros have been explored like this.

• The *in situ* missions. This third step is carried out by modules ('landers') that descend to the planet, providing information on its atmosphere, if any, and surface. The *in situ* results provide 'ground truth' which re-calibrates the remote-sensing techniques, and improves our knowledge even of the regions not visited. LUNOKHOD, MARS PATHFINDER and Galileo are examples, with the Cassini/Huygens probe to come soon. The Moon, Mars, Jupiter and arguably Eros have reached this stage; Titan will soon.

• The sample-return mission. The power and mass constraints on the instruments necessarily limit what can be discovered *in situ* about the surface material. The next stage

Solar system in figures

	Mercury	Venus	Earth	Mars	Jupiter	Saturn	Uranus	Neptune	Pluto
Mean distance from the Sun (10^6 km)	58	108	150	228	778	1426	2868	4494	5900
Mean distance from the Sun (AU)	0.39	0.72	1.00	1.52	5.20	9.54	19.18	30.06	39.44
Revolution period (year)	0.24	0.62	1.00	1.88	11.86	29.46	84.07	164.82	248.60
Orbital eccentricity	0.21	0.01	0.02	0.09	0.05	0.06	0.05	0.01	0.25
Inclination to ecliptic (°)	7.00	3.40	0	1.85	1.30	2.49	0.77	1.77	17.17
Rotation (day)	58.65	243.00 (R)	1.00	1.03	0.41	0.44	0.65 (R)	0.76	6.39 (R)
Diameter (km)	4878	12 102	12 756	6788	142 984	120 536	51 118	49 530	2390
Mass (Earth = 1)	0.06	0.82	1.00	0.11	317.89	95.18	14.54	17.15	0.002
Density (g cm^{-3})	5.43	5.24	5.52	3.94	1.33	0.69	1.27	1.64	1.80
Escape velocity (km s^{-1})	4.2	10.4	11.2	5.0	59.5	35.5	21.3	23.5	1.3

R = retrograde motion

is to return samples to terrestrial laboratories for virtually unlimited analysis. Examples of this type of mission are Luna 16, 20 and 24; some natural events have also transported material from other planets to the Earth (*see* METEORITE).

• The field-geology mission. The manned exploration of planets, limited by expense and practicality, makes it possible for trained scientists such as geologists to carry out studies with imagination and experience deployed to select samples (the only example is APOLLO 17).

Only the Moon has been explored anything like completely, and Pluto not at all.

Solar time

A time system based on the rotation of the Earth measured relative to the Sun; the time kept by a SUNDIAL. When corrected for irregularities in the solar motion, solar time is called 'mean solar time.' When mean solar time refers to Greenwich, according to international convention, it is UNIVERSAL TIME. *Contrast* SIDEREAL TIME.

Solar wind

One of the most enigmatic problems in solar-system research is how the Sun, with a surface temperature of only 5800 K, can heat up its atmosphere, the solar CORONA, to more than 1 million K. In fact, the corona is so hot that not even the Sun's enormous gravity can contain it. Part of it is continuously evaporating into interplanetary space as the solar wind. As a highly ionized magnetized PLASMA it dominates a huge volume around the Sun that is called the HELIOSPHERE. Only far beyond the outermost planets is a transition into the interstellar gas (*see* INTERSTELLAR MEDIUM) expected to occur.

The intricate structure of the corona is seen by ground-based observers during the rare occasions of a solar eclipse (*see* ECLIPSE AND OCCULTATION). The solar wind emerging from the corona is similarly inhomogeneous and creates a complicated three-dimensional shape of the plasma heliosphere. Interactions between outflowing streams of different speeds and solar transient phenomena cause further complications. Thus the solar wind as we see it from the Earth's orbit is characterized by an enormous variability in all its basic properties. This is what allows the solar wind to have a surprisingly large impact on the Earth (*see* AURORA).

It is now well established that there are two basic types of solar wind which differ markedly in their main properties. The fast solar wind emerges from magnetically open coronal holes which are representative of the inactive, or 'quiet,' Sun. The major coronal holes usually appear at high solar latitudes beyond 40–60° but the resulting solar wind expands significantly and fills up all the heliosphere except for the 40°-wide streamer belt close to the magnetic equator.

The slow solar wind originates from above the more active regions on the Sun. The slow wind shows variation, particularly in HELIUM content between maximum and minimum in the solar cycle. Note that there is no continuous transition between the fast and slow wind. Rather, the boundary layers between them are very thin close to the Sun and correspond in detail to the boundaries of the coronal holes. At present, existing theories on the production of the solar wind are highly controversial. This clearly illustrates that our understanding of solar-wind acceleration is not yet on firm ground, and further research is needed. *See also* ULYSSES MISSION.

Solstice

The point on the ECLIPTIC at which the Sun reaches its maximum declination north or south of the CELESTIAL EQUATOR. The greatest northerly declination corresponds to the summer solstice, the greatest southerly declination to the winter solstice. In the northern hemisphere the summer solstice occurs on June 21 or 22 and the winter solstice on December 21 or 22. In the southern hemisphere, where the seasons are reversed, the situation is exactly the opposite.

In the northern hemisphere at the time of the summer solstice, the north pole is tilted 23.45° toward the Sun. Because the Sun's rays are shifted northward by the same amount, the vertical noon rays are directly overhead at the Tropic of Cancer. Six months later, the south polar end of the Earth is inclined 23.45° toward the Sun. On this day of the summer solstice in the southern hemisphere, the Sun's vertical overhead rays progress to their southernmost position, the Tropic of Capricorn.

Sombrero Galaxy (M104) *See* MESSIER CATALOG

Somerville, Mary (1780–1872)

Scottish physicist and mathematician. With the early death of her husband and being of independent means, she took up mathematics, studying astronomy and dynamics. She carried out experiments on magnetism and the solar SPECTRUM. She wrote a popular book on the work of Pierre-Simon LAPLACE and Isaac NEWTON called *The Mechanism of the Heavens*, and an account of *The Connection of the Physical*

Sciences. Her discussion of a hypothetical planet causing the orbital perturbations of Uranus inspired John Couch ADAMS to predict the position of Neptune. Somerville and Caroline Herschel (*see* HERSCHEL FAMILY) were the first two women to write papers read to the RAS. Somerville College in the University of Oxford, England, was named in honor of her.

South African Astronomical Observatory

The South African Astronomical Observatory (SAAO) is the national facility for optical and infrared astronomy in South Africa. It has strong links with research groups worldwide through scientific collaboration and technological exchange. The SAAO contributes to the development of South Africa by providing training in a scientific and high-technology environment; by stimulating young people to follow careers in science and technology through a program educating schools and teachers about science; and by helping to create a culture of science and technology among all communities with a vigorous science-awareness program. SAAO's headquarters are in Cape Town (at the Royal Observatory, Cape of Good Hope, founded in 1820) and the telescope facilities are located at Sutherland in the Northern Cape. At present the main telescopes range in aperture from 1.9 to 0.5 m. The major new facility will be the South African Large Telescope (SALT) being constructed by South Africa with partners in Germany, Poland, the USA, New Zealand and the UK. The telescope, with a hexagonal mirror array 11 m across and based on the Hobby–Eberley Telescope at MCDONALD OBSERVATORY, Texas, USA, is scheduled for completion in December 2004.

Southern Pinwheel Galaxy (M83)

See MESSIER CATALOG

South Pole Aitken Basin

The largest and deepest multi-ringed impact structure on the MOON. It lies almost entirely on the far side, and extends for more than 2500 km from the south polar region to the 135-km-diameter crater Aitken. The impact that produced the basin occurred an estimated 4.1 billion years ago, near the beginning of the episode known as LATE HEAVY BOMBARDMENT. At its deepest point the South Pole Aitken Basin is nearly 13 km below the Moon's mean surface level. It was first revealed by the LUNAR ORBITER photographic reconnaissance of the mid 1960s, but its full extent was not appreciated until the Moon was revisited by spacecraft in the 1990s. The Galileo spacecraft en route to Jupiter photographed it in 1992 and showed it to have a darker coloration than its surroundings. Its size was established by the CLEMENTINE orbiter in 1994, which also found that the basin's floor is rich in titanium and iron oxides. This compositional anomaly may indicate that the impact reached a depth of 120 km, penetrating the crustal rocks completely and excavating metal-rich material from the mantle. For a photograph *see* ORIENTALE BASIN.

Space (or Spitzer) Infrared Telescope Facility (SIRTF)

NASA launched the SIRTF into space in August 2003. During its 2.5-year mission, SIRTF will obtain images and SPECTRA, radiated by objects in space between wavelengths of 3 and 180 μm. Most of this infrared radiation (*see* INFRARED ASTRONOMY) is blocked by the Earth's atmosphere and cannot be observed from the ground. Consisting of a 0.85-m telescope and three cryogenically cooled science instruments, SIRTF is the largest infrared telescope ever launched into space. It was renamed Spitzer Infrared Telescope Facility in 2003.

Because infrared is primarily heat radiation, the telescope must be cooled to near absolute zero (−273 °C) so that it can observe without interference from the telescope's own heat. Also, the telescope must be protected from the heat of the Sun and the infrared radiation put out by the Earth. To do this, SIRTF will carry a solar shield and will be launched into an Earth-trailing solar orbit.

SIRTF is the final mission in NASA's Great Observatories Program. SIRTF is also a part of NASA's Astronomical Search for ORIGINS PROGRAM, designed to provide information that will help us understand our cosmic roots, and how galaxies, stars and planets develop and form.

Space Interferometry Mission (SIM)

The NASA Space Interferometry Mission, scheduled for launch in 2009, will determine the positions and distances of stars several hundred times more accurately than any previous program (*see* ASTROMETRY). It will determine the distances to stars throughout the galaxy and probe nearby stars for Earth-sized planets. SIM will use optical interferometry, a technique pioneered by Albert MICHELSON. Light from two or more telescopes on a 10-m baseline will be combined as if they were pieces of a single gigantic telescope mirror. This technique is expected to achieve an accuracy of 1 microarcsec in a single measurement and will eventually lead to the development of telescopes powerful enough to take images of Earth-like planets orbiting distant stars and to determine whether these planets sustain life as we know it.

Space probe

An uncrewed spacecraft designed to investigate the interplanetary medium or conditions on or in the vicinity of another celestial body. A space probe travels along an ORBIT around the Sun or on a trajectory that takes it from the vicinity of the Earth to the neighborhood of its target (for example, a planet, comet or asteroid). The first probe to leave the vicinity of the Earth and enter an independent orbit around the Sun was LUNA 1, a Soviet probe that passed by the Moon at a range of 6000 km in 1959, and continued on into interplanetary space.

Space Telescope Science Institute (STScI)

The Space Telescope Science Institute is located on the Johns Hopkins University Homewood campus, Baltimore, Maryland, USA. It is responsible to NASA's GODDARD SPACE FLIGHT CENTER for the scientific operations of the HST, to maximize its scientific productivity and to serve the astronomical community in its operation. The institute solicits and reviews observation proposals and selects observations to be carried out. It schedules observations and assists guest observers. It also supports all spacecraft activities and provides the facilities and software to calibrate, analyze and archive

HST data. The STScI Guide Star Selection System provides reference stars and other bright objects so that the HST's fine-guidance sensors can point the telescope accurately.

The institute also played a key role in the early mission-concept studies for the Next Generation Space Telescope (now the JAMES WEBB SPACE TELESCOPE) scheduled for launch in 2011. It has also been selected by NASA to manage the science and mission operations for this successor to the HST.

Spacetime

A four-dimensional geometrical framework that unites the three dimensions of space and the dimension of time. A point in spacetime (identified by four coordinates) is called an event (or 'world-point'). The history of a particle (a line in four dimensions connecting all the events at which it is present) is called a world-line. The concept of spacetime was devised in 1907 by Hermann Minkowski (1864–1909). Whereas Isaac NEWTON regarded time as being independent from the three dimensions of space, Minkowski contended that they are intimately related. Four-dimensional spacetime was adopted by Albert EINSTEIN as the framework within which to develop his theory of GENERAL RELATIVITY. According to that theory, spacetime is curved in the presence of matter and the paths of rays of light and particles of matter are determined by the curvature of spacetime.

Space velocity

A measure of the true velocity in three dimensions of a star relative to the Sun, calculated from measurements of its RADIAL VELOCITY and tangential velocity (its proper motion – *see* ASTROMETRY).

Space weather and the solar-terrestrial connection

Radiation from the Sun and energy from the SOLAR WIND influence the Earth and its surroundings. The most significant solar variations cause long-term changes in the Earth's atmosphere and intense storms in space. Such storms, known as space weather, can damage spacecraft and power-generation networks.

Solar-induced variations in the Earth's atmosphere

The flux of solar energy, known as the solar constant, is indeed remarkably constant, although there is about an 8% increase in the radiation received at the Earth in January compared with June, owing to the eccentricity of the Earth's ORBIT. It is difficult to measure the solar constant from the ground because of the effects of the Earth's atmosphere, but space-based measurements began in the 1970s. Since then, the variation has been considerably less than 1% (1365 ± 2 W m^{-2}). It is estimated that a 1% increase in solar energy causes a 1-K increase in global temperature. Therefore the variation of solar energy is of minor importance in the recent global temperature rise, at the rate of about 0.5° per decade over 50 years. However, there are more than 2000 papers published in the refereed scientific literature describing meteorological phenomena that have a statistically significant relationship with the 11-year solar cycle. The connection between solar activity and meteorological weather is not understood.

The connection with space weather is clearer. Solar X-rays and ultraviolet (UV) radiation ionize some of the gases in the atmosphere, leading to the formation of the IONOSPHERE. Its strength depends on the flux of X-rays and UV, and on the solar ZENITH angle. During the day, the absorption of solar radiation causes the temperature of the upper layers of the atmosphere (the thermosphere) to rise by around 400 K, resulting in a large bulge on the day side of the Earth compared with the night side. A major consequence is that high-altitude winds blow at 100–400 m s^{-1} from the day side to the night side.

The intensity of the solar UV radiation increases on average by a factor of about 2.5 between sunspot minimum and maximum, and increases the electron concentration in the ionosphere by approximately the same factor (*see* SUNSPOT, FLARE AND ACTIVE REGION). Measurements of sunspot number and the intensity of the solar radio flux are often used to predict the state of the ionosphere for practical purposes of shortwave radio communication. However, they are at best only a guide to the general level of activity.

Solar flares last 1–100 min and release large amounts of UV energy. The biggest ones increase the ionization in the day-side ionosphere and profoundly affect radio-wave propagation. They may precede a coronal mass ejection (CME; *see* CORONA) from the Sun, which may travel to Earth as part of the solar wind. The Earth's MAGNETOSPHERE usually deflects the solar plasma. But under some conditions the plasma of a CME can be transferred into the magnetosphere. Its energy is redistributed in the magnetosphere by the effects of electric fields, currents, energetic particles and waves, which in turn have significant effects on the atmosphere and the magnetosphere, including AURORAE.

The occurrence of magnetic storms varies over the solar cycle. In general, storms are most frequent up to two years after sunspot maximum. However, the number of storms at each phase of the solar cycle has increased markedly over the last 100 years.

The effect of space weather on satellites

All Earth-orbiting satellites fly in the ionosphere and magnetosphere. Satellites are, of course, used for many practical purposes, especially telecommunications, navigation with the global positioning system (GPS), and Earth observations; they are also used for meteorological forecasting, hurricane track predictions and agricultural crop assessments.

Many satellites operate at geostationary orbit (altitude about 40 000 km), where the effects of energetic particles on spacecraft operations are very important. During storms, the particle fluxes of energetic electrons and protons increase enormously. This is particularly true in the outer radiation belts. The high fluxes damage spacecraft severely. The most energetic particles penetrate deep into the body of the spacecraft and into coaxial cables, changing the structure of the materials for the worse. They can lead to deep dielectric charging which sparks on discharge, resulting in damage to spacecraft components, spurious signals and occasionally to complete failure of some or all of the spacecraft, such as its computer memory. Solar cells are used to power satellites, and are also vulnerable to solar storms. A modest solar flare can reduce their efficiency by around 3% per event. Successive events affect the operational lifetime of the satellite.

Actions can be taken to reduce the effects, many of them incorporated into the design of the satellite. Shielding key elements and cables is the main technique, implemented during manufacture, but during the operational phase satellites may be oriented with less sensitive areas toward the radiation, to shield the more sensitive parts. Also operators try to avoid sensitive maneuvers during the storm.

Radio propagation

Guglielmo Marconi (1874–1937) first used shortwave radio communication via the ionosphere 100 years ago. Despite the advent of many new means of communication, short waves are still of great importance in many areas of the world, being cheap and, usually, reliable.

Free electrons in the ionosphere and magnetosphere affect the propagation of radio waves. Propagation conditions change markedly because of the effects of the Sun on timescales from seconds to a solar cycle. Even at the high frequencies used in satellite communications, small plasma irregularities cause the signal to fade rapidly. This can introduce errors into global positioning and can also distort radar images.

Spacecraft orbital decay

The orbits of all satellites below about 1500-km altitude decay, owing to atmospheric drag. The atmospheric density increases at times of sunspot activity, causing orbits to decay more rapidly, especially low-altitude orbits such as that of the ISS. This creates operational difficulties. There are side benefits, however, because the increase in atmospheric density de-orbits space debris, of which there are 10 000 pieces larger than 1 m, and millions smaller.

Human exposure to space weather

The exposure of humans to space weather is a significant problem. US Space Shuttle flights have been curtailed to prevent astronauts receiving high dosages of solar radiation. Space weather will limit human travel to Mars over the long (nine-month) journey. Although Earth's atmosphere is a good shield and the fluxes of particles at 15-km altitude are tiny compared with the fluxes in space, there is still a modest threat to the operation of high-flying aircraft. Again electronics may be vulnerable and the total radiation dosage for the crew, especially females of child-bearing age, is an issue. Such airplanes carry radiation monitors. If the radiation flux increases, the aircraft lowers its cruising altitude to reduce the effects. At present military aircraft are most vulnerable, although the current generations of passenger aircraft are pushing up toward 15 km and monitoring of the fluxes of energetic particles from geospace will be required.

Ground-based effects of space weather

During magnetic storms, very large currents flow horizontally in the ionosphere in the auroral oval. The magnetic variation from such currents is about 10% of the Earth's field at the latitude where these currents occur. These ionospheric currents induce currents in the sea, on land and in power lines. Transformers connected to the power lines can overheat and occasionally fail. The best-documented event occurred in March 1989 when the entire province of Quebec in Canada was blacked out for more than nine hours. The cost of the failure was put at $5 billion, similar to that of a hurricane. Likewise, currents induced in metallic pipelines induce economically damaging corrosion.

Predicting space weather

The accurate prediction of space weather will probably become more important, because we increasingly rely on satellite-based communications and navigation systems, and because the volume of high-flying commercial passenger air transport is growing. The prediction of space weather is in its infancy. Further scientific understanding of the solar-terrestrial system is essential before accurate predictions can be a reality, in particular the identification of the events on the Sun that lead to the severe storms. CME events are not the only kind.

The best forecasts of space weather so far achieved are based on solar wind and other measurements taken by satellites at the L_1 LAGRANGIAN POINT. These data give about a one-hour warning of space-weather conditions on Earth, which is insufficient.

Special Astrophysical Observatory, Russia

The Special Astrophysical Observatory (SAO) of the Russian Academy of Sciences was established in 1966 to manage the operation of the 6-m Bolshoi Azimuthal Telescope (BTA) and the 600-m aperture RATAN-600 radio telescope. The observatory is located in the North Caucasus. The 6-m telescope together with a few smaller instruments are at an alitude of 2.1 km. The radio telescope is about 20 km to the north while an administrative center, library, scientific laboratories and houses are located about halfway between the main instruments. Both telescopes were built in 1975, but regular observations started two years later. Today SAO is the main Russian center for ground-based astronomical observations. The BTA was the world's largest telescope until the beginning of the 1990s and was the first giant telescope to use an ALTAZIMUTH MOUNTING, thus becoming an example for later large telescopes.

Special relativity

The special theory of relativity, developed by Albert EINSTEIN in 1905, was a revolutionary change to Isaac NEWTON's concept of the nature of space and time, itself based on some of GALILEO GALILEI's ideas. According to Newton, space and time are separate and form an absolute framework in which events occur. According to special relativity, space and time are not absolute but are components of a single entity, SPACETIME. Special relativity formulates the laws of physics in inertial frames – that is, from the point of view of observers who are moving at constant speed relative to each other.

Special relativity is based on two postulates:

- Every law of nature has the same form in all inertial frames (the relativity principle);
- The speed of light in a vacuum, c, is the same in all inertial frames.

Because the speeds with which things move are usually, in normal experience, much less than the speed of light, Newton's concepts of space, time and gravity are usually good enough. When things move quickly special relativity is better. When things move quickly under the force of gravity, the theory of GENERAL RELATIVITY (also developed by Einstein) is the right one to use, but because gravity is a very weak force, special relativity is usually adequate.

Some of the implications of special relativity for astronomy are as follows:

- Simultaneity: observers in different frames will in general assign different times to a given event. Two events that are simultaneous in one frame of reference occur at different times according to observers in another frame. In astronomy, celestial events, like a double-star eclipse and the explosion of a NOVA in another galaxy, which happen on the same night for us may be at completely different times, separated by millions of years because of light-travel effects.
- Time dilation: a clock that is moving in a given frame runs slow compared with clocks that are stationary in that frame.

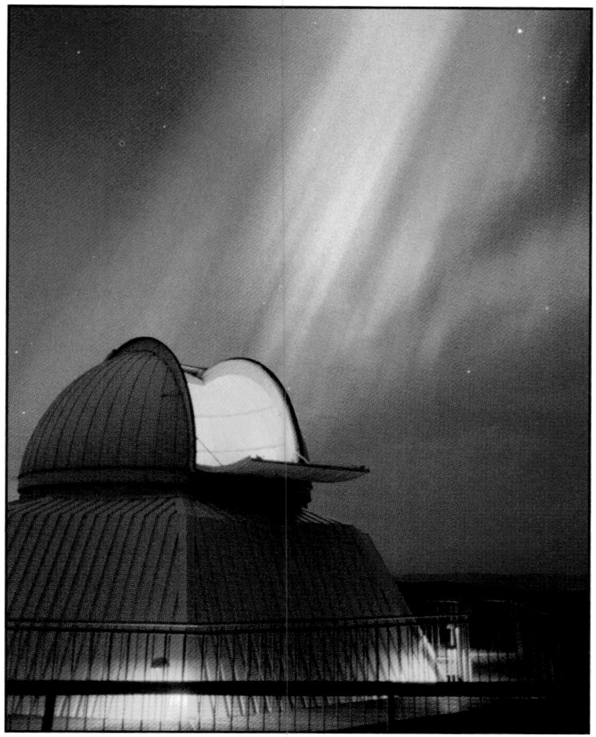

◀ ***Space weather*** *The visible form of space weather is the aurora, seen here in 2000 at the peak of the sunspot cycle, over Mount Mégantic Observatory in southern Quebec, located near the auroral oval around the magnetic north pole. Curtains of aurorae show different colors according to the conditions of the atmospheric atoms (green color comes from oxygen at 150 km altitude).*

Terrestrial clocks that are used to time the ticks of pulsars (*see* NEUTRON STAR AND PULSAR) run faster and slower depending on the annual speed of revolution of the Earth around the Sun, in its eccentric orbit. Supernovae (*see* SUPERNOVAE AND INTERSTELLAR MATTER) have a 'clock,' namely the time it takes for their light to fade from maximum. Because of time dilation, distant supernovae, which are receding with their parent galaxies in the expansion of the universe, fade more slowly than supernovae nearby.

- Length contraction: a body is shorter when measured in a frame in which the body is moving than it is in its own rest frame (the Lorentz contraction). Only the dimension in the direction of motion is contracted. The Lorentz contraction of one of the arms of the INTERFEROMETER in the MICHELSON–MORLEY EXPERIMENT to determine the speed of the Earth through the ether was the reason why the experiment gave a nil result and the ether was not detected.
- Relativistic beaming: if a source is moving at almost the speed of light, and emits particles or radiation more or less isotropically in its own rest frame, the radiation observed is highly beamed in the direction of the source's velocity. This makes the approaching side of the accretion disk orbiting a BLACK HOLE brighter than the receding side and is the reason why BLAZARS are so bright – we are looking down into a JET of material moving out toward us.
- Relativistic DOPPLER EFFECT: special relativity leads to a modification in the Doppler formula for the change in wavelength detected when the source of a light wave moves along the line of sight. Even motion across the line of sight leads to a Doppler effect in special relativity (the transverse Doppler effect). The star SS433 has fast jets of material that precess and change their angle relative to the line of sight. When, in the course of their twisting motion, the jets are passing across the line of sight, their spectral emissions nevertheless show a Doppler shift because of the transverse Doppler effect.
- Mass-energy relation: when a body's mass reduces by m, its energy changes by an equivalent amount $E = mc^2$. This underlies the whole of nuclear physics and is the reason why hydrogen fusion to helium, and other nuclear reactions, releases the energy radiated by the stars.
- Relativity and causality: an intriguing consequence of special relativity is that the time order of spatially separated events can be different in two frames. For cause-and-effect relations, however, the signal from the cause that triggers the effect can propagate no faster than the speed of light. In this case the cause precedes the effect in every frame. This argument is repeatedly used in astronomy to put a limitation on the dimensions of a source of light that is varying. If a cosmic source exhibits substantial variation in intensity over a time T, one can argue from causality that its dimensions cannot be greater than cT, because different parts of the varying structure must be able to communicate with each other to vary in synchronism. This argument was used to show how small the size of the central engine in a QUASAR is, because some vary over only days, and how pulsars are even smaller because they vary in less than a second.
- SUPERLUMINAL MOTION is observed in some extragalactic radio sources: material is ejected from a central source at a speed that seems to exceed that of light, sometimes by a factor as high as 10. However, there is an explanation entirely consistent with special relativity.

Spectral type *See* STELLAR SPECTRUM: CLASSIFICATION

Spectroheliograph/spectrohelioscope

An instrument that allows the Sun to be studied in light of one particular wavelength, by imaging the Sun photographically, electronically (with a spectroheliograph), or by direct vision (with a spectrohelioscope). The instrument consists of a SPECTROSCOPE or spectrograph, which isolates one spectral line (such as one of the HYDROGEN or calcium lines), combined with a mechanical or optical means of scanning the disk of the Sun. In the case of a spectrohelioscope, provided that the solar disk, or the region of interest, can be completely scanned at least 10 times a second, persistence of vision allows the observer to see a stationary monochromatic (single-wavelength) image. The instrument enables features such as PROMINENCES and FILAMENTS to be studied. The detail that can be seen increases as the width of the slit decreases.

Spectroscope and spectrograph

A device that enables the SPECTRUM of a light source to be viewed directly (with a spectroscope) or recorded by photographic or electronic means (with a spectrograph). A spectrograph contains four parts:

- an entrance aperture which lies at the focus of a telescope;
- a collimator which converts the cone of light emerging from the slit to a parallel beam;
- a prism or diffraction grating which disperses the parallel beam into a radiating bundle of beams, each beam at a given angle corresponding to a particular wavelength;
- a camera, which focuses the parallel beams into new images of the slit at a corresponding range of positions and records them on a detector, in the past a photographic emulsion but typically nowadays a CCD detector. Because each wavelength images the slit on the detector, the bright wavelengths appear as spectral lines.

The entrance aperture rejects sky background adjacent

to the object being studied; for example, a star. Otherwise the spectrum consists of an overlapping and hopelessly muddled set of spectra of each point in the focal plane of the telescope, and there is an underlying build-up of confusing background noise. The aperture can be a simple hole, but is often a slit that is narrower than the star image. This is so that the wavelength resolution in the final spectrum is determined by the width of the slit rather than by the distribution of light across the star (which might jiggle about at the telescope focus). The surface of the slit jaws is polished and tilted at an angle to the incident beam, so that an eye or television camera can be used to view the reflected image of the star on the slit. Where the star is very bright it may be possible to get away with using no entrance aperture at all (a slitless spectrograph).

To take many spectra at the same time (for example, of many galaxies in a cluster), a spectrograph can have many entrance apertures, their positions chosen so that the spectra produced do not overlap. One type uses multiple slits made by machining an aperture plate using mechanical tools or high-powered lasers so that the slits fit the image of the cluster or galaxy. Another type uses single optical fibers, positioned at the telescope focus so that each acquires the light from a separate target object. The output ends of the fibers are aligned to form a long slit, which feeds a conventional spectrograph. The fibers are accurately positioned in the focal plane by automated robotic positioning mechanisms. Multi-object fiber-fed spectrographs have been constructed that can obtain spectra for several hundred astronomical targets simultaneously. This efficiency more than compensates for the loss of telescope time used in reconfiguring the fibers between observations of successive clusters of galaxies.

A variation on this idea is an integral field unit, a tightly packed array of fibers that pick up the entire image of a galaxy and fan it out through a fish-tail of fibers that are aligned along the slit of a conventional spectrograph. Such a spectrum contains information about the motions, physical conditions and compositions of the entire galaxy at one hit.

The diameter of the collimated beam is 5–20 cm in a typical astronomical spectrograph. This size fills the diffraction grating, and uses it to best advantage; that is, best resolution. A larger diameter can be achieved by increasing the focal length of the collimator. This results in even higher resolution. But a larger grating is expensive or even impossible to make, and the camera lens to image the larger beam is bigger and of longer focal length. In such a case the camera is not only expensive but may flex as the telescope tracks and produce blurred pictures. This is why large spectrographs are fitted at the Coude Focus of a telescope (*see* COUDE TELESCOPE) or on the platform of a NASMYTH TELESCOPE.

Two key factors that characterize the performance of a spectrograph are dispersion and spectral resolution. A dispersion of 1 nm mm^{-1} corresponds to a change in wavelength of 1 nm over 1 mm of the imaged spectrum's length. Spectral resolution (R) is a measure of the instrument's ability to reveal fine detail in the spectrum and is defined by $R = \lambda/\Delta\lambda$, where λ denotes the wavelength of interest and $\Delta\lambda$ the smallest wavelength interval that can be distinguished within the spectrum. For example, at a wavelength of 500 nm, a spectral resolution of 500 would correspond to being able to distinguish two idealized spectral lines that were separated in wavelength by 1 nm. The dispersions of astronomical spectrographs range from about 100 to about 0.01 nm mm^{-1} and their resolutions from about 10 to about 100 000.

At radio wavelengths, spectroscopy with radio interferometers is carried out with correlators (*see* RADIO TELESCOPES AND THEIR INSTRUMENTS). Spectrographs on space satellites, high-flying aircraft and balloons have opened up new spectral windows. The desire to carry out spectroscopy at ever higher resolution on ever fainter objects drives the astronomers' passion for larger and larger telescopes, like the XMM-NEWTON OBSERVATORY X-ray satellite and the JAMES WEBB SPACE TELESCOPE.

Spectroscopic binary star *See* BINARY STAR

Spectroscopic parallax

If the spectral type of a star of apparent magnitude m is known (*see* STELLAR SPECTRUM: CLASSIFICATION; MAGNITUDE AND PHOTOMETRY)), including its luminosity class, then its absolute magnitude, M, is known (to limited accuracy). The star's distance in parsecs, d, is then calculated from the formula $m - M = 5 \log d - 5$ (*see* DISTANCE MODULUS). The derived PARALLAX, equal to the reciprocal of the distance in parsecs, is called the spectroscopic parallax.

Spectroscopy

A SPECTRUM is the wavelength or distribution of RADIATION emitted by any radiant source and spectroscopy is the study of spectra, with a SPECTROSCOPE AND SPECTROGRAPH. When the precision of the record is high the process is called spectrometry and the instrument is a spectrometer or spectrophotometer.

Radiation from astronomical sources spans the full spectrum from low-energy longwave radio emissions to highly energetic hard X-rays and gamma-rays, and spectrographs of one sort or another are used in each spectral window, not only the optical range. With the invention of spectroscopy during the development of ASTROPHYSICS, it became possible to determine the nature of the stars, and their velocities through space and orbital motions around previously unseen companion stars and planets. Spectroscopy distinguished the two primary types of nebulae (*see* NEBULAE AND INTERSTELLAR MATTER) – rarefied gases such as the Orion nebula, and unresolved stellar systems such as the Andromeda galaxy, and led to the realization that the universe was filled with galaxies extending far beyond the Milky Way. That view was further reinforced by Edwin HUBBLE and Milton HUMASON in 1929–30, when they discovered the expansion of the universe through spectroscopy of extragalactic nebulae.

The first concept of spectroscopy can be traced to Isaac NEWTON, who first noted that the dispersion of sunlight into its different colors by a prism showed the properties of sunlight itself. In the late 1800s, Johann Balmer (1825–98) recognized the HYDROGEN series in the absorption lines of the solar spectrum first seen by Joseph FRAUNHOFER. This hint from the heavens, that matter interacts with light in a very well-behaved manner, motivated the development of atomic physics. Each atom has a distinctive set of electronic levels, transitions between which emit light of an unambiguous signature for identification. Each transition has a particular energy and the light emitted or absorbed during the transition has a very precisely defined wavelength at a 'spectral line.' These make it possible to measure any shift in the rest wavelength due to the motion of the emitting source.

The various energy levels within an atom are populated

◄ ***Spectrographs*** *Two similar spectrographs for the VLT are shown here on the telescopes Kueyen (foreground) and Antu. Antu is visible through the open ventilation doors of Kueyen that keep Kueyen's temperature close to that of the night-time air, to mitigate any bad seeing caused by warm air trapped in the dome.*

according to the ambient physical conditions such as temperature and density. The relative strengths of the various atomic lines are therefore a means to identify the different kinds of stars (*see* STELLAR SPECTRUM: CLASSIFICATION) and the abundances of the elements in them.

The mere appearance of a spectrum can hint what astronomical object is creating it:

- emission lines: a heated rarefied gas (a nebula);
- emission lines from highly ionized elements (a gas ionized by energetic radiation, such as strongly ultraviolet light; for example, a PLANETARY NEBULA);
- wide emission lines: a gas rotating or expanding quickly, such as gas in a QUASAR, Seyfert Galaxy (*see* ACTIVE GALAXY), supernova or WOLF–RAYET STAR;
- continuous spectrum: a hot, dense surface (a star);
- continuous spectrum with more blue than red: a very hot star (for example a WHITE DWARF) or electrons spiraling in magnetic fields (SYNCHROTRON RADIATION);
- continuous spectrum with more red than blue: a very cool star (such as a red dwarf or red supergiant star) or a star reddened by interstellar material;
- an absorption spectrum: a surface overlaid by an absorbing gas (a star with an atmosphere);
- an absorption spectrum with wide spectral lines (a galaxy of many stars in motion, or a rapidly rotating star);
- an absorption spectrum with emission lines (a SPIRAL GALAXY or a BINARY STAR with gas lying between or around the stars);
- a red absorption spectrum with wide gaps: a cool star with an atmosphere containing molecules (for example, a red dwarf or red supergiant);
- doubled absorption spectrum: two stars; for example, a spectroscopic binary star.

There are a number of atomic lines that are of particular interest for astronomers. These include the Balmer lines in the optical and the 21-cm atomic hydrogen line, which are important in astronomy because hydrogen is by far the most abundant element in the universe. Radio spectral lines of interstellar molecules and the spectral lines of the commoner elements like carbon in the infrared are useful for studying hidden sources deeply embedded in dust.

Spectrum

In general terms, a spectrum (plural: spectra) is the distribution of intensity of ELECTROMAGNETIC RADIATION with wavelength. In the context of visible light, the spectrum is the band of rainbow colors produced when white light is passed through a dispersing element (*see* SPECTROSCOPE AND SPECTROGRAPH), which spreads light out according to wavelength. Such a spectrum is a continuous spectrum, or continuum, and is emitted by a hot solid body or hot gas under high pressure (in this case the Sun). A continuum is also emitted by the FREE-FREE RADIATION and SYNCHROTRON processes, but the form of the spectrum is different in each case.

A line spectrum is emitted by a gas under low pressure. In the simplest atom, HYDROGEN, there are a number of series of spectral lines because of the possible transitions between energy levels in the atom. For example, all the possible transitions down to the lowest energy level give rise to the LYMAN SERIES of emission lines at ultraviolet wavelengths, and all the possible transitions down to the second energy level (the first excited level) give rise to the Balmer series of lines in the visible spectrum (*see* HYDROGEN SPECTRUM). These lines are characteristic of hydrogen and hydrogen

Amateur spectroscopy

Amateur spectroscopy has seen a revival as of late. Pioneer spectroscopists in the nineteenth century like Angelo SECCHI and William HUGGINS observed spectra visually via large refractors under pristine skies. While the eye can see the brilliant solar spectrum and the spectra of a few brighter stars aided by a telescope, even low spectral DISPERSION greatly dilutes starlight. Photographic film can improve stellar penetration, but the supremely sensitive CCD (*see* CCDS AND OTHER DETECTORS) can snap spectra of the brightest stars in a fraction of a second, and down to magnitude 15 in a few minutes of exposure with a telescope in the 8- to 16-in aperture range, bringing the powerful tool of spectroscopy into the reach of amateurs, even on faint stars.

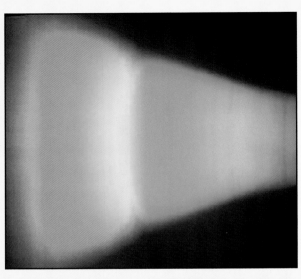

► ***Solar spectrum*** *Sunlight reflected off the surface of a compact disk is dispersed into a spectrum by the finely engraved lines on the disk.*

CD spectroscope

Most are familiar with the gorgeous blend of colors that emanate from a compact disk (CD) where the finely engraved lines of data act as the reflective DIFFRACTION GRATING, breaking up light into its constituent colors or wavelengths. Less appreciated is that the solar spectrum, complete with many dark absorption lines like sodium (D1 and D2 in yellow) and magnesium (B1, B2 and B3 in green), is revealed by the same CD when brought close to the eye with sunlight skimming its surface. This is best done in a darkened room with sunlight streaming onto the CD through a small aperture. (*Warning*: ensure that direct or reflected sunlight never enters the eye – only the colored spectrum!) The CD spectroscope (*see* SPECTROSCOPE AND SPECTROGRAPH) will also split the various emission lines in domestic fluorescent lamps and street lighting.

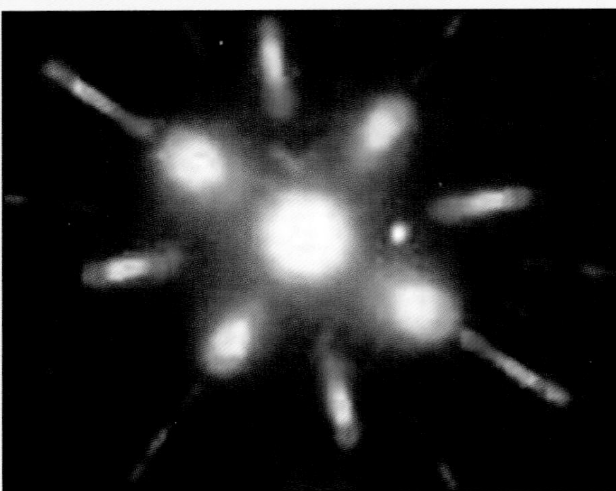

► ***Moon*** *White light is made up many colors as shown by this photo of the Moon, taken with a small transmission grating placed in front of the camera lens.*

The spectrum can be captured with a regular camera taking the eye's position above the CD, with the lens focused at infinity (*see* FOCUS), and a small lens aperture of *f*/16 to keep the spectrum lines sharp. The camera must be precisely aimed over the CD for success, and a digital camera with LCD or SLR-film camera will produce consistent results. The CD is really a party-piece and additional optics or a better grating (or prism) are need for serious work, but this need not be expensive.

Stellar spectrograph

Recording the night sky via a film or digital camera from a fixed tripod is very satisfying. Adding a prism before the lens in a lens hood converts the camera into an objective lens spectrograph, recording spectra of all the stars in the FIELD OF VIEW. However, the deviation in the light path due to the prism causes the recorded view to be offset, and this must be allowed for while pointing the camera.

▼ ***Spectra*** *of Nova Sgr 2002 with gamma Cas for comparison.*

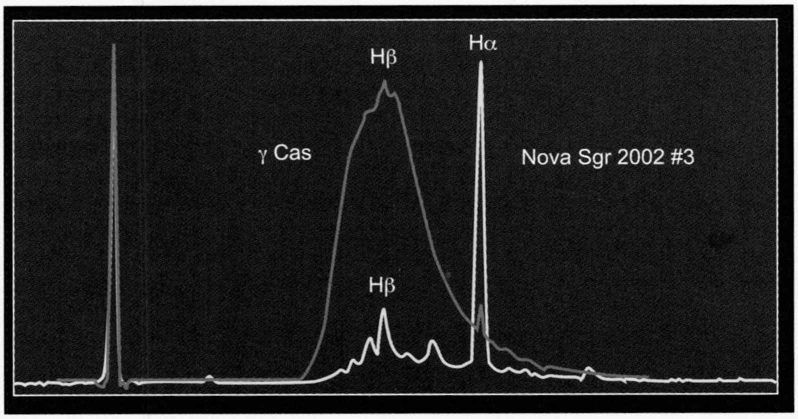

It is worth experimenting with a 45° prism from damaged binoculars. A better option is a 30° prism large enough to avoid vignetting (cropping) the lens at full aperture. The tiny spectrum of the brighter stars will be recorded, and resolution (the length of the spectrum) will be increased via a telephoto lens.

Stars begin to show a trail (through diurnal motion) after about 20 s of exposure with a standard 50-mm lens. This can be used to advantage by rotating the prism so dispersion is at a right angle to trailing (parallel to lines of declination in the sky). This broadens the spectrum into rectangles, making it easier to see the absorption lines crossing the spectrum. For maximum stellar penetration, the spectrograph should be piggybacked on an equatorially-driven mount with the longest exposure possible before skyfog spoils the results.

Building a solar or stellar spectrograph is relatively straightforward for those with wood- or metal-working skills to accurately support the optical components. There are many possible sources for the optical components, and the smaller industrial spectrographs can be adapted for amateur use. Due to the relatively small amateur demand, commercial astronomical spectroscopes are few in number, and tend to be expensive. The following are suitable for amateur telescopes greater than 15-cm aperture and less than 10-cm aperture for solar work.

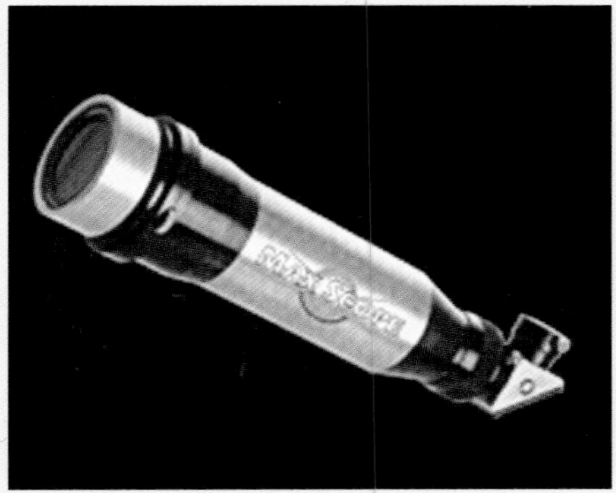

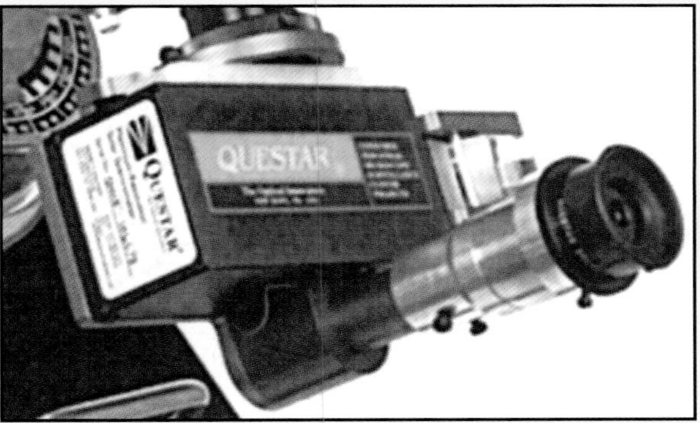

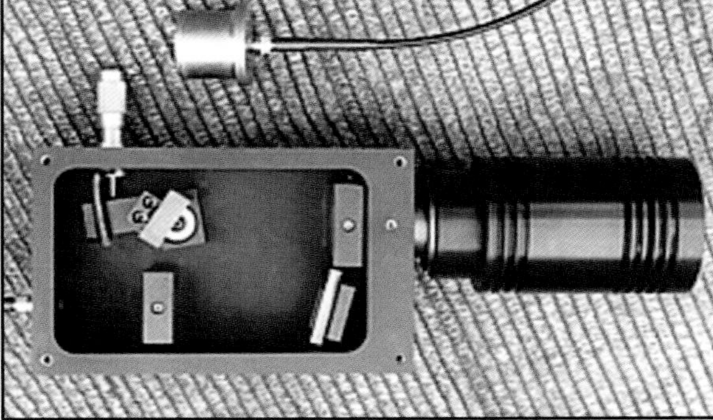

◀ ***Useful equipment*** *for amateur spectroscopy: (clockwise from far left) the Coronado SolarMax system, the Rainbow Optics Star spectrograph, the SBIG SGS spectrometer, the Sivo spectrometer, and the Questar Qmax spectrometer.*

Equipment

The SBIG SGS spectrograph attaches direct to a SCT backplate and is fully integrated around the ST–7/8 CCDs with a second guidance CCD to lock-on to the star for the duration of the exposure. It has two gratings (600 and 150 lines/mm) on a turntable and interchangeable entrance slits to swap resolution for bright and faint targets.

The Sivo spectrometer uses a fiber-optic feed to separate telescope from spectrograph so that no load is imposed onto the telescope. Most CCDs can be coupled. The fibers form a virtual slit at the spectrograph entrance, which has a single 1200 lines/mm grating for high dispersion and, as such, is best suited to bright targets.

The Rainbow Optics Star spectroscope comprises an efficient blazed transmission grating (200 lines/mm) in a standard 30-mm EYEPIECE-threaded mount. It screws directly into an eyepiece for visual use and a clip-on cylindrical lens broadens the spectrum. Attached before a photo-film or CCD detector (without cylindrical lens) in the convergent beam of the telescope, it forms an efficient, slitless, non-objective spectrograph of very low resolution, sufficient to classify stars, identify novae or supernovae by individual emission lines, or even study QUASARS.

Questar Qmax spectrometers couple to the Questar range of Maksutov telescopes for visual use at high resolution, and this is adequate to view the Zeeman splitting of sunspot lines (*see* ZEEMAN EFFECT). The special design renders the normally faint extremities of the spectrum (like the H and K lines in violet and A–C lines in deep red) readily visible to the eye.

Coronado SolarMaxes use a special INTERFERENCE filter before compact refractors that permit direct observation of the Sun's CHROMOSPHERE in the red light of hydrogen-alpha (Hα).

Such observations are also possible with a SPECTROHELIOSCOPE – a high-resolution spectroscope that scans the solar disk and is fully tunable to other lines in the Sun's spectrum like helium (D3 line), hydrogen-beta (Hβ) and the H and K lines of calcium.

Amateur projects

So much for the practicalities. There has been a jump in the quality of results produced by amateurs to match that in the equipment they are using. The most obvious project is to use stellar spectra to classify stars according to their spectral types (*see* STELLAR SPECTRUM: CLASSIFICATION), and this is done by identifying specific sets of absorption lines and comparing them with readily available comparison spectra. Another interesting project is to look for the presence of metallic elements (bearing in mind that in astronomical parlance a metal is anything except hydrogen and helium) in the atmospheres of the GIANT PLANETS. A particularly strong signal is detected from methane, but many other features can be identified with care. Not only is this interesting in itself, but it provides practice for observing similar features in the atmospheres of some unusual stars.

Deep-sky objects are also beginning to produce really good results, with PLANETARY NEBULAE being the most common targets. Measurement of intensity of spectral lines for these objects, together with a basic theoretical analysis, allows a temperature to be derived. At the most advanced end of the range of possible amateur projects lie the attempts being made to identify extrasolar planets through RADIAL VELOCITY measurements. At the time of writing, successful detection of one system, that around Tau Böotis, has been announced by an amateur group, the Spectrashift Team. Further development will be necessary, but we may well reach a point in the next couple of years when an amateur discovery of a planet around another star hits the headlines, something that would have been unthinkable just 10 years ago.

Maurice Gavin maintains a website that is devoted to amateur spectroscopy and contains links to other practitioners.

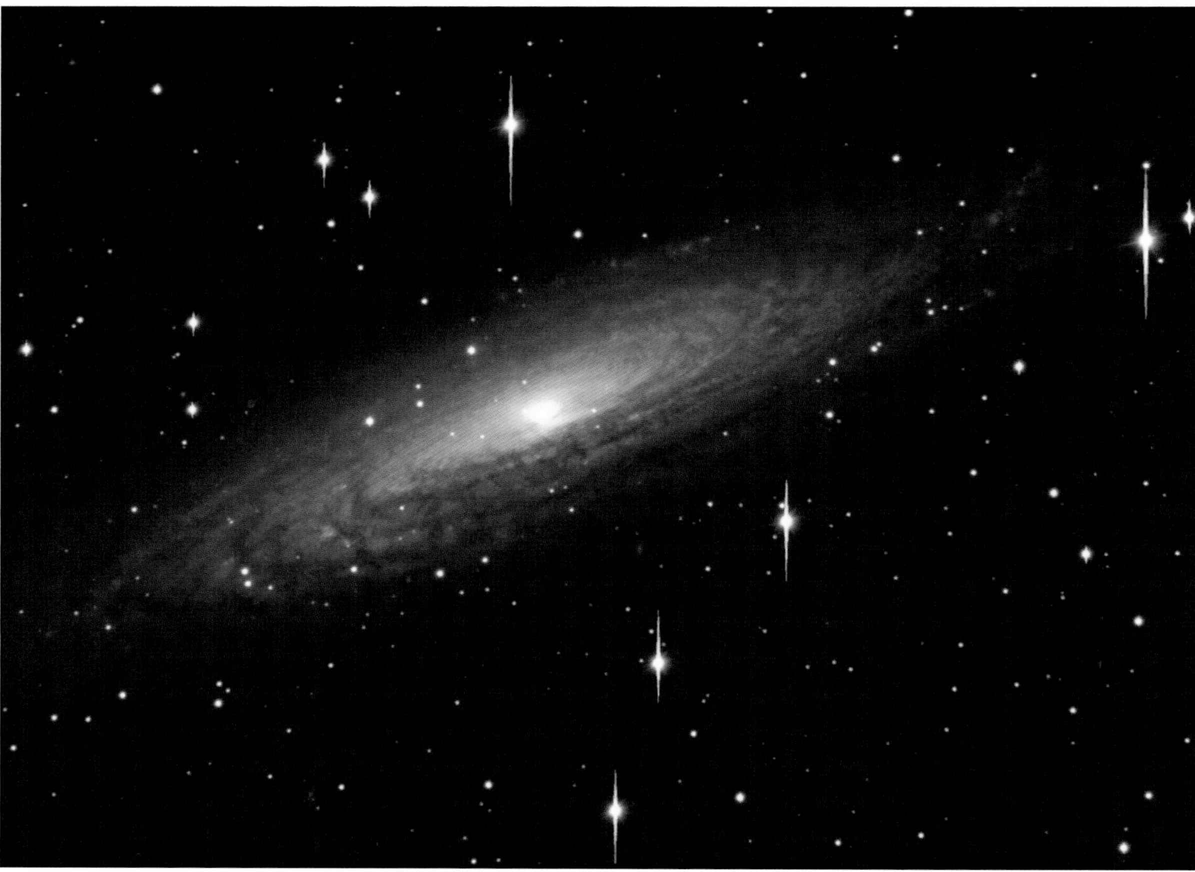

▶ ***Spiral galaxy*** *The tightly wound spiral arms of NGC 2613 resemble those of our own Milky Way Galaxy. NGC 2613 is seen obliquely, with the dark dust lanes of the nearer side silhouetted against the massed stars of its central regions. The galaxy is viewed through the individual stars of our own Galaxy, some of which were bright enough and made so much charge on the CCD detector of the Melipal telescope of the VLT that charge has leaked and smeared top to bottom over adjacent elements of the CCD.*

alone. Heavier elements have more complex spectra, and the spectra of molecules are complicated by vibrational and rotational energy states.

Dark-lined ABSORPTION SPECTRA are produced when atoms and molecules absorb radiation (from a background source) at the same wavelengths at which emission takes place by the mechanism described above. For example, a dark line of wavelength λ is produced when electrons of a given element absorb energy and make upward transitions from lower- to higher-energy orbits.

Spencer Jones, Harold (1890–1960)

English astronomer, chief assistant to Frank DYSON at the ROYAL OBSERVATORY, GREENWICH, and worked at the Royal Observatory at the Cape of Good Hope. He repeated David GILL's photographic survey of the southern sky (*Cape Photographic Catalogue*. He then organized the international project to determine the Earth-Sun distance by repeated measurements of the asteroid EROS during its close approach to the Earth in 1930–1, again repeating work on other asteroids by Gill. Spencer Jones returned to Greenwich and became ASTRONOMER ROYAL in 1933. Probably his most significant discovery was the irregularities in the Earth's rotation, which he proved by comparing observed and predicted positions of the planets. He replaced the concept of UNIVERSAL TIME, based on Earth's rotation, by ephemeris time, a system of time measurement eventually based on the frequency of oscillations in quartz then cesium atoms. Having experienced the clarity of South African skies, he was depressed by observing conditions in London, and set in train the removal of the Royal Observatory to the Sussex countryside.

Spherical aberration

The inability of a spherical lens or mirror to bring all parallel rays of light to the same focus. Rays reflecting from points close to the edge of the mirror are brought to a focus closer to the center of the mirror than rays reflected from points closer to the optical axis. A similar effect applies to rays of light passing through a simple lens, the surfaces of which have shapes that are parts of spheres.

Spherical aberration causes the image of a point source of light (such as a star) to be spread out into a disk. Between the focal positions for the central and peripheral rays there is a plane at which the smallest image, known as the circle of least confusion, occurs.

Spica

The star Alpha Virginis, a blue-white giant star, spectral type B1V, of apparent magnitude 0.98. Spica is a rotating ellipsoidal BINARY STAR, comprising two large, close components which are elongated toward each other by the strong gravitational attraction between them. Its magnitude varies very regularly between 0.95 and 1.05 over a period of 4.0146 days, as a result of the changing combined surface area of the component stars presented toward us. The primary component is also a regular pulsating VARIABLE STAR with a period of 0.17 days.

Spicules *See* SUN

Spiral arm

Lanes of gas, dust and hot young stars that spiral outwards from the central bulges of SPIRAL GALAXIES. The most prominent features of spiral arms are clusters and associations of O- and B-type stars (highly luminous, high-temperature stars) and H II REGIONS (luminous nebulae), both of which are indicative of recent star formation. At radio wavelengths spiral arms may be identified by measuring the distribution of neutral hydrogen clouds which radiate at a wavelength of 21 cm, and molecular clouds. Infrared observations reveal clouds of warm dust heated by new stars along spiral arms. Our Milky Way Galaxy's spiral pattern consists of two or more major arms and a number of shorter segments, one of which, the ORION ARM or 'spur,' contains the Sun. The Sagittarius arm (in the general direction of the constellation Sagittarius) lies closer to the galactic center than the Orion arm and the Perseus arm lies further out.

Spiral galaxy

A galaxy in which a central bulge of stars is surrounded by a flattened disk which contains a spiral pattern of hot young stars and luminous nebulae. Within the HUBBLE TUNING-FORK DIAGRAM system spiral galaxies denoted by 'S' are categorized by the size of the central bulge, the tightness of the spiral pattern and the patchiness in the spiral arms. The MILKY WAY GALAXY is a spiral galaxy.

Spitzer, Lyman, Jr (1914–97)

American astrophysicist, professor at the Princeton University faculty in New Jersey, where he founded the Princeton Plasma Physics Laboratory (PPPL). He researched spectral-line formation, stellar motions, star formation and the physics of the INTERSTELLAR MEDIUM. He showed that there must be at least two phases of material in the interstellar medium – high-temperature clouds around hot stars and cooler inter-cloud regions. He led the project to make and operate the ultraviolet astronomy satellite, COPERNICUS (OAO-3).

Spitzer Space Telescope

See SPACE INFRARED TELESCOPE FACILITY

Sporadic meteor

A METEOR that does not belong to an identifiable METEOR SHOWER. The METEOROIDS that produce sporadic meteors are part of the zodiacal dust cloud, which pervades the inner solar system, and are swept up by the Earth as it proceeds in its orbit. On average five sporadic meteors are visible from any given location on any clear night but the number varies with time of night and time of year. The seasonal variation has to do with the altitude of the Earth's APEX. The higher the apex is in the sky, the greater is the sporadic hourly rate.

Square of Pegasus

A square ASTERISM drawn through the bright stars Alpha, Beta and Gamma Peg together with the bright star Alpha And, which is a striking feature of the northern sky from late summer to early winter. The line joining the stars on the left of the square (Gamma Peg and Alpha And) with the bright stars Beta Cas and Polaris is approximately the line of zero (0) h right ascension.

SS433

A complex binary system within a supernova remnant (W50) in the constellation Aquila, which has proved to be of great astrophysical interest. It was discovered by the X-ray satellite ARIEL 5 in 1976. The unusual name derives from its inclusion in a catalog of emission-line stars published by C. Bonce Stephenson (1929–2001) and Nicholas Sanduleak (1933–1990) in 1977. As it is a VARIABLE STAR system it is also designated V1343 Aquilae.

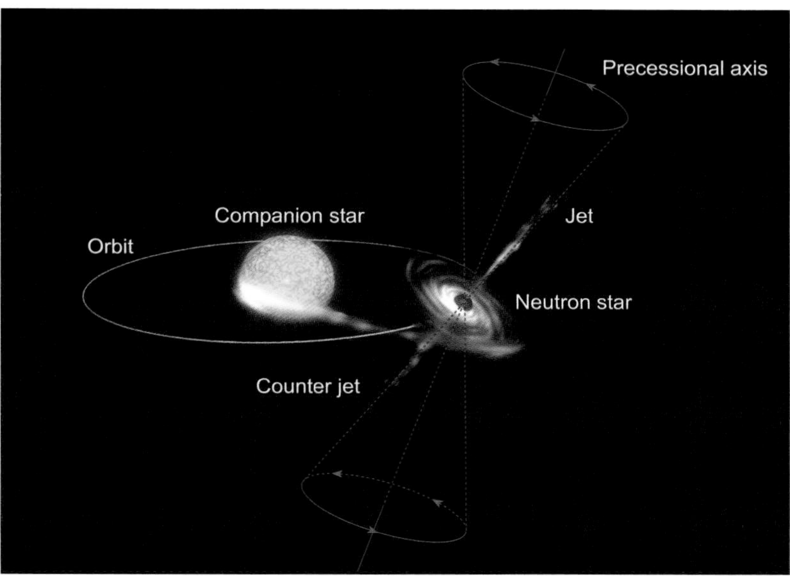

▲ ***SS433***
The neutron star feeds on its companion. Some of the matter fuels powerful jets which sweep the heavens like a rotating water-sprinkler.

In addition to the X-ray observations, SS433 has been studied at optical, radio and gamma-ray wavelengths. In its optical spectrum, in addition to a background continuum, there are strong emission lines due to hydrogen and helium. These lines have both a normal, broad, 'stationary' component, and also components that show Doppler red- and blueshifts alternately over a period of 164 days (*see* DOPPLER EFFECT). The stationary emission lines exhibit a period of 13.1 days.

This is a binary system, of orbital period 13.1 days, comprising a hot, massive star (estimated at 10–20 solar masses) and a neutron star (presumably the W50 supernova remnant) of perhaps 1.5–3 solar masses. A stellar wind from the massive star transfers matter through the LAGRANGIAN POINT, feeding an ACCRETION disk surrounding the neutron star. Not all of this matter reaches the surface of the neutron star, however, some being ejected at enormously high velocity (of the order of 80 000 km s^{-1}) via two opposed, narrow jets which PRECESS in a period of 164 days, sweeping out two conical volumes of space of about 40°.

The massive star, and possibly the accretion disk, contribute the background continuum. The stellar wind produces the broad component of the stationary emission lines, and the mass-transfer from the massive star to the accretion disk produces the narrow emission lines. The changing aspect of the accretion disk through the 164-day period explains the variation in total light from the system, and its periodic partial eclipses account for the lesser 13.1-day variation. The precessional motion of the jets explains the moving emission lines with their observed Doppler effects.

SS Cygni

SS Cygni (SS Cyg) is probably the best studied of the DWARF NOVA (DN) or U GEM subclass of cataclysmic variables (*see* VARIABLE STAR: CLOSE BINARY STAR). Like all cataclysmic variables SS Cyg is a BINARY STAR system consisting of a low-mass red dwarf star (the secondary; *see* RED DWARF AND FLARE STAR) and a WHITE DWARF star (the primary). The white dwarf, which was originally a massive MAIN SEQUENCE star of greater than 6.5 solar mass, now has a radius about equal to that of the Earth and a mass about 1.1 times solar mass. The secondary star is a main sequence star of spectral type K5 V and a mass of 0.7 times solar mass. The two stars are separated by about the diameter of the Sun and they orbit their common center of mass with a period of 6.6 hours.

Because of their proximity and the gravitational influence of the primary, the secondary star completely fills its ROCHE LOBE, causing matter to fall toward the primary forming an accretion disk around the white dwarf. Instabilities develop in the disk, causing semiperiodic outbursts, during which time the emitted light increases greatly – the signature of a DN-type of cataclysmic variable. SS Cyg has outbursts about every 50 days, which brighten the system from its normal quiescent level at magnitude 11.7 to a maximum level near 8.2, a factor of 25 increase in brightness.

SS Cyg was discovered in 1896 by Louisa Wells while she was examining Harvard College Observatory photographic plates. During the 1897 objective prism survey for the HENRY DRAPER CATALOG, a spectrum of SS Cyg was obtained. However, it was not until 1943 that a detailed study showed that SS Cyg is a spectroscopic binary.

Standard candle

Originally literally a candle made in a standard way, to be used as a reference source for PHOTOMETRY in order to measure candle-power. Now used figuratively by astronomers to refer to stars whose brightness is a known reference point and which can be used to determine distance; for example, CEPHEID VARIABLE STARS. 'Standard bomb' is a humorous extension of this terminology that is used to refer to a supernova of Type Ia, whose brightness at maximum light is a standard candle.

Star

Stars are self-luminous, self-supporting bodies in space such as the SUN, initially made mostly of hydrogen. Their masses are in the range that lies above planets and brown dwarfs, upwards from about 0.08 times the mass of the Sun, sufficiently massive to be able to burn hydrogen.

Stars are self-supporting because the inward gravitational pull of their material is balanced by internal pressure from the hot gas within (the 'hydrostatic equilibrium'). The pressure, and therefore density and temperature, of the material of the star increases toward its center. Over the whole of a star, except perhaps its atmosphere, the temperature is high enough that its atoms are ionized (*see* EXCITATION AND IONIZATION). In a central region of the star called the core, the densities, pressures and temperatures are so high that atoms are completely ionized.

Stars are self-luminous because, in their cores, individual nuclei collide and take part in nuclear reactions that release energy. The energy in the hot core flows to the cooler surface and radiates into space as light and heat. The pressure of the radiation flowing out through the material of the star also supports it against the force of gravity.

The hydrostatic equilibrium of most stars is, for long periods, stable: if the star heats up, the flow of energy increases, and the star expands, cools and stabilizes; if the star cools, the flow of energy decreases, and the star contracts, heats and stabilizes. For very massive stars, however, the rate of production of energy by nuclear reactions when the star first forms is so high that radiation pressure disrupts the outer parts of the star and it sheds mass, decreasing the energy flow and eventually finding stability. Stars in this unstable state show as LUMINOUS BLUE VARIABLE STARS. The maximum mass of a star is about 150 solar masses.

Nuclear burning

In MAIN SEQUENCE stars the source of thermonuclear energy is the fusion of light hydrogen nuclei into more massive helium nuclei, $4 \times {}^{1}H$ into $1 \times {}^{4}He$. The ^{4}He nucleus has a mass 3.97 times that of the proton, so mass equal to 0.03 times the mass of the proton has been converted into energy and liberated (*see* ELEMENTS, FORMATION OF).

Fusion of hydrogen to helium occurs through two channels, the proton-proton (PP) chain and the carbon-nitrogen-oxygen (CNO) cycle. The PP chain in its simplest version is a sequence of reactions in which two hydrogen nuclei make deuterium, and a third hydrogen nucleus is added to make ^{3}He. The first two reactions must occur twice in order to produce two ^{3}He nuclei, which fuse into one nucleus of ^{4}He, liberating two protons. In a star like the Sun, every second 600 million tonnes of hydrogen are converted into helium and 4 million tonnes of matter are transformed into energy.

In the CNO cycle, the same net process occurs through the operation of a cycle of intermediate stages involving carbon, nitrogen and oxygen nuclei. The CNO cycle happens at significantly higher temperatures than the PP chain and occurs in main sequence stars more massive than about 1.5 times the mass of the Sun.

When hydrogen fuel is exhausted the next step in nuclear fusion is the fusion in successive stages of three ^{4}He to ^{12}C (the triple-alpha process), which takes place in stars more than about 0.5 times the mass of the sun. At this stage the stars adjust themselves with a more contracted core, higher central temperatures and a larger size – they show as RED GIANT STARS. Following core helium exhaustion, stars exceeding about 10 solar masses ignite carbon and show as hotter giants and as supergiants. This progression of nuclear fusion sequence is terminated with silicon-burning and the formation of the most stable nuclei of iron.

While all this is happening, the star is adjusting its size and may be losing material into space like a SOLAR WIND, or, if it is a BINARY STAR, onto a nearby companion. Fusion may be turned off and the star may evolve to a PLANETARY NEBULA and WHITE DWARF. If a star is over about 6 solar masses, it will reach the stage at which its nuclear fuel is exhausted, so that the flow of energy and radiation dies away and can no longer support the star against gravity. Then it can no longer sustain itself in hydrostatic equilibrium and it collapses as a supernova of Type II (*see* SUPERNOVAE AND INTERSTELLAR MATTER).

Star formation

The formation of stars is a process that continues to the present day. Stars form from the gravitational collapse of the cold molecular interstellar material (gas and dust) that resides in structures known as interstellar molecular clouds. After a long search of these nearly opaque nebulae, the almost spherical infall process was finally identified in the late 1990s through radio studies in molecular lines of very young 'class 0' millimeter-wave sources. The first to be identified was B335, a Bok GLOBULE (*see* Bart BOK) with a new infrared-emitting star inside, only about 150 000 years old (*see* DARK NEBULA).

During the collapse and ACCRETION of infalling matter, a circumstellar disk forms around the accreting star, picking

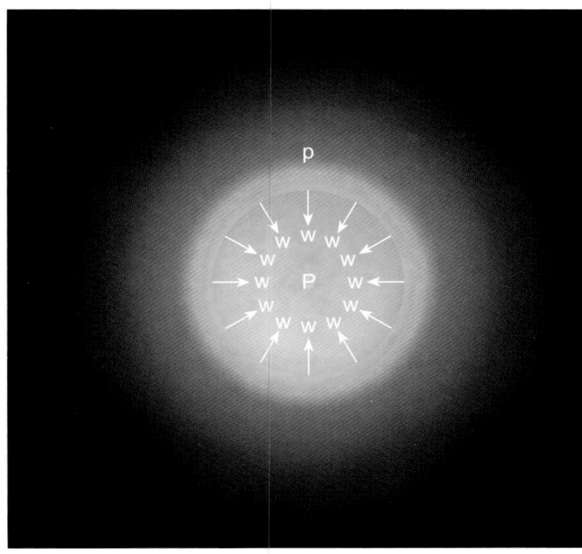

◀ ***Star***
The pressure at the center (P) of the star is higher than near the surface (p). The excess pressure of the interior supports the weight (w) of the layers of the star above.

up and amplifying the slow rotation of the molecular cloud. In many cases, two (or more) stars may form together as a DOUBLE STAR or MULTIPLE STAR. Other processes cause material to outflow as winds and JETS, which slows the rotation of the star and allows it to grow in mass ('class I sources'). Several such sources are seen in the nebula NGC 2024, the FLAME NEBULA.

As 1 million years pass, a thick disk forms ('class II sources' or proplyds, as found in the ORION NEBULA by the HST) and its matter is a reservoir for the formation of stellar and planetary companions (*see* PLANETS: ORIGIN). The new star starts to be seen as an optical star, like T TAURI STARS and Herbig–Haro objects (*see* George HERBIG). Eventually the disk thins as the dust sticks together in larger lumps and the star becomes fully visible ('class III sources'), like BETA PICTORIS.

From this collapse process, low-mass stars are more commonly generated than high-mass ones. The actual distribution of masses of the stars is called the IMF, first described by Edwin SALPETER.

Stellar atmospheres

The atmosphere of a star is the boundary between the stellar interior and the interstellar medium. It spans the layers from which photons can escape to the surrounding space. The atmosphere might be quite extended in space. However, the bulk of the radiation is emitted in a thin layer: the PHOTOSPHERE. The solar photosphere is about 600 km thick. Hotter stars have thicker photospheres, but in general the extent of the photosphere corresponds to about 1/1000 of the stellar radius.

The structure above the photosphere differs markedly between cool and hot stars. Cool stars (spectral type A7 and later on the main sequence) are characterized by increasing temperature in their outer layers, with a chromosphere (temperatures ranging from 7000 to 20 000 K), and above the chromosphere a hot, million-degree corona. Strong, fast winds typify the atmospheres of hotter main sequence stars (O- and early B-type). These atmospheres produce absorption lines (and sometimes emission lines) in the spectrum of the stars, which define their spectral type (*see* STELLAR SPECTRUM: CLASSIFICATION).

The most important parameter that describes the emergent spectrum is the temperature of the photosphere of the star. The density of the absorbing and emitting material above the photosphere also characterizes the spectrum. The density is set by the strength of gravity in the photosphere. Astronomers use surface gravity as the second fundamental parameter to describe the atmospheres of stars and their spectral types (*see* HERTZSPRUNG–RUSSELL DIAGRAM).

Further reading: Tayler R J (1994) *The Stars, Their Structure and Evolution* Cambridge University Press.

Starburst galaxy

A starburst galaxy is one undergoing a brief episode of intense star formation, usually in its central region. The massive stars in the burst generate most of the total luminosity of the entire galaxy. Starburst galaxies are fascinating objects in their own right and are the sites where roughly 25% of all the massive stars in the local universe are being formed. They offer unique laboratories for the study of the formation and evolution of massive stars, the effects of massive stars on the interstellar medium and the physical processes that were important in building galaxies and chemically enriching the intergalactic medium.

Star cluster *See* OPEN CLUSTER; GLOBULAR CLUSTER

Star diagonal

A device that fits in the draw-tube of a telescope to enable an object to be viewed at right angles to the direction in which the telescope is pointing. In such a device, light is reflected through a right angle by a flat mirror inclined at 45° to the optical axis or by total internal reflection in a prism. The resulting image is reversed left to right. It is often convenient to use a star diagonal when observing objects close to the ZENITH so the observer can avoid placing their head at an awkward angle.

A related device, the solar diagonal, uses an unsilvered mirror, or a thin prism called a Herschel wedge, to reflect only a little of the incoming solar radiation (typically about 5%), and so to reduce the amount of light and heat that reaches the eyepiece.

Stardust

Fourth mission in NASA's Discovery program, launched in February 1999. Stardust is the first US mission to a COMET and the first mission to bring material back from outside the Earth-Moon system. It arrived at its destination, Comet Wild-2, on January 2, 2004. It captured dust particles from the comet in aerogel and will return them to Earth. The sample return capsule is due to parachute onto the US military's Utah Test and Training Range on January 15, 2006.

Star of Bethlehem

In the story of the birth of Jesus Christ in the Gospel of Saint Matthew, the wise men follow a bright star which leads them to the birthplace. An attempted explanation of this 'star' is complicated by the fact that the actual date of birth is not certain. Most scholars believe it occurred between 8 and 4 BC. Possible explanations include a conjunction of Jupiter and Saturn in 7 BC, or of Jupiter, Saturn, and Mars in 6 BC, or of Jupiter and Venus in 3 BC. Other possibilities are appearances of COMETS in 5 BC and 4 BC, or a NOVA in 5 BC. Celebrations of Christ's birth date back to the third century and the date of December 25 follows the pre-Christian celebrations of the winter SOLSTICE in the north, when the days started to lengthen again.

Steady-state theory

A theory that postulates that the large-scale appearance of the universe does not change with time. The theory, devised in 1948 by Hermann Bondi (1919–), Thomas GOLD and Fred HOYLE, satisfies the 'perfect COSMOLOGICAL PRINCIPLE' that the universe is the same everywhere at all times. This theory avoided the necessity for a BIG BANG origin to the universe, but required the continuous creation of matter to maintain a constant uniform density in the universe at all times (where new galaxies formed as the older ones moved apart, maintaining the same average number of galaxies in each large volume of space). The steady-state theory is now of historic interest only. *See also* UNIVERSE: COSMOLOGICAL THEORY.

Stellar distribution

Our MILKY WAY GALAXY is a collection of stars moving in orbits through the Galaxy's gravitational field. A 'stellar distribution' is a description of the distribution (in real space and in velocity space) of the stars in our Galaxy. Different kinds of stars have different distributions. A knowledge of this, obtained from counts of stars of many kinds, leads to an understanding of the structure and dynamics of groups of stars in general and to a description of our Milky Way Galaxy in particular. It also leads to interesting conclusions about the origin and history of our Galaxy.

Stellar dynamics

Stellar dynamics describes the internal motions of star systems when considered as an aggregate of many point-mass particles whose mutual gravitational interactions determine their orbits. These particles may be taken to represent objects in small galaxy clusters with about 10^2–10^3 members, or in larger GLOBULAR CLUSTERS with 10^4–10^6 members, or in galactic nuclei with up to about 10^9 members, or in galaxies containing as many as 10^{12} stars. Under certain conditions, stellar dynamics can also describe the motions of galaxies in clusters, and even the general clustering of galaxies throughout the universe itself. This last case is known as the cosmological many-body problem.

The essential physical feature of all these examples is that each particle (whether it represents a star or an entire galaxy) contributes importantly to the overall gravitational field. In this way, the subject differs from celestial mechanics where the gravitational force of a massive planet or star dominates its satellite orbits. Stellar dynamical orbits are generally much more irregular and chaotic than the orbits of celestial mechanical systems.

Stellar evolution

See HERTZSPRUNG–RUSSELL DIAGRAM; STAR

Stellar nomenclature

Stars are identified in various ways. Most bright stars have proper names which are universally recognized – for example, Antares, Rigel and Vega. The brightest stars in each CONSTELLATION also have Bayer or Flamsteed names. Johann BAYER, in his star atlas *Uranometria* (Star Catalog) of 1603, gave names consisting of a lower-case Greek letter followed by the genitive constellation name; thus Antares became α Scorpii, Rigel is α Orionis and Vega is α Lyrae. The Bayer letters were allocated approximately in descending order of brightness within the constellation, but Bayer was also influenced by the detail of the constellation figure as drawn in his atlas, and also some stars have changed in brightness since Bayer's time. Thus Betelgeuse (α Orionis) is in fact less bright than Rigel (β Orionis). Fainter stars are designated by the numbers allocated to almost 3000 stars by the first ASTRONOMER ROYAL, John FLAMSTEED, in his star catalog *Historia Coelestis Britannica* (British Account of the Skies), which was published posthumously in 1725. Flamsteed allocated the numbers in order of increasing right ascension in the constellation. Thus the Flamsteed numbers for the three stars mentioned above are 21 Scorpii (Antares), 19 Orionis (Rigel) and 30 Lyrae (Vega). The constellation name is often abbreviated to three letters, using the convention agreed at the inaugural General Assembly in Rome in 1922 of the IAU. Thus our three typical bright stars are conveniently designated α Sco, β Ori and α Lyr.

The huge numbers of faint stars discovered since Flamsteed's time are identified by their numbers in various catalogs such as the *Bonner Durchmusterung* (Bonn Survey) of 1859–62. The *Bright Star Catalog* uses the same numbers as the *Harvard Revised Photometry* of 1908, and its stars are known by their HR numbers. Some stars have additional designations from catalogs compiled for a specific purpose; for example, the HENRY DRAPER CATALOG or the *General Catalog of Variable Stars*. A single star may thus have many designations, depending on the catalogs in which it is listed; hence Vega also has the numbers HR 7001, BD+38 °3238 and HD 172167, and many others.

Stellar population

A group of stars born at the same time and sharing the same initial composition. A GLOBULAR CLUSTER is an example of a simple stellar population, characterized by a single age and a single abundance mixture. The MILKY WAY is an example of a complicated mixture of many stellar populations, old and young, metal-poor and metal-rich.

In the Milky Way, the GALACTIC HALO is thought to have formed in the first third of the Galaxy's history and is composed of stars and clusters on randomly oriented, eccentric orbits so that the net effect is a large halo of roughly spherical shape, called POPULATION II. On the other hand, if one finds all the stars in the Galaxy younger than 1 billion years, virtually all are in the very thin GALACTIC DISK (POPULATION I). They have orderly, near-circular orbits with all stars rotating clockwise as seen from above the Galactic north pole. Wilhelm BAADE identified the two distinct stellar populations in the Milky Way around 1944. But by the 1960s it became evident that there were several populations in the Milky Way, rather than two distinct ones, with sub-components both in the disk and in the bulge.

It is easy to see that a stellar population is a very broad collective term for a sample of stars, ideally (but not necessarily) with well-defined statistical properties of age, chemical composition and kinematics. Several populations make up a galaxy, so it ought to be possible to construct a model of a galaxy from its constituent stellar populations. In fact, Allan SANDAGE (1986) noted: 'The main thrust and the ultimate aim of the work in stellar populations today is to achieve an understanding of the formation and development of galaxies with time.'

This work is called 'population synthesis,' in which a galaxy is thought of as the sum of its stellar populations. For example, one can model an irregular galaxy like the MAGELLANIC CLOUDS as star clusters of every age, *t*, from 1 million to 10 billion years. There may be more star clusters born at a few particular epochs than at others, and the fecund periods in the star-formation history are called starbursts. Astronomers make guesses at the star-formation history and try to build up a spectrum of the whole galaxy that fits the reality. For example, a galaxy might emit most of its light in the ultraviolet. What would fit is that the galaxy has many hot, massive stars, having recently undergone a massive starburst. Another galaxy might have a minority population of hot stars, as evidenced by weak ultraviolet radiation, but many cool stars, shown by strong titanium oxide spectral features in the red part of the spectrum, characteristic of RED GIANT STARS. Such a galaxy is obviously older, an ELLIPTICAL GALAXY perhaps.

The basic behaviors of a stellar population are that:

- the population becomes dimmer with age;
- the blue light from a population becomes dimmer, the red light brighter;
- the population becomes redder with age or increasing abundance;
- metallic lines increase in strength with age or abundance while the hydrogen Balmer lines decrease in strength.

Starting in the late 1960s, American astronomer Beatrice Tinsley refined the trial-and-error technique of population synthesis to something more definite. She constructed HERTZSPRUNG–RUSSELL DIAGRAMS of star clusters of different ages, or 'isochrones' – the locus of luminosities and temperatures at one instant in time for stars of all masses. An isochrone mimics a star cluster or a single-age stellar population. She then put together a collection of stellar spectra and computed a synthesized spectrum of a population by adding all the spectra of all the stars in an isochrone. The number of stars of each type came from the IMF as determined by Edwin SALPETER and the amount of heavy elements (metals) that each population contains is progressive with time, with metals made in the older populations of stars and recycled into the next generation.

Problems in applying this technique include dust in the observed galaxy, which affects the colors of starlight. The emission lines from H II REGIONS in a galaxy fill in the stellar absorption lines and make the Balmer absorption lines from the stars hard to measure. Finally, it is hard to see an old population underlying a much brighter young population – the youngest population shouts loudest and gets far more weight than the older ones.

Globular clusters are attractive targets for population synthesis because they are very homogeneous, and can be characterized by a single age and metal abundance. The first globular cluster system to be studied was the Milky Way's. It turns out that the Galaxy's globular clusters formed at various times during the first third of its life, some of them accreted from galaxies that have merged with the Milky Way. In other galaxies there are both metal-poor and metal-rich clusters, but for some reason ours is missing the younger, metal-rich globular clusters of most other galaxies.

Even with the HST, galaxies at high redshift are little more than spots of light, and astronomers' main tool to interpret them is population synthesis. This is complicated because the Doppler shift of the expanding universe moves their spectra to longer wavelengths, affecting the colors. At large look-back times the stars seen will be younger than their local counterparts, and therefore brighter and bluer. But the results show that star formation was much more active in the past, with most stars formed by the time the universe was only a few billion years old.

However, the results for local galaxies show not only a strong clump of old-appearing ages, but also a significant trail of galaxies toward young ages. This indicates that star formation is not just something that happened in the past then stopped. Indeed, the local DWARF GALAXIES show long periods of quiescence followed by successive bursts of star formation, triggered perhaps by the close approach of another galaxy, a near collision or even a merger.

▲ ***Cassiopeia*** *From Bayer's star atlas* Uranometria *(1723 edition), showing the Greek letters of stellar nomenclature (and, left, Tycho's Supernova).*

Stellar spectrum: classification

The SPECTRUM of a STAR carries a lot of information, principally the wavelengths and intensities of many different chemical elements that give rise to spectral lines, but also other qualities such as the sharpness of the lines. To organize the large set of observational data that a collection of possibly millions of stellar spectra represents, a sophisticated classification philosophy was developed by William MORGAN at the YERKES OBSERVATORY of the University of Chicago, Illinois, USA, and given the title 'the Morgan–Keenan (MK) process.' Interestingly, the astronomical development anticipated the thinking of Bavarian philosopher Ludwig Wittgenstein (1889–1951) on the subject of the use of specimens in classification. The unknown spectrum is compared with the known spectra of well-studied standard stars (specimens), using pattern-recognition techniques involving all the features in a stellar spectrum.

Angelo SECCHI, a Jesuit astronomer working in the 1860s at the VATICAN OBSERVATORY, was the first to provide a general classification scheme for stellar spectra. He looked at thousands of stars with a visual spectroscope and small telescopes (15–25 cm in aperture). The coolest stars dominate his classification scheme, forming two of his four classes, partly because they are among the apparently brightest stars in the visual region of the spectrum, and partly because of the striking appearance of their molecular bands. In modern terms, his four classes are:

- I – stars hotter than the Sun;
- II – stars like the Sun;
- III – stars cooler than the Sun with titanium oxide absorption bands in the yellow and red regions of the spectrum;
- IV – stars cooler than the Sun with carbon molecular absorption bands in the yellow and red regions of the spectrum.

Many other systems have been proposed.

By the end of the nineteenth century, Edward Pickering and others (*see* Edward PICKERING; William PICKERING) realized the importance of photographic spectra for classifying very large numbers of stars. The HENRY DRAPER (HD) project at Harvard College Observatory, Massachusetts, USA, was sponsored by the widow of Henry Draper and carried out primarily by Annie CANNON, who by the mid 1920s had classified more than a quarter of a million stars. The HD system, which corresponds more or less to the temperature scale of modern astronomy, was a one-dimensional scheme, originally based on the strength of the hydrogen lines, and types were assigned in alphabetical order (strong to weak hydrogen lines): A, B, C, and so on. Later, it was discovered that the hydrogen lines start weak in the hottest stars (O- and B-stars), reach a maximum strength in A-type stars, and fade with lower temperatures. The sequence of letters was changed to the order used today for the one dimension of temperature: OBAFGKMSRN. Also, some letters became redundant. There are various different mnemonics devised for (and by) students to help remember the sequence, but the most common is 'Oh, Be A Fine Girl (Guy), Kiss Me Sweetly Right Now.'

Henry Draper's niece, Antonia Maury (1866–1952), also at Harvard, identified a stellar spectrum that she classified with 'c' placed in front of the temperature class. Her important insight formed the basis of a second dimension, which relates to the LUMINOSITIES of stars and shows up in spectra because the large atmospheres of luminous stars are more rarefied than those of the smaller dwarfs. The spectral lines are sharper, and some spectral lines are subtly stronger or weaker than usual. 'c' turned out to represent supergiant stars.

In the Harvard system, the SRN types are in parentheses because these stars have temperatures that parallel the GKM stars, the spectra differing in chemical composition rather than temperature. Composition is a third dimension in the scheme.

The MK system

At the University of Chicago's Yerkes Observatory in 1943, Morgan, Philip Keenan (1908–2000) and Edith Kellman proposed a system of classification of stellar spectra known as the MKK system, which was developed into the MK system by Morgan and Keenan in 1953. The MK system works in this way:

- The Harvard sequence of the *Henry Draper Catalog* (OBAFGKMSRN) is retained for temperature. O-type stars are the hottest and most massive, whereas M stars are the coolest. The carbon (R and N) stars have more carbon than oxygen, compared with the G, K and M stars; S stars have similar temperatures to the M stars except that they show zirconium oxide (ZrO) absorption bands.
- Each letter is subdivided into tenths, 0–9, except for the O-type stars of which O3 is the hottest type. Not all subdivisions are used for each letter.
- A luminosity-related class is added, representing indirectly the absolute brightness of the star. Roman numerals are used to represent the luminosity class, where supergiants are I, giants are III and MAIN SEQUENCE stars (called dwarfs, although they are similar to the Sun) are V. Sometimes the subdivisions a, ab and b are used to subdivide the luminosity classification in the sense that a is brightest.
- There is no single third dimension representing abundance. Not only are there the SRN stars but also, in some stars, elements can be individually affected. Thus the abundance is noted by the element's symbol (iron, barium, carbon, CH, CN, and so on), followed by a number 1–5, which is positive for overabundance and negative for underabundance.
- There are several additional symbols, suffixed to the spectral classification. Rapidly rotating stars, with broad spectral lines, are designated n or nn, depending on the rotation rate. Stars with hydrogen lines in emission are labeled with the letter e. Hot, O-type stars with helium, carbon, nitrogen or oxygen in emission are marked with the letter f. Very hot stars showing by their broad emission bands that they are losing mass at a prodigious rate are classified as WOLF–RAYET STARS.
- If a star does not fit into the system at all, it is given the designation of the spectral type that matches it most closely and the letter p is attached, indicating a peculiar (and probably especially interesting) star. The peculiarity is then described in a note.

Typical examples of classifications are as follows.

- G2 V (the Sun's classification) indicates a star of moderate temperature and mass.
- M6 IIIe indicates a cool giant star with hydrogen emission.
- B0 Ib (close to the classification of the progenitor star of Supernova 1987a) represents a very hot, large massive star.
- K0 IIIb Fe_2, CH_1, CN-2 signifies a moderately cool giant star, slightly less luminous than most giants, with strong iron lines (Fe_2), a somewhat strong G-band from the CH molecule (CH_1) and weak CN.
- A5p (Cr, Eu) describes a moderately hot star with strong lines of chromium and europium.

The MK classifications are independent of the results of theory and all external influences. But after the classifications have been performed, it is useful to calibrate the system, and tie it to fundamental parameters such as temperature and absolute magnitude.

The MK system was defined in the blue-violet region of the spectrum, because that was the region recorded by old prism spectrographs on the blue-sensitive film-emulsions in the 1930s. It has been developed for the ultraviolet (UV), especially for the classification of hot, UV-emitting, O- and B-type stars observed by satellite, and the infrared, where it can be used advantageously for particular problems involving cool stars.

Like photographs, digital spectra obtained with CCDs can be compared by eye to obtain equivalent MK spectral types. They are ripe for powerful computer techniques to automate the pattern-recognition process. There are two successful ways to do this, which allow computers to rival humans. First, 'minimum metric distance' minimizes the discrepancy between the digital data and a set of the standard stars by a kind of least-squares process. Second, the neural-net approach constructs a network of parameters that relates every feature of a collection of standard star spectra by a system of weights

◀ ***Stephan's Quintet*** *The largest galaxy, NGC 7320, which lies just below center, is not associated with the others, but the small galaxy on the left edge has a similar redshift to them and seems to be an outlying member of the group.*

to the stars' spectral types. The network 'learns' how to recognize each of the standard star types, and applies this learning to data sets of millions of stars. In such cases, when new types of peculiarities are found, they are flagged for human consideration.

Further reading: Garrison R F (1984) *The MK Process and Stellar Classification* David Dunlap Observatory; Hearnshaw J (1986) *The Analysis of Starlight* Cambridge University Press.

Stephan's Quintet

A group of five galaxies (NGC 7317, 7318A, 7318B, 7319 and 7320) in the constellation Pegasus. Four appear to be gravitationally interacting, and show similar REDSHIFTS. NGC 7320 has a much lower redshift and is thus a foreground object not associated with the others. The group was discovered by Edouard Stephan in 1876.

Stereo comparator

A device that utilizes binocular vision to compare two images of the same area. The light paths are arranged so that the image seen in the left-hand eyepiece almost coincides with the image viewed in the right-hand eyepiece. If the images are well matched, any differences in the brightness or position of a particular object appear to stand out from the image plane. *See also* BLINK COMPARATOR.

Steward Observatory

The observatory, created in 1916, is the research arm of the Department of Astronomy of the University of Arizona (UA), USA. Its headquarters are in Tucson, Arizona, on the university campus. Its observational facilities are on Mount Graham (10-m Heinrich Hertz Submillimeter Telescope, or SMT; 1.8-m Vatican Advanced Technology Telescope (VATT) and the Large Binocular Telescope (LBT) with twin 8.4-m mirrors), Mount Hopkins (6.5-m Multiple Mirror Telescope, or MMT), Kitt Peak (2.3-m Bok) and the Catalina Mountains (2 × 1.5 m) in Arizona, and on Las Campanas, Chile (Magellan Project; 2 × 6.5 m). The LBT, Magellan, MMT, SMT and VATT are all collaborative projects. The Steward Observatory components of these facilities are used by scientists and students at the UA, Arizona State University and Northern Arizona University. The Steward Observatory is also involved in space astronomy research, currently including HST, ISO and SIRTF, and is participating in development programs for other space projects. The observatory also operates the Mirror Laboratory.

Stingray Nebula (Hen 1357)

A PLANETARY NEBULA in the constellation Ara. It is the youngest planetary nebula known, possibly only 200 years old.

Stony-iron meteorite

A METEORITE composed of roughly equal proportions of silicate (stony) material and metals (mostly nickel-iron). Stony-irons are the least common of the three main types of meteorite, accounting for less than 1% of all known specimens. The great majority of them fall into two main classes: the pallasites and the meso-siderites. Pallasites consist of a matrix of nickel-iron, the gaps in which are filled with olivine (magnesium-iron silicates). Meso-siderites resemble BRECCIAS: they consist of fragments of eucrite and howardite (varieties of achondritic material) mixed with nickel-iron. The largest known single fragment at 1.4 tonnes is the Huckitta meteorite (Australia, 1924), which is a pallasite. The fragments of the Brenham meteorite (Kansas, 1882), also a pallasite, that were collected had a total mass of 4.3 tonnes.

Stony meteorite

A METEORITE composed primarily of silicates, with only a small (typically 5%) metal content, mostly nickel-iron; also referred to simply as stones. Stones are by far the commonest meteorite, accounting for nearly 93% of all known specimens. There are two main classes of stony meteorite: CHONDRITES (96% of all stones) and ACHONDRITES (4%).

Strasbourg Astronomical Observatory

A research unit of the Université Louis Pasteur and the Centre national de la recherche scientifique (National Center of Scientific Research), or CNRS. It hosts teaching and research activities, the services of the Strasbourg Astronomical Data Center, and public outreach activities, with its planetarium. Founded in 1881, the Strasbourg Observatory currently carries out research in a wide variety of areas. The data center (known as CDS, Centre de données astronomiques de Strasbourg) is used by astronomers worldwide for its databases and services related to astronomical objects outside the solar system (*see* INFORMATION HANDLING IN ASTRONOMY).

Stratospheric Observatory for Infrared Astronomy (SOFIA)

A Boeing 747 SP aircraft modified to accommodate a 2.5-m reflecting telescope. SOFIA will be the largest airborne observatory in the world, and will make observations that are impossible for even the largest and highest of ground-based telescopes. The observatory is being developed and operated for NASA by industry experts led by the Universities Space Research Association. SOFIA will be based at NASA's Ames Research Center at Moffett Federal Airfield near Mountain View, California, and is expected to begin flying in the year 2004. Besides this contribution to scientific progress, SOFIA will be a major factor in the development of observational techniques and new instrumentation, and in the education of young scientists and teachers in the discipline of INFRARED ASTRONOMY.

String theory and superstring theory

String theory is a theory of FUNDAMENTAL PARTICLES and forces in which the basic entity is an exceedingly short one-dimensional structure rather than a point-like particle. These strings are envisaged as being about 10^{-36} m in length, different modes of vibration of the strings corresponding to different types of particle with different energies and masses. When the ideas of supersymmetry are applied to string theory, the outcome is superstring theory. Superstring theory requires a 10-dimensional SPACETIME, all but four of them (length, breadth, height and time) being hidden in the present-day universe. Superstring theory is one of the prime candidates for a 'theory of everything,' which embraces all of the forces and particles of nature.

Strömgren, Bengt (1908–87)

Danish astronomer who succeeded his father, Elis (1870–1947), as director of the Copenhagen University Observatory, but was also director of Yerkes and McDonald observatories in the USA. Strömgren calculated what happened when a hot star emitted ultraviolet light into the gas clouds around it, ionizing the hydrogen in a region (the H II REGION) now known as a Strömgren sphere. He created the Strömgren photometric system, which measures the brightness of stars in defined spectral bands as a quantitative counterpart to the Morgan–Keenan Spectroscopic Classifications System (*see* STELLAR SPECTRUM: CLASSIFICATION).

Strutt, John William [Lord Rayleigh] (1842–1919)

English scientist, Nobel prizewinner (1904) for the discovery of argon. He worked in many areas of physics, including electromagnetism and sound; the Rayleigh theory of the scattering of light was the first correct explanation of why the sky is blue.

Struve, Friedrich Georg Wilhelm (1793–1864)

Born in Germany, became professor of astronomy at Dorpat (now Tartu) University, Estonia, and director of the Dorpat Observatory. Using Joseph FRAUNHOFER's refractor he began his work on DOUBLE STARS. He became director of Pulkova (Pulkovo) Observatory near St Petersburg, Russia, which was constructed to his specifications by Tsar Nicholas I and became arguably the most productive observatory of the time in the world. He published a catalog in 1837 of more than 3000 binary stars. He carried out one of the first successful determinations of stellar distance, selecting VEGA for investigation because it had a large proper motion and was bright, thus indicating that it might be nearby. Although in 1837 he was the first to publish his PARALLAX results, his measurements were not believed, especially since he published a discordant value, twice as large, only three years later. The credit goes to Friedrich BESSEL and Thomas HENDERSON, who published the parallaxes of SIXTY-ONE CYGNI and ALPHA CENTAURI respectively in 1838 and showed that the measurements contained the expected patterns. This convinced astronomers that parallactic measurements had been reliably achieved. Friedrich was the great-grandfather of Otto STRUVE.

Struve, Otto (1897–1963)

Born in Kharkov, Ukraine. In an adventurous and varied career, he studied at Kharkov University, joined the White Russian army during the Revolution, and fled to Turkey at the time of their defeat. Struve emigrated to the USA, became director of the YERKES OBSERVATORY, near Chicago, Illinois, and founded the MCDONALD OBSERVATORY in Texas. He became director of the Leuschner Observatory in California and then the NRAO. He was a stellar spectroscopist, with achievements in close binaries, the INTERSTELLAR MEDIUM and stellar rotation. Struve was a fourth-generation astronomer. His father, Gustav Wilhelm Ludwig (1858–1920), uncle Karl Hermann (1854–1920), grandfather Otto Wilhelm (1819–1905) and great-grandfather Friedrich STRUVE were all directors of European observatories.

Subaru Telescope, Hawaii

The Subaru Telescope, Hawaii, is an 8.2-m optical-infrared telescope operated by the National Astronomical Observatory, Japan at the 4.2-km summit of Mauna Kea, Hawaii. It is one of the new-generation telescopes with an actively controlled large monolithic mirror, which became operational with

◀ ***Subaru Telescope*** *The view from inside the cylindrical enclosure on Hawaii's Mauna Kea mountain.*

first light in January 1999. The Subaru Telescope Base Facility at Hilo, Hawaii, has a large computer system and laboratories for developmental works. Subaru is the Japanese name for the Pleiades.

Subdwarf

A star that lies below the MAIN SEQUENCE in the HERTZSPRUNG–RUSSELL DIAGRAM and is consequently fainter than a dwarf star of the same spectral type. Subdwarfs are mainly old Population II objects which are metal-poor and with kinematics characteristic of the halo population. *See also* SDO STAR.

Subgiant

A star that has just moved away from the main sequence in the HERTZSPRUNG–RUSSELL DIAGRAM, having converted all the hydrogen in its core to helium. It has begun to burn hydrogen in a shell around the core. In a GLOBULAR CLUSTER the subgiant branch begins at the main sequence turnoff and extends up toward the region occupied by giants.

Sublimation

In space, solids melt into gases without being liquids. This is called sublimation. The ices that COMETS are made of sublimes into gas when the comet approaches the Sun. The same happens when solid carbon dioxide (dry ice) vaporizes, even on Earth.

Submillimeter Telescope Observatory

The observatory is a collaboration between the Max Planck Institute for Radio Astronomy (MPIfR) in Bonn, Germany, and STEWARD OBSERVATORY, Arizona, USA. It operates the 10-m Heinrich Hertz Telescope (HHT) for observations in the millimeter and submillimeter wavelength range. The observatory is located at an altitude of 3.2 km on top of Mount Graham, Arizona. The HHT is the first telescope to use a hot-electron bolometric mixer receiver for astronomical observations at 690 and 810 GHz.

Submillimeter Wave Astronomy Satellite (SWAS)

NASA Small EXPLORER mission (SMEX), launched December 1998. SWAS carries a 60-cm telescope to study submillimeter emissions from molecular clouds and star-forming regions. It found strong water emission lines in SPECTRA of all molecular-cloud targets. The SWAS Science Operations Center is located at the Harvard–Smithsonian Center for Astrophysics in Cambridge, Massachusetts, USA.

Sudbury Neutrino Observatory

The Sudbury Neutrino Observatory (SNO) studies NEUTRINOS from the Sun, supernovae and other astrophysical sources. The detector is situated 2000 m underground in an active nickel mine near Sudbury, Ontario, Canada and observes Cerenkov light from NEUTRINOS interacting with 1000 tonnes of ultra-pure heavy water (D_2O) surrounded by 7000 tonnes of light (ordinary) water shielding. Early neutrino experiments at a number of sites detected fewer neutrinos from the Sun than predicted by theory, leading to the so-called 'solar-neutrino problem.' The SNO is able to detect the three different types (or flavors) of neutrino. In September 2003 new results were announced from SNO confirming that 2/3 of electron-neutrinos from the Sun change into mu- or tau-neutrinos as they travel to the Earth. This is a further step in the process of proving that neutrinos have mass. The project involves more than 100 scientists from Canada, USA and UK and began observations in May 1999.

Suisei

Japanese mission to study COMET HALLEY. Launched in August 1985, Suisei approached to within 151 000 km of the comet on March 8, 1986, to observe its interactions with the SOLAR WIND. The hydrazine was depleted on February 22, 1991, and plans to extend the mission to visit other comets were cancelled. Suisei means 'comet' in Japanese.

Summer Triangle

The three bright stars, VEGA (Alpha Lyr), DENEB (Alpha Cyg) and ALTAIR (Alpha Aql) make a distinctive triangular ASTERISM visible in the northern sky in late summer and throughout the autumn.

Sun

The Sun is the nearest STAR, seen in greatest detail. A wide range of special instrumentation has been employed to study it, from the ground and from space. Of the space missions, those of particular interest have been the US SKYLAB satellite (1973–4), the Japanese YOHKOH mission (August 1991 to the present), the ESA–NASA SOHO mission (December 1995 to the present), and the US TRACE mission (April 1998 to the present).

The Sun as a star

The Sun is a gaseous sphere, which is in hydrostatic equilibrium; that is, it supports its weight by the pressure that it generates within. It emits light because it is hotter than the surrounding space. Treated as a star whose properties are more accurately known than most, the Sun is at an average position among other stars in its LUMINOSITY, radius, effective temperature, age, chemical composition, mass and rotation rate.

- The Sun's luminosity can be obtained from accurate spacecraft measurements. The solar luminosity (that is, the total radiant power in all wavelengths received at the top of the Earth's atmosphere) is 3.844×10^{26} W. This is equivalent to an absolute bolometric magnitude of +4.72.

- The diameter of the Sun (that is, of its photosphere) can be determined by direct measurement with a telescope or from the start and end times of total eclipses (*see* ECLIPSE AND OCCULTATION). Edmond HALLEY's 1715 measurement using this method gave the mean solar diameter as 1920 arcsec, almost as accurate as is known today, which translates into a linear value for the solar radius of 695 970 km. Despite claims made in the past, the measurements indicate a constant value over this period, and there seems to be a practically zero value for solar OBLATENESS (the excess of equatorial over polar diameters).

- The Sun's effective temperature, obtained from the solar luminosity as if it is a black body, is 5770 K. The equivalent spectral type is G2, luminosity class V. Its COLOR INDEX, as defined on the UBV photometric system, is given by B–V = +0.65 (*see* MAGNITUDE AND PHOTOMETRY).

- From the Sun's luminosity and spectral type (or equivalently the effective temperature, or color index), we may place the Sun's position, on the familiar HERTZSPRUNG–RUSSELL DIAGRAM, more or less in the middle.

- The Sun's age, determined from the age of the oldest METEORITES, is 4.6 billion years. This is midway between the ages of the youngest clusters (for example, the Pleiades at 60 million years) and the age of globular clusters, about 10 billion years.

- The chemical composition of the Sun is measured from high-resolution visible-light spectra of its photosphere. In broad terms, the composition is 91% hydrogen, 9% helium, and only 0.1% heavier elements (or 'metals'). The elements carbon, oxygen, nitrogen, neon and iron feature as some of the most abundant heavier elements, and the Sun is taken as a standard for POPULATION I.

- The Sun's mass can be very precisely determined from planetary motions using Kepler's laws (*see* CELESTIAL MECHANICS). It is 1.989×10^{30} kg. Stars in binary systems range from 0.1 to 50 times the solar mass.

- Because the solar magnetic field changes with a period of roughly 22 years, many features of solar activity, such as SUNSPOTS, FLARES AND ACTIVE REGIONS follow the SUNSPOT CYCLE. As a result, the Sun is a VARIABLE STAR. For older stars (1 billion years or more), mildly eruptive activity (as measured by numbers of active regions or propensity to flare) is related to rotational speed, because of the way in which magnetic fields are generated by dynamo action within the stars. The Sun rotates with a period that depends on latitude: 25 days at the equator and up to 33.5 days at latitude 80°. The corresponding rotational speed at the equator is 2 km s^{-1}. Below the surface the Sun's rate of rotation has been determined by HELIOSEISMOLOGY. Within the convection zone, the angular velocity has approximately the same latitude dependence as the surface. But at the base of the convection zone, around 30% of the way in, there is a transition region (of thickness about 10% of the solar radius) to a core that rotates like a solid body. The Sun rotates slowly compared with some other stars, because the planets, particularly Jupiter, have absorbed its angular momentum.

The solar interior

The internal structure of the Sun arises because of the way it supports its weight by high temperatures and pressures in its interior. As a result, nuclear energy is released by the fusion of hydrogen into helium, with the emission of NEUTRINOS. Energy is transported from the interior to the surface by radiation and convection.

The structure of the Sun has been calculated through the Standard Solar Model, a mathematical representation of the processes taking place based on defined knowledge or assumptions. It reproduces the basic properties of the Sun, even successfully challenging particle physics by demanding new properties for neutrinos. At the solar center, the temperature is 1.57×10^{7} K, density, 1.52×10^{5} kg m^{-3}, and pressure, 2.37×10^{16} Pa. The composition at the center of the Sun is 34% mass fraction of hydrogen and 64% of helium.

The PLASMA in the outermost 30% of the Sun is convectively unstable. Within the solar interior, the convection flow is turbulent and twisting. Eventually, magnetic flux emerges from the surface at sunspots when the flux is large enough.

Helioseismology has revolutionized our understanding of the solar interior. The convective motions of the outer regions of the Sun excite sound waves, which travel from the surface toward the interior, refracting until directed toward the solar surface again. This produces surface oscillations, which probe the regions through which the waves travel. The observations have led to modifications in the calculated basic physics of the gas and more accurate determinations of the abundances of chemical elements in the strata below the surface. Helium and the heavier elements settle toward the center of the Sun; this has been incorporated into the Standard Solar Model.

The solar atmosphere

There are three main regions of the solar atmosphere: the surface layer of the Sun (the PHOTOSPHERE); the overlying hotter and rarer CHROMOSPHERE with a temperature of about 10 000 K, lying at a mean height of 2000 km above the photosphere; and the much hotter (about 1 million K) and rarer CORONA, which can be seen out to several solar radii in a solar eclipse but stretches out to the Earth and beyond. The plasma that possesses temperatures between 10 000 and 1 million K comprises the solar transition region.

Many properties of the solar photosphere change systematically across the face of the Sun. FACULAE are small regions that are brighter than normal and 'filigree' are bright crinkle-shaped ribbons. CONVECTION exists on several scales, from 1 million to 15 million m. Over most of the solar surface the magnetic field is far from uniform, being concentrated into intense solar photospheric magnetic flux tubes at the edges of supergranule cells. In the interior of such cells, the magnetic fields are tangled by convection.

The chromosphere is even less uniform. A bright network

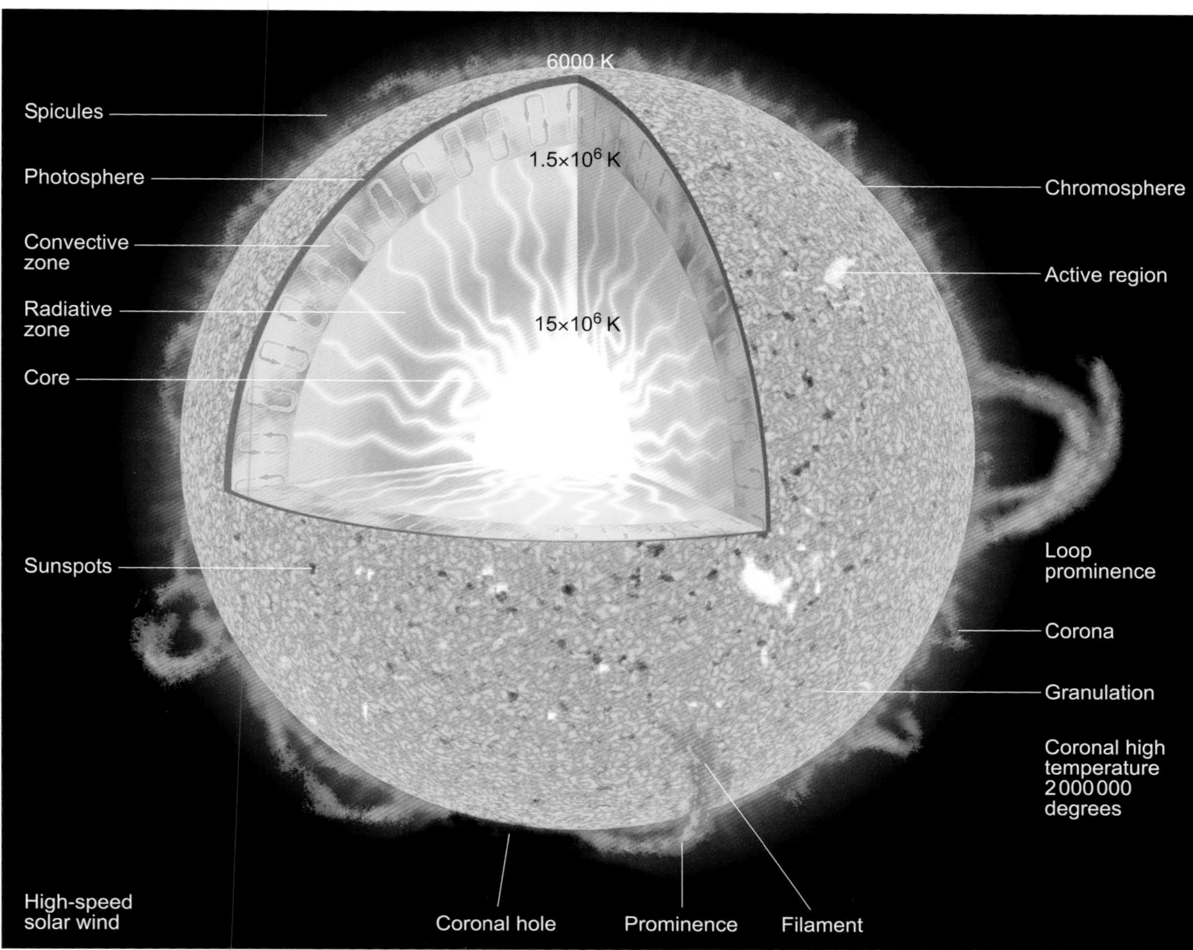

◀ ***Sun***
This cutaway reveals its changing features over 695 970 km from core to photosphere.

surrounds the supergranule cell boundaries. Around sunspots the magnetic field is up to 0.4 T and is associated with a brightening of the surrounding region (PLAGES). Transient brightenings occur near sunspots at the feet of surges ('Ellerman bombs'). Bipolar magnetic flux tubes break through the solar surface. Dark fibrils due to dense plasma along a canopy of magnetic flux tubes cross the interior of supergranule cells. Jets of plasma ('spicules') are ejected along flux tubes upwards at the network boundary. Solar chromospheric oscillations are observed and are related to one of the possible sources of the high temperatures in the chromosphere.

The solar transition region is highly dynamic and it is not easy to develop theories. There is a variety of flows and small high-velocity features and explosive events.

The corona possesses several different types of structure. CORONAL HOLES have magnetic fields that are open into interplanetary space. Polar plumes are dense rays in coronal holes outlining the magnetic field. Coronal loops have magnetic field lines that have both ends anchored to the solar surface. Small bright regions occur where magnetic fields are reconnecting. Streams of plasma (solar X-ray jets) are accelerated to several hundred kilometers per second by magnetic reconnection and emit X-rays. Large structures ('coronal streamers') are closed below a large fraction of a solar radius above the limb and are open above that height. The solar corona was traditionally seen only during an eclipse, but now can be observed by coronagraphs from the ground and direct by extreme ultraviolet and X-ray telescopes from space.

An important puzzle is how the corona is heated. Part of the mystery has been solved, because X-ray bright points are now known to be heated by magnetic reconnection.

The solar heliosphere

The SOLAR WIND is the outwards expansion of the corona from the Sun. Its composition and other properties are highly complex. Especially interesting is the behavior out of the ecliptic plane as measured by the only satellite to orbit in that region (*see* ULYSSES SPACE MISSION).

Neighboring streams that are flowing out from the Sun interact at their boundaries. Accelerated particles of many types produce bursts of radio waves. COSMIC RAYS may give rise to showers of electrons and other particles when they impact on the Earth's atmosphere ('extensive air showers').

The solar wind is rooted in the corona and is sensitive to solar activity. The comets, the planets, their satellites and their magnetospheres interact with the solar wind as they orbit the Sun. Of growing interest is the link between activity on the Sun and effects on the Earth's atmosphere – space weather (*see* SPACE WEATHER AND THE SOLAR-TERRESTRIAL CONNECTION). Heinrich Kreutz (1854–1907) discovered a family of comets that approach close to the Sun, some of which impact on it and alter the atmospheric composition. At the heliopause the solar wind connects with the rest of the Galaxy, and it is the Sun's outermost boundary at 100 AU.

Observing the Sun

Solar astronomy has long been a popular pastime among amateur astronomers. Compared to more traditional astronomical targets, the Sun is incredibly bright, and observing it doesn't require a large telescope as do many nighttime 'deep sky' objects. The detail on the Sun's disk is visible in white light, and its features are large enough that there is no difficulty resolving them with small instruments. In addition, the Sun provides convenient observing hours for those with day jobs, and solar observing is generally warmer and more comfortable than nighttime observing. And finally, for those with the enthusiasm and wherewithal to observe with narrow-band filters, the Sun offers a profusion of exotic detail.

For all of these reasons, amateur astronomers have observed the Sun for many generations. But the most important reason is probably its inherent attraction. The Sun is never the same from one day to the next, and users of small telescopes can often see changes occur from minute to minute. The constant variability of the Sun and the wide range of observable phenomena have led noted astronomer Brian Skiff to conclude the Sun is the most interesting object in the sky.

The first amateur solar observers in the modern sense were the first astronomers with telescopes. It took a generation or so from the time GALILEO GALILEI first looked at the Sun with his telescope before astronomy started to develop its professional class. Even long after Galileo, many groundbreaking solar observations were made by amateurs; for example, William Herschel (*see* HERSCHEL FAMILY) took the first important measurements of solar radiation and Richard CARRINGTON explained the Sun's DIFFERENTIAL ROTATION.

By the late nineteenth century, however, solar science was dominated by professionals, with amateurs mainly contributing sunspot counts. In modern times, amateur solar astronomy is largely a recreational pursuit, done for its own enjoyment and fascination. The popularity of solar-observing has given rise to a wide variety of solar-observing equipment – mainly filters, Sun finders and other accessories for telescopes.

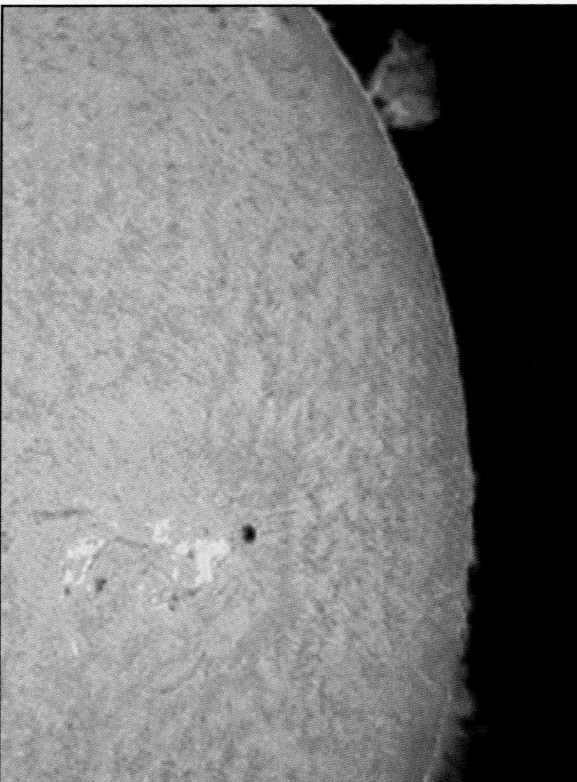

► ***Sun***
This photograph of a sunspot, prominences and a large filament was taken on November 17, 2002 from Divide, Colorado. It is a composite image in H-alpha, taken with a Canon D60, Coronado Solarmax 60 hydrogen-alpha narrowband filter, and a Televue 85. Image by Ginger Mayfield.

The most important issue in solar observing is *safety*. Solar astronomy is said to be the only form of observational astronomy which is inherently dangerous. Looking at the Sun through a telescope or binoculars without proper filtration can cause permanent blindness in a fraction of a second. It is important at all times to follow safe solar-observing practices, and follow all instructions that come with any solar-observing gear you use.

Most amateur astronomers use small astronomical telescopes for observing the Sun. These telescopes are generally fitted with pre-telescopic filters that attenuate the light of the Sun – including invisible but damaging infrared and ultraviolet light – to safe levels before it enters the telescope. Some observers do not use filters, but project the image of the Sun onto a white card using a telescope and EYEPIECE much like a slide projector. This is a safe practice only for certain designs of telescope – REFRACTORS and some REFLECTORS – because in some designs internal components can heat to such extremes that they catch fire.

Telescopic solar filters are available from a variety of companies. Most popular with amateurs today are filters made with Baader AstroSolar, which provide an optically excellent neutral white image for very little cost. Coated glass filters are also commonplace, although most budget-priced glass filters form an orange image of the Sun. Whichever is used, the filter should be purchased from a reputable manufacturer or dealer, and used strictly according to instructions. If a filter becomes damaged in any way, it should immediately cease to be used, at least until it can be repaired or replaced.

All of these filters are 'pre-telescopic'; that is, they are attached in front of the telescope, between it and the Sun. Most filters are available in a variety of different cells that will attach to most popular commercial telescopes. Generic filter material is available for those who have unusual or home-made telescopes, and many amateur astronomers opt to fabricate their own attachment cells. The only requirement is that the cell fit firmly on the telescope – a filter that can fall off or blow off in the wind is not acceptable.

One form of solar filter which was commonly included with inexpensive telescopes over the last several decades is inherently *unsafe*. No filter that screws into the telescope's eyepiece should ever be used. These filters can get extremely hot when exposed to the telescope's concentration of the Sun's light, and they have been known on occasion to shatter or crack while in use – with potentially blinding consequences for anyone using the telescope at the time. Solar filters of the eyepiece variety are best disposed of in the garbage.

When using a telescope for solar observing, it is important to be sure to cover any finderscopes or GUIDE TELESCOPES attached to the telescope. These smaller telescopes can still dangerously concentrate the Sun's rays, and people can inadvertently – or naively – look through them or pass their hair or hands close to them. This can cause burns and damage to vision.

Once the telescope is safely filtered, one must find and center the Sun in the field of view. Because finder scopes cannot be used, this can sometimes pose a challenge. Most seasoned solar observers look at the shadow of the telescope

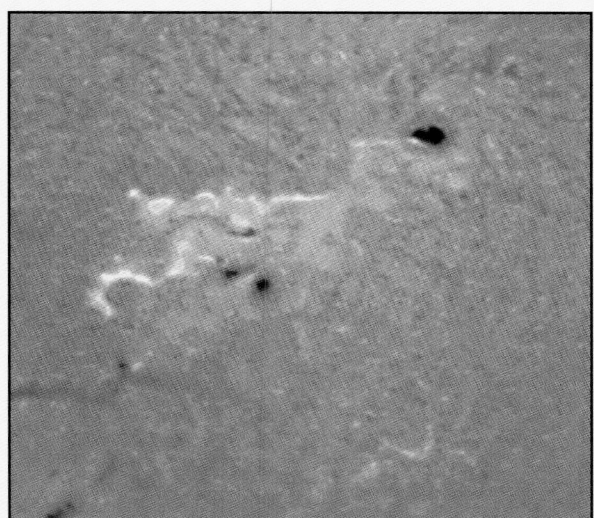

on the ground to get things centered up. When the telescope is pointed at the Sun, the shadow of the telescope will be at its smallest extent. One need merely look at the shadow and move the telescope around a bit, until it is clear what direction to go to minimize the shadow. At that point, the Sun will usually be within the field of a low-power eyepiece.

Some observers have trouble with the shadow method, however, and have created all manner of differently designed 'solar finders.' The general principle is to have two objects attached to the telescope. The first object, often looking like a pinhead, casts a small shadow onto the other object, which is typically a small card or piece of wood, marked so that when the shadow falls on the dot, the Sun is centered in the telescope. The simplest of these just uses a push-pin and an index card attached to a piece of foam which is tied to the telescope with string, but many more elaborate designs have been conceived by inventive observers.

Once the Sun is found, observing usually begins with low power. In a typical telescope at up to ×50 or so, the entire solar disk can be seen, and this gives the observer a broad overview of what is present on the Sun. Sunspots are the most conspicuous feature of the Sun in white light, and if any are present these should be very easily seen at low power as black splotches on the solar disk. Upon closer examination, the larger spots have a darker center, called the UMBRA, and lighter-gray outer regions, called the PENUMBRA. Sunspots are locations where convection of material in the Sun is slowed down by magnetic fields, giving the hot gases enough time at the surface to cool and darken.

Many observers like to observe individual sunspots and spot groups in more detail, and will increase magnification at this point in order to zoom in on individual features. At higher powers, assuming the atmosphere is steady, striations might be visible in the penumbra – and rarely, a spot might be crossed by a bright 'light bridge,' which is often seen to move over the course of 10 or 20 minutes.

The low-power view will also show solar FACULAE – broad, splotchy brighter areas on the solar disk, much larger than the sunspots. Often the faculae are easier to see closer to the LIMB of the Sun than in the center of the disk.

Beginners to solar observing often experience some difficulty with focusing the image of the Sun. In most parts of the world, the daytime atmosphere undergoes much more CONVECTION than is experienced at night. This manifests itself as astronomical SEEING distortions in the telescopic image caused by moving air currents. The effect has been likened to the difficulty of making out objects that are on the bottoms of swimming pools – if the surface of the pool is disturbed by waves, it can be very hard to make out objects under the water.

This astronomical seeing makes the limb of the Sun dance and sway, and the focusing difficulty comes in here. It is often possible to focus on an especially active layer of air, rendering the limb of the Sun apparently quite sharp – but clearly moving about. When this is the case, features on the Sun will often be a little fuzzy. Instead of looking at the limb while focusing, it is best to pick some solar feature, typically a sunspot, and focus on that until it is as sharp as possible.

Seeing also interferes with observing fine details on the Sun, of course. A good telescope of 100-mm APERTURE should resolve sunspots to the point that the penumbrae are striated, but frequently this isn't possible because of the atmosphere.

The characteristics of daytime seeing depend a great deal on the local climate and terrain. In the high desert of southeastern Arizona, the best times to observe the Sun are often within the 45 min after sunrise or the 45 min before sunset – at those times the atmosphere there frequently becomes very calm. In Ohio, however, the best time is often at noon. There is no fixed rule, and the only way to learn the characteristics of one's own site is to experiment.

Although amateur astronomers made virtually all of the important early scientific observations of the Sun, in modern times there is not nearly as much room for amateur contributions, unless the observer is prepared to acquire expensive instrumentation. One project that is of enduring scientific value, and that can be carried out with typical amateur observing equipment, is sunspot counts. The AAVSO is the North American clearinghouse for sunspot count observations.

For the vast majority of amateurs, however, solar observing is a recreational pastime. It is yet another target that can be viewed to satiate the amateur's hunger for observing the universe in all its splendor.

Tim Doyle is treasurer of the Huachuca Astronomy club of southeastern Arizona. He specializes in solar astronomy.

◀ ***Solar flare*** *close up on October 25, 2002. Image by Ginger Mayfield, using the equipment listed below.*

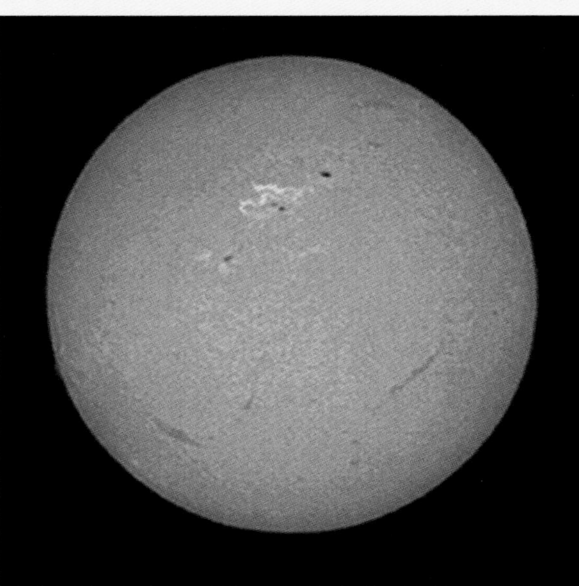

◀ ***Large flare*** *seen on October 25, 2002. Taken from Divide, Colorado with Canon D60, Coronado Solarmax 60 hydrogen-alpha narrowband filter, and Televue 85. Image by Ginger Mayfield.*

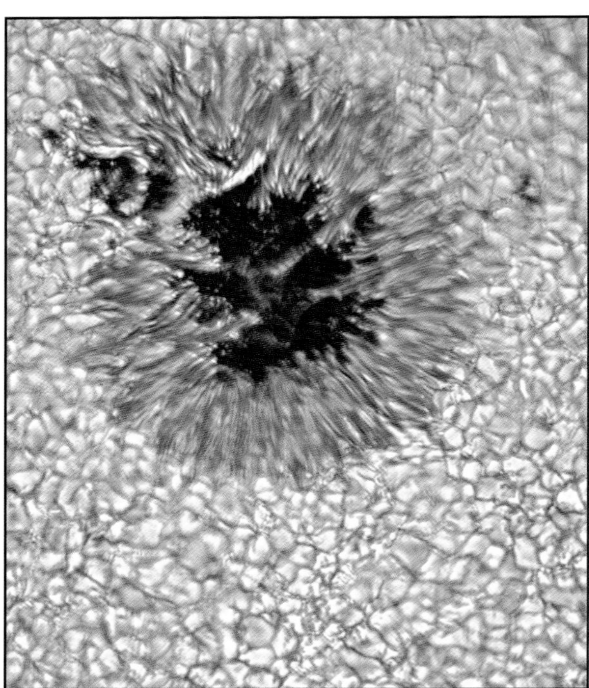

▶ ***The Sun's*** *granulated surface surrounds a sunspot. This very high resolution picture shows that the penumbra of the sunspot is actually individual streaks, formed by the shearing of the granulations into the central dark area, or umbra.*

Further reading: Phillips K J H (1995) *Guide to the Sun* Cambridge University Press; Stix M (1999) *The Sun, an Introduction* Springer.

Sundial

One of three types of timekeeping device based on the shadow cast by any fixed vertical object. As the day progresses, the Sun moves across the sky, causing the shadow of the object to move. Depending on the type of device, the object measures apparent solar time with some accuracy. From about 3500 BC the first device was probably a vertical stick known as a gnomon. By the eighth century BC, more precise devices were in use in Egypt. One example has a base inscribed with a scale of six time divisions fixed in an east-west direction. The design is still in use in some parts of Egypt.

Around 300 BC the hemispherical sundial, or hemicycle, is evident. It was made of a cubical block of stone or wood with a hemispherical opening cut into it. A pointer was fixed at the center of the opening. The path traveled by the shadow of the pointer was approximately a circular arc, which varied according to the seasons. These variations were inscribed on the internal surface of the hemisphere, and each was divided into 12 equal divisions or hours. Because the length of the day varied according to the season, these hours also varied in length from season to season and even from day to day, and were consequently known as temporary hours. There are records of the dial still being used during the tenth century AD.

The Greeks developed and constructed very sophisticated sundials. One version used the surface of a conic section for greater accuracy. The Romans also used sundials with temporary hours. The Arabs attached much importance to sundials and derived the principles and design from the Greeks. They increased the variety of designs available, and simplified the processes of design and construction by using trigonometry. With the advent of mechanical clocks in the early fourteenth century, sundials with equal hours gradually came into general use, evolving into the development of MEAN SOLAR TIME. *See also* TIME.

Sunflower Galaxy (M63) *See* MESSIER CATALOG

Sunspot cycle

The periodic increase and decrease in the number of sunspots and sunspot groups on the surface of the Sun (*see* SUNSPOT, FLARE AND ACTIVE REGION). The number of spots and groups reaches a maximum, on average, every 11 years. At the intervening sunspot minima, the solar disk may be devoid of spots for weeks on end. On average, the rise to maximum activity takes four to five years, and the subsequent decline to the next minimum about six to seven years. The level of activity in successive cycles can vary substantially.

At the beginning of each cycle, spots start to appear at latitudes of about 30° (occasionally as high as 40°) north (+) and south (–) of the solar equator. As the cycle advances, the bands of sunspot activity migrate toward the equator. At solar maximum, the average latitude at which spots occur is about ±15°, and by the end of the cycle, around ±8°. By now the first spots of the next cycle may be appearing at latitudes of 30°–40°. This cyclic variation in the mean latitudes of sunspots was pointed out in 1861 by Friedrich Spörer (1822–95), and is known as Spörer's law. When the latitude of each sunspot is plotted against time over a series of 11-year sunspot cycles, the distribution of spot positions resembles butterfly wings, so this is known as the butterfly diagram. A diagram of this kind was first plotted in 1904 by the English astronomer Edward MAUNDER.

In each spot pair, the spot that is ahead, in the sense of the direction in which the Sun rotates, is called the leader, and its companion, the follower. Throughout a complete cycle, from one minimum to the next, all of the spot pairs and groups in the northern hemisphere have the same polarity pattern, and those in the southern hemisphere have the opposite. For example, in one particular cycle, all the leaders in the northern hemisphere will have positive polarity and all the followers negative, whereas all the leaders in the southern hemisphere will have negative polarity and the followers positive. At the end of that cycle, the polarity pattern reverses, so that in the subsequent cycle all northern-hemisphere leaders will have negative polarity and all southern-hemisphere leaders, positive. The complete magnetic cycle consists of two successive 11-year sunspot cycles and is, therefore, 22 years long.

During the earlier part of each cycle, the leading spot in each pair (or the leading area of net magnetic polarity in each group) has the same magnetic polarity as the net polar polarity in the hemisphere within which it is located. As each spot or group decays, the polarity of the follower preferentially diffuses toward the pole. The cumulative effect eventually reverses polarity at the solar poles. This usually occurs around the time of solar maximum, but considerable variations occur and, while the piecemeal reversal is taking place, the Sun may for a time have the same net polarity at both poles.

The sunspot cycle is part of an overall solar cycle whereby all forms of solar activity, including sunspots, PLAGES, PROMINENCES, flares and coronal mass ejections (CMEs), together with the shape, extent and structure of the CHROMOSPHERE and CORONA, undergo cyclic variations with a period of about 11 years. The numbers, sizes and energies of prominences, flares and CMEs mirror the increase and

◀ ***Solar flare*** *of complex magnetic fields over a sunspot and active region, as viewed by the TRACE satellite.*

decrease in sunspot numbers and the corona is brighter, more extensive and more symmetrical around solar maximum than at times of minimum activity.

In addition to the 11-year cycle, there is evidence to suggest that sunspot numbers, and levels of solar activity as a whole, undergo longer-term modulations over periods of 80 years and more. Furthermore, records imply that there have been prolonged periods of enhanced and depressed activity in the past. In particular, solar activity appears to have remained at an unusually low level between 1645 and 1715, a period that is known as the MAUNDER MINIMUM.

Sunspot, flare and active region

Sunspots, flares and active regions are lumped together under the term 'solar activity,' referring to a wide range of transient solar phenomena that vary in complex ways with the SUNSPOT CYCLE. Sunspots are dark regions in the photosphere where large (1 T) magnetic flux tubes break through the surface. Active regions surround sunspot groups and have magnetic fields of one tenth of this. PROMINENCES are huge flux tubes up in the CORONA, containing a vertical sheet of PLASMA at temperatures of 10 000 K. Prominences are referred to as FILAMENTS when observed on the disk. Quiescent prominences can last for many months but can be shorter. Solar FLARES are enormous releases of magnetic energy, related to the eruption of prominences from active regions. Before the flare a prominence starts to rise slowly. The flare quickly develops in a few minutes, with a rapid acceleration of nonthermal electrons, sometimes producing gamma-rays. During the main phase, the energy release continues for several hours. In addition, a flare produces different kinds of radio emission and occasionally gives a continuum brightening in the photosphere, as discovered by Richard CARRINGTON.

Coronal mass ejections (CMEs) are large eruptions of mass associated with prominence eruptions, either from active regions or nearby. CMEs produce a wave-like disturbance on the solar surface that propagates from the ejection site. Solar flares often produce small ejections of mass that are called surges.

Sunspots often start as small dark patches called sunspot pores. The central part of the sunspot is dark and is called the umbra. The umbra is slightly depressed so it changes its appearance as a sunspot rotates across the solar disk (the Wilson effect). The umbra is surrounded by an annulus (the penumbra) with radial striations, since the magnetic field is horizontal. The penumbra exhibits a radial outflow (the Evershed effect).

Sun-synchronous polar orbit

The special case of the orbit of a satellite around the Earth in which it travels over the north and south poles as the planet turns below it. In a sun-synchronous orbit the satellite passes over the same part of the Earth at roughly the same local time each day. This makes communication with ground stations convenient. *See also* INFRARED IMAGING SURVEYOR.

Sunyaev–Zeldovich effect

The Sunyaev–Zeldovich (SZ) effect is a spectral effect on the COSMIC MICROWAVE BACKGROUND caused by interaction with the ELECTRONS in the hot gas inside a cluster of galaxies. This can manifest itself as a small (around 0.5 μK) dip in the temperature of the background in the direction of the cluster. Although it was first proposed in 1970, about 20 years of observational effort were required before it was reliably detected. Many tens of clusters have now been detected in the SZ effect, and early predictions of its usefulness for COSMOLOGY are starting to bear fruit. Its REDSHIFT independence makes it an excellent tool for investigating the evolution of structure with redshift. It is hoped that future ground-based telescopes and some space missions will be able to use it to detect large numbers of clusters. These include the PLANCK SURVEYOR mission, due for launch by ESA in 2007, which could detect tens of thousands of clusters with the SZ effect.

Supercluster; the Local Supercluster

Galaxies like to be with other galaxies. Gravity pulls them together and they form clusters of galaxies. In turn, clusters of galaxies act as nodal points in the filamentary structure of the universe, creating superclusters.

The LOCAL GROUP OF GALAXIES is an example of a typical environment for galaxies. In this cluster are two giant spirals, the ANDROMEDA GALAXY or M31 (the biggest) and the Milky Way, and two intermediate-size galaxies, Triangulum or M33 and the LARGE MAGELLANIC CLOUD. In total there are 40 known or suspected members but most of these are dwarfs. The Local Group is, itself, part of a much bigger supercluster. First, at 3–4 megaparsecs (10–13 million l.y.) away are the neighboring Sculptor, IC 342–Maffei and M81 groups. These groups and others string along the Coma–Sculptor Cloud, a branch filament toward the Virgo Cluster (*see* VIRGO CLUSTER OF GALAXIES), by far the most prominent nearby cluster, and usually taken as the center of the Local Supercluster. The Virgo cluster at a distance of 16 megaparsecs (50 million l.y.) is the central, densest part of the Local Supercluster. Galaxies in a spherical region within 7 megaparsecs have decoupled from the expansion of the universe and are falling back onto the cluster. North of Virgo, the band extends to the smaller Ursa Major cluster where one path of a bifurcation leads to the Coma–Sculptor cloud. South of Virgo, the band extends toward the Centaurus cluster. The flow of galaxies indicated by the COSMIC MICROWAVE BACKGROUND dipole anisotropy is in this general direction. Viewed from afar, the historical Local Supercluster would probably be considered an appendage of a larger Hydra–Centaurus–Virgo supercluster. Recent observations in the obscured regions of the Milky Way plane suggest that the Abell 3627 cluster (*see* ABELL CLUSTER) may be a particularly important component of this supercluster.

Supergiant

The largest and most luminous stars known (*see* HERTZSPRUNG–RUSSELL DIAGRAM), indicated by luminosity classes I and II (*see* STELLAR SPECTRUM: CLASSIFICATION). Only the most massive stars can evolve to become supergiants, and their lifetimes are short compared with less massive stars; consequently, they are very rare. They may be of spectral types O to M and have absolute magnitudes of between about −5 and −12. Many supergiants are variable (*see* WOLF–RAYET STAR, CEPHEID VARIABLE STAR, LUMINOUS BLUE VARIABLE), and they may be progenitors of some types of supernovae.

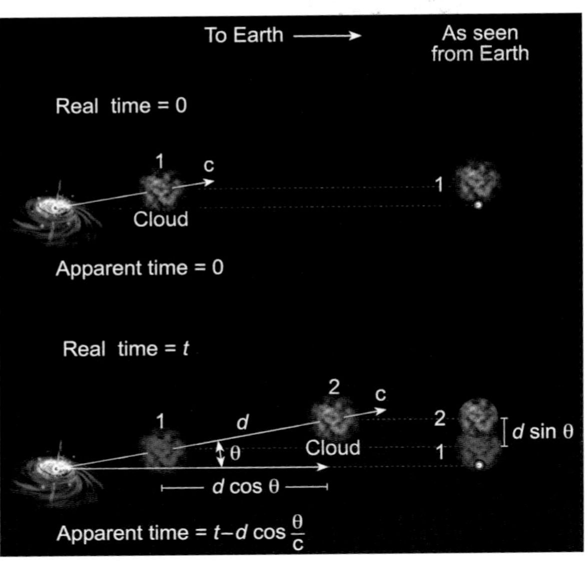

▶ ***Superluminal motion*** *Gas cloud moves from position 1 to position 2, a length d in time t at a speed near to c, so t = d/c. Light from position 2 has a shorter distance to travel to Earth so the image of 2 arrives at Earth after a shorter time t − d cos θ. In this image, the gas cloud seems to move a distance d sin θ in that time, so its apparent speed is d sin θ / (t − d cos (θ/c)) and is bigger than c.*

Superior planet

A collective term for the MAJOR PLANETS Mars, Jupiter, Saturn, Uranus, Neptune and Pluto, whose orbits lie outside that of the Earth, as opposed to the INFERIOR PLANETS, whose orbits lie inside the Earth's. *Compare* INNER PLANET.

Super-Kamiokande

Super-Kamiokande is a detector for NEUTRINO ASTRONOMY, located 1 km underground in Kamioka Mine in Japan. The detector consists of 51 000 tonnes of ultra-pure water which fills a 40-m tall, 40-m diameter, cylindrical stainless-steel tank. Cerenkov radiation (radiation produced by charged particles traveling in a medium faster than the speed of light in that medium) is produced by neutrino interactions in the water and is viewed with 11 146 photomultiplier tubes mounted on the inner surface of the tank. The detector is used in cooperation with about 120 physicists from 20 collaborating institutes in Japan and the USA. Super-Kamiokande found evidence for neutrino masses in 1998. The COSMIC RAY interactions in the atmosphere produce muon-neutrinos and electron-neutrinos. Super-Kamiokande has found that almost half of muon-neutrinos change to tau-neutrinos when they travel long distances, such as a diameter of the Earth. This phenomenon, called neutrino oscillation, happens only when neutrinos have masses; each 'flavor state' of neutrinos is a mixture of different mass states.

Another objective of Super-Kamiokande is a precise measurement of solar neutrinos produced by nuclear reactions deep within the core. The observed flux of solar neutrinos is about half that expected from the theories and the deficit may be due to neutrino oscillations. Super-Kamiokande is also searching for proton decays. If found, the GRAND UNIFIED THEORY, which unifies strong, electromagnetic and weak inteactions, will be proved.

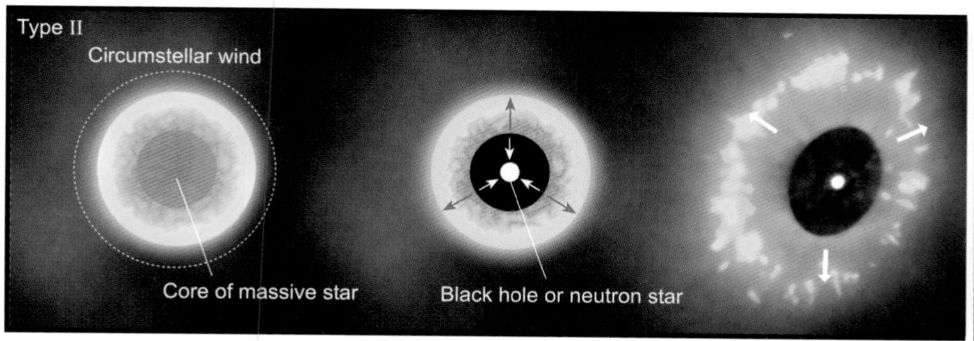

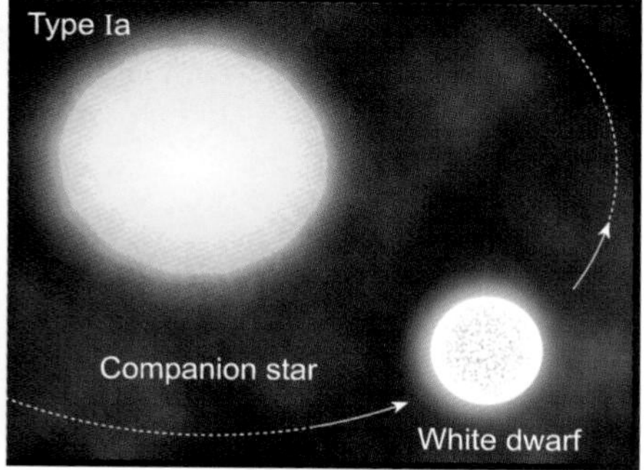

Superluminal motion

Superluminal motion describes a systematic change in the appearance of an astronomical object, when the speed of the change appears to exceed that of light (c). This phenomenon was recognized in 1971 in radio observations by very long baseline interferometry (VLBI) of the QUASAR 3C 279. It is an exciting observation because it seems to show that one of the fundamental postulates of SPECIAL RELATIVITY is wrong.

There are 66 RADIO GALAXIES for which speeds have been measured in this way, and well over half are superluminal. In most of them, the motion is along the length of a pair of JETS, which transport energy from the central engine outwards. We usually detect one jet only, or one jet is much brighter than the other. In all the cases, what is seen is a change of angle on the sky of some moving object over time. The angular speed is the change of angle divided by time, and the inferred speed of motion is calculated by multiplying by the distance to the source.

To produce an apparent speed faster than c requires that:

- the true speed of material in the jet approaches c;
- the direction of the jet is almost in the line of sight toward us.

The small angle between the jet and the line of sight is the reason why we see superluminal motion in one-sided jets – the approaching jet is made much brighter than the other by relativistic beaming, like a BLAZAR.

Superluminal motion is explained by the theory of special relativity as follows. At a time $t = 0$, a quasar nucleus ejects a blob of PLASMA traveling at v, nearly the speed of light, along a jet at an angle to the line of sight. Both the nucleus and the blob emit radio waves. The blob emits radio waves at times t_1 and t_2, picked up by VLBI images, which show the blob moving outwards from the nucleus. However, the time interval between the arrival at Earth of radio waves from the two events is much less than $t_1 - t_2$, because in the second image the blob is much closer to the Earth and its image has a shorter distance to travel to get here. This is the essence of the phenomenon: the time between the images is compressed and the apparent speed as the blob goes from one image to the next is increased. The apparent speed is an overestimate of the true speed. If the true speed was close to c in the first place, the apparent speed can be superluminal, depending on how closely the jet points to us.

Not all superluminal sources are quasars or AGN. A few objects in the Galaxy also show superluminal motion. An example is GRS 1915+105, which is a strong X-ray source as well as a radio source. Galactic superluminal sources are called 'mini blazars.'

Supermassive black hole *See* BLACK HOLE

Supernova

A catastrophic explosion at the end of the life of certain stars. In the explosion, a mass often much greater than that of the Sun is ejected at up to 10% of the speed of light. The corresponding energy is comparable to the gravitational binding energy of the exploding star, and the star is disrupted. The explosion ejects heavy elements that seed the gas of the surrounding INTERSTELLAR MEDIUM. The heavy elements are those formed in the progenitor star and others formed in the explosion. This enriched gas then condenses to form new stars, planets and living things. Some supernovae may produce stellar remnants such as neutron stars (*see* NEUTRON STAR AND PULSAR) or BLACK HOLES. Supernovae also produce gaseous shells as their EJECTA sweep outward into the interstellar medium. This not only helps to form new stars, it can blow matter entirely out of host galaxies in some circumstances. The shells and stellar remnant (if any) are known as 'supernova remnants.' Because of their great brightness, supernovae can be used to determine DISTANCES and the properties of the universe.

Historical supernovae

Historical records, particularly the careful data recorded by Chinese astronomers, show that seven or eight supernovae have exploded during the last 2000 years in the portion of the Galaxy not obscured from us by dust, about 15% of the area of the GALACTIC PLANE. The youngest of the expanding supernova remnants correspond to the historical supernovae, but we also observe the effects of several hundred older supernova remnants that have exploded over the last 1 million years or so.

The supernova of 1006 was the brightest supernova ever to be recorded, brighter than Venus and perhaps as bright as a quarter moon. However, the supernova of 1054 is by far the best-known. This explosion produced the rapidly expanding shell of gas that modern astronomers identify as the CRAB NEBULA. The supernova of 1572 was observed by Tycho BRAHE (*see* TYCHO'S SUPERNOVA). Tycho's student, Johannes KEPLER, in his turn witnessed the explosion of a supernova in 1604 (*see* KEPLER'S SUPERNOVA). Kepler and Tycho took careful data by which we deduce that they were the kind of supernova that modern astronomers label Type Ia. The powerful radio source, CASSIOPEIA A, is the remnant of a supernova that must have occurred about 1680. No very bright optical outburst was seen, but a faint star was recorded by John FLAMSTEED.

Supernova types

All supernovae observed since 1680, and hence all supernovae seen by modern astronomers, have been in other galaxies. Supernovae occur roughly once per century

▲ Supernova
A Type II (or Ib or Ic) supernova takes place in a massive star, which may have lost a little (or some or much) of its outer layers into a circumstellar nebula. Its central core collapses to a neutron star (or black hole if the progenitor star is massive), releasing energy that lifts off the progenitor star's mantle. Fragments of the mantle streak outwards and plow into the circumstellar nebula, outlining a spherical cavity that becomes visible as a supernova remnant. A Type Ia supernova is a similar collapse of a white dwarf, a very similar object to the central core of a massive star.

Discovering supernovae

▶ Supernova 2000P *discovery image by Robin Chassagne. It was found on the night of March 7/8, 2000 at magnitude 14.1, and confirmation was provided by the Observatoire du Pic-du-Midi on March 9.*

One of the attractions of serious amateur astronomy for many observers is the fact that valuable contributions to science can still be made. Nowhere are these contributions more obvious than in the field of SUPERNOVA hunting and observation. A dedicated effort has been made to improve communication between the serious amateur and the professional, and, in a few cases, the distinction has started to become blurred.

The simplest and cheapest way to observe supernovae, and perhaps to discover one, is to observe other galaxies directly through an EYEPIECE, using a backyard telescope. Especially during the 1980s and early 1990s, studies of bright supernovae found visually by amateurs provided basic information about the nature of supernovae; this work was foundational for modern research in COSMOLOGY. Supernovae brighter than magnitude 15 are also generally more scientifically significant than the fainter ones, and this is a continuing feature.

Any form of astronomical observation has certain minimum requirements. Observing supernovae by eye requires access to a REFLECTING TELESCOPE with a larger than, say, 20-cm APERTURE, and equipped with both low- and high-powered eyepieces. Access to a dark-sky site is also needed (*see* DARK SKIES AND GOOD OUTDOOR LIGHTING), together with maps showing the positions of galaxies and some photos or charts showing the normal appearances of these galaxies. A 31-cm telescope will generally reveal a star of magnitude 15.5 from a good, dark site.

Websites, together with lists issued by the Central Bureau for Astronomical Telegrams, contain information about supernovae that have recently been discovered. Providing the supernova is bright enough and is in a galaxy available in the night sky, its apparent magnitude can be recorded and reported, allowing the light curve to be compiled.

Any amateur with this equipment can also try to discover a new supernova. The idea is simply to observe as wide a range of galaxies as possible, steadily learning their normal

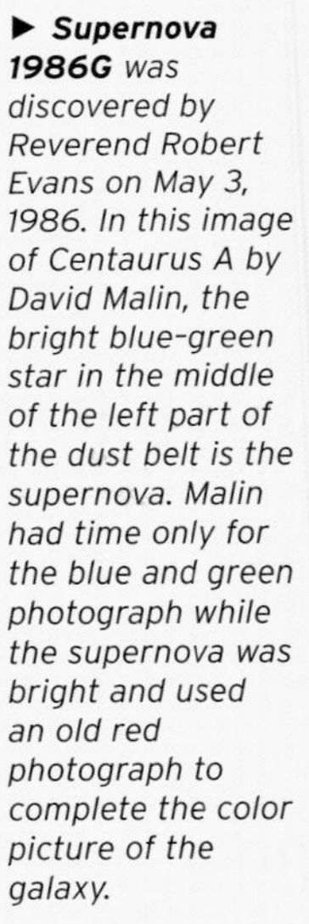

▶ Supernova 1986G *was discovered by Reverend Robert Evans on May 3, 1986. In this image of Centaurus A by David Malin, the bright blue-green star in the middle of the left part of the dust belt is the supernova. Malin had time only for the blue and green photograph while the supernova was bright and used an old red photograph to complete the color picture of the galaxy.*

appearance, so that the charts and photos do not have to be consulted each time. Training the memory is important, so that the positions of galaxies, and their appearances, can be memorized. This makes it possible to see and check more galaxies in the time you have available to observe. In turn, this increases your chances of success.

Generally, a good 31-cm telescope will reveal supernovae of all types in galaxies within 30 000 000 l.y. of the Earth, and brighter supernovae up to about 80 000 000 l.y. distant. The more distant galaxies should be checked twice a month, whereas the closest galaxies should be checked as regularly and as often as possible, as supernovae in these galaxies are our best sources of information and should be found as quickly as possible after explosion.

Confirming your discovery

When a star has been found that is not on the charts and photos of the galaxy under review, it is necessary to have the discovery checked to make sure it is publicly visible (not imaginary), and is not some other type of object. If it becomes clear that a new supernova has been discovered, then a CCD picture or ordinary photograph must be taken, either by the observer or some suitably equipped friend. This will provide an exact position of the new star with offsets from the nucleus of the galaxy, and a more accurate estimate of the brightness of the new star. All observations and details about the galaxy, the new star, positions, reference materials, telescope sizes, times and locations should then be e-mailed to the Central Bureau for Astronomical Telegrams in Boston.

Professional astronomers actively involved in supernova research will be contacted by the Bureau and asked to obtain SPECTRA of the new star, which provides the final confirmation of the new discovery. After that, the discovery will be officially announced, and research upon it will progress further.

These latter stages of a supernova discovery are shared, regardless of the method of discovery. It is true to say, however, that few discoveries are now made by the methods described above and are instead the product of very modern and often highly automated imaging through large telescopes.

The first and perhaps most important decision to be made regards the selection of a target catalog; a compromise must be reached between covering a large number of galaxies and obtaining longer exposure images of each (and hence detecting fainter supernovae). In any case, electronic catalogs allow data from a large number of catalogs to be easily combined. The choice of area to study may be influenced by the location of the observer; for example, southern-hemisphere observers may concentrate on the region south of –20° declination to avoid competition from northern-hemisphere observers. As with visual observations, nearer galaxies are surveyed more often than distant ones, with a typical night involving the imaging of up to 400 galaxies.

Automated system

A typical system will be almost completely automated, utilizing computer control software to point the telescope and obtain the image. The use of CCDs allows the image to be compared with a reference almost immediately after observation, offering a huge advantage over photographic film for which developing was often left until the next day. Using this method, 200 images an hour can be checked. This system can be implemented easily and relatively cheaply using commercially available telescopes such as the LX range produced by Meade. The CCD cameras used are often also commercially available; improvement in performance often depends on artificial cooling, which reduces the thermal noise that interferes with a 'clean' image. A typical system can produce a camera temperature of –10°C, even in warm external conditions.

However one observes, it is clear that the amateur influence on this important branch of astrophysics will continue for many years yet.

▲ ***Meade LX200***
Robin Chassagne's telescope under its rolling shelter. The camera is an SBIG ST7E CCD imaging camera.

Rev. Robert Evans is a Protestant clergyman (Uniting Church in Australia), recently retired from parish responsibilities. Since SN1981A, he has discovered 35 supernovae visually, mostly using backyard telescopes of 25-cm to 41-cm aperture, as well as four other supernovae and a comet on UK Schmidt films in 1996.

Robin Chassagne's St. Clotilde Observatory is on La Réunion Island, in the Indian Ocean east of Madagascar. He started searching for supernovae three years ago and has discovered 12 to date.

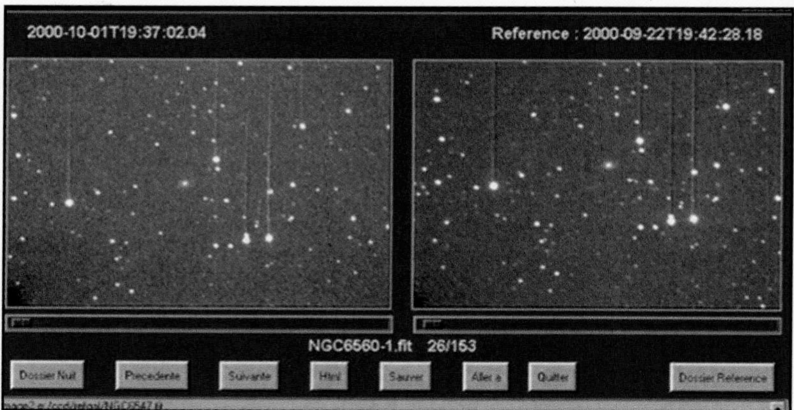

▼ ***SnVisu***
Programs like SnVisu can be very useful in analyzing data. Here, the image of what is being observed is compared against its reference; 200 fields per hour can be checked.

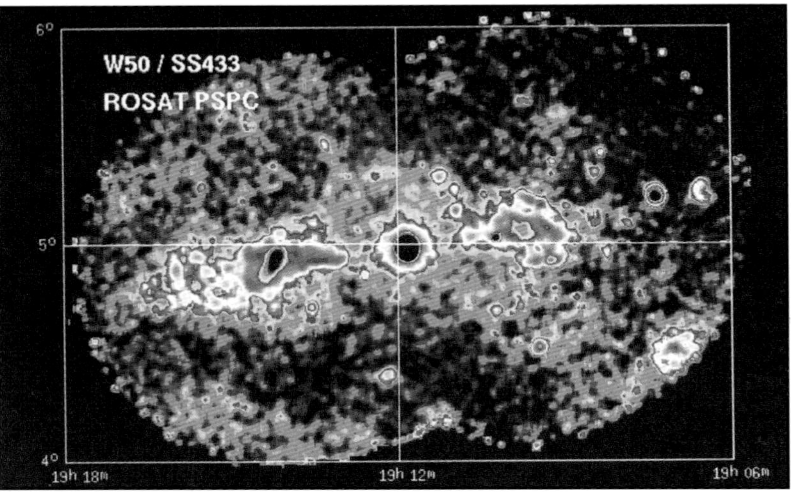

▲ ***Supernova remnant*** *W50 is centered on the star SS433 in this picture made by the ROSAT X-ray satellite. SS433's two fast-moving jets have collided with material inside the supernova remnant to make bright plumes to its left and right. The brighter left-hand jet is pointing toward us.*

for SPIRAL GALAXIES like the Milky Way. Astronomers observe a huge number of galaxies at great distances. Currently about 100 supernovae are discovered in them each year.

Astronomers classify supernovae by their SPECTRUM and by their LIGHT CURVE, the pattern of rapid brightening and slower dimming of each event. There are two basic types, Type I and Type II. The spectra of Type I supernovae reveal no detectable HYDROGEN, which is surprising since hydrogen is the most common element in the universe. Type II do have hydrogen spectral lines. This simple fact gives a clue to the progenitor stars of supernovae.

• Type Ia. Some Type I supernovae, Type Ia, appear in all kinds of galaxies, but avoid the arms of spiral galaxies. It follows that Type Ia supernova are explosions of older, longer-lived stars, since no stars formed recently in ELLIPTICAL GALAXIES. Type Ia supernovae show characteristic elements in their spectrum, such as magnesium, silicon, sulfur, calcium and iron. Their light curves have a quick (two weeks) initial rise to a peak and a slower decay lasting months (as recorded for their supernovae by Tycho and Kepler). All Type Ia supernovae are very similar.

Observationally, most WHITE DWARFS are near their maximum mass of about 1.4 solar masses, which is known as the Chandrasekhar limit (*see* Subrahmanyan CHANDRASEKHAR). White dwarfs are old stars and almost identical. The similarity of Type Ia supernovae points to an origin in exploding white dwarfs. One star in an orbiting pair evolves to a white dwarf, gathers mass from its companion, surpasses the Chandrasekhar limit, and explodes.

• Type II. By contrast with Type I supernovae, Type II supernovae have normal abundances, including a normal complement of hydrogen. No one has ever discovered a Type II supernova in an elliptical galaxy. They occur mostly in the arms of spiral galaxies, so the stars that make Type II supernovae are massive stars. The light curve of a Type II supernova shows a rise to peak brightness in a week or two, then a period of a month or two when the light output is nearly constant. The optical luminosity then drops suddenly over a few weeks and more slowly over months. This light curve is consistent with an explosion in the core of a RED GIANT STAR with a massive, extended envelope.

• Types Ib and Ic. These are two other less common varieties of hydrogen-deficient supernovae. They are probably closely related to each other and to Type II supernovae, because Types Ib and Ic are also associated with massive stars. Type Ib have helium in their spectrum, but Type Ic do not. Both Type Ib and Ic show oxygen, magnesium and calcium at later times. Their composition is similar to the core of a massive star that has been stripped of its hydrogen, and in the case of Type Ic, most of the helium.

Type II and Type Ib/c supernovae thus represent the explosion of massive stars that have evolved from the MAIN SEQUENCE to red giants. Neutron stars form in the Galaxy at the same rate that Type II and Type Ib/c supernovae occur and at the rate that stars exceeding about 10 solar masses die (*see* NEUTRON STAR AND PULSAR). This does not prove that these three are related through cause and effect, but it is a nearly universal working hypothesis. Some massive stars may explode without any stellar remnant and some form black holes. Perhaps stars of 6–10 solar masses produce no stellar remnant and stars over 50 solar masses produce a black hole.

Supernova remnants

A supernova remnant (SNR) is the aftermath of a supernova, including interstellar gas swept up in a shock wave, the expanding stellar debris and, in some cases, a compact stellar remnant. Some SNRs, like the Crab Nebula, are powered by pulsars while others show no sign of a neutron star or black hole. The kinetic energy of the exploding material is transformed by a strong shock wave into thermal energy and cosmic-ray energy. As the SNR ages, the shock slows down, and the shell of shocked gas merges back into the interstellar medium. During the course of this evolution, an SNR can be observed as an expanding shell of hot gas containing many solar masses and emitting radiation across the entire electromagnetic spectrum, from radio to gamma-rays, with details depending on its age.

SNRs are often observed as shells of hot gas a few parsecs to tens of parsecs in size. They produce strong X-rays and radio emission, and infrared emission from heated interstellar dust grains.

For the first 1000 years or so of its life, the SNR is strongly influenced by the nature of the explosion. The simplest SNRs are produced by Type Ia supernovae, appearing as nearly circular, hollow shells in X-ray and radio emission, like Tycho's Supernova.

In a few young Type II SNRs, fragments of the core of the progenitor star are visible. They appear as knots of material completely devoid of hydrogen and helium. The prototype is Cassiopeia A (Cass A), which ejected tens of solar masses of material from the progenitor star that has undergone nuclear burning. Some material shows pure oxygen emission, some oxygen mixed with sulfur, and some carbon, argon and calcium as well. The Cass A knots travel at around 6000 km s^{-1}. Cass A also shows 'quasi-stationary flocculi,' which are other knots, relatively slow moving. They are rich in helium and nitrogen, which are material shed by the SN progenitor during an earlier red giant phase of its evolution.

If a Type II supernova leaves a neutron star, and it is a strong pulsar, it dominates the SNR for many years. The Crab Nebula is an irregular system of rapidly expanding filamentary emission and an amorphous cloud of polarized SYNCHROTRON emission, powered by the Crab pulsar. The filaments occupy a thick shell expanding at 700–1800 km s^{-1}. SNRs similar to the Crab include 3C58 and CTB87. 3C58 may be the remnant of a supernova explosion in the year 1181.

◀ ***Supernova 1987A*** *(right) and the same region of sky (left) before the supernova. In March 1987, a month after its discovery, the supernova reached magnitude 3, easily visible to the eye, then slowly faded.*

After an SNR has swept up much more interstellar gas than the mass of the stellar ejecta, the character of the explosion itself becomes unimportant. The shock expansion slows and a bubble of hot, X-ray-emitting gas remains inside. Older SNRs generate radio emission by compressing the cosmic-ray electrons and magnetic field already present in the interstellar medium.

Supernova 1987a

Supernova 1987a was a Type II supernova that occurred in the Large Magellanic Cloud about 150 000 l.y. from Earth. It was discovered at magnitude 5 through photography by Ian Shelton and by eye by Oscar Duhalde on February 24, 1987.

A flash of NEUTRINOS, lasting 12 s and from the direction of SN1987A, occurred on February 23, 1987, detected through their interaction with protons in large underground tanks of water at the Kamiokande II experiment in Japan and the Irvine–Michigan–Brookhaven (IMB) experiment in Ohio, USA. The details of the 20 neutrinos confirmed that a neutron star formed. However, no neutron star has since been detected. It may be a weak pulsar (slowly spinning or with weak magnetic field), or it may have been a temporary stage on the way to form a black hole.

Gamma-rays from the decay of cobalt-56 in SN 1987A were directly detected by the SMM. They came from the radioactive decay of about 0.07 solar masses of nickel-56 produced during the explosion (*see* ELEMENTS, FORMATION OF).

The progenitor star of Supernova 1987A was Sk-69°202, the only progenitor star of a supernova about which anything is known. It was spectral type B2 Ia, with visual magnitude 12. Thus, against expectation, SN 1987A exploded as a rather compact blue supergiant instead of a red giant. The reasons for this are still not clear; the progenitor star might have had a binary companion that was consumed when the progenitor became a red giant.

SN 1987A is surrounded by nested rings of gas formed by the progenitor before the explosion. They may be related to the binary interaction or some stellar-wind phenomenon. The ejecta from the supernova have begun to collide with the innermost ring, which will eventually shred them into quasi-stationary flocculi.

Supernovae and the universe

Because they are so bright, and 'standard bombs,' always reaching the same absolute brightness (*see* STANDARD CANDLE), Type Ia supernovae have been used to determine distances. This information suggests that the universe is accelerating its expansion (*see* DARK ENERGY AND THE COSMOLOGICAL CONSTANT).

Further reading: Wheeler J C (2000) *Supernovae, Gamma-ray Bursts and Adventures in Hyperspace* Cambridge University Press.

Supersymmetry

A set of rules that attempts to link matter particles (fermions, such as QUARKS and LEPTONS) and force-carrying particles (GAUGE BOSONS) into a single framework, or 'family.' It requires that each type of particle has a supersymmetric partner. For example, a PHOTON (which is a boson) has a fermion counterpart (the photino), an electron (fermion) is partnered by a selectron (boson), a quark by a squark, and so on. The theory therefore predicts the existence of a wealth of hitherto unknown (and undiscovered) particles. When the ideas of supersymmetry are applied to gravity, the resulting theory – supergravity – predicts the existence of one or more types of gravitinos (fermions) to partner the graviton (the hypothesized force-carrying particle for gravitation).

Surface gravity

The surface gravity, *g*, (or acceleration due to gravity) of a body is the acceleration toward the center of that body due to its gravitational field. It is equal to $G M/R^2$ where G is the Gravitational constant and M and R are the mass and radius of the body respectively. It is thus independent of the mass of the accelerated object. On Earth the value of g is 9.81 m s^{-2}, but it can vary from place to place on the Earth's surface because of different distances to the Earth's center and also because it is affected by local deposits of light or heavy material. The weight of an object on the surface of a planet is equal to the mass of that object times the surface gravity of that planet.

Surveyor missions

Series of seven NASA automated MOON landers, launched 1966–8 and designed to send back images and other data on potential APOLLO landing sites near the lunar equator. Surveyor 1 was the first spacecraft to make a controlled soft landing on another world. Surveyors 5–7 carried out *in situ* soil analyses. Surveyor 6 performed the first lift-off from the lunar surface. Surveyors 2 and 4 were failures. Parts of Surveyor 3 were returned to Earth by Apollo 12 astronauts.

Synchronous orbit

An orbit in which the period of REVOLUTION of an orbiting body is the same as the rotation period of the body it is orbiting. In the solar system, the only natural instance of a synchronous orbit is that of Pluto's satellite, Charon. Some artificial satellites are put into synchronous orbit around the Earth; such orbits are termed geosynchronous. If in addition such an orbit is circular and in the Earth's equatorial plane, it is termed geostationary, since from a point on the Earth's surface a satellite in such an orbit appears to keep approximately the same position in the sky. Geosynchronous and geostationary orbits are used for navigation and communications satellites. *See also* SYNCHRONOUS ROTATION.

Synchronous rotation

The ROTATION on its axis of a celestial object in the same time as it takes to orbit another object, also called captured rotation. In other words, the orbiting body's rotation and REVOLUTION periods are identical. Synchronous rotation results from tidal forces between the two bodies. Most of the major satellites in the solar system, the Moon included, have synchronous rotation and always keep the same face turned toward their parent planet. In the Moon's case LIBRATION means that in the course of time rather more than half of the Moon's surface is visible from the Earth. Pluto's satellite Charon is unique in having both synchronous rotation and a SYNCHRONOUS ORBIT: its period of revolution, and the rotation periods of both planet and satellite, are identical. From either body, the other is motionless in the sky. Synchronous rotation can also occur in BINARY STARS with near-circular orbits.

Synchrotron radiation

Radiation emitted by ELECTRONS moving at very high speeds in magnetic fields. An electron follows a spiral path around a magnetic field line and, at relativistic (close to the speed of light) velocities, emits radiation in a narrow cone in the direction of its motion. Synchrotron radiation has a characteristic SPECTRUM and generally exhibits high polarization. The synchrotron process is an important source of radiation from astronomical sources. For example, synchrotron radiation is observed at radio wavelengths from electrons spiraling through Jupiter's magnetic field. In the CRAB NEBULA synchrotron radiation generated by electrons moving in the magnetic field associated with the supernova remnant is observed at radio wavelengths and in the visible and ultraviolet. Synchrotron radio emission has also been detected from other supernova remnants in our Galaxy and from extragalactic objects such as RADIO GALAXIES and QUASARS.

Synodic period

The time interval between two successive similar alignments of two celestial bodies. In the case of a planet, the synodic period may be taken to be the mean time interval between, say, two successive OPPOSITIONS or two successive CONJUNCTIONS. It is this period that determines the times at which particular planets will be visible in the night sky. The Moon's synodic period is the time between successive recurrences of the same phase such as between one full moon and the next. *See also* SIDEREAL PERIOD.

Syrtis Major

A prominent triangular feature on the equator of Mars, dark by contrast with its surroundings. It takes its name from the historical name for the larger of two quicksands off the North African coast. Syrtis Major is prominent in telescopic views of the planet, and was in fact the first feature of Mars ever to be recorded, in a sketch made by Christiaan HUYGENS in 1659. It coincides with a volcanic plateau which bears the official name Syrtis Major Planum (changed in 1982 from Syrtis Major Planitia). A low-relief shield volcano on the plateau is probably responsible for the dark material that covers the region; winds shift this material around, causing changes in the appearance of Syrtis Major as seen from Earth and creating dune fields.

Syzygy

The approximate alignment of three celestial bodies. Examples are the Earth, Sun and Moon at new Moon and full Moon, and the Earth, Sun and a planet when the planet is at OPPOSITION or CONJUNCTION.

Tarantula Nebula (NGC 2070)

An EMISSION NEBULA in the Large Magellanic Cloud, in the constellation Doradus (*see* MAGELLANIC CLOUDS). At 0.5° across it is far larger than any emission nebula visible in our Galaxy. It is energized by several young, hot O- and B-type stars (*see* STELLAR SPECTRUM: CLASSIFICATION), the brightest of which is designated R136. The Tarantula Nebula gets its name from its spidery appearance.

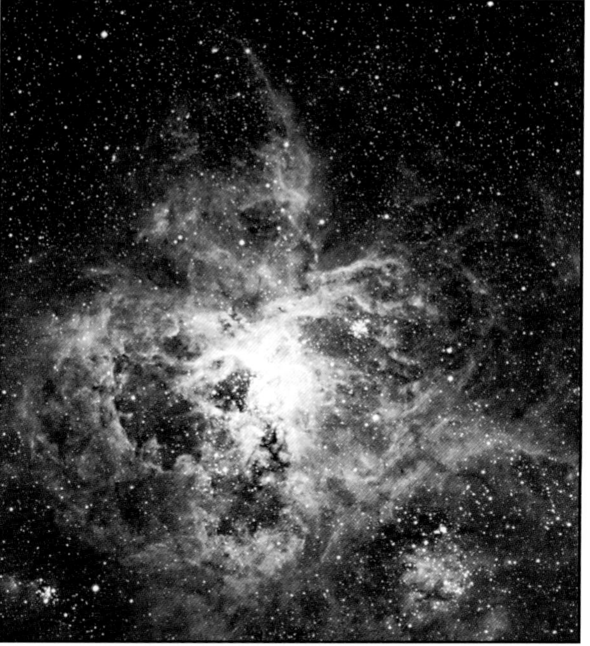

◀ ***Tarantula Nebula*** *This object is the only extra-galactic emission nebula that can be seen with the unaided eye. It is a faint patch of light at the eastern end of the Large Magellanic Cloud, 160 000 l.y. distant. A small telescope reveals narrow spindly tendrils of glowing gas, which have been likened to the legs of a spider.*

Tartu Observatory

This observatory in Estonia accommodates the northernmost 1.5-m telescope in the world. It is located about 20 km south-west of Tartu in the village of Tõravere. The observatory was originally founded in 1808 as part of Tartu University and it gained worldwide renown under the leadership of Friedrich STRUVE (director of the observatory from 1820 to 1839). In 1824 the observatory was equipped with a 23-cm Fraunhöfer telescope, which was then the world's largest REFRACTOR. In 1958 building began at a new observing site in Tõravere, and in 1964 the new observatory was opened. Major observing facilities include the 1.5-m telescope (installed in 1975) equipped with a Cassegrain spectrograph, and the 0.6-m telescope (installed in 1998) with a CCD photometer.

Taurids

A METEOR SHOWER with two RADIANTS in the constellation Taurus, at DECLINATION +14° and +22°. The METEOR STREAM consists of debris ejected from the now feebly active COMET ENCKE. The Taurid stream has spread out over a broad swathe of the inner solar system, and it takes the Earth six weeks during October to December to pass through it, with low but steady rates of slow, bright METEORS. The Earth also passes through the stream between June and July, producing the daytime Beta Taurid shower observed by radar.

Taurus *See* CONSTELLATION

Taylor, Joseph (1941–)

Radio astronomer, born in Philadelphia, Pennsylvania, USA. Won the Nobel Prize for Physics in 1993 with Russell HULSE 'for the discovery of a new type of pulsar, a discovery that has opened up new possibilities for the study of gravitation' (*see* NEUTRON STAR AND PULSAR). From an interest in radio as a boy, he went on to become a professor at the University of Massachusetts, Amherst, and with Hulse, his research student, searched with the Arecibo radio telescope for pulsars (*see* ARECIBO OBSERVATORY). With this telescope, they discovered the Hulse–Taylor Pulsar, PSR 1913+16, a binary pulsar that on prolonged study showed general relativistic effects, including loss of energy by gravitational radiation. The pulsar confirmed Albert EINSTEIN's theory of GENERAL RELATIVITY. Taylor helped found the FIVE COLLEGE RADIO ASTRONOMY OBSERVATORY.

Tectonics

The term for processes that shape the surface features of a planet (or satellite) through forces generated in its crust by heating or cooling in its interior. Tectonic forces occur in planets that have undergone DIFFERENTIATION and have developed a molten mantle beneath the crust. Global heating causes the crust to expand, creating tensional forces in the crust; global cooling will cause it to contract, creating compressive forces. Lunar RILLES are examples of tectonic features caused by expansion. On Mercury, extensive LOBATE SCARPS (long cliff-like features) result from compression. Wrinkle ridges, found on most of the terrestrial planets, are also compressive features. Localized heating in the mantle that produces upwelling can raise the crust above; on Venus this created the mountainous region MAXWELL MONTES. The Earth's crust consists of slowly drifting plates, driven by CONVECTION currents in the mantle. Plate tectonics, as this process is known, may also have operated on Venus. Tectonic processes have also shaped the surfaces of EUROPA and GANYMEDE, and of some of the satellites of Saturn and Uranus, including ENCELADUS and ARIEL.

Teide Observatory

The Observatorio del Teide, at 2.4 km above sea level on the island of Tenerife in the Canary Islands, contains several solar telescopes (the diameter of the largest exceeds 90 cm) operated by different countries, as well as helioseismological instrumentation, radio telescopes to study the COSMIC MICROWAVE BACKGROUND, various optical telescopes, a 1.55-m infrared telescope and an optical ground station for communication with satellites and the cataloging of space debris.

Tektite

A small, rounded piece of natural glass that is green, black or brown and typically centimeter-sized. Tektites are mainly silica with some metal oxides. Once believed to have a volcanic origin or to be peculiar METEORITES, they are formed by the impact on Earth of a kilometer-sized ASTEROID, which melts terrestrial rock and splatters molten droplets into the atmosphere (tektite comes from the Greek *tektos*, meaning molten). During their flight, they rapidly cool from a temperature of 2000 K, but their spinning flight through the atmosphere melts their surface and shapes the tektites into disk shapes or tear-shaped droplets. Tektites are found in up to seven main locations, known as strewn fields (*see* table). The tektites found in a strewn field have nothing in common geologically with their surroundings, and in three cases there is an identifiable impact structure.

Microtektites are microscopic tektites. Some are associated

Strewn fields of tektites

Strewn field	Age (millions of years)	Impact structure (diameter)
Australasia, including Southeast Asia	0.8	-
Ivory Coast	1	Bosumtwi Crater, Ghana (10.5 km)
Irgiz, near Aral Sea	1.07	-
Mauritania & West Africa	3.5	-
Moldavia & Central Europe	15	Reiskessel, Germany (24 km)
Libya	28.5	-
North America (Texas, Georgia, Martha's Vineyard, Cuba)	35	Chesapeake Bay, USA (90 km)

with the known strewn fields. Others are found across the world in layers deposited at the time of the Chicxulub meteor crater (*see* CHICXULUB BASIN). Yet others are impact glass found close to smaller meteor craters.

Telescope
See REFLECTING TELESCOPE; REFRACTING TELESCOPE. *See also* practical astronomy feature Telescopes, p. 420 .

Telescopium *See* CONSTELLATION

Telluric planet *See* TERRESTRIAL PLANET

Temperature scales
Three different scales for describing temperature are in common use:
- The Celsius scale was devised in 1742 by the Swedish astronomer Anders Celsius (1701–44). One Celsius degree (1 °C) is one-hundredth part of the difference in temperature between the freezing point and the boiling point of water (at standard sea-level atmospheric pressure). Because it is based on dividing that temperature range into 100 parts, it is also known as the centigrade scale. Temperatures below 0 °C are given negative values.
- The Kelvin, or absolute, temperature scale was named after the British physicist Lord Kelvin (William THOMSON). One Kelvin (1 K) is equal in magnitude to one degree Celsius. The zero of the Kelvin scale is ABSOLUTE ZERO, the temperature at which, classically, the motion of atoms would cease completely and which, therefore, is the lowest possible temperature. Absolute zero (0 K) corresponds to a temperature of –273.15 °C. Because the Kelvin is equal in magnitude to the Celsius degree, a temperature expressed in Kelvin is equal to the temperature in degrees Celsius plus 273.15. The Kelvin is the SI unit of temperature (the unit of temperature in the International System of Units).
- Although obsolescent, the Fahrenheit scale, which was devised by the German instrument-maker Gabriel Fahrenheit (1686–1736), is still used in some contexts. The freezing point of water is 32 °F, and the boiling point of water 212 °F. To convert a temperature in °C to a temperature in °F, multiply by 9, divide by 5 and add 32.

Tenma (Astro-B)
Second Japanese X-ray satellite, launched in February 1983. Tenma detected X-rays in the range 2–60 keV. It also carried a transient source monitor. Tenma ceased operations in 1984. Its name means 'Pegasus' in Japanese. *See also* ASTRO.

Terminator
The line marking the boundary between the day and night sides of a planet or moon. As the relative positions of the Earth, the Sun and the object shift, the proportion of the object's sunlit side visible from the Earth changes and the terminator moves across the object's disk. The terminator thus determines the object's PHASE. The Moon's rough terrain gives its terminator a jagged appearance. Sometimes, elevated features just to the night side of the terminator are visible, obliquely illuminated by the Sun.

Terrestrial planet
A collective term for the major planets Mercury, Venus, Earth and Mars, which are all 'Earth-like' or 'telluric' in that they are about 10 000 km in diameter; have densities of roughly 5000 kg m^{-3}; have a rock/metal composition with a solid surface; and possess only a small number of satellites or none at all.

Tethys
A mid-sized icy satellite of SATURN, discovered by Giovanni Cassini in 1684 (*see* CASSINI DYNASTY). Its diameter is 1060 km, and its orbital distance from Saturn is 295 000 km. The parts of Tethys imaged during the VOYAGER missions show a variety of terrains with different crater densities. The heavily cratered side contains a large impact feature: the 440-km diameter basin Odysseus, a 'relaxed' structure that has signs of an inner mountain ring. On the hemisphere opposite Odysseus is a lightly cratered plain cut by Ithaca Chasma, a huge valley system of multiple faults over 100 km across at its widest and around 1000 km long. The formation of Ithaca Chasma may be linked with the impact that produced Odysseus. The plains are a consequence of resurfacing in the distant past.

Thales of Miletus (624–546 BC)
The first known Greek philosopher, he was also a scientist, mathematician and engineer. Thales was born in Miletus, Asia Minor (now Turkey). He is believed to have taught the Greek philosopher Anaximander. Thales apparently wrote a book on navigation, in which he defined the constellation Ursa Minor and used it as a navigation aid. He is credited with predicting an eclipse of the Sun in 585 BC, although it is not known how he did this. Nevertheless, the eclipse stopped the war between the Medes and the Lydians, according to Herodotus. Thales determined the height of pyramids by measuring the length of their shadow at the moment when a man's shadow is equal to his height. He taught that 'all things are water,' which may seem an unpromising hypothesis but was the first expression of the idea of explaining different phenomena from a common underlying cause, and foreshadowed scientific theory. Like the Egyptians, Thales believed that the Earth was a flat disk floating on an infinite ocean, explaining earthquakes as the shaking of the floating 'boat.'

Tharsis Bulge
A large raised area in the northern hemisphere of MARS, also known as the Tharsis Ridge. It takes its name from the ancient Spanish town known formerly as Tartessus. Tharsis measures 2105 km across at its greatest dimension, and contains the planet's most prominent volcanoes. In a 1000-km chain from northeast to southwest lie Ascraeus Mons (rising to 26 km above Mars's average surface level), Pavonis Mons (18 km) and Arsia Mons (20 km), all of which are about 150 km

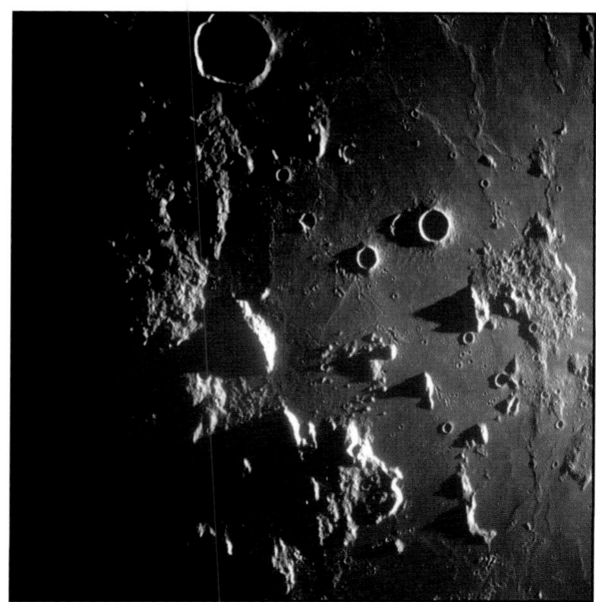

◄ ***Terminator*** *This image, taken from Apollo 12 in lunar orbit, shows the lunar terminator, the line separating daylight from darkness. The low Sun angle causes objects near the terminator to appear high and sharp. The crater Gambart, 25 km in diameter, is on the terminator at the north (upper) part of the frame.*

across, while 1000 km to the northwest is the solar system's largest volcano, OLYMPUS MONS. These peaks apart, the highest point on the Tharsis Montes is the complex, fractured terrain named Noctis Labyrinthus, with an elevation of 11 km. This is part of an extensive fracture system aligned approximately on the center of Tharsis; other prominent fractures are Ceraunius Fossae to the north and Claritas Fossae in the south. From the east of the Tharsis region, the giant canyon system of Valles Marineris extends eastward. It may be that the whole region was uplifted by an upwelling in the underlying mantle, possibly as a consequence of the event that created the impact basin Hellas on the opposite side of the planet.

Tharsis Montes

The main mountain and volcanic peak on the THARSIS BULGE.

Thémis (Téléscope héliographique pour l'étude du magnétisme et des instabilités solaires)

A heliographic telescope for the study of magnetic instabilities on the Sun, built by a Franco-Italian collaboration at Teide Observatory in Tenerife (Canary Islands, Spain) at an altitude of 2456 m. Thémis is a 90-cm-diameter Ritchey–Chrétien altazimuthal telescope for observing the Sun's high (spatial, spectral and polarimetric) resolution.

Thermal emission *See* FREE-FREE RADIATION

Thomson, William (Baron Kelvin of Largs) (1824–1907)

Born in Belfast, Ireland, he became professor of physics at Glasgow, Scotland, studied thermodynamics, and proposed an absolute scale of temperature (measured by the Kelvin TEMPERATURE SCALE). In parallel with Hermann von HELMHOLTZ Thomson formulated a theory of stellar evolution, in which a star radiated the energy that it released by progressive gravitational contraction. From this theory, the lifetime of a star (the time over which its gravitational potential energy is radiated from its surface) is the Kelvin–Helmholtz timescale, but it proved to be too short to be compatible with the geological age of the Earth.

Three-body problem

The problem of calculating the motions of three bodies moving under their mutual gravitational attraction, like the Sun, Earth and Moon. In the general three-body problem, three bodies of arbitrary mass start from an arbitrary initial configuration and move purely under the inverse-square force of attraction. There is no general analytical solution to this problem and the orbits are chaotic (*see* CHAOS; N-BODY PROBLEM). The astrophysically interesting three-body systems are:

- The orbits of COMETS and ASTEROIDS, particularly the TROJAN ASTEROIDS, and similar resonant orbits in the solar system.
- Stable hierarchical systems, examples of which are the numerous triple stars observed in the Galaxy (*see* DOUBLE STARS).
- Systems of stars (of comparable masses and initial distances) that break up after some dynamical evolution. Such configurations are typical in star-forming regions, like the TRAPEZIUM.
- Scattering of single stars off of close binary stars (*see* VARIABLE STAR: CLOSE BINARY STAR). These events are important in the dynamics of star clusters.

In the last two cases, as a rule, close approaches of two of the bodies occur while the third body is at a greater distance. Often the third body goes so far away that the other two bodies form a BINARY STAR and the third orbits the other two. At other times all bodies stay close to each other and perform repeated close encounters at frequent intervals. This type of motion is called 'interplay.' When one of the members of a temporary binary is replaced by another body, the event is called 'exchange.' Finally, one of the bodies may leave the binary permanently, in which case we have an 'escape'; this process forms RUNAWAY STARS.

Tide

Everyone who has lived near the ocean is aware of the variation in the level of the ocean water, called the tide, that is caused by the Moon. On most days there are two high tides and two low tides, occurring on average about 50 min later than they did the day before. The solid body of the Earth also experiences a tide, with an amplitude of about 30 cm. Each planet or satellite in the solar system experiences tidal distortions due to all the other objects attracting the planet or satellite gravitationally. The Earth tide due to the Sun is not as large as that due to the Moon, but it is quite noticeable when one considers that the highest and lowest tides occur when the Sun and Moon are nearly lined up.

The tidal deformation of a body is caused by differences in the gravitational forces exerted by the disturbing body on different parts of the disturbed body. As the tide oscillates, tidal energy is converted into heat, and causes slow changes in the orbits and rotations of many objects in the solar system. Tidal forces also affect stars, star clusters, galaxies and clusters of galaxies.

The Moon's gravitational attraction causes the Earth's oceans to take up an ellipsoidal shape. The ocean tide rises and falls at a given point on a coast because the Earth rotates under the Moon. The tidal bulge tends to stay more or less aligned with the Moon, so for our fixed position on the rotating Earth we are carried through alternate high and low points of the tidal deformation.

The consequence of the dissipation of tidal energy is that the Earth's rotation slows down, while the Moon is accelerated in the direction of its orbital motion and recedes. The rate at which Earth's rotation is slowing is such that 400 million

Telescopes

Optical types

All telescopes for amateur and professional astronomy fall into one of two basic categories: REFRACTORS and REFLECTORS. The former uses a main lens or OBJECTIVE LENS to gather and FOCUS the incoming light. Reflectors use a concave MIRROR. These two classes each contain many variations, including hybrid designs that combine elements of both refractors and reflectors. The most common form of refractor (popular in APERTURES sized between 60 and 150 mm) uses a doublet lens, formed by two lens elements, one made of crown glass and one made of flint glass. The different refractive properties of the two glass types work in tandem to cancel out the principle flaw of single-lens refractors, CHROMATIC ABERRATION, or false color. Lenses with this flaw do not bring all colors to the same focus, producing an image surrounded by out-of-focus blue and red light.

Even the best achromatic lenses do not eliminate false color completely. Refractors deemed 'apochromatic' reduce false color to negligible levels through the use of exotic low-DISPERSION glasses in two-, three-, or even four-element designs. Though expensive and limited to apertures below 8 in, 'apo' refractors can provide the finest optical performance of any telescope design. However, for amateurs seeking greater aperture at lower cost, the reflector is the best choice.

▶ *Starter scopes* *Meade's DS-2070 AT (left) and ETX-70 AT (right) are typical starter scopes. They are 70-mm achromatic refractors with Go To computer control. They can give the beginner an automated tour of the sky.*

All reflectors use a concave mirror finished with a thin aluminized coating that serves as the actual reflective surface. The simplest reflector (the Newtonian telescope), a design invented by Isaac NEWTON in 1668, employs a parabolic PRIMARY MIRROR (*see* PARABOLA) and a small, flat secondary mirror to direct light out of the side of the tube at the top. Newtonian reflectors with apertures as large as 600 mm remain extremely popular among amateur astronomers today because of the design's ability to deliver large aperture at low cost (*see* NEWTONIAN TELESCOPE).

A principal drawback is the need to occasionally adjust the tilt of the mirrors to keep the optics lined up or 'collimated' (*see* COLLIMATION). Newtonians, especially those with fast FOCAL RATIOS, also provide fields marred by COMA, an ABERRATION that spreads stars at the edges of the field into seagull shapes. Nevertheless, when fitted with first-class optics and mated to a Dobsonian style of mount, a large Newtonian provides the amateur with views of dim and distant objects unequalled by any other type of portable telescope.

All variations of the alternative reflector design first devised by Guillaume Cassegrain in 1672 use a convex secondary mirror to direct light back down the tube to exit through a hole in the primary mirror. The classical CASSEGRAIN TELESCOPE uses a parabolic primary mirror and a hyperbolic secondary mirror (*see* HYPERBOLA). The RITCHEY–CHRÉTIEN TELESCOPE design employs mirrors with more complex, aspherical surfaces to provide a wider aberration-free field than other Cassegrains, making this design popular among amateurs conducting long-focal-length imaging with film and CCD cameras (*see* CCDS AND OTHER DETECTORS).

Since it was introduced in the 1970s, the single most popular telescope design for amateurs has been the SCHMIDT–CASSEGRAIN TELESCOPE. This hybrid type adds a thin lens-like corrector plate at the front of the tube that cancels the spherical aberration of the fast spherical primary mirror. This combination, most popular in the 8-in size, provides wide aperture in a compact and portable tube assembly. The principal drawback is the large secondary mirror, typically 35–38% of the diameter of the primary mirror. A secondary obstruction over 20% begins to introduce a degradation of contrast due to the added DIFFRACTION created by the central obstruction. However, Schmidt–Cassegrains are still capable of providing sharp images of all types of celestial objects, making them good portable instruments for those interested in all types of observing and imaging.

A mirror-lens variation invented in the 1940s, the MAKSUTOV TELESCOPE employs a steeply curved corrector lens that works in tandem with the primary mirror to provide nearly aberration-free images. Most commercial Maksutov–Cassegrains in the popular 90- to 180-mm range of sizes use a separate secondary mirror, while in the design introduced by John Gregory in 1957, the secondary mirror is an aluminized spot on the inside surface of the corrector lens. Maksutov–Cassegrains are known for their sharp, refractor-like optics. Their principal disadvantages are their slow focal ratios (a drawback for deep-sky imaging) and, in large apertures, the long wait required for the optics to cool off and perform at their best on chilly nights.

Commercial manufacturers have recently introduced 6- to 10-in Schmidt and Maksutov versions of Newtonian reflectors, designs with front-element correctors to provide a wider and flatter field than is possible with a classic mirror-only Newtonian. Maksutov-Newtonians, in particular, can

▶ *8-in Schmidt–Cassegrain* *The Meade LX-90 8-in telescope is computer-controlled and fork-mounted. It is a popular scope for keen amateurs, giving high resolution visual observing of deep-sky objects. With proper accessories it is capable of CCD imaging and astrophotography.*

rival an aporefractor for sharp, high-contrast image quality over a wide field, but can suffer from the long cool-down time of Maksutov–Cassegrains.

Go To telescopes

One of the biggest changes in the way that amateurs have used their telescopes in the last decade has been the advent of widely available computer control. Indeed, all of today's common optical types of telescopes can be purchased with computerized mounts that can slew automatically to any of thousands of celestial targets. Called 'Go To telescopes,' these models have quickly become a prime choice for the aspiring amateur astronomer.

Their advantages are many; not only do they make it easy to locate targets, they can follow objects over many minutes or hours without need for any polar alignment, as required by traditional EQUATORIAL MOUNTS. For fork-mounted Makustov– and Schmidt–Cassegrain telescopes, for example, this means that the telescope does not need to be tilted over on a polar-aligned wedge. Instead, it can be mounted directly to the tripod as an 'altazimuth' telescope (*see* ALTAZIMUTH MOUNTING), making for a sturdier unit that is simpler to set up and less prone to the effects of shaking and vibration.

All Go To telescopes require an input of the current date and time, and the telescope's latitude and longitude on Earth. Some units acquire this data through a built-in receiver that picks up signals from orbiting Global Positioning System (GPS) satellites. Less elaborate models require the user to input that data. Then, the telescope needs to be pointed to and synchronized on two stars. This allows the telescope to match its internal database of object positions to the real sky. Once calibrated in this fashion, the user need only call up the object of interest from the keypad, or initiate one of the preset guided tours. The computer-driven motors take over to slew the telescope over to the object of interest and begin tracking to keep it centered in the field, all at the push of a button.

Reflecting telescopes

Reflecting telescopes remain the most popular type among amateur astronomers, and for good reason: they provide the maximum aperture for the lowest cost. This is particularly true of the simplest class of reflectors, the Newtonian. For this reason alone, Newtonians are the telescopes of choice for amateur astronomers wishing to track down faint deep-sky objects such as distant galaxy clusters (*see* GALAXY CLUSTER AND GROUP), or simply to see as much structure as possible in showpiece objects such as the ORION NEBULA. For deep-sky viewing, sheer aperture is the most important consideration. Through a large-aperture (11 in and larger) reflector, views of bright nebulae and galaxies begin to look like photographs, rather than like blurry patches of light.

However, such views require transporting the telescope to a dark-sky site. For the required portability, serious deep-sky observers often opt for a large Newtonian with a truss-pole tube that breaks apart into smaller components and that is cradled in a simple Dobsonian-style mount. With such a design, even a large 24-in telescope can fit into a sport utility vehicle, van or station wagon, a feat impossible with a traditional solid-tube reflector and equatorial mount.

However, Newtonians are not just deep-sky 'light buckets.' When fitted with first-class optics, a Newtonian can provide stunning views of the Moon and planets. This is particularly true of high-quality 11-in and larger Newtonians. Their generous apertures will often provide views of planetary detail that can rival, if not exceed, a large apochromatic refractor at much less cost. Dobsonian owners who wish to track targets can place their mounts on so-called Poncet or equatorial platforms. These swiveling tables can provide up to an hour of motorized tracking, a great aid when observing at high magnifications.

Those pursuing imaging with film or CCD cameras often choose a reflector with a more compact Cassegrain configuration, perhaps a Ritchey–Chrétien or Schmidt–Cassegrain. The rear-mounted focuser facilitates the attachment and balancing of heavy cameras and guiding accessories. And, in large telescopes, the camera is more conveniently placed (in a Newtonian it ends up at the top of a tall tube).

Reflectors of any type do have their disadvantages. The mirrors (usually two) need to be collimated from time to time. If the mirrors shift out of position, the result can be aberrated star images and blurry planetary detail. With a Newtonian telescope, both the primary and secondary mirrors can require adjustment; with a Cassegrain type, often only the secondary mirror need be adjusted. Commercially available collimation sighting tools (instructions on collimation procedures are usually included) allow for coarse adjustments of mirror-centering and tilt. Final adjustments can be performed by looking at a star at high power and tweaking the mirrors to ensure a symmetrical star image.

◀ 6-in Schmidt–Newtonian *Meade's 6-in LXD55 on an equatorial mount with Go To computer control gives access to hundreds of deep-sky objects for the amateur stepping up from a starter scope.*

Exposed mirror surfaces can also require cleaning from time to time. However, with all reflectors, frequent and careless cleaning can scratch the delicate coatings. A little dust is far less detrimental. When cleaning is required, one should use a mild solution of distilled water and a few drops of liquid soap. Moistened cotton balls should be used to gently swab the mirror, with no more than the weight of the cotton balls for pressure. It should then be rinsed with distilled water and stood on its edge to dry so that the water runs off without spotting the surface. With a little care, a reflector can perform well for decades.

An amateur astronomer since the 1970s, Alan Dyer is a frequent contributor of telescope reviews to *Sky & Telescope* and *SkyViews* magazines, and is co-author with Terence Dickinson of *The Backyard Astronomer's Guide*.

Refracting telescopes

The refracting telescope was invented by Hans LIPPERHEY in 1608 and first used astronomically by GALILEO GALILEI in 1610. Until the first achromatic lens was designed in 1729 by Chester Moor Hall (1704–71) and was made by George Bass (died 1769) in 1733, astronomers developed telescopes of enormously long focal length in order to reduce the chromatic aberration of singlet lenses. Improvements in lens design and glass technology have resulted in a wide variety of refractors being available to the modern amateur astronomer. These range from inexpensive achromats to apochromats of exceptional quality and correspondingly exceptional cost. Today's mid-range refractors match the quality of those used by our forebears in the nineteeth century, the 'heyday of the refractor.'

▶ 6-in achromatic refractor *The Meade LXD55 6-in refractor has higher contrast than a reflector of the same aperture. The mounting is a German-type equatorial mount.*

The advantages of the refractor over conventional reflectors include:

- no obstruction in the light path to degrade the image;
- they are less affected by temperature changes;
- the fully enclosed tube reduces air currents, leading to steadier images;
- the optical components rarely require collimation – an especially important consideration if the telescope is intended to be portable;
- it is easier to effectively baffle stray internal reflections that would otherwise degrade the image;
- a smaller optical tube for equivalent image quality;

The disadvantages include:

- chromatic aberration (which is, however, reduced to almost imperceptible levels in high-quality apochromats);
- expense: a refractor is generally more expensive than a reflector of equivalent aperture;
- aperture: there are very few amateur refractors of over 6-in aperture.

A good refractor will give images of very high contrast and detail, and, consequently, is the preferred instrument of lunar and planetary observers, as these observers strive to see as much detail as possible. The optimal magnification for planetary observation is approximately equal to the diameter of the aperture measured in millimeters. Refractors are also favored by DOUBLE STAR observers, for whom image contrast is of primary importance; they may use magnifications more than twice the measurement of the aperture (again, in millimeters). Many deep-sky observers appreciate this high contrast, and consider it to be adequate compensation for the increased light-gathering of larger reflectors.

Owing to their use at high magnifications, large refractors are best mounted equatorially, thus enabling easy tracking of the target object. A German equatorial mount, supported by a pier, is the mount of choice because of the length of the optical tube. The height of the EYEPIECE above the ground varies a great deal, so either observing steps or a variable-height observing chair are desirable for comfortable viewing. These considerations do not apply to smaller 'grab it and go' refractors, which may be mounted more simply and whose variation in eyepiece height is correspondingly smaller.

The essential accessory for a refractor is a STAR DIAGONAL. This is a prism or mirror that is placed between the focuser and the eyepiece, and which reflects the light through 90°. Without a star diagonal, observing near the ZENITH is extremely uncomfortable unless the observer is reclining.

Other desirable accessories depend upon the uses to which the refractor is put. Color filters can enhance certain lunar and planetary features; a minus-violet filter will reduce the effects of chromatic aberration; and proprietary correctors can minimize the aberrations of achromats, giving them near-apochromatic optical properties.

Stephen Tonkin is the author of *AstroFAQs* and *Amateur Telescope Making*, and editor of *Practical Amateur Spectroscopy* and *Astronomy with Small Telescopes*.

Software for amateur telescopes

In 1970, an amateur might dream of access to a small, state-of-the-art institutional telescope run by a $150 000 minicomputer with only 4 kilobytes of memory and nothing more elegant than assembly-language programming. By the mid 1980s, small numbers of amateurs were building computer-operated telescopes thanks to hardware advances such as, very importantly, the personal computer. The rise of the Internet allowed software and project knowledge to be widely shared. Today, amateurs look to both remote-control and robotic operation, using software from a variety of vendors.

Telescope software falls into three categories:

- positional encoders, called computer-aided pointing;
- motors for tracking and slewing, called computer-operated or 'Go To' Telescopes;
- integrating software, used to connect encoders and motors to cameras, filter wheels and focusers, thus allowing coordinated control.

The most popular types of positional sensors are optical wheel encoders. These small (5-cm diameter), lightweight, inexpensive (around $60) devices convert rotary motion into digital pulses by passing light through an optical wheel that is marked with opaque spokes. The microprocessor decodes each passing of a spoke into four events. This is called quadrature decoding. The best encoders are those with the highest number of pulses per revolution. Because of the necessity to service a high rate of encoder pulses, encoder software nearly always resides in a dedicated microprocessor. Tangent Instruments marketed the first encoder interface box, and it is used as the basis for a number of commercially available units.

Commercial units vary in capability, size of the object database and cost. While most can be used in conjunction with planetarium software running on a PC, laptop or hand-held computer, most of the time their built-in LCD screens

will be used. Consequently, ease of use coupled with capability should be the guiding factors when considering the purchase of an encoder interface box.

Integrating software such as the Astronomy Common Object Module (ASCOM), shepherded by Bob Denny, aims to give a framework for software-controlled devices so that they can bridge to, and interface with, each other. The bridging can be as simple as a Windows-based Excel script, or as complicated as a full-blown application written in C++. The script can command a sequence of events such as 'go to a star,' 'focus on the star,' 'autoguide on star,' 'open shutter for five minutes,' 'close shutter' and then 'download image'.

Computer control of telescopes has certainly advanced since the pioneering days, with even full-scale robotic telescopes being within the reach of today's advanced amateur. Both off-the-shelf and customized equipment can be utilized to make the experience of viewing the sky easier than ever before.

Mel Bartels is an amateur astronomer for whom asteroid 17823 has been named. He is a professional systems manager/analyst/programmer who as a hobby writes real-time code for computer-controlled telescopes.

Observatories

Whatever the nature of your telescope, you are more likely to enjoy it if it is permanently set up. Even portable equipment can be time-consuming to unpack, set up and align. If your telescope has been stored indoors you will have to wait while it stabilizes to the outdoor temperature before you can start serious observing. Many amateurs choose to house their telescopes in an observatory with power, catalogs, star charts and perhaps a computer and Internet access. These observatories can be built in backyards or on rooftops.

The observatory must be large enough and have an appropriate aperture for the telescope to move around the whole sky, and have room for the observer at the eyepiece with a few friends looking on. The traditional dome design consists of a fixed cylindrical base of appropriate height topped with a hemispherical roof (or dome). The dome has a shuttered aperture and rotates to allow the telescope access to the sky. More cost-effective structures include a clamshell-shaped dome or a shed with a roll-off roof, both of which open wide to the whole sky. All types of structure are available commercially, although some astronomers choose to build their observatories themselves.

Domes provide shelter from the wind and any street lights or security lights. However, a structure which opens to reveal the whole sky cools to the ambient temperature more quickly and is thus better for the seeing.

Margaret Penston

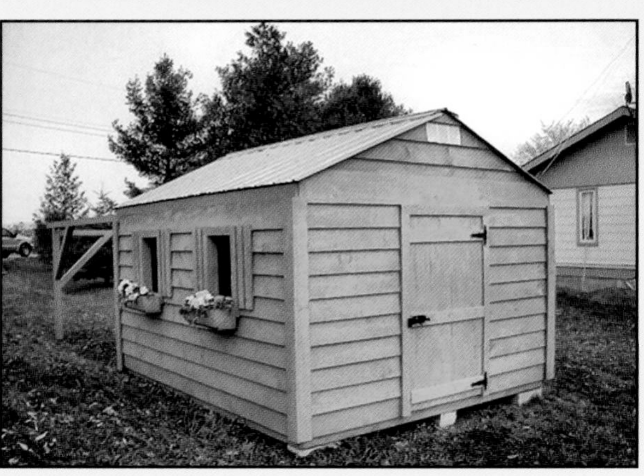

◀ ***Observatories*** *(Top, middle) Sirius Observatories of Australia's commercial 2.3-m dome for home use. (Bottom, left and right) The SkyShed roll-off observatory from NorthSheds, Canada.*

years ago, in the Devonian period, the day was 22 hours long. The Moon is receding at about 3.8 cm per year.

If a rapidly spinning planet is very massive compared to a satellite that orbits in a close circular orbit, the satellite is quickly slowed to SYNCHRONOUS ROTATION by large tides raised on the satellite by the planet. Where the two bodies are of comparable mass they lock into rotation synchronously with their revolution, as with many close binary stars.

If the tidal forces on a body are strong enough it will disintegrate. This happens if a satellite comes within the ROCHE LOBE of a planet. In such a case the debris may at first form a chain of pieces like COMET SHOEMAKER–LEVY 9 (which was broken up by Jupiter in this way), and eventually a PLANETARY RING. On a much larger scale, tidal forces between colliding galaxies produce streams of stars, dust and gas trailing from their parent galaxy (*see* GALAXIES, COLLIDING).

Time

There are many fascinating philosophical and deep scientific questions about the nature of time (*see* GENERAL RELATIVITY; SPECIAL RELATIVITY). This article is about the simpler process of determining time with clocks. For the purposes of astronomy, 'clocks' measure intervals of time that are shorter than a day (hours, minutes, seconds) or not much longer. CALENDARS are clocks that measure intervals of time longer than a day.

A clock delivers a repeated marker, such as the tick of an oscillating spring, and counts them. To use a clock to time the interval between two events, you read the clock when the events occur. All clocks are imperfect – even if two clocks agree at a given moment they will progressively depart from each other. The history of timekeeping has been a series of technical advances in choosing natural clocks or making artificial ones, with better and better accuracy, the latest clock showing up the imperfections of the earlier ones. At any period of history people have come together to choose and agree on standard clocks as the best available for the desired purpose.

For agricultural and domestic purposes, and for not very demanding pre-industrial purposes such as travel on foot or horseback, the standard clock used to be the SUNDIAL, showing SOLAR TIME. As the Earth rotates on its axis relative to the Sun, sunlight casts a shadow of a stick (a gnomon) onto a graduated scale (sundial) and shows the time, from day to sunny day. When the Sun is at CULMINATION the time is conventionally defined as 12 noon. Originally, and even now in some religious communities, the period between sunrise and sunset was divided into 12 hours from 6 a.m. to 6 p.m., even though the length of the time of daylight is manifestly variable from day to day throughout the year. Normal practice is to divide the period from noon to the next noon into 24 hours. Two sundials separated from east to west show times that differ by an amount equal to the difference of longitude of the two places, and could be as much as 12 hours different. For this reason time shown by a sundial is called 'local solar time.'

At night, the Earth can be observed to be rotating relative to stars, and it is possible with a telescope fixed to a point in a north-south plane to observe 'sidereal time,' as stars successively transit the MERIDIAN. Because in a year the Earth makes one more rotation relative to the stars than it does relative to the Sun, the sidereal day is on average 365.25/(365.25 + 1) times as long as a solar day (the sidereal day is 23 hours 56 min of solar time). Furthermore,

Radio-transmitted UTC time signals

Station	Location	Frequency
JJY	Japan	40, 60 kHz
RTZ	Irkutsk, Russia	50 kHz
MSF	Rugby, England	60 kHz
WWVB	Fort Collins, Colorado, USA	60 kHz
RBU	Moscow, Russia	66.66 kHz
BPC	China	68.5 kHz
HBG	Prangins, Switzerland	75 kHz
DCF77	Mainflingen, Germany	77.5 kHz
Loran-C		100 kHz
WWV	Fort Collins, Colorado, and	
WWVH	Kaui, Hawaii	2.5, 5, 10, 15, 20 MHz

astronomers found that solar time is irregular through the year, due to the tilt of the Earth's axis and the ECCENTRICITY of the Earth's orbit around the Sun, which causes the Earth to accelerate and decelerate in its orbit. Solar time corrected for these departures from uniformity is called MEAN SOLAR TIME, and the difference between local mean solar time and local solar time is called the EQUATION OF TIME (the difference ranges up to about 16 min).

Time can be read from a sundial and corrected for the equation of time to an accuracy of about 1 min. Technical advances in the seventeenth and eighteenth centuries produced mechanical clocks (pendulum clocks, spring operated clocks – *see*, for example, John HARRISON and Thomas TOMPION) that were able to read finer intervals of time (a second or so), and telescopes that could observe the transits of stars to comparable accuracy (*see* John FLAMSTEED). Mechanical clocks were kept in time with the telescopic observations to produce a uniform scale of sidereal time, which was converted to local mean solar time by calculation. Individual nations carried out this process at their national or subsidiary observatories (at Washington, Greenwich, Edinburgh, Paris, Cadiz, Cape Town and so on). Once each day time was transferred by a signal (a gunshot or the fall of a 'time ball' from a post) from the observatory to mechanical clocks on ships, and transferred around the world. The problem of knowing where you are and finding out the time can be inverted to knowing the time and finding out where you are. The history of time, clocks and star positions is directly connected through the requirements of NAVIGATION. This is true even in the twenty-first century with the US GPS (Global Positioning System), Russian Glonass (Global Navigation Satellite System) and European Galileo satellite navigational systems, and with radio networks such as LORAN (Long Range Navigation System).

With the advent of the railroad and travel that was timed to the minute, particularly with the east-west rail systems in the USA, differences of local mean solar time had a significant effect on the scheduling of trains. The International Meridian Conference in Washington, DC, in 1884 agreed a system of zones of 'standard time' based on the local mean solar time at Greenwich ('Greenwich Mean Time,' GMT), with any country or group of countries conventionally able to offset its time from GMT in a given region (such as a group of states) for a given season (in winter or in summer – daylight saving time) by a whole number of hours (or, exceptionally, half-hours). The time zones across North America are (from east to west) Atlantic (GMT –4 hours), Eastern (–5), Central (–6), Mountain (–7), Pacific (–8), Yukon (–9) and

◀ ***Time ball*** *The ball was raised up the mast, and at a fixed time each day allowed to fall. This was the first method for providing accurate time information to the general public.*

Alaskan/Hawaiian (–10) Standard Times. Across Europe, from west to east lie GMT, Central European Time (+1), and Eastern European Time (+2). The term Greenwich Mean Time has been replaced in scientific usage by the term UNIVERSAL TIME (UT).

Because of the TIDES, the day, as measured in UT, is lengthening; indeed there are smaller seasonal fluctuations within any given year. As a result UT is not the most accurate timescale, however convenient it is for regulating human existence. The motion of the planets was for a period used as a more accurate timescale (Ephemeris Time, ET, now called Terrestrial Dynamical Time, TDT, or Barycentric Dynamical Time, BDT). Time is at present derived from atomic clocks. One standard clock in common use (1 billion are produced annually) is a quartz crystal oscillator of particular shape, which can potentially split time into microseconds. The SI second, the unit of time of the International System of Units, has been defined since 1967 in terms of a hyperfine transition of the cesium atom, and the most accurate standard clocks now realize it with a relative uncertainty of a few parts in 10^{15} (approximately a microsecond over a millennium). This makes time the most accurately measurable physical quantity.

International Atomic Time (known from its French initials as TAI) is based on more than 200 atomic clocks distributed worldwide, coordinated since 1988 by the International Bureau of Weights and Measures in Paris. Clocks that contribute most significantly include ones at the US NAVAL OBSERVATORY in Washington, DC. TAI is used in the most critical astronomical measurements (for example, the pulses from pulsars). The accuracy demanded by these observations is so high that the terrestrial clocks have to be corrected for subtle effects from special and general relativity (due to the motion of the Earth around the solar system) in order to produce a truly uniform, regular time system.

Universal Time (Coordinated) (UTC) is derived from TAI, using a system of 'leap seconds' inserted when necessary into the timescale at the end of December and at the end of June to force UTC to agree with UT within a second. UTC is the basis of all legal timescales and is broadcast as radio time signals, from, for example, the US shortwave stations WWV in Fort Collins, Colorado, and WWVH in Kaui, Hawaii; MSF in Rugby, England; and the LORAN-C navigational network operated worldwide by the US Coast Guard. A quartz crystal clock (such as a good wristwatch) can easily be kept at UTC by listening to these transmissions, with an accuracy better than 1 s. UTC is suitable to record time-critical astronomical observations, such as of variable stars. It can be related directly to TAI when necessary (for example, to relate data accurately to the second).

There is discussion about replacing UTC with the time kept by the GPS system of satellites, which is used for navigation, and therefore disseminated across the world with high accuracy. Soon a new generation of clocks is expected to provide an accuracy of 10^{-17} or less, in the laboratory or onboard satellites. When this happens it may be necessary to reconsider both the definition and the way of realizing TAI, and new time-transfer techniques will need to be developed to compare the new clocks.

Titan

Saturn's satellite Titan is the only satellite in the solar system to have a significant ATMOSPHERE. It is so thick that Titan's surface has never been seen (it is covered by a haze).

Titan is approximately half the size of Mercury and is made of a mixture of water ice and rock. The ice contains greater amounts of frozen trapped gases than that on other icy satellites, because it is so distant from the Sun. The crust of the solid body of Titan is expected to be water ice.

The satellite travels on an almost circular prograde orbit in the ecliptic plane, with a period of almost 16 days. Its orbit is significantly eccentric (0.028), and this ECCENTRICITY is too large to be due to perturbations by the existing bodies in their present locations. It may be due to the relatively recent impact of a large comet.

Titan's atmosphere originated from the primordial solar nebula. It was originally gaseous material trapped in the ice accreted by the satellite. The gases were later released by heat generated from the decay of short-lived radioactive elements in Titan's rocks. In the cold environment, at more than 1 billion km from the Sun, Titan has preserved its atmosphere for more than 4.5 billion years. Mercury, the Moon, GANYMEDE and CALLISTO became too warm to do this.

None of the above explains why Titan has so much more atmosphere than Mars or the Earth. Perhaps Titan's atmosphere was frozen at the time when the atmospheres on Mars and and the Earth were depleted by heavy bombardment (*see* LATE HEAVY BOMBARDMENT). Alternatively, ammonia, methane and carbon monoxide may be escaping slowly through cracks or 'cryovolcanoes' on the satellite.

Titan's atmosphere is molecular nitrogen, with a few percent of methane; 0.1% is lightweight hydrogen gas, which escapes to space rapidly because of Titan's low gravity but is being continuously replenished. Titan moves in and out of Saturn's MAGNETOSPHERE but has no significant magnetic field itself, so that IONS and ELECTRONS from Saturn interact directly with its upper atmosphere, sweeping its upper layers away. The atmosphere contains some simple hydrocarbons, such as methane, acetylene, ethylene and ethane, and some that are more complex: diacetylene (C_2H_4), methylacetylene (CH_4), propane (C_2H_2) and monodeuterated methane (C_2H_6). Ultraviolet (UV) light from the Sun converts methane into more complex hydrocarbon molecules and hydrogen gas.

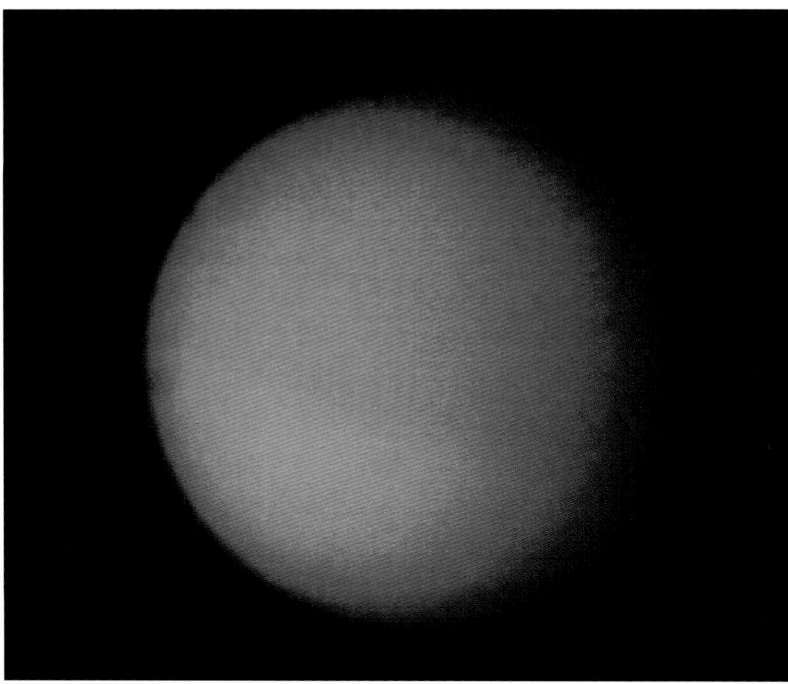

▲ ***Titan***
Titan's thick cloudy atmosphere is mostly nitrogen, like Earth's, but contains higher percentages of 'smog-like' chemicals such as methane and ethane. The smog may be so thick that it actually rains 'gasoline-like' liquids. Because of its thick cloud cover, Titan's actual surface properties remain unknown.

The hydrocarbons combine with nitrogen to form nitriles. All these organic chemicals remain on Titan, condensing at low altitudes. Water was recently detected in Titan's atmosphere.

The haze around Titan extends up to an altitude of 300 km. The AEROSOLS are mostly condensed ethane, which forms clouds, haze or near-surface mist. There are thin, patchy clouds of methane ice crystals. The Sun's UV light acts on the methane and ethane to produce polymers; these condense to form oily droplets and a brownish tarry organic sludge (called 'tholins' – the word was first used by Carl SAGAN and comes from a Greek word for 'muddy'). The aerosols rain or drift like snow onto the surface, into lakes of liquid methane, ethane and nitrogen at –180 °C, seeping into subsurface reservoirs. Some of the tholin molecules are prebiotic, similar to those produced in experiments on the origin of life on Earth. They are the precursors of organic molecules and hence amino acids and, ultimately, the first living cells in the Earth's oceans (*see* EXOBIOLOGY AND SETI).

Recent observations from space (with the HST), and from the ground (using the new techniques of ADAPTIVE OPTICS), have shown that Titan's surface has some dark and bright areas. It is not covered with uniform oceans. It is covered mainly with water ice and organic deposits, with frosty relief (perhaps mountains or craters) covered with methane snow, their shores beaten by the waves produced in vast hydrocarbon lakes. When the Huygens probe drops from the Cassini spacecraft and impacts the surface, it may bounce, splash or squelch on hard ground, lakes or swamps, depending where it lands (*see* CASSINI/HUYGENS MISSION).

Further reading: Burns J A and Matthews M S (eds) (1986) *Satellites* University of Arizona Press; Coustenis A and Taylor F (1999) *Titan: the Earth-like Moon* World Scientific.

Titania

A mid-sized satellite of Uranus, discovered by William Herschel in 1787 (*see* HERSCHEL FAMILY). Its diameter is 1580 km and it orbits at a distance of 191 000 km. The part of Titania imaged by VOYAGER 2 consists of a large plain covered with small craters and crossed by large canyons where the crust has faulted. The most extensive of these canyons is the Messina Chasmata system, which is nearly 1500 km long and 100 km across at its widest point. The 326-km diameter Gertrude is the only large crater. The lack of large craters may indicate that Titania has been resurfaced in the past. The surface resembles that of ARIEL, another Uranian satellite, and the processes that operated there are probably those that have shaped Titania's surface.

Titius, Johann Daniel (1729–96)

Astronomer, born in Konitz, Germany (now Poland). Titius became professor of physics at Wittenberg and built the first lightning conductor. He discovered the relation between the distances of the planets, and interpolated this into the translation he had prepared of a book by the naturalist Charles Bonnet. Titius's discovery was popularized by Johann Bode and is known as BODE'S LAW or the Titius–Bode law.

Tokyo, University of, Institute of Astronomy

Founded in 1987, the Institute of Astronomy, University of Tokyo is located at Ohsawa, Mitaka, Japan, 30 km west of central Tokyo. It operates the Kiso Observatory, which has a 1.05-m Schmidt telescope. The MAGNUM (Multicolor Active Galactic Nuclei Monitoring) telescope is a 2-m optical/infrared telescope on Haleakala, Hawaii, used for determining distances to remote ACTIVE GALAXIES from multicolor monitoring of their variability. It will eventually be operated remotely. There is also a remotely operated 1.2-m submillimeter telescope on Mount Fuji, Japan, which is observing the emission of the neutral carbon atom (CI) at 492 and 809 GHz toward various molecular clouds.

Tombaugh, Clyde William (1906–97)

Born near Streator, Illinois, USA, Tombaugh became interested in astronomy as a boy and, on the basis of some observations that he sent to the LOWELL OBSERVATORY, was hired as an astronomer there to search for PLANET X. Photographing 65% of the sky, he discovered several star clusters, two comets, more than 100 asteroids, dozens of clusters of galaxies, and the planet Pluto. At White Sands Missile Range, New Mexico, USA he worked on V-2 rockets captured from German forces in World War II, and became professor at New Mexico State University. There he confirmed the rotation period of Mercury on its axis, determined the vortex nature of Jupiter's GREAT RED SPOT, and looked for small, natural Earth satellites.

Tompion, Thomas (c.1639–1713)

Clockmaker, born in Northill, Bedfordshire, England. Tompion made the clocks for the ROYAL OBSERVATORY, GREENWICH, the first to keep Greenwich Mean Time and demonstrate the uniformity of the Earth's rotation. The clocks had 4-m pendulums with a 2-s swing, and were the reason why the observing room (the Octagon Room at Flamsteed House) was designed by Christopher Wren with such high ceilings. After John FLAMSTEED's death, the clocks were remade with conventional meter-long pendulums and sold for domestic purposes, but one was recently restored to its original location.

TRACE (Transition Region and Coronal Explorer)

A NASA Small-Explorer (SMEX) satellite developed to study the impact of magnetic fields on the solar outer atmosphere.

The instrument observes the solar surface and the hotter overlying domains (*see* CORONA; TRANSITION REGION) with high angular resolution, equivalent to 725 km on the Sun. The TRACE observatory was launched in April 1998 from Vandenberg Air Force Base on a Pegasus XL launch vehicle. The spacecraft is in a polar orbit roughly over the day-night TERMINATOR. This Sun-synchronous orbit allows TRACE to see the Sun without interruptions for approximately nine months of the year. It carries a 30-cm aperture CASSEGRAIN TELESCOPE to image wavelengths at visible, ultraviolet and extreme-ultraviolet wavelengths. The TRACE images show structural details and rapid changes in the Sun that are stunning to see.

Transient lunar phenomena (TLP)

See Observing the Moon, p. 276; *see also* LUNAR TRANSIENT PHENOMENA

Transit

The passage of a celestial body across the disk of a much larger one. The term is applied to passages of Mercury and Venus across the face of the Sun, passages of the satellites across the faces of Jupiter and the other giant planets, and passages of exoplanets (*see* EXOPLANET AND BROWN DWARF) across the face of their parent star.

Transits of the INNER PLANETS were first predicted reliably by Johannes KEPLER. From these predictions, Pierre Gassendi (1592–1655) became the first to observe a transit of Mercury in 1631 and, eight years later, Jeremiah HORROCKS made the first observation of a transit of Venus. These two planets are seen in transit when near one of their nodes at inferior CONJUNCTION. Mercury is at its ascending node in November, and its descending node in May; these occasions coincide with inferior conjunction at intervals of 7, 13 and 46 years (ascending node) and 13 and 46 years (descending node). Transits of Venus are much rarer. They can happen near June 7 (descending node) or December 8 (ascending node), and recur in pairs, 8 years apart, at intervals of 105.5 years and 121.5 years. The fifth and sixth transits since the one Horrocks observed in 1639 occur on June 8, 2004, and June 5/6, 2012. From the eighteenth century transits afforded an important means of determining the scale of the solar system (*see* DISTANCE).

The four Galilean satellites of Jupiter frequently transit the planet's disk. A satellite's shadow is sometimes visible crossing the planetary disk in a 'shadow transit.' The term 'transit' is also used to indicate the passage of an atmospheric or surface feature across a planet's face.

Transition region

The solar transition region comprises the PLASMA between the CHROMOSPHERE and the CORONA. In both of these regions the temperature is fairly uniform. The transition region, by contrast, is believed to be characterized by a steep temperature rise from a chromospheric temperature of slightly less than 10^4 K to coronal temperatures of about 10^6 K. Within the transition region occur explosive events (frequent, very small bursts). Thirty to forty thousand such events take place on the Sun at any time, and each lasts for about a minute. The events give an impression of explosion because highly supersonic motions are nearly instantaneously produced within a small area, but they are not isotropic outflows (as explosions would be), but are actually explosive jets or rapidly spinning disks. The events occur when magnetic energy is converted to kinetic energy by reconnection of the Sun's magnetic field in the transition region, and are probably the larger of an enormous number of events that heat the solar corona.

▲ ***Transit*** *Jupiter's satellite Io transits above the cloud tops in this image from the spacecraft Cassini. It was captured on January 1, 2001, two days after Cassini's closest approach. The image is deceptive: there are 350 000 km, roughly 2.5 Jupiters, between Io and Jupiter's clouds. Io is only the size of our Moon.*

Trans-neptunian object

An object found beyond Neptune in the Kuiper Belt (*see* OORT CLOUD AND KUIPER BELT).

Trapezium

The unstable multiple star Theta-1 Orionis (*see* DOUBLE STAR), in the heart of the ORION NEBULA. The Trapezium is visible when viewed directly with a moderately powerful telescope and on short-exposure photographs of the nebula. It consists of four stars arranged in a pattern resembling an irregular trapezium. Identifying the individual stars can be confusing: they were formerly designated Theta-1 Ori A, B, C and D in order of increasing RIGHT ASCENSION; in the *Hipparcos Catalogue*, however, they are lettered in order of decreasing brightness (*see* HIPPARCOS). Clockwise from the southernmost component they are now designated A (formerly C), C (formerly D), D (formerly B) and B (formerly A). The faintest component, D, an eclipsing BINARY STAR with a period of 6.5 days, is also known as BM Orionis. B was recently discovered to be an eclipsing binary also, with a period of 65.4 days. All four components are hot blue-white stars of spectral types O and B (*see* STELLAR SPECTRUM: CLASSIFICATION).

Triangulum, Triangulum Australe *See* CONSTELLATION

Triangulum Galaxy (M33)

A nearby SPIRAL GALAXY in the constellation of Triangulum, otherwise known as M33 (*see* MESSIER CATALOG). It is a face-on spiral that lies at a distance of 2 500 000 l.y. (slightly further away than the much larger ANDROMEDA GALAXY) and is a member of the LOCAL GROUP OF GALAXIES. With a diameter of 40 000 l.y. and a population of 15 billion stars, it is about half the size, and has about one-tenth of the mass, of the Milky Way Galaxy. It is an Sc galaxy and has an AGN.

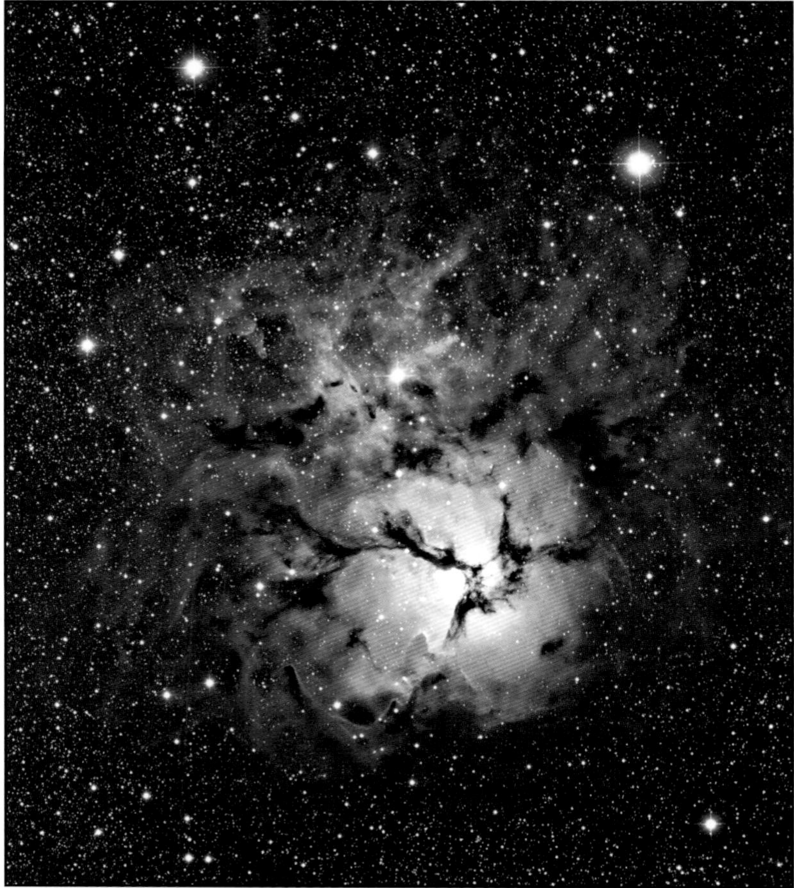

▲ ***Trifid Nebula** (M20) This nebula beautifully contrasts two types of nebula, emission and reflection, in one hybrid object. The red coloration is typical of an emission nebula, its light primarily being that emitted by excited hydrogen atoms. Blue coloration indicates a reflection nebula, in which the dust particles scatter more blue light than red.*

Trifid Nebula (M20)

A nebula in the constellation Sagittarius. It is of magnitude 9, with quite a high surface brightness, and measures 29 × 27 arcmin. It surrounds the multiple star HN 40, the light of whose brightest members energize the nebula (*see* DOUBLE STARS). The Trifid gets its name from dark lanes that trisect it.

Triton

The largest satellite of NEPTUNE, discovered in 1846 by William LASSELL, a few weeks after the discovery of Neptune itself.

Trojan asteroid

A member of one of two groups of ASTEROIDS whose orbital elements are similar to those of Jupiter. One group orbits ahead of Jupiter, clustered around the L_4 LAGRANGIAN POINT, while the other group orbits behind the planet, around the L_5 Lagrangian point. The L_4 group, which is more numerous, is sometimes called the Achilles group, after the first Trojan to be named, (588) ACHILLES. The L_5 group is sometimes called the Patroclus group, after the second Trojan discovery, (617) Patroclus. Achilles and Patroclus are both around 150 km in diameter, and were both found in 1906. The largest Trojan is (624) Hektor, at 225 km.

The names given to Trojans were originally those of heroes from Homer's writings on the Trojan wars, but too many have been found for this to continue. As of June 2003 there were 1016 known Trojans in Jupiter's L_4 group, and 602 in the L_5 group. By analogy with Jupiter's Trojan asteroids, asteroids at the L_4 and L_5 points of the orbits of other planets are also called Trojan asteroids. Mars has six Trojans, the first of them (5261) Eureka, and all except one are at Mars's L_5 point. There is one Trojan at the L_4 position of Neptune.

Perturbations by other planets cause Jupiter's Trojans to oscillate about the Lagrangian points. An individual Trojan will drift back and forth in longitude by 15° to 20° either side of its Lagrangian point with a period of up to 200 years. According to computer simulations Trojans at the L_4 and L_5 positions of the Earth's and Venus's orbit are possible, but none are known.

Trumpler, Robert Julius (1886–1956)

Astronomer, born in Zürich, Switzerland. He worked at the LICK OBSERVATORY in California, USA, where he showed that distant clusters of stars are systematically too faint for their size, as if dimmed by an obscuring medium whose effects accumulate with distance. This was the first convincing evidence of the presence of an INTERSTELLAR MEDIUM in our Galaxy, consistent with the indications from Heber CURTIS's recognition that SPIRAL GALAXIES seen edge on had obscuration along their central plane.

T Tauri star

A T Tauri star is a type of PRE–MAIN SEQUENCE STAR. T Tauri stars are of low to intermediate mass with ages of typically a few million years. At this stage in their evolution the newly collapsed stars have become optically visible but are still surrounded by gas and dust from their formation, typically in a disk. They are cool, luminous stars that exhibit emission lines, excess infrared and ultraviolet, variability, and are sometimes associated with JETS, molecular outflows, winds or X-ray emission. *See also* PLANETS: ORIGIN.

Tucana *See* CONSTELLATION

Tully–Fisher relation

This relates the rotational velocity of a SPIRAL GALAXY to its LUMINOSITY. The Tully–Fisher relation was originally described in 1974 by Chantal Balkowski and collaborators, but it was developed fully as a distance indicator by Brent Tully and Richard Fisher in 1976. A spiral galaxy rotates under the influence of its own gravity and so the rotational speed is connected with its mass. A more massive galaxy also contains a larger number of stars and is therefore more luminous.

The Tully–Fisher relation has become one of the most widely used methods of measuring extragalactic DISTANCES. It is so accurate that it shows the expansion of the universe, and the deviations of the motion of individual galaxies from uniform outflow caused by local concentrations of matter in clusters of galaxies. The primary features identified in such maps are:

- galaxies within 10 megaparsecs of the LOCAL GROUP OF GALAXIES;
- the VIRGO CLUSTER OF GALAXIES;
- the GREAT ATTRACTOR located at a distance of about 60 megaparsecs in the direction of the Hydra–Centaurus cluster complex (*see* UNIVERSE).

Tunguska event

A midair explosion on the morning of June 30, 1908, over a remote forest region of eastern Siberia that caused widespread devastation. The few eyewitnesses to the explosion, near the Podkamennaya Tunguska River, reported seeing a FIREBALL as bright as the Sun. The explosion itself was heard up to 1000 km away, and seismographic stations recorded pressure waves that went twice around the planet. The forest canopy burned for over two days. Trees and reindeer in the central 1000-square-km area beneath the blast were incinerated, and up to 50 km away trees were felled by the shock wave. Dust from the explosion blanketed the area and was carried around the world by atmospheric currents, causing colorful sunsets. It was 20 years before the first scientific research team, led by Estonian scientist Leonid Kulik (1883–1942), reached the area. They found no main impact crater or meteoritic fragments; later expeditions, however, recovered dust particles preserved in tree resin and found them to be consistent in composition with STONY METEORITES. It seems, then, that a stony meteorite with a diameter of about 50 m fragmented and vaporized at an altitude of about 8 km. An alternative explanation, widely believed by Russian scientists, is that the event was the explosion of a COMET.

Tuorla Observatory

The observatory lies at an altitude of 53 m and is located 12 km from Turku in Finland. It is the site of the headquarters of the NORDIC OPTICAL TELESCOPE. Founded in 1952, early research activities included setting up a geodetic triangulation network for measuring distances; optical work; zenith tube studies of polar motion; and detections and orbital calculations of minor planets and comets. The main instrument today is a 1-m Dall–Kirkham REFLECTING TELESCOPE, used for CCD photometry of QUASARS. Other instruments include a 0.7-m SCHMIDT TELESCOPE, a 0.6-m RITCHEY–CHRÉTIEN TELESCOPE and a 0.39-m zenith telescope. Tuorla also contributes to the VLBI facility at the Metsähovi Radio Research Station of the Technical University of Helsinki.

Twilight glow

The scattered light before sunrise and after sunset that partially illuminates the sky. This relatively steady glow is part of the overall phenomenon called AIRGLOW.

Tycho

A prominent young crater in the MOON's southern uplands, and the hub of the Moon's most striking ray system. Its diameter is 102 km, and it is named for the astronomer Tycho BRAHE. The crater is one of the youngest features of the Moon, perhaps just 100 million years old. Its ray system, which extends for up to 1500 km in all directions, is most prominent at full Moon, when the fresh EJECTA from which it is composed reflect sunlight back toward the Earth, and the rays are seen to extend over much of the Moon's nearside. The crater itself has terraced walls (caused by partial slumping of sections of the inner rim, which rise to 4.5 km), and a pronounced central peak. For over 80 km beyond the rim there is a continuous deposit of ejecta (an ejecta blanket) in all directions. Rock melted by the impact that created Tycho flowed down the outside to solidify in pools, and down the inside of the rim in rivulets toward the crater floor. Material ejected on higher trajectories produced the rays and also many secondary craters.

Tycho Brahe *See* BRAHE, TYCHO

Tycho's Supernova

A supernova remnant in the constellation Cassiopeia, 7.7° north of Alpha Cas (*see* SUPERNOVAE AND INTERSTELLAR MATTER). It suddenly appeared as a brilliant naked-eye star in November 1572 and reached a maximum apparent magnitude of –3.5. Until its disappearance 16 months later, it was studied extensively by the Danish astronomer Tycho BRAHE, who described its early appearance as follows: 'Initially, the new star was brighter than any other fixed star, including Sirius and Vega. It was even brighter than Jupiter...It kept approximately the same brightness for almost the whole of November. On a clear day it could be seen at noon.' (From *De stella nova*, On the New Star, 1573.)

In 1952 observations made by Robert Hanbury-Brown at JODRELL BANK OBSERVATORY identified the supernova location with the radio source 3C 10 (also known as Cassiopeia B). Hanbury-Brown mapped the detailed structure of the radio source, following which Rudolph MINKOWSKI detected optical remnants with the 200-in Hale telescope at the PALOMAR OBSERVATORY. The remnant of Tycho's supernova has also been identified as an X-ray source.

The detailed records kept by Tycho Brahe enabled a LIGHT CURVE to be constructed, which indicates that the object was a Type 1 supernova. Recent measurements suggest that the remnant is probably between 8000 and 10 000 l.y. distant, in which case it reached an absolute magnitude at maximum of about –16. This is equivalent to a luminosity approaching 250 million times that of the Sun.

The faint remnants of the expanding gas shell now have a radius of 3.7 arcmin, equivalent to an actual diameter of 17–22 l.y., and a rate of expansion of 6000–7500 km s^{-1}. This is the highest expansion rate deduced for any object, and is about 10 times that of the CRAB NEBULA, also a young supernova remnant. For an illustration *see* STELLAR NOMENCALTURE.

Tychonic system

The world system proposed by Tycho BRAHE. Unable to accept the doctrine of Nicolaus COPERNICUS that the Earth moves around the Sun, he put forward the view, later disproved by Johannes KEPLER, that the planets move around the Sun, but the Sun and Moon move around the Earth. The theory explained the observed variations of the phases of Venus, for which the PTOLEMAIC SYSTEM had no explanation.

Tycho star catalogs

The HIPPARCOS space ASTROMETRY mission determined the position and brightness of 2.5 million stars to high precision. The data, for all stars above magnitude 11 and containing many stars to magnitude 12.5, were published in the Hipparcos and Tycho Catalogues (1997–2000). The *Millennium Star Atlas* presents the data on 1 million stars in a form suitable for amateur astronomers.

The positions of the stars are accurate to about 1 milliarcsec, and magnitudes are accurate to a few hundredths of a magnitude. PARALLAXES are given for the nearer stars, and these measurements form the steps from the solar system to the stars on the ladder of DISTANCE in astronomy. Proper motions are given, based on the positions measured by Hipparcos compared to archival data. DOUBLE STARS and VARIABLE STARS have been identified. The survey for these is complete and uniform over the sky to an extent that will only be superseded by a future astrometric satellite.

U Gem star

U Geminorum (U Gem) was the first of the DWARF NOVAE to be discovered and its name is often used to describe the whole group. The star itself falls into the SS Cygni (UGSS) subclass of dwarf novae (*see* VARIABLE STAR: CLOSE BINARY STAR). Dwarf novae brighten by factors of hundreds to thousands, sometimes in just a few hours. The eruptions recur quasi-periodically on intervals of weeks to years, with durations from a few days to a few weeks. U Gem stars, like all cataclysmic variables, are BINARY STAR systems. They consist of a WHITE DWARF star, which is the primary, and a low-mass red dwarf star, which is the secondary. The red dwarf is losing mass to the white dwarf through the inner LAGRANGIAN POINT (L_1). This gas forms an accretion disk surrounding the white dwarf, and this is usually the major source of radiation in the system. Another source of light is a bright spot formed where the stream collides with the outer edge of the disk. The other two subclasses of DN are the Z Cam and SU UMa stars. The Z Cam stars have similar outburst patterns with the addition of occasional extended standstills about 0.7 magnitude below maximum brightness. The SU UMa stars sometimes have super-outbursts that are brighter and of a longer duration than an ordinary outburst.

Uhuru (SAS-1/Explorer 42)

NASA satellite, launched from the San Marco platform off the Kenyan coast in 1970, and the first satellite dedicated to X-RAY ASTRONOMY. Uhuru completed an all-sky X-ray survey and studied individual sources, discovering X-ray binaries, including Hercules X-1 and Centaurus X-1, and confirming the variability of Cygnus X-1. Uhuru means 'freedom' in Swahili. *See also* EXPLORER; SMALL ASTRONOMY SATELLITE.

UKIRT

The UK Infrared Telescope (UKIRT) is the world's largest telescope dedicated solely to observations at infrared wavelengths of 1–30 μm. It is sited in Hawaii near the summit of Mauna Kea at an altitude of 4.194 km above sea level. Officially opened in October 1979, UKIRT is a 3.8-m classical CASSEGRAIN TELESCOPE with a primary mirror which is about half as thick as those of conventional telescopes of its era.

UK Schmidt Telescope (UKST)

The UKST is a standard SCHMIDT TELESCOPE with a very wide-angle field of view. It was designed as a copy of the Palomar Observatory Oschin telescope (*see* REFLECTING TELESCOPE) to photograph 6.6 × 6.6° areas of the night sky on photographic plates 356 × 356 mm square. This 1.2-m telescope was commissioned in 1973, and from then until 1988 it was operated by the ROYAL OBSERVATORY, EDINBURGH. It then became part of the AAO. The UKST's initial task was the first deep, blue-light photographic survey of the southern skies, which was completed in the 1980s. It has since undertaken many other survey projects in different colors and in the near infrared. About 75% of the time on the telescope is now spent carrying out an all-southern-sky galaxy redshift survey using a multi-object spectroscopy system called 6dF (six degree field). This uses up to 150 fibres on interchangeable field-plate units, which replace the original photographic plateholders of the telescope.

Ultra-luminous infrared galaxy (ULIRG)

First detected by IRAS in the early 1980s, ULIRGs emit copious amounts of infrared radiation, 100 times brighter than our Milky Way Galaxy. The infrared comes from a large amount of dust, which absorbs and re-radiates the light of hot, newborn stars in the galaxies. The dust itself has been made from carbon and silicate material distributed by the many SUPERNOVAE of Type II in the galaxies. The starbursts were precipitated from the interstellar material in the galaxies by collisions between and mergers with other galaxies.

Ultraviolet astronomy

The study of the electromagnetic radiation emitted by celestial bodies in the ultraviolet (UV) wavelength range. The original meaning of UV is the portion of the spectrum beyond the shortest wavelength seen by human beings, that is, beyond the violet region, and with a wavelength of less than about 400 nm. The region of the spectrum between 400 and 320 nm is transmitted by our own atmosphere and is accessible to detectors such as photographic emulsions and some CCDs, so it can be studied by conventional techniques of ground-based astronomy. For astronomers, this region therefore falls within the optical, or OptIR, spectrum. The Earth's ozone layer shields us from the radiation coming from space with wavelengths shorter than 320 nm. The UV spectral range for astronomers thus lies from 320 to 6 nm, the conventional start of the X-ray region.

There are at least three reasons for being interested in UV astronomy. Firstly, most radiation emitted by stars whose temperature exceeds 10 000 K falls in the UV region. Such stars are astrophysically active, and form the key component of POPULATION I. Secondly, the UV wavelength range contains the resonance spectral transitions, the most intense spectral lines from the most common atoms, ions and molecules. Finally, when studying galaxies at high REDSHIFT, astronomers make measurements in the OptIR region of electromagnetic radiation, which left the galaxies as UV light. To understand the light from distant galaxies it is necessary to understand the UV radiation from nearby ones.

History of UV astronomy

UV astronomy started when it became possible, by means of balloons, rockets and satellites, to carry astronomical telescopes above the atmosphere. Observations with rocket-borne payloads started in the late 1940s, with the first, unguided experiments flown by the US Naval Research Laboratory (NRL) on board captured German V2 rockets (*see* ROCKETS IN ASTRONOMY). They recorded the intensity and spectral distribution of the UV emission of the Sun. Simple, single-band UV photometers, flying on unstabilized US Aerobee rockets, provided the first UV observations of stars and nebulae in the 1950s, including about 50 early-type (hot) stars. It became possible to target individual stars from 1965, when rocket-borne telescopes could be pointed, using three-axis stabilized mounts.

The first of NASA's OAO satellites was launched in 1966, with OAO-2 in 1968 and OAO-3 (the Copernicus satellite) in 1972 providing breakthroughs in UV astronomy, including luminous hot stars and interstellar matter. In 1972 the TD-1 satellite and from 1974 to 1978 the Netherlands Astronomical Satellite (ANS) made surveys of stars in the UV range.

UV astronomy was expanded enormously by the outstanding success of the IUE, launched in 1978, a joint project of the USA, ESA, and the UK. It lasted for more than a decade and left an archive that is still productive. The EUVE (1992–2001) was the first orbiting observatory to focus on

◀◀ ***Omega Nebula*** *UKIRT obtained this infrared image of M17 with its innovative imaging spectrometer, UIST. The image is a true-color image but, since they would be invisible to the eye, its infrared colors have been shifted into the optical region. Bright stars in M17's central cluster cause nearby gas (lower right) to emit infrared radiation (colored blue-white). Longer wavelength infrared radiation (colored orange) penetrates through the dust cloud in the top-left half of the picture from background stars.*

the EUV spectrum. The X-ray astronomy satellite ROSAT (1990–9) carried instruments that pushed its spectral response capability into the EUV, while at the other end of the UV spectrum, HST has near-UV capability, as envisaged in the original concept by Lyman SPITZER. FUSE was launched in 1999 and, after a number of close calls, is still (2003) in operation. GALEX is the Galaxy Evolution Explorer, with a much delayed launch in 2003, and carries a UV telescope intended over three years to survey UV emission from young galaxies. Several spectrographs have been carried into space and operated more than once from the Space Shuttle, including the German/US ORFEUS instrument (Orbiting and Retrievable Far and Extreme Ultraviolet Spectrometer).

Techniques

Although UV radiation is neither transmitted nor reflected by optical components as efficiently as visible light and infrared wavelengths, UV telescopes are basically the same as ground-based optical reflecting telescopes. The EUV waveband uses techniques that are, however, more similar to those of X-RAY ASTRONOMY.

As far as detectors are concerned, CCDs are capable of recording UV photons, and, together with microchannel plate detectors (MCPs), have replaced photographic emulsions. MCPs consist of a thin disk of a lead oxide glass crossed by many microscopic channels which act as individual photomultipliers and make the detector as a whole an image intensifier. Photon-counting MCP detectors like MAMA (Multi-Anode Microchannel Array) were used in the Space Telescope Imaging Spectrograph (STIS) on HST.

Achievements of UV astronomy

UV spectroscopy provides the tool for investigating the composition of interstellar gas by analyzing the absorption spectrum of common elements toward bright stars. UV techniques identified the UV emitters in globular clusters and elsewhere in POPULATION II, including eruptive variables, the mysterious BLUE STRAGGLERS, WHITE DWARFS and PLANETARY NEBULAE. The most intriguing and fruitful studies have been of hot stellar populations in galaxies, where spaceborne, UV-sensitive facilities offer ways to recognize active sites of star formation at different scales (from individual stars to spiral arms, nuclear rings and star-forming complexes). UV observations become even more key when a galaxy hosts a starburst, a major event of star formation dominating its energetic output. In giant elliptical galaxies and the spheroidal components of spirals, UV astronomy recognized the unexpected presence of hot components in their stellar populations, a phenomenon that depends sensitively on the chemical composition.

UV astronomy techniques have been particularly important in the study of ACTIVE GALAXIES because the light from the hot material orbiting the central black holes emits UV copiously. UV observations have been at the core of prolonged, generally multi-wavelength, spectral monitoring campaigns for REVERBERATION MAPPING the distribution of the components of the nucleus of the active galaxies, and so on.

Ulugh Beg (1394–1449)

Astronomer, born in Timurid, Persia (now Iran), the grandson of the Asian conqueror Timur, and under his father's patronage built a large, well-equipped observatory in the city of Samarkand, directed by Ali-Kudschi. Ulugh Beg and the staff, including the brilliant al-Kashi (1390–1450), observed and published the positions of the Sun, the Moon, the planets and 992 stars, revealing errors in PTOLEMY's data.

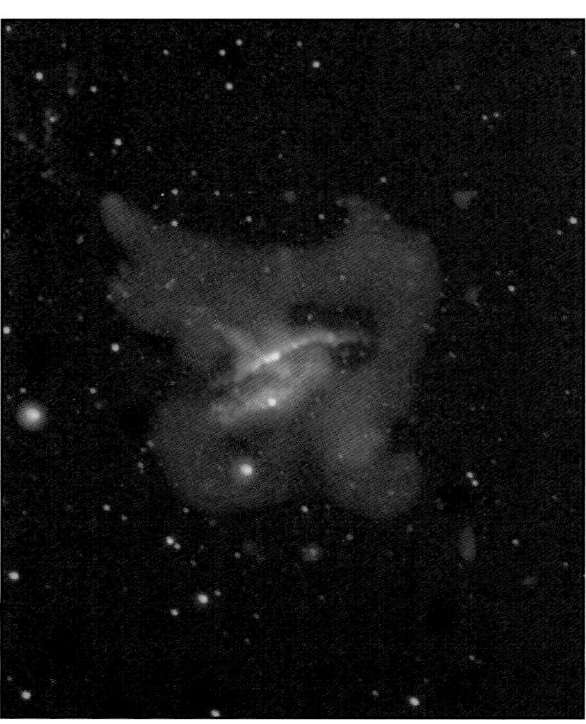

▶ Centaurus A *This picture is a combination of the GALEX far-UV image (colored blue), the GALEX near-UV image (colored green) and an image taken by NASA's Chandra Observatory (colored red), which measures the X-ray jets coming out of this galaxy. The UV light patches at the upper left are from hot newly born stars formed by the X-ray jets.*

Ulysses mission

Ulysses, a joint ESA/NASA mission, is the first space mission in a polar orbit around the Sun, and the first to make observations in the areas of the HELIOSPHERE over the polar regions at high heliolatitudes. The objective of the Ulysses space mission, launched by the Shuttle Discovery on October 6, 1990, is the exploration of the dependence on heliolatitude of the heliospheric medium and its relationship with the Sun's coronal structure and dynamics.

Ulysses's unique orbit was achieved by first targeting the spacecraft to Jupiter and using gravity assist (the maneuver that uses the gravitational potential of a planet to modify the orbit of a spacecraft during a close flyby) to deflect the trajectory of Ulysses out of the ecliptic plane. The Jupiter flyby took place in February 1992, and since then Ulysses has been in a nearly polar orbit around the Sun. Its orbital plane makes an angle of about 80° with the solar equatorial plane, its APHELION at 5.4 AU and its PERIHELION at 1.4 AU. The orbital period of Ulysses is 6.2 years. Ulysses is equipped with a comprehensive payload to make observations in the solar wind. Instruments to measure the speed, density, temperature and composition of the solar-wind plasma are complemented by instruments to measure the magnetic field, radio and plasma waves, as well as energetic particles from the Sun and COSMIC RAYS.

Umbra

(1) The dark inner region of a shadow cast by an object illuminated by a light source of finite angular size, such as the Sun. For someone standing within this part of the shadow the light source is completely hidden. The penumbra is the outer part of the shadow which is only partly illuminated. In a solar eclipse (*see* ECLIPSE AND OCCULTATION), the shadow of the Moon falls on the Earth. A total eclipse is seen by an

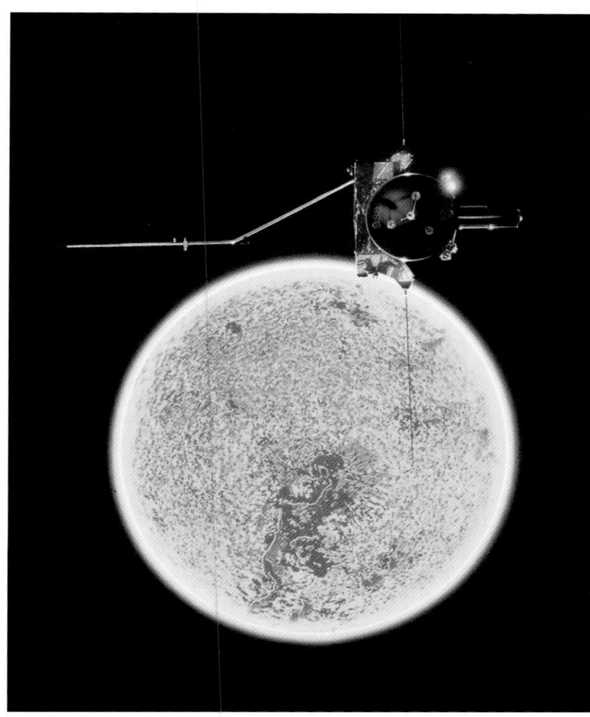

observer within the umbra whereas a partial eclipse is observed from the penumbral region.

(2) The terms 'umbra' and 'penumbra' are also used to describe the dark core and light outer areas in a sunspot (*see* SUN; SUNSPOT CYCLE).

Umbriel

A mid-sized satellite of Uranus, discovered by William LASSELL in 1851. Its diameter is 1170 km and it orbits the planet at a distance of 266 000 km. The satellite was not well imaged by VOYAGER 2; an ancient, cratered surface was visible, broken only by the bright-floored crater Wunda (diameter 160 km). Umbriel appears to be a primitive, undifferentiated body: one that has never melted internally and segregated into a core and mantle, and has remained largely geologically inactive. *See also* DIFFERENTIATION.

Uncertainty principle

Also known as Heisenberg's uncertainty principle, it was proposed in 1927 and named for the German physicist Werner Heisenberg (1901–76). The principle states that it is impossible to know simultaneously the precise position and momentum of a SUBATOMIC PARTICLE.

Universal Time (UT)

A measurement of TIME used for civil timekeeping and based on the SUN's daily apparent motion. It is the MEAN SOLAR TIME of the Greenwich meridian at 0° longitude. UT replaced Greenwich Mean Time in 1928. For purposes requiring a precision of 1 s or greater, different categories of UT were defined in 1955 by the IAU. These are known as UT0, UT1 and UT2, the differences being due to the effects of irregularities in the Earth's motion. Coordinated Universal Time (UTC) is the international basis of civil and scientific time, based on a 24-hour clock, and is used for broadcast time signals. Zero (0) hours UTC is midnight at Greenwich. UTC is obtained from an atomic clock that is set to remain within ±0.90 s to UT. To do this it is sometimes necessary to insert a leap second, and in this way the solar time that is indicated by UT is synchronized with atomic time. Leap seconds are added only at the end of June or the end of December in any year. Since 1972 when the system was introduced, there have been 32 leap seconds added so that international atomic time (TAI) is 32 s ahead of UTC.

◀ ***Ulysses*** *Named for the legendary Greek hero, the spacecraft Ulysses charts the polar regions of the Sun.*

Universe

COSMOLOGY is the study of the entire universe, and cosmological theories attempt to explain its nature and evolution (*see* UNIVERSE: COSMOLOGICAL THEORY). Clearly, to do so one must know what the universe consists of. We can directly investigate only the very local universe (the laboratory, the Earth and the solar system out to the heliopause). For the rest, of course by far most of the universe, we rely on sensing radiation from its components – essentially what we can see, or otherwise detect by telescopes. What we see as we look out into space are stars, galaxies and gas in a mostly empty-looking void – the average density of the universe that we see is measured in units of the mass of one hydrogen atom per cubic meter. The air we breathe has a density that is one million, million, million, million times more. With this in mind, it is easier to understand why twentieth-century scientists began to appreciate that cosmology was about the properties of space and time. To determine the properties of the SPACETIME continuum, astronomers used galaxies like surveyor's stakes. As an example, HUBBLE'S LAW of the recession of the galaxies can be interpreted to show that space is expanding.

Galaxies, however, do more than merely map space and time: the matter in the universe provides the gravitational force under the influence of which the universe evolves, together with effects produced by dark energy, described by the cosmological constant (*see* DARK ENERGY AND THE COSMOLOGICAL CONSTANT). Thus the distribution of matter in space is a key to understanding the past, present and future of the universe. The description of the distribution is uncertain because we see only ordinary, 'baryonic' matter, and there is evidence that there is more DARK MATTER than matter made of BARYONS. The situation is further complicated because it seems there is more than one kind of dark matter. But, to an approximation, dark matter broadly follows the distribution of baryonic matter.

Most baryonic matter in the universe is in the form of the HYDROGEN clouds in the INTERGALACTIC MEDIUM. It is difficult to determine the distribution of the hydrogen clouds because we detect them only in directions to the rather rare bright QUASARS where they show themselves as absorption lines of different redshift in the spectra of the quasar. It is easier to study the distribution of stars and galaxies, and most of what follows is about them. We must bear in mind that galaxies are just a small fraction of what is there.

The cosmological principle and clustering of galaxies

A guiding principle of cosmology has been that if one looks out to large enough distances, then the universe is homogeneous and isotropic. This assumption is referred to as the cosmological principle. Present observations suggest that the cosmological principle is valid but only on distance scales over 500 million l.y. The most directly visible representation of the principle is the distribution on the sky of over 2700 GRBs that originated in very distant galaxies. Their distribution in the sky is completely uniform.

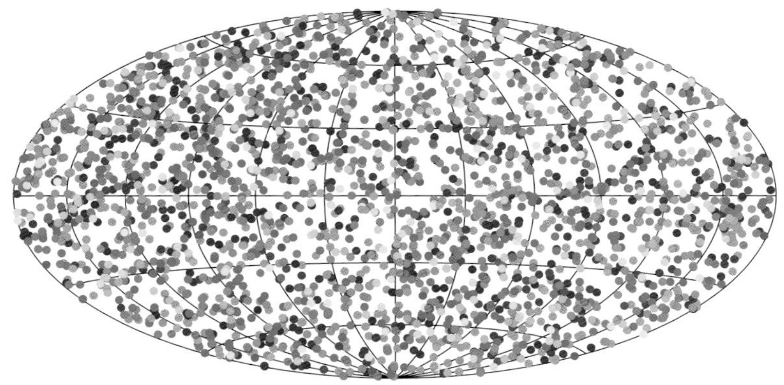

▲ ***Gamma-ray bursts*** *This plot of 2704 GRBs, which come from distant galaxies, shows uniform distribution in the sky, confirmation that the universe is homogeneous and isotropic. The data came from BATSE.*

On smaller scales, we do see non-uniformity, and the distribution of galaxies is highly non-random: when one galaxy is found, the likelihood of finding a nearby companion galaxy is significantly higher than chance would predict.

Our Milky Way Galaxy is a member of a group (the Local Group) of a score of galaxies spread over a region about 3 million l.y. in extent. The Local Group contains a total mass of up to 10 million million solar masses. Groups like this are relatively common: on average, there is approximately one group for each 60 000 cubic million light years volume. Clusters of galaxies are comparable in size, spanning 3 million–15 million l.y., but are considerably more massive and rare than galaxy groups. Clusters contain hundreds of galaxies and have a mass of at least 100 million million solar masses (the richest systems can be 50 times more massive than that). Although quite rare, clusters of clusters (known as superclusters) have been detected. These systems can contain in excess of 100 000 million million solar masses, spread across regions 30 million–100 million l.y. in extent.

Structures on scales in excess of 100 million l.y. were not discovered until wide-area, three-dimensional galaxy surveys were first undertaken in the early 1980s. They revealed an amazing bubble-like network of voids, sheets of galaxies and clusters, extending for hundreds of millions of light years. Crossing right through the frothy distribution of galaxies within 500 million l.y. is one striking feature called the GREAT WALL, perhaps the largest structure known.

Voids

Striking features of the early maps of the distribution of matter in the universe were voids, whose existence had been unsuspected. They are large regions, empty of galaxies, extending for tens of millions of light years. They were first identified in 1978, when Stephen Gregory and Laird Thompson commented that there were regions nearly devoid of galaxies to the foreground of a supercluster that extends from the Coma cluster to the cluster Abell 1367. Several other empty regions were found within the next few years and, in 1981, the Boötes void was reported by Robert Kirshner, August Oemler, Paul Schechter and Stephen Shectman. This void extends over roughly 30 million cubic light years. Over the years since, there have been searches to see whether there is anything to be found in the Boötes void. A few small galaxies have turned up. Still, it is an empty place.

Cosmic flows

The irregular structures form a complex of gravitational attraction forces, which pull galaxies that lie in their vicinity. The gravitational pull accelerates galaxies and clusters so that they depart from Hubble's law. In a given region, all the galaxies may be moving together in the direction of the attractor. This is called the cosmic flow.

The first observational study of a cosmic flow came in 1976, with the discovery that the temperature of the COSMIC MICROWAVE BACKGROUND (CMB) is warmer in one direction and cooler in the opposite direction by a few millidegrees. This is due to motion of our Galaxy relative to the CMB, motion that is assumed to be caused by an imbalance in the gravitational tug of the various structures which surround it. The COBE satellite measured the motion as 620 km s^{-1} toward the direction which has galactic coordinates l = 271°, b = +29° in the constellation Hydra. The Local Group itself is moving due, in part, to the gravitational pull of the Virgo Cluster, which lies about 50 million l.y. away. But its mass is sufficient to explain only about 40% of the velocity of our Galaxy relative to the CMB and the velocity vector does not point to Virgo. Additional structures at larger distances must, therefore, play a role. In the mid 1980s it was discovered that a vast number of galaxies nearby to us within a sphere nearly 300 million l.y. in diameter appear to share a common motion toward a massive aggregate of matter, dubbed the GREAT ATTRACTOR. The Great Attractor is a moderately rich supercluster lying 120 million–150 million l.y. from our Galaxy. The Great Attractor itself may be moving, suggesting that the origin of the flow is due, in part, to structure on scales larger than 100 million l.y.

The origin of structure

Cosmologists call the irregular distribution of galaxies, clusters of galaxies, superclusters, voids and filaments 'structure.' How did structure originate from the mêlée of the BIG BANG? Most theories of the formation of structure in the universe start with initial fluctuations and grow them through gravitational interaction of each dense clump with others. The growing clumps interact with both visible matter and dark matter. In particular, the results are sensitive to the type of dark matter, 'cold' or 'hot,' a nomenclature coined by Dick Bond, referring to how fast it moves, slowly or nearly at the speed of light, as it would if it were made up of elementary particles of low mass.

There is a fundamental difference in the way in which galaxies are predicted to form in hot and cold dark-matter models. If dark matter is hot, it can stream from the over-dense regions into adjacent under-dense areas and fill them up. In the case of cold dark matter, on the other hand, streaming is not important and density fluctuations persist.

In hot dark-matter calculations, the first structures that form are flat, pancake-like objects of supercluster scale. They must somehow fragment later to form galaxies. Computer simulations of this process carried out in 1981 by Marc Davis, Simon White and Carlos Frenk showed that galaxies formed after about half the age of the universe had gone by. However, we know that there is a large population of galaxies that are only 10% of the age of the universe. The alternative, a universe dominated by cold dark matter, proved to be much more successful, as shown, for example, in a series of computer simulations by Davis, George Efstathiou, White and Frenk. For a period of time, the cold dark matter (CDM) model of the formation of structure was regarded as the standard model. Astronomers are currently (2004) attempting to investigate the effects of adding a cosmological constant (the Λ-CDM model), and complicating the dark matter by making it a little 'warm.'

Further reading: Fairall A (1998) *Large Scale Structures in the Universe* John Wiley & Sons and Praxis; Ferris T (1988) *Coming of Age in the Milky Way* Doubleday.

Universe: cosmological theory

COSMOLOGY, the branch of astronomy that deals with studies of the large-scale structure of the UNIVERSE, requires knowledge about the most remote objects and the extremes of physics. Despite this, cosmology has emerged as a very successful and important branch of science.

Modern cosmology began when Isaac NEWTON attempted to apply his theory of gravitation to the universe, on the assumptions that it was homogeneous and isotropic, but static. He realized that the universe would be unstable, collapsing in on itself because of the inward pull of gravity.

The advent of the general theory of relativity in 1915 offered a possible resolution. Only two years after he proposed the theory, Albert EINSTEIN made a bold attempt to apply it to the entire universe. But like Newton, Einstein found that a static universe was unstable. He introduced the cosmological constant (*see* DARK ENERGY AND THE COSMOLOGICAL CONSTANT), which acted like a repulsive force that increased with distance. The universe that emerged was closed: finite but unbounded.

However, Willem DE SITTER in the same year demonstrated that the model was not unique. De Sitter found that the universe could be empty but expanding. Between 1922 and 1927, Alexandr FRIEDMANN and Georges LEMAÎTRE determined theories of the expanding universe for which the cosmological constant was not required. In 1929 Edwin Hubble arrived at what is today known as HUBBLE'S LAW, showing that the universe is expanding. It was later realized that in his 1927 paper Lemaître had predicted a linear velocity–distance relation of this kind.

The Big Bang

The theory of the BIG BANG hinges on the supposition that, at a time $t = 0$, the universe came into existence in a singular event (*contrast* STEADY-STATE THEORY). No physical description of the original event is possible, and physical theories examine the behavior of the universe from about 10^{-43} s after $t = 0$, when the universe was compacted into about the size of a grapefruit. 10^{-43} s is the Planck time, the time it would take a photon traveling at the speed of light to cross a distance equal to the Planck length, or the scale at which classical ideas about gravity and spacetime cease to be valid, and quantum effects dominate. The Planck time is the quantum of time, the smallest measurement of time that has any meaning within the laws of physics as we understand them at present.

One of the early attempts to go close to the Big Bang epoch was made in the late 1940s by George GAMOW, with his collaborators Ralph Alpher (1921–) and Robert Herman (1914–97). They worked out the physics of the universe at t = 1–4 min old, when it was full of radiation at a very high temperature – the cosmic fireball. While the universe was hot, the matter in the universe was at first elementary particles. It became electrons, protons and neutrons, which 'cooked' to become 75% hydrogen (including deuterium) and 24% helium, according to calculation. The first of the elements were thus created in the first four minutes or less (*see* ELEMENTS, FORMATION OF). The standard Big Bang cosmology has successfully met the challenge to demonstrate how nucleons and leptons evolved out of more primordial particles. The abundance of helium and deuterium in the universe is close to the calculated value.

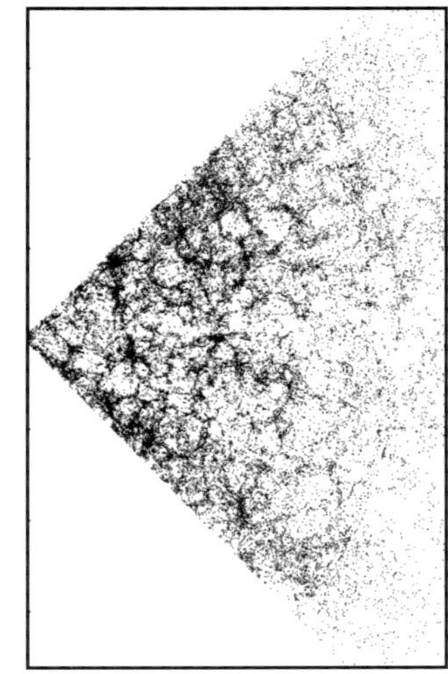

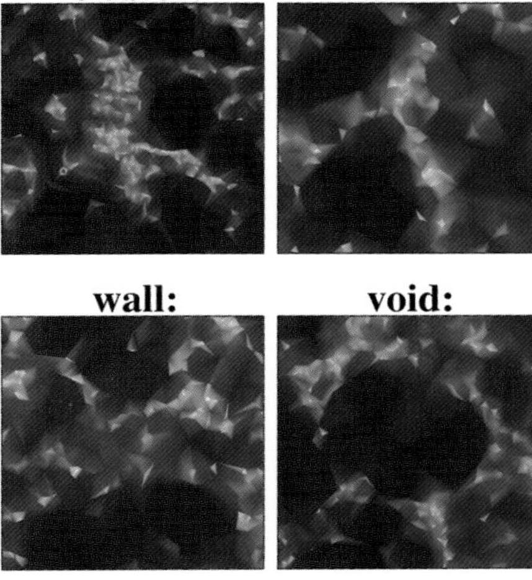

▲ ***Universe*** *The large-scale structure of the universe is revealed by the Anglo-Australian Telescope's 2-degree field (2df) survey of galaxies. Each galaxy in a thin strip of the sky extending over 100° of the sky is plotted at the correct distance according to its redshift. The map (left) extends from the Earth (at the apex of the wedge) to a distance of about 5 billion l.y. The plots reveal typical structures in the figures at right. Concentrations of galaxies are called clusters, and regions where they are absent are called voids. Voids and clusters are connected by sheets of galaxies or by a web of thin filaments.*

In 1965 the COSMIC MICROWAVE BACKGROUND (CMB) was discovered, the relic of the fireball, surviving in the form at which it became free of the effects of matter in the universe at t = 380 000 years. COBE and the WMAP have made accurate measurements of the CMB's properties, including its inhomogeneities or fluctuations from place to place.

Inflation

A second major challenge of the Big Bang theory in cosmology has been to demonstrate how the large-scale structures in the universe formed. How did a hot dense universe of elementary particles evolve into a cold rarefied universe of clusters of galaxies?

Alan Guth, André Linde and others introduced the concept of 'inflation' to start the solution to the problem. The basic idea of the inflationary Big Bang theory is the following. The universe was almost infinitely hot at $t = 0$, but its temperature dropped with time. During this process, for a very brief period, the universe inflated at an exponential rate between $t = 10^{-43}$ and 10^{-36} s. The universe was in a kind of unstable vacuum-like state (a state with large energy density, but without elementary particles). The unstable state generated a force that had the same effect as the cosmological constant first introduced by Einstein. The force drove the universe outward exponentially during inflation. At the end of inflation, the vacuum-like state decayed, and the subsequent evolution of the universe could be described by the standard Big Bang theory.

Inflation was necessary to resolve several difficult problems of the standard Big Bang theory. In combination with the modern theory of elementary particles, the standard Big Bang theory predicted the existence of a large amount of superheavy stable particles carrying magnetic charge: magnetic monopoles. According to the standard Big Bang theory, monopoles should appear at the very early stages of the evolution of the universe, and they should now be as abundant as protons. They are not; in fact, not one has been found.

The standard Big Bang theory was unable to answer other questions. One of these pertained to the geometry of the universe. The universe can be described as closed or open, depending on its extent, which in turn depends on the density of energy within it. There is a value of the energy

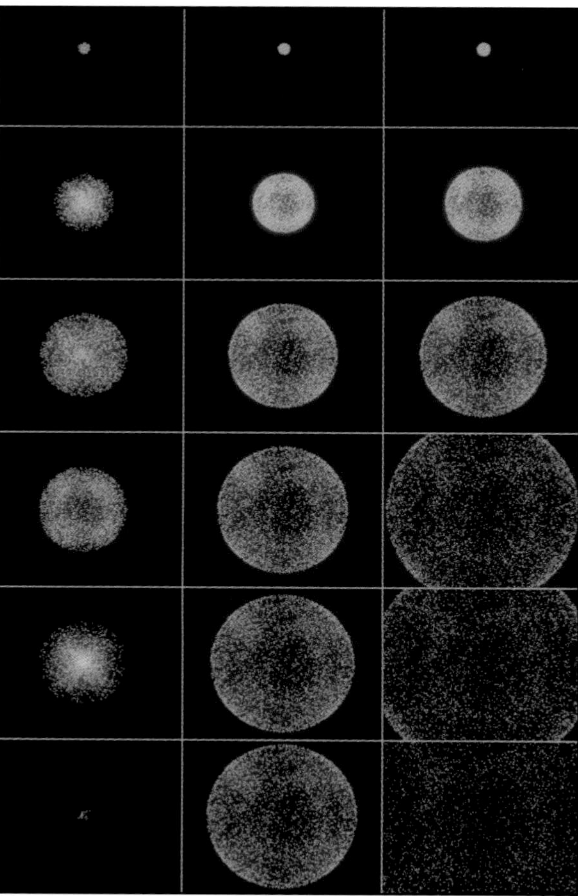

▶ ***The fate of the universe*** *(Left) The density of the universe is greater than a critical value; it stops expanding, contracts under gravity, ending in the 'Big Crunch.' (Center) If the density is neither too great nor too little, the expansion gradually slows down — the 'flat' universe. (Right) If less than the critical value, the universe expands forever.*

density of the universe that forms a boundary between the two situations, and in this case the universe is said to be flat. If the density of the universe is ρ and the density that causes the universe to be flat (the critical density) is ρ_0, then astronomers use the symbol Ω for the density of the universe in terms of the critical density, $\Omega = \rho / \rho_0$. The universe is closed if $\Omega > 1$, it is flat if $\Omega = 1$ and it is open if $\Omega < 1$. The value of Ω changes in an evolving universe. Observational data suggest that now Ω is close to 1. However, according to the Big Bang theory, this requires that very early in the Big Bang the value of Ω should coincide with 1 with an accuracy of 10^{-60}. A slightly higher value of Ω would lead to a closed universe, which would immediately collapse; a slightly smaller value would lead to an open universe, which by now would be practically empty. The need for this incredible fine-tuning of initial conditions in the standard Big Bang theory is the essence of the flatness problem.

Astronomers sought a reason for Ω to be exactly 1. They found it in inflation. By blowing up the universe by a huge factor, inflation ensures that the curvature of space becomes indistinguishably close to zero (the 'flat' case). The observable universe is a tiny portion of the whole, and space appears flat in much the same way as the surface of the Earth appears flat if we examine only a tiny portion of it.

Solving a difficult cosmological problem by a rapid stretching of the universe created another difficulty. If space stretched so much, all inhomogeneities would be stretched away. If there were no inhomogeneities left, there would be nothing to grow into galaxies and other structures.

Fortunately, while removing existing inhomogeneities, inflation created new ones. The basic mechanism can be understood as follows. According to quantum-field theory, empty space is not entirely empty. It is filled with quantum fluctuations of all types. If these fluctuations average over time to zero, then space seems to us empty and can be called the vacuum. In the inflationary universe the vacuum structure is more complicated. The first quantum fluctuations freeze in the inflationary expansion and are stretched by the expansion of the universe. Newer and newer fluctuations become stretched and freeze on top of each other. They do not disappear on being averaged over time.

These inflationary perturbations are responsible for the subsequent appearance of galaxies and formation of the large-scale structure of the universe. The distribution of the perturbations is quite specific and takes a mathematical form discovered in quite different contexts by Carl Friedrich GAUSS, now called the gaussian distribution. Other theories of the origin of the fluctuations give distributions of different shapes (non-gaussian). There are observations of the structure in the universe that support the idea that the initial fluctuations were gaussian, so this supports the inflationary theory of the Big Bang.

The cosmological constant

The universe is still accelerating outwards, although not at all as fast as it did when inflation ruled. According to observations by astronomers of distant SUPERNOVAE of Type Ia, and according to analysis of the fluctuations in the CMB as measured by WMAP, the universe was expanding slower in the past than now. Astronomers characterize this acceleration with a value for the cosmological constant (Λ) of 0.7.

The fate of the universe

Inflation requires that the universe should be extremely flat, $\Omega = 1 \pm 0.0001$. Observational data are consistent with this. If there was no cosmological constant, and $\Omega > 1$, the universe will expand forever, if $\Omega < 1$ gravity will win over the expansion and the universe will fall back in on itself in the Big Crunch, as Newton and Einstein worried. Given that $\Omega = 1$, the universe is poised between a situation where gravity or expansion wins. In this case, the cosmological constant has the deciding role and makes the universe accelerate out, expanding forever.

Parameters of the universe

As the observational details about the universe become more and more focused, the Big Bang cosmology becomes more and more defined. The ages of stars in very old globular clusters are close to the age of the universe, 14 billion years, according to WMAP's observations of the CMB fluctuations. WMAP and observations of distant supernovae show that there is a cosmological constant, $\Lambda = 0.7$, such that dark energy provides 70% of the energy content of the universe, with DARK MATTER providing 25% and baryonic matter about 5%, with the critical density of the universe, ρ_0, having a value of the mass of five atoms of hydrogen per cubic meter. Observations confirm that the calculated 24% of baryonic matter is helium. Several lines of attack, including theories of the growth of structure, agree that $\Omega = 1$.

It used to be said that cosmology was uncertain although cosmologists were never in doubt; we ought to remember this warning and avoid overconfidence. Nevertheless, in spite of themselves, astronomers are saying to each other that cosmology may have started to become an exact science.

Further reading: Narlikar J V (1993) *Introduction to Cosmology* Cambridge University Press; Peacock J A (1999) *Cosmological Physics* Cambridge University Press.

Universe: one or many?

The universe is the sum of everything of which we can be aware. But it may be (a) that our universe is finite, yet unbounded; (b) that the accessible universe is only a small part of a much larger entity, most of which we cannot observe; or (c) that there exist other universes of which we are not aware (that is, our universe is one of many universes that form, in total, the 'multiverse'). *See also* ANTHROPIC PRINCIPLE.

Uppsala Astronomical Observatory

Uppsala University, home of the observatory, was founded in 1477, and preserved lecture notes show that astronomy was taught in the 1480s. Among well-known professors at the observatory were Anders CELSIUS, Anders Angström (1814–74), Gunnar MALMQUIST, Erik HOLMBERG and Bengt Westerlund (1921–). Current research is characterized by the combination of theory and observations with the observational material coming from large ground and space telescopes. At the observatory, about 70 km north of Stockholm, there is a 0.3-m double refractor from 1893 currently used mainly for public shows and at Kvistaberg, 50 km south of Uppsala, there is a 1964 1.4-m Schmidt telescope.

Uranus

The first of the non-naked-eye planets to be discovered, by William Herschel (*see* HERSCHEL FAMILY). Technically, at sixth magnitude, Uranus is marginally visible with the unaided eye. However, it was not distinguished from background stars until Herschel's telescopic observations. Uranus is more than twice the distance of Saturn from the Sun; its discovery gave great impetus to searches for even more distant planets (Neptune and Pluto).

Uranus is the third largest of the planets in the solar system, slightly larger than Neptune. However, because of its lower density (1.27 g cm^{-3} compared with 1.64 g cm^{-3} for Neptune), Uranus ranks fourth among the solar system planets in mass. Its characteristic blue-green appearance is due to a layer of clouds of methane ice in its upper atmosphere, which is not present in the atmospheres of Jupiter or Saturn. Prior to the flyby of Uranus by VOYAGER 2 in 1989, little was known about the physical characteristics of Uranus.

Uranus has a ring system (*see* PLANETARY RING) and many satellites. Nine of the rings were detected as they occulted distant stars (*see* ECLIPSE AND OCCULTATION), and the tenth was discovered by Voyager 2. Prior to 1986, only five moons of Uranus were known. Herschel discovered the satellites OBERON and TITANIA, while UMBRIEL and ARIEL were discovered by William LASSELL, and MIRANDA by Gerard KUIPER. Almost 40 years on from Kuiper's discovery, 10 more satellites (Cordelia, Ophelia, Bianca, Cressida, Desdemona, Juliet, Portia, Rosalind, Belinda and Puck) were discovered by Voyager 2 in 1986. All of the moons discovered by Voyager 2 orbit Uranus closer to the planet than the five classical satellites. In 1999, Erich Karkoschka found a previously undiscovered moon with images from Voyager 2 made in 1986. Between 1997 and 2003, further satellites were discovered by Brett Gladman and his associates. They are more than 10 times as far from the planet as any of the previously known regular satellites, with orbits that are retrograde, highly inclined and/or eccentric.

The orbits of the five classical Urananian satellites provided the first evidence of the unusual tilt of the equator of Uranus relative to its orbital plane. In fact, excluding Miranda and the irregular satellites, all of the Uranian satellites orbit within 0.4° of Uranus's equator. The tilt of the rotation axis of the prograde rotating pole of Uranus is 98°. Alternatively, by IAU convention, which specifies the north pole to be that rotation pole which lies north of the ecliptic plane, the tilt

Uranus

Semimajor axis of the orbit	19.191 AU
Orbital eccentricity	0.046
Inclination to ecliptic	0.77°
Sidereal period	83.7474 years
Equatorial diameter	51 118 km
Equatorial diameter relative to Earth's	4.007
Polar diameter	49 946 km
Flattening	0.023
Mass relative to Earth's	14.54
Mean density	1.27 g cm^{-3}
Surface gravity relative to Earth's	0.906
Escape velocity	21.28 km s^{-1}
Rotation period	17.24 hours
Inclination of equator versus orbital plane	97.92°
Atmosphere	hydrogen (83%) helium (15%) methane (2%)

Uranus's regular satellites

Name	D (km)	P (days)	i (°)	e	d (km)
Cordelia	49 792	0.3350	0.08	0.000	26
Ophelia	53 764	0.3764	0.10	0.001	32
Bianca	59 165	0.4346	0.19	0.001	44
Cressida	61 767	0.4636	0.01	0.000	66
Desdemona	62 659	0.4737	0.11	0.000	58
Juliet	64 358	0.4931	0.07	0.001	84
Portia	66 097	0.5152	0.06	0.000	110
Rosalind	69 927	0.5585	0.28	0.000	54
Belinda	75 255	0.6235	0.03	0.000	68
Puck	86 006	0.7618	0.32	0.000	154
Miranda	129 847	1.4135	4.22	0.027	472
Ariel	190 929	2.5204	0.31	0.0034	1158
Umbriel	265 979	4.1442	0.36	0.0050	1170
Titania	436 273	8.7059	0.14	0.0022	1578
Oberon	583 421	13.4632	0.10	0.0008	1522

D = distance from planet; P = period of revolution; i = inclination of orbit to equatorial plane of Uranus; e = orbital eccentricity; d = diameter

Uranus's irregular satellites

Name and Designations			D (km)	R	d (km)
XVI	Caliban	S/1997 U 1	7 000 000	21.9	60
XX	Stephano	S/1999 U 2	8 000 000	24.1	20
		S/2001 U 1	8 500 000	faint	small
XVII	Sycorax	S/1997 U 2	12 000 000	20.4	120
XVIII	Prospero	S/1999 U 3	16 000 000	23.2	30
XIX	Setebos	S/1999 U 1	17 500 000	23.3	30

D = distance from planet; R = magnitude in red light; d = diameter

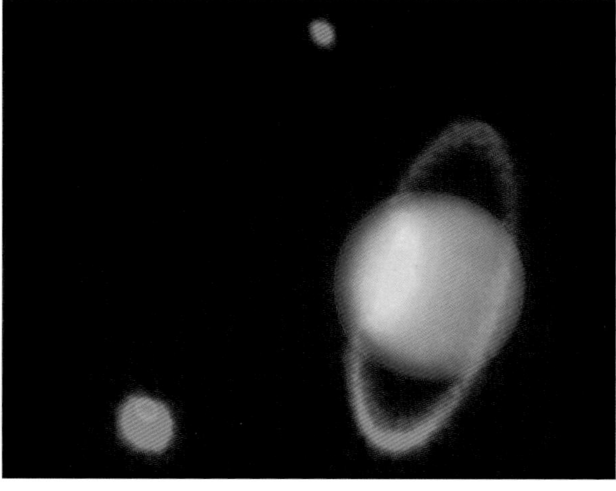

► ***Uranus*** *The Subaru telescope shows the ring system and the satellites Miranda (top) and Ariel (left).*

is designated as 82° with retrograde rotation. Much speculation has surrounded this unusual rotation of Uranus. One popular theory is that, during its early history, Uranus had a more normal tilt, but during the latter stages of formation the planet was struck by an Earth-sized PLANETESIMAL, resulting in the present extreme tilt. The only other planets to exhibit more than a modest tilt are Venus and Pluto, which are much smaller in mass.

This orientation means that, part of the time, the Sun will be nearly overhead at Uranus's north or south pole. Because the rotational axis stays fixed in direction while Uranus orbits around the Sun, one-quarter of an orbit later the Sun will be over Uranus's equator. Therefore the differences between the solar heating during Uranus's summer and winter are much more extreme than on Earth, and the planet is much colder overall because of its much greater distance from the Sun. Despite these extremes of heat input, the temperature extremes are small, apparently because the atmosphere can equalize temperature differences by transporting heat.

The low reflectivity of Uranus at the red end of the spectrum led spectroscopist Rupert Wildt (1905–76) to propose in 1932 the presence of methane (CH_4) in the atmospheres of each of the gas-giant planets. Their low densities led to the conclusion that their atmospheres were mostly hydrogen (H_2), and helium (He) was detected in the atmosphere of Uranus by Voyager 2 in 1986. Voyager 2 provided not only high-resolution images of Uranus, the ring system and the satellites, but also confirmed that the planet has a highly tilted and offset magnetic field (*see* PLANETARY MAGNETOSPHERE) and measured the rotation period of its interior.

Composition and structure

Uranus is composed primarily of the light elements hydrogen and helium, and even the somewhat heavier compounds, CH_4, ammonia (NH_3) and water (H_2O), are composed largely of hydrogen. Uranus and Neptune may be smaller than Jupiter and Saturn as a consequence of the solar nebula being less dense at these greater distances from the Sun. Perhaps the growth of Uranus and Neptune therefore lagged behind that of Jupiter and Saturn.

The water in Uranus's upper atmosphere is exogenic; that is, from an external source, presumably micrometeoroids.

There is a methane cloud deck in the atmosphere near the 1-bar pressure level, where the temperature is near 80 K. This is colder than the atmospheres of Jupiter and Saturn, which is why they have no methane clouds.

Unlike the other three gas-giant planets, and for unknown reasons, Uranus has little or no heat escaping from its interior to warm the upper regions of the planet. It may or may not have a molten rocky core out to a distance of about 5000 km from the center. It may be that Uranus's interior is cooler than those of the other gas giants, perhaps because the collision that tipped Uranus over also stirred it enough to let its excess heat out.

With little or no discernible internal heat source, Uranus might be expected to have little atmospheric weather. All atmospheric features are symmetric around the rotation axis and lie in zones.

Further reading: Miner E D (1998) *Uranus: The Planet, Rings and Satellites* Wiley

Urey, Harold Clayton (1893–1981)

American chemist, 1934 Nobel prizewinner for Chemistry 'for his discovery of heavy hydrogen' or deuterium. At the University of Chicago, Illinois, USA he worked on the origin of the elements, their abundances in stars, and the origin of the planets, including the chemical properties of the Earth.

Ursa Major, Ursa Minor *See* CONSTELLATION

US Naval Observatory

The oldest astronomical observatory in the USA, and the oldest continuously operating scientific institution in the US government. Founded in 1830 as a Depot of Charts and Instruments for rating chronometers and maintaining navigational instruments, by 1844 it had become the first national observatory of the USA, analogous to the ROYAL OBSERVATORY, GREENWICH in the UK. The observatory's headquarters, in Washington, DC, include the historic 0.66-m refractor used to discover the two moons of Mars in 1877. The observatory's largest telescope, a 1.55-m astrometric reflector, is located in Flagstaff, Arizona. The observatory provides the national time service for the USA, determines the precise positions and motions of celestial bodies, measures Earth rotation parameters including polar motion, and produces a variety of almanacs for use by astronomers, navigators and the general public.

Ussher, James (1581–1656)

The Archbishop of Armagh and Primate of All Ireland, Ussher was a churchman and a scholar. He correlated Middle Eastern and Mediterranean histories with Jewish genealogies of the Old Testament, and with the resulting chronology established the year of creation as 4004 BC.

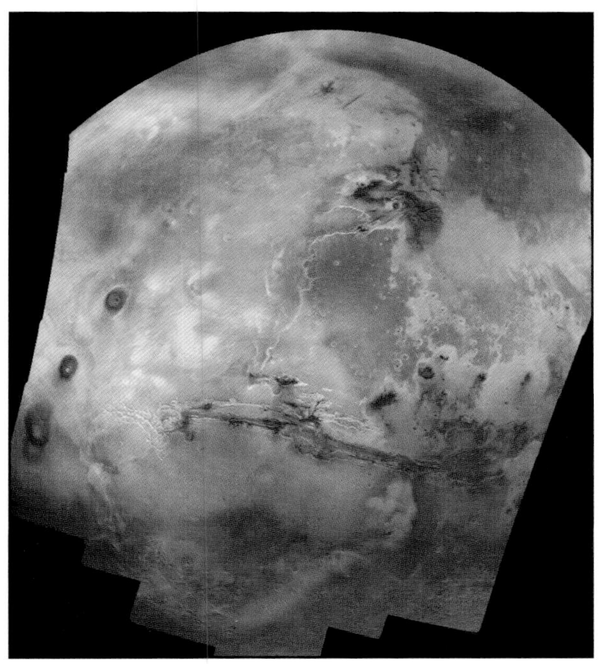

◀ ***Valles Marineris*** *The largest canyon in the solar system cuts a wide swathe across the face of Mars. The grand valley extends over 3000 km, spans as much as 600 km across, and delves over 6 km deep. By comparison, the Earth's Grand Canyon in Arizona, USA is 800 km long, 30 km across, and 1.8 km deep.*

Valles Marineris

A complex system of canyons on MARS, also known as Mariner Valley, stretching for a total of 4128 km in the east-west direction just south of the equator, and reaching depths of over 6 km. This length is equivalent to the distance across the USA. It is named for the Mariner probes (*see* MARINER MISSIONS) which returned the first close-up images of the planet's surface. Individual sections of the system are termed 'chasmata.' At its western end the system abuts the faulted area at the east of the THARSIS BULGE known as Noctis Labyrinthus. At the eastern end the main system opens out into the so-called chaotic terrain of the region known as Margaritifer Sinus.

Valles Marineris was created largely by faulting, but other forces have also been at work. The deep branching valleys running into the southern edge of Ius Chasma suggest erosion by groundwater, while at the eastern end are teardrop-shaped islands suggestive of flowing water. Close-up views of some areas clearly reveal the presence of layered sediments, which could only have been deposited under water.

Van Allen, James (1914–)

American magnetospheric physicist who discovered the VAN ALLEN BELTS that surround Earth. After war-time service, van Allen used V-2 rockets for high-altitude experiments, and later the Aerobee, when supplies of the V-2 ran out. During the International Geophysical Year (1957/8), the first US satellite, EXPLORER 1, carried a micrometeorite detector and a COSMIC RAY experiment designed by Van Allen. Data from Explorer 1 and Explorer 3 (launched March 26, 1958) revealed the existence of a doughnut-shaped region of charged particle radiation trapped by Earth's magnetic field. Later in 1958, Pioneer 3 (*see* PIONEER MISSIONS) led to the discovery of a second radiation belt.

Van Allen belt

Two toroidal (doughnut-shaped) regions surrounding the Earth that contain trapped charged particles, discovered by James VAN ALLEN.

van de Hulst, Hendrik (1918–2000)

Dutch astronomer, he became professor at LEIDEN OBSERVATORY. While a student (under Jan OORT), van de Hulst showed that, as a result of a spontaneous change of direction of spin from the electron orbiting a neutral hydrogen atom, strong emission would arise from hydrogen in the radio spectral region. Although of low probability, there are so many neutral hydrogen atoms in space that the cumulative effect is significant. The prediction of radio emission at a wavelength of 21 cm (1420 MHz) was confirmed by van de Hulst, Jan Oort (1900–92) and C Alex Muller in 1951.

van de Kamp, Peter [Piet] (1901–95)

Dutch astronomer who became director of the Sproul Observatory and a professor at Swarthmore College, both in Pennsylvania, USA. He worked on the astrometric measurement of proper motions of stars across the sky (*see* ASTROMETRY) and discovered oscillations in their paths. He interpreted the deflections as being caused by the revolution of planetary systems around the stars. The oscillations have not been confirmed, and many astronomers think that they were instrumental effects caused by the telescopes and measurement techniques.

van Maanen, Adriaan (1884–1946)

Dutch astronomer, who became a member of the Mount Wilson staff (*see* MOUNT WILSON OBSERVATORY). He tried to measure the rotation of spiral nebulae as a means of establishing their distances. His apparent detection of rotation misleadingly suggested that the nebulae were relatively nearby. Through its proper motion (*see* ASTROMETRY) and PARALLAX, he discovered the white dwarf VAN MAANEN'S STAR.

van Maanen's Star

Star in the constellation of Pisces which, although faint, is very important, as it is the nearest readily observable WHITE DWARF. The white dwarfs SIRIUS B and PROCYON B are both closer, but their proximity to their much brighter companions makes them very difficult to study as individual stars.

Van Maanen's Star, known also as Wolf 28 and HIP 3829, was discovered in 1917 by Adriaan VAN MAANEN to have the large proper motion of 2.978 arcsec per year. From its PARALLAX, it is only 14.4 l.y. distant. Its apparent magnitude of 12.37 indicates that it has a very low intrinsic luminosity, with an absolute magnitude of 14.1. It has been estimated that its diameter may be no more than 12 500 km, which is comparable to that of the Earth.

Van Vleck Observatory

Situated on the campus of Wesleyan University in Middletown, Connecticut, USA, the observatory was built in the years 1914–6 and named in memory of John van Vleck (1853–1904), professor of natural science at Wesleyan throughout much of the nineteenth century. The primary instrument of the observatory is a refracting telescope of 0.5-m aperture and a focal length of 8.41 m. Since 1992 it has been mainly used to determine trigonometric PARALLAXES and proper motions (*see* ASTROMETRY) of many hundreds of faint nearby stars. In the last few decades the van Vleck astrometric program has been one of the leaders in obtaining ground-based parallaxes to respond to problems in stellar astrophysics (*see* HIPPARCOS). In 1971, a 0.6-cm reflecting telescope was obtained from the estate of Richard Perkin (?–1969), founder of the Perkin–Elmer Corp, a leading US optoelectronics company.

Variable star

All stars are variable. They form from the interstellar medium, burn hydrogen, evolve into RED GIANT STARS or SUPERGIANTS and die. While they do this they change their brightness and their temperature and move along tracks in the HERTZSPRUNG–RUSSELL DIAGRAM. But usually the timescale for these changes is very long, perhaps millions or even billions of years: in fact some stages of the lives of less massive stars last longer than the age of the universe.

Variable stars are stars whose light varies on short enough times for the variability to have been observed – from milliseconds to millennia. Hundreds of thousands of such stars have been discovered. They are listed in the General Catalogue of Variable Stars (GCVS). Many such stars have names like 'RR Lyrae,' with two letters and the genitive of the constellation name, or 'V1974 Cygni,' with the letter V, a number and the constellation; such stars first appeared in a catalog of variable stars, discovered by a survey set up for the purpose. Others may have been known as stars under another name before their variability was discovered, like Delta Cephei. Others have special designations like 'Mira,' 'SN1987A,' 'PSR 0542+35' and 'Cygnus X-1.'

Both the GCVS and the AAVSO classify variable stars into their many types. The classification system has evolved for more than a century. In its simplest form a classification might be described in terms of a prototype star like RR Lyrae or Cepheid (abbreviated from Delta Cephei). AAVSO and the GCVS use observational characteristics when the underlying mechanism in the variable star is unknown but work toward a generic taxonomy as the various sorts of variable star become understood. There are the following overall categories:

- Eruptive variable stars (*see* VARIABLE STAR: ERUPTIVE): the variability is caused by flares or the ejection of a shell;
- Pulsating variable stars (*see* VARIABLE STAR: PULSATING): the variability is caused by radial or non-radial pulsation;
- CATACLYSMIC VARIABLE STARS: the variability is caused by explosions on the star;
- X-ray stars: they show variable X-ray emission from a neutron star or black hole in a binary.

In all the above cases the variability is intrinsic to the star. If the variability is caused by a change of viewing angle, it is called extrinsic variability, as in the following two cases:

- Eclipsing variable stars: the variability occurs when one component passes in front of the other. (For the last three categories, *see* VARIABLE STAR: CLOSE BINARY STAR.)
- Rotating variable stars (*see* VARIABLE STAR: ROTATING): the variability is caused by starspots coming into view at the front of the star or by the star's changing shape, as it rotates.

Within these overall categories AAVSO and the GCVS recognize specific types, and give them an alphabetic code usually of one to five capital letters. The code is sometimes subdivided into less important classes based on observational differences, and the system is sometimes inconsistent with modern understanding. And of course, the modern understanding may not be accurate and so may change.

Variable star: close binary star

Variability of the light from a BINARY STAR system often occurs, usually where the stars are 'close binaries'; the opposite is 'wide' (*see* DOUBLE STARS). The light from the system may change because the system rotates and presents different faces toward us or, in the case of an ECLIPSING BINARY STAR, because one star passes periodically behind the other. In addition, the stars may vary because of some intrinsic instability in the component stars, as in CATCLYSMIC BINARY STARS or X-ray stars.

Classification of close binaries by current physical state

Binaries can be classified in terms of the masses and evolutionary states of the two stars. ROCHE LOBES are the zones in a binary star system around each star such that material in the zone belongs to that star. If a star fills its Roche Lobe, material can be pushed onto the other or leave the system entirely. The classes are as follows:

- D: detached binaries are binaries where neither star fills its lobe;
- SD: semi-detached binaries have one lobe-filling star (often a RED GIANT STAR), and material spilling through the point where the Roche Lobes touch, onto the other star or into a disk orbiting around it;
- K: in contact-binaries, both stars fill their Roche Lobes;
- if material has overflowed the lobes and surrounds both stars, we speak of a common envelope binary like Z ANDROMEDAE.

Classification of close binaries by evolutionary phase

Assigning binary systems to their place in a scenario for how systems evolve or their place in astronomical phenomena is uncertain, but provides the most physical understanding. These classes are often named for a prototype:

- EW: W Ursae Majoris variables have components that almost touch, and so have very short periods (less than 1 day).
- EB and EA: BETA LYRAE and ALGOL variables. In Beta Lyr mass transfer is taking place from the more massive, lobe-filling star to its companion. This mass transfer has just started and is happening rapidly. In Algol stars the stars are almost spherical and the less massive star is the one that fills the Roche Lobe, because it has transferred so much material onto the other. The 'Algol paradox' recognizes that this is unusual in that the less massive star appears to have evolved further than the more massive star.
- RS: RS CANUM VENATICORUM stars have a detached subgiant component, and fast rotation, almost but not quite synchronized with the orbital periods. This produces 'photometric waves' in the light curve that are caused by solar-like activity phenomena, such as dark spots or bright PLAGES, that modulate the star's brightness at the beat period between the star's rotation period and the binary's orbital period.
- A CATACLYSMIC BINARY STAR shows various types of non-destructive explosions in a binary system that contains a WHITE DWARF.
- V 471 Tauri stars are detached pairs of white dwarf and red dwarf (*see* RED DWARF AND FLARE STAR), expected to evolve into cataclysmic variables;
- N: NOVAE and recurrent novae have outbursts resulting from explosive hydrogen-burning – thermonuclear runaways – on the surfaces of white dwarfs that are accreting hydrogen-rich material from red dwarf companions.
- UG: U Geminorum stars or DWARF NOVAE like SS CYGNI are binary stars that brighten by factors of hundreds to thousands, sometimes in just a few hours. An accretion disk is usually the major source of radiation in the system. Another source of light is a bright spot formed where the stream collides with the outer edge of the disk.
  - UGSS: SS CYGNI-type stars increase in brightness by 2–6 magnitudes in 1–2 days, returning to their original

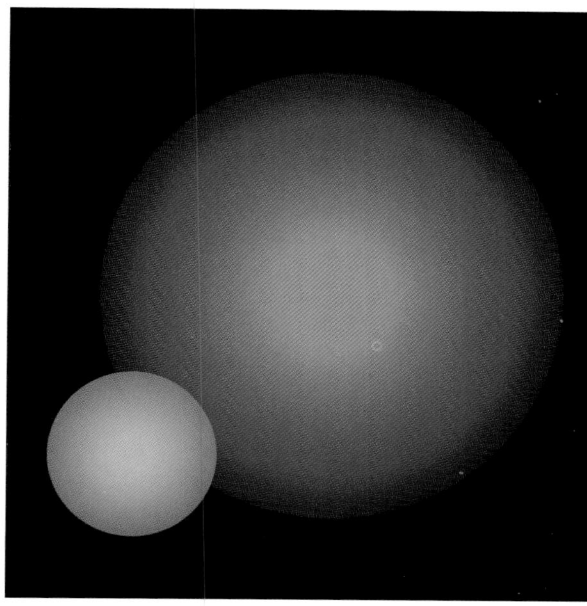

◀ ***Eclipsing binary star*** *Artist's representation of a red giant and hot blue dwarf in close orbit. As the small blue star transits the large red one, the red light from the star will reduce by a fraction. As it passes behind, the blue star will be fully eclipsed and its blue light will be almost extinguished.*

brightness in a few days. The cycle is repeated with periods between 10 and several thousand days.

- UGSU: SU Ursae Majoris stars have two distinct kinds of outbursts: one is faint, frequent, and short, with a duration of 1–2 days; the other (a 'super-outburst') is bright, less frequent, and long, with a duration of 10–20 days. During super-outbursts, small periodic modulations ('superhumps') appear.
- UGZ: Z Camelopardis stars show cyclic variations, interrupted by intervals of constant brightness called 'standstills.' These standstills last the equivalent of several cycles, with the star stuck at a near constant brightness approximately one-third of the way from maximum to minimum.

- ZAND: Z Andromedae stars are close binaries where a hot component actually orbits inside the extended envelope of its cool giant companion.
- Symbiotic stars: literally two objects of quite different kinds, living close together in interdependence. One star is a red giant of various sorts. The second is a white dwarf (in two cases a neutron star; *see* NEUTRON STAR AND PULSAR), or even a normal main sequence star. There is an essential hot light source, which gives characteristic spectral properties to surrounding gas. It is either the white dwarf star itself or a hot accretion disk surrounding the companion.
- AM: AM HERCULIS variables are like novae, except that the magnetic field of the white dwarf is strong, preventing the formation of an accretion disk, and channeling the gas direct to the surface of the white dwarf. The magnetic fields of the white dwarf and its companion are locked and the white dwarf rotates on its axis at the same rate as the revolution of the binary. Also known as 'polars' because the white dwarf has a bright magnetic pole where the accretion strikes.
- DQ HERCULIS variables are like AM Her stars, except the magnetic field is weaker, there is an incomplete accretion disk, and the rotation rate is not locked. Also known as 'intermediate polars,' because the magnetic field is intermediate between strong and insignificant.
- AM CVN STARS are apparently a binary system which has two white dwarfs in orbit, one, of helium, feeding a helium accretion disk around the other.
- SN Ia: Type Ia SUPERNOVAE are the collapse of a white dwarf to a neutron star, because of accretion of material from a companion. In principle neutron stars could likewise collapse to BLACK HOLES.
- X: X-ray binary stars. A donor star provides material to a compact accreting star that can be a white dwarf (like DQ Herculis), neutron star (like HERCULES X-1) or black hole (like CYGNUS X-1). The gas is heated to X-ray emitting temperatures. The donors may be of high mass (in which case their light dominates the system) or low mass (in which case it is possible more readily to study the accretion disk, accretion stream, place where they collide, and so on). The light and the X-rays from such stars are often both highly variable.

The following types of variable star are hypothesized to be in binary systems:

- Dwarf carbon stars: a so-called CH star has a spectrum showing it has a carbon-rich atmosphere (*see* CH STAR AND BARIUM STAR). It seems to have been a normal star with its hydrogen atmosphere contaminated with carbon that it has accumulated by mass transfer from an evolved red giant companion, with plenty of carbon after helium-burning. The mass transfer would have turned the red giant into a white dwarf star.
- Millisecond pulsars with weak magnetic fields seem to have been old, slowly rotating neutron stars that have been spun up (made to rotate faster) by the impact of falling material from a companion. Some are definitely binaries with white dwarf or neutron star companions.
- SNIa and SNIc. Type Ib and Ic supernovae are apparently from massive progenitors, like Type II supernovae, but with their hydrogen-rich envelopes entirely stripped off, presumably by mass donation in a close binary.

Eclipsing binaries

An eclipsing binary star is a binary system oriented such that the stars of an orbiting pair periodically pass in front of one another. Bright eclipsing stars such as Algol and EPSILON AURIGAE excite the interest of observers, and many are within reach of binoculars.

Both stars are eclipsed in most eclipsing binaries, although usually one of the alternating eclipses is much more prominent. The larger the light loss (eclipse depth) the more likely that the orbit is edge-on and the larger the ratio of star size to orbit size. Total eclipses (entire star covered) and annular eclipses (a ring of surface not eclipsed) happen when the stars are unequal sizes. For circular orbits, the star with higher average surface brightness (ordinarily higher surface temperature) has the deeper eclipse. Where the orbit is eccentric, the eclipses are usually unequally spaced. LIMB DARKENING may cause an eclipsed star to be significantly fainter at its limb, so the eclipse starts slowly.

Further reading: Kallrath J and Milone E F (1999) *Eclipsing Binary Stars, Modeling and Analysis* Springer.

Variable star: eruptive

The variability of eruptive variable stars is caused by flares or the ejection of a shell, as a result of rapid rotation, unstable strong magnetic fields (often caused by the rapid rotation), strong stellar winds and/or low surface gravity.

- IN: Orion variable stars are PRE–MAIN SEQUENCE STARS and other young stars in or near nebulae, like the ORION NEBULA. They have irregular variations in their light of perhaps several magnitudes. They are in the last stages of evolution toward being a stable main sequence star. The FLARES are caused by rapid rotation (in which case the stars are called IN(YY) stars), unstable surface magnetic fields or other surface eruptions. There are several subspecies indicated by one or more suffixes that denote spectral differences (for example, INA and INB stars). In particular, INT stars are T TAURI STARS (which have intense iron emission lines). RW Aurigae stars are stars of these classes, and are not now recognized as a distinct group.

- FU: FU Orionis stars show a slow rise of many magnitudes over many months to a maximum lasting for several years, followed by the development of an emission spectrum. These stars are all associated with REFLECTION NEBULAE and are connected in an evolutionary line with Orion variable stars.
- GCAS: Gamma Cassiopeiae is the prototype of a class of rapidly rotating blue giants that occasionally eject a ring of matter from their equator, accompanied by irregular brightness changes, which in Gamma Cas itself are readily visible to the naked eye.
- RCB: R CORONAE BOREALIS STARS are hydrogen-poor, helium and carbon-rich stars that irregularly fade by many magnitudes for weeks or months. These stars can disappear in just a few hours, as a result of the ejection of an almost opaque shell of carbon dust.
- SDOR: S Doradus stars are massive LUMINOUS BLUE VARIABLES.
- UV: UV Ceti stars or flare stars are red dwarf stars showing flares that brighten the star for only a few minutes by many magnitudes.
- WR: WOLF–RAYET STARS have broad emission lines of carbon and nitrogen, and a massive and unstable stellar wind.

Variable star: pulsating

A star is in balance between the force of gravity and its internal pressure, caused both by gas pressure and radiation pressure. The pressure depends on the chemical composition of the star. In particular the radiation pressure depends on the opacity of the material; the more opaque it is, the more the radiation is trapped and the harder it can push on the star's structure. The opacity of the star's chemical material depends in turn on how many electrons the material has released. This depends on the temperature of the material and how many electrons are knocked off the ions. There can be a situation in which the material, usually helium, is ionized, its opacity is high, and the material is pushed out. The star expands, it cools, and its material recombines and becomes transparent. Radiation is released and the pressure reduces, so the star shrinks. It becomes hotter, ionizes the material, becomes more opaque and expands again. This mechanism is what drives CEPHEID VARIABLE STARS and RR LYRAE STARS to pulsate. Since the mechanism is quite precise in terms of the structure of the star, such stars occupy a particular region of the HERTZSPRUNG–RUSSELL DIAGRAM that is called the INSTABILITY STRIP.

There is a wide variety of other pulsating stars whose brightness varies. In all cases, the star can vibrate, like a musical instrument, in different modes, throbbing with a 'fundamental' note as the pulsation carries at sound-speed through the star, or at harmonics. In fact, a star can sometimes vibrate in more than one frequency as the sound wave carries through or round the star in different ways, in which case the multiple modes may 'beat' one with the other (common in Alpha Cygni stars, Beta Cephei stars, Cepheids and RR Lyrae stars). The LIGHT CURVE can be very complicated, particularly of non-radial pulsations (*see* ASTROSEISMOLOGY).

Pulsating variable stars fall into the following categories recognized by the General Catalogue of Variable Stars (GCVS) and the AAVSO:

- ACYG: Alpha Cygni stars, named for DENEB, are supergiants with multiple non-radial pulsations showing very small variations with periods of weeks. The modes beat with each other.
- BCEP: Beta Cephei stars (formerly known also as Beta Canis Majoris stars) are blue stars with small variations and periods of less than 1 day, caused by radial pulsations. Multiple periods are common. BCEPS is a subgroup with much smaller, faster variations.
- CEP: Cepheid variable stars are named for DELTA CEPHEI. They are radially pulsating white to yellow giants with variations of up to 2 magnitudes and periods from 1 to over 100 days. Cepheids show a period-luminosity (P-L) relation, as initially discovered by Henrietta LEAVITT, but the P-L relations for the two main subgroups are different (*see* DISTANCE).
- CW: W Virginis stars are POPULATION II Cepheids found in the halos of galaxies and some globular clusters. They are 0.7–2 magnitudes fainter than POPULATION I Cepheids of the same period, and can be recognized from subtle differences in the shape of their light curves, and by their different spectral features.
- DCEP: delta Cephei stars or 'classical' Cepheids are young Population I stars, the ones whose P-L relation was first identified in the magellanic clouds.
- DSCT: DELTA SCUTI STARS are Population I pulsating stars.
- M or LPV: MIRA stars are named for Omicron Ceti and are RED GIANT STARS with large variations (2.5–11 magnitudes or more). The period of the variability ranges from 80 to over 1000 days and is the reason for the alternative name for this class, of Long Period Variable (LPV). The light variations do not repeat exactly from one cycle to the next and some Mira variables have multiple periods.
- PVTEL: PV Telescopii stars are pulsating helium-rich supergiants with small variations. They may be related to the WOLF–RAYET STARS.
- RR: RR LYRAE STARS (formerly known as 'cluster Cepheids') are radially pulsating stars of the galactic halo commonly found in GLOBULAR CLUSTERS, with periods of about 1 day and variations of up to 2 magnitudes. A suffix 'a', 'b' or 'ab' (for example, RRa) indicates that the star has an asymmetric light curve and vibrates in the fundamental mode, while RRc stars have symmetric light curves and vibrate in the first overtone. Some RR Lyrae stars show the Blazhko Effect, which is periodic variations in period and light curve.
- RV: RV Tauri stars are radially pulsating yellow to red supergiants with alternating primary and secondary minima, with variations of up to 4 magnitudes and periods of 30–150 days.
- SR: semiregular variable stars are related to Mira variable stars. Like Mira stars, SR variables are red giants, but they have irregular periods and light curves. There are subgroups A to D, denoting how periodic they are and what their spectral type is.
- SXPHE: SX Phoenicis stars are Population II subdwarf stars resembling Delta Scuti variable stars, with periods of about 1 hour (often multiple periods) and variability of about 0.5 magnitude.
- ZZ: ZZ CETI STARS are non-radially pulsating white dwarfs with small, rapid variations (periods of less than 30 minutes).

See also ASTROSEISMOLOGY.

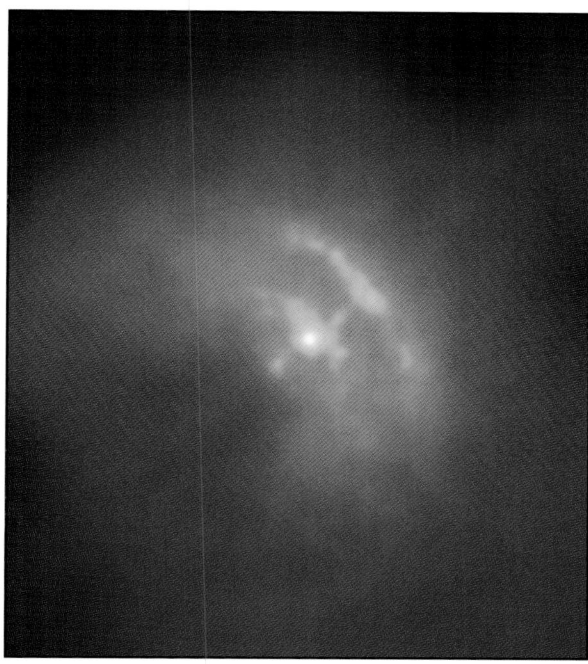

◀ ***Vela pulsar*** *The image shows a dramatic bow-like structure at the leading edge of the nebula that surrounds the Vela pulsar (the bright X-ray star at the center). The swept-back appearance of the nebula is due to the motion of the pulsar through the Vela supernova remnant, toward the upper right corner.*

Variable star: rotating

The light from a rotating star can vary if the star is spotty, is heated on one side more than the other, or is not symmetric around its rotation axis (pointy, due to a tide raised by a companion). Some stars in binary systems vary due to tides and mutual heating, collectively known as 'proximity effects.' 'Ellipsoidal variation' is variability due to the ellipsoidal figure of a tidally distorted star. Mutual heating produces the 'reflection effect.' *See* VARIABLE STAR: CLOSE BINARY STAR. Rotating variable stars fall into the following categories:

- ACV: Alpha-2 CVn stars are white stars showing small variations due to large 'starspots' generated by intense magnetic fields (*see* MAGNETIC FIELDS IN STARS);
- BY: BY Dra stars are red dwarfs with large 'starspots';
- ELL: elliptical stars are tidally distorted close binary stars showing ellipsoidal variation;
- FKCOM: FK Com stars are rotating yellow-orange giants with non-uniform surface brightness;
- PSR: rotating NEUTRON STARS AND PULSARS;
- R: binary stars with a reflection effect, having a large cooler component illuminated by a hotter companion;
- RS: RS CANUM VENATICORUM stars are close binaries in which one member, rotating non-synchronously with the binary system, has patchy chromospheric activity and starspots;
- SXARI: SX Ari stars are high temperature helium-rich Alpha-2 CVn stars

Variable stars, observing

See practical astronomy feature overleaf.

Vatican Observatory

One of the oldest astronomical institutes in the world, it began with the reformation of the CALENDAR in 1582. At the Roman College, Father Angelo SECCHI first classified stars according to their spectra. With these rich traditions, Leo XIII in 1891 formally founded the Vatican Observatory on a hillside behind the dome of St Peter's Basilica. In 1935 Pius XI provided a new location for the observatory at the papal summer residence at Castel Gandolfo. In 1981 the observatory founded a second research center in Tucson, Arizona. In 1993 the observatory, in collaboration with the STEWARD OBSERVATORY, completed the construction of the Vatican Advanced Technology Telescope (VATT), which has pioneered the new technology of creating large, lightweight, stable mirrors in a rotating furnace.

Vega

The star Alpha Lyrae, the only bright star in the constellation. Its name, formerly Wega, derives from the Arabic 'al-Waki,' 'the Swooping (or Falling) Eagle,' hence the form Alvaka used on some seventeenth-century celestial globes. Because of PRECESSION of the equinoxes, it was the Pole Star about 12 000 years ago, and will be again 14 000 years hence. Fairly close at 25.3 l.y., Vega is the fifth brightest star in the heavens, with an apparent magnitude of 0.03. With its high northern declination it dominates the skies in northern latitudes on clear summer nights, forming the Summer Triangle ASTERISM with the bright stars ALTAIR and DENEB. The IRAS survey showed it to be surrounded by a disk of gas and dust, which may eventually form a planetary system.

VEGA mission

A combined spacecraft mission to VENUS and COMET HALLEY, launched by the USSR at the end of 1984. The mission consisted of two identical spacecraft VEGA 1 and VEGA 2. VEGA is an acronym from the words 'Venus' and 'Halley' ('Galley' in Russian spelling). The basic design of the spacecraft was the same as had been used many times to deliver Soviet landers and orbiters to Venus. In June 1985, these spacecraft successfully delivered the first balloons into the venusian atmosphere and also landers, providing the first direct measurements of the properties of Venus's atmosphere and surface. After this the VEGA 1 and VEGA 2 cometary probes were directed to Comet Halley, which they encountered on March 6 and March 9, 1986.

Vela *See* CONSTELLATION

Vela pulsar

The pulsar PSR 0833-45, discovered in 1968. With a period of only 9.3 ms, it has one of the fastest pulse-rates known, implying that it is one of the youngest pulsars. This is borne out by an observed deceleration in its pulse-rate of 10.7 ns per day, which sets an upper limit to the time elapsed since it was formed in a supernova explosion of about 11 000 years ago. In 1977 its pulsations were also recorded in visible light, making it one of the first optical pulsars to be confirmed. It is surrounded by extensive nebulosity more than 5° across, also a product of the supernova (*see* VELA SUPERNOVA REMNANT), and is associated with a moderately strong radio source, Vela X; its distance has been estimated at between 1300 and 1600 l.y. The Vela pulsar and supernova remnant lie within a much older and larger supernova remnant, the Gum Nebula. Observations with NASA's CHANDRA X-RAY OBSERVATORY show that the neutron star (*see* NEUTRON STAR AND PULSAR) is enveloped in a cloud of high-energy particles emitting X-rays as they spiral around magnetic field lines. The X-ray image of the nebula resembles a gigantic cosmic crossbow. This is thought to show a shock wave caused by matter rushing away from the neutron star. This cloud, or nebula, is embedded in a much larger cloud produced by the supernova and has a swept-back, cometary shape because of its motion through the larger cloud.

Observing variable stars

There are many research programs where contributions made by amateurs are crucial. A particular example is the long-term monitoring of variable stars in support of space missions. Amateur variable-star associations, such as the AAVSO, the BAA and the Royal Astronomical Society of New Zealand, have made important historical contributions in coordinating such observations. As one example, the HIPPARCOS program contained many thousands of variable stars, some showing very large and unpredictable variations. The monitoring helped ESA predict the observing time needed by the satellite and was also used to complete the light curves drawn using the very accurate, but sporadic, Hipparcos photometric measurements.

Observations can be made visually, using a CCD detector, or (less commonly) photographically. Whatever the method, measurements are made by comparing the star you are observing with standard comparison stars in its vicinity.

Observing variable stars by eye

To monitor naked-eye variables, it is necessary to be able to be able to see a fainter companion star than the one being monitored, so in practice this means that for this method the star should not be expected to drop below about magnitude 4. There are 34 variable stars which have a range of at least 0.4 magnitude and become brighter than visual magnitude 4.0, and so can be monitored with the naked eye. These include ECLIPSING BINARY STARS, CEPHEID VARIABLE STARS, semiregular red variables, a few long-period Mira variable stars, and the recurrent nova T Coronae Borealis. Of these, 24 do not fall below magnitude 5, and therefore remain visible.

As your observations are going to be incorporated into a worldwide database, it is worth spending some time planning your observing program. There are many possibilities. You will obtain most satisfaction if the stars you observe match your interests and so you should find out as much as you can about the different types of variable star and decide where your contributions will be most valuable. You must also consider the equipment you will use and the darkness of your observing site to fit with the expected range of brightness of your star(s). The timescale of the star's variability is also to be taken into consideration as this affects how often you should plan to observe it.

Stars in the magnitude range 4 to 9 are commonly observed with 7 × 50 or 10 × 50 binoculars, and the observing program includes several thousand stars. At the faint end of the range, the binoculars need a firm mount to keep the image steady. Below magnitude 9, a telescope is necessary.

Before making any observations you should allow your eyes to become very well adapted to the dark. This can take about 15 min but on your first night you will have plenty to do in this time. Try to approach each observation with a fresh mind. Do not be tempted to try to remember your last observation of the same star or what others have observed.

You will have the official star chart that is provided by the organization which will receive your observations. It is essential to use the comparison stars selected by the coordinator of the program if you want your observations to be accepted into the database. While you are observing, the charts need to be protected from the elements so that they do not get damp from the dew, blown away by the wind or covered in mud if they fall on the ground. To avoid

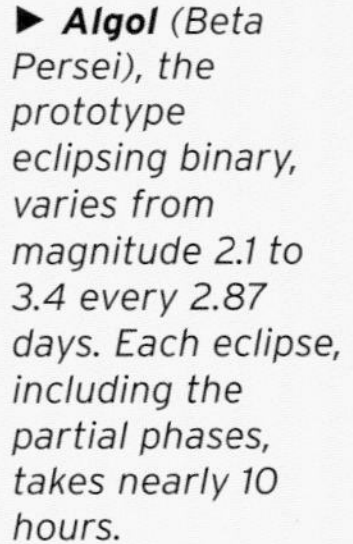

► ***Algol*** *(Beta Persei), the prototype eclipsing binary, varies from magnitude 2.1 to 3.4 every 2.87 days. Each eclipse, including the partial phases, takes nearly 10 hours.*

►► ***Beta Lyrae*** *is a different type of eclipsing binary. Its components are so close together that they are distorted into ellipsoids by each other's gravity. As the system revolves in its 12.94-day orbital period, there is continuous change at all phases of its light curve.*

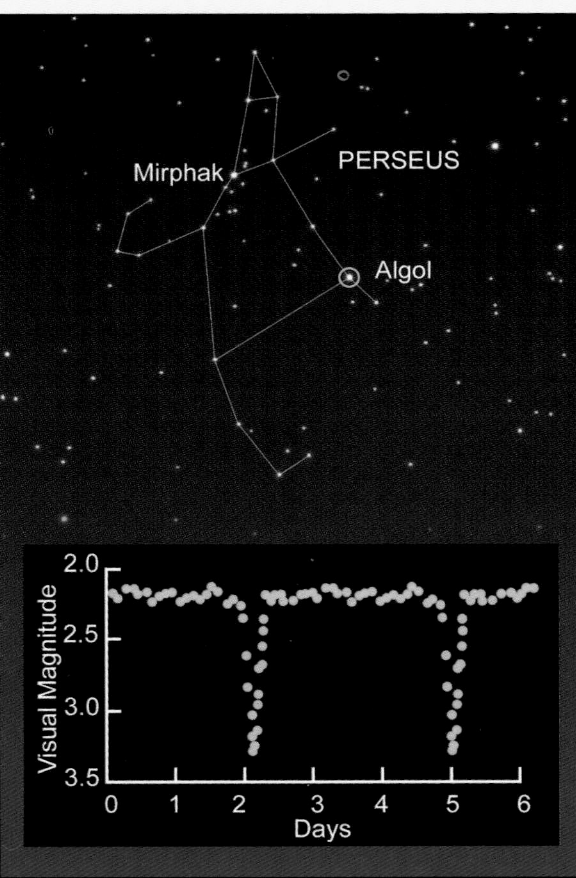

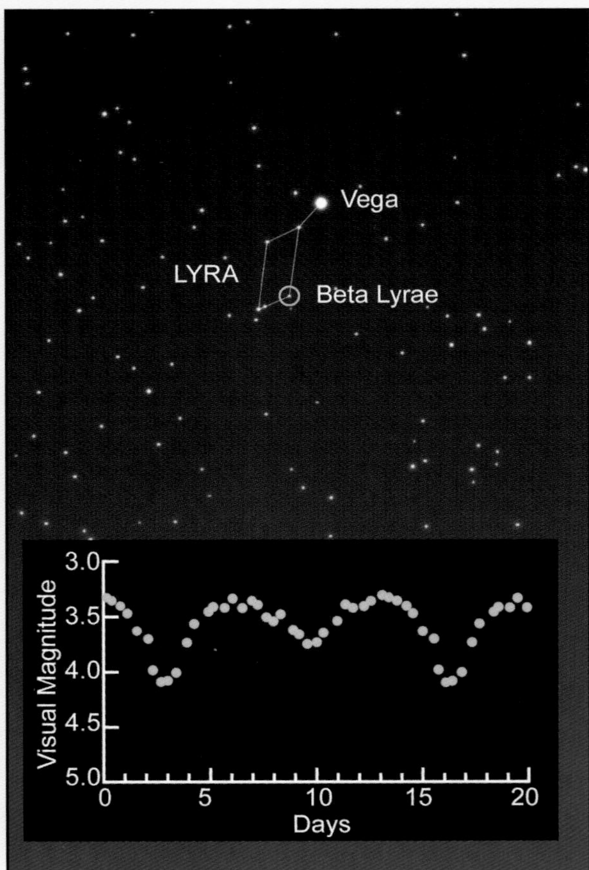

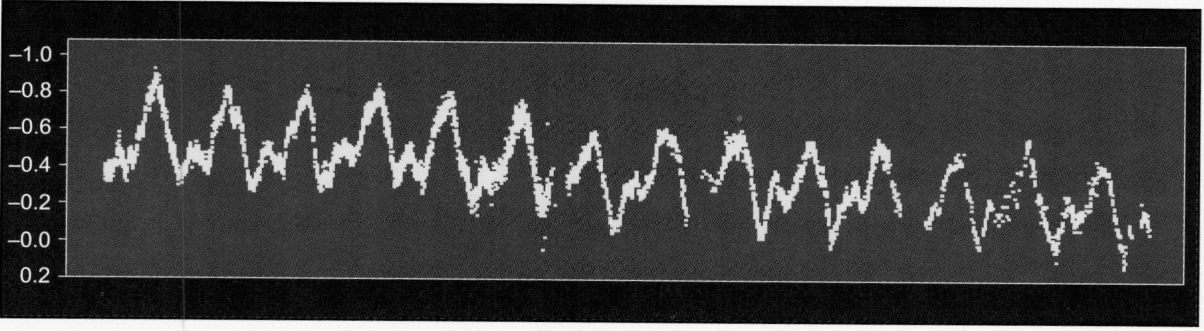

◀ ***Superhumps of WZ Sge.*** *The light curve spanning 19 hours is constructed from observations reported to VSNET from different longitudes.*

such disasters consider keeping your charts in transparent sleeves in a ring-binder.

It hardly needs saying that you must be very careful to identify the field correctly. How big an area does your chart cover and how does this compare with your observing field of view? Are any of the bright stars on other charts that you are familiar with? Check the orientation of the chart. If you are in the northern hemisphere facing toward the south, a corresponding chart will have north at the top of the chart and east to the left. This is as you will see it with the naked-eye or binoculars, but a telescope may invert the image. Look for a distinctive pattern of stars and 'star-hop' to your variable. Are you sure of the identification? Mistakes are often made at this stage.

Your chart will identify the comparison stars either with identifying letters (A, B, C, and so on) or with numbers (67, 94, 101 and so on). The numbers correspond to the magnitudes of the comparison stars without the decimal points, which are omitted to avoid confusion with the dots of stars. Once you have identified the variable you should find a couple of the comparison stars that most closely bracket the variable. Your observations then consist of interpolating between these stars; if, for example, your variable star 'V' is one-third of the way between the brightness of, say A (magnitude 6.2) and B (magnitude 6.8), then the magnitude of V is 6.4.

Experienced observers can achieve an accuracy of about 0.1 magnitude and this must be the target of anyone who wants their observations to be taken seriously. However, everyone's eyes respond slightly differently to different colors of light, and when making a comparison between stars of different colors it is quite common for there to be a difference between different observers. One observer may consistently estimate a star's brightness as several tenths of a magnitude brighter or fainter than other observers. However, once this is known the 'personal equation' can be applied by the coordinator.

Margaret Penston

◀ ***Mira****, Omicron Ceti, is a pulsating giant star, a long-period red variable, the prototype for Mira variables. At a typical peak Mira reaches magnitude 3.4. Mira's period of 332 days means that its maxima come one month earlier each succeeding year, but the exact date is never predictable.*

Observing eruptive variables

Observation of eruptive variables requires a telescope. Nightly visual monitoring of selected cataclysmic variables (CVs), particularly CVs with rare outbursts, is still a very powerful method of detecting outbursts of these objects. As a reminder, the fourth historical outburst of WZ Sge, 23 years since the last one, was first detected in July of 2001 by a high-school amateur observer Tomohito Ohshima, Japan, with a 78-mm telescope. This observation was quickly relayed to the appropriate organizing network, and has led to historic worldwide campaigns, including X-ray satellite and HST observations. Those observers who can reach magnitude 14 or so can make a significant contribution in this field. The most important factors are patience and quick communication of outburst detections to the professional community, by reporting to the Variable Star Network (VSNET) or similar organizations. Monitoring of outbursts with CCDs can have similar scientific impact, and can reach fainter magnitudes (*see* CCDS AND OTHER DETECTORS).

With the advent and wide availability of CCDs, VSNET and the Center for Backyard Astrophysics have developed a worldwide network of observers, with telescopes distributed all over the globe to watch continuously on selected CVs or other objects of timely interest. These organizations mainly focus on superhumps of SU UMa-type DWARF NOVAE or other CVs, short-term variation of black-hole binaries, novae, supernovae, GRBs and other unusual eruptive variable stars. Satellite triggers of high-energy photons (X-ray and gamma-ray) also invoke prompt optical follow-up observations.

Observation of eruptive variables in this modern style has thus opened a new window through which we may observe the universe. This field of astronomy (sometimes called 'transient object astronomy') is now one of the fields where amateur astronomers can play an essential role.

Because outbursts of CVs have less color variation than in other variable stars, time-resolved, continuous photometry to reveal the periodicity does not usually require filtered photometry (*see* MAGNITUDE AND PHOTOMETRY; CAMERA AND FILTER). The instruments can thus be minimal: a telescope and a CCD. Typical CCD observations of outbursting dwarf novae require a sequential acquisition of many CCD images in more than several continuous hours with short exposure times (typically tens of seconds, sometimes shortened to a few seconds when the detection of short-term variations is very important).

For accurate and highly efficient photometry of CVs, small FIELD OF VIEW (a few tens of arcminutes) and low spatial resolution (2 arcsec/pixel) are recommended to reduce the readout time. The use of non–ABG-type CCDs is highly recommended. The observer should take dark frames (it is often sufficient to take them before and after the runs when the CCD temperature is stable) and flat fields, as in usual CCD observations. The observer should also choose a set of unsaturated comparison stars and make differential photometry. Several commercial packages like, for example, AIP4WIN have a function to perform photometry of sequential images automatically.

The selection of targets is the most important factor for successful CCD photometry of eruptive variables. Two of the most promising targets include super-outbursting SU UMa-type dwarf novae (including rarely outbursting WZ Sge-type dwarf novae) and outbursting black-hole binaries. These phenomena usually last for a short time, and so regular access to electronic alerts (such as those from VSNET mailing lists) is essential for a successful observation. You should observe as long as the outburst continues, and possibly even longer, since such objects frequently show one or more post-outburst rebrightenings. A typical light curve obtained from such an observation would show superhumps in a super-outbursting SU UMa-type dwarf nova. Eclipses may be associated with such a light curve (*see* ECLIPSE AND OCCULTATION).

Period analyses of these phenomena (superhumps, eclipses) will provide essential scientific information about the physics of CVs and ACCRETION disks. Period analyses can be done with a popular algorithm, Phase Dispersion Minimization (PDM), which is implemented in some light-curve-generating programs. This job can be done by the amateur on their own data, but submitting these observations to the worldwide organizations leading such a campaign (like the VSNET Collaboration) is an excellent idea, as an individual observer's data can then be combined with others' data, and can be made available to a greater audience. The results of amateur-professional collaborations are continually being published in professional journals.

Short-term variation of black-hole binaries can be observed in a similar manner to CVs. These objects are sometimes calm and sometimes violently variable, to timescales of seconds or even less! These variations are even more unpredictable than in CVs, and regular access to electronic alerts is even more essential. With the detection of these short-term variations, one can 'see' the variations arising from the accretion disk or flow close to a black hole. This is a novel and quickly developing field of astronomy, pioneered by a leading amateur-professional collaboration.

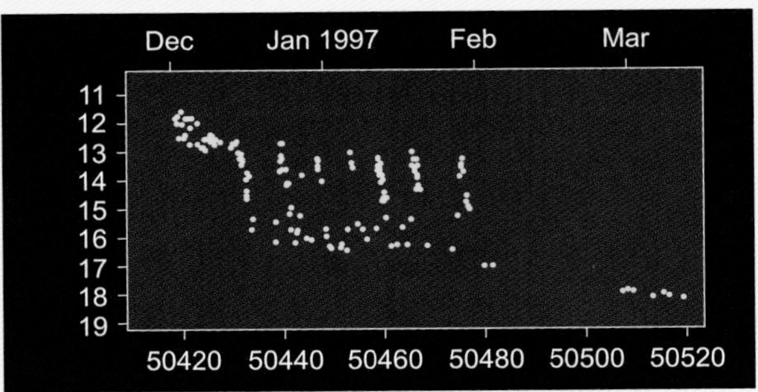

► ***Post-outburst rebrightenings*** *in EG Cnc. This phenomenon was discovered by a amateur-professional collaboration in VSNET.*

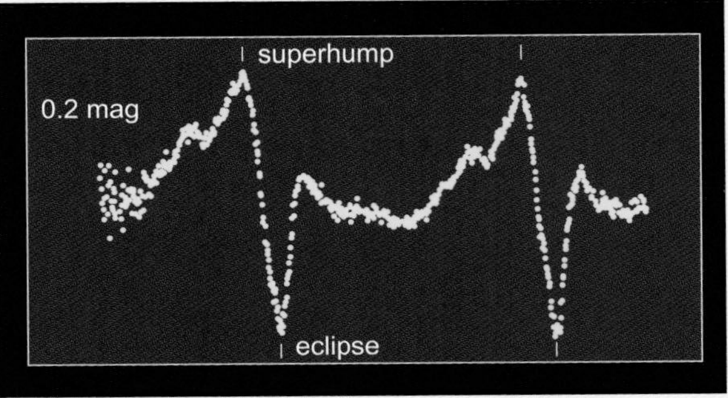

► ***Superhumps and eclipses*** *in an eclipsing SU UMa-type dwarf nova IY UMa.*

11.4
11.6
11.8
12.0
12.2
0.02 day

► ***Light curve*** *of a violent variation*

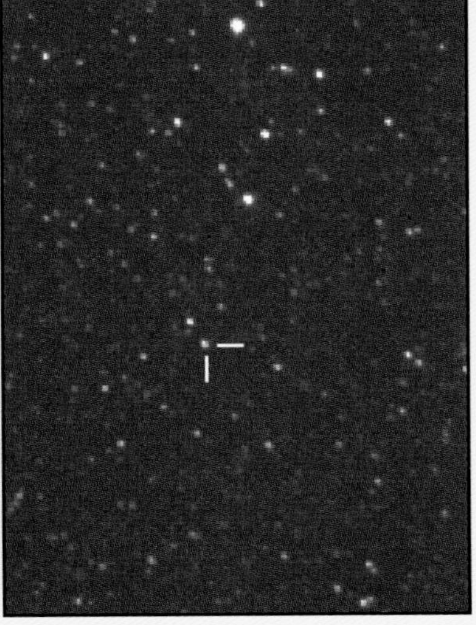

▼ ***Violent variation*** *in a black-hole binary (microquasar) V4641 Sgr. (Left) Seen 30 s before an optical flare. (Right) Seen during the flare.*

Taichi Kato is a lecturer in the astronomy department at Kyoto University, Japan. He is a member of VSNET, an international variable-star observing network.

Vela supernova remnant

A SUPERNOVA remnant in the constellation Vela, extending to nearly 5° in diameter, and consisting of material expelled by a supernova an estimated 11 000 years ago. The core of the supernova remains as the VELA PULSAR.

Venera missions

Russian (Soviet) spacecraft missions to Venus launched in the period 1961–83 (*see* VEGA MISSION). Venera missions V1–V8 were the so-called first generation of the spacecraft. The spacecraft mass varied from 900 to 1200 kg and each was launched by the three-stage Soyuz missile. Mission V-1, launched in February 1961, was the first interplanetary spacecraft, achieving a 100 000-km flyby of Venus. All of V-3 to V-8 reached the planet surface, with V-7 achieving the first soft landing. The second generation of Venera spacecraft had a mass of 5000 kg and they were launched by a Proton rocket. Those missions were successful and reliable, and provided the basis of our current knowledge of Venus, along with ground-based observations and the US MARINER 2, 5 and 10, PIONEER Venus and MAGELLAN spacecraft.

Venus

A TERRESTRIAL PLANET, an INNER PLANET, an INFERIOR PLANET and the second planet from the Sun. At a distance of 0.72 AU, it is always close to the Sun, either as a MORNING STAR or as an EVENING STAR. The distance from Venus to the Earth varies from 1.72–0.27 AU, the closest that any planet approaches (apart from NEOs). The distance of Venus has thus provided the basis for the scale of the solar system, both by measurements of its TRANSITS across the Sun and by RADAR ASTRONOMY. As seen from the Earth, Venus's apparent diameter varies from 10 (superior conjunction) to 64.5 arcsec (inferior conjunction) and shows PHASES, like the Moon. Venus is covered by white clouds and reflects 75% of the sunlight incident on it. This light, combined with Venus's nearness, makes it the brightest object in the sky, after the Sun and the Moon, but its brightness is very variable owing to the combined effects of the change of distance and change of phase.

Venus is the most Earth-like of the (other) terrestrial planets in size and overall properties. However, now that we know Venus better, we find many more differences from the Earth than was believed only 50 years ago. In particular, Venus has a very high surface temperature (hot enough to melt lead) and a very dense atmosphere (93 times that of Earth). The main atmospheric constituent is carbon dioxide, instead of nitrogen. Venus shows no evidence for plate tectonics in recent history and has no satellites. The surface of Venus is essentially of volcanic origin. Its clouds are white and opaque, but are not water droplets, and the surface is invisible except for the highest mountain areas imaged in infrared radiation during Venus's night time.

The rotation of Venus is nearly synchronous. It rotates slowly in the retrograde direction (*see* RETROGRADE MOTION). Because of the combined effects of the slow retrograde rotation of Venus and its orbital motion around the Sun, the solar day (time between two sunrises) on Venus corresponds to 117 Earth days. The eccentricity of the orbit and the inclination of the planetary rotation axis are both very small and there are no major seasonal changes.

Earth-based investigations

Low-contrast markings on Venus have been seen since about 1666. In 1761, Mikhail Lomonosov (1711–65) observed a halo around Venus that he interpreted as being caused by an atmosphere. In 1974, the polarization of light reflected from the clouds gave a value for the refractive index of the particles making up the clouds, which are droplets of concentrated sulfuric acid (H_2SO_4), as confirmed by space missions. Measurements of the radio emission of the planet in the late 1950s first indicated that the temperature at the surface of Venus was much higher than that on Earth. The rotation of the planet was measured by radar astronomy, which also provided the first radar images of portions of the surface at resolutions of 5–20 km.

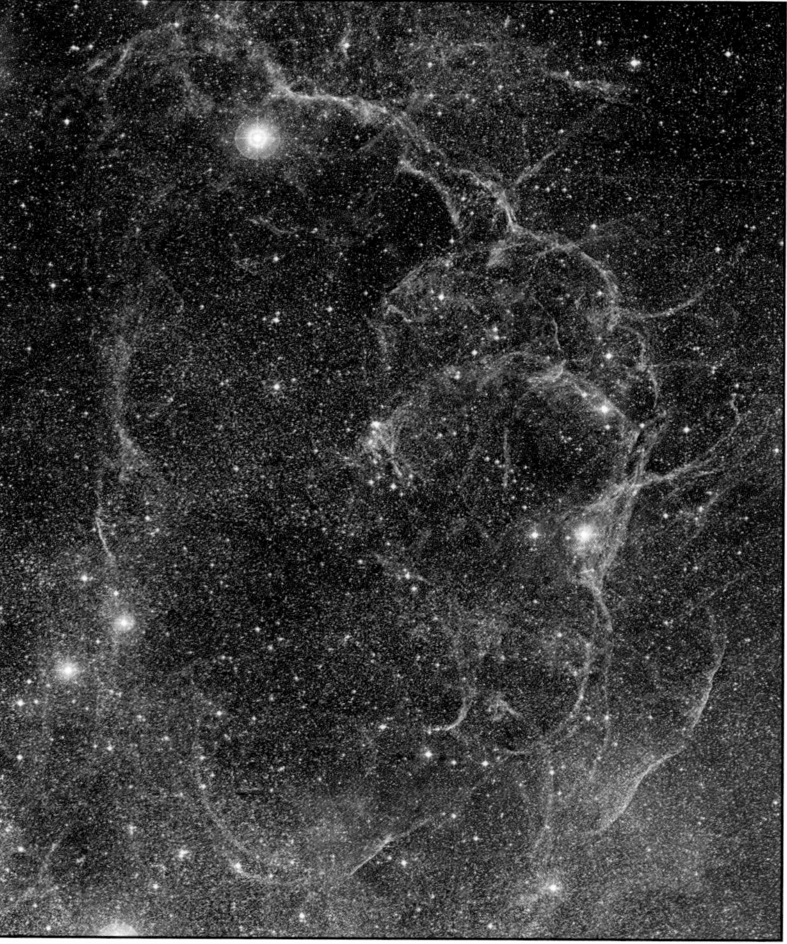

▲ ***Vela supernova remnant*** *About 120 centuries ago, an inconspicuous star in the constellation of Vela brightened by about 100 million times to rival the Moon as the brightest object in the night sky. This photograph shows a portion of the northwestern quadrant of an expanding nebulous shell that is now centered on the Vela pulsar, the site and stellar remnant of the explosion.*

Space missions

The most favorable periods for the launch of a spacecraft to Venus occur at times of inferior conjunction, when it is closest to the Earth, which happen about every 19 months. The first flybys were by the MARINER MISSIONS, starting in 1962. From 1967–92 Venera and Pioneer probes explored the planet, its atmosphere and its surface (*see* PIONEER MISSIONS; VENERA MISSIONS). A major step in the exploration of Venus was achieved by the US MAGELLAN radar mapping spacecraft which, between 1990 and 1992, mapped 98% of the surface at resolutions between 120 and 300 m. It also provided altimetric, radiometric and gravity data. The latest spacecraft investigations were fleeting visits in February 1990 by the Galileo spacecraft, on its way to the planet Jupiter (*see* GALILEO MISSION TO JUPITER), and in April 1998 by Cassini, bound for Saturn (*see* CASSINI/HUYGENS MISSION).

▲ ***Venus***
This image, created using data from the Magellan orbiter, uses simulated color to enhance the appearance of small-scale structures. The simulated hues are based on color images recorded by the Venera 13 and 14 landing craft. Maxwell Montes, the planet's highest mountain at 11 km above the average elevation, is the bright feature in the lower center of the image.

Venus	
Mean distance from Sun (km)	108.2 million
Mean distance from Sun (AU)	0.72333
Eccentricity	0.00677
Sidereal period (d)	224.701
Equatorial diameter (km)	12 1044
Equatorial diameter (relative to Earth's)	0.95
Sidereal rotation period (d)	243 retrograde
Inclination	177.3°
Mass (kg)	4.8689×10^{24}
Mass (relative to Earth's)	0.95
Density (relative to water)	5.24
Escape velocity (km s^{-1})	10.36
Escape velocity (relative to Earth's)	0.93
Albedo	0.65
Surface temperature	740 K

The atmosphere

Venus has a very hot, dense atmosphere, with carbon dioxide (CO_2) the major constituent. Minor components are nitrogen (N_2), water vapor (H_2O), sulfur dioxide (SO_2), sulfur monoxide (SO), carbon monoxide (CO), carbonyl sulfide (COS), sulfuric acid (H_2SO_4), oxygen (O_2), hydrogen chloride (HCl), hydrogen fluoride (HF), argon (Ar), neon (Ne) and krypton (Kr). One of the striking characteristics of Venus is its dryness. At its high temperature, liquid water does not exist at the surface. Even in the atmosphere, only very small amounts of water vapor are found. The deuterium/hydrogen ratio is very high (about 150 times that on Earth). The noble gases neon, argon and krypton are much more abundant on Venus than on Earth.

The high surface temperature is at the same time expected and surprising. Its proximity to the Sun means that it receives twice the solar energy that the Earth does. But the clouds are highly reflective and only about 25% of the solar flux penetrates into the atmosphere. The high surface temperature is due to a very efficient runaway greenhouse effect but, although the ATMOSPHERE of Venus contains gases that are strong infrared absorbers (carbon dioxide, sulfur dioxide, water vapor, carbonyl sulfide), they are not enough to produce such an extreme effect. The clouds of water droplets are responsible, because, like greenhouse glass, they efficiently absorb infrared wavelengths from the surface. However, they diffuse downward those of the incoming solar photons that they do not reflect to space. Their properties also explain the result, surprising at the time, that the Venera landers in the 1970s were able to photograph the surface in natural light.

Venus has an ionosphere, first detected by Mariner 5 in 1967, at a height of about 150 km. It is produced primarily by solar extreme ultraviolet radiation acting on the atmospheric gases to produce mostly ionized oxygen and carbon dioxide. At solar maximum, even the night-side ionosphere can sometimes be significant, the oxygen ions having moved around the planet from the day-side.

Clouds cover Venus in three layers between 48 and 68 km. The total mass of the clouds is relatively small. The venusian clouds are more like the hazes on Earth, with droplets about 2 μm in size. The clouds are made mostly of concentrated H_2SO_4, with about 75% by weight sulfuric acid and 25% water. They resemble terrestrial acid rain (caused by industrial pollution). Their composition is probably the result of the chemical combination of sulfur dioxide with water. Chemical reactions with the surface should be removing sulfur quickly from the atmosphere. To explain the permanent clouds of sulfuric acid, there must be a source of sulfur (continuous or episodic): volcanism. Wind speeds on Venus decrease from 100 m s^{-1} at the cloud-tops to close to 1 m s^{-1} near the surface.

Like other terrestrial planets, Venus lost its primitive atmosphere. It acquired a secondary atmosphere, outgassed from inside and brought by cometary and meteoritic impacts. On Earth, the atmosphere changed further, from biological activity, but this never happened on Venus. Venus may have had the same amount of water as the Earth, but lost it through the atmosphere's high temperature. Water molecules dissociate in the upper atmosphere, and hydrogen then escapes, with deuterium escaping more slowly because it is heavier, and thus raising the deuterium/hydrogen ratio to a very high value. It is not known why the noble gases are more abundant on Venus.

Surface of Venus

There are about 1000 impact CRATERS on Venus, which is not many compared with Mercury, Mars or the Moon. About half the craters have been formally assigned names, all after famous women in history or common female names. The largest crater, Mead, has a diameter of 270 km. The smallest ones have diameters of about 1.5 km. The absence of smaller craters and the fact that craters on Venus seem to come in groups are both probably caused by the dense atmosphere, which has broken the larger METEROIDS and completely burnt the smaller ones. The paucity of craters generally suggests that the surface of Venus is very young, no older than 500–800 million years.

The surface of Venus has an elevation between –1 and

+2.5 km below or above its reference mean radius of 6051.8 km.

The major features of the highlands are:

• Regio (plural: regiones): topographically high regions, often hosting large shield volcanoes;
• Terra (terrae): cover vast areas and have variable topographic relief, as continents do on Earth;
• Planum (plana): Lakshmi Planum is the only planum recognized on Venus. It is a plateau 3–4 km high, bordered by mountainous ridges.

The lowlands are called:

• Planitia (planitiae): these are topographic low-lying regions, generally the most featureless on Venus in terms of tectonic and volcanic structures.

The surface is dominated by the plains, such as the Atalanta Planitia, which cover about a quarter of the surface and are depressed by 1–2 km. The upland or rolling plains constitute 65% of the surface and have elevations of 0–2 km. About 8% of the surface is made of highlands with elevations of 2–12 km.

The lowland plains are generally very smooth (dark at radar wavelengths), whereas the upland or rolling plains are a little more rough and possess many small landforms such as scarps, ridges, troughs, hills, channels and so on. The highlands are dominated by two continent-sized features: Aphrodite Terra and Ishtar Terra. The highlands can be subdivided into three groups: the volcanic rises, the crustal plateaux and Ishtar Terra.

Thousands of volcanoes have been identified on Venus, from kilometer-size vents to broad shields hundreds of kilometers aross, with numerous long lava flows and central calderas. Anemones are smaller volcanic edifices, also characterized by flows radiating outward, but often in bilateral fashion from a central graben or fissure. Individual volcanoes and other volcanic features are found everywhere. Their deposits and other consequences of volcanism have been largely preserved on Venus, in contrast with Mars and the Earth. The plains are apparently completely covered by volcanic deposits, which arise from two main categories of volcanic process. Very significant flows come from large vent regions during repetitive eruptions, which create large volcanic edifices, flow fields and calderas. There are additionally smaller regions (some 100–200 km in diameter) characterized by the presence of many small edifices, including so-called shield fields, and by relatively limited lava flows.

The longest of the various classes of channels observed on Venus are called canali. They have remarkably constant width along their great length, which may exceed 500 km (up to 6800 km for Baltis Vallis). They are 1–3 km wide, and less than 50 m deep. They were formed by highly fluid lava. In contrast, the pancake domes and festoon flows correspond to high-viscosity lavas.

The largest tectonic forms (100–1000 km) on Venus are:

• Tessera (plural: tesserae): a terrain network consisting of two or more directions of linear ridges and troughs. They are largely compressional in origin and are the oldest part of the crust;
• Chasma (chasmate): a broad trench or linear zone consisting of a parallel arrangement of troughs or valleys bounded by fault scarps, caused by stretching;
• Mons (montes): large highland provinces are termed montes (for example, Maxwell Montes, Danu Montes, Akna Montes and Freyja Montes), descriptive of their mountain range-like appearance.

There are numerous other tectonic surface features on Venus that have been formed by compression or by extension: ridges, valleys, mountain belts, rift valleys and so on.

Unique to Venus are the forms known as coronae, which are distributed across the surface in hundreds, most with diameters of less than 300 km. They seem to be collapsed domes over large magma chambers. Arachnoids are circular to elliptical structures that consist of a central dome or depression surrounded by an extensive network of radial and concentric linear features, and are apparently uncollapsed coronae.

Somewhat unexpectedly, the Magellan images showed abundant wind streaks. They are widespread and generally oriented with the downwind direction toward the equator. Eolian (wind deposit) activity causes parabolic deposits in the lee of crater walls, which are centimeters thick and made up of fine debris.

Internal structure

Earth and Venus are similar in structure. Venus has differentiated, with a crust, a mantle and a core. Although the relatively large value of the mean density of Venus strongly suggests that the planet has an iron core (about 3200 km in radius), Venus lacks a magnetic field and it apparently does not have a large liquid-iron core like Earth. The mantle is some 2800 km thick. The crust of Venus is about the same thickness as the Earth's. Volcanism is the result of activity in the mantle, with long-lived plumes causing sustained eruptions that create the large volcanoes, and short-lived plumes creating the coronae and arachnoids.

Further reading: Bougher S W, Hunten D M and Phillips R J (eds) (1997) *Venus II* University of Arizona Press; Hunten D M, Colin L, Donahue T M and Moroz V I (eds) (1983) *Venus* University of Arizona Press.

Venus, observing

See practical astronomy feature Observing Mercury and Venus, p. 254.

Very Large Array (VLA)

The NRAO's VLA, 80 km west of Socorro, New Mexico, USA, is one of the world's premier RADIO ASTRONOMY facilities,

▼ ***Very Large Array*** *The VLA is composed of 27 individual antennae arranged in a Y pattern. Four times each year, the VLA antennae are moved into new configurations by a transporter that moves along dual sets of railroad tracks. In their closest configuration (about 1 km wide) the VLA is able to image large portions of the sky. In its largest configuration (about 36 km wide) the VLA is able to hone in on the fine details of astronomical objects.*

offering researchers a unique combination of resolving power, sensitivity and observational flexibility. Dedicated in 1980, the VLA includes 27 × 25 m diameter dish antennae, arranged in a Y pattern, that work together as a single aperture-synthesis interferometric imaging system.

Capable of observing at frequencies from 74 MHz to 50 GHz (non-continuous) the VLA provides resolution ranging from 15 arcmin to 0.05 arcsec. Its versatility has allowed it to serve a wide range of research specialties, including planetary, solar, stellar and galactic astronomy; as well as cosmology. It has made important contributions to the study of both galactic and extragalactic relativistic jets, the Milky Way's central region, galactic structure, dynamics and evolution (*see* GALAXY), SUPERNOVA remnants and transient events such as supernovae and GRBs (*see* GAMMA-RAY ASTRONOMY).

Very Large Telescope (VLT)

See EUROPEAN SOUTHERN OBSERVATORY

Very Long Baseline Interferometry (VLBI)

A technique used by radio astronomers (*see* RADIO ASTRONOMY) to achieve very high angular resolution imaging of celestial radio sources, by combining the signals from widely separated radio telescopes. It uses the principle of the Michelson Interferometer as applied to radio wavelengths, achieving resolution greater than that by single-aperture telescopes at any wavelength (*see* INTERFEROMETER). In aperture synthesis, the same technique is extended to an assemblage of many telescopes which can be combined in pairs; an array of N suitably distributed telescopes will yield $N(N-1)/2$ pairs, allowing sources to be mapped with an instrument whose effective diameter is the length of the maximum interferometer spacing. An example is the VLA in New Mexico. By observing as the Earth rotates, the effective spacing of the telescopes changes and so the density of samples is further increased.

VLBI is a further extension of the technique, made possible by the exquisite control that modern electronics offers for time measurement. Rather than being transmitted from the telescopes to the receiver, the amplified signals are recorded on wide-band magnetic tape recorders together with signals from a highly stable clock (usually a hydrogen maser frequency standard) for time control. The tapes are then shipped to a central processing facility, where the received signal is now a string of discrete samples at known times. The two data streams are then cross-correlated numerically, giving the fringe amplitude and phase. Several VLBI arrays are in operation: the VLB Array (VLBA) of the NRAO with its processor in Socorro, New Mexico, USA; the European VLBI Network (EVN), with a processor in Dwingeloo, the Netherlands (*see* JOINT INSTITUTE FOR VLBI IN EUROPE); and the Southern Hemisphere Array, with its processor in Narrabri, New South Wales, Australia. There is also a worldwide network, a collaboration between all these VLBI arrays and other national radio telescopes.

There is a surprising variety of scientific problems that VLBI can address. QUASARS and the nuclei of ACTIVE GALAXIES all have structures in the milliarcsec range or smaller. Pulsars and GRB sources are even smaller, requiring baseline lengths far longer than anything currently feasible. Molecular maser sources are also of milliarcsec size, and these occur in association with star-forming regions and cool, evolved stars such as the Mira variables. Three general classes of observation can be identified: the mapping of milliarcsec structures, the measurement of relative source positions to achieve higher astrometric accuracy, and the use of known sources as references to study geophysical phenomena such as continental drift and polar wandering.

Vesta

Minor planet (4) Vesta is the third-largest main-belt ASTEROID, with a mean diameter of 529 km, and the fourth asteroid to be discovered. It was found in 1807 by the German astronomer Heinrich OLBERS and named for the goddess of fire and the hearth in Roman mythology. Vesta is the sole 'intact' asteroid that may have undergone heating at temperatures able to produce a complete planetary-type DIFFERENTIATION. Telescope observations at high angular resolution, from space or from the ground, show that Vesta underwent a large-impact event 4.5 billion years ago. This discovery supports the idea that Vesta is the parent body of a group of small Vesta-like asteroids and is possibly the source of a particular type of METEORITE (the basaltic achondrite meteorites) collected on Earth.

Viking mission

Spacecraft mission to MARS whose main purpose was to search for life. Viking 1 and 2 were two identical spacecraft, each carrying an orbiter (VO-1 and VO-2) and a lander (VL-1 and VL-2). They were launched on August 20, 1975, and September 9, 1975, and arrived at Mars the following summer. At Mars the orbiters searched for safe and scientifically interesting sites from which the landers would operate. The landing site chosen for VL-1 was the western slope of Chryse Planitia (the Plains of Gold), while VL-2 settled down at Utopia Planitia.

Each lander had a sampling arm with a scoop for returning samples to selected instruments for analysis and also meteorology sensors and a seismometer. The landers had a two-way communication link with Earth and a one-way (transmit) link with the orbiters. While the primary scientific objective was to search for life, other objectives were to image the surface and moons, determine the composition of the atmosphere and surface, monitor the weather and climate, and detect marsquakes. Over 50 000 images were returned by the Viking orbiters covering the entire planet at a resolution of 200 m and large areas at resolutions as high as 7.5 m. For almost two marsian years, the orbiter cameras monitored weather patterns, the advance and retreat of the polar caps, dust-storm activity and the continual redistribution of fine-grained material on the surface. They returned pictures of enormous volcanoes, a vast equatorial canyon system, numerous impact craters, polar layered deposits and fluvial features suggesting catastrophic flooding and possible climate change.

The primary mission was to last 90 days for each lander. However, the mission was repeatedly extended as long as the spacecraft remained healthy and returned useful data. One by one the components ceased to function but the mission only finally ended on November 13, 1982 when contact was lost with VL-1. At the end of the mission, the consensus was that no compelling evidence for life had been found.

Vilnius University Observatory

Astronomical observatory of Vilnius University, Lithuania, founded in 1753. In 1831 the university was closed and until 1881 the observatory was managed by the Academy of

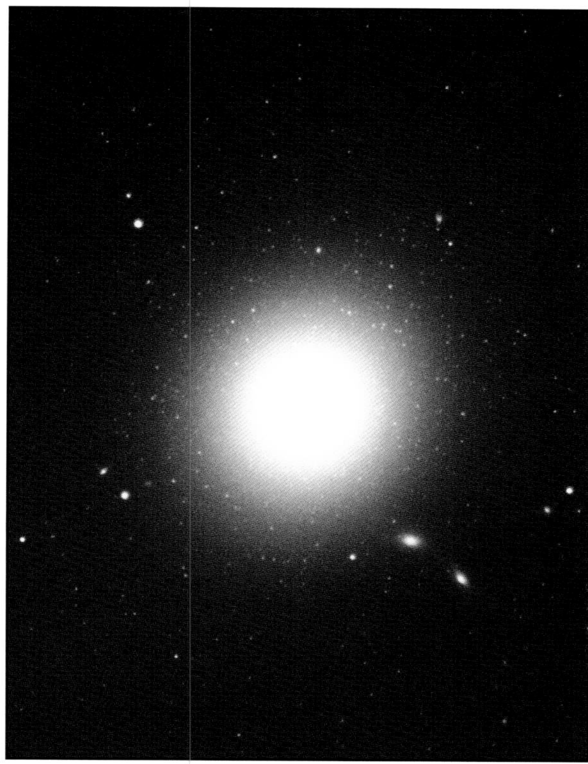

◀ **Virgo A** *(M87) A giant elliptical galaxy in the Virgo cluster, and a strong radio source. Over 500 globular clusters have been detected in visible light and many are visible here, looking like fuzzy stars near the galaxy.*

Sciences of St Petersburg, Russia. The observatory resumed its operation in 1919 when Vilnius University was reopened. Activities of the observatory were again interrupted by Nazi occupation (1941–4) and resumed after World War II. Its main instruments are a 60-cm reflector located at Moletai Observatory, 80 km north of Vilnius, and a 48-cm reflector located at the MAIDANAK OBSERVATORY site in Uzbekistan, Central Asia.

Virgo *See* CONSTELLATION

Virgo A (M87)

The brightest radio source in the constellation of Virgo. Virgo A, otherwise known by the catalog number 3C 274 and M87 in the MESSIER CATALOG, is a giant elliptical galaxy, and an ACTIVE GALAXY, located at a distance of some 50 million l.y. in the Virgo cluster of galaxies. While some of the radio emission comes from two lobes (unusually) contained within the optical galaxy, most of the radio output comes from a jet that emerges from the core of the galaxy and stretches out some 8000 l.y. into one of the two lobes. The jet radiates at all wavelengths from X-RAY to radio. Like the lobes, it emits SYNCHROTRON RADIATION. The motion of bright knots in the jet indicates that they are traveling outwards at about half the speed of light and implies that the electrons themselves must be traveling within the jet at speeds at least as high, if not higher, than this. With a mass of at least 10^{12} solar masses, M87 is an exceptionally massive galaxy. On very deep optical images it is seen to cover an area of more than the diameter of the full Moon, corresponding to a linear extension of 500 000 l.y. Observations of the rates at which stars and gas clouds revolve within its central core, and of the way that stars appear to be concentrated at its center, imply that M87 contains a compact massive object, most probably a BLACK HOLE, of about 3 billion solar masses.

Virgo cluster of galaxies

The nearest and best-studied rich cluster of galaxies (*see* GALAXY CLUSTER AND GROUP), lying at a distance of about 55 million l.y. in the constellation of Virgo. In three-dimensional (3D) space, the cluster constitutes the nucleus of the Local Supercluster of galaxies, in whose outskirts our MILKY WAY GALAXY is situated. As early as 1784, Charles MESSIER noted a concentration of 'nebulae' in Virgo. Of the 109 famous 'Messier objects,' 16 are, in fact, member galaxies of the Virgo cluster. However, only in the 1920s, following Edwin HUBBLE's proof of the extragalactic nature of those 'nebulae,' was Messier's group understood as a self-gravitating system of hundreds of galaxies, and the first systematic investigations of the cluster were carried out by Harlow SHAPLEY and others. In the 1930s Sinclair Smith and Fritz ZWICKY studied the dynamics of the cluster. They showed that the dynamical mass of Virgo, estimated by using the so-called virial theorem, was much larger than the mass inferred by integrating the light of all the galaxies in the cluster and multiplying by a mass-to-light ratio such as the average of stars in the solar neighborhood. This was the first clear detection of DARK MATTER. The cluster has over 2000 members, most of which are very faint dwarf galaxies. It is, and always has been, one of the most important stepping stones for the cosmological distance scale. Much of the debate on the value of the HUBBLE CONSTANT boils down to a debate on the mean distance of the Virgo Cluster, with quoted distances ranging from 14 to 22 megaparsecs. The giant elliptical galaxy M87 (*see* VIRGO A) is considered to be the dominant galaxy in the cluster and it also lies close to the physical center.

VIRGO Gravitational-wave Interferometer

A collaboration between Italian and French research teams to construct an interferometric gravitational-wave detector at Cascina, near Pisa, Italy (*see* GRAVITATION, GRAVITATIONAL LENSING AND GRAVITATIONAL WAVE). The VIRGO Gravitational-wave Interferometer relies on a technique called laser interferometry to measure with immense accuracy the minute changes in distance induced by gravitational waves from astronomical sources (*see* INTERFEROMETER). VIRGO consists of a laser interferometer made of two 3-km-long orthogonal arms. A split laser beam travels along the arms. By using multiple reflections the effective optical length of each arm is increased to 120 km. VIRGO will be sensitive to gravitational waves at frequencies from 10–6000 Hz. It should be able to detect radiation produced by SUPERNOVAE and by the coalescence of binary systems in the Milky Way and other galaxies. VIRGO uses high-power ultra-stable lasers and high-reflectivity mirrors and the optical components are isolated from seismic noise by a 10-m-high elaborate system of compound pendulums. The 6-km-long, 1.2-m-diameter evacuated tube through which the light beam passes will be one of the largest vacuum vessels in the world. The construction of VIRGO was completed in June 2003. At the end of the commissioning phase, it will run day and night listening to all gravitational signals, which may arrive at any time and from any part of the universe.

Virtual observatory

See INFORMATION HANDLING IN ASTRONOMY

VLBI Space Observatory Program (VSOP)

See HALCA

Void

Extensive REDSHIFT surveys in recent years have shown that 70% of galaxies are found in groups or clusters that are probably gravitationally bound. In turn, the groups of galaxies lie along filaments that interconnect. The largest clusters of galaxies tend to be found at the intersections of several filaments. Even the 30% of galaxies that are not in bound groups lie, almost entirely, within the filaments, which themselves occupy only a small fraction of space. Most of space has no observable matter. These empty regions are the voids. *See also* UNIVERSE.

Volans *See* CONSTELLATION

Voyager missions

Launched from Cape Canaveral, Florida, USA, in 1977 on August 20 (Voyager 2) and September 5 (Voyager 1), these two hardy spacefarers visited the planets Jupiter, Saturn, Uranus and Neptune. Although their last planetary encounter was in 1989, they are continuing to collect useful data about the outer solar system and are expected to do so into the foreseeable future as they travel toward interstellar space.

The two Voyager spacecraft were built with identical engineering designs and identical scientific payloads, and benefited from the results of the smaller Pioneer 10 and 11 spacecraft (*see* PIONEER MISSIONS). For example, neither Pioneer spacecraft encountered the expected problems crossing the asteroid belt, but they did experience very high levels of radiation in the vicinity of Jupiter; this led to a campaign to build the Voyager engineering and scientific hardware to be resistant to damage from these higher-radiation fields.

Closest approach to Jupiter occurred on March 5, 1979 (Voyager 1) and July 9, 1979 (Voyager 2). Voyager discovered the rings of Jupiter and three new satellites (Metis, Adrastea and Thebe) and images were obtained of eight of Jupiter's 16 then known moons. Voyager revealed active volcanoes on IO and information on the sizes, masses and surface detail of the other moons.

Closest approach to Saturn occurred on November 12, 1980 (Voyager 1) and on August 25, 1981 (Voyager 2). Four new satellites (Pan, Atlas, Prometheus and Pandora) of Saturn were discovered and images of all 18 then known satellites of Saturn were obtained. Enormous amounts of detail in the rings were revealed. Voyager also showed that TITAN was the solar system's second largest satellite, and the only one with a substantial atmosphere. In 2004 Titan will be visited by the Huygens probe on the NASA/ESA CASSINI/HUYGENS MISSION.

Closest approach to Uranus occurred on January 24, 1986 (Voyager 2). Ten new satellites (Cordelia, Ophelia, Bianca, Cressida, Desdemona, Juliet, Portia, Rosalind, Belinda and Puck) were added to the five previously known and the first detailed photographs of the rings of Uranus were obtained.

Closest approach to Neptune occurred on August 25, 1989 (Voyager 2). Six new satellites (Naiad, Thalassa, Despoina, Galatea, Larissa and Proteus) were added to Neptune's prior repertoire of two and the first images of the ring system were obtained. Triton was discovered to have nitrogen ice, extremely cold temperatures (32 K), and geyser-like plumes spewing dust through cracks in the icy surface.

On October 1, 1989, after completion of the planetary flybys, the mission was extended and renamed as the Voyager Interstellar Mission (VIM). As of May 2003, the spacecraft were at a distance from the Sun of 88.4 AU (Voyager 1) and 70.2 AU (Voyager 2) traveling at 3.6 and 3.3 AU per year respectively. Voyager 1 is traveling out of the ecliptic plane in the north in the general direction of the solar apex. Voyager 2 is traveling out of the ecliptic plane to the south. Both are heading toward the outer boundary of the solar system in search of the heliopause, where the influence of the Sun ends and interstellar space begins. Sometime in the next 10 years, the two spacecraft should come to an area known as the 'termination shock' where the solar wind slows to about 150 000 km h^{-1}. A decade or two later the Voyagers should cross the heliopause. They have enough electrical power and thruster fuel to operate at least until 2020. By that time, Voyager 1 will be 118.3 AU from the Sun and Voyager 2 will be 43.4 AU.

Both Voyager spacecraft carry a greeting to any form of life, should that be encountered. The message is carried by a phonograph record, a 30-cm gold-plated copper disk containing sounds and images selected to portray the diversity of life and culture on Earth. The contents of the record were selected for NASA by a committee chaired by Carl SAGAN who assembled numerous images and a variety of natural sounds. To this they added musical selections from different cultures and eras, and spoken greetings in 55 languages.

Vulcan

The name given to a hypothetical planet believed by Urbain LEVERRIER to exist within the orbit of Mercury. Leverrier first studied the orbit of Mercury in the early 1840s and managed to explain the greater part of the discrepancy between its calculated and observed positions (the advance of its PERIHELION) as being caused by gravitational perturbations by the other planets. He returned to the problem in 1859, having in the meantime successfully predicted the existence of Neptune. Now he predicted an intramercurial planet or asteroid belt as the cause of Mercury's irregularities. Shortly after, a physician and amateur astronomer, Edmond Lescarbault, reported the observation of a small body in transit across the Sun's disk. Leverrier was convinced the observations were genuine and announced the existence of a new planet which he named Vulcan. Subsequent sightings were reported but never confirmed. Vulcan is now known not to exist and the observed discrepancy in position is exactly that predicted by the theory of GENERAL RELATIVITY proposed by Albert EINSTEIN.

Vulpecula *See* CONSTELLATION

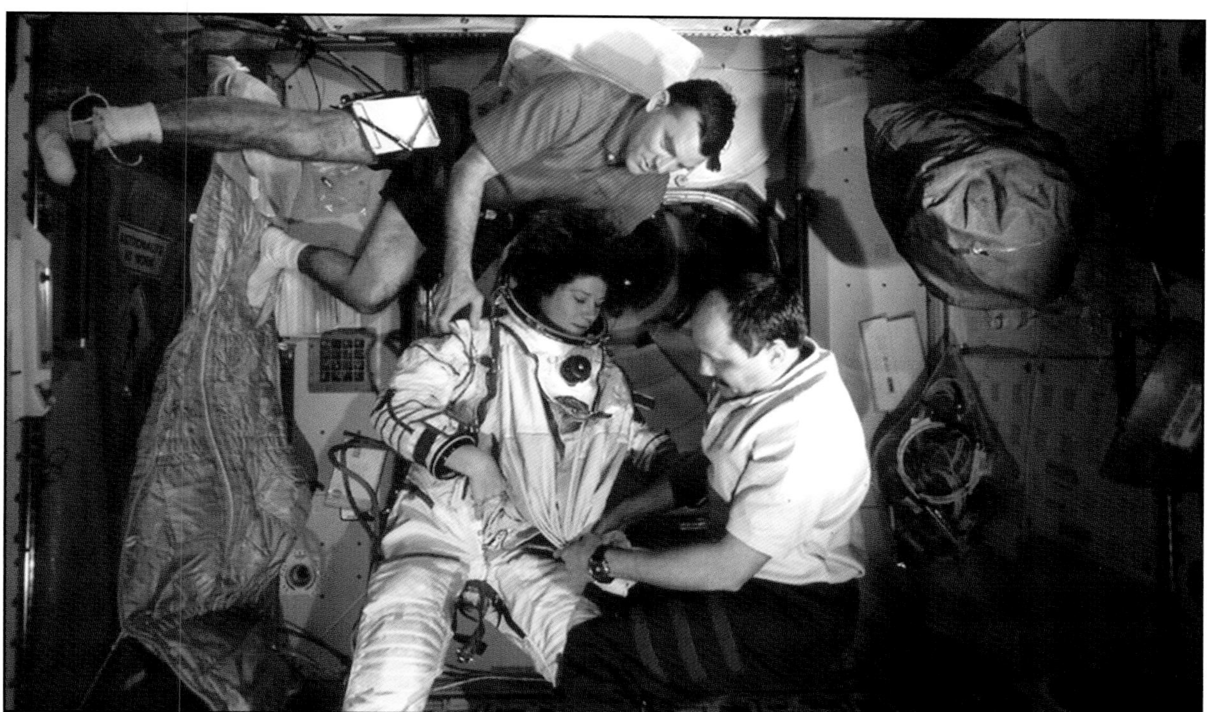

◀ ***Weightlessness*** *Astronaut Susan Helms donning a Russian EVA (extravehicular activity) suit inside the US laboratory Destiny on board the ISS, assisted by the other members of the Expedition 2 crew. This was the second crew to live aboard the ISS.*

Warner & Swasey Observatory

Located at Washburn University in Topeka, Kansas, USA, and home of the Warner & Swasey 29-cm REFRACTING TELESCOPE. Built in the late nineteenth century, the telescope was displayed at the 1912 World's Fair, then acquired by Washburn. Crane Observatory was built on campus to house the telescope. The Warner & Swasey survived a tornado in the 1960s. During telescope refurbishment, which was completed in 1998, all of the original parts were retained.

Wavelength

The distance between two successive crests of a wave motion. Usually, in transverse waves (waves with points oscillating at right angles to the direction of their advance), wavelength is measured from crest to crest. In longitudinal waves (waves with points vibrating in the same direction as their advance), it is measured from compression to compression. The term is applied to ELECTROMAGNETIC RADIATION, which is regarded as a wave motion. For example, blue light has a wavelength of about 440 nm and red light about 700 nm. X-rays have wavelengths of the order of 10^{-10} m and radio waves of the order of meters. Wavelength is usually denoted by the Greek letter lambda (λ); it is equal to the speed (v) of a wave train in a medium divided by its FREQUENCY (f): $\lambda = v/f$.

Webb Society

An international society of amateur and professional astronomers specializing in the observation of DOUBLE STARS and deep-sky objects. It was founded in 1967 and is named in honor of the Reverend Thomas William Webb (1807–85), an eminent amateur astronomer whose classic *Celestial Objects for Common Telescopes* has been an inspiration to amateur astronomers.

Weight

The force experienced by a body resting on, for example, the surface of a planet. A person standing on the Earth's surface experiences weight because the surface on which he is standing resists the effect of the force of GRAVITY, which otherwise would accelerate that person toward the center of the Earth. In other words, the weight of a body depends upon the gravitational force to which it is subjected. On the surface of any planet the weight F is equal to the mass of the body times the SURFACE GRAVITY. By convention, mass and weight at the surface of the Earth are numerically equal, so to find the weight of an object on another planet, say, we have to multiply the weight on the Earth by M/R^2 where M and R are the ratios of the planet's mass and radius to those of the Earth.

Weight of a body on different surfaces

Surface	Weight of body
Earth	100 kg
Moon	16 kg
Mars	38 kg
Jupiter	264 kg
Sun	2790 kg
white dwarf	30 000 000 kg

Weightlessness

A body that is falling in a vacuum is subject to only one force, the force of gravity . Such a falling object will speed up, or accelerate, with the 'acceleration due to gravity' that is usually denoted by 'g,' and the body will then be said to be in free-fall. Objects in a state of free-fall are said to be weightless. On the surface of the Earth the acceleration due to gravity (or SURFACE GRAVITY) is equal to 9.81 m s^{-2}.

Astronauts circling the Earth in the ISS are also weightless, but this is not, as some people think, because there is no gravity in space. The Earth's gravitational field is proportional to $1/R^2$ (where R is the distance to the center of the Earth)

▲ ***Whirlpool Galaxy*** *(M51) At only 20 million l.y. from Earth and fully 60 000 l.y. across, the Whirlpool Galaxy is one of the brightest and most picturesque galaxies in the sky, visible to anyone with a good pair of binoculars. Astronomers speculate that its spiral structure is primarily due to gravitational interaction with a smaller galaxy just off the top of this image.*

and at about 400 km above the surface maintains 88.8% of its strength at the surface. Isaac NEWTON devised a thought experiment in which a cannonball is fired with increasing velocity from the top of a mountain. The faster the cannonball is fired the further it will travel before falling to the Earth. Eventually, with sufficient speed the cannonball will orbit the Earth and never fall to the ground, but it is still experiencing the force of gravity and is in a state of continuous free-fall. Because orbiting spacecraft like the ISS are kept in orbit around the Earth by gravity, the spacecraft and a man inside it both fall at the same rate, and the astronaut seems weightless.

Westerbork Synthesis Radio Telescope

An APERTURE SYNTHESIS INTERFEROMETER in Hooghalen, the Netherlands, that consists of a linear array of 14 antennae arranged on a 3-km east-west strip. The array works by combining the signal from all the antennae and simulates a 3-km aperture telescope.

The antennae are equatorially mounted 25-m dishes with a FOCAL RATIO of 0.35. Ten of the telescopes are fixed, 144 m apart, while two nearby dishes are movable along a 300-m track and two others are on a 180-m track at a distance of 1.5 km. In the array, the baselines extend from 36 m to 3 km. The pointing accuracy of the dishes is 15– 20 arcsec, and the surface accuracy is of the order of 1.7 mm.

The array can routinely operate at wavelengths of 92, 49, 21, 18, 13, 6 and 3 cm, and change frequency in less than a minute. The telescope, used by astronomers from the Netherlands and from many other countries, also participates in the European (EVN) and Global VLBI networks.

Whipple, Fred Lawrence (1906–)

American astronomer who became director of the Smithsonian Astrophysical Observatory in Cambridge, Massachussetts, USA. While still a graduate student he helped compute the ORBIT of newly discovered Pluto. Using a new method of photography from two separated wide-angle cameras, he triangulated on METEOR tracks and determined their orbits. He deduced that nearly all are made up of bits from COMETS. He proposed the 'dirty snowball' model for comets, suggesting that comets have icy cores inside layers of dirt. This was confirmed in 1986 when the GIOTTO spacecraft flew close to and imaged COMET HALLEY. Whipple tracked artificial satellites to determine the shape of the Earth.

Whirlpool Galaxy (M51)

A face-on spiral galaxy located at a distance of some 20 million l.y. in the constellation of Canes Venatici. M51 (also known as NGC 5194), with its two well-defined arms that spiral out from its relatively small central bulge, is a classic example of a 'grand design' spiral. It is classified as an Sc galaxy in the Hubble Classification scheme (*see* GALAXY). With a diameter of 65 000 l.y. and a mass of about 5×10^{10} solar masses, it is somewhat smaller than the Milky Way Galaxy, although it is several times more luminous, its spiral arms laden with bright young clusters and H II regions.

M51 has a smaller, fainter companion (NGC 5195), which lies at the end of one of the larger galaxy's spiral arms. NGC 5195 appears to be orbiting around M51 in a period of about 500 million years. Tidal interaction between the two during their last close encounter, which took place some 70 million years ago, probably played a large part in establishing the bold spiral pattern in M51 and stimulating a vigorous bout of star formation within it.

M51 was the first galaxy to be recognized as having a spiral shape. This discovery was made in 1845 by William Parsons (*see* ROSSE, THIRD EARL OF AND FOURTH EARL OF) with the aid of the 1.8-m telescope at Birr Castle, Ireland which, at that time, was the largest telescope in the world.

White dwarf

Stars whose masses range from about 0.08 to about 8 solar masses end up as white dwarfs. Such stars are the vast majority (95%) of stars formed in our Galaxy. White dwarfs are typically of 0.5–1 solar mass, while their size is more akin to that of a planet. Because they are compact, they have high densities and strong surface gravities.

The first white dwarf to be identified was inferred in 1844 by Friedrich BESSEL, the great German mathematician and astronomer, who measured the proper motion of SIRIUS. Its irregular path led him to suggest the presence of an unobserved, solar-mass companion. The companion, Sirius B, was first seen by Alvan Clark (*see* CLARK FAMILY) in 1862.

The first spectrum of Sirius B, secured by Walter ADAMS in 1914, showed that the white dwarf was comparable to its companion in temperature. Since A and B were both at the same distance but the white dwarf was faint, it had to be small. Sirius B is about 10 000 times smaller than Sirius A.

How could such a dense star withstand the tendency to collapse under the influence of its gravitational field? The answer to that question came in 1925, when Ralph Fowler (1889–1944) applied the newly developed principles of quantum mechanics to stars. He showed that, in white dwarf stars, the density is high enough for the gas of free electrons to become 'degenerate.' Electrons are said to be degenerate

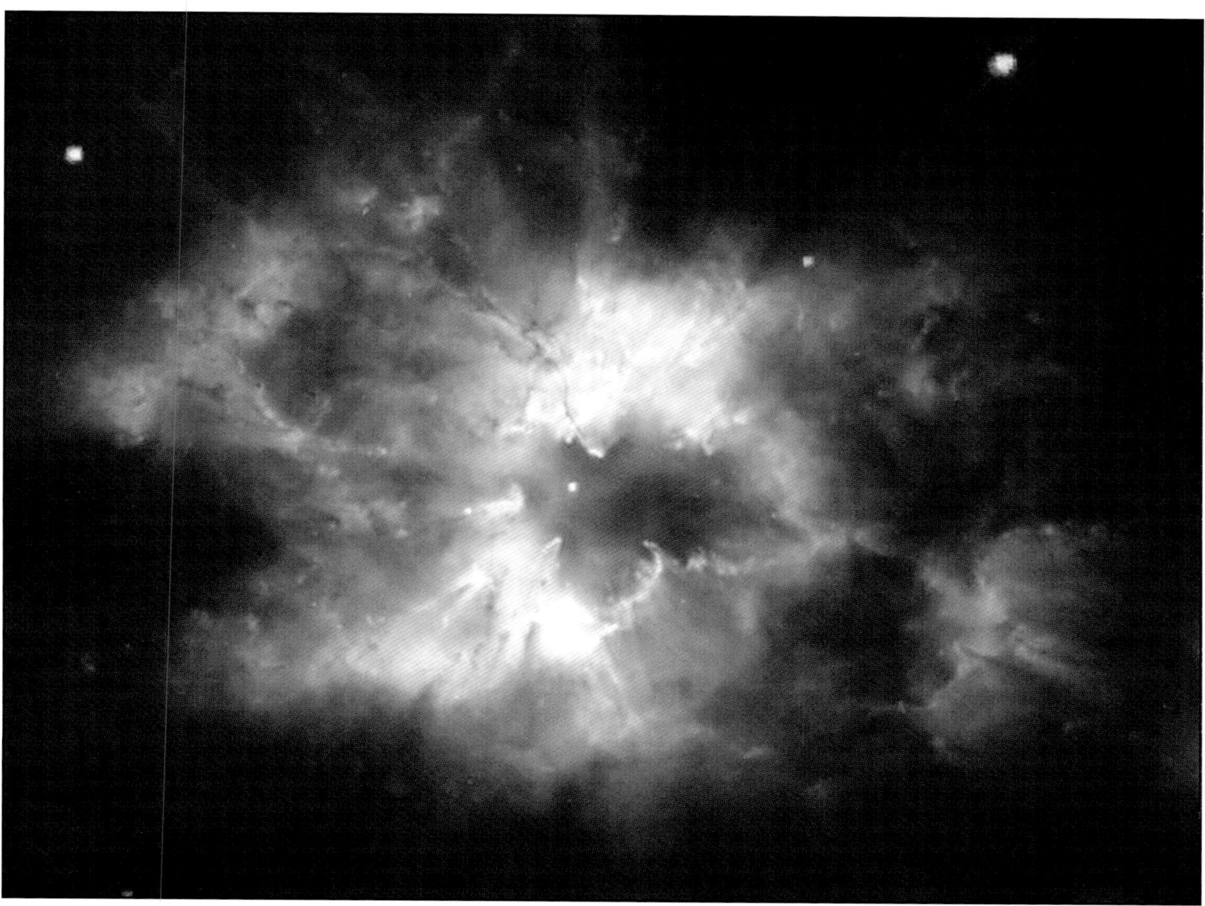

◀ ***White dwarf*** *Like a butterfly, a white dwarf star begins its life by casting off a cocoon that enclosed its former self. In this analogy, however, the Sun would be a caterpillar and the ejected shell of gas would become the prettiest of all! The cocoon shown here, the planetary nebula designated NGC 2440, contains one of the hottest white dwarf stars known, which can be seen as the bright dot near the photo's center. Our Sun will eventually become a white dwarf, but not for another 5 billion years. It is making its surrounding cocoon with the solar wind.*

when a majority of them occupy the lowest possible energy states available. This occurs when the electrons are packed close to each other. Because of the PAULI EXCLUSION PRINCIPLE, no more than two electrons (with oppositely directed spins) can occupy the same energy state. The electrons retain kinetic motion even when cooled to zero temperature. In a white dwarf, the pressure generated by this kinetic motion prevents the gravitational collapse of the star. The first detailed stellar models appropriate to white dwarfs were calculated in the 1930s by Subrahmanyan CHANDRASEKHAR (winner of the Nobel Prize in 1983).

Electron degeneracy is directly responsible for the curious relationship between the mass and the radius of a white dwarf: the more massive the star, the smaller its size. This means that there is a limiting mass above which a white dwarf cannot exist. This is known as the Chandrasekhar mass, and is about 1.4 times the mass of the Sun.

The masses of white dwarfs can be estimated from:

- their spectral characteristics;
- their influence on a companion star, if they are in a binary system;
- and, for roughly three dozen DA stars, the GRAVITATIONAL REDSHIFT of their Balmer lines (*see* HYDROGEN SPECTRUM). (The star under scrutiny has to belong to an open cluster or a binary system with a well-determined radial velocity, so that the gravitational redshift can be separated from the redshift caused by the physical motion of the star.)

White dwarfs are typically 0.6 solar masses.

Because white dwarfs are faint, most of the 2000 confirmed white dwarfs are within 1000 l.y. of the Sun, although some are known in GLOBULAR CLUSTERS located at 10 000 l.y. or more. Traditionally, however, white dwarfs have been culled from samples of objects showing significant PROPER MOTION, and thus located relatively nearby the Sun. More recently, large numbers of hot white dwarfs have been detected by searches for blue stars.

Their distribution in the Galaxy is consistent with that of an old disk population, within some 1000 l.y. of the Galactic Plane. In a typical 30-l.y. cube there are five white dwarfs.

About three-quarters of white dwarfs have a spectrum dominated by the Balmer series of hydrogen. These objects are termed DA stars; D signifying white dwarf and A being the spectral type, as in the classification of stellar spectra (*see* STELLAR SPECTRUM: CLASSIFICATION). Their temperatures range from above 100 000 K to 4000 K.

Other white dwarfs have optical spectra dominated by helium. The hottest ones have ionized helium and temperatures above 45 000 K and are called DO stars. Between 30 000 and 12 000 K, helium in white dwarfs is neutral; these are the DB stars. Below 12 000 K, the temperature is too cool for helium to show, although hydrogen would if there was any there; the spectrum is then featureless, and is termed DC.

DQ white dwarf stars have spectra characterized by carbon, generally in molecules. The DZ stars show lines of heavy elements other than carbon; for example, calcium, magnesium or iron.

The spectrum of a white dwarf is related to the composition of its atmosphere. The hydrogen-line DA stars have atmospheres where hydrogen is the dominant element, to the near complete exclusion of any other. The so-called non-DA stars, which encompass objects of the DO, DB, DC, DQ and DZ spectral types, all have helium-dominated atmospheres.

The gravity of white dwarfs is so strong that they are highly stratified, like the Earth, with the heavier elements in the middle. The apparent chemical purity of white dwarfs is understood today as the result of this gravitational settling. All elements heavier than the dominant atmospheric constituent rapidly sink into the deep atmospheric layers of the stars and remain out of sight. In fact, this settling mechanism is so efficient that the presence of any element heavier than hydrogen in the atmospheres of DA stars, or helium in those of non-DA stars, constitutes a puzzle because it seems to mean that these stars have no hydrogen at all. It is generally believed that the immediate progenitors of most white dwarfs are nuclei of PLANETARY NEBULAE, themselves the products of stars of intermediate and low mass. Hydrogen is the most abundant element in the universe and must have been the most abundant element in the progenitor stars, so all of it has been lost in the formation of a non-DA white dwarf.

Most of the mass of a typical white dwarf is contained in a core made of the products of helium-burning, namely, for typical stars that turn into white dwarfs, carbon and oxygen. The process of mass-loss in white dwarf progenitors, which have a wide range of initial masses, seems to be focused well enough to produce white dwarfs whose masses are all the same.

A typical white dwarf has a carbon-oxygen core surrounded by a thin, helium-rich envelope itself surrounded by a hydrogen-rich layer. The outer layers are very thin, with very much less than 1% of the mass of the star, but they are extremely opaque to radiation and they regulate the energy outflow from the star. A white dwarf's spectrum is produced by less than one trillionth of the mass of the star.

As former nuclei of planetary nebulae, most white dwarfs are born extremely hot. Their nuclear energy sources are depleted, and gravitational energy can no longer be tapped efficiently, as they cannot contract any further. They cool, each white dwarf radiating its thermal reservoir. At first they cool by radiating neutrinos, then by radiating ultraviolet radiation. Ultimately, the reservoir of thermal energy becomes depleted, and the star cools further and gets dimmer. It now radiates light. At a certain temperature the white dwarf crystallizes, its constituent nuclei rearranging themselves into a regular lattice. The white dwarf abruptly disappears from sight and eventually becomes a BLACK DWARF.

With the HST and 8–10 m ground-based telescopes, it is possible to study faint white dwarfs in open and globular clusters. This exciting development has led to a renewed interest in white dwarf cooling calculations because the sudden crystallization of white dwarfs provides a good indicator of the age of, and distance to, its parent star cluster.

Apart from white dwarfs described above which form the vast majority, there are further families of white dwarfs:

- Magnetic white dwarfs (there are nearly 50 known) have magnetic fields that range from 10 T all the way to 10 000 T. They are known in X-ray binary stars.
- Variable white dwarfs, or pulsators, are known among the hot DB and the DA stars (*see* ASTROSEISMOLOGY).
- A white dwarf may be made by exposing the core of a star that has not progressed to helium burning (for example, by some event in a binary system which strips off the outer layers of a hydrogen-burning star). Some helium white dwarfs may therefore be, not carbon-oxygen cores with helium atmospheres, but made of helium all through (as in AM CVN STARS).
- As proposed in 1973 by John Whelan (1945–81) and Icko Iben, SUPERNOVAE of Type Ia are believed to be the result of the collapse of a white dwarf pushed over the Chandrasekhar limit by accreting mass from a companion, or by the merger of two white dwarfs in a binary system.

Whole Earth Telescope

A worldwide network of astronomers interested in short-period irregular VARIABLE STARS, established in 1986 by astronomers from the University of Texas, USA. The headquarters moved to Iowa State University in 1997. Typically twice a year, stars (WHITE DWARFS, DELTA SCUTI STARS and CATACLYSMIC VARIABLES) are chosen for a coordinated global time-series photometry campaign at about 10 observatories worldwide. The target objects are visible from the night side of the planet 24 hours a day and have characteristic periods ranging from minutes to roughly an hour. The extended time series allows analysis to probe the interiors of the target objects using the technique of ASTROSEISMOLOGY.

Widmanstätten pattern

A characteristic roughly hexagonal pattern of intersecting lines that appears on the surface of an octahedrite, a type of iron METEORITE, when it is sectioned, polished and etched with acid. The Austrian mineralogist Aloys von Widmanstätten (1753/4–1849) discovered the pattern in 1804. It is formed by the intergrowth of two nickel-iron alloys under the conditions of slow cooling that pertained in the solidifying core of an asteroidal parent body that had undergone DIFFERENTIATION, and is found only in meteorites. The two alloys are kamacite, with a low nickel content, and taenite, which is richer in nickel.

Wilcox Solar Observatory

Observatory at Stanford University, California, USA that measures the Sun's large-scale synoptic magnetic and velocity fields with the goal of understanding solar variability and how it affects our terrestrial environment. With more than a 22-year solar cycle (*see* SUNSPOT CYCLE) of spectrograph observations since 1975, observatory staff investigate the solar interior, photosphere, corona, wind and cycle. The observatory was rededicated in honor of its first director, John Wilcox (1925–83).

Wild Duck Cluster (M11) *See* MESSIER CATALOG

Wilkinson Microwave Anisotropy Probe (WMAP)

Launched in June 2001, WMAP is a NASA satellite named for David Wilkinson (1935–2002), a pioneer in the study of cosmic background radiation; *see* COSMIC MICROWAVE BACKGROUND (CMB). The satellite has accurately measured the CMB temperature all over the sky with high angular resolution and sensitivity (angular resolution of 0.3°, relative sensitivity of 20 μK). It uses two microwave radiometers to measure the temperature difference between two points on the sky. WMAP observes the sky from an orbit about the L_2 Sun-Earth LAGRANGIAN POINT, 1.5 million km from Earth. This vantage point offers an exceptionally stable environment for observing in INFRARED ASTRONOMY. WMAP has measured the anisotropy of the CMB with much finer detail and greater sensitivity than its predecessor COBE. These measurements reveal the size, matter content, age, geometry and fate of the universe. WMAP data, combined with other measurements about the universe, including clusters of galaxies (*see* GALAXY CLUSTER AND GROUP), the INTERGALACTIC MEDIUM and SUPERNOVAE, indicate that:

- the universe is 13.7 billion years old with a margin of error of close to 1%;
- the first stars ignited 200 million years after the BIG BANG;
- the CMB is from 379 000 years after the Big Bang;
- the content of the universe is 4% matter made of BARYONS, 23% DARK MATTER and 73% dark energy (*see* DARK ENERGY AND THE COSMOLOGICAL CONSTANT);
- dark energy behaves more like a cosmological constant than QUINTESSENCE;
- the expansion rate of the universe (HUBBLE CONSTANT) is 71 km s^{-1} Mpc^{-1} (with a margin of error of about 5%);
- the universe will expand forever.

William Herschel Telescope

See ISAAC NEWTON GROUP OF TELESCOPES

Wilson, Olin Chaddock (1909–94)

American astronomer and spectroscopist, who became a staff member at MOUNT WILSON OBSERVATORY. He studied stellar chromospheres and stellar activity cycles (*see* STAR), showing by intensive analysis of the H and K lines of ionized calcium that other stars besides the Sun have cycles of activity. With Manali BAPPU, he found a means of determining luminosity, and thus distance, of stars from the widths of the emission in these two lines that comes from the chromosphere (the 'Wilson–Bappu effect'). He studied spectra of nebulae (*see* NEBULAE AND INTERSTELLAR MATTER), eclipsing stars (*see* ECLIPSE AND OCCULTATION), WOLF–RAYET STARS and PLANETARY NEBULAE.

Wilson, Robert Woodrow (1936–)

American radio astronomer, and Nobel prizewinner for physics in 1978 with Arno PENZIAS 'for their discovery of COSMIC MICROWAVE BACKGROUND radiation.' He was interested in radio as a boy and was drawn to radio astronomy by working with John BOLTON at the California Institute of Technology, USA, mapping the Milky Way. He joined Bell Laboratories at Crawford Hill, New Jersey, USA, where with Penzias he shared a small allowance given to the laboratory for radio astronomy projects. With new millimeter-wave receivers at 100–120 GHz they discovered unexpectedly large amounts of carbon monoxide in a molecular cloud behind the ORION NEBULA, enabling them to use isotopic spectral line ratios as a probe of nucleogenesis (*see* ELEMENTS, FORMATION OF). With a large radio telescope (the Holmdel horn) and a new, sensitive low-noise receiver they discovered the cosmic microwave background radiation.

WIMP

A 'Weakly Interacting Massive Particle' and one of the possible constituents of DARK MATTER. *Contrast* MACHO.

Wind mission

NASA satellite, part of NASA's Global Geospace Science program and the International Solar Terrestrial Physics program (ISTP). It was launched in November 1994 to study the solar wind from both inside and outside the Earth's magnetosphere. Initially it followed a double-lunar swingby orbit with APOGEE of 80–250 Earth radii and PERIGEE of 5–10 Earth radii. In this orbit, lunar gravity-assists maintained apogee over Earth's day hemisphere. Later the spacecraft was inserted into a 'halo' orbit at the sunward Sun-Earth LAGRANGIAN POINT (L_1) to measure the solar wind, magnetic fields and particles, and provide a 1-hour warning to other ISTP spacecraft of changes in the solar wind. Since October 1998, it has been placed in a 'petal' orbit that takes it out of the ecliptic plane. This allows the spacecraft to sample regions of interplanetary space and of the magnetosphere that have never before been studied.

Wise Observatory

Observatory in Mitzpe Ramon, Israel, dedicated in 1971. It is owned and operated by Tel Aviv University, and has a well-equipped 10-m telescope. The large percentage of clear nights at its desert site and its unique longitude have made the observatory particularly useful for long-term monitoring projects (for example, REVERBERATION MAPPING of QUASARS and ACTIVE GALAXIES), and as a part of global monitoring networks. A highlight published in 1999 was the first detection, via gravitational microlensing, of a planet orbiting a binary star in the system known as MACHO-97-BLG-41 (*see* MACHO).

WIYN Observatory

Located at Kitt Peak in Arizona, USA the WIYN Observatory (pronounced 'win') is owned and operated by a consortium of the universities of Wisconsin, Indiana, Yale and by NOAO. NOAO, which operates the other telescopes of the KITT PEAK NATIONAL OBSERVATORY, provides most of the services. The 3.5-m WIYN telescope, which was completed in 1994, is the second largest telescope on Kitt Peak. Many innovative design features have resulted in a compact lightweight instrument (42 000 kg compared to the 340 000 kg of the 4-m Mayall telescope on the site) with superb image quality. WIYN is equipped with the latest instruments for astronomical SPECTROSCOPY and imaging (*see* SPECTROSCOPE AND SPECTROGRAPH). A multiple-object spectrograph employing optical fibers allows the simultaneous observation of the spectra of 100 objects. The imaging cameras employ highly sensitive arrays of electronic detectors. In 2001 the WIYN consortium also took over the operational responsibility for the historic 0.9-m telescope on Kitt Peak.

W M Keck Observatory

Located on the island of Hawaii, USA, the observatory operates the world's two largest optical/infrared TELESCOPES, each with a primary mirror 10 m in diameter, near the 4200-m summit of Mauna Kea. Made possible through grants totaling more than $140 million from the W M Keck Foundation, the observatory is operated by the California Institute of Technology, the University of California and NASA, which joined the partnership in October 1996. The Keck I telescope began scientific observations in May 1993, whilst Keck II began in October 1996. Over 400 astronomers per year are involved with observations from the Keck telescopes, which are carried out from the administrative facility at Waimea, Hawaii via a fiber-optic link to the summit of Mauna Kea.

The primary mirrors of the Keck telescopes each consist of 36 × 1.8-m diameter hexagonal segments, aligned to a tolerance of 2.5×10^{-6} cm under computer control.

The Keck INTERFEROMETER is a NASA-funded project that combines the light from both telescopes as an 85-m baseline

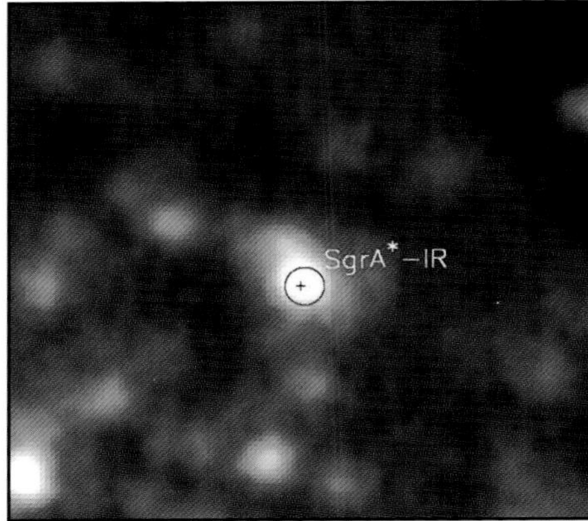

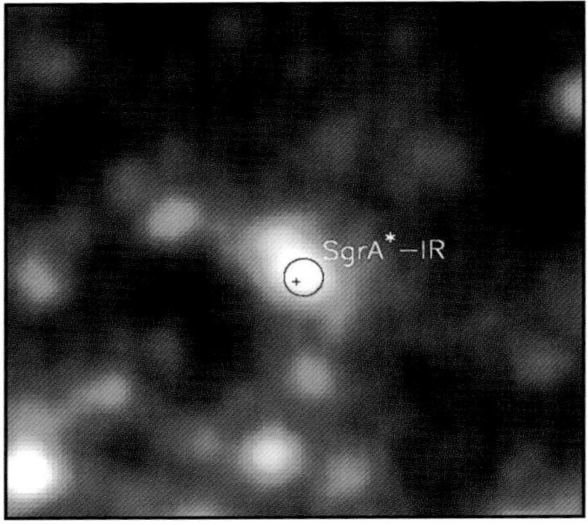

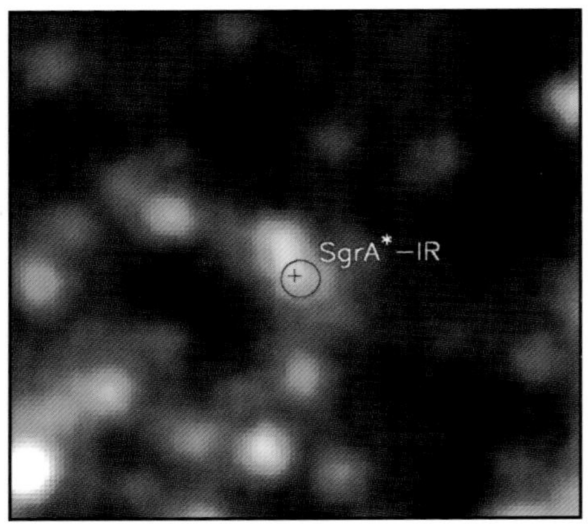

▲ ***Sagittarius A**** *Infrared light from plasma falling onto the supermassive black hole at the center of the Milky Way, viewed by the W M Keck telescope on June 10, 16 and 17, 2003 (left to right). The location of the black hole is marked with a cross and the newly detected infrared source is encircled and labeled as SgrA*-IR. The brightness variations reveal that violent events occur almost continually.*

infrared interferometer. This is the ground-based component of NASA's 'origins' program that addresses fundamental questions about the formation of galaxies, stars and planetary systems, as well as the prevalence of planetary systems around other stars, and the formation of life on Earth.

The first test observation obtained by linking the two Keck 10-m telescopes was made on March 12, 2001 on the star HD 61294, a faint star in the constellation Lynx. In July 2003 observations of a dusty disk around the T TAURI star DG Tau were reported. To phase the two telescopes properly, adaptive optics on both telescopes remove the distortion caused by the Earth's atmosphere. In addition, the optical system in the tunnel adjusts the light path to within 2.5×10^{-6} cm.

Major discoveries from this young observatory include: the discovery of several planetary systems around other stars; the identification of GRBs as being at cosmological distances; the discovery of the most distant objects in the universe; the measurement via SUPERNOVA observations of the apparent acceleration of the universe.

Wolf, Johann Rudolf (1816–93)

Swiss astronomer, who became professor of astronomy at the University of Bern, Switzerland and director of the Bern Observatory, then professor of astronomy in Zürich, where he founded an observatory. He devised a system now known as 'Wolf's sunspot numbers,' used to quantify solar activity by counting sunspots and sunspot groups, and used it to confirm the sunspot cycle discovered by Heinrich Schwabe (1789–1875) and measure its period at 11 years. He also co-discovered with Edward Sabine (1788–1883) its connection with geomagnetic activity.

Wolf, Max[imilian] Franz Joseph Cornelius (1863–1932)

German astronomer who founded and became the first director of the Königstuhl Observatory at the University of Heidelberg, Germany. He took widefield photographs of the MILKY WAY GALAXY and counted stars of different brightnesses, plotting the results in a Wolf diagram of number versus magnitude to prove the existence of clouds of obscuring dust. He showed that the spiral nebulae (*see* SPIRAL GALAXY) have ABSORPTION SPECTRA typical of stars, rather than EMISSION SPECTRA from gas. He pioneered the use of photography to discover hundreds of ASTEROIDS.

Wolf–Rayet nebula

Nebulosity surrounding a WOLF–RAYET STAR. Wolf–Rayet stars are of around 10 solar masses and have very high surface temperatures, up to about 40 000 K. This gives them powerful stellar winds, up to 2000 km s^{-1}, and an enormous rate of mass loss. Material is usually ejected in the form of a spherical shell or ring (the term 'Wolf–Rayet bubble' is sometimes used), and the accumulating envelope from successive ejection episodes constitutes the nebula. Examples of Wolf–Rayet nebulae are NGC 2359, surrounding the star HD 56925, and NGC 6888, surrounding the star MR 102.

Wolf–Rayet star

A small class of peculiar stars first identified in 1867 by Charles Wolf (1827–1918) and Georges Rayet (1839–1906). The spectra of Wolf–Rayet (W–R) stars show broad emission lines (*see* EMISSION SPECTRUM), and this makes them easy to identify by spectroscopic observations, even at large distances. (The spectra of most stars are dominated by narrow absorption lines.) W–R stars are extremely massive, high-luminosity stars descended from O-type stars and are probable progenitors of Type II SUPERNOVAE.

W–R stars are divided into three broad spectroscopic classes (WN, WC and WO) based on the emission lines present in their spectrum. WN stars show emission lines predominantly of helium and nitrogen, although emission due to carbon, silicon and hydrogen can readily be seen in some of these objects. In contrast, the spectra of WC stars are dominated by carbon and helium emission lines with hydrogen and nitrogen emission absent. WO stars, which are much rarer than either WN or WC stars, are similar to WC stars except that oxygen lines are more prevalent, and there is a tendency to exhibit lines arising from atomic species of higher ionization.

In the Milky Way Galaxy, POPULATION I W–R stars are primarily located, as are O-type stars, in the spiral arms and near H II REGIONS. Their masses range from an uncertain lower limit of about 5 to greater than 60 solar masses while surface temperatures range from about 25 000 to greater than 100 000 K. Because of their spatial association with O stars, and their peculiar surface abundances, W–R stars are generally believed to be descended from O stars.

Approximately 220 W–R stars are known in our Galaxy but this number is certainly incomplete. Most are hidden from our view by dust and estimates of the total number range from 1000 to 2000. The rarity of W–R stars is due to the IMF (INITIAL MASS FUNCTION), which favors the production of low-mass stars, and the short evolutionary lifetime of W–R stars which is only a few hundred thousand years. Their rarity belies their importance. All stars more massive than approximately 25 solar masses (and of similar composition to the Sun) pass through a W–R phase and over the lifetime of a galaxy, W–R stars (and their progenitors) have an important influence on the dynamics and chemical evolution of the INTERSTELLAR MEDIUM.

W–R stars are expected to end their life via a supernova explosion. As the supernova ejecta expands, it will interact with the complex circumstellar environment that reflects the previous mass-loss history of the progenitor star.

The basic model of a W–R star is of a hot star that is losing mass at a great rate via a continuous stellar wind traveling at speeds of up to 3000 km s^{-1}. The star typically loses 10^{-5} solar mass per year or more (*see* WOLF–RAYET NEBULA). The observed emission lines originate from this extended atmosphere that can extend beyond 10 stellar radii.

Approximately 50% of W–R stars occur in binaries, which is a number comparable with O stars; all confirmed binary systems consist of a W–R and an O or B star (*see* BINARY STAR). Such systems are extremely useful. First and foremost they allow a direct determination of stellar masses independent of evolutionary models. Secondly the companion can be used to probe the structure of the W–R stellar wind. Most W–R stars are X-ray sources and some are strong and/or variable X-ray emitters.

W–R stars lie in the upper part of the HERTZSPRUNG–RUSSELL DIAGRAM, alongside other classes of massive luminous stars. These include Of stars, blue supergiants, red supergiants, LBVs and WN/Of stars. One of the goals of massive star evolution is to understand the links between the various classes of objects and the distribution of massive stars between the different classes.

W–R stars are moderately easy to detect in external galaxies owing to their strong emission lines. Extensive surveys of both the LARGE MAGELLANIC CLOUD (LMC) and the SMALL MAGELLANIC CLOUD (SMC) have resulted in the discovery of 134 W–R stars in the LMC, and nine in the SMC. The difference in the number is believed to be due to a combination of the star formation rates and the lower metallicity of the SMC (which inhibits W–R production). W–R stars have also been found in many Local Group galaxies and in many galaxies exhibiting extensive star formation (often called 'starburst galaxies'). Indeed some galaxies are termed 'W–R galaxies' if they exhibit strong W–R features in their integrated spectra. The presence of W–R stars in these galaxies immediately provides an age determinant. The galaxy has to be older than about 2 million years for the most massive O stars formed in the burst to have had sufficient time to evolve into W–R stars. They also provide an upper limit of about 7 million years, because after this time all massive stars that pass through a W–R stage will have done so. Both age limits are metallicity dependent.

Wormhole

A hypothetical shortcut, or 'tunnel' that in principle may link the interior of a BLACK HOLE to another universe or to another location in our universe. During the 1930s Albert EINSTEIN and Nathan Rosen (1909–95) showed that the sharply curved SPACETIME of the interior of a black hole may open out again into another spacetime (another universe). The hypothetical connection between these two regions of spacetime came to be known as an Einstein–Rosen bridge. An alternative interpretation is that the bridge, or tunnel, links two different regions in the spacetime of our own universe. More recently the term 'wormhole' has been used to describe a spacetime tunnel of this kind. Although it has been speculated that wormholes could be used to facilitate virtually instantaneous interstellar travel, in practice it seems likely that even if wormholes do exist they will be too small and too short-lived (and too physically hazardous) to be utilized in this way.

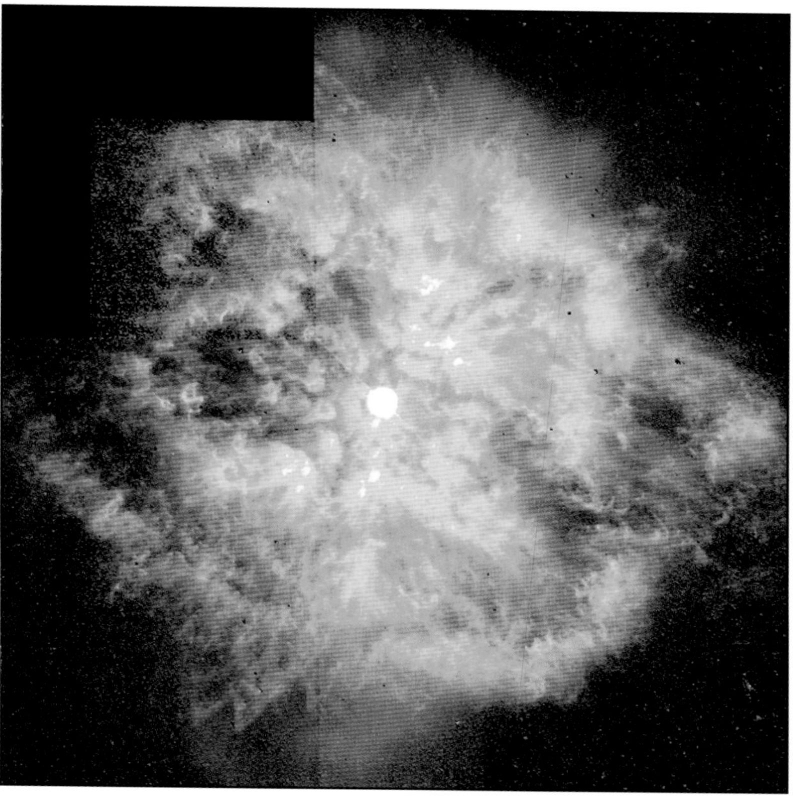

▲ ***Wolf–Rayet star*** *Resembling an aerial fireworks explosion, this dramatic picture of the energetic star WR124 reveals it to be surrounded by hot blobs of gas that are being ejected into space at speeds of over 160 000 km h^{-1}. The blobs may result from the furious stellar wind, which does not flow smoothly into space but has instabilities that make it clumpy.*

Wrinkle ridge

A long ridge on the surface of a planetary body. Wrinkle ridges were first identified on the Moon, where they are often associated with RILLES. As the lunar maria solidified (*see* MARE), tensile forces in the outer regions opened up the faults that produced rilles, while compressive forces nearer the center pushed up the surface to form wrinkle ridges. Lunar wrinkle ridges are typically several hundred meters high and several hundred kilometers long. On Venus wrinkle ridges are common features on the plains, where they extend for 10–50 km. The alignment of many of them suggests that they are associated with the compressive forces that uplifted the northern upland region of APHRODITE TERRA.

XMM-Newton Mission
ESA's X-ray Muliple Mirror (XMM) Mission was launched by an Ariane-5 rocket in December 1999 and later named in honor of Isaac NEWTON to mark the association of the mission with Newton's work on both spectroscopy and gravity. Its nominal mission was for two years but this has been extended for a further four years (until 2005). XMM-Newton orbits the Earth in a 48-hour elliptical orbit with PERIGEE altitude 7000 km and APOGEE at 114 000 km. It is the biggest scientific satellite ever built in Europe, with a total length of 10 m and a span of 16 m when the solar arrays are deployed. It covers a spectral range of 0.1–12 nm (12–0.1 keV) with a resolution of 5 arcsec (full width half-maximum) at all wavelengths.

X-rays are not easily focused because they pass straight through most materials, but they may be captured by grazing-incidence reflections. The XMM-Newton telescopes each contain 58 high-precision gold-plated concentric mirrors, nested to offer the largest collecting area possible. There are three main instruments on board the spacecraft; the European Photon Imaging Cameras (EPIC), the Reflection Grating spectrometers (RGS) and the Optical Monitor (OM). The OM is a 30-cm telescope co-aligned with the main X-ray telescope, enabling coincident optical images to be obtained. In addition, XMM-Newton is equipped with a particle detector to measure the radiation levels in the Earth's radiation belts and from solar flares (*see* SUN), radiation that can perturb the sensitive CCD detectors of the main science instruments. XMM-Newton (ESA's second Cornerstone mission) and NASA's CHANDRA X-RAY OBSERVATORY were both launched in 1999 and have complimented each other in producing exciting observations of the X-ray universe. XMM-Newton has a larger collecting area than Chandra and can observe fainter objects, but Chandra has higher resolution and can see finer detail.

X-ray
Energetic part of the ELECTROMAGNETIC SPECTRUM with wavelengths between 0.01 and 10 nm. Cosmic X-rays are usually described in terms of photon energies, a wavelength of 10 nm corresponding to an energy of 80 eV and a wavelength of 0.01 nm to 8×10^5 eV (80 keV). Alternately 1 eV is equivalent to a frequency of 2.418×10^{14} Hz or a wavelength of 1240 nm.

X-rays from cosmic sources are absorbed high in the Earth's atmosphere. Even the most energetic (shortest-wavelength) X-ray photons fail to penetrate much closer to the ground than an altitude of about 40 km.

X-rays were discovered in 1895 by the German physicist Wilhelm Konrad Röntgen (1846–1923).

X-ray astronomy
Astronomy discovered by observing X-RAYS, part of the ELECTROMAGNETIC SPECTRUM of radiation. X-rays lie between the extreme ultraviolet and gamma-rays, with wavelengths between 10 and 0.01 nm and energies between 0.1 and 100 keV. The waveband is subdivided into hard X-rays (more energetic) and soft X-rays (less energetic), at about 10 keV. X-ray astronomy is an achievement of the space age, because the Earth's atmosphere is completely opaque at photon energies beyond the UV region. Today, we know more than 150 000 X-ray sources in the sky and they include almost all astrophysical species – from the nearby COMETS to the most distant QUASARS at the edge of the universe, from the tiny neutron stars (*see* NEUTRON STAR AND PULSAR) to the clusters of galaxies (*see* GALAXY CLUSTER AND GROUP) and superclusters, which are the largest physical formations in the COSMOS. Some species radiate most of their power in X-rays, for instance BLACK HOLES or neutron stars accreting matter from a binary companion, as well as SUPERNOVA remnants and single cooling neutron stars.

Many X-ray objects shine because they are hot, with temperatures of millions to billions of kelvins. Another mechanism that commonly gives X-rays is SYNCHROTRON RADIATION, from extremely energetic electrons spiraling in magnetic fields. Less well known is the 'inverse Compton effect,' which occurs when low-energy photons interact with hot electrons and pick up some of their energy in the scattering process. This happens when millimeter wave photons from the 2.7-K cosmic background radiation (*see* COSMIC MICROWAVE BACKGROUND) interact with gas of a million K in clusters of galaxies (the SUNYAEV–ZELDOVICH EFFECT). In any case, the emission of X-rays points to extreme physical conditions in the source region.

◄◄ *Bode's Galaxy* *This ultraviolet (UV) image was obtained by the Optical Monitor on XMM-Newton in April 2001. The image was formed from three 1000-s exposures taken with different UV filters and an X-ray image (red). The galaxy has a mini-quasar at its nucleus and hot stars in its spiral arms.*

History
In 1949 the first X-rays from the solar corona were detected by Herb Friedman (1916–2000) using a Geiger counter on a V-2 rocket. In 1962 there followed the discovery by Riccardo GIACCONI and Bruno Rossi (1905–93) of the first X-ray source outside the solar system: SCORPIUS X-1 (*see* ROCKETS IN ASTRONOMY). The instruments used in the first rocket experiments were Geiger counters, with little spectral resolution. The rockets scanned the sky by spinning. The first discoveries of X-ray astronomy were totally unexpected, with rockets sent aloft to look at other things – Sco X-1 was found in the background of a rocket sent up to detect X-rays from the Moon. Most of the few dozen sources like this found during the rocket era are powered by the infall of matter into the deep gravitational potential well of a neutron star or a black hole.

The rocket experiments had shown that the spectra of compact X-ray sources are quite hard and could be studied using balloon-borne instruments. The atmosphere is transparent above 40 km for X-ray energies above 20 keV. A highlight of these activities was that the CRAB NEBULA was identified as a bright X-ray source by watching the source Taurus X-1 as it was occulted by the Moon, establishing its position. This proved that some supernova remnants are X-ray sources.

Most significant progress in X-ray astronomy came with the advent of satellite observatories. Their ancestor, the first satellite entirely devoted to X-ray astronomy, was UHURU. Launched in 1970, it was a spinning spacecraft, which performed the first all-sky survey and located 339 objects, mostly X-ray binaries (*see* BINARY STAR) and supernova remnants , showing a strong clustering near the galactic plane. It also discovered a distribution of SEYFERT GALAXIES and clusters of galaxies. It determined the positions of many sources sufficiently accurately that many were identified with optical and radio sources, including Cygnus X-1, a spectroscopic binary which contained an X-ray source whose mass turned out to be larger than a neutron star could be. Thus it had to be a black hole.

After Uhuru, an impressive series of X-ray-sensitive satellites with increasing capability was launched, including the EINSTEIN OBSERVATORY, EXOSAT, KVANT, GRANAT, ROSAT, ASCA, BEPPOSAX, the CHANDRA X-RAY OBSERVATORY and the XMM-NEWTON

MISSION. Imaging of X-rays is possible in different ways. Early solar observations used pinhole cameras. X-ray telescopes use paraboloidal-hyperboloidal mirrors that reflect X-rays at grazing incidence (otherwise the X-rays penetrate into the mirror material and do not reflect). The first X-ray satellite carrying such a telescope was the Einstein Observatory (1979). Chandra and XMM have huge, nested grazing-incidence mirrors with almost optical image quality. Two very successful recent missions include the RXTE, optimized to look at raidly varying X-ray sources, and BeppoSAX, which discovered the X-ray afterglows of GRBs, leading to their identification in very distant galaxies.

Further reading: Charles P and Seward F (1995) *Exploring the X-ray Universe* Cambridge University Press.

X-ray Evolving Universe Spectroscopy mission (XEUS)

The potential follow-up to ESA's highly successful XMM-Newton mission, with a launch date sometime after 2014. XEUS will consist of two spacecraft – one carrying the mirror, the other carrying the detectors – formation-flying 50 m apart. It is designed to search for the first massive BLACK HOLES that formed in the early universe, over 10 billion years ago. An enormous free-flying X-ray telescope, XEUS would orbit much lower than previous ESA satellites for X-RAY ASTRONOMY. But by orbiting close to the ISS, XEUS can be brought alongside every few years and have new mirrors and features added. By being serviceable, XEUS can last up to 25 years. XEUS will operate in the energy range 0.05–30 keV.

X-ray spectrum

The part of the ELECTROMAGNETIC SPECTRUM extending from about 10 to 0.01 nm, bounded at the long WAVELENGTH end by the extreme ultraviolet and at the short wavelength end by gamma-rays.

Year

The period of the Earth's revolution around the Sun, or of the apparent motion of the Sun on the ECLIPTIC. It may be defined in a number of ways, each of which leads to a slightly different value:

- Sidereal year (*see* SIDEREAL TIME): the time interval during which the Sun apparently completes one revolution of the CELESTIAL SPHERE relative to the stars (which, for this purpose, are regarded as being fixed in space). This is equal to the revolution period of the Earth around the Sun as measured relative to the stars, and is equivalent to 365.2564 mean solar days.
- Tropical year: the time interval between two successive passages of the Sun through the vernal EQUINOX. Its length is 365.2422 mean solar days, about 20 min shorter than the sidereal year. The difference arises because of the effects of PRECESSION. As this definition of the year is related to the recurrence of the seasons, the term 'year,' if unqualified, is generally taken to mean 'tropical year.'
- Anomalistic year: the interval between two successive passages of the Earth through the perihelion of its orbit which, because of a slow change in the position of PERIHELION, is not quite the same as the sidereal year. Its length is 365.2596 mean solar days.
- Gregorian calendar year: this is the value of the year adopted for CALENDAR purposes, and is equal to 365.2425 mean solar days. For practical purposes it can be taken as equal to the tropical year (the difference amounts to 0.0003 mean solar days).

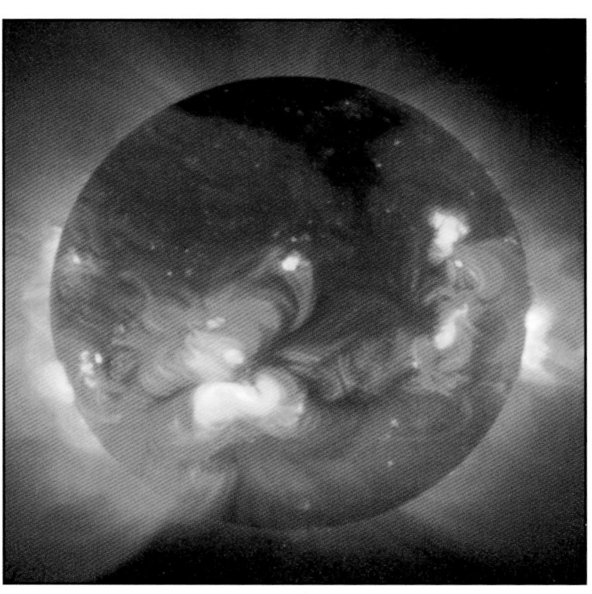

► ***Yohkoh mission*** *This photo, captured by the Soft X-ray Telescope on board the orbiting Yohkoh satellite, captures a rare event: the passage or transit of Mercury in front of the Sun. Transits of Mercury (and Venus) were historically used to discover the geometry of the solar system and to map Earth itself.*

Yerkes Observatory

Observatory, completed in 1897, lying 334 m above sea level in Williams Bay, Wisconsin, USA. It is a research branch of the department of astronomy and astrophysics of the University of Chicago. It was financed by Charles Tyson Yerkes, a Chicago transportation tycoon, but the inspiration behind its construction was George HALE. The showpiece of the observatory was the 1-m REFRACTOR, the world's largest telescope in 1897, and still the largest refracting telescope ever built. The other telescopes are a 1-m REFLECTOR which is used for ADAPTIVE OPTICS studies, a 0.6-m reflector and an 18-cm Schmidt camera for widefield photography.

Recent research at Yerkes includes: measuring the velocities and distances of the furthest star clusters within the Milky Way to better determine the mass of our Galaxy; spectroscopic measurements of lithium abundances; spectra of the dust disk around Beta Pictoris; and studies of the properties of distant galaxies.

Yohkoh mission

Satellite, originally Solar-A, which was launched by Japan's ISAS on August 30, 1991 for observations of solar flares and the solar corona (*see* SUN) in X-ray and gamma-ray wavelengths. During its 10-year operation (until a power failure in mid-December 2001) the satellite experienced almost the entire solar cycle. Yohkoh (meaning 'sunbeam') carried two X-ray telescopes: the hard X-ray telescope had imaging capability above 40 keV that gave new information on where and how energetic electrons are accelerated in solar flares. The soft X-ray telescope was a grazing-incidence X-ray (0.5–5 nm) telescope to observe the solar corona and solar flares with high spatial (3 arcsec) resolution. Two other instruments onboard were the Bragg crystal spectrometer to observe the iron, calcium and sulfur lines from flare PLASMAS and the wide-band spectrometers to observe flare spectra from 5 keV to 10 MeV (*see* SPECTROSCOPY).

Yohkoh revolutionized our understanding of the solar corona and the behavior of magnetized plasmas in general. It showed for the first time how the Sun dynamically relaxes its magnetic energy, buoyant as a result of the subsurface dynamo mechanism, by the process of magnetic reconnection. The formation of the solar corona with frequent sporadic energy releases including solar flares, solar wind and coronal mass ejections is a consequence and manifestation of the energy release process. This suggests that magnetic energy conversion through magnetic reconnection is likely to be a common occurrence in the cosmos.

Z Andromedae (Z And)

The prototype of the symbiotic class of variable star. Symbiotic stars are close binary systems consisting of a RED GIANT STAR and a hot blue star both embedded in nebulosity. They are one of the classes of CATACLYSMIC VARIABLES which, as their name implies, undergo occasional violent outbursts. In 1901 Williamina FLEMING first noted that the star had a peculiar spectrum but this M-type VARIABLE STAR with a 2-magnitude range in visual brightness, received more attention in the 1920s when John Plaskett (1865–1941) reported a peculiar class-A spectrum with strong nebular lines on spectra near maximum light. Frank Hogg later noted titanium oxide (TiO) bands on Plaskett's spectra. The simultaneous presence of A-type features with the TiO bands observed in much cooler M-type stars led Hogg to speculate that Z And and a handful of similar 'stars with combination spectra' might be a new type of (binary) stellar system. Roughly a dozen of these systems were known in the 1940s, when Paul MERRILL coined the term 'symbiotic star' for the class.

As in all symbiotic stars, the behavior of Z And can be divided into two states: eruption and quiescence. During quiescence, the optical spectrum resembles an M-type giant, with a strong red continuum and deep TiO absorption bands. This continuum peaks in the near-infrared, where carbon monoxide (CO) absorption bands dominate the spectrum. In addition to the M-type stellar photosphere, optical data show a weak blue continuum and intense emission lines from a variety of ionized species.

Z And consists of a cool red giant and a WHITE DWARF companion, orbiting each other with a period of 759 days. From the orbital motion, masses of about 2 solar masses and 0.5–1.0 solar mass are obtained. The brightness and the spectrum of Z And vary on very short timescales but there are also occasional outbursts where the system brightens by about 2 magnitudes and slowly fades, the whole process lasting about 7 years. Material flows from the red giant in a stellar wind and forms an accretion disk around the white dwarf companion. The gas drifts inward through the disk and falls onto the white dwarf. Nuclear reactions occur, in a process similar to that in other cataclysmic variables, when the material reaches a critical mass, producing the outburst in optical brightness. It is not understood why these outbursts, which are repeated every 10–20 years, can occur so frequently. Calculations imply a recurrence time of 100 years of more.

Zeeman effect

The splitting of a spectral line into two, three or more components, which occurs when the source of that line lies within a magnetic field. This phenomenon is named for Dutch physicist Pieter Zeeman (1865–1943), who discovered the effect in the laboratory, in 1896. The separation of the components of a line is proportional to the strength of the magnetic field and the number of components, and the polarization of the light in each component depends on the orientation of the field to the observer's line of sight. The Zeeman effect enables the strength and orientation of magnetic fields (for example, the magnetic fields in sunspots) to be measured. Where the components are too close together to be resolved into separate lines, the line appears broader than would be the case in the absence of a magnetic field (this phenomenon is called Zeeman broadening).

The Zeeman effect occurs because each of an atom's orbiting electrons has a small magnetic field (or magnetic moment). When the atom is placed in a magnetic field, the electrons can align themselves at certain discrete angles to the magnetic field (the orientations are quantized), each of which corresponds to a marginally different energy level. Consequently, each energy level of the atom is split into two or more closely spaced sub-levels, and more transitions (movements of an electron from one level to another) are then possible, each transition corresponding to a spectral line (or a component of a line).

Zeldovich [Zel'dovich, Seldowitsch], Yakov Borisovich (1914–87)

Russian physicist who worked at the Institute of Chemical Physics in Leningrad (later in Moscow), playing a significant role in the development of Soviet nuclear and thermonuclear weapons. In the 1960s he worked on astrophysics and cosmology, including the theory of BLACK HOLES, the formation of galaxies and clusters, and the large-scale structure of the universe. He identified the SUNYAEV–ZELDOVICH effect of a 'shadow' in the COSMIC MICROWAVE BACKGROUND caused by intervening electrons in clusters of galaxies. He developed ASTROPARTICLE PHYSICS in the cosmological theory of the BIG BANG and started to develop a QUANTUM THEORY of gravity.

Zenith

The point on the CELESTIAL SPHERE that is vertically above an observer on the Earth's surface. It is 90° distant from any point on the horizon. The point 180° opposite (directly underneath) the zenith is the nadir.

Zenith distance

The angular distance, measured along a great circle on the CELESTIAL SPHERE, between the ZENITH and a celestial object. The zenith distance of a celestial object is equal to 90° minus the object's altitude.

Zenithal hourly rate

A measure of the activity of a meteor shower. It is defined as the number of meteors that would be seen by a single 'ideal' observer in a cloudless, perfectly dark sky if the radiant were at the ZENITH.

Zero-age main sequence

The locus of points in the HERTZSPRUNG–RUSSELL DIAGRAM where newly formed stars lie during the stable hydrogen-burning phase of their evolution. This is seen as a diagonal line across the diagram with massive stars being the most luminous and burning at the highest temperature (seen as blue in color) while the smallest stars burn much less brightly and are cool and red. In a cluster of stars of the same age the most massive stars evolve first away from the top of the zero-age main sequence.

Zero gravity

A term that is used to describe the state of WEIGHTLESSNESS or free-fall.

▲ ***Zodiacal light*** *Sometimes the sky itself seems to glow. Usually, this means you are seeing a cloud reflecting sunlight or moonlight. Or you could be seeing the combined light from the billions of stars that compose our Milky Way Galaxy. If, however, the glow appears triangular and near the horizon, after sunset, or before sunrise, you may be seeing zodiacal light, which is sunlight reflected by tiny dust particles orbiting in our solar system.*

Zodiac

The band of CONSTELLATIONS around the CELESTIAL SPHERE extending about 8 degrees on either side of the ECLIPTIC. In ASTROLOGY the zodiac is divided into into 12 equal signs, each 30 degrees long. The 12 well-known constellations are Aries, Taurus, Gemini, Cancer, Leo, Virgo, Libra, Scorpius, Sagittarius, Capricornus, Aquarius and Pisces. However, because of the effects of precession and the redefinition of constellation boundaries, the constellation Opiuchus now also lies on the ecliptic and a number of other constellations (for example Cetus, Orion and Sextans) lie within the band of the zodiac.

Zodiacal light

A faint cone-shaped glow in the night sky stretching along the ECLIPTIC, alternatively known as the 'counterglow' or 'gegenschein.' Given a dark sky and the absence of moonlight, it is visible at all times from the tropics. From temperate latitudes it is best seen about an hour and a half before sunrise in the fall or the same time after sunset in the spring, for at these times the ecliptic makes its greatest angle with the HORIZON. The zodiacal light is caused by sunlight scattered by interplanetary dust particles in the plane of the ecliptic.

Zodiacal star

Stars located within 8° of the ECLIPTIC; that is, within the zodiac. There are 3539 stars listed in the Zodiacal Catalog (ZC) of apparent magnitude 8.5 and brighter. They are the only stars that can be occulted by the Moon: observations of lunar occultations (*see* ECLIPSE AND OCCULTATION) are valuable as a check on the Moon's position.

Zond mission

Series of eight Soviet deep-space missions. Launched 1964–70. Zond 1 was a failed Venus flyby. Zond 2 was a failed MARS mission. Zond 3 (launched July 1965) conducted a lunar flyby. Zonds 4–8 were part of the test program for a Soviet manned lunar mission. Zond 5 (launched September 1968) was the first spacecraft to successfully circumnavigate the Moon and return to Earth. Zond means 'probe.'

Zwicky, Fritz (1898–1974)

Swiss physicist, born in Bulgaria, who became professor at the California Institute of Technology, USA. He researched galaxies and produced a comprehensive catalog of them. He had an all-inclusive approach to astronomy, which suggested that if something was physically possible then it existed somewhere in the universe. He called this 'morphological astronomy.' In 1934 he predicted the existence of neutron stars (*see* NEUTRON STAR AND PULSAR) and BLACK HOLES, formed by SUPERNOVAE (a word he coined). His studies of the dynamics of galaxies showed the existence of DARK MATTER decades before this was generally accepted.

ZZ Ceti star

The generic name for pulsating WHITE DWARFS of spectral type DA, which have a pure hydrogen outer-layer composition. This name is equivalent to DAV (for variable DA white dwarfs), a term that is frequently used in the literature.

Once on the white dwarf sequence, the stars cross three more instability strips as they cool down, which we recognize as distinct groups:

- the pulsating PG 1159 STARS (also called GW Vir stars), which are the nuclei of PLANETARY NEBULAE, at an effective temperature between 150 000 and 80 000 K;
- the variable DB white dwarfs (DBV) at about 25 000 K (*see* DBV PULSATING STAR);
- the variable DA (ZZ Ceti stars) at about 12 000 K.

The number of pulsating white dwarfs (which undergo NON-RADIAL PULSATIONS) is quite small: only four PG1159, seven DB and 29 DA white dwarfs are known to pulsate at present. The ZZ Ceti form the largest group because the hydrogen-atmosphere white dwarfs (DA) are the most numerous and because the cooling timescale increases with decreasing luminosity and the ZZ Ceti define the coolest instability strip, so more stars will be found in a given range of temperature. However, because of the difficulty of discovering these intrinsically faint stars, the population of pulsating white dwarfs may in fact constitute the largest group of variable stars in our Galaxy.

The mean mass of a ZZ Ceti star is the same as the mean mass of white dwarfs, namely 0.6 solar mass. They are found in a narrow instability strip in the HERTZSPRUNG–RUSSELL DIAGRAM extending from about 12 400 K on the blue edge to about 11 200 K on the red edge. There are no stable stars in this region and all DA white dwarfs crossing the ZZ Ceti instability strip should become pulsators. As a consequence, the properties derived for ZZ Ceti from ASTROSEISMOLOGY are presumably applicable to all DA white dwarfs.

The ZZ Ceti stars cool from the blue to the red edge of the instability strip and a clear relation between pulsation periods and effective temperature, T_e, is observed with pulsation periods increasing as T_e decreases. There is also a clear tendency for ZZ Ceti of longer periods to have larger amplitude, except close to the red edge where amplitudes fall abruptly to small values. This shows that the pulsation excitation mechanism stops being efficient at a given T_e (for a given total mass) and this defines the red edge.

Picture credits

The publishers have made every effort to obtain permission to use copyright material and to avoid errors or omissions. We would welcome any such errors or omissions being brought to our attention, so that we may correct them in any future editions.

Abbreviations: *t* = top, *b* = bottom, *c* = center, *l* = left, *r* = right

viii, **6**, **46**, **65*t***, **81**, **83**, **86*br***, **87**, **98**, **113**, **130**, **163*l***, **163*c***, **163*r***, **167**, **180*r***, **190**, **197**, **215**, **230**, **271**, **286**, **332**, **355**, **362**, **377**, **412*b***, **415**, **417**, **428**, **447**, **451**, Anglo-Australian Observatory / David Malin Images; **1**, Nigel Sharp, National Optical Astronomical Observatories / National Solar Observatory; **7**, **156**, **158**, **409**, TRACE (Stanford–Lockheed Institute for Space Research, NASA); **8*t***, STS-59 crew, NASA; **8*b***, Air Force Maui Optical Station; **9**, **27**, **34**, **99**, **102**, **103**, **171**, **186**, **191*l***, **191*r***, **192**, **194**, **293*b***, **314**, **334**, **399**, **425**, Royal Astronomical Society; **11*t***, Stefan Binnewies; **13**, Malin / Caltech; **15*t***, Smithsonian Institution; **15*b***, Courtesy of D D Dixon (University of California, Riverside) and W R Purcell (Northwestern University); **16**, **48**, **55*t***, **69**, **214**, **344**, **348**, Image courtesy of NRAO / AUI / NSF; **17**, **149**, **313**, **320**, **367**, **374**, **419**, NSSDC / NASA; **18**, **120**, **203**, **358**, Copyright Calvin J Hamilton; **19**, digital image © 2002 The Museum of Modern Art / Scala, Florence; **21**, **66**, **86*bl***, **92**, **93**, **154**, **202**, **254*b***, **376**, **464**, Akira Fujii / David Malin Images; **24*b***, **173**, **217**, **253**, **273*t***, **288**, **289**, **325**, Courtesy of NASA / JPL / Caltech; **26*t***, **341**, Whipple Museum of the History of Science, University of Cambridge; **28**, Courtesy of the Archives, California Institute of Technology; **31**, Kamioka Observatory, ICRR (Institute for Cosmic Ray Research, University of Tokyo); **45*t***, **58**, AIP Emilio Segrè Archives; **47**, **125**, **147**, **162**, **166**, **235*t***, **287**, **336**, **343**, **391**, **394**, European Southern Observatory; **54**, NASA Marshall Center; **55*bl***, **55*br***, NASA, the Hubble Heritage Team (AURA / STScI) and M Rich (USLA); **56**, Lowell Observatory Archives; **57**, J F Sepinsky *et al.* (Villanova University), NASA; **59**, Inventory no. 59738, Oxford Museum of the History of Science; **61**, Malin / IAC / RGO; **62**, David Messier; **65b**, Courtesy Harvard College Observatory; **67**, Photograph by R Pelisson, SaharaMet; **70**, Lawrence Berkeley National Laboratory / Roy Kaltschmidt; **77**, H. Hammel (SSI), WFPC2, HST, NASA; **79*l***, Photograph by Dorothy Davis Locanthi, courtesy of AIP Emilio Segrè Visual Archives, Dorothy Davis Locanthi Collection; **79*r***, Fred Baganoff (MIT), Mark Morris (UCLA), *et al.*, CXC, NASA; **80**, Image by V L Sharpton, Courtesy: Lunar & Planetary Institute; **82**, Andrew Fruchter (STScI) *et al.*, WFPC2, HST, NASA; **95**, **105**, **382**, SOHO (ESA & NASA); **96**, Prof. Mark Bailey, Director of Armagh Observatory; **97**, **433**, **460**, ESA; **106**, NASA/WMAP Science Team; **109**, Malin/Pasachoff / Caltech; **110**, Charles Danforth, FUSE; **116**, David Dunlap Observatory, University of Toronto; **117**, **143*t***, **327**, **432**, **448**, Courtesy of NASA / JPL / Caltech; **124**, **229**, Robert Reeves; **128**, NASA, ESA and the Hubble Heritage Team (STScI / AURA); **129**, Leiden University; **138**, Photo contributed by the Max-Planck-Institut für Radioastronomie, copyright: Ingenieurbüro Aerocart, Bonn; **139**, Einstein Observatory / HEASARC / NASA ; **145**, NASA - Goddard Space Flight Center Scientific Visualization Studio; **146**, NASA/CXC/SAO; **155**, Faulkes Telescope Project; **172**, EGRET Team/NASA; **178**, B Balick (University of Washington) *et al.*, WFPC2, HST, NASA; **182**, Randall L Ricklefs / McDonald Observatory; **183**, W N Colley and E Turner (Princeton University), J A Tyson (Bell Labs, Lucent Technologies) and NASA; **185**, Jeff Hester and Paul Scowen (Arizona State University), and NASA; **187**, **339**, National Oceanic and Atmospheric Administration; **196**, Ronald Bentley; **200*t***, **200*b***, **366**, **453**, NASA; **204**, Robin Rees / Canopus Publishing Limited; **205**, E Karkoschka (University of Arizona) and NASA; **206**, Infrared Processing and Analysis Center, Caltech / JPL. IPAC is NASA's Infrared Astrophysics Data Center; **207**, IRAS; **211**, STS-108 Crew / NASA; **212**, Photo courtesy of the Isaac Newton Group of Telescopes, La Palma; **213**, James Clerk Maxwell Telescope, Mauna Kea Observatory, Hawaii; **222**, **414**, ROSAT Mission / Max-Planck-Institut für extraterrestrische Physik (MPE); **225**, **240**, **300**, **303**, **307**, **315**, **337**, **401**, NOAO / NSF / AURA; **226**, A Caulet (ESO) / NASA; **228**, Harvard College Observatory, courtesy of AIP Emilio Segrè Visual Archives; **233**, NASA / JPL / Northwestern University; **235*b***, Norman Lockyer Observatory Society; **237**, Atlas Image courtesy of 2MASS / Umass / IPAC-Caltech / NASA / NSF; **238**, Courtesy Lund Observatory; **239**, J Shalf, Y. Zhang (UIUC) *et al.*, GCCC; **245**, NASA, Lunar Orbiter 4; **247*l***, **247*r***, NASA / JPL / Malin Space Science Systems; **250**, ESA, D Ducros; **252**, David P Anderson; **257**, Tom Muxlow & Alan Pedlar, Jodrell Bank Observatory, University of Manchester; **263**, Courtesy of Meteor Crater North Arizona; **268**, Chen Huang-Ming; **273*b***, Photo courtesy of the Maria Mitchell Association; **280**, NASA / LARC; **293*t***, National Trust photolibrary / Derrick Witty; **295**, Photo copyright Scott Tucker; **311**, J Bally, D Devine, R Sutherland, D Johnson (CITA), HST, NASA; **316**, Painting of Hale Telescope by J D Cremi courtesy Caltech archives; **318**, From *Ring of Truth: Doubt* by PBS, courtesy of AIP Emilio Segrè Visual Archives, Physics Today Collection; **321**, Axel Mellinger; **326**, Matt Bobrowsky / NASA; **328**, Apollo 16 / NASA; **333**, Alan Stern (Southwest Research Institute), Marc Buie (Lowell Observatory), NASA and ESA; **335**, NASA / STScl / AURA; Cambridge University; **342**, Scott Croom for the 2dF QSO Redshift Survey team; **347**, M Cobb/Cassicorp; **349**, Keel, Ledlow & Owen, STScI, NRAO / AUI / NSF, NASA; **351**, CSIRO; **356**, Griffith Observatory / Anthony Cook; **357**, NASA / JHUAPL; **359**, Walter Nowotny, Walter Koprolin, University of Vienna, Austria; **361**, NASA / Goddard Space Flight Center; **363**, David Davidson / Birr Castle; **364**, AIP Emilio Segrè Visual Archives, Margaret Russell Edmondson Collection; **365**, NASA / Penn State / G. Garmire *et al.*; **371**, Bruce Balick, Jason Alexander, Arsen Hajian, Yervant Terzian, Mario Perinotto, Patrizio Patriarchi, NASA; **375**, *Through Rugged Ways to the Stars*, by Harlow Shapley, New York: Charles Scribner's Sons, 1969, courtesy AIP Emilio Segrè Visual Archives, Shapley Collection; **386**, Allan Chapman; **389**, Sebastien Gauthier; **403**, **438**, Copyright © Subaru Telescope, National Astronomical Observatory of Japan. All right reserved; **408**, T Rimmele (NSO), M Hanna (NOAO) / AURA / NSF; **420**, **421**, **422**, Brockhurst, Clarkson & Fuller Ltd. / Meade; **426**, Voyager 2 / NASA; **427**, NASA / JPL / University of Arizona; **430**, Joint Astronomy Centre; **434**, BATSE team; **435*l***, **435*r***, Willem Schaap, Rien Weygaert; **436**, Julian Baum; **439**, Viking Project, USGS, NASA; **441**, Dr Ian Short; **443**, NASA / SAO / CXC; **449**, Image courtesy of NRAO / AUI and photographer Kelly Gatlin, digital composite Patricia Smiley; **458**, A Ghez *et al.*, UCLA / W M Keck Observatory; **454**, N Scoville, T Rector *et al.*, Hubble Heritage Team, NASA; **455**, NASA and the Hubble Heritage Team (AURA/STScI); **459**, Yves Grosdidier, Anthony Moffat, Gilles Joncas, Agnes Acker and NASA; **462**, ISAS, Yohkoh Project.

Unless otherwise stated, all artworks are courtesy of James Symonds.

Contributors

Ambroise, Bruno
Anderson, S B
Anderson, Scott
Arlot, Jean-Eudes
Arnaboldi, Magda
Arpigny, Claude
Aschwanden, Markus J
Ashtekar, Abhay
Asplund, Martin
Audouze, Jean
Avrett, E
Ayres, Thomas R
Bacon, Roland
Bagenal, F
Bahcall, John
Bai, T
Bailes, Matthew
Baker, D
Baliunas, S
Balogh, A
Bai, Taeil
Barnes, Joshua
Barucci, Antonella
Bastian, T S
Batten, Alan H
Baugh, Carlton
Beckers, J
Beer, Hermann
Beer, J
Bely, Pierre
Benedict, George F
Benest, Daniel
Bentley, Robert D
Benz, A
Berger, Mitchell
Bernstein, Rebecca
Bessell, Michael S
Beutler, G
Bézard, Bruno
Bibring, Jean-Pierre
Biermann, Peter L
Biesecker, Douglas
Binggeli, Bruno
Binney, James
Bizouard, Christian
Blaauw, A
Blandford, Roger
Bland-Hawthorn, Joss
Blitz, Leo
Bockelee-Morvan, Dominique
Bochsler, P
Bogdan, T
Böker, Torsten
Boland, Wilfried
Bolton, C T
Bond, H
Bond, Peter
Borovicka, Jiri
Boucher, Claude
Bougeret, J L
Bouwens, Rychard
Brack, Andre
Bradley, Paul
Brandenburg, Axel
Brandt, Peter N
Bregman, J
Brekke, P
Broughton, Peter
Brown, A G A
Brown, Robert
Bruhweiler, F C
Brunner, Robert J
Brush, Stephen G
Bunce, E J
Burgess, David
Burke, Bernard F
Burlaga, Leonard F
Buson, Lucio M
Buta, Ronald J
Caldwell, Nelson
Canfield, R
Capetti, A
Capitaine, Nicole
Cargill, P
Carlberg, Raymond G
Carlsson, M
Carney, Bruce
Carr, Bernard
Carroll, Sean M
Cattaneo, Fausto
Chapman, Allan
Chapman, Gary
Charles, Philip A
Charlton, Jane
Chassefiere, Eric
Chiosi, Cesare
Christensen-Dalsgaard, Jørgen
Chun, Leung Min
Churchill, Chris
Chu, You-Hua
Clementini, Gisella
Cliver, Edward W
Coates, Andrew
Cole, Trevor
Colless, Matthew
Combes, Francoise
Coradini, Marcello
Cornilleau-Wehrlin, Nicole
Couch, Warrick
Courtin, Regis
Courvoisier, T
Coustenis, Athena
Cowley, S W H
Cranmer, Steven R
Cravens, Thomas
Crawford, David L
Croft, Steven K.
Crovisier, Jacques
Cruise, A M
Crutcher, Richard
Culhane, Len
Dalgarno, Alex
Dappen, W
Davies, Roger L
Davis, John
de Bergh, Catherine
de Blok, Erwin
de Bruijne, J H J
de Zeeuw, Tim
DeForest, Craig
Dehant, V
del Toro Iniesta, Jose Carlos
Delsemme, Armand
Dere, Ken
Deupree, Robert G
DeVorkin, David H
Dick, Steven J
Djorgovski, S George
Dopita, Mike
Dorman, Ben
Dougherty, M K
Dowling, T E
Draine, Bruce T
Dreizler, Stephan
Drilling, John S
Drossart, Pierre
Dumas, Christophe
Dunkin, Sarah
Duric, Neb
Durrant, C J
Duvall, T
Dyer, John W
Dyson, J E
Eckart, Andreas
Eggleton, Peter
Ehlers, J
Ekers, Ron
Elliot, James
Ellis, John
Ellsworth, Robert W
Encrenaz, Thérèse
Englmaier,
Engvold, Oddbjörn
Erard, Stéphane
Esposito-Farèse, Gilles
Esser, Ruth
Fabian, Andy
Feast, Michael
Ferguson, Harry
Ferrari, Cecile
Ferriz-Mas, Antonio
Fontaine, Gilles
Forbes, Duncan A
Forbes, T G
Forni, Olivier
Fox, Nicola J
Franco, J
Franx, Marijn
Fraser, George
Freeland, Sam
Freedman, Wendy
Freeman, Ken
Frenk, Carlos S
Friis-Christensen, E
Fulchignoni, Marcello
Fulle, Marco
Fuselier, Stephen A
Fusi-Pecci, Flavio
Gaizauskas, V
Gallagher, Jay
Garrison, Robert F
Geissler, Paul
Giampapa, Mark S
Giavalisco, Mauro
Gingerich, Owen
Glazebrook, Karl
Godin, Sophie
Golub, L
Goodman, Jordan A
Goodrich, Robert
Gough, Douglas
Grant, Edward
Gray, Richard O
Green, Paul
Greife, U
Grevesse, Nicolas
Griest, Kim
Gurnett, D
Habbal, S Rifai
Haberle, Robert M
Häfner, Reinhold
Hagenaar, Hermance J
Hamann, Fred
Handy, Brian N
Hanson, Robert
Hansteen, Viggo
Harrington, J Patrick
Harrison, R
Harvey, J
Harvey, K

Hauschildt, Peter H
Heck, André
Heckman, T
Heggie, Douglas C
Heintz, Wulff D
Heiselberg, Henning
Henoux, J C
Herbert, Floyd
Herbst, Eric
Herring, David
Hewett, P
Heyvaerts, J
Hickson, Paul
Higdon, James L
Hillerbrand, Wolfgang
Hillier, John
Hines, Dean
Hinshaw, Gary
Hjellming, R
Ho, Luis C
Ho, Paul
Hodapp, Klaus-Werner
Høg, Erik
Hollweg, Joe
Hood, Alan
Hoogerwerf, R
Horanyi, M
Hornig, Gunnar
Hoskin, Michael
Hough, James H
Hough, Jim
Howard, R F
Howell, Steve B
Hoyng, Peter
Hubeny, Ivan
Huchra, John
Hudson, Mary
Hughes, D W
Ibata, Rodrigo
Iglesias, C A
Innes, D
Irwin, Mike
Jacoby, B A
Janes, Kenneth
Jarrell, Richard A
Jeffery, C Simon
Jenkins, Edward B
Jerjen, Helmut
Jokipii, J R
Jones, Alexander
Jordan, Stefan
Kahler, S
Kaifu, Norio
Kaitchuck, Ronald
Kaler, James B
Kamide, Y
Karovska, Margarita
Katgert, P
Kelley, David
Kenyon, S
Keppens, Rony
Kesteven, Mike
King, Michael D
Kjeldseth-Moe, O
Klapdor-Kleingrothaus, H V
Klemaszewski, J
Klemola, A R
Kolb, Edward
Kormendy, John
Kosovichev, A
Koutchmy, S
Kovalevsky, Jean
Kozyra, Janet
Krasnopolsky, Vladimir
Krisciunas, Kevin
Kriss, Gerard A
Kudritzki, Rolf-Peter
Kuhn, J
Kunow, H
Kurtz, D W
Kurucz, Robert L
Kwitter, Karen
Kwok, Sun
Lada, Charles J
Lamers, Henny J G L M
Landstreet, John D
Langevin, Yves
Lanz, Thierry
Lanzetta, Kenneth M
Lasenby, Anthony
Lasher, Lawrence E
Laskar, Jacques
Latham, David W
Le Quéau, Dominique
Lean, Judith
Lebreton, Jean-Pierre
Leckrone, David
Lee, M A
Leer, E
Lellouch, Emmanuel
Lemaire, P
Leonard, Peter
Lepping, R P
Leschiutta, S
Lesser, Michael
Lester, M
Leung, Chun Ming
Liebert, Jim
Linde, Andrei
Lites, B
Livingston, William
Lloyd Evans, Thomas
Lockwood, Mike
Lopez, Carlos E
Low, B C
Lowman, P D
Louarn, Philippe
Luhmann, J G
Lunine, Jonathan
Luppino, Gerard
Luu, Jane
Lyne, Andrew
Ma, C
Macchetto, D
Machado, M
Machetel, Philippe
Madau, Piero
Maddison, Ron
Madore, Barry F
Malin, David
Malkan, Matthew Arnold
Maltby, Per
Mandzhavidze, N
Margon,Bruce
Marsch, Eckart
Martens, P C H
Martin, Sara F
Marvel, Kevin B
Mason, Helen
Massey, Philip
Masson, Philippe
Mateo, Mario L
Mathews, Lynn
Mathieu, R
Matson, Dennis L
Matthaeus, W
McAlister, Harold A
McCarthy, D D
McCray, Richard
McEwen, Alfred S
McFarland, John
McIntosh, Patrick S
McKibben, R Bruce
McLean, Ian
McMullin, Ernan
McNally, Derek
Meadows, A J
Melrose, D B
Mercier, Raymond
Mermilliod, Jean-Claude
Meyer, Michael R
Mihos, Chris
Mikkola, Seppo
Milkey, Bob
Miner, Ellis D
Moffatt, Keith
Mohr, Rob
Moore, Patrick
Moore, Ronald L
Moreno-Insertis, F
Moroz, Vassili I
Mörzer Bruyns, Willem F J
Mosser, Benoit
Mosqueira, Ignacio
Mould, Jeremy
Mowlavi, Nami
Mozurkewich, David
Mukai, Toshifumi
Mullan, Dermott
Müller, Ewald
Muller, Richard
Murdin, Paul
Murray, Carl
Najita, Joan
Narlikar, Jayant V
Neckel, Heinz
Neubauer, F
Niemeyer, Jens C
Nomoto, Ken´ichi
Nordlund, Åke
Nordtvedt, Kenneth
Norris, John
Norris, R
Olive, Keith A
Oliver, Ramon
Onsager, Terrance
Osmer, Patrick S
Ossendrijver, Mathieu
Owocki, Stanley
Parker, Eugene N
Parker, Joel Wm
Parker, P
Parmar, Arvind
Parnell, Clare E
Pasachoff, Jay M
Peacock, A
Peacock, J A
Peale, Stanton J
Perozzi, Ettore
Perron, Claude
Perryman, Michael A C
Pesce, J E
Peters, Geraldine J
Peterson, Bradley M
Peterson, Bruce A
Petit, G
Philip, A G Davis
Phillips, Kenneth J H
Phillips, Thomas G
Pierce, Michael

Pillet, Valentin Matinez
Pingree, David
Platais, Imants
Postman, Marc
Poulet, Francois
Pounds, Ken
Prantzos, Nikos
Prialnik, Dina
Priest, Eric
Proctor, M
Pulkkinen, Tuija I
Qibin, Li
Rabin, D
Raffelt, Georg
Raga, Alejandro C
Ramaty, R
Rauch, Michael
Raulin, Francois
Raychaudhury, Somak
Raymond, John C
Reames, D
Rector, Travis A
Reid, Mark
Reid, Paul B
Reiff, Patricia
Reynolds, Ron
Richardson, E Harvey
Ridpath, Ian
Roberts, B
Robertson, Gordon
Roddier, Francois
Rodger, A
Rodonò, Marcello
Rogers, F J
Rolfs, Claus
Roques, Françoise
Rosner, Robert
Roueff, Evelyne
Rowan, Sheila
Ruggles, Clive L N
Russell, C T
Rust, D
Rutten, Robert J
Saffer, Rex
Saha, Prasenjit
Sahade, Jorge
Saliba, George
Sanchez, Francisco
Sandage, Allan
Sartori, Leo
Saslaw, William
Sasselov, Dimitar
Sauval, A Jacques
Savage, Blair
Savonije, Gerrit Jan
Schindler, K
Schmidt, Gary
Schmidt, Wolfgang
Schmieder, Brigitte
Schmoldt, Inga
Schneider, Donald P
Schneider, Jean
Schneider, Stephen E
Scholer, Manfred
Schreier, E
Schrijver, Carolus J
Schubert, Gerald (Jerry)
Schüssler, M
Schutz, Bernard F
Schwehm, Gerhard
Schwenn, Dr Rainer
Sears, Derek W G
Seidelmann, P Kenneth
Sekii, T
Shara, Mike
Sharples, Ray M
Shectman, Steve
Shellard, E Paul
Shibata, Kazunari
Shine, Richard
Shortridge, Keith
Shustov, Boris
Silk, Joseph I
Sim, Helen
Simnett, George
Simon, George W
Sinclair, Mal
Sion, Edward M
Sivin, Nathan
Slavin, James A
Smith, Craig H
Smith, Eric P
Smith, Robert W
Snyders, Conway W
Soderblom, David
Soffel, M H
Solanki, Sami K
Spencer, John R
Spruit, Hendrik C
Stahler, Steven
Starrfield, Sumner
Steffen, Matthias
Steigman, Gary
Stein, Robert
Steiner, Oskar
Stenflo, Jan
Sternberg, Amiel
Stetson, Peter B
Stevenson, David
Stix, Michael
Stocke, John
Strobel, Darrell F
Strom, Robert
Sutherland, Will
Svestka, Zdenek F
Swank, Jean
Swerdlow, Noel M
Sykes, Mark
Tandberg-Hanssen, E
Tapping, Ken
Taylor, Fred
Terrell, Dirk
Thielemann, Friedrich-K
Thomas, John H
Thomas, Nicolas
Thomas, Pierre
Thompson, Barbara
Thorsett, Stephen E
Thuillier, Gerard
Title, Alan
Tody, Doug
Tolstoy, Eline
Toomre, J
Totsuka, Yoji
Treumann, Rudolf A
Trimble, Virginia
Truemper, Joachim
Tsuneta, S
Tull, Robert G
Tully, R Brent
Turck-Chieze, Sylvaine
Turner, Michael S
Tyson, J Anthony
Ulmschneider, P
Ulrich, Marie-Helene
Unwin, Stephen C
Valtonen, M
van Altena, William
van Ballegooijen, A
van der Klis, M
van der Kruit, Piet
Vauclair, Gerard
Vauclin, Michel
Vazquez, M
Verbunt, F
Voelkel, James R
von Steiger, Rudolf
Vrtilek, S
Wallerstein, George
Walterbos, Rene A M
Wang, Haimin
Warner, B
Webb, D
Weber, Fridolin
Weekes, Trevor
Weidenschilling, S J
Weiss, N O
Weissman, Paul
Weistrop, Donna
Werner, Klaus
Wesemael, François
West, Robert A
Wheeler, J Craig
White, Stephen
Whittle, M
Wielebinski, Richard
Wilkes, Belinda
Williams, D A
Williams, Gareth V
Willingale, Richard
Wilson, Curtis
Wilson, Robert E
Wilson, Thomas L
Wilson, Warwick
Winget, Donald E
Winter, Klaus
Witt, Adolf
Wöhl, H
Wolf, Richard
Woo, Richard
Wood, Matt
Woodruff, John
Worthey, Guy
Wright, Edward L
You-Hua Chu
Young, R E
Zacharias, Norbert
Zahn, Jean-Paul
Zank, Gary P
Zarka, P
Zepf, Steve
Zirker, Jack
Zuckerman, Ben M

Acronyms and abbreviations

2dF or 6dF	2- or 6-degree field spectroscopy
AAO	Anglo-Australian Observatory
AAVSO	American Association of Variable Star Observers
ADS	Astrophysics Data System
AGN	active galactic nucleus
ALEXIS	Array of Low Energy X-ray Imaging Sensors
ALMA	Atacama Large Millimetre Array
APM	Automatic Plate Measuring machine
AURA	Association of Universities for Research in Astronomy
BAA	British Astronomical Association
BATSE	Burst and Transient Source Experiment
CCD	charge-coupled device
CDS	Centre de Données astronomiques de Strasbourg (Strasbourg astronomical data center)
CERN	European Organization for Nuclear Research
CGRO	Compton Gamma Ray Observatory
CHARA	Center for High Angular Resolution Astronomy
COBE	Cosmic Background Explorer
COS-B	Celestial Observation Satellite
ESA	European Space Agency
ESO	European Southern Observatory
EUVE	Extreme Ultraviolet Explorer
EXOSAT	European X-ray Observatory Satellite
FIRST	Far-Infrared and Submillimetre Telescope
FUSE	Far-Ultraviolet Spectroscopic Explorer
GONG	Global Oscillation Network Group
GPS	Global Positioning System
GRB	gamma-ray burst(er)
HA	hour angle
HAO	High Altitude Observatory
HD	Henry Draper Catalog
HDF	Hubble Deep Field
HEAO	High Energy Astrophysical Observatory
HETE	High Energy Transient Experiment
HR diagram	Hertzsprung–Russell diagram
HST	Hubble Space Telescope
IAC	Instituto de Astrofísica de Canarias (Canary Islands Institute of Astrophysics)
IAR	Instituto Argentino de Radioastronomía (Argentine Institute for Radioastronomy)
IAU	International Astronomical Union
IC	Index Catalogue
ICE	International Cometary Explorer
IMF	initial mass function
ING	Isaac Newton Group of Telescopes
INTEGRAL	International Gamma-Ray Astrophysics Laboratory
IRAM	Institut de Radioastronomie Millimétrique (Institute of Millimeter-Wave Astronomy)
IRAS	Infrared Astronomy Satellite
IRTF	Infrared Telescope Facility
IRTS	Infrared Telescope in Space
ISAS	Institute of Space and Astronautical Sciences
ISEE	International Sun-Earth Explorer
ISO	Infrared Space Observatory
ISS	International Space Station
IUE	International Ultraviolet Explorer
JCMT	James Clerk Maxwell Telescope
JIVE	Joint Institute for VLBI in Europe
JPL	Jet Propulsion Laboratory
LBV	Luminous Blue Variable
LF	Life Finder
LINER	Low Ionization Nuclear Emission Region
M	Messier Catalog number
MMT	Multiple Mirror Telescope
NASA	National Aeronautics and Space Administration
NEO	near-Earth object
NGC	New General Catalogue of Clusters and Stars
NOAA	National Oceanic and Atmospheric Administration
NOAO	National Optical Astronomy Observatory
NRAO	National Radio Astronomy Observatory
NVSS	NRAO VLA Sky Survey
OAO	Orbiting Astronomical Observatory
OWL	Overwhelmingly Large Telescope
ROSAT	Röntgen Satellite
RXTE	Rossi X-ray Timing Explorer
SAFIR	Single Aperture Far Infrared Observatory
SAMPEX	Small Anomalous and Magnetospheric Particle Explorer
SAS	Small Astronomy Satellite
SETI	search for extraterrestrial intelligence
SIM	Space Interferometry Mission
SIMBAD	Set of Identifications, Measurements and Bibliography for Astronomical Data
SIRTF	Space Infrared Telescope Facility
SMART	Small Mission for Advanced Research in Technology
SMM	Solar Maximum Mission
SOFIA	Stratospheric Observatory for Infrared Astronomy
SWAS	Submillimeter Wave Astronomy Satellite
TAI	international atomic time
TPF	Terrestrial Planet Finder
TRACE	Transition Region and Coronal Explorer
UKST	UK Schmidt Telecope
ULIRG	ultra-luminous infrared galaxy
UT	universal time
UTC	coordinated universal time
VLA	Very Large Array
VLBI	very long baseline interferometry
VLT	Very Large Telescope
VSOP	VLBI Space Observatory Program
WIMP	weakly interacting massive particle
WMAP	Wilkinson Microwave Anisotropy Probe
XEUS	X-ray Evolving Universe Spectroscopy Mission

Units and conversions

Numerical

We follow US usage and scientific notation, for example:
1 million = 1 000 000 = 10^6
1 billion = 1 thousand million = 1 000 000 000 = 10^9
1 trillion = 1 000 000 000 000 = 10^{12}
1 billionth = 10^{-9}

kilo (k) = 1000 (10^3) times
mega (M) = 1 million (10^6) times
giga (G) = 1 billion (10^9) times
milli (m) = 1 thousandth (10^{-3})
micro (μ) = 1 millionth (10^{-6})
nano (n) = 1 billionth (10^{-9})

Units

As usual in science, metric units are generally used.

Telescope diameters: we follow the IAU convention for professional telescopes and use metric units except where US/Imperial units have become the historic descriptor. We also follow common practice and use US/Imperial units to describe larger amateur telescopes.

Units are identified in exponential notation: e.g. speed is in kilometers per second, written km s^{-1}. Here are some conversions:

Length

1 meter (m) = 100 centimeters (cm) = 39.37 inches (in)
1 kilometer (km) = 1000 m = 0.62 mile (mi) = 3281 feet (ft)
1 millimeter (mm) = 10^{-3} m
1 micron (μm) = 10^{-6} m = 10 000 ångstroms (Å)
1 nanometer (nm) = 10^{-9} m = 10 ångstroms (Å)
1 ångstrom (Å) = 10^{-10} m = 10^{-8} cm
1 mile (mi) = 1.609 km
1 inch (in) = 2.54 cm
1 astronomical unit (AU) = 149 597 870 691 m = 150 million km (approx.)
1 light year (l.y.) = 9.46×10^{12} km
1 parsec (pc) = 3.26 l.y. = 3.09×10^{13} km
1 megaparsec (Mpc) = 10^6 (1 million) parsecs

Volume

1 liter (L) = 10^{-3} cubic meter (approx.).
1 milliliter (mL) = 1 cubic centimeter (approx.) = 1 cm^3 or 1 cc
1 US gallon = 3.78 L.

Time

The scientific unit of time is the second (s). Conventionally, 60 s = 1 minute (min), 60 min = 1 hour (h), 24 h = 1 day (d), 365.25 d = 1 year (y). But for the astronomical definitions, *see* TIME.

Note that 1 year = 365.25 d
= 60 × 60 × 24 × 365.25 s
= 3.16×10^7 s
= $\pi \times 10^7$ s (approx.)

Mass

1 kilogram (kg) = 1000 grams (g) = 2.21 pounds (lb)
1 tonne = 1000 kg =0.984 US ton
The rest mass of a particle can also be expressed in units of energy (for example in electron-volts), through the relation $E = m c^2$
1 electron-volt (eV) = 1.6022×10^{-19} J
= ($1.6022 \times 10^{-19} / c^2$) kg
= ($1.6022 \times 10^{-19} / (299\,792\,458)^2$) kg
= 1.7827×10^{-36} kg
Mass of a proton = 1.6726×10^{-27} kg
Mass of proton in energy units = ($1.6726 \times 10^{-27} / 1.7827 \times 10^{-36}$) eV
= 938.2 MeV (*see* MASS)

Energy

1 joule (J) = 1 kg m^2 s^{-2} = 1/1054 BTU (approx).
1 electron-volt (eV) = 1.6022×10^{-19} J (*see* ELECTRON-VOLT)

Power

Power is the rate at which energy is transferred; for example the radiation of light.
1 kilowatt (kW) = 1000 J s^{-1}.
The luminosity (power output) of the Sun is 3.84×10^{26} W

Speed

1 km s^{-1} = 2237 miles per hour (mph) = 0.62 mi s^{-1}.
The speed of light in a vacuum (c) = 299 792 458 m s^{-1} = 3×10^8 m s^{-1}
The 'speed of light' is also conventionally (and incorrectly) referred to as the 'velocity of light.'

Magnetic field

1 tesla (T) = 10 000 gauss

Pressure

1 newton (N) = 1 kg m s^{-2}
1 pascal (Pa) = 1 N m^{-2} = 101 325 standard atmospheres
1 lb per square inch = 6.895 kPa

Density

1000 kg m^{-3} = 1 tonne m^{-3} = 1 gram per cubic centimeter = density of water (approx.)

Angle

1 degree = 1° = 60 arcminutes (arcmin) or minutes of arc = 60'
1 arcmin = 1' = 60 arcseconds (arcsec) or seconds of arc = 60"
1 hour of Right Ascension (RA) = 15° cos (Dec) = 15° at the celestial equator
1 minute of RA = 15' cos (Dec) = 15' at the celestial equator
1 second of RA = 15" cos (Dec) = 15" at the celestial equator
1 radian = 180° / π

Temperature (*see* TEMPERATURE SCALES)

The freezing point of water is 0 degrees Celsius (°C) = 273.15 Kelvin (K) = 32 degrees Fahrenheit (°F)
The size of the unit 1 °C = 1K = 1.8 °F
To convert Celsius (°C) to Farenheit (°F): (°C × 1.8) + 32 = °F
To convert Fahrenheit (°F) to Celsius (°C): (°F – 32) / 1.8 = °C

Frequency

1 hertz (Hz) = 1 cycle per second